Molecular Pathomechanisms and New Trends in Drug Research

Molecular Pathomechanisms and New Trends in Drug Research

Edited by
György Kéri and István Toth

Taylor & Francis
Taylor & Francis Group
LONDON AND NEW YORK

First published 2003
by Taylor & Francis
11 New Fetter Lane, London EC4P 4EE

Simultaneously published in the USA and Canada
by Taylor & Francis Inc,
29 West 35th Street, New York, NY 10001

Taylor & Francis is an imprint of the Taylor & Francis Group

Typeset in Baskerville by
Newgen Imaging Systems (P) Ltd, Chennai, India
Printed and bound in Great Britain by
St Edmundsbury Press, Bury St Edmunds, Suffolk

British Library Cataloguing in Publication Data
A catalogue record for this book is available from the British Library

Library of Congress Cataloging in Publication Data
A catalog record for this book has been requested

ISBN 0-415-27725-6

To Dr Maria Szekerke
who dedicated her life for antitumor drug research

"Sickness comes from lost integrity
Sickness comes from lack of inner peace
The unity of spirit, soul, body and mind
Will make you happy and shine."

Contents

Contributors

Bharat B. Aggarwal, Professor of Medicine (Biochemistry), Chief, Cytokine Research Section, Department of Molecular Oncology, University of Texas, M.D. Anderson Cancer Center, 1515 Holcombe Boulevard, Box 143, Houston, Texas 77030-4095
E-mail: aggarwal@audumla.mdacc.tmc.edu

Stefan Barlage, Institute of Clinical Chemistry and Laboratory Medicine, University of Regensburg, Germany

Chris J. Barry, Department of Molecular Ophthalmology, Centre of Ophthalmology and Visual Science, University of Western Australia, Perth, Australia

Pál I. Bauer, Semmelweis University, Budapest, Department of Medical Biochemistry, 1088 Budapest, Puskin u. 9., 1444 Bp. 8, POB 262, Hungary
E-mail: PIBauer@Puskin.sote.hu

Veronika Bender, Department of Rheumotology, Clinical Sciences, Westmead Hospital, Westmead, Australia

Martin J. Berg, Nathan Kline Institute, Center for Neurochemistry

Chris Birch, Victorian Infectious Diseases Reference Laboratory, 10 Wreckyn Street North Melbourne, 3051, Victoria, Australia
E-mail: chris.birch@nwhcn.org.au

Hana Borsukova, Institute of Clinical Chemistry and Laboratory Medicine, University of Regensburg, Germany

Lindsay Brown, University of Queensland, Physiology and Pharmacology, Brisbane, Qld 4072, Australia
E-mail: l.brown@mailbox.uq.edu.au

Andrew Constanti, School of Pharmacy, University of London, 29–39, Brunswick Square, London WC1N 1AX, UK
E-mail: andy.constanti@ulsop.ac.uk

Péter Csermely, Semmelweis University, Budapest, Department of Medicinal Chemistry, Molecular Biology and Pathobiochemistry, 1444 Bp. 8. POB 260, Hungary, 1088 Budapest, Puskin utca 9
E-mail: CSERMELY@puskin.sote.hu

Balázs Döme, First Institute of Pathology and Experimental Cancer Research, Semmelweis University, Budapest, Hungary

Nicholas Drinnan, 3 Hi-Tech Court, Brisbane Technology Park, Eight mile Plains Qld 4113, Brisbane, Australia
E-mail: NDrinnan@alchemia.com.au

István Ember, Department Preventive Medicine, Faculty of Medicine, University of Pécs, Hungary, 7624 Pécs, Szigeti út 12, Hungary
E-mail: Ember@pubhealth.pote.hu

Anna Erdei, Department of Immunology, Eotvos University, 1117 Budapest Pazmany S. 1/c
E-mail: Anna.Erdei@freemail.hu

Csaba Ertsey, Semmelweis University, Budapest, School of Medicine, Department of Neurology, 1083 Budapest, Balassa u. 6, Hungary

David P. Fairlie, Centre for Drug Design and Development, Institute for Molecular Bioscience, University of Queensland, Brisbane, Qld 4072, Australia
E-mail: D.Fairlie@imb.uq.edu.au

András Falus, Department of Genetics, Cell and Immunobiology, Semmelwei University, Budapest, Nagyvárad tér 4., H-1089 Budapest Hungary
E-mail: faland@net.sote.hu

Árpád Furka, Eötvös Loránd University, organic Chemistry, P.O. Box 32, Budapest 112, H-1518, Hungary
E-mail: afurka@szerves.chem.elte.hu

János Gergely, Department of Immunology, Eotvos University, 1117 Budapest Pazmany S. 1/c
E-mail: Janos.Gergely@freemail.hu

Michael Good, Cooperative Research Centre for Vaccine Technology, Division of Infectious Diseases and Immunology, Queensland Institute of Medical Research, 300 Herston Road, Herston, Brisbane, Qld 4029, Australia
E-mail: michaelG@qimr.edu.au

Péter Halász, National Institute of Psychiatry and Neurology, 1021 Budapest, Hüvösvölgyi út 116, Hungary
E-mail: h12197hal@ella.hu

T. Harkány, University of Groningen, The Netherlands

Ferenc Hudecz, Department of Organic Chemistry, Eötvös University, H-1518 Budapest 112, POB 32, Hungary
E-mail: hudecz@szerves.chem.hu

Richard Anthony Hughes, Head, Drug Design Laboratory, Department of Pharmacology, University of Melbourne, Grattan Street, Victoria 3010
E-mail: r.hughes@pharmacology.unimelb.edu.au

András Illés, Semmelweis University, Budapest, Department of Medical Biochemistry, 1088 Budapest, Puskin u. 9., 1444 Bp. 8, POB 262, Hungary

Ilona Jelencsik, Semmelweis University, Budapest, School of Medicine, Department of Neurology, 1083 Budapest, Balassa u. 6, Hungary

Tamás Kardon, Semmelweis University Department of Medical Chemistry, Molecular Biology and Pathobiochemistry

Devarajan Karunagaran, The Rajiv Gandhi Centre for Biotechnology, Thiruvananthapuram, Kerala, India

Michael J. Kelso, Centre for Drug Design and Development, Institute for Molecular Bioscience, University of Queensland, Brisbane, Qld 4072, Australia
E-mail: M.Kelsko@imb.uq.edu.au

Erzsébet Kénesi, Semmelweis University, Budapest, Department of Medical Biochemistry, 1088 Budapest, Puskin u. 9., 1444 Bp. 8, POB 262, Hungary

György Kéri, Department of Medicinal Chemistry, Semmelweis University, Budapest, H-1444. Bp 8. POB 260., 1088 Puskin u.9. Budapest, Hungary
E-mail: Keri@Puskin.sote.hu

György M. Keserű, Computer Assisted Drug Discovery, Gedeon Richter Ltd., POB 27, H-1475 Budapest, Hungary
E-mail: gy.keseru@richter.hu

István Kiss, Department Preventive Medicine, Faculty of Medicine, University of Pécs, Hungary, 7624 Pécs, Szigeti út 12, Hungary
E-mail: Ember@pubhealth.pote.hu

György Kóczán, Research Group of Peptide Chemistry, Hungarian Academy of Sciences, Eötvös L. University, Budapest, Hungary

László Kopper, Semmelweis University, Budapest, 1st. Institute of Pathology and, Experimental Cancer Research, 1085 Budapest, Üll i út 26
E-mail: KOPPER@korbl.sote.hu

István Kövesdi, EGIS Pharmaceuticals, Budapest, Hungary

Péter Krajcsi, Semmelweis University, Budapest, Department of Medical Biochemistry, 1088 Budapest, Puskin utca 9, H-1444 Bp. 8, POB 262, Hungary
E-mail: Krajcsi@Puskin.sote.hu

Alexander Levitzki, Wolfson family Professor of Biochemistry, Department of Biological Chemistry, The Alexander Silverman Institute of Life Sciences, Director, The Institute for Advanced Studies, The Hebrew University of Jerusalem, Jerusalem 91904, Israel
E-mail: levitzki@vms.huji.ac.il

Raymund Machovich, Semmelweis University, Budapest, Department of Medical Biochemistry, 1088 Budapest, Puskin u. 9., 1444 Bp. 8. POB 266, Hungary
E-mail: MR@Puskin.sote.hu

Ildikó Magdó, Computer Assisted Drug Discovery, Gedeon Richter Ltd., POB 27, H-1475 Budapest, Hungary
E-mail: i.magdo@richter.hu

József Mandl, Department of Medicinal Chemistry, Molecular Biology and Pathobiochemistry, Semmelweis University, Budapest, 1444 Bp. 8. POB 260, Hungary, 1088 Budapest, Puskin utca 9
E-mail: Mandl@Puskin.sote.hu

Zoltán Marcsek, Department of Molecular Cell Biology, National Institute of Chemical Safety, "Jozsef Fodor" National Center for Public Health, 1097 Gyali ut 2/6 Budapest, Hungary
E-mail: Marcsek@microbi.hu

Neville Marks, New York University, Department of Psychiatry, Orangeburg, NY, USA

Ross P. McGeary, University of Queensland, School of Pharmacy, Qld 4072, Brisbane, Australia
E-mail: ross@pharmacy.uq.edu.au

György Mészáros, Semmelweis University, Budapest, Department of Medicinal Chemistry, 1444 Bp. 8. POB 260, Hungary, 1088 Budapest, Puskin utca 9

Gábor Mikala, Semmelweis University, Faculty of Health Sciences, Department of Internal Medicine and Geriatrics, 35 Szabolcs St, Budapest, Hungary H-1135
E-mail: gmikala@helios.hiete.hu

Béla Molnár, Second Department of Medicine, Faculty of Medicine, Semmelweis University, Budapest, Hungary

Greg Monteith, University of Queensland, School of Pharmacy, Qld 4072, Brisbane, Australia
E-mail: greg@pharmacy.uq.edu.au

Zoltán Nagy, Institute of Psychiatry and Neuorology, National Stroke Centre, 1021 Hüvösvölgyi út 116, Hungary
E-mail: nagy@opni.hu

Gábor Náray-Szabó, Department of Organic Chemistry, Eötvös University, H-1518 Budapest 112, POB 32, Hungary
E-mail: naray@para.chem.elte.hu

Colleen Olive, Cooperative Research Centre for Vaccine Technology, Division of Infectious Diseases and Immunology, Queensland Institute of Medical Research, 300 Herston Road, Herston, Brisbane, Qld 4029, Australia
E-mail: colleenO@qimr.edu.au

László Örfi, Semmelweis University, Department of Pharmaceutical Chemistry, Hőgyes E. u. 9., 1092 Budapest, Hungary
E-mail: orlasz@hogyes.sote.hu

Sándor Paku, First Institute of Pathology and Experimental Cancer Research, Semmelweis University, Budapest, Hungary

B. Penke, Department of Medical Chemistry, 6720 Szeged, Dóm tér 8, Hungary
E-mail: penke@ovrisc.mdche.u-szeged.hu

Gyula Poór, National Institute of Rheumatology and Physiotherapy, Budapest, Hungary

László Prónai, Second Department of Medicine, Faculty of Medicine, Semmelweis University, Budapest, Hungary

Piroska E. Rakóczy, Deputy Director (research), Lions Eye Institute, 2 Verdun Street, University of Western Australia, WA, 6009 Australia
E-mail: rakoczy@cyllene.uwa.edu.au

Tracie Ramsdale, 3 Hi-Tech Court, Brisbane Technology Park, Eight mile Plains Qld 4113, Brisbane, Australia
E-mail: TRamsdale@alchemia.com.au

Judit Reményi, Research Group of Peptide Chemistry, Hungarian Academy of Sciences, Eötvös L. University, Budapest, Hungary

Balázs Sarkadi, Department of Cell Metabolism, National Institute of Hematology and Immunology, 1113 Budapest, Daróczi út 24, Hungary
E-mail: B.Sarkadi@ohvi.hu

Gerd Schmitz, Institute for Clinical Chemistry, University Hospital, Franz-Josef-Strauss-Allee 11, 93053 Regensburg, Germany
E-mail: gerd.schmitz@klinik.uni-regensburg.de

Wei-Yong Shen, Department of Molecular Ophthalmology, Centre of Ophthalmology and Visual Science, University of Western Australia, Perth, Australia

Maree Smith, University of Queensland, School of Pharmacy, Qld 4072, Brisbane, Australia
E-mail: maree@pharmacy.uq.edu.au

Michael S. Starr, School of Pharmacy, University of London, 29-39 Brunswick Square, London WC1N 1AX, UK
E-mail: mike.starr@ntlworld.com

Gergely Szakács, Research Group of Peptide Chemistry, Hungarian Academy of Sciences, Budapest, Hungary

Béla Szende, Semmelweis University, Budapest, 1st Institute of Pathology and Experimental Cancer Research, 1085 Budapest, Üll i út 26
E-mail: BSzende@korbl.sote.hu

János Szolcsányi, Pécs University Sacully of Medicine, Department of Medical Chemistry, 7624 Pécs, Szigeti út 12, Hungary
E-mail: Szolcs@apacs.pote.hu

József Tímár, National Institute of Oncology, Department Tumor Progression, 1122 Budapest, Ráth György utca 7-9, Hungary
E-mail: Jtimar@Oncol.hu

Michael Torzewski, Institute of Clinical Chemistry and Laboratory Medicine, University of Regensburg, Germany

István Toth, University of Queensland, School of Pharmacy, Qld 4072, Brisbane, Australia
E-mail: i.toth@pharmacy.uq.edu.au

Edward J. Triggs, University of Queensland, School of Pharmacy, Qld 4072, Brisbane, Australia
E-mail: T.Triggs@pharmacy.uq.edu.au

Zsolt Tulassay, Semmelweis University, Budapest, II.Department of Medicine, 1088 Budapest, Szentkirályi utca 46
E-mail: Tulassay@Bel2.sote.hu

Axel Ullrich, Director, Max-Planck-Institut für Biochemie, Am Klopferspitz 18A, 82152 Martinsried, Germany
E-mail: ullrich@biochem.mpg.de

István Vályi-Nagy, Semmelweis University, Faculty of Health Sciences, Department of Internal Medicine and Geriatrics, Budapest, Hungary

Szilveszter E. Vizi, Institute of Experimental Medicine, Hungarien Academy of Sciences, 1083 Budapest, Szigony u. 43, Hungary,
E-mail: Esvizi@koki.hu

Allan Wong, University of Queensland, School of Pharmacy, Qld 4072, Brisbane, Australia
E-mail: allan@pharmacy.uq.edu.au

Ichiro Yahara, Department of Cell Biology, the Tokyo Metropolitan Institute of Medical Science and CREST, Japan Science and Technology Corporation, Tokyo, Japan

Eugene R. Zabarovsky, Department of Chromosome3 Sequencing and Function, Microbiology and Tumor Biology Center, Karolinska Institute, Stockholm, Sweden

Acknowledgments

The editors wish to thank all the authors for their dedicated hard work and also for their patience. The editors express their gratitude for Dr. Zsolt Szegedi for his very important technical assistance in preparation of the unified electronic version of the text, figures and tables and also his major contribution in the cover page design. The contribution and support of Tracy Breakell, Matthew Honan and Jemma Nissel is also highly appreciated.

Aims and scope

Knowledge of the basic pathomechanisms of human diseases is essential to every student and professional, engaged in drug research and development. In recent years a breakthrough has occurred in the field of pathobiochemistry and very broad perspectives have opened up for drug research. According to recent predictions, at the beginning of this century, Pharmaceutical Industry (including biomedical research, pathobiochemistry, molecular biology and combinatorial chemistry) and Informatics will be the most dramatically developing scientific and industrial areas. Due to the increasing amount of research having been carried out on the molecular pathomechanisms of diseases, a great amount of knowledge have accumulated about cancer, AIDS, atherosclerosis, neurodegenerative disorders (like Alzheimer, Parkinson etc.). This book aims to introduce the pathomechanisms of various diseases, emphasising the recent results and techniques, including the new pathologically relevant target molecules and rate limiting steps considering the therapeutic aspects. The book is aimed for students studying chemistry, biochemistry, pharmacy or medicine and researchers from the same field including pharmaceutical scientists and it will be a good read for everybody who are interested in the pathomechanism of diseases and in the new perspectives of therapies.

Part I

Introduction

Chapter 1

Introduction

A breakthrough in modern drug research

György Kéri and István Toth

In recent years dramatic developments in molecular biology, pathobiochemistry and bioinformatics have greatly increased our understanding of the molecular pathomechanisms of various diseases. With the help of functional gene analysis (genomics), functional protein analysis (proteomics), molecular diagnostics, and other molecular biological techniques it has become possible to understand in more detail the molecular mechanisms which contribute to complex pathological states, opening up novel therapeutic possibilities.

Molecular pathomechanisms of diseases are usually very complex; their complete understanding still demands a lot of effort. The first question is how do we define the term "pathomechanism"? The greek word "patho" means disease, so pathomechanism means disease-mechanism, while in molecular terms we must clarify the differences in the molecular mechanisms of the normal and the pathological state. This question will lead us to the more philosophical one: what is the normal state, when do we function properly? Without getting too philosophical it is worthwhile reminding ourselves of an old Chinese saying:

> *The unity of spirit, soul, body and mind*
> *Will make you happy and shine*

In other words, the normal state is when we function properly, think properly, feel properly and act properly, or in a more simple way when we have peace of mind because we have a meaningful life, we know what we want and we do it.

In contrast, in a pathological state due to physical, mental or emotional problems (or a mixture of all of them), we do not function properly, and in most cases these problems cannot be localized to a particular organ or area of the body, as the pathological state usually means a systemic problem. If certain parts of the organism do not fulfill their task or duty, this influences or inhibits the proper functioning of the whole organism, resulting in system mal-function. For example, cancer is the result of genomic changes, which usually develops from a single transformed cell (monoclonal origin). However whether such a transformed cell can become a fully fledged malignant tumor very much depends on the systemic response, on the intra- and intercellular (and may be even interpersonal) relationships and communication. This systemic view is also true for pathological states caused by external factors, for example for viral or bacterial diseases, where the pathogens take over and reprogram the communication channels of the host cell in order to serve the intruder optimally.

Recently, it has become evident that, in most cases, intra- or intercellular communication disorders form the background to complex pathomechanisms. Thus modern drug research has focused on signal transduction therapy, with the primary task of understanding the relevant molecular pathomechanism.

The aim of this book is to present a relatively broad picture about molecular pathomechanisms and new trends in drug research, with a special emphasis on signal transduction problems and potential therapeutic strategies.

Many molecular pathomechanisms are the result of an intracellular or intercellular communication disorder, while a series of genomic changes can be the cause and the consequence of these communication disorders. In a healthy organism, normal cells fulfill their duties, do not send or receive false messages and are strongly controlled by the external messages of the communication network. On the other hand, for example, cancer cells generate a false, mimicked proliferation signal for themselves via oncogenes and other genomic changes. Whether this communication failure is the result of environmental factors and/or external messages (generating changes at the genomic level), or originates in the genetic program is still a question, and can be answered only on a case-by-case basis. However we have to consider that cells, like human beings, live in a well organized society and in a given ecosystem, which, to a certain extent, determines their receptivity and responsiveness as well as the systemic response for the various carcinogenic agents and effects. In other words, carcinogenic compounds can be carcinogenic in a given *in vitro* system or in a given organism, but the same agent can have different effects in different systems, depending, of course, on the extent of the effect. Clearly changes at the genomic level are critical steps during carcinogenesis, however the manifestation of these genomic changes and the system response depends very much on the communication state and responsibility of the system.

The processes of cellular growth and differentiation as well as the maintenance of specialized functions show a remarkable degree of coordination and it has been clearly demonstrated that this involves intercellular communication, rather then relying entirely on intracellular programing. In the pathological state, the normally interdependent system controls are uncoupled and certain cellular functions or malfunctions are stimulated in such a way as to result in further damage-causing signals, or often in the growth of the malfunctioning cells. Proliferation of infected, damaged or malfunctioning cells is very often a key factor in the generation of the pathological state, not only in cancer and infectious diseases but also in inflammation or autoimmune-related diseases like arteriosclerosis, arthritis, or certain inflammation-related neurodegenerative diseases. Inflammation has been found to be a determinatory pathological cause of many chronic diseases, where proliferation of immune cells due to false signaling turned out to be a critical factor in the generation of the disease.

In the above mentioned cases, disease manifestation starts when these malfunctioning or transformed cells diversify, and variants with altered properties arise in the population. To survive and function in a competitive environment, such variants must have selective growth and communicative properties and other competitive advantages over other cells. The surviving malfunctioning cells must have specific signal transduction pathways turned on, with which all the feedback effects and inhibitory actions of the microenvironmental parameters cannot interfere. Since communication disorders represent a major cause of pathological states and most of the recently identified validated target molecules of drug research are signal transduction related macromolecules, most of the pathomechanisms and drug research areas described in this book relate to signal transduction.

This book describes several general strategies and tactics and related results for molecular pathomechanism-based target selection and validation. This is an area where the most dramatic progress has occurred in recent years. A series of genomic and proteomic approaches together with bioinformatic tools have been developed and used to identify novel molecular pathomechanism-related drug discovery targets. Functional genomics (not only in the human genome project but also in model organisms, pathogens and experimental animals) and proteomics in conjunction with an information-intensive "knowledge base" approach for comparative genomics and proteomics has generated a large amount of data for pathological target selection and validation. Dominant negative mutants, antisense, ribozyme and antibody methods and other modern molecular biological techniques have been used successfully for molecular target validation. For example cDNA microarrays and two-dimensional gel electrophoresis unmask the expression of genes with unassigned or unexpected functions, while depletion of mRNA with ribozymes or neutralization of proteins with intracellular antibodies or small molecular inhibitors enable the investigators to select and validate relevant target molecules for certain diseases.

Most of the potential novel molecular targets for signal transduction therapy can be grouped into the following categories: growth factor-, hormone-, cytokine receptor targets, tyrosine kinases and serine/threonine kinase signal transduction pathway targets; cell cycle targets; apoptosis-related targets; extracellular matrix targets, immune cell receptor targets, angiogenesis and metastasis targets; and cell life-span targets. On the other hand, because of the recent emphasis on the systemic nature of cellular and pathological states, genome-wide expression monitoring and analysis of genomic networks has become a novel tool of functional genomics. This method is aimed at identifying groups of coregulated genes and discovering genes expressed differentially in distinct situations. Such a global (genome-wide) view of "gene function" in the regulation of the dynamic relationship between proliferation, differentiation and apoptosis can provide new insights into cellular homeostasis.

Computational biology and bioinformatics is becoming increasingly important in pathomechanism-based drug target selection and validation. An integrated "knowledge base" approach is required to understand the interrelationship between genetic information, transcription and translation, the existing phenotype and cellular function and how this is modified by environmental factors. Cell biology, molecular medicine and genomic technologies are inseparable for defining and validating new molecular targets.

Recently developed pharmacogenomic approaches provide additional research tools for modern drug research elucidating the genetic polymorphisms in drug-metabolizing enzymes, transporters, receptors, and other drug targets. These have been linked to interindividual differences in the efficacy and toxicity of many medications, thus providing novel target molecules and putting more emphasis on novel drug delivery and formulation approaches. Drug delivery has also become a key issue in modern drug research, not only to improve bioavailability of the drug molecules by utilizing passive and active transport systems, but to target the drug to its active site. In addition, rational drug design, combinatorial chemistry and high-throughput screening against purified molecular and cellular targets have emerged as powerful techniques for target oriented drug discovery.

Without an exhaustive description of all known molecular pathomechanisms, a series of representative diseases have been discussed in this book. The examples have been selected on the basis of pathological and epidemiological significance and on the basis of the present trends in modern drug research. The diseases discussed provide examples of signal transduction therapy and the systemic nature of the molecular pathomechanisms of the diseases, indicating that rate limiting steps or key signaling elements can become validated targets of modern drug research, reminding us of the saying:

The sea is in the drop and the drop is in the sea.

Part 2

Pathomechanisms and molecular target finding

Chapter 2

Drug discovery based on functional genomics

Target selection and validation

Gábor Mikala and István Vályi-Nagy

Contents

Discovery and development of therapeutically useful drugs requires the concentration of resources. Traditionally, models of human disease are utilized for drug screening purposes. Most frequently, these are animal or cell-based screening models representing a well-defined disease phenotype. Drug candidates are tested in these models in order to identify compounds that alter the disease phenotype. The majority of the therapeutically available drugs used in the Western world were developed along these "traditional" avenues. The main advantage of this screening method is that it uses a full biological system, and compounds passing this test are highly likely to be useful as therapeutic agents. Parallel to this, serious toxicity can also be ruled out. The major disadvantage of this approach rests in its inability to pinpoint target molecules, therefore, one cannot predict and dissect mechanisms of action and side effects. As a consequence, there are still a number of widely used drugs with obscure molecular mechanism of action.

Another "traditional" approach to drug selection has been aided by developments in protein purification and in molecular biology. Here, a molecular target of drug action is identified first, a molecule that is predicted to have a rate-limiting role in the disease development or phenotype. Once a target molecule is defined, potent and effective small molecules may be identified that have a well-defined mechanism of action and *in vitro* activity. Unfortunately, toxicity, metabolism and poor bioavailability are problems frequently encountered in the process of development of candidates with this approach. Moreover, since this procedure starts with a target molecule that plays a pivotal role in the disease process, correct identification and proper evaluation of the target "receptor" is essential for eventual success. Clearly, both "canonical" methods of drug screening have led to the development of many life-saving and life-improving medicines, however, they are also inherently responsible for the very high and costly attrition rate in drug development. Should it be possible to screen a large number of candidate "receptors" and drug molecules against one another, cost containment for drug development may be found.

The advent of the genomics era and development of numerous new screening procedures have the potential to provide the raw material for a revolution in pharmaceutical research. The human (and human pathogenic organism) genome projects have given rise to a tremendous number of target genes for screening, and combinatorial chemistry has

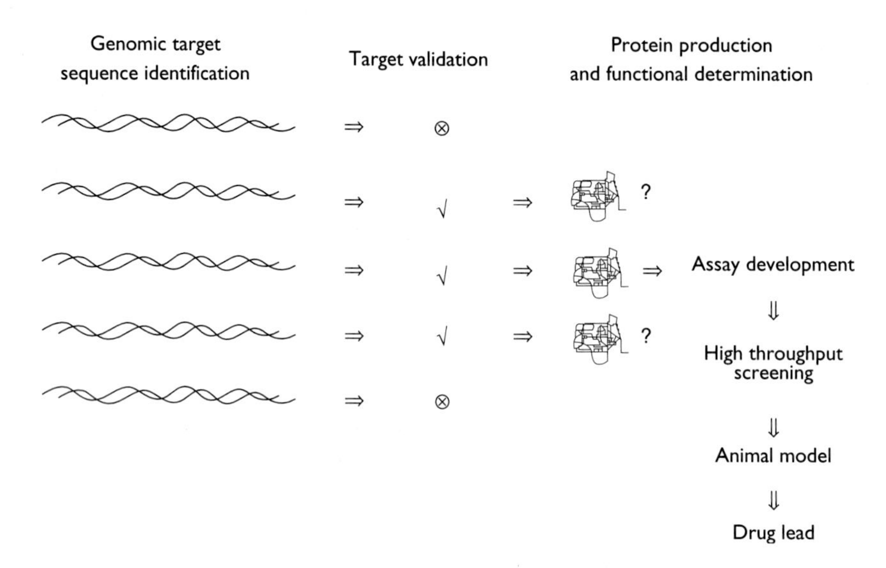

Figure 2.1 Classical drug discovery process based on functional genomics. In the genomic era much of the drug discovery process is still done in the traditional manner, target selection and validation (and frequently detailed functional determination) are placed upstream of assay development and drug screening. The major disadvantage of this time-tested approach is that it cannot be scaled up effectively to handle the large numbers of candidate drug target genes provided by genome projects.

the potential to provide countless compounds to test against them. Unfortunately, the complexity of both approaches is such that homing in on the right drug-receptor pair remains quite a challenge. In this chapter the authors attempt to introduce the reader to some of the revolutionary techniques that help to select and validate the molecular targets for drug development in the genomics era (Figure 2.1).

2.1 Genomics-based drug target selection

The first step in genome-wide DNA sequence analysis has been the identification and tagging of each gene (expressed sequence tags – ESTs) in the human genome. At the same time, the genomes of several important microorganisms that are pathogenic to humans were sequenced. Both approaches provide a plethora of genetic information that is expected to deliver approximately 10,000 new therapeutic targets. Unfortunately, it has been difficult to pinpoint the most important drug-receptor candidates since more than three quarters of the proteins predicted to exist based exclusively on their corresponding cDNA sequence have no known function. The scientific fields of functional genomics and proteomics have developed out of necessity to handle this mass of data. Proteomics (in analogy to genomics that aims to provide sequence information on nucleic acids of a species, the genome) seeks to provide functional information for all proteins, the proteome. EST database mining, expression analysis and proteomics technologies are the most important new avenues in drug target discovery of the genomics era (Jones, DA *et al.* 1999, Johnston *et al.* 1999).

2.1.1 EST database mining based on homology searches

There are huge databases that contain EST information on the human genome. There are two main methods of utilizing these databases in the discovery process. First, it is possible to identify new members of previously known target protein families. This goal may be reached by simply searching in the databases for new sequences that possess structural features known to be characteristic for a certain class of proteins. In the simplest case this structural feature is a certain sequence to be found in the cDNA (by reverse translation). This approach has been particularly useful in identifying new ion channel subunits or proteins that regulate apoptosis or inflammation. New findings include isoforms of the much sought-after T-type calcium channel or novel cytokines related to tumor necrosis factor (TNF). Searching of databases using a universally available Basic Local Alignment Search Tool (BLAST) (Altschul *et al.* 1990) and the amino acid sequence of another calcium channel or TNF, respectively, helped to identify these novel molecules. Follow-up cloning and expression studies have confirmed that the identified targets are really the T-type calcium channel (and several distinct isoforms) or TNF-like proteins. At this point the "conventional" molecular-based drug discovery processes may take over and utilize the new protein as a drug screening target as in the case of the new ion channel. Alternatively, as in the case of the new TNF-analogues, though new pharmacological intervention points to inflammatory pathways may exist, more pre-drug-development research is needed to identify the correct targets among them for further drug discovery.

2.1.2 Differential tissue expression revealed by EST databases

The other route available for use of EST databases in the selection of possible targets is through the use of differential gene expression as revealed by tissue specific EST databases. EST databases currently encompass several different human tissues. Since the ideal drug target is expressed only in a single tissue or cell type – i.e. the diseased one to be modified by the drug under development – selection of a target molecule specifically expressed in that particular tissue is especially promising for further drug development. Theoretically, these tissue-restricted genes provide drug targets with great specificity and a reduced number of potential side effects. An example for this type of target selection approach is the identification of cathepsin K (an osteoclast-specific protease) as a promising drug target (Drake *et al.* 1996). Cathepsin K was discovered in a human osteoclast EST library through homology matches to known cathepsin, and further experimentation proved the molecule to be exclusively expressed in osteoclasts. Since pronounced bone matrix protein degradation was thought to be characteristic to the bone disease osteoporosis, inhibitors for cathepsin K were produced and tested in *in vitro* and *in vivo* models of bone resorption.

2.1.3 Expression analysis of diseased tissues for target identification

Database minings through homology searches and differential tissue expression are powerful tools for target selection, however, they have one crucial drawback. They could only be used effectively for selection of targets that have certain features that place them into a category of homologous proteins with known function. Unfortunately, most of the possible targets have no known structural relative. On the other hand, in certain settings it may not be crucial to have a known homologue of a target gene product.

Expression microarrays – or DNA-chips – are novel molecular tools that allow comprehensive analysis of RNA expression in different cells or tissues (e.g. healthy versus diseased) (Schena *et al.* 1998, Harrington *et al.* 2000). Though the methodology is detailed elsewhere in this book, it is useful to outline the essence of this new tool here. Expression microarray analysis involves quantitative hybridization of a large set of DNA probes (currently in the order of thousands of genes) with the total cDNA derived from the tissue to be analyzed. The tool is called the DNA chip as the probe sequences are fixed to glass slides to form microscopic arrays consisting of a single kind of probe per each coordinate on the slide. The characteristics of the hybridization are set in such a manner as to reveal the absolute abundance of every message level to be evaluated with the chip. DNA microarray technology can measure even poorly expressed genes, therefore comprehensive expression analysis can be achieved. If one checks a large number of genes with unknown function and compare two tissues (e.g. healthy or diseased, control or drug treated, etc.), it is possible to identify expression differences useful for a lead in drug target selection. Alternatively, the method can be used to group structurally unrelated drugs to classes of common mode of action, or link compound specificity and toxicity to the intended molecular target versus other identified targets.

An example in which microarrays were used to establish transcriptional profiles in yeast (a useful eukaryotic model cell system) treated with inhibitors of Cdc28p (a cyclin-dependent kinase) has been reported recently. In this study, genome-wide transcriptional perturbations were analyzed in the presence of two structurally distinct inhibitors of Cdc28p and a third structurally related inactive compound. It was established that both active compounds induced similar changes in the expression pattern while the structurally related inactive compound yielded little effect (Grey *et al.* 1998).

Though expression microarrays represent a formidable new tool towards characterizing gene expression in different tissues, the method has in inherent drawback. Expression microarrays stand as "closed system" gene expression profiling platforms that means they can only quantify the expression of a predetermined set of genes. Expression profiling by them may never be complete, therefore research must not rely solely on microarrays otherwise the drug development process may miss important targets. Fortunately, an alternative exists, a somewhat complementary approach to Serially Analyze Gene Expression (SAGE). SAGE is a sequence-based approach that aims to comprehensively identify which genes are expressed and to simultaneously quantify their level of expression (Madden *et al.* 2000). The catalog of gene expression thus obtained for a given cell type or tissue is called a transcriptome. SAGE is able to generate transcriptomes though exact identification and precise quantification of mRNA molecules and still do it in an unbiased and fairly efficient manner. The theoretical basis of the procedure is described in Figure 2.2. The end product of a SAGE analysis is a library/database of 14 bp tags, and this library/database of information that is characteristic for the given cell type or tissue analyzed may be dissected by analytical software. There is a public repository of human SAGE data (SAGEmap, http://www.ncbi.nlm.nih.gov/SAGE), this online resource provides tools for viewing and analyzing data from a compilation of >40 SAGE libraries (mostly data collected on human cancer cell lines).

Since interpretation of SAGE data is very complex, the procedure has been best suited for the discovery of transcriptional changes due to a single or a few gene changes, in other words for novel pathway elucidation. An example of global transcription profiling for pathway elucidation is the use of SAGE to evaluate the effects of the tumor suppressor gene APC (Adenomatous Polyposis Coli), a gene that is responsible for monitoring growth of the epithelial cells in the large

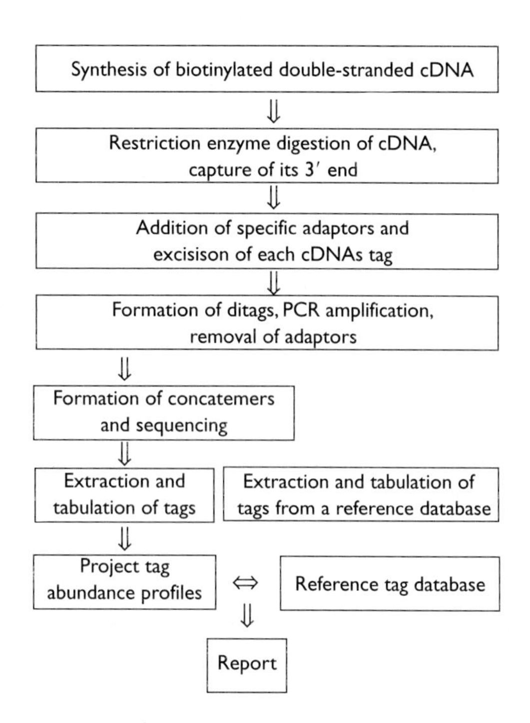

Figure 2.2 Flowchart of serial analysis of gene expression. The process may be divided to a pre-sequencing experimental part and sequence analysis.

intestine. Previously, it was known that APC regulates cell growth by altering the activity of the β-catenin/Tcf-4 transcription factor complex. Transcription profiling by SAGE drew attention to the c-Myc oncogene that, as it turned out, is regulated by β-catenin/Tcf-4 (Madden *et al.* 2000).

2.1.4 Proteomic methods in target identification

Microarray analyses and SAGE are powerful methods to analyze the expression of target genes, however, it is well known that mRNA expression is frequently unrelated to protein expression and may be related even less to protein activity. Protein fingerprinting of the tissues analyzed is a means of addressing certain aspects of this shortcoming. Proteomics aims to comprehensively describe cellular proteins (the proteome) and their change in disease (Edwards *et al.* 2000). One main branch of current proteome analysis aims to achieve this goal through fractionation of cellular components by two-dimensional gel electrophoresis followed by extraction and mass spectrometry identification of individual proteins. Alternatively, reverse-phase liquid chromatography coupled to two-dimensional mass spectrometric analysis may be used for this purpose. Though, in theory, mass spectrometry has the potential to analyze and identify every protein of the cell, in practice it has been difficult to identify proteins in low abundance. The approach may be best used to compare diseased and normal tissues or two differentially treated cell lines, etc. Comparing protein profiles in drug-responsive versus unresponsive cell lines will help to identify drug targets and mechanisms as well as locating new molecular target pathways. Nevertheless, one must never forget that powerful methodologies they may be, but transcriptional and proteomic technologies have a common shortcoming in that they are unable to discriminate between the true molecular cause of the disease and the consequences of such a change.

An example of such a proteome-based methodology used to assess the effects of certain antineoplastic drugs has been reported. In this study, the authors applied proteomic approaches utilizing 60 different cell lines and several experimental anticancer compounds, and revealed correlations between the pharmacological efficacy and specific protein profiles (Myers *et al.* 1997).

2.1.5 Critical role of bioinformatics in target identification

Identification of the proper targets for drug discovery, as noted in the previous paragraphs, is an initial key step in drug development based on functional genomics. The identification of candidate drug-target proteins (i.e. that have potential binding sites for small-molecule drugs that alter their function and possibly their structure) in the sequence or expression databases has been facilitated substantially by bioinformatics solutions of three-dimensional (3D) protein structures. This point is critical since the 3D structure and function of a protein may not be predicted simply on the basis of its amino acid sequence (the information provided by structural genomics). This limitation has been a major drawback for the full exploitation of data emerging from human and pathogen/model organism genome programs.

One way to tackle this problem has already been mentioned: most proteins may be catalogued to be members of a limited number of protein families sharing essentially similar structures and functions. Therefore, if the sequence of an uncharacterized protein can be shown to be related to a known protein, it is likely that the two proteins belong to the same family. As a result, they may have similar structural, functional, and possibly drug-receptor properties. This way, our limited knowledge of protein structures may be used to advance the recognition of proper drug targets, provided that a match between sequences can be found.

This approach already has far-reaching consequences with regards to a few drugs available today. During the development of anti-HIV drugs, it was recognized that HIV proteinase is in part homologous to renin and pepsin (aspartic proteinases of well known structure and a great number of available inhibitors). New protease inhibitors based on this structural knowledge could be developed reasonably quickly and these inhibitors are now crucial elements in the battle against HIV infection. Yet another possible target of anti-HIV drugs is integrase; identification and lead drug selection of this viral enzyme was initiated based on its partial 3D homology to the well characterized avian sarcoma virus integrase (Chen *et al.* 2000).

Unfortunately, searching genomic databases for structural homology between proteins with relatively simple search procedures may have only limited value. For identification of many drug targets hidden in databases much more sophisticated programs are needed. The main reason for this is that while there are a great number of proteins related in their 3D structure, there is little to no similarity in their amino acid sequences. Conventional search techniques that simply compare sequences are unable to detect many potential structural similarities, and therefore may be insufficient to pinpoint important drug targets. Database mining programs based on the "threading technique", that combine knowledge of sequence with already known 3D structures, will enable matches between even the most distantly related proteins to be identified (Jones *et al.* 1992). On the other hand, these powerful programs may locate some matches that may prove later to be false, a reason for much concern in teams dealing with drug target identification.

2.2 Genomic-based drug target validation

The drug target identification approaches provided by functional genomics produce an unreasonable number of target molecules for drug screening processes. In order to advance further in the drug discovery process, an additional step providing technical feasibility and biological significance of the protein target under study is often required, known as target validation. This additional functional information about the candidate drug-target protein may be gained by demonstrating the effects of the elimination of or significant reduction of its level on physiological properties thought to be relevant to the disease process. There are currently four avenues generally used to gain additional information about the potential drug-target protein (Table 2.1). In specialized settings, further alternative validation techniques may also exist.

2.2.1 Gene disruption techniques

Targeted gene disruption or ablation techniques use molecular biological methodology to alter the genome of experimental species and study the effect of the "knockout". From the viewpoint of target validation one needs to determine whether the drug target to be validated is in the human genome or in the genome of a pathogenic microorganism. In the latter case gene disruption and analysis of the null-mutant is an extremely important method of data collection that is presently unrivalled. The goal of a null-mutant may be achieved in a few months, a time frame that is acceptable for pharmaceutical development. This approach also opens up the potential to investigate the importance of the gene for survival or pathogenicity of the microorganism. Such studies have resulted in the estimation that, on average, one third of the genes present in a microbial genome are essential for viability and even more for pathogenicity (Johnston 1999). Unluckily, mutant strains of pathogens may not supply information concerning the susceptibility of a particular gene product to functional antagonism by a potential drug molecule.

The applicability for drug target validation of mammalian knockouts (mostly mice) is disappointing. Gene ablation is difficult to achieve in mammalian cells and to generate knockout

Table 2.1 Features of generally applicable target validation approaches

Approach	*Information required*	*Information source*	*Selectivity*	*Human/pathogen organism*	*Speed/efficiency*	*Limitations*
Gene disruption	Genomic sequence	Literature	High	Model mammalian organism + Pathogen organism +++	Slow Good	Technical proficiency, rodent model, compensatory mechanisms
Antisense	Partial cDNA	cDNA database	Variable	Human cell line +++ Pathogen organism +++ Mammalian organism +/−	Fast	Stability of oligos, half-life of target protein
Ribozymes	Partial cDNA	cDNA database	Good	Cell lines +++	Fast	Metabolic stability of ribozymes
Antibodies and intrabodies	Protein/peptide sequence	Amino acid sequencing or nucleotide sequence translation	Good	Cell lines +++	Moderate	Antibody engineering and expression

animals from targeted cell lines requires a minimum of 12–18 months and formidable expertise. Such a time frame and expense to create and evaluate a knockout animal makes this model poorly suited for drug target validation. Moreover, since the "end-product" animal is nearly invariably the mouse, a rodent with several fundamental physiological differences from the human, it may not always be an appropriate model for human physiology and pathophysiology. Furthermore, knockout animals frequently have phenotypes that are ill-suited for pharmacological analysis. Examples are the embryonically lethal or developmentally fully compensated (null) phenotypic configurations. Taken together, knockout mice are better suited for academic purposes, for the evaluation of basic biological phenomena. Drug target validation in mammalian subjects requires methods that allow modulation of selected genes from the prefered human background which can be done in a timely manner.

2.2.2 *Antisense methods*

Antisense technologies rely on the presence of an RNA or short DNA (or, more recently, peptide-backboned nucleic acid, PNA) molecule that is complementary to an mRNA species. Upon hybridization of the two species, the mRNA cannot direct the production of the encoded protein properly (Taylor *et al.* 1999). There are several key points in the lifecycle of an mRNA molecule that are altered by the presence of an antisense analogue, Figure 2.3 gives an overview of these effects. Nonetheless, successful antisense inhibition of target protein production in cell lines depends on numerous variables which are at times difficult to predict in advance. These include mRNA and protein turnover rate, secondary structure of the target mRNA, access of the antisense molecule to its target, biological half-life of the antisense analogue etc. Moreover, the maximum level of inhibition achieved by antisense technologies rarely reaches 90% on the protein level, thus one always needs to calculate with the potential effect of leftover functional proteins remaining in the cell.

The main advantages of antisense inhibition technology are its ease of use, and its ability to produce data in a timely manner fairly inexpensively. This method has now been in use for more than a decade. In certain settings, the technology may be used not only in cell culture systems but scaled up for use in whole animals. On the other hand, one must always bear in mind that some mRNA molecules remain poor targets for antisense inhibition.

2.2.3 *Ribozymes*

Ribozyme technology may be considered to be an outgrowth of antisense technologies that takes advantage of novel discoveries in catalysis by RNA molecules. Ribozymes are catalytically proficient RNA enzyme molecules that are able to specifically recognize target mRNA species and cleave

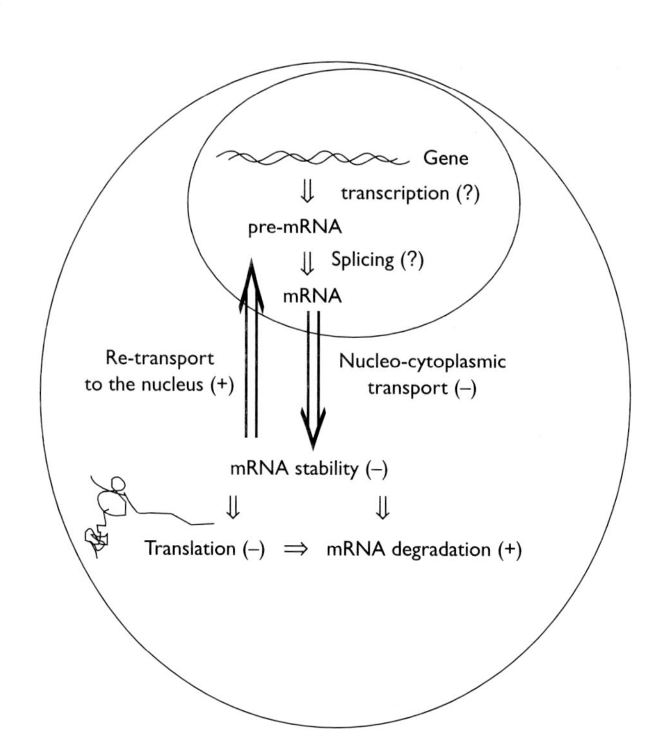

Figure 2.3 Inhibition mechanisms by antisense technologies. Antisense RNA and oligonucleotides (and peptide-nucleic acids, PNAs) are known to specifically influence several key steps in the lifecycle of their target mRNA. The processes influenced include transcription (mostly by PNAs), splicing (questionable), nucleo-cytoplasmic transport, retransport of double-stranded RNA to the nucleus (mostly by long antisense RNA), mRNA stability (dramatically decreased by ribozymes) and, probably most significantly, translation (inhibited by all classes of antisense structures).

them. They share the advantage with conventional antisense technologies that only a limited sequence of the mRNA molecule is sufficient to design them, however, depletion of the target mRNA by ribozymes generally offers a more reliable tool for inhibiting production of a target protein. Effective depletion of the target mRNA species depends on the metabolic stability of the cell as well as the cellular delivery and catalytic efficiency of the ribozyme. Currently, these parameters are still incompletely characterized, and thus subject to empirical improvement. Elimination or depletion of the target mRNA species is the key step to inhibition of target protein production, however, the level and rate of target protein depletion remains a function of protein half-life.

The most commonly used ribozymes are termed hammerhead ribozymes. Their name is based on the structure of the original viroid hammerhead-structured ribozymes. They contain a catalytic core region with highly conserved sequence of bases and this "enzyme-type" region is flanked by two hybridization sequences that are responsible for correct recognition of the target mRNA. We now know of more than a hundred different effective ribozymes. Rules of effective ribozyme generation may be extrapolated from existing data (Zhao *et al.* 1998).

Ribozymes offer experimental advantage over the more simple antisense approach since their inhibition of target protein levels commonly exceeds 90% (the average level of inhibition by antisense oligos may be estimated to be in the 50% range). Controlled expression of ribozyme structures is a highly useful tool for delineating the role of a certain protein in model systems and thus provide an important tool for drug validation.

2.2.4 *Antibodies and intrabodies*

Gene ablation technology is generally very time-consuming and labor-intensive and antisense and ribozyme methods have quite unpredictable effects on target protein levels especially those with long half-lives. Therefore, the need emerged to introduce techniques that directly inhibit or neutralize the protein target. The fact that highly specific antibodies produced by the immune system are frequently able to neutralize or at least modulate the function of proteins, and the availability of molecular biological techniques to produce such antibodies, have led to the development of gene constructs that produce intracellular antibodies (intrabodies). These intracellularly expressed single-chain antibodies (scFvs) are able to immunologically neutralize macromolecules within the living cell itself. Structurally, these intrabodies are scFvs with a variable domain of an immunoglobulin heavy chain linked to a variable domain of the light chain. They can be produced using various recombinant methods. It is also possible to attach subcellular localization signals to the intrabodies and therefore specifically inhibit target proteins in a particular cellular compartment. Intrabodies have a further use in dissecting molecular cascades. Since their target epitopes are usually short peptides within the target protein it is possible to use the method to target certain regions or splice variants of a protein in a biological process. While intrabodies represent an extremely valuable tool for the study of intracellular proteins, one should always remember that conventional antibodies applied to the surface of cells, or microinjected into cells, may also provide a viable experimental alternative to their use, especially if the target epitope resides facing the exterior of the cell.

An elegant example of the use of the intrabody expression method to clarify the role of a protein was demonstrated by Wright and colleagues. An adenovirus encoding anti-erbB-2 scFv was used to study the role of erbB-2 in breast cancer. erbB-2 is a protooncogene encoding a protein that, upon activation by ligand binding, causes activation of the Ras signaling pathway and cell growth. The anti-erbB-2 scFv was able to induce apoptosis and antiproliferative effects in certain cell lines but not in others (Wright *et al.* 1997). This is due to the differential utilization of the erbB-2 pathway of growth regulation by the different cell lines. Moreover, this technology has helped to identify that the COOH-terminal cytoplasmic domain of the erbB-2 protein is required for apoptosis.

2.2.5 *Expression of dominant negative alleles*

The target validation methods detailed in the previous section could theoretically be used for any given target and do not require detailed knowledge of the function of the target protein product. In certain settings, especially in the case of targets identified as members of a previously known protein family, a particularly powerful method for the study of the effects of the lack of target protein function can be used. Dominant negative alleles are genes that, upon expression, inhibit the function of their normal counterparts. Most frequently, these alleles exist in multimeric proteins. The presence of a dominant negatively-acting protein may inhibit the correct multimerization of its counterpart and effectively inhibits its function provided that multimerization is essential for normal function.

An example of the use of dominant negative expression technologies to characterize a protein of unknown function may be found in the characterization of novel multimeric ion channels. Voltage-sensitive potassium channels function as tetramers and only a very short sequence within the protein is required for tetramerization which also determines the correct specificity of molecular aggregation. Expression of excess tetramerization signal polypeptides in cells results in the effective and selective removal of target functional channel tetramers. Similarly, the catalytic component of telomerase, hTERT, has been found to be a valid target of small molecule inhibitors using both antisense and dominant-negative validation techniques (Kelland 2000).

2.2.6 *Validating protein targets with peptides*

Combinatorial peptide libraries, such as those expressed in phage display libraries, have served as a rich source of ligands for a large number of targets (Kay *et al.* 1998). In bacteriophage display systems, peptides are fused to one of the capside proteins and displayed in a way as to be accessible for molecular interactions. These libraries may be screened by affinity selection if a sufficient amount of the protein target is available. The selected phage that expresses the peptide with affinity to the target protein may be amplified and the sequence of the displayed peptide determined. In certain settings, these peptides may be utilized as target validation tools in the drug discovery process. For instance, if a peptide introduced to a cell is found to inhibit a molecular interaction critical for the role of the target protein, biologically significant effects may emerge. Most frequently, relevant disrupted molecular interactions turn out later to be interprotein interactions or enzyme-substrate interactions. Target validation with peptides is faster than "knockout" experiments though more time-consuming than antisense-based methodologies, however, they offer significant potential advantages over other technologies. One advantage is that a peptide with affinity to its target will probably interfere with only one or a few functions of a protein target as opposed to depletion techniques that aim to eliminate the whole protein with all its functions from the cell. The other potential advantage is that existence of the peptide ligand offers a promising route towards finding a small molecular drug which could act on the target protein.

Delivery of the peptide used for target validation to its protein target can be achieved simply by cellular microinjection. To bypass the tediousness of this procedure, several new technologies have recently emerged which enable peptides to cross cell membranes. A notable example is linkage of a peptide to the Tat protein of human immunodeficiency virus type 1, a protein that can translocate across the plasma membrane and reach the nucleus, and which has the ability to retain this function in hybrid structures (Vives *et al.* 1997). Another approach is the use of lipophilic membrane-coated vesicles that can fuse with the plasma membrane of target cells and deliver their content to the cytoplasm.

2.2.7 Alternatives to the conventional genomics-based drug discovery process

As mentioned previously, the genome projects provided such a plethora of possible molecular targets for drug research that the existing drug discovery process was overwhelmed. Target validation and approximate functional determination of a candidate protein is costly and time consuming. Therefore, the need to couple lead drug discovery with target identification and validation procedures in order to focus the drug discovery process on the best targets is emerging.

2.2.8 The ligand-directed drug discovery process

The ligand-directed drug discovery process aims to provide early discovery of biologically active ligands that act on the identified target (Lenz *et al.* 2000). As depicted in Figure 2.4A, the process is initiated by identification of a target protein. As opposed to the canonical drug discovery process, target validation is postponed. Rather, recombinant expression of the target protein and high-throughput affinity selection of libraries of small molecules are the next steps in development. In the selection process molecules are identified that bind with high affinity and specificity to each (at this point still unvalidated) target protein. Generally, peptide display libraries are used for screening and therefore peptide ligands of the target proteins are identified. Target validation uses low-throughput biological assays and chiefly uses the identified (peptide) ligand as the main validation tool. As a result, a validated drug target is produced as well as a lead drug molecule known to modulate the target effectively. Thus, proof is provided for the amenability of the protein target for effective drug modulation.

This latter feature is a potentially crucial advantage from the viewpoint of drug development. Moreover, the screening process frequently turns up multiple ligands for a target molecule, sometimes with different biological activities, an attractive feature which may aid understanding of a potentially multifunctional target protein. Yet another advantage of using biologically active compounds as filters for narrowing down the potential drug targets in the discovery process is that the system is suitable for scaling up to handle the large number of targets provided by genomic research. The main disadvantage of this discovery pathway is the use of peptides as ligands. Peptides are not ideal drug types as they are biologically unstable and usually membrane impermeable, therefore, they are viewed by the industry as hits to be reengineered to develop useful drug leads. Unfortunately, there is no simple and reliable way to do so. The most useful approach is to use the biologically active peptides in ligand-displacement high-throughput assays to discover small molecule ligands that inhibit their binding to their target.

2.2.9 High-throughput automated ligand identification

The discovery of ligands for a target protein is considered paramount in the ligand-directed drug-discovery pathway. The drug industry prefers small molecular ligands as drug leads and fortunately sensitive, high-throughput approaches have recently been incorporated into the discovery process.

An example for this approach is depicted in Figure 2.4B. In this robotized approach, a library of hundreds to thousands of compounds is incubated with a target protein in free solution. Ligands, if they exist in the mixture, can bind to their receptor. Size exclusion chromatography is then applied to this mixture to separate the protein and its bound ligands from the unbound library members. Dissociation of the ligand and its concentration is performed by reverse phase HPLC. The concentrated free ligand is then fed to a mass spectrometer for structural identification (Lenz *et al.* 2000).

2.2.10 Chemical ligands for RNA targets

A much neglected yet intriguing area of drug discovery is among drugs that bind to RNA and not protein molecules. The most important initial targets are bacterial RNA molecules. The secondary structure of bacterial RNA is highly conserved and there are sufficient differences from mammalian RNA species (Ecker *et al.* 1999). Moreover, two important drug classes of antibiotics, macrolides and aminoglycosides, are known to act upon binding to bacterial RNA.

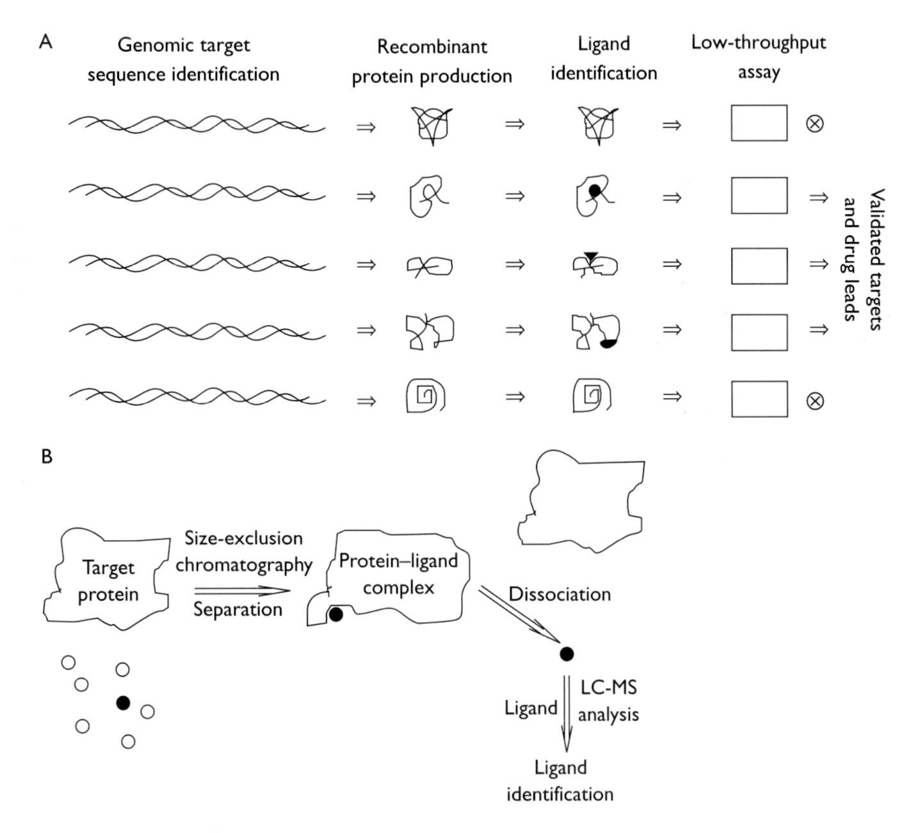

Figure 2.4 (A) The ligand-directed drug discovery pathway. In this process, recombinant protein production for identified drug targets and *in vitro* identification of a therapeutically active ligand (most commonly a peptide selected from a library) precedes detailed functional analysis and formal drug target validation. The high affinity ligand identified during the process is then used to validate the drug target in biological assays. The pathway thus simulaneously provides a drug target with a lead drug structure for downstream development of a small molecule drug. (B) High throughput automated ligand identification. A multi-thousand library mixture produced by combinatorial chemistry is screened by size exclusion chromatography (removal of unbound molecules from the target with a bound drug), followed by reverse-phase HPLC and high resolution mass spectrometry (for ligand identification). The libraries of molecules are designed in such a manner that the mass of each potential ligand would identify its composition.

Nevertheless, the possibility also exists for finding RNA molecules with sufficient secondary structure to serve as targets in mammalian cells.

The chemical ligand-based discovery process and ligand identification are both applicable to this field. In this process, the target RNA is incubated with a chemical library for ligand selection by affinity. Subsequently, the mixture (the associated drug molecules together with the target) is analyzed by high-resolution mass spectrometry. Since the mass of the RNA molecule and that of each ligand is known, the combined molecular weight identifies the bound ligand.

2.3 High-throughput drug screening

As emphasized earlier in this chapter the drug discovery process based on functional genomics has been made possible by developments in modern molecular biology, protein biochemistry, and the genome projects. One must keep in mind that many of the technologies used for drug target validation and lead drug selection involve high-throughput screening methods. This is especially evident in the chemical-based drug discovery pathway. Therefore, it is necessary to provide a quick insight into some of the high-throughput drug screening methods used today.

High-throughput screening was originally defined as a procedure that allows the screening of 100 to 1000 compounds per day. As automatized technologies develop, some current technologies allow the collection of 100,000 data points per day. Nevertheless, high-throughput screening is more than just numbers, it represents a special way of handling the screening work. Assays should be designed to have the smallest number of steps and smallest volumes of material involved to maximize throughput and minimize reagent need, yet they should still produce highly reliable data. Some examples of novel high-throughput screening methods to identify molecular interactions (Knight 2000) are shown in Figure 2.5 and are described in the subsequent sections.

2.3.1 *Scintillation proximity assay*

This technology is an outgrowth of the conventional radioligand binding assays, designed to eliminate the filtration step involved in that procedure. The target protein is bound to special beads that also contain scintillant chemicals (molecules that convert the energy of β-radiation to photons). A tritiated (^{3}H) lead drug is usually used to analyze displacement. The need for filtration is removed by taking advantage of the very short path-length of the β-particle emitted by ^{3}H in water; only those drug molecules that are bound to the target protein can activate the scintillant, while those in solution are too distant to do so effectively (Bosworth *et al.* 1989). The method is highly reliable, but can be used only if a high affinity ligand for the site has already been identified (such as a peptide in the drug-based discovery pathways).

2.3.2 *Fluorescence resonance energy transfer*

Fluorescence resonance energy transfer is a method that relies on the phenomenon that two different fluorescent labels placed in close proximity (<10 nm) influence the behavior of each other. Most commonly, one fluorophore is excited by a photon of certain energy and this excitation is transferred by resonance to the other fluorophore that emits a photon of a characteristic wavelength. The two fluorescent labels are sometimes located at the opposite ends of a single

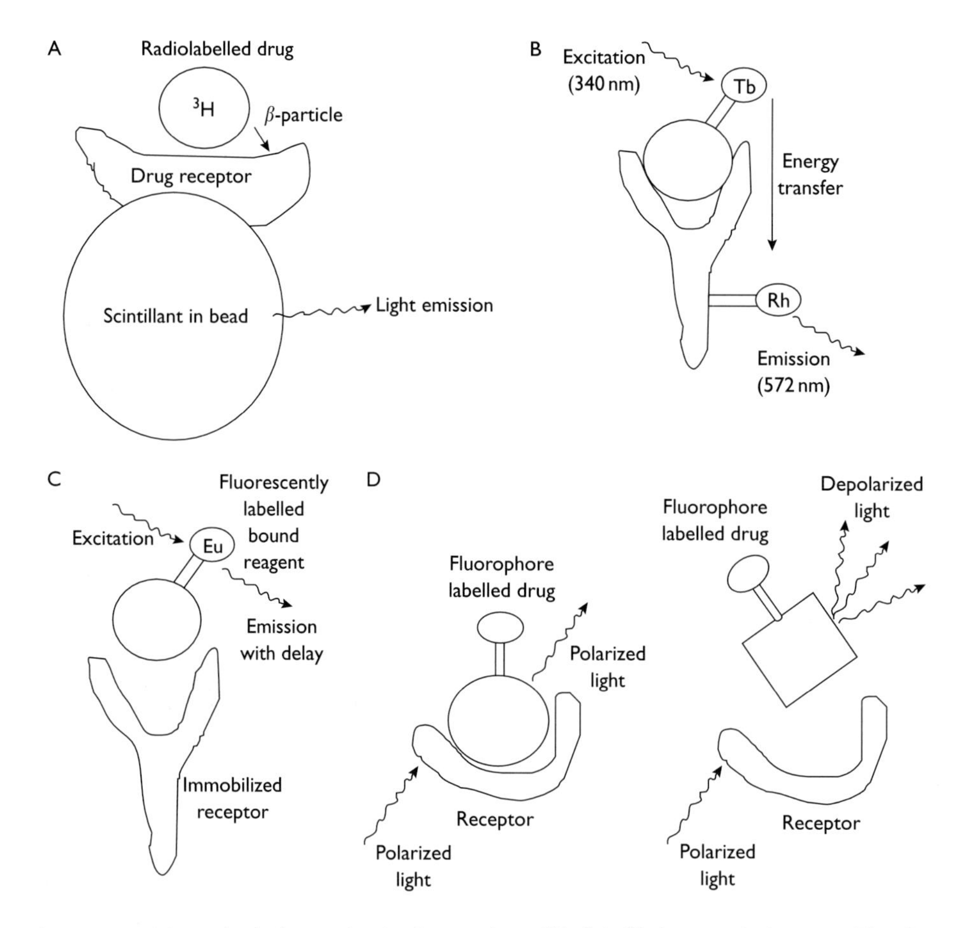

Figure 2.5 High-throughput screening methods for molecular interactions. (A) Scintillation proximity assay. The drug receptor to screen against is attached to a resin bead that contains scintillant molecules. Binding of a radioactively labeled ligand to the receptor enables activation of the scintillant by released β-radiation, molecules in solution are unable to do so effectively. (B) Fluorescence resonance energy transfer. Two different fluorescent labels, if placed in a close proximity, can transfer excitation energy to one another by resonance. Emission of a photon of characteristic energy by the second labels is indicative of binding. (C) Time-resolved fluorescence. A ligand tagged with a fluorescent lanthanide may be used in binding reactions and the delay in fluorescent emission by these tags provides an assay with very low background and high sensitivity. (D) Fluorescence polarization. Binding of a fluorescently tagged ligand to its much larger receptor prevents the ligand from rotating freely and the degree of depolarization falls.

molecule. In such a setting, the system is sensitive to enzymatic modification (such as phosphorylation) or cleavage and may be used as an enzyme assay. Alternatively, the two fluorophore-labeled compounds may be drug and receptor. The disadvantage of using this method for binding reactions is that small molecules often may not be labeled with fluorescent tags without significant alteration in their properties.

2.3.3 *Time-resolved fluorescence*

This technique is based on the unique fluorescent properties of lanthanide ions. These ions, particularly europium and terbium, re-emit fluorescent light of a characteristic wavelength with a delay after accepting excitation energies. This long lifetime of fluorescence makes it possible to design a system with significantly reduced background fluorescence; emitted light is measured only after the background fluorescence had leveled off. The method is particularly amenable for binding assays performed with large biomolecules, as it also suffers from the problem that the large size of the fluorescent tag makes it inappropriate for labeling small molecules.

2.3.4 *Fluorescence polarization*

This technology is based on the fact that immobilized fluorescent molecules, when excited with polarized light, emit polarized light. This is not the case with molecules free in solution, as they can rotate freely and the emitted light, therefore, is depolarized. The assay is based on binding of a small fluorescently tagged molecule to a larger molecule; upon binding the small molecule loses its ability to rotate freely and the degree of light depolarization falls. As with all fluorescent methods described here, the size of the fluorescent tag puts a limitation on the use of the method for small molecule ligands.

2.3.5 *Reverse yeast two-hybrid system*

As opposed to radioactive or fluorescent methods as described above, this is a microbially-based screening procedure that allows the screening of a large number of drug molecules. It is useful for the identification of membrane permeable drugs that disturb protein-protein interaction.

The yeast two-hybrid system has been used for nearly a decade to identify protein-protein interactions by making use of fast genetic screening and selection. The two-hybrid system is based on the fact that transcription activation of a gene occurs via binding of a transcription factor to a specific upstream DNA sequence and activation of RNA polymerase II (Figure 2.6). The DNA binding and activating functions of the transcription factor are located on two domains, the DNA-binding and activator domains which are physically separable. Protein-protein interactions can reconstitute functional transcriptional activation provided they bring these two domains into close physical proximity. If the gene regulated in this way confers selective growth advantage to yeast, colonies can be identified that have protein interactions which bring together the activator and DNA binding domains of the transcription factor.

In the reverse system, protein domains that interact are placed in a two-hybrid setting. A selection system is chosen so that interaction of the protein domains confers a lethal phenotype on the yeast (negative selection). This system makes it possible to screen many compounds at the same time: only those yeast colonies where protein-protein interactions were

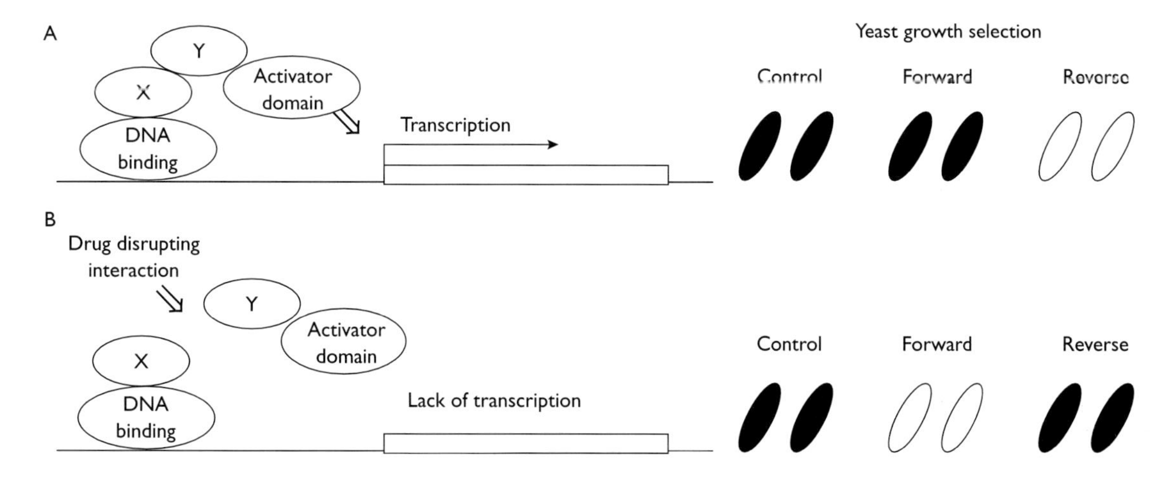

Figure 2.6 Forward and reverse yeast two-hybrid systems. Growth of yeast on selective media is used to analyze two-hybrid interaction in the nuclear transcription system. (A) In the forward two-hybrid system, protein-protein interactions are identified based on the transcriptional activation of a reporter gene required for yeast growth. (B) In the reverse two-hybrid system, proteins known to interact are placed into the transcriptional activatory machinery of a toxic reporter gene. The yeast colony will grow only if a small membrane-permeant molecule selectively disturbs the protein-protein interaction upstream of the toxic gene.

specifically disturbed by the drug molecule present will grow (Vidal *et al.* 1999). It is important to emphasize that disturbance of protein-protein interaction occurs in the biological system of a eukaryote, thus cell permeability and cytotoxicity are already included as parameters in the screen. It is predicted that this system and other systems in which drug modulation of protein-protein interaction is studied will produce many important therapeutic agents in the future.

2.4 Concluding remarks

Functional genomics holds the key to many new drugs still to be discovered. Better understanding and improvement of drug target selection, their validation, and successful use of the drug screening procedures will accelerate the discovery process. It is predicted that within a decade the pharmaceutical industry and mankind will enjoy the consequences of these emerging technologies.

References

Altschul, S.F., Gish, W., Miller, W., Myers, E.W., and Lipman, D.J. (1990) Basic local alignment search tool. *J. Mol. Biol., 215, 403–410.*

Bosworth, N., and Towers, P. (1989) Scintillation proximity assay. *Nature, 341, 167–168.*

Chen, I.J., Neamati, N., Nicklaus, M.C., Orr, A., Anderson, L., Barchi, J.J., Kelley, J.A., Pommier, Y., and MacKerell, A.D. (2000) Identification of HIV-1 integrase inhibitors via three-dimensional database searching using ASV and HIV-1 integrases as targets. *Bioorg. Med. Chem., 8, 2385–2398.*

Drake, F.H., Doods, R.A., James, I.E., Connor, J.R., Debouck, C., Richardson, S., Lee-Rykaczewski, E., Coleman, L., Rieman, D., Barthlow, R., Hastings, G., and Gowen, M. (1996) Cathepsin K, but not cathepsins B, L, or S, is abundantly expressed in human osteoclasts. *J. Biol. Chem., 271, 12511–12516.*

Ecker, D.J., and Griffey, R.H. (1999) RNA as a small-molecule drug target: Doubling the value of genomics. *Drug Discovery Today, 4, 420–429.*

Edwards, A.M., Arrowsmith, C.H., and Des Pallieres, B. (2000) Proteomics: New tools for a new era. *Modern Drug Discovery, 3, 35–44.*

Gray, N.S., Wodicka, L., Thunnissen, A.M., Norman, T.C., Kwon, S., Espinoza, F.H., Morgan, D.O., Barnes, G., LeClerc, S., and Meijer, L., (1998) Exploiting chemical libraries, structure, and genomics in the search for kinase inhibitors. *Science, 281, 533–538.*

Guild, B.C. (1999) Genomics, target selection, validation, and assay considerations in the development of antibacterial screens. *Annu. Rep. Med. Chem., 34, 227–236.*

Harrington, C.A., Rosenow, C., and Retief, J. (2000) Monitoring gene expression using DNA microarrays. *Curr. Opin. Microbiol., 3, 285–291.*

Johnston, D.A., Blaxter, M.L., Degrave, W.M., Foster, J., Ivens, A.C., and Melville, S.E. (1999) Genomics and the biology of parasites. *Bioessays, 21, 131–147.*

Jones, D.A., and Fiztpatrick, F.A. (1999) Genomics and the discovery of new drug targets. *Curr. Opin. Chem. Biol., 3, 71–76.*

Jones, D.T., Taylor, W.R., and Thornton, J.M. (1992) A new approach to fold recognition. *Nature, 358, 86–89.*

Kay, B.K., Kurakin, A.V., and Hyde-DeRuyscher, R. (1998) From peptides to drugs via phage display. *Drug Discovery Today, 3, 370–378.*

Kelland, L.R. (2000) Telomerase inhibitors: Targeting the vulnerable end of cancer? *Anticancer Drugs, 11, 503–513.*

Knight, T. (2000) High throughput screening. *Drug Discovery World, Summer 2000, 33–38.*

Lenz, G.R., Nash, H.M., and Jindal, S. (2000) Chemical ligands, genomics and drug discovery. *Drug Discovery Today, 5, 145–156.*

Madden, S.L., Wang, C.J., and Landes, G. (2000) Serial analysis of gene expression: From gene discovery to target identification. *Drug Discovery Today, 5, 415–425.*

Myers, T.G., Anderson, N.L., Waltham, M., Li, G., Buolamwini, J.K., Scudiero, D.A., Paull, K.D., Sausville, E.A., and Weinstein, J.N. (1997) A protein expression database for the molecular pharmacology of cancer. *Electrophoresis, 19, 647–653.*

Pal, K. (2000) The keys to chemical genomics. *Modern Drug Discovery, 3, 47–55.*

Powell, K. (2000) Target discovery and drug design: Extracting the value from genomics. *Drug Discovery World, Summer 2000, 26–30.*

Schena, M., Heller, R.A., Theriault, T.P., Konrad, K., Lachenmeier, E., and Davis, R.W. (1998) Microarrays: Biotechnology's discovery platform for functional genomics. *Trends Biotechnol., 16, 301–306.*

Skolnick, J., Fetrow, J.S., and Kolinski, A. (2000) Structural genomics and its importance for gene function analysis. *Nature Biotechnology, 18, 283–287.*

Taylor, M.F., Wiederholt, K., and Svedrup, F. (1999) Antisense oligonucleotides: A systematic high-throughput approach to target validation and gene function determination. *Drug Discovery Today, 4, 562–567.*

Vidal, M., and Endoh, H. (1999) Prospects for drug screening using the reverse two-hybrid system. *Trends Biotechnol., 17, 374–381.*

Vives, E., Brodin, P., and Lebleu, B. (1997) A truncated HIV-1 Tat protein basic domain rapidly translocates through the plasma membrane and accumulates in the nucleus. *J. Biol. Chem., 272, 16010–16017.*

Werbovetz, K.A. (2000) Target-based drug discovery for malaria, leishmaniasis, and trypanosomiasis. *Curr. Med. Chem., 7, 835–860.*

Wright, M., Grim, J., Deshane, J., Kim, M., Strong, T.V., Siegal, G.P., and Curiel, D.T. (1997) An intracellular anti-erbB-2 single-chain antibody is specifically cytotoxic to human breast carcinoma cells overexpressing erbB-2. *Gene Ther., 4, 317–322.*

Zhao, J.J., and Lemke, G. (1998) Rules for ribozymes. *Mol. Cell. Neurosci., 11, 92–97.*

Chapter 3

Mapping of the human genome

Zoltán L. Marcsek and Eugene R. Zabarovsky

Contents

The genome is often referred as the "blueprint of the human being", knowledge of its structure will revolutionize medical practice and biological research into the 21st century and beyond. "Changes that will have effects comparable to those of the Industrial Revolution and the Computer-based Revolution are now beginning. The next great era, a genomics revolution, is in an early phase" (1).

The human genome is the complete set of genetic, biochemical and structural instructions used by a cell or organism. The genome contains all information processed (which can be processed or may never be processed) in any cell of an organism during its entire lifetime, in the case of humans, from the formation of the zygote, through the different phases and steps of its development, through aging till its death. It is the collection of all biological codes physically represented by the cellular deoxyribonucleic acid (DNA). DNA is a very large molecule, identical copies of it are present in the nucleus of all cells of the organism. When a cell enters division, mitosis, the stretched out lengths of DNA packed in the nuclei become visible under the microscope as tightly coiled threads of DNA and associated protein molecules, organized into structures called chromosomes. In the time between cell divisions the chromosomes are not visible, they are not so condensed, and consist of the "interphase" chromatin.

3.1 Coordinated genome research

The Human Genome Program (HGP) was conceived in the mid-1980's and was widely discussed within the scientific community and public press over the last half of that decade. The Human Genome Initiative was proposed to the Congress of the United States in 1988 by James D. Watson, initiating a worldwide coordinated research activity to find all information coded and stored in the DNA of humans and several other test organisms. In the United States, the DOE initially, and the NIH soon thereafter, were the main research agencies within the US government responsible for developing and planning the project. By 1988, the two agencies were working together, a relationship that was formalized by the signing of a Memorandum of Understanding to "coordinate research and technical activities related to the human genome." The initial planning process culminated in 1990 with the publication of a joint research plan. The HGP started in 1990 as a 15-year program. One of its main coordinators is Dr. Francis Collins, director of the National Human Genome Research Institute (NHGRI) at the NIH. Soon, the Human Genome Organization (HUGO) spread to Europe (HUGO Europe) and Japan (HUGO Pacific) to coordinate the research performed in other countries outside of the US.

The HGP is an international research program designed to construct detailed genetic and physical descriptions of the human genome, to determine the complete nucleotide sequence of human DNA, to localize the estimated 50,000–100,000 genes and pseudogenes (and gene-like structures not expressed) within the human genome, and to perform similar analyses on the genomes of several other organisms used extensively in research laboratories as model systems. The scientific products of the HGP will comprise a resource of detailed information about the structure, organization and function of human DNA, information that constitutes the basic set of inherited "instructions" for the development and functioning of a human being. The number of real genes expressing at any phase of human differentiation in any tissues is now known to be around 30,000–35,000. The different tissues in the body express information across a wide range, between less than 2000 (in specialized and terminally differentiated white blood cells) and more than 20,000 (in the placenta).

Successfully accomplishing the ambitious goals of HGP will demand the development of a variety of new technologies. It will also necessitate advanced means of disseminating information widely to scientists, physicians, and others in order that the results may be rapidly used for the public good. Improved technology for genetic and biomedical research will thus be another important product of the HGP by characterization of the genomes of human and selected model organisms through complete mapping and sequencing of their DNA, development of technologies for genomic analysis, examination of the ethical, legal, and social implications of human genetic research, and by training scientists who will be able to utilize the tools and resources developed through the HGP to pursue biological studies that will improve human health. From the inception of the HGP, it was clearly recognized that acquisition and use of such genetic knowledge would have momentous implications for both individuals and society and would pose a number of policy choices for public and professional deliberation.

The HUGO program aims at a detailed understanding of the organization of the human genome, to enable a molecular definition of the basis for normal function as well as genetic disorders affecting human cells and the organism as a whole. An intermediate aim in this program was the construction of a high-resolution physical map for the human genome. At the beginning of this work, much attention was paid to developing techniques for obtaining long-range restriction maps of individual chromosomes or parts of them. In parallel, the mapping efforts can be directed to define some genetic aspects of human diseases, especially cancer.

The initial research plan set out specific goals for the first five years (1990–1995) of what was projected to be a fifteen-year research project and focused the efforts of the research community on the most important initial objectives. Because progress was more rapid than anticipated, the 1990 plan was updated in 1993 by extending the initial goals and scope of genome research. The new goals were also a joint NIH-DOE effort and covered Fiscal Years 1994–1998. The project was originally planned to last 15 years, but effective resource and technological advances have accelerated the expected completion date to 2003.

Several types of genome maps have already been completed, and a working draft of the entire human genome sequence was announced in June 2000, with analyses published in February 2001. The Human Genome Working Draft Sequence was published February 15 and 16, 2001 in Nature (Vol. 409. issue 6822) and Science (Vol. 291, No. 5507).

3.2 Structural organization of the genome

The 3 billion bp in the human genome are organized into 24 distinct, physically separate microscopic units called chromosomes. All genes are arranged linearly along the chromosomes. The nucleus of most human cells contains 2 sets of chromosomes, one set given by each parent. Each set has 23 single chromosomes – 22 autosomes and an X or Y sex chromosome. (A normal female will have a pair of X chromosomes; a male will have an X and Y pair.) Chromosomes contain roughly equal parts of protein and DNA; chromosomal DNA contains an average of 150 million bases (Table 3.1). DNA molecules are among the largest molecules now known. In the somatic cells all information is held by 46 double-stranded DNA molecules. After fertilization the zygote incorporates information from both the oocyte and the sperm, each from the mother and from the father, respectively. The genetic material from the oocyte contains the maternal information while DNA from the sperm carries the paternal information. The two sets of haploid parental genomes will include all genetic information available for the organism for its entire life.

Chromosomes are condensed before cell division and in this form they can be seen under a light microscope and, when stained with certain dyes, reveal a pattern of light and dark bands reflecting regional variations in the amounts of A and T vs. G and C. Differences in size and banding pattern allow the 24 chromosomes to be distinguished from each other, an analysis called a karyotype. A few types of major chromosomal abnormalities, including missing or extra copies of a chromosome or gross breaks and rejoinings (translocations), can be detected by microscopic examination; Down's syndrome, for example, in which an individual's cells contain a third copy of chromosome 21, is routinely diagnosed by karyotype analysis. Most changes in DNA, however, are too subtle to be detected by this technique and require molecular analysis. These subtle DNA abnormalities (mutations) are responsible for many inherited diseases such as cystic fibrosis and sickle cell anemia or may predispose an

Table 3.1 Sizes of the individual human chromosomes

Chromosome 1	263 million bases
Chromosome 2	255 million bases
Chromosome 3	214 million bases
Chromosome 4	203 million bases
Chromosome 5	194 million bases
Chromosome 6	183 million bases
Chromosome 7	171 million bases
Chromosome 8	155 million bases
Chromosome 9	145 million bases
Chromosome 10	144 million bases
Chromosome 11	144 million bases
Chromosome 12	143 million bases
Chromosome 13	114 million bases
Chromosome 14	109 million bases
Chromosome 15	106 million bases
Chromosome 16	98 million bases
Chromosome 17	92 million bases
Chromosome 18	85 million bases
Chromosome 19	67 million bases
Chromosome 20	72 million bases
Chromosome 21	50 million bases
Chromosome 22	56 million bases
Chromosome X	164 million bases
Chromosome Y	59 million bases

(data from: N.E. Morton. Parameters of the human genome. Proc. Natl. Acad. Sci. USA 88: 7474–7476, 1991.)

individual to cancer, major psychiatric illnesses, and other complex diseases.

Human DNA contains two different types of sequences. The first type is the collection of sequences which are unique in the genome, i.e. occur only in one copy per parental allele. The other group contains the so-called repetitive sequences which occur many times (sometimes many thousand copies) in the genome. They are sized very differently, the repeated unit length may vary from a single base to several kilobases. Sometimes repeats intrude into the coding regions. The most frequently occurring repeat is the Alu-type repeat, another is the LINE. Microsatellites are relatively short DNA fragments (usually less than 100 bp) that contain runs of tandemly repeated DNA with repeat unit 1–5 bp (such as A, AC, AG, AAC, AAAG etc.). It is estimated that they are very numerous (5×10^5) and uniformly distributed throughout the human genome with an estimated average spacing of about 6 kbp. Microsatellites can be highly polymorphic (more than 10 alleles of the same locus) and polymorphism usually increases with the number of repeats. Microsatellites with less than 10 copies are usually not polymorphic. Since microsatellites are short, they can be analyzed quickly using PCR with primers flanking each locus. Different alleles can be resolved using denaturing gel electrophoresis. Thus microsatellites may be the most useful polymorphic markers (see later) since they are the most plentiful and have many alleles.

The importance of repetitive DNA sequences are not clear, especially their extremely high representation: approximately 40% of the human genome consists of repetitive elements.

3.3 Functional organization of the genome

How is the information used in the cells, in higher organized biological systems such as man? The information is held in the nucleus by the genome (genotype) but not all information is used during the organism's lifetime, only a part of it is realized to form all the properties that are manifest in the organism (phenotype). Behind any property some biochemical or structural, qualitative or quantitative process is found: what is working, how (much) is it working? A very early philosophical definition of life was given by Friedrich Engels (1820–1895), he defined life as the existence of different modes of proteins, although, at that time no information was available about the biological functions of proteins nor about the existence of enzymes, the biological catalysers of metabolic processes. Now it is clear that almost all cellular functions are performed by enzyme-based biocatalytic systems. Enzymes are protein structures, polymers of their building blocks, the amino acids, and their construction information is coded in the DNA. Those parts of the DNA which contain code for protein synthesis are called genes. Each DNA molecule contains many genes (Table 3.2) – the basic physical and functional units of heredity. A gene is described in the DNA by its code, a specific sequence of nucleotide bases, three nucleotides for each amino acid. The nucleotide sequences carry the information required for constructing proteins, enzymes for essential biochemical reactions as well as the structural components of cells and tissues.

Human genes vary widely in length, often extending over thousands of bases, but only about 3% of the genome is known to include the protein-coding sequences (exons) of genes. The vast majority of DNA in the human genome – 97% – has no known function. Interspersed within many genes are intron sequences, which have no coding function. The balance of the genome is thought to consist of other noncoding regions, whose functions are obscure. All living organisms are composed largely of proteins; humans can synthesize 30,000–35,000 different kinds. Proteins are large, complex molecules made up of long chains of amino acid subunits. Twenty different kinds of amino acids are usually found in proteins. Within the gene, each specific sequence of three DNA bases (codons) directs the cell's protein-synthesizing machinery to add specific amino acids. For example, the base sequence ATG codes for the amino acid methionine. Since 3 bases code for 1 amino acid, the protein coded by an average-sized gene (3000 bp) will contain 1000 amino acids. The genetic code is thus a series of codons that specify which amino acids are required to make up specific proteins. The

Table 3.2 Genes and pseudogenes on the human chromosomes

Chromosome	*Genes*	*Pseudogenes*
1	975	44
2	601	72
3	467	26
4	378	18
5	476	25
6	614	57
7	587	34
8	323	20
9	362	21
10	301	15
11	646	47
12	535	8
13	152	16
14	492	108
15	300	26
16	351	25
17	633	22
18	149	9
19	603	33
20	215	6
21	218	45
22	352	62
X	662	34
Y	47	15

Up-to-date information is available at many WEB sites, for e.g.: http://www.ncbi.nlm.nih.gov, http://www-shgc.stanford.edu/ or http://www.tigr.org/tdb/humgen/humgen.html

protein-coding instructions from the genes are transmitted indirectly through messenger ribonucleic acid (mRNA), a transient intermediary molecule similar to a single strand of DNA. For the information within a gene to be expressed, a complementary RNA strand is produced (a process called transcription) from the DNA template in the nucleus. This mRNA is moved from the nucleus to the cellular cytoplasm, where it serves as the template for protein synthesis. The cell's protein-synthesizing machinery then translates the codons into a string of amino acids that will constitute the protein molecule for which it codes. In the laboratory, the mRNA molecule can be isolated and used as a template to synthesize a complementary DNA (cDNA) strand, which can then be used to locate the corresponding genes on a chromosome map. This strategy is described in the section on physical mapping.

In higher organisms DNA contains sequential information without overlaps. The human genome is recorded in the DNA in a specific manner; while bacterial DNA contains the code in several single genes without interruption, in the human (and also in most eukaryotic organisms) the code is fragmented, expressed information is coded in exons and variable non-coding sequences are found between (introns). On the chromosomes, control sequences and intergenic regions with largely unknown function are located between genes.

3.4 Genomic maps

3.4.1 Linkage maps

A genetic linkage map shows the relative locations of specific DNA markers along the chromosome. Any inherited physical or molecular characteristic that differs among individuals is easily detectable in the laboratory is a potential genetic marker. Markers can be expressed DNA regions (genes) or DNA segments that have no known coding function but whose inheritance pattern can be followed. DNA sequence differences are especially useful markers because they are plentiful and easy to characterize precisely.

On the genetic map, distances between markers are measured in terms of centimorgans (cM), named after the American geneticist Thomas Hunt Morgan. Two markers are said to be 1 cM apart if they are separated by recombination 1% of the time. A genetic distance of 1 cM is roughly equal to a physical distance of 1 million bp (1 Mb). One goal of the genome project is to develop a high-resolution genetic map containing all genetic markers.

The value of the genetic map is that an inherited disease can be located on the map by following the inheritance of a DNA marker present in affected individuals (but absent in unaffected individuals), even though the molecular basis of the disease may not yet be understood nor the responsible gene identified. Genetic maps have been used to find the exact chromosomal location of several important disease genes, including cystic fibrosis, sickle cell disease, Tay-Sachs disease, fragile X syndrome, and myotonic dystrophy.

Genetic mapping resolution has been increased through the application of recombinant DNA technology, including *in vitro* radiation-induced chromosome fragmentation and cell fusion (joining human cells with those of other species to form hybrid cells) to create panels of cells with specific and varied human chromosomal components. Assessing the frequency of marker sites remaining together after radiation-induced DNA fragmentation can establish the order and distance between the markers. Because only a single copy of a chromosome is required for analysis, even non-polymorphic markers are useful in radiation hybrid mapping.

3.4.2 Physical maps

A physical map not only provides a revolutionary organizational framework for all genomics research results, such as linkage maps, DNA markers, genes, STSs (site-tagged sequences), ESTs (expressed site-tags), repeat elements, etc., but also paves a genomic "freeway" for structural, functional and evolutionary genomics research. Integrative physical mapping is central to genomics research. One aspect is physical mapping of the genome, the reconstruction of genomes from large-insert, ordered DNA libraries (Figure 3.1).

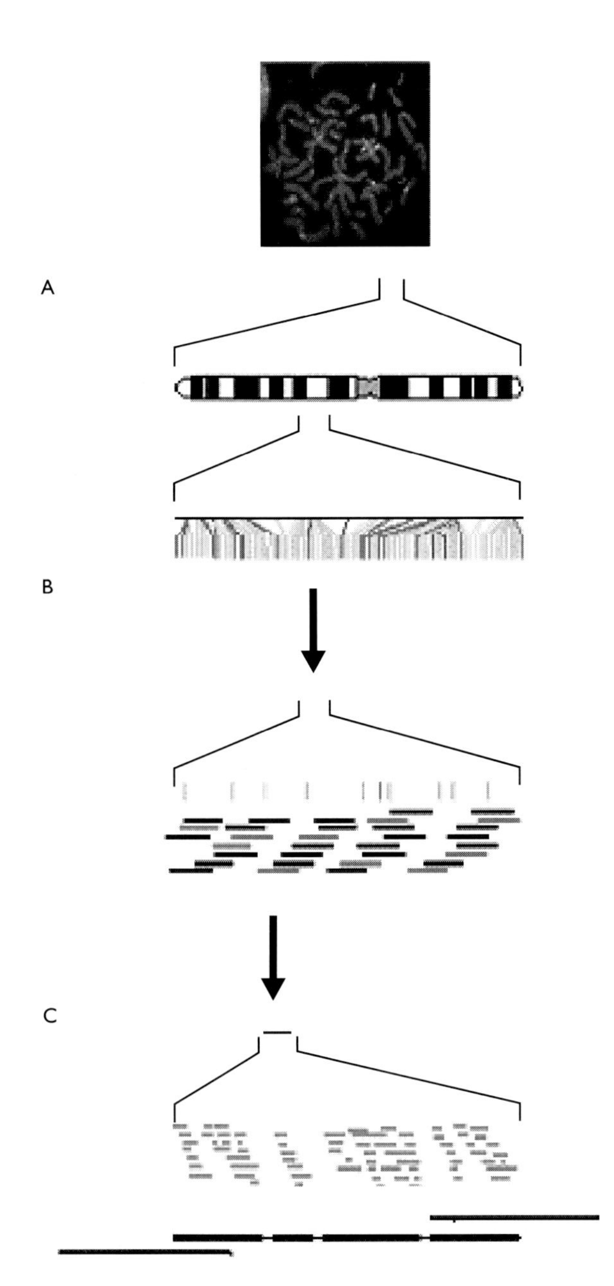

Figure 3.1 Strategy for sequencing the human genome by mapped clones. (A) Construction of maps on the basis of ordered genes, ESTs, STSs provides long-range map, (B) Physical mapping of overlapping clones, (C) Sequencing and reconstitution of the original chromosomal sequence.

Different types of physical maps vary in their degree of resolution. The lowest-resolution physical map is the chromosomal (sometimes called cytogenetic) map, which is based on the distinctive banding patterns observed by light microscopy of stained chromosomes. A cDNA map shows the locations of expressed DNA regions (exons) on the chromosomal map. The more detailed cosmid contig map depicts the order of overlapping DNA fragments spanning the genome. A macrorestriction map describes the order and distance between enzyme cutting (cleavage) sites. The highest-resolution physical map is the complete elucidation of the DNA base pair sequence of each chromosome in the human genome.

By determining the location in the genome from which the particular DNA fragment was derived, it acquires the property of a DNA marker. Such DNA markers are prerequisites for physical and genetic mapping of the genome. DNA markers are also important in the diagnosis of genetic diseases. DNA markers can be divided into several different classes depending on the way in which the markers were selected among the fragments of genomic DNA. Examples of such classes are: anonymous, micro- and minisatellites, restriction fragment length polymorphism (RFLP) markers, *NotI* linking clones, ESTs, STSs etc.

3.4.3 Markers

Markers must be polymorphic to be useful in mapping; that is, alternative forms must exist among individuals so that they are detectable among different members in family studies. Polymorphisms are variations in DNA sequence that occur on average once every 300 to 500 bp. Variations within exon sequences can lead to observable changes, such as differences in eye color, blood type, and disease susceptibility. Most variations occur within introns and have little or no effect on an organism's appearance or function, yet they are detectable at the DNA level and can be used as markers. Examples of these types of markers include (1) restriction fragment length polymorphisms (RFLPs), which reflect sequence variations in DNA sites that can be cleaved by DNA restriction enzymes (see box), and (2) variable numbers of tandem repeat sequences, which are short repeated sequences that vary in the number of repeated units and, therefore, in length (a characteristic easily measured). The human genetic linkage map is constructed by observing how frequently two markers are inherited together. Two markers located near each other on the same chromosome will tend to be passed together from parent to child. During the normal production of sperm and egg cells, DNA strands occasionally break and rejoin in different places on the same chromosome or on the other copy of the same chromosome (i.e. the homologous chromosome). This process (called meiotic recombination) can result in the separation of two markers originally on the same chromosome. The closer the markers are to each other – the more "tightly linked" – the less likely a recombination event will fall between and separate them. Recombination frequency thus provides an estimate of the distance between two markers.

Cloned DNA fragments, located at a specific site on a chromosome can be used for physical and genetic mapping.

Different types of markers are used. Many markers represent randomly cloned DNA fragments whose functions or specific features are not known. Other DNA markers may contain sequence of a known gene or any expressed sequences with unknown function (EST), CpG islands (frequently associated with genes) or recognition sites for rare cutting restriction enzymes convenient for long-range mapping, etc. Markers can also be polymorphic, i.e. they have slightly different sequences in different individuals. Polymorphic markers are extremely important for construction of genetic maps and in mapping human disease genes.

To clone markers, different types of vectors and different types of libraries can be used. The most common methods are: whole-genome, shotgun cloning and sequencing using large-insert vectors (PACs, BACs, YACs); optical mapping; PFGE; and chromosome jumping. Chromosome jumping has been helpful for isolating genes involved in cystic fibrosis, acute nonlymphocytic and myeloid leukemias, and mapping chromosomal regions containing tumor suppressor genes involved in lung, breast, renal, and other cancers. Lambda-based vectors and different kinds of genomic libraries are commonly used for this purpose. Many different variants of lambda-based vectors, which combine features of different cloning vehicles (plasmids, M13 and P1 phages), have been created for this purpose. The use of each vector is usually limited to a specific task; the construction of general genomic libraries which contain all genomic DNA fragments, or special genomic libraries which contain only particular subsets of genomic DNA fragments. Among these special libraries, NotI linking and jumping libraries have particular value for physical and genetic mapping of the human genome.

Whole-genome sequencing will only be performed for selected organisms. However, analysis of disease genes in other species may be extremely useful for understanding the fundamental processes leading to disease development. For organisms for which whole-genome maps and sequences are not available, only two methods exist to move along the chromosomes: chromosome walking and chromosome jumping. Moving along the chromosome can be very important for cloning particular genes of interest, e.g. genes that cause disease or confer disease resistance. Thus, chromosome jumping can be used for two purposes: to move along the chromosome towards a gene of interest, and to construct whole-genome physical maps.

In a chromosomal map, genes or other identifiable DNA fragments are assigned to their respective chromosomes, with distances measured in base pairs. These markers can be physically associated with particular bands (identified by cytogenetic staining) primarily by *in situ* hybridization, a technique that involves tagging the DNA marker with an observable label (e.g. one that fluoresces or is radioactive). The location of the labeled probe can be detected after it binds to its complementary DNA strand in an intact chromosome. As with genetic linkage mapping, chromosomal mapping can be used to locate genetic markers defined by traits observable only in whole organisms. Because chromosomal maps are based on estimates of physical distance, they are considered to be physical maps. The number of base pairs within a band can only be estimated. Until recently, even the best chromosomal maps could be used to locate a DNA fragment only to a region of about 10 Mb, the size of a typical band seen on a chromosome. Improvements in fluorescence *in situ* hybridization (FISH) methods allow orientation of DNA sequences that lie as close as 2 to 5 Mb. Modifications to *in situ* hybridization methods, using chromosomes at a stage in cell division (interphase) when they are less compact, has increased map resolution to around 100,000 bp. Further banding refinement might allow chromosomal bands to be associated with specific amplified DNA fragments, an improvement that could be useful in analyzing observable physical traits associated with chromosomal abnormalities.

cDNA map

A cDNA map shows the positions of expressed DNA regions (exons) relative to particular chromosomal regions or bands. (Expressed DNA regions are those transcribed into mRNA). The cDNA can then be mapped to genomic regions. Because they represent expressed genomic regions, cDNAs are thought to identify the parts of the genome with the most biological and medical significance. A cDNA map can provide the chromosomal location for genes whose functions are currently unknown. For disease-gene hunters, the map can also suggest a set of candidate genes to test when the approximate location of a disease gene has been mapped by genetic linkage techniques.

The individual (and thus allele-specific) variation in the number of tandemly repeated (VNTR) DNA sequences in a particular locus can be used as marker. Minisatellites are DNA fragments of 200–2,000 bp that contain several copies (from a few to near 50) of 15–60 bp repeats. All these repeats share a 10–15 bp core sequence. When DNA from different individuals is digested with a restriction enzyme that does not cut into these repeats, the length of the fragments produced will depend on the number of repeats at the locus. Since the size of the minisatellites and repeats is big enough, it is possible to discriminate between different alleles using ordinary non-denaturing gel electrophoresis and Southern blot analysis.

Electrophoretic techniques are used to separate the fragments according to size into different bands, which can be visualized by direct DNA staining or by hybridization with DNA probes of interest. The use of purified chromosomes separated either by flow sorting from human cell lines or in hybrid cell lines allows a single chromosome to be mapped. The development of pulsed-field gel electrophoretic (PFGE) methods has improved the mapping and cloning of large

DNA molecules. While conventional gel electrophoretic methods separate pieces less than 40 kb in size, PFGE separates molecules up to 10 Mb, allowing the application of both conventional and new mapping methods to larger genomic regions.

The PFGE method is a powerful technique that is very useful for long range mapping. But this method cannot be used without cloning, because it is only possible to obtain information about fragments already cloned, it is not possible to obtain new ones. To this end several approaches have been suggested that combine the advantages of PFGE and gene cloning. The best known are YAC cloning and chromosome jumping techniques (see later). These approaches do not exclude each other; using them together will greatly facilitate achievement of the final aim – construction of a physical map for the whole human genome and obtaining contigs of cloned human DNA fragments covering part of or the entire genome.

3.5 High resolution mapping using libraries

A genomic library is a collection of recombinant vectors which contains DNA fragments representing the genome of a particular organism. Genomic libraries can be either general, containing DNA fragments covering the whole genome, or special, containing only specific genomic fragments which differ in certain parameters. They can be CG rich, contain only particular size fragments of DNA obtained after digestion with a particular restriction enzyme, or contain specific repeats etc. Important types of special genomic libraries are jumping and linking libraries.

YAC

Technological improvements now make the cloning of large DNA pieces possible, using artificially constructed chromosome vectors that carry human DNA fragments as large as 1 Mb. These vectors are maintained in yeast cells as artificial chromosomes (YACs). (For more explanation, see DNA Amplification.) Before YACs were developed, the largest cloning vectors (cosmids) carried inserts of only 20 to 40 kb. YAC methodology drastically reduces the number of clones to be ordered; many YACs span entire human genes. A more detailed map of a large YAC insert can be produced by subcloning, a process in which fragments of the original insert are cloned into smaller insert vectors. Because some YAC regions are unstable, large capacity bacterial vectors (i.e. those that can accommodate large inserts) are also being developed.

YAC cloning is now widespread and many laboratories have used this technique to obtain large cloned DNA fragments from different eukaryotic genomes. The construction of YAC contigs is not free from problems – certain regions of the human genome are difficult to clone by YAC, certain YAC clones may be artificial hybrids containing DNA fragments from different chromosomes, and rearrangements of DNA can occur in YAC clones.

BAC

Bacterial artificial chromosome (BAC) represents the state-of-the-art technology for such large-insert DNA library development. It has been demonstrated that BAC libraries are invaluable and desirable genetic resources for all kinds of modern structural, functional and evolutionary genomics research. Genome-wide, as well as regional, physical maps of the human genome from BACs and/or binary BACs have been developed to minimize duplication of effort in different laboratories constructing genomic DNA libraries, to prevent contamination of the libraries during manipulation and applications, and to maximize and accelerate genomics research (Figure 3.2).

PAC

P1-derived artificial chromosome (PAC) A vector used to clone fragments of 100 to 300 kb insert size (average at 150 kb) in Escherichia coli cells.

Jumping and linking libraries

Long-range mapping using chromosome jumping creates physical maps with large holes, and large-insert cloning vectors are needed to fill these holes. A big advantage in using chromosome jumping and linking clones is that they are small-insert vectors and mapping with these clones can be completely automated.

Jumping and linking libraries can be used independently for construction of a long range restriction map using PFGE. The first methods developed by Poutska and Lehrach for the construction of jumping libraries were laborious, expensive, and complicated. To make a representative library one needed to package 40–150 ug of vector arms and use a supF marker with all its disadvantages. This might be the reason why these types of genomic libraries are not so widespread although the creation of such libraries was suggested quite long time ago.

cDNA libraries contain information on all expressed genes for a particular state of a cell, cell line, tissue or organism. They are synthesized by reverse transcriptase from the isolated messenger RNAs, converted to double-stranded DNA and cloned in an appropriate vector.

A number of strategies can be used to reconstruct the original order of the DNA fragments in the genome. Many approaches make use of the ability of single strands of DNA and/or RNA to hybridize – to form double-stranded segments by hydrogen bonding between complementary bases. The extent of sequence homology between the two strands can be inferred from the length of the double-stranded

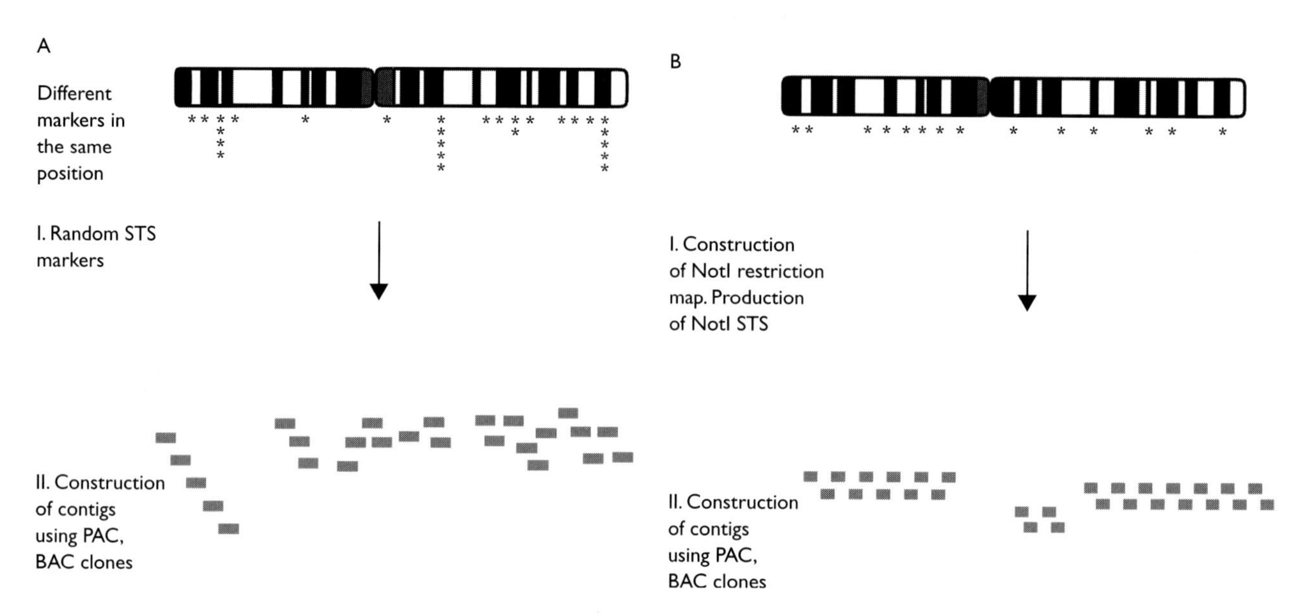

Figure 3.2 Two strategies to map the human genome. (A) Random STS strategy. (B) The strategy based on the NotI sites (CpG islands) as framework markers.

segment. Fingerprinting uses restriction map data to determine which fragments have a specific sequence (fingerprint) in common and therefore overlap. Another approach uses linking clones as probes for hybridization to chromosomal DNA cut with the same restriction enzyme.

3.5.1 Cloning approaches

For cloning DNA markers from a particular chromosome one method is based on the use of fluorescence-activated cell sorters (FACS). FACS operates by rapid analysis of suspended particles, single cells or even chromosomes. For chromosome sorting, metaphase chromosomes are isolated from cells blocked in metaphase with colcemid. A suspension of chromosomes, stained with fluorescent dyes (usually Hoechst 33258 and chromomycin A3) passes through the focus of a laser beam that excites the DNA-bound fluorescent dyes. Photomultipliers are used to monitor the emissions from each chromosomal particle and a computer-based system evaluates the fluorescence signal of each individual chromosome. Equally-sized droplets are formed by ultrasonic dispersion and the droplets containing the desired chromosome as indicated by the fluorescence measurement are deflected from the main stream by an electric field and collected in a tube (Figure 3.3). DNA isolated from sorted chromosomes can be used for construction of genomic libraries. Chromosome sorted libraries are currently available for many chromosomes. The most important characteristic property of the library is the purity of the sorted chromosomes (how many recombinant clones will contain inserts from other chromosomes).

Another approach is based on the use of hybrid cell lines. To obtain such somatic cell hybrids, human cells are fused

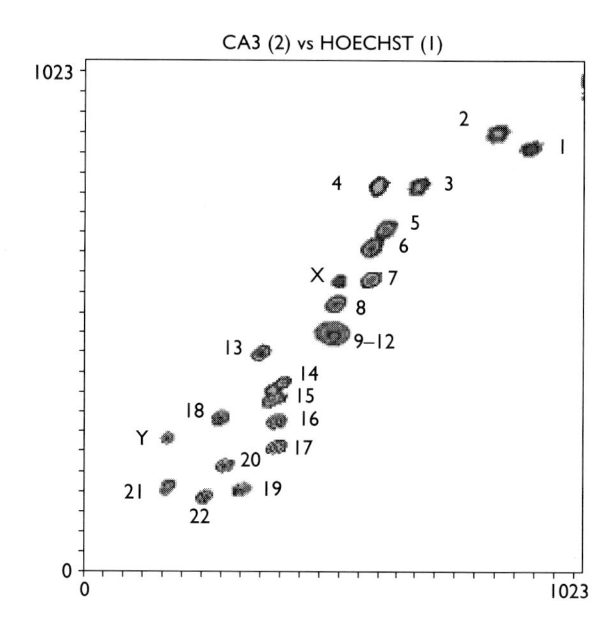

Figure 3.3 Separation of human chromosomes on the basis of their relative fluorescence by flow sorting. Chromomycin A3 signals are plotted versus Hoechst 33258 and separated chromosomes are collected as shown (see Colour Plate I).

with rodent cells (e.g. mouse or hamster). When the resulting hybrid cells are grown in culture, there is a progressive loss of human chromosomes until only one or a few of them are left. At this step some of the segregant hybrid cells can be quite stable. In a modification of this technique, human cell line is transfected with plasmid vector or with a retroviral vector which contains selectable marker (e.g. gpt or neo). Transfected clones are screened for those that contain only a single integrated plasmid per cell (and thus contain a single marked chromosome). These transformants are micronucleated by prolonged colcemid treatment and the microcells, containing only one chromosome, are produced using special techniques. The microcells are fused to mouse cells and the resulting human-mouse microcell hybrids, containing the single marked chromosome, are isolated by growth in selective medium. Hybrid cells containing only fragments of human chromosomes can be produced using human chromosomes with translocations and deletions, or the fragments can be produced experimentally by irradiation of a hybrid with complete chromosomes (radiation hybrids). Such hybrid cell lines are very useful not only for constructing genomic libraries but also for physical mapping of already isolated DNA markers and genes. Another advantage of somatic hybrid cells is that DNA is available in large amounts and it can be used for different purposes. Human specific clones can be isolated from the library by hybridization to total human DNA.

Region-specific libraries may be constructed by the use of chromosome microdissection. Using this approach it is possible to physically remove the chromosomal region of interest and to clone the minute quantities of microdissected DNA by a microcloning procedure. Spreads of human chromosomes are made and stained using standard cytogenetic techniques. DNA from an individual band is then cut from the chromosome using ultrafine glass needles or isolated with the help of a laser. In the latter case, all other chromosomes are destroyed by the laser and intact DNA is present only in the chromosome of interest. DNA obtained from only a few (2–20) chromosomes is enough to construct a region specific library. This DNA is amplified using PCR and cloned in plasmid or lambda vectors. It is possible to divide all methods suggested for amplification and cloning of microdissected DNA into two groups. In the first group, chromosomal DNA is usually digested with frequently cutting enzymes, like MboI, and ligated to the linker adaptor. These adaptor molecules serve as a binding site for primers in a PCR amplification and may contain recognition sites for restriction enzymes. Amplified DNA is then digested (e.g. with MboI) and cloned into the vector. In another scheme, chromosomal DNA serves as a direct substrate for PCR amplification. Usually such amplification is performed in several steps: one or two are non-specific and performed under low-stringency annealing conditions to permit random annealing of primer to chromosomal DNA. The last step of amplification is performed under high-stringency conditions, in which the primers anneal only to the PCR products generated in the first step. Amplified DNA is then digested with restriction enzyme and cloned into the vector. The biggest problem with this technique is connected with PCR. Any impurity will lead to unpredictable results. The average size of the inserts is usually in the range 200–600 bp which is not convenient for most purposes. Also these inserts are random, without any linkage to genes or polymorphic sequences. About 50% of the clones are usually useless since they contain repetitive sequences such as Alu repeats. Nevertheless, this method makes it possible to obtain DNA sequences from a specific region of a chromosome and these sequences can also serve as probes to isolate cosmid, YAC or other clones from that region.

3.5.2 *Cloning vectors in genome research*

Cosmids are essentially plasmids that contain the *cos* region of phage lambda responsible for packaging of DNA into phage particles. Any DNA molecule containing this region and sized between 37.7 kb and 52.9 kb, can be packaged into phage particles and introduced to the E. coli host cells. The advantages of cosmids are their easy handling (similar procedures as working with plasmids) and large cloning capacity. Since the plasmid body is usually small (3–6 kb), large DNA molecules (46–49 kb) can be cloned in cosmid vectors. Problems can arise if cosmids are used for construction of genomic libraries from DNA containing numerous repeats. In this case cosmid molecules become unstable and are subject to rearrangement in the E. coli cell.

Phasmids are lambda phages that have an inserted plasmid. They have the same basic features as lambda phage vectors but inserted foreign DNA fragments can be separated from the body of phage DNA and converted into plasmid form. After conversion the cloned DNA fragment will exist as a recombinant plasmid. There are two main methods for such a conversion: biological and enzymatic. In the first case, the phasmid vector will contain signal sequences bordering the cloned DNA fragment and plasmid body in the phage vehicle. These signal sequences (e.g. from P1, lambda, M13 phages) can be recognized by specific proteins (e.g. M13) and the cloned DNA fragment together with plasmid body will be cut out.

In some cases it is more convenient to work with a genomic library in plasmid than in lambda phage form. The construction of a representative genomic library directly into a plasmid vector has several drawbacks and difficulties. But all these problems can be solved easily with the help of phasmid and diphasmid vectors. Often a genomic library is constructed in lambda phage, and then the whole library is converted to plasmid form.

Diphasmids are even newer types of vectors which can combine the advantages of phages (l and M13) and plasmids, i.e. the three main types of vectors used in molecular cloning. They can be divided into two classes: (1) diphasmids that can replicate as phage lambda (a further improvement of phasmids); (2) diphasmids that are incapable of replication as phage lambda, i.e. a cosmid capable of being packaged into phage M13 particles.

Hyphages represent another type of lambda based vector. They are constructed from M13 vectors with a built-in cos site of lambda. As these vectors have the main features of

M13 vectors they can be obtained in single stranded form. Their distinctive feature is that they can be packaged efficiently into lambda phage-like particles. This decreases the chance of recovering non-recombinant vectors and enables the construction of a representative genomic library in single strand form.

3.5.3 Phage lambda-based methods

Among other vectors used for construction of genomic libraries (YACs, BACs, P1), phage lambda-based vectors are still the most popular. The reason is that the genetics and features of both lambda phage and E. coli (the host for lambda phage) are well known. The two basic features which make cloning work with lambda-based vectors convenient are: (1) lambda phage has genes that are not critical for its replication and can be replaced by any foreign DNA fragment of a given size. The size of the phage DNA that can be packaged into viable phage particles is limited to between 37.7 and 52.9 kbp. This means that it is possible to biologically regulate the size range of the cloned DNA fragment; (2) The easiest way to obtain a maximal proportion of recombinant molecules is to perform ligation at high concentration of vector and inserted (genomic) DNA since elevated concentration of DNA facilitates intermolecular ligation instead of self-ligation of vector molecules. In this case the main product is long DNA chains (>200 kbp) containing many copies of vector and genomic DNA fragments. Extremely efficient *in vitro* systems are used for packaging lambda DNA into phage particles (10^9 plaque forming units per μg of DNA) to produce viable phages. Extensive modifications of lambda phage-derived vectors have been developed which combine the features of different vector systems.

Representation is one of the most important features of a genomic library. A representative genomic library means that every genomic DNA fragment will be present in at least one of the recombinant phages of the library. In practice, however, this is difficult to achieve. Some genomic fragments are not clonable because of the strategy used for construction of the genomic library. For example, if the maximal cloning capacity of the vector is 18 kb and EcoRI digestion is used to construct the library, no genomic EcoRI fragments bigger than 18 kb will be present in the library. In some cases, genomic DNA fragments can suppress growth of the vector or the host cell so that its cloning can be restricted to specific vector systems.

An important reason for decreased representation is the different replication potential of different recombinant phages. Most researchers work with amplified libraries. The ligated DNA molecules are packaged into the lambda phage particles and plated on a lawn of E. coli cells (usually many Petri dishes are used for such plating). Then all lambda phage particles are eluted from the plaques and mixed with the liquid eluted from the Petri dishes. Glycerol or DMSO is added to the eluate and aliquotes are kept at $-76°C$. This procedure is called amplification of the library. The advantage of this procedure is that a library can be kept for many years and can be used for many experiments. However, each recombinant phage present in the library gives a single plaque at the first plating, and since different recombinant phages have different growth potential, sometimes differences of 100 times in the abundance of clones can result after amplification. This means that in the amplified library some of the phages are present 100 times more often than others. In this case, to recover all recombinant phages obtained after packaging into lambda phage particles, one needs to plate 100 times more phages than were obtained after the original plating. In reality, this is difficult to achieve and means that after amplification some recombinant phages are virtually lost from the library.

How is the representation of the library estimated? The library is considered to be representative if, after the first plating (before amplification), it contains a number of recombinant clones containing genomic DNA fragments equal to 7–10 genome equivalent. For example, human genomic DNA contains approximately 3×10^9 bp. If the vector contains on average 15 kb inserts, the representative library should contain 1.4–2.0×10^6 recombinant clones.

Another important characteristic feature of a library is the percentage of recombinants. If a library contains 100% of recombinants, one needs to obtain 1.5×10^6 phages to get representation of the human genome. In the case of a library with 50% of recombinants, one needs to work with 3×10^6 phages to get the same result. On the other hand, genetic selection (to increase the proportion of recombinants) usually decreases the representation of the library. For most purposes, if a library contains more than 80% of recombinant phages, it is better to omit the genetic selection procedures. To calculate the percentage of recombinants genetic (Spi) selection, as in the case of lambda-EMBL-based vectors, can be used. Another selection relies on blue-white color identification (e.g. lambda-Charon series). A third class of vectors has both genetic selection and blue-white color identification (lambda SK4, lambda SK6). The three commonly used methods to construct genomic libraries are described below.

Genomic libraries can also be constructed in cosmids. The absence of selection against nonrecombinant vector and the potential to package concatemers of only cosmid fragments into phage particles makes preventing self-ligation of vector fragments even more important. Many similar approaches have been suggested, one of which prevents the formation of vector-concatemers by partial filling-in. A similar effect can be achieved by dephosphorylation or by digestion at the first step with AccI and SmaI instead of EcoRI and HindIII. SmaI produces blunt ends and AccI gives sticky ends with only two protruding base pairs. The ligation of these ends will be far less effective than for BamHI and Sau3AI sticky ends (four protruding base pairs).

3.5.4 *Alu-PCR as a tool to clone markers from specific regions of the chromosomes*

The human genome consists of at least 50% of different kinds of repetitive sequences. Among them about 50% are repeats randomly distributed in the human genome – different kinds of short (SINE) and long (LINE) interspersed repeats. The most abundant among them is the Alu repeat family. Alu-repeats are present at $0.5–1.0 \times 10^6$ copies per genome with an estimated average spacing of about 4 kb. Alu repeats have a length of about 300 bp and consist of two homologous units. Related repeats also exist in other mammals. Different members of this family from the same species usually have homology of 80–90% but are only about 50% identical in different species. These repeats have conserved and variable regions. It is possible to find conserved sequences that are species specific. These conserved sequences can be used as primers for PCR, to specifically amplify human sequences in the presence of nonhuman DNA. These features are the basis for using Alu-repeats for isolation of human chromosome specific sequences from hybrid cell lines containing human and nonhuman DNA sequences. Moreover, if a hybrid cell line contains only a short piece of human chromosome, the Alu-PCR approach can be used for isolation of markers specific for a defined region of the chromosome. In the case of two hybrid cell lines, one of which contains complete human chromosome (HCL1) and the other carries the same chromosome but with deletion (HCL2), there is the potential to obtain markers specific for the deletion. This method may be called the differential Alu-PCR approach for obtaining DNA markers. There are usually two variations to this approach.

In the first variation, Alu-PCR is done using DNA from both cell lines and the products of the reactions are separated by agarose gel electrophoresis. Some bands present in products from HCL1 will be absent among the products from HCL2. These bands can be excised and cloned, giving markers localized in the deletion. The disadvantage of this approach is that Alu-PCR usually results in a large number of products. The products have a very complex pattern and look like a smear on the gel. Different approaches to solve this problem have been suggested: use more specific primers (for only a subset of the Alu-repeats) or genomic DNA digested with restriction enzyme, or use hybrid cell lines that contains only small pieces of human chromosomes.

The second variation is mainly used in connection with sources which contain only a limited amount of human material: YAC clones and radiation hybrid cell lines containing small pieces of human chromosome. The YACs can, for example, be used for Alu-PCR and products of the PCR reaction used as a probe to screen genomic libraries (e.g. in cosmids). The hybridization pattern reveals which cosmids are present in one YAC, which in other YACs and which are present in one but absent in another. Such an approach is also useful for mapping.

Alu-PCR and subtractive procedures can be used to clone CpG islands from defined regions of the chromosomes. The Alu-PCR approach is used successfully for cloning DNA markers, but the disadvantage is that it results in cloning of small DNA fragments (500 bp) between Alu sequences. Alu sequences are distributed in a random fashion and are not linked with genes or other kind of markers. An obvious suggestion for making Alu-PCR more useful for mapping is to use not just genomic DNA from different sources, but linking libraries constructed from these sources. This modification has at least two advantages: using isolated probes, it is easy to clone a parental linking clone (e.g. *NotI*) which is a natural marker on the chromosome convenient for linkage with other markers. Furthermore, linking clones are located in CpG rich islands which are associated with genes. In this scheme, linking libraries are constructed from different hybrid cell lines containing either whole or deleted human chromosomes. Then total DNA isolated from these libraries can be used for Alu-PCR. But in this case every PCR product (either discrete bands or total product) is used as a probe to isolate linking clones from the defined region of the chromosomes. Genomic subtractive methods represent potentially powerful tools for identification of deleted sequences and cloning region specific markers. This approach has given rewarding results in less complex systems, like yeast or cDNA libraries, but the great complexity of the human genome has generated serious problems. These problems can be overcome by reducing the complexity of the human genomic sequences. Two approaches have been suggested to achieve this aim. In one (representational difference analysis) only a subset of genomic sequences (e.g. BamHI fragments less then 1 kb) is used for subtractive procedures and this approach will result in cloning of random sequences. In the other, NotI or XhoI (SalI) linking libraries are used instead of whole genomic DNA (Figure 3.4). The complexity of the NotI-linking library is at least 100 times less than the complexity of the whole human genome. It is approximately equal in complexity to the yeast genome. Since this approach is not linked with Alu repeats, it allows isolation of NotI-linking clones, which are unavailable for cloning using Alu-PCR.

3.5.5 *Mapping with jumping and linking clones*

Chromosome jumping is principally different from chromosome walking. In chromosome walking (Figure 3.5), moving along the chromosome is achieved by cloning all genomic sequences between the start and the final point, i.e. the gene of interest. DNA fragments that cannot be cloned in the particular vector or host create problems. One way to overcome these problems is to jump over the regions containing these fragments. Chromosome jumping can be performed using jumping and linking libraries.

The two main types of jumping libraries are the rare cutting enzyme (e.g. *NotI*) jumping library and the hopping

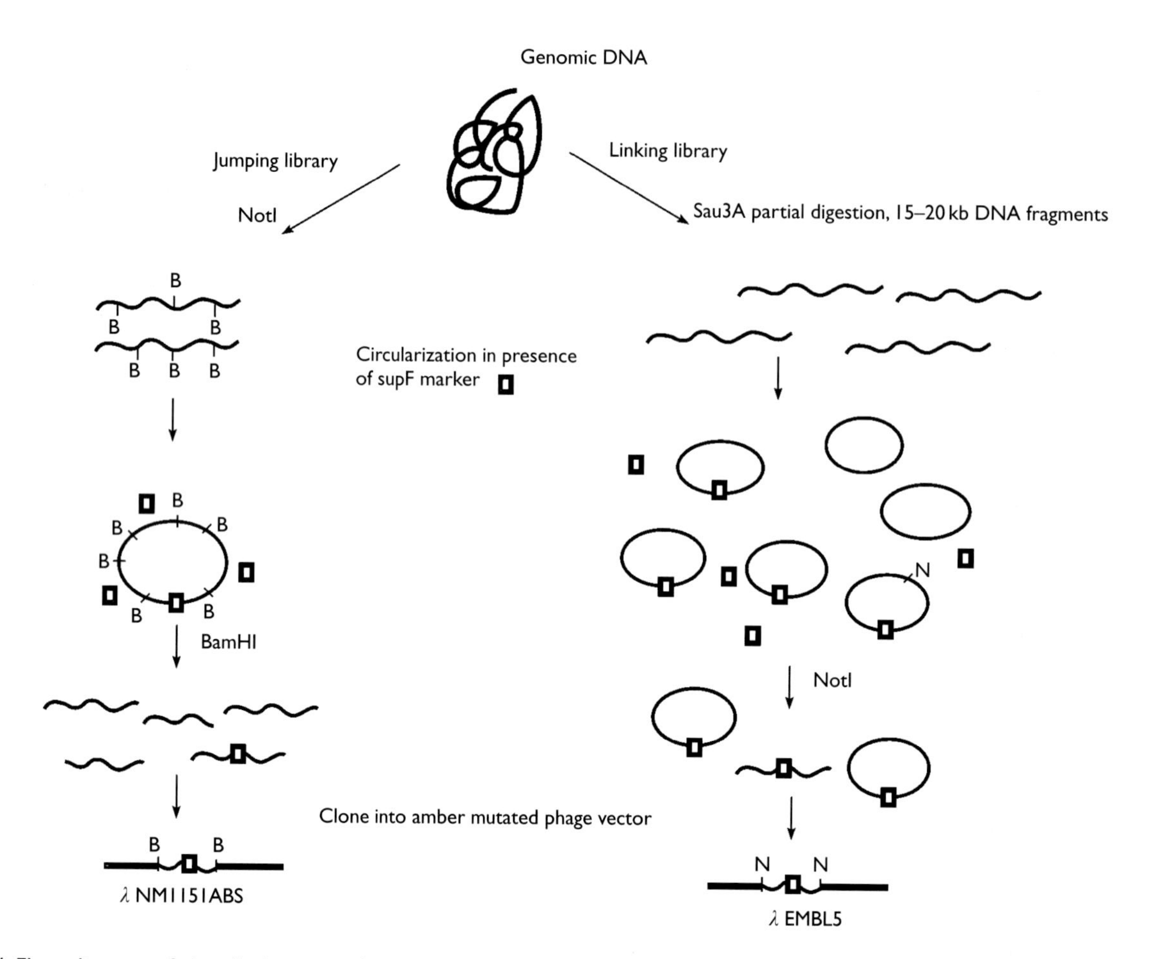

Figure 3.4 Flow diagram of the cloning procedure to construct NotI jumping and linking libraries using supF marker, in lambda vector. B – BamHI, N – NotI.

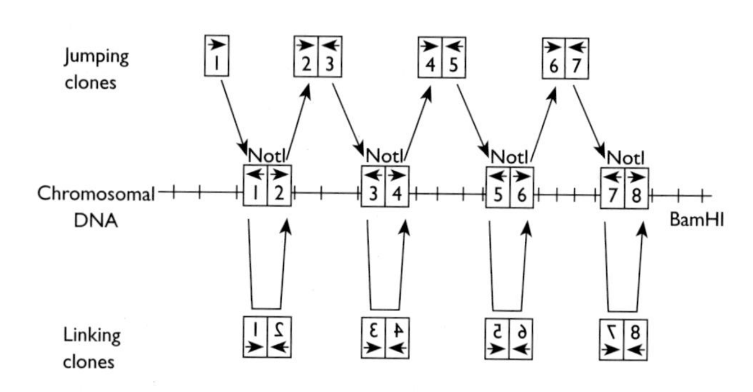

Figure 3.5 General scheme of computer-assisted large-scale mapping, using NotI linking and jumping libraries. DNA fragments containing NotI sites are located between two BamHI sites. Arrow heads show the sequenced part of the DNA fragments. In the chromosomal DNA, the NotI sites are positioned at the junction between the numbered boxes. *Bam*HI sites are indicated by small vertical arrows.

library. The main difference between these libraries is that the jumping library is made by complete digestion with *NotI* whereas the hopping library is made by partial digestion with *EcoRI* or *BamHI*. Using a *NotI* jumping library, it is possible to jump over a long distance, but a jump can only be initiated from a *NotI* recognition site. In screening *NotI* jumping libraries, clones corresponding to jumps over more than 1000 kb have been isolated. Using a hopping library, it is possible to start jumping from practically any point, but over a shorter distance (50–150 kb).

The basic principles of jumping and linking library construction are described later in the chapter. In the case of jumping libraries, only the ends of large DNA fragments are cloned. *NotI* jumping clones contain DNA sequences adjacent to neighboring restriction sites and *NotI* linking clones contain DNA sequences surrounding the same restriction site. Using these two types of libraries, it is possible to move quickly along the chromosome because the human genome contains less than 10,000 recognition sites for this enzyme. *NotI* linking and jumping clones covering the whole human genome can be placed on one high-density filter. In this case, screening can be done quite rapidly. Another way to perform chromosome jumping is using the sequencing approach. More than 10,000 human *NotI* linking clones have been sequenced; therefore, chromosome jumping can be performed by sequencing *NotI* jumping clones and then finding the respective *NotI* linking clone among the *NotI* linking sequences.

A shotgun sequencing method that may be used for whole-genome mapping has been suggested for *NotI* linking and jumping libraries constructed in lambda SK15 and lambda SK22. This approach is based on the following considerations: (1) essentially the same vectors and scheme are used to construct *NotI* linking and jumping libraries; and (2) the *NotI* sites in both libraries are available for sequencing with standard sequencing primers, and sequence information for the 800–1000 bp surrounding each *NotI* site can be easily obtained by automated sequencing. Plasmid DNA can be isolated automatically and thousands of clones sequenced. The linear order of the *NotI* clones can then be established using a computer program. This approach to *NotI* linking and jumping library construction is illustrated in Figure 3.5. The numerous repeats in the mammalian genome are not expected to create serious problems for this approach, as a 500 bp sequence will uniquely identify a *NotI* clone. A more detailed map (100–300 kb resolution) can be constructed using this approach and by using other enzymes that cut more frequently, like *XmaIII*, *SalI*, and *XhoI*.

In the case of hopping library construction, a number of clones (at least 1,000,000) should be obtained to cover the whole genome. It is more laborious to work with hopping libraries.

3.5.6 *Linking and jumping libraries*

For long range mapping and cloning of large stretches of genomic DNA, the two most widely used methods are construction of contigs using chromosome walking (e.g. YAC cloning) and chromosome jumping. The YAC cloning technique is now used by many laboratories. Still, this approach is not devoid of problems and drawbacks. These problems could be diminished using jumping/linking libraries. Moreover, jumping and linking libraries can be used independently for construction of a long range restriction map using PFGE. Jumping clones contain DNA sequences adjacent to neighboring *NotI* sites and linking clones contain DNA sequences surrounding the same restriction site.

The two best known types of jumping libraries are the *NotI* jumping library and the "general" jumping (hopping) library. The basic principle of both methods is to clone only the ends of large DNA fragments rather than continuous DNA segments, as in YAC clones. Internal DNA is deleted by controlled biochemical techniques. The main difference is that in the first type of library, complete digestion with a rare cutting enzyme (the *NotI* is the most popular) is used and the second is based on a partial digestion with a frequently cutting enzyme with subsequent isolation of DNA fragments of desired size. Using the first type of library it is possible to jump over long distances (>200 kb) but only from certain starting points containing the recognition site for the rare cutting enzyme. Using the hopping library it is possible to start jumping from practically any point and over a defined but shorter distance (<200 kb). Only the first type of jumping libraries can be used in conjunction with linking libraries to create genomic maps.

There are two major approaches to construction of *NotI* jumping and linking libraries. In the first method (Figure 3.4) jumping libraries are constructed as follows: DNA of high molecular mass, isolated in low-melting agarose, is completely digested with *NotI*. Then the DNA is ligated at very low concentration in the presence of dephosphorylated plasmid containing a marker (supF gene), and is encircled, trapping the supF gene, which acts a marker to select clones which contain the ends of a long fragment. The large circular molecules are digested with another enzyme (that has no recognition site in the plasmid) into small fragments, each of which is cloned in a vector phage carrying amber mutations. Recombinant phages containing the plasmid with the two terminal fragments are selected in a E. coli strain lacking the suppressor gene.

The linking library can be constructed in different ways. In the original (and still the most popular) protocol, the genomic DNA is partially digested with Sau3A and size selected to obtain 10–20 kb fragments. The DNA is then diluted and encircled in the presence of supF marker plasmid. The circular products are digested with NotI, ligated into a NotI digested suppressor-dependent vector (NotEMBL3A) and plated on a suppressor negative host.

One modification is based on the use of a genomic library in a circular form (e.g. plasmid). DNA isolated from the total library is digested with NotI and separated using specific conditions for PFGE in which linear and circular DNAs segregate into two discrete zones by size. The zone containing linear DNA fragments is cut out, the DNA is eluted, self-ligated and introduced into E. coli.

In another approach, DNA from a total genomic library is digested with NotI, and a selectable marker (e.g. resistance to the antibiotic) is inserted into recombinants containing this site. Such recombinants are then selected by their resistance to the antibiotic.

The most important drawback is that all these methods to construct linking libraries exploit different strategies and vectors compared to the construction of jumping libraries. It means that some fragments present in one library (e.g. jumping) will be absent in another (e.g. linking), which creates serious problems for their use in mapping.

An integrated approach for construction of jumping and linking libraries is outlined in Figure 3.4. The most important feature here is that the same vectors and protocol are used for construction of both libraries. For the linking library, genomic DNA is completely digested with *BamHI*. Subsequently, the DNA is self-ligated at a very low concentration (without a supF marker), to yield circular molecules as the main product. In order to eliminate any linear molecules, the sticky ends are partly filled-in with the Klenow fragment in the presence of dATP and dGTP. Since the Klenow fragment has exonuclease activity all the *BamHI* sticky ends are neutralized and nearly all ends generated during random DNA breakage become unavailable for ligation. The DNA is then cut with NotI and cloned in lambda SK4, lambda SK17 and lambda SK22 vectors.

The same strategy is applied for the construction of a *NotI* jumping library. One initial step is added: genomic DNA is fully digested with NotI and self-ligated at a very low concentration. The subsequent steps are the same as for the linking library. An obvious difference between this approach and the previous ones is that this procedure combines a biochemical selection for NotI jumping fragments with improved ligation kinetics during the preparation of the libraries.

Chromosome jumping has been used to construct *NotI* maps of human chromosomes 3, 11, and 21, and to map and analyze genes involved in cystic fibrosis, Alzheimer's disease, amyotrophic lateral sclerosis, paraganglioma 2, other genetic diseases, and different leukemias and solid tumors.

A linking library is a collection of cloned genomic DNA fragments that contain internal sites for rare cutting enzymes. A linking library in combination with PFGE is sufficient to construct a physical chromosomal map. Linking libraries can be constructed using any restriction enzyme, but the most well known and useful are *NotI* linking libraries, and as the principles to construct all linking libraries are the same, only *NotI* linking libraries will be described here. In some studies, *NotI* boundary libraries, containing clones with only one half of the *NotI* linking clones, have been used. Linking libraries have been useful for isolating many genes including oncogenes and tumor suppressor genes and mapping the chromosomal regions containing tumor suppressor genes involved in lung, breast, renal, and other cancers.

When linking clones are used as probes for hybridization with a PFGE genomic blot, each clone reveals two DNA fragments, adjacent in the genome. In theory, it is possible to order the *NotI* sites within a single linking library, but in reality, this is not feasible as many fragments are the same size. To solve these ambiguities, different linking libraries constructed with different rare cutting enzymes should be used and a restriction map can be created just as in ordinary restriction fragment analysis. The use of jumping and linking libraries in a complementary fashion simplifies this approach. By successive screening of the two libraries, it is possible, in principle, to jump from clone to clone (–jumping–linking–jumping– etc.) and to construct an ordered map without using PFGE. However, establishing the order of linking and jumping clones using the hybridization approach has some limitations due to the high level of background hybridization between unrelated *NotI* clones. Another way to establish this order is shotgun sequencing of *NotI* linking and jumping clones.

When linking clones are used to probe a PFGE genomic blot, each clone should reveal two DNA fragments, and they must be adjacent in the genome. In principle, with just a single library and digest, one should be able to order the rare cutting sites, although it will not generally be possible to distinguish between two fragments of the same size. To resolve such ambiguities it is important to use several different libraries, each for a particular enzyme and overlap the resulting patterns just as in ordinary restriction fragment analysis.

The first difficulty with the schemes outlined above arises due to repetitive sequences in genomic DNA. This difficulty can be minimized in several ways but ultimately some regions of mammalian genomes will still contain too many repetitive sequences to be mapped easily. Secondly, different vectors and protocols are used for the construction of these libraries, so jumping and linking libraries are not really complementary, and genomic DNA fragments present in one library might be absent in another for different reasons: different cloning capacity of the vectors, different fragment sizes when different restriction enzymes have been used to construct jumping and linking libraries, etc. Thirdly, jumping libraries would not be representative due to the small cloning capacity of the vectors used. This is also obvious from the fact that the largest jump isolated is 450 kb while the average size of the *NotI* fragments is estimated to be about 500–1000 kb. Fourthly, application of these schemes for the whole human genome would be a very laborious task and the small cloning capacity of the vectors used would not permit isolation of a sufficiently representative set of *NotI* jumping clones specific for a particular chromosome using, for instance, human-rodent hybrid cell lines. Finally, different *NotI* linking clones have been shown to hybridize with each other due to homologous CG rich stretches of the DNA. This complicates the cloning procedure and leads to the isolation of non-related clones.

For linking libraries, genomic DNA is completely digested with *BamHI*. (Libraries can also be constructed with *BglII*, *PstI* etc., but not with frequently cleaving enzymes like *Sau3A* or infrequently cutting enzymes like *XhoI*). Subsequently, the DNA is self-ligated at a very low concentration (without a supF marker), to yield circular molecules as the main product. In order to eliminate any remaining linear molecules, the sticky ends are partly filled-in using the Klenow fragment in the presence of dATP and dGTP. This way, all the *BamHI* sticky ends are neutralized and since the Klenow fragment also has exonuclease activity, nearly all ends generated upon random DNA breakdown become unavailable for ligation. The DNA is then digested with *NotI* and ligated to lambda SK17 and lambda SK22 "arms", produced by the combined digestion with *NotI* and *EcoRI.* The ligation product is packaged *in vitro* into phage lambda particles. The phage particles are then used to infect E. coli cells in which only recombinant phages can grow. The same strategy can be applied to the construction of a NotI jumping library. One initial step may be added: genomic DNA is first completely digested with NotI and self-ligated at a very low concentration. The subsequent steps are the same as for the linking library (Myers *et al*, 2002; Poustka and Lehrach, 1988).

This strategy has the following advantages, compared to previous approaches:

1 The same scheme and the same vectors are used for the construction of jumping and linking libraries. The use of two vectors instead of one widens the size-range of cloned fragments (0.2–19 kb);
2 It is possible to use complete BamHI digestion instead of partial Sau3A digestion, because of the larger cloning capacity of SK17 and SK22 vectors;
3 The supF marker is not used. This simplifies the procedure;
4 The partial filling-in of sticky ends makes it possible to clone only the products of regular ligation, i.e. circular molecules;
5 The application of biochemical selection (cleavage of phage DNA by two restriction enzymes) and genetic selection yields almost 100% recombinants;
6 The library can be converted into plasmid and single-stranded forms;
7 The NotI linking and jumping clones are suitable for direct sequencing using standard sequencing primers, which is very useful for the STS-approach (STS-sequence tagged site);
8 After encircling and partial filling-in, the genomic DNA can be used to prepare any other linking library simply by digesting with the appropriate enzyme. Dephosphorylation or other additional operations are not required.

The procedure is very simple, does not take much time and labor, requires no DNA fractionation or dephosphorylation, and is effective.

As a consequence of these modifications only 1–2 ug of vector arms need be packaged to obtain a representative jumping library. Previous protocols called for large amounts of vector arms (40 ug) and required specially prepared packaging extracts. The efficiency of the new approach for construction of NotI jumping libraries is evident from the fact that even before biological counter-selection in the P2-lysogenic E. coli strain NM646, a significant proportion (about 50%) of the phages were recombinant, whereas previous methods gave only about 0.1% recombinants. These libraries also contain jumps over larger distances than described previously. Using procedures described above we have constructed numerous jumping and linking libraries. These pilot experiments have, however, revealed some limitations of the modified procedures.

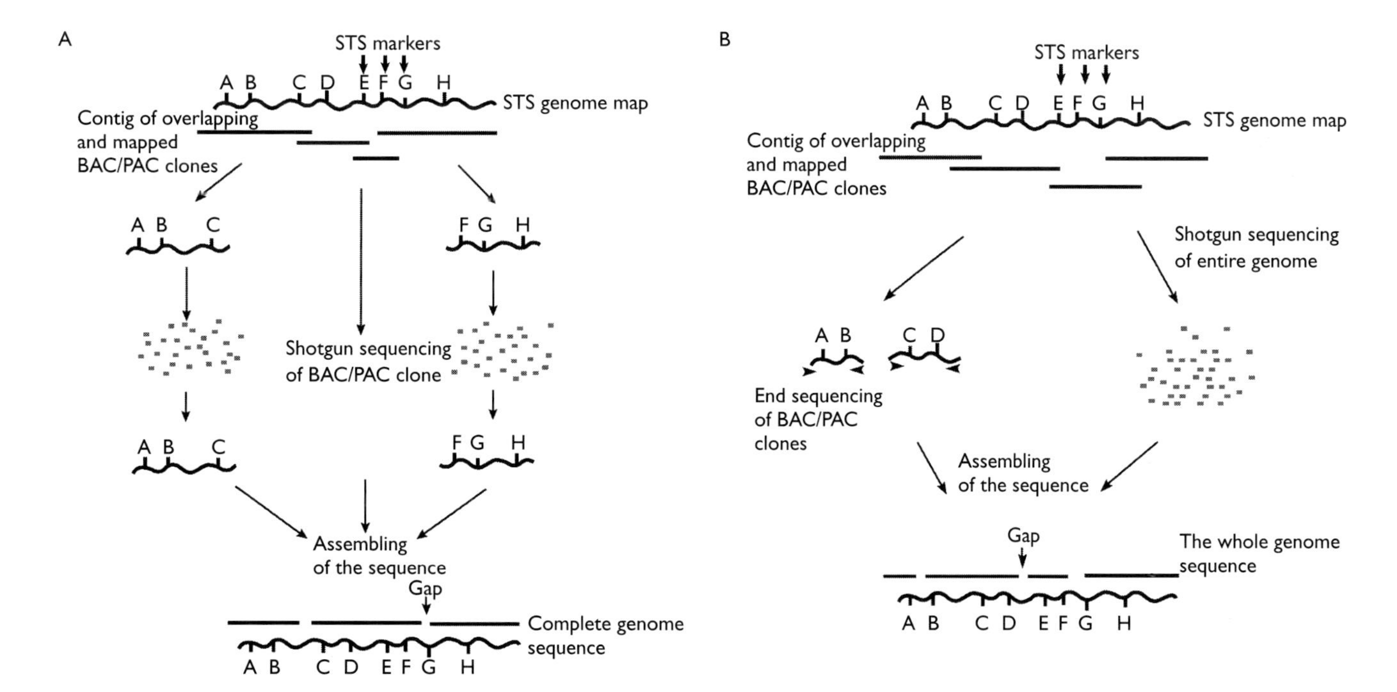

Figure 3.6 Shotgun approaches to genome sequencing. (A) Approach based on shotgun sequencing of BAC/PAC clones. (B) The whole genome shotgun sequencing scheme.

The same DNA and vectors are used and essentially the same scheme is employed to construct NotI linking and jumping libraries (Figure 3.6). These two types of libraries contain DNA fragments of the same physical structure, but arranged differently. As a consequence of the scheme for construction of the libraries, the NotI sites are, in both cases, adjacent to direct and reverse sequencing primers in the vector arms. The libraries can easily be transmitted to plasmid form and sequencing reactions can be performed, without recloning. Sequence information about 500–600 bp surrounding each NotI site can thus be obtained easily by automated sequencing. Since the efficiency for construction of the libraries is high, it is easy to obtain sets of *NotI* clones for a specific chromosome or for a whole genome, increasing the representation of the library. Inserts in the libraries are large enough to permit isolation of chromosome specific clones from mouse-human microcell hybrid cell lines by hybridization with total human DNA.

3.5.7 A shotgun sequencing strategy for genome mapping

For example, human chromosome 3 contains 200–300 *NotI* sites. Random sequencing of 600–1000 clones from each (linking and jumping) library will give information about a considerable fraction of all these *NotI* sites. Subsequently the linear order of the *NotI* clones on chromosome 3 can be established using a computer program. The resulting map, however, does not take into account the physical distance between the *NotI* sites. This additional information can be obtained using a combination of PFGE and in situ hybridization on metaphase chromosomes. Obviously, the same approach can be used to construct a NotI map for the entire human genome.

The problem of numerous repeats in the human genome is not expected to create serious problems. It is known that *NotI* sites are located preferentially in the 5′ regulatory regions of the genes. It is therefore reasonable to expect that unique sequences are located near *NotI* sites. Some of them may be quite short if a repeat is located near a *NotI* site, but even a 20 bp sequence is likely to uniquely identify a sequence in the human genome. Moreover, the sequence data provide a means to discriminate between different occurrences of the same class of repeats. This is not possible using other mapping approaches (hybridization or restriction enzyme mapping). This approach will also yield STS information for *NotI* sites.

After completing the NotI restriction map, a more detailed map (100–300 kbp resolution) can be constructed using *XmaIII*, the recognition site of which is nested within the *NotI* site. Other enzymes (SalI, *XhoI*) can also be used to create linking/jumping libraries with the SK17 and SK22 vectors.

Special lambda vectors (lambda SK4, lambda SK15 and lambda SK22) for the construction of jumping and linking libraries have been created using an integrated scheme. Genomic DNA is completely digested with *Eco*RI or with another 6-bp cutter. Then the DNA fragments are self-ligated and the sticky ends of the remaining linear molecules are partly filled-in with the Klenow fragment in the presence of dATP. This way, all the sticky ends are destroyed and practically all the ends originating from linear molecules become unavailable for ligation. The resulting DNA sample is digested with *NotI* and ligated to the vector arms. The phage particles are then grown in an E. coli strain in which only recombinant phages can replicate. These libraries can be converted into plasmid forms. The *NotI* linking clones are suitable for sequencing using standard primers and this sequence information can be used for generation of sequence tagged sites (STS). It has also been demonstrated that sequencing these clones is a very efficient method of gene isolation because even short nucleotide sequences (approximately 500 bp from each side) flanking *NotI* sites are sufficient to detect genes. This approach has been used for isolating and mapping many new genes. The *NotI* sites are linked to genes because practically all *NotI* sites are localized in CpG islands, which are tightly associated with genes.

A jumping library is a collection of cloned DNA fragments that contains the ends of large genomic DNA fragments. Chromosome jumping can be performed using jumping and linking libraries. In principle, chromosome jumping can be done without linking libraries, with the help of general libraries containing DNA fragments generated by partial digestion with restriction enzymes or physical shearing. However, the use of linking clones makes it significantly easier to move along the chromosome. In this case, linking clones link two jumping clones and jumping clones jump from one linking clone to another. Jumping clones are also useful for constructing PFGE maps and for gene isolation. Jumping libraries have been helpful for isolating genes involved in cystic fibrosis and acute nonlymphocytic and myeloid leukemias, and mapping chromosomal regions containing tumor suppressor genes involved in lung, breast, renal, and other cancers.

Different types of jumping libraries have been suggested, the most well known being the *NotI* jumping library and the hopping library. The basic principle for constructing jumping libraries is to generate clones that contain only the ends of large DNA fragments, rather than a continuous DNA segment. The DNA between the two ends of the fragments is deleted with restriction enzymes, and the clones containing only the ends of the large DNA molecules are enriched by genetic or biochemical selection. Four main techniques have been suggested for constructing jumping libraries.

The first method involves the following steps: DNA with a very high molecular mass is isolated in agarose blocks and is either completely (for the NotI jumping libraries) or partially (for the hopping libraries) digested. Then the DNA is ligated at very low concentration in the presence of a marker (supF gene), and is encircled, trapping the supF gene, which works as a selective marker to detect clones that contain the ends of a long DNA fragment. The resulting DNA sample is digested with EcoRI or BamHI into small fragments, each of which is cloned in a lambda phage vector carrying amber mutations, e.g. lambda Ch3Adlac or lambda 1151ABS. Only the recombinant phages containing the supF gene can grow in selective E. coli strains. Moving along the chromosome can then be achieved by successive hybridization screening of jumping and linking libraries.

The technique described above for the construction of jumping libraries has several drawbacks. The first problem is the presence of repetitive sequences in genomic DNA and the high CG content in *NotI* clones. Thus, different *NotI* clones hybridize with each other. This complicates the cloning procedure and leads to the isolation of non-related clones. Second, different vectors and protocols have been used for the construction of jumping and linking libraries, so DNA fragments present in one library might be absent in another for different reasons, e.g. the small cloning capacity (0–12 kb) of the vectors used. Low representation of the libraries is also apparent from the fact that the largest jump isolated is 450 kb, while the average size of the *NotI* fragments is estimated to be about 500–1000 kb. It is important to note that construction of jumping libraries according to this scheme is very laborious and expensive. The use of the supF selective marker itself creates intrinsic problems because for different sized genomic fragments, different concentrations of the supF marker should be used, and the selective strength of the supF marker is dependent on flanking sequences.

3.5.8 Integrating experimental data in to a genetic map

Linking and jumping clones can be used to construct a physical map of chromosomes. In theory, a linking library is sufficient to construct a physical chromosomal map. When linking clones are used to probe a PFGE genomic blot (*NotI* digested genomic DNA), each clone should reveal two DNA fragments, which are adjacent in the genome. Thus in principle, one should be able to order the rare cutting sites with just a single library and one digest, although it will not generally be possible to distinguish between two fragments of the same size. To resolve such ambiguities it is important to use several different libraries, each for a particular enzyme and to overlap the resulting patterns just as in ordinary restriction fragment analysis. To accomplish this for the whole human genome will be a very laborious work. But for small stretches of the genome containing 5–10 NotI sites, this approach can be efficient.

In the shotgun sequencing approach for long-range genome mapping, hybridization technique is replaced by sequencing. In the case of individual chromosomes, the main points of the shotgun sequencing strategy are the following: (1) Jumping and linking libraries are constructed using the same scheme and the same vectors; (2) The NotI sites in both libraries are adjacent to direct and reverse sequencing primers available for sequencing reactions without any recloning procedure; (3) An average human chromosome contains about 150 NotI sites. Random sequencing of 1000 clones from each (linking and jumping) library will give sequence information (300–500 bp) about a considerable fraction of all these NotI sites. Subsequently, the linear order of the NotI clones on a chromosome can be established using a computer program. Even a 20 bp sequence is likely to uniquely identify a sequence in the human genome. The sequence data provide a means to discriminate even between different instances of the same class of repeats.

This approach (using CpG islands as landmarks for genome mapping) should lead to identification of new genes and join genetic and physical mapping. The approach can be used in a complementary fashion for creating YACs and cosmid contigs and results in generation of new STS. Since the average size of the inserts in SK vectors is greater than 6 kb, practically every linking clone will contain some microsatellites.

In summary, while several different strategies are available to obtain and use DNA markers for identifying and mapping DNA sequences in complex organisms, no single system is likely to suffice for obtaining a complete and accurate map of the human genome. Rather, a combination of different approaches and vector systems are needed to corroborate data from different sources.

Creating contig maps involves ordering the cloned chromosomal fragments. The ordered fragments form contiguous DNA blocks (contigs). Currently, the resulting "library" of clones varies in size (several megabases). An advantage of this approach is the accessibility of these stable clones to other researchers. Contig construction can be verified by FISH, which localizes cosmids to specific regions within chromosomal bands. Contig maps thus consist of a linked library of small overlapping clones representing a complete chromosomal segment. While useful for finding genes localized to a small area (under 2 Mb), contig maps are difficult to extend over large stretches of a chromosome because all regions are not clonable. DNA probe techniques can be used to fill in the gaps, but they are time consuming.

CpG islands are powerful markers for genome mapping. Although human DNA is highly methylated, stable unmethylated sequences (about 1% of the genome) have been observed in human chromosomal DNA. Such sequences occur as discrete "islands", usually 1–2 kb long, that are dispersed in the genome. They are usually called CpG (rich) islands because they contain more then 50% of C + G (the human genome contains on average about 40% C + G). Their distinctive feature is the presence of a CpG pair at predicted frequency, whereas elsewhere in the genome it is present at a frequency of less than 25%. Altogether, there are about 30,000 islands in the haploid genome (the average spacing is about one per 100 kb). It is now clear that the majority (if not all) CpG islands are associated with genes.

Since one of the main goals of the human genome project is the isolation of all genes and the construction of a transcriptional, gene map, it is clear that markers located in the CpG islands have an additional value for physical mapping. It has been shown that recognition sites for many of the rare cutting enzymes are closely associated with CpG islands. For example, at least 82% of all *NotI* and 76% of all *XmaIII* sites are located in the CpG islands. More than 20% of CpG island-containing genes have at least one *NotI* site in their sequence, while about 65% of those genes have *XmaIII* site(s). Summarizing the data for all genes (with or without CpG islands), we can conclude that approximately 12% of all well characterized human genes contain *NotI* sites, and 43% of them have *XmaIII* sites. For human genome mapping this means that by sequencing DNA fragments containing NotI sites it is possible to tag up to 1/5 of all expressed genes. The recombinant clone containing a NotI (or other rare cutting enzymes) recognition site is called a linking clone.

The ultimate physical map of the human genome is the complete DNA sequence – the determination of all base pairs on each chromosome. The completed map will provide biologists with the ultimate template for studying human biology and will enable medical researchers to begin to unravel the mechanisms of inherited diseases. Much effort continues to be spent locating genes; if the full sequence were known, emphasis could shift to determining gene function. The Human Genome Project is creating research tools for 21st-century biology, when the goal will be to understand the sequence and functions of the genes residing therein. Achieving the goals of the Human Genome Project will require substantial improvements in the rate, efficiency, and reliability of standard sequencing procedures. While technological advances are leading to the automation of standard DNA purification, separation, and detection steps, efforts are also focusing on the development of entirely new sequencing methods that may eliminate some of these steps.

Obtaining a substantially uniform representation of the genome using recombinant DNA clones was itself a problem, up to a few years ago. The problem was solved by the DOE-supported development of the more stable and larger recombinant BACs (bacterial artificial chromosomes). To support the contig building requirements of sequencers, sequence tag connectors (STCs) for the BACs are now being generated. The BACs whose STCs overlap an already sequenced region are candidate clones for extension of the sequence.

Large-insert, ordered BAC libraries in forms of high density colony filters, duplicated library, clones and DNA pools are essential tools of genomic research as well as for structural and functional mapping. In the late 90's genome-wide BAC and binary BAC physical map contigs became available that are ready for large-scale genome sequencing and human cell transformation. The BAC and binary BAC contigs are integrated with the developed genetic maps. New large-insert BAC and binary BAC libraries and genome-wide sequence-ready physical maps of human genome from ordered BAC and/or binary BAC libraries have also been developed.

Because coding sequences of genes represent most of the potentially useful information content of the genome (but are only a fraction of the total DNA), some investigators have begun partial sequencing of cDNAs instead of random genomic DNA. In addition to providing unique markers, these partial sequences also identify expressed genes. This strategy can thus provide a means of rapidly identifying most human genes. Other applications of the EST approach include determining locations of genes along chromosomes and identifying coding regions in genomic sequences.

Starting maps and sequences is relatively simple; finishing them will require new strategies or a combination of existing methods. After a sequence is determined using the methods described above, the task remains to fill in the many large gaps left by current mapping methods. One approach is single-chromosome microdissection, in which a piece is physically cut from a chromosomal region of particular interest, broken up into smaller pieces, and amplified by PCR or cloning. These fragments can then be mapped and sequenced by the methods described previously. Chromosome walking, one strategy for filling in gaps, involves hybridizing a primer of known sequence to a clone from an unordered genomic library and synthesizing a short complementary strand (called "walking" along a chromosome). The complementary strand is then sequenced and its end used as the next primer for further walking; in this way the adjacent, previously unknown, region is identified and sequenced. The chromosome is thus systematically sequenced from one end to the other. Because primers must be synthesized chemically, a disadvantage of this technique is the large number of different primers needed to walk a long distance. Chromosome walking is also used to locate specific genes by sequencing the chromosomal segments between markers that flank the gene of interest.

3.5.9 Large-scale ordered EST and cDNA clones of the genome in the form of microarrays

After mapping is completed, the next step is to determine the base sequence of each of the ordered DNA fragments. The ultimate goal of genome research is to find all the genes in the DNA sequence and to develop tools for using this information in the study of human biology and medicine. Improving the instrumentation and techniques required for mapping and sequencing – a major focus of the genome project – will increase efficiency and cost-effectiveness. Goals include automating methods and optimizing techniques to extract the maximum useful information from maps and sequences. A genome map describes the order of genes or other markers and the spacing between them on each chromosome. Human genome maps are constructed on several different scales or levels of resolution. At the coarsest resolution are genetic linkage maps, which depict the relative chromosomal locations of DNA markers (genes and other identifiable DNA sequences) by their patterns of inheritance. Physical maps describe the chemical characteristics of the DNA molecule itself.

Impact of the human genome project

The atlas of the human genome will revolutionize medical practice and biological research into the 21st century and beyond. All human genes will eventually be found, and accurate diagnostics will be developed for most inherited diseases. In addition, animal models for human disease research will be more easily developed, facilitating the understanding of gene function in health and disease. Researchers have already identified single genes associated with a number of diseases, such as cystic fibrosis, Duchenne muscular dystrophy, myotonic dystrophy, neurofibromatosis, and retinoblastoma. As research progresses, investigators will also uncover the mechanisms for diseases caused by several genes or by a gene interacting with environmental factors. Genetic susceptibilities have been implicated in many major disabling and fatal diseases including heart disease, stroke, diabetes, and several kinds of cancer. The identification of these genes and their proteins will pave the way to more-effective therapies and preventive measures. Investigators determining the underlying biology of genome organization and gene regulation will also begin to understand how humans develop from single cells to adults, why this process sometimes goes awry, and what changes take place as people age. New technologies developed for genome research will also find myriad applications in industry, as well as in projects to map (and ultimately improve) the genomes of economically important farm animals and crops. While human genome research itself does not pose any new ethical dilemmas, the use of data arising from these studies presents challenges that need to be addressed before the data accumulate significantly. To assist in policy development, the ethics component of the Human Genome Project is funding conferences and research projects to identify and consider relevant issues, as well as activities to promote public awareness of these topics.

Once the neighborhood of a gene of interest has been identified, several strategies can be used to find the gene itself. An ordered library of the gene neighborhood can be constructed if one is not already available. This library provides DNA fragments that can be screened for additional polymorphisms, improving the genetic map of the region and further restricting the possible gene location. In addition, DNA fragments from the region can be used as probes to search for DNA sequences that are expressed (transcribed to RNA) or conserved among individuals. Most genes will have such sequences. Then individual gene candidates must be examined. For example, a gene responsible for liver disease is likely to be expressed in the liver and less likely in other tissues or organs. This type of evidence can further limit the search. Finally, a suspected gene may need to be sequenced in both healthy and affected individuals. A consistent pattern of DNA variation when these two samples are compared will show that the gene of interest has very likely been found. The ultimate proof is to correct the suspected DNA alteration in a cell and show that the cell's behavior reverts to normal.

The current human genetic map has about 550,000 markers, or 1 marker spaced every 6 thousand bp; practically no genes lie between each pair of markers. Higher-resolution genetic maps and even sequences have been made in regions of particular interest. New genes can be located by combining genetic and physical map information for a region. The genetic map basically describes gene order. Rough information about gene location is sometimes available also, but these data must be used with caution because recombination is not equally likely at all places on the chromosome.

The degree of difficulty in finding a disease gene of interest depends largely on what information is already known about the gene and, especially, on what kind of DNA alterations cause the disease. Spotting the disease gene is very difficult when disease results from a single altered DNA base; sickle cell anemia is an example of such a case, as are probably many major human inherited diseases. When disease results from a large DNA rearrangement, this anomaly can usually be detected as alterations in the physical map of the region or even by direct microscopic examination of the chromosome. The location of these alterations pinpoints the site of the gene. Identifying the gene responsible for a specific disease without a map is analogous to finding a needle in a haystack. Actually, finding the gene is even more difficult, because even close up, the gene still looks like just another piece of hay. However, maps give clues on where to look; the finer the map's resolution, the fewer pieces of hay to be tested.

The current map of the human genome comprises approximately 350,000 clones (either as "bacterial-" or "P1-derived" artificial chromosomes: "BAC" or "PAC" clones, respectively). These clones have been assembled into 1500 contigs using a similar fingerprinting principle to that employed for the *C. elegans* genome project, and are all anchored to the GeneMap. 30,000 of the clones were selected for the tiling path, and sequenced by random shotgun method to provide the assembled unfinished sequence as the "working draft". Each clone is then taken through the process of directed finishing, as with the *C. elegans* project, to provide the reference sequence at an accuracy of >99.99%. The status of the project can be assessed based on the amount of sequence that is available and comparing this figure to the estimated size of the genome (3,200,000,000 bases). On this basis, 90%

of the human genome is available on the internet, and 20% is finished reference sequence (as of 26th June 2000). Another measure of completeness has been obtained by examining the genome sequence for gene content. The sequence contains 95% of all known genes, and also 97% of all genes that are implicated in human disease.

References

Adams, MD, Kelley, JM, Gocayne, JD, Dubnick, M, Polymeropoulos, MH, Xiao, H, *et al.* (1991) Complementary DNA sequencing: Expressed sequence tags and human genome project. *Science. 252(5013): 1651–1656.*

Allikmets, RL, Kashuba, VI, Pettersson, B, Gizatullin, R, Lebedeva, T, Kholodnyuk, ID *et al.* (1994) *Not*I linking clones as tools to join physical and genetic mapping of the human genome. *Genomics, 19: 303–309.*

Brown, TA (1999) *Genomes.* John Wiley & Sons, NewYork.

Collins, FS (1988) *In Chromosome Jumping in Genome Analysis: A Practical Approach.* Davis, KE (ed.) (1988), IRL Press, Oxford, England, pp. 73–94.

Myers, EW, Sutton, GG, Delcher, AL, Dew, IM, Fasulo, DP, Flanigan, MJ, *et al.* (2000) *Science 287, 2196–2204.*

Poustka, A, Lehrach, H (1988) Chromosome Jumping: A long range cloning technique. In *Genetic Engineering Principles and Methods*, Setlow, JK (ed.) Brookhaven National Laboratory, Plenum Press New York and London, vol. 10, pp. 169–193.

(1998) *Science, 279: 2019–2020.*

Smith, CL, Lawrence, SK, Gillespie, GA (1987) Strategies for mapping and cloning macroregions of mammalian genomes. In *Methods in Enzymology*. Gottesman, MM (ed.) San Diego, Academic Press, vol. 151, pp. 461–489.

Watson, JD, Gilman, M, Witkovski, J, Zoller, M (1992) *Recombinant DNA* (second edition). Scientific American Books, Inc., New York.

Zabarovsky, ER (1996) Cloned DNA markers. In *The Encyclopedia of Molecular Biology. Fundamentals and Applications.* Meyers, R (ed.), VHS Publishers Inc., New York, pp. 66–78.

Chapter 4

Molecular diagnosis of signal transduction related diseases

István Ember and István Kiss

Contents

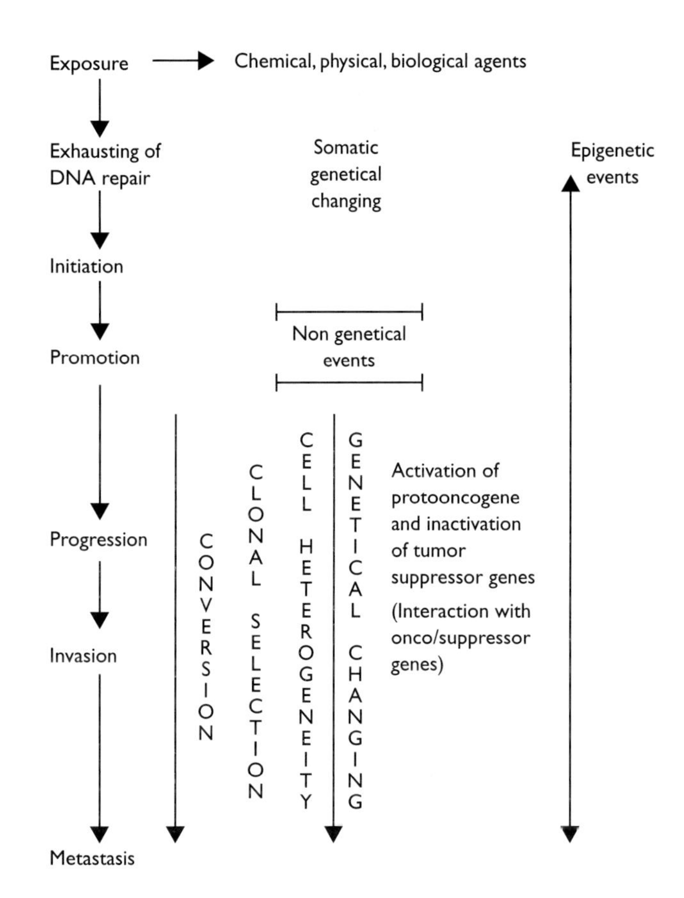

Figure 4.1 Scheme of multistage carcinogenesis at gene level.

4.1 Introduction

It has long been known that cancer development includes genetic components. Our most recent understanding is that cancer is a genetically based disease, which is induced in 85–90% of cases by environmental factors (Doll and Peto 1983, Macload and Ford, 1991). Some decades ago viruses were considered to be significant external causes of cancer, but later environmental aspects were implicated (Ember 1993). Discovery of oncogenes and tumor suppressor genes has brought a breakthrough in cancer research, serving as a solid base for the step by step unfolding of the molecular events from the very first alteration to the tumor formation (Figure 4.1). Rapid advances in the field of molecular biology ensured the further development of oncogene and suppressor gene research, unfolding new aspects of carcinogenesis. Comparing molecular changes (affecting the oncogenes and tumor suppressor genes) with the morphological changes resulted in a focus on molecular and predictive approaches to environmentally-induced carcinogenesis. Since 1941 carcinogenesis (especially chemical carcinogenesis) has been considered as a two-step process of initiation–promotion (Berenblum 1979), but since the early eighties, the five-step model has been widely accepted (Harris 1985). In this model we are able to match specific onco/suppressor changes to these stages, and the model is able to explain the whole process of cancer formation, independently from the immediate causative agent (Figure 4.1). However, this model cannot be mechanistically applied to all cases, since there is room for individual variation.

As a consequence, onco/suppressor gene alterations are to be measured at population level, and evaluated with epidemiological-statistical methods. Recognition of the need for population level studies and epidemiological evaluation has lead to the formation of a new field of sience: molecular epidemiology (or molecular and predictive epidemiology) (Harris 1985, Hulka *et al.* 1990, Schulte and Perera 1993). Molecular epidemiology has grown from traditional epidemiology, with the incorporation of molecular biological methods and techniques (Perera 1995) (Figure 4.2). One of its basic concepts and ideas was the introduction of "molecular biomarkers", and connecting them to stages of carcinogenesis. In the early stages onco/suppressor gene alterations may be present without any sign of morphological change or clinical symptoms, preceding them by months, years or even decades (Figure 4.1). This opens up the possibility of early and effective prevention, using early biomarkers, including primary, secondary and tertiary prevention as well. By applying molecular epidemiology, it is possible to create onco/suppressor gene level screening strategies for the healthy (primary prevention) or clinically healthy (secondary prevention) population and for cancer patients (tertiary prevention) (Figure 4.3).

The first predictive-sequential model was created for colorectal carcinogenesis by Fearon and Vogelstein in 1990 (Figure 4.4). To date, there are numerous similar models for other tumor types (e.g. head and neck, thyroid, gastric, pancreas, hepatocellular, breast, lung, kidney cancer, melanoma and different leukemias and lymphomas (Yamamoto 1995). Using these models at population level, the early stages of carcinogenesis could be recognized (without micro and macromorphological pictures and clinical symptoms) enabling effective prevention (and sometimes chemoprevention, Figures 4.1 and 4.3).

We have focused on onco/suppressor genes, which are involved in the carcinogenesis process as markers of the effect of exposure (Figure 4.1). The majority of oncogenes are also involved in signal transduction pathways. All these discoveries have immediate practical significance: since the oncogene and suppressor gene changes follow the model-sequence in most cases, they can be used in early recognition of precancerous lesions, early tumors, and also for diagnostic purposes (molecular pathology).

The incorporation of the new molecular, onco/suppressor gene level screening and diagnostic methods do not solve certain problems, and several new questions are generated.

4.2 Problems

1 Unfortunately the cure is often lacking proper therapeutic methods – except that of very early diagnosis and certain type of tumors.
2 If the molecular methods are applied as diagnostic tools – molecular pathology – then much more precise, but not earlier, diagnosis can be established than with traditional methods.

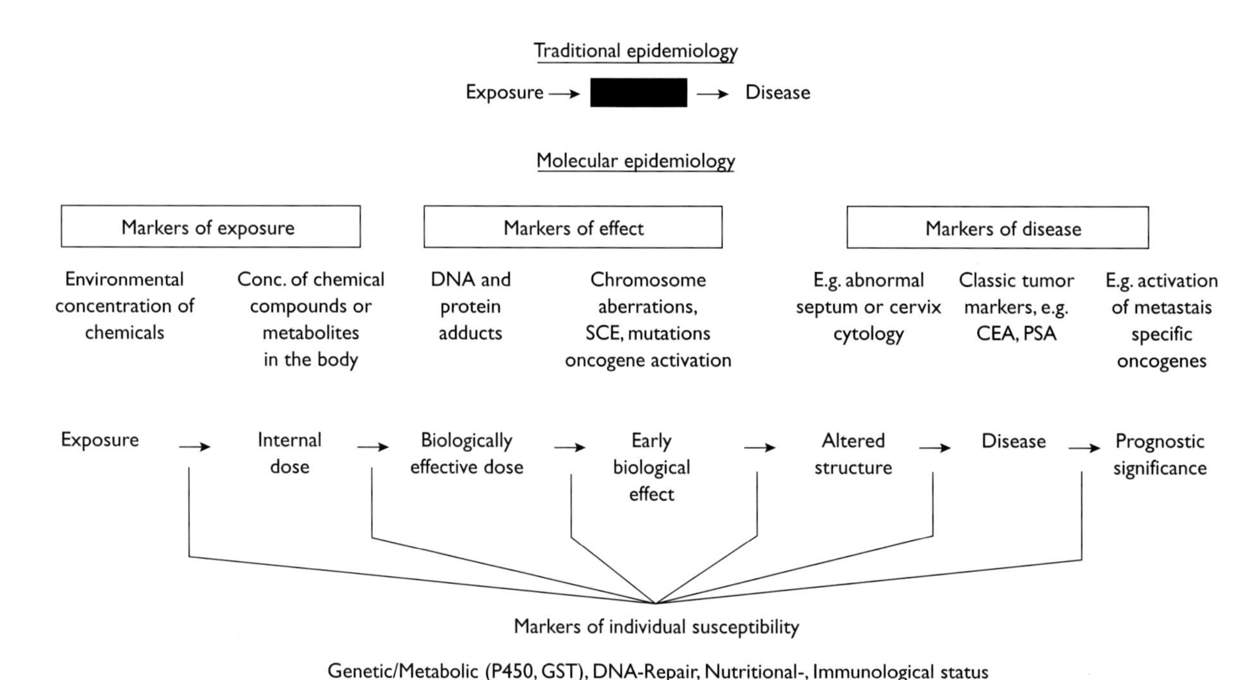

Figure 4.2 Traditional and molecular epidemiology (modified from Schulte).

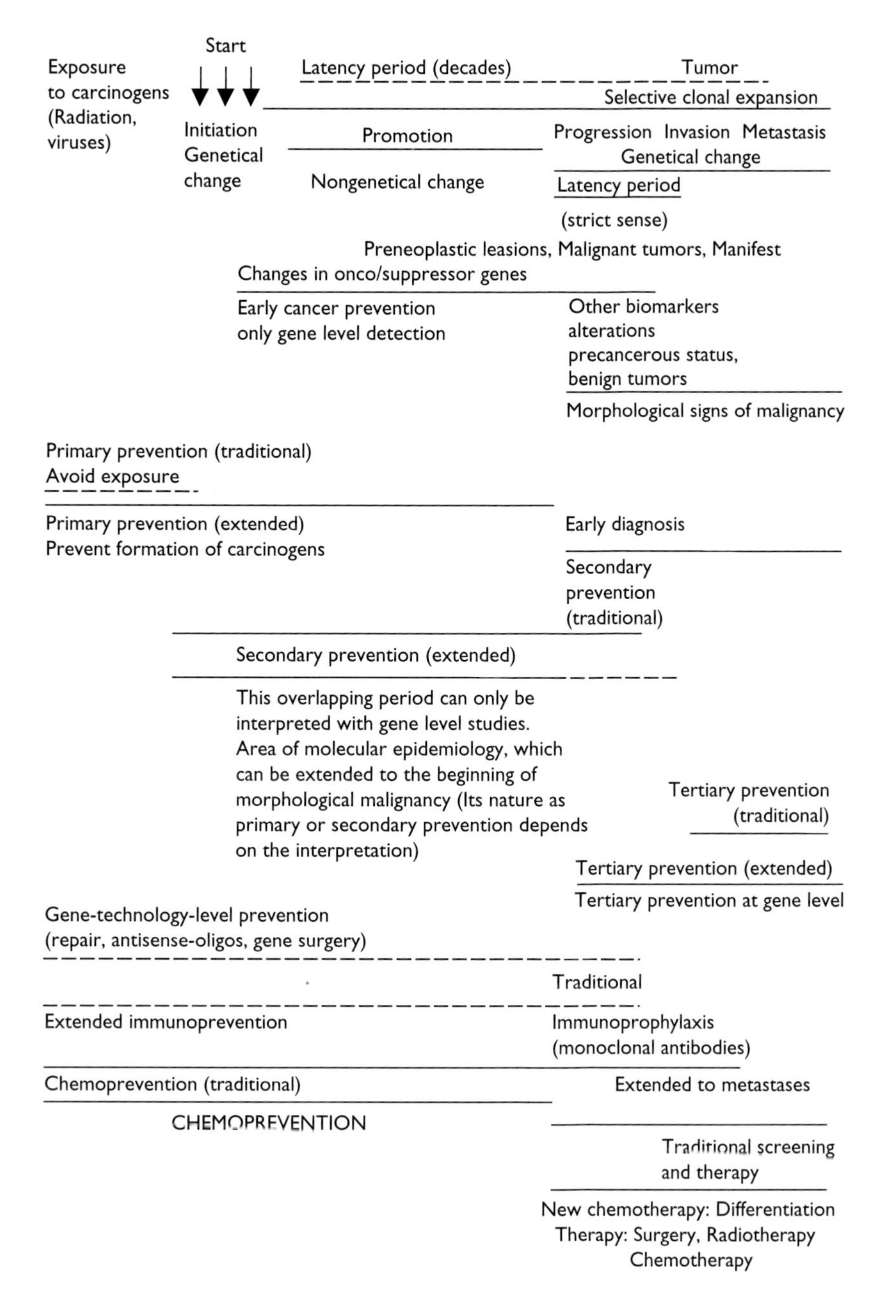

Figure 4.3 Changes in level of prevention, due to incorporation of molecular biological methods, Ember (1996).

3 There is an increased risk of false negative and false positive results, particularly because most of these molecular markers work much better at population level than at individual level.
4 Carrying out molecular genetic examinations without having a solid therapeutic method raises ethical problems.

Considering all the above issues, these types of molecular screening and diagnostic tools are still useful methods for cancer treatment and prevention. What are their main advantages?

4.3 Advantages

1 Their applicability at population level. When using appropriate epidemiological methods, these onco/suppressor gene studies give useful and informative results, without the previous uncertainty experienced at individual level. This raises the possibility of carrying out large

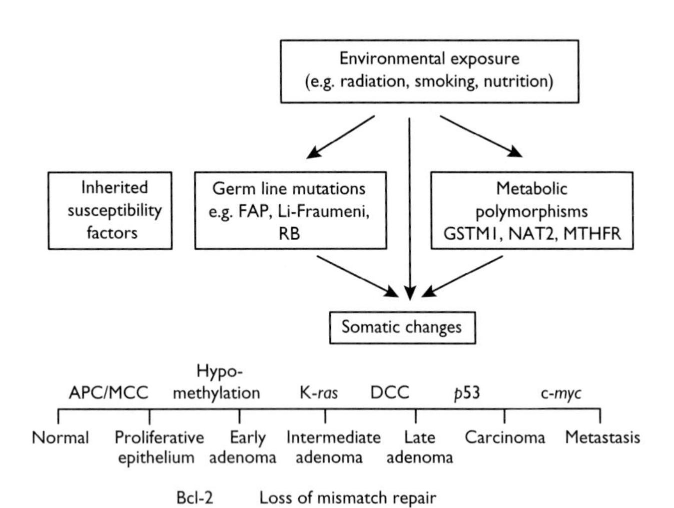

Figure 4.4 Multistep model of human colorectal carcinogenesis.

preventive studies and checking the results at molecular level. The potential for continuous gene-level monitoring provides the researcher with practical results. Similarly, new therapeutic protocols can also be evaluated.

2 Early genetic biomarkers make intervention possible before the appearance of clinical or even preclinical signs of the disease. This is best ensured by chemoprevention, an area of cancer prevention that has undergone rapid development over recent years.
3 Genetic markers can, in certain cases, be used to identify a high-risk, exposed, and/or genetically altered population. The high-risk population can be examined further or screened regularly to ensure the early diagnosis of cancer.
4 The alterations of oncogenes and/or tumor suppressor genes can be used to signal or monitor carcinogenic exposure, and thus help the effective primary prevention by ceasing the exposure.

Table 4.1 Functions of proto-oncogene products in the cell

Group I	*Growth factors*	N-ras	Membrane-associated G-protein GTP/GDP transducer
Sis	PDGF B-chain growth factor	gsp	Mutant activated form of Gs GTP/GDP transducer
int-1	Growth factor?	gip	Mutant activated form of Gi
int-2	FGF-related growth factor	*Class 2*	*Cytoplasmic protein-serine kinases*
hst(KS3)	FGF-related growth factor	atk	Cytoplasmic kinase
Group II: Protein-tyosine kinases		raf/mil	Protein-serine kinase
Class 1	*Receptor type*	B-raf	Raf-like kinase
ros	Membrane-associated receptor-like protein	pim-1	Protein-serine kinase
erbB	Truncated EGF receptor	mos	Cytostatic factor
HER2/c-erbB-2/neu	Receptor-like protein (EGF)	cot	Protein-serine kinase
fms	Mutant CSF-1 receptor	*Class 3*	*Adaptors*
met	Soluble truncated receptor-like protein	crk	SH2/SH3-containing protein
trk	Soluble truncated receptor-like protein GF receptor	vav	SH2/SH3-containing protein
kit	Truncated stern cell receptor protein	cbl	Proline-rich adaptor?
sea	Membrane-associated truncated receptor-like protein	*Group IV*	*Nuclear transcription factors*
ret	Truncated receptor-like protein (cell surface)	myc	Sequence-specific DNA-binding protein nuclear GF
Class 2	*Non-receptor type*	N-myc	Sequence-specific DNA-binding protein? nuclear GF
src	Membrane-associated protein-tyrosine kinase	L-myc	Sequence-specific DNA-binding protein? nuclear GF
yes	Membrane-associated Src-like kinase	myb	Sequence-specific DNA-binding protein
fgr	Membrane-associated Src-like kinase	fos	Component of AP-1 transcription factor
lck	Membrane-associated Src-like kinase	jun	Sequence-specific DNA-binding protein; component of AP-1, nuclear GF
fyn	Membrane-associated Src-like kinase	erbA	Dominant negative mutant of thyroxine (T3) receptor
fps/fes	Cystoplasmic protein-tyrosine kinase	rel	Dominant negative mutant of $NF_{K}B$-related protein
abl	Cystoplasmic/nuclear protein-tyrosine kinase	ets	Sequence-specific DNA-binding protein
Group III	*Transducers of extracellular signals*	ski	Transcription factor?
Class 1	*GTP-binding/GTPase*	evi-1	Transcription factor?
H-ras	Membrane-associated G-protein GTP/GDP binding	qin	Transcription factor?
K-ras	Membrane-associated G-protein GTP/GDP transducer		

Modified from Yamamoto (1995).

5 The gene level follow-up of the course of a disease may help in evaluating therapeutic responses, and in choosing the adequate therapy.
6 Certain genetic markers – even at individual level – may be specific to a tumor subtype or stage (e.g. the analysis of point mutation pattern of Ha-ras or p53 genes could be specific to chemical carcinogens, or to the tumor which will develop later). Table 4.1 shows some specific genetic alterations induced environmentally.

The methods and biomarkers applied in molecular epidemiology are diverse. The markers can be specific for a certain tumor type, subtype or stage, and may be used for individual level diagnostic or prognostic purposes. On the other hand, a lot of these biomarkers are not specific enough for individual level application. These are used at population-level and are particulary suitable for identifying high-risk population (early limiting biomarkers).

Since numerous oncogenes and tumor suppressor genes play a role in signal transduction pathways, applying the described molecular epidemiological framework to genes participating in signal transduction is an important issue in cancer molecular epidemiology. It is of great advantage that the functions of several genes in the signal transduction pathways are relatively well described, which allows us to theoretically substantiate the experimental results, and connect the theory with practical applications. The main preventive efforts are being targeted at very early diagnosis. Early diagnosis at gene level may have a role in primary prevention, and in targeting chemopreventive interventions, while early diagnosis in the traditional sense helps to provide effective and early treatment. However, oncogene and tumor suppressor gene alterations are found throughout human carcinogenesis from initiation to metastasis formation, and may be monitored with molecular biological techniques (Figure 4.3).

4.4 Onco and suppressor genes as biomarkers

As Figures 4.1–4.3 illustrate, activation of oncogenes and/or inactivation of tumor suppressor genes are key events in carcinogenesis. A single genetic change never leads to complete carcinogenesis independently from the presence of an external carcinogenic – triggering – agent. Alterations of several oncogenes and/or tumor suppressor genes are needed for cancer formation. Protooncogenes and tumor suppressor genes are part of the normal genome, playing important roles in cell functions in the regulation of cell differentiation, cell cycle control, apoptosis, and signal transduction processes. Expression of these genes is generally specific to cell cycle, cell types, tissues, organs, developmental status, and age. Their ectopic activation – spatially or timely – could be pathogenic. Unfortunately certain oncogene and tumor suppressor gene alterations are not specific to cancer, they might be present in chronic or acute inflammation, benign tumors and several other diseases. On the other hand, a particular change is not necessarily present in all cancer patients, so in most cases the molecular level diagnosis must be based on several biomarkers. Therefore, analysis of population-level data is indispensable for validation of these markers. Description of step by step genetic changes during human carcinogenesis and working out multistep models for different tumor types increases the applicability of oncogene/tumor suppressor gene markers.

There is firm evidence, based on molecular genetic research, that genetic damage is involved in tumorigenesis. This damage may include point mutation, deletion, amplification, translocation or inversion. The changes in oncogenes are generally manifest as dominant effects: oncogene activation promotes cell proliferation. Today the number of known oncogenes exceeds 80, several of them are frequently overexpressed in different tumors, at RNA and protein levels too. In contrast to oncogenes, the suppressor gene exerts a negative regulation on cell proliferation.

Tumor suppressor gene alterations are considered to be recessive effects, because inactivation of both alleles is necessary for the manifestation of functional disturbance. This is the reason why "loss of heterozygosity" for tumor suppressor genes is often found in cancer. For tumor manifestation, cooperation between oncogene and tumor suppressor gene dysfunction is needed. The main functional categories of oncogenes and tumor suppressor genes are listed in Table 4.1 and Table 4.2. The tables clearly illustrate that the vast majority of oncogenes and tumor suppressor genes function as part of signal transduction pathways or in very strong interaction with them. Thus, disturbed onco/suppressor gene function can lead to dysfunction of signal transduction pathways. The location of oncogene products in the cell are shown in Figure 4.5.

The first step in these pathways is ligand-receptor binding (in carcinogenesis mitogen signals are particularly important). Most of these receptors have intracellular tyrosine kinase activity. Dimerization of receptor molecules induce the kinase activity, helping each other with interactive phosphorylation. Tyrosine kinases are able to initialize phosphorylation cascades, including serine/threonine kinases as well. This cascade may be involved in activation of other different oncoproteins. The protein kinase cascade leads to the activation of mitogen activated protein kinases (MAP kinases), which may activate transcription factors such as the c-jun or c-fos proteins. On the other hand, phosphatase enzymes inhibit the activity of these proteins. Obviously, protein kinase and phosphatase activities correlate strongly with cell proliferation or tumor promotion.

Not all proteins involved in signal transfer have protein activity. A different signal transduction mechanism involves G proteins. The most important G protein-related signal

Table 4.2 Suppressor genes involved in signal transduction

Gene	*Chromosomal localization*	*Cellular localization*	*Action*	*Organ or tissue origins of tumor*
DCC (deleted colon cancer)	18q21.3	Cell membrane	Cell adhesion and signal transduction	Colon, stomach, mouth, esophagus, uterine, blood, testis, prostate, pancreas
NF-1	17q11	Cell membrane-associated	Regulates p21 (ras) signaling pathway	Nerve, bone, skin
BRCA1	17q21	Nucleus	Transcription factor	Breast, ovary, prostate
MST1	9p21–22	Nucleus	CDK inhibition regulates cell-cycle	Skin, esophagus, pancreas, nose, lung, bladder, kidney, brain, blood, soft tissue
WT-1	11p13	Nucleus	Transcription factor regulates gene activity	Bladder, eye, sarcoma, lung, ovary, soft tissue, Wilms' tumor
p53	17p13.1	Nucleus	Transcription factor regulates gene activity and cell-cycle	Breast, colon, bladder, ovary, brain, testis, skin, soft tissues, kidney, blood, lung, esophagus, stomach, liver, prostate
RB-1	13q14	Nucleus	E2F inhibition	Eye, bone, breast, lung, esophagus, prostate, bladder, kidney, cervix

Modified from Gao & Honn (1995).

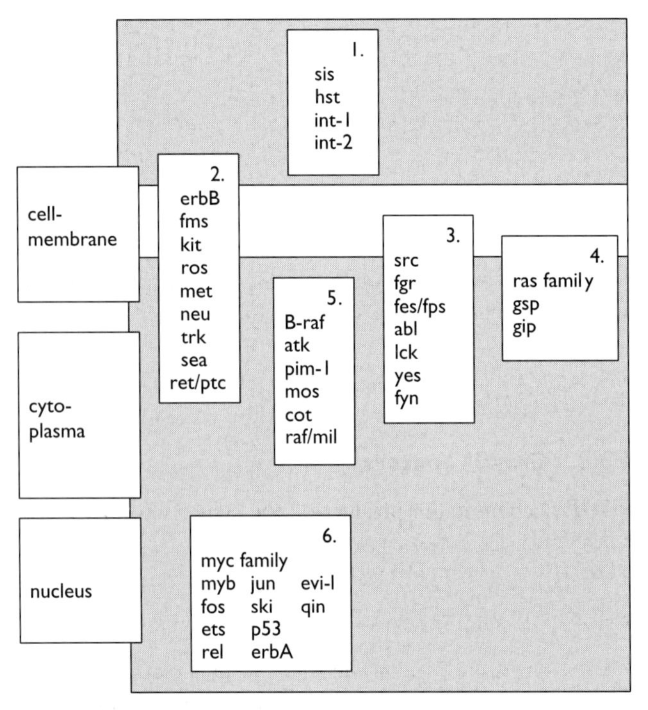

Figure 4.5 Localization of oncogene products.

transduction pathways involve activation of adenylate cyclase or phospholipase C. These enzymes catalyze formation of intracellular secondary messenger molecules, which have amplifier roles in intracellular signaling. The effect of adenylate cyclase is transmited through cAMP and protein kinase A, while phospholipase C is a key enzyme in inositol-phopholipid metabolism. It hydrolyzes the phosphatydilinositol-biphosphate to diacylglycerol and inositol-triphosphate. Inositol-triphosphate binds to Ca-ion containing vesicles in the cytoplasm, consequently Ca-ions are released into the cytosol, and bind to several Ca-dependent proteins (e.g. calmodulin). Thus it regulates the activity of different enzymes, e.g. Ca-dependent protein kinases, through which the signal can reach further transcriptional factors. The other metabolite, diacylglycerol, a molecule containing three C atoms remains at the cell membrane, and activates protein kinase C, which is a "pacemaker" molecule in mitogen signal transfer. It is important to note that promoters (e.g. phorbol-esters) are generally strong activators of protein kinase C. The activation does not lead to genetic changes, but alters the signal transduction mechanism. This makes protein kinase activity a useful molecular biomarker for promoter effect of different carcinogenic chemicals.

After initiation and promotion, carcinogenesis reaches an irreversible phase, but for progression, invasion and metastasis formation further genetic changes are necessary. Growth factors, cytokines, and receptors are produced by protooncogenes, and are important factors in paracrine regulation. Their contribution to carcinogenesis may be through paracrine or autocrine regulation.

Table 4.3 Some examples of disturbance of transcriptional factors in human tumors

Gene mistake	*Breakpoint*	*Gene product*	*Clinical manifestation*
t (15,17)	15q22 PML; 17q12–21 RARa	TF	Acute promyelocytic leukemia
PML	Contain Zn-fingerprint domain, DNA bounder protein family member		
RAR	Together with retinoic acid could bind to promoters of different genes		
t (8;14)	8q24 c-myc; 14q32 Ig heavy chain		
t (2;8)	2p12 Ig kappa chain	Gene fusion	Burkitt-lymphoma; ALL (L3)
t (8;22)	22q11 Ig lambda chain		
c- myc	Translocation on Ig locus, overexpression in heterodimer form bound to DNA (CANNTG)		
t (1;14)	1p32 tal-1; 14q11 TCR α/δ	TF	T-ALL
tal-1	Gene product (phosphoprotein)		
t (7;19)	7q35 TCR β chain; 19 lyl-1	TF	T-ALL
lyl-1	Break on 19 chromosome		

Some other examples could be described: Ttg-1 – T-ALL; E2A, PBX1 – pre-B-ALL in childhood; Hlf – ALL; Hox11 – T-ALL.

Activation of the erb-B oncogene is a good example of these types of regulatory disfunctions: the altered receptor protein produces a continuous intracellular signal, without requiring the presence of an external transmitter molecule. Another example is the TGF β receptor (type II): its point mutation may lead to resistance to TGF β. Considering that TGF β is able to activate proteolytic enzymes, influence the formation of matrix proteins, and has an inhibitory effect on certain immune functions, it might affect tumor formation as well.

End-points of the different signal transduction pathways are nuclear transcriptional factors: proteins with the ability to recognize and bind to specific DNA sequences. These proteins are able to promote or inhibit transcription of the affected genes, thus playing the role of transcriptional regulators of oncogenes and tumor suppressor genes as well.

Functionally they can be divided into two major groups, those participating in the control of differentiation (this group is not involved in human carcinogenesis), and those activating genes required for cell proliferation (this group is important in carcinogenesis). Certain hormone receptors also act as transcriptional factors, such as the steroid receptors (e.g. estrogen receptor positivity is a bad prognostic sign in certain types of breast cancers, through resistance to chemotherapy).

Gene expression regulation involves hundreds of different genes. An important gene group is the group of "early response genes", consisting of about 100 members, with several oncogenes (c-jun, c-fos, c-myc, c-myb, etc.) among them. They have an important role in G0/G1/S phase transition. Early response genes are activated very shortly after mitogen stimuli, and the "late early genes" are targets of early response genes themselves. These regulatory signs activate other cell cycle regulator genes through different cascades.

Several oncogenes are involved in signal transduction, there are even theories that define oncogenesis as an error in the cellular information flow, which is particularly true for the promotion stage of carcinogenesis.

Figure 4.3 summarizes the molecular events during malignant transformation and tumorigenesis. Several protooncogene products play important roles in signal transduction pathways, and recently tumor suppressor genes have also been studied from this point of view. Protooncogenes involved in signal transduction can be divided into at least four categories: growth factor encoding genes, growth factor receptors and receptor associated protein kinases, transducers of extracellular signals (these can be further divided into three groups: G proteins, Ser/Thr kinases, adaptor molecules) and nuclear transcription factors (see Tables 4.2 and 4.3).

4.5 Examples

4.5.1 *Growth factors*

PDGF β chain is the product of the c-sis oncogene. It has a proliferative effect on certain cell types.

4.5.2 *Growth factor receptors*

Normally, these receptors bind the extracellular ligand, and transmit the signal through the cell membrane. In oncogenesis, their decreased function (lower affinity to the ligand, no binding at all, or no phosphorylation) may be important for inhibitory regulation, while their overactivity (producing signals without receptor-ligand binding) might cause very strong autocrine or paracrine proliferation. The classic example is the EGF receptor (Erb-B2 oncogene) mutation, which may encode a truncated receptor protein, with signal overproduction. It is frequently found in breast cancers, where it has a prognostic significance.

4.5.3 *Src oncogene encodes a membrane associated tyrosine kinase protein*

Its mutations are often found in diverse tumors. Under normal conditions the protein phosphorylates the next member of the cascade, and then loses activity. The mutated form, however, remains active, and produces the proliferative signal. The mechanism is similar to the bcr-abl fusion, which is a frequent event in chronic myelocytic leukemia. The regulatory sequence is lost, and continuous signal is produced in these cases.

4.5.4 *The G proteins are very important membrane associated, signal transducers*

The most significant family of the G proteins is the ras gene family. Their point mutations are present in a large number of human tumors, but other genetic alterations (amplification, overexpression) can also be found in some cases. Presence of genetic alterations in the ras genes is often a good prognostic biomarker, which can be studied with Kaplan-Meier survival curves (Figure 4.6). illustrating the difference between survival of head and neck cancer patients with or without ras overexpresion (a point mutation is supposedly behind this). The Ha-ras gene may be involved in a wide variety of tumors, e.g. pancreatic cancer, soft tissue sarcomas, skin and oral cavity cancers, and in chemically-induced tumors, while Ki-ras mutations can often be found in lung cancer and in different gastrointestinal tumors.

4.6 Sampling

In cancer molecular epidemiology and molecular pathology, sampling is crucial. For nucleic acid-based investigations we

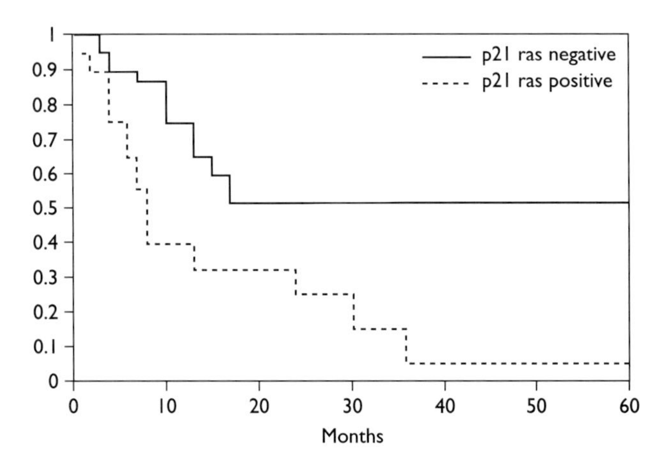

Figure 4.6 Survival of patients with squamous head and neck cancers grouped by ras p21 expression (Kaplan-Meier).

can obtain samples from the target organ itself, or we can use "surrogate" tissues when samples cannot be taken from the target organ. Surrogate tissues are generally biological fluids, most frequently blood, but in certain cases sputum, saliva, stool, pancreatic juice, or bronchoscopic fluid can also be used. Blood is often used as a surrogate tissue to detect point mutations derived from tumor cells in the circulatory system and to determine DNA adduct levels, but recently gene expression or amplification studies have also been performed, and point mutations are detected from cells in the peripheral blood (white blood cells or drifting tumor cells). Dot blot, slot blot or Southern blot hybridization is used to study gene amplification, while dot blot, slot blot and Northern blot are the choices for gene expression investigations. Detection of point mutations generally involves a polymerase chain reaction (PCR) amplification, and then RFLP, SSCP, hybridization with allele specific oligonucleotides, or DNA sequencing. Comparative genomic hybridization is still commonly used in research, as is fluorescent in situ hybridization (FISH). In addition to the DNA- and RNA-based investigations, oncoprotein expression is also often determined.

Certain genes have mutational hot spots – parts with very frequent occurence of mutations. In the ras gene family these hot spots are codons 12, 13, 61, and less frequently 59, 62.

Cytoplasmic kinases do not have as important roles as oncogenes, but some, e.g. protein kinase C, may be a target of certain carcinogenic agents. Phorbol-myristate-acetate (TPA, a strong promoter) receptors interact with PKC, causing the activation of the kinase. Several other tumor promoters acting on protein kinase C have since been demonstrated and this effect is now used in promoter tests, e.g. blocking of intracellular communication with Lucifer yellow (Pásti 1987, Ács and Parragh 2000). Since traditional carcinogen tests are only capable of detecting initiator or complete carcinogenes, application of promoter tests is very important for evaluation of potentially carcinogenic (promoter) compounds.

Nuclear transcription factors stand at the end of the signal transduction cascade. These proteins are responsible for regulating gene transcription required for cell proliferation. If the transcription factors do not require phosphorylation to bind to DNA, and stop repression of certain genes, this may lead to tumor formation. However, in most cancers the overexpression of these genes (e.g. c-jun, c-fos) is a consequence of an earlier dysfunction in the signal transduction cascade. Transcription factors are generally involved in hematological malignancies and their application in molecular diagnosis is useful for following the results of the therapy (Table 4.3). The most commonly affected oncogenes from this group are the myc oncogenes. Amplification and overexpression of the c-myc gene is frequently found in several tumor types. The myc protein can immortalize cultured cells, and it also has a role in apoptosis regulation. N-myc is also worth to mention; its amplification is typical in neuroblastomas, and there is a strong correlation between the presence of amplification (even the copy number) and poor prognosis of the disease.

When discussing nuclear factors participating in oncogenesis, one cannot omit the p53 tumor suppressor gene from the list. p53 alterations are found in about 90% of human tumors – of course this does not prove a causative role of p53 in all these tumors. The most important role of the gene is the inhibition of cell proliferation following genetic damage, preventing the cells from entering the S phase. If the damage can be repaired, p53 allows the cell to enter the S phase, but if the repair is not possible, it starts the apoptosis program, preventing the proliferation of genetically damaged cells. The main expressional regulation of p53 is at the protein level, through post-translational modification. Typical genetic events leading to p53 inactivation are point mutations and deletions. Like the ras genes, p53 also has certain mutational hot spots, but the mutational spectrum is much wider than that of ras genes. Most of the mutations fall within the area of exons 5–8 (codons 130–290).

We have described the potential for using oncogenes and tumor suppressor genes in revealing precancerous stages and in early diagnosis. They can be especially important in the case of a good anamnesis (a known carcinogenic exposure) (e.g. ras expression status and connection with alcohol consumption and smoking). Field (1991) investigated the effect of smoking on p53 expression, withouth clinical manifestation of any tumours (Figure 4.7). He found very strong correlation between smoking habits and p53 over expression, showing the predictive value of molecular epidemiological investigations.

In our own studies, onco/suppressor gene expression showed good correlation with exposure to ethylene oxide. Gene expression was significantly higher among exposed people than in the control group, and after decreasing the exposure, gene expression dropped to normal levels. Hence the usefulness of such molecular epidemiological methods can be seen for the recognition of high risk groups (exposed population), primary prevention of cancer and in following the fate (gene monitoring) of persons (Ember *et al.* 1999).

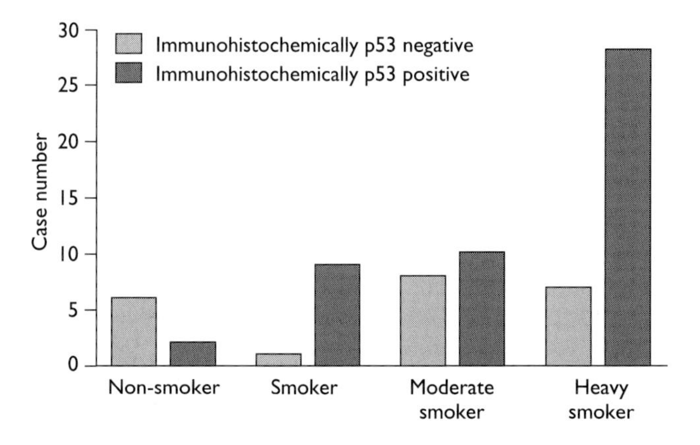

Figure 4.7 Correlation between p53 and head and neck cancers.

The routine application of onco/suppressor gene level molecular biological methods is continuously increasing, but introduction of new methods requires a strong molecular biological, ethical and epidemiological background. Scientifically, strongly established primary and secondary preventional and interventional programs in model populations (high risk people) provide the best approach for the validation and specification of these markers.

4.7 Methods

According to the molecular epidemiological stages of disease development (Figure 4.1), the first stage in which chemical compounds are measured in the human body is the internal dose level. Internal dose, however, has clinical significance only in very rare cases. It is applied in primary prevention, mainly in occupational health, using several different analytical methods to detect the very diverse scale of compounds that humans are environmentally or occupationally exposed to.

Biologically effective dose is also a marker of exposure, so from this point of view it can be treated in a similar manner as molecular biomarkers of internal dose. For carcinogenesis, the most significant biomarkers of biologically effective dose are the DNA-adducts since their presence indicates contact between the genetic material and the possibly carcinogenic compound. In the genesis of other diseases, protein-adducts also play a very important role, since the pathomechanism of several diseases is based on inhibition or disturbance of certain cellular functions by external agents (e.g. mercury inhibits SH-enzymes). The usual methods for detection of adducts are: high performance liquid chromatography (HPLC) sometimes combined with isotopic postlabelling or immunoassay, mass spectroscopy, and gas chromatography.

For physicians, the most interesting group of biomarkers is that with the so called early biological effect. In carcinogenesis this covers a wide range of markers such as different chromosome aberrations, sister chromatid exchange, point mutations, activation of oncogenes (mutation, amplification, over expression), inactivation of tumor suppressor genes (mutation, deletion). Deletion is often detected by loss of heterozygosity. Amplifications are investigated by Southern blotting, gene expressions are examined by Northern blotting and immunocytochemistry using antibodies against the oncoprotein. These genetic alterations are important not only in carcinogenesis, but also in the genesis of numerous other diseases. PCR based methods can be used to detect point mutations. With new PCR methods it is even possible to detect point mutations occuring in only 1 in 10^5 cells. This is achieved by enrichment of the mutant alleles in the specimen by digesting the wild type alleles with restriction endonucleases. The method gives an excellent opportunity to study biologically relevant early changes at gene level. The standard method for detection of point mutations is DNA sequencing. Today the dideoxy termination method is mainly used for DNA sequencing, which involves adding dideoxy nucleotides (which stop the elongation) to an *in vitro* DNA synthesizing system, and detecting the length of synthesized fragments by high resolution gel electrophoresis. Recently electrophoresis and fragment detection has been performed in automatic DNA sequencer units, which ensures precise reading of fragment lengths and the reliability of the investigation. Sequencing, however, is a time consuming and expensive molecular biological method, and efforts are being made to replace it with simpler, faster and cheaper methods. Single-stranded conformational polymorphism can be used instead of sequencing in certain cases. It is based on the different migrating velocity of DNA fragments even when differing by only one single base. The base change causes conformational changes in the DNA, and the altered conformation can be detected in gel electrophoresis by the altered migration speed. In order to detect the change, single-stranded fragments have to be electrophoresed, so the electrophoresis is performed under alkaline conditions. The fragment under investigation cannot be too long, as the effect of a single base change would then be so small that the difference in migration speed would not be detectable. If the site and type of the mutation is known, and it is located within a recognition site of a restriction endonuclease, restriction fragment length polymorphism (RFLP) can be applied. It is based on digestion with the appropriate restriction endonuclease and electrophoretic detection, whether the digestion took place or not. If the restriction site is not altered (no mutation), the enzyme will digest and two fragments appear in the gel, while in the case of mutation, the recognition site is altered and consequently no digestion occurs so only one fragment is visible in the gel (in some cases the situation is reversed: mutated alleles exhibit the restriction site and the wild type alleles do not). RFLP is a simple and effective method for studying known mutations. Unfortunately it cannot be used if the mutation does not fall within the recognition site of a restriction endonuclease. In addition to the methods mentioned above, many other common molecular biological techniques can be used to detect point mutations, some of which are still in the development and validation phase.

DNA amplification is classically studied by Southern blotting. It includes enzymatic digestion of the isolated DNA, gel electrophoresis, transfer of the fragments to a nylon or nitrocellulose membrane (=blotting), and hybridization with a labelled (radioactive isotope, dye, chemiluminescence) gene probe. A high detection signal indicates high numbers of the studied gene in the cells (gene amplification). A simplified version of the Southern blot is the dot blot (or slot blot) investigation, where the isolated DNA is immediately bound to the membrane without digestion. It is a much faster method than the Southern blot, and can be applied in large scale human epidemiological studies or to control efficiency of population level intervention. Care must be taken to control the specificity, because unspecific binding of the probe might occur.

RNA expression is examined in a similar manner, but the starting material is RNA instead of DNA. The method is called Northern blot, and in certain cases it might also be replaced by dot blot or slot blot hybridization.

Cytogenetic studies give information on chromatide and chromosome level, generally detecting abnormalities affecting larger DNA fragments. These studies are traditionally used to investigate chromosomes under a light microscope after dying or applying different chromosome banding techniques (G-banding). Visible alterations in number, shape, dying of the chromosomes, loss of arms, translocations (may be present due to DNA damaging agents), or hereditary genetic defects, can be specific for the disease or causative agent. Banding techniques are able to detect relatively small changes, based on the presence of chromosome bands (regions with different DNA condensation take up the dye differently, and have different shades). Sister chromatide exchange (SCE) studies the presence and rate of exchange of DNA fragments between chromatides. Mutagenic and carcinogenic compounds generally increase the rate of sister chromatide exchange. It is measured by labelling only one of the chromatides with fluorescent dye, after a period of time the presence of the dye can be detected in the other chromatide under the microscope. Another mutagenicity test used in human investigation is the Comet assay. The Ames test and micronucleus test also detect mutagenicity, but the investigations are carried out in nonhuman primates and cell cultures.

With fluorescent in situ hybridization (FISH) small genetic aberrations (down to gene level) can also be studied, and the other advantage of this method is good visualization and localization of the disorders. FISH is, as a matter of fact, a hybridization of fluorescently labelled probes to metaphasic or interphasic cells, thus enabling visualization of the location of nucleic acid sequences in the genome, in the chromosomes. Minor translocations, deletions, and other defects affecting only a small region are, in most cases, not detectable by traditional methods, but can be studied with fluorescent in situ hybridization. It is particularly useful for the study of cell or tissue specificity of certain genetic alterations, e.g. intratumoral heterogeneity.

Comparative genomic hybridization detects differences between genetic material of cells – normal and altered cells (e.g. tumor cells). DNA of the tumor cells (or other type of genetically altered cells) is hybridized with DNA of normal cells, where the nucleic acids are labelled with different fluorescent dyes. This arrangement is suitable for detecting any fragment losses, but also duplications, amplifications, and several structural changes. The hybridization is carried out in the interphasic stage of the recipient cells, so chromosomal localizations are clearly and precisely indicated. Comparative hybridization is particularly useful when several abnormalities (or not exactly described, not well characterized alterations) exist in the studied cells, which would make FISH very difficult or impossible.

Application of the molecular biological methods is not confined to the stage of early biological effect. These techniques can be used to study the following stages of exposure during the progression of the disease (altered structure/function, clinical disease and prognostic significance), if chromosome or gene level markers must be investigated, but this is a task for clinical laboratory investigation.

References

Ács, P., Paragh, Gy. (2000) A proteinkináz C, mint lehetõség a rákkutatásban specifikus gátlószerek elõállítására. *Orvosi Hetilap, 141, 1767–1771.*

Berenblum, I. (1979) In: Griffin, C.A., Show, R.C. (eds.), *Carcinogenesis: Identification and mechanism of action.* Raven Press, New York, USA, pp. 1–35.

Doll, R., Peto, R. (1983) *The causes of cancer,* Oxford University Press, Oxford, England.

Ember, I. (1993) *Környezetünk és a rák,* REEH Alapítvány, Budapest, Hungary (in Hungarian).

Ember, I., Kiss, I., Málovics, I. (1998) Oncogene and tumour suppressor gene expression changes in persons exposed to ethylene oxide. *European Journal of Cancer Prevention, 7, 167–168.*

Fearon, E.R., Vogelstein, B. (1990) A genetic model for colorectal tumorigeneis. *Cell, 61, 759–767.*

Field, J.K. (1992) Molecular and genetic analysis reveals chromosoma deletions, genetic alterations and overexpression of oncogenes/onco-suppressorgenes in the progression of head and neck cancer. In: Spandidos, D.J. (ed.), *Current perspectives on molecular and cellular oncology.* AI Press LTD, London. UK. pp. 119–152.

Gao, X., Honn, K.V. (1995) Recessive oncogenes: Current status. *Pathol Oncol Res. 1, 7–23.*

Harris, C. (1985) *Biochemical and molecular epidemiology of cancer.* Alan R Liss, Inc., New York, USA.

Hulka, B., Wilcosky, T.C., Griffith, J.D. (1990) *Biological markers in epidemiology,* Oxford University Press.

Jove, R. (ed.), STAT signalling. (2000) *Oncogene, 19, 2.*

Kiss, I., Ember, I., (1997) *Molekuláris epidemiológia.* Magyar Gasztroenterológiai Társaság, MEDICOM-Glaxo, Budapest, Hungary.

Kopper, L. (1993) Tumorbiológia. In: Ádány, R., Ember, I., Kásler, M., Kopper, L., Thurzó, L. (eds.), *Az onkológia alapjai.* Medicina, Budapest, Hungary, (in Hungarian) 1997, pp. 1–30.

Macdonald, F., Ford, J. (1991) *Oncogenes and suppresor genes,* Bios Scientific Publishers, UK.

Pasti, G., Lacal, J.C., Warren, B.S., Aaronson, S.A., Blumberg, P.M. (1986) Loss of mouse fibroblast cell response to phorbolesters restored by microinjected protein kinase C. *Nature, 324, 375–377.*

Perera, F.P. (1995) Molecular cancer epidemiology: A prevention of cancer. *Env Health Perspectives, 103, 233–236.*

Schulte, P.A., Frederica, P., Perera, F.P. (1993) *Molecular epidemiology-principles practices.* Academic Press, Harcourt Brake et Company.

Yamamoto, T. (1995) Oncogenes and suppresor genes. In: Kurzrock, R., Talpaz, M. (eds.), *Molecular biology in cancer medicine.* Martin Dunitz, London, UK, pp. 98–113.

Part 3

Common pathway and general mechanism

Chapter 5

Signal transduction therapy

Alexander Levitzki

Contents

5.1 Introduction

Over the past 15 years the molecular basis of numerous diseases has been discovered. The pathological states of these diseases are frequently derived from aberrations in either **intra**cellular or **inter**cellular signaling pathways. This is particularly true of proliferative diseases such as cancers, atherosclerosis, psoriasis and of inflammatory conditions including sepsis, rheumatoid arthritis, autoimmune diseases and tissue rejection. These findings have refocused medical research on seeking out new modalities for disease management. The new paradigm shift focuses on designing therapeutic modalities aimed at restoring normal signaling or bringing about the demise of the diseased cells without harming normal neighboring cells. Global genomic and proteomic approaches will further assist this new focus. This chapter will discuss the emerging patterns of this shift and examples of emerging therapies based on new molecular understanding.

Cancer cells respond to a variety of growth signals as do normal cells. Growth factors produced either by neighboring cells (paracrine stimulation) or by the tumor cells themselves (autocrine stimulation), interact with specific receptors on the cell surface or sometimes within the cell. Activating mutations in the receptor or in one of the downstream signal propagation elements induce persistent signaling. Inactivating mutations of tumor suppressor genes such as p53 or Rb, or the complete deletion of such genes, eliminates negative growth regulation signals. Table 5.1 lists examples of proto-oncoproteins, oncoproteins and tumor suppressors that play key roles in numerous cancers. The table is not intended to be a comprehensive list of all oncogenes or signal transducers known to be involved in cancer, but shows the types of molecules involved. Clearly, in cases where the oncoprotein represents an activated version of the proto-oncoprotein or where the proto-oncoprotein is overexpressed, it is desirable to inhibit its activity. Where inactivation of the gene results in weakened activity of a tumor suppressor gene product, like p53, it is desirable to enhance its activity. The frequent involvement of receptor tyrosine kinases (RTKs) and cellular tyrosine kinases (CTKs) in cancer put these molecules at the forefront of this field. In the case of receptor

Table 5.1 Targets for signal transduction therapy

Biochemical target	*Relevant disease*
Growth factors, cytokines	Cancers, inflammatory diseases, psoriasis
Protein tyrosine kinases (PTKs)	Cancers and other proliferative diseases Restenosis, psoriasis, papilloma, etc. Inflammatory conditions
Ser/Thr protein kinases	Cancers, other proliferative conditions
Anti-apoptotic proteins	Cancers
Estrogen and Androgen receptors	Breast and prostate cancers
Unique genetic make-up of the cancer cell	Modalities of gene therapy of cancer
Ras	Cancers, restenosis
Cell cycle proteins	Cancers

tyrosine kinases, major pathways of signal transduction have recently been deciphered and are now clear candidates for interception. Figure 5.1 represents current knowledge of the RTK-Ras-Raf-1 pathway, the RTK-PLCγ-PKC, the RTK-PI-′3 kinase pathway and the cell cycle machinery where the points of possible interception are depicted. The recent finding that certain mitogenic G protein coupled receptors (GPCRs) transmit their signals through PTKs, broadens the scope of signal interception therapy and antagonists against certain GPCRs and can now be considered valid lead agents for new anti-cancer and anti-proliferative drugs.

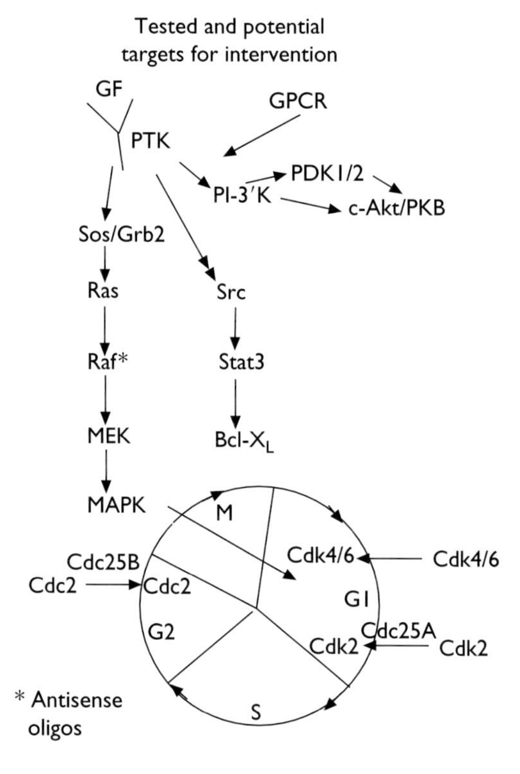

Figure 5.1 Targets for signal transduction therapy: a partial list.

5.1.1 The targets

The targets can be categorized according to their biochemical activity. Table 5.1 summarizes the biochemical targets currently being explored for drug development.

5.2 Receptor tyrosine kinases as targets for small molecules

Receptor tyrosine kinases, (RTKs), is a growing family of proteins that signal cells through tyrosine phosphorylation reactions. Some of these receptors, such as EGFR Her-2/neu and PDGFR are overexpressed in many tumors and/or are persistently activated by autocrine loops. In the most common and severe cancers, receptor over expression and autocrine loops have been demonstrated. The hypothesis that certain RTKs drive tumor growth can be tested by checking the consequences of the expression of dominant negative mutant receptors in transformed cells. Since RTKs must dimerize as an initial step for signal transduction, an inactive receptor, when expressed, can generate inactive dimers thus producing a "dominant negative" effect. Indeed, dominant negative PDGF or PDPFR inhibit growth of tumors expressing PDGF/PDGFR autocrine loops and dominant negative Flk-1 inhibits tumor growth which is, of course, dependent on intensive blood supply mediated by the Flk-1/KDR receptor. The first step of RTK signaling, subsequent to receptor dimerization, is receptor trans-autophosphorylation. This induces receptor activation which enables it to phosphorylate exogenous substrates and recruit adaptor molecules and enzymes to its autophosphorylated sites through SH2 domains, such as Grb2, PLC-γ, P I3′-kinase and SHP1 phosphatase (Figure 5.1). These two events allow signaling to propagate through a number of signaling pathways (Figure 5.1). The repertoire of these activated signaling pathways depends on the cell and each step is, in fact, a potential candidate for interception.

Thus, RTK signaling can theoretically be inhibited through blocking ligand binding, receptor dimerization, RTK activity or recruitment of signaling molecules. The *in vivo* effect of dominant negative growth factors and their receptors actually validates RTKs as targets for drugs. Thus far, successful modality is arrived at through inhibition of the PTK catalytic reaction and antibodies acting against the extracellular domain of the receptor.

5.2.1 Protein tyrosine kinase inhibitors

PTKs comprise over 80% of all known oncoproteins and proto-oncoproteins that play key roles in cancers. The development of **tyr**osine **phos**phorylation **in**hibitor**s** (tyrphostins) was initiated in the late 1980s in response to this fact (Yaish *et al.*, 1988). In 1991 the first successful *in vivo* utilization of a **tyr**osine **phos**phorylation **in**hibitor was reported (Yoneda *et al.*, 1991). This study demonstrated for the first time that tyrphostin RG 13022, an inhibitor of the EGF receptor kinase, exhibits efficacy *in vivo*, suppressing the growth of human squamous cell carcinoma in nude mice. Furthermore, in combination with an antibody of the EGFR extracellular domain (mAb 108), RG 13022 acts synergistically to inhibit tumor growth *in vivo*. Since the publication of this pioneering study (Yoneda *et al.*, 1991), a number of PTK inhibitors (tyrphostins) have shown great efficacy in pre-clinical animal models and have entered clinical trials (see discussion below and Table 5.3). With the exception of AG 490 (Meydan *et al.*, 1996b) and AG 556 (Novogrodsky *et al.*, 1994; Vanichkin, 1996; Sevransky *et al.*, 1997; Lopez-Talvera *et al.*, 1997; Brenner *et al.*, 1998) which are effective *in vivo*, all the other PTK inhibitors (tyrphostins) currently being developed are ATP competitors (ATP mimics). These are EGFR kinase inhibitors, VEGFR kinase inhibitors, FGFR kinase inhibitors and PDGFR kinase inhibitors. The high intracellular concentration of ATP (~1 mM) requires the use of high doses of the agent, which is likely to produce side effects and may therefore limit the efficacy of ATP-mimics (see also below).

5.2.1.1 Design and synthesis of protein tyrosine phosphorylation inhibitors

Kinetic analysis of the mode of EGFR kinase shows that ATP and the substrate bind independently to the kinase domain and no sequential binding occurs (Posner *et al.*, 1994). This property simplifies the kinetic analysis on the inhibitors' mode of inhibition of PTKs (Posner *et al.*, 1994). One of the most surprising findings on the selectivity of inhibitors discovered to date, is the extent of selectivity found in ATP-competitive inhibitors (Levitzki and Gazit, 1995; Buolamwini, 1999; Garcia-Echeverria *et al.*, 2000). For example, quinoxalines and quinazolines are highly selective inhibitors of PDGFR kinase (Kovalenko *et al.*, 1994; Gazit *et al.*, 1996) and EGFR kinase respectively (Ward *et al.*, 1994; Osherov and Levitzki, 1994 and see below). Further analysis of the tyrosine kinase inhibition mode reveals that the affinity of the inhibitor and its mode of binding to the kinase domain depends on whether the kinase is activated or in its basal inactive conformation. Two examples illustrate this point: Activated Abl kinase like $p210^{Bcr\text{-}Abl}$ and $p185^{Bcr\text{-}Abl}$ possess different affinities to both substrate and inhibitors (tyrphostins) as compared to the proto-oncogenic form $p140^{c\text{-}Abl}$. For example, $P210^{Bcr\text{-}Abl}$ and $p185^{Bcr\text{-}Abl}$ are inhibited by the tyrphostin AG 957 with Ki values of 0.75 μM and 1.5 μM respectively as compared to Ki = 10.0 μM for the cellular wild type $p140^{c\text{-}Abl}$ (Anafi *et al.*, 1992).

In all these cases AG 957 is competitive with the substrate and non-competitive with ATP (Anafi *et al.*, 1992). AG 957 is effective in purging Ph^{+} cells with good selectivity against normal blood cells (Carlo-Stella *et al.*, 1999). The difference between c-Abl and Bcr-Abl is not found in the kinase domain, since the Bcr sequence is fused in-frame with the intact kinase domain. It would therefore appear that tethering the Bcr sequence upstream to the c-Abl alters the conformation of the kinase domain such that it binds the inhibitor more tightly. The same pattern of behavior is observed vis-à-vis the substrates: the oncogenic forms exhibit lower Km values towards the substrates as compared to the proto-oncoprotein towards the substrates and inhibitors as compared to the proto-oncogenic form of the protein (Anafi *et al.*, 1992). In the case of PDGFR, kinase activation of the receptor also leads to changes in the structure of the kinase domain although the situation is more complex: upon activation the mode of inhibition of the selective inhibitor AG 1296 (or AG 1295) is altered. While the inhibitor is competitive *vis-à-vis* ATP in the inactive form of the receptor, it binds with higher affinity and becomes *mixed competitive vis-à-vis* ATP subsequent to receptor activation by PDGF (Kovalenko *et al.*, 1997). These two examples suggest general pattern of behavior for PTKs (Levitzki and Böhmer, 1998). This finding also points to the necessity of examining potential PTK inhibitors not only as blockers of PTK autophosphorylation, which is performed routinely, but also as blockers of the PTK action on exogenous substrates.

With the advance of X-ray crystallography and the ability to determine the three dimensional structure of the kinase with the bound inhibitor, drug design has become more precise and rational. The three-dimensional structures of three tyrosine kinases complexed with a kinase inhibitor have already been solved: (1) The structure of the FGF receptor with selective and non-selective inhibitors has already been published (Mohammadi *et al.*, 1997); (2) The structure of the c-Abl kinase domain, in its inactive conformation, with the inhibitor STI-571 has recently been solved (Schindler *et al.*, 2000); (3) The Src kinase Hck has been crystallized in its inactive form with the inhibitor PP1 complexed with it

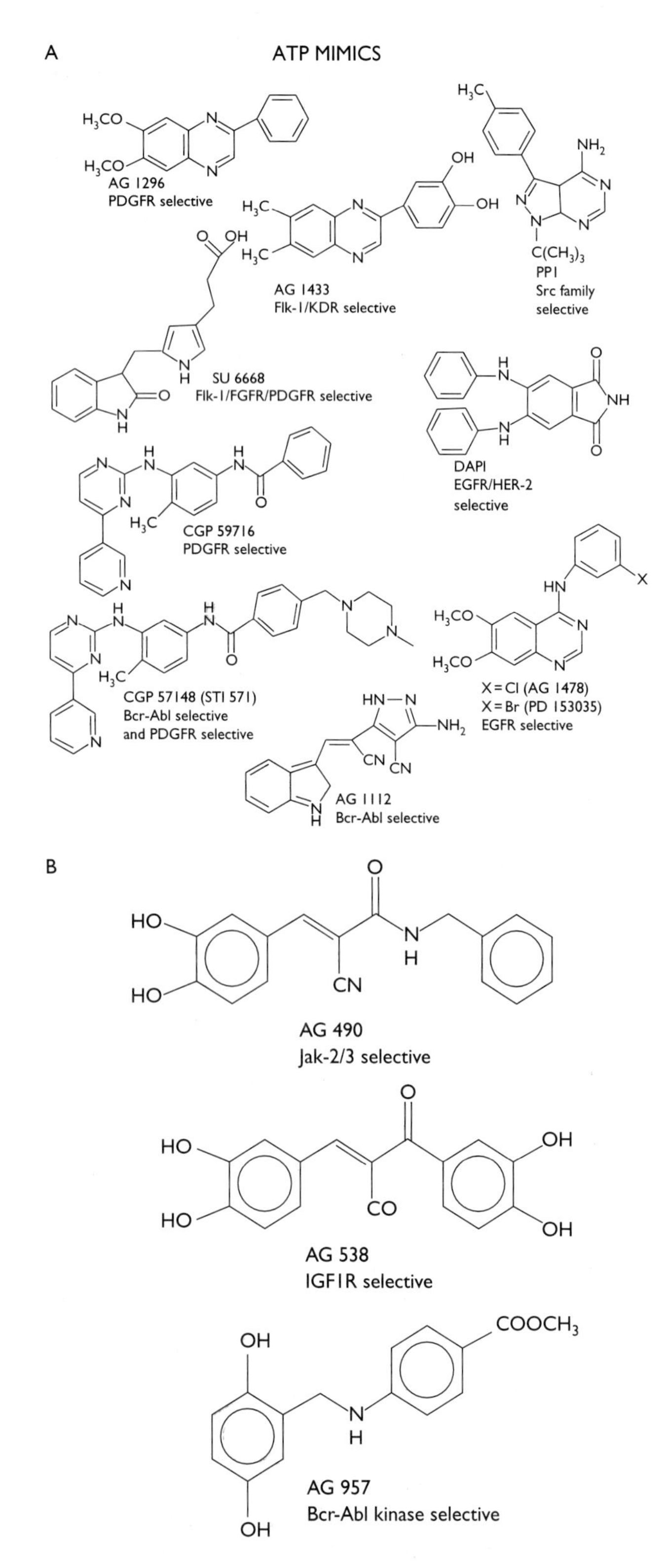

Figure 5.2 Pharmacophores utilized for PTK inhibitors. (A) ATP mimics, (B) Substrate mimics.

(Schindler *et al.*, 1999). The Src family member Lck in its active form has been crystallized with the inhibitor PP2 (Zhu *et al.*, 1999). Several laboratories, including our own, are currently using these structures as guides for designing more selective Src kinase inhibitors. Since pp60$^{c\text{-}Src}$ is found to be activated in many human malignancies and appears to play a key role in these tumors (Levitzki, 1996a), it follows that a Src kinase inhibitor could prove an effective anti-cancer drug for numerous cancers. Similarly, the availability of the insulin receptor kinase structure in both its inactive as well as active forms (Hubbard *et al.*, 1994) complexed with APPNHP and a peptide substrate (Hubbard, 1997) allows us to conduct an educated search for inhibitors of IGF-1R kinase, which is highly homologous to the insulin receptor kinase. One compound, AG 538, was identified and found to inhibit IGF1-R kinase with competitive kinetics against the substrate and non-competitive kinetics against ATP (Blum *et al.*, 2000). The aim is now to determine whether a compound can be developed that discriminates between an insulin receptor kinase and an IGF-1 receptor kinase. Figure 5.2 summarizes the main pharmacophores that have proven themselves effective PTK inhibitors with no significant effects on Ser/The kinases. Figure 5.3 depicts the evolution of tyrphostins and how the "drift" to ATP-mimics occured. It should be noted that the "drift" from one-ring tyrphostins to two-ring tyrphostins has generated more ATP-mimics than substrate-mimics. (Figure 5.2; Posner *et al.*, 1994).

5.2.2 *ATP mimics vis-à-vis substrate mimics*

We have frequently argued that the best potential PTK inhibitors would be those compounds that compete for the substrate binding site within the kinase binding domain. Substrate competitive compounds should be less toxic than ATP mimics since they bind to those domains at the kinase site that are less conserved than the substrate binding domains. ATP mimics are likely to bind to many other targets which normally bind ATP and other nucleotides. It is reasonable to assume that ATP mimics will be somewhat toxic, especially if administered in high doses. The need for high doses of PTK inhibitors that are ATP mimics is due to their need to compete with high intracellular concentrations of ATP (1 to 2 mM). Thus even compounds which possess nanomolar affinities towards the ATP binding site only affect their targets in intact cells at the μM concentration range. At this concentration range ATP mimics are likely to bind to other targets as well, thus inducing undesirable effects. Indeed tyrphostins, such as AG 490 which blocks Jak-2 and AG 556 which possesses anti-inflammatory properties, have been shown to be non-toxic *in vivo* (see above). The main problem with these compounds is that they possess hydroxyl groups which are metabolized quite fast. It is therefore essential to develop metabolically stable "substrate scaffolds", a rather difficult task which we have recently overcome

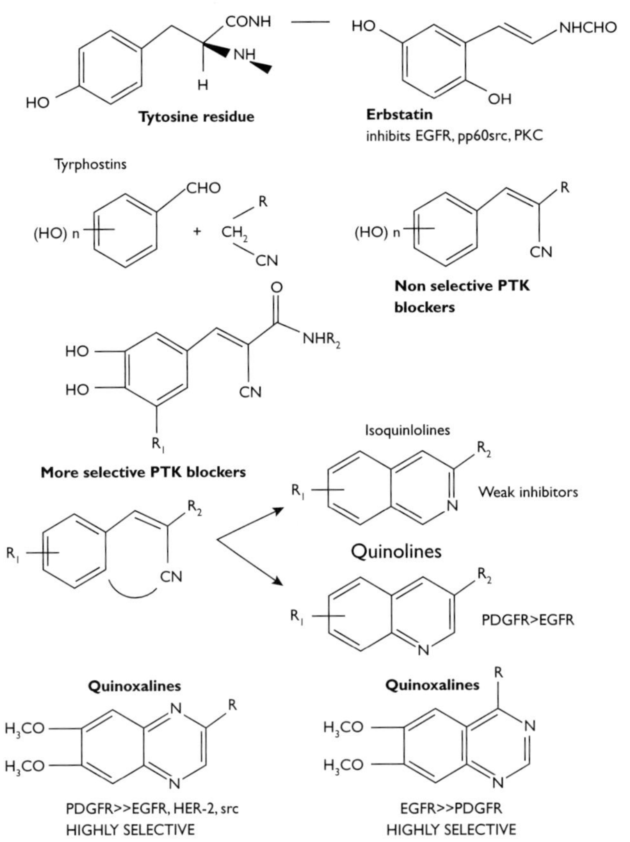

Figure 5.3 The evolution of tyrphostins. Initially tyrphostins were designed as substrate competitive inhibitors. With the closure of a second ring by incorporating the nitrogen of the nitrilo group (CN) into a second ring, ATP mimics rather than substrate mimics were obtained. Initially these compounds were generated deliberately. Later they were also found through random searches by other investigators.

(Blum *et al.*, unpublished). One unexplored direction is the development of bi-substrate inhibitors which covers areas of the ATP binding site as well as the substrate binding domain. If successfully developed, it is likely that these compounds will exhibit very high affinities towards the binding site of the specific kinase against which they will be developed. This will allow for the use of reduced doses thus diminishing side effects. Some bi-substrate EGFR kinase inhibitors have already been identified (Posner *et al.*, 1994) but their affinity is still not high enough and needs to be optimized.

Over the past decade, the PTK inhibitors favored by most investigators have been ATP mimics that compete with ATP at the binding site and are non-competitive inhibitors with the substrate. Some compounds act as "mixed" competitive inhibitors, namely they bind simultaneously with ATP and/or substrate but decrease the affinity of ATP and the substrate to their respective sites. In the case of the PDGFR we find that the quinoxaline AG 1296 is competitive with ATP in the non-activated form of PDGFR and mixed competitive with ATP in the non-activated form (Kovalenko *et al.*, 1997). STI 571 (CGP 57148), which is an ATP competitor has been administered to patients at a dose of 300–800 mg/day with only minor adverse effects. This dose is, however, unusually low compared to doses of 20 to 100 mg/kg of PTK inhibitors used in *in vivo* experiments and in recent clinical trials aimed at treating solid tumors. These high doses reflect the relatively low efficacy of these compounds *in vivo* despite the fact that their IC50 values towards their molecular targets such as the EGFR, VEGFR-2/Flk-1 etc., are within the low nanomolar concentration range. When examining the efficacy of the ATP competitive inhibitors in cellular assays it is observed that nanomolar compounds act on cells in the high sub-micromolar to micromolar concentration range. For example, quinazolines that bind to the EGFR with Ki of certain nanomolars inhibit EGFR autophosphorylation in intact cells at the sub-micromolar to the low micromolar concentration range. Similarly, PP1 and PP2, which inhibit Src kinase with IC50 values of 20–100 nM (Hanke *et al.*, 1996), block Src activity in cells in the 5 to 40 μM concentration range (Karni *et al.*, 1999; Karni and Levitzki, 2000). The *in vivo* effects of PP1 are notable only at fairly high doses of 20 mg/kg (He *et al.*, 2000). It seems that the high doses required *in vivo*, at least partly reflect the competitive relationship between the intracellular millimolar concentrations of ATP and the administered drug. It is noteworthy that drugs such as β-adrenergic blockers are administered at doses about 100-fold lower. In this case a drug possessing an affinity in the nanomolar concentration range only has to compete with up to 100 nM (0.0001 mM) concentration of the endogenous ligand adrenalin or noradrenalin. Thus β-blockers can be administered in this case at doses of 1.0 mg/kg or less and be highly effective. This is true for other receptor-directed drugs where the endogenous ligand is present in low concentrations within body fluids. Targeting an intracellular protein is problematic when competing against high concentrations of substrates such as ATP. Moreover, the selectivity of newly developed compounds are only tested against a limited number of PTKs and Ser/Thr kinases, where the number of PTKs falls in the 150 range and the number of Ser/kinases is in the 600 range. It has already been observed that the so called selective Src family kinase inhibitor PP1 is, in fact, equipotent as a PDGFR kinase inhibitor (Waltenberger *et. al.*, 1999). Similarly the Novartis Bcr-Abl kinase inhibitor, CGP57148/STI 571 (Druker *et al.*, 1996) already approved for the treatment of chronic myeloid leukemia (CML), is just as potent against PDGFR kinase and c-kit. This type of behavior is likely to be encountered frequently. Short-term treatment regimens should be designed in order to minimize negative side effects. Our own long-term goal should be to design and synthesize substrate competitive PTK inhibitors, a highly difficult task. In one case, IGF-1 receptor kinase substrate competitive inhibitors were designed on the basis of the three-dimensional insulin receptor structure. These compounds inhibit IGF-1R signaling in intact cells (Blum *et al.*, 2000) and attempts are currently being made to improve potency and selectivity. An important challenge in this case is to generate inhibitors of IGF-1R which do not block insulin receptor kinase.

5.2.3 *How many tyrphostins do we need?*

The approximate number of PTKs is estimated at around 150. In every cancer a number of active PTKs appear to play a role in promoting cell proliferation and/or anti-apoptotic pathways. A few of PTKs show up in many cancers and as a result inhibitors for these kinases can be utilized in many types of cancers. These **broad spectrum PTK inhibitors** can qualify as **universal** PTK targets (Levitzki, 1996b) since they may well prove effective in the treatment of numerous cancer types. Many PTKs are activated in only a few types of cancer. Table 5.2 lists examples of universal and selective targets which include PTKs.

5.3 Antibodies to receptor tyrosine kinase

The success of Herceptin among a subgroup of patients with advanced metastatic breast cancer in which Her-2/neu overexpression was involved (Stebbing *et al.*, 2000), presents a strong case for antibody therapy. The success of Herceptin is based on pioneering work conducted on anti-Her-2 antibodies. This was aimed at the extracellular domain which inhibits the growth of breast cancer cell lines overexpresssing Her-2 (Drebin *et al.*, 1985; Stancovski *et al.*, 1991; Shepard *et al.*, 1991; Harwerth *et al.*, 1993; Kern *et al.*, 1993; Lewis *et al.*, 1993). Anti-EGFR antibodies are also currently undergoing clinical trials. The early observations of Yoneda *et al.* (1991) showed that monoclonal antibodies are able to synergize with tyrphostin RG 13022 to inhibit the growth of EGFR overexpressing tumors *in vivo*. The efficacy of anti EGFR C225 as a blocker against EGFR-driven tumors *in vivo* (Naramura *et al.*, 1993; Fan *et al.*, 1993) has been recreated in a clinical setting. These antibodies are currently in clinical trials for head and neck cancers. Although anti-CD20 (Rituxan) is not effective against an RTK there is evidence that its active mechanism involves an intracellular PTK (Shan *et al.*, 2000).

Table 5.2 Examples of universal and selective targets

Universal targets	*Selective targets*
VEGF/Flk-1(KDR)	PDGF/PDGFR
IGF-1/IGF-1R	EGF/EGFR
$p60^{c\text{-}Src}$	JAK-2
	FGF/FGFR
Shc/Grb2 SH2 domains	CSF-1R
Sos SH3 domain	Her-2/neu
Raf-1	
MAPK	
MEK(MAPKK)	
Cdk2/Cdc2/Cdk4-6	
Cdc25 phosphates	
PKB/c-Akt	
PDK1	

Universal targets are activated in many cancers whereas the selective ones in fewer cancers.

5.4 Inhibition of growth factor binding by growth factor

5.4.1 *Antagonists*

Scientists have not yet managed to design growth factor antagonists *in vivo* but this option is still being explored. The validity of this approach is exemplified by the case of Suramin as an anticancer agent. It is thought that Suramin competes with growth factors by either binding to them directly or competing for receptor binding. Suramin is, in fact, being used in the clinic (Eisenberger *et al.*, 1993; Van Rijswick *et al.*, 1993). It may serve as a lead compound for novel growth factor antagonists, but so far no effective new compounds have been reported from this family. Medical biotechnology companies have made some headway in this direction.

5.5 Growth factor-toxin and antibody-toxin chimeras

Chimeras between EGF and a bacterial (or plant) toxin can be used to kill tumor cells that overexpress the growth factor receptor. Similarly, antibody-toxin chimeras can also be used. While undergoing internalization, the receptor carries with it either the ligand-toxin chimera or the antireceptor antibody-toxin chimera. This principle is also useful for immunotoxins based on cancer antigens which are not themselves receptors but which internalize the antibody subsequent to its binding (for review see Pai and Pastan, 1993). Clinical trials with immunotoxin conjugates are being conducted.

5.6 Ser/Thr kinase inhibitors

A number of Ser/Thr kinases play key roles in pathophysiology. These enzymes have also become targets for therapy.

5.6.1 *Cyclin dependent kinase inhibitors*

Cdks, such as Cdk2 and Cdk4, have attracted the attention of drug designers due to their pivotal role in the regulation of the cell cycle (Kaubisch *et al.*, 2000). Flavopiridol is the most widely evaluated Cdk inhibitor. This compound is a strong inhibitor of both Cdk2 and Cdk4. It induces cell cycle arrest in G1 and is not dependent on the p53 or Rb status of the cell. The compound also induces cell cycle arrest in G2, presumably because it inhibits Cdk1 (Cdc2). Endeavors in combinatorial chemistry have led to the discovery of Purvalanol B (IC50 = 6 nM) (Gray *et al.*, 1998), a purine analog, which is a highly selective and potent inhibitor of the Cdks. This compound is an ATP mimic and could therefore suffer similar difficulties to those of ATP mimics for PTKs (see above). Time will tell if these compounds will prove effective in a clinical setting.

5.6.2 The Raf-MEK/MAPK module

So far one inhibitor, the compound PD 98059, has been reported as being selective for the Raf-MEK/MAPK pathway. This compound is an allosteric inhibitor which binds to MAPK and inhibits its activation through MEK. Although this compound has reportedly inhibited cell growth and reversed Ras transformation, its efficacy in animal models has not yet been reported. However, a more potent inhibitor, PD 184352 has shown great efficacy *in vivo* as an inhibitor of metastatic colon cancer (Sebolt-Leopold *et al.*, 1999). This finding strongly suggests that the inhibition of Ras in the MAPK pathway is most probably important for numerous scenarios since this pathway is a key factor in the invasiveness of the tumor.

5.6.3 PKB/Akt

In recent years it has been established that PKB/c-Akt plays a pivotal role in enhancing the anti-apoptotic robustness of cancer cells. PTEN deletions and overexpression of PKB/c-Akt in highly malignant tumors have been reported (DiCristofano and Pandolfi, 2000). Furthermore, an important component of IGF-1/IGF-1R signaling is mediated via the PI-3′kinase/PDK1/c-Akt pathway. These considerations point to PKB/c-Akt as a target for developing novel anti-cancer agents. A recently developed PKB/c-Akt inhibitor Aktstatin NL 71–101 (Reuveni *et al.*, 2002) induces apoptosis in Panc1 and Ovcar3 cells in which the enzyme is overexpressed and is constitutively active. These findings appear to validate PKB/c-Akt as a legitimate target for cancer therapy (Reuveni *et al.*, 2001). This inhibitor has already lead to a novel PKB/Akt inhibitor in the 100 nm range (unpublished).

5.6.4 Other Ser/Thr kinases

Enzymes such as PDK1 and p70S6 kinase can also be considered for drug development since these enzymes are most probably involved in the enhancement of oncogenic pathways. Rapamycin, which blocks the activation of S6 kinase via mTor, is currently in clinical trials for cancer treatment (Hosol *et al.*, 1999).

5.6.5 Protein kinase C isozymes

A number of PKC isozymes have been implicated in cancer and serve as targets for anti-cancer drug design. Bryostatin showed some efficacy (Kaubisch and Schwartz, 2000; Pendville *et al.*, 1993; Basu, 1993) against melanoma (Schachter *et al.*, 1991) but has not seriously moved into the clinical trial stage. The multiplicity of PKC isozymes on the one hand and the obscurity that still surrounds our understanding of the role played by these isozymes on the other, has made drug development in this area slow.

5.6.6 Cdc25 phosphatases

Cdc25 isozymes are dual-specificity phosphatases which promote the transitions from G1 to S and G2 to M by dephosphorylation of Cdk2 and cdk1/Cdc2 respectively (Figure 5.1). Cdc25A is also a target gene for Myc which accounts for part of the oncogenic activity of this well known oncoprotein. The overexpression of Cdc25 is a tumor hallmark and identification of this enzyme in tumors indicates poor prognosis (Hernandez *et al.*, 2000; Aihara *et al.*, 2000). Thus, these enzymes also qualify as universal targets for anti-cancer drug development. There have been a few reports on attempts to generate such inhibitors (Peng *et al.*, 2000) but none has so far been tested *in vivo*.

5.6.7 Ras protein

Since oncogenic Ras is the hallmark of over 30% of all tumors and more than 80% of colon and pancreatic tumors it is considered a prime target for anti-cancer drug development. The major biochemical reaction currently targeted is the farnesylation of Ras which is required for membrane anchoring which is essential to its biological activity. Farnesyl transferase inhibitors (FTIs) have been developed by a number of laboratories and have shown efficacy in cellular assays as well as *in vivo*. FTI is expected to have great effect on the anchoring of H-Ras (Kohl *et al.*, 1993; Reuveni *et al.*, 1997, 2000) but not on K-Ras where anchoring is achieved through geranyl-granylation and the poly-basic sequence of the c-terminal domain. Since oncogenic K-Ras is the more abundant of the Ras proteins involved in human cancer, it makes no sense that FTIs are effective against most tumors. It is likely that the observed inhibitory effects of FTIs are mediated by farnesylation inhibition of other targets. It is, in fact, difficult to establish an exact target since approximately 20 farnesylated proteins exist. Due to the conceptual difficulty underlying the use of FTIs, other avenues should be explored to target Ras. In this vein Kloog *et al.* (Elad *et al.*, 1999) designed and synthesized S-trans, trans-farnesylthiosalicylic acid (FTS) which blocks docking of Ras proteins. This promising strategy has already proven efficacious *in vivo*. Another approach which may yet prove useful is based on the finding that by modifying GTP the full hydrolytic activity of the oncogenic version of the Ras protein can be restored (Ahmadian *et al.*, 1999; Kosloff *et al.*, 2000). This type of "substrate-assisted catalysis" reinstates oncogenic Ras sensitivity to GAP. Thus, in principle, there is the possibility of finding a small molecular weight nucleophile which would restore the ability of oncogenic Ras to hydrolyze GTP.

Another approach is to utilize dominant negative Ras constructs which block the GDP/GTP exchange on the Ras protein by trapping the exchanger. Dominant negative RasN17, when applied locally to balloon-injured areas, succeeded in inhibiting the restenosis process (Indolfi *et al.*, 1995) suggesting the utility of such constructs in local gene therapy.

5.6.8 *Transcription factors*

In contrast to the rapid development of enzyme inhibitors and receptor antagonists, there has been no development of small molecules that inhibit the action of transcription factors. The extremely high affinity of transcription factors to DNA makes this task highly challenging. In some cases, such as Stat 3 which is activated by tyrosine phosphorylation which leads to dimerization and migration to the nucleus, small molecules aimed at preventing dimerization could, in principle, be designed, though again this would be a difficult task.

5.6.9 *Steroid and androgen antagonists*

Breast cancer is often estradiol-dependent while prostate cancer is androgen-dependent in its early stages. These qualities mark these two diseases as targets for signal transduction therapy based on the design and synthesis of anti-estrogens and anti-androgens respectively. Attempts have been made to antagonize the estradiol by generating antagonists (Dreicer and Wilding, 1992). Tamoxifen (ICI 46, 474) and Raloxifen are estradiol-partial antagonists and have demonstrated efficacy in patients (Litherland and Jackson, 1990; Love and Koroltchouk, 1993). However, the pure estrogen antagonists ICI 164,384 and ICI 182,780 proved much more efficacious in cell culture tumor models (Wakeling *et al.*, 1992) and more recently in limited clinical trials (DeFriend *et al.*, 1994). ICI 182,780 induces over 80% inhibition of estrogen-dependent MCF-7 cell growth in nude mice, in contrast to 40% inhibition achieved by tamoxifen under identical conditions. The steroid antagonist ICI 164,384 is similar in action to ICI 182,780 although its IC_{50} is 13 nM compared to $IC_{50} = 0.2$ nM for ICI 182,780 (Wakeling *et al.*, 1992). It has also been shown recently that human breast cancer cells which are resistant to tamoxifen and 4-hydroxytamoxifen retain their sensitivity to ICI 182,780 (Brunner *et al.*, 1993). ICI 182,780 and ICI 104,384 inhibit the nucleocytoplasmic shuttles of the estrogen receptor, whereas tamoxifen does not (Dauvois *et al.*, 1993). The estrogen receptor shuttles between the cytoplasm and the nucleus in the absence of hormone, while in the presence of hormone it is targeted to bind to the receptor binding domain of the DNA. The steroid receptor antagonists ICI 187,780 and ICI 164,384 disrupts nuclear entry and reduces the overall level of the receptor. The net result is a complete shut off of estrogen receptor function. Such a complete shut-off is not accomplished by tamoxifen, probably because this is a partial rather than a pure antagonist. Indeed, it has recently been shown that tamoxifen-resistant variants of human breast cancer cells are completely inhibited by ICI 182,780 (Hu *et al.*, 1993). It is therefore likely that ICI 182,780 will eventually replace tamoxifen in the clinic.

5.6.10 *P53 as a target*

In many tumors p53 is mutated and functionally inactive. Numerous studies have shown that the mutations nullify the ability of p53 to bind to DNA due to a conformational transition of the protein. It was therefore argued that small molecules which "sway" the conformation to its biologically active state may restore its function. Such compounds have not yet been discovered. However, another ingenious approach has been developed: a small molecular weight p53 inhibitor, when administered to patients, protects normal cells from chemotherapy and radiation therapy-induced apoptosis, while allowing tumor cells devoid of functional p53 to suffer extensive damage and death (Komarov *et al*; 1999).

5.7 Signal transduction gene therapy

The identification of signal transducers as an element crucial to the process of tumorigenesis immediately brings to mind modalities of gene therapy. Thus, for example, one can reverse the transformed phenotype of v-Src-transformed cells by expressing antisense RNA under an inducible promoter (Karni *et al.*, 1999). Many scenarios may be envisaged where dominant negative constructs could inhibit tumor growth *in vivo* (Millawer *et al.*, 1994). Antisense constructs or ribozyme constructs can be expressed in the diseased tissue, to suppress a particular oncogene or proto-oncogene. It is also possible to envisage the expression of tumor suppressors in tumors harboring inactive mutant p53. The use of antisense RNA can also be envisaged in other diseases, such as restenosis/atherosclerosis. In this case it can be shown that proliferation of smooth muscle cells can be greatly inhibited by antisense oligonucleotides complementary to the RNA messenger of cell proliferating nuclear antigen (PCNA) (Speier and Epstein, 1992). Local application of RasN17 (dominant negative) to a balloon-injured area prevents stenosis in rats (Indolfi *et al.*, 1996). Encouraging results were recently obtained with the Onyx 015 viral vector for head and neck cancer (McCormick, 1999). This adenovirus lacks the AdA gene product, that is the target native to p53, and arrests virus growth upon binding. Thus Onyx 015 can only grow in cells which lack p53 or harbor inactive mutant p53. While clinical success has been achieved from local application for head and neck cancer, targeting the construct still remains the central issue and so far precludes systemic application. Another approach is the double stranded RNA (dsRNA) killing strategy (DSKS). The principle of this approach runs as follows: The enzyme double stranded RNA dependent protein kinase, PKR, is usually activated by double stranded RNA molecules 30–80 nucleotide long that are themselves normally activated by unique dsRNA molecules generated by invading RNA viruses. Activation of PKR leads to cell death, thus combating the viral infection. Since many cancer cells possess unique species of mRNAs due to chromosomal aberrations, antisense RNA molecules complementary to these sequences can be expressed in cancer cells. This selectively activates PKR in these cells, inducing them to apoptose. This hypothesis was successfully tested in glioma cells that express a mutated form of the EGFR. A lentiviral construct possessing a sequence coding for an antisense (AS) RNA 39 nucleotide long, complementary to the unique sequence at the junction of the truncated EGFR RNA has been constucted. The AS sequence encompasses the unique sequence generated by the fusion of exon 1 with exon 8 due to the deletion of the exons 2 to 7. The AS construct can bind only to 20 nucleotides in exon 1 and 8 in the native EGFR RNA. This length is insufficient to trigger the activation of PKR. In cells that harbor the truncated EGFR, PKR is activated, leading to cell death (Shir and Levitzki, 2002). This strategy can, in principle, be implemented for numerous cancers, leukemias and lymphomas since they all possess unique RNA species.

5.8 Combination therapies

The malignant tumor is an evolving entity and usually quite heterogeneous. It is comprised of different types of cells with different signaling networks. Different stages of the tumor express different repertoires of oncogenes and proto-oncogenes and exhibit a variety of chromosomal aberrations and deletions. Thus, in order to successfully combat tumor progression or even induce tumor regression, combinations of drugs are necessary. Indeed, though very few novel anti-cancer drugs have been developed since the 1960s, the medical community recognizes that different drug cocktails are effective against different cancers. Cytotoxic drugs by themselves may, however, have reached maximal optimization in the clinic and novel agents are needed. Small organic signal interceptors and bio-reagents such as antibodies and immunotoxins seem to fit this demand. We have shown that tyrphostins that effectively block EGFR synergize with antibodies against the extracellular domain of the receptor to inhibit the growth of human squamous cell carcinoma (Yoneda *et al.*, 1991). The combination of the EGFR kinase inhibitor AG 1478 with CDDP is effective against glioma which harbors the mutated EGFR, both in tissue culture (Nagane *et al.*, 1998) and *in vivo* (Nagane *et al.*, 2001). Experiments performed on human cell lines derived from lung cancer, which over express $p185^{Her\text{-}2}$ show that cytotoxic drugs synergize with Her-2 selective tyrophostins such as AG 825. This agent synergizes with cis platin (CDDP), doxorubicin and etoposide (Tsai *et al*, 1996). Tyrphostins, which are more effective against EGFR than Her-2, do not synergize with these cytotoxic compounds but act only in an additive fashion. Also, cell lines derived from the same tumors, but which do not overexpress Her-2, are not inhibited in a synergistic manner by the Her-2 selective compound. These results (Tsai *et al.*, 1996) strongly suggest that tyrphostins can be extremely useful as components of drug cocktails together with cytoxic drugs or anti-Her-2 antibodies such as Herceptin. Synergism has also been observed between anti-EGFR antibodies and cytoxic drugs (Bruny *et al.* 2000) and between Herceptin and cytotoxic drugs (Burris, 2000). In IL-6 dependent multiple myeloma it was shown that the disease is driven by the persistent activation of Jak2. The Jak2 kinase inhibitor AG 490 with the Fas agonistic antibody CH-11 were found to induce massive cell death, a combination now considered for clinical development (Catlett-Falcone *et al.*, 1999). Interestingly, the combinaion of the Jak2 inhibitor with IL-12 was also found to be very effective in the IL6/Jak2 driven leukemia (Burdeyla *et al.*, 2001).

5.9 Induction of apoptosis in cancer cells as a future modality

Most anticancer drugs are aimed at *proliferating* cells but the tumor mass always possesses a large number of quiescent undividing cells that escape chemotherapy. Once there is a break in therapy these cells take off. Apoptosis-inducing drugs that kill non-dividing cells by inducing them to die should be developed. In several cases it has been found that the synergy between a signal interceptor and a cytotoxic agent leads to enhanced apoptosis. For example, the PKC blocker bryostatin 1 and the cytotoxic drug 1-β-D-orabinofurcnosylcytosine cause enhanced apoptosis in HL-60 cells (Lin *et al.*, 1992). Similarly, the tyrphostin AG1112 induces K562 cells towards terminal differentiation (Anafi *et al.*, 1993). Similarly, AG 957 a potent Bcr-Abl kinase inhibitor (Anafi *et al.*, 1992) synergizes with Fas ligand to induce apoptosis in Ph^+ cells (Carlo-Stella *et al.*, 1999). AG 490 induces massive apoptosis by itself in pre-B ALL (Meydan *et al.*, 1996a) but is only effective in IL-6 dependent multiple myeloma in the presence of the agonistic anti-Fas antibody CH-11 (Catlett Falcone *et al.*, 1999). It seems that the stress signaling in pre-B ALL cells is more intense so that once the anti-apoptotic shield provided by Jak-2 is blocked, the already enhanced pro-apoptotic signals take over and kill the cells. In the case of IL-6 dependent multiple myeloma it seems that stress signaling is less intense and thus the removal of the Jak-2 anti-apoptotic shield is insufficient to induce cell death. The cells are, however, left more vulnerable to pro-apoptotic agents. If this working hypothesis is a general paradigm of all cancer cells one would expect to find that cancer cells possess intensified stress signaling at the same time as possessing robust anti-apoptotic shields. This does indeed seem to be the case since the examination of transformed cells showed that p38 and Jnk are highly sensitized to stress signaling as compared to the parent un-transformed cells (Benhar *et al.*, 2000, 2002). Thus, as a general strategy, the status of stress signaling as well as identification of the exact nature of the enhanced proliferative and ant-apoptotic elements should be sought in every type of cancer. This will determine the exact nature of the signaling modifiers "cocktail".

5.10 Immune suppression

Modulation of the immune system is beneficial for a whole host of pathophysiological conditions. Autoimmune diseases, tissue rejection, and other inflammatory responses, such as sepsis, are mediated by signaling of immune cells. Modulation of such signals may be of therapeutic value and therefore this field has become an active area of investigation. The drugs cyclosporine A(CsA), FK506 and rapamycin have revolutionized organ transplantation. These drugs are used as immunosuppressive drugs since they intercept T-cell mediated immune response. CsA, a lipid soluble cyclic undecapeptide of fungal origin, is an extremely potent immunosuppressive drug (Sigal and Dumont, 1992). CsA blocks the G_0 to G_1 transition of the T-cell by blocking Ca^{2+} signaling which leads to production of IL-2 and the activation of the Il-2/Il-2 receptor autocrine loop. It was found that CsA binds to cyclophilin and the complex binds to the Ca^{2+}/calmodulin-dependent protein phosphatase (type 2B) known as calcineurin, thus inhibiting phosphatase activity essential for signal transduction (see below). FK506, a bacterial product, binds to FKBP (FK506 binding protein), which, when complexed with the drug, also inhibits calcineurin (Sigal and Dumont, 1992; Schreiber and Crabtree, 1992). Cyclophilins and FK506 are immunosuppressant proteins known collectively as immunophilins (Sigal and Dumont, 1992; Schreiber, 1991). This inhibition of calcineurin results in the blocking of the dephosphorylation and Ca^{2+} dependent nuclear transport of the cytoplasmic subunit NF-AT. These events are required for nuclear transcription and activation of the T-cells (Schreiber and Crabtree, 1992) including the formation of the cytokine interleukin-2 (IL-2). In summary, these drugs actually switch off gene transcription by inhibiting calcineurin which is a key signaling phosphatase.

Table 5.3 Signal trunsduction inhibitors which exhibit efficacy *in vivo*

Specific inhibitor	*Molecular target*
Protein tyrosine kinases	
RG 13012, AG 1478, ZD-1839, CP-358,774	EGF receptor
GW 2016	EGFR, Her-2
AGL 2043	PDGF receptor
SU 6668	VEGF receptor / Flk-1, FGF receptor, PDGF receptor
AG 490, AGL 2010	Jak-2
CGP 57148/STI 571, AG 957	Bcr-Abl kinase
Ser/Thr Kinases	
CGP 60474, Flavopiridol	Cdk2/4
CGP 41251, UCN-01	Protein Kinase C
ISIS-5132	Raf expression
NL-71–101	Protein Kinase B / c-Akt
Ras	
SCH 66336, HR-12	Farnesylation of Ras proteins
FTS (S-trans, trans-farnesylthiosalicylic acid)	Anchoring of Ras proteins
Dominant negative N17 Ras	Inhibition of Ras activation
ISIS-2503 (antisense)	Ras expression
Anti-receptor antibodies	
Herceptin	Her-2/c-ERB-2
C225	EGF receptor
Riruxan, Bexxar	CD 20 (Non-Hodgkin B lymphomas)
Receptor antagonists	
Tamoxiphen, ICI 182,780	Estrogen receptor
Flutamide	Androgen receptor
Anti-appoptotic proteins	
G 3139 (antisense)	Bcl-2
Transcription factor	
Onyx 015	Cancer cells with no/mutant p53

5.10.1 Inflammatory signals

Inflammatory signals are mediated by different types of immune cells and, in principle, modes of signal intervention can be designed. PTKs are involved in inflammatory signals although the initial events may not involve the activation of PTKs (Williams *et al.*, 1992). We therefore decided to examine whether tyrphostins can inhibit inflammatory signals elicited for example by LPS. Indeed, PTK blockers of the tyrphostin family have recently been demonstrated to be highly effective in blocking the activation of mouse macrophages by lipopolysaccharide (LPS) and the production of TNFα which is an important mediator of inflammatory responses (Novogrodsky *et al.*, 1994). Mice can be protected from LPS toxicity induced mortality by this class of tyrphostins (Novogrodsky *et al.*, 1994). This class of compounds also protects mice (Vanichkin *et al.*, 1996) and dogs (Servansky *et al.*, 1997) against sepsis, rats against cirrhosis-like syndrome (Talavera *et al.*, 1997) and mice against experimental autoimmune encephalytis (EAE, Brener *et al.*, 1998).

The protective effect of this class of tyrphostins appears to result from their ability to block $p56^{Lck}$, $p59^{Hck}$ and stress kinase activation induced by LPS binding to CD14 (Weinstein *et al.*, 1991; Stefanova *et al.*, 1993). The detailed stages involved in signal transduction from LPS binding to CD14 to TNFα production have yet to be identified. Since rheumatoid arthritis as well as other inflammatory conditions appear to be mediated by PTKs it is to be expected that PTK blockers will find broad usage in modulating immune responses. It is, in fact, surprising that this has not yet occurred.

5.11 Summary

Impressive progress has been made since the late 1980s when the concept of signal transduction emerged. Table 5.3 summarizes the signaling modifiers demonstrated to possess efficacy *in vivo* and in cells.

References

Ahmadian, M.R., Zor, T., Vogt, D., Kabsch, W., Selinger, Z., Wittinghofer, A., Scheffzek, K. (1999) Guanosine triphosphatase stimulation of oncogenic Ras mutants. *Proc Natl Acad Sci USA, 96(12), 7065–70.*

Anafi, M., Gazit, A., Gilon, C., Ben-Neriah, Y., Levitzki, A. (1992) Selective interactions of transforming and normal abl proteins with ATP, tyrosine-copolymer substrates, and tyrphostins. *J Biol Chem, 267(7), 4518–23.*

Anafi, M., Gazit, A., Zehavi, A., Ben-Neriah, Y., Levitzki, A. (1993) Tyrphostin-induced inhibition of p21 Obcr-Abl tyrosine kinase activity induces K562 to differentiate. *Blood, 82(12), 3524–9.*

Antoniades, H.N., Galanapoulos, T., Neville-Goldin, J., O'Hara, C.J. (1992) Malignant epithelial cells in primary human lung carcinomas coexpress *in vivo* platelet-derived growth factor (PDGF) and PDGF receptor mRNAs and their protein products. *Proc Natl Acad Sci, 89, 3942–46.*

Basu, A. (1993) The potential of protein kinase C as a target for anticancer treatment. *Pharmacol Ther, 59(3), 257–80.*

Benhar, M., Engelberg, D., Levitzki, A. (2000) Enhanced ROS production in oncogenically transformed cell potentiates JNK and p38 MAP kinase activation and sensitization to genotoxic stress. *Submitted.*

Benhar, M., Engelberg, D., Levitzki, A. (2002) Ros stress-activated kinases and stress signaling in cancer. *EMBO Rep, 3, 420–5.*

Besalga, J., Norton, L., Masui, H., Papdilla, A., Caplar, K., Miller, W.H., Mendelsohn, J. (1993) Antitumor effects of doxorubian in combination with anti-EGF receptor monoclonal antibodies. *J. Natl Cancer Inst, 85, 1327–33.*

Bishop, J.M. (1991) Molecular themes in oncogenesis. *Cell, 64(2), 235–48.*

Blum, G., Gazit, A., Levitzki, A. (2000) Substrate competitive inhibitors of IGF-1 receptor kinase. *Biochemistry, in press.*

Bos, J.L. (1989) ras oncogenes in human cancer: a review published erratum appears in Cancer Res 1990 Feb 15;50(4):1352]. *Cancer Res, 49(17), 4682–9.*

Bowman, T., Yu, H., Sebti, S., Dalton, W., Jove, R. (1999) Signal transducers and activators of transcription: novel targets for anticancer therapeutics. *Cancer Control, 6(5), 427–35.*

Brenner, T., Poradosu, E., Soffer, D., Sicsic, C., Gazit, A., Levitzki, A. (1998) Suppression of experimental autoimmune encephalomyelitis by tyrphostin AG-556. *Exp Neurol, 154(2), 489–98.*

Brunner, N., Frandsen, T.L., Holst-Hansen, C., Bei, M., Thompson, E.W., Wakeling, A.E., Lippman, M.E., Clarke, R. (1993). MCF7/LCC2: a 4-hydroxytamoxifen resistant human breast cancer variant that retains sensitivity to the steroidal antiestrogen ICI 182, 780. *Cancer Res, 53(14), 3229–32.*

Buchdunger, E., Zimmermann, J., Mett, H., Meyer, T., Muller, M., Druker, B.J., Lydon, N. B. (1996) Inhibition of the Abl protein-tyrosine kinase *in vitro* and ill vivo by a 2-phenylaminopyrimidine derivative. *Cancer Res, 56(1), 100–4.*

Buolamwini, J.K. (1999) Novel anticancer drug discovery. *Curr Opin Chem Biol, 3(4), 500–9.*

Buolamwini, J.K. (2000) Cell cycle molecular targets in novel anticancer drug discovery. *Curr Pharm Des, 6(4), 379–92.*

Burdelya, L., Catlett-Falcone, R., Levitzki, A., Coppola, D., Sun, J., Sebti, S., Dalton, W., Jove, R., Yu, H. (2001) Interleukin-12 enhances the antitumor effect of the tyrosine kinase inhibitor AG-490 in a murine tumor model. *In press.*

Burris, H.A., 3rd. (2000) Docetaxel (Taxotere) in HER-2-positive patients and in combination with trastuzumab (Herceptin). *Semin Oncol, 27(2 Suppl 3), 19–23.*

Carlo-Stella, C., Regazzi, E., Sammarelli, G., Colla, S., Garau, D., Gazit, A. (1999) Effects of the tyrosine kinase inhibitor AG957 and an anti-Fas receptor antibody on CD34+ chronic myelogenous leukemia progenitor cells. *Blood, 93(11), 3973–82.*

Dauvois, S., White, R., Parker, M.G. (1993) The antiestrogen ICI 182780 disrupts estrogen receptor nucleocytoplasmic shuttling. *J Cell Sci, 106(Pt 4), 1377–88.*

DeFriend, D.J., Howell, A., Nicholson, R.I., Anderson, E., Dowsett, M., Mansel, R.E., Blarney, R.W., Hundred, N.J., Robertson, J.F., Saunders, C., *et al.* (1994) Investigation of a new pure antiestrogen (ICI 182780) in women with primary breast cancer. *Cancer Res, 54(2), 408–14.*

Di Cristofano, A., Pandolfi, P.P. (2000) The multiple roles of PTEN in tumor suppression. *Cell, 100(4), 387–90.*

Dive, C., Evans, C.A., Whetton, A.D. (1992) Induction of apoptosis–new targets for cancer chemotherapy. *Semin Cancer Biol, 3(6), 417–27.*

Drebin, J.A., Link, V.C., Stem, D.F., Weinberg, R.A., Greene, M.I. (1985) Down-modulation of an oncogene protein product and reversion of the transformed phenotype by monoclonal antibodies. *Cell, 41(3), 697–706.*

Drebin, J.A., Link, V.C., Weinberg, R.A., Greene, M.I. (1986) Inhibition of tumor growth by a monoclonal antibody reactive with an oncogene-encoded tumor antigen. *Proc Natl Acad Sci USA, 83(23), 9129–33.*

Dreicer, R., Wilding, G. (1992) Steroid hormone agonists and antagonists in the treatment of cancer. *Cancer Invest, 10(1), 27–41.*

Druker, B.J., Tamura, S., Buchdunger, E., Ohno, S., Segal, G.M., Fanning, S., Zimmermann, J., Lydon, N.B. (1996) Effects of a selective inhibitor of the Abl tyrosine kinase on the growth of Bcr-Abl positive cells. *Nat Med, 2(5), 561–6.*

Ducruet, A.P., Rice, R.L., Tamura, K., Yokokawa, F., Yokokawa, S., Wipf, P., Lazo, J.S. (2000) Identification of new Cdc25 dual specificity phosphatase inhibitors in a targeted small molecule array [In Process Citation]. *Bioorg Med Chem, 8(6), 1451–66.*

Eisenberger, M.A., Reyno, L.M., Jodrell, D.I., Sinibaldi, V.J., Tkaczuk, K.H., Sridhara, R., Zuhowski, E.G., Lowitt, M.H., Jacobs, S.C., Egorin, M.J. (1993) Suramin, an active drug for prostate cancer: interim observations in a phase I trial [see comments] [published erratum appears in J Natl Cancer Inst 1994 Apr 20;86(8):639–40]. *J Natl Cancer Inst, 85(8), 611–21.*

Elad, G., Paz, A., Haklai, R., Marciano, D., Cox, A., Kloog, Y. (1999) Targeting ofK-Ras 4B by S-trans, trans-farnesyl thiosalicylic acid. *Biochim Biophys Acta, 9(3), 228–42.*

Evans, C.A., Owen-Lynch, P.J., Whetton, A.D., Dive, C. (1993) Activation of the Abelson tyrosine kinase activity is associated with suppression of apoptosis in hemopoietic cells. *Cancer Res, 53(8), 1735–8.*

Evans, C.A., Lord, J.M., Owen-Lynch, P.J., Johnson, G., Dive, C., Whetton, A.D. (1995) Suppression of apoptosis by v-ABL protein tyrosine kinase is associated with nuclear translocation and activation of protein kinase C in an interleukin-3-dependent haemopoietic cell line. *J Cell Sci, 108(pt 7), 2591–8.*

Fan, Z., Baselga, J., Masui, H., Mendelsohn, J. (1993) Antitumor effect of anti-epidermal growth factor receptor monoclonal antibodies plus cis-diamminedichloroplatinum on well established A431 cell xenografts. *Cancer Res, 53(19), 4637–42.*

Fearon, E.R., Vogelstein, B. (1990) A genetic model for colorectal tumorigenesis. *Cell, 61(5), 759–67.*

Fischel, J.L., Barbe, V., Berlion, M., Formento, P., Berrile, J., Bizzari, J.P., Milano, G. (1993) Tamoxifen enhances the cytotoxic effects of the nitrosourea fotemustine. Results on human melanoma cell lines. *Eur J Cancer, 16, 2269–73.*

Garcia-Echeverria, C., Traxler, P., Evans, D.B. (2000) ATP site-directed competitive and irreversible inhibitors of protein kinases. *Med Res Rev, 20(1), 28–57.*

Gazit, A., Yaish, P., Gilon, C., Levitzki, A. (1989) Tyrphostins I: synthesis and biological activity of protein tyrosine kinase inhibitors. *J Med Chem, 32(10), 2344–52.*

Gazit, A., Osherov, N., Posner, I., Yaish, P., Poradosu, E., Gilon, C., Levitzki, A. (1991) Tyrphostins. 2. Heterocyclic and alpha-substituted benzylidenemalononitrile tyrphostins as potent inhibitors of EGF receptor and ErbB2/neu tyrosine kinases. *J Med Chem, 34(6), 1896–907.*

Gazit, A., Osherov, N., Posner, I., Bar-Sinai, A., Gilon, C., Levitzki, A. (1993) Tyrphostins. 3. Structure-activity relationship studies of alpha-substituted benzylidenemalononitrile 5-S-aryltyrphostins. *J Med Chem, 36(23), 3556–64.*

Gazit, A., App, H., McMahon, G., Chen, J., Levitzki, A., Bohmer, F.D. (1996) Tyrphostins. 5. Potent inhibitors of platelet-derived growth factor receptor tyrosine kinase: structure-activity relationships in quinoxalines, quinolines, and indole tyrphostins. *J Med Chem, 39(11), 2170–7.*

Gibbs J.B. (2000) Mechanism-based target identification and drug discovery in cancer research. *Science, 287, 1969–73.*

Gray, N.S., Wodicka, L., Thunnissen, A.M., Norman, T.C., Kwon, S., Espinoza, F.H., Morgan, D.O., Barnes, G., LeClerc, S., Meijer, L., Kim, S.H., Lockhart, D.J., Schultz, P.G. (1998) Exploiting chemical libraries, structure, and genomics in the search for kinase inhibitors. *Science, 281(5376), 533–8.*

Groundwater, P.W., Solomons, K.R., Drewe, J.A., Munawar, M.A. (1996) Protein tyrosine kinase inhibitors. *Prog Med Chem, 33, 233–329.*

Hanke, J.H., Gardner, J.P., Dow, R.L., Changelian, P.S., Brissette, W.H., Weringer, E.J., Pollok, B.A., Connelly, P.A. (1996) Discovery of a novel, potent, and Src family-selective tyrosine kinase inhibitor. Study of Lck- and FynT-dependent T cell activation. *J Biol Chem, 271(2), 695–701.*

Harwerth, I.M., Wels, W., Schlegel, J., Muller, M., Hynes, N.E. (1993) Monoclonal antibodies directed to the erbB-2 receptor inhibit *in vivo* tumour cell growth. *Br J Cancer, 68(6), 1140–5.*

He, H., Hirokawa, Y., Levitzki, A., Maruta, H. (2000) An anti-Ras cancer potential of PP1, an inhibitor specific for Src family kinases: *in vitro* and *in vivo* studies. *The Cancer Journal, 6(4), 243–8.*

Henrikson, M., Funa, K., Hartman, M., Claesson-Welsh, L., Heldin, C.H., Westermark, B., Nister, M. (1992) Platelet-derived growth factor and its receptor in human glioma tissue: expression of messenger RNA and protein suggests the presence of autocrine and paracrine loops. *J Neurol Sci, 111(2), 119–133.*

Hernandez, S., Hernandez, L., Bea, S., Pinyol, M., Nayach, I., Bellosillo, B., Nadal, A., Ferrer, A., Fernandez, P.L., Montserrat, E., Cardesa, A., Campo, E. (2000) cdc25a and the splicing variant cdc25b2, but not cdc25B1, -B3 or -C, are over-expressed in aggressive human non-Hodgkin's lymphomas. *Int J Cancer, 89(2), 148–52.*

Hosol, H., Dilling, M.B., Shikata, T., Liu, L.N., Shu, L., Ashmun, R.A., Gennain, G.S., Abraham, R.T., Houghton, P.J. (1999) Rapamycin causes poorly reversible inhibition of mTOR and induces p53-independent apoptosis in human rhabdomyosarcoma cells. *Cancer Res, 59(4), 886–94.*

Hubbard, S.R., Wei, L., Ellis, L., Hendrickson, W.A. (1994) Crystal structure of the tyrosine kinase domain of the human insulin receptor [see comments]. *Nature, 372(6508), 746–54.*

Hubbard, S.R. (1997) Crystal structure of the activated insulin receptor tyrosine kinase in complex with peptide substrate and ATP analog. *EMBO J, 16(18), 5572–81.*

Indolfi, C., Avvedimento, E.V., Rapacciuolo, A., Di Lorenzo, E., Esposito, G., Stabile, E., Feliciello, A., Mele, E., Giuliano, P., Condorelli, G., *et al.* (1995) Inhibition of cellular ras prevents smooth muscle cell proliferation after vascular injury *in vivo* [see comments]. *Nat Med, 1(6), 541–5.*

Kaubisch, A., Schwartz, G.K. (2000) Cyclin-dependent kinase and protein kinase C inhibitors: a novel class of antineoplastic agents in clinical development [In Process Citation]. *Cancer J Sci Am, 6(4), 192–212.*

Kern, J.A., Torney, L., Weiner, D., Gazdar, A., Shepard, H.M., Fendly, B. (1993) Inhibition of human lung cancer cell line growth by an anti-p185HER2 antibody. *Am J Respir Cell Mol Biol, 9(4), 448–54.*

Kohl, N.E., Mosser, S.D., deSolms, S.J., Giuliani, E.A., Pompliano, D.L., Graham, S.L., Smith, R.L., Scolnick, E.M., Oliff, A., Gibbs, J.B. (1993) Selective inhibition of ras-dependent transformation by a farnesyltransferase inhibitor [see comments]. *Science, 260(5116), 1934–7.*

Komarov, P.G., Komarova, E.A., Kondratov, R.V., Christov-Tselkov, K., Coon, J.S., Chernov, M.V., Gudkov, A.V. (1999) A chemical inhibitor of p53 that protects mice from the side effects of cancer therapy [see comments]. *Science, 285(5434), 1733–7.*

Kosloff M, Tsaffrir Z, Selinger Z. (2000) Substrate-assisted catalysis: implications for biotechnology and drug design. *Drug Development Research, 50.*

Kovalenko, M., Gazit, A., Bohmer, A., Rorsman, C., Ronnstrand, L., Heldin, C.H., Waltenberger, J., Bohmer, F.D., Levitzki, A. (1994) Selective platelet-derived growth factor receptor kinase blockers reverse sis-transformation. *Cancer Res, 54(23), 6106–14.*

Kovalenko, M., Ronnstrand, L., Heldin, C.H., Loubtchenkov, M., Gazit, A., Levitzki, A., Bohmer, F.D. (1997). Phosphorylation site-specific inhibition of platelet-derived growth factor beta-receptor autophosphorylation by the receptor blocking tyrphostin AG1296. *Biochemistry, 36(21), 6260–9.*

Larsen, A.K. (1993) Suramin: an anticancer drug with unique biological effects. *Cancer Chemother Pharmacol, 32(2), 96–8.*

Ledwith, B. (1992) Reversal of transformed phenotypes by antisense fos. *Ann NY Acad Sci, 660, 124–35.*

Levitzki, A. (1992) Tyrphostins: tyrosine kinase blockers as novel antiproliferative agents and dissectors of signal transduction. *Faseb J, 6(14), 3275–82.*

Levitzki, A., Gazit, A. (1995) Tyrosine kinase inhibition: an approach to drug development. *Science, 267(5205), 1782–8.*

Levitzki, A. (1996a) SRC as a target for anti-cancer drugs. *Anticancer Drug Des, 11(3), 175–82.*

Levitzki, A. (1996b) Targeting signal transduction for disease therapy. *Curr Opin Cell Biol, 8(2), 239–44.*

Levitzki, A., Bohmer, F.D. (1998) Altered efficacy and selectivity of tyrosine kinase inhibitors of the activated states of protein tyrosine kinases. *Anticancer Drug Des, 13(6), 731–4.*

Lewis, G.D., Figari, I., Fendly, B., Wong, W.L., Carter, P., Gonnan, C., Shepard, H.M. (1993) Differential responses of human tumor cell lines to anti-p185HER2 monoclonal antibodies. *Cancer Immunol Immunother, 37(4), 255–63.*

Lopez-Talavera, J.C., Levitzki, A., Martinez, M., Gazit, A., Esteban, R., Guardia, J. (1997) Tyrosine kinase inhibition ameliorates the hyperdynamic state and decreases nitric oxide production in cirrhotic rats with portal hypertension and ascites. *J Clin Invest, 100(3), 664– 70.*

Love, R.R., Koroltchouk, V. (1993). Tamoxifen therapy in breast cancer control worldwide. *Bull World Health Organ, 71(6), 795–803.*

Mauro, A., Bulfone, A., Turco, E., Schiffer, D. (1991) Coexpression of platelet-derived growth factor (PDGF) B chain and PDGF B-type receptor in human gliomas. *Childs Nerv Syst, 7(8), 432–6.*

McConnick, F. (1999) Cancer therapy based on p53. *Cancer J Sci Am, 5(3), 139–44.*

Meydan, N., Grunberger, T., Dadi, H., Shahar, M., Arpaia, E., Lapidot, Z., Leeder, J.S., Freedman, M., Cohen, A., Gazit, A., Levitzki, A., Roifman, C.M. (1996). Inhibition of acute lymphoblastic leukaemia by a Jak-2 inhibitor. *Nature, 379(6566), 645–8.*

Millauer, B., Shawver, L.K., Plate, K.H., Risau, W., Ullrich, A. (1994) Glioblastoma growth inhibited *in vivo* by a dominant-negative Flk-1 mutant. *Nature, 367(6463), 576–9.*

Mohammadi, M., McMahon, G., Sun, L., Tang, C., Hirth, P., Yeh, B.K., Hubbard, S.R., Schlessinger, J. (1997) Structures of the tyrosine kinase domain of fibroblast growth factor receptor in complex with inhibitors. *Science, 276(5314), 955–60.*

Mohammadi, M., Froum, S., Hamby, J.M., Schroeder, M.C., Panek, R.L., Lu, G.H., Eliseenkova, A.V., Green, D., Schlessinger, J., Hubbard, S.R. (1998) Crystal structure of an angiogenesis inhibitor bound to the FGF receptor tyrosine kinase domain. *Embo J, 17(20), 5896–904.*

Nagane, M., Levitzki, A., Gazit, A., Webster, K., Cavanee, Huang, S. (1998) Drug resistance of human glioblastoma cells conferred by a tumor-specific mutant epidermal growth factor receptor through modulation of Bcl-XL and caspace-3-like protease. *Proc Natl Acad Sci, 95, 5724–29.*

Nagane, M., Narita, Y., Mishima, K., Levitzki, A., Burgess, A.W., Webster, K., Cavenee, W.K., Su Huang, H.-J. (2001) Human glioblastoma xenografts overexpressing a tumor-specific mutant EGF receptor sensitized to cisplatin by the AG1478 tyrosine kinase inhibitor. *J Neurosurgery, 95, 472–9.*

Naramura, M., Gillies, S.D., Mendelsohn, J., Reisfeld, R.A., Mueller, B.M. (1993) Therapeutic potential of chimeric and murine anti-(epidermal growth factor receptor) antibodies in a metastasis model for human melanoma. *Cancer Immunol Immunother, 37(5), 343–9.*

Novogrodsky, A., Vanichkin, A., Patya, M., Gazit, A., Osherov, N., Levitzki, A. (1994) Prevention of lipopolysaccharide-induced lethal toxicity by tyrosine kinase inhibitors. *Science, 264(5163), 1319–22.*

Osherov, N., Levitzki, A. (1994) Epidermal-growth-factor-dependent activation of the src-family kinases. *Eur J Biochem, 225(3), 1047–53.*

Pai, L., Pastan I. (1993) The use of immunotoxins for cancer therapy. *Eur. J. Cancer, 29a, 1606–9.*

Pai, L.H., Pastan, I. (1994) Immunotoxins and recombinant toxins for cancer treatment. *Important Adv Oncol, 3–19.*

Parrizas, M., LeRoith, D. (1997) Insulin-like growth factor-l inhibition of apoptosis is associated with increased expression of the bcl-xL gene product. *Endocrinology, 138(3), 1355–8.*

Peng, H., Xie, W., Kim, D.I., Zalkow, L.H., Powis, G., Ottemess, D.M., Abraham, R.T. (2000) Steroidal derived acids as inhibitors of human Cdc25A protein phosphatase. *Bioorg Med Chem, 8(2), 299–306.*

Posner, I., Engel, M., Gazit, A., Levitzki, A. (1994) Kinetics inhibition by tyrphostins of the tyrosine kinase activity of the epidermal growth factor receptor and analysis by a new computer program. *Mol Pharmacol, 45(4), 678–83.*

Prendville J, Crowther D, Thatcher N, Woll, P.J., Fox, B.W. (1993) A phase 1 study of intravenous bryostatinl in patients with advanced cancer. *J Cell Physiol, 155(2), 301–12.*

Reuveni, H., Livnah, N., Geiger, T., Klein, S., Ohne, O., Cohen, I., Benhar, M., Gellerman, G., Levitzki, A. (2002) Towards a PKB inhibitor: Modification of a selective PKA inhibitor by rational design. *Biochemistry*, in press.

Ross, R.C. (1993) The pathogenesis of atherosclerosis: a perspective for the 1990s. *Nature, 302, 801–9.*

Schachter, L.M., Esa, A.H., May, S., Laulis, M.K., Pettit, G.R., Hess, A.D. (1991) Successful treatment of murine melanoma with bryostatin 1. *Cancer Res, 51(2), 682–7.*

Schindler, T., Sicheri, F., Pico, A., Gazit, A., Levitzki, A., Kuriyan, J. (1999) Crystal structure of Hck in complex with a Src family-selective tyrosine kinase inhibitor. *Mol Cell, 3(5), 639–48.*

Schreiber, S.L. (1991) Chemistry and biology of the immunophilins and their immunosuppressive ligands. *Science, 251(4991), 283–7.*

Schreiber, S.L., Crabtree, G.R. (1992) The mechanism of action of cyclosporin A and FK506. *Immunol Today, 13(4), 136–42.*

Schreiber, S.L. (1992) Immunophilin-sensitive protein phosphatase action in cell signaling pathways. *Cell, 70(3), 365–8.*

Sebolt-Leopold, J.S., Dudley, D.T., Herrera, R., Van Becelaere, K., Wiland, A., Gowan, R.C., Tecle, H., Barrett, S.D., Bridges, A., Przybranowski, S., Leopold, W.R., Saltiel, A.R. (1999) Blockade of the MAP kinase pathway suppresses growth of colon tumors *in vivo*. *Nat Med, 5(7), 810–16.*

Sevransky, J.E., Shaked, G., Novogrodsky, A., Levitzki, A., Gazit, A., Hoffinan, A., Elin, R.J., Quezado, Z.M., Freeman, B.D., Eichacker, P.Q., Danner, R.L., Banks, S.M., Bacher, J., Thomas, M.L., 3rd, Natanson, C. (1997) Tyrphostin AG 556 improves survival and reduces multiorgan failure in canine Escherichia coli peritonitis. *J Clin Invest, 99(8), 1966–73.*

Shan, D., Ledbetter, J.A., Press, O.W. (2000) Signaling events involved in anti-CD20-induced apoptosis of malignant human B cells. *Cancer Immunol Immunother, 48(12), 673–83.*

Shepard, M.H., Lewis, G.D., Sarup, J.C., Fondby, B.M., Maneval, D., Mardenti, J., Figari, I., Kotts, C.E. (1991) Monoclonal antibody therapy of human cancer: Taking the Her2 proto-oncogene to the clinic. *J Clin Immunol, 11(9), 117–27.*

Shir, A., Levitzki, A. (2002) Selective inhibition of glioma growth by tumor specific activation of double-stranded RNA dependent protein kinase – PKR. *Nature Biotechnology*, in press.

Sigal, N.H., Dumont, F.J. (1992) Cyclosporin A, FK-506, and rapamycin: pharmacologic probes of lymphocyte signal transduction. *Annu Rev Immunol, 10, 519–60.*

Slamon, D.J., Godolphin, W., Jones, L.A., Holt, J.A., Wong, S.G., Keith, D.E., Levin, W.J., Stuart, S.G., Udove, J., Ullrich, A., *et al.* (1989) Studies of the HER-2/neu proto-oncogene in human breast and ovarian cancer. *Science, 244(4905), 707–12.*

Speir, E., Epstein, S.E. (1992) Inhibition of smooth muscle cell proliferation by an antisense oligodeoxynucleotide targeting the messenger RNA encoding proliferating cell nuclear antigen. *Circulation, 86(2), 538–47.*

Stancovski, I., Hurwitz, E., Leitner, O., Ullrich, A., Yarden, Y., Sela, M. (1991) Mechanistic aspects of the opposing effects of monoclonal antibodies to the ERBB2 receptor on tumor growth. *Proc Natl Acad Sci USA, 88(19), 8691–5.*

Stebbing, J., Copson, E., O'Reilly, S. (2000) Herceptin (trastuzamab) in advanced breast cancer. *Cancer Treat Rev, 26(4), 287–90.*

Stefanova, I., Corcoran, M.L., Horak, E.M., Wahl, L.M., Bolen, J.B., Horak, I.D. (1993) Lipopolysaccharide induces activation of CD14-associated protein tyrosine kinase p53/56lyn. *J Biol Chem, 268(28), 20725–8.*

Takemasa, I., Yamamoto, H., Sekimoto, M., Ohue, M., Noura, S., Miyake, Y., Matsumoto, T., Aihara, T., Tomita, N., Tamaki, Y., Sakita, I., Kikkawa, N., Matsuura, N., Shiozaki, H., Monden, M. (2000) Overexpression of CDC25B phosphatase as a novel marker of poor prognosis of human colorectal carcinoma. *Cancer Res, 60(11), 3043–50.*

Tsai, C., Levitzki, A., Li-Hwa, W., Kuo- Ting, C., Cheng, C.-C., Gazit, A., Perng, R.-P . (1996) Enhancement of chemosensitivity by tyrphostin AG825 in high-p185neu expressing non-small cell lung cancer cells. *Cancer Research, 56, 1068–74.*

Van Rijswick, R.E., Cuitovic, E., Wagstaff, J., Myers, C.E. (1993) Workshop on suramin with emphasis on prostate cancer. *Eur J Cancer, 29A, 298–300.*

Vanichkin, A., Patya, M., Gazit, A., Levitzki, A., Novogrodsky, A. (1996) Late administration of a lipophilic tyrosine kinase inhibitor prevents lipopolysaccharide and Escherichia coli-induced lethal toxicity. *J Infect Dis, 173(4), 927–33.*

Wakeling, A.E., Dukes, M., Bowler, J. (1992) A potent specific pure antiestrogen with clinical potential. *Cancer Res, 51(15), 3867–73.*

Waltenberger, J., Uecker, A., Kroll, J., Frank, H., Mayr, U., Bjorge, J. D., Fujita, D., Gazit, A., Hombach, V., Levitzki, A., Bohmer, F.D. (1999) A dual inhibitor of platelet-derived growth factor beta-receptor and Src kinase activity potently interferes with motogenic and mitogenic responses to PDGF in vascular smooth muscle cells. A novel candidate for prevention of vascular remodeling. *Circ Res, 85(1), 12–22.*

Ward, W.H., Cook, P.N., Slater, A.M., Davies, D.H., Holdgate, G.A., Green, L.R. (1994) Epidermal growth factor receptor tyrosine kinase. Investigation of catalytic mechanism, structure-based searching and discovery of a potent inhibitor. *Biochem Pharmacol, 48(4), 659–66.*

Williams, W.V., Von Feldt, J.M., Ramanujam, T., Weiner, D.B. (1992) Tyrosine kinase signal transduction in rheumatoid synovitis. *Semin Arthritis Rheum, 21(5), 317–29.*

Yaish, P., Gazit, A., Gilon, C., Levitzki, A. (1988) Blocking of EGF-dependent cell proliferation by EGF receptor kinase inhibitors. *Science, 242(4880), 933–5.*

Yoneda, T., Lyall, R., Alsine, M., Pearsons, P., Spada, A., Levitzki, A., Zilberstein, A., Mundy, G.R. (1991) The antiproliferative effects of tyrosine kinase inhibitor tyrphostin on a human squamous cell carcinoma *in vitro* and in nude mice. *Cancer Research, 51, 4430–4435.*

Zhu, X., Kim, J.L., Newcomb, J.R., Rose, P.E., Stover, D.R., Toledo, L.M., Zhao, H., Morgenstem, K.A. (1999) Structural analysis of the lymphocyte-specific kinase Lck in complex with non-selective and Src family selective kinase inhibitors. *Structure Fold Des, 7(6), 651–61.*

Chapter 6

Heat shock proteins

Péter Csermely and Ichiro Yahara

Contents

6.1 Introduction

When a cell experiences environmental stress, it stops, or at least slows down most of its original functions, such as transport processes, DNA, RNA and protein synthesis. However, there is a peculiar set of proteins, called stress proteins, which are preferentially expressed under these, restrictive conditions. The archetypal stress response is a sudden rise in the outside temperature, called heat shock. The heat shock response was first discovered by Feruccio Ritossa (1962), who observed an enlargement of special sections of *Drosophila melanogaster* chromosomes (heat shock puffs) after heat treatment of the flies. Later it became clear that these chromosome segments encoded a special class of proteins, heat shock proteins. Heat shock proteins (Hsp-s) are induced by a large variety of stimuli besides heat shock itself. Table 6.1 lists a few examples of environmental stresses leading to the expression of heat shock, or other stress proteins.

Table 6.1. Effects inducing the synthesis of stress proteins. (For details please consult the following references: Lindquist (1986), Welch (1992) and Feige *et al.* (1996).)

Heat shock	Too much Ca^{2+} inside the cell
Cold shock	Viral infection
UV radiation	Bacterial products
Electrosmog	Parasite toxins
Amino acid analogues	Acute phase reaction
Alcohol	Overload of the endoplasmic reticulum
Heavy metal ions	Phagocytosis
Arsenite	Hormonal effects
Too much oxidation	Increased cell proliferation
Too much reduction	Cell differentiation
Too little glucose	Increased blood pressure
Too little ATP	Too little exercise
"Feeding" of cell cultures	Too much exercise
Osmotic shock	Mental stress

Heat shock proteins play an essential role in the etiology of numerous diseases, with a rapidly increasing role in clinical practice (Latchman, 1991; Welch, 1992; Jindal, 1996; van Eden and Young, 1996; Feige *et al.*, 1996). Their function is necessary for the homeostasis of the living cell, and becomes especially important in disease, when our cells have to cope with a stressful environment. Our increasing knowledge about the function and biochemistry of heat shock proteins gives us more and more tools to characterize them in patients, enabling us to monitor and improve the status of this intracellular defense system during infections, diseases and aging.

6.2 Heat shock proteins, molecular chaperones

Heat shock proteins form an ancient, primary system for "intracellular self-defense". Why are heat shock proteins beneficial? How does this defense system save and, in fact, extend our life? Most of the heat shock proteins are molecular chaperones. Chaperones have been defined as "proteins that bind to and stabilize an otherwise unstable conformer of another protein – and, by controlled binding and release, facilitate its correct fate *in vivo*: be it folding, oligomeric assembly, transport to a particular subcellular compartment, or disposal by degradation" (Hartl, 1996). The majority of chaperones prevent the aggregation of "sticky" protein folding intermediates. By binding to their targets, chaperones are acting as "collectors" of damaged proteins. Two classes of chaperones which are especially effective in completing this job are the small heat shock proteins and the 90 kDa heat shock proteins (the Hsp90 family, Table 6.2) may thus be considered as "recycling-concerned dustmen" of cells, binding their targets and keeping them in a folding competent state until the whole cell recovers, and becomes able to provide the energy for the refolding process. Members of the 90 kDa chaperone family bind to various peptides both *in vitro* and *in vivo* (Menoret *et al.*, 1999), which may extend their "collector/dustman" role to free intracellular peptides (Baranyi, Juhasz, Csermely, unpublished observations). These residual peptides, which escape from the major cytoplasmic proteolytic apparatus, the proteasomal system, may seriously interfere with signaling processes (Blum *et al.*, 2000), and therefore pose a great threat to cellular function.

A class of chaperones (in yeast: the 100 kDa heat shock proteins) is able to desaggregate proteins from loose protein aggregates (Glover and Lindquist, 1998). Other classes of chaperones (mainly the 60 and 70 kDa heat shock proteins, Table 6.2) rescue misfolded proteins from folding traps, giving them renewed chance for spontaneous folding. Chaperones do not determine the tertiary structure of proteins, but help them find their structure more efficiently. This mechanism increases the yield, but not the rate, of protein folding. Only a few chaperones behave as true catalysts by increasing the speed of folding. These special chaperones, peptidyl prolyl cis/trans isomerases and protein disulfide isomerases, are therefore better named "folding catalysts" (Hartl, 1996). All these mechanisms have paramount importance following stress, when the cell has to refold damaged proteins and re-establish its original structure.

Chaperone protein monomers never work alone. Most chaperones form oligomers and/or are helped by co-chaperones. A special multimeric chaperone machine is the eukaryotic "foldosome", which governs the folding of various signaling kinases and nuclear hormone receptors (Pratt and Toft, 1997). Refolding of proteins requires energy. Therefore it is not surprising that most chaperones bind and hydrolyze ATP. ATP cleavage induces a conformational change in chaperone structure, which is an essential part of the folding assistance. However, chaperones are rather poor ATP-ases. The inefficient and slow hydrolysis of ATP enables them to provide enough time to the target protein to re-fold. The rate of ATP hydrolysis and exchange of the product ADP to another ATP is regulated by several co-chaperone molecules. Co-chaperones may also influence binding and dissociation of target molecules as well as direct the targets, or the whole chaperone complex to various destinations inside the cell.

Figure 6.1 shows the "working cycles" of the two major classes of molecular chaperones, Hsp60 and Hsp70. Hsp60 binds its target to the internal cavity of the oligomeric protein. In most cases the client protein is completely isolated from the outside world by the acquisition of an Hsp10-heptamer behaving as a "cap". Due to this insulation of the target, Hsp60-type chaperone machines are often called as Anfinsen-cages after Christian Anfinsen, a Nobel-laureate

Table 6.2 Major families of molecular chaperones

Chaperone families	*Prokaryotic family members*	*Major functions of chaperone families*
Small heat shock proteins (e.g. Hsp27*)		Prevent the aggregation of other proteins, by collecting protein "garbage", act as "dustmen" of cells
Hsp60 family	GroEL (co-chaperone: GroES)	Assistance in protein folding and re-folding
Hsp70 family	DnaK (co-chaperones: DnaJ, GrpE)	Assistance in protein folding and re-folding
Hsp90 family	HtpG	Stabilize substrate proteins and maintain their active, or inactive state, prevent the aggregation of other proteins, by collecting protein "garbage", act as "dustmen" of cells
Hsp100 family	ClpA	Desaggregation of proteins
Protein disulfide isomerases	DsbA	Assistance in gradual oxidation of secreted proteins, promotion of correct disulfide-bridge formation, reorganization of disulfide bridges
Peptidyl prolyl cis/trans isomerases (rotamases, FKBP-s, cyclophilins, parvulins)		Catalysis of the cis/trans isomerization of peptide-bonds besides proline residues to set the correct conformation of protein segments

*The "Hsp" abbreviation stands for "heat shock protein", numbers refer to the molecular weight in kDa.

pioneer in protein-folding studies. The protected environment of the Hsp60-cavity prevents the aggregation of the client protein completely (Hartl, 1996). Binding and hydrolysis of ATP induces large conformational changes in Hsp60 (see http://www.cryst.bbk.ac.uk/~ubcg16z/cpn/chaperone.html), which help to loosen the hydrophobic core of the target protein in a partial unfolding process. During this multi-directional pulling of the target, water may enter its hydrophobic core and facilitate reorganization (Csermely, 1999). Hydrolysis of ATP and binding of a new target on the other side of the chaperone-machine releases the Hsp10 cap, and liberates the target protein. For most unfolded or damaged proteins a single round of Hsp60-assisted folding is not enough, and they are recycled through multiple unfolding-refolding cycles. The above mechanism has been deciphered by studying the *E. coli* Hsp60, GroEL and its co-chaperone, GroES. Substrate specificity of eukaryotic Hsp60 chaperones tends to be restricted to actin and tubulin, and details of the folding mechanism may differ (Bukau and Horwich, 1998).

Hsp70 binds the target proteins utilizing the help of its co-chaperone, Hsp40 (in *E. coli*: DnaK and DnaJ, respectively Figure 6.1B). During Hsp70 ATP hydrolysis some proteins acquire their native conformation, while others require the successive action of the Hsp60-machine. Both Hsp70 and Hsp40 have many isoforms, which enables members of these families to perform highly specialized functions. In contrast to Hsp60, which surrounds its target, Hsp70 binds a small peptide-segment of the target protein. Among the Hsp70-bound 7 amino acids, 3–4 are usually hydrophobic. The Hsp70-bound peptide segment adopts a highly extended conformation, which requires considerable unfolding of local secondary structures (Bukau and Horwich, 1998).

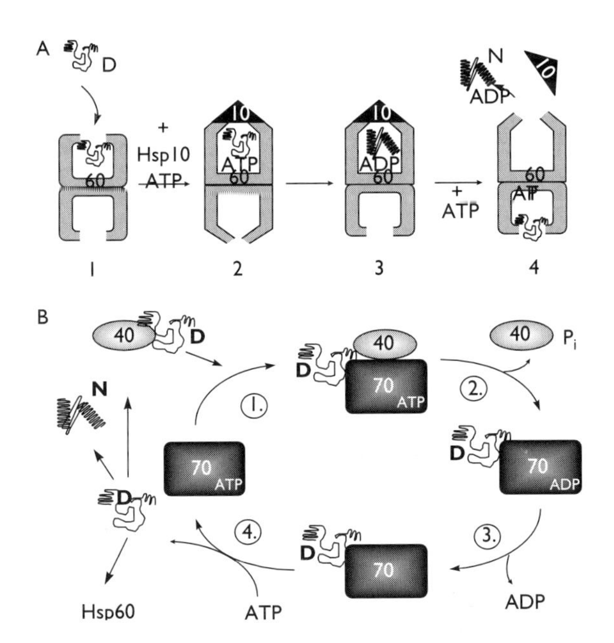

Figure 6.1 A, B Catalytic cycle of the 60 kDa (*panel A*) and 70 kDa (*panel B*) molecular chaperones. Abbreviations: 60, Hsp60; 10, Hsp10 – a co-chaperone of Hsp60; 70, Hsp70; 40, Hsp40 – a co-chaperone of Hsp70; D, denatured protein; N, native protein.

6.3 Synthesis of heat shock proteins

Chaperones (such as heat shock proteins) are ubiquitous, highly conserved proteins which probably played a major role in the pre-biotic evolution of modern enzymes (Csermely, 1997). Chaperones are vital throughout the whole lifetime of our cells. However, they are needed even more after environmental stress, which induces protein damage. In eukaryotic organisms the expression of heat shock protein messenger RNA-s is mediated by a family of transcription factors, called heat shock factors. Heat shock factor 1 (HSF-1) plays a major role in heat shock response, while other members of the family are activated after prolonged stress, or participate in processes such as embryonic development, or cell differentiation. In resting cells HSF-1 is complexed with various heat shock proteins, such as with Hsp70, or with Hsp90. After stress, damaged proteins become abundant and liberate the heat shock factor from its Hsp70/Hsp90 complexes. This process sets the stage for the trimerization, nuclear translocation and phosphorylation of HSF-1, which are all pre requisites for its binding to the special nucleotide segments, called heat shock elements, in the promoter region of heat shock protein genes. All these steps are modulated by numerous co-chaperones of the major heat shock proteins, Hsp70 and Hsp90, and most probably by other proteins as well (Morimoto, 1999; Figure 6.2). The nucleosomal structure of the DNA-segment containing the heat shock element is reorganized by a special protein-machine, called the GAGA-factor (Tsukiyama *et al.*, 1994). Interestingly, many heat shock protein genes recruit an active DNA-dependent RNA polymerase II even in the absence of heat shock factor. This "pausing polymerase" transcribes a small segment of the gene, but becomes arrested by its binding to the initial complex of TATA-binding general transcription factors. Binding of the heat shock factor-trimer to the heat shock element sets the polymerase free, which can proceed to complete the transcription of the heat shock RNA. During stress, all the subsequent steps of protein synthesis (RNA splicing, nuclear export and translation itself) are blocked. Heat shock RNA-s developed various strategies to circumvent these problems. Primary transcripts (such as Hsp70 RNA) usually do not contain introns, or the open reading frame encoding the protein itself begins after the intron and the initialization may proceed from the intron as well (e.g. Hsp90, see Csermely *et al.*, 1998). Recognition of heat shock RNA-s also utilizes special routes avoiding those translational initialization factors, which became inactivated during stress.

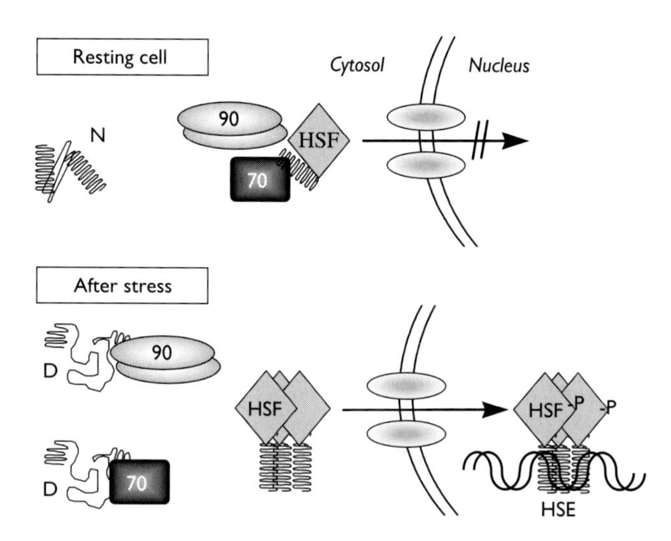

Figure 6.2 Induction of heat shock factors during stress. Abbreviations: 70, Hsp70; 90, Hsp90; D, denatured protein; HSE, heat shock element – a binding site for heat shock factor in the promoter region of heat shock protein genes; HSF, heat shock factor; N, native protein. "-P" represents the phosphorylation of heat shock factor during stress.

Another set of stress proteins, glucose regulated proteins, reside in the endoplasmic reticulum (ER) and are expressed when the ER experiences stress. ER stress can be provoked by any circumstances, which hinder folding and maturation of secreted proteins. Thus changes of redox conditions, calcium concentration, deprivation of the glycosylating apparatus from its building blocks, monosaccharides, as well as the "ER-overload", when more secretory proteins are synthesized than the amount which can be handled by the ER folding apparatus, all lead to an ER stress response. The endoplasmic reticulum uses a unique mechanism for stress-signaling. A transmembrane receptor protein, called IRE, binds unfolded proteins in the lumen of the ER. Occupation of yeast IRE leads to the activation of its endonuclease activity in the cytosol. The endonuclease is specific for the primary RNA of the transcription factor, Hac. With the coordinated action of an RNA-ligase Hac-RNA undergoes a nonconventional splicing event. Only the spliced Hac mRNA can be translated. The Hac protein binds to the promoter region of ER-chaperones, and activates their transcription. The human substitute of Hac is called XBP1 (Kaufman, 1999; Yoshida *et al.*, 2001).

6.4 Heat shock proteins and signal transduction

Almost all chaperone classes are involved in various signaling events. Small heat shock proteins become phosphorylated by stress-kinases, and increase the amount of reduced glutathione in the cytoplasm (Arrigo, 1998). Similarly, Hsp70 protects cells against oxidative stress, and inhibits stress kinases and apoptosis (Gabai *et al.*, 1998). Other chaperones, such as mitochondrial Hsp60 (which is liberated after the disruption of mitochondrial membrane), promote apoptosis by activating caspases (Samali *et al.*, 1999). Another mechanism, the peptidyl prolyl cis/trans isomerase, Pin1, is required for the activation of several key modulators of the cell cycle.

Hsp90 has a special role among "signaling-chaperones". As we mentioned before, Hsp90 is a key organizer of several cytoplasmic complexes. It binds to steroid receptors and to several serine and tyrosine kinases including the Src, Raf, focal adhesion kinases and protein kinase CK-II, or cyclin-dependent kinases-4, 6 and 9 (Buchner, 1999; Csermely *et al.*, 1998; Miyata and Yahara, 1992; Pratt and Toft, 1997). Hsp90 helps these proteins to reach their fully signaling-competent form, as well as regulating their association with other proteins and membranes. Figure 6.3 shows the maturation of the steroid receptor. Chaperones bind to the *de novo* synthesized receptor protein in a sequential manner. First Hsp70 is attached with its co-chaperones, Hsp40 and Hip. In the next step, Hsp90 associates with the "bridging" chaperone, Hop. Finally, peptidyl prolyl cis/trans isomerases from the FK506-binding class join the complex helping to direct the receptor to its final destination in the cell. The small co-chaperone, p23 stabilizes the complex preventing its premature dissociation. When the steroid hormone binds to its receptor, the receptor dissociates from Hsp90 and becomes translocated to the nucleus resulting in the dimerization-coupled binding of the receptor-hormone complex to the steroid response element DNA-region. Chaperones like Hsp90 may also play a role in the termination of the steroid response promoting the recycling of the receptor to the

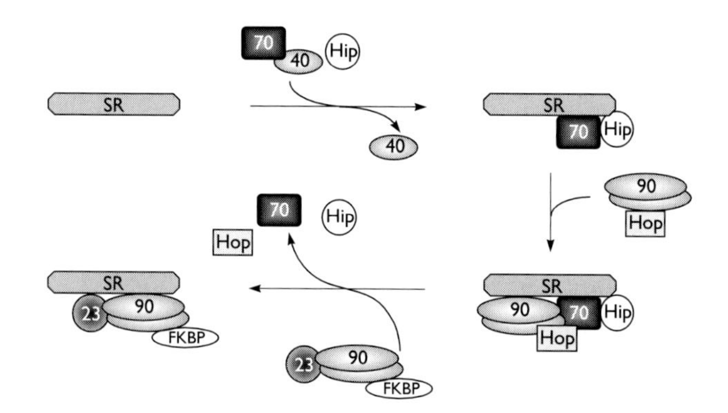

Figure 6.3 Involvement of the Hsp90-organized foldosome in the folding of steroid receptors. Abbreviations: 23, a co-chaperone of Hsp90; 40, Hsp40 – a co-chaperone of Hsp70; 70, Hsp70; 90, Hsp90; FKBP, FK506-binding protein, a peptidyl prolyl cis/trans isomerase binding to Hsp90; Hip, a co-chaperone of Hsp70; Hop, a chaperone, which connects Hsp70 with Hsp90; SR, steroid receptor.

cytoplasm. The above mechanism gives an overview of activation of glucocorticoid receptors. Activation of other steroid receptors differ in the extent of chaperone involvement in the maturation/translocation of the receptor, and in the termination of the response.

A similar Hsp90-organized foldosome helps to acquire the activation-competent state of several protein kinases. The major difference from the steroid-parenting foldosome is that, in case of kinase-folding, the foldosome contains Cdc37, which is a kinase-specific "destination-tag" on the folding complex. Recent data also indicated the involvement of Hsp90 in the maturation of NO-synthase, telomerase and viral reverse transcriptase (Buchner, 1999; Csermely *et al.*, 1998; Pratt and Toft, 1997).

6.5 Heat shock proteins and disease

6.5.1 *Chaperones as general helpers of cell survival*

Molecular chaperones are responsible for the "conformational homeostasis" of cellular proteins. When the homeostasis of the host organism is perturbed, an increased capacity of the chaperones is highly advantageous. Many of the perturbations (such as alcohol, other poisons, sunburn, anxiety, etc.) may induce the synthesis of these chaperone proteins *per se*, but in case of bacterial and viral infections the developing fever also helps this process. Several common drugs, such as aspirin, also promote the induction of heat shock proteins (Jurivich *et al.*, 1992).

Ischemia and the consecutive oxidative damage of reperfusion are also common perturbations in higher organisms. Since Currie *et al.* (1988) have shown that the induction of molecular chaperones, most notably Hsp70, may prevent damage to cardiac muscle by both ischemia and reperfusion, molecular chaperones are actively being investigated as possible tools in the treatment of heart attack or stroke. Indeed, transgenic animals, where Hsp70 has been constitutively expressed either in heart or brain, are much less likely to develop heart attack or stroke. Induction of heat shock proteins is also beneficial in transplanted organs, where moderate heat treatment reduces transfer-damage and the risk of organ rejection (Perdrizet *et al.*, 1993).

6.5.2 *Chaperones in aging and in neurodegenerative diseases*

Aging is frequently described as a consequence of an impaired function of repair processes (immune system, DNA-repair, elimination of free radicals, etc.). Molecular chaperone-catalyzed refolding of damaged proteins may well be one of these crucial repair processes. In agreement with this hypothesis, aged organisms contain an increased amount of misfolded proteins, and the induction of Hsp70 is impaired in both aged rats and humans (Söti and Csermely, 2002). The increased amount of damaged proteins in aged organisms causes a chaperone overload which may contribute to the development of polygenic diseases such as atherosclerosis, diabetes and cancer (Csermely, 2001b). On the contrary, better induction of heat shock proteins leads to an increased life expectancy in yeast, *Drosophila* or *C. elegans* (Tatar *et al.*, 1997).

Protein damage becomes especially dangerous when it affects neuronal cells, which, generally, cannot renew themselves by multiple mitotic events. In most of the neurodegenerative diseases, such as in Alzheimer's disease, in Parkinson's disease, in Huntington's disease, in Wilson's disease, in Alexander's disease and in prion-related human syndromes, nerve cells develop massive protein aggregates. These inclusion bodies usually contain various heat shock proteins, such as ubiquitin-tags, the small heat shock protein, Hsp27, Hsp70 and Hsp90 (Mayer *et al.*, 1991). Heat shock protein co-aggregation may reflect the fight of these chaperones against the aggregation process. In accordance with this view, over expression of Hsp70, or other heat shock proteins protects *Drosophila*-s from neurodegeneration in Huntington disease-like polyglutamine-induced aggregation.

6.5.3 *Chaperones and the immune response*

Molecular chaperones are one of the most conserved proteins in living organisms (Lindquist, 1986). Invading bacteria experience major changes in their environment when entering their host. These changes and the activation of defense mechanisms (depletion of nutrients, pH changes, osmotic changes, digestive enzymes, peroxides, superoxides and an increase in temperature) induce numerous heat shock proteins in bacteria, among which some are also expressed on the bacterial surface. Because of their conservative structure, these bacterial heat shock proteins, especially the bacterial homologue of Hsp70 become a common recognition signal, and therefore provoke a general, high-capacity immune response (van Eden and Young, 1996). There are at least two dozen infectious diseases in which immune responses to heat shock proteins have been reported, including tuberculosis, leprosy, legionnaire's disease, Chagas's disease, lyme disease, chlamidyal infections and Q fever.

In some unfortunate cases (such as in rheumatoid arthritis, in lupus erythematosus, in multiple sclerosis and in insulin dependent diabetes mellitus, IDDM) certain proteins of the host organism resemble some epitopes of these bacterial heat shock proteins. In these patients the common, antibacterial immune response attacks the cells bearing these host-proteins, and a severe autoimmune response develops. Vaccination with modified epitopes of a bacterial Hsp70 homologue diminish, and in some cases prevent the development of the disease (van Eden and Young, 1996). Some recent reports raise the possibility that expression of human Hsp60 on the surface of epithelial cells may be one of the initial events of arterial plaque development.

6.6 Therapeutic approaches

6.6.1 *Chaperone induction*

As we have mentioned earlier, if an organism experiences stress, the induction of heat shock proteins is beneficial for its survival. Fever and some common drugs, such as aspirin, help to induce heat shock protein expression. Other general methods, such as sauna, or hyperthermia can also be efficient. However, due to the pleiotropic effects of chaperones, in most cases an organ-specific induction of special heat shock proteins is sought. Though recent data revealed the complexity of the regulation of the heat shock response at the molecular level (Morimoto, 1999), which may expose new targets for therapeutic interventions, we are still far from an efficient "stress-pill", which would reduce the risk of stroke, or heart attack. Moreover, the expression- and induction-pattern of various heat shock proteins seems to be rather complex. Each individual, but most probably each cell at each point of its history, has a unique pattern of heat shock protein level and transcription. This "stress-status" may be a complex readout of the individual "stress-history" of the given cell. This complexity makes a general therapeutic intervention even more difficult. As an additional note of caution we must emphasize that artificially high levels of heat shock proteins are extremely cytotoxic (Feder *et al.*, 1992). Therefore limitless chaperone-induction may not always be advantageous. Recently a non-toxic drug candidate has been described, which behaves as a co-inducer of various heat shock proteins (Vígh *et al.*, 1997). Such molecules behave as "smart drugs" affecting only those cells which have already been damaged.

6.6.2 *Chaperone inhibition*

Chaperones help survival, regardless of whether the host cells are "normal" or malignant. Tumor cells experience a lot of stress, such as hypoxia or glucose-limitation. Various chemotherapy, phototherapy, radiation therapy and hyperthermia protocols increase heat shock protein induction and consequent tumor cell survival even further. Under these conditions the inhibition of chaperones would be highly beneficial. When thinking about chaperone-inhibition based, signaling-related therapeutic approaches we find two major difficulties:

1 chaperones have many general functions in cells
2 there are only a few compounds which bind to one or another chaperone specifically.

Fortunately in several chaperone functions, chaperone-isoforms, or even chaperones from different classes, may substitute each other, and the number of chaperone-specific pharmacological agents is increasing.

As we have mentioned earlier, the most important signaling-chaperone is Hsp90. It has numerous distinct, and overlapping binding sites on its surface. Hsp90 contains at least two chaperone-sites, one in its N-terminal domain, and another in the C-terminal domain. The highly charged connecting bridge after the N-terminal domain binds protein kinase CK-II, and there are additional binding sites for calmodulin, peptidyl prolyl cis/trans isomerases and other co-chaperones. Hsp90 forms dimers, binds ATP in both its N-terminal and C-terminal domains and changes its conformation after ATP addition (Minami *et al.*, 1994; Csermely *et al.*, 1998; Maruya *et al.*, 1999; Söti *et al.*, 2002).

This abundance of binding-sites and the conformational flexibility of Hsp90 suggests the presence of a large variety of active surfaces on Hsp90, which may all be potential binding sites for various pharmacological compounds. Moreover, different occupancy and conformation may expose different binding surfaces of this chaperone. Indeed, the number of Hsp90-binding drugs is steadily increasing. Figure 6.4 shows an overview of the Hsp90-binding pharmacological agents identified so far. Geldanamycin, and an unrelated antibiotic, radicicol, occupy the ATP-binding pocket of the N-terminal domain of Hsp90. This ATP-binding site is rather unique, does not resemble the nucleotide binding site of protein-, or other kinases. The only distant homologs are some DNA-binding bacterial topoisomerases. Therefore geldanamycin and radicicol seem to be specific agents for the modification of the function of the 90 kDa chaperones (Hsp90, and its homologs in the endoplasmic reticulum and mitochondria). Treatment of cells with these cell-permeable drugs induces the dissociation of Hsp90-complexes and the subsequent degradation of most Hsp90 client proteins by the proteasome.

Recently other compounds, such as the coumarine-derivative novobiocin and cisplatin, have been found which bind to the C-terminal domain of Hsp90. Their binding site overlaps with

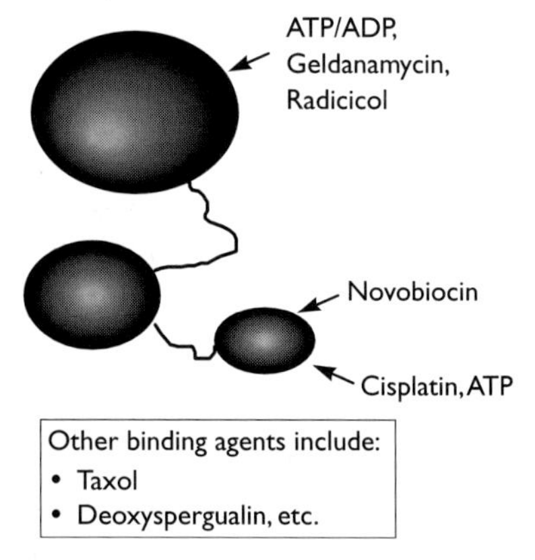

Figure 6.4 Various drugs and drug-candidates binding to the 90 kDa molecular chaperone, Hsp90. The Figure shows the domain structure of Hsp90 and lists those drugs which bind to the N-terminal (top) domain, or to the C-terminal (bottom) domain of Hsp90. Both domains contain a binding site for target proteins and peptides.

the C-terminal ATP binding site (Söti *et al.*, 2002). The exact binding site of another widely applied anti-tumor agent, taxol, on Hsp90 has not been identified yet (Byrd *et al.*, 1999; Itoh *et al.*, 1999a; Marcu *et al.*, 2000; Neckers *et al.*, 1999). It will be an exciting task to explore whether some of these compounds specifically affect the efficiency of the N-terminal, or C-terminal chaperone site of Hsp90. Various Hsp90 conformations and/or complexes may be stabilized by one, or another of these drugs, and therefore different compounds (or their combination) may inhibit different functions of Hsp90.

Another possible target, where chaperone-inhibition may help, is immune-suppression. Indeed, widely used immune-suppressants, such as cyclosporin, or FK506 bind to different classes of peptidyl prolyl cis/trans isomerases, and block various signaling pathways in T lymphocytes. Other chaperones are also targets of immune-suppressive drugs: Hsp70 and Hsp90 bind deoxyspergualin (Nadeau *et al.*, 1994), while Hsp60 binds mizobirine (Itoh *et al.*, 1999b).

6.6.3 *Chaperones as anti-cancer or anti-viral vaccines*

As we have discussed before, tumor cells experience a lot of stress (hypoxia, nutrient deprivation, etc.). Stressful conditions induce heat shock protein synthesis, and several types of cancer cells expose heat shock proteins on their surface (Multhoff and Hightower, 1996). These cells are recognized and killed by a special class of T lymphocytes, the $\gamma\delta$-cells. There are ongoing clinical trials, where tumor cells are exposed to various types of stresses to prime them for a subsequent immune attack.

At the beginning of the eighties, several research groups isolated molecular chaperones (Grp94 and Hsp90) as tumor-specific transplantation antigens, i.e. surface proteins, which provoked a highly specific immune response against a certain type of tumor. When painstaking research efforts made it clear that neither the primary structure, nor major post-translational modifications were different in chaperones coming from normal versus malignant tissue, researchers in the field were puzzled. Pramod Srivastava first suggested that the differences may come from the peptides, which are carried by the chaperones. In a 1994 hypothesis paper he proposed the existence of a relay-type mechanism, where various cytoplasmic chaperones, such as Hsp70 and Hsp90 as well as their counterparts in the endoplasmic reticulum, such as Grp94, give each other the peptides and help their presentation to the MHC-I complex, specialized to present the "self" antigens to the immune system. MHC-I molecules induce a cytotoxic T-cell response, which is a fast and local immune attack, against a cell expressing "false" self-antigens (e.g. after a viral infection). This may significantly enhance the efficiency of the MHC-II immune response, which is based on helper T lymphocytes, and generates a slow and general immune attack. According to Srivastava's hypothesis,

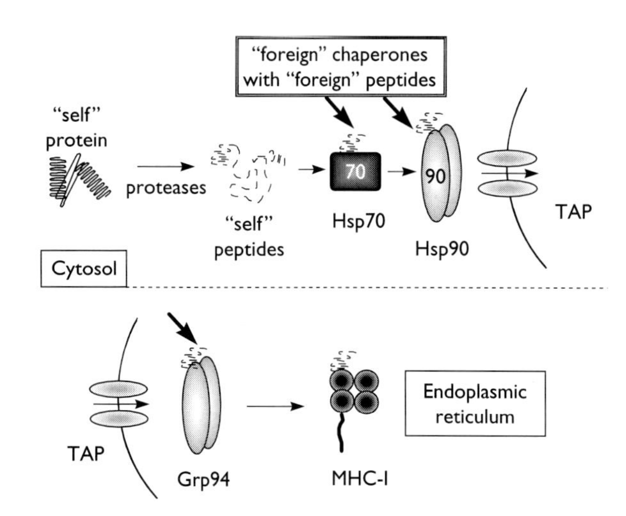

Figure 6.5 Heat shock proteins may enable the presentation of "foreign" peptides (antigens) to the MHC-I complex specialized to present "self" antigens. TAP represents the protein responsible for peptide transport through the endoplasmic reticulum membrane and Grp94 is a homologous protein to Hsp90 in the endoplasmic reticulum.

chaperones carrying tumor-peptides may enter the MHC-I pathway of macrophages and dendritic cells (Figure 6.5) and activate a cytotoxic immune response against a "foreign" antigen. Later studies proved that each major step of the original hypothesis was correct (Srivastava *et al.*, 1998). Chaperone-induced "cross-priming" (channeling of the MHC-II peptides to the MHC-I presenting pathway) is an interesting phenomenon by itself, but its real importance lies in the clinical application.

Srivastava's discovery made it possible to induce an anti-tumor immune response by vaccination. Isolation of peptide-chaperone complexes from the primary tumor and administration of the complex to the dendritic cells as a vaccine provokes an efficient anti-tumor immune response and is the subject of ongoing clinical trials (Srivastava *et al.*, 1998). In the future the establishment of common important peptide-antigens in certain type of tumors, such as prostate cancer, may alleviate the need for time consuming and costly "personalized vaccines". Similar approaches may be used to provoke an anti-viral immune response by the help of viral-specific peptide-chaperone complexes.

6.7 Final remarks

Molecular chaperones were necessary for the establishment of life on Earth, for the bursts of activity during evolution, and are crucial for the protection of our own life against proteotoxic stress. Chaperones are one of the most abundant proteins in our cells, and may help to maintain and remodel the structure of the cytoplasm in eukaryotes (Csermely *et al.*,

1998; Pratt and Toft, 1997; Csermely, 2001a). These pleiotropic effects make chaperones an attractive target for pharmacological interventions, but also create difficulty for the researcher looking for proper compounds and methods which specifically modify one or another chaperone function. Recent developments in Hsp90-related pharmacology and the establishment of several clinical methods which utilize the role of chaperones in peptide presentation make this field an exciting area for future clinical studies. Even in our own private life we may take a lesson from chaperones: small, but frequent challenges lead to a modest induction of heat shock proteins, which increases our fitness and (at least in animal studies) prolongs life expectancy.

Acknowledgments

We would like to apologize to those authors whose important primary work could not be cited due to length limitations. Work in the authors' laboratory was supported by research grants from CREST, ICGEB, Hungarian Science Foundation (OTKA-T37357), Hungarian Ministry of Social Welfare (ETT-21/00) and the Volkswagen Foundation (I/73612). Authors are thankful to Tamas Schnaider (Semmelweis University, Department of Medical Chemistry) for providing Figures 6.1 and 6.3.

References

Arrigo A.P. (1998) Small stress proteins: Chaperones that act as regulators of intracellular redox state and programmed cell death. *Biol. Chem. 379, 19–26.*

Blum, J.H., Dove, S.L., Hochschild, A. and Mekalanos, J. (2000) Isolation of peptide aptamers that inhibit intracellular processes. *Proc. Natl. Acad. Sci. USA 97, 2241–2246.*

Buchner, J. (1999) Hsp90 & Co. – a holding for folding, *Trends Biochem Sci. 24, 136–141.*

Bukau, B. and Horwich, A.L. (1998) The Hsp70 and Hsp60 chaperone machines, *Cell 92, 351–366.*

Byrd, C.A., Bornmann, W., Erdjument-Bromage, H., Tempst, P., Pavletich, N., Rosen, N., Nathan, C.F. and Ding, A. (1999) Heat shock protein 90 mediates macrophage activation by Taxol and bacterial lipopolysaccharide. *Proc. Natl. Acad. Sci. USA 96, 5645–5650.*

Currie, R.W., Karmazyn, M., Kloc, M. and Mailer, K. (1988) Heat-shock response is associated with enhanced postischaemic ventricular recovery. *Circ. Res. 63, 395–397.*

Csermely, P. (1997) Proteins, RNA-s, chaperones and enzyme evolution: A folding perspective. *Trends Biochem. Sci. 22, 147–149.*

Csermely, P. (1999) The "chaperone-percolator" model: A possible molecular mechanism of Anfinsen-cage type chaperone action. *BioEssays 21, 959–965.*

Csermely, P. (2001) A nonconventional role of molecular chaperones: involvement in the cytoarchitecture. *News in Physiol. Sci. 15, 123–126*; Chaperone-overload as a possible contributor to "civilization diseases": atherosclerosis, cancer, diabetes. *Trends in Genetics 17, 701–704.*

Csermely, P., Schnaider, T., Söti, Cs., Prohászka, Z. and Nardai, G. (1998) The 90-kDa molecular chaperone family: Structure, function and clinical applications. A comprehensive review. *Pharmacol. Therap. 79, 129–168.*

Feder, J.H., Rossi, J.M., Solomon, J., Solomon, N. and Lindquist, S. (1992) The consequences of expressing hsp70 in Drosophila cells at normal temperatures. *Genes Dev. 6, 1402–1413.*

Feige, U., Morimoto, R.I., Yahara, I. and Polla, B. (eds., 1996) *Stress Inducible Cellular Responses*, EXS vol. 77, Birkhauser Verlag, Basel.

Gabai V.L., Meriin, A.B., Mosser, D.D., Caron, A.W., Rits, S., Shifrin, V.I. and Sherman, M.Y. (1998) Hsp70 prevents activation of stress kinases. A novel pathway of cellular thermotolerance. *J. Biol. Chem. 272, 18033–18037.*

Glover, J.R. and Lindquist, S. (1998) Hsp104, Hsp70, and Hsp40: A novel chaperone system that rescues previously aggregated proteins. *Cell 94, 73–82.*

Hartl, F.-U. (1996) Molecular chaperones in cellular protein folding. *Nature 381, 571–580.*

Itoh, H., Ogura, M., Komatsuda, A., Wakui, H., Miura, A.B., Tashima, Y. (1999a) A novel chaperone-activity-reducing mechanism of the 90-kDa molecular chaperone, Hsp90. *Biochem. J. 343, 697–703.*

Itoh, H., Komatsuda, A., Wakui, H., Miura, A.B., Tashima, Y. (1999b) Mammalian Hsp60 is a major target for an immunosuppressant Mizobirine. *J. Biol. Chem. 274, 35147–35151.*

Jindal, S. (1996) Heat shock proteins: Applications in health and disease. *Trends Biotechnol. 14, 17–20.*

Jurivich, D.A., Sistonen, L., Kroes, R.A. and Morimoto, R.I. (1992) Effect of sodium salicylate on the human heat shock response. *Science 255, 1243–1245.*

Kaufman, R.J. (1999) Stress signaling from the lumen of the endoplasmic reticulum: Coordination of gene transcriptional and translational controls. *Genes Dev. 13, 1211–1233.*

Latchman, D.S. (1991) Heat shock proteins and human disease. *J. Royal Coll. Physic. London 25, 295–299.*

Lindquist, S. (1986) The heat shock response. *Annu. Rev. Biochem. 55, 1151–1191.*

Marcu, M.G., Schulte, T.W. and Neckers, L. (2000) Novobiocin and related coumarins and depletion of heat shock protein 90-dependent signalling proteins. *J. Natl. Canc. Inst. 92, 242–248.*

Maruya, M., Sameshima, M., Nemoto, T. and Yahara, I. (1999) Monomer arrangement in HSP90 dimer as determined by decoration with N and C-terminal region specific antibodies. *J. Mol. Biol. 285, 903–907.*

Mayer, R.J., Arnold, J., Laszlo, L., Landon, M. and Lowe, J. (1991) Ubiquitin in health and disease. *Biochim. Biophys. Acta 1089, 141–157.*

Menoret, A., Peng, P. and Srivastava, P.K. (1999) Association of peptides with heat shock protein gp96 occurs *in vivo* and not after cell lysis. *Arch. Biochem. Biophys. 262, 813–818.*

Minami, Y., Kimura, Y., Kawasaki, H., Suzuki, K. and Yahara, I. (1994) The carboxy-terminal region of mammalian HSP90 is required for its dimerization and function *in vivo. Mol. Cell. Biol. 14, 1459–1464.*

Miyata, Y. and Yahara, I. (1992) The 90-kDa heat shock protein, HSP90, binds and protects casein kinase II from self-aggregation and enhances its kinase activity. *J. Biol. Chem. 267, 7042–7047.*

Morimoto, R.I. (1999) Regulation of the heat shock transcriptional response: Cross talk between a family of heat shock factors, molecular chaperones, and negative regulators. *Genes Dev. 12, 3788–3796.*

Multhoff, G. and Hightower, L.E. (1996) Cell surface expression of heat shock proteins and the immune response. *Cell Stress Chaperones 1, 167–176.*

Nadeau, K., Nadler, S.G., Saulnier, M., Tepper, M.A. and Walsh, C.T. (1994) Quantitation of the interaction of the immuno-suppressant deoxyspergualin and analogs with Hsc70 and Hsp90. *Biochemistry 33, 2561–2567.*

Neckers, L., Mimnaugh E. and Schulte, T.W. (1999) Hsp90 as an anti-cancer target. *Drug Resistance Updates 2, 165–172.*

Perdrizet, G.A., Kaneko, H., Buckely, T.M., Fishman, M.S., Pleau, M., Bow, L. and Schweizer, R.T. (1993) Heat shock recovery protects renal allografts from warm ischemic injury and enhances HSP72 production. *Transpl. Proc. 25, 1670–1673.*

Pratt, W.B. and Toft, D.O. (1997) Steroid receptor interactions with heat shock protein and immunophilin complexes, *Endocrine Rev. 18, 306–360.*

Ritossa, F. (1962) A new puffing pattern induced by temperature shock and DNP in Drosophila. *Experientia 18, 571–573.*

Samali, A., Cai, J., Zhivotovsky, B., Jones, D.P. and Orrenius, S. (1999) Presence of a pre-apoptotic complex of pro-caspase-3, Hsp60, and Hsp10 in the mitochondrial fraction of Jurkat cells. *EMBO J. 18, 2040–2048.*

Söti, Cs. and Csermely, P. (2002) Chaperones come of age. *Cell Stress Chaperones 7, 186–190.*

Söti, Cs., Rácz, A. and Csermely, P. (2002) A nucleotide-dependent molecular switch controls ATP binding at the C-terminal domain of Hsp90: N-terminal nucleotide binding unmasks a C-terminal binding pocket. *J. Biol. Chem. 277, 7066–7075.*

Srivastava, P.K., Menoret, A., Basu, S., Binder, R.J. and McQuade, K.L. (1998) Heat shock proteins come of age: Primitive functions acquire new roles in an adaptive world. *Immunity 8, 657–665.*

Tsukiyama, T., Becker, P.B. and Wu, C. (1994) ATP-dependent nucleosome disruption at a heat-shock promoter mediated by binding of GAGA transcription factor. *Nature 367, 525–532.*

van Eden, W. and Young, D.B. (eds., 1996) *Stress Proteins in Medicine,* Marcel Dekker Inc., New York.

Vígh, L., Literáti, P.N., Horváth, I., Török. Z., Balogh, G., Glatz, A., Kovács, E., Boros, I., Ferdinándy, P., Farkas, B., Jaszlits, L., Jednákovits, A., Koranyi, L. and Maresca, B. (1997) Bimoclomol: A nontoxic, hydroxylamine derivative with stress protein-inducing activity and cytoprotective effects. *Nature Med. 3, 1150–1154.*

Welch, W.J. (1992) Mammalian stress response: Cell physiology, structure/function of stress proteins, and implications for medicine and disease. *Physiol. Rev. 72, 1063–1081.*

Yoshida, H., Matsui, T., Yamamoto, A., Okada, T. and Mori, K. (2001) XBP1 mRNA is induced by ATF6 and spliced by IRE1 in response to ER stress to produce a highly active transcription factor. *Cell 107, 881–891.*

Chapter 7

Transcription factors as targets for drug development

Devarajan Karunagaran and Bharat B. Aggarwal

Contents

7.1 Introduction

Transcription factors are genetic switches that regulate the expression of various genes. Because dysregulated expression of a gene may lead to the development of a disease, transcription factors have been used for drug development. For instance nonregulated activation of transcription factor NF-κB has been implicated in inflammation, rheumatoid arthritis, autoimmune diseases, tumorigenesis, and viral replication; it is therefore considered an ideal target for drug development. This review discusses the structure of transcription factors, their binding to DNA, gene activation, and drugs that can suppress them.

Transcription is an important regulatory event in the process of gene expression. Transcription factors are proteins that regulate transcription by binding specific sequences present within the promoter, enhancer, or other regulatory regions of DNA or even RNA. Transcription factors can interact directly with RNA polymerase or through other transcription factors. Eukaryotic transcription factors are nearly always positive regulators of transcription, although negative regulation is seen occasionally. Hundreds of transcription factors with functionally separable domains essential for DNA-binding and activation have been identified and characterized in several organisms (Latchman, 1995).

A brief description of transcription clarifies the importance of transcription factors. RNA polymerase II is mainly involved in the formation of mRNA in eukaryotes, and the basal transcriptional complex contains a TATA box, an AT-rich sequence found 30 bases upstream of the start site of transcription. The TATA box interacts with TBP (TATA-binding protein) and a complex of proteins known as TFIID (transcription factor D for RNA polymerase II) and other proteins. Prokaryotic transcription factors bind DNA at sites upstream of the TATA box, whereas in eukaryotes an array of such sites may be present upstream or downstream of the RNA start site of a gene. A binding site that responds to specific signals is known as a response element (Lewin, 2000). Table 7.1 lists some of the well-known binding sites and the corresponding transcription factors. This chapter deals with the general structure of most transcription factors and then describes NF-κB and AP-1, the two most important transcription factors, as targets for drug development. Most transcription factors consist of a DNA-binding domain (DB) and an activation domain (AD). The factor binds to the DNA through DB and causes gene activation through the AD. The following are specific features of each domain.

Table 7.1 Transcription factors and their binding sites

Factors	*Binding sites*
GR	GGTACANNNTGTTCT
Sp1	GGGCGG
AP-1/TRE	TGACTCA
NF-κB	GGGPuNNPyPyCC
C/EBP	TGTGGAAAG
AP-2	CCCCAGGC
CREB/ATF	TGACGTCA
OCT-1	ATTGCAT
OCT-2	ATTTGCAT
SRF	GATGTCCATATTAGGACATC
HSF	CTNGAATNTTCTAGA

7.2 DNA binding domains

A number of structural motifs have been identified in the DNA binding domains of transcription factors which are important in protein-DNA or protein-protein interactions and other regulatory functions. These are grouped as helix-turn-helix, zinc finger, helix-loop-helix (HLH), and leucine-zipper motifs.

7.2.1 *Helix-turn-helix motif*

In the helix-turn-helix motif, a short stretch of amino acids forms an α-helical structure that is followed by a β-turn and another α-helical region. The second helix lies partly within the major groove of DNA, where it can make specific contact with the bases of the DNA. The homeotic gene *fushi tarazu* (ftz) encodes a protein that contains a helix-turn-helix motif and binds specifically to the sequence TCAATTAAATGA. When this gene is introduced into *Drosophila* cells along with a marker gene containing this sequence, transcription of the marker gene is increased. This up-regulation is dependent upon the binding of the Ftz protein to this sequence in the promoter of the marker gene. The product of another homeotic gene, *engrailed*, binds to the same site as that of ftz and prevents activation by ftz (Latchman, 1995).

7.2.2 *Zinc finger motif*

The zinc finger motif is a repeating unit of 30 amino acids 12 of which make up a loop of conserved leucine, phenyl alanine and several basic amino acids. The loop projects from the surface in the form of a finger and is anchored at its base by the conserved cysteine and histidine residues, which directly coordinate an atom of zinc. The tips of the finger can directly contact DNA (major groove) through their basic amino acids, and successive fingers bind on opposite sides of the helix. Zinc finger proteins are important in controlling development in vertebrates, and they interact with helix-turn-helix-containing proteins. Some proteins contain a multi-cysteine zinc finger motif in which four cysteines replace the two cysteine-two histidine structure of the conventional finger in binding zinc and which are separated by a linker region containing 15–17 variable amino acids (Mitchell and Tjian, 1989).

7.2.3 *Helix-loop-helix (HLH) and the leucine zipper motifs*

Transcription factors belonging to the HLH family are involved in regulation of neurogenesis, myogenesis, cell proliferation and differentiation, cell lineage determination, sex determination, and other essential processes. Over 240 HLH proteins have been identified to date in organisms ranging from yeast to man. The HLH structure is different from the helix-turn-helix, and the two helices are separated by an intervening non-helical loop. The dimerization motif contains about 50 amino acids and produces two amphipathic α-helices (containing all the charged amino acids on one side of the helix) separated by a loop of variable length. HLH proteins include the Myc, MyoD and Id family of transcription factors. A motif mainly of basic residues permits HLH proteins to bind to a consensus hexanucleotide E-box (CANNTG). The bHLH motif was first identified in murine transcription factors E12 and E47 (Massari and Murre, 2000). Additionally, some basic HLH proteins contain a "leucine zipper dimerization motif" characterized by heptad repeats of leucine residues that occur immediately C-terminal to the bHLH motif.

The leucine zipper was identified first in C/EBP (CCAAT/enhancer binding protein) (Landschultz *et al.*, 1988a; 1988b) and has been found to mediate very specific interactions between many transcription factors owing to the distribution of charged amino acid residues within the leucine zipper domain. Dimerization of proteins brings two basic domains into juxtaposition, each of which makes an equal contribution to DNA site recognition and binding (Lewin, 2000). A number of transcription factor families contain the leucine zipper domain including AP-1, cAMP response element binding protein family (CREBs; also called activating transcription factors ATFs) (Hai *et al.*, 1989), the c/EBP family (Landschultz *et al.*, 1988b; Williams *et al.*, 1991), TEF/DBP (Drolet *et al.*, 1991), the Maf/Nrl family (Kataoka *et al.*, 1994a), and NF-E2 (Andrews *et al.*, 1993). There are many instances of cross-talk between families, whereby heterodimers formed between members of different families bind to composite or unique elements that can only be recognized by certain protein combinations (Lewin, 2000).

7.3 Activation domains and their mechanisms of action

Transcription factors have one or more activation domains containing 30 to 100 amino acids. Different combinations of activation and DNA-binding domains from unrelated species were found to activate reporter genes, and these "domain-swapping" experiments demonstrated the functional independence of DNA-binding and activation domains. Several transcription factors contain a number of acidic amino acids, glutamine, or proline in their activation domains: acidic activation domains are present in GAL4, GCN4, and GR; glutamine-rich portions are present in the activation domains of Sp1, OCT-1, OCT-2, AP-2 and many other transcription factors; proline-rich motifs were identified in SRF, AP-2, Jun, OCT-2 and CTF (Mitchell and Tjian, 1989). Acidic activation domains may form α-helical or antiparallel β sheets and have the ability to stimulate transcription initiation by interacting with TFIID or RNA polymerase II (Hahn, 1993; Mitchell and Tjian, 1989). Some acidic domains were shown to stabilize a complex with TBP on promoters from yeast and mammalian cells (Horikoshi *et al.*, 1988). Many genes are known to function without a TATA box and still require TBP for transcription (Weis and Reinberg, 1992). Specific associations between regulatory regions of active genes and nuclear scaffold proteins are known to occur (Gasser and Laemmli, 1986). Histone acetylation is said to contribute to the transcriptional activation process by disrupting the repressive chromatin structure and facilitating the sequestration of the basal transcriptional machinery (Wolffe and Pruss, 1996). GCN5p stimulates the action of acidic activators in yeast such as GCN4, and GAL4-VP16 as a coactivator of their transcriptional complex interacts directly with the acidic activation domain of VP16 and with TBP (Barlev *et al.*, 1995). During HIV infection, the viral protein Tat enhances transcriptional initiation by binding to RNA (Greenblatt *et al.*, 1993). General mechanisms by which the activation domains function are shown in Figure 7.1.

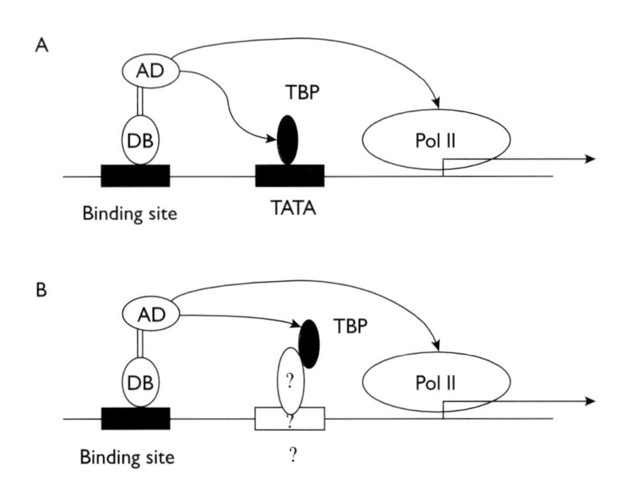

Figure 7.1 General mechanisms of action of activator domains. (A) The activator domain of a transcription factor interacts with RNA polymerase II either directly or through TBP, stabilizing and activating the basal transcriptional machinery. (B) In a TATA-less promoter the activator domain can still interact with TBP, which presumably interacts with other proteins that bind DNA.

7.4 Regulation of transcription factors

There are several ways by which the level and activity of transcription factors may be regulated. Like any other protein, the synthesis of a transcription factor is also controlled at the levels of transcription and translation during protein synthesis. Many transcription factors regulate their own gene expression (autoregulation). Splicing, transport to the cytoplasm, and degradation of mRNA are some of the important regulatory events. Selection of alternative start sites may generate distinct protein isoforms. After protein synthesis, transporting the protein to the nucleus is again a regulatory step. Masking of the nuclear localization signal (NLS) by another protein or by phosphorylation may block the nuclear translocation of certain transcription factors. DNA binding, dimerization, and trans-activation may be affected by post-translational modifications or ligand binding. A number of transcription factors form stable dimers with each other, and the number of possible combinations of dimers capable of interacting with specific sequences/sites is enormous. This also increases the complexity of issues such as the mechanisms that regulate the formation of specific homo and heterodimers among the particular transcription factor family members, the regulation of their transcriptional and DNA-binding activities and most important, the role played by each specific transcription factor in cell and organismal physiology and in development. Knockout mouse strains deficient in individual transcription factors have helped to understand part of the complexity (Calkhoven and Geert, 1996).

7.5 AP-1 (activated protein-1)

Transcription factor AP-1 (activated protein-1) was first described as a DNA-binding activity in nuclear extracts from HeLa cells that specifically recognized the enhancer elements of SV40 and the human metallothionein IIA gene (Lee *et al.*, 1987a). The AP-1 consensus sequence (TGACTCA/TGAGTCA) known as TRE (TPA-responsive element) is present in all vertebrate cell types, and recognition occurs only by dimeric complexes consistent with its palindromic nature (Lee *et al.*, 1987b; Angel *et al.*, 1987). Two families of proteins, Jun and Fos, dimerize and bind DNA at specific AP-1 sites. The Jun family consists of c-Jun (the cellular counterpart of v-Jun), JunB, and JunD. The Fos family includes c-Fos, the cellular counterpart of v-Fos, Fos-related antigen-1 (Fra-1), Fra-2, and FosB and its naturally truncated form FosB2 (Karin *et al.*, 1997). Heterodimeric complexes of

Jun and Fos bind to the AP-1 site with much greater affinity than the Jun homodimer, and Fos cannot form homodimers (Halazonetis *et al.*, 1988). Jun proteins can also form heterodimers with ATF-2 and bind the CRE sequence 5′ TGACGTCA3′. This sequence is also recognized by the CREB and ATF families of bZIP transcription factors, which contain both the basic amino acid and leucine zipper regions (Wisdom, 1999). Many other transcription factors recognize DNA motifs similar to AP-1 sites, or AP-1 half sites, or large elements with an AP-1 consensus sequence at their core (Montminy *et al.*, 1986). Maf proteins (v-Maf and c-Maf) and the neural retina-specific gene product Nrl can form heterodimers with c-Jun or c-Fos that bind to TRE, CRE, and an asymmetric DNA sequence consisting of juxtaposed AP-1 and Maf half sites (Karin *et al.*, 1997). The Maf-related proteins (MafB, MafF, MafG and MafK) form heterodimers with Fos but not with Jun. On the basis of their interaction with c-Jun, two novel bZIP proteins were also recently identified as Jun-dimerizing partners (Karin *et al.*, 1997).

7.5.1 AP-1 target genes

Addition of serum to quiescent fibroblasts increased the expression of about 300 genes (Iyer *et al.*, 1999). A large number of these show kinetics compatible with being AP-1 target genes, and a database scan for available promoter sequence indicates the presence of a number of AP-1 sites within 600 bp of the start site of transcription (Wisdom, 1999). The IL-2 gene was one of the first cellular genes shown to have an AP-1 site within its promoter (Angel *et al.*, 1987). Subsequently, AP-1 has been shown to play an integral role in the regulation of the IL-2 gene, either alone or in combination with other transcription factors such as the NFAT and octamer proteins (Oct-1 and Oct-2) (Foletta *et al.*, 1998). There are two near-consensus AP-1 sites within the IL-2 promoter to which various dimeric combinations of Fos and Jun family members have been shown to bind. However, the proximal AP-1 site was observed to mediate transcriptional activation of IL-2 promoter constructs (Jain *et al.*, 1992). This proximal site is also the primary target site for the activation by PKC. Five NFAT sites have been characterized in the IL-2 promoter, four of which permit cooperative NFAT and AP-1 binding. Hence, AP-1 acts to stabilize the interaction of NFAT with the DNA, thereby enhancing the induction of IL-2. In activated T-cells, a variety of Fos and Jun proteins functionally interact with NFAT and promote transactivation of IL-2 reporter constructs. In primary T-cells, Fra-1/JunB heterodimers predominated. In stimulated Jurkat leukemia and EL4 lymphoma cells, c-Jun, JunD, FosB, and JunB were found to cooperatively interact with Oct proteins at the proximal Oct site and increase IL-2 promoter activity. The regulation of IL-4 in leukocytes is thought to primarily involve NFAT, and AP-1 has been shown to interact with NFAT at a composite site within the IL-4 promoter. Cooperative and coordinate binding of both transcription factors was found to be necessary for full activity of the IL-4 promoter (Foletta *et al.*, 1998). Promoters of IL-5, IL-3, and GM-CSF contain sequences similar to NFAT and AP-1 sites (Karlen *et al.*, 1996). The IL-9 promoter contains an AP-1 like site, and mutational analysis demonstrated that AP-1 will enhance IL-9 promoter activity. Under certain stimulatory conditions in various T-cell preparations, the CD28RE (CD28-repressive element) site can bind NF-κB family members or it can act as a composite site with flanking NFAT and AP-1 binding sites with which NFAT/AP-1 and c-Rel plus c-Fos/c-Jun proteins can, respectively, interact (Foletta *et al.*, 1998). AP-1 complexes have been shown to trans-activate the kappa light chain gene promoter and the Ig heavy chain (IgH) gene. Both Th1 and Th2 cells express different profiles of AP-1 activity and may account for the specific cytokine expression patterns of these cells. A c-Jun/ATF-2 protein combination was found to cooperate with NFAT and bind a CRE site, leading to the activation of the TNF-α gene in calcium-stimulated T-cells (Foletta *et al.*, 1998). In LPS-stimulated THP-1 monocytic cells, however, c-Jun-containing complexes interacted with the NF-κB constituents p50 and p65 and synergistically enhanced the induction of the TNF-α promoter. A c-Jun/ATF-2 heterodimer and other CREB/ ATF proteins were also shown to interact with and promote IFN-γ promoter constructs. The IL-8 gene has an AP-1-like binding site within its promoter, to which a c-Fos/JunD heterodimer may bind in Jurkat cells. A c-Jun/ATF complex binds and induces the macrophage inflammatory protein 1β (MIP-1β) promoter function. IL-1β stimulation of endothelial cells appears to induce expression of monocyte chemoattractant protein-1 (MCP-1) gene via the synergistic activation of AP-1 and NF-κB. AP-1 and Ets sites are required for the activation of heparin binding epidermal growth factor, urokinase tissue plasminogen activator, and type I collagenase genes (Foletta *et al.*, 1998). Families of transcription factors known to interact with AP-1 are listed in Table 7.2.

7.5.2 Activators of AP-1

A wide variety of stimuli such as growth factors (PDGF, EGF, IGFs, nerve growth factor), peptide hormones, interleukins, hematopoietic cytokines, neurotransmitters,

Table 7.2 Transcription factors known to interact with AP-1

Family of transcription factors	*AP-1 interacting factors*
CREB/ATFs	CREB, ATF-1, ATF-3, ATF-4, CREBP1
NF-κB	P65
bHLHzip proteins	MyoD, USF
Maf	c-Maf, MafB, MafF, MafG, MafK, Nr1
Others	GR, p300, TBP, NFAT

12-O-tetradecanoylphorbol-13-acetate (TPA), UV, and many factors that alter the cellular oxidation/reduction status activate AP-1 proteins (Ryder and Nathans, 1988; Birrer *et al.*, 1992; Abate *et al.*, 1991; Angel and Karin, 1991). The genes encoding AP-1 proteins are often among the first genes (immediate-early genes, i.e, genes whose transcription is rapidly induced, independent of *de novo* protein synthesis, following cell stimulation) to be transcribed after stimulation of a cell. The most rapid, highly transient induction is exhibited by c-Fos, while induction of other Fos genes such as Fra-1 is somewhat slower and longer lasting (Angel and Karin, 1991; Hai and Curran, 1991). In 3T3 cells, c-Jun transcription is stimulated by TPA and partially inhibited by cAMP, while JunB transcription is stimulated by cAMP (Chiu *et al.*, 1989). In most cells, JunD expression is constitutive, but in HeLa cells its expression is induced only by the simultaneous activation of both PKC and PKA (Chiu *et al.*, 1989). Activation of AP-1 sites is low in nonstimulated cells, and rapid activation occurs in response to growth factors such as PDGF or oncogenically activated Src, Ras and Raf proteins (Binetruy *et al.*, 1991; Smeal *et al.*, 1991; 1992). In fibroblasts, Jun and Fos proteins are involved in controlling cell proliferation and are activated by a variety of mitogenic growth factors (Angel and Karin, 1991). Alternatively, in other cell types such as the pheochromocytoma cell line PC12 and the myeloid leukemia cell line U937, induction of c-Jun and c-Fos by nerve growth factor (PC12) or TPA (U937 cells) leads to cessation of cell division and differentiation (Szabo *et al.*, 1991). In addition, different Jun and Fos family members can have opposing effects on AP-1-mediated trans-activation (Chiu *et al.*, 1989). In myeloid and lymphoid cells, phosphorylation of c-Jun occurs after stimulation by TPA (Pulverer *et al.*, 1991) but not in fibroblasts and epithelial cells (Boyle *et al.*, 1991); these differences may be due to different modes of Ha-Ras regulation in lymphoid cells versus fibroblasts (Downward *et al.*, 1990). This type of differential response may aid the formation of different types of dimers at different time intervals after stimulation. This may also explain the observed repression of activation of c-Jun by JunB (Chiu *et al.*, 1989). In mouse fibroblast mutant for either c-Jun or c-Fos and FosB, cyclin D expression and G1 progression in response to serum or peptide growth factors are impaired (Wisdom, 1999). In addition, microinjection of neutralizing antibodies to either Fos or Jun proteins inhibits cell cycle progression of fibroblasts following exposure of cells to serum growth factors (Kovary and Bravo, 1991). There is a rapid induction of AP-1 activity after T cell activation (Serfling *et al.*, 1995).

7.5.3 Regulation of AP-1

Like most transcription factors, Fos and Jun proteins possess functionally independent DNA-binding and transcriptional activation domains. Figure 7.2 shows the sequence of the bZIP region of AP-1. AP-1 is mainly regulated by its phosphorylation at specific sites (Karin *et al.*, 1997) and the redox potential of cells (Abate *et al.*, 1990). The activation domain of Jun family proteins is regulated by the JNK family of MAP kinases (Karin, 1995). C-Jun is phosphorylated at two sites within its amino-terminal activation domain and three that are next to its carboxy-terminal DNA-binding domain (Karin and Smeal, 1992). Phosphorylation of serine residues (Ser 63 and Ser 73) of the activation domain of c-Jun increases its ability to stimulate transcription. This phosphorylation also enhances the interaction between c-Jun and the CBP/p300 family of coactivators and transcriptional activation. Conformational changes in the activation domain are thought to be important for the trans-activation by c-Jun (Karin and Smeal, 1992). The docking site for JNK on c-Jun is physically separable from the phosphorylation sites, and the substrate binding site on JNK is distinct from the active site (Kallunki *et al.*, 1994). It has been shown that phosphorylation of two or three c-terminal sites next to its DNA-binding domain negatively regulates c-Jun DNA binding and AP-1 activity (Boyle *et al.*, 1991).

The carboxy-terminal phosphorylation sites of c-Jun were identified as Thr 231, Ser 243, and Ser 249 (Boyle *et al.*, 1991). These sites are present amino-terminally to the basic region of c-Jun, and another site present in this region, Thr 239, does not appear to be phosphorylated *in vivo* (Karin and Smeal, 1992). Two of the sites, Thr231 and Ser 249, are phosphorylated by CKII (Karin and Smeal, 1992), whereas the third site, Ser 243, is phosphorylated *in vitro* by the ERK kinases (Karin and Smeal, 1992). *In vitro* phosphorylation of Thr 231 and Ser249 by CKII inhibits the DNA-binding activity of c-Jun in a reversible manner, and injection of peptides

c-Jun MESQERIKAERKRMRNRIAASKSRKRKL**ERIAR**LEEKVKTLKAQNSELASTANMLREQVAQL

C-Fos LSPEEEEKRRIRRERNKMAAAKSRNRR**RELTD**TLQAETDQLEDEKSALQTEIANLLKEKEKL

Basic region Spacer Leucine zipper

Figure 7.2 Sequence of the bZIP region of the AP-1 subunits c-Jun and c-Fos. This figure shows the basic, spacer and the leucine zipper regions. The spacer region is shown in bold letters.

that inhibit CKII activity induces the expression of an AP-1-dependent reporter (Karin and Smeal, 1992). Proximity of phosphorylation sites to the basic region suggests that inhibition is due to electrostatic repulsion between phosphates on c-Jun and those of the DNA backbone. However, some studies showed that phosphorylation of c-Jun did not produce a significant change in dimerization, DNA binding, or *in vitro* transcription activity (Baker *et al.*, 1992; Bannister *et al.*, 1993). CKII is a constitutive protein kinase (Karin and Smeal, 1992), whereas the ERKs exhibit inducible activity (Boulton *et al.*, 1991). Cell stimulation with TPA coincides with rapid dephosphorylation of two of the carboxy-terminal sites (Boyle *et al.*, 1991; Binetruy *et al.*, 1991; Smeal *et al.*, 1991), suggesting the existence of a phosphatase. Presumably, the role of the phosphatase is to increase DNA-binding activity by relieving the inhibitory effect of phosphorylation by CKII. Replacing Ser243 of c-Jun by Phe increases the trans-activation potential of c-Jun (Boyle *et al.*, 1991).

A primary mechanism regulating c-Fos transcriptional induction in response to a diverse spectrum of stimuli is a ras-dependent signal cascade involving ERK-2 MAP kinases (Wisdom, 1999). The cis elements mediating c-fos induction are a cAMP response element (CRE), serum response element (SRE), and TRE. The CRE responds to neurotransmitters (cAMP as second messenger) to activate protein kinase A (PKA) and to polypeptide hormones (Ca2+ as second messenger) to activate calmodulin-dependent kinases (Karin, 1995). SRE responds to growth factors, cytokines, and other stimuli that activate MAP kinases (Treisman, 1992), and a Sis-inducible enhancer mediates induction by stimuli that activate the JAK group of protein kinases (Karin, 1995). The serum response factor (SRF) recognizes and binds SRE to recruit the ternary complex factor (TCF), which cannot bind to the SRE by itself (Treisman, 1992). Following mitogenic stimulation, MAP kinases rapidly phosphorylate Elk-1, one of several candidate TCFs (Treisman, 1994). The sites at which Elk-1 is phosphorylated are within its activation domain and are conserved in other candidate TCFs such as SAP-1 (Treisman, 1994). ERK activation leads to elevated AP-1 activity via *c-fos* induction, and c-Fos translocates to the nucleus to combine with pre-existing c-Jun (Wisdom, 1999). Increased stability results in higher levels of AP-1 DNA binding activity because it shifts the equilibrium towards dimer formation essential for DNA binding. A novel serine/threonine kinase has been identified that phosphorylates c-Fos in lymphocytes after stimulation by various surface receptors such as CD3 (Nel *et al.*, 1994). This kinase, called Fos kinase, appears to be activated by PKC and is similar to a nerve growth factor-regulated kinase in PC 12 cells (Tabata *et al.*, 1989). JNK also increases expression of Fos proteins through its ability to phosphorylate the ternary complex factors Elk-1 and SAP-1 (Janknecht and Hunter, 1997). Mechanisms of activation of AP-1 are shown in Figure 7.3.

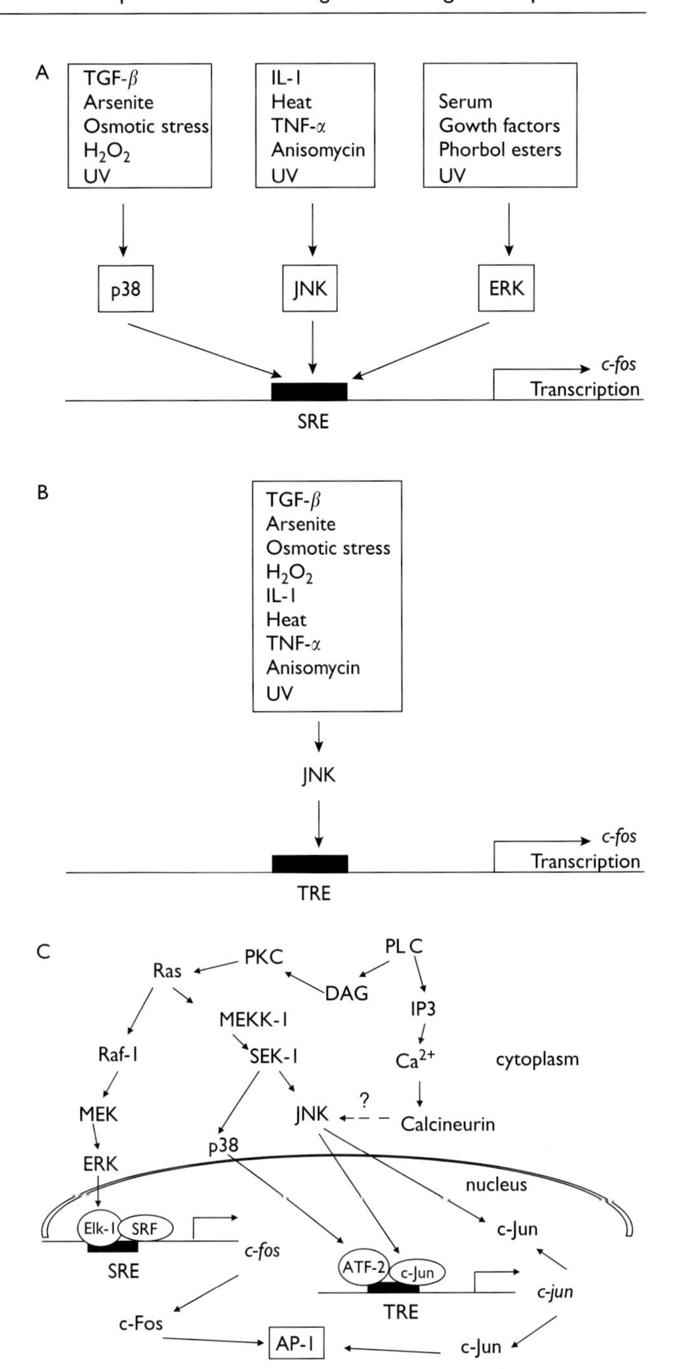

Figure 7.3 Mechanisms of activation of Fos and Jun. (A) Induction of c-fos is mediated by three cis elements, SRE, SIE and CRE. Signals through ERK, JNK and p38 reach SRE. (B) When JNK-mediated signals reach TRE, they induce c-Jun transcription. (C) Extracellular factors/agents inducing Fos/Jun shown in A and B signal through several pathways such as the Ras-MAP kinase cascade, PKC, and calcium. Dashed arrow with question mark indicates an unknown pathway. PKC, protein kinase-C; PLC, phospholipase C; MEK, MAP kinase/ERK; MEKK-1, MAP kinase-1/ERK kinase-1; SEK-1, stress-activated protein kinase; JNK, Jun amino-terminal kinase; ERK, extracellular signal regulated kinase.

Redox regulation appears to be another attractive mechanism to regulate AP-1 activity, and the oxidation and reduction state of the c-Fos/c-Jun protein complex is reported to be an important modulator of DNA-binding activity *in vitro* (Abate *et al.*, 1991). Redox regulation is mediated by a conserved motif of lysine, cysteine, and arginine (KCR), located in the basic DNA-binding domain of both c-Fos and c-Jun (Abate *et al.*, 1991) and in several other transcription factors (Kerppola and Curran, 1995). *In vitro* DNA-binding activity of c-Fos and c-Jun is enhanced via mutations of these conserved cysteines to serines in the basic DNA binding domains of c-Fos and c-Jun (c-Fos-C154S and c-Jun C262S) (Gius *et al.*, 1999). Naturally occurring cysteine-to-serine mutations for both c-Fos and c-Jun strongly enhance their transforming ability (Gius *et al.*, 1999). Chemical reduction of c-Fos and c-Jun also activates DNA binding. A cellular nuclear protein called redox factor-1 (Ref-1) can substitute for reducing conditions and activate the DNA-binding of c-Fos/c-Jun. Unlike NF-κB, AP-1 is strongly induced by some antioxidants. In HeLa cells the DNA-binding and transcriptional activities of AP-1 are induced following treatment with pyrrolidine dithiocarbamate (PDTC), butylated hydoxyanisole, and N-acetyl cysteine (NAC). AP-1 DNA-binding activity increases during hypoxia; following reoxygenation, however, NF-κB is activated presumably due to the rapid formation of oxidants suggesting that NF-κB responds to oxidation while AP-1 to reduction (Gius *et al.*, 1999).

7.5.4 AP-1 inhibitors as potential targets for drug development

Transcription is an important regulatory stage in the expression of cytokines, cell adhesion molecules, erosive metalloproteinases, collagenases, and stromelysin, which mediate inflammation. Inhibitors of pleiotropic transcription factors are very likely to inhibit multiple aspects of complex diseases. The anti-inflammatory effects of steroids, retinoids, and a variety of antirheumatic drugs are mediated through the inhibition of pro-inflammatory transcription factors such as AP-1 and NF-κB. Inhibitors of AP-1 have important implications in the treatment of inflammatory diseases and cancer. Apart from drugs and chemicals that inhibit AP-1, dietary compounds also have profound effects on AP-1 activity. In T-cells, Jun proteins may also be calcium regulated and are sensitive to the immunosuppressant cyclosporin A, which inhibits the phosphatase calcineurin (Ullman *et al.*, 1993). Glucocorticoids inhibit AP-1 activity, and binding of GR to Jun and Fos causes mutual inhibition of GR and AP-1 DNA binding. In some cell types DNA binding of AP-1 was unaltered by dexamethasone, although AP-1-mediated transcription was inhibited. These results may be explained by known competition between GR and AP-1 for binding to CREB binding protein (CBP), which is involved in both transcriptional complexes but does not alter direct DNA binding of either GR or AP-1 (Handel, 1997). Retinoic acid receptor (RAR)-α can antagonize AP-1's interaction with the AP-1/Oct binding site (Foletta *et al.*, 1998). A class of less toxic retinoids, heteroarotinoids, inhibited AP-1-driven transcription by activating RAR and RXR receptors (Zacheis *et al.*, 1999). DHEA (dehydroepiandrosterone) and its analogs inhibited DNA binding of AP-1 and proliferation of airway smooth muscle and may be useful for treatment of asthmatic airway remodeling (Dashtaki *et al.*, 1998). The immunosuppressant FK506 selectively inhibits the expression of early T-cell activation genes. Immunosuppressant drugs such as anisomycin did not block AP-1 induction in Jurkat cells, whereas they were able to inhibit AP-1 induction in primary T-cells (Foletta *et al.*, 1998). A mutant c-Jun protein is known to function as a transdominant inhibitor to quench endogenous AP-1-mediated transcriptional activation (Brown *et al.*, 1994). Such naturally occurring AP-1 inhibitors that bind to Jun proteins and inactivate the AP-1 complex have been reported in different cell types (Chen *et al.*, 1996). Such inhibitors of transcription factors may be capable of inhibiting growth factor-induced proliferation of cells and may ultimately be useful in identifying which transcription factors are most important in regulating cell growth. The nuclear factor Yin-Yang 1 (YY1) may block the binding of a c-Jun homodimer through the interaction of YY1 with a regulatory site that overlaps the AP-1 site within the IFN-7 promoter (Foletta *et al.*, 1998).

Several natural compounds have been identified that could inhibit the activity of AP-1. For instance green tea polyphenols, which reduce tumor growth rate and/or metastatic capacity in several animal studies (McCarty, 1998), inhibit cell growth, MAP kinase pathway and AP-1 activity in H-ras-transformed cells (Chung *et al.*, 1999), hinting at the possibility of polyphenols in diet having some use in cancer therapy. Additional measures for downregulating AP-1 activity, may include fish oil, heparin, and the drug LY 290181 (McCarty, 1998). Cannabinol decreased IL-2 production in EL4 T-cells through a transient inhibition of AP-1 and a sustained inhibition of NFAT (Yea *et al.*, 2000). Curcumin is known to inhibit AP-1 activity and HNE (4-hydroxynonenol), a peroxidation product of omega-6-polyunsaturated fatty acid, decreased c-Fos expression induced by growth factors (Xu *et al.*, 1997; Kreuzer *et al.*, 1998). Aspirin is known to inhibit AP-1 activity induced by UV irradiation (Ma *et al.*, 1998). Gold (I) thiolates significantly inhibit AP-1 DNA binding by forming disulphide bonds with the cysteine residues present in the DNA binding domains of Jun and Fos. D-Penicillamine inhibits AP-1 DNA-binding by a similar mechanism (Handel, 1997). However, these thiol-reactive drugs may react with many other cysteine residues, and an escalation of dose to enhance their effect may lead to toxicity limiting their clinical use. Gold and penicillamine have adverse effects, including rashes, proteinuria, leukopenia, and thrombocytopenia, that are probably consequences of their reactivities with thiols.

Each of the AP-1 family members differs in its ability to trans-activate or repress transcription, and experiments in cultured cells have shown that each protein has distinct effects on cell proliferation and oncogenic conversion. The function of any individual AP-1 component will depend on the state of post-translational modification of that protein, and also on which potential partners are available for dimerization (Foletta *et al.*, 1998). Many cytoplasmic transforming proteins including Src and Ras activate AP-1, due to their ability to activate ERK and JNK signaling pathways. Serum, peptide growth factors, and TPA induced large increases in the expression of c-Jun and c-Fos mRNA and in AP-1 trans-activating activities in human breast cancer cells (Chen *et al.*, 1996). C-Fos mutant cells are partially defective in the transforming response to activated src alleles, and c-Jun mutant cells are resistant to transformation by activated ras (Wisdom, 1999). In addition, the formation of invasive skin cancers induced by the combination of an activated ras transgene and topical application of phorbol esters is markedly impaired in c-Fos null mice (Foletta *et al.*, 1998). AP-1 inhibition can also impede angiogenesis and tumor invasiveness (McCarty, 1998). It may be possible to titrate the dose of an inhibitor within a dynamic range to achieve a therapeutic and subtoxic response. Many therapeutic drugs against a particular disease may show a common mode of action and yet have quite different adverse effects. Thus the adverse effects of each drug cannot be mediated by the common mode of action. A specific inhibitor of AP-1 or NF-κB, Z-A should be an effective therapeutic agent with minimal adverse effects. At present there are no selective antagonists of AP-1 or NF-κB available for human therapy.

7.6 Nuclear factor (NF)-κB

Rel/NF-κB was first identified in 1986 as a nuclear factor bound to an enhancer element of the immunoglobulin kappa light chain gene (Sen and Baltimore, 1986). It was considered to be a B-cell transcription factor in the beginning, but now it is known to constitute a family of ubiquitous proteins. NF-κB proteins contain a Rel homology (RH) domain (DNA-binding domain/dimerization domain) with a nuclear localization sequence and are conserved from Drosophila to man. Class I proteins include p50, p52, p100, and p105. Multiple copies of ankyrin repeats are present in p100 and p105 and proteolytic cleavage of p100 forms p52 and that of p105 forms p50. They form dimers with class II proteins (c-Rel, Rel B and RelA/p65), which exclusively contain C-terminal activation domains. RelB forms only heterodimers, whereas all other members can form both homo and heterodimers. The term "NF-κB" refers to the most common heterodimer formed between RelA and p50. Dimeric NF-κB transcription factors bind to the 10-base-pair consensus site GGGPuNNPyPyCC where Pu is purine, Py is pyrimidine, and N is any base. The individual dimers have distinct DNA binding specificities for a collection of related κB sites (Gilmore, 1999; Baldwin, 1996; Ghosh and Kopp, 1998).

7.6.1 Signaling pathway

NF-κB proteins are sequestered in the cytoplasm by another family of inhibitory proteins, the IκBs. Several IκBs have been identified, including α, β, and ϵ IκBs. They are expressed in a tissue-specific manner and have distinct affinities for individual Rel/NF-κB complexes. IκBs contain six or more ankyrin repeats, an N-terminal regulatory domain, and a C-terminal domain that contains a PEST motif. IκBs bind to NF-κB dimers and sterically block the function of their nuclear localization sequences (NLSs), thereby causing their cytoplasmic retention. Most agents that activate NF-κB mediate the phosphorylation-induced degradation of IκB. Upon receiving a signal, phosphorylation of IκBα takes place on two conserved serine residues (S32 and S36) in the N-terminal regulatory domain. Another member of the IκB family, Bcl-3, however, stimulates transcription after interacting with p50 and p52 subunits of NF-κB. The kinases which phosphorylate IκBs are termed IκB kinases (IKKs), and several members of this complex have now been characterized (IKKα, IKKβ and IKKγ). Mutation analysis revealed that IKKβ and not IKKα mediates proinflammatory signals. Once phosphorylated, the IκBs, still bound to NF-κB, almost immediately undergo a second post-translational modification, polyubiquitination. The major ubiquitin acceptor sites in human IκBα are lysines 21 and 22. Protein ubiquitination occurs through E1 ubiquitin activating enzyme, E2 ubiquitin conjugating enzyme, and E3 ubiquitin protein ligases. After ubiquitination, IκBs are degraded in 26 S proteasomes, leading to the release of NF-κB dimers to translocate into the nucleus (Karin, 1999; Gilmore, 1999; Baldwin, 1996; Ghosh and Kopp, 1998). This general mechanism of activation of NF-κB is depicted in Figure 7.4. In contrast, activation of NF-κB in response to UV radiation is accompanied by IκB degradation but not phosphorylation on the N-terminus of IκBα (Li and Karin, 1998). Hypoxia or pervanadate treatment stimulates the phosphorylation of IκBα at tyrosine 42, but other IκBs do not have a tyrosine at this position (Imbert *et al.*, 1996). Phosphorylation on Ser-276 by the catalytic subunit of protein kinase A (PKAc) can contribute to the intrinsic transcriptional capacity of the p65 subunit of NF-κB. PKAc was also found associated with NF-κB and IκB in the cytoplasm and was able to phosphorylate p65 only after IκB degradation (Mayo and Baldwin, 2000). A site-directed mutant of p65 (Ser-276 to Ala) is still phosphorylated at Ser-529 in response to TNF, suggesting that multiple physiological stimuli modulate p65 through distinct phosphorylation sites to control transcriptional activity. RelA (C-terminus) has been shown to interact with basal transcriptional apparatus proteins like TBP, TFIIB and TAF 105 as well as

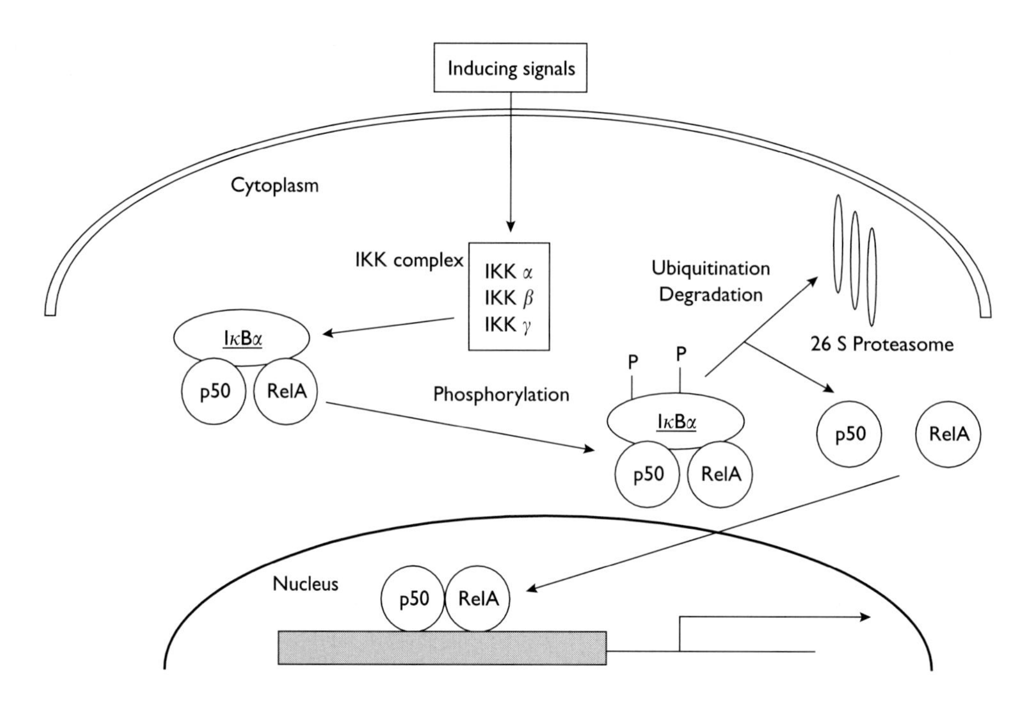

Figure 7.4 NF-κB signaling pathway. Divergent stimuli converge on IKK complex which then phosphorylates IκB at 2 N-terminal serines, signaling for its ubiquitination. These events lead to the degradation of IκB by 26S proteasome, releasing the dimers of NF-κB to enter the nucleus, bind DNA and activate transcription of target genes. As an autoregulatory mechanism, to limit its own activation, NF-κB induces the synthesis of IκB (not shown).

coactivators such as CBP and p300 though the actual role of these interactions is not clear (Mayo and Baldwin, 2000).

7.6.2 Activators and target genes of NF-κB

NF-κB is activated by many divergent stimuli including proinflammatory cytokines such as TNF-α, interleukin-1, T- and B-cell mitogens, bacteria and lipopolysaccharide, viruses, viral proteins, double-stranded RNA, and physical and chemical stresses. Accumulation of proteins in the endoplasmic reticulum (ER), a form of cellular stress as a result of viral infection, drugs interfering with ER function, mutant proteins with folding defects, and transient transfection of proteins, activates NF-κB (Pahl, 1999). Cellular stress ultimately leads to apoptosis, and a number of stimuli that activate NF-κB, such as TNF-α, ionizing radiation and chemotherapeutic agents, also induce apoptosis (Wang *et al.*, 1996). Inducers of NF-κB are listed in Table 7.3.

NF-κB is considered a central mediator of the immune response since it promotes the expression of over 150 target genes, most of which encode immunoregulatory proteins such as chemokines, cytokines, major histocompatibity complex (MHC) molecules, and immunoreceptors. Numerous target genes are selectively regulated by the transcriptional activation potential of different homo- and heterodimer combinations. In addition, variations in the NF-κB consensus sequence to which the subunits bind and cooperativity between different transcription factor families and NF-κB/Rel contribute to the differential specificity of gene activation. Many viral promoters contain NF-κB sites, and these may help to promote viral transcription and maintain chronic infection. Apart from immunoregulation, NF-κB also plays an equally important role in activating stress response proteins such as nitric oxide synthase (iNOS) and cyclooxygenase-2 (COX-2). NF-κB has both pro- and anti-apoptotic activities in different cell types and accordingly its target genes appear to encode proteins having opposite effects on apoptosis (Gilmore, 1999). Some of the target genes of NF-κB are shown in Table 7.4.

7.6.3 NF-κB as a potential target for drug development

There has been great interest in the modulators of the NF-κB signaling pathway since it is involved in a variety of physiological and pathological processes. These compounds have provided an insight into the complex molecular control of immunoregulatory expression. They have a potentially promising role in suppressing inflammatory responses and offer a pharmacological basis for interfering with the course of many diseases. NF-κB activation is inhibited by several unrelated compounds such as peptide aldehydes, phytochemicals, steroid hormones, antioxidants and specific protease inhibitors. An inhibitor may block any one or more

Table 7.3 Inducers of NF-κB activity

Bacteria or their products
Helicobacter pylori
Mycobacterium tuberculosis
Lipopolysaccharide (LPS)
Exotoxin B
Viruses or their products
HIV-1:Tax, gp 160
Influenza virus
HTLV-1:Tax 1
Double-stranded RNA viruses
HCV: core protein
HBV: HBx
Cytokines
Interleukins (1, 2, 12, 15, 17, 18)
TNF
Lymphotoxin
THANK
VEGI
CD29L, CD30L, CD40L, 4-1BBL, OX40L, RANKL
Growth factors
Serum
EGF
Insulin
PDGF
TGF-α
NGF
Stress conditions
UV and ionizing radiation
Hyperosmotic shock
Shear stress
H2O2
Pervanadate
Chemotherapeutic compounds
Cisplatin
Daunorubicin
Etoposide
Tamoxifen
Taxol
Chemical agents
Phorbol ester
N-Methyl-D-aspartate
Okadaic acid
Calcium ionophores

Table 7.4 Proteins whose genes contain NF-κB binding sites

Class	*Examples*
Cytokines/chemokines	TNF-α, interleukins (1, 2, 6, 8, 11, 12), interferon-γ, IP-10
Immunoreceptors	CD 23, CD 48, T-cell receptor β chain
Cell adhesion proteins	E-Selectin, ICAM-1, VCAM-1
Acute phase proteins	Angiotensinogen, C4b binding protein, complement factors B and C4, C-reactive protein, tissue factor-1, urokinase type plasminogen activator
Stress response proteins	Angiotensin II, COX-2, MnSOD, iNOS
Cell surface receptors	Platelet activator receptor-1, neuropeptide YY-1 receptor, bradykinin B1-receptor
Apoptosis regulatory proteins	Bcl2 homologs (A1, Bcl-XL, Nrl 3), Fas, FasL, IAPs, IEX-1L
Growth factors	G-CSF, GM-CSF, VEGF
Transcription factors	c-Myb, c-Myc, JunB, c-Rel, NF-κB p100, NF-κB p105, IκBα, p53
Viruses	HIV-1, CMV, Adenovirus, SV-40

Abbreviations used:
GR, Glucocorticoid receptor; SRF, Serum response factor; AP-1, Activator protein-1; AP-2, Activator protein-2; TPA, 12-O-tetradecanoylphorbol-13-acetate; NF-κB, Nuclear factor kappaB; CREBs, cAMP response element binding proteins; ATFs, Activating transcription factors; HLH, Helix-loop-helix; HSF, Heat shock factor; Oct-1, Octamer-1; C/EBP, CCAAT/enhancer binding protein; TNF-a Tumor necrosis factor-a EGF, Epidermal growth factor; PDGF, Platelet-derived growth factor; LPS, Lipopolysaccharide; TBP, TATA binding protein; ICAM, Intercellular adhesion molecule; VCAM, Vascular cell adhesion molecule; VEGF, Vascular endothelial growth factor; G-CSF, Granulocyte colony stimulating factor; GM-CSF, Granulocyte macrophage colony-stimulating factor; COX-2, Cyclo oxygenase-2; iNOS, Inducible nitric oxide synthase; TF-1, Tissue factor-1; NFAT, Nuclear factor of activated T-cells; HTLV-1, Human T-cell leukemia virus type 1; CRE, cAMP response element; SRE, Serum response element; SIE, sis-inducible element
PKA, Protein kinase A; TCF, Ternary complex factor; NAC, N-Acetyl cysteine; PDTC, Pyrrolidine dithiocarbamate; ROIs, Reactive oxygen intermediates.

steps in the signaling pathway such as the incoming signals that activate NF-κB, signaling cascade, translocation of NF-κB into the nucleus, DNA binding of the dimers, and/or interactions with the basal transcriptional machinery. Since NF-κB is activated by oxidative stress, antioxidants act as NF-κB inhibitors, among which are curcumin, NAC, PDTC, and vitamin E derivatives. Many of these agents act by scavenging the ROIs formed during activation of NF-κB. Antioxidants appear to act at the early stages of NF-κB activation by blocking phosphorylation and/or degradation of IκBα. Owing to their ability to block the activation of NF-κB, antioxidants such as PDTC may have clinical application in the treatment of septic shock and certain retinal neovascular diseases (Beauparlant and Hiscott, 1996; Epinat and Gilmore, 1999). Proteasome inhibitors inhibit the chymotrypsin-like activity of the proteasome complex. Calpain inhibitor I is a cysteine protease inhibitor and is less potent than MG132 and MG115. Molecules like lactacystin irreversibly block proteasome activity by acylating a threonine residue in the active site of the subunit X of the mammalian proteasome. Some, serine protease inhibitors also act as proteasome inhibitors and block the phosphorylation and degradation of IBα. Retaining the NF-κB dimers in the cytoplasm by preventing the phosphorylation and degradation of IκBα is an effective way to inhibit NF-κB and this can be achieved by several signaling molecules such as NO, estrogen, oxidized LDL, 4-hydroxynonenal, and prostaglandin A.

Several natural compounds from plants are also known to downregulate the activation of NF-κB. For instance, many phytochemicals known for their anti-inflammatory and anti-tumor activities also inhibit NF-κB activity, among which are sanguinarine, emodin, and sesquiterpene lactones (Epinat and Gilmore, 1999). Nuclear translocation of the RelA subunit and degradation of IκBα induced by TNF are prevented by IL-13 and human chorionic gonadotropin (Manna and Aggarwal, 1998; 2000c). IL-10 and IL-13 both suppress nuclear localization of NF-κB and increase IκBα mRNA expression (Ehrlich *et al.*, 1998; Lentsch *et al.*, 1997). In addition, fungal metabolites such as gliotoxin also block the degradation of IκBα. African swine fever virus encodes a protein similar to IκB which can bind RelA and inhibit NF-κB activation. NF-κB limits its own synthesis by inducing the synthesis of IκBα (Epinat and Gilmore, 1999). The anti-cancer agents genistein and erbstatin suppress TNF-α induced NF-κB activation in myeloid U937 cells but not in MCA-101 fibrosarcoma cells and in human alveolar epithelial A439 cells (Natarajan *et al.*, 1998; Epinat and Gilmore, 1999). Treatment of fetal rat cortical neurons, which exhibit constitutive NF-κB activity with β-amyloid peptide decreases NF-κB activity and increases IκBα mRNA and protein whereas exposure of astroglial cell cultures to β-amyloid peptide activates NF-κB (Bales *et al.*, 1998). Atrial natriuretic peptide can prevent NF-κB activity associated with rat liver damage during reperfusion (Gerbes *et al.*, 1998). Similarly leflunomide, a drug approved for the treatment of rheumatoid arthritis, also mediates its effects through suppression of NF-κB (Manna *et al.*, 1999c, 2000d).

Evidence that NF-κB is involved in many inflammatory diseases and in oncogenesis has led a number of investigators to determine if NF-κB is a specific target of pharmaceutical and dietary compounds known to prevent disease. Aspirin and sodium salicylate specifically inhibit NF-κB nuclear translocation by preventing IκBα phosphorylation and degradation (Pierce *et al.*, 1996). Recently, Yin *et al.* (1998) have shown that aspirin can directly bind to and inhibit the kinase activity of IKKβ by reducing its ability to bind ATP. Other non-steroidal anti-inflammatory drugs such as ibuprofen, defereoxamine and tepoxaline also inhibit NF-κB (Epinat and Gilmore, 1999). Several dietary chemopreventive compounds, including curcumin (Singh and Aggarwal, 1995; Kumar 1998b), silymarin (Manna and Aggarwal, 1999a), emodin (Kumar *et al.*, 1998a), capsaicin (Singh *et al.*, 1996), genistein & erbastatin (Natarajan *et al.*, 1998), caffeic acid phenylether ester (CAPE) (Natarajan *et al.*, 1996), sanguinarin (Chaturvedi *et al.*, 1997), oleandrin (Manna *et al.*, 2000a), β-lapachone (Manna *et al.*, 1999a) and resveratrol (Manna *et al.*, 2000b), are known to block NF-κB activation. These agents suppress NF-κB activation by different mechanisms. While some agents (e.g. curcumin, silymarin, emodin, and capasaicin) suppress IκBα phosphorylation and degradation, others (e.g. CAPE) modifies NF-κB protein in a manner that it can no longer bind to the DNA (Natarajan *et al.*, 1996). Other polyphenols (e.g. resveratrol) block the phosphorylation of p65 subunit of NF-κB thus suppressing its transcriptional activity (Manna *et al.*, 2000b). These studies suggest that NF-κB is a functionally relevant target of chemopreventive drugs and dietary compounds that prevent cancer (Mayo and Baldwin, 2000). Cyclosporin A inhibits NF-κB activity by preventing its nuclear translocation in activated T-cells (McCaffrey *et al.*, 1994) and inhibits the inducible degradation of IκBα and IκBβ. Another immunosuppressant, FK506, can block specific NF-κB activation involving c-Rel nuclear translocation (Sen *et al.*, 1995). Another approach to inhibiting NF-κB has been to use mutant forms of IκBs called superrepressors in which the serine phosphorylation sites (32 and 36) of IκBα have been mutated and thus cannot be phosphorylated or degraded (Mayo and Baldwin, 2000). Some of the well-known inhibitors of NF-κB and their sites of action are shown in Figure 7.5.

7.6.4 Role of NF-κB in pathogenesis

NF-κB regulates the transcription of a very large number of genes, and it is not surprising that dysregulation of NF-κB is observed in a wide range of human disorders, including cancers, neurodegenerative disorders, ataxia telangiectasia, arthritis, asthma, inflammatory bowel disease, and several other inflammatory conditions (Gilmore *et al.*, 1996; Luque and Gelinas 1997; Grilli and Memo, 1999). Retroviruses encoding v-rel are oncogenic in avians and rel genes are prone to

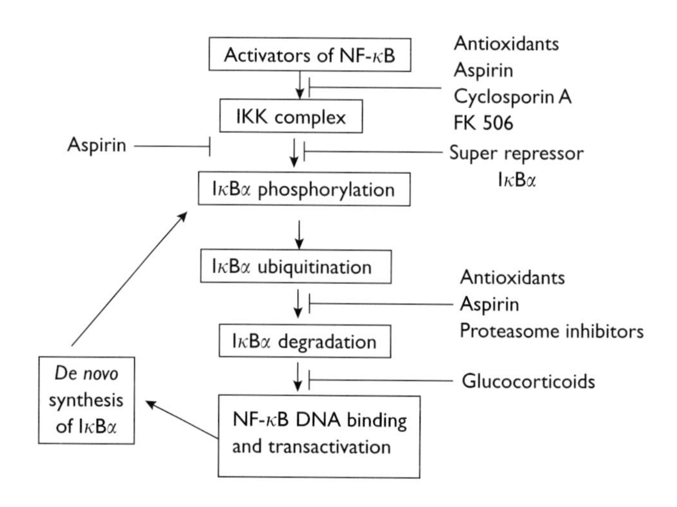

Figure 7.5 Inhibitors of NF-κB activation target its signaling pathway. Some inhibitors like immunosuppressants and antioxidants block both the phosphorylation and degradation of IκBα. Proteasome inhibitors block the degradation of IκBα whereas others like glucocorticoids inhibit nuclear binding and transactivation.

rearrangements and translocations (Luque and Gelinas, 1997). In certain lymphomas bcl-3 and nf-κb2 genes are known to be translocated (Baldwin, 1996). Mutations in iκbα have been observed in Hodgkin's lymphoma (Cabannes *et al.*, 1999), and blocking iκbα by antisense mechanisms appears to induce oncogenic transformation (Beauparlant *et al.*, 1994). Overexpression of p50 subunit of NF-κB has been noticed in cell lines derived from cancers of the lung, prostate, breast, bone and brain (Bours *et al.*, 1994; Mukhopadhyay *et al.*, 1995). Antisense to relA blocks tumorigenesis induced by Tax (Kitajima *et al.*, 1992). Interestingly, constitutive expression of NF-κB has been reported in tumors of breast, ovary, colon, pancreas, thyroid, prostate, bladder and skin (Rayet and Gelinas, 1999). A recent report shows that human breast tumors accumulate activated NF-κB complexes consisting of p50, p52 and Bcl-3 rather than p65 (Mayo and Baldwin, 2000). HT1080 fibrosarcoma cells exposed to ionizing radiation, TNF-α or daunorubicin exhibited enhanced activation of NF-κB and inhibition of NF-κB dramatically enhanced apoptosis in response to radiation or daunorubicin (Wang *et al.*, 1996). Fibrosarcoma and colorectal tumors grown in nude mice were induced to undergo apoptosis when infected with an adenovirus expressing a modified form of IκBα along with systemic delivery of CPT-11 chemotherapy (Mayo and Baldwin, 2000).

Stable transfection of iκbα did not yield enhanced cytotoxicity in response to chemotherapy despite the ability of these agents to activate NF-κB (Benties-Alj *et al.*, 1999). Transient inhibition of NF-κB using adenoviral delivery of iκbα in some of these cells had an enhanced apoptotic response to CPT-11, and stable transfectants are presumably the variants having NF-κB-independent anti-apoptotic pathways (Mayo and Baldwin, 2000). Inhibition of NF-κB in the skin via expression of the superrepressor form of IκBα led to squamous cell carcinomas and increased apoptosis (Mayo and Baldwin, 2000). Activation of NF-κB by LPS may have a role in the development of septic shock, since NF-κB activates transcription of the iNOS genes known to be involved in septic shock (MacMicking *et al.*, 1995). Autoimmune diseases such as systemic lupus erythromatosus (SLE) may also involve activation of NF-κB. In Alzheimer's disease, under chronic conditions, the amyloid β peptide causes production of ROIs and indirectly activates gene expression through κB sites (Behl *et al.*, 1994). The influenza virus protein hemagglutinin activates NF-κB, and this activation may contribute to viral induction of cytokines and to some of the symptoms associated with flu (Pahl and Baeuerle, 1995). The oxidized lipids from LDL associated with atherosclerosis activate NF-κB, which then activates other genes (Berliner *et al.*, 1995). Interestingly, mice that are susceptible to atherosclerosis exhibit NF-κB activation when fed an atherogenic diet (Liao *et al.*, 1994). Another important contributor to artherosclerosis is thrombin, which stimulates the proliferation of the vascular smooth muscle cells through the activation of NF-κB (Nakajima *et al.*, 1994). A truncated form of IκBα has been shown to protect AT (ataxia telangiectasia) cells, which express constitutive levels of an NF-κB-like activity, from ionizing radiation (Jung *et al.*, 1995). From these studies it is quite clear that abnormal activation or expression of NF-κB is associated with a wide variety of pathological conditions.

7.7 Conclusion

From this description it is clear that transcriptional factors play a major role in the pathogenesis of various diseases. The suppression of the activity of these factors will downregulate the expression of genes that cause the disease. Several extracellular stress stimuli have been identified that lead to activation of these transcription factors. The pathways that lead to the activation of these transcription factors are, however, poorly understood. Multiple pathways exist that could activate a single transcription factor. The design of specific inhibitors of the activation pathway could provide a disease-specific drug. Another limitation is that the same transcription factor may regulate multiple genes, some of whose products may be harmful and others useful. Thus inhibitors of transcription factors may also exhibit harmful effects. Nevertheless, transcription factors represent ideal targets for drug development for various diseases.

Acknowledgment

This research was conducted in part by The Clayton Foundation for Research.

References

Abate, C., Luk, D., and Curran, T. (1991) Transcriptional regulation by Fos and Jun *in vitro*: interaction among multiple activator and regulatory domains. *Mol. Cell. Biol. 11, 3624–3632*

Andrews, N. C., Erdjument-Bromage, H., Davidson, M. B., Tempst, P., and Orkin, S. H. (1993) Erythroid transcription factor NF-E2 is a haematopoietic specific basic leucine zipper protein. *Nature 362, 722–728*

Angel, P., Imagawa, M., Chiu, R., Stein, B., Imbra, R. J., Rahmsdorf, H. J., Jonat, C., Herrlich, P., and Karin, M. (1987) Phorbol ester-inducible genes contain a common cis element recognized by a TPA modulated trans-acting factor. *Cell 49, 729–739*

Angel, P., and Karin, M. (1991) The role of Jun, Fos and the AP-1 complex in cell-proliferation and transformation. *Biochem. Biophys. Acta 1072, 129–157*

Baker, S. J., Kerppola, T. K., Luk, D., Vandenberg, M. T., Marshak, D. R., Curran, T., and Abate, C. (1992) Jun is phosphorylated by several protein kinases at the same sites that are modified in serum-stimulated fibroblasts. *Mol. Cell. Biol. 10, 4694–4705*

Baldwin, A. S. (1996) The NF-κB and IκB proteins: new discoveries and insights. *Annu. Rev. Immunol. 14, 649–681*

Bales, K. R., Du, Y., Dodel, R. C., Yan, G. M., Hamilton-Byrd, E., Paul, S. M. (1998) The NF-kappaB/Rel family of proteins mediates A beta-induced neurotoxicity and glial activation. *Brain Res. Mol. Brain Res. 57, 63–72*

Bannister, A. J., Cook, A., and Kouzarides, T. (1991) *In vitro* DNA binding activity of Fos/Jun and BZLF1 but not C/EBP is affected by redox changes. *Oncogene 7, 1243–1250*

Barlev, N. A., Candau, R., Wang L., Darpino P., Silverman N., and Berger S.L. (1995) Characterization of physical interactions of the putative transcriptional adaptor, ADA2, with acidic activation domains and TATA-binding protein. *J. Biol. Chem. 270, 19337–19344*

Beauparlant, P., and Hiscott, J. (1996) Biological and biochemical inhibitors of the NF-κB/Rel proteins and cytokine synthesis. *Cytokine & Growth Factor Reviews 7, 175–190*

Beauparlant, P., Kwan, I., Bitar, R., Chou, P., Koromilas, A., Sonenberg, N., and Hiscott, J. (1994) Disruption of IκB-α regulation by antisense RNA expression leads to malignant transformation. *Oncogene 9, 3189–3197*

Behl, C., Davis, J., Lesley, R., and Schubert, D. (1994) Hydrogen peroxide mediates amyloid β protein toxicity. *Cell 77, 817–827*

Bentires-Alj, M., Hellin, A. C., Ameyar, M., Chouaib, S., Merville, M. P., and Bours, V. (1999) Stable inhibition of nuclear factor kappaB in cancer cells does not increase sensitivity to cytotoxic drugs. *Cancer Res. 59, 811–815*

Binetruy, B., Smeal, T., and Karin, M. (1991) Ha-Ras augments c-Jun activity and stimulates phosphorylation of its activation domain. *Nature 351, 122–127*

Birrer, M. J., Alani, R., Cuttitta, F., Preis, L. H., Sabich, A. L., Sanders, D. A., Siegfried, J. M., Szabo, E., and Brown, P. H. (1992) Early events in the neoplastic transformation of respiratory epithelium. *J. Natl. Cancer Inst. Monogr. 13, 31–37*

Bours, V., Azarenko, V., Dejardin, E., and Siebenlist, U. (1994) Human RelB (I-Rel) functions as a kappa B site-dependent transactivating member of the family of Rel-related proteins. *Oncogene 6, 1699–1702*

Boyle, W. J., Smeal, T., Defize, L. H., Angel, P., Woodgett, J. R., Karin, M., and Hunter, T. (1991) Activation of protein kinase C decreases phosphorylation of c-Jun at sites that negatively regulate its DNA-binding activity. *Cell 64, 573–584*

Brown, P. H., Chen, T. K., and Birrer, M. J. (1994) Mechanism of action of a dominant-negative mutant of c-Jun. *Oncogene 3, 791–799*

Cabannes, E., Khan, G., Aillet, F., Jarrett, R., and Hay, R. (1999) Mutations in the IκBα gene in Hodgkin's disease suggest a tumour suppressor role for I kappa B alpha. *Oncogene 18, 3063–3070*

Calkhoven, C. F., and Geert, A. B. (1996) Multiple steps in the regulation of transcription factor level and activity. *Biochem. J. 317, 23–36*

Chainy, G. B. N., Manna, S. K., Chaturvedi, M. M., and Aggarwal, B. B. (2000) Anethole blocks both early and late cellular responses transduced by tumor necrosis factor: effect on NF κB, AP-1, JNK, MAPKK and Apoptosis. *Oncogene 19, 2943–2950*

Chaturvedi, M., Kumar, A., Darnay, B., Chainy G. B. N., Agarwal, S., and Aggarwal, B. B. (2000) Sanguinarine (Pseudochelerythrine) is a potent inhibitor of NF-kB activation, IkBa phosphorylation, and degradation. *J. Biol. Chem. 272, 30129–30134*

Chen, T. K., Smith, L. M., Gebhardt, D. K., Birrer, M. J., Brown, P. H. (1996) Activation and inhibition of the AP-1 complex in human breast cancer cells. *Mol. Carcinog. 15, 215–226*

Chiu, R., Angel, P., and Karin, M. (1989) JunB differs in its biological properties from, and is a negative regulator of c-Jun. *Cell 59, 979–986*

Chung, J. Y., Huang, C., Meng, X., Dong, Z., and Yang, C. S. (1999) Inhibition of activator protein 1 activity and cell growth by purified green tea and black tea polyphenols in H-ras-transformed cells: structure-activity relationship and mechanisms involved. *Cancer Res. 59, 4610–4617*

Dashtaki, R., Whorton, A. R., Murphy, T. M., Chitano, P., Reed, W., Kennedy, T. P. (1998) Dehydroepiandrosterone and analogs inhibit DNA binding of AP-1 and airway smooth muscle proliferation. *J. Pharmacol. Exp. Ther. 285, 876–883*

Downward, J., Graves, J. D., Warne, P. H., Rayter, S., and Cantrell, D. A. (1990) Stimulation of p21ras upon T-cell activation. *Nature 346, 719–723*

Drolet, D. W., Scully, K. M., Simmons, D. M., Wagner, M., Chu, K. T., Swanson, L. W., and Rosenfeld, M. G. (1991) TEF, a transcription factor expressed specifically in the anterior pituitary during embryogenesis, defines a new class of leucine zipper proteins. *Genes Dev. 5, 1739–1753*

Ehrlich, L. C., Hu, S., Peterson, P. K., and Chao, C. C. (1998) IL-10 down-regulates human microglial IL-8 by inhibition of NF-kappaB activation. *Neuroreport. 9, 1723–1726*

Epinat, J-C., and Gilmore, T. D. (1999) Diverse agents act at multiple levels to inhibit the Rel/NF-κB signal transduction pathway. *Oncogene 18, 6896–6909*

Foletta, V. C., Segal, D. H., and Cohen, D. R. (1998) Transcriptional regulation in the immune system. *J. Leuk. Biol. 63, 139–152*

Gasser, S. M., and Laemmli, U. K. (1986) Cohabitation of scaffold binding regions with upstream/enhancer elements of three developmentally regulated genes of D. melanogaster. *Cell 46, 521–530*

Gerbes, A. L., Vollmar, A. M., Kiemer, A. K., Bilzer, M. (1998) The guanylate cyclase-coupled natriuretic peptide receptor: a new target for prevention of cold ischemia-reperfusion damage of the rat liver. *Hepatology 5, 1309–1317*

Ghosh, S., May, M. J., Kopp, E. B. (1998) NF-kappa B and Rel proteins: evolutionarily conserved mediators of immune responses. *Annu. Rev. Immunol. 16, 225–260*

Gilmore, T. D. (1999) The Rel/NF-κB signal transduction pathway: introduction. *Oncogene 18, 6842–6844*

Gilmore, T. D., Koedood, M., Piffat, K. A., and White, D. W. (1996) Rel/NF-kappaB/I-kappaB proteins and cancer. *Oncogene 13, 1367–1378*

Gius, D., Botero, A., Shah, S., and Curry, H. A. (1999) Intracellular oxidation/reduction status in the regulation of transcription factors NF-i-synthetic powdered diet at 0.03125% level for the whole lifespan of Swiss mice starting from 6 weeks of age. This treatment resulted in the development of tumors of the cecum in 22% of females akdB and AP-1. *Toxicol. Lett. 106, 93–106*

Greenblatt, J., Nodwell, J. R., and Mason, S. W. (1993) Transcriptional antitermination. *Nature 364, 401–406*

Grilli, M., and Memo, M. (1999) Nuclear factor-kappa B/Rel proteins: a point of convergence of signaling pathways relevant in neuronal function and dysfunction. *Biochem. Pharmacol. 57, 1–7*

Hahn, S. (1993) Structure (?) and function of acidic transcription activators. *Cell 72, 481–485*

Hai, T., and Curran, T. (1991) Fos/Jun and ATF/CREB cross family dimerisation alters DNA binding specificity. *Proc. Natl. Acad. Sci. USA 88, 3720–3724*

Hai, T., Liu, F., Coukos, W. J., and Green M. R. (1989) Transcription factor ATF cDNA clones: an extensive family of leucine zipper proteins able to selectively form DNA-binding heterodimers. *Genes Dev. 3, 2083–2090*

Halazonetis, T., Georgopoulos, K., Greenberg, M. E., and Leder, P. (1988) c-Jun dimerises with itself and with c-Fos, forming complexes of different DNA binding affinities. *Cell 55, 917–924*

Handel, M. L. (1997) Transcription factors AP-1 and NF-κB: where steroids meet the gold standard of anti-rheumatic drugs. *Inflamm. Res. 46, 282–286*

Horikoshi, M., Carey, M. F., Kakidani, H., and Roeder, R. G. (1988) Mechanism of action of a yeast activator: direct effect of GAL4 derivatives on mammalian TFIID-promoter interactions. *Cell 54, 665–669*

Imbert, V., Rupec, R. A., Livolsi, A., Pahl, H. L., Traenckner, E. B., Mueller-Dieckmann, C., Farahifar, D., Rossi, B., Auberger, P., Baeuerle, P. A., and Peyron, J. F. (1996) Tyrosine phosphorylation of I kappa B-alpha activates NF-kappa B without proteolytic degradation of I kappa B-alpha. *Cell 86, 787–798*

Iyer, R., Eisen, M. B., Ross, D. T., Schuler, G., Moore, T., Lee, J. C. F., Trent, J. M., Staudt, L. M., Hudson, J., Boguski, M. S., Lashkari, D., Shalon, D., Botstein, D., and Brown, P. O. (1999) The transcriptional program in the response of human fibroblasts to serum. *Science 283, 83–87*

Jain, J., Valge-Archer, V. E., and Rao, A. (1992) Analysis of the AP-1 sites in the IL-2 promoter. *J. Immunol. 148, 1240–1250*

Janknecht, R., and Hunter, T. (1997) Convergence of MAP kinase pathways on the ternary complex factor Sap-1-sensitizing and irritating properties of anise oil are reported (324), studies have shown the cancer chemopreventive potential of anetholead. *EMBO. J. 16, 1620–1627*

Jung, M., Zhang, Y., Lee, S. M., and Dritschilo, A. (1995) Correction of radiation sensitivity in ataxia telangiectasia cells by truncated IkBα. *Science 268, 1619–1621*

Kallunki, T., Su, B., Tsigelny, I., Sluss, H. K., Derijard, B., Moore, G., Davis, R., Karin, M. (1994) JNK2 contains a specificity-determining region responsible for efficient c-Jun binding and phosphorylation. *Genes Dev. 8, 2996–3007*

Karin, M. (1995) The regulation of AP-1 activity by mitogen-activated protein kinases. *J. Biol. Chem. 270, 16483–16486*

Karin, M. (1999) How NF-tumor and liver were significantly elevated in animals fed anethole trithione. This study suggested that the inhibition of AOM-induced colon carcinogenesis by anethole trithione may be askdB is activated: the role of the IκB kinase (IKK) complex. *Oncogene 18, 6867–6874*

Karin, M., Liu, Z-G., and Zandi, E. (1997) AP-1 function and regulation. *Curr. Opin. Cell Biol. 9, 240–246*

Karin, M., and Smeal, T. (1992) Control of transcription factors by signal transduction pathways: the beginning of the end. *Trends Biochem. Sci. 17, 418–422*

Karlen, S., D'Ercole, M., and Sanderson, C. J. (1996) Two pathways can activate the interleukin-5 gene and induce binding to the conserved lymphokine element 0. *Blood 88, 211–221*

Kataoka, K., Noda, M., and Nishizawa, M. (1994a) Maf nuclear oncoprotein recognizes sequences related to an AP-1 site and forms heterodimers with both Fos and Jun. *Mol. Cell. Biol. 14, 700–712*

Kataoka, K., Fujiwara, K. T., Noda, M., and Nishizawa, M. (1994b) Maf B, a new Maf family transcription activator that can associate with Maf and Fos but not with Jun. *Mol. Cell. Biol. 14, 7581–7591*

Kerppola, T. K., Curran, T. (1991) Transcription factor interactions: basics on zippers. *Curr. Op. Struct. Biol. 1, 71–79*

Kovary, K., and Bravo, R. (1991) The jun and fos protein families are both required for cell cycle progression in fibroblasts. *Mol. Cell. Biol. 11, 4466–4472*

Kreuzer, T., Grube, R., Wutte, A., Zarkovic, N., and Schaur, R. J. (1998) 4-Hydroxynonenal modifies the effects of serum growth factors on the expression of the c-fos proto-oncogene and the proliferation of HeLa carcinoma cells. *Free Radic. Biol. Med. 25, 42–49*

Kumar, A., Dhawan, S., and Aggarwal, B. B., (1998a) Emodin (3-methyl-1,6,8 trihydroxyanthraquinone) inhibits the TNF-induced NF- kB activation, IkB degradation and expression cell surface adhesion protein in human vascular endothelial cells. *Oncogene 17, 913–918*

Kumar, A., Dhawan, S., Hardegen, N. J., and Aggarwal, B. B. (1998b) Curcumin (Diferuloylmethane) inhibition of TNF-mediated adhesion of monocytes to endothelial cells by suppression of cell surface expression of adhesion molecules and of nuclear factor-kB activation *Biochem. Pharmacol. 55, 775–783*

Landschultz, W. H., Johnson, P. F., and McKnight, S. L. (1988a) The leucine zipper: a hypothetical structure common to a new class of DNA binding proteins. *Science 240, 1759–1764*

Landschultz, W. H., Johnson, P. F., Adashi, E. Y., Graves, B. J., and McKnight, S. L. (1988b) Isolation of a recombinant copy of the gene encoding C/EBP. *Genes Dev. 2, 786–800*

Latchman, D. (1995) *Gene regulation: a eukaryotic perspective*. Chapman and Hall, London, UK

Lee, W., Haslinger, A., Karin, M., and Tjian, R. (1987a) Activation of transcription by two factors that bind promoter and enhancer sequences of the human metallothionein gene and SV40. *Nature 325, 368–372*

Lee, W., Mitchel, P., and Tjian, R. (1987b) Purified transcription factor AP-1 Interacts with TPA-inducible enhancer elements. *Cell 49, 741–752*

Lentsch, A. B., Shanley, T. P., Sarma, V., and Ward, P. A. (1997) *In vivo* suppression of NF-kappa B and preservation of I kappa B alpha by interleukin-10 and interleukin-13. *J. Clin. Invest. 100, 2443–2448*

Lewin, B. (2000) *Genes VII.* Oxford University Press Inc. New York

Li, N., and Karin, M. (1998) Ionizing radiation and short wavelength UV activate NF-kappaB through two distinct mechanisms. *Proc. Natl. Acad. Sci. USA. 95, 13012–13017*

Liao, F., Andalibi, A. I., Qiao, J. H., Allayee, H., Fogelman, A., and Lusis, A. (1994) Genetic evidence for a common pathway mediating oxidative stress, inflammatory gene induction, and aortic fatty streak formation in mice. *J. Clin. Invest. 94, 877–884*

Luque, I., and Gelinas, C. (1997) Rel/NF-kappa B and I kappa B factors in oncogenesis. *Semin. Cancer Biol. 8, 103–111*

Ma, W. Y., Huang, C., Dong, Z. (1998) Inhibition of ultraviolet C irradiation-induced AP-1 activity by aspirin is through inhibition of JNKs but not ERKs or P38 MAP kinase. *Int. J. Oncol. 3, 565–568*

Mac Micking, J., Nathan, C., Hom, G., Chartrain, N., Fletcher, D., Trumbauer, M., Stevens, K., Xie, Q-W., Sokol, K., Hutchinson, N., Chen, H., and Muddgett, J. (1995) Altered responses to bacterial infection and endotoxic shock in mice lacking inducible nitric oxide synthase. *Cell 81, 641–650*

Manna, S. K., and Aggarwal, B. B. (1998) IL-13 suppresses TNF-induced activation of nuclear factor-kappa B, activation protein-1, and apoptosis. *J. Immunol. 161, 2863–2872*

Manna, S. K., Mukhopadhyay, A., Van, N. T., and Aggarwal, B. B. (1999) Silymarin suppresses TNF induced activation of nuclear transcription factor-k B, c-Jun N-terminal kinase and apoptosis, *J. Immunol. 163, 6800–6809*

Manna, S., Gad, Y., Mukhopadhyay, A., and Aggarwal, B. B. (1999) Suppression of tumor necrosis factor-activated nuclear transcription factor-kB, activator protein-1, c-Jun N-terminal kinase and apoptosis by b-lapachone. *Biochem. Pharmacol. 57, 763–774*

Manna, S. and Aggarwal B. B. (1999) Immunosuppressive leflunomide metabolite (A77 1726) blocks TNF-dependent nuclear factor-κB activation and gene expression. *J. Immunol., 162: 2095–2102*

Manna, S. K., Sah, N. K., Newman, R. A., Cisneros, A., and Aggarwal, B. B. (2000) Oleandrin suppresses activation of nuclear transcription factor-kB, activator protein-1, and c Jun N-terminal kinase. *Cancer Research 60, 3838–3847*

Manna S. K, Mukhopadhyay A., and Aggarwal B. B. (2000) Resveratrol suppresses TNF induced activation of nuclear transcription factors NF-kB, activator protein-1, and apoptosis: potential role of reactive oxygen intermediates and lipid peroxidation. *J. Immunol. 164, 6509–6519*

Manna S. K, Mukhopadhyay A, and Aggarwal B. B (2000) Human chorionic gonadotropin suppresses activation of nuclear transcription factors NF-kB and activator protein-1 induced by tumor necrosis factor. *J. Biol. Chem. 275, 13307–13314*

Manna S. K, Mukhopadhyay A. and Aggarwal B. B (2000) Leflunomide suppresses TNF induced cellular responses: effects on NF-κB, AP-1, JNK, and apoptosis. *J. Immunol. 164, 5962–5969*

Mitchell, P. J., and Tjian, R. (1989) Transcriptional regulation in mammalian cells by sequence specific DNA binding proteins. *Science 245, 371–378*

Massari, M. E., and Murre, C. (2000) Helix-loop-helix proteins: regulators of transcription in eukaryotic organisms. *Mol. Cell. Biol. 20, 429–440*

Mayo, M. W., and Baldwin, A. S. (2000) The transcription factor NF-Resveratrol is a polyphenol (trans-3, 4′, 5-trihydroxystilbene) found in various fruits and vegkďB: control of oncogenesis and cancer therapy resistance. *Biocehim. Biophys. Acta 1470, M55–M62*

McCaffrey, P. G., Kim, P. K., Valge-Archer, V. E., Sen, R., Rao, A. (1994) Cyclosporin A sensitivity of the NF-kappa B site of the IL2R alpha promoter in untransformed murine T cells. *Nucleic Acids Res. 22, 2134–2142*

Montminy, M. R., Severino, K. A., Wagner, J. A., Mandel, G., and Goodman, R. H. (1986) Identification of a cyclic -AMP-responsive element within the rat somatostatin gene. *Proc. Natl. Acad. Sci. USA 83, 6682–6686*

Mukhopadhyay, T., Roth, J. A., and Maxwell, S. A. (1995) Altered expression of the p50 subunit of the NF-kappa B transcription factor complex in non-small cell lung carcinoma. *Oncogene 11, 999–1003*

Nakajima, T., Kitajima, I., Shin, H., Takasaki, I., Shigeta, K., Abeyama, K., Yamashita, Y., Tokioka, T., Soejima, Y., and Maruyama, I. (1994) Involvement of NF-κB activation in thrombin-induced human vascular smooth muscle cells. *Biochem. Biophys. Res. Commun. 204, 950–955*

Natarajan, K., Manna, S. K., Chaturvedi, M. M., and Aggarwal, B. B. (1998) Protein tyrosine kinase inhibitors block tumor necrosis factor-induced activation of nuclear factor-kappaB, degradation of IkappaBalpha, nuclear translocation of p65, and subsequent gene expression. *Arch. Biochem. Biophys. 352, 59–70*

Natarajan, K., Singh, S., Burke, Jr. T. R., Grunberger, D., and Aggarwal, B. B. (1996) Caffeic acid phenethyl ester (CAPE) is a potent and specific inhibitor of activation of nuclear transcription factor NF kB. *Proc. Natl. Acad. Sci. USA 93, 9090–9095*

Nel, A. E., Taylor, L. K., Kumar, G. P., Gupta, S., Wang, S. C., Williams, K., Liao, O., Swanson, K., and Landreth, G. E. (1994) Activation of a novel serine/threonine kinase that phosphorylates c-Fos upon stimulation of T and B lymphocytes via antigen and cytokine receptors. *J. Immunol. 152, 4347–4357*

Pahl, H. L. (1999) Activators and target genes of Rel/NF-explain resveratrol's activity against carcinogenesis, inflammation, and cell growth modulation, all dependent on the produkďB transcription factors. *Oncogene 18, 6853–6866*

Paul, H., and Baeuerle, P. (1995) Expression of influenza virus hemagglutinin activates NF-κB. *J. Virol. 69, 1480–1484*

Pierce, J. W., Read, M. A., Ding, H., Luscinskas, F. W., Collins, T. (1996) Salicylates inhibit I kappa B-alpha phosphorylation, endothelial-leukocyte adhesion molecule expression, and neutrophil transmigration. *J. Immunol. 156, 3961–3969*

Pulverer, B. J., Kyriakis, J. M., Avruch, J., Nikolakaki, E., and Woodgett, J. R. (1991) Phosphorylation of c-Jun mediated by MAP kinase. *Nature 353, 670–674*

Rayet, B., and Gelinas, C. (1999) Aberrant rel/nfkb genes and activity in human cancer. *Oncogene 18, 6938–6947*

Ryder, K., Lau, L., and Nathans, D. (1989) A gene activated by growth factors is related to the oncogene v-jun. *Proc. Natl. Acad. Sci. USA 85, 1487–1491*

Sen, R., and Baltimore, D. (1986) Multiple nuclear factors interact with the immunoglobulin enhancer sequences. *Cell 46, 705–716*

Sen, J., Venkataraman, L., Shinkai, Y., Pierce, J. W., Alt, F. W., Burakoff, S. J., and Sen, R. (1995) Expression and induction of nuclear factor-kappa B-related proteins in thymocytes. *J. Immunol. 154, 3213–21*

Serfling, E., Avots, A., and Neumann M. (1995) The architecture of the interleukin-2 promoter: a reflection of T lymphocyte activation. *Biochim. Biophys. Acta. 1263, 181–200*

Singh, S. and Aggarwal, B. B., (1995) Activation of transcription factor NF-kB is suppressed by curcumin (Diferulolylmethane) *J. Biol. Chem. 270, 24995–25000*

Singh, S., Natarajan, K., and Aggarwal, B. B. (1996) Capsaicin (8-methyl-N-vanillyl-6 nonenamide) is a potent inhibitor of nuclear transcription factor activation by diverse agents. *J. Immunol., 157, 4412–4420*

Smeal, T., Binetruy, B., Mercola, D. A., Birrer, M., and Karin, M. (1991) Oncogenic and transcriptional cooperation with Ha-Ras requires phosphorylation of c-Jun on serines 63 and 73. *Nature 354, 494–496*

Smeal, T., Binetruy, B., Mercola, D., Grover-Bardwick, A., Heidecker, G., Rapp, U. R., and Karin, M. (1992) Oncoprotein-mediated signalling cascade stimulates c-Jun activity by phosphorylation of serines 63 and 73. *Mol. Cell. Biol. 8, 3507–3513*

Szabo, E., Preis, L. H., Brown, P. H., and Birrer, M. J. (1991) The role of jun and fos gene family members in 12-O-tetradecanoylphorbol-13-acetate induced hemopoietic differentiation. *Cell Growth Differ. 10, 475–482*

Tabata, T., Takase, H., Takayama, S., Mikami, K., Nakatsuka, A., Kawata, T., Nakayama, T., and Iwabuchi, M. (1989) A protein that binds to a cis-acting element of wheat histone genes has a leucine zipper motif. *Science 245, 965–967*

Treisman, R. (1992) The serum response element. *Trends Biochem. Sci. 17, 423–426*

Treisman, R. (1994) Ternary complex factors: growth factor regulated transcriptional activators. *Curr. Opin. Genet. & Dev. 4, 96–101*

Ullman, K. S., Northrop, J. P., Admon, A., and Crabtree, G. R. (1993) Jun family members are controlled by a calcium-regulated, cyclosporin A-sensitive signaling pathway in activated T lymphocytes. *Genes Dev. 7, 188–196*

Wang, C. Y., Mayo, M. W., Baldwin, A. S. Jr. (1996) TNF- and cancer therapy-induced apoptosis: potentiation by inhibition of NF-kappaB. *Science. 274, 784–787*

Weis, L., and Reiberg, D. (1992) Transcription by RNA polymerase II initiator directed formation of transcription-competent complexes. *FASEB J. 6, 3300–3309*

Williams, S. C., Cantwell, C. A., and Johnson, P. F. (1991) A family of C/EBP-related proteins capable of forming covalently linked leucine zipper dimers *in vitro*. *Genes Dev. 5, 1553–1567*

Wisdom, R. (1999) AP-1: one switch for many signals. *Exp. Cell Res. 253, 180–185*

Wolffe, A. P., and Pruss, D. (1996) Targeting chromatin disruption: transcription regulators that acetylate histones. *Cell 84, 817–819*

Xu, Y. X., Pindolia, K. R., Janakiraman, N., Chapman, R. A., Gautam, S. C. (1998) Curcumin inhibits IL1 alpha and TNF-alpha induction of AP-1 and NF-κB DNA-binding activity in bone marrow stromal cells. *Hematopathol. Mol. Hematol. 11, 49–62*

Yea, S. S., Yang, K. H., and Kaminski, N. E. (2000) Role of nuclear factor of activated T-cells and activator protein-1 in the inhibition of interleukin-′2 gene transcription by cannabinol in EL4 T-cells. *J. Pharmacol. Exp. Ther. 292, 597–605*

Yin, M. J., Yamamoto, Y., and Gaynor, R. B. (1998) The anti-inflammatory agents aspirin and salicylate inhibit the activity of I(kappa)B kinase-beta. *Nature 396, 77–80*

Zacheis, D., Dhar, A., Lu, S., *et al.* (1999) Heteroarotinoids inhibit head and neck cancer cell lines *in vitro* and *in vivo* through both RAR and RXR retinoic acid receptors. *J. Med. Chem. 42, 4434–4445*

Chapter 8

Apoptosis in various diseases

Béla Szende

Contents

Apoptosis, together with cell proliferation, physiologically controls the homeostasis of various cell populations, which can be imbalanced by intrinsic or extrinsic factors. The imbalance may result in pathological conditions characterised by either too much or insufficient apoptosis.

8.1 Pathological conditions characterised by apoptosis at insufficient rate

8.1.1 Carcinogenesis

When DNA damage reaches the "no repair" stage, the cells may continue reproducing the damage which results in the first step of carcinogenesis, or they may commit suicide, i.e. undergo active, programmed cell death. This phenomenon was first demonstrated in hepatic carcinogenesis by Schulte-Hermann *et al.* (1990) and Bursch and co-workers (1984). It also became evident that phorbol esters, which are well known tumor-promoters, exert their enhancing effect, at least partially, by suppressing apoptosis. Withdrawal of the promoter results in increased apoptosis in precancerous lesions and the net cell loss produced in this way may even cause regression of preneoplasia (Bursch *et al.*, 1985). The role of wild type p53 in the induction of apoptosis is widely proven (Canman and Kastan, 1997). Mutations of p53 cease the pro-apoptotic activity of this gene product and facilitate the process of carcinogenesis (Donehower *et al.*, 1992). Other gene products, like the anti-apoptotic bcl_2 and the pro-apoptotic bax, also influence carcinogenesis but the up-regulation or down-regulation of p53, bcl_2 and bax is variable in different types of neoplasia (Trump and Berezesky, 1998). Moreover, wild type p53 plays a major role in regulation of bcl_2 and bax, as well as another important pro-apoptotic gene, APO-1 (fas, CD-95). It has also become evident that mutation of p53 which is equivalent to loss of pro-apoptotic activity, usually occurs in a later stage of tumor development, i.e. in the stage of promotion (Grasi-Kraupp *et al.* in press). Inhibition of cell death may result in increased genomic instability and stabilies aneuploidy. Data from studies on ultraviolet (UV) light-induced skin carcinogenesis suggest that loss of p53 function due to UV light induced mutation decreases the apoptotic potential of keratinocytes in an earlier rather than later stage of carcinogenesis (Krajewski *et al.*, 1995). These statements are of course only valid in cases of p53-dependent apoptosis.

Recent experimental studies by Schulte-Hermann and his group (1995) have shown that single initiated cells possess the potential to proliferate under the influence of growth promoting stimuli. A considerable number of these cells undergo apoptosis, probably due to "spontaneous" deprivation of growth factors required for their survival. Researchers are still investigating whether this mechanism is also valid, for instance, in human liver carcinogenesis induced by viruses and enhanced by aflatoxin B and/or alcohol intoxication.

8.1.2 Decreased apoptotic activity in tumor tissue

It has been presumed that malignant neoplasms are characterised by low apoptotic activity compared to highly proliferating normal tissues, for instance the cells of the Lieberkühn's crypts of the intestinal mucosa. In other words, the balance between mitotic and apoptotic rate in tumor tissue is shifted in favour of mitosis, which results in a constant growth of the tumor mass.

According to our studies on various types of malignant neoplasms, the apoptotic index in tumor tissues of untreated patients varies considerably according to organs of origin and histological structure. In prostate carcinomas (Szende *et al.*, 1996) the rate of apoptosis is relatively low in highly differentiated tumors and increases with de-differentiation (Figure 8.1). Very low apoptotic activity was observed in mesopharyngeal (Kraxner *et al.*, 1997) and oral mucosal squamous cell carcinomas. Similarly, squamous cell carcinomas of the lung contain few apoptotic tumor cells. Adenocarcinomas of the lung exhibit higher apoptotic activity (Figure 8.2) and the highest apoptotic index can be found in small cell carcinomas of the lung (Demoura *et al.*, 1999). The apoptotic index in peripheral blood lymphoblasts was low in almost all of the 50 acute lymphoid leukaemia cases investigated by our group (Schuler and Szende, 1997). In gliomas of the central nervous system the tumors of histologically low grade malignancy showed a low apoptotic rate, which increased parallel with increasing malignancy; the highest apoptotic index was found in glioblastoma multiforme (Sipos *et al.*, 1998). Neuroblastomas, however, show very low apoptotic activity; in contrast, primitive peripheral neuroendocrine tumors (PNET) have a relatively high apoptotic index. It is noteworthy that when the retinoic acid receptor (RAR) was determined immunohistochemically, the ratio of RAR positive cells varied in parallel with the ratio of apoptotic cells, both in neuroblastomas and in PNET-s (Farid *et al.*, 2000). Papillary, follicular and even anaplastic carcinomas of the thyroid gland are characterised by very low apoptotic activity, whereas medullary thyroid carcinomas show an increased apoptotic index. An almost invariably low number of apoptotic tumor cells was observed in colorectal adenocarcinomas (Farczádi *et al.*, 1999) and the apoptotic index is similarly low in various histological subtypes of ductal mammary carcinomas.

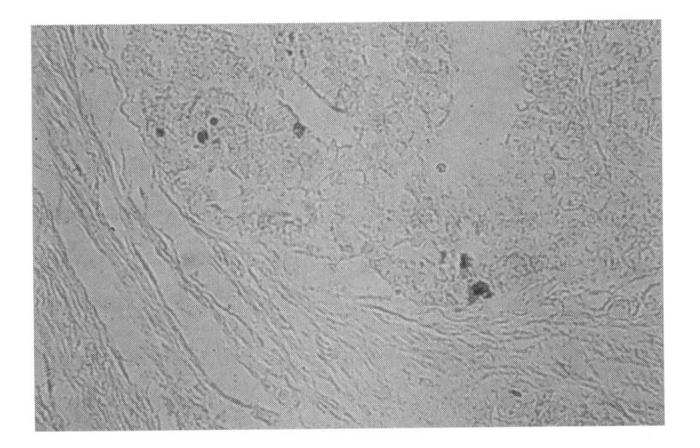

Figure 8.1 Apoptotic tumor cells in a poorly differentiated prostate carcinoma, TUNEL reaction, ×400 (see Colour Plate II).

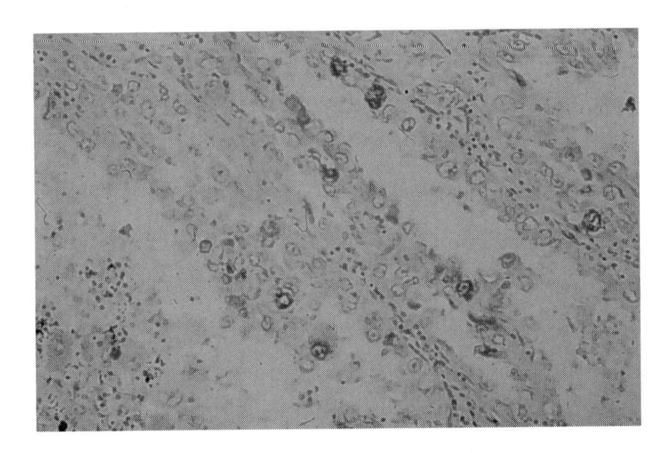

Figure 8.2 Adenocarcinoma of the lung. Note numerous apoptotic tumor cells (brown colour), TUNEL reaction, ×400 (see Colour Plate III).

The intensity of apoptotic activity in untreated tumors does not regularly correlate with the clinical outcome of the disease. Good correlation was found between the low apoptotic rate and poor clinical prognosis in the case of oral, mesopharyngeal and pulmonary squamous cell carcinomas as well as neuroblastomas. On the other hand, the high apoptotic rate observed in glioblastoma multiforme, PNET, high grade prostate carcinoma and small cell cancer of the lung is inconsistent with the intensive growth and propagation of these tumors. Clinical outcome of colon carcinomas is in statistically significant relation to apoptotic activity in the tumor tissue. In the case of acute lymphoid leukaemia, thyroid carcinoma and mammary carcinoma no such correlation could be detected (Kraxner *et al.*, 1997; Demoura *et al.*, 1999; Schuler and Szende, 1997; Sipos *et al.*, 1998; Farczádi *et al.*, 1999).

The contradictions outlined above should be considered by taking into account the proliferative activity of tumor cells, measurable either by determination of the mitotic index or by immunohistochemical demonstration of proliferation markers like Ki67 or PCNA (proliferating cell nuclear antigen). Proliferative activity was measured in our studies of colorectal, thyroid, mesopharyngeal, mammary and prostate carcinomas. The results show that cell proliferation co-exists with apoptosis and in slowly growing tumors the low apoptotic rate was accompanied by low mitotic rate, whereas a relatively high apoptotic index was exceeded by high mitotic activity e.g. in glioblastoma multiforme or poorly differentiated prostate carcinoma (Szende *et al.*, 1996; Kraxner *et al.*, 1997; Demoura *et al.*, 1999; Schuler and Szende, 1997; Farid *et al.*, 2000; Sipos *et al.*, 1998; Farczádi *et al.*, 1999).

Apoptosis in tumor tissue can be induced by hormone, radiation or cytostatic therapy and this poses one of the most interesting questions regarding the prognosis of neoplastic disease and prediction of therapeutical efficacy. It has been supposed that activation of the fas-fas L system may be involved in the mechanism of apoptosis caused by the above mentioned agents (Golstein *et al.*, 1995). Another theory suggests that DNA damage itself is the trigger of apoptosis (Lowe *et al.*, 1994). Histological follow-up of human tumors by repeated biopsies during antitumor therapy can provide valuable data in this respect. Because of technical difficulties and ethical considerations, only a few systematic studies have been published on changing apoptotic activity in human neoplasms after treatment by hormones, cytostatic agents or ionising radiation. Clinico-pathological investigations of our group showed that apoptosis occurs in malignant neoplasms and it can be enhanced by hormone, cytostatic or radiation therapy. The degree of enhancement can be determined individually and the data obtained from several individuals may help to create a *new predictive factor* in various tumors. The following human malignancies were investigated in this respect: acute lymphoid leukaemia (ALL) (50 cases) (Garami *et al.*, 1999), prostate carcinoma (30 cases) (Szende *et al.*, 1999), rectal carcinoma (16 cases) (Farczádi *et al.*, 1999), mesopharyngeal carcinoma (15 cases) (Kraxner *et al.*, 1997) and gliomas (12 cases) (Sipos *et al.*, 1998). Apoptosis (TUNEL reaction), mitosis, p53 (Figure 8.3), bcl2 and Ki-67 expression was determined in the tumor tissue before and during treatment of the same patients.

Changes in apoptotic ratio in the first 24 hours after glucocorticoid monotherapy are of predictive value in ALL (Table 8.1 and 8.2). Similarly, good correlation between apoptotic activity and clinical outcome was observed in prostate carcinomas treated with androgen ablation (Figures 8.4 and 8.5).

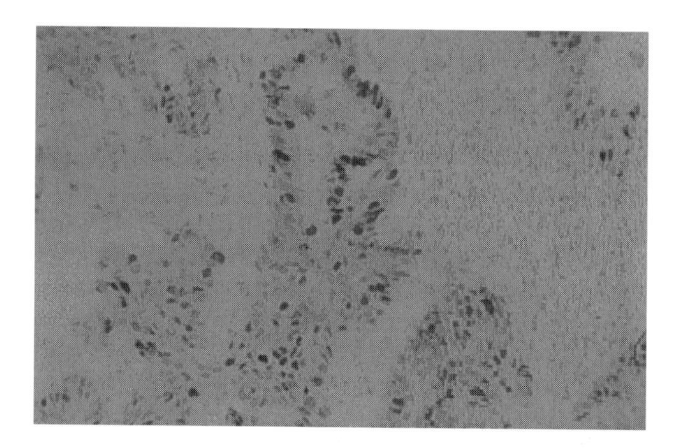

Figure 8.3 Colon biopsy containing both normal and tumor tissue. The tumor cells are p53 positive, the normal glands are p53 negative. Immunoperoxidase for p53, ×300 (see Colour Plate IV).

Table 8.1 Correlation between apoptotic index and effect of prednisolone

	Increased AI after 6 hours	*Non-increased AI after 6 hours*
Prednisolone + at day 8	39	4
Prednisolone − at day 8	3	4

Table 8.2 Correlation between apoptotic index and clinical course

	Increased AI after 6 hours	*Non-increased AI after 6 hours*
Remission	17/21	4/21
Relapse	2	4

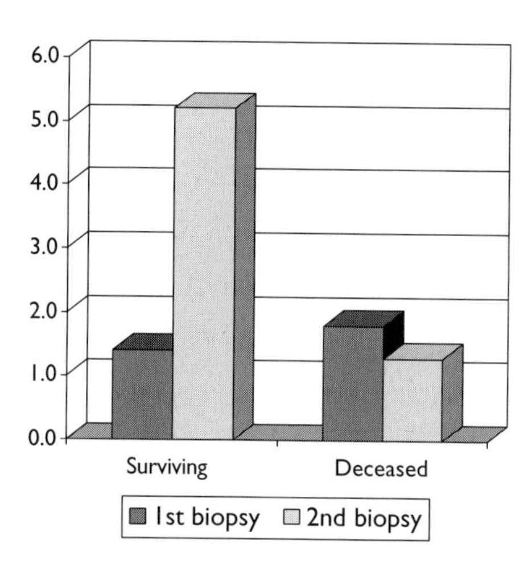

Figure 8.4 Apoptosis index of prostate carcinomas of surviving and deceased patients.

Increase in apoptotic ratio after short-term cytostatic therapy of rectal carcinoma correlated with a favorable prognosis (Figure 8.6). Mesopharyngeal carcinomas are both radiation- and chemotherapy-resistant, which is reflected in their poor response regarding apoptotic activity. The most malignant glioma, glioblastoma multiforme, is characterised by a paradoxically high ratio of spontaneous apoptosis. The predictive value of changes in apoptotic index upon the effect of various types of therapy should be considered separately according to malignant tumors of specific organs and specific histological structure of the tumors.

8.2 Apoptosis in autoimmune diseases

Autoimmune diseases, like systemic lupus erythematosus (SLE), scleroderma, dermatomyositis, rheumatoid arthritis,

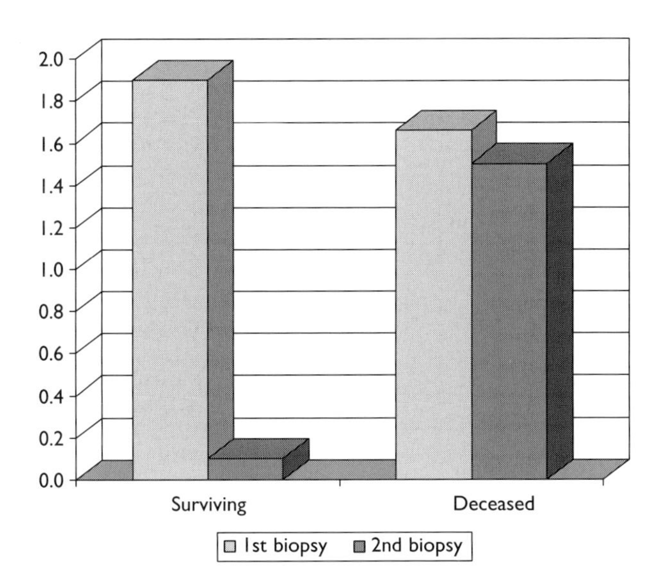

Figure 8.5 Mitosis index of prostate carcinomas of surviving and deceased patients.

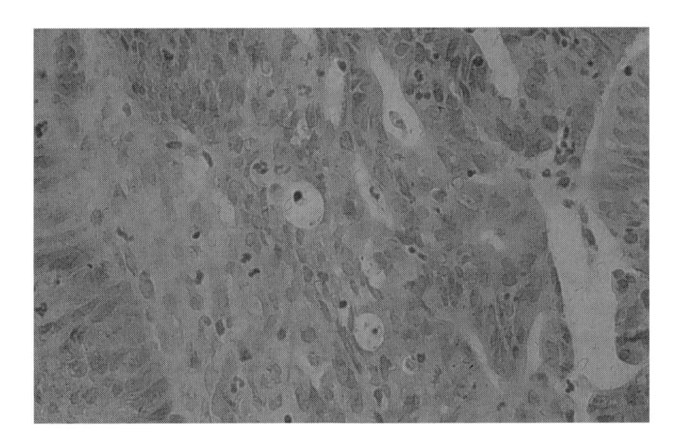

Figure 8.6 Apoptotic tumor cells in a colon tumor biopsy, taken from a patient pre-treated with 5-fluorouracyl, H and E, ×400 (see Colour Plate V).

Sjögren's syndrome, Hashimoto's thyroiditis, Graves' disease, autoimmune arteritis, autoimmune hepatitis and others, are related to abnormalities of the immune system (Tan, 1994; Budd, 1998). The negative selection of self-reactive lymphocytes occurs via apoptosis. It has been shown experimentally, and also in humans, that a genetic defect causing mutations in the genes for fas or fas L results in failure of apoptosis and in accumulation of lymphocytes which have retained the capability to attack cells of the same organism. The T-cell growth cytokine IL-2, which promotes both cell proliferation and apoptosis, may also be failing or be defective, thus leading to the development of autoimmunity (Su *et al.*, 1994).

The tissues involved in various autoimmune diseases are heavily infiltrated by T-lymphocytes, which present fas L to fas-expressing parenchymal cells, and initiate apoptosis. This phenomenon can clearly be observed in various forms of autoimmune thyroid diseases, in parotid glands and lacrimal glands of Sjögren's disease, in the synovial membrane in rheumatoid arthritis, as well as in autoimmune hepatitis (lupoid hepatitis) and autoimmune orchitis (Huang *et al.*, 1994).

Therapeutic attempts to influence apoptosis in autoimmune diseases have not yet been published. Of course, any therapeutic measure (glucocorticoids, immunosuppressive agents) which decrease the number and activity of T lymphocytes, could prevent or minimize apoptosis induced by these cells.

8.3 Apoptosis in infectious diseases

The body of knowledge on apoptosis caused by infectious agents is steadily increasing. First of all, hepatitis B and C viruses are known to act on hepatocytes by inducing the expression of fas on the cell surface. Lymphocytes which provide fas L infiltrate the liver and initiate apoptosis resulting in the formation of Councilman bodies. In fact, this was the first disease to draw attention of pathologists to active, programmed cell death (Kerr *et al.*, 1972).

The most thoroughly investigated viral disease based on induction of apoptosis by virus is AIDS. The onset of clinical symptoms in AIDS is preceeded by a considerable latency period, during which a gradual decrease in number of $CD4^+$ T lymphocytes can be observed. According to a series of studies, these lymphocytes undergo apoptosis far exceeding the physiological level. It has also been shown that the majority of the dying lymphocytes are not infected by virus (Badley *et al.*, 1997). The fas L binding mechanism for apoptosis is thought to play a role in the death of HIV-infected lymphocytes and a protein, gp12, produced by the HIV-infected cells, may interact with non-infected $CD4^+$ T lymphocytes, thus initiating apoptosis. Attempts have been made to inhibit $CD4^+$ T cell apoptosis before the onset of clinical symptoms of AIDS in order to prolong the incubation period. For this purpose the *in vitro* and *in vivo* observed anti-apoptotic action of oligopeptide, which are considered as active centres of thymopoietin, may be used (Dénes *et al.*, 1987).

The range of other viruses reportedly able to induce apoptosis includes adenoviruses, herpes viruses, other retroviruses and the influenza virus (Yoshida *et al.*, 1995). It should be mentioned that a number of viruses cause apoptosis *in vitro* when added to cultured cells. Recently apoptosis has been reported as an effect of lymphocytic infiltration mimicking autoimmune inflammation in the heart muscle and synovial membranes after infection by Borrelia burgdorferi. One of the newly observed phenomena in Lyme disease is the interaction between lymphocytes and heart muscle cells and synovial cells including apoptosis (Mukasa *et al.*, 1995).

8.4 Apoptosis in neurodegenerative diseases

Increasing evidence shows that neuronal apoptosis can result from a wide range of insults such as trophic insufficiency, excitatory amino acids, metamphetamine and others (Tatton and Chalmers-Redman, 1996). Apoptosis can be identified in neuronal cells in culture or in tissue sections by demonstration of nuclear and cytoplasmic shrinkage (fluorescent dye methods; flow cytometry) or by cleavage of nuclear DNA (polyacrylamide gel-electrophoresis to identify the "ladder" pattern of oligonucleosomal DNA fragmentation; terminal deoxynucleotidyl-transferase-mediated dUTP-x nick-end labelling, known as the TUNEL method).

Apoptosis contributes to neuronal loss in human neurodegenerative diseases, such as Parkinson's disease, Alzheimer's disease and amyotrophic lateral sclerosis (ALS). Anglade *et al.* (1995) reported apoptotic cell death of nigral dopaminergic neurons in Parkinson's disease. Cotman and Anderson (1995) suggested a potential role for apoptosis in neurodegeneration in Alzheimer's disease. Su *et al.* (1994) gave immunohistochemical evidence of apoptosis in Alzheimer's disease. This finding correlates with the DNA damage and apoptosis described by Anderson *et al.* (1996), who also showed co-localisation of apoptosis with c-Jun, using immunohistochemistry.

Lassmann *et al.* (1995) evaluated apoptotic cell death in Alzheimer's disease by in situ end-labelling of fragmented DNA. The same was done by Dragunow *et al.* (1995) in Alzheimer's disease (temporal lobes) and Huntington's disease (striatum). Yoshiyama *et al.* (1994) found that apoptosis-related antigen Le (Y) and nick-end labelling are positive in spinal motor neurons in amyotrophic lateral sclerosis. Moreover, according to the work of Muller *et al.* (1992), the AIDS protein gp120 of HIV-1 induces apoptosis in rat cerebral cortical cell cultures.

The monoamino oxydase B inhibitor (−)-deprenyl (Selegiline, Jumex) which has been introduced in the therapy of neurodegenerative diseases, prevents neuronal apoptosis in various experimental models. We have also shown that this effect is caspase-3 dependent and needs metabolic alteration in order to achieve efficacy. The active metabolite is still unknown (Magyar *et al.*, 1998, Magyar and Szende 1999).

Apoptotic death of neurons has been reported also in case of Down's syndrome and in Huntington's chorea (Portera-Cailliau *et al.*, 1995). The peripheral, so-called necrobiotic zone of liquefaction necrosis in cerebral infarction undergoes damage caused by ischemia-reperfusion and results in neuronal as well as glial apoptosis. Similarly, we found strikingly large numbers of apoptotic cells around the necrotic areas of cerebral leukomalacia in pre-term infants (Figure 8.7).

The retina, a peripheral outlet of the central nervous system, also presents examples of apoptosis under certain pathological conditions. Mutation in the light-sensitive protein rhodopsin is seen in dominantly inherited retinal degeneration, which is characterised by apoptosis of the imperfect photoreceptor cells (Papermaster and Nir, 1994). The more frequent age-related macular degeneration may also be a consequence of retinal apoptosis (Bird, 1995).

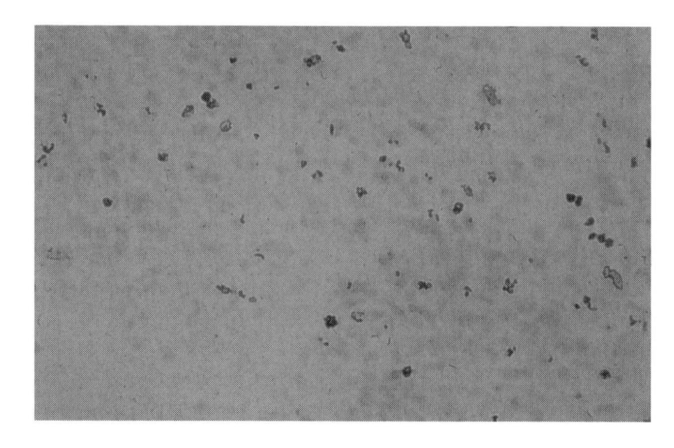

Figure 8.7 Apoptotic cells in periventricular leukomalacia in a pre-term infant, Neurotacs, ×200 (see Colour Plate VI).

8.5 Apoptosis in liver diseases

Apoptosis of liver cells during hepatitis B and C infections and in autoimmune hepatitis has been discussed above. Alcohol intoxication may also lead to single cell death of hepatocytes consistent with apoptosis (Goldin *et al.*, 1993). Acetaldehyde, the final product of ethanol metabolism may be the causative agent since experimental data point to the importance of aldehyde radicals in the initiation of apoptosis (Szende *et al.*, 1999). Cholestasis, either due to mechanical obstruction of the bile ducts or as a primary disease, causes apoptosis of biliary epithelial cells, and also of hepatocytes (Patel *et al.*, 1994). After liver transplantation the damage to bile duct epithelium due to acute or chronic rejection reaction is via apoptosis of the epithelial cells by the killer mechanism of T lymphocytes (Nawaz and Fennell, 1994).

8.6 Apoptosis in renal diseases

Apoptosis of mesangial and endothelial cells has been reported in mesangiocapillary glomerulonephritis (Buttyan and Gobé, 1997), due to immune mechanisms. Endothelial apoptosis may also contribute to the vascular lesions observed in kidney arterioles in hypertension (Hamet *et al.*, 1995). The tubular epithelial cells of the kidney may also undergo apoptosis as a consequence of ischemia followed by reperfusion (Gobé *et al.*, 1990). This condition usually occurs in cadaveric renal transplants. We have observed both apoptosis and necrosis of renal tubular epithelial cells shortly after kidney transplantation. The rate of apoptosis correlated well with

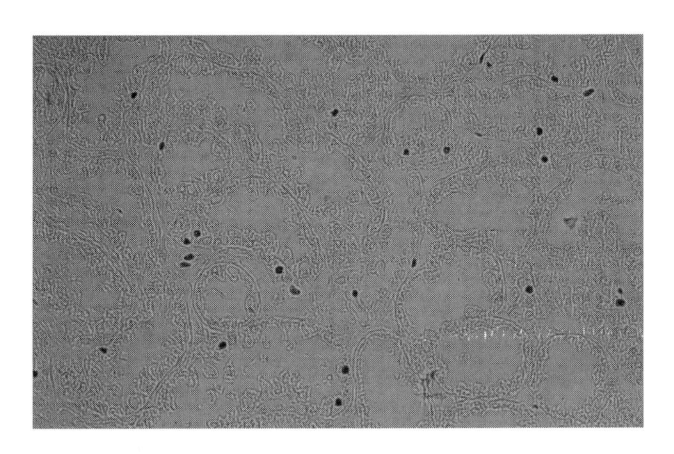

Figure 8.8 Apoptotic cells in a renal biopsy taken from a transplanted kidney 30 minutes after transplantation, TUNEL reaction, ×200 (see Colour Plate VII).

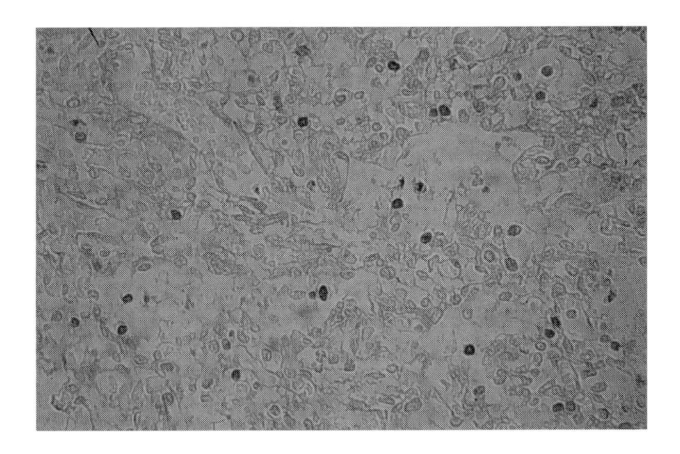

Figure 8.9 Apoptotic pneumocytes in alveolar walls in a case of bronchopulmonary dysplasia, TUNEL reaction, ×200. (see Colour Plate VIII).

delayed renal function after transplantation (Figure 8.8). In experimental studies of others and of our group, reperfusion after temporary clamping of renal artery in rats resulted in extensive tubular epithelial cell apoptosis, which could be diminished by the preventive administration of Ca channel blockers (Toronyi *et al.*, 1999), as well as the monoamino oxydase B inhibitor, Deprenyl (Selegiline).

8.7 Apoptosis in pulmonary disorders

Some data point to the occurrence of apoptosis of type II pneumocytes in adult respiratory distress syndrome (ARDS) (Bardales *et al.*, 1996). According to our investigations this phenomenon can also be observed in the lung of pre-term infants suffering from phase II or III bronchopulmonary dysplasia (Figure 8.9). This alteration develops after intensive oxygen therapy in infant respiratory distress syndrome (IRDS). These circumstances may suggest a role for oxygen free radicals in the induction of apoptosis.

8.8 Apoptosis in cardiovascular diseases

The role of apoptosis in the development of congestive cardiomyopathy and in chronic cardiac failure of different origins has widely been discussed (Umansky and Tomei, 1997). Recently, apoptosis of myocardial cells in myocardial infarction has become one of the major points of interest in cardiac pathology. The so-called necrobiotic zone, between the areas of coagulation necrosis and the normal myocardium, contains an immense amount of heart muscle cell nuclei, showing all the morphological criteria of apoptosis, including TUNEL positivity. These cells, in a later stage of the process, contribute to the necrotic area of infarction. Ischemia-reperfusion, the oxygen paradox is also considered a causative agent in this situation (Gottlieb *et al.*, 1994). Little is known about the possible therapeutic measures to prevent apoptosis of heart muscle cells, thus decreasing the extent of myocardial necrosis.

References

Anderson, A. J., Su, J. H. and Cotman, C. W. (1996) DNA damage and apoptosis in Alzheimer's disease: Colocalization with c-Jun immuno-reactivity, relationship to brain area and the effect of postmortem delay. *J. Neurosci., 16, 1710–1719.*

Anglade, P., Michel, P., Marquez, J. *et al.* (1995) Apoptotic degeneration of nigral dopaminergic neurons in Parkinson's disease. *Abs. Soc. Neurosci., 21, 1250.*

Badley, A. D., Dockrell, D. and Paya, C. V. (1997) Apoptosis in AIDS. In S. H. Kaufmann, (ed.) *Apoptosis. Pharmacological implications and therapeutic opportunities.* Advances in Pharmacology, Vol. 41., Academic Press, San Diego, pp. 271–294.

Bardales, R. H., Su, X., Schaefer, R. F. and Su Ming, H. (1996) Apoptosis is a major pathway responsible for the resolution of type II pneumocytes in acute lung injury. *Am. J. Pathol., 149, 845–852.*

Bird, A. C. (1995) Retinal photoreceptor dystrophies: LI Edward Jackson Memorial Lecture. *Am. J. Ophthalmol., 119, 543–562.*

Budd, R. C. (1998) Apoptosis in autoimmunity. In R. A. Lockshin, Z. Zakeri, and J. L. Tilly (eds.), *When cells die. A comprehensive evaluation of apoptosis and programmed cell death.* Wiley-Liss, New York, pp. 279–288.

Bursch, W., Lauer, B., Timmermann-Trosiener, I., Barthel, G., Schuppler, J. and Schulte-Hermann, R. (1984) Controlled death (apoptosis) of normal and putative preneoplastic cells in rat liver following withdrawal of tumor promoters. *Carcinogenesis, 5, 453–458.*

Bursch, W., Taper, H. S., Lauer, B., and Schulte-Hermann, R. (1985) Quantitative histological and histochemical studies on the occurrence and stages of controlled cell death (apoptosis) during regression of rat liver hyperplasia. *Virch. Arch. Cell Pathol., 50, 153–166.*

Buttyan, R. and Gobé, G. (1997) Apoptosis in the mammalian kidney: Incidence, effectors, and molecular control in normal development and disease states. In S. H. Kaufmann, (ed.) *Apoptosis. Pharmacological implications and therapeutic opportunities*. Advanced in Pharmacology, Vol. 41. Academic Press, San Diego, pp. 369–381.

Canman, C. E. and Kastan, M. B. (1997) Role of p53 in apoptosis. In S. H. Kaufmann, (ed.) *Apoptosis. Pharmacological implications and therapeutic opportunities*. Advanced in Pharmacology, Vol. 41. Academic Press, San Diego, pp. 429–452.

Cotman, C. W. and Anderson, A. J. (1995) A potential role for apoptosis in neurodegeneration and Alzheimer's disease. *Mol. Neurobiol., 10, 19–45.*

Demoura, S., Kogan, E., Szende, B. and Paltsev, M. (1999) Apoptosis in neuroendocrine tumors of the lung. *Virchows Archiv, 435, 268.*

Dénes, L., Szende, B., Hajós, Gy., Szporny, L. and Lapis, K. (1987) Therapeutic possibilities of thymopoietin fragments/TP-3 and Tp-4/based on experimental animal models. *Drugs. Exptl. Clin. Res., 13, 279–287.*

Donehower, L. A., Harvey, M., Slagle, B. L., McArthur, M. J., Montgomery, C. A., Butel, J. S., and Bradley, A. (1992) Mice deficient for p53 are developmentally normal but susceptible to spontaneous tumors. *Nature, 356, 215–221.*

Dragunow, M., Faull, R. L., Lawlor, P., Beilharz, E. J., Singleton, K., Walker, E. B. and Mee, E. (1995) In situ evidence for DNA fragmentation in Huntington's disease striatum and Alzheimer's disease temporal lobes. *Neuroreport, 6, 1053–1057.*

Farczádi, E., Szántó, J., Kaszás, J., Benyó, I., Bodnár, Zs., Szlobodnyik, J. and Szende, B. (1999) Changes in apoptotic and mitotic activity in rectal carcinoma after short-term cytostatic therapy as possible predictive factors. *Neoplasma, 46, 219–223.*

Farid, P., Babosa, M., Hauser, P., Schuler, D. and Szende, B. (2000) Spontaneous apoptosis and retinoic acid receptor incidence in neuroblastomas and PNET-s. *Ped. Hematol. Oncol. In press.*

Garami, M., Bocsi, J., Kovács, G., Csóka, M., Schuler, D. and Szende, B. (1999) Apoptotic ratio as an early prognostic factor in acute lymphoblastic leukemia in children. *Medical Ped. Oncol., 33, 344.*

Gobé, G. C., Axelsen, R. A. and Searle, J. W. (1990) Cellular events in experimental unilateral ischemic renal atrophy and in regeneration after contralateral nephrectomy. *Lab. Invest., 63, 770–779.*

Goldin, R. D., Hunt, N. C., Clark, J. and Wickramasinghe, S. N. (1993) Apoptotic bodies in a murine model of alcoholic liver disease: Reversibility of ethanol-induced changes. *J. Pathol., 171, 73–76.*

Golstein, P., Marguet, D. and Depraetere, V. (1995) Homology between reaper and the cell death domains of Fas and TNFR1. *Cell, 81, 185–186.*

Gottlieb, R. A., Burleson, K. O., Kloner, R. A., Babior, B. M. and Engler, R. L. (1994) Reperfusion injury induces apoptosis in rabbit cardiomyocytes. *J. Clin. Invest., 94, 1621–1628.*

Grasi-Kraupp, B., Lübeck, G., Wagner, A., Löw-Baselli, A., de Gunst, M., Waldhör, T., Moolgavkar, S. and Schulte-Hermann, R. (in press) Quantitative analysis of tumor initiation in rat liver: Role of cell replication and cell death (apoptosis). *Carcinogenesis, in press.*

Hamet, P., Richard, L, Dam, T. V., Teiger, E., Orlov, S. N., Gaboury, L., Gossard, F. and Tremblay, J. (1995) Apoptosis in target organs of hypertension. *Hypertension, 26, 642–648.*

Huang, L., Soldeville, G., Leeker, M., Flavell, R. and Crispe, N. (1994) The liver eliminates T cells undergoing antigen-triggered apoptosis *in vivo. Immunity, 1, 741–749.*

Kerr, J. F. R., Wyllie, A. H. and Currie, A. R. (1972) Apoptosis: A basic biological phenomenon with wide-ranging implications in tissue kinetics. *Br. J. Cancer, 26, 239–257.*

Krajewski, S., Blomqvist, C., Franssila, K., Krajewska, M., Wasenius, V. M., Niskanen, E., Nordling, S. and Reed, J. C. (1995) Reduced expression of proapoptotic gene BAX is associated with poor response rates to combination chemotherapy and shorter survival in women with metastatic breast adenocarcinoma. *Cancer Res., 55, 4471–4478.*

Kraxner, H., Tamás, L., Járay, B., Ribári, O., Szentirmay, Z. and Szende, B. (1997) Search for prognostic factors in head and neck cancer. *Acta Otolaryngol, 527, 145–149.*

Lassmann, H., Bancher, C., Breitschopf, H., Wegiel, J., Bobinski, M., Jellinger, K. and Wisniewski, H. M. (1995) Cell death in Alzheimer's disease evaluated by DNA fragmentation in situ. *Acta Neuropathol. Ber., 89, 35–41.*

Lowe, S. W., Bodis, S., Bardeesy, N., McClatchey, A., Remington, L., Ruley, H. E., Fisher, D. E., Jacks, T., Pelletier, J. and Housman, D. E. (1994) Apoptosis and the prognostic significance of p53 mutation. *Cold Spring Harbor* Symp. *Quant. Biol., 59, 419–426.*

Magyar, K., Szende, B., Lengyel, J., Tarczali, J. and Szatmáry, I. (1998) The neuroprotective and neuronal rescue effects of (-d)-deprenyl. *J. Neural. Transm., 52, 109–123.*

Magyar, K. and Szende, B. (1999) The neuroprotective and neuronal rescue effect of deprenyl. In R. G. Cameron and G. Feuer (eds.), *Apoptosis in neurodegenerative diseases*. Chapter 18, Springer Verlag, Berlin, pp. 457–472.

Mukasa, A., Hiromatsu, G., Matzusaki, R., O'Brien, R., Borne, W. and Nomoto, K. (1995) Bacterial infection of the testis leading to autoagressive immunity triggers apparently opposed responses of $\alpha\beta$ and $\gamma\delta$ T cells. *J. Immunol., 155, 2047–2056.*

Muller, W. E., Schroder, H. C., Ushijima, H., Dapper, J. and Bormann, J. (1992) gp120 of HIV-1 induces apoptosis in rat cortical cell cultures: Prevention by memantine. *Eur. J. Pharmacol., 226, 209–214.*

Nawaz, S. and Fennell, R. H. (1994) Apoptosis of bile duct epithelial cells in hepatic allograft rejection. *Histopathology, 25, 137–142.*

Papermaster, D. S. and Nir, I. (1994) Apoptosis in inherited retinal degenerations. In E. Mihich, R. H. Schimke (eds.), *Apoptosis, Pharmacological implications and therapeutic opportunities.*, Plenum Press, New York, pp. 15–30.

Patel, T., Bronk, S. F. and Gores, G. J. (1994) Increases of intracellular magnesium promote glycodeoxycholate-induced apoptosis in rat hepatocytes. *J. Clin. Invest., 94, 2183–2192.*

Portera-Cailliau, C. P., Hedreen, J. C., Price, D. L. and Koliatsos, V. E. (1995) Evidence for apoptotic cell death in Huntington's disease and excitotoxic animal models. *J. Neurosci., 15, 3775–3787.*

Schuler, D. and Szende, B. (1997) Apoptosis and acute lymphocytic leukemia in children. Challenges and opportunities in pediatric oncology. *Annals of the New York Academy of Sciences, 824, 28–37.*

Schulte-Hermann, R., Timmermann-Trosiener, I., Barthel, G., and Bursch, W. (1990) DNA synthesis, apoptosis and phenotype expression as determinants of growth of altered foci in rat liver during phenobarbital promotion. *Cancer Research, 50, 5127–5135.*

Schulte-Hermann, R., Bursch, W. and Grasi-Kraupp, B. (1995) Active cell death (apoptosis) in liver biology and disease. In J. L. Boyer and R. K. Ockner (eds.), *Progress in liver diseases*, Saunders, Philadelphia, pp. 1–35.

Sipos, L., Szegedi, Zs., Fedorcsák, I., Áfra, D. and Szende, B. (1998) Apoptosis and p53 expression in human gliomas. *Pathol. Oncol. Res., 4, 267–270.*

Su, J. H., Anderson, A. J., Cummings, B. J. and Cotman, C. W. (1994) Immunohistochemical evidence for apoptosis in Alzheimer's disease. *Neuroreport, 5, 2529–2533.*

Su, X., Zhou, T., Wu, J., Jope, R. and Mountz, J. D. (1994) Dephosphorylation of a 65 kD protein delivers a signal for Fas-mediated apoptosis. *FASEB J., 8, 218–222.*

Szende, B., Lübben, T., Romics, I. and Vass, L. (1996) Apoptosis in untreated and hormone-treated prostate cancer of various histological types. Pathol. Oncol. Res., 2, 239–241.

Szende, B., Tyihák, E., Trézl, L., Szöke, É., László, I., Kátay, Gy. and Király-Véghely, Zs.(1999) Formaldehyde generators and captures as influencing factors of mitotic and apoptotic processes. Acta Biol. Hung., 49, 323–329.

Szende, B., Romics, I., Torda, I., Bély, M., Szegedi, Zs. and Lovász, S. (1999) Apoptosis, mitosis, p53, bcl2, Ki-67 and clinical outcome in prostate carcinoma treated by androgen ablation. Urol. Int., 63, 115–119.

Tan, E. M. (1994) Autoimmunity in apoptosis. *J. Exp. Med., 179, 1083–1086.*

Tatton, W. G. and Chalmers-Redman, R. M. E. (1996) Modulation of gene expression rather than monoamine oxidase inhibition: (-)-Deprenyl-related compounds in controlling neurodegeneration. *Neurology, 47, S171–183.*

Toronyi, É., Hamar, J., Perner, F. and Szende, B. (1999) Prevention of apoptosis reperfusion renal injury by calcium channel blockers. *Exp. Toxic. Pathol., 51, 209–212.*

Trump, B. F. and Berezesky, I. K. (1998) The reactions of cells to lethal injury: Oncosis and necrosis – the role of calcium. In R. A. Lockshin, Z. Zakeri and J. L. Tilly (eds.), *When cells die. A comprehensive evaluation of apoptosis and programmed cell death.* Wiley-Liss, New York, pp. 57–96.

Umansky, S. R. and Tomei, L. D. (1997) Apoptosis in the heart. In Kaufmann, S. H. (ed.) *Apoptosis. Pharmacological implications and therapeutic opportunities.* Advanced in Pharmacology, Vol. 41. Academic Press, San Diego, pp. 383–407.

Yoshida, H., Sumichika, H., Hamano, S., He, X., Minamishima, Y., Kimura, G. and Nomoto, K. (1995) Induction of apoptosis of T cells by infecting mice with murine cytomegalovirus. *J. Virol., 69, 4769–4775.*

Yoshiyama, Y., Yamada, T., Asanuma, K. and Asaki, T. (1994) Apoptosis related antigen, Le(Y) and nick end labeling are positive in spinal motor neurons in amyotrophic lateral sclerosis. *Acta Neuropathol. (Berl), 88, 207–211.*

Chapter 9

Caspases

The regulators and executioners of the apoptotic process

Pál I. Bauer, Erzsébet Kénesi and András Illés

Contents

Apotosis or programmed cell death is a genetically programmed process that is essential for tissue homeostasis during human development and adult life. Its malfunctions or erroneous regulation may lead to serious pathological conditions like cancer, neurodegenerative diseases, immunological problems etc. and is reviewed elsewhere in this book. In certain cases apoptosis is artificially induced by drugs to remove excess cells (e.g. cancer treatments). Characterised previously only by morphological criteria, the biochemical background of changes occuring during apoptosis (cell blebbing, chromatin condensation, apoptotic body formation, DNA degradation, proteolytic degradation etc.) have been more or less understood in the past ten years, though the fine details of both the apoptotic process and its regulation are still missing (Wyllie, 1997; Schwartz, 1993).

In this review we discuss current knowledge about the caspase system, the structures of its members, the process and regulation of caspase activation and the substrates and inhibitors of caspases (Budihardjo, 1999; Earnshaw, 1999; Fadell, 1999; Nicholson, 1999; Steinnicke, 1999; Ekert, 1999; Villa, 1997; Cohen, 1997 and Porter, 1999).

9.1 Procaspases and the structure of caspases

Procaspases are mostly found in the cytosol of cells (Ceretty, 1992). Their amounts and the rate of their synthesis show cell type dependence. Until now 14 types of procaspases were isolated and described, 12 of them in humans (Thornberry *et al.*, 1998). In addition to the murine types, caspases have been isolated from *Caenorhabditis elegans, Drosophila melanogaster* and from *Xenopus laevis* too (Fraser, 1997). Caspases can be sorted into three subfamilies. Members of the caspase-1 family (caspase-1, -4, -5, -11, -12, -14) are primarily involved in regulation of the inflammation process (Nicholson, 1999). On the other hand, members of the caspase-3 group (caspase-3, -6, -7, -8, -9, -10) are involved in the control and execution of the apoptotic process. The cuckoo's egg is caspase-2 which is involved in the apoptotic process but structurally is more related to the caspase-1 family (Harvey, 1997) and so represents solely the caspase-2 family. The recently isolated caspase-14 is involved in the developmental process. The role of the other caspases is not known at this time. The phylogenetic family tree of caspases is shown in Figure 9.1.

The matured caspases are heterotetramers (homodimers of heterodimers), consisting of two identical large and small subunits (Rotonda, 1996). In the proenzyme the N-terminal part contains a prodomain with different length. In certain procaspases (caspase-2, -8, -9, -10) it also contains specific association domains responsible for their participation in the activation complexes (Cohen, 1997). While the death effector domains (DEDs) serve as anchors to signaling adaptor molecules and are present in caspase-8 and in caspase-10 in duplicates, the caspase recruitment domains (CARDs) are present in caspase-1, -2, -4 and -9 and mediate protein-protein interactions during the caspase activation process (Chou, 1998). To carve out the large and the small subunits from the proforms, specific peptide bonds are split at the carboxyl groups of aspartate residues. Between the two subunits

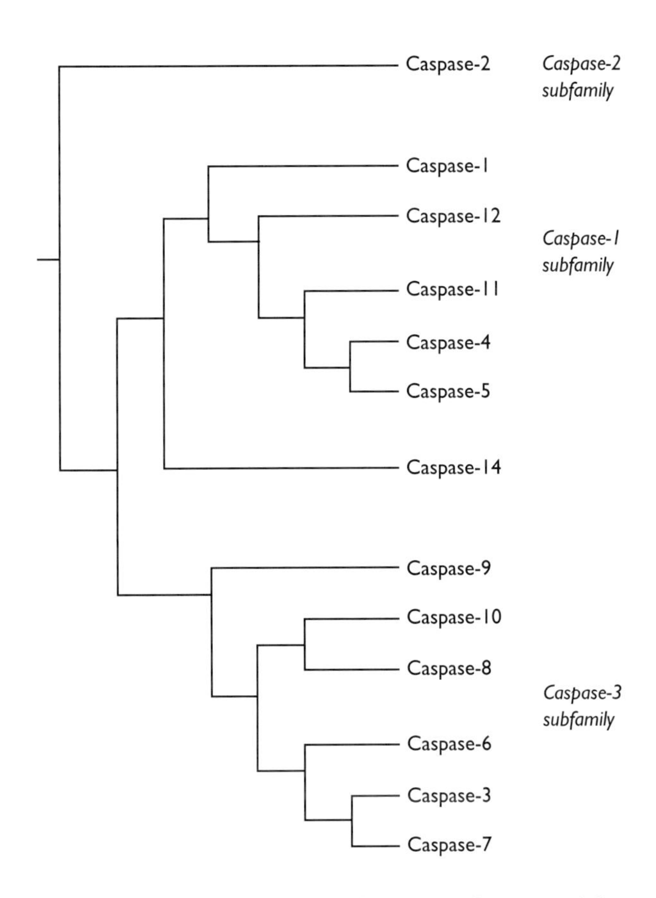

Figure 9.1 The phylogenetic tree of caspases. Constructed from the known DNA sequences.

a linker region is found whose length varies between the different procaspase isotypes. A specific feature of the procaspases is that the zymogen form itself might show a small amount of enzymatic activity. When the proenzymes are clustered into the activation complexes, the first active enzyme is liberated by this remnant enzyme activity yielding to a cascade type of activity increase. The process is experimentally described for procaspase-8 and is called the "close proximity activation model" (Muzio, 1998). The active caspases are also able to activate caspases further downstream. This process is still not known in full detail. Hirata (1998) showed that in Fas-Ligand (FasL) induced apoptosis in Jurkat cells, caspase-8 is activated first which is then able to activate caspase-7 and -3. Caspase-3 in turn can activate caspase-6. Which of the other caspases and how are they activated is not known (Hirata, 1998). On the other hand, when caspase activation proceeds through the cytochrome C driven pathway, the first caspase activated is caspase-9. Six other caspases are activated (caspase-2, -3, -6, -7, -8, -10) while caspase-1, -4 and -5 remain in the precursor form. Immunodepletion of caspase-3, -6 and -7 abolished the activation of the others. Caspase-3 is able to activate caspase-2, -6, -8 and -10 (Slee, 1999). Based on these experiments a branched, hierarchical pathway of caspase activation was constructed. It is worth mentioning that in addition to caspases, Granzyme B is the only protease splitting at aspartate residues and which is able to induce apoptosis *in vitro* when applied together with porine (Yang, 1998). Artificial activation of apoptosis can be achieved when proteases are overexpressed in cells.

As shown for caspase-1 and -8 by analysing X-ray crystallographic data and for caspase-3 and -8 using computer aided estimation processes, the active centres are found on opposite ends of the enzyme dimers and are located on the large subunits. The small and large subunits of a heterodimer derive from different zymogen molecules during the maturation process (Walker, 1994). All caspases contain a conserved QACR(Q/G)G sequence at the active site. At the center of this sequence is the cys285 (caspase-1 based numbering) and the nucleophilicity of the thiol group is very much enhanced by the proton bridge formed with the imidazole ring of his237. The tetrahedral intermediate formed during splitting of a peptide bond is stabilized by gly238. As we already mentioned, caspases are very specific to cut at the carboxy site of aspartates. This aspartate (P1 position) is firmly anchored into the S1 subsite, a pocket structure, and is fastened there by both hydrogen bonding and by ionic forces. It is interesting to note that the P2 and P3 positions of substrates have limited influence on substrate specificity of the enzyme, contrary to the P4 position. While in the caspase-1 family the hydrophobic residue (tyrosine) interacts with the shallow hydrophobic S4 binding pocket, in caspase-3 the S4 subsite is a narrow, deep hydrophilic cave into which the acidic side group of aspartate at the P4 position is firmly attached (Nicholson, 1999; Walker, 1994).

9.2 Activation of procaspases

In the past five years a tremendous amount of knowledge has emerged about the apoptotic process. A plethora of physiological and pathological inducers became known. The majority of the apoptosis-inducing compounds express their effect through the activation of caspases. Two fields have been much explored to date, the death ligand-receptor induced activation of caspases and the cytochrome C mediated process of caspase activation, and these are shown in Figure 9.2.

9.2.1 *Activation through cell surface death receptors*

The cell surface death receptors belong to the TNFR/NGF receptor superfamily and induce apoptosis after binding ligands to their cognate receptors (Schulze-Osthoff, 1998). The cytokine TNF-α induces apoptosis by binding to its receptors TNFR55 or TNFR75. Most probably this is not the physiological function of TNF-α: it is probably involved in the inflammation process. The TNF $-/-$ mouse shows an

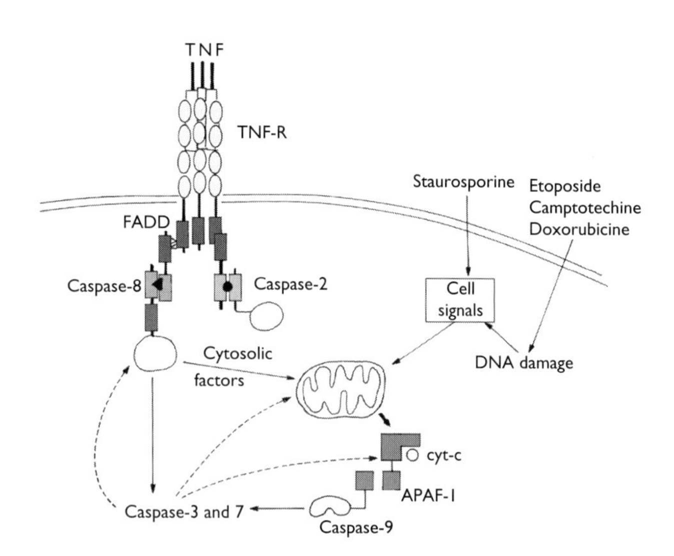

Figure 9.2 The receptor dependent and independent activation of caspases. The activation of caspase-3 is shown both through the receptor dependent (e.g. TNRF/TNRFR) and receptor independent (e.g. staurosporine, camptotechine etc.) pathways.

increased susceptibility towards microbial infections and reduced inflammatory response to bacterial endotoxine (Wallach *et al.*, 1997). On the other hand, the FAS/CD95/APO-1 ligand and its receptor's physiological function is the elimination of activated lymphocytes and the removal of virus-infected or cancerous cells (Ashkenazi, 1999). Recently two other ligand/receptor couples have been isolated. TRAIL-R1/APO-2/DR4 and TRAIL-R2/DR5 (TNF related apoptosis inducing ligand) are expressed mostly in tumour cells and are involved in the apoptotic elimination of cancer cells from the body. TRAIL-R3/DcR1 and TRAIL-R4/DcR2 are decoy receptors with truncated intracellular structures (French, 1999). TRAMP/APO-3 (TNF-receptor related apoptosis mediating protein) is similar to TNFR and is expressed in T lymphocytes. Recently it has also been shown that osteoprotegerin (a protein involved in osteoclast metabolism) belongs to the same family of proteins, binds to the TRAIL receptor and may be involved in the apoptotic process too (Emery, 1998).

The extracellular part of all the receptor proteins contains a cysteine-rich domain, a stretch of 40 amino acids containing six cysteines repeated 2–6 times per molecule (Schulze-Osthoff, 1998). The intracellular C-terminal part of the receptor molecules are unique, but both TNFR1 and FAS contain a segment of 90 amino acids which are similar in sequence and are designated as death domains (DD). Deletion of this part from the TNFR1 molecule abolishes its apoptosis-inducing capacity. When the ligands bind to their receptors they induce oligomerization (trimerization) of the receptors. The ligands recognize their receptor targets through a common structural motif which contains antiparallel beta sheets arranged into jelly-roll structures. While the receptors are type I transmembrane proteins, the ligands are type II. The extracellular trimerization is followed by intracellular decondensation of the receptors. The DD domains not only strengthen this process but also serve as adaptors, to recruit other proteins having similar DD structures. Using various biochemical methods different DD-containing proteins were isolated. Treating cells with APO-1 antibodies resulted in receptor trimerization and the formation of the death inducing signaling complex (DISC) isolated by co-immunoprecipitation (Boldin, 1995). Adapter FADD (FAS associated death domain protein) has been isolated and it contains, besides the DD domain, DED (death effector domain) motifs too. Other components of the isolated DISC complex are the FLASH protein and procaspase-8 (known also as FLICE-1) (Martin, 1998). Binding of procaspase-8 to FADD enhances its proteolytic activity and the further autocatalytic process releases the active caspase-8 molecule, which after dissociating from the DISC complex, can activate procaspases further downstream (Muzio, 1998).

Other DD containing proteins, TNF receptor associated death protein (TRADD) and receptor interacting protein (RIP) were also isolated. TRADD is bound most pronouncedly after TNF/TNFR1 interaction and is able to lure other DD-containing proteins like FADD and RIP. While the former can recruit procaspase-8, the latter is able to activate the transcription factor NF-κB (Cyrns, 1998). It is worth mentioning that in the case of CD95 another adaptor molecule RAIDD (RIP associated/Ced-3 homologous death domain protein) is able to bind the DD parts of the receptor. The C-terminus of RAIDD contains CARD (caspase recruitment domain) motifs and through these domains procaspase-2 (also containing a CARD association part) can be recruited and finally activated (Cyrns, 1998). The TRAIL/TRAILR system functions on the same basis. Although the pattern of adaptors is still questionable its DISC complex contains procaspase-10 instead of procaspase-8 (Pan, 1997).

9.2.2 *Activation through mitochondrial cytochrome C release*

When apoptosis is induced by UV radiation, staurosporine or the DNA damaging cisplatin or doxorubicin, or the topoisomerase poisons camphtotecine or etoposide are applied, caspase activation follows a different route. This activation is not mediated by the surface receptor system. Biochemical analysis has shown that mitochondria play an important role in this process and that besides the regulation of the ATP level mitochondria produce (release) a soluble factor which is able to induce DNA and protein degradation in cell free extracts (Liu, 1996). This protein is cytochrome C. It forms a multimeric complex with APAF-1 (apoptosis activating factor-1) in a process which involves the hydrolysis of dATP or ATP (Bossy-Wetzel, 1998). In the absence of ATP the oligomerization process is aborted. Cytochrome C has to contain the porphyrin ring, but the redox state of the complexed iron is irrelevant. The N-terminal part of the APAF-1 molecule contains a CARD domain which is able to form a complex with the prodomain of procaspase-9. Other procaspases (procaspase-1 and -2) are also able to bind but are not activated. Besides the nucleotide binding motif at the C-terminus of APAF-1, a dozen WD-40 repeats are also found which mediate protein-protein interactions involved in the release of activated caspase-9. The activation of procaspase-9 is autocatalytic. When released from the activation complex, caspase-9 is able to activate downstream procaspase-3 and -7.

Besides cytochrome C the mitochondria also releases AIF (apoptosis inducing factor), a 57kDa protein present in the intermembrane space. AIF is homologous to bacterial NAD-oxidoreductase and after its release it is translocated to the nucleus where it causes chromatin condensation and fragmentation of genomic DNA into 50 kDa size pieces. AIF also mediates the release of cytochrome C from mitochondria. However, its nuclear effect is not mediated by caspases, because caspase inhibitors do not block or suspend its nuclear effects. The fine details of AIF action are not known (Susin, 1999).

9.3 Regulation of caspase activity

To prevent accidental induction of the apoptotic process, the regulation of caspase activity is controlled at different levels and using different strategies. Major efforts are focused on learning the different regulatory steps, because both the activation and the inhibition of apoptosis have tremendous medical relevance. Much was learned from viruses that have different ways to protect themselves and keep the infected cells alive, delaying the apoptotic process.

9.3.1 Regulation of the amount of zymogen at the transcriptional/translational level

Procaspase mRNAs and zymogens were found in most of the cells investigated. Different cell lines contain vastly different amounts of zymogen. While cells of haemopoetic origin contain high levels of procaspase-3, neurons contain only a small amount of procaspase-3 though apoptosis plays an important role in the control of innervated neuron number. On the other hand it has been shown that in apoptotic neurons the procaspase-3 mRNA level is upregulated. To discover the importance of regulation at the synthesis level many more experiments have to be carried out. Generally, we do not even know which sets of procaspases are used in different cell lines both *in vivo* and *in vitro* (Earnshaw, 1999).

The different procaspases/caspases are compartmentalized in certain cell types. While procaspase-9 is cytoplasmic in many tumour cell lines, it has also been found in the intermembrane space of liver mitochondria together with procaspase-2 and these are released into the cytosol when the transition pore of the mitochondria opens by various apoptotic stimuli. Moreover, in neurons, procaspase-9 resides in the mitochondria and is translocated to the nucleus by death-inducing signals (Susin, 1999). Procaspase-3 is accepted to be mostly cytosolic though in certain cell lines it is found in the mitochondria. Contrary to this, the mainly cytoplasmatic procaspase-7 is translocated into mitochondria and microsomes when activated during apoptosis. Immunostaining and substrate availability studies indicate that capase-3 and -1 can be found in the nucleus and it is also known that PARP, topoisomerases, DNA-dependent protein kinase etc. are all substrates of caspase-3 and are located exclusively in the nucleus of cells. Moreover, nuclear localization signals have been found in the prodomain of caspases-1 and -2. The compartmentalization of different caspases might contribute to the fact that in different cell lines different sets of proteins might be digested during the apoptotic process.

9.3.2 Regulation of caspase activation

9.3.2.1 Regulation at the level of the death receptor

To induce trimerization of the ligands besides the natural substrates, either membrane bound or proteolitically liberated, different methods can be applied. Antibodies generated against the receptors can activate the receptor pathway. Artificial switches have been created in the form of crosslinkers with proper spacing and were shown to induce the activation of caspases (MacCorkle, 1998). On the other hand, to avoid accidental activation of apoptosis different processes have been developed by nature. In the case of TNFR1, decoy adapters have been developed to avoid clumping of the cytoplasmic tails of the receptors through the death domains without ligand activation. The silencer of death domain (SODD) repressor protein masks the DDs in the receptor by binding to them thus preventing the erroneous association of DD domains followed by the binding of adapters and procaspases (Yiang, 1999). As we mentioned earlier in the case of TRAIL receptor, decoy receptors (DcR1 and DcR2) are used. Although both have the extracellular cysteine rich domains, their intracellular part is different. DcR1 does not have an intracellular part at all and it is truncated in the case of DcR2, so both are unable to carry out physiological roles. The upregulation of DcR1–2 may protect cells from the cytotoxic action of APO2L/TRAIL. In contrast, the DcR1 level decreases in stimulated T-cells.

In the late lytic phase, herpes virus infected cells were found to be resistant to CD95 (Schulze-Osthoff, 1998). From these cells a protein called v-FLIP (viral FLICE inhibitory protein) was isolated as the causative agent. Later its cellular homologue, c-FLIP, was also found. Its short form contains two N-terminal DED-motifs which are involved in association of caspase-8 and -10. The longer splice variant contains a C-terminal part which resembles caspases but the active site cysteine is replaced by a tyrosine. Both forms of c-FLIP serve as decoys for procaspase binding and activation (Irmler, 1997).

When the constituents of the DISC complex were analyzed they were shown to contain proteins with different levels of phosphorylation. This observation opened up a new pathway for regulation. By metabolic labeling it was shown that after the induction of apoptosis a couple of caspase species displayed phosphorous labeling and simultaneously showed decreased enzymatic activities (Martins, 1998). Cardone (1988) was able to show experimentally that procaspase-9 can be phosphorylated by Akt-kinase, concomittantly loosing its ability to be activated (Cardone, 1998). More experiments are needed to further prove the existence and significance of these pioneering results.

9.3.2.2 Regulation at the level of cytochrome C release. The Bcl-2 system

To avoid unwanted apoptosis the components of apoptosome are carefully separated. The cytoplasmic APAF-1 and procaspase-9 are kept away from cytochrome C present in the intermembraneous space of mitochondria. Cytochrome C is synthesized outside the mitochondria and the apoenzyme is constructed from its components and is imported separately into the mitochondria by cytochrome C haem lyase present in the intermembrane space (Kranz, 1998). Because the holoenzyme itself is not apoptotic, accidental apoptosis is avoided. When mitochondria are triggered by some apoptotic stimuli (decrease of membrane potential, ROS, etc.) cytochrome C is released. The mechanism of release is still not known but three major routes are pictured. Kroemer's group supports the opening of the mitochondrial transition pore (Petit, 1996). The TP is a megachannel located at the contact points, having proteins as porin, the ADP-ATP translocator or the benzodiazepine receptor. Its modus operandi is unknown, but the pore openers are inducers of apoptosis and its inhibitors, such as bongkrekic acid or cyclosporin A, inhibit the onset of apoptosis (Petit, 1996). Another model was introduced by Van der Heiden (1997) who supports the idea that water enters through the TP causing the matrix space to balloon, rupturing the outer membrane so that cytochrome C spills out (Van der Heiden *et al.*, 1997). The fact that no broken mitochondrial membranes were shown in electron microscopic pictures published to date argues against this model. The third model is based on the fact that the antiapoptotic Bcl-xL and the proapototic Bax proteins both form channels in artificial membranes. These channels (for cationic compounds in the case of Bcl-xL and for anionic ones when Bax is applied) might not be involved directly in the release of cytochrome C but may influence the opening of the TP (Antonsson, 1997). It is worth mentioning that Bcl-2 can exert its antiapoptotic action even when cytochrome C is injected into the cytoplasm of the cell. The flagship of this apoptosis regulatory protein family is Bcl-2. Bcl-2 was discovered as an oncogene causing B-cell malignancies and the gene is activated by a 14→18 chromosome translocation. It has been discovered that its oncogenicity is connected to its ability to prolong cell survival (Yang, 1996). Using yeast, two hybrid experiments and co-immunoprecipitation techniques, a large family of proteins became known having either anti-apoptotic (Bcl-2, Bcl-xL, Bcl-w, Mcl-1 and A1) or proapoptotic (Bax, Bad, Bak, Bim, Bid, Bik) properties (Li, 1999). In analysing the structures of these proteins, four homology groups (BH1–4) have been found. The BH1 and BH2 motifs are thought to be responsible for membrane anchoring, and the pore forming ability of the proteins. BH1 and BH3 are involved in protein-protein interactions. BH3 is known as the minimal proapototic domain, because it can be found in all proapototic members of the family, some of them lacking any of the other motifs (Wang, 1998). A rheostate model was first proposed by Korsmeyer (Yang, 1996) based on their associative properties, i.e. that these proteins are able to form homo and hetero dimers. Dominance of the Bax-Bax dimer formation is lethal for the cell, but when Bax-Bax dimer formation is kept under control by heterodimer formation with anti-apoptotic members of the family, the cell survives (Yang *et al.*, 1996). The survival/death decision is based on binding equilibria between the members which is a function of the conditions. Recently the theory has been expanded, and considers as part of the regulatory process any reactions, which either by physical or by chemical means, influence the binding capacity of any of the pro- or antiapoptotic family members (Porter 1999). From the point of view of cell survival, the family of anti- and proapototic proteins of the Bcl-2 family are shown in Figure 9.3.

Proteolysis is a form of regulation. Bcl-2 can be cleaved by caspases and its anti-apoptotic behaviour turns into

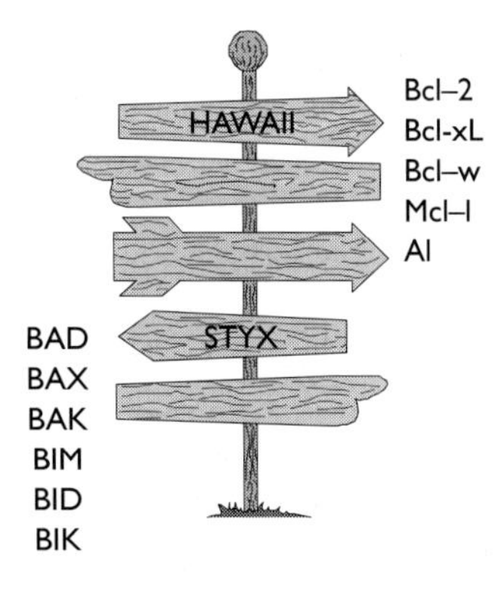

Figure 9.3 The Bcl-2 family of proteins involved in the regulation of the apoptotic process. The anti-apoptotic and proapoptotic members of the Bcl-2 protein family are shown from the point of view of cell survival.

proapoptotic, probably because its membrane anchoring part is removed but it is still able to form dimers in the cytoplasm (Hsu, 1997). The proapoptotic molecule Bid is a substrate for caspase-8, and its cleavage which also happens *in vivo*, enhances cytochrome C release nearly a thousandfold (Luo, 1998).

The potential role of phosphorylation of caspases has already been discussed as part of the regulatory circuit. However, in addition to caspases, other molecules might also be phophorylated. It has been shown that Bcl-2, when phosphorylated, loses its ability to protect cells from apoptosis. Taxol and other microtubule-targeting drugs induce Bcl-2 phosphorylation and, as a consequence, it is released from the microtubules with a concomitant abrogation of its antiapoptotic function (Srivastava, 1999).

In certain cell types receptor connected PI-3-kinase activity is a survival signal for the cell. Through a cascade of phosphorylation steps the proapoptotic Bad is finally phosphorylated. The modified form is sequestered by a 14-3-3 protein into the cytoplasm (Dotter, 1997). When cellular calcium concentration increases it might activate calcineurin (a calcium calmodulin dependent phosphatase) which releases Bad. The free Bad travels to the mitochondrial membrane, heterodimerizes with Bcl-xL and, by shifting the equilibrium, promotes apoptosis (Wang, 1997). Another method of regulation is the membrane anchored Bcl-xL binding the APAF-1-procaspase-9 complex to the mitochondrial membrane. However, when apoptosis is induced, proapoptotic proteins associate with Bcl-xL, releasing the APAF-1-procaspase-9 complex into the cytoplasm which, together with the released cytochrome C, might initiate the caspase activation cascade (Pan, 1998).

The reactions shown here represent only the tip of the iceberg. More has to be learned about the regulation of the apoptotic process, in order to develop drugs which influence only the apoptotic role of these proteins and leaves their other functions in the cell unchanged.

9.4 The inhibitors of caspases

The earliest known inhibitors of caspases are viral proteins (Krajcsi, 1999). The cowpox virus-derived CrmA protein is a serpin-like molecule. Upon cleavage it binds strongly to the active center of the enzyme. It inhibits caspase-1 (K_i = 10 pM), caspase-8 ($K_i = 1\ \mu M$) and caspase-3 ($K_i = 1\ \mu M$). It inhibits apoptosis induced by the receptor pathway (TNF-α, CD95, anoikis etc.) but is ineffective when cell death is induced by ionizing radiation, DNA damaging drugs, or by staurosporin. The role of CrmA could be the prevention of inflammation during viral infections. The p35 protein derives from baculovirus and acts by making the cleaved product dissociating extremely slowly from the surface of the enzyme. It blocks the apoptotic process induced either by TNF and CD95 binding, by DNA damaging drugs, ceramides or growth factor withdrawal (Ekert, 1999).

From homology searches, first in baculovirus and later in mammalian cells, proteins able to inhibit apoptosis were isolated. They all contain specific motifs called BIR (baculovirus IAP repeats) a tandem repeat of 70 amino acids. The copy number varies between 1 and 3. Some of the proteins contain a RING-finger motif at the C-terminus, which is a rather hydrophobic, asymmetrical zinc-finger structure (Miller, 1999). To date five human IAP-s have been isolated: c-IAP-1, c-IAP-2, XIAP, NAIP and survivin (Ekert, 1999). cIAP-1 and cIAP-2 are good active site inhibitors of caspase-3 and -7. They have also been shown to inhibit the activation of procaspase-9 by binding to it. Certain IAP-s have CARD domains and through these also inhibit the activation of procaspase-9 by competing with APAF-1. IAP-s can be considered as safeguards against erroneously activated caspases.

During the intensive research for inhibitors of caspases, inhibitory compounds with small molecular weights were also found. From their sequence it appears that they act as peptide inhibitors against both caspase-1 and caspase-3 families. Active groups were attached to YVAD and DEVD tetrapeptide. While aldehydes (Garcia-Calvo *et al.*, 1998) and beta-homo-aldehydes (Bajusz, 1999) are reversible inhibitors, chloromethyl-, flurometbyl- and acyloxymethyl-ketones and epoxydes are irreversible inhibitors with high specificity (Margolin, 1997).

It is interesting to note that NO is also able to inhibit caspases through S-nitrosylation (Dimmeler, 1997). C-NO compounds (nitroso-benzamide etc.) are also able to inhibit/delay apoptosis induced by different drugs through the inhibition of caspase activities (Mihalik, 1999). Isocyanates were shown to block the enzymic activity of caspase-3 (Petak, 1998). Zinc ions were shown as inhibitors of caspase-3 by reacting with the active center thiol group (Perry, 1997). Phenylarsine oxide also inhibited both apoptosis and caspases.

In the case of small molecular weight inhibitors, extension of their short half-life (NO, isocyanates, etc.) and their delivery to the specific target sites are both subjects of intensive research.

9.5 Caspase substrates

When a computer aided search was carried out using the Swiss Protein Bank looking for proteins containing DEVD, DEXD, DXVD, DXXD sequences, more than two hundred caspase-3 substrate candidates were found (PI Bauer, unpublished). The question arises, are they all important in apoptosis or are most of them only evolutionary bystanders? Since the discovery of PARP splitting only one-third of the proteins have been shown as substrates and are proteolyzed during apoptosis. But returning to the original question, what is the function of their degradation, is it a cause or a causative? The PARP protein clearly represents this dilemma. The first protein was formed to be cut into 85 and 29 kDa fragments, a process which became the hallmark of apoptosis (Kaufmann, 1993). Later it was shown that PARP is not cut in all cases and other proteases are also responsible for its degradation. Then by using PARP −/− cell lines it has been shown that most inducers induce apoptosis with nearly the same concentration both in the +/+ and −/− cells (Leist, 1997). This observation argues against any role of PARP in apoptosis. On the other hand, it has been shown that poly(ADP-ribose)polymerase activity increases in the early stage of apoptosis and the transcription of PARP mRNA is upregulated (Rosenthal, 1997). Are these results meaningless artefacts, or are we looking at the wrong targets? This debate can be applied to almost all of the proteins listed in Table 9.1. The number of caspase substrates analyzed with hair-splitting accuracy, enough to express a final sentence, is very limited. One of them is ICAD which, when degraded, liberates the

Table 9.1 Caspase substrates

Caspase substrate	*Cutting site/sequence*	*Caspase responsible*	*References*
Proteins participating in the apoptotic process			
procaspase-1	D103, D119, D294/WFKD; D316/FEDD		See text in this review
procaspase-2	D152, D316/DQQD; D330/EESD		See text in this review
procaspase-3	D9, D28, D175/IETD		See text in this review
procaspase-4	D104, D270/WEVD; D289/LEED		See text in this review
procaspase-5	D121, D311/WRVD; D330/LEAD		See text in this review
procaspase-6	D23, D179/DIID; D193/TEID		See text in this review
procaspase-7	D23, D198/IQAD		See text in this review
procaspase-8	D210, D218, D374/VETD; D384/LEMD		See text in this review
procaspase-9	D130, D330/DQLD		See text in this review
procaspase-10	D219, D372/IEAD		See text in this review
procaspase-11			See text in this review
procaspase-12			See text in this review
procaspase-13	D289/LEED		See text in this review
procaspase-14			See text in this review
ICAD	DEPD	3	Liu et al., 1998, Cell, 94; 491
Bcl-2	DAGD	?	Hsu et al., 1997, Blood, 89; 4470
Bcl-x_L	HLAD	1, 3	Clem et al., 1998, PNAS US, 95; 554
Bid	LQTD	8	Luo et al., 1998, Cell, 94; 481
Bax	FIQD	?	Sakahira et al., 1998, Nature, 391; 96
$FLIP_L$	LEVD	3, 8, 10	Irmlev et al., 1997, Nature, 388; 190
Proteins involved in signal transduction			
pro-interleukin 1β	FEAD	1	Thornberry et al., 1992, Nature, 356
pro-interleukin 16	SSTD	3	Zhang et al., 1998, JBC, 273; 1144
pro-interleukin 18	LESD	1	Gu et al., 1992, Science, 275; 206
Ras GTP-ase activating protein	DTVD	3	Widmann, 1998, JBC, 273; 7141
D4-inhibitor of GDP dissociation	DELD	3	Na et al., 1996, JBC, 271; 11209
cytosolic phospholipase A2	DELD	3	Wissing et al., 1992, PNAS US, 99
protein phosphatase 2A	DEQD	3	Santoro, 1998, JBC, 273; 13119
Stat1	MELD	3	King, 1998, JBC, 273; 8699
Cbl	?	?	Widmann et al., 1998, JBC, 273
NF κB, p50, p65	?	3	Ravi et al., 1998, Cancer Res., 58
IκB	DRHD	3	Barkett et al., 1997, JBC, 272; 29419
DCC	LSVD	3	Mehlen et al., 1998, Nature, 395; 801
steroid response element	SEPP; DEPD	3	Wang et al., 1996, EMBO J., 15
Raf1	?	?	Widmann et al., 1998, JBC, 273
Akt1	?	?	Widmann et al., 1998, JBC, 273
calpstatin	ALDD; LSSD; ALAD	1, 3, 7	Wang et al., 1998, ABB, 356; 187
PKCδ	DMQD	3	Tewari et al., 1995, JBC, 270; 18738
PKCτ	DEVD	3	Gorhing et al., 1997, EMBO J., 16

(Continued)

Table 9.1 (Continued)

Caspase substrate	*Cutting site/sequence*	*Caspase responsible*	*References*
PKC-related	DITP	3	Erhardt *et al.*, 1997, JBC, 272; 15049
PKN	LGTD	3	Takahasi *et al.*, 1998, PNAS US
Ca/calmodulin	PAPD	3	McGinnis, 1998, JBC, 273; 19993
PITSLRE	YVPD	3	Lahti *et al.*, 1995, Mol. Cell Biol., 15
p21 activated	SHVD	3, 8	Rudel *et al.*, 1997, Science, 276; 1571
MEKK1	DTVD	3	Deak *et al.*, 1998, PNAS US, 95
MST1 and 2	DEMD; DELD	?	Lee *et al.*, 1998, Oncogene, 16; 3029
focal adhesion kinase	DQTD;VSWD	3, 6, 7	Datta, 1997, JBC, 272; 20317
Wee1 kinase	?	3, 7	Zhou, 1998, PNAS US, 95; 6785
Proteins involved in cell cycle regulation and DNA repair			
PARP	DEVD	3, 7, 9	Kaufmann *et al.*, 1993, Cancer Res.
DNA dependent kinase	DEVD	3	C-Rosen *et al.*, 1995, J Exp. Med.
topoisomerase I	DDVD	3	Shao *et al.*, 1997, JBC, 272; 31321
replication factor	DEVD	3	Ubeda *et al.*, 1997, JBC, 272; 19562
Rb	DEAD	3	Janicke *et al.*, EMBO J., 15; 6969
cdc 27	?	3	Zhou *et al.*, 1998, PNAS US, 95
RAD 51	DLVD	3	Flygare *et al.*, 1998, FEBS Lett.
heat shock TF	?	3	Zhang, 1999, Cell Growth Diff., 10
APC*	DNID	3	Browne *et al.*, 1998, Cell Death Diff.
p21 waf1	DHVD	3, 7	Levkan *et al.*, 1998, Mol. Cell, 1; 553
p27 kip1	DPSD	3, 7	Levkan *et al.*, 1998, Mol. Cell, 1; 553
Cytoplasmic proteins			
β-catenin	?	3, 6, 8	Herren *et al.*, 1998, Mol. Biol. Cell
γ-catenin	?	?	Schmeiser *et al.*, 1998, FEBS Lett.
cytokeratin	VEVD	3, 6, 7	Ku *et al.*, 1992, JBC, 272; 33197
β-spectrin	DEVD	3	Janicke *et al.*, 1998, JBC, 273; 15540
fodrin	DETD; DSLD	3	Cyrns *et al.*, 1996, JBC, 271; 31277
gas-2	SRVD	?	Brancolini, 1995, EMBO J., 14; 5179
gelsolin	DQTP	3	Kothakota *et al.*, 1992, Science, 278
huntingtin	DSVD	3, 7	Goldberg, 1992, Nature Genet., 13
DPAP	DSLD	3	Miyashita, 1997, JBC, 272; 29238
cadherin E	?	?	Schmeiser *et al.*, 1999, Cell Death
cadherin P	?	?	Schmeiser *et al.*, 1999, Cell Death
Nuclear and membrane bound proteins			
lamin A	VEVD	6	Oberhammer *et al.*, 1994, J. Cell
lamin B	VEVD	3, 6	Orth *et al.*, 1996, 271; 20977
NuMa	?	3	Weaver *et al.*, 1996, J. Cell Sci., 109
HnRNP C_1 and C_2	?	3	Waterhouse *et al.*, 1996, JBC, 271
SnRNP U1	DGPD	3	C-Rosen *et al.*, 1996, J. Exp. Med.
mdm2	DVPD	3, 6, 7	Erhardt *et al.*, 1997, JBC, 272; 15049
presenilin-1	ARQD	?	Kim *et al.*, 1997, Science, 277; 373
presenilin-2	DSYD	3, 8	Kim *et al.*, 1997, Science, 277; 373
kinectin	?	7	Machleidt *et al.*, 1998, FEBS Lett.
LAP2	?	3	Buendina *et al.*, 1999, J. Cell Sci.
Nup153	?	3	Buendina *et al.*, 1999, J. Cell Sci
AMPA receptor	?	?	Chan, 1999, J. Neuro Sci., 57; 315

nuclear localization signal of CAD protein which is mainly responsible for the ladder type degradation of DNA, another leitmotif of apoptosis (Sakahira, 1998).

We tried to group the caspase split proteins according to their role or physical location. Of course these artificial classifications can never be foolproof, and sometimes a protein belongs to more than one category, in this case grouping according to biochemical role was preferred. Table 9.1 lists the caspase substrate proteins discovered during the past ten years.

References

Antonsson, B., Conti, F., Ciaratta, A.H., Montesuit, S., Lewis, S., Martinou, I., Bernasconi, L., Bernard, A., Hermond, J.J., Mazzei, G., Maundrell, K., Gambale, F., Sadoul, R. and Martinou, J.G. (1997) Inhibition of bax channel forming activity by Bcl-2. *Science, 277, 370–372.*

Ashkenazi, A. and Dixit, V.M. (1999) Apoptosis control by death and decoy receptors. *Current Opinion in Cell Biology 11, 255–260.*

Bajusz, S., Fauszt, I., Nemeth, K., Barabas, E., Juhasz, A., Patthy, M. and Bauer, P.I. (1999) Peptidyl β-homo-aspartals (3-amino-4-carboxybutyraldehydes): new specific inhibitors of caspases. *Biopolymers, 51, 109–118.*

Boldin, M.P., Varfalomeev, E.E., Panzer, Z., Mett, I.L., Carmonis, J.H. and Wallach, D. (1995) A novel protein that interacts with the death domain of Fas/APO-1 contains a sequence motif related to the death domain. *J. Biol. Chem., 270, 7795–7798.*

Bossy-Wetzel, E., Newmeyer, D.D. and Green, F.R. (1998) Mitochondrial cytochrome c. Release in apoptosis occurs upstream of DEVD-specific caspase activation and independently of mitochondrial transmembrane depolarisation. *EMBO J., 17, 37–49.*

Budihardjo, I., Oliver, H., Lutter, M., Luo, X. and Wang, X. (1999) Biochemical pathways of caspase activation during apoptosis. *AR. Cell. Dev. Biol., 68, 251–306.*

Cardone, M.H., Roy, N., Stennicke, H.R., Salvesen, G.S., Franke, F., Stanbridge, E., Friesch, S. and Reed, J.C. (1998) Regulation of cell death protease cspase-9 by phosphorylation. *Science, 282, 1318–1321.*

Chou, J.J., Matsuo, H., Duan, H. and Wagner, G. (1998) Solution structure of the RAIDD/CARD and model for CARD/CARD interaction in caspase-2 and caspase-9 recruitment. *Cell, 94, 171–180.*

Cohen, G.M. (1997) Caspases: the executioners of apoptosis. *Biochem. J., 326, 1–16.*

Corcle, M., Freeman, K. and Spencer, D.M. (1998) Synthetic activation of caspases: artificial death switches. *PNAS Sci. US., 95, 3655–3660.*

Cyrns, V. and Yuan, J. (1998) Protease to die for. *Genes Dev., 12, 157–170.*

Dattta, S.R., Dudek, H., Tao, X., Masters, S., Fu, H., Gotoh, Y. and Greenberg, M.E. (1997) Akt phosphorylation of Bad couples survival signals to the cell-intrinsic death machinery. *Cell, 91, 231–241.*

Dimmeler, S., Rippmann, V., Weiland, U., Haendeler, J. and Zeikar, A.M. (1997) Angiotensin II induces apoptosis of human endothelial cells. Protective effect of nitric oxide. *Circ. Res., 81, 970–976.*

Earnshaw, W.C., Martins, L.M. and Kaufmann, S.H. (1999) Mammalian caspases: structure, activation, substrates and functions during apoptosis. *AR. Biochem., 68, 383–424.*

Ekert, P.G., Silke, J. and Vaux, D.L. (1999) Caspase inhibitors. *Cell Death and Differ., 6, 1081–1086.*

Emery J.G., McDonell, P., Brigham-Bute, M., Deen, K.C., Lyn, S., Silverman, C., *et al.*, (1998) Osteoprotegerin is a receptor for the cytotoxic ligand TRAIL. *J. Biol. Chem., 273, 14363–14367.*

Fadell, B., Zhivotovsky, B. and Orrenius, S. (1999) All along the watchtower: on the regulation of apoptosis regulators. *FASEB J., 13, 1647–1657.*

Fraser, A.G. and Evan, G.I. (1997) Identification of a Drosophila Melanogaster ICE/CED-3 related protease. *EMBO J., 16, 2805–2813.*

French, L.E. and Tschopp, J. (1999) The TRAIL to selective tumor death. *Nature Med., 5, 146–147.*

Garcia-Calvo, M., Peterson, E.P., Leiting, B., Ruel, R., Nicholson, D.W. and Thornberry, N.A. (1998) Inhibition of human caspases by peptide-based and macromolecule inhibitors. *J. Biol. Chem., 273, 32608–32613.*

Harvey, N.L., Butt, A.J. and Kumar, S. (1997) Functional activation of Nedd2/ICH-1 (caspase-2) is an early process in apoptosis. *J. Biol. Chem., 272, 13134–13139.*

Hirata, M., Takahashi, A., Kobayashi, S., Yosehara, S., Sawai, H., Okazaki, T., Yamamoto, K. and Sasada, M. (1998) Caspases are activated in a branched protease cascade and control distinct downstream processes in Fas-induced apoptosis. *J. Exp. Med., 187, 587–600.*

Hsu, C.A., Rishi, J.K., Su, L.K., Gerald, T.M., Dawson, M.I., Schiffer, C., *et al.*, (1997) Retinoic acid induced apoptosis in leukaemia cells through a retinoic acid nuclear receptor independent pathway. *Blood, 89, 4470–4479.*

Irmler, M., Thome, M., Hahne, M., Schneider, P., Hofmann, K., Steiner, V., *et al.*, (1997) Inhibition of death receptor signals by cellular FLIP., *Nature, 388, 190–195.*

Kaufmann, S.H., Desnoyers, S., Ottaviano, Y., Davidson, N.E. and Poirier, G.G. (1993) Specific proteolytic cleavage of PARP an early marker of chemotherapy-induced apoptosis. *Cancer. Res., 53, 3976–3985.*

Krajcsi, P. and Wold, S.M. (1999) Viral proteins that regulate cellular signalling. *J. Gen. Virol., 79, 1323–1335.*

Kranz, R., Lill, R., Goldman, B., Bonnard, G. and Merchard, S. (1998) Molecular mechanisms of cytochrome c biogenesis: three distinct systems. *Mol. Microbiol., 29, 383–396.*

Kumar, S. (1999) Mechanisms mediating caspase activation in cell death. *Cell Death Differ., 6, 1060–1066.*

Leist, M., Single, B., Kunstle, G., Volbracht, C., Hentze, H. and Nicotera, P. (1997) Apoptosis in the absence of PARP. *Biochem. Biophys. Res. Com., 233, 518–522.*

Li, H. and Yuan, J. (1999) Dechiphering the pathways of life and earth. *Current Opinion in Cell Biology, 11, 261–266.*

Liu, X., Kim, C.M., Yang, J., Jemmerson, R. and Wang, X. (1996) Induction the apoptotic program in cell free extracts: requirement for dATP and cytochrome c. *Cell, 86, 147–157.*

Luo, X., Budihardjo, I., Zou, H., Slaughter, C. and Wang, X. (1998) Bid a Bcl-2 interacting protein, mediates cytochrome C release from mitochondria in response to activation of cell surface death receptors. *Cell, 94, 481–490.*

Margolin, N., Raybuck, S.A., Wilson, K.P., Chen, W.Y., Fox, T., Gu, Y. and Livingston, D.J. (1997) Substrate and inhibitor specificity of interleukin-beta-converting enzyme and related caspases. *J. Biol. Chem., 272, 7223–7228.*

Martin, D.A., Siegel, R.M., Zheng, L. and Lenardo, M.J. (1998) Membrane oligomerization and cleavage activates the csapase-8. *J. Biol. Chem., 273, 4345–4349.*

Martins, L.M., Kottke, T.J., Kaufmann, S.H. and Earnshaw, W.C. (1998) Phosphorylated forms of activated caspases are present in the cytosol of HL-60 cells during etoposide induced apoptosis. *Blood, 92, 3042–3049.*

Mihalik, R., Bauer, P.I., Petak, I., Krajcsi, P., Marton, A., Kun, E. and Kopper, L. (1999) Interaction of cytoidal drugs and the inhibition of caspase-3 by 3-nitrosobenzamide. *Int. J. Cancer, 82, 875–879.*

Miller, L.K. (1999) An exegesis of IAP: salvation and surprises from BIR motifs. *TICS., 9, 323–328.*

Muzio, M., Stockwell, B.R., Stennicke, M.R., Salvesen, G.S. and Dixit, V.M. (1998) An induced proximity model for caspse-8 activation. *J. Biol. Chem., 273, 2926–2930.*

Nicholson, D.W. (1999) Caspase structure, proteolytic substrates and function during apoptosis. *Cell Death and Differ., 6, 1028–1042.*

Pan, G., O'Rourke, K. and Dixit, V.M. (1998) Caspase-9, Bcl-xL and Apaf-1 form a tertiary complex. *J. Biol. Chem., 273, 5841–5845.*

Pan, G., O'Rourke, K., Chinnaiyan, A.M., Gentz, R., Ebner, R., Ni, J. and Dixit, V.M. (1997) The receptor for the cytotoxic ligand TRAIL. *Science, 276, 111–113.*

Perry, D.K., Smyth, M.J., Stennicke, M.R., Salvesen, G.S., Duriez, P., Poirier, G.G. and Hannun, Y.A. (1997) Zinc is a potent inhibitor of the apoptotic protease caspase-3. A novel target for zinc in the inhibition of apoptosis. *J. Biol. Chem., 272, 18530–18533.*

Petak, I., Mihalik, R., Bauer, P.I., Suli-Vargha, H., Sebestyen, A. and Kopper, L. (1998) BCNU is a caspase-mediated inhibitor of drug induced apoptosis. *Cancer. Res., 58, 614–618.*

Petit, P.X., Susin, S.A., Zamzami, N., Mignotte, B. and Kroemer, G. (1996) Mitochondria and programmed cell death: back to the future. *FEBS Lett., 963, 7–13.*

Porter, A.G. (1999) Protein translocation in apoptosis. *TICS., 9, 394–401.*

Rosenthal, D.S., Ding, R., Simbulan-Rosenthal, C.M., Vaillancourt, J.P., Nicholson, D.W. and Smulson, M. (1997) Intact cell evidence for the early synthesis and subsequent late apopain mediated suppression of PARP during apoptosis. *Exp. Cell. Res. 232, 313–321.*

Sakahira, H., Enari, M. and Nagata, S. (1998) Cleavage of CAD inhibitor in CAD activation and DNA degradation during apoptosis. *Nature, 391, 96–99.*

Schulze-Osthoff, K., Ferrari, D., Los, M., Wesselberg, S. and Peter, M.E. (1998) Apoptosis signaling by death receptors. *Eur. J. Biochem., 254, 439–459.*

Schwartz, L.M. and Osborne, B.A. (1993) Programmed cell death, apoptosis and killer genes. *Immunology Today, 14, 582–590.*

Slee, E.A., Hartz, M.T., Kluck, R.M., Wolf, B.B., Casiano, C.A., Newmayer, D.D., *et al.*, (1999) Ordering the cythochrome c.- initiated caspase cascade: hierarchical activation of caspases -2, -3, -6, -7, -8 and -10 in a caspase-9 dependent manner. *J. Cell Biol., 144, 281–292.*

Srivastava, R.X., Mi, Q.S., Hardwick, J.M. and Longo, D.L. (1999) Deletion of the loop region of Bcl-2 completely blocks paclitaxel induced apoptosis. *PNAS. Sci. US., 96, 3775–3780.*

Stennicke, H.R. and Salvesen, G.S. (1999) Catalytic properties of caspases. *Cell Death and Differ., 6, 1054–1059.*

Susin S.A., Lorenzo, H.K., Zamzami, N., Marzo, I., Sroe, S.A., Bothers, G.H., *et al.*, (1999) Molecular characterisation of mitochondrial apoptosis-inducing factor. *Nature, 397, 441–446.*

Susin S.A., Lorenzo, M.K., Zamzami, N., Marzo, I., Brenner, C., Larochette, N., *et al.*, (1997) Mitochondrial release of caspase-2 and -9 during the apoptotic process. *J. Exp. Med., 189, 381–394.*

Thornberry, N.A. and Lazebnik, Y. (1998) Caspases: enemies within. *Science, 281, 1312–1316.*

Vander Heiden, M.G., Chandel, N.S., Williamson, E.K., Schumacher, P.T. and Thompson, C.B. (1997) Bcl-xL regulates the membrane potential and volume homeostasis of mitochondria. *Cell, 91, 627–637.*

Villa, P., Kaufmann, S.H. and Earnshaw, W.C. (1997) Caspases and caspase inhibitors. *TIBS., 22, 388–393.*

Walker, N.P., Talenian, R.V., Brady, K.D., Dang, L.C., Bump, N.J., Ferenz, C.R., *et al.* (1994) Crystal structure of the cysteine protease interleukine-1-beta converting enzyme: a $(P_{20}/P_{10})_2$ homodimer. *Cell, 78, 343–352.*

Wallach, D., Boldin, M., Varfolomeev, E., Beyaert, R., Vandenabeele, P. and Fiers, W. (1997) Cell death induction by receptors of the TNF family: towards a molecular understanding. *FEBS Lett., 410, 96–106.*

Wang, G.H., Pathan, N., Ethel, I., Krajewski, S., Yamaguchi, Y., Shibasak, F., *et al.* (1999) Ca^{2+} induced apoptosis through calcineurine dephosphorylation of Bad. *Science, 284, 339–343.*

Wang, K., Gross, A., Waksman, G. and Korsmeyer, S.J. (1998) Mutagenesis of the BH3-domain of Bax identifies residues critical for dimerisation and killing. *Mol. Cel. Biol., 10, 6083–6089.*

Wyllie, A.H. (1997) Apoptosis: an overview. *British Medical Journal, 53, 451–466.*

Yang, E. and Korsmeyer, S.J. (1996) Molecular thanatopsis: a disclosure on the Bcl-2 family and cell death. *Blood, 88, 386–401.*

Yang, X., Stennicke, H.R., Wang, B., Green, D.R., Janicke, R.U., Srinivasan, A., *et al.* (1998) Granzyme B mimics apical caspases. Description of a unified pathway for transactivation of execution caspase-3 and -7. *J. Biol. Chem., 273, 34278–34283.*

Yiang, Y., Woronicz, J.D., Liu, W. and Goeddel, D.V. (1999) Prevention of constitutive TNF receptor signalling by silencer of death domain. *Science, 283, 543–546.*

Chapter 10

The role of drug metabolism enzymes in signal transduction processes

József Mandl

The regulation of cellular functions occurs mainly via chemical signals. The entire signal transduction process is essentially determined by the availability of the signal. Chemical signals can be represented by various molecules, among them by "specialized" signal molecules: e.g. hormones, neurotransmitters. The majority of these signals are small molecules of non-protein nature (molecular weight = 250 ± 200). Several foreign compounds, xenobiotics, can mimic their various biological effects. Both endogenous and exogenous compounds are substrates of drug metabolism enzymes (DME). Drug metabolism (biotransformation) was previously considered to be the main route of detoxification. It is through this pathway that small endogenous compounds and xenobiotics get eliminated. Recently this view has been changed and broadened. Based on the metabolic role of various DME, drug metabolism is also responsible for the conversion of a series of endogenous chemical signals, thus, it can also be called signal metabolism (Figure.10.1) (Nebert, 1994). Therefore, the actual expression of DME, which is dependent on various determinants, such as metabolic state, hormonal state, redox homeostasis etc., affects signal transduction through the supply of chemical signal molecules and also by altering their biological half-life. Vertebrates adapt to adverse chemical environments by upregulating batteries of DME, also called xenobiotic metabolizing enzymes.

Drug metabolism was defined previously as that part of the intermediary metabolism which is not related to the formation of ATP. This was generally accepted in spite of the fact that the various reactions of different phases of drug metabolism are ATP-consuming processes. This definition cannot be maintained anymore. It has turned out that several DME participate in intermediary metabolism pathways and their regulation is in accordance with the control of metabolism especially at the level of gene expression. Thus, there is a strong interrelationship between biotransformation and intermediary metabolism, which influences signal transduction through the metabolism of the signal molecules.

In the course of biotransformation substrates/drugs are converted mainly from lipid soluble compounds into water soluble molecules. The whole process results in the inactivation of the drugs. Biotransformation consists of three phases: first the preparatory phase, after which the product is generally suitable for conjugation. Following the 2nd conjugatory phase, the product can be transported through the (plasma) membrane, which is the 3rd transport phase. Thus, the first and second phase are catalysed by DME and the third is mediated by transporters. The main organ of biotransformation in the organism is the liver. DME are expressed mostly in hepatocytes, however they are also involved in several other cellular functions and in different physiologic processes in the liver and also in extrahepatic tissues. DME are mainly bound to the endoplasmic reticulum membranes within the cells, but some are located in other organelles (e.g. in mitochondrion) (Jeffery, 1993).

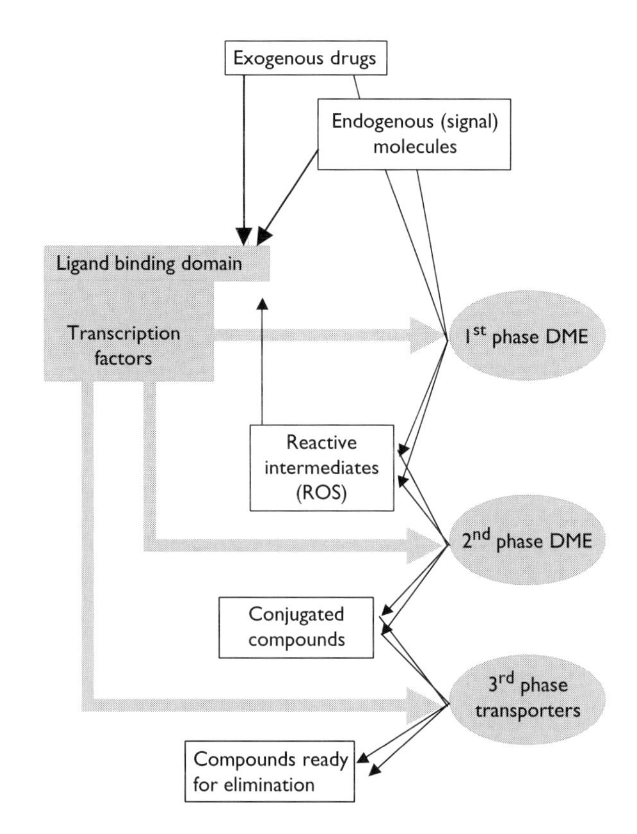

Figure 10.1 Scheme of the role of transcription factors in drug metabolism. ROS: reactive oxygen intermediates.

In the first phase substrates are usually oxygenated, but can also be reduced or hydrolyzed. First phase DMEs are

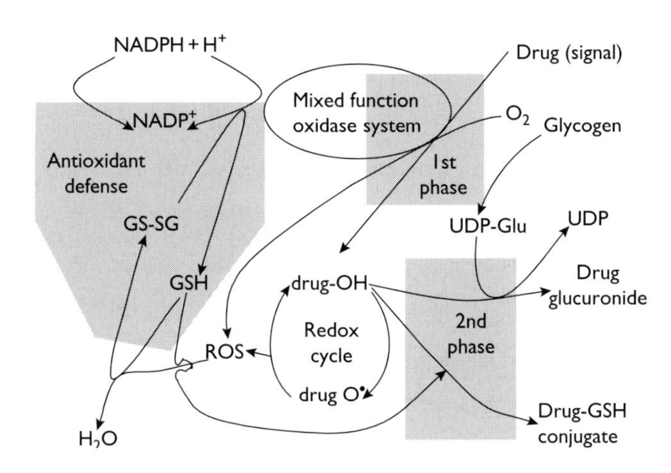

Figure 10.2 Connection between drug metabolism and antioxidant defense. Reactive oxygen species (ROS) formation in biotransformation. GSH: reduced glutathione, GS-SG: oxidized glutathione, UDP-Glu: UDP-glucuronic acid.

mainly mixed function oxygenases (Figure 10.2). Mostly NADPH is required for their action. Thus, as an NADPH-consuming process the 1st preparatory phase of biotransformation influences the reducing equivalent supply especially in hepatocytes. The 1st phase is also an oxygen-consuming process, which can be responsible for up to 30% of the hepatic oxygen consumption. Oxidoreductases (mixed function oxidase system) catalyzing the first phase of drug metabolism belong to the superfamily of cytochrome P 450 enzymes (P450), one of the largest superfamilies of enzyme proteins. There are hundreds of P450 isozymes and several tens of thousands of substrate molecules. The products of the *CYP1-CYP4* P450 gene families participate in human drug metabolism (Coon *et al.*, 1992; Werk-Reichart and Feyereisen, 2000).

In the 2nd, conjugation, phase different DME (gene families) are involved: UDP-glucuronosyl transferases, glutathione transferases, sulfotransferases etc. The bulk of the drugs are inactivated by glucuronidation, catalyzed by different UDP-glucuronosyl transferase isozymes. UDP-glucuronosyl transferases participating in drug metabolism belong to two gene families (*UGT1, UGT2*) (Mackenzie 1995). They need UDP-glucuronic acid for their actions, therefore, glucuronidation is ultimately a glucose- (glycogen) consuming process. Sulfation is also an energy dependent process. Another important type of conjugation is the glutathione conjugation. For conjugation with reduced glutathione, acetyl CoA is also required. As a glutathione-consuming process this method of conjugation may alter the intracellular antioxidant reserve as well. There are also other minor forms of drug conjugation (methylation, amino acid conjugation etc.), which can also be involved in formation or degradation of chemical signals (Mandl *et al.*, 1995).

The "youngest" phase of biotransformation is the 3rd phase. The various families of transporters involved in this phase have been discovered recently. It has also been shown that Dubin Johnson syndrome, a long known form of hyperbilirubinemia can be classified as a 3rd phase disease (the expression of a multidrug resistance protein (MRP2), an ABC transporter responsible for the transport of conjugated bilirubin is altered, underlying this illness) (König *et al.*, 1999).

The majority of drugs are inactivated in the first phase of biotransformation. However, in some cases first phase DME might also cause activation of the drug and consequently the second phase enzymes inactivate these drugs. In the course of the first phase reactive, activated, electrophilic intermediates can be formed. They are both direct and indirect sources of reactive oxygen species (ROS) production and can lead to pathologic processes. Some reactive intermediates can be bound to bases constituting DNA or protein-building amino acids, thus forming adducts. This way cytotoxic, carcinogenic processes can be started depending on the actual activity of the first phase DME. There are some rare exemptions, where drug conjugation leads to activation of the drug. Finally the product of the conjugation is a substrate of a transporter; in this way the conjugated molecule can be eliminated from the cell and ultimately from the organism (Mandl *et al.*, 1995).

DME are inducible. Upregulation of DME batteries is an effective way of adapting to the adverse chemical environment, namely to drug exposure, by decreasing the biological half-life of the drugs or other compounds. The rate of induction depends on the gene families. The inducibility of *CYP1* isozymes can be more than a hundred fold, while UDP-glucuronosyl transferases can usually be induced 2–3 fold. This variable inducibility is related to their physiological role and also their induction state during development. The induction of the DME are examples of a coordinated induction. DME (and sometimes transporters) involved in different biotansformation phases are induced together with intermediary metabolism enzymes (e.g. NADPH generating pentose phosphate pathway enzymes upon addition of phenobarbital, a "classic" DME inducer). The drug dose required for the drug action is also determined by the induction state of the DME. As emphasized earlier, foreign drugs often imitate effects of endogenous compounds; xenogenic molecules are analog of them, as substrates or as ligand agonists or antagonists. Several drugs have been discovered as "pure synthetic" molecules and were later found to be plant mediators or their derivatives (salycilates etc.). The physiologic substrates of several DME are unknown, however, it is supposed that they have a real, original biological function, other than the xenogenic compound metabolism. Thus DME is a site of interaction of endogenic and exogenic metabolism. It is noteworthy to mention that this is not the only interference point.

Drugs exert their pharmacologic effects through receptors (Figure 10.1). These receptors are frequently transcription factors, therefore, drugs are considered as their ligands. Thus, the receptor-transcription factors serve as another point of interference in the metabolism of endogenous and xenogenic compounds. DME batteries are driven by these receptors/transcription factors. This is why drug metabolism is often called effector-ligand metabolism. As part of a (usually negative) feedback, DME influence the concentration of signal molecules. It is well known, that several, mainly first phase DME are essential elements in synthetic pathways of various signal molecules (e.g. steroid hormones, thyroid hormones, eicosanoids). At the same time inactivation of signal molecules are functions of several first and second phase DME (retinoids, steroids, thyroid hormones etc.).

Signal transduction processes mediated by various intracellular receptors are of special interest. There are some (orphan) receptors with no (known) ligands, and some receptors with only xenogenic ligands, but at the same time their biological functions are vital. Their pathology was recognized before their physiologic significance, as their endogenous ligand has not yet been found.

At the cellular level energy transformation needs electron donors and electron acceptors. Electron donors are the various food molecules: glucose, fatty acids etc. The ultimate electron acceptor is oxygen. Intracellular/nuclear receptors, transcription factors can be categorized by different aspects. Nuclear hormone receptors of the steroid/thyroid/retinoid gene superfamily are implicated in energy homeostasis (Mangelsdorf *et al.*, 1995). As an example peroxisome proliferator-activated receptor (PPAR) isoforms with retinoid ligands are involved in metabolic regulation of fatty acid metabolism, e.g. in starvation (Escher and Wahli, 2000). Thyroid, retinoid and steroid hormones regulate many metabolic processes and multiple cross-talk exists among the various pathways under normal and pathologic conditions; they also influence signal transduction routes of other hormones (e.g. insulin). However besides the gene expression of enzymes in intermediary metabolism, many first and second phase DME (products of CYP2, CYP4 gene subfamilies, UGT2 isozymes etc.) are regulated at the gene expression level by these nuclear receptors.

Other transcription factors are also connected to oxygen metabolism and intracellular reactive oxygen species (ROS) formation. ROS is also produced in biotransformation and there is an interrelationship between ROS formation and drug metabolism (Figure 10.2). Recently several transcription factors have been shown to be reactive to redox changes; alterations in intracellular ROS (e.g. NO, hydrogen peroxide, superoxide) levels affect signal transduction pathways. There are many papers on the relation of mechanisms of oxygen sensing and transcription factors (Semenza, 1999a). In addition, there are several reports on DNA binding of these transcription factors to corresponding response elements. The EPRE (electrophile response element, also called ARE, antioxidant response element) was found to be a DNA motif responsible for the upregulation of genes induced by electrophiles. The electrophiles might include various molecules, such as hydrogen peroxide, or some CYP1A1 substrate/product planar aromatic hydrocarbons (PAH, see later). Moreover, it is supposed that some of these transcription factors (e.g. eukaryotic PAS proteins) act as sensors of environmental signals (Gu *et al.*, 2000).

Aryl hydrocarbon (Ah) receptor (AHR) is a good example of one of the many open questions and controversies on the character, inducibility and physiologic role of receptors/transcription factors involved in drug/xenobiotic metabolism and on their participation in signal transduction (Fernandez-Salguero *et al.*, 1995; Rowlands and Gustafsson, 1997). AHR is a transcription factor, and is a bHLH-PAS protein (basic (b), helix-loop-helix (HLH), and PAS are domain structures) (Crews and Fan, 1999). The PAS domain is a multi-functional interaction domain. It is responsible for substrate/ligand binding, therefore it acts as a ligand binding environmental sensor and signal transducer. It forms heterodimers with another bHLH-PAS protein, ARNT (the AHR nuclear translocator). The AHR-ARNT heterodimer then binds to the dioxin response element (DRE, also called XRE xenobiotic response element) and activates transcription of downstream target genes (Hankinson, 1995). Induction of P450 enzymes, such as CYP1A1 and CYP1A2 occurs in response to exposure to PAH or dioxins (Whitlock, 1999). Furthermore, other DME such as CYP1B1, glutathione S transferase Ya subunit and quinone oxidoreductase and a UDP-glucuronosyltransferase (*UGT1a6*) are products of target genes of the so called AHR gene battery (Nebert, 1994). Highly toxic PAHs are widely distributed in the environment and also accumulate as by-products of industrial technologies. They are inducers of these CYP isozymes, and AHR mediates the adaptive response to PAH. The most potent agonists of this signal transduction system are halogenated aromatic compounds, such as dioxins, the most powerful carcinogenic agents. Much is known about the mechanism and recent biological, toxicological importance of this PAH/dioxin signal transduction, but its physiologic agonist and its original physiologic function is unknown. The implication of ARNT and DRE mediated gene expression in various toxic mechanisms and most aspects of AHR mediated toxic response also need further confirmation and experimental support (Gu *et al.*, 2000). At present it is suggested that the AHR signal transduction pathway is an adaptive one, that is especially important in a developing organism due to endogenous toxicants. On the other hand DME, products of *CYP* genes of AHR battery can catalyze the formation of intermediates, which can generate an oxidative stress response (Whitlock, 1999).

ARNT has, in fact, been found to be is identical to HIF-1β, which is involved in a parallel cellular signal transduction pathway. It is needed to form a heterodimer with HIF-1α, essential to hypoxia signal transduction. Thus, ARNT is a common component of (at least) two (but probably more) signal transduction mechanisms (Rowlands and Gustafsson, 1997; Maltepe *et al.*, 1997). The hypoxia response pathway is essential in adaptation to low oxygen tension. It can stimulate systemic (e.g. transcriptional upregulation of the gene encoding erythropoietin stimulating erythropoiesis, or of tyrosine hydroxylase, the rate limiting enzyme of catecholamine synthesis), local (upregulation of transcription of genes encoding various angiogenic growth factors, such as vascular endothelial growth factor (VEGF) etc. in tumors) and cellular (transcriptional activation of genes encoding glycolytic enzymes etc.) responses. To our knowledge no ligand of ARNT has been described yet (Semenza, 1999b).

Genetic polymorphism of DME is a known phenomenon. The polymorphism of certain first phase enzymes (e.g. CYP2D, CYP2C isozymes) and several second phase enzymes (e.g. N-acetyl transferases, UGT1A1) have been investigated and used in planning individual drug therapy. There are major differences among different populations. Necessity of DME screening examinations has been reported for various diseases and for several drug molecules. It might determine the dosage, adverse effects of various drugs and can also cause differences in signal transduction mechanisms (Jeffery, 1993; ISSX Proceedings 1998).

In summary, efficacy and dose of drugs or concentrations of chemical signals can be influenced by dramatic changes of activity and expression of various DME. On the other hand there is also a great variability in the transcription factors mediating the drug effects. Thus signal transducing processes are modified by the actual induction state of DME and by the receptors, which mediate the regulation of their expression.

Acknowledgments

Thanks are due to the Hungarian Academy of Sciences, Ministry of Health, and Ministry of Education for their support.

References

Coon, M.J., Ding, X., Pernecky, S.J., Vaz, A.D.N. (1992) Cytochrome P450: progress and predictions. *FASEB J. 6, 669–73*

Crews, S.T., Fan, C-M. (1999) Remembrance of things PAS: regulation and development by bHLH-PAS proteins. *Curr. Opin. Genet. Dev. 9, 580–87*

Escher, P., Wahli, W. (2000) Peroxisome proliferator-activated receptors: insight into multiple cellular functions. *Mut. Res. 448, 121–38*

Fernandez-Salguero, P., Pineau, T., Hilbert, D.H., McPhail, T., Lee, S.S.T., Kimura, S., Nebert, D.W., Rudikoff, S., Ward, J.M., Gonzales, F.J. (1995) Immune system impairment and hepatic fibrosis in mice lacking the dioxin-binding Ah receptor. *Science 268, 722–26*

Gu, Y.Z., Hogenesch, J.B., Bradfield, C.A. (2000) The PAS superfamily: sensors of environmental and developmental signals. *Annu. Rev. Pharmacol. Toxicol. 40, 519–61*

Hankinson, O. (1995) The aryl hydrocarbon receptor complex. *Annu. Rev. Toxicol. 35, 307–40*

ISSX Proceedings (1998) Vol. 13.

Jeffery, F.C. ed. Human Drug Metabolism. From Molecular Biology to Man CRC Press, Boca Raton, 1993

König, J., Nies, A.T., Cui, Y. Leier, I., Keppler, D. (1999) Conjugate export pumps of the multidrug resistance protein (MRP) family: localisation, substrate specificity and MRP2-mediated drug resistance. *Biochem. Biophys. Acta 1461, 377–394*

Mackenzie, P.I. (1995) The UDP glucuronosyltransferase multigene family. *Rev. Biochem. Toxicol. 11, 29–72*

Maltepe, E., Schmidt, J.V., Baunoch, D., Bradfield, C.A., Simon, M.C. (1997) Abnormal angiogenesis and responses to glucose and oxygen deprivation in mice lacking the protein ARNT. *Nature 386, 403–7*

Mandl, J., Bánhegyi, G., Kalapos, M., Garzó, T. (1995) Increased oxidation and decreased conjugation of drugs in the liver caused by starvation. *Chem. Biol. Interactions 96, 87–101*

Mangelsdorf, D.J., Thummel, C., Beato, M., Herrlich, P., Schütz, G., Umesono, K., Blumberg, B., Kastner, P., Mark, M., Chambon, P., Evans, R.M. (1995) The nuclear receptor superfamily: The second decade. *Cell 83, 835–39*

Nebert, D.W. (1994) Drug metabolizing enzymes in ligand-modulated transcription. *Biochem. Pharmacol. 47, 25–37*

Nebert, D.W., Roe, A,L., Dieter, M.Z., Solis, W.A., Yang, Y., Dalton, T.P. (2000) Role of the aromatic hydrocarbon receptor and (Ah) gene battery in the oxidative stress response, cell cycle control and apoptosis. *Biochem. Pharmacol. 59, 65–85*

Rowlands, C., Gustafsson, J.A. (1997) Aryl hydrocarbon receptor-mediated signal transduction. *Crit. Rev. Toxicol. 27, 109–134*

Semenza, G.L. (1999a) Perspectives of oxygen sensing. *Cell 98, 281–84*

Semenza, G.L. (1999b) Regulation of mammalian homeostasis by hypoxia-inducible factor 1. *Annu. Rev. Cell. Dev. Biol. 15, 551–78*

Werck-Reichhart, D., Feyereisen, R. (2000) Cytochrome P450: a success story. *Genome Biology 1(6): reviews 3003.1–3003.9*

Whitlock, J.P.Jr (1999) Induction of cytochrome P4501A1. *Annu. Rev. Pharmacol. Toxicol. 39, 103–25*

Chapter 11

Recent developments in calcium signaling

Implications for disease and drug development

Greg Monteith

Contents

Until very recently, our knowledge of the way cytosolic intracellular free calcium ($[Ca^{2+}]_{CYT}$) is regulated in mammalian cells was relatively simple. We knew that increases in $[Ca^{2+}]_{CYT}$ could trigger a diverse array of physiological responses such as neurotransmitter and hormone release and muscle contraction, however it is only recently that the true complexity of calcium regulation has been appreciated. In this chapter, recent advances in our understanding of Ca^{2+} regulation will be discussed, particular attention will be paid to the way Ca^{2+} can increase to a greater degree in specific subcellular domains, and the possible significance of this in disease. The current targets for direct modification of Ca^{2+} transport in the cell will be outlined briefly, with those targets currently without selective pharmacological inhibitors highlighted.

11.1 Understanding Ca^{2+} regulation in the good old days – when life and calcium regulation was simpler

The key event in cellular signaling in many cell types is an increase in $[Ca^{2+}]_{CYT}$. Ca^{2+} can act as a cellular signal because, despite high (>1 mM) Ca^{2+} levels in the extracellular space, the level of $[Ca^{2+}]_{CYT}$ in resting cells is maintained at a level of approximately 100 nM. On stimulation of a variety of cells, $[Ca^{2+}]_{CYT}$ increases, this is mediated via the release of Ca^{2+} from intracellular stores (such as the sarcoplasmic/endoplasmic reticulum (S/ER)) and/or the influx of Ca^{2+} across the plasma membrane. These two events are represented in Figure 11.1 and a typical "calcium transient" is shown in Figure 11.2, such Ca^{2+} transients are often

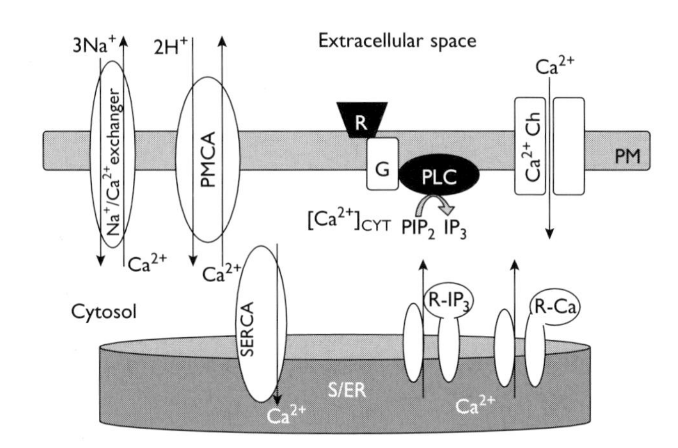

Figure 11.1 Bulk cytosolic free Ca^{2+} homeostasis in mammalian cells. The figure shows a simplified view of intracellular Ca^{2+} regulation, without reference to the possible compartmentalization of the Ca^{2+} signal (shown in Figure 11.3). Resting $[Ca^{2+}]_{CYT}$ levels are maintained by the active transport of Ca^{2+} across the plasma membrane via the PMCA. Ca^{2+} can be elevated via release from the intracellular (S/ER) Ca^{2+} store or via Ca^{2+} influx across the plasma membrane. The subsequent decline in $[Ca^{2+}]_{CYT}$ is mediated via Ca^{2+} efflux (PMCA or the Na^+/Ca^{2+} exchanger) or Ca^{2+} re-sequestration into the S/ER, mediated by SERCA. Abbreviations: Ca^{2+} Ch – plasma membrane Ca^{2+} influx channel; $[Ca^{2+}]_{CYT}$ – cytosolic free Ca^{2+}; G – G protein; IP_3 – ionositol-1,4,5-trisphophate; PIP_2 – phosphatdylinositol-4,5-bisphosphate; PLC – phospholipase C; PM – plasma membrane; PMCA – the plasma membrane Ca^{2+}-ATPase; R – receptor; R-Ca – Ca^{2+} sensitive Ca^{2+} channel; R-IP_3 – IP_3-sensitive Ca^{2+} channel; S/ER – the sarcoplasmic/endoplasmic reticulum; SERCA – the sarcoplasmic/ endoplasmic reticulum Ca^{2+}-ATPase.

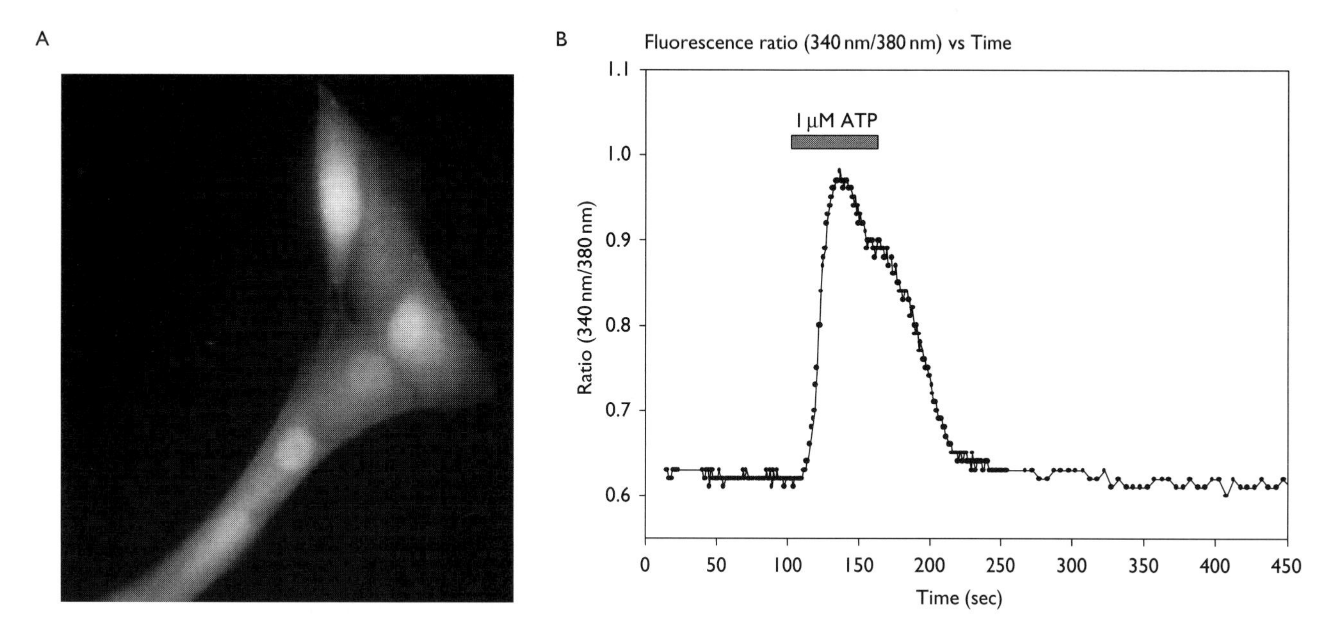

Figure 11.2 A "typical" $[Ca^{2+}]_{CYT}$ transient in a stimulated mammalian cell line loaded with the Ca^{2+}-sensitive probe Fura-2. Panel 2A is an image of a group of human MDA-MB-231 mammary gland epithelial cells loaded with the Ca^{2+}-sensitive probe Fura-2 (4 μM Fura-2 AM, 30 min 37°C), cells were excited at 340 and 380 nm and fluorescence observed at 520 nm, the 340 nm image is shown. Cells were stimulated with 1 μM ATP, a typical Ca^{2+} transient which resulted from ATP stimulation is shown in Panel 2B. An increase in the 340/380 ratio, corresponds to a relative increase in $[Ca^{2+}]_{CYT}$ (unpublished data, H. Rhee and G.R. Monteith, 2000).

quantified using Ca^{2+}-sensitive fluorescent probes and fluorescence digital imaging. The release of Ca^{2+} from intracellular stores is activated by the interaction of inositol 1,4,5 trisphosphate (IP_3) or Ca^{2+}, with channels on the the S/ER. IP_3 is generated in the cytosol via the cleavage of PIP_2 through hormone-induced phospholipase C (PLC) activation (see Figure 11.1). Ca^{2+} can also augment its own signal via a Ca^{2+}-induced Ca^{2+} release mechanism whereby Ca^{2+} activates ryanodine/caffeine sensitive channels on the S/ER. The influx of Ca^{2+} also occurs through channels but these are located on the plasma membrane and can be voltage sensitive or receptor operated, there are numerous classes of Ca^{2+} channels which differ in their electrophysiological properties (e.g. T-type, L-type, Q-type) or agonist sensitivity. The expression of these individual classes of Ca^{2+} channels is cell type dependent which is consistent with their selective roles in cell regulation. For Ca^{2+} to be an effective cellular signal, its signal, of course, must be able to be turned off. This is mediated via the reduction of $[Ca^{2+}]_{CYT}$ after cell stimulation. Although processes such as Ca^{2+} release from stores and Ca^{2+} influx will eventually cease after stimulation, the cell also has pathways to directly remove Ca^{2+} from the cytosol. These processes involve either the re-sequestration of Ca^{2+} into intracellular stores or the extrusion of Ca^{2+} across the plasma membrane. Re-sequestration of Ca^{2+} into the intracellular stores (the S/ER) is mediated via an active transport system – the S/ER Ca^{2+} ATPase (SERCA). The extrusion of Ca^{2+} across the plasma membrane is mediated via either the Na^+/Ca^{2+} exchanger, which is driven by the Na^+ and Ca^{2+} gradients across the plasma membrane and is membrane-potential sensitive, or via a "P-type" ATPase – the plasma membrane Ca^{2+} ATPase (PMCA). The transport systems important in the sequestration of Ca^{2+} into intracellular stores and Ca^{2+} extrusion are illustrated in Figure 11.1 (Carafoli, 1987).

11.2 Understanding Ca^{2+} regulation today – more complex, but more exciting

We now appreciate that Ca^{2+} homeostasis is far more complex than a global increase in intracellular Ca^{2+} followed by a reduction in $[Ca^{2+}]_{CYT}$ mediated via SERCA, PMCA or the Na^+/Ca^{2+} exchanger. Much smaller Ca^{2+} signaling events have now been resolved. They arise from the transient opening of ryanodine-sensitive Ca^{2+} channels of the S/ER.

It is now apparent that the golgi apparatus, the nucleus, mitochondria, the subplasmalemmal space as well as the S/ER can undergo distinct changes in Ca^{2+} during cell stimulation. As will be discussed later in this chapter, it may be the regulation of Ca^{2+} in these compartments which is abnormal in disease states and which is most affected by pharmacological regulators of Ca^{2+} transporters.

Differential regulation of Ca^{2+} in the nucleus has been hypothesized for some time, and recent experiments now support this, nuclear Ca^{2+} may be important in the triggering of apoptosis and the regulation of gene transcription (Gerasimenko, 1996; Rogue and Malviya, 1999). Another organelle that now appears to be involved in Ca^{2+} homeostasis is the golgi. Experiments using an aequorin (a protein capable of Ca^{2+} sensitive luminescence) targeted to the golgi, have demonstrated that the golgi serves as an IP_3-sensitive intracellular Ca^{2+} store (Pinton *et al.*, 1998). Indeed, HeLa cells release Ca^{2+} from the golgi after histamine stimulation, and then re-sequester Ca^{2+} via a mechanism involving a "conventional" thapsigargin-sensitive SERCA and possibly a novel thapsigargin-insensitive Ca^{2+} sequestration mechanism (Pinton *et al.*, 1998), which may be similar to the novel thapsigargin-insensitive pathway in the S/ER (Golovina and Blaustein, 1997).

Mitochondria represent an organelle that has been the subject of some controversy in regard to Ca^{2+} homeostasis. However, it is now clear that Ca^{2+} accumulation by mitochondria plays an important role in the regulation of global and local $[Ca^{2+}]_{CYT}$, and that excessive mitochondrial Ca^{2+} is likely to be a trigger for cell death (Ichas and Mazat, 1998; Hajnoczky *et al.*, 1995; Hajnoczky *et al.*, 1999; Boitier *et al.*, 1999).

The localization of Ca^{2+} signaling is not confined just to organelles and Ca^{2+} sparks. The space just below the plasma membrane has also been proposed as an area where Ca^{2+} changes differently to Ca^{2+} levels in the bulk cytosol. Differential Ca^{2+} regulation of subplasmalemmal Ca^{2+} ($[Ca^{2+}]_{spm}$) has been demonstrated using plasma membrane targeted aequorin and novel lipophillic Ca^{2+} sensitive dyes (Marsault *et al.*, 1997; Etter *et al.*, 1996). The maintenance of this Ca^{2+} gradient beneath the plasma membrane may involve the close association between intracellular Ca^{2+} stores and the localization of Ca^{2+} influx, release and efflux pathways (Blaustein *et al.*, 1998). Perhaps the first and best known example of a non-uniform change in $[Ca^{2+}]_{CYT}$ is a Ca^{2+} wave. In contrast to a homogenous increase in $[Ca^{2+}]_{CYT}$ after physiological stimuli (e.g. an IP_3 generating hormone), an increase in $[Ca^{2+}]_{CYT}$ is first manifest in a confined region of the cell, the increase in Ca^{2+} is then propagated throughout the cell like a wave. This wave can also be propagated across individual cells, such as a cultured monolayer of glia. The rate of the calcium wave varies between cell type, type of stimuli and the degree of stimulus. Table 11.1 shows just a few examples of the differences in the rates of Ca^{2+} waves in different cell types. As well as Ca^{2+} waves, cells may also undergo Ca^{2+} oscillations, the frequency of which is likely to modulate gene transcription (Li *et al.*, 1998). Hence, Ca^{2+} regulation is highly complex, and it is this complexity which allows Ca^{2+} to control a variety of different cellular functions. Figure 11.3 is a simple representation of our current understanding of the intricacies of Ca^{2+} homeostasis in mammalian cells.

Table 11.1 Variability in the rates of Ca^{2+} wave propagation and type of stimuli

Cell type	*Activator*	*Rate of $[Ca^{2+}]_{CYT}$ wave*
Rat lung capillary endothelium	Spontaneous	5 μm/s (18)
Retinal glial cells	ATP	25.3 μm/s (19)
Rat cultured astocytes	Noradrenalin	10.9 μm/s (20)
Blowfly salivary gland	IP_3	13.8 μm/s (21)

11.3 Ca^{2+} and disease – when a precise regulator of cell signaling is disrupted

The vital role of Ca^{2+} in the contraction of cardiac and vascular smooth muscle cells, has made it a focus of research in cardiovascular disease, indeed altered $[Ca^{2+}]_{CYT}$ regulation has been observed in some cells from hypertensive animals (Sada *et al.*, 1990). The effectiveness of L-type Ca^{2+} channel blockers such as nifedipine in the treatment of hypertension is indicative of the importance of Ca^{2+} in the regulation of vascular tone (Hutri-Kahonen *et al.*, 1999). Furthermore, abnormal Ca^{2+} regulation has also been implicated in heart failure and may result in altered regulation of key Ca^{2+} regulated proteins such as calcineurin (Lim and Molkentin, 1999). Excessive levels of $[Ca^{2+}]_{CYT}$ have also been implicated in diseases associated with neuronal cell death such as stroke (ischemia) and Alzheimer's disease (Choi, 1995; Mark *et al.*, 1995). The trigger for neuronal cell death is often excessive $[Ca^{2+}]_{M}$, which opens the mitochondrial transition pore to a level which allows the release of cytochrome C, which subsequently leads to cell death (Ichas and Mazat, 1998). Indeed drugs which block the opening of the mitochondrial transition pore, such as cyclosporin A, are neuroprotective in some *in vitro* models (Khaspekov *et al.*, 1999). Abnormal Ca^{2+} and mitochondria regulation has also been observed in Trisomy 16 mice, a Down's syndrome animal model (Schuchman *et al.*, 1998). Future studies are likely to continue to focus on the regulation of Ca^{2+} in cells from disease states, and may begin to assess Ca^{2+} levels in subcellular compartments such as mitochondria, the E/SR, golgi, nucleus and the subplasmalemmal space. These studies will most likely rely on advanced fluorescence digital imaging techniques (e.g. confocal and multi-photon microscopy) and novel Ca^{2+}-sensitive fluorescence probes. Ca^{2+} sparks are of higher amplitude in cardiac myocytes from hearts with hypertrophy (Shorofsky *et al.*, 1999), and altered amplitude and frequency of fundamental Ca^{2+} release events are likely in other diseases. Some of the diseases where alterations in Ca^{2+} homeostasis have been implicated are listed in Table 11.2.

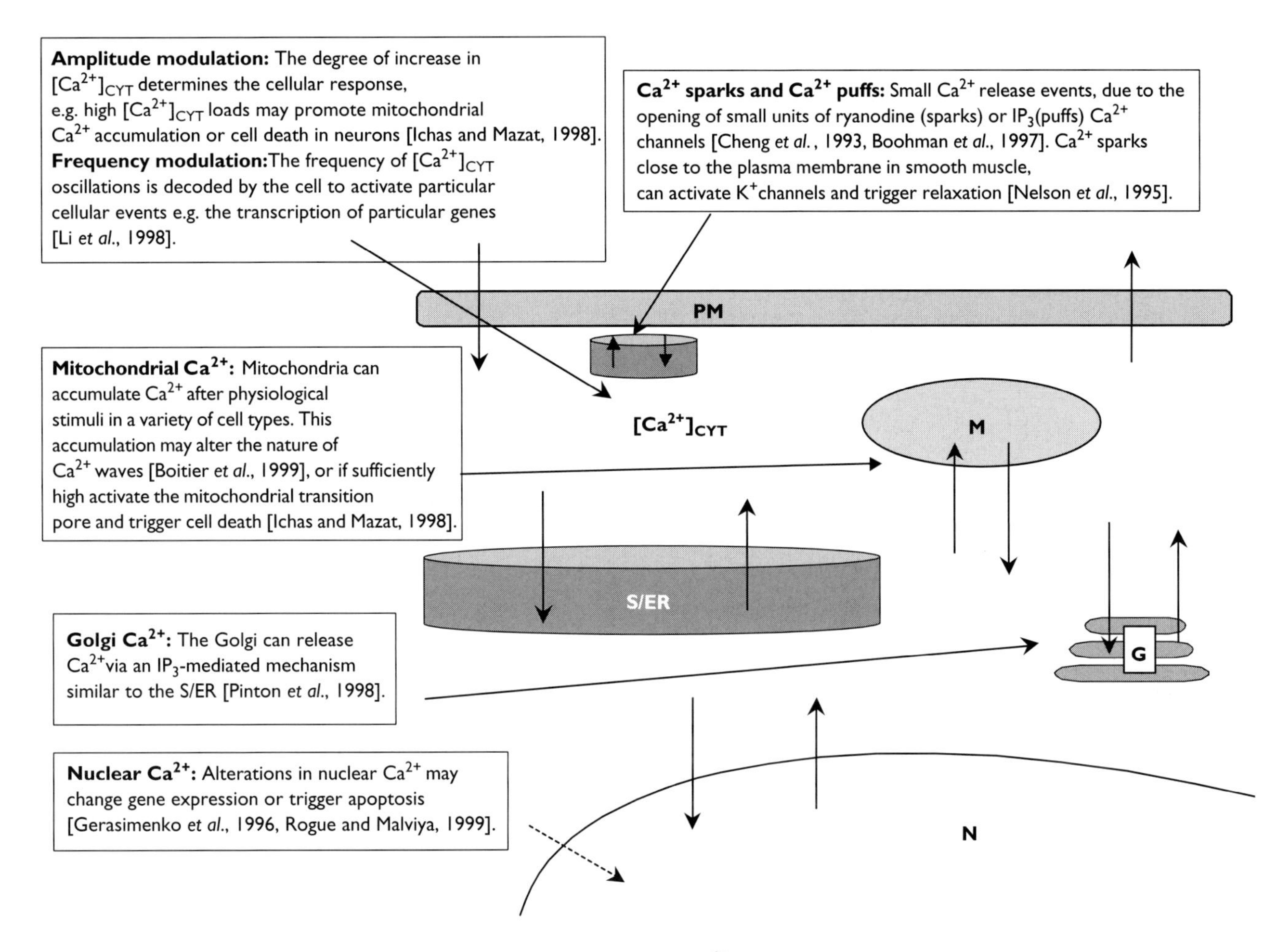

Figure 11.3 Calcium complexities. Illustrates some of the complexities in Ca^{2+} signaling, note the role of non-S/ER intracellular organelles in Ca^{2+} regulation and the localization of the Ca^{2+} signal. Arrows denote Ca^{2+} fluxes across membranes.

11.4 Pharmacological modulators of Ca^{2+} transporters

Perhaps the best known pharmacological regulators of Ca^{2+} are plasma membrane Ca^{2+} channel blockers. Table 11.3, shows just a few examples of plasma membrane Ca^{2+} channel blockers and the classes of plasma membrane Ca^{2+} channels they block. It can be seen that Ca^{2+} channel blockers include toxins isolated from venoms as well as small organic molecules. Perhaps one of the most interesting Ca^{2+} channel blockers is (+)-BAYK8644, since its enantiomer (−)-BAYK8644 is a Ca^{2+} channel opener. This is an encouraging sign that small molecules, which open other Ca^{2+} channel sub-types may be developed in the future. As described above, Ca^{2+} channels are also present on the S/ER. There are two general classes of these Ca^{2+} store channels, those channels sensitive to ryanodine and caffeine and those which are IP_3-sensitive. IP_3-sensitive Ca^{2+} channels can be inhibited by heparin and activated by analogues of IP_3 (Li *et al.*, 1998). As discussed above, the accumulation of Ca^{2+} into the S/ER can be mediated via Ca^{2+} pumps known as SERCA. SERCA is inhibited by a tumor promoting sesquiterpene lactone thapsigargin, as well as cyclopiazonic acid (Golovina and Blaustein, 1997).

Table 11.2 Some of the conditions where deregulation of Ca^{2+} homeostasis may be involved

Condition/disease state
Glutamate-mediated neuronal cell death (Choi, 1987)
Alzheimer's disease (Mark *et al.*, 1995)
Stroke and hypoxic ischemic brain damage (Choi, 1995)
Cardiovascular diseases including hypertension and heart failure (Sada *et al.*, 1990; Hutri-Kahonen *et al.*, 1999; Lim and Molkentin, 1999; Shorofsky *et al.*, 1999)
Complications in diabetes (Voitenko *et al.*, 1999)

Table 11.3 Pharmacological modulators of plasmalemmal Ca^{2+} channels

Class of Ca^{2+} Channel modulator	*Examples*
L-type Ca^{2+} channel blockers	Nimodipine, Nitrendipine, Nicardipine, Nifedipine, Verapamil, Diltiazem, (+)-BAYK8644, Calciseptine (peptide) from Black Mamba.
L-type Ca^{2+} channel openers	(−)-BAYK8644
T-type Ca^{2+} channel blockers	Flunarizine, Amiloride, Mibefradil
N-type Ca^{2+} channel blockers	ω-conotoxin GVIA, ω-conotoxin SVIB, ω-conotoxin MVIIA
P-type Ca^{2+} channel blockers	ω-agatoxin TK (from funnel web spider)
Q-type Ca^{2+} channel blockers	ω-conotoxin MVIIAC

11.5 The Ca^{2+} transporters without selective pharmacological modulators – targets waiting for a drug

Until only very recently there have been no small molecules which could inhibit Ca^{2+} efflux. As described above, Ca^{2+} efflux can be mediated in mammalian cells via the PMCA and the Na^+/Ca^{2+} exchanger. The Na^+/Ca^{2+} exchanger is unique due to its ability to contribute to Ca^{2+} influx via operation in "reverse mode" (Blaustein and Lederer, 1999). Recently, 2-[2-[4-(4-nitrobenzyloxy)phenyl]ethyl]isothiourea methanesulfonate, also known as KB-R7943 has been shown to more effectively inhibit Ca^{2+} influx mediated via the Na^+/Ca^{2+} exchanger, than Ca^{2+} efflux (Iwamoto *et al.*, 1996; Hoyt *et al.*, 1998). This compound has been used to demonstrate that Ca^{2+} inflow, mediated via the Na^+/Ca^{2+} exchanger is one of the mechanisms by which glutamate increases $[Ca^{2+}]_{CYT}$ in cultured neurons (Hoyt *et al.*, 1998). Inhibition of Ca^{2+} influx via this mechanism may be neuroprotective, however, initial studies suggest that Ca^{2+} influx via the Na^+/Ca^{2+} exchanger is not the primary mode of glutamate mediated excitotoxicity in cultured forebrain neurons (Hoyt *et al.*, 1998). An inhibitor of Ca^{2+} efflux mediated via the Na^+/Ca^{2+} exchanger, the "forward mode", may increase myocardial contractility and have therapeutic potential, and may become the focus of rational drug design.

Despite the presence of selective inhibitors of other "P-type" ATPases (e.g. ouabain: Na^+ pump; omeprazole: H^+ pump; thapsigargin: SERCA) there is no selective small molecule inhibitor of the PMCA. Inhibitors which have been used include La^{3+}, eosin and eosin analogs, however these agents are not selective and in the case of La^{3+}, can inhibit Na^+/Ca^{2+} exchange at higher concentrations and, in the case of eosin, inhibit other ATPases (Shimizu *et al.*, 1997). An inhibitor of PMCA would not only have great value in studies examining the role of PMCA in cellular Ca^{2+} homeostasis and physiology, but have therapeutic potential in the modulation of muscle contractility and cellular responsiveness. Furthermore, given that a reduction in PMCA activity and excessive $[Ca^{2+}]_{CYT}$ have both been implicated in some disease states, an activator of PMCA may be a useful agent for the treatment of hypertension and other diseases. Indeed, the therapeutic potential of a PMCA activator, is evident by the recent finding that platelets isolated from patients with hypertension have a greater degree of tyrosine kinase-mediated PMCA phosphorylation, which may contribute to abnormal platelet Ca^{2+} regulation in hypertension (Blankenship *et al.*, 2000).

The recent studies that have characterized Ca^{2+} transport in intracellular organelles, such as the golgi, nucleus and mitochondria may help identify future drug targets. For example, to maintain a Ca^{2+} gradient between the nucleus and the cytosol an active Ca^{2+} transport system to extrude Ca^{2+} from the nucleus is required, and this enzyme may be a novel therapeutic target, inhibition or activation of such a protein is likely to affect gene transcription and proliferation. The golgi's proposed novel active transport system for the sequestration of Ca^{2+} which is distinct from SERCA and PMCA (Pinton *et al.*, 1998), may be another therapeutic target for drug development. Inhibition of golgi Ca^{2+} sequestration may alter cellular responsiveness to IP_3 generating agonists and modulate the cellular organization and modification of other proteins. The proteins associated with mitochondria Ca^{2+} regulation are perhaps the most logical target for the development of therapeutic agents, due to the important role of mitochondrial Ca^{2+} in cell death (Ichas and Mazat, 1998).

11.6 Conclusions

In this chapter, the regulation of intracellular Ca^{2+} has been described. Particular attention has been paid to drawing the reader's attention to recent examples of the intricacies of Ca^{2+} signaling in mammalian cells. Many of the recently characterized pathways involved in cellular Ca^{2+} homeostasis involve differential Ca^{2+} regulation in sub-cellular organelles, in particular the nucleus, mitochondria and golgi. The Ca^{2+} transport pathways in these organelles are potential targets for the development of new therapeutic agents, as are the Ca^{2+} transporters that have been well characterized but are still without small molecules which can modulate them.

References

Blankenship, K., Dawson, C., Aronoff, G., Dean, W. (2000) Tyrosine phosphorylation of human platelet plasma membrane Ca^{2+}-ATPase in hypertension. *Hypertension, 35, 103–107.*

Blaustein, M., Juhaszova, M., Golovina, V. (1998) The cellular mechanism of action of cardiotonic steroids: a new hypothesis. *Clin. Exp. Hypertens., 20, 693–701.*

Blaustein, M., Lederer, W. (1999) Sodium/calcium exchange: its physiological implications. *Physiol. Rev., 79, 763–854.*

Boitier, E., Rea, R., Duchen, M.R. (1999) Mitochondria exert a negative feedback on the propagation of intracellular Ca^{2+} waves in rat cortical astrocytes. *J. Cell Biol., 145, 795–808.*

Bootman, M., Niggli, E., Berridge, M., Lipp, P. (1997) Imaging the hierarchical Ca^{2+} signalling system in HeLa cells. *J. Physiol. (Lond), 499, 307–314.*

Carafoli, E. (1987) Intracellular calcium homeostasis. *Annu. Rev. Biochem., 56, 485–433.*

Cheng, H., Lederer, W., Cannel, M. (1993) Calcium sparks: elementary events underlying excitation-contraction coupling in heart muscle. *Science, 262, 740–744.*

Choi, D. (1987) Ionic dependence of glutamate neurotoxicity. *J. Neurosci., 7, 369–379.*

Choi, D. (1995) Calcium: still center-stage in hypoxic-ischemic neuronal cell death. *Trends in Neuroscience, 18, 58–60.*

Etter, E.F., Minta, A., Poenie, M., Fay, F.S. (1996) Near-membrane [Ca^{2+}] transients resolved using the Ca^{2+} indicator FFP18. *Proc. Natl. Acad. Sci. USA, 93, 5368–5373.*

Gerasimenko, O.V., Gerasimenko, J.V., Tepikin, A.V., Petersen, O.H. (1996) Calcium transport pathways in the nucleus. *Pflugers Arch., 432, 1–6.*

Golovina, V.A., Blaustein, M.P. (1997) Spatially and functionally distinct Ca^{2+} stores in sarcoplasmic and endoplasmic reticulum. *Science, 275, 1643–1648.*

Hajnoczky, G., Hager, R., Thomas, A.P. (1999) Mitochondria suppress local feedback activation of inositol 1,4,5-trisphosphate receptors by Ca^{2+}. *J. Biol. Chem., 274, 14157–14162.*

Hajnoczky, G., Robb-Gaspers, L., Seitz, M., Thomas, A. (1995) Decoding of cytosolic calcium oscillations in the mitochondria. *Cell, 82, 415–424.*

Hoyt, K., Arden, S., Aizenman, E., Reynolds, I. (1998) Reverse Na^{+}/Ca^{2+} exchange contributes to glutamate-induced intracellular Ca^{2+} concentration increases in cultured rat forebrain neurons. *Mol. Pharmacol., 53, 742–749.*

Hutri-Kahonen, N., Kahonen, M., Wu, X., Sand, J., Nordback, I., Taurio, J., Porsti, I. (1999) Control of vascular tone in isolated mesenteric arterial segments from hypertensive patients. *Br. J. Pharmacol., 127, 1735–1743.*

Ichas, F., Mazat, J-P. (1998) From calcium signaling to cell death: two conformations for the mitochondrial permeability transition pore. Switching from low- to high-conductance state. *Biochim. Biophys. Acta, 1366, 33–50.*

Iwamoto, T., Watano, T., Shigekawa, M. (1996) A novel isothiourea derivative selectively inhibits the reverse mode of Na^{+}/Ca^{2+} exchange in cells expressing NCX1. *J. Biol. Chem., 271, 22391–22397.*

Khaspekov, L., Friberg, H., Halestrap, A., Viktorov, I., Wieloch, T. (1999) Cyclosporin A and its nonimmunosuppressive analogue N-Me-Val-4-cyclosporin A mitigate glucose/oxygen deprivation-induced damage to rat cultured hippocampal neurons. *Eur. J. Neurosci., 11, 3194–3198.*

Li, W-H., Llopis, J., Whitney, M., Zlokarnik, G., Tsien, R.Y. (1998) Cell-permeant caged $InsP_3$ ester shows that Ca^{2+} spike frequency can optimize gene expression. *Nature, 392, 936–941.*

Lim, H., Molkentin, J. (1999) Calcineurin and human heart failure. *Nat. Med., 5, 246–247.*

Lipp, P., Niggli, E. (1998) Fundamental calcium release events revealed by two-photon excitation photolysis of caged calcium in guinea-pig cardiac myocytes. *J. Physiol. (Lond), 508.3, 801–809.*

Mark, R., Hensley, K., Butterfield, D., Mattson, M. (1995) Amyloid beta-peptide impairs ion-motive ATPase activities: evidence for a role in loss of neuronal Ca^{2+} homeostasis and cell death. *J. Neurosci., 15, 6239–6249.*

Marsault, R., Murgia, M., Pozzan, T., Rizzuto, R. (1997) Domains of high Ca^{2+} beneath the plasma membrane of living A7r5 cells. *EMBO J., 16, 1575–1581.*

Nelson, M.T., Cheng, H., Robart, M., Santana, L., Bonev, A., Knot, H., Lederer, W. (1995) Relaxation of arterial smooth muscle by calcium sparks. *Science, 270, 633–637.*

Newman, E.A., Zahs, K.R. (1997) Calcium waves in retinal glial cells. *Science, 275, 844–847.*

Pinton, P., Pozzan, T., Rizzuto, R. (1998) The Golgi apparatus is an inositiol 1,4,5-trisphosphate-sensitive Ca^{2+} store, with functional properties distinct from those of the endoplasmic reticulum. *EMBO J., 17, 5298–5308.*

Rogue, P.J., Malviya, A.N. (1999) Calcium signals in the cell nucleus. *EMBO J., 18, 5147–5152.*

Sada, T., Koike, H., Ikeda, M., Sato, K., Ozaki, H., Karaki, H. (1990) Cytosolic free calcium of aorta in hypertensive rats. Chronic inhibition of angiotensin converting enzyme. *Hypertension, 16, 245–251.*

Schuchmann, S., Muller, W., Heinemann, U. (1998) Altered Ca^{2+} signaling and mitochondrial deficiencies in hippocampal neurons of trisomy 16 mice: a model of Down's syndrome. *J. Neurosci., 18, 7216–7231.*

Shimizu, H., Borin, M.L., Blaustein, M.P. (1997) Use of La^{3+} to distinguish activity of the plasmalemmal Ca^{2+} pump from Na^{+}/Ca^{2+} exchange in arterial myocytes. *Cell Calcium, 21, 31–41.*

Shorofsky, S.R., Aggarwal, R., Corretti, M., Baffa, J., Strum, J., Al-Seikhan, B., Kobayashi, Y., Jones, L., Wier, W., Balke, C. (1999) Cellular mechanisms of altered contractility in the hypertrophied heart: big hearts, big sparks. *Circ. Res., 84, 424–34.*

Voitenko, N., Kostyuk, E., Kruglikov, I., Kostyuk, P. (1999) Changes in calcium signaling in dorsal horn neurons in rats with streptozoticin-induced diabetes. *Neuroscience, 94, 887–890.*

Yagodin, S., Holtzclaw, L., Russell, J. (1995) Subcellular calcium oscillators and calcium influx support agonist-induced calcium waves in cultured astrocytes. *Mol. Cell. Biochem., 159/150, 137–144.*

Ying, X.Y., Minamiya, Y., Fu, C.Z., Bhattacharya, J. (1996) Ca^{2+} waves in lung capillary endothelium. *Circ. Res., 79, 898–908.*

Zimmerman, B., Walz, B. (1999) The mechanism mediating regenerative intercellular Ca^{2+} waves in the blowfly salivary gland. *Embo J., 18, 3222–3231.*

Chapter 12

T-cell receptors and coreceptors; cytokine receptors

András Falus

Contents

12.1 Types of T-cell receptor (TCR)

T lymphocytes express antigen specific receptors (T-cell receptors-**TCR**). These $\alpha\beta$ or $\gamma\delta$ TCRs are generated by gene rearrangement and their diversity is further increased by other genetic modifications. T-cells with $\alpha\beta$ TCRs recognize antigenic peptides presented together with products of the major histocompatibility complex (MHC), while $\gamma\delta$ T-cells see conformation antigens reminding to B-cell recognition pattern.

12.2 The structure of the α-, β-, γ- and δ-chains

All four chains have a domain structure and belong to the immunoglobulin supergene family. The $\alpha\beta$ TCRs (Figure 12.1) and $\gamma\delta$ TCRs are **heterodimers linked covalently by S-S bonds**. (Non-covalently associated forms of $\gamma\delta$ TCRs also occur.) It is possible that not one, but two neighbouring $\alpha\beta$ heterodimers form the functionally active TCR on the surface of the T-cell. The constant part of the TCR-chains in the carboxyl terminal is encoded by C-genes, while their variable part is coded by the variable genes V, D, J (β and α) and V, J (α and γ). The homology between the V, C parts of TCR as well as the domains of the immunoglobulin H-chain is significant. The homology is especially considerable in the regions around the disulphide bonds in the chain, that is in the regions around the cystein amino acids. The β-flat structure of the N-terminals is also similar to the variable parts of immunoglobulin.

Without the carbohydrate content the size of the TCRs (α: 40–50 kDa, β: 40–45 kDa, γ: 36–55 kDa, δ: 40–60 kDa) and their structure is very similar. It is interesting that in every case there are two polypeptide pairs in the heterodimer, where one of them contains a D-region and the other

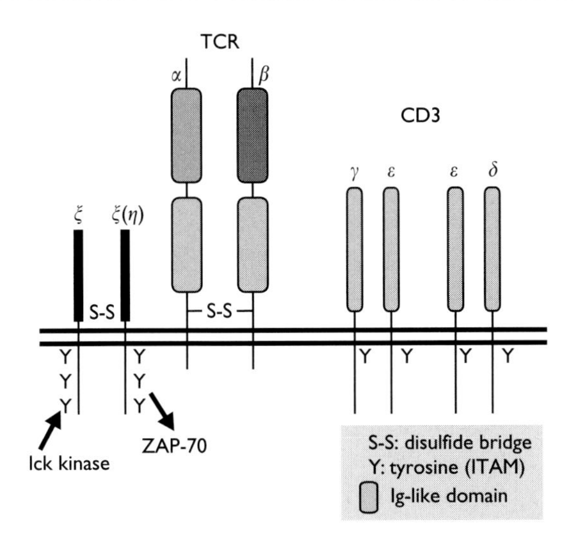

Figure 12.1 The structure of the T-cell receptor (TCR) and CD3. The TCR is a heterodimer of an α- and a β-chain (having a domain structure and belonging to the immunoglobulin super gene family) bound together by a disulphide bond. In CD3 there is a ζ–ζ dimer linked together by a disulphide bond beyond the 1-1 γ-, δ- and 2 ε-chains. (In the minority of T-cells there is a ζ–η dimer in instead of the ζ–ζ dimer.) We can find phosphorylable serine on the γ- and the δ-chain and tyrosine amino acid on the ζ-chain. The complex binds to a lck-type tyrosine kinase intracellularly and its task is to activate a kinase (ZAP-70) linked on the tyrosine-kinases (ITAMs).

one does not. Each TCR-chain is glycosylated, much more than the chains of immunoglobulins. Their other common characteristic is that all four TCR-chains contain a lysine and a valine (plus an arginine in the α- and β-chains) on the rather conservatively structured transmembrane part. The amino acid(s) of positive charge presumably play a role in binding the CD3-chains of negative charge which contain an asparagine acid. There is a "hinge-region-like" (H) structural element in all four chains. The very short cytoplasmic tail (5–12 amino acids) is also a common peculiarity (von Boehmer, 1997).

12.2.1 The chains of the CD3-complex

The very short cytoplasmic tail of the $\alpha\beta$ and $\gamma\delta$ TCR-chains already suggests that intracellular signal transduction can only happen with the help of other molecules. The general characteristic of the **TCR-complex is that it is "anchored" to the plasma membrane of the T-cells by a molecule-complex called CD3**. The chains of the CD3-complex (γ, δ, ε, ζ and η) occur in the stochiometric rate characteristic of the T-cell's developmental state. (The γ- and δ-chains of CD3 are only "namesakes" of the TCR's $\gamma\delta$-chains.) The γ- and δ-CD30-chains are glycosylated, the other four are not. The structure of the γ-, δ- and

Table 12.1 Molecular features of chains of CD3 complex

CD3 chain	*Function*	*kDa*	*Phosphorylation*	*Multimeric form*
γ	signal transduction	25–28	P-Ser	–
δ	signal transduction	20	P-Ser	–
ε	signal transduction	20	P-Tyr	–
ζ	signal transduction	16	P-Tyr	$\zeta 2$*
η	signal transduction	21 (?)	P-Tyr	$\zeta\eta$**

Notes
* 90% of TCRs contain CD3 $\zeta\zeta$ dimer.
** 10% of TCR contain CD3 $\zeta\eta$, in a part of gut associated T cells γ chain of FcεRI occurs.

ε-chains – presumably formed by gene-duplication – shows great homology. Extracellularly each contains an immunoglobulin domain-like element (they are members of the immunoglobulin supergene-family). Their gene, in man, is located on chromosome 11.

The asparagine acid, ensuring the negative charge, can be found on the cytoplasm part of all three. They link to the positively charged amino acid of the TCR-chains. The ζ- and η-chains have different structures, but they are homologous with each other. Their extracellular (9 amino acids) and transmembrane segments – containing the asparagine acid – are totally identical, however their long (113 and 155 amino acids) cytoplasmic parts differ from each other.

The ζ-chain is special in that it is detectable on other cells and receptors (e.g. in the Fcγ receptors of the NK-cells) as well. There is no ζ-chain among the molecules linking to the surface immunoglobulin molecules in the antigen-receptor complex of B-cells. Moreover, the γ-polypeptides of FcεRI are located in the CD3-complex of the special T-cells (for instance in intraepithelial lymphocytes, IEL) found in the intestines, instead of the $\zeta\eta$-dimer.

The six components of the CD3-complex surround the $\alpha\beta$-chain as a ring. Biophysical measurements show dimers of the $\alpha\beta$-chains of TCR occur on T-cells, and decamer-like (2α-, 2β- and 6 CD3-chains) complexes form a functional unit. Intracellularly, several adapter-proteins link to this complex directly. Table 12.1 presents a few biochemical and functional data of the CD3-complex's chains.

12.3 T-cell receptor (TCR) genes

The basic principle that one gene codes one polypeptide chain has recently required amendment, since it became evident in the middle of the sixties that only the variable part of the immunoglobulin's heavy and light chains show significant diversity. The segments determining the isotype are much more homogenous. Therefore, several genes must code one immunoglobulin-chain. Later, direct proof was found that, at the DNA-level, the matured T- (and B-) cells differ from the cells of, embryonic tissue and from other cells which do not produce immunoglobulins. In other words, in the course of the T- (and B-) cells' ontogenesis, the embryonic (germ-line) DNA goes through somatic recombination, known as **rearrangement**. This way the remote genes can get close and functional immunoglobulin chains can be synthesized. Recently, similar proof has come to light regarding the TCR genes and their products (von Boehmer *et al.*, 1998).

12.3.1 The genomic organization of TCR genes

In humans the α-chain and δ-chain gene is located on chromosome 14 (Figure 12.2), while the gene of the β- and γ-chain are found on chromosome 7. Similarly to immunoglobulin genes, every TCR gene is made of V (variable), J (joining) and C (constant) regions. In addition, the β-chain gene also contains D (diversity) gene-segments. On the gene-segment of the β-chain we can find two constant genes (Cβ1, Cβ2), each having four exons. At the 5′ end of constant β-genes there are 6–7 J and 1–1 D gene-segments. Compared to these, the 70–100 Vβ gene-segments, which can be ranged in several families according to their nucleotide-homologies – are found in 5′ orientation, although there have been cases when a Vβ gene was found in 3′ direction from the Cβ genes.

On the α-chain gene-segment there is single C gene made of four exons, and 60–90 J-sections have been described in the 5′ direction from it. The 50–100 Vα gene-segments found in the 5′ direction can also be found in more (at least 10) families.

A special insertion can be observed concerning δ-chain genes, since these are found inside the gene region of the α-chain between the Vα and Jα regions of the δ-chain gene-segment. The special arrangement is conserved phylogenetically; there is a similar genomic arrangement in mice. Four Vδ, 3 J, 2 D and 1 C segment can be found on the δ-chain gene-segment. The structure of the gene-segment coding the γ-chain is similar to the β gene-region. Two J segments were detected next to the Vγ segment; they contain 5 J and 2 C segments altogether. As opposed to what was observed at Cα and Cβ genes, the characteristic of Cγ genes (followed by functional results) is that the two Cγ genes differ significantly from one another.

12.3.2 The rearrangement of TCR genes

As with the expression of immunoglobulin genes, somatic rearrangement is a precondition for the expression of TCR genes (Figure 12.3). First the β gene-segment is rearranged. As the first step a Dβ- and a Jβ-segment move next to each other. Then the DJ-segment aligns beside a Vβ. At this point the primary mRNA, which includes another intron between the VDJ- and the C-segments, is transcribed. In the case of the β-chain, it is purely chance which Cβ ends up in the final mRNA and it is of no functional consequence either.

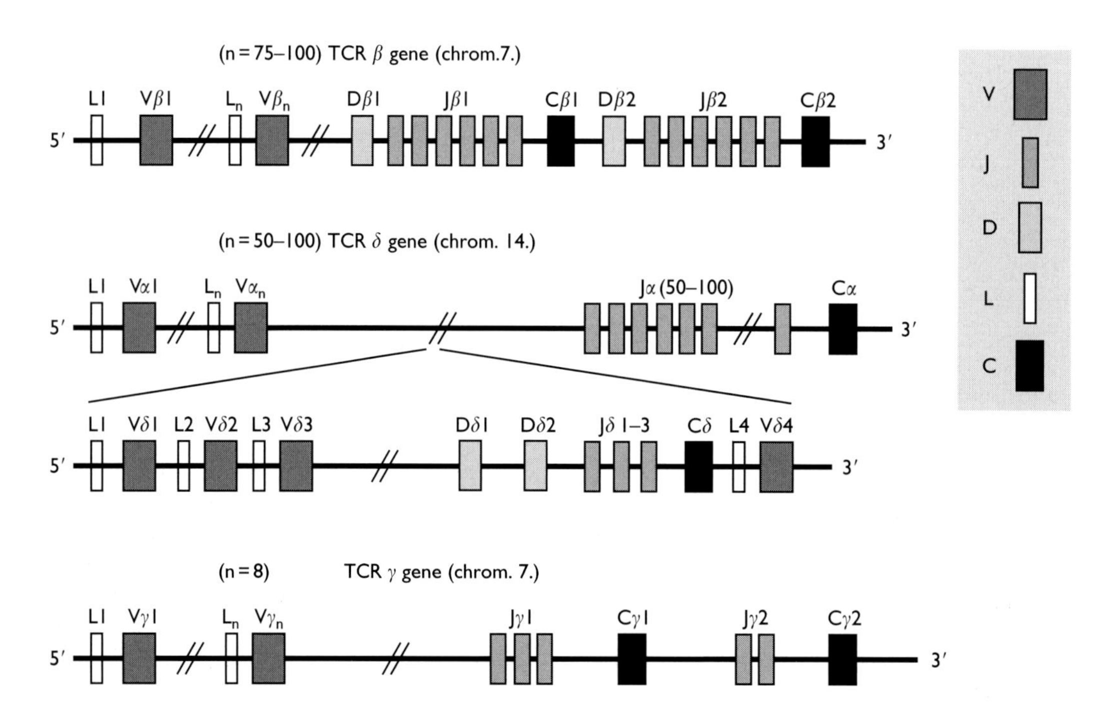

Figure 12.2 The genomic organization of T-cell receptor (TCR) genes (α, β, γ, δ). The signal (L), variable (V), joining (J), diversity (D) and constant (C) gene-segments.

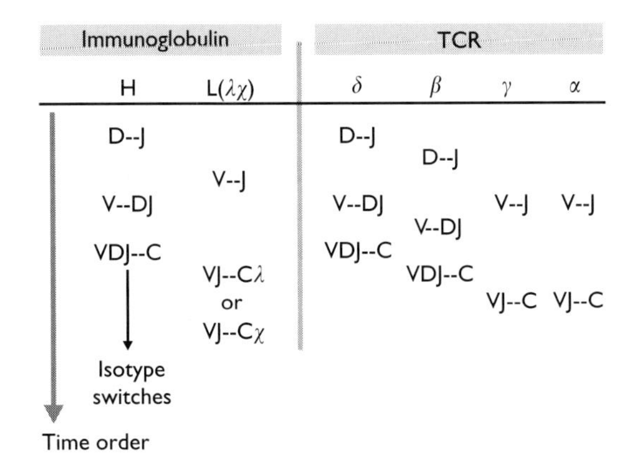

Figure 12.3 The rearrangement order of TCR genes.

Recently, the possibility of the pre-TRC's production and its biological role has been investigated. The rearrangement of the α-chain's gene is started and stimulated by the functional rearrangement of the β-chain. The rearrangement takes place in a similar manner to the β-gene, the difference is that there is no D-segment here. Certain results indicate multiple α-chain rearrangement, which means further variation.A number of **enhancer elements** have been found in both gene-segments. Some of them have negative ("silencer") effects and display these effects only in non T- or $\gamma\delta$ T-cells.

In both the α- and the β-chain of the TCR, the phenomenon called allelic exclusion – which inhibits the second chromosome rearrangement – is effective. Thus, in the T-cell only one α- and one β-chain is rearranged to form a functional TCR. If the rearrangement of one of the chains is not functional, the rearrangement of the appropriate segment may take place on the other chromosome. The rearrangement of the γ- and the δ-chains arises likewise. According to some opinions, the very first motif is the rearrangement of the δ-genes. It is an interesting phenomenon that, although the much fewer germ-line genes in these two chains result in less combination, yet, because of the extremely high linkage diversity and numerous inaccurate, but successful rearrangements, theoretically more $\gamma\delta$-TCR molecules can be formed than $\alpha\beta$-TCR or immunoglobulin (Sleckman *et al.*, 1998). The rearrangement of TCR-genes begins in the bone marrow and ends in the thymus.

12.3.3 RAG-1, RAG-2 recombinases

Molecularly the TCR gene rearrangement takes place by cutting out the DNA-segments between the neighboring gene-segments and by ligating the elements that get into close proximity. The process is catalyzed by recombinase enzymes specific to lymphoid lines. These enzymes – expressed both in the B- and T-precursors (RAG-1 and RAG-2) – are cell-specific, meaning that they are only expressed in lymphoid cells, presumably they are the same enzymes in B and T lymphocytes, and they recognize specific nucleotide-segments. These greatly conserved target segments can be detected on the 3′ end of the V genes, on the 5′ ends of the J genes and on both ends of the D genes. The nucleotide-motifs recognized by the recombinases are made of 7 or 9 bases. These are separated by a (non-conserved) "spacer" part made of 12 or 23 bases (Figure 12.4). The enzyme splits on the bottom of the noose, and as a result of the ligation, the V and J genes get close to each other in the primary structure as well. Similar processes happen with the genes of the H-chain and with the genes of the TCR-chain too.

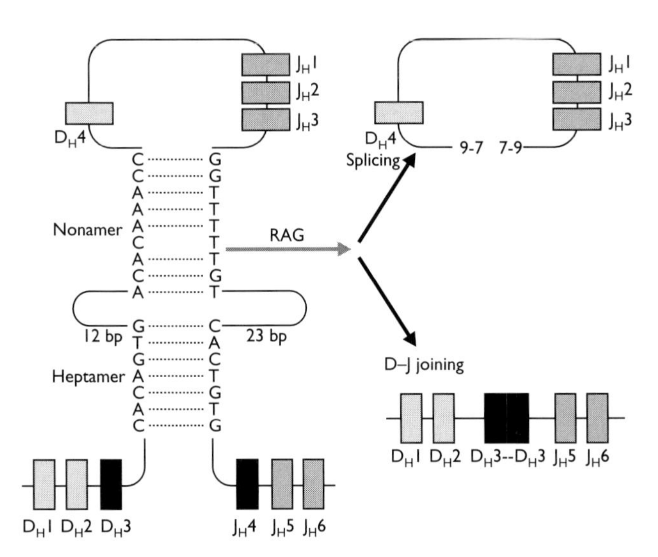

Figure 12.4 Gene-rearrangement signal.The rearrangement signal-sequence is made of a hepta- and nona-nucleotide of conserved structure. There are 12 or 23 (not conserved) base pairs located between them. During rearrangement the two hepta- and 2 nona-nucleotides (having symmetrical arrangement) link and cutting out of the segment bordered by the rearrangement signals takes place. As a result two variable gene-segments (DH3 and JH4 in this example) move close to each other and fuse.

The genetic lack or lesion of recombinases leads to the lack of matured B- and T-cells. This happens in the case of SCID ("severe combined immuno-deficiency syndrome") mice, too. Besides Ig and TCR genes, we currently do not know of any other genes that rearrange somatically and where the process is the condition for the functioning of the genes. According to evolutionary analyses, the phylogenetic ancestor of the current rearrangement processes was an ancient transposon-like mobile genetic element. Now a days researchers have succeeded in proving the somatic gene rearrangement of antigen receptors (Ig, TCR) in every higher living organism from the cartilaginous fish (e.g. shark).

The whole process reminds us very much of the general DNA-repair mechanism, but the RAG-enzymes and the accessory factors connected to them (e.g. Ku 70/80) are specific to the rearrangement of antigen receptor genes. New data indicate the participation of *DNA-dependent protein-kinases*, that is protein-phosphorylation during which the DNA-segment, with a shape reminiscent of a hairpin, becomes accessible for nucleases.

12.4 The origin of TCR gene diversity; T-cell repertoire

A number of factors participate in establishing the complete pool of TCR molecules, as well as T-cells, and in determining the size of the actual repertory:

1 The germ-line genes: the number differs in every species. Although the number of V genes is lower on the α- and β-chains of the TCR than on the appropriate immunoglobulin gene-segment, the much higher number of J genes compensates for this difference.
2 The combination variations: the values determined by the number of V, D and J genes mean a number of possible variations of the hypervariable regions.
3 The linkage diversity, which comes from two sources:

   a From the incorrect DNA-rearrangement between the ends of the V-J segments of the light chain and the ends of the V-D-J gene-segments of the heavy chain. This brings more variations into the system in the case of the TCR than in the case of Ig genes.
   b Nucleotides can be added to the VJ- and VDJ-complexes by the co-operation of the terminal desoxyribo-nucleotydil-tranferase (TdT) (N-region diversity). In the case of immunoglobulins, this process only takes place in the H-chain, but occurs in all four chains with TCRs. The special feature of Dβ and the Dδ genes is that, in their case all three reading windows produce usable translation products. Nevertheless, the direct structural examination of numerous immunoglobulins indicates that, in spite of the theoretical possibility not all reading frame patterns appear in the repertory.

4 The somatic mutations.
5 The free combinations of the α/β-chains increases the diversity significantly.

In the course of ontogenesis the chance of the T-cell repertory's regeneration decreases. For reasons that are not completely known yet, the size of the repertory is smaller during the non-thymus-dependent T-cell maturation.

12.5 The expression of the TCR and the CD3-complex

The expression of the elements of TCR and CD3 complex have a mutual effect on each other. In the immature thymocytes transcription of CD3 γ, δ and ε genes precedes the expression of the TCR β- and α-chains. Following this, the γ-, δ- and ε-chains are glycosylated and associate with the TCR-$\alpha\beta$ heterodimer (appearing meanwhile) in the endoplasmic reticulum. All five chains are produced in great surplus, in a much bigger amount than that which appears on the plasma membrane. However, the ζ- and the η-chain is formed in limited amount and it can be shown that the association of the $\zeta\zeta$ (more rarely $\zeta\eta$) dimers is a condition for the TCR $\alpha\beta$-$\gamma\delta\varepsilon$ complex to reach the cell surface. The ζ-chain is practically always in a phosphorylated state: an intracellular tyrosine-kinase, ZAP-70 plays an important role in its operation. The intracellular ω-chain has a role in the CD3-complex's linking to the endoplasmic reticulum. Essentially, present knowledge shows a similar mechanism at work in the linkage between the $\gamma\delta$–TCR and the CD3-complex.

12.6 Peripheral T-cell subpopulations and T-cell polarization

T-cells that escape to the periphery carry a number of so-called accessory surface components besides the $\alpha\beta$- or $\gamma\delta$-antigen receptors and the CD3-complex. The common characteristic of these non-polymorphic, invariable molecules (also used as markers) (e.g. CD2, CD4, CD8, CD28, LFA-1, etc.) is that they react with the surface ligands of other cells (e.g. APCs, target cells). As "**co-stimulators**", they further increase the intensity of the cell-cell interactions and transmit signals towards the interior of the cell.

Many accessory molecules have already been mentioned as adhesion proteins. According to present conceptions, the accessory (costimulatory) molecule pattern and their interactions with neighboring cells is decisive regarding the "fate" of the T lymphocyte of certain TCR-specificity. This is the "micro-milieu" that currently characterizes the cell.

12.6.1 *αβ TRC/CD3 cells*

In the circulation, 90–99% of T lymphocytes are $\alpha\beta$ TCR/CD3 cells. As a result of differing cytokine production routes (Figure 12.5) several further subgroups (Th0, Th1 and Th2) can be distinguished among Th-cells which mainly carry CD4 surface antigen (about two thirds of the peripheral T-cells are CD4+, less than one third of them are CD8+). However, according to our current knowledge, we are closer to the truth if we talk about **T-cell-polarization** instead of sharply isolated subgroups. Here, the Th1- and Th2-cells are found at the two ends of the Th- (and Tc-) cell spectrum. In many cases the separation is almost "dogmatic", since "Th1-" and "Th2-"like cytokines participate equally in a number of normal and pathogenic immune processes. Nevertheless it is true that Th1-cells mainly tend to produce IL-2, IFNγ, IL-12 and lymphotoxin (LT, TNFβ) while Th2-cells produce IL-4, IL-5, IL-6, IL-9, IL-10 and IL-13. The two types of Th-cells influence the different immunological

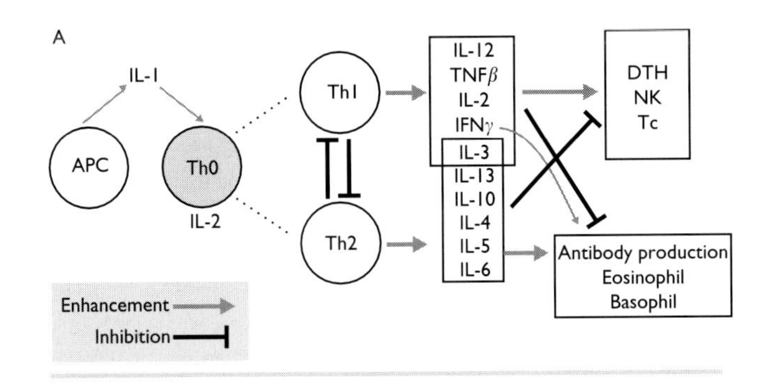

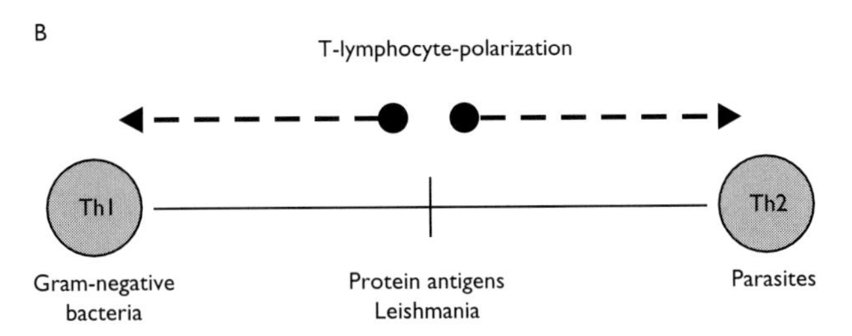

Figure 12.5 The cytokine pattern (a) and effects (a, b) of Th1- and Th2-cells. IFNγ of Th1-origin inhibits functioning of Th2-cells; IL-10 and IL-4 of Th2-cell origin inhibit functioning of Th1-cells. Cytokines of the Th1-cells usually stimulate the cell-mediated immune response and inhibit the humoral immune response (except IFNγ, which stimulates IgG-production). The Th2-cytokines stimulate the humoral and inhibit the cell-mediated immune response. DTH: delayed type hypersensitivity; APC: antigen-presenting cell; NK: natural killer cell; Tc: cytotoxic T-cell.

functions in different (helper or inhibiting) directions through their cytokines. The T-cell polarization (that is, the dissimilar cytokine pattern) is found among CD8 positive cytotoxic T-cells and among natural killer (NK) cells, as well.

In the peripheral blood CD8-antigen is detectable on the majority of Tc-cells. The so-called "suppressor cells" used to be considered members of this group, but their existence as a separate cell group with suppressor function is under debate.

Both the monomer CD4 and the heterodimer CD8 are **co-receptors**, therefore they react with the constant, non-variable part of MHC Class I (CD8) and MHC Class II (CD4) molecules. In the course of this reaction, their cytoplasmic part is phosphorylated and both molecules connect with a $p56^{lck}$ tyrosine-kinase. The name co-receptor indicates that, following the CD4/MHC Class II, CD8/MHC Class I and the specific (TCR-MHC/peptide) interactions, additional signals head towards the interior of the cell. Similarly, non-antigen specific, independent cell-cell relations and signal transduction processes occur between other cell surface molecule partners, as well. For instance, the interactions between CD2/LFA-3, the CD28/B7 and the CD40/CD40 ligand (gp39) are such (Figure 12.6). It has also become clear that IL-16 is a ligand, e.g. for CD4-coreceptor; it causes the

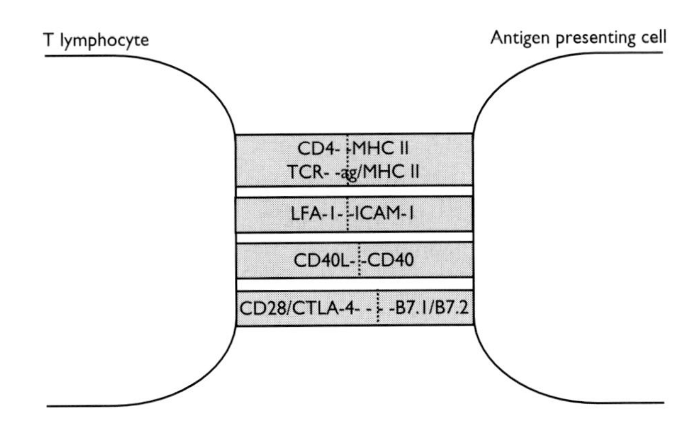

Figure 12.6 Costimulatory signals with intercellular relations between the T lymphocyte and antigen presenting cell. Antigen specific interaction (TCR–MHCII/ag) must be completed by action of co-receptor (CD4–MHCII) and costimulations (CD40–CD40L, CD28/CTLA4–B7.1/B7.2).

chemotaxis of CD4+ cells. Thus, according to our present knowledge, a number of non-antigen specific costimulatory intercellular relations (which increase the bond and activate the connecting cells) have an essential role beyond the antigen specific relation. Not only does activation not develop in their absence, but anergy evolves.

The kinetics of accessory relations is also important. For instance, the TCR-MHC Class II/peptide, the CD4-MHC Class II and the LFA-1-ICAM-1 relation develops first between the recognizing T-cell and the antigen-presenting B lymphocyte. This is followed by the CD40 ligand-CD40 and by the connection of CD28/CTLA-4-B7.1/B7.2 (CD80/ CD86) membrane proteins. Only then follows the IL-4-production of the T-cell and this stimulates the antigen-presenting neighboring B lymphocyte.

Besides this, soluble effects in the form of cytokines are also realized among cells participating in the interaction (e.g. IL-1 and IL-4). We will also touch upon their importance.

12.6.2 γδ TCR/CD3 cells

About 1–3% of the CD3+ cells found in the peripheral lymphatic glands carry γδ-TCR. However, the rate of γδ-T-cells is suprisingly high in the skin, in the mucous membrane of the sexual organs, in the placenta and in the tissues of the digestive and respiratory systems. The variance of γδ-cells exceeds that of αβ-T-cells and of B-cells. This is in spite of the fact that the V gene "usage" of γδ-T-cells is limited. Although there is significant debate about the function, MHC-restriction and target-specificity of these cells, it has been shown that there are many parasites (e.g. Plasmodium), phylogenetically rather conservative antigens mainly from infectious microbes (e.g. heat-shock proteins) and microbial superantigens among the antigens recognised by γδ-cells. In their recognition pattern, γδ-T-cells remind us of B lymphocytes rather than "classic" $\alpha\beta$-T-cells. In other words: $\gamma\delta$-T-cells recognize the conformation of antigens (that is their native, non-processed structure) and not the structure determined by the sequence. As effector cells, $\gamma\delta$-T-cells show cytotoxicity and cytokine production as well. The keratinocyte growth factor is especially interesting, certifying the role of skin-associated $\gamma\delta$-T-lymphocytes in the regulation of the dermal cell compound. The recirculation of intraepithelial T-cells (either $\alpha\beta$- or $\gamma\delta$-cells) is significantly smaller than that of $\alpha\beta$- and $\gamma\delta$-cells found in other parts of the organism.

12.7 Activation of T-cells, coreceptors

So far, we have discussed the molecular participants of the interaction between T-cells and antigens. This chapter deals with the results of the interaction, the activation and clonal accumulation of T-cells. In the course of activation, T-cells undergo division regulated in an autocrine way and perform their relative effector functions (Th-cells produce cytokines, Tc-cells destroy the target cells carrying the appropriate antigen). Moreover, antigen-specific memory T-cells are formed. During the activation of T-cells, multiple signal transduction occurs after the antigen-specific event. Genes are activated temporarily, cell surface and secreted proteins coded by these genes are produced, and the mitosis of T-cells begins.

According to present knowledge, tyrosines of the cytoplasmic part (conserved segments made of 26 amino acids, immune-receptor-tyrosine based activation element, ITAM, 10 occur in the T-cell) undergo covalent modification. The **phosphorylation** of a number of intracellular proteins (CD3 ζ, ε and γ in the T-cell, then ZAP-70, phospholipase-C and numerous intracellular adapter-proteins) participating directly or indirectly can be considered to be a key moment. This process occurs in antigen-receptor complexes (TCR in the case of the T-cell, BCR in B lymphocytes) and in Fc-receptors (in FcεRI and several FcγR). More and more adapter-proteins of related function have been recognised. These are protein-molecules appropriate for building macromolecular complexes of different composition. During this process **protein-kinases** phosphorylate and modify the **transcription proteins** (connected to DNA), which link to enhancer elements of genes during the regulation of gene expression and alter transcription. Modification includes oligomerization and/or disposal from endogenous inhibitors. The ITAM-connected stimulating events are inhibited by molecular processes related to ITIM-motifs represented by 13 amino acids (Garcia *et al.*, 1999).

12.7.1 Early phosphorylation events in T-cell activation

The interactions of T-cells with contact and soluble proteins help T-cells to reach the activated state in a determined temporal order. These processes are related to characteristic intracellular signal transduction processes. The CD4 molecule is linked with p56^{lck} protein-kinase; the TCR/CD3 complex is mainly related to ZAP-70 (also belonging to the src-gene-family). The stimulus (B7.1 or B7.2) of the CD28 chains activates PI 3-kinase (Onishi *et al.*, 1998).

In the course of the process (Figure 12.7), the ζ-chain of the CD3-complex is phosphorylated first, presumably by the lck-kinase linked with the CD4-molecule. Then, the ZAP-70-kinase (binding tyrosine-P and having a SH2-domain) links to the ζ-chain, which is phosphorylated on its right tyrosine (a different antigen-peptide would cause inhibition by inducing "disturbed" ζ-phosphorylation). After the linkage,

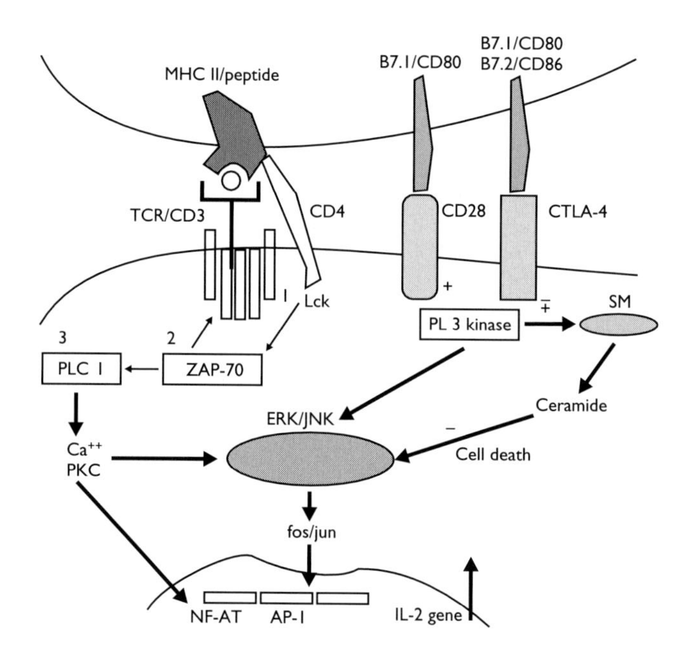

Figure 12.7 The early events of T-cellular signal transduction. After TCR/CD3-MHC/peptide interaction the lck protein-tyrosine-kinase (linked to the CD4 intracytoplasmatically) first phosphorylates the CD3-ζ-chains (1). Another protein-tyrosine-kinase, ZAP-70 (containing Sh2-domains) binds to the phosphotyrosines of the phosphorylated CD3-ζ-chains (2). The activation (phosphorylation) leads to the γ-activation of phospholipase-C (3), which participates in the activation of the JUN/ERK complex through protein-kinase C (PCK). JUN/ERK receives the other signal from CD28 (that got into a complex with the B7.1) through the PI-3 kinase. These two signals are required for the total activation (and jun/fos production) of JNK and ERK. The CTLA-4-B7.2 relation (whose affinity is a lot higher than the CD28-B7 relation) regulates PL-3-kinase. The jun/fos forms the AP-1 transcription factor; it is necessary for transcription of the IL-2 gene together with NF-AT. The sphyngomyelinase (SM) enzyme hinders the process through the ceramide in an apoptotic manner.

ZAP-70 phosphorylates phospholipase-C (PLC-γ1), which is activated by this. Then, phosphatidil-inositol-4,5-diphosphate (PIP_2) hydrolysis takes place and it dissolves into diacylglycerol (DAG) and inositol-1,4,5-triphosphate (IP_3). IP_3 binds to its intracellular receptors and mobilizes the intracellular Ca^{++}, consequently the intracellular Ca^{++}-level increases. Following the effect of DAG and Ca^{++} the protein-kinase-C (PCK) enzyme is activated and translocated. Ca^{++} forms a complex with calcineurin, which (among others) activates one of the transcription factors necessary for IL-2 synthesis (NF-AT). Besides this, PKC activates other cytoplasmic proteins (e.g. Vav), which is one of the components of JNK, a mitogen-activated protein-kinase (MAPK). This is one of the sources of the jun and fos oncogenes' activation. They activate AT-1, the other transcription factor of IL-2.

The activation of PKC (with phorbolesters) and the increase in Ca^{++} level (Ca-ionophores) can also be attained artificially. Together these also lead to the activation of the T-cell (without antigen stimulus).

Besides the signal from TCR presentation mediated by the CD3-complex, a second signal is also necessary. This is mainly ensured by another membrane-protein-membrane-protein, the CD28-B7(CD80/86) interaction, during which, CD28 binds and activates the phospho-inositol-3-kinase (P1 3-kinase) and another protein-kinase. Activation of JNK now becomes complete; i.e. activation of jun and fos is necessary for the AP-1 transcriptional factor. B7 (B7.2) (supposedly having a different structure) links to (CD28-analog) CTLA-4, which is expressed later in time, in fact with significantly greater affinity than CD28. The CTLA-4-B7.1, B7.2 (CD80, CD86) interaction is a similar process to that of CD28. According to certain data, it participates in developing the peripheral T-cell tolerance by inhibiting JNK activity. However, it might be also possible that inhibition takes place by other mechanisms. According to the newest data, the effect of CTLA-4 is not necessarily negative. Thus, the CD28/CNTL-4 B7.1/B7.2 (CD80/CD86) interaction constitutes one of the most important costimulation regulating motifs of IL-2 synthesis, that is of T-cell proliferation. The expression of the B7-molecule is decisive in the process. This is regulated by numerous effects of natural immunity (see chapter 18). Low costimulation (e.g. because of low B7 expression) leads to anergy, that is, to immune tolerance. Thus, in essence, regulation of the TCR-MHC/peptide takes place by costimulation; its lack (only TCR-peptide/MHC connection) calls forth a specific incapability to respond (anergy).

There are numerous other known costimulation relations. Which T-cell population responds and how depends on their local, current pattern. Previously we thought that the production of costimulatory cytokines close to the T-cell (IL-1) was of critical significance. It is thought that they may participate in the fine regulation of the process (Auron *et al.*, 1998), in addition to events causing the intracellular pH of T-cells to elevate, the inflow of Na^+ and the outflow of K^+ to increase and quabain-sensitive ATP-ase activity to intensify.

Among the accessory proteins, CD45 of tyrosine-phosphatase activity occurs in very significant amount (about 10%). This indicates a reversible and complex, process regulation. Thus, beyond the phosphorylation events, more and more attention has been drawn to phosphatases, especially to the tyrosine-phosphatases (PTP-ases). The two most well-known PTP-ases are the receptor-like transmembrane CD45 and the non-receptor type cytoplasmic SHP (or HCP). Both proteins are expressed in all hemopoietic cells their role is in the negative and positive regulation of cell-activation by removing tyrosine-phosphate. Today, we think that phosphatases (binding to the phosphorylated tyrosine Y-P and containing an SH2 domain) have some importance in the development of intracellular, temporary macromolecular complexes, which are formed during lymphocyte activation.

12.7.2 Gene activation and expression during T-cell activation

Minutes and hours after the TCR's antigen binding, a number of new mRNAs coding for proto-oncogens, cytokines and cytokine receptors appear in the T-cells. Among the proto-oncogenes, the most important are c-myc, c-fos and c-jun.

C-myc is necessary for cell division, c-fos and c-jun for transcription of the IL-2 gene (Figure 12.8). Cyclosporin A and FK-506 – already used in therapy – also inhibit T-cell activation at this level. They bind to the cyclophylin of the cytoplasm and hinder the Ca^{++}-dependent activation of calcineurin, a transactivation factor necessary for IL-2 transcription.

Transcription of IL-2 and IFNγ genes starts within an hour of antigen binding. According to current knowledge, during transcription of IL-2, a specific T-cell growth factor, at least four (NFAT, NFkB, Oct-1 and AP-1) transcription factors bind to the specific enhancer part of IL-2 genes. At the same time the protein units of the AP-1 molecule are the products of the jun and fos oncogene proteins. The fact that two signals are necessary for T-lymphocyte activation attains molecular sense at this level.

12.8 Cytokine receptors and associated molecules

Cytokines react with surface receptors with differing distribution. There are cytokines (IL-1, TNF), whose receptors are detectable on many types of cell. Others (e.g. IL-2, IL-4) are able to bind to much fewer cells (primarily to lymphocytes). Figure 12.9 presents the "module"-like structure of the most frequent cytokine receptors. Overlaps are also observed between receptors belonging to the immunoglobulin super-gene family (A) and to the so-called "hematopoietic cytokine receptor family" (B) (containing four cysteins and a pentapeptide,

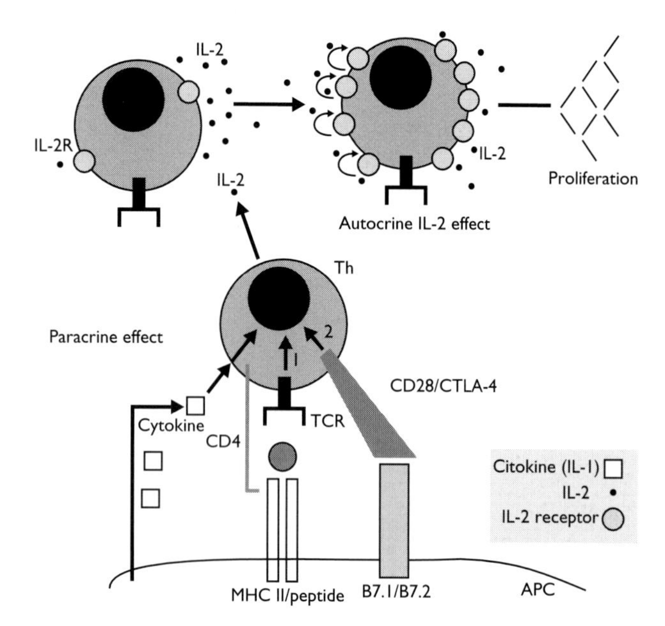

Figure 12.8 Generation of IL-2 in T lymphocytes. The peptide/MHC Class II-TCR (1) and the CD28/CTLA4-B7.1/B7.2 (2) interactions evoke induction of jun and fos oncogenes. The effect of the local cytokine (IL-1) adds to this. The FOS and JUN transcriptional factors link to each other and form AP-1. It binds to the enhancer of the IL-2 gene and starts IL-2 mRNA transcription. The activated T-cells produce IL-2 that binds to the (initially small number of) IL-2 receptors of neighboring cells. Further synthesis results in and secretion of IL-2 molecules results in clonal multiplication of the T-cells.

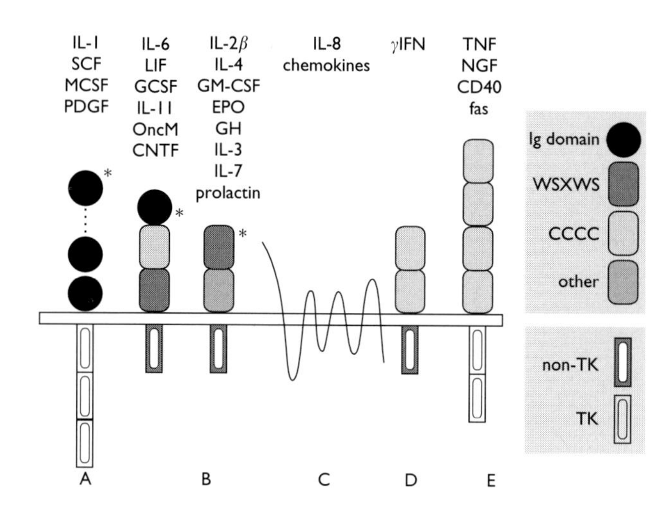

Figure 12.9 Modules and families of cytokine receptors. In the case of the so-called "hemopoietic" cytokine receptor gene family (A, B) there is an immunoglobulin-like domain, WSXWS (tryptophane-serine-X-tryptophane-serine) and motifs containing four cysteins on the extracellular part of the molecule. The IL-8 receptor (C) contains a number of trans membrane elements, too. The cysteins, in particular, are characteristic of IFNγ (D) and TNF receptor (E) gene families. In the cytoplasmic part of the molecules we can find tyrosine-kinase (TK) (A, E) and other segments too. Various combinations of the elements illustrated may occur.

the WSXWS, that is, triptophane-serine-X-triptophane-serine motif). The IL-8 receptor and other chemokines (C) – which can be grouped into the rhodopsine gene-family and have seven trans-membrane regions – display a completely different, serpentine-like structure. The IFNγ (D), which contains eight and four cysteins on its extra- and intracellular sections, and the TNF-receptor I and II, which have 6-6 cysteins in the four extracellular domains, are different again. The receptor of nerve growth factor (NGF), Fas on the cells that induce apoptosis, CD27, CD30 and CD40 also belong here (Onishi *et al.*, 1998, Moutoussamy *et al.*, 1998). In terms of their structure, cytokine receptors may be divided into "families".

12.8.1 Hematopoietic cytokine-receptor gene family

This is the largest group we know today. It consists of mixed elements (of fibronectin "III-type" and "immunoglobulin C2"). It contains characteristic domains too, which are made of about 100 amino acids, Cys-X-Try motifs and 3 Cys amino acids. Their characteristic element is the WSXWS (Try-Se-X-Try-Se motif).

A common "β"-chain can also be found in this cytokine receptor family (Table 12.2); these are the gp130, the gp140 (KH-97) and the γ-chain of the IL-2 receptor. In certain cases the cytokine's binding creates receptor-complexes of different stochiometry (e.g. with IL-6 receptors, the binding of IL-6 causes the homodimerization of gp130). These chains (gp130, gp140, γ-chain) themselves do not bind the cytokine or they do it with small affinity, however they significantly increase the ligand's binding affinity, that is, the biological effect of cytokines on cells. The gene knockout accomplishments indicate the decisive role of the associated chains. Gp130-deficient mice die of cardiac failure in the embryonic age. The reason for this is that cardiotrophin also acts through gp130, and the myocardium atrophies if the cardiotrophin signal (through gp130) is missing. The fact that the development of T and B lymphocytes decreases significantly when the γ-chain is missing, also indicates the decisive role of the associated chains.

In this cytokine receptor family receptors can also be found in soluble form. We know the soluble form of IL-4, IL-5, IL-7, IL-9, GM-CSF, G-CSF, GH, LIF, IL-6 and CNTF receptors.

12.8.2 Interferon receptor gene family

The IFN α/β, IFNγ and IL-10 receptors belong here.

Table 12.2 Common chains in cytokine receptors

β chain	*Cytokine receptor*
gp130	IL-6
	IL-11
	Oncostatin M
	LIF
	CNTF
	G-CSF (?)
	IL-12 (*)
	Cardiotrophin-1
γ-C	IL-2
	IL-4
	IL-7
	IL-9
	IL-13
	IL-15(**)
gp140	IL-3
	IL-5
	GM-CSF

Notes
? debated.
* IL-12 receptor is an gp130 homologous molecule.
** IL-15 receptor on T lymphocytes consists of the other two (α, β) components of IL-2 receptor.

12.8.3 "Immunoglobulin" receptor family

Most cytokine receptors contain an Ig-domain (IL-1, IL-6, FGF, PDGF, M-CSF and SCF receptors), but these cytokine receptors can be grouped primarily into this family. Thus, it is evident that certain cytokine receptors belong to several families.

12.8.4 Protein-tyrosine-kinase receptor family

These receptors carry tyrosine-kinase activity on their cytoplasmic section. The EGF, NGF and BDNF receptors belong here.

12.8.5 NGF receptor family

The recurrence of parts which are rich in 3-4 cysteins is structurally characteristic of this cytokine receptor family. This is where NGF, TNF (receptor-I), CD27, CD30, CD-40, OX-40, 4-1BB, Fas receptors and other molecules belong.

12.8.6 Receptor family connected to G-protein

The cytokine receptors of this group are in direct connection with G-protein. This is the rhodopsine-like receptor family ("serpentine"); their members contain seven trans-membrane domains. Chemokines and the receptors of IL-8 belong here. It has recently been discovered that chemokine receptors have important co-receptor functions in HIV infection.

12.9 Signal transduction processes caused by cytokines

Cytokine binding to the receptor activates different kinds of second messenger systems. IL-8 and other chemokines, as we have already mentioned, bind to "serpentine" receptors with characteristic, repeated transmembrane domains. This structure (together with many other hormone receptors) directs signal transduction through the G-protein, as already shown for IL-8 and other chemokines.

Src-type kinase enzymes ($p56^{lck}$, $p59^{fyn}$) (in the case of the IL-2-receptor) made of three chains (α: p55, Tac, β: p75), JAK1/JAK3 tyrosine-kinases and stat3/stat5 transcription factors participate in signal transduction via IL-2 (Figure 12.10). With other cytokines (e.g. SCF which binds to c-kit oncogenes and IL-6), particular types of tyrosine-kinases are activated. Probably the best known are the JAK (Janus-kinases) which occur in interferons and in many other cytokines. These kinases have conserved structure and belong to the src family. Their peculiarity is the fact that the kinase domain itself also contains a tyrosine, whose phosphorylation significantly increases the activity of the enzyme. The use of JAK enzymes shows typical differences in certain cytokines (Table 12.3).

Among the effects of the cytokines, three transcriptional factors may be emphasized: NF-kb, which occurs frequently in immunological processes, C/EBP (cAMP-dependent "enhancer"-binding protein) of Lucien-"zipper" structure and the stat ("signal transduction and activation of transcription") molecule. The main feature of stat proteins of conserved structure is that they contain a SH2 ("sacra homology") domain, which is able to bind the phosphorylated tyrosine. This is why stat proteins are able to "dock" on phosphorylated receptor-chains. Stat molecules also have tyrosine which can be phosphorylated, thus, there is a possibility for stat molecules to dimerize later (with SH2-tyrosine-P binding). This dimerization enables the stat transcriptional factors to bind to DNA. Stat proteins also contain a SH3-region, they can bind to intracellular elements that are rich in proline (PPPLP) and to other "adapter"-like elements. This enables the development of intracellular protein-complexes of temporary size and composition. According to latest research, protein kinase C- (or protein kinase A) dependent phosphorylation of stat through serine is also necessary for the stat-dimer to migrate into the cell nucleus. This fact also indicates the importance of the "dialogue" between different signal transduction mechanisms (receptors connected to G-protein and tyrosine-kinase). The lower part of table 12.3 presents the stat-protein usage of certain cytokine-induced signal pathways.

Again, gene knockout experiments have shown the biological importance of stat proteins. For instance the cells of stat1-deficient mice are not able to react to interferons. The ras oncogene also participates in certain signal transduction path ways of cytokines; MAP-kinase has central importance here (Yamauchi *et al.*, 1998).

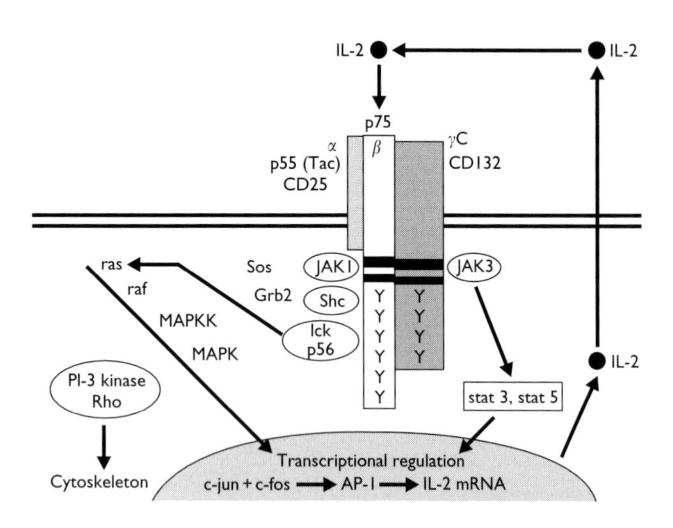

Figure 12.10 Signal transduction of the IL-2 receptor. Among the three chains of the IL-2 receptor (α, β and γ) the β-chain belonging to the hemopoietic cytokine receptor gene family contains structural elements, which, by activating lck p56 src-kinase via the ras/RAF/MAPK pathway, leads to generation of c-jun and c-fos, resulting in AP-1, necessary for IL-2 production. The β- and the γ-chains also bind tyrosine-kinases (JAK3); they activate stat3 and stat5 transcription factors. Activation of the PI-3-kinase system is linked to the cytoskeletal system.

Table 12.3 Usage pattern of JAKs and stats

Cytokine family	*JAK*	*stat*
IFN family	Jak1, Jak2, Tyk2	stat1, stat3
gp130 family	Jak1, Jak2, (Tyk2)	stat3 (stat1)
γ-C family	Jak1, Jak3	stat3, stat5, (stat6)
gp140 family	Jak2	stat5
GH family	Jak2	stat5
RTK family	Jak1, Jak2, Tyk2	stat1, stat3, stat5

JAK: Janus kinase, IFN: interferon, gp: glycoprotein, GH: growth hormone, EPO: erythropoietin, PRL: prolactin, THR: thrombopoietin, RTK: receptor tyrosine kinase.

Usually, transcription of a given cytokine is formed by the simultaneous action of 2–5 transcriptional factors. For example, in the case of IL-4, the transcriptional factor can bind to 5 enhancer elements, while IL-3/GM-CSF can bind to 4 enhancer elements. We will discuss the regulation of IL-2 synthesis in the section on T lymphocyte activation. As with signal transduction of TCR and BCR structures, several signal transducing systems are also activated through cytokine receptors. In addition to stat- and src-kinase, the ras-dependent MAP-kinase system must be considered here as well (Heinrich *et al.*, 1998, Nelms *et al.*, 1999, Clements *et al.*, 1999). Figure 12.11 presents the signal transduction scheme for cytokine receptors.

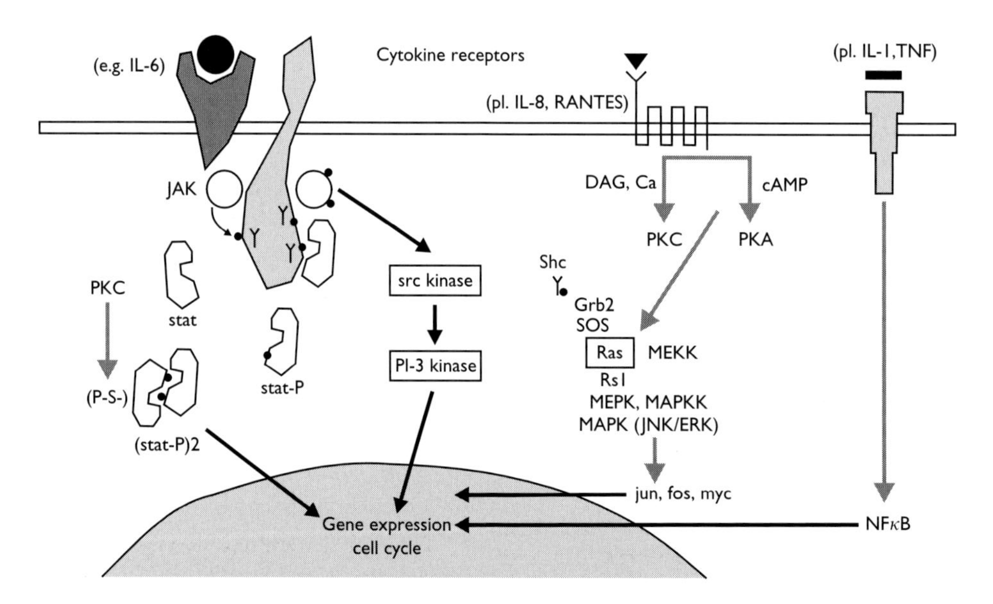

Figure 12.11 The general signal transduction scheme of cytokine receptors. Upon binding of cytokine to cytokine receptors several types of intracellular signal transducing systems are coactivated. The stat-proteins phosphorized by JAK/TYK tyrosine-kinases are able to enter the nucleus and bind to the DNA in homo- or heterodimerized form. The src-kinases activate transcription factors through PI-3 kinase, while the ras-raf system modifies gene expression through MAP-kinase. Activation of cytokine receptors that contain seven transmembrane domains and connect to G-protein (related to generation of cAMP and release of Ca^{++}) is required for the final activation (by serine-phosphorylation) of particular stat-dimers. The NF-kB transcription factors (representing a separate class) can also bind to the stat-proteins. JAK: Janus-kinase, MAPK: MAP-kinase.

12.10 Soluble cytokine receptors – agonists and inhibitors

Almost every cytokine receptor can be detected in soluble form. Two processes can lead to the formation of soluble cytokine receptors (Figure 12.12): alternative "splicing" and surface proteolytic digestion. The majority of them – competing with the receptor bound to the membrane – have inhibitory effects (e.g. soluble TNF-α receptor).

TNF-α is quite important as it frequently occurs attached to the cell membrane, affecting the TNF receptor of the neighboring cell in a "juxtacrine" manner. In some cases, a significant increase was observed in the effects of IL-6 and LIF (for soluble IL-6 receptor and soluble LIF-receptor). Since the common β-chain of IL-6 and LIF-receptors, the gp130, is detectable on practically every cell, this means that, in the presence of enough IL-6/LIF and soluble IL-6/LIF receptors, every cell is able to bind IL-6/LIF. A similar interesting regulation is presumed in the case of CNTF which affects neuronal cells.

Thus the *presence of soluble cytokine receptors* close to the cell indicates a very *important controlling system*. It "slows down" the effect of cytokines on cells that express cytokine receptor themselves and at the same time, in some cases (e.g. IL-6/LIF), it "sensitizes" those cells which do not have cytokine-receptors to cytokine.

12.11 T-cell lymphomas

Pathological aspects of T-cell physiology include many disorders with abnormal activity or appearence of T-cells. Here, we give a summary on the currently accepted list of T-cell malignancies (Table 12.4.). These diseases are characterized by a blockade of T-cell ontogeny at particular differentiation stages. Most forms of T-cell neoplasms involve leukemias with relatively mature T-cells. In some cases the real molecular background of the disease is still undiscovered and only clinical classification can be applied. In some, tissue specificity (e.g. hepatosplenic $\gamma\delta$ T-cell, enteropathy-type intestinal or cutaneous T-cell lymphomas), in other cases the proposed etiological factor (e.g. mycosis fungoides) or cytokine/cytokine receptor pattern is predominant (Jaffe *et al.*, 1999).

In the future, scientific classification will be made on the basis of surface adhesion molecules (determining tissue location) and gene expression (actual mRNA composition) pattern using either FACS analysis and differentiation display/ cDNA chip systems. These results will alter the strategy and possibilities of therapeutic approaches, too.

Table 12.4. Proposed World Health Organization (WHO) classification for T-cell neoplasms

Precursor T-cell lymphoblastic leukemia/lymphoma
Mature T-cell and NK-cell neoplasms
T-cell prolymphocytic leukemia
T-cell large granular lymphocytic leukemia
NK-cell leukemia
Extranodal NK/T-cell lymphoma, nasal-type (angiocentric lymphoma)
Mycosis fungoides
Sézary syndrome
Angioimmunoblastic T-cell lymphoma
Peripheral T-cell lymphoma
Adult T-cell leukemia/lymphoma (HTLV1)
Systemic anaplastic large cell lymphoma (T- and null-cell types)
Primary cutaneous anaplastic large cell lymphoma
Enteropathy-type intestinal T-cell lymphoma
Hepatosplenic gamma-delta T-cell lymphoma

(with help of Dr Klára Gadó)

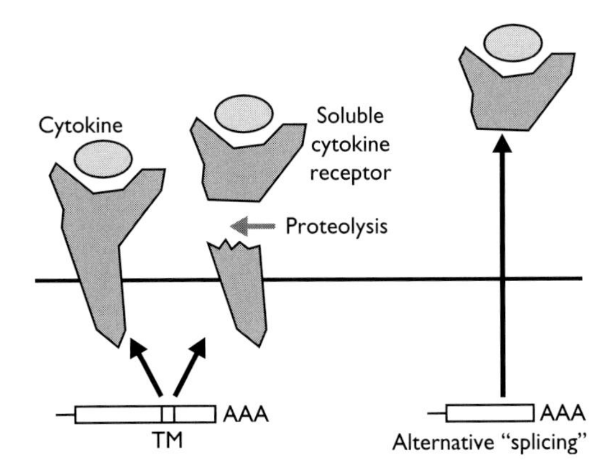

Figure 12.12 Formation of soluble cytokine-receptors. Soluble cytokine-receptors may develop by surface proteolysis or by alternative "splicing" resulting in mRNA that does not contain a transmembrane region.

References

Auron PE (1998) The interleukin 1 receptor: ligand interactions and signal transduction. *Cytokine Growth Factor Rev 9, 221–37*

von Boehmer H, Fehling HJ (1997) Structure and function of the pre-T cell receptor. *Annu Rev Immunol 15, 433–52*

von Boehmer H, Aifantis I, Azogui O, Feinberg J, Saint-Ruf C, Zober C, Garcia C, Buer J (1998) Crucial function of the pre-T-cell receptor (TCR) in TCR beta selection, TCR beta allelic exclusion and alpha beta versus gamma delta lineage commitment. *Immunol Rev 165, 111–9*

Clements JL, Boerth NJ, Lee JR, Koretzky GA (1999) Integration of T cell receptor-dependent signaling pathways by adapter proteins. *Annu Rev Immunol 17, 89–108*

Garcia KC, Teyton L, Wilson IA (1999) Structural basis of T cell recognition. *Annu Rev Immunol 17, 369–97*

Guthridge MA, Stomski FC, Thomas D, Woodcock JM, Bagley CJ, Berndt MC, Lopez AF (1998) Mechanism of activation of the GM-CSF, IL-3, and IL-5 family of receptors. *Stem Cells 16, 301–13*

Heinrich PC, Behrmann I, Müller-Newen G, Schaper F, Graeve L (1998) Interleukin-6-type cytokine signalling through the gp130/ Jak/STAT pathway. *Biochem J 334, 297–314*

Jaffe T (1999) World Health Organization classification of neoplastic diseases of the hematopoietic and lymphoid tissues. *Am J Clin Pathol. 111 (Suppl.1): S8–S12*

Moutoussamy S, Kelly PA, Finidori J (1998) Growth-hormone-receptor and cytokine-receptor-family signaling. *Eur J Biochem 255, 1–11*

Nelms K, Keegan AD, Zamorano J, Ryan JJ, Paul WE (1999) The IL-4 receptor: signaling mechanisms and biologic functions. *Annu Rev Immunol 17, 701–38*

Onishi M, Nosaka T, Kitamura T (1998) Cytokine receptors: structures and signal transduction. *Int Rev Immunol 16, 617–34*

Sleckman BP, Bassing CH, Bardon CG, Okada A, Khor B, Bories JC, Monroe R, Alt FW (1998) Accessibility control of variable region gene assembly during T-cell development. *Immunol Rev 165, 121–30*

Yamauchi T, Ueki K, Tobe K, Tamemoto H, Sekine N, Wada M, *et al.* (1998) Growth hormone-induced tyrosine phosphorylation of EGF receptor as an essential element leading to MAP kinase activation and gene expression. *Endocr J 45, S27–31*

Chapter 13

Neuro-immune responses in different diseases

Role of tumor necrosis factor-α and IL-1

E. Sylveszter Vizi

Contents

Over the past two decades it has become known that the effect of the central nervous system (CNS) on immune responses is not only mediated via the hypothalamo-pituitary-adrenal axis (HPA), but also by direct neuronal control (Besedovsky and Del Rey, 1996; James *et al.* 1996; Elenkov *et al.*, 2000). The sympathetic branch of the autonomic nervous system (ANS) is able to regulate blood flow and cytokine production. Besides the primary (thymus, bone marrow) and secondary lymphoid organs (spleen, tonsils, and lymph nodes), many other tissues involved in immune responses are heavily influenced by incoming chemical signals via norepinephrine (NE) derived from varicose axon terminals of the sympathetic nervous system (Vizi, 1998; Elenkov *et al.*, 2000). In addition to NE released from nonsynaptic varicosities of noradrenergic terminals (Vizi, 2000), circulating catecholamines (epinephrine, NE) are able to influence the production of pro- and anti-inflammatory cytokines by different immune cells (Table 13.1) (Mire-Sluis and Thorpe, 1998).

Around the turn of the century, W.B. Coley observed that treatment with bacterial extracts was as effective as inoculation of live bacteria in cancer treatment (cf. Beyaert and Fiers, 1995). Shear and colleagues provided evidence that *Serratia marcescens* extracts induced hemorrhagic tumor necrosis when administered to mice with transplanted sarcomas (Shear *et al.*, 1943). Later, the active component of these extracts was identified as lipopolysaccharide (LPS) derived from the cell wall of Gram-negative bacteria. In addition, when a serum factor named "*cachectin*" was purified and partially sequenced, it turned out to be identical with TNF (cf. Rothwell, 1995). As far as the origin of cytokines is concerned, there are several cells able to produce them. Macrophages are essential for host defense and play an important role in orchestrating immune responses (Mills *et al.*, 2000). They are the first cells to receive signals. During phylogeny, lymphocytes take over the duties of macrophages using their recognition structures. This shift is very important. T or other types of lymphocytes control the type of immune response. Th1 immune responses are associated with TNF-α and IFN-γ and Th2 responses with IL-4, IL-5 and IL-10 (Table 13.1). IL-12 secreted by myelomonocytic cells is an important factor promoting the development of Th1-cells.

Table 13.1 Major pro- and antiinflammatory cytokines

Proinflammatory cytokines	*Source*
IL-12	APCs
TNF-α	APCs
IL-1	APCs, fibroblasts, endothelium
IFN-γ	Th1 and NK cells
Antiinflammatory cytokines	*Source*
IL-4	Th2 cellls
IL-10	APCs, Th2 cells
TGF-β	Different cells

APCs, antigen-presenting cells; NK, natural killer cell; Th, helper lymphocyte.

13.1 Biological activities

13.1.1 Activation of the hypothalamic-pituitary-adrenal (HPA) axis

Endotoxin, IL-1 and TNF-α activate the HPA axis (Dunn, 2000). The site of their action is in the hypothalamus through the production of corticotropin-releasing factor resulting in an increase of ACTH and corticosterone secretion. Our results (Elenkov *et al.*, 1992) showed that lesions of the hypothalamic paraventricular nucleus did not completely block the effect of endotoxin (lipopolysaccharide, LPS) on plasma ACTH levels in the late phase of the response. Because the lesions of the median eminence totally abolished the effect of LPS, it is conceivable that the median eminence might be a target for the effect of LPS and/or related cytokines.

13.1.2 Neurodegeneration

It has been shown that TNF released from different immune cells (Figure 13.1) activates IL-1 production and induces inflammation, fever, necrosis or apoptosis, i.e. swelling, destruction and lysis of the cell and cell shrinkage. Inflammation and subsequent cytotoxicity can also be mediated by reactive oxygen species (ROS), such as peroxides and nitric oxide (NO), which can be produced by TNF-α (Hoffman and Weinberg, 1987; Schulz *et al.* 1995). TGF-β released from immune cells acts in a different way (Figure 13.1), it deactivates macrophages and reduces the release of ROS and NO. TNF-α has also been shown to be mitogenic for a number of normal cells. There is now considerable evidence to suggest that specific cytokines, particularly IL-1, are involved directly in neurodegeneration and neuronal death in the CNS (Rothwell, 2000). Most forms of neuronal death are associated with excitotoxicity, i.e. excessive release of glutamate and subsequent activation of NMDA and AMPA receptors (Vizi *et al.*, 1998). There is some evidence that IL-1β is the predominant form of IL-1 induced by brain insults. It is likely that IL-1 interacts in some way with this cascade to modify glutamate release, reuptake or actions (Rothwell, 2000).

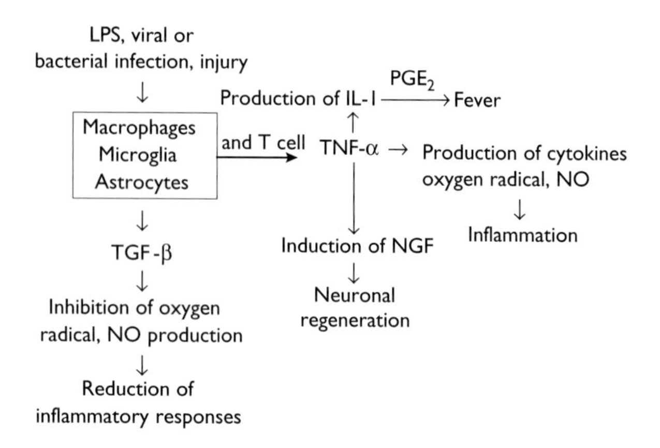

Figure 13.1 Bacterial and/or viral infection increases cytokine production by different immune cells. TNF-α produces IL-1, which, in fact, results in fever via PGE_2 release and other proinflammatory cytokines and inflammation of the cells. During septic shock the inflammatory immune response can be influenced by catecholamines released from nonsynaptic varicosities of noradrenergic neurons (Elenkov *et al.*, 2000). TGF-β production reduces the inflammatory responses.

13.1.3 Antiviral activity

TNF-α has an antiviral effect against several viruses. In contrast, TNF-α has been shown to increase the infection rate and replication of human immunodeficiency virus (HIV).

13.1.4 Antitumor activity

TNF-α and IFN-γ have been shown to produce either additive or synergistic antitumor activity (Brouckaert *et al.*, 1986; Marquet *et al.*, 1987). A number of studies have shown that TNF-α limits experimental metastasis. An interesting observation is that cytotoxic drugs, such as cyclophosphamide, doxorubicin, adriamycin and staurosporine increase the antitumor activity of TNF-α (Beyaert *et al.*, 1993b).

13.2 Cytokines in diseases and therapy

The release of cytokines by immunocompetent cells may be responsible for the development of different pathological conditions. The reproduction of TNF-α is the first-line reaction of the organism to microbial, parasitic, viral, or mechanical stress.

13.2.1 Sepsis

The most common cause of death in surgical intensive care units is septic shock (Song *et al.*, 1993; McCarthy Pastores *et al.*, 1996). Sepsis, septic shock, and the resulting multiple organ dysfunction syndrome remain leading causes of morbidity and mortality in critically ill and trauma patients (Parrillo *et al.*, 1990; Beal and Cerra, 1994). It is now generally accepted that sepsis is a systemic inflammatory response in the host organism evoked by the production and secretion of endogenously synthesized mediators, including proinflammatory cytokines (Bone, 1991; Bone *et al.*, 1992). In addition, other endogenous molecules such as prostaglandins, leukotrienes, reactive oxygen metabolites, platelet activating factor (PAF), and NO have also been implicated as potentially toxic mediators produced during sepsis (Bone, 1991; Natanson *et al.*, 1995). Recent clinical trials in patients with sepsis and septic shock have focused on novel therapies directed at different elements of the inflammatory cascade, including monoclonal antibodies to endotoxin and TNF-α, the administration of IL-1 receptor antagonist, NO, and products of the lipoxygenase and cyclooxygenase pathway (Natanson, 1994; Fisher *et al.*, 1994; Abraham *et al.*, 1995; Dhainaut *et al.*, 1994; 1995). Until now the preliminary results of these expensive clinical trials failed to be promising, some have even been disappointing.

It is increasingly recognized that the inflammatory immune response during sepsis can be modulated by catecholamines released by the SNS, as well as by the administration of adrenergic and dopaminergic receptor agonists and phosphodiesterase inhibitors utilized in the therapy of septic shock (McCarthy Pastores *et al.*, 1996). These agents can alter pro- and antiinflammatory cytokine production from immune cells through the activation of adrenergic and dopaminergic receptors located on these cells, possibly through changes in cAMP metabolism. Availability of selective receptor antagonists (e.g. α_2-adrenoceptor antagonists, Elenkov *et al.*, 2000), although not routinely available or utilized, offers further hope for immunomodulation through this indirect mechanism. Mortality from sepsis has not changed significantly over the past few decades: in many ICUs less than fifty percent of the patients with septic shock survive.

13.2.2 Autoimmune disorders

Rheumatoid arthritis, multiple sclerosis and insulin-dependent diabetes mellitus are diseases which seem to have little in common. All three are inflammatory disorders (Wekerle, 1998). TNF-α is involved in a number of autoimmune and inflammatory conditions (rheumatoid arthritis, inflammatory polyarthritis). TNF-α plays a role in demyelination and oligodendrocyte toxicity in multiple sclerosis (Selmaj and Raine, 1988; Selmaj *et al.*, 1991). Patients with active psoriasis often have increased TNF-α and IL-6 levels in the psoriatic plaques, and even in their plasma (Nickoloff *et al.*, 1991). Corticosterone inhibits production, and down-regulates transcription and translation of TNF-α. In accordance with this, the Lewis strain rat (with low corticosterone response to stress) is particularly susceptible to such chronic inflammatory models as arthritis and experimental allergic emcephalomyelitis.

13.2.3 Traumatic brain injury

Traumatic brain injury in humans (Table 13.2) is associated with immunosuppression and is often accompanied by infectious complications. Brain tumor resection frequently results in localized brain damage and inflammation. The sympathetic activation (stress) resulting in catecholamine release is presumably due to acute decompression, brain stem manipulation, lesion and irritation during the neurosurgical procedure (Figure 13.2).

Interleukin-10 not only downregulates MHC class II expression on monocytes but also inhibits monocyte production of proinflammatory cytokines, including IL-1, IL-2, IL-6, IL-8, TNF-α and IFN-γ (de Waal Malefyt *et al.*, 1991; Fuchs *et al.*, 1996). Woiciechowsky and colleagues (1998) demonstrated that catecholamines trigger the release of IL-10 from unstimulated monocytes without costimulation via β-adrenoceptors (Suberville *et al.*, 1996). Iatrogenic brain injury is associated with sympathetic activation, which results in catecholamine release i.e. increased release of NE, and activation of β-adrenergic receptors that are coupled to the cyclic AMP-protein kinase A signal transduction pathway. Protein kinase A induces release of preformed IL-10 from monocytes. Using a rat model of acute brain injury, propranolol, a β-adrenoceptor antagonist, was found to prevent the increase of IL-10 plasma levels.

Table 13.2 Diseases known to induce cytokine production in the brain

Alzheimer's disease
Brain injury, surgical intervention
Down's syndrome
Scrapie and Creutzfeldt-Jakob disease
Multiple sclerosis and experimental allergic encephalomyelitis
Stroke
Parkinson's disease

Stress → Increased sympathetic outflow
↓
NE/E release is increased
↓ via β_2-adrenoceptors
Increase in IL-10 production
↓
Immunosuppression
↓
Infection

Figure 13.2 Effect of stress or sympathetic activation on IL-10 production. Note the effect of norepinephrine (NE) and/or circulating epinephrine (E) is mediated via β_2-adrenoceptors. This is how traumatic brain injury results in immune suppression, and subsequently, infections.

If we can understand the pathomechanisms and, etiology of these diseases, we will be able to design better and more specific therapies. Reduction of sympathetic overaction by drug therapy (β-adrenoceptor antagonist) may be a possible means for prevention of fatal infections and complications.

13.2.4 Cachexia

Cachexia is usually associated with chronic diseases such as cancer, arthritis and AIDS, and acute disease such as severe injury or sepsic conditions (Tisdale, 1997). Cytokines are primary mediators of cachexia through actions in the periphery and the CNS.

13.3 Role of sympathetic innervation in diseases related to pro- and antiinflammatory cytokines

During the past two decades, evidence from various experimens has shown that the immune system is regulated by signaling from both the central and autonomic nervous systems (Madden *et al.*, 1995). The role of the CNS in the regulation of immune reactions arising from within the brain, or in response to peripheral immune stimuli, has been extensively studied and reviewed (Reichlin, 1993; Chrousos, 1995; Michelson *et al.*, 1994; Wilder, 1995). Of note, several hormones and peptides under CNS control exert immunoregulatory effects. In particular, hormones of the hypothalamic-pituitary-adrenal axis (e.g. glucocorticoids, corticotropin, β-endorphin, corticotrophin-releasing hormone, and arginine vasopressin) have either stimulatory or inhibitory effects on the immune response during antigenic or inflammatory challenge (Reichlin, 1993; Chrousos, 1995).

The ANS relays information from the CNS to visceral target tissues through the sympathetic and parasympathetic nervous systems (Madden *et al.*, 1995; Cohen *et al.*, 1994). Neural regulation of immune function through the SNS is modulated by the release of NE from the sympathetic nerve endings and epinephrine from the adrenal medulla (Elenkov *et al.*, 2000). The major lymphoid organs, (i.e. spleen, lymph nodes, thymus and intestinal Peyer's patches) are extensively innervated by noradrenergic sympathetic nerves (Felten *et al.*, 1992; Vizi *et al.*, 1995). NE and other catecholamines have a wide variety of direct effects on immune cells, particularly monocytes and lymphocytes (Madden *et al.*, 1994; 1995; Carlson *et al.*, 1989). These neurotransmitters regulate the functions of immune cells through interactions with cell surface adrenergic and dopamine receptors (Elenkov *et al.*, 1995; 1996; Haskó *et al.*, 1995; Szelényi *et al.*, 2000). Both the adrenergic and dopamine receptors are members of the superfamily of guanosine triphosphate-binding protein (G-protein) coupled receptors (Linder and Gilman, 1992). These receptors are coupled intracellularly to the adenylate cyclase complex which, when activated, leads to increased levels of cAMP (Bourne *et al.*, 1974). The manipulation of cAMP metabolism through adrenergic (β_2) and dopaminergic receptors, or through phosphodiesterase inhibition, is the crucial element of the immunomodulatory potential of vasoactive pharmacological agents used for circulatory support.

Since all primary and secondary lymphoid organs are innervated by sympathetic division of the autonomic nervous system and when these are activated by stress or other stimuli, these neurons release micromolar quantities of NE into the vicinity of T lymphocytes and macrophages (Felten *et al.*, 1987; Shimizu *et al.*, 1994). β_2-adrenoceptors expressed on the lymphocyte plasma membrane bind NE and signal the cellular interior via $G_{\alpha S}$ subunit of a heterotrimeric G protein linked to the adenylyl cyclase-cAMP-protein kinase A (PKA) signaling cascade (Kammer, 1988; Kobilka, 1992; Landmann, 1992).

Lymphoid organs represent the primary site of HIV pathogenesis. Long before the onset of clinical illness, high amount of virus and large numbers of infected CD4 T lymphocytes are present. It has been shown (Cole *et al.*, 1998) that NE can accelerate HIV-I replication by up to 11-fold. The effect of NE on viral replication is mediated via the β-adrenoceptor-adenylyl-cyclase-cAMP-PKA signaling cascade confirming that cAMP can increase HIV-I replication (Nokta and Pollard, 1992; Chowdhury *et al.*, 1993) at concentrations ranging from 10^{-8} to 10^{-5} M. This effect could be reproduced by protein kinase A activators (forskolin) and inhibited by β-adrenoceptor antagonists (sotalol, propranolol). The effect of NE on cytokine production can also be involved, since NE via activation of β_2-adrenoceptors present on the surface of the immune cells reduces the production of proinflammatory (TNF-α, IL-I, IFN-γ, Il-21) and increases that of antiinflammatory (IL-10) cytokines (Vizi, 1998). These data suggest that the activation of SNS neurons innervating lymphoid tissue is able to influence HIV disease pathogenesis (Cole *et al.*, 1998).

Increase of cAMP level by NE (10 μM) increases HIV-I replication: this is a β-receptor-mediated effect (Figure 13.3), sotalol and propranolol β-adrenoceptor blocking agents block it. This, in fact, provides an attractive therapeutic approach to the treatment of HIV-I replication.

Salbutamol, a β_2-adrenoceptor agonist is a potent suppressor of collagen-induced arthritis and IL-12 release (Malfait *et al.*, 1999). Daily administration of 200 μg of

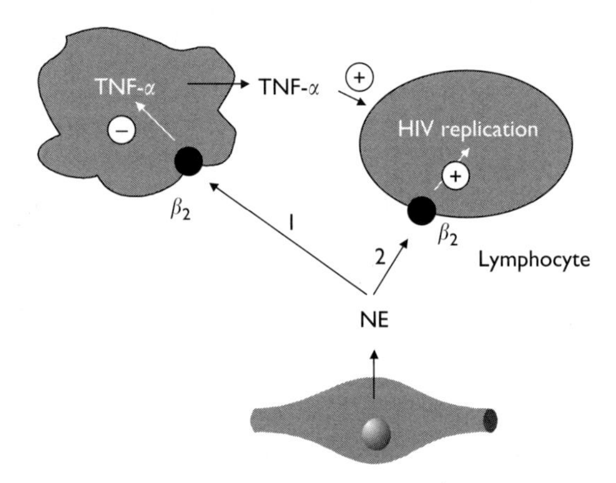

Figure 13.3 Cartoon of the effect of sympathetic activation on HIV replication. Note that increased norepinephrine (NE) release reduces the production of TNF-α, thereby inhibiting HIV replication. In addition, NE directly increases cAMP level via β_2-adrenoceptors, and subsequently HIV replication in the lymphocytes, therefore, β_2-adrenoceptor blocking agent will prevent this direct effect.

salbutamol protected male mice from joint damage as assessed by histology.

References

Abraham, E., Wunderink, R., Silverman, H., Perl, T.M., Nasraway, S., Levy, H., *et al.* (1995) Efficacy and safety of monoclonal antibody to human tumor necrosis factor alpha in patients with sepsis syndrome. A randomized, controlled, double-blind, multicenter clinical trial. TNF-alpha MAb Sepsis Study Group. *JAMA, 273, 934–941.*

Beal, A.L., Cerra, F.B. (1994) Multiple organ failure syndrome in the 1990s. Systemic inflammatory response and organ dysfunction. *JAMA, 271, 226–233.*

Besedovsky, H.O., Del Rey, A. (1996) Immune-neuro-endocrine interactions: facts and hypothesis. *Endocrine Rev, 17, 64–102.*

Beyaert, R., Fiers, W. (1995) Tumor necrosis factor and lymphotoxin. In *Cytokines*, Academic Press Ltd., p. 335.

Beyaert, R., Heyninck, K., De Valck, D., Boeykens, F., van Roy, F., Fiers, W. (1993b) Enhancement of tumor necrosis factor cytotoxicity by lithium chloride is associated with increased inositol phosphate accumulation. *J Immunol, 151, 291–300.*

Bone, R.C. (1991) The pathogenesis of sepsis. *Ann Intern Med, 115, 457–469.*

Bone, R.C., Balk, R.A., Cerra, F.B., Dellinger, R.P., Fein, A.M., Knaus, W.A., Schein, R.M., Sibbald, W.J. (1992) Definitions for sepsis and organ failure and guidelines for the use of innovative therapies in sepsis. The ACCP/SCCM Consensus Conference Committee. American College of Chest Physicians/Society of Critical Care Medicine. *Chest, 101, 1644–1655.*

Bourne, H.R., Lichtenstein, L.M., Melmon, K.L., Henney, C.S., Weinstein, Y., Shearer, G.M. (1974) Modulation of inflammation and immunity by cyclic AMP. *Science, 184, 19–28.*

Brouckaert, P.G., Leroux-Roels, G.G., Guisez Y., Tavernier J., Fiers, W. (1986) *In vivo* anti-tumour activity of recombinant human and murine TNF, alone and in combination with murine IFN-gamma, on a syngeneic murine melanoma. *Int J Cancer, 38, 763–769.*

Carlson, S.L., Brooks, W.H., Roszman, T.L. (1989) Neurotransmitter-lymphocyte interactions: dual receptor modulation of lymphocyte proliferation and cAMP production. *J Neuroimmunol, 24, 155–162.*

Chowdhury, M.I., Koyanagi, Y., Horiuchi, S., Hazeki, O., Ui, M., Kitano, K., Golde, D.W., Takada, K., Yamamoto, N. (1993) cAMP stimulates human immunodeficiency virus (HIV-1) from latently infected cells of monocyte-macrophage lineage: synergism with TNF-alpha. *Virology, 194, 345–349.*

Chrousos, G.P. (1995) The hypothalamic-pituitary-adrenal axis and immune-mediated inflammation. *N Engl J Med, 332, 1351–1362.*

Cohen, N., Ader, R., Felten, D.L. (1994) Psychoneuroimmunology. In L.H. Sigal and Y. Ron, (eds.), *Immunology and Inflammation: Basic Mechanisms and Clinical Consequences*, New York, McGraw-Hill, pp. 465–494.

Cole, S.W., Korin, Y.D., Fahey, J.L., Zack, J.A. (1998) Norepinephrine accelerates HIV replication *via* protein kinase A-dependent effects on cytokine production. *J Immunol, 161, 610–616.*

de Waal Malefyt, R., Abrams, J., Bennett, B., Figdor, C.G., de Vries, J.E. (1991) Interleukin 10(IL-10) inhibits cytokine synthesis by human monocytes: an autoregulatory role of IL-10 produced by monocytes. *Exp Med, 174, 1209–1220.*

Dhainaut, J.F., Tenaillon, A., Le Tulzo, Y., Schlemmer, B., Solet, J.P., Wolff, M., *et al.* (1994) Platelet-activating factor receptor antagonist BN 52021 in the treatment of severe sepsis: a randomized, double-blind, placebo-controlled, multicenter clinical trial. BN 52021 Sepsis Study Group. *Crit Care Med, 22, 1720–1728.*

Dhainaut, J.F., Vincent, J.L., Richard, C., Lejeune, P., Martin, C., Fierobe, L., *et al.* (1995) CDP571, a humanized antibody to human tumor necrosis factor-alpha: safety, pharmacokinetics, immune response, and influence of the antibody on cytokine concentrations in patients with septic shock. CPD571 Sepsis Study Group. *Crit Care Med, 23, 1461–1469.*

Dunn, A.J. (2000) Effects of the IL-1 receptor antagonist on the IL-1- and endotoxin-induced activation of HPA axis and cerebral biogenic amines in mice. *Neuroimmunomodulation, 7, 36–45.*

Elenkov, I.J., Haskó, G., Kovács, K., Vizi, E.S. (1995) Modulation of lipopolisaccharide-induced tumor necrosis factor-α production by selective α- and α-adrenergic drugs in mice. *J Neuroimmunol, 61, 123–131.*

Elenkov, I.J., Kovács, K., Bertók, L., Vizi, E.S. (1992) Lipopolisaccharide is able to bypass corticotrophin-releasing factor in affecting plasma ACTH and corticosterone levels: evidence from rats with lesion of the paraventricular nucleus. *J Endocrinol, 133, 231–236.*

Elenkov, I.J., Wilder, R.L., Chrousos, G.P., Vizi, E.S. (2000) The sympathetic nerve – an integrative interface between two "supersystems": the brain and the immune system. *Pharm Rev 52, 595–638.*

Elenkov, I.J., Papanicolaou, D.A., Wilder, R.L., Chrousos, G.P. (1996) Modulatory effects of glucocorticoids and catecholamines on human interleukin-12 and interleukin-10 production: Clinical implications. *Proc Ass Am Phys, 108, 1–8.*

Felten, D.L., Felten, S.Y., Bellinger, D.L. (1992) Noradrenergic and peptidergic innervation of lymphoid organs. In J.E. Blalock (ed.), *Chemical Immunology: Neuroimmunoendocrinology*, Basel, Karger, pp. 25–48.

Felten, D.L., Felten, S.Y., Carlson, S.L., Olschowka, J.A., Livnat, S. (1985) Noradrenergic and peptidergic innervation of lymphoid tissue. *J Immunol, 135, 755s–765s.*

Felten, D.L., Felten, S.Y., Bellinger, D.L., Carlson, S.L., Ackerman, K.D., Madden, K.S., Olschowki, J.A., Livnat, S (1987) Noradrenergic sympathetic neural interactions with the immune system: structure and function. *Immunol Rev, 100, 225–260.*

Fisher, C.J. Jr., Dhainaut, J.F., Opal, S.M., Pribble, J.P., Balk, R.A., Slotman, *et al.* (1994) Recombinant human interleukin 1 receptor antagonist in the treatment of patients with sepsis syndrome. Results from a randomized, double-blind placebo-controlled trial. Phase III rhIL-1ra Sepsis Syndrome Study Group. *JAMA, 271, 1836–1843.*

Fuchs, A.C., Granowitz, E.V., Shapiro, L., Vannier, E., Lonnemann, G., Angel, J.B., *et al.* (1996) Clinical, hematologic, and immunologic effects of interleukin-10 in humans. *J Clin Immunol, 16, 291–303.*

Haskó, G., Elenkov, I.J., Kvetan, V., Vizi, E.S. (1995) Differential effect of selective block of α_2-adrenoceptors on plasma levels of tumor necrosis factor-α, interleukin-6 and corticosterone induced by bacterial lipopolisaccharide in mice. *J Endocrinol, 144, 457–462.*

Hoffman, M., Weinberg, J.B. (1987) Tumor necrosis factor-alpha induces increased hydrogen peroxide production and Fc receptor expression, but not increased Ia antigen expression by peritoneal macrophages. *J Leukoc Biol, 42, 704–707.*

James, E.G., Kendal, D.D. and Kendal, M.D. (1996) Peripheral and central neural mechanisms for immune regulation through the innervation of immune effector sites. In J.A. Marsh and M.D. Kendal, (eds.) *The Physiology of Immunity*, CRC Press, pp. 103–127.

Kammer, G.M. (1988) The adenylate cyclase-cAMP-protein kinase A pathway and regulation of the immune response. *Immunol Today, 9, 222–229.*

Kobilka, B. (1992) Adrenergic receptors as models for G protein-coupled receptors. *Annu Rev Neurosci, 15, 87–114.*

Landmann, R. (1992) Beta-adrenergic receptors in human leukocyte subpopulations. *Eur J Clin Invest, 1, 30–36.*

Linder, M.E., Gilman A.G. (1992) G proteins. *Sci Am, 267, 56–61, 64–65.*

Madden, K.S., Felten, S.Y., Felten, D.L., Hardy, C.A., Livnat, S. (1994) Sympathetic nervous system modulation of the immune system. II. Induction of lymphocyte proliferation and migration *in vivo* by chemical sympathectomy. *J Neuroimmunol, 49, 67–75.*

Madden, K.S., Sanders, V.M., Felten, D.L. (1995) Catecholamine influences and sympathetic neural modulation of immune responsiveness. *Annu Rev Pharmacol Toxicol, 35, 417–448.*

Malfait, A.M., Malik, A.S., Marinova-Mutafchieva, L., Butler, D.M., Maini, R.N., Feldmann, M. (1999) The beta2-adrenergic agonist salbutamol is a potent suppressor of established collagen-induced arthritis: mechanisms of action. *J Immunol, 162, 6278–6283.*

Marquet, R.L., Ijzermans, J.N., De Bruin, R.W., Fiers, W., Jeekel, J. (1987) Anti-tumor activity of recombinant mouse tumor necrosis factor (TNF) on colon cancer in rats is promoted by recombinant rat interferon gamma; toxicity is reduced by indomethacin. *Int J Cancer, 40, 550–553.*

McCarthy Pastores, S., Hasko, G., Vizi, E.S., Kvetan, V. (1996) Cytokine Production and its Manipulation by Vasoactive Drugs. *New Horizons, 4, 252–264.*

Michelson, D., Gold, P.W., Sternberg, E.M. (1994) The stress response in critical illness. *New Horiz, 2, 426–431.*

Mills, C.D., Kincaid, K., Alt, J.M., Heilman, M.J., Hill, A.M. (2000) M-1/M-2 Macrophages and the Th1/Th2 paradigm. *J Immunol, 164, 6166–6173.*

Mire-Sluis, A., Thorpe, R. (eds.), (1998) *Cytokines*, Academic Press.

Natanson, C., Hoffman, W.D., Suffredini, A.F., Eichacker, P.Q., Danner, R.L. (1994) Selected treatment strategies for septic shock based on proposed mechanisms of pathogenesis. *Ann Intern Med, 120, 771–783.*

Natanson, C., Hoffman, W.D., Parrillo, J.E. (1995) Septic shock and multiple organ failure. In J.E. Parrillo and R.C. Bone, (eds.), *Crtitical Care Medicine. Principles of Diagnosis and Management*, St. Louis, Mosby, pp. 355–375.

Nickoloff, B.J, Karabin, G.D., Barker, J.N., Griffiths, C.E., Sarma, V., Mitra, *et al.* (1991) Cellular localization of interleukin-8 and its inducer, tumor necrosis factor-alpha in psoriasis. *Am J Pathol, 138, 129–140.*

Nokta, M.A., Pollard, R.B. (1992) Human immunodeficiency virus replication: modulation by cellular levels of cAMP. *AIDS Res Hum Retroviruses, 8, 1255–1261.*

Parrillo, J.E., Parker, M.M., Natanson, C., Suffredini, A.F., Danner, R.L., Cunnion, R.E., Ognibene, F.P. (1990) Septic shock in humans. Advances in the understanding of pathogenesis, cardiovascular dysfunction, and therapy. *Ann Intern Med, 113, 227–242.*

Reichlin, S. (1993) Neuroendocrine-immune interactions. *N Engl J Med, 329: 1246–1253.*

Rothwell, N.J. (1995) *Immune responses in the Nervous System*. Biol Scientific Publishers, Oxford.

Rothwell, N.J. (1998) Cytokines – killers in the brain? *J Physiol, 514, 3–17.*

Rothwell, N.J. (2000) Show them how it's really done. *Nature, 405: 621.*

Schulz, J.B., Matthews, R.T., Jenkins, B.G., Ferrante, R.J., Siwek, D., Henshaw, D.R., *et al.* (1995) Blockade of neuronal nitric oxide synthase protects against excitotoxicity *in vivo*. *J Neurosci, 15, 8419–8429.*

Selmaj, K.W., Raine, C.S. (1988) Tumor necrosis factor mediates myelin and oligodendrocyte damage *in vitro*. *Ann Neurol, 23, 339–346.*

Selmaj, K., Raine, C.S., Cannella, B., Brosnan, C.F. (1991) Identification of lymphotoxin and tumor necrosis factor in multiple sclerosis lesions. *J Clin Invest, 87, 949–954.*

Shear, M.J., Turner, F.C., Perrault, A. (1943) Chemical treatment of tumors. V. Isolation of the hemorrhage producing fracion from *Serratia mercescens* culture filtrate. J Natl Cancer Inst, 4, 81–97.

Shimizu, N., Hori, T., Nakane, H. (1994) An interleukin-1 beta-induced noradrenaline release in the spleen is mediated by brain corticotropin-releasing factor: an *in vivo* microdialysis study in conscious rats. *Brain Behav Immun, 8, 14–23.*

Song, S.K., Karl, I.E., Ackerman, J.J., Hotchkiss, R.S. (1993) Increased intracellular Ca2+: a critical link in thepathophysiology of sepsis? *Proc Natl Acad Sci USA, 90, 3933–3937.*

Szelényi, J., Kiss, J.P., Vizi, E.S. (2000) Differential involvement of sympathetic nervous system in the modulation of TNF-α production by α_2 and β-adrenoceptors in mice. *J Neuroimmunol, 103, 34–40.*

Suberville, S., Bellocq, A., Fouqueray, B., Philippe, C., Lantz, O., Perez, J., Baud, L. (1996) Regulation of interleukin-10 production by beta-adrenergic agonists. *Eur J Immunol, 26, 2601–2605.*

Tisdale, M.J. (1997) Cancer cachexia: metabolic alterations and clinical manifestations. *Nutrition, 13, 1–7.*

Vizi, E.S. (1998) Receptor-mediated local fine-tuning by noradrenergic innervation of neuroendocrine and immune systems. *Ann NY Acad Sci, 851, 388–396.*

Vizi, E.S. (2000) The sympathetic nerve – an integrative interface between two "supersystems": the brain and the immune system. *Pharmacol Reviews, 52, 63–89.*

Vizi, E.S., Mike, Á. and Tarnawa, I. (1996) 2,3-Benzodiazepines (GYKI 52466 and analogs): negative allosteric modulators of AMPA receptors. *CNS Drug Reviews, 2, 91–126.*

Vizi, E.S., Orso, E., Osipenko, O.N., Hasko, G., Elenkov, I.J. (1995) Neurochemical, electrophysiological and immunocytochemical evidence for a noradrenergic link between the sympathetic nervous system and thymocytes. *Neuroscience 68, 1263–1276.*

Wekerle, H. (1998) The viral triggering of autoimmune disease. *Nature Med, 4, 770–771.*

Wilder, R.L. (1995) Neuroendocrine-immune system interactions and autoimmunity. *Annu Rev Immunol, 13, 307–338.*

Woiciechowsky, C., Asadullah, K., Nestler, D., Eberhardt, B., Platzer, C., Schoning, B., *et al.* (1998) Sympathetic activation triggers systemic interleukin-10 release in immunodepression induced by brain injury. *Nature Med, 4, 808–813.*

Chapter 14

Regulation of B lymphocyte function by crosstalk between receptors with stimulatory and inhibitory activity

János Gergely and Anna Erdei

Contents

14.1 Introduction

Recognition and the transfer of information between immune cells are mediated by membrane bound receptors, by the crosstalk between various receptors expressed on the same cell and by direct and/or indirect interaction of various immunocompetent cells. Activation of these cells is mainly induced by receptors belonging to the family of *multisubunit immune recognition receptors* (MIRR), the members of which are characterized by a complex heterooligomeric structure in which ligand binding and signal transducing functions are segregated into distinct receptor substructures (Figure 14.1). Members of the MIRR receptor family, such as the antigen receptors of T lymphocytes (TCR – T-cell receptor), that of the B lymphocytes (BCR – B-cell receptor) and certain types of receptors interacting with the Fc parts of immunoglobulins (FcRs – Fc receptors) (Cambier *et al.*, 1993) transduce signals by very similar mechanisms (Lin *et al.*, 1994).

Due to interactions with counter receptors on partner cells and/or co-crosslinking of various receptors on the same cell, several other receptors may be affected simultaneously in addition to the members of the MIRR family. The involvement of other receptors (co-receptors) may alter the quality and magnitude of the responses induced by MIRR receptors: e.g. the co-crosslinking of type 2 complement receptors (CR2) and BCR enhances (Tedder *et al.*, 1997), while that of BCR and FcγRIIb inhibits BCR-mediated activation of B-cells (reviewed in Gergely *et al.*, 1996). The control of B lymphocyte function is essential for the development of the humoral immune response.

To find targets for potential therapeutic substances that may influence the signaling cascade/s associated to a receptor, one has to know in detail the signaling pathways induced by the receptor itself and those activated by the corresponding co-receptor/s. The advantage of this approach is that it might be possible to interfere with cellular events independently from the initial stimuli. One must also consider that the functional consequences of the receptor–co-receptor cooperation may be a quantitative issue, i.e. it may depend on the number of both types of receptors involved in the "crosstalk". To illustrate receptor interactions of this sort, this chapter focuses on BCR–FcγRIIb "crosstalk" inhibition, and BCR–CD19/CD21/81 complex interaction, enhancing BCR-induced activation of B-cells.

14.2 Tyrosine-based activation and inhibition motifs of MIRR-family receptors

IMMUNORECEPTOR TYROSINE-BASED ACTIVATION MOTIF (ITAM)

The B-cell antigen receptor complex (BCR) consists of the membrane-bound immunoglobulin molecule (mIg), representing the ligand binding unit of the receptor, and of the signaling unit, composed of the noncovalently associated disulphide-linked heterodimer of Igα and Igβ chains (Reth *et al.*, 1991). Each of the signal transducing subunits contain two copies of a conserved sequence module in the cytoplasmic tail (Flaswinkel *et al.*, 1995), termed *immunoreceptor tyrosine-based activation motif* (ITAM) that endows the antigen receptors with specific binding sites for nonreceptor PTKs. The

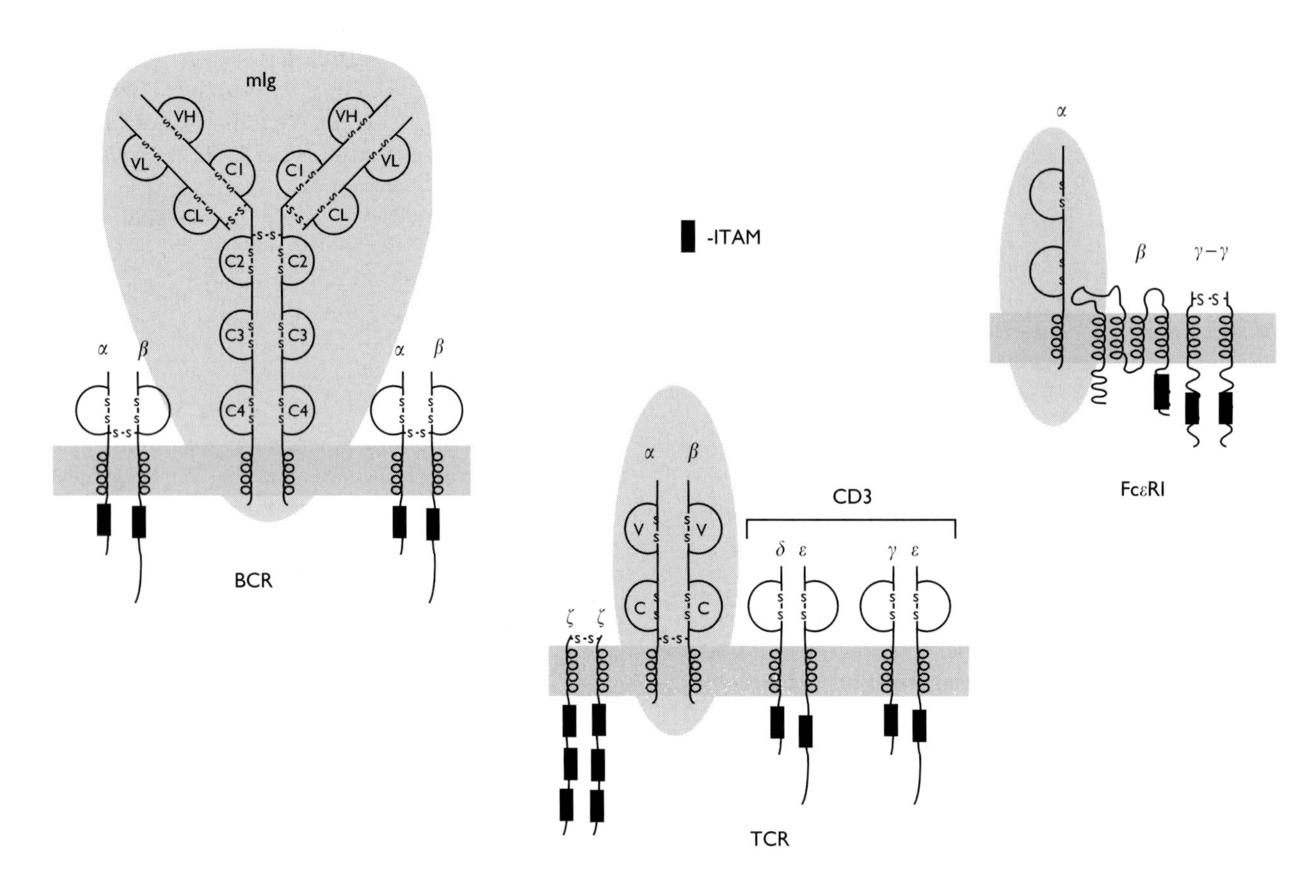

Figure 14.1 Multichain immune recognition receptors. Antigen receptors on B-cells (BCR), T-cells (TCR) and the high affinity IgE-binding Fc-receptor (FcεRI) contain ligand-binding polypeptides (shaded background) and ITAM-bearing signaling chains.

following structural and functional features characterize the ITAM:

- ITAMs are composed of a twice-repeated YXXL sequence flanking 7 variable residues: D/E X7 D/E X_2 ***YXXL/I*** X_7 ***YXXL/I***.
- Upon receptor aggregation, ITAMs rapidly become tyrosine-phosphorylated by src PTKs;
- P-ITAMs provide docking sites for Src-homology (SH2) domains of cytosolic PTKs of the syk family;
- P-ITAM-bound PTKs (syk or ZAP-70) become phosphorylated by src kinases or transphosphorylated by each other, and once activated, they induce downstream signaling pathways (including phosphorylation, recruitment of signaling and adaptor molecules, activation of GTP-binding molecular switches, second messenger generation and gene transcription).

IMMUNORECEPTOR TYROSINE-BASED INHIBITION MOTIF (ITIM)

In FcγRIIb-mediated inhibition of B-cell activation, a highly conserved 13 amino acid region has been described within the intracytoplasmic tail of the murine FcγRIIb, containing a single YXXL motif. The tyrosine and leucine within this motif seem to be a critical element in the FcγRIIb mediated inhibitory function (Miettinen *et al.*, 1992; Daeron *et al.*, 1993; Muta *et al.*, 1994). Recently it has been found that, in addition to the FcγRIIb, a whole family of ITIM-bearing negative co-receptors exists (Vivier *et al.*, 1997; Daeron, 1996). The following structural and functional features characterize the ITIM:

- A single YXXL;
- the mandatory Y is followed, at position Y+3, by a L or V;
- the tyrosine is generally preceded by a small hydrophobic residue at the −2 position;
- phosphorylation of the tyrosine residue leads to the recruitment of SH2-containing phosphatases of two distinct groups: protein tyrosine phosphatases SHP-1 and SHP-2; polyinositol phosphate phosphatase SHIP;
- the association of ITIM-bearing receptors with phosphatases and the dephosphorylation of PTK substrates and/or that of PIP3 represent critical steps in their inhibitory function;
- The ITIM in human FcγRIIB: A E N T I T ***Y S L L*** M H P.

14.3 Early events in BCR-mediated signal transduction

It has been suggested that in BCR-stimulated B-cells a pre-formed BCR transducer complex (consisting of kinases, phosphatases and their substrates and adaptor proteins which connect physically and functionally to these elements) (Wienands *et al.*, 1996; Wienends *et al.* 1998) maintains signal specificity and avoids interference with other receptors that are not stimulated. The first step in BCR-mediated activation of B-cells is the ligation and aggregation of the receptor complexes, resulting in the immediate tyrosine phosphorylation of several cellular protein components, such as Src-PTKs Lyn, Fyn and Blk, and followed within minutes by activation of the non-Src related kinases, Btk and Syk (reviewed in Campbel, 1999, Figure 14.2). The membrane-proximal tyrosines of the ITAMs in Igα and Igβ are the first targets of phosphorylation by activated Lyn, which can then bind tandem SH2 domains of Syk (a structural homolog of ZAP-70 of T-cells) (Kurosaki *et al.*, 1994). It has been shown that the phosphotyrosine phosphatase CD45 subtly mediates early signaling events such as tyrosine phosphorylation and Ca^{2+} mobilization (Brown *et al.*, 1994; Katagiri *et al.*, 1995). CD45 is physically associated with Igα-Igβ and with Lyn, which is a selective *in vivo* substrate for CD45. The interaction of Syk kinase with the phosphphorylated ITAMs of Igα and Igβ results in cross-phosphorylation and activation of several src family PTKs and in their enhanced binding to the signaling unit of the BCR complex. These early events of the

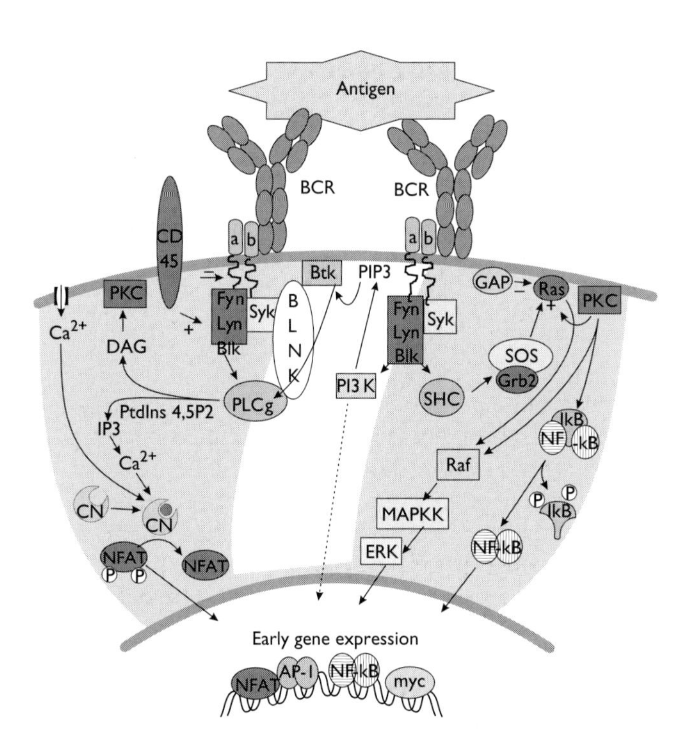

Figure 14.2 Intracellular signaling events after crosslinking B-cell surface immunoglobulins by antigen.

BCR complex-mediated signaling processes are coupled to at least three signal transduction pathways, the activation of which leads to transcriptional activation of a number of immediate early genes and, finally, to B-cell proliferation and differentiation or apoptosis of the cells. These pathways are:

(1) *The phosphatidylinositol (PI)-specific phospholipase C (PLC) pathway* (Coggeshall *et al.*, 1992). The crosslinking of BCR complexes and the activation of Btk and Syk result in phosphorylation, activation and membrane translocation of PLCγ1 and PLCγ2. The translocation as well as the phosphorylation of PLCγ on at least two tyrosine residues are required steps for receptor-mediated PI hydrolysis. These PLC isoforms convert phosphatidylinositol 4,5-bisphosphate (PIP2) to two intracellular second messengers: diacylglycerol (DAG) and inositol 1,4,5-trisphosphate (IP3). DAG remains in the membrane and induces membrane translocation and activation of the serine/threonine kinase PKC. By engaging its receptor on the endoplasmic reticulum, IP3 activates the release of intracellular Ca^{2+} stores leading to Ca influx from the extracellular space. This first phase of Ca^{2+} influx occurs very early, within seconds of BCR stimulation (Bootman *et al.*, 1995; Sugawara *et al.*, 1997), while the second phase occurs later, mediated by the entry of Ca^{2+} from extracellular sources. Several of the Ca^{2+}-activated calmodulin and PKC regulated effectors have been implicated in regulation of gene expression.

(2) *p21ras activation.* BCR ligation also induces activation of the ras proto-oncogene product p21ras, which binds to guanine nucleotides. The active, GTP-bound p21ras triggers a kinase cascade including the serine/threonine kinase Raf-1, which in activated form, phosphorylates and activates MAP-(mitogen activated protein) kinases. The latter play a role in regulation of gene expression via phosphorylation of transcription factors including c-jun, c-fos and c-myc, which are involved in regulation of several cellular processes (Lazarus *et al.*, 1993; Kawauchi *et al.*, 1994; Hashimoto *et al.*, 1998). p21ras can be activated via a PKC independent pathway as well (Harwood *et al.*, 1993). During this process Shc (a highly conserved SH2 domain and a collagen-like sequence containing protein) plays an important role (Saxton *et al.*, 1994). After its phosphorylation Shc interacts with the adapter protein Grb-2 that in turn reacts with the guanine nucleotide exchange factor for Ras, mSos. Signaling by BCR stimulates phosphorylation of Shc and Sos and induces the formation of membrane-associated complexes containing Shc, Grb-2, Sos and a 145/kDa protein (Smit *et al.*, 1996).

(3) The *PI3-kinase pathway* is involved in signaling mechanisms leading to cell proliferation, survival and cell motility (Yamanashi *et al.*, 1992). The PI3-Kα-mediated phosphorylation of its regulatory subunit p85 decreases the kinase activity of the enzyme (reviewed in Bondeva *et al.*, 1998). The lipid products of PI3-Ks activate certain calcium-independent PKCs and bind to a subset of SH2 domains

modifying their actions. PtdIns-3,4,5-P3 was found to bind the C-terminal SH2 domain of PLCγ and the generation of PtdIns-3,4,5-P3 by receptor-associated PI3-K causes an increase in IP3 production and intracellular calcium release (Rameh *et al.*, 1998). In addition the lipid products of PI3-K can bind and/or stimulate pleckstrin homology (PH) domain containing proteins including the antiapoptotic protein kinase AKT. Furthermore, the p110 catalytic subunit of certain PI3-Ks binds the GTP-binding protein Ras, and consequently PI3-K is a direct effector of Ras (Franke *et al.*, 1997 and reviewed in Toker *et al.*, 1997).

Recently it has been shown that linker or adapter proteins provide mechanisms by which receptors can amplify and regulate downstream effector proteins. The cytosolic adapter protein *BLNK* (B-cell linker protein) represents a central linker molecule that bridges the BCR-associated kinases with several signaling pathways that link the BCR-associated Syk with PLCγ, the Vav guanine nucleotide exchange factor, and the Grb2 and Nck adapter proteins. Syk-mediated tyrosine phosphorylation of BLNK is one of the earliest events in BCR-induced cell activation (Fu *et al.*, 1998; Wienands *et al.*, 1998) and provides docking sites for these molecules that, in turn, allow the phosphorylation and/or activation of their respective signaling pathways (Fu *et al.*, 1998). BLNK possesses a phosphotyrosine-binding SH2 domain containing six tyrosines in the context of YXXP which, if phosphorylated, may interact with the SH2 domains of Vav, Nck, PLCγ and Grb2. In addition, the proline-rich domain of BLNK binds to the two SH3 domains of Grb2 which mediates Grb2-Sos complex recruitment. Finally, phosphorylation of the adaptor complex results in its translocation from the cytosol to the plasma membrane.

In addition to BLNK, several other adaptor proteins are also involved in the activation of signaling events following BCR-clustering. The multisubstrate docking protein *Gab1* contains an amino-terminal PH domain as well as multiple tyrosine residues which are potential binding sites for proteins that contain SH2 or phosphotyrosine binding (PTB) domains. Phosphorylated Gab1 binds to the SH2 domains of Shc, PI3-K and protein-tyrosine phosphatase, SHP-2 (Ingham *et al.*, 1998).

Finally, it is worth mentioning the regulatory linker $\alpha 4$ protein, which binds to and regulates the catalytic activity of a serine/threonine phosphatase (PPA2). It has been found that rapamycin disrupts the $\alpha 4$-PP2A complex and this points to the possibility that this regulatory linker protein is a component of the rapamycin-sensitive signaling pathway (Inui *et al.*, 1998).

14.4 CD22

CD22 is a lectin-like B-cell-specific membrane protein, the cytoplasmic domain of which contains at least one ITAM and at least three ITIM-like motifs. Consequently the molecule is a positive and negative regulator of BCR-induced cell activation. CD22 is a member of the sialoadhesin subclass of the immunoglobulin superfamily, its N-terminal Ig-domains bind Siaα2–6Galβ1–4GlcNAc, contained in glycoconjugates on the surface of several cell types, such as activated endothelial cells, erythrocytes and monocytes. The molecule is involved in cell adhesion, required for normal antibody responses to thymus-dependent antigens, regulates the lifespan of mature B lymphocytes and plays a role in regulating the threshold of BCR signaling. CD22 reacts preferentially with T lymphocytes, suggesting that the function of this protein on B-cells is restricted mainly to the secondary lymphoid organs where T-cells provide the necessary second signals. The engagement of CD22 by T-cell surface molecules has proved to be a potent costimulatory signal for B lymphocytes. In CD22-deficient mice the resting B-cells resemble chronically stimulated B-cells.

The ITIMs of CD22 are phosphorylated on their tyrosine residues upon ligation of BCR and its cytoplasmic domain binds proteins containing SH2 domains. Syk co-immunoprecipitates with CD22 either via the SH2 domains of the latter, or by interaction with α/β signaling chains of the mIg complex. Tyrosine-phosphorylated CD22 then associates with phosphatase SHP-1, a putative negative regulator of the B lymphocyte (Sato *et al.*, 1996; Nitschke *et al.*, 1997; Tedder *et al.*, 1997; Nadler *et al.*, 1997).

14.5 FCγRIIb-mediated negative regulation of ITAM-dependent B-cell activation

The feedback suppression of B-cells as a consequence of negative signaling induced by immune complexes containing the antigen and the secreted specific antibody is well documented (reviewed in Coggeshall, 1998; Gergely *et al.*, 1999). This form of negative signaling is mediated by co-ligation of BCR and the IgG-binding FcγRIIb, and can be studied using an *in vitro* model (Phillips *et al.*, 1984) in which anti-Ig antibodies of IgG isotype are used to co-crosslink the two receptors on B-cells (Figure 14.3).

FcγRs (with the exception of the PI-linked FcγRIIIb) are integral membrane proteins. FcγRs, which mediate activation, signal to cells possessing one or several ITAMs in the cytoplasmic tail of their signal transduction subunits, and are expressed on a wide range of cells. In contrast, the FcγRIIb constitutively expressed on B-cells possess ITIM in their intracytoplasmic domain and mediate inhibitory signals to BCR-activated cells.

The single chain FcγRII receptors are encoded by three structurally distinct similar genes (FcγRIIA, IIB and IIC) resulting in six different transcripts (FcγRIIa and a', b1, b2, b3 and c (Warmerdam *et al.*, 1990)). There is a structural difference between the cytoplasmic tail of the ITIM-bearing

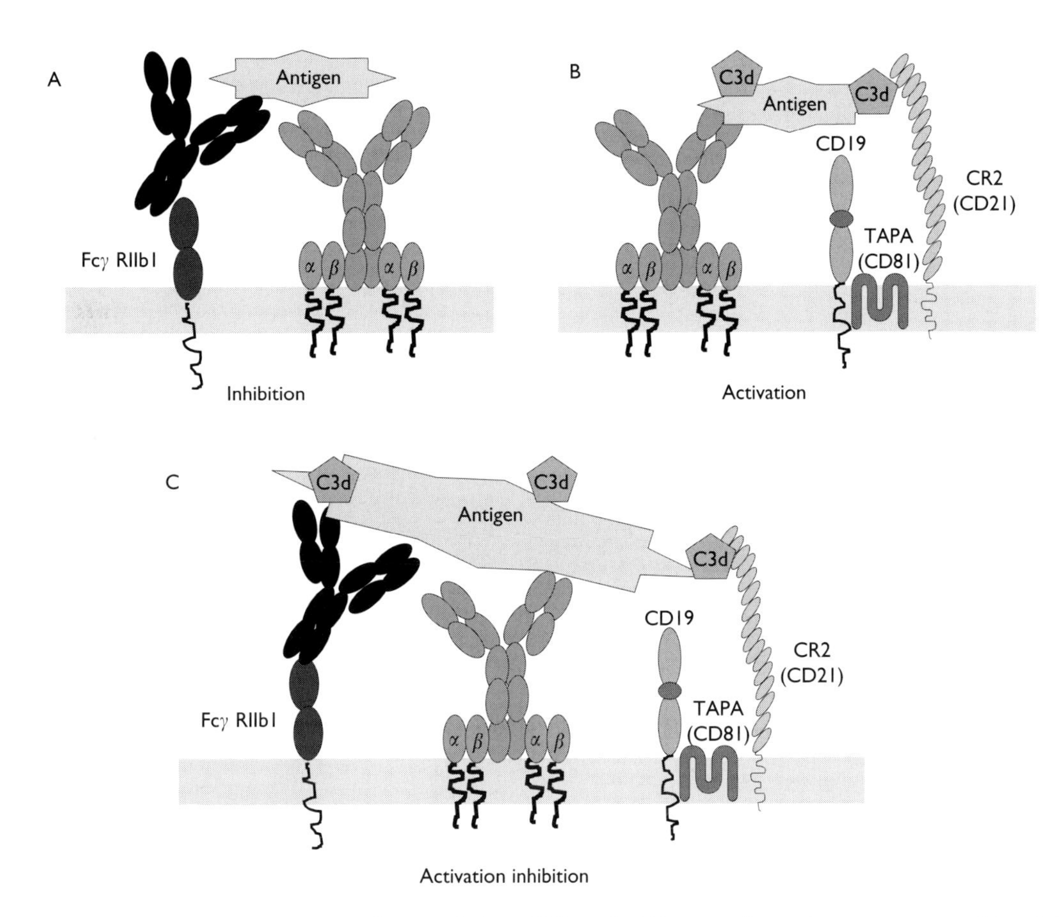

Figure 14.3 Regulation of B lymphocyte activation by crosslinking BCR with different cell membrane structures. Crosslinking BCR with FcγRIIb1 results in inhbition (A), crosslinking BCR with the CD21, CD19-CD81 complex leads to activation (B), while co-crosslinking of the three cell membrane structures results in inhibition of activation.

alternatively spliced isoforms FcγRIIb1 and b2. The cytoplasmic part of b1 is longer (47 amino acids in murine and 19 in human) than that of b2, and due to this diversity the FcγRIIb2 mediates internalization but does not cause capping, while the b1 isoform induces capping but does not internalize (Miettinen *et al.*, 1992). Human B-cells express both FcγRIIb1 and b2, while murine B-cells express only FcγRIIb1. Following cell activation an alteration in number, ligand binding capacity and release from the cell membrane of the FcγRII expressed on human resting B-cells has been observed (Sarmay *et al.*, 1991; Gergely *et al.*, 1992). Cell activation induces the splicing of FcγRIIb1 mRNA, while the alternative splicing of FcγRIIb2 mRNA is downregulated and results in increased surface expression of FcγRIIb1 (Sarmay *et al.*, 1995). Co-clustering of BCR and FcγRII is known to inhibit Ca^{2+} influx in murine B-cells without affecting the release of Ca^{2+} from intracellular stores (Choquet *et al.*, 1993). In contrast, in human B-cells co-crosslinked with BCR, both ITIM-bearing FcγRII isoforms (b1 and b2) inhibit both the Ca^{2+} influx from the extracellular space and its release from the endoplasmic reticulum pool (Koncz *et al.*, 1998).

INHIBITORY SIGNALING BY ITIM-BEARING FCγRIIb IN B-CELLS

The co-ligation of BCR and FcγRIIb on the surface of B-cells results in tyrosine phosphorylation of FcγRII-ITIM which recruits the SH2 domain-containing phosphatases, the protein tyrosine phosphatases SHP1 and SHP2 (D'Ambrosio *et al.*, 1995; D'Ambrosio *et al.*, 1996) and the polyphosphoinisitol 5′-phosphatase, SHIP (Chacko *et al.*, 1996; Ono *et al.*, 1996; Sarmay *et al.*, 1997; Nakamura *et al.*, 2000) to the co-clustered receptors. These result in the translocation of SH2 domain-containing regulatory cytoplasmic molecules to the cell membrane. The association of these phosphatases with the phosphorylated ITIM (P-ITIM) suggest that negative signaling following BCR-FcγRIIb co-clustering might be a consequence of dephosphorylation of specific PTK substrates.

Phosphorylation of FcγRIIb ITIM as well as interaction of P-ITIM with phosphatases SHIP and SHP-2 are among the earliest events in BCR-FcγRIIb coligation-induced inhibition of B-cell activation. Due to BCR aggregation, a number of signaling molecules associated with FcγRIIb are phosphorylated. Since Lyn, Shc, SHP-2 and PI3K were found to associate with both P-ITIM and BCR-coaggregated FcγRIIb (Sarmay *et al.*, 1999), the significance of these molecules in the early phase of FcγRIIb-mediated inhibition of B-cell activation seems to be obvious. In the negatively regulated B-cells, Lyn is responsible for ITIM phosphorylation (Nishizumi *et al.*, 1998; Sarmay *et al.*, 1999). Most likely the next step is the recruitment of Shc, SHP-2, SHIP and PI3K (p85) as a consequence of the interaction of their SH2 domain with the P-ITIM of FcγRIIb as well as with tyr-P residues of some of the interacting molecules. Recently the participation of the docking protein Gab-1 in the molecular complex was also demonstrated (Koncz *et al.*, 2001). In addition it was shown that SHIP and Shc are released from the P-ITIM-recruited molecular complexes as a consequence of the bound phosphatase activity (Koncz *et al.*, 1999). Dephosphorylation of these components can be attributed to SHP-2, i.e. this protein tyrosine phosphatase may play a critical role in the FcγRIIb-dependent inhibition process. Dephoshorylation of Shc is particularly important in the chain of events of FcγRIIb-mediated negative regulation since it leads to inhibition of the ras/MAPK pathway (Sarmay *et al.*, 1996).

Recently, a number of new members of the ITIM-bearing receptor family were described and the ITIM-concept of negative regulation of ITAM-induced activation of immune competent cells was extended to nonhemopoietic cells as well (Vely *et al.*, 1997). It is worth mentioning that one can find several contradictory data concerning the mechanism of ITAM-induced cell activation and ITIM-mediated inhibition. However, comparing such observations one has to consider that in various species and cells a different selection of PTKs is available for phosphorylation of the receptors in question. It is also evident that various selections of signaling molecules exist in the immediate proximity of the receptors, and consequently the composition of transducer complexes may be different. Similarly, in the microenvironment of ITIM-bearing receptors different phosphatases are available for interaction with P-ITIM, i.e. depending on the species and cells, SHP-1, SHP-2 or SHIP might be the key phosphatases that initiate the chain of inhibitory events. These differences may explain the incidental differences and one has to emphasize that all these must be considered when relevant target molecules are searched.

14.6 Positive regulation of B-cell activation mediated by the CD19-CD21-CD81 complex

Involvement of complement, particularly activation fragments of component C3 in triggering and modulating the acquired immune response has long been recognized (Pepys, 1974). The role of opsonization in various forms of complement or complement receptor deficiencies have been demonstrated (reviewed in Carroll and Fischer, 1997; Heyman, 1994), especially in primary T-dependent immune responses. The interplay between complement and acquired immunity depends very much on the covalent deposition of C3-fragments – such as C3b and C3d – on the surface of pathogens/antigens, which readily interact with cellular receptors expressed on various cells mediating versatile functions (for review see Fearon, 1998; Prodinger *et al.*, 1999).

Complement receptors type 1 (CR1 or CD35) and type 2 (CR2 or CD21) expressed by B-lymphocytes are single-chain polypeptides belonging to the family of molecules containing Short Consensus Repeat (SCR) units (Reid *et al.*, 1986). These proteins are encoded by the *regulators of complement activation* gene cluster on chromosome 1 (Fearon *et al.*, 1995). While the role of CR1 on B-cells is not yet understood, CR2 expressed by these cells are known to be essential for the development of the humoral immune response (Carroll, 1999).

CR2 is expressed in higher density on mature B lymphocytes and on follicular dendritic cells, while it is present in lower amounts on T lymphocytes and on epithelial cells. The receptor for the Epstein-Barr virus, human CD21, is also thought to be involved in infection by this pathogen. The extracellular portion of CR2 is composed entirely of SCRs, modules of about 60 amino acids stabilized by two disulfide-bonds. It is the two N-terminal, membrane-distal SCRs which are essential for ligand binding. The SCRs are linked by flexible sequences of approximately 8 amino acids. The glycosylated 145 kDa cell membrane protein expressed on B-lymphocytes contains 15 SCRs in its extracellular portion, followed by a transmembrane region of 24 residues and a cytoplasmic tail of 34 amino acids. This intracellular sequence is not involved in signal transduction, however CR2 forms a non-covalently linked complex via its transmembrane region with CD19 and CD19-associated CD81 (Tedder *et al.*, 1997; Tsubata, 1999).

As mentioned earlier, to modulate B-cell activation through mIg, B lymphocytes integrate signals from at least three membrane proteins: FcγRIIb1, CD19 and CD22 (Doody *et al.*, 1996). While FcγRIIb1 – along with cell membrane molecule CD22 – mediate negative signals, CD19, appearing in a complex with CD21 and CD81 (or TAPA1), is a positive regulator of B-cell activation (Figure 14.3). The glycoprotein, CD19 is present on the surface of B-cells throughout their development until the plasma cell stage. Crosslinking with mIg results in phosphorylation of tyrosine residues in the cytoplasmic tail of CD19 followed by association with signaling molecules such as Vav, PI3-kinase, lyn and fyn. Recently it has been shown that coligation of CD19 and mIgM have a synergistic effect on phosphorylation of MAP kinases erk-2, JNK and p38. The only known ligand for CD19 is CD21, the receptor for complement fragment C3d, and also for EBV. C3 fragments are generated upon activation of the complement cascade, and the ligand for CD21 is

C3d, attached covalently to the antigen. This complex has the capacity to crosslink the CD19-CD21 complex to mIg on antigen specific B-cells (Figure 14.3). It has been shown in elegant experiments, that coligation of the CD19-CD21-CD81 complex with mIg reduces the amount of antigen required for B-cell activation by 10- to 1000-fold (Dempsey *et al.*, 1996). The immunogenicity of hen egg lysosyme (HEL) coupled to 1 to 3 copies of C3d was shown to act both *in vitro* and *in vivo*. Antigen-C3d complexes are able to co-ligate mIg and CR2 in the subset of B-cells that recognize the antigen in the immune complex via their BCR. This event has been shown to decrease the affinity threshold required for signaling (Fearon and Carter, 1995) thereby increasing the efficiency of activation of antigen specific B-cell clones.

14.7 Conclusion and future perspectives

Our increased understanding of the signal transduction pathways implicated in the BCR mediated B lymphocyte activation and the regulation of this activation process by the CD19-CD21-CD81 complex, CD22 and/or FcγRIIb has opened the door for the search for original drugs to use in treating diseases connected to disorders of signaling mechanisms. Although potent and selective modulators of the signaling pathways leading to the inhibition of B lymphocyte activation or to suspension of ITIM-dependent inhibitory events are not known yet, substances influencing the signaling mechanisms in other cells resulting in inhibition of immune mediated events have already been described. For example, the p38 MAP kinase is one of the most studied signaling molecules in this regard, and its relevance in disease states has been established primarily through the use of p38 MAP kinase inhibitors, such as the pyridinylimidazole compound SB 203580. The use of this specific inhibitor has led to the dissection of pathways where various substrates of p38 are involved. It is known that the regulation of cytokine biosynthesis in many different cell types is through activation of p38 MAP kinase and it has been demonstrated that p38 MAP kinase inhibitor decreased cytokine production in a number of cells (reviewed in Lee *et al.*, 2000). The role of rapamycin in influencing signal transduction in lymphocytes through binding to the catalytic subunit of protein phosphatase PPA2 was mentioned above (Inui *et al.*, 1998). The modulation of neutrophil responses to CD32 by potent inhibitors of PP1 and PPA2 serine/threonine phosphatases has also been shown. All these recent findings point to possible cross-talk between serine/threonine and tyrosine phosphorylation (Rollet-Labelle *et al.*, 2000). Another type of therapeutic approach is based on targeting drugs to CD19-associated tyrosine kinases. In these experiments CD19-specific monoclonal antibodies conjugated to tyrosine kinase inhibitors such as genistein were found to promote the survival of mice with B lymphomas (Uckun *et al.*, 1995).

The discovery of novel drugs specifically influencing the function of signaling molecules and improvement of targeting techniques by better utilization of monoclonal antibodies and/or molecular constructs with similar function will lead to new successful therapeutic interventions in the near future.

References

Bondeva, T., Pirola, L., Bulgarelli-Leva, G., Rubio, I., Wetzker, R., and Wymann, M.P. (1998) Bifurcation of lipid and protein kinase signals of PI3Kγ to the protein kinases PKB and MAPK. *Science 282, 293–296.*

Bootman, M.D., and Berridge, M.J. (1995) The elemental principles of calcium signaling *Cell 83, 675–678.*

Brown, V.K., Ogle, E.W., Burkhardt, A.L., Rowley, R.B., Bolen, J.B., and Justement, L.B. (1994) Multiple components of the B cell antigen receptor complex associate with the protein tyrosine phosphatase, CD45. *J. Biol. Chem. 269, 17238–17244.*

Cambier, J.C., Bedzyk, W., Campbell, K., Chien, N., Friedrich, J., Harwood, A., *et al.* (1993) The B-cell antigen receptor: structure and function of primary, secondary, tertiary and quaternary components. *Immunol. Rev. 132, 85–106.*

Campbell, K.S. (1999) Signal transduction from the B cell antigen-receptor. *Curr. Opin. Immunol. 11, 256–264.*

Carroll, M.C. (1999) Role of complement receptors CD21/CD35 in B lymphocyte activation and survival. *Curr. Top. Microbiol. Immunol. 246, 63–68.*

Carroll, M.C. and Fischer, M.B. (1997) Complement and the immune response. *Curr. Opin. Immunol., 9, 64–69.*

Chacko, G.W., Tridandapani, S., Damen, J.E., Liu, L., Krystal, G., and Coggeshall, K.M. (1996) Negative signaling in B lymphocytes induces tyrosine phosphorylation of the 145-kDa inositol polyphosphate 5-phosphatase, SHIP. *J. Immunol. 157, 2234–2238.*

Choquet, D., Partiseti, M., Amigorena, S., Bonnerot, C., Fridman, W.H., and Korn, H. (1993) Cross-linking of IgG receptors inhibits membrane immunoglobulin-stimulated calcium influx in B lymphocytes. *J. Cell. Biol. 121, 355–363.*

Coggeshall, K.M. (1998) Inhibitory signaling by B cell FcγRIIb. *Curr. Opin. in Immunol. 10, 306–312.*

Coggeshall, K.M., McHugh, J.C., and Altman, A. (1992) Predominant expression and activation-induced tyrosine phosphorylation of phospholipase Cγ2 in B lymphocytes. *Proc. Natl. Acad. Sci. USA 89, 5660–5664.*

Daeron, M. (1996) Building up the family of ITIM-bearing negative coreceptors. *Immunol. Lett. 54, 73–76.*

Daeron, M., Malbec, O., Latour, S., Bonnerot, C., Segal, D.M., and Fridman, W.H. (1993) Distinct intracytoplasmic sequences are required for endocytosis and phagocytosis via murine fcγRII in mast cells *Int. Immunol. 5. 1393–1401.*

D'Ambrosio, D., Fong, D.C., and Cambier, J.C. (1996) The SHIP phosphatase becomes associated with Fc gammaRIIB1 and is tyrosine phosphorylated during "negative" signaling. *Immunol. Lett. 54, 77–82.*

D'Ambrosio, D., Hippen, K.L., Minskoff, S.A., Mellman, I., Pani, G., Siminovitch, K.A., and Cambier, J.C. (1995) Recruitment and activation of PTP1C in negative regulation of antigen receptor signaling by Fc gamma RIIB1. *Science 268, 293–297.*

Dempsey, .P.W., Allison, M.E.D., Akkaraju, S., Goodnow, C.C., and Fearon, D.T. (1996) C3d of complement as a molecular adjuvant: bridging innate and acquired immunity. *Science, 271, 348–350.*

Doody, G.M., Dempsey, P.W., and Fearon, D.T. (1996) Activation of B lymphocytes: integrating signals from CD19, CD22 FcγRIIb1. *Curr. Opin. Immunol. 8, 378–382.*

Fearon, D.T. (1998) The complement system and adaptive immunity. *Sem. Immunol. 10, 355–361.*

Fearon, D.T., and Carter, R.H. (1995) The CD19/CR2/TAPA-1 complex of B lymphocytes: linking natural to acquired immunity. *Annu. Rev. Immunol. 13, 127–150.*

Flaswinkel, H., Barner, M., and Reth, M. (1995) The tyrosine activation motif as a target of protein tyrosine kinases and SH2 domains. *Semin. Immunol. 7, 21–27.*

Franke T.F., Kaplan D.R., Cantley L.C., and Toker A. (1997) Direct regulation of the Akt proto-oncogene product by phosphatidylinositol-3,4-bisphosphate. *Science 275, 665–668.*

Fu C., Turck C.W., Kurosaki T., and Chan A.C. (1998) BLNK: a central linker protein in B cell activation. *Immunity 9, 93–103.*

Gergely, J. and Sarmay, G. (1996) FcγRII-mediated regulation of human B cells. *Scand. J. Immunol. 44, 1–10.*

Gergely, J. and Sarmay, G. (1992) B-cell activation-induced phosphorylation of FcγRII: A possible prerequisite of proteolytic receptor release. *Immunol. Rev. 125, 5–19.*

Gergely, J., Pecht, I., and Sarmay, G. (1999) Immunoreceptor tyrosine-based inhibition motif-bearing receptors regulate the immunoreceptor tyrosione-based activation motif-induced activation of immune competent cells. *Immunol. Lett. 68, 3–15.*

Harwood, A.E. and Cambier, J.C. (1993) B cell antigen receptor cross-linking triggers rapid protein kinase C independent activation of $p21^{ras}$. *J. Immunol. 151, 4513–4522.*

Hashimoto, A., Okada, H., Jiang, A., Kurosaki, M., Greenberg, S., Clark, E.A., and Kurosaki, T. (1998) Involvement of guanosine triphosphatases and phospholipase C-γ2 in extracellular signal-regulated kinase, c-Jun NH2-terminal kinase, and p38 mitogen-activated protein kinase activation by the B cell antigen receptor. *J. Exp. Med. 188, 1287–1295.*

Heyman, B. (1994) The role of complement receptors in the regulation of the immune response. In: Erdei A. (ed.), *New Aspects of Complement Structure and Function*, R.G.Landes Co. Austin, pp. 59–72.

Ingham, R.J., Holgado-Madruga, M., Siu, C., Wong, A.J., and Gold, M.R. (1998) The Gab 1 protein is a docking site for multiple proteins involved in signaling by the B cell antigen receptor. *J. Biol. Chem. 273, 30630–30637.*

Inui, S., Sanjo, H., Maeda, K., Yamamoto, H., Myamoto, E. and Sakaguchi, N. (1998) Ig receptor binding protein 1 (α4) is associated with rapamycin-sensitive signal transduction in lymphocytes through direct binding to the catalytic subunit of protein phosphatase 2A. *Blood 92, 539–546.*

Katagiri, T., Ogimoto, M., Hasegawa, K., Mizuno, K., and Yakura, H. (1995) Selective regulation of Lyn tyrosine kinase by CD45 in immature B cells. *J. Biol. Chem. 270, 27987–27990.*

Kawauchi, K., Lazarus, A.L., Rapaport, M.J., Harwood, A., Cambier, J., and Delovitch, T.L. (1994) Tyrosine kinase and CD45 tyrosine phosphatase activity mediate $p21^{ras}$ in activation in B cells stimulated through the antigen receptor. *J. Immunol. 152, 3306–3316.*

Koncz, G., Gergely, J., and Sarmay, G. (1998) FcγRIIb inhibits both B cell receptor- and CD19-induced Ca^{2+} mobilization is FcγR-transfected human B cells. *International Immunology 10, 141–146.*

Koncz, G., Pecht, I., Gergely, J., and Sarmay, G. (1999) Fcγ receptor mediated inhibition of human B cell activation: the role of SHP-2 phosphatase. *Eur. J. Immunol. 29, 1980–1989.*

Koncz, G., Toth, G.K., Bokonyi, G., Keri, G., Pecht, I., Medgyesi, D., Gergely, J., and Sarmay, G. (2001) Co-clustering of Fcgamma and B cell receptors induces dephosphorylation of the Grb2-associated binder 1 docking protein. *Eur. J. Biochem. 268, 3898–3906.*

Kurosaki, T., Takata, M., Yamanashi, Y., Inazu, T., Taniguchi, T., Yamamoto, T., and Yamamura, H. (1994) Syk activation by the Src-family tyrosine kinase in the B cell receptor signaling. *J. Exp. Med. 179, 1725–1729.*

Lazarus, A.H., Kawauchi, K., Rapoport, M.J., and Delovitch, T.L. (1993) Antigen-induced B lymphocyte activation involves the p21ras and ras. GAP signaling pathway. *J. Exp. Med. 178, 1765–1769.*

Lee, J.C., Kumar, S., Griswold, D.E., Underwood, D.C., Votta, B.J., and Adams, J.L. (2000) Inhibition of p38 MAP kinase as a therapeutic strategy. *Immunopharmacology, 47, 185–201.*

Lin, C-T., Shen, Z., Boros, P., and Unkeless, J.C. (1994) Fc receptor-mediated signal transduction. *J. Clin. Immunol. 14, 1–13.*

Miettinen, H.M., Matter, K., Hunziker, W., Rose, J.K., and Mellman, I. (1992) Fc receptor endocytosis in controlled by a cytoplasmic domain determinant that actively prevents coated pit localization *J. Cell. Biol. 116, 875–888.*

Muta, T., Kurosaki, T., Misulovin, Z., Sanchez, M., Nussenzweig, M.C., and Ravetch, J.V. (1994) A 13-amino-acid motif in the cytoplasmic domain of FcγIIB modulates B-cell receptor signalling. *Nature 368, 70–73.*

Nadler, M.J., McLean, P.A., Neel, B.G., and Wortis, H.H. (1997) B cell antigen receptor-evoked calcium influx is enhanced in CD22-deficient B cell lines. *J. Immunol. 159, 4233–4243.*

Nishizumi, H., Horikawa, K., Mlinaric-Rascan, I., and Yamamoto, T. (1998) A double-edged kinase Lyn: a positive and negative regulator for antigenreceptor-mediated signals. *J. Exp. Med. 187, 1343–1348.*

Nitschke, L., Carsetti, R., Ocker, B., Kohler, G., and Lamers, M.C. (1997) CD22 is a negative regulator of B-cell receptor signalling. *Curr. Biol. 7, 133–43.*

Ono, M., Bolland, S., Tempst, P., and Ravetch, J.V. (1996) Role of inositol phosphatase SHIP in negative regulation of the immune system by FcγRIIB. *Nature, 383, 263–266.*

Pepys MB. (1974) Role of complement in induction of antibody production *in vivo*. Effect of Cobra Venom Factor and other C3-reactive reagents on thymus-dependent and thymus-independent antibody response. *J. Exp. Med. 140, 126–145.*

Phillips, N.E. and Parker, D.C. (1984) Cross-linking of B lymphocyte Fcγ receptors and membrane immunoglobulin inhibits antiimmunoglobulin induced blastogenesis. *J. Immunol. 132, 627–632.*

Prodinger, W.M., Würzner, R., Erdei, A., Dierich, M.P., (1999) Complement, In: Paul, W.E. (ed.) *Fundamental Immunology*, Lippincott-Raven Publishers, Phialdelphia, pp. 967–995.

Rameh, L.E., Rhee, S.G., Spokes, K., Kazlauskas, A., Cantley, L.C., and Cantley, L.G. (1998) Phosphoinositide 3-kinase regulates phospholipase Cgamma-mediated calcium signaling. *J. Biol. Chem. 273, 23750–23757.*

Reid, K.B.M., Bentley, P.I., Cambell, R.D., Chung, L.P., Sim, R.B., Kristensen, T., and Tack, B.F. (1986) Complement system proteins which interact with C3b or C4b. *Immunol. Today. 7, 230–234.*

Reth, M., Hombach, J., Wienands, J., Campbell, K.S., Chien, N., Justment, L.B., and Cambier, J.C. (1991) The B-cell antigen receptor complex. *Immunol. Today 12, 196–201.*

Rollet-Labelle, E., Gilbert, C., and Naccache, P.H. (2000) Modulation of human neutrophil responses to CD32 cross-linking by serine/threonine phosphatase inhibitors: cross-talk between serine/threonine and tyrosine phosphorylation. *J. Immunol. 164, 1020–1024.*

Sarmay, G., Koncz, G., and Gergely, J. (1996) Human type II Fcγ receptors inhibit B cell activation by interacting with the $p21^{ras}$-dependent pathway. *J. Biol. Chem. 271, 30499–30504.*

Sarmay, G., Koncz, G., Pecht, I., and Gergely, J. (1997) Fcγ receptor type IIb induced recruitment of inositol and protein phosphatases to the signal transductory complex of human B.cell. *Immunol. Lett. 57, 159–164.*

Sarmay, G., Rozsnyai, Z., Szabo, I., Biro, A., and Gergely, J. (1991) Modulation of Fc gamma receptor type II expression on activated human B lymphocytes. *Eur. J. Immunol. 21, 541–549.*

Sarmay, G., Rozsnyay, Z., Koncz, G., Danilkovich, A., and Gergely, J. (1995) The alternative splicing of human FcγRII mRNA is regulated by activation of B cells with mIgM cross-linking, interleukin 4, or phorbolester. *Eur. J. Immunol. 25, 262–268.*

Sarmay. G., Koncz, G., Pecht, I., and Gergely, J. (1999) Cooperation between SHP-2 phosphatidyl 3-kinase and inositol 5 phosphatase in the FcγRIIb mediated B cell regulation. *Immunol. Lett. 68, 25–34.*

Sato, S., Miller, A.S., Inaoki, M., Bock, C.B., Jansen, P.J., Tang, M.L., and Tedder, T.F. (1996) CD22 is both a positive and negative regulator of B lymphocyte antigen receptor signal transduction: altered signaling in CD22-deficient mice. *Immunity 5, 551–62.*

Saxton, T.M., van Oostveen, I., Bowell, D., Aebersold, R., and Gold, M.R. (1994) B cell antigen receptor cross-linking induces phosphorylation of the $p21^{ras}$ oncoprotein activators SHC and mSOS1 as well as assembly of complexes containing SHC, GRB-2, mSOS1, and a 145-kDa tyrosine-phosphorylated protein. *J. Immunol. 153, 623–633.*

Smit L; van der Horst G; and Borst J. (1996) Sos, Vav, and C3G participate in B cell receptor-induced signaling pathways and differentially associate with Shc-Grb2, Crk, and Crk-L adaptors. *J. Biol. Chem. 271, 8564–8569.*

Sugawara, H., Kurosaki, M., Takata, M., and Kurosaki, T. (1997) Genetic evidence for involvement of type 1, type 2 and type 3 inositol 1,4,5-trisphospahate receptors in signal transducing throug the B-cell antigen receptor. *EMBO J. 16, 3078–3088.*

Tedder, T.F., Inaoki, M., and Sato, S. (1997) The CD19-CD21 complex regulates signal transduction thresholds governing humoral immunity and autoimmunity. *Immunity 6, 107–118.*

Tedder, T.F., Tuscano, J., Sato, S., and Kehrl, J.H. (1997)CD22, a B lymphocyte-specific adhesion molecule that regulates antigen receptor signaling. *Annu. Rev. Immunol. 15, 481–504.*

Toker A., and Cantley L.C. (1997) Signalling through the lipid products of phosphoinositide-3-OH kinase. *Nature 387, 673–676.*

Tsubata, T., (1999) Co-receptors on B lymphocytes. *Curr. Opin. Immunol., 11, 249–255.*

Uckun, F.M., Evans, W.E., Forsyth, C.J., Waddick, K.G., Ahlgren, L.T., Chelstrom, L.M., Burkhardt, A., Bolen, J., and Myers, D.E. (1995) Biotherapy of B cell precursor leukemia by targeting genistein to CD19-associated tyrosine kinases. *Science, 267, 886–891.*

Vély, F.; and Vivier E. (1997) Conservation of structural features reveals the existence of a large family of inhibitory cell surface receptors and noninhibitory/activatory counterparts. *J. Immunol. 159, 2075–2077.*

Vivier, E. and Daeron, M. (1997) Immunoreceptor tyrosine-based inhibition motifs. *Immunology Today 18, 286–291.*

Warmerdam, P.A.M., van de Winkel, J.G.J., Gosselin, E.J., and Capel, P.J.A. (1990) Molecular basis for a polymorphism of human Fcρ receptor II (CD32). *J. Exp. Med. 172, 19–25.*

Wienands, J., Larbolette, O., and Reth, M. (1996) Evidence for a preformed transducer complex organized by the B cell antigen receptor. *Proc. Natl. Acad. Sci. USA 93, 7865–7870.*

Wienands, J., Schweikert, J., Wollscheid, B., Jumaa, H., Nielsen, P.J., and Reth, M. (1998) SLP-65: A new signaling component in B lymphocytes which requires expression of the antigen receptor for phosphorylation. *J. Exp. Med. 188, 791–795.*

Yamanashi, Y., Fukui, Y., Wongsasnat, B., Kinoshita, Y., Ichimori, Y., Toyoshima, K., and Yamamoto, T. (1992) Activation of *src*-like protein tyrosine kinase *lyn* and its association with phosphatidylinisitol 3-kinase upon B cell antigen receptor-mediated signaling. *Proc. Natl. Acad. Sci. USA 83, 1118–1122.*

Part 4

Drug discovery

Chapter 15

Combinatorial chemistry

Árpád Furka

Contents

15.1 Introduction

Pharmaceutical research is considered to be a slow process. Generally, thousands of new compounds need to be prepared to find a new drug and their conventional one-by-one synthesis as well as their one-by-one testing was very tedious, time consuming and expensive. When one compares the production of new compounds – before the eighties – to automated industrial manufacturing of other items and the extensive application of production lines, the backwardness of synthetic technologies becomes quite obvious. This situation was changed dramatically by new combinatorial synthetic methods introduced in the eighties. These methods initiated a revolution, and radically changed our theory and practice in designing and preparing new substances for pharmaceutical research and other applications. The new way of thinking expanded very quickly and contributed to the foundation of a rapidly growing new scientific field, Combinatorial Chemistry. In addition to the new synthetic methods, combinatorial chemistry also comprises the new and very efficient high throughput screening (HTS) methods. This chapter, however, is devoted only to description of the synthetic methods and the deconvolution strategies.

The synthetic methods applied in combinatorial chemistry are used to prepare either large series of individual compounds or mixtures of large numbers of compounds. Both mixtures and series of compounds are termed libraries. The synthesis methods themselves can be classified into two categories:

- real combinatorial synthetic methods and
- parallel synthetic methods.

The product of a parallel synthesis is a library containing discrete compounds. Some real combinatorial methods produce mixtures, others make individual compounds. Some can be used in the synthesis of both mixtures and individual compounds. In parallel synthesis each product is usually prepared individually in a separate reaction vessel, consequently needing more synthetic steps than the number of compounds produced. As will be shown later the real combinatorial synthesis is organized so as to allow production of a very large number of compounds in a very limited number of synthetic steps.

First, parallel synthetic methods will be described, these most resemble conventional procedures. These are followed by the real combinatorial methods. When a combinatorial synthesis leads to a mixture, a special strategy needs to be followed that enables the user to identify the bioactive components of the library. These deconvolution strategies will also be described.

15.2 Parallel synthesis

Parallel synthesis was first introduced by Geysen *et al.*, 1984. Their "multipin" method was originally used for the parallel synthesis of arrays of peptides. This method, which is still in use, was soon followed by the introduction of another parallel procedure, the "teabag" method also developed for preparation of peptides (Houghten *et al.*, 1985). The parallel approach of preparation of peptide and non-peptide libraries culminated in the development of automatic synthesizers capable of producing several hundreds of individual compounds in a single run.

15.2.1 The multipin method

The multipin method (Geysen *et al.*, 1984) was developed as a simple and efficient tool to help preparation of series of peptide epitopes. The apparatus developed for this purpose is shown in Figure 15.1. The multipin apparatus has a block of 96 wells serving as reaction vessels (B and C) and a cover plate with mounted polyethylene rods fitting into the wells (A). The first amino acids of the peptides are attached to the ends of polyethylene rods (pins) covered with derivatized polyacrilic acid (marked in black in A). The solvents and coupling reagents are added to the wells. Depending on the number of the coupling step, and the sequence of the peptides to be prepared on the pins fitting into the wells, a solution of protected amino acid is also added to each well. The peptides formed on the pins are immersed into solutions. The sequence of the peptide attached to a pin depend on the protected amino acids added step by step to the wells. The peptides are screened after deprotection without cleaving them from the pins. In the parallel process a single – and different – peptide forms on each pin. Since the sequences of protected amino acids fed into each well are known, the sequences of the peptides forming on the corresponding pins are also known. These sequences can be identified by the location (row and column) of the pins in the cover.

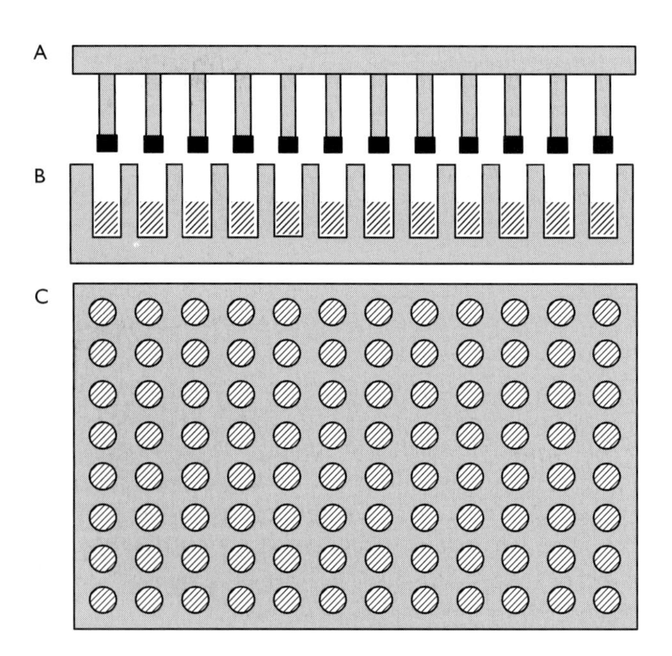

Figure 15.1 The multipin apparatus. (A) Side view of the cover with pins, (B) Side view of the reaction wells, (C) top view of the reaction block with wells.

The multipin method is still used and the multipin apparatus is a commercially available product (CHIRON Technologies, Australia). Among many applications the multipin procedure has been used, for example, by Ellman *et al.* (1992) in pioneering preparation of organic libraries by parallel synthesis. More than 10,000 derivatives of 1,4-benzodiazepines were constructed from 2-aminobenzophenones, amino acids and alkylating agents (Figure 15.2). The Fmoc protected 2-aminobenzophenenones were first attached to an acid labile linker (L) then through the linker to the pins (P). After removal of the protective group it was coupled with a protected amino acid (1). This was followed by the removal of the Fmoc protective group and cyclization (2), then by alkylation of the ring nitrogen to introduce R^4 (3). Finally the product was cleaved from the support (4).

15.2.2 The teabag method

A different version of parallel synthesis was developed by Houghten *et al.* (1985) for preparing arrays of peptides (Figure 15.10). The beads of the solid support were enclosed in permeable plastic bags (A) then placed for coupling into a reaction vessel containing the solution of amino acid and the coupling reagent. All operations, including removal of protective groups, couplings, washings and even the cleavages

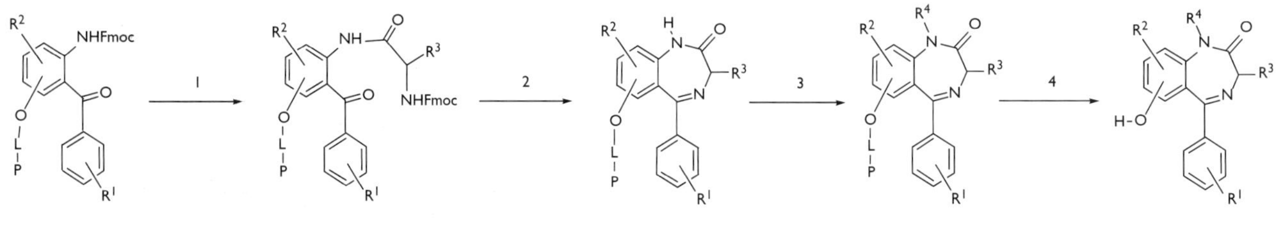

Figure 15.2 Synthesis of 1,4-benzodiazepines. P: pin, L: linker.

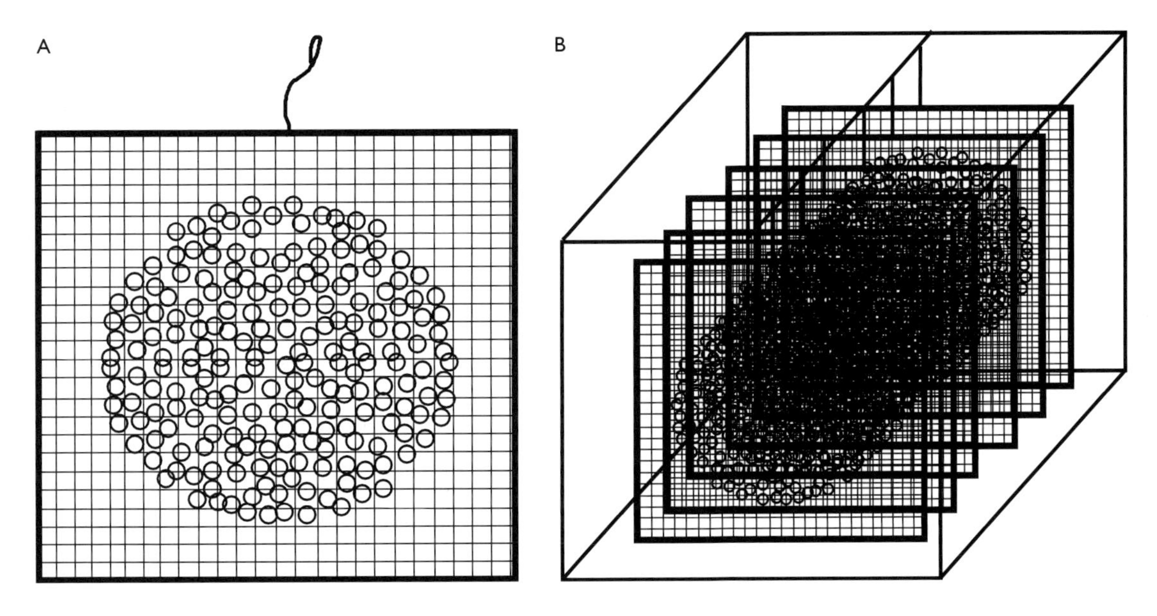

Figure 15.3 The teabag method.

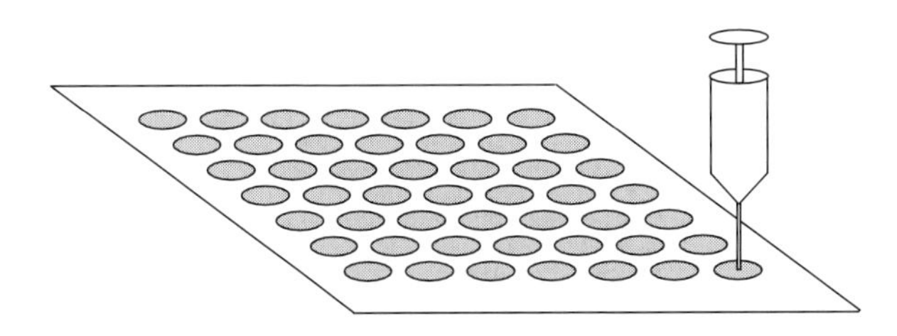

Figure 15.4 The SPOT technique.

were performed on solid supports enclosed in bags. This procedure has a significant advantage. All those bags which needed attachment of the same amino acid (e. g. alanine) were grouped together, placed into the same reaction vessel (B), and the coupling could be done in a single operation. The method is still in use in the laboratory of the inventor.

15.2.3 The spot technique

The SPOT method was introduced by Frank *et al.*, (1992). This method was also developed for preparing peptide arrays. The synthesis is carried out on cellulose paper membrane serving as anchors for the first amino acids of the sequences to be prepared. Small droplets of solutions of protected amino acids dissolved in low volatility solvents and coupling reagents are pipetted onto predefined positions of the membrane (Figure 15.4). An apparatus for spotting has also been constructed and made commercially available. An array of as many as 2000 peptides can be made on an 8×12 cm paper sheet. The peptides can be screened on the paper after removal of the protective groups. The method was also used to make mixtures in the spots.

15.2.4 Other manual devices for parallel synthesis

DeWitt *et al.* (1993) also developed a manual device for parallel synthesis. This apparatus was designed for the synthesis of small organic molecules. The solid support was placed in porous tubes immersed into vials containing solutions of reagents, which diffused into the tubes. The temperature of the reaction mixtures could be controlled by heating or cooling the reaction block.

Another inexpensive device was described by Meyers *et al.*, (1995). Beckman polypropylene deepwell plates were modified by drilling a small hole in the bottom of each well. A porous polyethylene frit was fitted into the bottom of the wells to allow removal of the solutions and solvents by vacuum. At the bottom, a rubber gasket prevented the leakage of the wells during the reactions. However these devices did not gain wide application.

15.2.5 Automatic parallel synthesizers

Like many other important developments in combinatorial chemistry, automation also began in the field of peptide chemistry. Introduction of the solid phase peptide synthesis procedure by Merrifield (1963) opened the possibility for automation and the first solid phase synthesizer was also developed in his laboratory. The new technology made it possible to carry out multi-step synthesis without the need for purification of the intermediates. The advent of combinatorial chemistry opened a flourishing era for automatic synthesizers. As a consequence of automation, parallel synthesis became the most extensively used method in combinatorial chemistry. Application of automatic machines has many advantages. Only two of them are mentioned here:

- Unlike humans, they can work 24 hours a day;
- Unlike humans they are not supposed to make errors.

These two features alone make the automatic synthesizers indispensable. Experience using the synthesizers made construction of organic synthesizers with heating and cooling possible. Because many organic reactions are moisture sensitive, the content of the reaction vessels needs to be isolated from the environment. Nowadays quite a few manufacturers offer such products. The automatic machines are exemplified by a top performing organic synthesizer produced by Advanced ChemTech, Louisville, USA.

15.2.5.1 Benchmark 384 HTS

The synthesizer is demonstrated in Figure 15.5. It is a computer-controlled machine that can prepare 384 different organic compounds in a single run. It has four blocks, each containing 96 reaction wells that are covered by a septum and have a filter at the bottom to keep the solid supported in the vessels while the solutions are removed by vacuum. The blocks can be shaken, heated up to 150 °C or cooled down to −70 °C. The machine also has a monomer rack where the solutions of monomers are stored. In addition to the reaction blocks and the monomer rack, containers for solutions of the reagents are also found on the tabletop. The bottles containing the solvents, however, are placed outside the tabletop. The machine has two needle like probes that can penetrate the septum and transfer the solutions from the monomer rack and the reagent bottles into the reaction wells. The solvents and liquids for washings are also delivered by the probes. The synthesizer is driven by flexible Windows® operating software.

Among other favorite automatic machines worthy of mention and available at the same company is a peptide synthesizer (ACT Model 396 Omega) and an organic synthesizer (Benchmark 496 Omega).

15.3 Combinatorial synthetic methods

Combinatorial synthetic methods work according to the combinatorial principle, which means that they enable the preparation of compound libraries which comprise all molecular structures that can be theoretically deduced from the monomers used in the synthesis as building blocks. At the same time, these methods are very efficient. The enormous potential they offered pharmaceutical chemists shocked the scientific community and a new way of thinking began to prevail. The organization of pharmaceutical research underwent a dramatic change. New companies were founded to exploit this potential and combinatorial groups were formed at many big companies. The three most important methods were introduced between 1988 and 1991.

One of these methods is the split-mix procedure, originally called "portioning-mixing", invented by Furka *et al.* (1988a, 1988b and 1991). This method was originally developed to enable the user to prepare millions of new peptides but later was successfully used to synthesize organic libraries too. It embodies the combinatorial principle, and "combinatorial thinking", the theoretical basis of the method, proved to be so fruitful that it is used, for example, even in biology (Khmelnitsky *et al.*, 1996) or material science (Xiang *et al.*, 1995; Hsieh-Wilson *et al.*, 1996). Even the libraries of

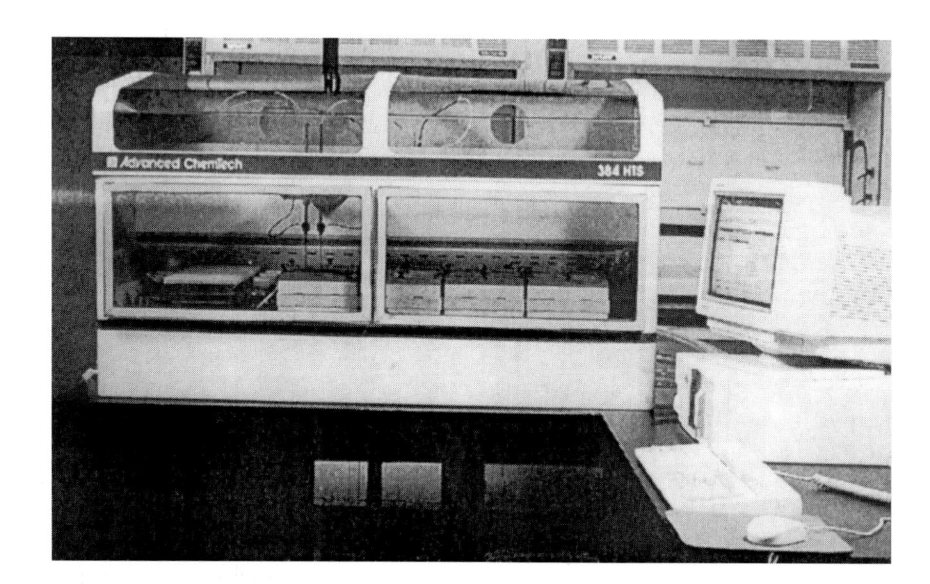

Figure 15.5 The Benchmark 384 HTS automatic parallel synthesizer.

individual compounds made by automated parallel synthesis are constructed following combinatorial principles.

The biological method of preparation of peptide libraries was introduced by three different groups in the same year (Scott and Smith, 1990; Cwirla *et al.*, 1990; Devlin *et al.*, 1990). This method, which is also based on combinatorial principles, can be used to prepare as many as billions of peptide sequences, each attached to the outer end of the coat proteins of a phage. Only the natural L-amino acids can be used as monomers in composition of the libraries. Very effective methods are available to determine the bioactive sequences.

Light-directed, spatially addressable parallel chemical synthesis was developed by Fodor *et al.* (1991). Arrays of about 1000 individual peptides were prepared on the surface of glass slides. The key technology applied in the synthesis was photolithography, which is regularly used to make computer chips. Although the synthesized array may be composed of any collection of individual sequences, the method is most effective when the library is constructed according to combinatorial principles.

15.3.1 Split-mix (portioning-mixing) synthesis

A combinatorial oligomer library containing all sequences that can be deduced from 3 different ("white", "gray" and "black") monomers could be synthesized by conventional methods following the branches of the combinatorial tree demonstrated in Figure 15.6. The sequences of the library

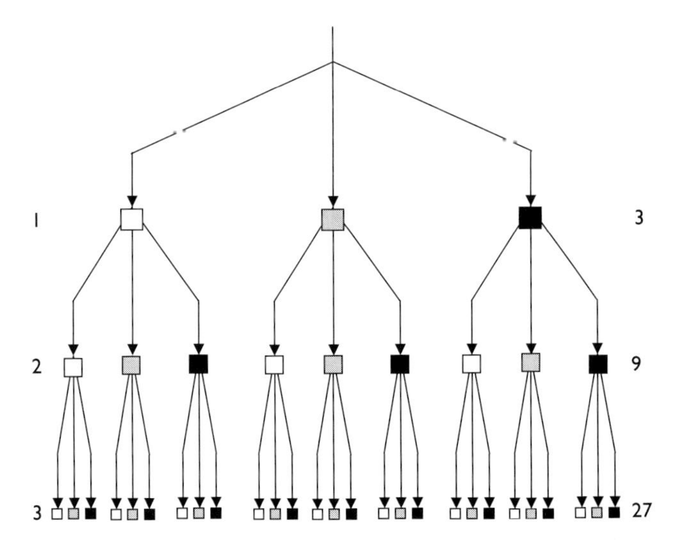

Figure 15.6 The combinatorial tree. Each square represents a reaction vessel and their white, gray and black colors symbolize different monomers. The numbers on the left side indicate the number of the reaction step and the order of branches. The numbers on the right side show the number of reaction vessels, the number of individual reaction cycles and the number of products.

members can be read by moving from the origin along the branches (e.g. white-white-white and black-black-black on the left and right sides, respectively). The combinatorial tree displays a very important rule, the combinatorial distribution rule:

> Each product formed in a given reaction step of a combinatorial process must be distributed into samples then each sample reacted with one of the monomers of the next reaction step.

The split-mix method (Furka *et al.*, 1988a, 1988b, 1991) realizes the combinatorial distribution rule by mixing the products after each reaction step then distributing the mixture into equal portions.

The method developed for preparing peptide libraries is based on Merrifield's solid phase procedure (Merrifield 1963). Each coupling cycle of the solid phase synthesis is replaced by the following simple operations:

- Dividing the solid support into equal portions;
- Coupling each portion individually with only one of the amino acids;
- Mixing the portions.

The method is exemplified by the synthesis of a dipeptide library on solid support using only three different amino acids, which are represented in Figure 15.7 by white, gray and black circles. In the first coupling cycle the amino acids are coupled to equal portions of the resin and the final product – after recombining and mixing the portions – is the mixture of the three amino acids bound to resin. In the second cycle, this mixture is again divided into three equal portions and the amino acids are individually coupled to these mixtures. In each coupling step, three different resin-bound dipeptides are formed, so the end product – after mixing – is a mixture of 9 dipeptides. The divergent, vertical and convergent arrows indicate portioning, coupling (with one kind of amino acid) and combining-mixing, respectively. If the synthesis is continued, a further portioning, coupling and mixing step leads to the formation of a mixture of 27 resin-bound tripeptides (Figure 15.8A). And, if an additional coupling step is carried out, the end product comprises 81 tetramers (Figures 15.8B,C,D).

The split-mix synthesis has several key features that are crucial for the use of this method in drug discovery or other types of application.

15.3.1.1 Formation of all possible sequence combinations

By examining the di-, tri- and tetrapeptide libraries in Figures 15.7 and 15.8 it becomes quite clear that these libraries contain all sequences that can be deduced from the three ("white", "gray" and "black") amino acids. No more

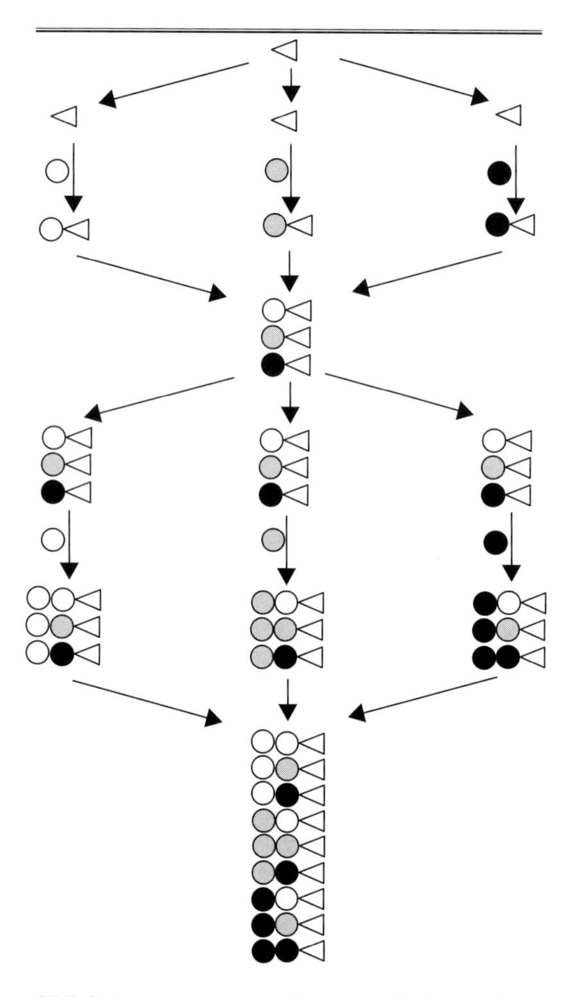

Figure 15.7 Scheme of the split-mix synthesis. The triangles represent the solid support, and the white, gray and black circles are amino acids or other kinds of monomers.

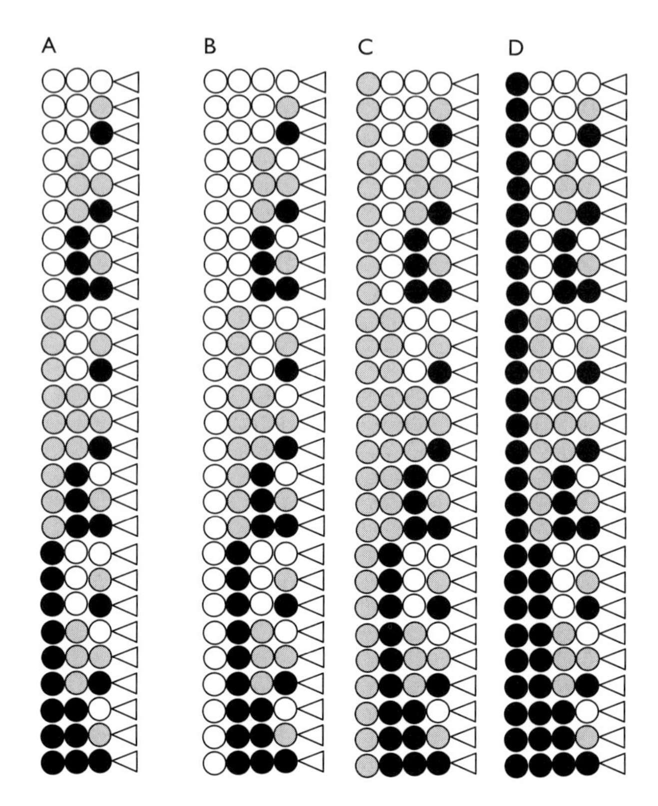

Figure 15.8 Trimer (A) and tetramer (B, C, D) libraries constructed from three different monomers. Triangles: solid support, white, gray and black circles: monomers.

sequence combinations of the white, gray and black circles can be deduced than those found in the figures. That means that the consecutive execution of three simple operations (portioning, coupling and mixing) ensures, with mathematical accuracy, the formation of all possible sequence combinations of amino acid or other kinds of building blocks used in the synthesis. The combinatorial principle embodied in the synthesis captured the imagination of many researchers all over the world and had a profound effect on the development of the field. The combinatorial nature of the product of the split-mix synthesis is also reflected in its name "combinatorial library".

15.3.1.2 Efficiency

Figures 15.7 and 15.8 also show that starting with a single substance (the resin, used as the solid support), the number of compounds is tripled after each coupling step: first $3 \times 1 = 3$ $(=3^1)$ resin-bound amino acids, then $3 \times 3 = 9$ $(=3^2)$ resin-bound dipeptides, then $3 \times 9 = 27$ $(=3^3)$ resin-bound tripeptides and finally $3 \times 27 = 81$ $(=3^4)$ tetrapeptides are formed. This means that the number of products increases exponentially as synthesis proceeds. If 20 different amino acids are used, the number of peptides in each coupling step is increased by a factor of 20. The total number of synthesized peptides can be expressed by a simple formula 20^n, where n is the number of amino acid residues in the peptides. Even if only a manual procedure is used in the synthesis, one coupling step with each of the 20 amino acids can be easily executed every day. As a consequence, in 2, 3, 4 or 5 days 400, 8,000, 160,000 or 3,200,000 peptides, respectively, can be made. Or to put it in another way, a chemist working with this method can produce in a single week more compounds than were produced in the whole previous history of chemistry. Such efficiency had never been dreamed of before the introduction of the method and explains the explosion in the field of combinatorial chemistry.

15.3.1.3 Formation of compounds in one to one molar ratio

Libraries are most often prepared in order to find biologically active substances among new products. In the identification process, or screening, the goal is to find the biologically most effective component. Serious problems may arise in screening mixtures of compounds that are not present in equal quantities. A low activity component, for example, if it is present in a large amount, may show a stronger effect than a highly active component present in lower quantity. It is important, therefore, to prepare libraries in which the constituents are present in equal molar quantities. The split-mix method was designed to comply with this requirement. Before each round of couplings, the resin is thoroughly mixed then divided into homogenous equal portions. This ensures that the previously formed peptides or other compounds are present in equimolar quantities in each portion. Since couplings with the different monomers are executed on spatially separated samples, it is possible to use appropriate chemistry to drive each coupling reaction to completion regardless of the reactivity of the different amino acids or other kinds of monomers. Since each peptide is quantitatively transformed into an elongated new one, as a result, both the numbers of peptides originally present and their equimolar ratio is preserved in every portion at each step.

15.3.1.4 *Formation of individual compounds*

The split-mix procedure has another intrinsic feature, which plays an important role in screening and gives a unique character to the method: on any individual bead of the solid support, only one kind of peptide or other compound is formed. This may seem surprising at first glance, but becomes quite understandable upon closer examination. In Figure 15.9, the fate of a randomly selected bead is followed in a three step coupling process. The bead in every coupling step meets only a single amino acid. This is the amino acid which is used for coupling in the reaction vessel into which the given bead was randomly transferred. Thus, only this single amino acid is coupled to all of its free sites. This amino acid in the first, second and third coupling step is the "gray", "white" and "black" one, respectively, so the bead ends up with the gray-white-black sequence (read in order of coupling).

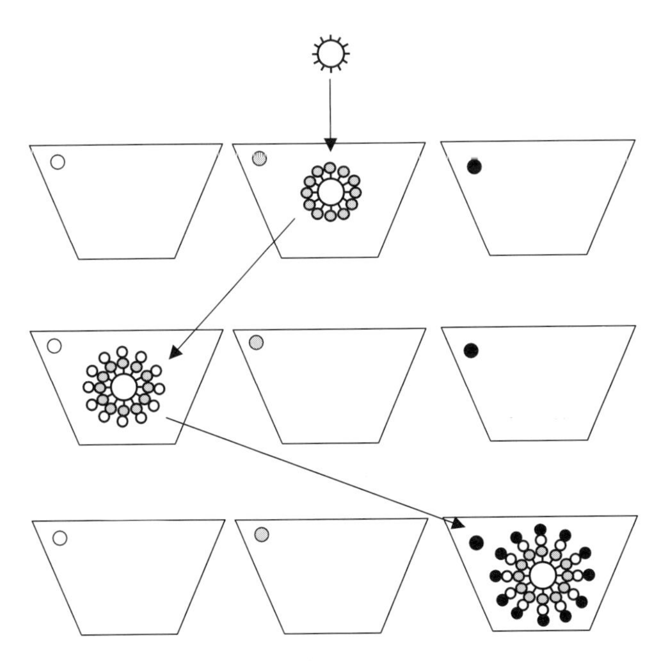

Figure 15.9 Formation of a single compound on each bead. Large circle: resin bead, white, gray and black smaller circles: monomers (e.g. amino acids).

In the split-mix synthesis, like in other solid phase procedures, the beads behave very much like tiny reaction vessels, which do not interchange their contents with the others. Each of the millions of these reaction vessels preserves its content until the end of the synthesis, when they become containers of a single substance. If peptides are produced, their identity can be determined by automatic sequencing (Lam *et al.*, 1991). It is acceptable to sacrifice a fraction of the total quantity for this purpose. This means that the PM synthesis is in fact a parallel procedure, with unprecedented efficiency, but leading to individual compounds. This feature of PM synthesis enables three different means of screening:

- Doing binding experiments with the individual compounds tethered to the beads;
- Cleaving the product from a single bead and testing it as an individual compound;
- Pooling the beads before cleavage then screening with a solution of a mixture.

15.3.1.5 *Applicability of the PM method in the synthesis of "organic" libraries*

Peptides are not preferred drug candidates because of their high susceptibility to enzymatic degradation. The ideal drug leads are small organic compounds. The split-mix method is fully applicable in the synthesis of such organic libraries. Both sequential type and cyclic libraries can easily be prepared if the reaction conditions for solid phase are well developed. The synthesized libraries are often screened as individual substances released from beads. Since determination of the structure of the various organic compounds is not as simple as sequencing peptides, the beads are usually encoded. The building blocks of the encoding tags are attached to the beads in parallel with coupling the organic building blocks of the library. Different types of encoding have been described in the literature, such as encoding with sequences (Nielsen *et al.*, 1993; Nikolaiev *et al.*, 1993; Kerr *et al.*, 1993), binary encoding (Ohlmeyer *et al.*, 1993) and encoding with fluorescent colloids (Battersby *et al.*, 2000).

When encoding by sequences, the encoding tags are either peptides or oligonucleotides. Their sequences encode both the identities of organic reagents coupled to the bead and the order of their coupling. In the binary encoding system the coding units are halobenzenes carrying a varying length hydrocarbon chain attached to the beads through a cleavable spacer. It is characteristic of this labeling technique that the coding units do not form a sequence. It is simply their presence which codes for the organic building blocks and their position.

15.3.1.6 The liquid phase split-mix synthesis

Han *et al.* (1995) described a liquid phase variant of the split-mix method. The synthesis is carried out on polyethylene glycol monomethyl ether (MeO-PEG) support. This polymer is soluble in the course of the reaction, and the homogeneous phase is advantageous for coupling, but can be precipitated allowing excess reagents to be washed out. Only mixtures can be made by this method.

15.3.2 Mixed monomer synthesis

As shown by Geysen *et al.* (1986) combinatorial peptide libraries can be prepared by using mixtures of amino acids in the acylation step of solid phase synthesis. Although the method is even more efficient than the split-mix procedure, it has a serious disadvantage: the 1 to 1 molar ratio of the compounds formed cannot be assured due to differences in the coupling rates of amino acids. These differences can, in part, be compensated by proper adjustment of the concentrations of amino acids in the coupling mixtures (Rutter *et al.*, 1991). The method has been applied mainly in preparation of peptide libraries but non-peptide libraries have also been synthesized (Winter *et al.*, 1997). Because mixtures of reactants are used in couplings, at the end of the synthesis all library components are present on every bead. This means that only mixtures can be prepared by this method.

15.3.3 Light-directed, spatially addressable combinatorial synthesis

The light-directed method enables the preparation of an array of peptides or other types of oligomers on the surface of a glass slide (Fodor *et al.*, 1991). The surface of the glass is functionalized with aminoalkyl groups protected by photolabile 6-nitroveratryloxycarbonyl (Nvoc) groups. The amino acids used in the synthesis are also protected by Nvoc groups. In the example demonstrated in Figure 15.10, nine dipeptides are synthesized from amino acids A, G and K. Before each coupling step one (or more) area of the slide is irradiated through a mask to remove the protective group, then submitted to coupling with the protected form of the indicated amino acid. Coupling occurs only in the irradiated area. In Figure 15.10, the irradiated areas are white and those shadowed by the mask are gray. After completing the **a** through **f** cycles of irradiation and coupling, nine dipeptide sequences are found in locations indicated in **g**. The efficiency of the synthesis resembles that of the split-mix procedure. The method has, however, a unique and very important feature:

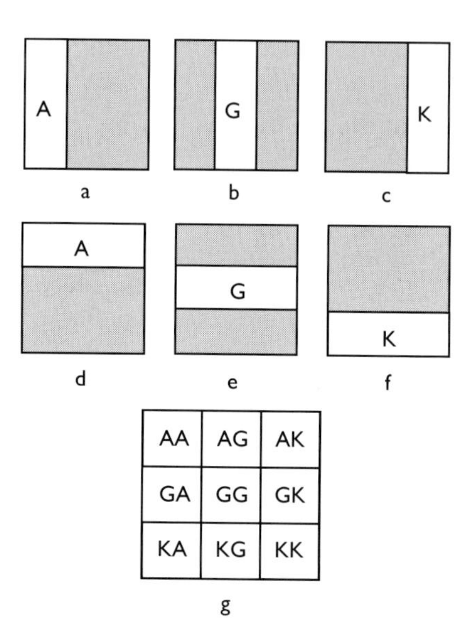

Figure 15.10 Light directed synthesis of nine dipeptides.

the sequences are defined by their locations on the slide. It is also worth mentioning that the quantity of the compounds is low, unsuitable for cleavage in preparation of a real mixture. The technology has been applied very successfully in manufacturing oligonucleotide chips (Fodor, 1997).

15.4 Deconvolution methods

Libraries prepared by application of real combinatorial synthetic methods – except the light-directed method – are used in screening experiments either as soluble mixtures or as unknown discrete compounds cleaved from or tethered to individual beads of the solid support. The task in deconvolution is to identify the substance with the desired property. Deconvolution methods can be classified into two groups: (i) deconvolution of mixtures cleaved from support and (ii) deconvolution of tethered libraries. Libraries prepared by parallel synthetic methods do not need deconvolution.

15.4.1 Mixtures cleaved from the solid support

As already mentioned, components of libraries produced by the split-mix method are discrete compounds. If the libraries are cleaved from the solid support, however, mixtures are formed. The products of the mixed monomer synthesis are always mixtures. Initially, finding an active component in a mixture of thousands or millions of structurally related compounds seemed to be akin to finding a needle in a haystack. Later, however, reliable methods were developed to solve the problem. All these methods are based on preparation and screening of properly designed partial libraries.

15.4.1.1 The iteration method

The principle of the iteration method was first described in an unpublished document (Furka, 1982). Experimental realization was demonstrated by deconvolution of a multicomponent peptide library prepared by the mixed monomer method (Geysen *et al.*, 1986). Application to libraries prepared by the split-mix method was published by Houghten *et al.* (1991).

The principle of this method is outlined in Figure 15.11, demonstrating a simple example for determination of a bioactive sequence in three stages. First, a tripeptide library is synthesized by applying the same three ("white", "gray" and "black") amino acids in all three coupling cycles. Before mixing the resin portions in the first and second coupling cycles, a small sample is taken out from all portions for later use.

In the first stage of the iteration process, the amino acid occupying the N-terminal position (coupling position 3) is determined. After the last coupling cycle of the library synthesis, the samples are not mixed. They are cleaved and screened separately (A). The components of the three samples differ only in the amino acid occupying coupling position 3 (left) and this makes its identification possible. If, after cleavage, the sample marked by + shows activity in screening, then the position 3 (N-terminal) amino acid is the "black" one.

In the second stage of the process, this "black" amino acid is then coupled to each of the three samples taken before mixing in the dipeptide stage of synthesis. The products of couplings differ only in the amino acid occupying coupling position 2 (B). The three products are cleaved from the resin, then separately screened to identify the amino acid occupying coupling position 2 in the active peptide. If the sample marked by + carries the activity, this assigns the "gray" amino acid to the second position of the active tripeptide.

The third stage of iteration returns to the samples taken after the first coupling cycle of the synthesis (C). First the "gray" then the "black" amino acid is coupled to each of the unmixed sample, since it is already known that these two amino acids occupy the second and third positions, respectively, in the active compound. Each of the products contains a single tripeptide. Determination of the activity of the samples after cleavage identifies the coupling position 1 (C-terminal) amino acid ("white"). The final sequence, starting from the N-terminus, is black-gray-white.

The method described above can be used for deconvolution of libraries made by the split-mix procedure. In somewhat modified form, however, it can be applied to screening of libraries synthesized by the mixed monomer method too. Many successful experimental results show that a previously unthinkable task can be accomplished by iterative synthesis and screening of partial libraries. The need for iterative

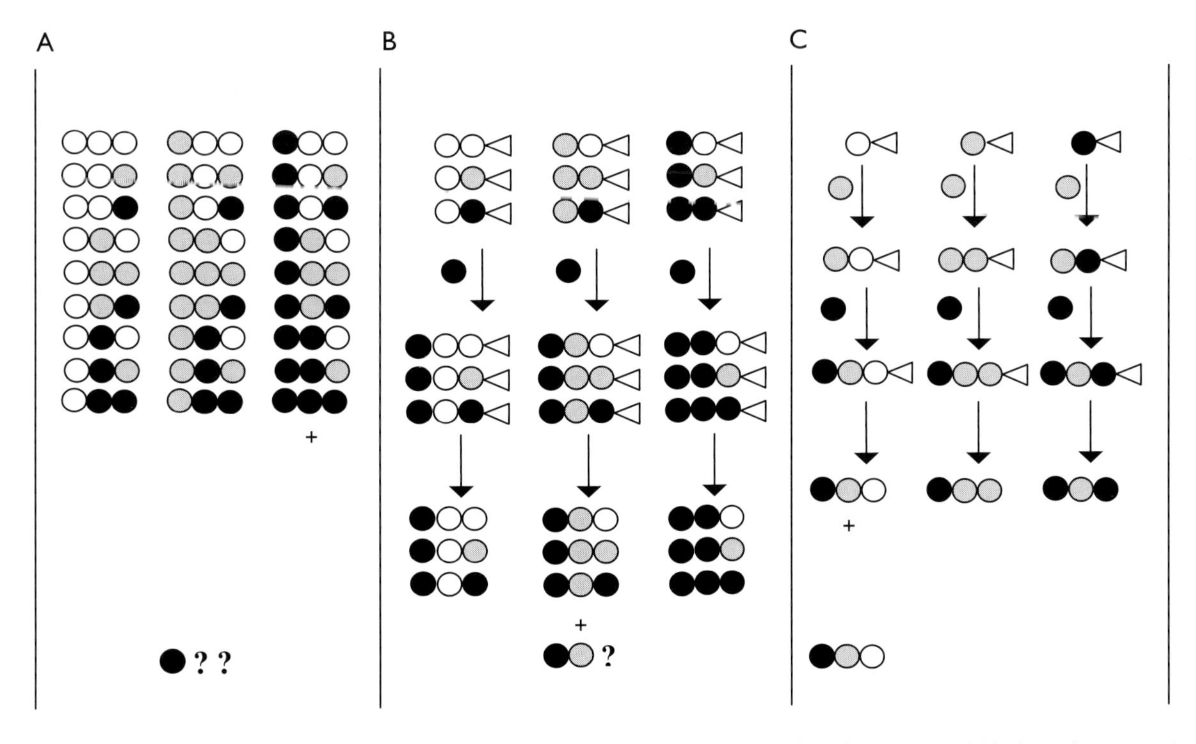

Figure 15.11 Deconvolution by iteration. The triangles represent the solid support, and the white, gray and black circles are amino acids. (A) coupling position 3, (B) coupling position 2, (C) coupling position 1.

synthesis and screening can be eliminated by properly designed and pre-prepared sets of partial libraries (see below).

15.4.1.2 Positional scanning

The sets of pre-prepared partial peptide libraries and their applicability in screening were introduced by two groups (Furka *et al.*, 1993; Sebestyén *et al.*, 1993 and Pinilla *et al.*, 1993). In all peptides present in a partial library (first order sub-library) used in positional scanning, one position is occupied by the same amino acid while any other amino acid may occur in all other positions. Such a library can be prepared by omitting the portioning operation in one selected coupling position and using a single amino acid in coupling. In all other coupling positions normal split-mix steps are executed.

Figure 15.12 shows all possible first order sub-libraries (B–J) that can be deduced from a full library (A). In H, for example, all C-terminal positions are occupied by the "black" amino acid. If the C-terminal position in the bioactive peptide also happens to be occupied by the "black" residue, this sub-library is expected to show activity in screening. Since this sub-library comprises all possible sequences with the "black" C-terminal, the bioactive peptide must be present in the sub-library. If the bioactive peptide has a "gray" or "white" C-terminal, the H sub-library is inactive since the bioactive peptide is not present.

If the sub-libraries marked with + are active in screening (C, G and H), the sequence of the bioactive peptide is gray-white-black. The method can be applied to screening libraries prepared by both the split-mix and mixed monomer method. In deconvolution of a tripeptide library prepared from 20 amino acids, for example, 60 sub-libraries have to be synthesized and screened. For screening of hexapeptides, a set of 120 sub-libraries is needed. Once the set of sub-libraries has been prepared in sufficient quantity, it can be used in many different screening experiments.

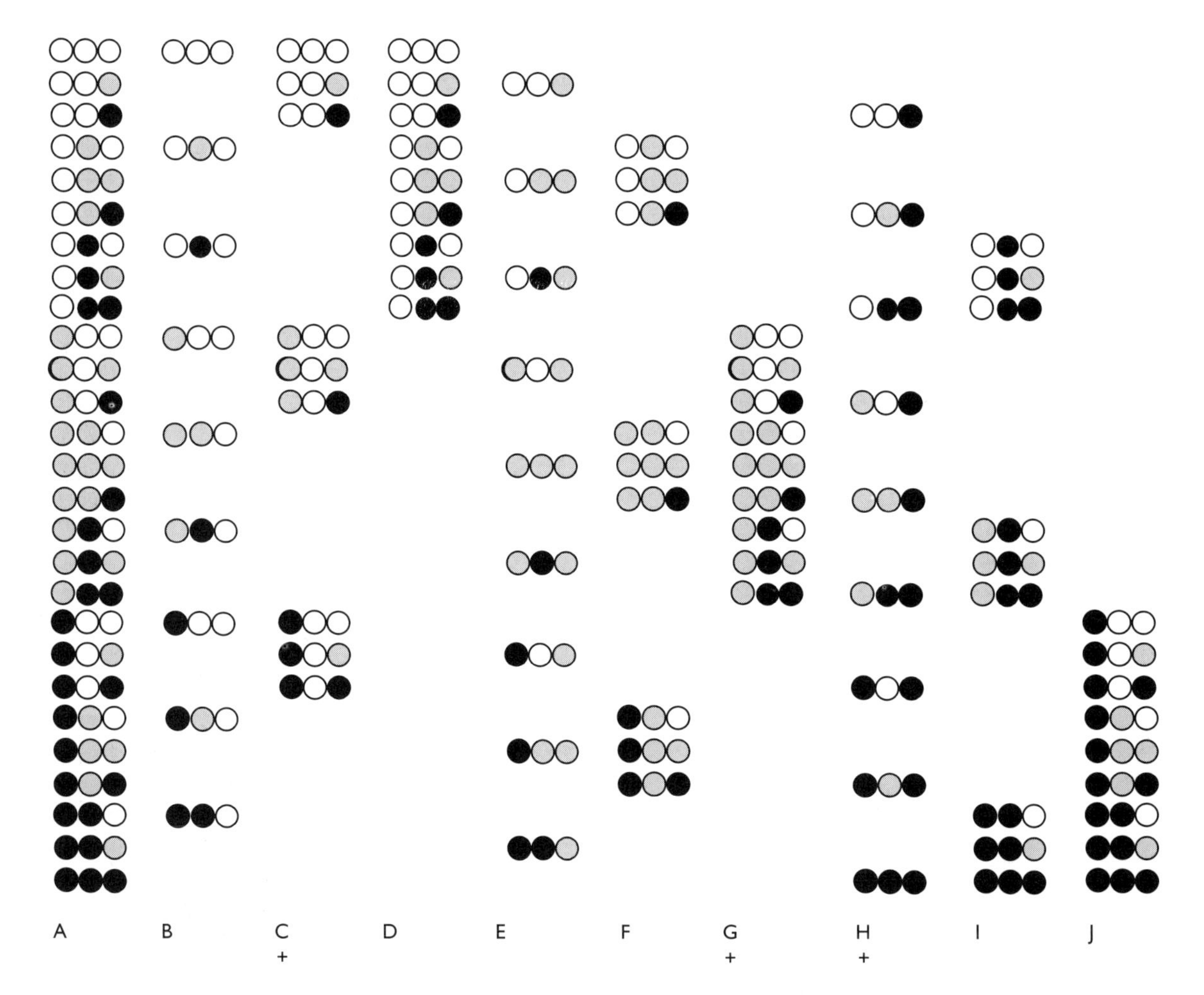

Figure 15.12 Positional scanning. (A) full library, (B)–(J): a full set of first order sublibraries.

15.4.1.3 Determination of amino acid composition of bioactive peptides. Omission libraries

Omission libraries developed by Furka *et al.*, (1998) enable the determination of the amino acid composition of bioactive peptides and, as a consequence, reduce the time and cost of deconvolution. Omission libraries can be prepared by omitting one amino acid in all positions during library synthesis. Composition of omission libraries (B, C, D) and their relation to the full library (A) are demonstrated in Figure 15.13. It can be seen in B, for example, all "white" containing peptides are missing. If the "white" amino acid happens to be present in the active component of the full library, the "white" (B) omission library is expected to be inactive in screening since the active peptide is missing from the mixture. If an omission library proves to be active in screening, it means that the omitted amino acid is not present in the active peptide or it is not essential for the activity. This feature allows the amino acid composition of bioactive peptides to be determined by screening a full set of omission libraries. The number of omission libraries in a full set is the same as the number of amino acids used in the synthesis of the full library. If the B and C omission libraries are inactive and only D is active, then the active tripeptide is composed from the "white" and "gray" amino acids.

The advantages of using omission libraries are illustrated by a simple example. If the full tripeptide library is composed from 20 amino acids it contains 8000 peptides. Only 20 omission libraries have to be prepared (or bought) and tested (instead of the 60 sub-libraries needed in positional scanning) in order to determine the amino acid composition. By varying the three (or less) amino acids in all three positions, a library can be created containing only 27 peptides yet still comprising the bioactive sequence. These peptides can even be prepared by parallel synthesis to identify the active one or positional scanning can be carried out with the first order sub-libraries of this very simple library.

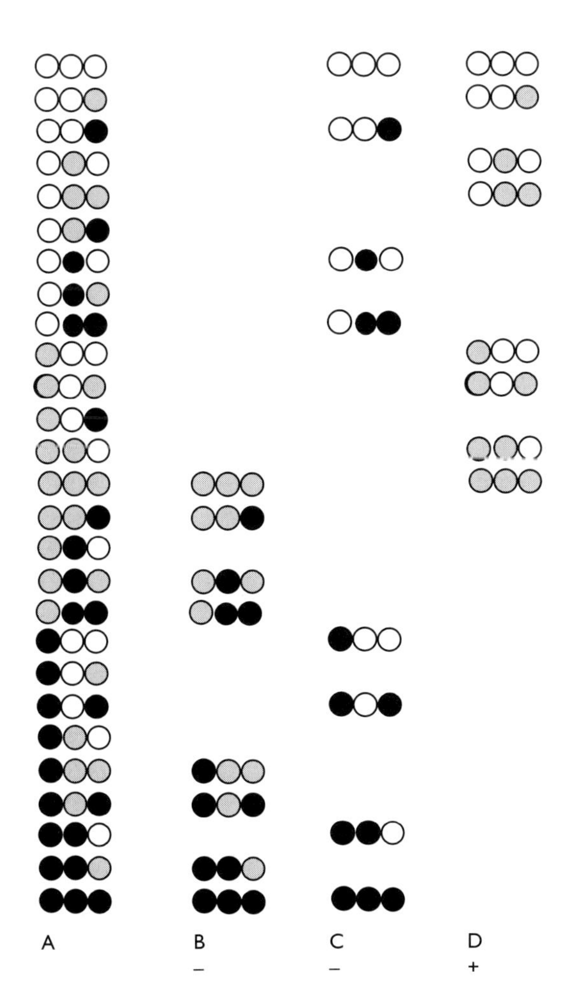

Figure 15.13 Omission libraries. A: full library, B, C and D: "white", "gray" and "black" omission libraries.

15.4.2 Tethered libraries

Tethered libraries prepared by the split-mix method have an enormous advantage: they contain individual compounds enclosed in beads as containers. Several deconvolution methods take advantage of this fact. The first method was developed by Lam *et al.* (1991) to determine the bioactive sequence in peptide libraries tethered to the support. In their procedure the beads are mixed with the solution of the target protein carrying a color label. The beads binding the target can be distinguished by color (Figure 15.14). They are picked out manually then sequenced after removal of the attached protein.

Another approach in deconvolution is to cleave the compound from individual beads and use it in the individual compound screening solutions. One of these procedures, developed by Ohlmeyer *et al.* (1993), is used for deconvolution of libraries of small organic compounds. The libraries are prepared by applying the binary encoding technique and using a photolabile linker which allows a two stage release of the organic substance. After portions of beads are distributed

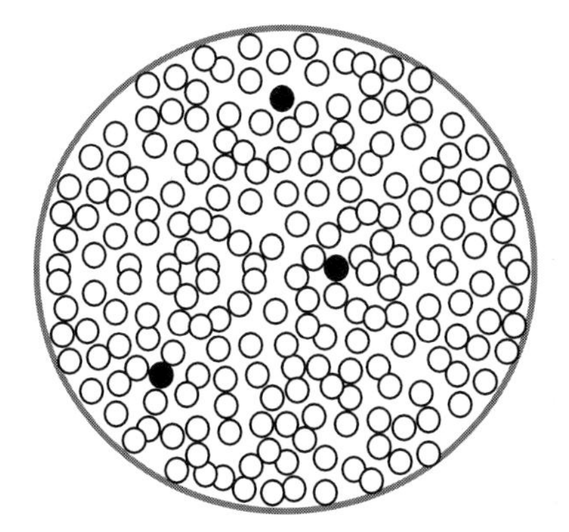

Figure 15.14 Identification of beads carrying active peptide sequences.

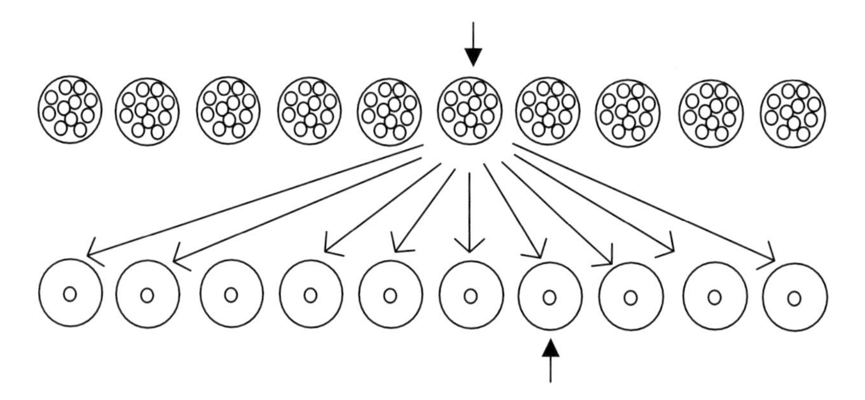

Figure 15.15 Two stage identification of beads carrying active organic molecule. The arrow marks the vessel containing the active substance.

into small vessels (Figure 15.15), the first portions of the substances are released by irradiation. The content of each vessel is then screened. If one of them proves to be active (marked by an arrow in the figure), the beads are re-distributed into vessels, each containing a single bead. After releasing the second portion of the substances, a second screening identifies the bead carrying the active substance (marked by an arrow). Finally the encoding molecules are released from the "active" bead and determined by electron capture gas chromatography thus identifying the structure of the organic molecule responsible for the biological activity.

15.5 Combinatorial synthesis on macroscopic solid support units

The parallel synthetic methods, while being very slow and expensive, have two advantages over the split-mix method:

- The products of the parallel synthesis are known individual compounds and no special deconvolution process is needed for their identification;
- The products can be prepared in relatively large – multi milligram – quantities.

There have been efforts in the last few years to modify the split-mix method to eliminate the disadvantages and preserve the high efficiency.

15.5.1 Radiofrequency encoding of capsules containing resin

The synthesis suggested by two groups (Moran *et al.*, 1995; Nicolaou *et al.*, 1995) follows the split-mix pattern. The solid support units are permeable capsules containing resin. Between two synthetic steps the units are pooled then sorted according to the combinatorial principle. To enable proper sorting, the capsules need to be encoded. In the radiofrequency encoding method suggested by the two groups, an electronic chip, a small bar, is enclosed besides the resin in the permeable capsules (Figure 15.16A). The Nicolaou version has been developed into a commercial product at IRORI. The key operation in the synthetic process is the automated sorting. Figures 15.16B–G demonstrate the components of the automatic sorter. The capsules are placed into a vibratory bowl (B) then pass through a solenoid gate (C) where their code is read by an antenna (D) and, based on the code, the computer (F) determines which vessel (G) the capsule must be delivered to for the next synthetic step. The delivery is executed by the X-Y movement of the delivery mechanism. After sorting, the capsules collected in each vessel are reacted with a different reagent. The automatic sorter can sort 10,000 capsules in 10 hours. IRORI have also developed a manual version of sorting. Chiron applies radiofrequency encoding. In their system the small bar, that is the radiofrequency transponder, is attached to their crowns and sorting is carried out manually.

15.5.2 String synthesis

It has been shown by Furka *et al.* (2000), that in split synthesis encoding becomes unnecessary if the spatial relation between each of the macroscopic solid support units can be followed in every phase of the synthetic process. This can be accomplished if (i) the unlabeled support units are arranged into spatially ordered groups, (ii) the relative spatial arrangement of the units can be maintained during the chemical reactions, (iii) the support units are redistributed between the reaction steps according to a predetermined pattern and (iv) the synthetic history of each support unit is traced by a computer.

All these conditions have been fulfilled in the "String Synthesis" published by Furka *et al.* (2000). As an example, a library of 125 tripeptides were synthesized using Chiron crowns as solid support units and 5 amino acids in each coupling step (I,F,L,V,G – E,F,W,Y,S – E,F,W,Y,S). The crowns were stringed as shown in Figure 15.17A. The 5 strings were

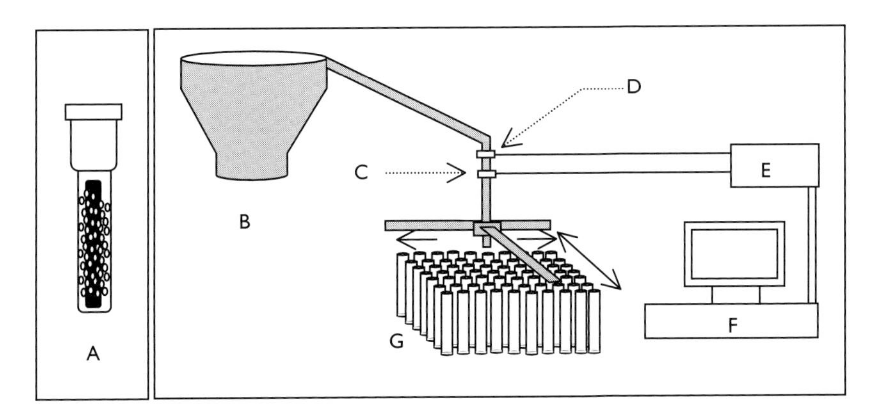

Figure 15.16 The IRORI automatic sorter. (A) permeable capsule containing resin and a transpoder, (B) vibratory bowl, (C) solenoid gate, (D) antenna, (E) control unit, (F) computer, (G) vessel.

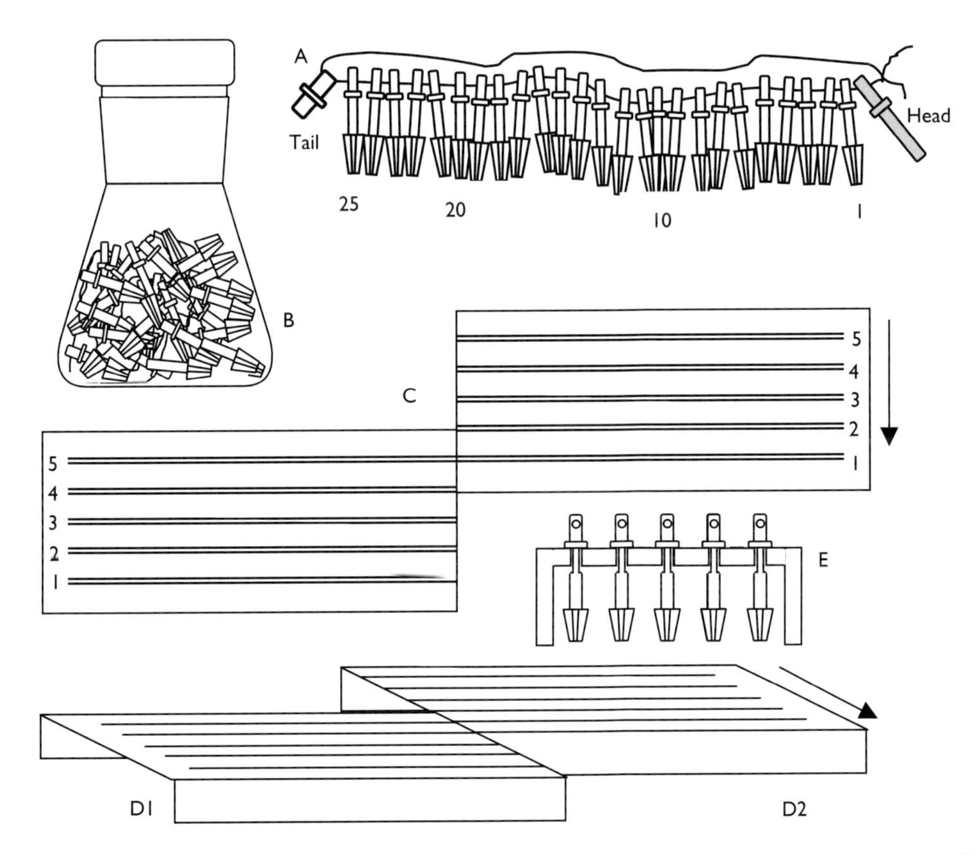

Figure 15.17 String synthesis. (A) Stringed Chiron crowns, (B) stringed crowns in flask, (C) top view of manual sorter, (D1,D2) side view of manual sorter (D1: source tray, D2: destination tray), (E) crowns in the slots of the sorter.

then placed into flasks for couplings (B). Before the next coupling step, the crowns were sorted according to a predetermined pattern (semi-parallel sorting) complying with the combinatorial distribution rule. A very simple manual device was used (C, D1, D2) for sorting. The stringed crowns were inserted into the slots of the source tray of the device (D1) then, after removal of the string, the crowns hanging in the slots (E) were pushed – according to the predetermined pattern – into the slots of the destination tray (D2), stringed again and resubmitted to coupling. Sorting is exemplified in Figure 15.18 by showing stages 1, 5 and 9 from both the 9 stage first sorting and the 45 stage second sorting.

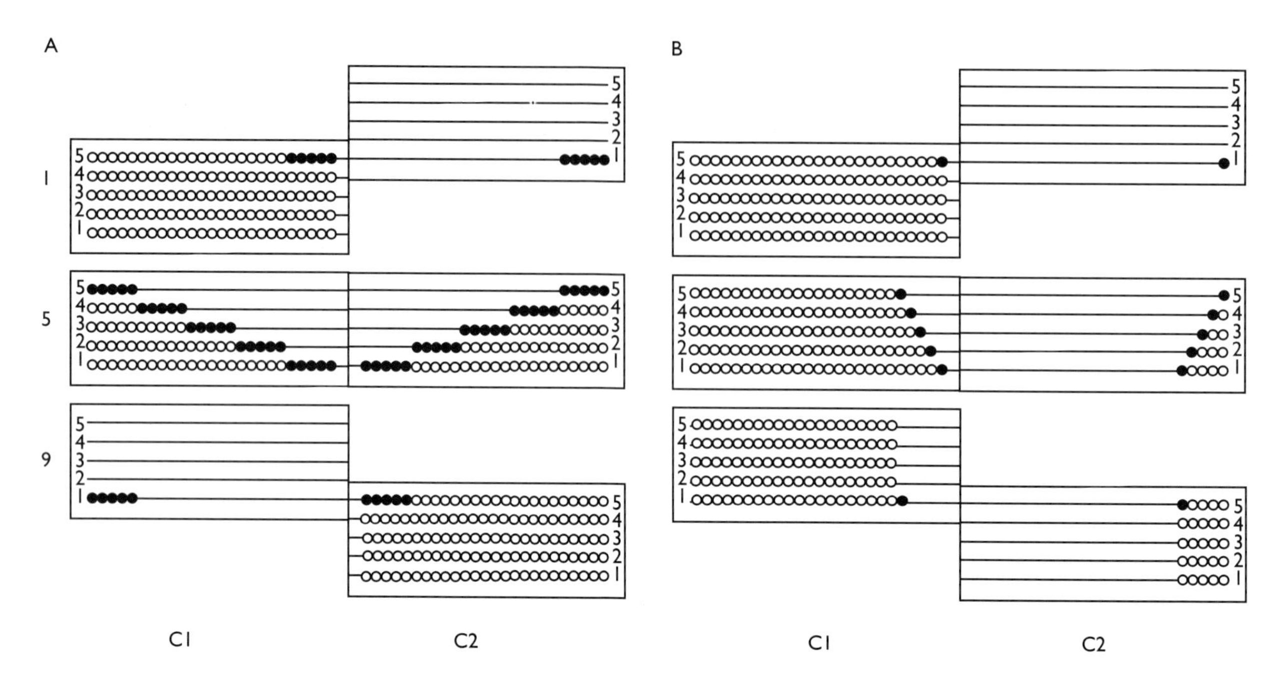

Figure 15.18 Sorting in string synthesis. (A) first sorting, transferring 5 crowns from a slot, (B) second sorting, transferring 1 crown from a slot, C1 source trays, C2 destination trays. The destination tray was gradually moved from the position of Step 1 into position for Step 9, then during second sorting, it was returned to position 1 four times. In each position transfers occured from all source slots that were aligned with destination slots. The original and destination positions of crowns are marked by black.

Table 15.1 Position of tripeptides on strings

Position	*String 1*	*String 2*	*String 3*	*String 4*	*String 5*
1	EEI-*	FEI-*	WEI-*	YEI-*	SEI-*
2	EFI-*	FFI-*	WFI-*	YFI-*	SFI-*
3	EWI-*	FWI-*	WWI-*	YWI-*	SWI-*
4	EYI-*	FYI-*	WYI-*	YYI-*	SYI-*
5	ESI-*	FSI-*	WSI-*	YSI-*	SSI-*
6	EEF-*	FEF-*	WEF-*	YEF-*	SEF-*
7	EFF-*	FFF-*	WFF-*	YFF-*	SFF-*
8	EWF-*	FWF-*	WWF-*	YWF-*	SWF-*
9	EYF-*	FYF-*	WYF-*	YYF-*	SYF-*
10	ESF-*	FSF-*	WSF-*	YSF-*	SSF-*
11	EEL-*	FEL-*	WEL-*	YEL-*	SEL-*
12	EFL-*	FFL-*	WFL-*	YFL-*	SFL-*
13	EWL-*	FWL-*	WWL-*	YWL-*	SWL-*
14	EYL-*	FYL-*	WYL-*	YYL-*	SYL-*
15	ESL-*	FSL-*	WSL-*	YSL-*	SSL-*
16	EEV-*	FEV-*	WEV-*	YEV-*	SEV-*
17	EFV-*	FFV-*	WFV-*	YFV-*	SFV-*
18	EWV-*	FWV-*	WWV-*	YWV-*	SWV-*
19	EYV-*	FYV-*	WYV-*	YYV-*	SYV-*
20	ESV-*	FSV-*	WSV-*	YSV-*	SSV-*
21	EEG-*	FEG-*	WEG-*	YEG-*	SEG-*
22	EFG-*	FFG-*	WFG-*	YFG-*	SFG-*
23	EWG-*	FWG-*	WWG-*	YWG-*	SWG-*
24	EYG-*	FYG-*	WYG-*	YYG-*	SYG-*
25	ESG-*	FSG-*	WSG-*	YSG-*	SSG-*

The computer predicted – and experimentally verified – result of the two semi-parallel sortings and the three couplings is demonstrated in Table 15.1. String synthesis provides cheap, fast and reliable sorting and offers the user the advantages of both parallel and split-mix synthesis.

Acknowledgement

The author is grateful for the grant FKFP/0149/2000.

References

Battersby, B.J., Bryant, D., Meutermans, W., Matthews, D., Smythe, M.L., Trau, M. (2000) Toward larger chemical libraries: encoding with fluorescent colloids in combinatorial chemistry. *J. Am. Chem. Soc. 122, 2138–2139.*

Bunin, B.A. and Ellman, J.A. (1992) A general and expedient method for the solid phase synthesis of 1,4-benzodiazepine derivatives. *J. Amer. Chem. Soc. 114, 10997–10998.*

Câmpian, E., Peterson, M.L., Saneii, H.H. and Furka, Á. (1998) Deconvolution by omission libraries. *Bioorganic & Medicinal Chemistry Letters 8, 2357–2362.*

Cwirla, S.E., Peters, E.A., Barrett, R.W. and Dower, W.J. (1990) Peptides on phage: A vast library of peptides for identifying ligands. *Proc. Natl. Acad. Sci. USA 876378–6382.*

Devlin, J.J., Panganiban, L.C. and Devlin, P.E. (1990) Random peptide libraries: A source of specific protein binding molecules. *Science 249, 404–406.*

DeWitt, S.H., Kiely, J.K., Stankovic, C.J., Schroeder, M.C., Cody, D.M.R. and Pavia, M.R. (1993) "Diversomers": An approach to nonpeptide, nonoligomeric chemical diversity. *Proc. Natl. Acad. Sci. USA 90, 6909–6913.*

Fodor, S.P.A. Laboratory Automation News 2, 50 (1997).

Fodor, S.P.A., Read, J.L., Pirrung, M.C., Stryer, L., Lu, A.T. and Solas, D. (1991) *Science 251, 767.*

Frank, R. (1992) Spot synthesis: An easy technique for the positionally addressable, parallel chemical synthesis on a membrane support. *Tetrahedron 48, 9217–9232.*

Furka, Á. (1982) Study on possibilities of searching for pharmaceutically useful peptides. http://szerves.chem.elte.hu/Furka/

Furka, Á. (1994) Sub-Library Composition of Peptide Libraries. Potential Application in Screening. *Drug Development Research 33, 90–97.*

Furka, Á. Redistribution in combinatorial synthesis. A theoretical approach. *Comb. Chem. & HTS* In print.

Furka, Á., Christensen, J. W., Healy, E., Tanner, H.R. and Saneii, H.J. (2000) The string synthesis. A spatially addressable split procedure *J. Comb. Chem.* In print.

Furka, Á., Sebestyén, F. (1993) Peptide sub-library kits. *PCT application WO 93/24517.*

Furka, Á., Sebestyén, F., Asgedom, M. and Dibó, G. (1991) General method for rapid synthesis of multicomponent peptide mixtures. *Int. J. Peptide Prot. Res. 37, 487–493.*

Furka, Á., Sebestyén, F., Asgedom, M. and Dibó, G. (1988) Cornucopia of peptides by synthesis. In *Highlights of Modern Biochemistry, Proc. 14th Internat. Congr. Biochemistry*, VSP. Utrecht, The Netherland, Vol. 5, p. 47.

Furka, Á., Sebestyén, F., Asgedom, M. and Dibó, G. (1988) More peptides by less labour. *Proc. 10th Internat. Symp. of Medic. Chem.*, Budapest, Hungary, 1988, p. 288, Abstract P-168.

Furka, Á., Sebestyén, F. and Câmpian, E. (1994) In R.S. Hodges, J.A. Smith (Eds) *Peptides. Chemistry, Structure and Biology*, ESCOM, Leiden, p. 986.

Geysen, H.M., Meloen, R.H. and Barteling, S.J. (1984) Use of peptide synthesis to probe viral antigens for epitopes to a resolution of a single amino acid. *Proc. Natl. Acad. Sci. USA 81, 3998–4002.*

Geysen, H.M., Rodda, S.J. and Mason, T.J. (1986) A priori delineation of a peptide which mimics a discontinuous antigenic determinant. *Mol. Immunol. 23, 709–715.*

Han, H., Wolfe, M.M., Brenner, S. and Janda, K.D. (1995) Liquid-phase combinatorial synthesis. *Proc. Natl. Acad. Sci. USA 92, 6419–6423.*

Houghten, R.A., Pinilla, C., Blondelle, S.E., Appel, J.R., Dooley, C.T. and Cuervo, J.H. (1991) Generation and use of synthetic peptide combinatorial libraries for basic research and drug discovery. *Nature 354, 84–86.*

Houghten, R.A. (1985) General method for the rapid solid-phase synthesis of large numbers of peptides: Specificity of antigen-antibody interaction at the level of individual amino acids. *Proc. Natl. Acad. Sci. USA 82, 5131–5135.*

Hsieh-Wilson, L.C., Xiang, X.D. and Schultz, P.G. (1996). Lessons from the immune system system: From catalysis to material science. *Account. Chem. Res. 29, 164–170.*

Kerr, J.M., Banville, S.C. and Zuckermann, R.N. (1993) Encoded combinatorial peptide libraries containing non-natural amino acids. *J. Amer. Chem. Soc. 115, 2529–2531.*

Khmelnitsky, Y.L., Michels, P.C., Dordick, J.S. and Clark, D.S. (1996) In I.M. Chaiken and K.D. Janda (Eds) *Molecular Diversity and Combinatorial Chemistry*, American Chemical Society, Washington, DC, p. 144.

Lam, K.S., Salmon, S.E., Hersh, E.M., Hruby, V.J., Kazmierski, W.M. and Knapp, R.J. (1991) A new type of synthetic peptide library for identifying ligand-binding activity. *Nature 354, 82–84* and its correction, *Nature 360, 768.*

Merrifield, R.B. (1963) V Solid phase peptide synthesis. I. The synthesis of a tetrapeptide. *J. Am. Chem. Soc. 85, 2149–2154.*

Meyers, H.V., Dilley, G.J., Durgin, T.L., Powers, T.S., Winssinger, N.A., Zhu, H. and Pavia, M.R. (1995) Multiple simultaneous synthesis of phenolic libraries. *Molecular Diversity 1, 13–20.*

Moran, E.J., Sarshar, S., Cargill, J.F., Shahbaz, M., Lio, A., Mjalli, A.M.M. and Armstrong, R.W. (1995) *J. Am Chem. Soc., 117, 10787–1788.*

Nicolaou, K.C., Xiao, X.-Y., Parandoosh, Z., Senyei, A. and Nova, M.P. (1995) *Angew. Chem. Int. Ed. Engl. 36, 2289–2291.*

Nielsen, J., Brenner, S. and Janda, K. D. (1993) Synthetic methods for the implementation of encoded combinatorial chemistry. *J. Am. Chem. Soc. 115, 9812–9813.*

Nikolaiev, V., Stierandova, A., Krchnak, V., Seligmann, B., Lam, K.S., Salmon, S.E. and Lebl, M. (1993) Peptide-encoding for structure determination of nonsequenceable polymers within libraries synthesized and tested on solid-phase supports. *Peptide Res. 6, 161–170.*

Ohlmeyer, M.H.J., Swanson, R.N., Dillard, L.W., Reader, J.C., Asouline, G., Kobayashi, R., Wigler, M. and Still, W.C. (1993) Complex synthetic chemical libraries indexed with molecular tags. *Proc. Natl. Acad. Sci. USA 90,10922–10926.*

Pinilla, C., Appel, J.R. and Houghten, R.A. (1993) In Schneider, C.H. and Eberle, A.N. (Eds) *Peptides 1992*, ESCOM, Leiden, p. 65.

Rutter, W.J. and Santi, D.V. (1991) General method for producing and selecting peptides with specific properties. *U.S. Pat. 5,010,175.*

Scott, J.K. and Smith, G.P. (1990). Searching for peptide ligands with an epitope library. *Science 249, 386–390.*

Sebestyén, F., Dibó, G. and Furka, Á. (1993) In Schneider, C.H. and Eberle, A.N. (Eds) *Peptides 1992*, ESCOM, Leiden, 1993, p. 63.

Winter, E.A. and Rebek, J. Jr. (1997) In S.R. Wilson and A.W. Czarnik (Eds) *Combinatorial Chemistry. Synthesis and Application*, John Wiley & Sons, New York, NY, p. 95.

Xiang, X.D., Sun, X., Briceno, G., Lou, Y., Wang, K.A., Chang, H., Wallace-Freedman, W.G., Chen S.W. and Schultz, P.G. (1995). A combinatorial approach to materials discovery. *Science 268, 1738–1740.*

Chapter 16

Lead search, selection and optimization in silico (virtual) screening

László Őrfi and István Kövesdi

Contents

16.1 Introduction

Discovery of active substances (new hits and leads) and optimization of lead molecules via synthetic modifications (i.e. increasing activity, enhancing selectivity, decreasing toxicity, optimizing ADME (absorption, distribution, metabolism, excretion) properties are basic tasks of medicinal chemistry.

Historically, drugs were inorganic substances and natural products, they came from folklore traditions or were found accidentally by fortunate discoveries. Penicillin is a good example of this. The era of the chemical industry (beginning in the last century) brought a great number of new bioactive compounds. Developments in separation methods enabled the isolation of active components of natural extracts. Products of soil microorganisms have been studied systematically since 1940, some of them are still applied as drugs while others were selected as leads for further optimization. Antibiotics, like erythromycin, streptomycin, neomycin, gentamycin, and tetracyclines are all of microbial origin. The search for new leads originally started from known drug structures, taking into account their main and side effects. Systematic synthetic modifications of the known drug structures were aimed at producing more active, more selective, less toxic, etc. and patentable (therefore more profitable) drugs (Figure 16.1 – the drug discovery paradigm). Small modifications in the original structure can greatly affect biological activity as well as the side effects. Synthetic modifications of morphine resulted in several more selective drugs where the ratios of the different effects (analgesic effect, constipation, respiratory depression, dependence liability) were changed. In the nalorphine molecule the substitution of the N-methyl group of morphine with an allyl group gave a partial agonist, which has some morphine-like activity but at higher concentrations antagonizes the effects of morphine. The systematic simplification of the original morphine structure was also succesful: the analgesic pethidine has better oral absorption and fewer gastrointestinal side effects. Methadone does not even contain the piperidine ring yet it still shows good oral analgesic activity and a long half time in the body.

The parallel development of new drugs from the same lead structures by different research groups resulted in "me-too" drugs. Such drugs are more or less analog copies of an original drug molecule, owned by other pharmaceutical companies. "Me-too" drugs can also be considered as the result of an optimization process: some of the newer drugs can usually be much more effective and selective than their

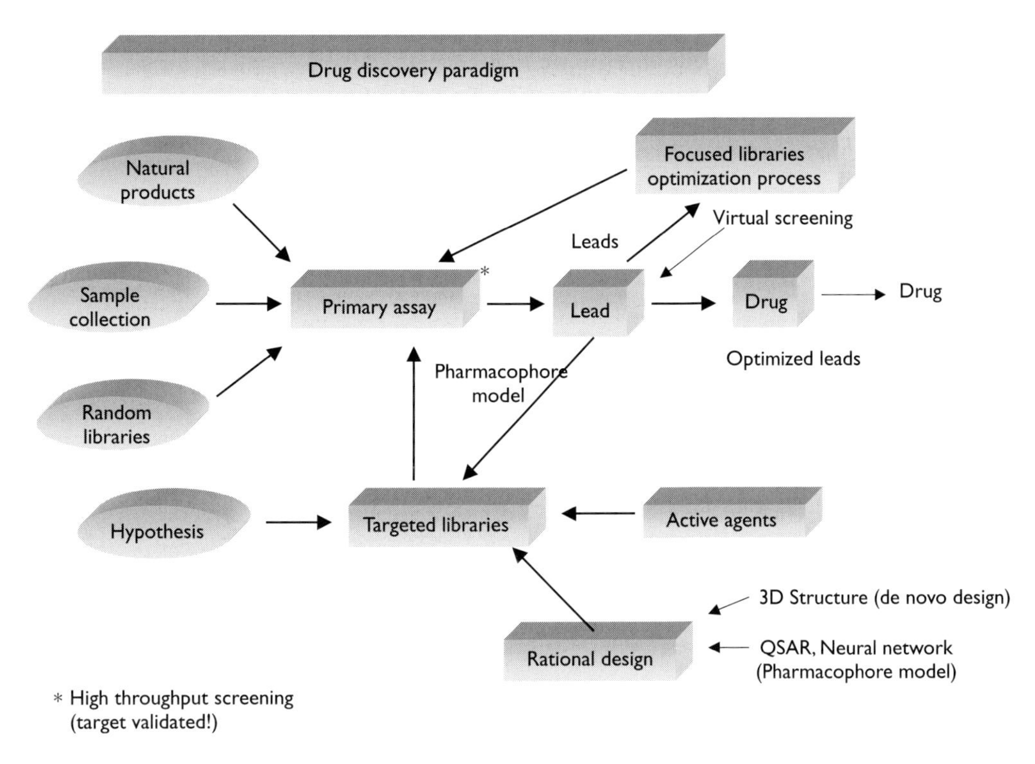

Figure 16.1 Drug discovery paradigm.

predecessors. One can find good examples for this among the penicillins.

Original, new compounds, derived from either natural sources or synthesis, are generally treated as potential (and original) drugs, therefore their pharmacological applicability is systematically studied in all possible therapeutic fields. There is also the opposite approach, where a validated biological target is selected first, then a great number (several thousands) of compounds are randomly screened in an experimental *in vivo* or *in vitro* model.

Molecular biology and genomics research generates large numbers of potential target molecules. Advances in automation, miniaturization, and informatics (application of *high-tech* methods) have resulted in **high-throughput screening** and **combinatorial chemistry** methods providing much more powerful tools for molecular pharmacology than ever before. Thousands to tens of thousands of *in vitro* assays can be performed by robots daily. Combinatorial chemistry laboratories are able to produce an incredible number of compounds, the number of synthesizable compounds is still much greater than the available synthetic capacity. Random screening of millions of compounds and their required follow up is also a great challenge. In order to prevent the overload of development resources, medicinal chemists should select the most promising molecules for synthesis, testing and for further development in a careful and rational way. This task can be completed only with the aid of computer assisted drug design (CADD) and bioinformatics. Originally, synthetic work and biological testing were separated. Medicinal chemists synthesized and derivatized small molecules ad hoc and it took a long time to obtain the biological data. In contrast, the present trend is for an integrated, iterative process, where fast feedback exists between bioassays and synthesis, helping the dynamic optimization of structures (Figure 16.2). Different approaches are used successfully in CADD depending on the availability of the 3D structures of the target receptor or the biologically active compounds. With the ongoing progress in analytical chemistry and biochemistry, many validated target molecule structures have been determined.

16.2 Binding site structure-based methods

Docking softwares were the first binding site structure-based computer methods applied for the selection of possible leads. ***DOCK*** software searched for 3D ligands in databases and selected possible leads by matching them in the binding cavity using their geometry data (1).

In ***de novo drug design*** bioactive molecule structures are generated by incremental construction within a computer-simulated model of a receptor binding site of a macromolecule. ("Receptor structure-based" approach.) Receptor structures can be obtained directly from X-ray and NMR data, or can be modeled by means of known protein structures having the

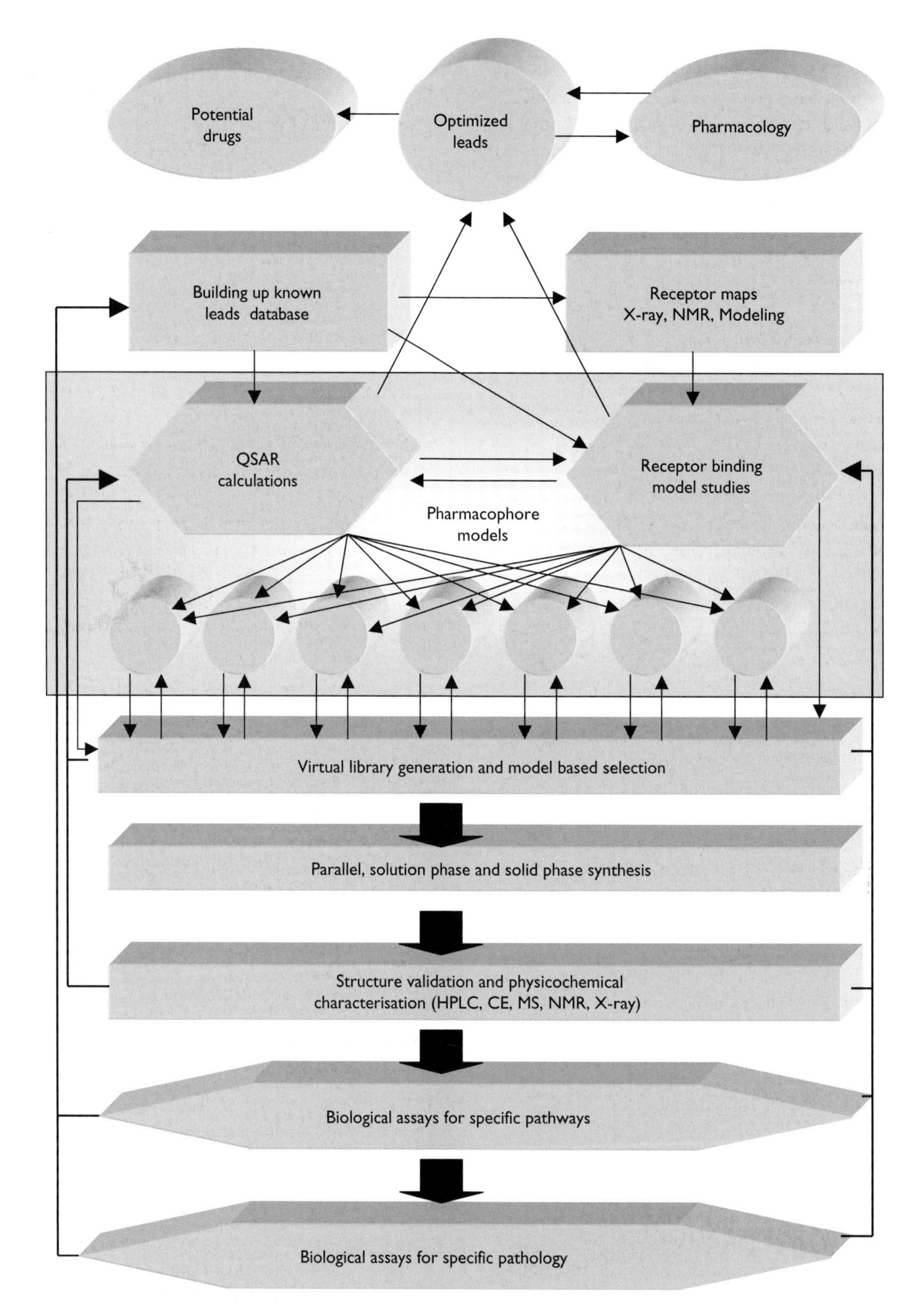

Figure 16.2 Scheme for modern drug development.

same (or similar) sequence ("homology modeling"). One of the first de novo design programs, ***LUDI*** was developed by Böhm, at BASF (2). The program automatically maps the geometry of the interacting groups of the user defined binding site (H-acceptors, H-donors, and hydrophobic areas), then searches 3D structure databases for potential ligands. All possible conformers and orientations are tested for every molecule. Possible leads are selected by the number of potential interactions, while ligands sterically overlapping with the receptor site are ignored and the remaining structures are ranked by a **scoring function**.

FlexX (3) has been developed for automated docking of a large number of flexible compound structures into a rigid binding site model. Basic, rigid fragment(s) of the potential lead is positioned in favorable directions in the binding site. Then the ligands' low energy conformations can be built up step-by-step, automatically or manually from fragments.

These softwares can select molecules showing the best fitting and highest binding energy with the help of **scoring functions** (like the scoring function of LUDI or CScore of Tripos (St. Louis, USA)). There are different types of scoring functions, using different weights for the interactions in the predictions. Therefore, scoring functions should be validated on the given receptor-ligand model by experimental binding data.

LeapFrog, de novo design software of Tripos, can create ligand structures from scratch in a receptor binding site or in a CoMFA model. It can also be used for optimizing lead structures in a stepwise manner via small structural changes of an existing ligand. Potential ligands are optimized for optimal binding energy by comparing the binding energy of the new ligand structures to that of the intermediate precursors.

Captopryl, an ACE (angiotensin-converting enzyme) inhibitor antihypertensive drug was the first successful drug designed by de novo design method. The 3D structural information of the closely related Carboxypeptidase A inhibitor complex was used in the binding site model (4).

Functionalities required for practical applications of de novo design programs in medicinal chemistry were recently summarized by Kubinyi (5):

- Searches in large 3D databases for potential ligands;
- Consideration of conformational flexibility (at least of the ligand);
- Option to create new ligands, or to modify existing leads by fusion of groups, fragments and rings;
- A scoring function appropriate for evaluation and sorting of the hits.

(Target based drug design methods are reviewed in detail in Chapter 19 of this book.)

16.3 Ligand structure-based methods

In order to design a ligand which fits into the binding site it is essential to have the validated 3D structure of the target receptor. Unfortunately, in the overwhelming majority of cases, biological targets are poorly characterized and there are not enough reliable data for receptor based modeling. Still, if there are sufficient biological activity data (on the given target) for a number of compounds, ligand structure-based methods can be applied, even in the absence of structural data about the binding site.

Structure-activity relationships (SAR) have been analyzed since the early days of systematic drug research. Researchers tried to map qualitatively the core structures and their optimal substituents required for the given biological activity in a ***pharmacophore model***. The pharmacophore model approach correlates the structures and properties of the bioactive molecules in a quantitative structure activity relationship (QSAR) study. The resulting ***pharmacophore*** or ***pharmacophore pattern*** is a purely abstract concept rather than a chemical functionality or structural skeleton (6). A pharmacophore is really an ensemble of steric (structural, topological), physico-chemical and electronic features that is required for the desired biological activity. QSAR studies in most cases resulted in QSAR equations (mathematical "pharmacophore model"), describing the given biological activity as a function of measurable physico-chemical and/or calculated molecular descriptors.

16.4 General considerations for test compound selection

Every pharmaceutical company has its own empirical rules for excluding useless test compounds in the very early stages of drug development in order to reduce costs. Some of these rules are evident, other rules are calculated by means of mathematical statistics from known drug properties. Several properties are used in ranking compounds: degree of synthetic difficulty, chemical stability, melting point, water solubility, lipophilicity, metabolic liability, pK_a (acidic dissociation constant), *in vitro* membrane permeability etc.

16.5 Undesired (reactive) chemical groups

Since drugs are mainly for systemic use, compounds possessing obviously reactive – therefore toxic – chemical groups should be excluded from the test compound sets. Usual undesirable chemical groups are listed in Table 16.1.

16.6 Drug likeness

"Drug likeness" is an important criteria for the selection and prioritization of compounds from compound libraries for biological testing. Many attempts have been made to determine molecular properties that make a compound "drug like". A strict statistical analysis of all molecular properties of known drugs would answer this question. Useful filters are

Table 16.1 Examples of undesirable chemical groups in potential drug molecules

Chemical group	*Structure*	*Chemical group*	*Structure*
Acid anhydride	–(C=O)–O–(C=O)–	Nitrite	–O–N=O
Acid halide	–(C=O)–Hal	Nitroso	–C–N=O
Aldehyde	–CH–CH=O	Ozonide	O–O, O
Azide	$-N=N^{+}=N^{-}$	P-Phosphorous	–P–P–
Carbodiimide	–N=C=N–	Perchlorate	$-ClO_4$
Diazonium	$-N^{+}\equiv N$	Periodate	$-IO_4$
Epoxide	O	Peroxide	–O–O–
		Phosphine	PC_3, PH
Hetero-halide	–N–Hal, –O–Hal, S–Hal	Phosphonium salt	$P^{+}C_4$
Isocyanate	–N–C=O	Quaternary ammonium	$N^{+}C_4$
Isocyanide	$-C-N^{+}\equiv C^{-}$	Sulfonyl chloride	$-SO_2Cl$
N-Phosphorous	–N–P–	Unusual atoms	Not: C, H, N, O, S, P, F, Cl, Br, I
Nitrate	$-O-NO_2$		

very simple, applying substructure searches for toxic groups (see Table 16.1), calculating molecular weights and conformational freedom (number of rotatable bonds). Their main goal is to filter out the definitely toxic, non-drug like substances.

There are other, more accurate methods comparing knowledge bases containing drugs and (presumed) non-drugs: artificial neural network programs (ANN, see in detail later) were trained to successfully distinguish – better than 80% hit rate – between drugs and non-drugs, using 1D, 2D molecular descriptors and a Bayesian neural network (7), or atomic type descriptors and a feedforward neural network (8). A genetic algorithm-based approach has been used at GlaxoSmithKline (Stevenage, Hertfordshire, UK) to filter out compounds prior to high throughput screening (9).

16.7 Rule-of-five ("Lipinski's Rules")

Rule-of-five is probably the best known method for predicting intestinal absorption, the condition for oral administration of a drug. Lipinski and coworkers at Pfizer (Groton, NJ, USA) analyzed 2245 drugs reaching Phase II trials and built up these rules (10). If any two of the following rules are satisfied for a compound, they indicate a potential absorption problem:

- Molecular weight > 500;
- Number of hydrogen bond acceptors > 10;
- Number of hydrogen bond donors > 5;
- Calculated logP > 5.0 (if ClogP (11) is used) or > 4.15 (if MlogP (12) is used).

In the rules, any oxygen and nitrogen atoms are defined as **H-bond acceptors**, and NH or OH groups are defined as **H-bond donors**. **LogP** is the logarithm of calculated octanol/water partition coefficient, and it is used as a measure of lipophilicity.

Molecular weight (160–480) and logP (ALOGP = −0.4–+5.6) ranges were later modified by Ghose *et al.* (Amgen, Thousand Oaks, CA, USA) (13). (It is obvious that for the different drug classes the ranges can vary). They have also analyzed the relative abundance of structural units in the drug database and found the following order: ***Benzene ring >> tertiary amine > carboxamide = alcohol > aliphatic heterocyclic ring > aromatic heterocyclic ring***.

16.8 Blood brain barrier (BBB) penetration

Although CNS (central nervous system) drugs must pass through the BBB (blood brain barrier), this could be a serious drawback for drugs targeting other organs, because of CNS side effects. Brain penetration can be characterized by the ratio of the steady state concentrations of the drug molecule: $\log BB = \log(C_{brain}/C_{blood})$. Experimentally determined logBB values are in the range of about −2.0 to +1.0. Abraham *et al.* have studied the logBB data of approximately 60 compounds.(14) They found good correlation of logBB with five solute descriptors: increasing molecular volume and molar refraction increased partitioning of compounds in the brain, while dipolarity/polarizability and hydrogen-bonding and acidity/basicity decreased it (15). PLS analysis of topological descriptors by Luco (16) provided an 18 parameter equation applicable for prediction of logBB. Recently a simple two-variable equation was reported by Clark (17) containing polar molecular surface area (PSA) and calculated ClogP (11) values:

$$\log BB = -00148 * PSA + 0.152 \times ClogP + 0.139$$
$$(r = 0.887,\ n = 55)$$

The polar surface area is the area contribution of the N, O, and S atoms or of hydrogen atoms attached to N, O, and S atoms.

16.9 Application of QSAR methods in lead selection and optimization

Rational drug research applies sophisticated methods to chemical structure-biological activity correlation (QSAR) studies (MLR, PLS, ANN, CoMFA, etc.).

16.10 Narrow scope and limitations of the most popular methods

MLR (multivariate linear regression) is a fast method but limited to linear and pseudo-linear modeling. MLR determines the linear relation between the matrix of explanatory variables and the matrix of responses. Some conventional MLR software packages cannot handle situations where the number of molecules is either smaller or larger than the number of explanatory variables. Our MLR implementation does not have such limitations. It always gives a unique solution with the smallest Frobenius norm. However, in the case of correlated inputs/outputs and/or limited observations, MLR methods usually fail to give a model which is robust to noise and which does not overfit.

PLS (partial least squares regression) is an extension of MLR. PLS is based on factor analysis fundamentals and is mainly used in overdetermined cases where the number of variables is larger than the number of compounds. The models obtained in PLS are linear, even in cases of application of advanced variable selection methods such as genetic algorithm or simulated annealing.

The number of explanatory variables may run into thousands, whereas the number of compounds rarely exceeds 100. In this situation conventional statistical methods like MLR are vulnerable to overfitting. Linear regression by PLS is designed to avoid this problem. PLS reduces the explanatory data to a small number of components, or linear combinations, which are strongly correlated with the responses. The first PLS component is a trend vector of the responses in the space of the explanatory variables. The next component is the trend within a subspace orthogonal to the first; and so on. Most QSAR calculations entail enough redundancy that the major risk is that an unrecognized chance correlation misdirects experimental work. PLS filters out any chance correlations but with a price of a very small, usually acceptable, risk of overlooking a correct correlation.

CoMFA (comparative molecular field analysis) became one of the most popular methods for QSAR. It uses multivariate statistical methods for correlating shapes and properties of structures with their biological activity. Bioactive conformation of compounds are aligned and superimposed according to the supposed binding to the receptor. This method also assumes great similarity between the structures, otherwise finding their best superposition becomes problematic. CoMFA compares the 3D steric and electrostatic fields generated for the molecules and selects the correlating features with biological activity. It correlates molecular properties to biological activities by:

a calculating steric and electrostatic (and optionally lipophylic) potentials around the molecules, and then
b applying the PLS method to the data sets.

Application of ANN (artificial neural networks). In all cases, the relationship between biological activity and physicochemical properties and structure is naturally nonlinear. Recently, a conceptually different approach, neural network methodology, has been shown able to recognize complex relations between structural or physicochemical features of molecules and their biological activities.

Most of the published QSAR findings are valid only for a limited number of compounds showing strong structural similarity to each other. Several positive ANN attempts have been made to detect "drug-likeness" (see above) or to predict biological activity spectra of molecules, however these experiments provided only qualitative (e.g. matching keyword) results. **ANN** and **NPLS** (nonlinear partial least squares) can be used successfully for recognition of nonlinear correlations in QSAR (18).

Most applications of neural networks in chemistry use a fully connected three-layer, feed-forward computational neural network with back-propagation training. The schematic architecture of this typical neural network is shown in Figure 16.3. The basic processing unit, represented by a circle, is the neuron which receives one or more inputs and produces an output. The input usually takes values from the descriptors. These inputs are commonly called, and sketched as, input neurons in the input layer, though in a sense that is a misnomer. No processing is done by an input neuron. They all produce an output equal to their single input. The input neurons are only a semantic construct to suggest that they pass their input toward each hidden neuron. Unlike the hypothetical input neurons, hidden layer neurons and output layer neurons are very real. Each of the hidden and output neurons accept inputs, sum them and produce an output. At each processing neuron, every input has an associated weight that modifies the strength of each input connected to that neuron. The processing neuron simply sums all the inputs and calculates an output which should be forwarded to all other neurons in the next layer or it is displayed to the outer world. Principally, the neural networks proceed as follows:

1 each input descriptor value is multiplied by the connection weight;
2 the products are summed up at each hidden unit neuron, where a non-linear transfer function is applied; and
3 the output of each hidden unit neuron is multiplied by the connection weight, summed up at the output layer neurons, and the result interpreted.

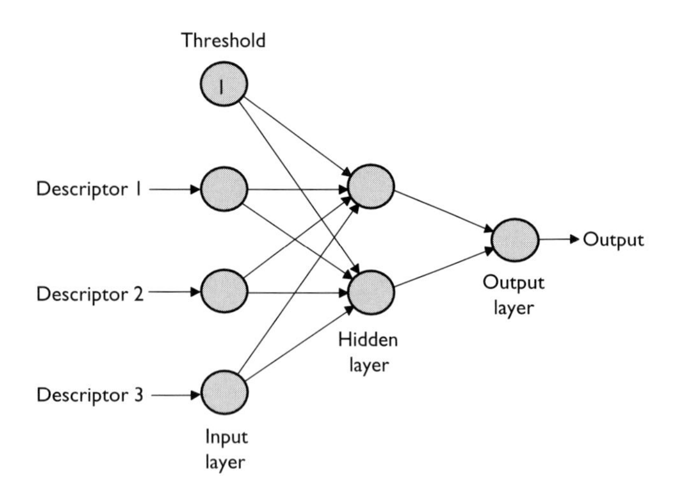

Figure 16.3 Schematic view of an artificial neural network

There is a special, so-called bias neuron in the input layer. Its output is always one and its connection weights to the non-linear hidden neurons set the switching thresholds of those non-linear neurons. Neural networks are not explicitly pre-programed for making solutions; rather they are trained through examples. During the training process values of the weights are adjusted to make the output of the network close to the expected output.

In respect to the performance of a network, two mathematical issues need to be considered: the representation power of the network, and the training algorithm. The first relates to the ability of a neural network to represent a desired function. Since a neural network is built up from a set of standard functions, it can only approximate the desired function. Therefore, even in the case of an optimal set of weights, the error of approximation can never reach the value of zero.

Fully connected, three-layer, feed-forward computational neural networks with non-linear transfer function in the hidden layer have produced excellent performance in many applications of fitting and reproducing almost any non-linear hypersurface, due to the universal approximation theorem. The theorem says that these types of networks can approximate any functions with a finite number of discontinuities to arbitrary precision. As discussed above, most QSAR methods are based on MLR or PLS analysis. Therefore, these approaches can only capture linear relationships between molecular characteristics and functional properties. In contrast, neural networks can recognize highly non-linear relationships between structural or physicochemical descriptors and biological activities or any other molecular features. This inherent feature of non-linearity makes neural networks particularly well suited to treatment of generally non-linear SARs.

A neural network learns by passing the data through repeatedly and adjusting its connection weights to minimize error, e.g. the difference between predicted *versus* actual biological activities. The method of weight adjustment is known as the training algorithm. There are now various algorithms in use, the most common is the back propagation of errors. Although it is not the fastest method in terms of training, it has a very useful convergence property. Namely, if the number of input descriptors are greater than the number of hidden neurons – a carefully selected network architecture usually has fewer hidden neurons than input descriptors – convergence of the network to a global optimum is always ensured by back propagation.

Some important practical features of neural networks should still be considered. They can learn everything, apparently, without any limitation, and this ability might be a source of overfitting the data. To avoid this, first, like other QSAR methods, the experimental error of measured data, which should be predicted or represented by the neural network calculations, must be defined. Any model derived from measured data with a fitting error less than the experimental error is meaningless. A validation process must evaluate the competence of a neural network. Generally, the known cases are divided into two disjointed sets. One is the training set, the other is the validation set. In many respects, a proper validation process is more important than proper training. The latter, more general method is suitable for the recognition of existing relationships between data, even in cases where the other procedures failed (e.g. underdetermined cases).

The **general form of a QSAR relationship** is:

$$f(b_i, \ldots, z_i) = A_i$$

i.e. the biological activity of the "i" molecule (A_i) can be approximated from a (linear or non-linear) function of a significant set of the corresponding theoretically or experimentally determined molecular descriptors (b_i, z_i).

Successful applications of many different kinds of calculated descriptors for QSAR studies have been recorded (see Table 16.2). The experimental determination of physicochemical properties (e.g. logP, pK_a, dipole moment etc.) for thousands of compounds is a time-consuming and expensive procedure. Obtaining calculated descriptors is cheaper, faster, and their reliability are comparable to experimental biological data.

The **3D low energy structural data** of conformers of compounds can be obtained from quantum chemical (*ab initio*) or semi-empirical (e.g. *AM1, PM3*) calculations. The exact calculation of data for only one hundred molecules in this way would need unbelievably long computer time or extremely high performance. Therefore many methods apply simple, standardized transformation of 2D structures into 3D using experimental datasets and/or theoretically calculated data (e.g. the popular **Concord** (48) or **Corina** (49)). These 3D structures could be far from the energy minimized conformations and represent only one conformation from the

Table 16.2 Calculated molecular descriptors used in QSAR studies

Descriptor	*No. of available descriptors*[a]	*Ref.*
Molecular mass[b]	1	
Molecular volume, solvent extended volume[b]	2	19, 22, 36
Molecular surface, solvent accessible surface, solvent extended surface[b]	3	19, 21, 22
Globularity[b]	1	20
WHIM descriptors of atomic mass, position, electronegativity, localized charge, atomic polarizability contribution, atomic electro topological index, pi functionality. Moments and T A V K combinations were used[b]	$7 \times 7 = 49$	31
Polarizability[b]	1	23, 24
Dipole moment[b]	1	25
Hildebrand solubility parameter[b]	1	30
LogP[b]	1	26
Unsaturation number[b]	1	
Degree of chemical bond rotational freedom[b]	1	27
Wiener index[b]	1	32
Randics index[b]	1	33
HDSA1, HDSA2, HASA1, HASA2 hydrogen bond (HB) descriptors[b]	4	34
Gravitational index[b]	1	34
Topological electronic index[b]	1	34
QN, QO, QNO, QTOT Bodor charge descriptors for logP[b]	4	35
Min., max. and average of electrostatic potential (ESP) on the vdw surface[b]	3	23
Histogram of ESP distribution on the vdw surface (8 cells)[b]	8	23
Min., max. and average of molecular lipophylicity potential (MLP) on the vdw surface[b]	3	23
Histogram of MLP distribution on the vdw surface (8 cells)[b]	8	23
Number of specified atom types[b]	35	
Min. max. and average of localized charge on any atom type[b]	95	23, 25
Electrostatic HB basicity and acidity, max. plus summed values[b]	4	29
HOMO, LUMO (AM1)	2	29
Auto correlation functions of atomic mass, position, electronegativity, localized charge, atomic polarizability contribution, atomic electro-topological index, pi functionality, logP contribution and of any atom type[b]	$(35+8)\ (\times 6) = 258$	26
Pair correlation functions of atomic mass, position, electronegativity, localized charge, atomic polarizability contribution, atomic electro-topological index, pi functionality, logP contribution and of any atom type[b]	$(903)\ (\times 6) = 5418$	26
3D MoRSE codes of atomic mass, position, electronegativity, localized charge, atomic polarizability contribution, atomic electro-topological index, pi functionality, logP contribution and of any atom type from 0 to 8 angstroms^{-1} in 16 steps.	$(903)\ (\times 16) = 14440$	43

Note

a The number of calculated descriptors are shown for 35 atom types.

b Here we calculated with pair correlations from 1 angstrom to 7 angstroms in 6 steps. We used 11 atom types in this example: H (lypophylic, HB don.), HB don. (O, N), HB acc. (O, N), C (sp^3, sp^2), N (sp^3, sp^2), O (sp^3, sp^2), halogens.

possible dozens but they are still applicable for comparison of compounds because all of the structures are derived by the same standard rules. Many of the descriptors listed in the table above can be calculated with satisfactory precision from even 2D (or connectivity) data.

QSAR correlations can be published in the literature even at a correlation coefficient value of 0.3. There is an acceptably low risk that such a correlation is merely a chance correlation, in any event it is far from those that yield reliable predictions.

QSAR analysis can also be performed by the simultaneous, automatic application of PLS, MLR and ANN algorithms to achieve an optimal (maximal) Q^2 or (minimal) SEP value for the given model.

$\mathbf{Q^2}$ is the **crossvalidated correlation coefficient**, in other words it is the "goodness of estimation". This parameter focuses on predictive ability instead of fitting capability. Therefore it is more strict than the r^2 "correlation coefficient" which was used widely in QSAR studies previously.

An extended formula for calculation of Q^2:

$$Q^2 = 1 - \frac{\text{press}}{\overline{press}} = 1 - \frac{\Sigma(calc - \exp)^2}{\Sigma(\overline{\exp} - \exp)^2}$$

calc = calculated value
exp = experimental value
$\overline{\exp}$ = mean of those experimental values that were used to generate the model
press = predictive residual sum of squares.

SEP is the **standard error of prediction**. The classic expression for SEP is:

$$\text{SEP} = \sqrt{\frac{press}{m}} = \sqrt{\frac{\Sigma(calc - \exp)^2}{m}}$$

m = number of molecules.

Statistically a more reliable estimate of SEP is

$$\text{SEP} = \sqrt{\frac{press}{m - n}}$$

where n = number of parameters
but the above expression is valid only if: $m > n$.

The first step of a QSAR study is to calculate the molecular descriptors for each molecule for the model generation and select the significant descriptors by ranking them according to the ratio of the normalized contribution (e.g. %) of the descriptors to the output. In order to make the different kinds of descriptors comparable with each other, their values (x_i) should be preprocessed (scaled) by standard normalization (50).

$$x_{scaled} = (x_i - x_{mean}) / x_{standard\ dev.}$$

The scaled variable (x_{scaled}) has zero mean and the standard deviation (SD) is one. This process is also known as *autoscaling* or *regularization*.

16.11 The importance of the descriptors

The importance (="significance") of a given descriptor can be defined and calculated in different ways, taking into account the effect of the descriptor on the output value (52, 53, 54). The importance of the descriptors is not only ranked but can also be normalized by taking the most important descriptor as 100%. Another key step in the generation of an optimal QSAR model is a stepwise, statistically not self-referencing cross validation. When the validated statistical parameters, e.g. Q^2 of the optimal QSAR model are satisfactory, e.g. $Q^2 > 0.4$, the model can be used for the reliable prediction of biological activity/properties of existing or virtual libraries of molecules. This way potential drug molecules can be selected from large databases where the selection is based upon all structural information we have.

16.11.1 Validation of QSAR models

QSAR models should be validated by the recently used, most accepted cross validation methods. The data should be randomly split (split-half, leave-n-out, leave-one-out, shuffle), before building the model into work set and external validation sets. The work set (used for descriptor selection and model building) should be further randomly separated into an ensemble of training sets and test sets. The external validation set does not take part in the model generation process.

The models can be generated successively: the best model results from the analysis of the training set/test set ensemble validated by the first validation set. The reproductive details of this validation (the number of structures, which structures, predicted values, Q^2 values, SEP, etc.) are stored for each step. Then the data of the validation set will be merged with the data of the working set and a new optimization (recognition) starts. The procedure should be repeated with each validation set and the results summarized. The increase of Q^2 values (or monotone decreasing SEP values) indicates a statistically robust model which converges toward the optimal model that can be obtainable on the basis of the available information. Randomly fluctuating, low Q^2, and high SEP values indicate that even the optimal model obtained from the existing dataset cannot be used for prediction due to lack of (or not sufficiently good quality) data. Non-self-referencing, iterative validation means that a validation set can be used for validation only once in the same model building process and its molecules are never "seen" by the model before the validation.

The optimal pharmacophore model, generated by this method, specifies value intervals (ranges) for the descriptors needed for the description of the relationship therefore the "pharmacophore model" can be fitted on diverse molecular structure sets as well. We can find the significant (important) descriptors, if any, and the correlation function between these descriptors and between the biological activity automatically. Then we have clear-cut statistical measures of the best predictive correlation in our dataset. The basic assumption, however, is that similar molecules tend to have similar biological activity. The key point here is that our method can find similarity patterns in the space of calculated abstract or experimentally measured descriptors for largely differing chemical structures. In other words the scope of the term "similarity" is expanded to the realm of very different chemical structures. This method also automatically indicates whether an optimal model could be obtained from the existing dataset or if more data are necessary.

16.11.2 Use of the QSAR method for lead selection and optimization (prediction)

Test compound structures can be real or virtual and they can come from different sources: Dictionary of Drugs, ACD (Available Chemicals Directory), Chemical Abstracts Service, Beilstein Crossfire, combinatorial companies' databases, etc. providing synthesizable (=real) compounds. CombiLibMaker (TRIPOS), MDL Project and other structure generators give virtual compounds based on virtual reactions on selected core structures. The structure library data (from any source) should be transfered into a unified database in 2D and/or 3D structural format. Then, following selection of an acceptable pharmacophore model (2D or 3D) from the model database by the user, the software calculates descriptors and predicts the desired values for the library and stores the calculated values in the database. Finally the potential lead molecules are ranked.

16.11.2.1 Representative examples

In this study we used the 3D structures available for all of the compounds obtained previously using the Concord module of Tripos SYBYL program system (48). The 2D and 3D chemical structures, along with the activity data, were stored in MDL ISISBASE format (51). In every model optimization we allowed the program to use 3D holistic descriptors. Only 11 atom types were taken into consideration. Validations were performed as "true validations" (or "real validations") using external validation sets, randomly selected from the datasets before model building (the validation sets were not used in the default model building).

16.11.3 Tumor dihydrofolate reductase (DHR) inhibitors

As an example we can analyze the classical dihydrofolate-reductase inhibitors dataset studied by Hansch *et al.* (MLR, PLS, ANN models). Hansch utilized his QSAR approach in the analysis of 256 4,6-diamino-1,2-dihydro-2,3-dimethyl-1-(X-phenyl)-s-triazines which were tested against tumor dihydrofolate reductase (38). These data became a test set for several QSAR studies (45, 46, 47). The $\log(1/IC_{50}) = pIC_{50}$ values were reproduced or predicted.

It is interesting to note that the original article contains two pairs of identical compounds among the 256 (namely compounds no.112, 202 and compounds no. 186, 188), i.e. different IC_{50} values are listed for the same structures. None of the following publications mentioned this, but used and printed the original data. In each identical pair we excluded the higher activity compounds from our study. Therefore we performed our calculations with 254 DHR inhibitors only. 240 molecules were randomly selected randomly for the work set and 14 molecules for the validation set. The validation parameters of the optimized models are shown in Table 16.3. For the sake of comparison, we performed leave-one-out cross validation with the best NN model. Even the

Table 16.3 QSAR model data for DHR inhibitors

Model	*MLR*	*PLS*	*NN*
Maximum average Q^2 of monitoring validations	0.499 (average of 32 values)	0.503 (average of 32 values)	0.712 (average of 8 values)
Q^2 of final validation	0.553	0.648	0.661
Model parameters	23 parameters	15 parameters, 14 components	5 hidden neurons, 140 parameters, $\rho = 1.71$ (at 240 compounds)
Common descriptors that appear in at least 2 optimized models	Volume, degree of rotational freedom, 1st lipophylicity moment, Wiener index, electronegativity-vdw volume pair correlation, vdw volume – pi functionality pair correlation, vdw volume – electro-topological index pair correlation		

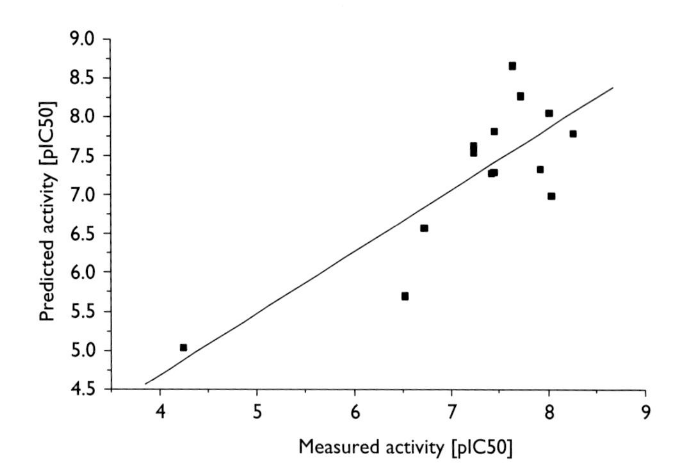

Figure 16.4 Linear regression for DHR-MLR pharmacophore model validation:

Y = A + B*X

Parameter	*Value*		*Error*
A	1.50909		1.20714
B	0.7933		0.16438
R	*SD*	*N*	*P*
0.81238	0.59263	14	4.14846E-4

Parameter Descriptions for Linear Regression
A: Intercept value and its standard error.
B: Slope value and its standard error.
R: Correlation coefficient.
P: value – Probability (that R is zero).
N: Number of data points.
SD: Standard deviation of the fit.

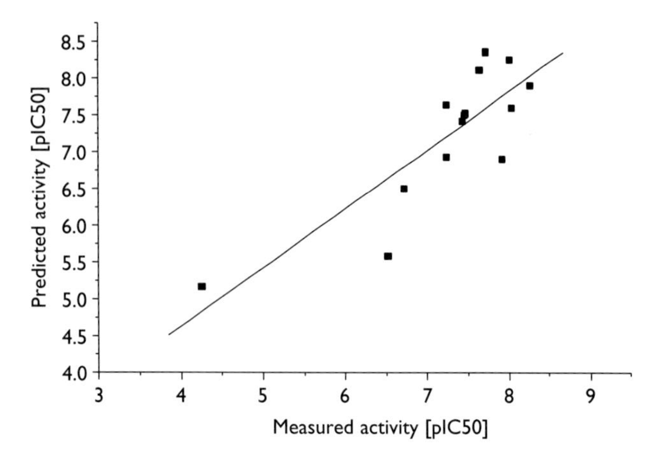

Figure 16.5 Linear regression for DHR-PLS pharmacophore model validation:

Y = A + B*X

Parameter	*Value*		*Error*
A	1.44249		1.09908
B	0.79614		0.14967
R	*SD*	*N*	*P*
0.83798	0.53957	14	1.82479E-4

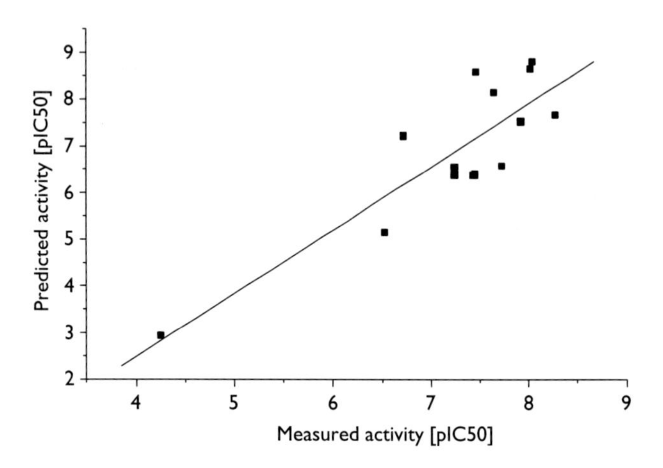

Figure 16.6 Linear regression for DHR-ANN pharmacophore model validation:

Y = A + B*X

Parameter	*Value*		*Error*
A	−2.91689		1.69773
B	1.35248		0.23119
R	*SD*	*N*	*P*
0.86046	0.83347	14	<0.0001

leave-one-out cross validated $Q^2 = 0.855$ value compared well with the best R^2 values of fitting found in the literature with NN, MLR and MLR models (45), where the corresponding figures were 0.850, 0.494 and 0.773, respectively. The R^2 of fitting of our cross validated NN model was 0.910 for these compounds.

16.11.3.1 Demonstration of a single external validation of DHR pharmacophore models

The models were optimized via series of training set–test set selections, trainings and validations cycles. (The maximum averages of Q^2 values are given in the tables.) We visualizd below the validation data of the final models with a single test set which was excluded from the model buildings (Figures 16.4–16.6).

References

1 Kuntz, I.D. (1992) *Science 257 1078–1082.*
2 Böhm, H.J.J. (1992) *Comput.-Aided Mol. Design 6 61–78.*
3 Rarey, M., Kramer, B., Lengauer, T., Klebe, G. J. (1996) *Mol. Biol. 261 470–489; http://cartan.gmd.de/flexx.*
4 Wermuth, C. G. (1996) The Practice of Medicinal Chemistry *Academic Press, London.*
5 Kubinyi, H. (1998) Chimica Oggi/Chemistry Today, Oct. 17–22.
6 Wermuth, C.-G.; Ganellin, C.R.; Lindberg, P.; Mitscher, L.A. (1998) *Pure and Applied Chemistry 70, 1129–1143.*

7 Ajay, A., Walters, W. P., Murcko, M. A. J. (1998) *Med. Chem. 41, (18): 3314–24.*
8 Sadowski, J., Kubinyi, H. J. (1998) *Med. Chem. 41, (18): 3325–29.*
9 Clark, D.E., Pickett, S.D. (2000) *Drug Discovery Today 5, (2) 49–58.*
10 Lipinski, C.A., Lombardo, F., Dominy, B.W., Feeney, P.J. (1997) *Advanced Drug Delivery Reviews, 23, p.3–25.*
11 Daylight Information Systems, Mission Viejo, CA, USA.
12 Moriguchi I., Hirono, S., Liu, Q., Nakagome, Y., Masushita, Y. (1992) *Chem. Pharm. Bull. 40 127–130.*
13 Ghose, A. K., Viswanadhan, V. N., Wendoloski, J. J. (1992) *J. Comb. Chem. 1 55–68.*
14 Abraham, M. H., Weathersby, P. K. (1994) *J. Pharm. Sci. 83(10):1450–6.*
15 Platts, J. A., Butina, D., Abraham, M. H., Hersey, A. (1999) *J. Chem. Inf. Comput. Sci. 39, 835–845.*
16 Luco, J. M. (1999) *J. Chem. Inf. Comput. Sci. 39, 396–404.*
17 Clark D. E. (1999) *J. Pharm. Sci. 88, 815–821.*
18 Kövesdi, I., Dominguez-Rodriguez, M. F., Örfi, L., Náray-Szabó, G., Varró, A., Papp, J. G., Mátyus, P. (1999) *Med. Res. Rev. 19(3) 249–69.*
19 Connolly, M. L. (1985) *J. Am. Chem. Soc., 107, 1118–1124.*
20 Meyer, A. Y. (1986) *J. Chem. Soc. Rev., 15, 449–474.*
21 Iwase, K., Komatau, K., Hirono, S., Nakagawa, S., Moriguchi, I. (1985) *Chem. Pharm. Bull., 1985, 33, 2114–2121.*
22 De Bruijn, J., Hermkens, J. (1990) *J. Quant. Struct-Act. Relat., 9, 11–21.*
23 Breindl, A., Beck, B., Clark, T., Glen, R. C. (1997) *J. Mol. Model., 3, 142–155.*
24 Miller, K. J. (1990) *J. Am. Chem. Soc., 112, 8533–8542.*
25 Mortier, W. J., van Genechten, K., Gasteiger, J. (1985) *J. Am. Chem. Soc., 107,* 829.
26 Broto, P., Moreau, G., Vandycke, C. (1984) *Eur. J. Med. Chem., 19, 71–78.*
27 Andrews, P. R., Craik, D. J., Martin, J. L. (1984) *J. Med. Chem., 1984, 27, 1648–1657.*
28 Gaillard, P., Carrupt, P., Testa, B., Boudon, A. (1994) *J. Comput.-Aided Mol. Des., 8, 83–86.*
29 Cronce, T. D., Famini, G. R., De Soto, J. A., Wilson, L. Y. (1998) *J. Chem. Soc. Perkin Trans. 2., 2, 1293–1301.*
30 Fedors, R. F., Van Krevelen, D. V., Hoftyzer, P. J. (1986) *In CRC Handbook of Solubility Parameters and Other Cohesion Parameters, CRC Press, New York.*
31 Todeschini, R., Grammatica, P. (1997) *Quant. Struct.-Act. Relat. 16, 120–125.*
32 Wiener, H. (1947) *J. Am. Chem. Soc., 69, 2636–2641.*
33 Randic, M. (1975) *J. Am. Chem. Soc., 97,* 6609–6615.
34 Katritzky, A. R., Lobanov, V. S., Karelson, M. (1998) J. *Chem. Inf. Comput. Sci.* 38, 28–41.
35 Bodor, N., Huang, M. J., Harget, A. (1994) *J. Mol. Struct. (Theochem), 309, 259–266.*
36 Bodor, N., Buchwald, P. (1997) *J. Phys, Chem., 101, 3404–3412.*
37 3DNET, Ver. 1.0 (2000) *VICHEM Ltd., Budapest, Hungary.*
38 Hansch, C., Fujita, T. (1964) *J. Am. Chem. Soc., 86, 1616–1620.*
39 Hansch, C., Silipo, C. (1975) *J. Am. Chem. Soc. 97, 6849–6861.*
40 Hornik, K., Stinchcombe, M., White, H. (1989) *Neur. Net., 5, 359–366.*
41 Hartman, E., Keeler, J. D., Kowalski, J. M. (1990) *Neur. Comp., 2,* 210–215.
42 Hornik, K., Stinchcombe, M., White, H., Aurer, P. (1994) *Neur. Comp., 9, 1262–1275.*
43 Schuur, J., Selzer, P., Gasteiger, J. (1996) *J. Chem. Inf. Comput. Sci., 36, 334–344.*
44 Wold, H. (1966) In *Research Papers in Statistics, 1966, Wiley, New York.*
45 Andrea, T. A., Kalayeh, H. (1991) *J. Med. Chem., 34, 2824–2836.*
46 So, S.-S., Richards, W. G. (1992) *J. Med. Chem., 35, 3201–3207.*
47 King, R. D., Muggleton, S., Lewis, R. A., Sternberg, M. J. E. (1992) *Proc. Natl. Acad. Sci. USA, 89, 11322–11326.*
48 Rusinko, A., Skell, J. M., Balducci, R., McGarity, C. M., Pearlman, R. S., Univ. of Texas, Austin, TX and Tripos, St. Louis, MO USA (1988) *CONCORD 6.0, 1992, TRIPOS Associaties Inc., St. Luis, Missouri.*
49 Gasteiger, J., Rudolph, C., Sadowski, J. (1990) *Tetrahedron Comput. Methodol. 3 537.*
50 Kowalsky, B. R., Bender, C. F. (1972) *J. Am. Chem. Soc. 94, 5632.*
51 ISIS/Base, Ver. 2.2.1 (1999) *MDL Information Systems Inc. San Leandro CA.*
52 Masters, T. (1996) *In: Practical Neural Network Recipes in C++, Academic Press, Boston.*
53 Despagne, F., Massart, L. (1996) *Chemom. Intell. Lab. Syst., 40 145.*
54 Andersson, F. O., Aberg, M., Jacobson, S. P. (2000) *Chemom. Intell. Lab. Syst., 51 61.*

Chapter 17

The development of carbohydrate based therapeutics

Nicholas Drinnan and Tracie Ramsdale

Contents

17.1 The roles of carbohydrates

Carbohydrates constitute one of the three main classes of biopolymers and stand as the least exploited class in regard to therapeutic development. It is well recognized that certain cell-stage specific carbohydrate sequences play a key role in intercellular recognition and serve as cell surface markers. Glycoconjugates are involved in such biomolecular processes as cell-cell contact as well as cellular, bacterial and viral adhesion.

Carbohydrates also assist in improving protein solubility as well playing a critical role in protein folding (1). In fact, glycosylation is one of the most frequently occuring post-translational modifications made to proteins (2). There are three major classes of glycoconjugates; glycoproteins, proteoglycans and glycolipids. Some of the more common types of glycoforms will be outlined below.

17.1.1 N-Glycans

Included in the class of glycoproteins are the *N*-glycans. The *N*-glycans have a common pentasaccharide core structure with differentiation arising from the formation of different antennaries, up to penta-antennaries, which branch from mannose residues in the core structure (Figure 17.1). *N*-Linked glycoproteins are characterized by a β-glycosidic linkage between a GlcNAc residue and a δ-amide nitrogen atom of an asparagine residue (3). All *N*-glycans are synthesised via a common starting material, a lipid bound tetradecasaccharide. Most trimming and differentiation, including re-glycosylation, of the *N*-glycans occurs in the golgi apparatus.

Highly branched and sialylated *N*-linked oligosaccharides are expressed in increased amounts on transformed cells and are implicated in enhanced tumorigenicity and metastatic capability of tumor cells (4). Such neoglycosylation has, for example, the effect of enabling tumorigenic cells to evade immune destruction.

The enzymes called glycosyltransferases that catalyze the formation of new glycosidic linkages, and the glycosidases that effect the reverse process, are both targets for the inhibition of cellular oligosaccharide synthesis. It has been suggested that the inhibition of either glycosyltransferases or glycosidases could prevent the formation of cancer associated aberrant glycoforms. For example, the action of compounds such as swainsonine and deoxynojiromycin is to effect the trimming process that occurs in the golgi, inhibiting further glycosylation and resulting in the loss of *N*-acetyllactosamine antennae, preventing, for example, neo-sialylation (5).

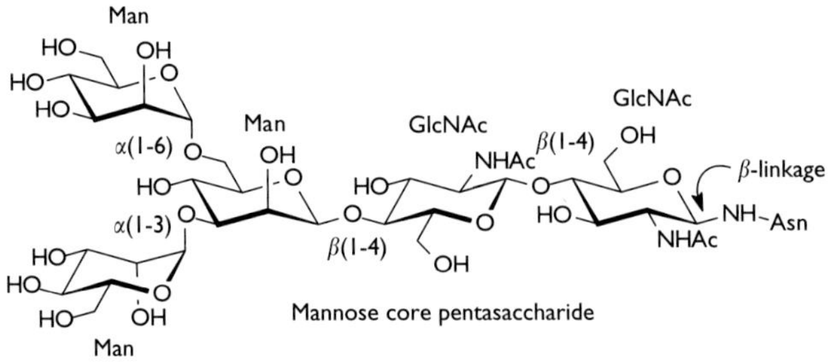

Mannose core pentasaccharide

Figure 17.1

17.1.2 O-Glycans

O-Glycans occur on soluble, secreted and membrane-bound glycoproteins and proteoglycans. *O*-Linked glycoproteins are characterized by an α-linked *N*-acetylgalactosamine (GalNAc) or *N*-acetylglucosamine (GlcNAc) (Figure 17.2), or β-linked xylose conjugated to the hydroxyl group of a serine or threonine (6). The synthesis of *O*-linked glycans occurs in the golgi apparatus by the sequential addition of saccharide units by glycosyltransferases. *O*-Glycans can be classified into different subgroups depending on the terminal sugar and the amino acid to which they are coupled. One sub-class of *O*-glycans are mucins which are characterized by a *N*-acetyl-galactosamine residue coupled to a serine or threonine. Mucins are high molecular weight glycosylated proteins that form a major part of a protective biofilm on the surface of epithelial cells. An example of the function of mucins is to act as decoys for infectious agents. The invading microorganism or parasite, in an attempt to cross the mucosal membrane, recognizes a cognate oligosaccharide sequence of a mucin and upon binding to this sequence is effectively swept away by ciliary action (4).

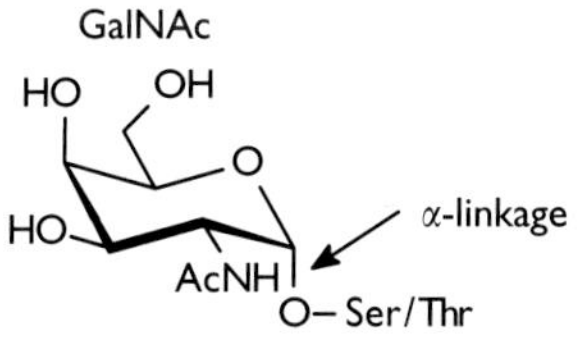

GalNAc

α-linkage

O–Ser/Thr

Tn antigen, *O*-linked oligosaccharide core structure

Figure 17.2

17.1.3 Glycolipids

Glycolipids are comprised of an oligosaccharide conjugated to a lipid by means of an inositol or sphingosine moiety (Figure 17.3). The lipidic chain is usually anchored in the cellular membrane whilst the carbohydrate terminus is exposed on the cell surface in the extracellular matrix. In this way, the oligosaccharide moiety can act as an anchoring site for the attachment of proteins, as do the class of glycolipids known

$(CH_2)_{12}CH_3$

R = Long fatty chain

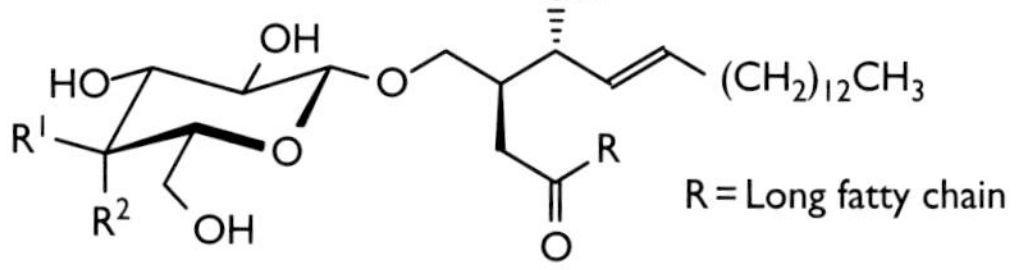

For galactoceramides, R^2 = OH and R^1 = H
For glucoceramides, R^2 = H and R^1 = OH

Figure 17.3

as glycophosphatidyl inositol (GPI) anchors. Another type of glycolipid, the gangliosides, are thought to be critical in the development of nervous tissue. Differences in chemical composition of membrane glycolipids have been found in various transformed cells in comparison with their untransformed counterparts (7). Modifications to glycolipids have been seen in melanoma systems, highly metastatic lung cancers, and in chemically induced pre-cancerous and cancerous liver cells.

17.1.4 Proteoglycans

Another class of glycoconjugates, the proteoglycans (Figure 17.4), are characterized as a protein with a high molecular weight carbohydrate component. The carbohydrate moieties of the proteoglycans, known as glycosaminoglycans, take part in a wide range of biological functions. Proteoglycans are found in the jelly-like intercellular cement, which fills the space between the cells of most tissues. They are also present in cartilage, tendons and skin, and in the synovial fluid, the secretion that lubricates skeletal joints. Growth factors bind with a high degree of specificity to this extracellular matrix, which thereby regulate growth factor activity (8). In proteoglycans, the *O*-glycosidic linkage occurs between the serine residue of the polypeptide chain and a xylose residue of the oligosaccharide. The highly charged carbohydrate polymer heparin, is an example of a proteoglycan of considerable significance. It is generated by certain types of cells, which are especially abundant in the lining of arterial blood vessels. Heparin is an extremely powerful inhibitor of blood clotting and aids in the prevention of clot formation in circulating blood. Its use as an agent for the treatment of arterial and venous thrombosis will be discussed later.

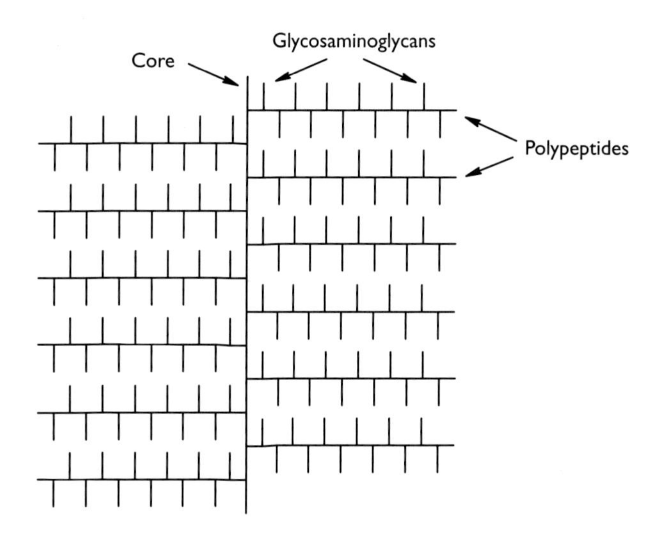

Figure 17.4

17.2 Carbohydrate therapeutics – the challenge

As mentioned earlier, carbohydrates have numerous biological roles, they potentiate cancer, inflammatory, infective, and thrombotic disease states. By and large there are many inherent problems in developing therapeutics based on carbohydrates. The medicinal chemist considers carbohydrates too complex for process development, and too hydrophilic for good bioavailability. Carbohydrates are generally considered to be orally inactive and metabolically unstable. Moreover, many carbohydrate receptor bindings are weak, with dissociation constants in the millimolar range. This weak binding is generally prescribed to the lack of charged and hydrophobic groups on carbohydrate epitopes. It is true that not all carbohydrate epitopes are of this nature, many cell surface carbohydrates are sialylated or are composed of uronic acid moieties providing charged groups. Unfortunately the incorporation of charged moieties presents a further synthetic challenge, although it does alleviate some of the concern surrounding low binding affinities inherent in carbohydrates (9).

The following discussion details some of the recent developments made towards establishing carbohydrates as therapeutics, as well as commenting on the historical progression of some existing carbohydrate-based therapeutics.

17.2.1 Antithrombosis – a historical perspective

Vascular thrombosis is a cardiovascular disease indicated by the partial or total occlusion of a blood vessel by a clot containing blood cells and fibrin. In arteries it results predominantly from platelet activation and leads to heart attack, angina or stroke, whereas venous thrombosis results in inflammation and pulmonary emboli (10). The coagulation of blood is the result of a cascade of events employing various enzymes collectively known as activated blood-coagulation factors (11). Heparin, a powerful anticoagulant has been used since the late 1930's in the treatment of thrombosis. In its original implementation, tolerance problems were noted and so reduced dosage was suggested to reduce bleeding and improve efficacy. In the early 1970's clinical trials did indeed indicate acceptable tolerance was obtainable, whilst still preserving antithrombotic activity. Unfractioned heparin (UFH) is primarily used as an anticoagulant for both therapeutic and surgical indications, and is usually derived from either bovine lung or porcine mucosa. Amongst the modern uses of unfractioned heparin are the management of unstable angina, an adjunct to chemotherapy and anti-inflammatory treatment, and as a modulation agent for growth factors and treatment of hemodynamic disorders.

In the late 1980's, the development of low molecular weight heparins (LMWHs) led to improvements in antithrombotic therapy. LMWHs are derived from UFH by such processes as; chemical degradation, enzymatic depolymerization and γ-radiation cleavage. This class of heparins have recently been used for treatment of trauma-related thrombosis. Of particular interest is the fact that their relative effects on platelets are minimal compared to heparin, providing an immediate advantage when treating platelet compromised patients. The degree of depolymerisation of UFH can be controlled to obtain LMWH of different lengths. Dosage requirements for the treatment of deep vein thrombosis (DVT) are significantly reduced when employing LMWH as opposed to UFH, although in general the efficacy of both therapeutics seems to be comparable. In addition, LMWH can be effective as an alternative therapeutic for patients who have developed a sensitivity to UFH (12). Unfortunately there has been some recent concern about the use of LMWH due to the perceived potential for cross-species viral contamination as a result of the animal source of the parent UFH.

One way of avoiding the possibility of cross-species contamination, is to prepare heparins by chemical synthesis. This method would also provide the opportunity to develop second generation heparins and heparinoids, that can be tailored to target particular biological events in the blood coagulation cascade. Studies have shown (13) that the significant biological event in preventing thrombosis is the binding of a pentasaccharide sequence (14) of heparin to heparin cofactor antithrombin III (ATIII). A derivative of this pentasaccharide fragment has been prepared by total synthesis (15), Figure 17.5, and has completed phase III clinical trials.

17.2.2 Anti-inflammatories

In response to injury or inflammation, damaged tissue releases cytokines, which trigger the expression of P-selectin followed by E-selectin on the endothelium. These selectins recognize a terminal tetrasaccharide fragment called sialyl-Lewis^x (sLe^x), which precipitates various protein-protein interactions. The culmination of this cascade results in extravasation of leukocytes into the epithelium, giving rise to inflammation (16). Blocking the initial sLe^x/selectin interaction at an early stage in the inflammation cascade is seen as a method for treating acute and even chronic inflammatory diseases. Employing sLe^x itself as a therapeutic has several

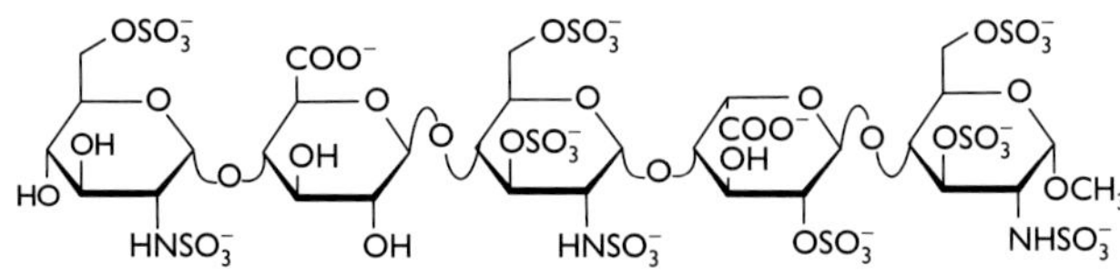

Figure 17.5

drawbacks; it has a weak binding affinity for selectins, it is orally inactive, and is unstable in the blood. What sLe^x does provide is a model for the development of second generation therapeutics. On the basis of NMR studies of the solution conformation of sLe^x, and its conformation when bound to either E- or P-selectin, the functional groups critical for binding have been established. On this basis, a rational approach to mimetics of the tetrasaccharide can be undertaken with the aims of (1) improving binding affinity, (2) improving bioavailability, and (3) improving stability *in vivo*. A number of mimetics of sLe^x have been synthesised. Figure 17.6 gives an example of one particular fucopeptide (**II**), that compares well with sLe^x (**I**) with regard to binding affinity yet has a much simplified structure (17).

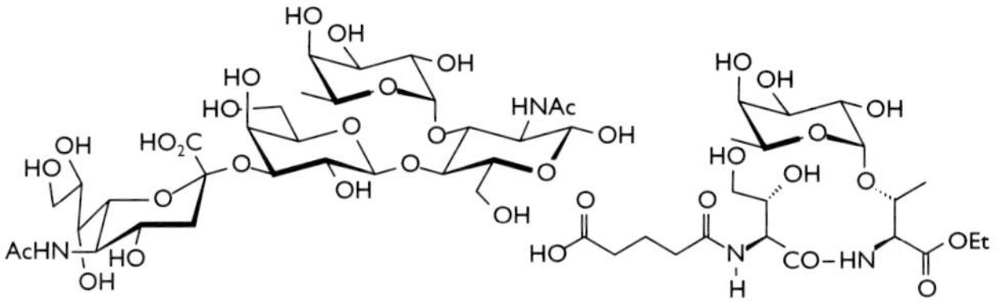

Figure 17.6

In an alternative rationalized approach towards sLe^x mimetics, researchers from Glycomed Inc. searched the Fine Chemicals Directory database against a solution phase derived pharmacophoric profile of sLe^x. The resulting 'hits' were tested as inhibitors of selectin binding. A natural product, glycyrrhizin (Figure 17.7), was found to block selectin binding to sLe^x (18). Consequently derivatives were designed and synthesized, based on the glycyrrhizin structure, resulting in compounds that displayed *in vivo* anti-inflammatory properties.

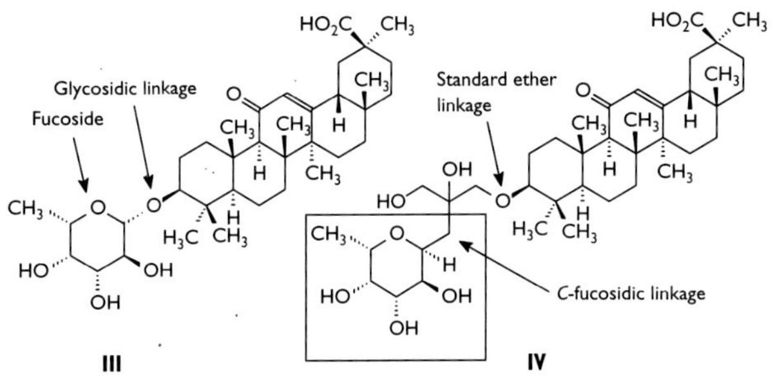

In an effort to further simplify the structure and improve efficacy, the glucuronic acid dimer of glycyrrhizin was substituted with galactose, fucose and mannose moieties. The fucose conjugate (**III**) (Figure 17.8) retained full activity against P-selectin and partial activity against L-selectin. More significantly the conjugate showed an increased ability, relative to glycyrrhizin, to block swelling and neutrophile

Figure 17.7

Figure 17.8

influx in an arachidonic acid stimulated inflammation model in mice. Finally a *C*-fucoside (**IV**) (Figure 17.8) was synthesized eliminating the problem of the labile glycosidic fucosyl linkage, and in turn providing a closer mimic of the spacing between functionalities as found on the sLe^x structure. Structure (**IV**) was shown to be more potent than glycyrrhizin against E- and L-selectin and of equal efficiency against P-selectin. The *in vivo* studies showed that the *C*-glycoside (**IV**) was of equal efficacy to the fucosyl derivative (**III**).

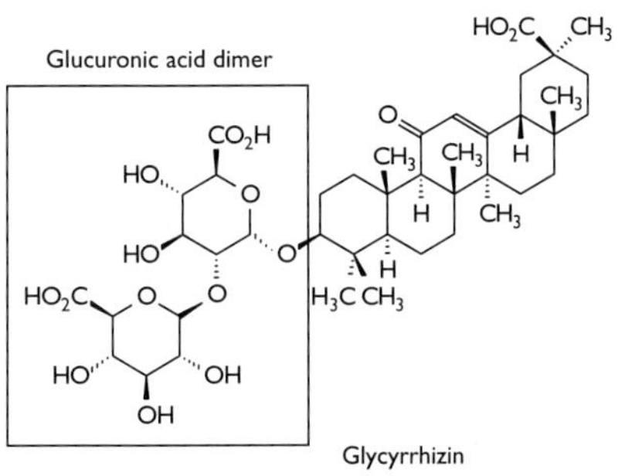

17.2.3 Antibiotics

In recent years, multi-drug resistance (MDR) and the emergence of new pathogenic strains of bacteria have become more prevalent, driving the need for the discovery of new antimicrobial drug therapies. In addition, the dissection of bacterial genomes continues to unearth potential new drug targets, further fueling the race to discover the next generation of anti-infectives.

There are two main classes of antibiotics, those that affect cell wall biosynthesis, and those that interact with bacterial RNA resulting in errors in protein biosynthesis. For example, penicillins are a group of antibiotics that act by interfering with the synthesis and crosslinking of mucopeptides essential for the formation and integrity of the bacterial cell wall (19). Cell wall biosynthesis may also be affected by targeting the muramyl cascade of enzymes. In this case, the interruption of cell wall biosynthesis proceeds through the action of inhibitors based on substrates of the muramyl enzyme cascade. This will be discussed in more detail shortly.

The aminoglycosides are an example of a structurally diverse family of antibiotics that affect the fidelity of protein synthesis by their interaction with bacterial RNA, resulting in truncation and miscoding in protein biosynthesis (20). More detailed discussion of this class of antibiotics will be dealt with in the following section.

17.2.3.1 RNA binders

Aminoglycosides are currently the only well characterized class of small molecules binding to RNA (21). Due to the

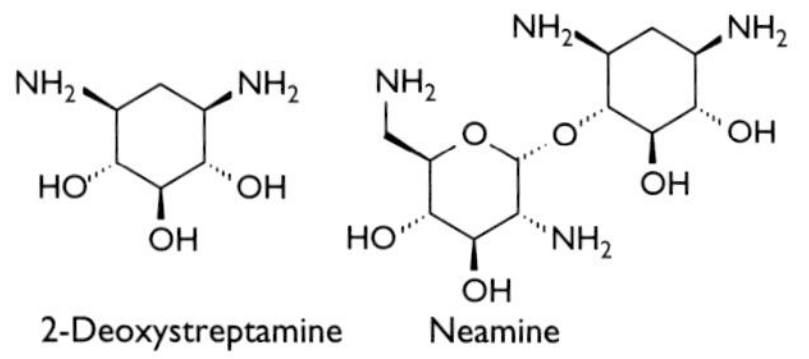

Figure 17.9

emergence of prokaryotic resistance against aminoglycosides, some serious side effects (eg. ototoxicity and nephrotoxicity), poor bioavailability, and bio-stability of these aminoglycosides, new structures are required. Because of the bulk and overt structural complexity of some of the aminoglycosides, one of the requirements of the synthesis of aminoglycoside mimetics is to identify smaller, less bulky structures with only those functionalities present that are required for optimal activity. The majority of these antibiotics contain the *meso*-1,3 diaminocyclitol 2-deoxystreptamine glycosylated at the 4 position and in some cases at the 5- or 6- position. Several aminoglycosides also share a common pseudo-disaccharide core 4-*O*-(2,6-diamino-2,6-dideoxy-α-D-glucopyranosyl)-2-streptamine, commonly known as neamine Figure 17.9. These structures present themselves as possible starting points for concerted combinatorial libraries aimed at improving biological uptake and half-life, and enhancing parasitotrophism whilst decreasing host trophism of the aminoglycosides.

17.2.3.2 Bacterial cell wall biosynthesis

All unicellular bacteria contain a rigid cell wall, which is associated with a diverse range of functions (22). Structure function studies of the components of bacterial cell walls were first published in 1965 by Strominger, who examined the cell wall of *Staphylococcus aureus* (23). Since then numerous other bacterial cells have been investigated. Although the ratio and composition of lipids, glycolipids, glycoproteins and proteins differs from genus to genus, there is remarkable homogeneity in that all genera exhibit a similar but not identical polymeric unit which acts as the basis of the cell wall structure. The common repeating unit is a network of polysaccharide chains covalently linked to each other via small polypeptide chains.

The polysaccharide unit is a repeating heterodimer of N-acetyl-glucosamine (β1-4) N-acetyl-muramic acid. The peptide components consist of two parts:

1. a tetrapeptide (composed of both D and L amino acids) attached to the N-acetyl-muramic acid via a C3 lactic acid side chain, and
2. a pentaglycine which cross-links the lysine side chain of one N-acetyl-muramic acid with the terminal alanine of another N-acetyl-muramic acid. The latter peptide is added in the final step in the biosynthesis of cell walls and is the site of action of the penicillin antibiotics.

Schematically, cell wall synthesis can be depicted as in Scheme 1, in which the *N*-acetyl glucosamine first forms a β1-4 linkage to the muramyl pentapeptide. This heterodimer is extended to form a linear polysaccharide whose chains are subsequently cross-linked to form the rigid cell wall structure.

The ultimate precursor of the heterodimers is UDP-*N*-acetyl-D-glucosamine, which serves as a donor for half of the heterodimer, as well as being converted to UDP-murNAc. UDP-murNAc is then converted in a cascade of events to the muramyl peptide donor (24). Inhibition of any of the enzymes involved in this process, or in the synthesis of the heterodimer, are attractive targets for new antibiotics, particularly in the arena of MDR anti-infectives (25).

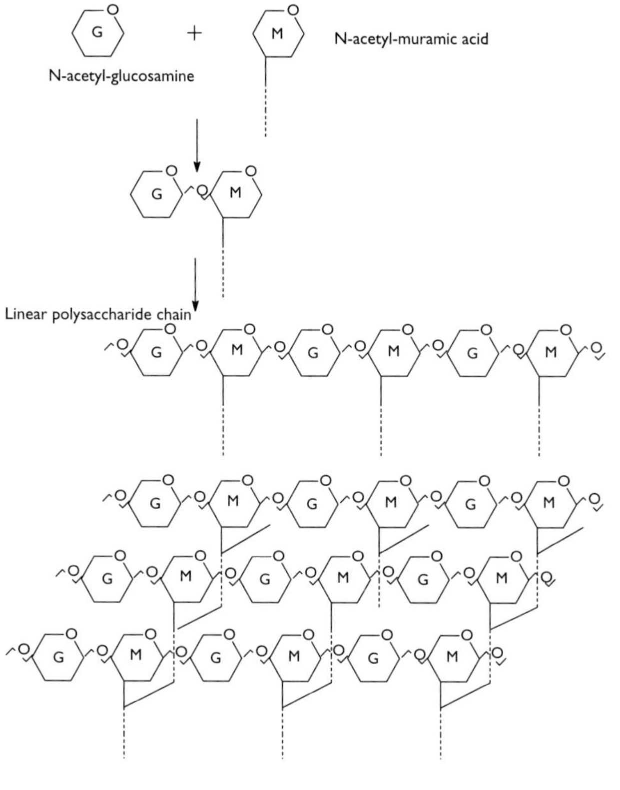

Scheme 1

17.2.4 Carbohydrate anti-cancer vaccines

Cell surface glycosylation patterns can be classified by stages of embryonic and adult growth. For example in fetal erythrocytes, cell surface poly-*N*-acetyllactosamine chains are unbranched. The conversion to adult erythrocytes usually occurs within one year of birth and is characterized by branching through the conjugation of further *N*-acetyllactosamine moieties. Similarly, the appearance of cell surface polysialic acid chains in adults is restricted to mucosal membranes, neural cell adhesion molecules (N-CAM) and on the α-subunit of sodium channels in the brain. Malignant cells are commonly characterized by the appearance of large or unusual oligosaccharide motifs on their cell surface which distinguish them from their normal cell counterparts (26). Characteristic in cancer growth is the "switching on" of the genes relevant to the synthesis of the embryonic glycoforms. For example, fetal N-CAM in several human fetal tissues has a high polysialic acid content that is correspondingly re-expressed by neoplasms. In tumor development, the effect of this cell surface sialylation varies, eg. masking tumor cells from binding interactions, decreasing the tumor cell susceptibility to immune defense, producing larger tumor emboli and increasing metastatic cell adherence to vascular endothelium.

Because a distinction can be made between these two glycoforms, an opportunity exists for the development of therapeutics to treat cancer based on interruption of embryonic glycoform biosynthesis mechanisms. For example, mucins are produced with an altered pattern of glycosylation in many cancers (27). A typical alteration is oligosaccharide chain shortening leading to the exposure of carbohydrate determinants such as the Tn antigen (Figure 17.2). The Tn structure in normal cells is usually cryptic, ie. it is usually further glycosylated in the construction of complex *O*-linked glycans (28). Neoglycosylation by sialyltransferases leads to the formation of the sialyl-Tn (STn) structure (Figure 17.10). This terminal sialic acid moiety effectively precludes any further glycosylation leading to clusters of the STn antigen on human carcinoma cells.

As an atypical carbohydrate form, STn can serve as a target for antibodies. Development of the vaccine Theratope was based on the STn antigen. This vaccine was synthesized by coupling the STn antigen to a crotyl linker functionality, which allowed attachment of the disaccharide to keyhole limpet hemocyanin (KLH), an immunogenic carrier system. Currently in phase III clinical trials, Theratope reliably induces anti-STn antibody responses and appears to induce T-cell mediated responses in many individuals (29).

Similarly, ganglioside-KLH conjugates GMK and MGV were developed as anti-cancer vaccines. Gangliosides are predominantly sialic acid containing glycosphingolipids that are anchored in the cell wall by lipophilic ceramide chains. They are characteristically overexpressed on tissues of neuroectodermal origin, particularly in tumors such as melanomas, sarcomas, neuroblastomas, astrocytomas, and small cell lung

STn antigen

Figure 17.10

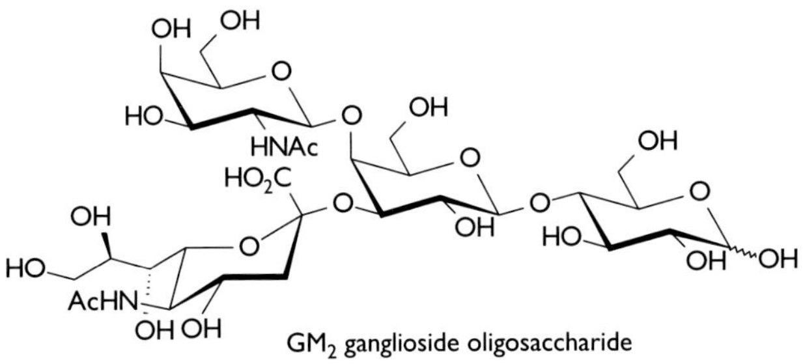

GM_2 ganglioside oligosaccharide

Figure 17.11

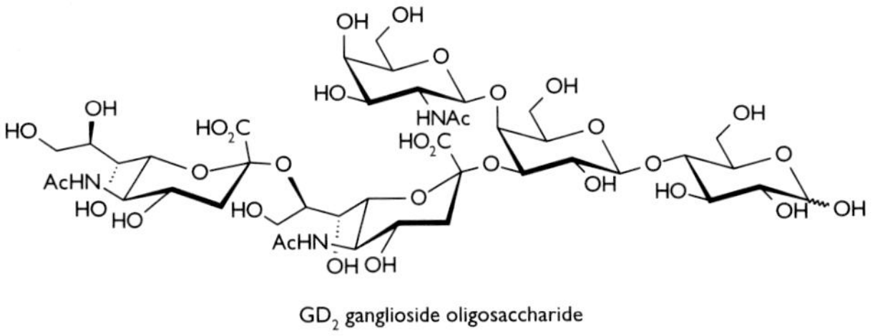

GD_2 ganglioside oligosaccharide

Figure 17.12

cancer (30). The vaccine GMK, a conjugate comprising the ganglioside GM_2 (Figure 17.11) coupled to KLH is currently in phase III clinical trials, for the treatment of breast cancer, melanoma, ovarian cancer and prostate tumor.

MGV, a multi-ganglioside vaccine comprised of both GM_2 and GD_2 (Figure 17.12) is currently in phase II clinical trials for the treatment of colorectal tumor, lung tumor, lymphoma, melanoma, neoplasm, nervous system tumor, sarcoma and stomach tumor.

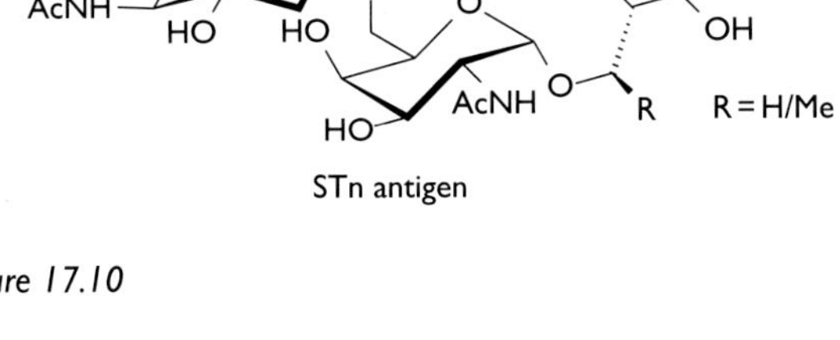

17.3 Future potential for carbohydrates in drug discovery

In order to fully explore the potential of carbohydrate based therapeutics, it is necessary to have rapid access to a wide variety of structures. This section is primarily concerned with advances that have been made in the synthesis of carbohydrate based therapeutics through employing *combinatorial chemistry* techniques as well as detailing some of the chemical advances that have been made with regard to synthetic carbohydrate-based building blocks.

17.3.1 Combinatorial chemistry

The partnership of combinatorial chemistry and the development of carbohydrate based therapeutics is one of recent origin. The aims of combinatorial carbohydrate chemistry can be described as follows:

i To employ carbohydrates as scaffolds in the synthesis of carbohydrate like structures;

ii To employ carbohydrates as scaffolds for the generation of mimetics of non-carbohydrate structures; and

iii To utilise carbohydrates as scaffolds for structural and functional diversity.

17.3.1.1 Carbohydrates as scaffolds in rational design

PROOF OF CONCEPT

In 1996, a proof of concept library of aminosugars was synthesized, employing both combinatorial and solid phase chemical techniques. The library was based on Gal-β-1,3-GalNAc a natural ligand of a *Bauhinia Purpurea* lectin (31). The lectin is known to bind to carbohydrates on the surfaces of erythrocytes causing them to agglutinate, and for this reason was considered a good model system for cell adhesion proteins and other carbohydrate binding proteins that recognize cell surface carbohydrates. The library was synthesized in a stepwise manner, six sugar monomers were introduced to a resin via a glycosylation reaction providing 6 batches of resin. Each of these resin bound sugars was then further glycosylated with twelve different glycosyl donors providing 72 batches of resin. Twenty different acyl groups were introduced to each of these batches of resin at which stage all resins were recombined for the final deprotection step. The library as a composite was then screened against biotin-labeled *Bauhinia purpurea* lectin. Identification of any binding was then accomplished by exposing the resin to streptavidin-linked alkaline phosphatase followed by staining. The library, which contained approximately 1300 di- and trisaccharides, provided two ligands that bound more tightly than the known ligand.

STRUCTURE-ACTIVITY IN AMINOGLYCOSIDE LIBRARIES

A recent study by Wong and co-workers (32) has probed the effect of varying the substitution point of the amino functionality on the glucopyranosyl moiety of neamine to provide neamine type pseudo-disaccharides in the hope of elucidating a new direction for aminoglycoside antibiotic synthesis (Table 17.1). Testing indicated that compound **1**, neamine itself, was the most suitable pseudodisaccharide core for further development. The initial screening involved studying the interactions of aminoglycosides with small model prokaryotic RNA sequences. Although compounds **3** and **6** displayed good binding affinity in one of the screens, only neamine displayed any reasonable antibiotic activity in the Kirby-Bauer disk assay (33).

Subsequently the 5-hydroxy position of the cyclitol moiety of neamine was substituted with diamine, triamine, amino alcohol and aromatic groups to gauge their efficacy as replacements for saccharide substitution (kanamycin, neomycin B analog) at the same position. Screening results implied that there is only a weak correlation between RNA binding affinity and antibacterial activity. Several of the analogues bound more tightly than neamine but failed to exhibit the same antibiotic activity. This discrepancy between strength of binding and antibiotic activity may be due to the use of incomplete pieces of bacterial RNA employed in the binding assays. Two of the analogs, **12** and **13**, displayed similar antibiotic activity to the parent neamine (Figure 17.13).

Table 17.1

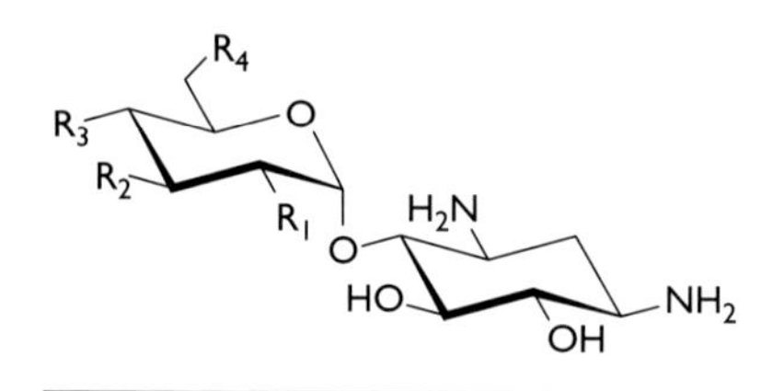

	R_1	R_2	R_3	R_4
1	NH_2	OH	OH	NH_2
2	OH	OH	OH	OH
3	NH_2	OH	OH	OH
4	OH	NH_2	OH	OH
5	OH	OH	NH_2	OH
6	OH	OH	OH	NH_2

Figure 17.13

ANTIBIOTICS BASED ON CELL WALL BIOSYNTHESIS INHIBITORS

In an effort to discover novel antibacterial agents, Intercardia Research Labs developed a disaccharide library based on the decomposition product of the known antibiotic moenomycin A. The moenomycins are a family of natural products known to inhibit the synthesis of bacterial cell wall peptidoglycan

Figure 17.14

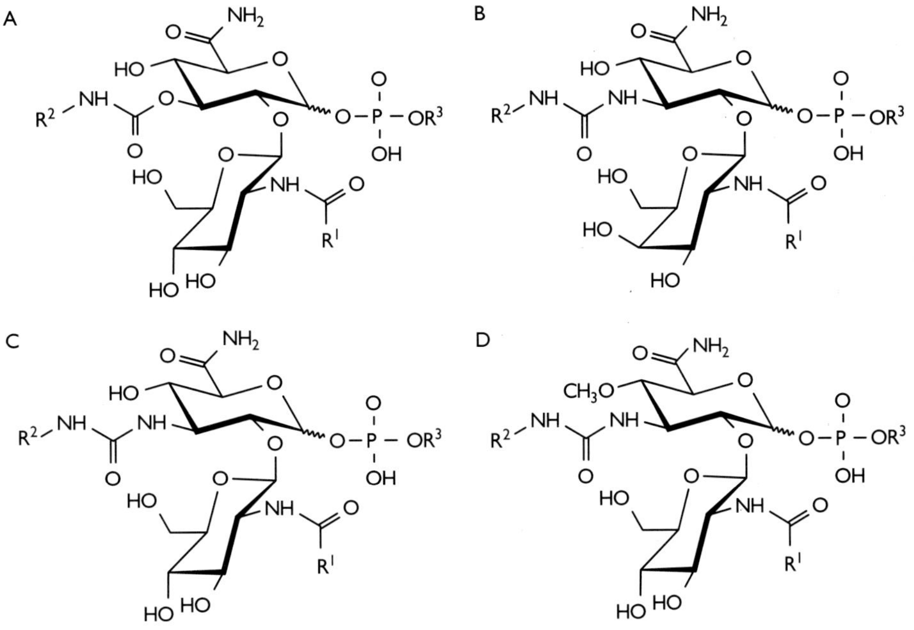

Figure 17.15

through inhibition of transglycosylase. It was shown that activity in inhibition of cell wall biosynthesis was retained by the post decomposition disaccharide core structure. To this end, the synthesis of a library of novel bacterial transglycosylase inhibitors was embarked upon. Solid phase synthesis was the methodology of choice. Sites linked to the activity of the disaccharide had been identified and included the carbamate at C3, the amide at C2′, and the phosphoglycerate at moiety C1 (Figure 17.14). Four disaccharide intermediates **A–D** were prepared with slight variations in structure and functionality. By coupling these blocks to a solid phase via the C6 amide of the disaccharide, and varying substitution at the positions indicated in Figure 17.14, a library of 1300 disaccharides was prepared (Figure 17.15).

The library was screened for both inhibition of cell wall biosynthesis and bacterial growth. Although the parent moenomycin A was still the most potent inhibitor of cell wall biosynthesis, several structurally simpler novel disaccharides were identified that inhibited both cell wall biosynthesis and bacterial growth. These results gave strong support to the proposition that the entire pentasaccharide structure is not required for target recognition and efficacy. In addition, the active disaccharides were shown to be effective against a strain of bacteria naturally resistant to moenomycin and showed similar efficacy to the clinically employed antibiotic vancomycin (34).

THE MURAMYL CASCADE

Other amino function containing carbohydrate-based structures also show potential as antibiotics. As has been mentioned, it has been demonstrated that inhibition of the muramic acid enzyme cascade also leads to interrupted cell wall biosynthesis of bacterial pathogens.

Glucosamine, a readily available sugar constitutes the core structure of UDP-MurNAc and therefore presents itself as an ideal scaffold for a muramic acid based combinatorial library. The crystal structure of the Mur D (UDP-*N*-acetylmuramoyl-L-alanine:D-glutamate ligase) enzyme, co-crystallised with the UDP-MurNAc substrate (35), indicates that the 4-, or 6-hydroxyl groups of the substrate are very close to the wall in the enzyme binding pocket, suggesting little likelihood of successful substitution at these hydroxyl groups, although there has been some indication that the substrate has a degree of motility (36). In (Figure 17.16), the four positions around a glucosamine scaffold where structural variation is most appropriate are indicated.

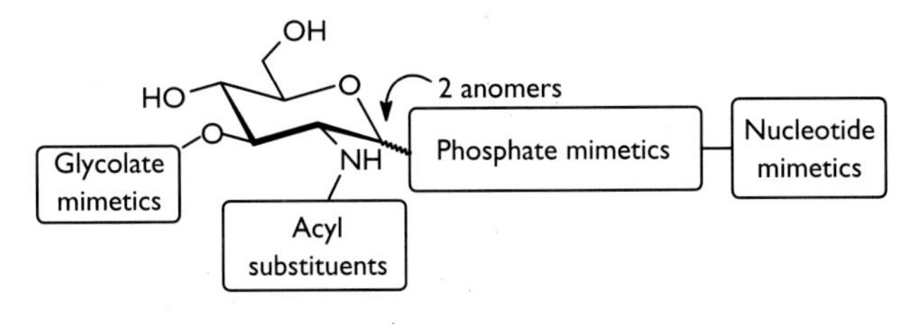

Figure 17.16

It would also appear from the crystal structure (Figure 17.17) that nucleotide or base mimetics need to be planar and have aromatic character to successfully fit into the narrow binding region in the enzyme. The linking arm needs to be composed of at least one charged or strongly dipoled functional group to act as a phosphate mimetic. There is a relative amount of freedom in substitution of various acyl substituents onto the sugar nitrogen, as the spatial requirements in that region of the binding pocket are not so constrained. Finally, a carboxylate mimetic or some similar charged functionality, arranged in an appropriate spatial relationship to the scaffold core, needs to be substituted to C3 of the glucosamine moiety.

The first effective inhibitors of the D-glutamic adding enzyme (Mur D) of the muramyl cascade were reported by Tanner and co-workers in 1996 (37). Similarly inhibitors of the Mur E enzyme were synthesized in 1998 (38). Inhibitors of both enzymes included a phosphinic acid moiety as part of the peptidic arm (Figure 17.18). The rationale behind incorporation of the phosphinic acid moiety was to mimic a tetrahedral transition state intermediate formed during the bio-conjugation of a diaminopimelic acid chain. The Mur D and Mur E inhibitors synthesized displayed reasonable activity, both in the low μM range.

Further inhibitors of the Mur D enzyme were synthesized in 1998 by researchers at Merck (39). Compounds were synthesized based on a similar strategy to Tanner's work, with compounds containing the phosphinite transition state mimic. The major difference was the inclusion of the glucosamine moiety which, in the previous examples, had been replaced by an alkyl chain. The inclusion of the carbohydrate moiety, as theorized, increased the activity of the inhibitor by three orders of magnitude to give activity in the low nM range (Figure 17.19).

17.3.1.2 *Carbohydrates as scaffolds for the generation of mimetics*

From a drug discovery perspective, carbohydrates are well suited as templates. Each sugar represents a three-dimensional scaffold to which a variety of substituents can be attached,

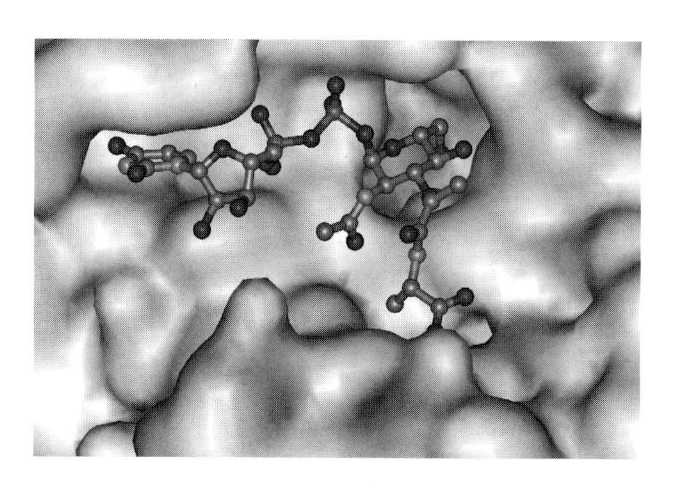

Figure 17.17 (See Colour Plate IX.)

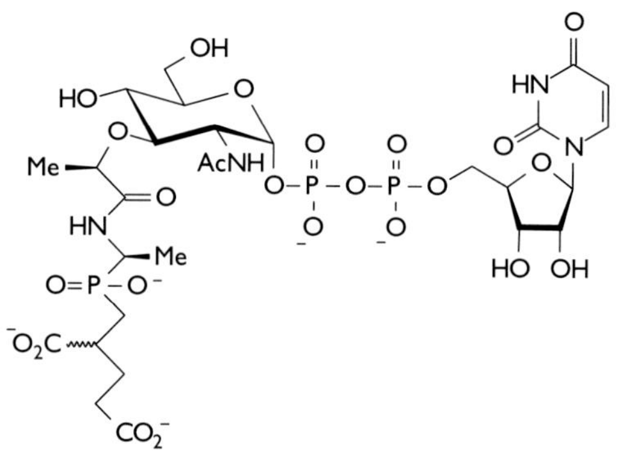

Figure 17.19

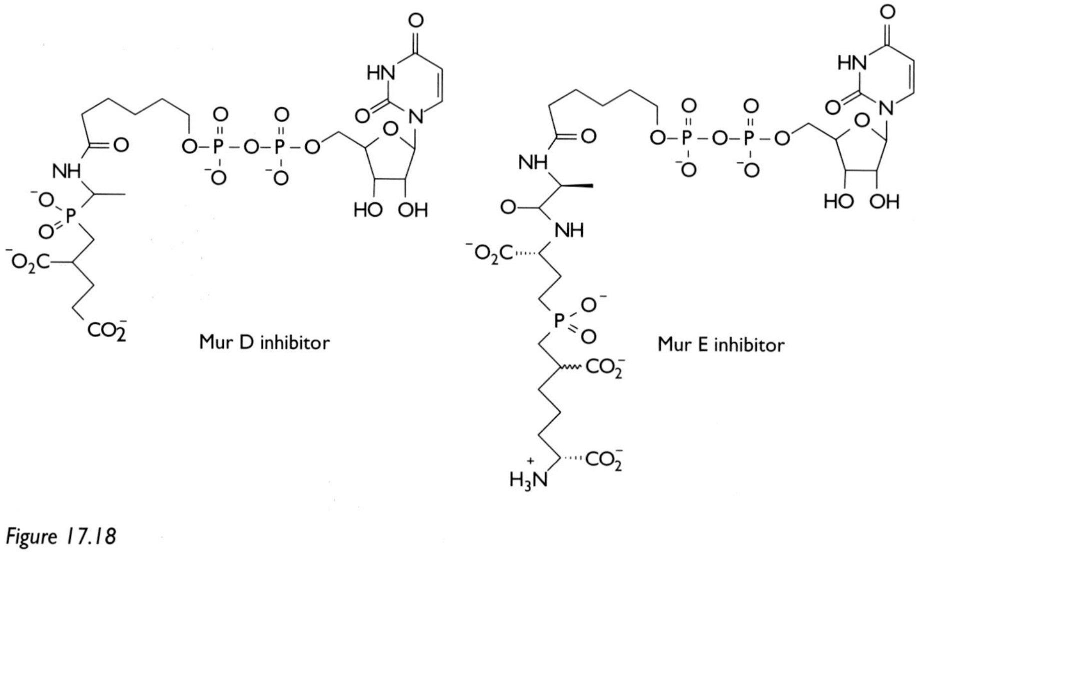

Mur D inhibitor

Mur E inhibitor

Figure 17.18

usually via a hydroxyl group, but occasionally a carboxyl or amino group. By varying the substituents, their relative position on the sugar, and the type of sugar to which the substituents are coupled, numerous highly diverse structures are obtainable. An important feature to note with carbohydrates is that molecular diversity is achieved not only in the type of substituents, but also in the three dimensional presentation. The different stereoisomers of carbohydrates that naturally occur (glucose, galactose, mannose etc.) (Figure 17.20), offer the inherent structural advantage of providing alternative presentation of substituents.

The first example of a combinatorial approach employing carbohydrate chemistry, was a symposium report on the design and synthesis of a peptidomimetic using a glucose scaffold in the early 1990's (40). The results revealed that the glucose based structures (**VI**) and (**VII**) (Figure 17.21), designed as mimetics of the potent somatostatin (SRIF) agonist (**V**) (41), acted as agonists at low concentration, and at high concentration became the first known antagonists of SRIF. Although hardly a library, the results were unique. The synthesis of compounds (**VI–VIII**) demonstrated the potential for saccharide building blocks to act as scaffolds in structure-based design.

Although compound **VIII** was shown to be ineffective as a somatostatin agonist, it did exhibit high affinity and specificity for the substance P (SP) receptor, inhibiting SP-mediated inositol phosphate production. As the authors noted, it was interesting to see the effect such a small structural variation could have on biological activity.

Continuing in part the work commenced in the early 1990's, Nicolaou and co-workers began developing carbohydrate-based peptido-mimetics targeting integrins (42). Integrins, a major class of cell surface receptors, mediate many cell adhesion processes (cell-cell adhesion, cell-matrix adhesion, cell migration, etc.). For this reason they are considered major therapeutic targets in the effort to overcome diseases such as thrombosis, inflammation, tumor metastasis, osteoporosis and arthritis.

Most integrins recognize an Arg-Gly-Asp (RGD) sequence in ligands such as fibronectin, vitronectin and fibrinogen, each binding with different affinities to the individual integrin receptors. This selectivity has been prescribed to the different conformational presentation of the RGD segment in the ligand. Inhibition of binding of these integrins by antibodies or cyclic peptides has been shown to inhibit angiogenesis and induce tumor regression. One of the known integrin antagonists specific for the vitronectin binding $\alpha_v\beta_3$, is the peptide cRGDFV (**IX**, Figure 17.22). Through a process of rational design, with a structure of (**IX**) determined by NMR spectroscopy, and taking into account the importance of the spatial relation of the guanidine to the carboxylate functional groups, a number of mimetics were synthesized. The carboxyl function was introduced at either C1 or C2 while maintaining a guanidine side chain at C6 of varying lengths. A benzyl group was introduced as a phenylalanine isostere and similarly a methyl group introduced to mimic a valine side chain.

A set of nine structures was synthesized, exemplified by structure **X** (Figure 17.22). None of the structures synthesized displayed strong binding to $\alpha_v\beta_3$, and when tested in other integrin assays, the compounds displayed similar lack of activity. The chemical synthesis of nine different compounds with very few common intermediates requires a considerable amount of chemical effort. It was evident that in order to generate a number of different structures in a facile manner, a suite of building blocks was required that allowed the required chemistry to be achieved without protracted syntheses. One of the most noteworthy recent developments to this end has been the synthesis of universal building blocks.

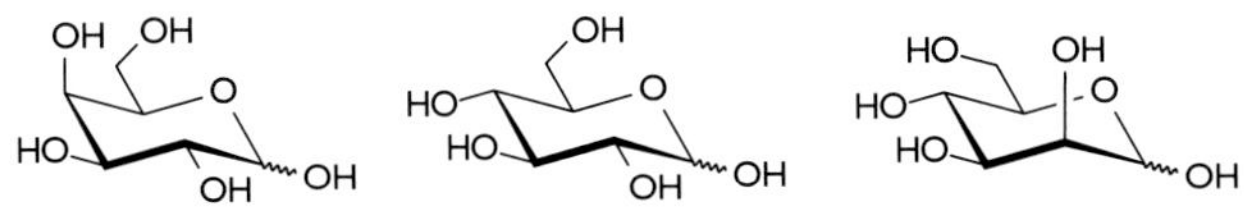

α,β-D-Galactose α,β-D-Glucose α,β-D-Mannose

Figure 17.20

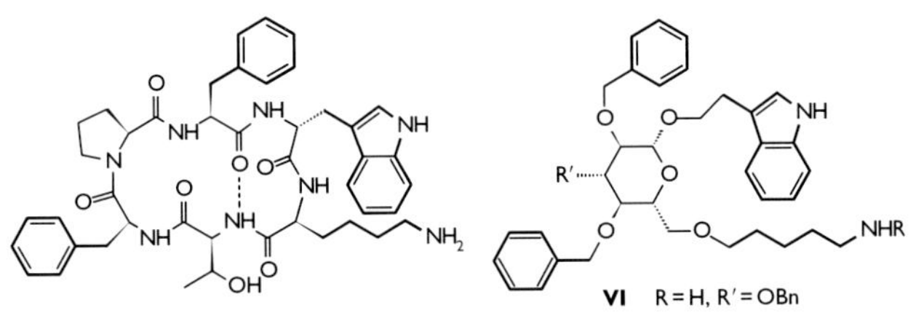

V

VI R = H, R′ = OBn
VII R = R′ = H
VIII R = Ac, R′ = OBn

Figure 17.21

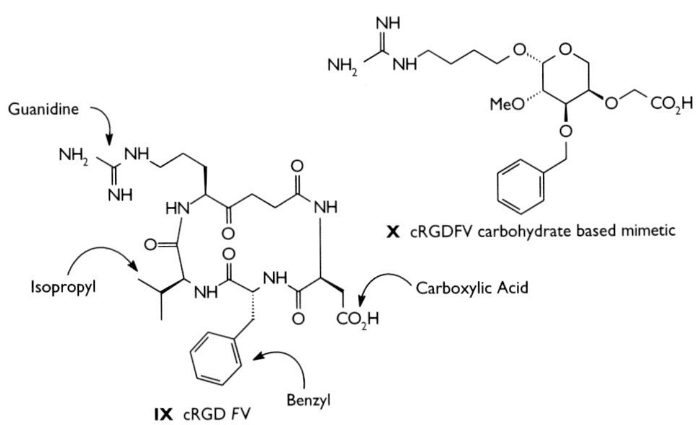

IX cRGD FV

X cRGDFV carbohydrate based mimetic

Figure 17.22

17.3.1.3 Carbohydrates in molecular diversity – universal building blocks

The idea of the universal building block is to have a monosaccharide that is differentially protected at each functional group on the saccharide core such that any one protecting group can be selectively removed without compromising any of the remaining protecting groups. Ideally this orthogonal protection strategy is to be achieved with no preference in order of the removal of protecting group.

CARBOHYDRATE LIBRARIES

The first example of a small, entirely carbohydrate-based library, was prepared by Wong and coworkers at The Scripps Research Institute (43) (Scheme 2). The key to the success of Wong's library was the employment of a quasi-universal building block. The only drawback of Wong's building block is that the chloroacetyl protecting group must be removed before the levulinoyl (lev) protecting group, as the chloroacetyl group is susceptible to cleavage under the conditions for removal of the lev group. This is a common problem and new protecting group chemistry is in constant demand to alleviate such problems. A typical procedure would see initial conjugation of the universal building block to a linking arm *via* glycosylation, so as to allow for later coupling of the oligosaccharide to a solid support or carrier protein. This would be followed by a sequence of orthogonal deprotections and glycosylations.

As Wong envisaged, if seven different donor sugars were employed with this one universal building block, each donor could glycosylate in one of four different positions, which, taking into account alpha and beta anomers, provides 56 disaccharides. Similarly, if we extend this to the trisaccharide level, meaning two protecting groups are sequentially deprotected and glycosylated, again with seven different donors sugars, we find that 1176 structures are generated. If we extend this to the pentasaccharide limit, i.e. five sequential deprotections and glycosylations, we find that as many as 38,416 pentasaccharide structures are possible. Finding a set of orthogonal protecting groups that is compatible with glycosylation chemistry is a reasonably straightforward procedure, the more challenging task lies with selection of a set of protecting groups that can withstand strong base or alkylation conditions. Such is the chemistry required to effect ether linkages for the introduction of alkyl chains as peptidomimetics.

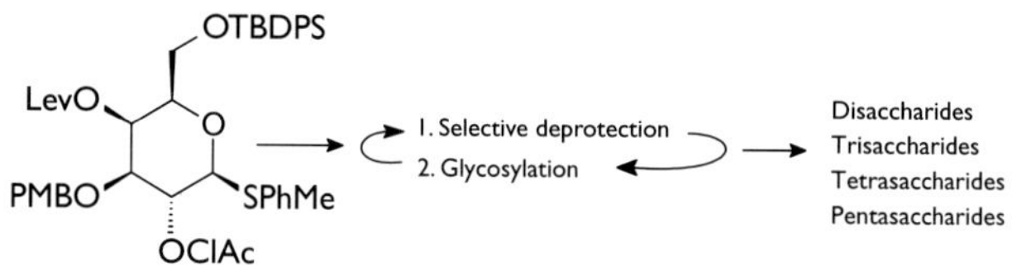

OTBDPS
LevO
O
PMBO
SPhMe
OClAc
1. Selective deprotection
2. Glycosylation
Disaccharides
Trisaccharides
Tetrasaccharides
Pentasaccharides

Scheme 2

DIVERSITY LIBRARIES

Since 1998 researchers in the group of Kunz (44) have developed a number of universal building blocks, generally coupling the building block to a solid support to effect the desired chemical transformations. The chemistry developed can be employed to achieve, like the work of Hirschmann and co-workers (45), the introduction of peptidomimetic side chains to carbohydrate scaffolds in an effort to produce glyco-based mimetics of cyclic peptides. Some of the building blocks synthesized by Kunz are displayed in Figure 17.23.

In general one of the five protecting groups on the carbohydrate scaffolds shown in Figure 17.24 is a protecting group modified as a linker, so as to allow coupling of the block to

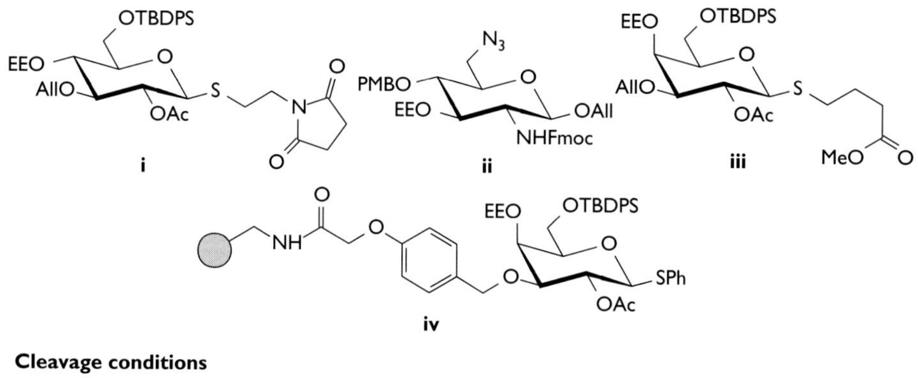

OTBDPS
EEO
AllO
O
S
OAc
N
O
O
i
PMBO
EEO
N_3
O
OAll
NHFmoc
ii
EEO
OTBDPS
AllO
O
S
OAc
iii
MeO
O
NH
O
O
EEO
OTBDPS
O
O
SPh
OAc
iv

Cleavage conditions

EE (Ethoxyethyl) cleave with R-OH, cyclohexene, Et_4NBr, **TBDPS** (*tert*-butyldiphenylsilyl) cleave with HF/pyridine or tetrabutylammoniumfluoride (TBAF), **Ac** (acetate) cleave with MeONa/MeOH, PMB (*p*-methoxybenzyl) cleave with cericammoniumnitrate (CAN), or DDQ, **All** (Allyl) cleave with Pd^0 reagents, Fmoc (fluorenylmethyloxycarbonyl) cleaves with 2^0 amines, eg. triethyl amine, piperidine, thioethers (eg. **SPh** (S-phenyl)), are activatable with thioglycoside promoters.

Figure 17.23

a solid support. The strategy is simple, remove a protecting group and effect coupling at the freed functionality with a peptidomimetic reagent. Remove another protecting group and couple with the next peptidomimetic, and so on.

Similar to the work of both Wong and Kunz, researchers at Alchemia have developed a number of orthogonally protected universal building blocks with the intention of employing carbohydrates as scaffolds for both molecular diversity, and the development of focussed libraries based on specific peptide sequences (46) (Figure 17.24).

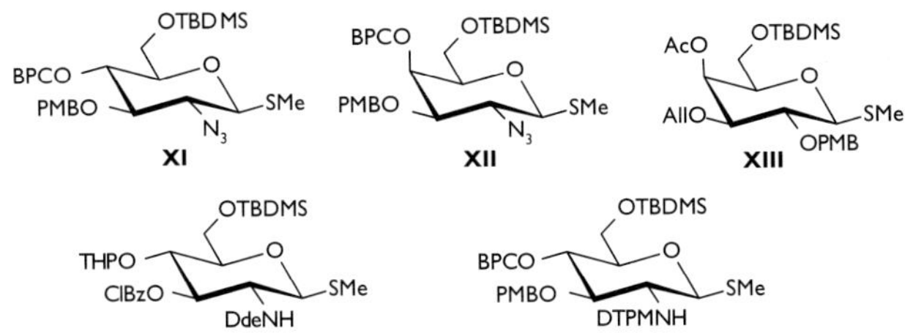

Cleavage conditions
BPC (biphenylcarbonyl) and **ClBz** (chlorobenzoyl) cleave with NaOMe/MeOH, **THP** (tetrahydropyranyl) cleave with TsOH/MeOH, $\mathbf{N_3}$ (Azido) reduce to amine with propanethiol-TFA/MeOH.

Figure 17.24

The amino function of structure **XIV** is protected by a hydrazine labile protecting group called Dde [*N*-1-(4,4-dimethyl-2,6-dioxocyclohexylidene)ethyl] and **XV** by another hydrazine labile protecting group, (1,3-dimethyl-2,4,6-(1H,3H,5H)-trioxopyrimidin-5-ylidene)methylamino (DTPM). The above building blocks can be employed in solution and solid phase synthetic schemes for the production of both carbohydrate libraries and peptidomimetic libraries. Initially, a library is synthesized using solid phase methodologies, with a glucosamine building block, with five different alkyl substituents. This potentially provides a library in excess of three thousand members. The series of alkylations must begin with the removal of the acyl protecting group, as this is susceptible to migration or cleavage under alkylation conditions (Scheme 3).

17.4 Conclusion

Carbohydrate-based drug discovery has, to some extent, been a series of forays into uncharted waters. The scientists leading the vanguard of research in glycobiology and carbohydrate chemistry are presently mapping the ground of what is likely to be one of the highest growth areas in therapeutic development in the years to come. The concerted efforts of organic chemists, medicinal chemists, biologists, biochemists

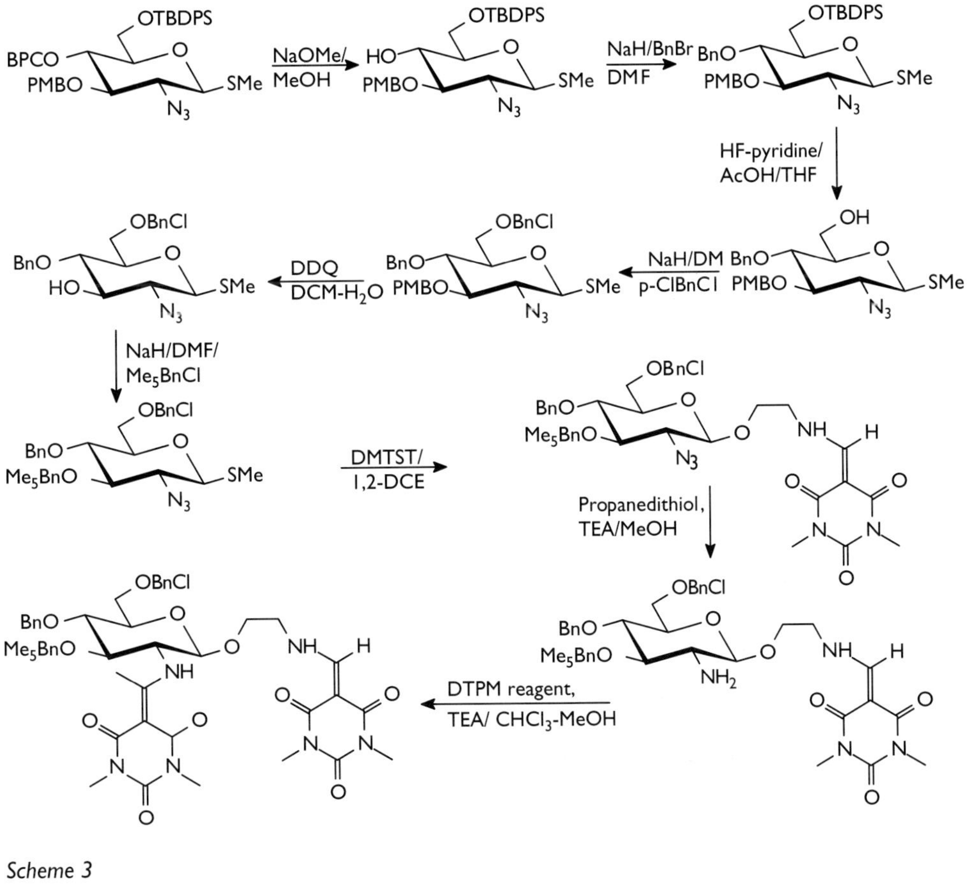

Scheme 3

and physical scientists, are all required to maintain momentum to see realisation of end-goals that will have a profound effect on the future treatment of disease states.

References

1 P. K. Qasba, *Carbohydrate Polymers*, 41, 2000, 293–309.
2 F. Dall'Olio, *J. Clin. Pathol: Mol Pathol.*, 1996; 49: M126–M135.
3 C-H. Wong, R L. Halcomb, Yoshitaka Ichikawa, and Tetsuya Kajimoto, *Angew. Chem. Int. Ed. Eng.*, 1995, *34*, 521–546.
4 A. Varki, *Glycobiology*, Vol. 3, No. 2, 1993, 97–130.
5 (a) J. M. Chrispeels, A. Vitale, *Plant Physiol.*, 1985, 78(4), 704–709, (b) P. E. Goss, A. Michael, J. P. Carver, J. W. Dennis, *Clin. Cancer. Res.*, 1995, 1(9), 935–944.
6 H. Brockhausen, J. Schutzbach, W. Kuhns, *Acta Anat.* 1998, 161, 36–78.
7 G. L. Nicholson, *Biochemica et Biophysica Acta*. 695, 1982, 113–176
8 P.K. Qasba, *Carbohydrate Polymers*, 41 2000 293–309.
9 C-H. Wong, Journal of the Chinese Chemical Society, 1999, 46, 271–281.
10 M. Petitou and C. A. A. van Boeckel, *Pure & Appl. Chem.*, 1997, Vol. 69, No. 9, 1839–1846.
11 P. Sinäy, *Nature*, 1 April 1999, Vol 398, 377–378.
12 J. Fareed, D. Hoppensteadt and R. L. Bick, Current Opinion in Cardiovascular, Pulmonary and Renal Ivestigational Drugs, 1999, 1(1), 40–55.
13 K.B. Wlasichuk *et al.*, *J. Biol. Chem.*, 1993, 268(19), 13971–13977.
14 J. Choay, M. Petitou, J. C. Lormeau, P. Sinäy, J. Fareed, *Ann. NY Acad. Sci.*, 1981, 370, 644–649.
15 Petitou, M., Duchaussoy, P., Lederman, I., Choay, J., Jacquinet, J. C., Sinäy, P., and Torri, G., *Carbohydrate Research*, 1987, 167, 67–75.
16 A. Etzioni, J. M. Harlan, S. Pollack, L. M. Phillips, R. Gershoni-baruch, and J. C. Paulson, *Immunodeficiency*, 1993, Vol. 4, pp. 307–308.
17 (a) C-H. Wong, T. F. J. Lampe, and G. Weitz-Schmidt, *Angew. Chem. Int. Ed.*, 1998, 37, No. 12, 1707–1711; (b) G. Thoma, F. Schwarzenbach, and R. O. Duthaler, *J. Org. Chem.*, 1996, *61*, 514–524; (c) K. Shibata, K. Hiruma, O. Kanie, and C.-H. Wong, *J. Org. Chem.*, 2000, *65*, 2329–2398; (d) C-C. Lin *et al*, *J. Am. Chem. Soc.*, 1996, 118, 6826–6840; (e) M. Buerke, A. S. Weyrich, Z. Zheng, F. C. A. Gaeta, M. J. Forrest, A. M. Lefer, *J. Clin. Invest.* 1994, 1140; (f) A. Giannis, *Angew. Chem. Int. Ed. Engl.* 1994, 33, 178; (g) X. Cheng, N. Khan, and D. R. Mootoo, *J. Org. Chem.*, 2000, *65*, 2544–2547.
18 B. N. Narasinga Rao, M. B. Anderson, J. H. Musser, J. H. Gilbert, M. E. Schaefer, C. Foxall, and B. K. Brandley, *The J. Bio. Chem.*, 269 3, 1994, 19663–19666.
19 *AMA Drug Evaluations* (Publishing Sciences Group, Inc., Little, Massachusetts, 1977), pg. 689.
20 C-H. Wong, *J. Chin. Chem. Society*, 1999, *46*, 271–281, U. Von Ahsen, H.F. Noller, *Science*, 1993, 260, 1506.
21 S. J. Sucheck, W. A. Greenberg, T. J. Tolbert, and C-H. Wong, *Angew. Chem. Int. Ed.*, 2000, 39, No. 6, 1080–1084.
22 A. H. Lehninger, *Principles of Biochemistry*, (Worth Publishers, Inc., 1982), pp. 292–294.
23 J. L. Strominger, *Ann. N. Y. Acad. Sci.*, 128(1) 1965, 59–61.
24 S. A. Hitchcock, C. N. Eid, J. A. Aikins, M. Zia-Ebrahimi, L. C. Blaszczak, *J. Am. Chem. Soc.*, 1998, *120*, 1916–1917.
25 H. C. Neu, *Science*, Vol. 257, 21 August 1992, 1064–1073.
26 S. J. Danishefsky, and J. R. Allen, *Angew. Chem. Int. Ed.*, 2000, *39*, 836–863, and references contained therein.
27 M. A. Morse, Current Opinions in Molecular Therapeutics, 2000, 2(4) 453–458.
28 T. Toyokuni and A. K. Singhal, *Chemical Society Reviews*, 1995, 231–242.
29 Drug report, iddb database, 27th September.
30 P. Livingstone, *Semin. Oncol.*, 1998, 25(6), 636–645.
31 R. Liang, L. Yan, J. Loebacg, M. G. Y. Uozumi, K. Sekanina, N. Horan, J. Gildersleeve, C. Thompson, A. Smith, K. Biswas, W. C. Still, D. Kahne, *Science*, Vol. 274, 29 Nov. 1996, 1520–1522.
32 C-H. Wong *et al*, *J. Am. Chem. Soc.* 1999, 121, 6527–6541.
33 (a) I. Phillips; D. Williams, in *Laboratory Methods in Antimicrobial Chemotherapy*; Garrod, L., Ed.; Churchill Livingstone Press: Edinburg, 1978; pp 31–40; (b) Bary, A. L. *The Antimicrobic Susceptibility Test: Principles and Practice*; Lea and Febiger: Philadelphia, PA, 1976.
34 P. Welzel, F. Kunisch, F. Kruggel, H. Stein, A. Ponty, H. Duddeck, *Carbohydr. Res.*, 1984, 126, C1–C5.
35 J. A. Betrand, G. Auger, E. Fanchon, L. Martin, D. Blanot, J. van Heijenoort, O. Dideberg, *EMBO J.*, 16, 1997, pp. 3416; structure available PDB-1UAG.
36 L. D. Gegnas, S. T. Waddell, R. M. Chabin, S. Reddy, and K. K. Wong, *Bioorganic & Medicinal Chemistry Letters*, 8, (1998), 1643–1648.
37 M. E. Tanner, S. Vaganay, J. van Heijenoort, and D. Blanot, *J. Org. Chem.*, 1996, 61, 1756–1760.
38 B. Zeng, K. K. Wong, D. L. Pompliano, S. Reddy, and M. E. Tanner, *J. Org. Chem.*, 1998, 63, 10081–10086.
39 L. D. Gegnas, S. T. Waddell, R. M. Chabin, S. Reddy, and K. K. Wong, *Bioorganic & Medicinal Chemistry Letters*, 8, (1998), 1643–1648.
40 K. C. Nicolaou, J. M. Salvino, K. Raynor, S. Pietranico, T. Reisine, R. M. Freidinger, R. Hirschmann, *Pept.: Chem., Struct. Biol., Proc. Am. Pept. Symp.*, 11th, 1990.
41 R. Hirshmann, K. C. Nicolaou, S. Pietranico, J. Salvino, E. M. Leahy, P. A. Sprengeler, G. Furst, A. B. Smith III, *J. Am. Chem. Soc.*, 1992, *114*, 9217–9218 and references therein.
42 K. C. Nicolaou, J. I. Trujillo, K. Chibale, *Tetrahedron*, Vol. 53, No. 26, 1997, 8751–8778.
43 C-H. Wong, X-S. Ye, and Z. Zhang, *J. Am. Chem. Soc.*, 1998, *120*, 7137–7138.
44 (a) H. Kunz, T. Wundberg, C. Kallus, T. Opatz, S. Henke, W. Schmidt, *Angew. Chem. Int. Ed.*, 1998, *37*, No. 18; (b) K. Kallus, T. Wundberg, W. Schmidt, S. Henke, H. Kunz, *Tet. Lett.*, 40, 1999, 7783–7786; (c) U. Hünger, T. Maidhof, O. Knöll, H. Kunz, *Poster Presentation, 20th International Carbohydrate Symposium, Hamburg-Germany*; (d) T. Opatz, C. Kallus, T. Wundberg, W. Schmidt, S. Henke, H. Kunz, *Poster Presentation, 20th International Carbohydrate Symposium, Hamburg-Germany*.
45 R. Hirschmann, K. C. Nicolaou, S. Pietramico, J. Salvino, E.M. Lealy, W. C. Shakepeare, P.S. Spengler, P. Hamley, A. B. Smith, T. Reisine, K. Raynor, C. Donaldson, W. Vale, L. Maechler, R. M. Freidinger, C. D. Strader, *J. Am. Chem. Soc.*, 1993, *115*, 12550.
46 John Papageorgiou, Laurent Bornaghi, Steven Taylor, Latika Singh, Joachim Seifert, Gyula Dekany, *Poster Presentation, 20th International Carbohydrate Symposium*, Hamburg-Germany.

Chapter 18

Introduction to structure and ligand-based drug design

György M. Keserű, Ildikó Magdó and Gábor Náray-Szabó

Contents

18.1 Introduction

Drug discovery is a challenging and extremely expensive game played by global companies all over the world. Since most of the companies are involved in several mergers creating pharma giants, the number of players decreases. One of the major reasons for resource integration is the increasing R&D cost of a new drug. The Pharmaceutical Manufacturer's Association reported an increase from \$4 million to over \$350 million during 34 years between 1962 and 1996. On the other hand, development time from the first synthesis to launch has almost quadrupled during this period. Rational drug design – in combination with experimental techniques – is a potential tool to increase efficiency in early phase development and therefore it has been widely applied in drug discovery programs.

In principle biological properties of ligands are related to structural features and represent the starting point of rational drug design. Strategies applied in computer assisted drug discovery (CADD) depend on the extent of structural information available on the ligands and the target. The conceptual frame of CADD and its relationship to molecular modeling is depicted in Figure 18.1.

Ligand-based design evaluates structural similarities and dissimilarities of active and inactive ligands, receptor-based design considers the structure of the macromolecular target as well. Since structural investigations on macromolecules became available in the mid-80s, early attempts on CADD were limited to quantitative structure–activity relationship (QSAR) studies. In the early 70s, molecules were treated as 2D objects and QSAR equations were derived using chemical and physicochemical descriptors including lipophilicities,

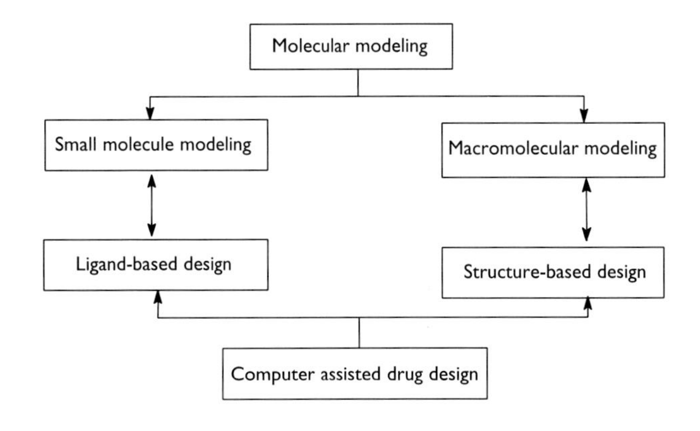

Figure 18.1 Conceptual basis of CADD.

substituent constants and quantum mechanical parameters. Second generation QSAR works from the mid-80s entered the third dimension, considering stereochemical features and several 3D descriptors as well. Dramatic developments in experimental techniques in structural biology allowed information to be obtained on macromolecular structures and initiated the so called direct approach of CADD. Analysis of the ligand–macromolecule interaction can be useful in determining the mechanism of action and helping to find new molecules with improved binding affinity. Although these approaches are conceptually different, their parallel use was found to be beneficial. A flowchart of CADD support is depicted in Figure 18.2.

After target identification, molecular biology is used to produce the purified protein. Then, the structural biology group crystallizes the protein or sets up preliminary NMR studies to check whether the protein is suitable for 3D structure elucidation. Meanwhile assay development is started and the lead discovery unit performs high throughput screening (HTS) tests using the corporate compound library to find hits for lead optimization. Structure-based design can also produce potential leads if structural investigations on the target macromolecule are successful. Otherwise ligand-based methods are used to facilitate the conversion of hits to viable leads. Finally, lead molecules are converted to clinical candidates using an integrated approach involving traditional medicinal chemistry, combinatorial chemistry and ligand and/or structure-based design. In this chapter we summarize basic principles of both ligand and receptor-based design through case studies and overview recent trends within the CADD discipline.

18.2 Ligand-based design

Although complementary principles serve as a theoretical basis for both ligand and structure-based design [1], if the receptor structure is not available, the only approach is to depend on a hypothesis based on known active ligands (ligand/analog-based design, also referred to as indirect design).

18.2.1 Pharmacophore model

Most qualitative models used by medicinal chemists to rationalize structure–activity relationships have been based

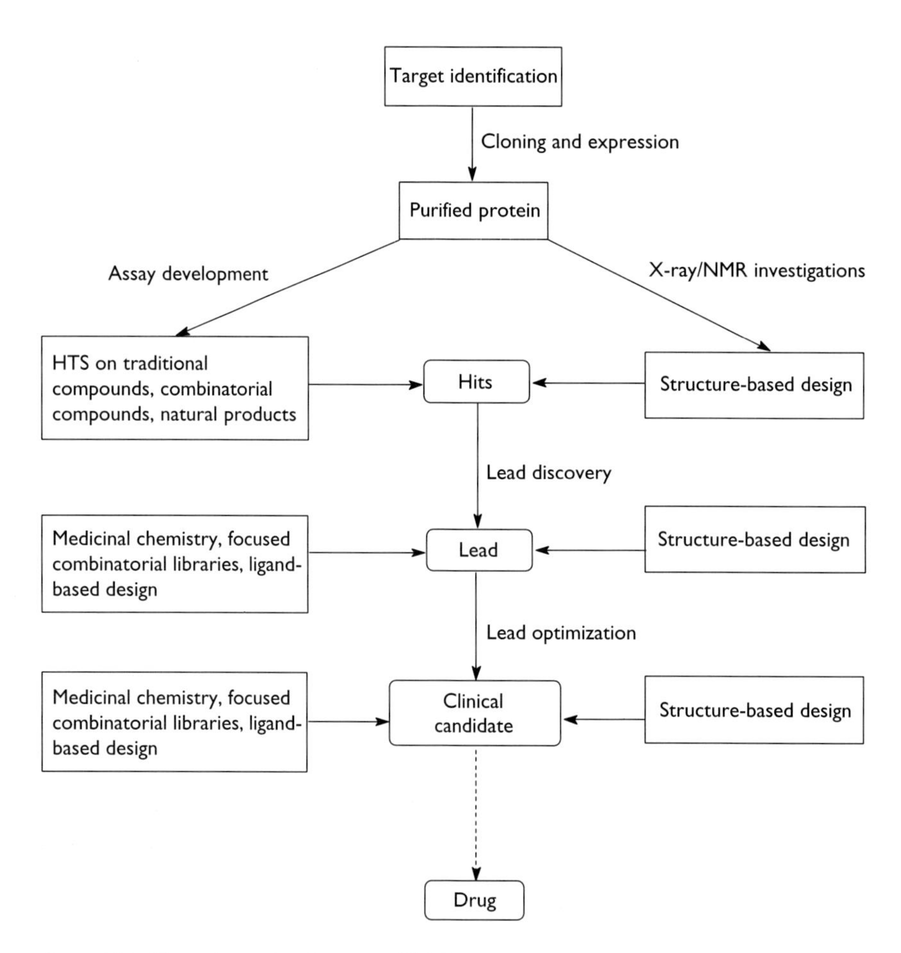

Figure 18.2 Flow chart of early phase CADD.

on the pharmacophore concept. Given sets of biologically active molecules, which produce activity by the same mechanism, are assumed to have the same pharmacophore. A pharmacophore may be defined as a particular arrangement of atoms or functional groups essential to produce a given biological response. These pharmacophoric groups are chosen based on their ability to participate in different intermolecular interactions, thus charged groups, hydrogen donors/ acceptors, polar groups, hydrophobic groups and aromatic rings are usually selected. Pharmacophores have been described as topologic (graph-theoretic or connectivity based) and topographic (geometric, 3D) patterns. Since molecules are (in most cases) not rigid, construction of a 3D pharmacophore model requires a conformational analysis to be performed as a first step to identifying the so-called bioactive conformations. One has to note that the bioactive conformer is not necessarily identical to the lowest energy conformer, but it cannot be a high-energy conformation. Identifying the bioactive conformer is one of the major tasks in molecular modeling, especially when the molecules have several rotatable bonds. It is helpful if rigid analogs exist, but it is often not the case.

Several programs have been developed for automatic generation of pharmacophoric models, e.g. APEX [2], DISCO [3], and GERM [4]. Pharmacophore mapping studies are not usually performed in isolation but are a prelude to 3D QSAR study, 3D database searching or de novo design.

18.2.2 Database searching

3D database searching in terms of geometric relations between pharmacophoric groups has become an important tool in the discovery of new active structures. Molecules with pharmacophore groups arranged spatially at appropriate positions are considered to have a higher probability of exerting activity.

A prerequisite for any 3D search is a chemical database that includes 3D structural information. The primary source of experimental structures for small organic molecules is X-ray crystallography, and these structures are collected in the Cambridge Structural Database (CSD) [5]. Commercial 3D databases are also available. There is much larger number of databases for which the 2D (connectivity) information is available. These 2D structures can be converted to 3D using different computational methods. The most commonly used method is CONCORD [6] which generates a single low-energy conformer of a small molecule. Several other programs are available for converting a 2D database to a 3D database, e.g. CORINA [7], WIZARD [8] and MIMUMBA [9]. The consideration of a single rigid conformation will certainly miss some molecules that could present the pharmacophore in a conformation other than that stored in the database. Conformational flexibility can be taken into account either by storing multiple conformations or performing conformational analysis at search time. Several algorithms have been developed to examine conformations during a search using distance geometry, systematic search, Monte Carlo or genetic algorithm techniques. The most commonly used program packages for 3D database searches are UNITY™ [10], MACCS-3D™ and Catalyst™ [11]. Flexible searching has the advantage of finding more hits, however it often requires more resources and time than is feasible.

The usefulness of a database search is the provision of a hit list for screening; it also provides the medicinal chemists with a variety of molecules that might inspire new synthetic directions.

18.2.3 QSAR

Pharmacophore mapping is of great value in generating chemical lead structures, however, an inherent limitation with pharmacophoric modeling is its inability to quantitatively describe the effect, it can only distinguish active from inactive compounds. In the process of optimizing a lead structure it is necessary to utilize the information from quantitative activity data and from structural properties in order to predict more active congeners. The biological activity can be regarded as a function of the physicochemical and structural properties of the ligand.

QSAR methods are used to attempt to correlate biological activities of compounds with their stereoelectronic properties. The aim is to produce a suitable model to predict the activities of novel compounds. Two different approaches can be used, 2D (traditional) and 3D QSAR.

18.2.3.1 2D QSAR

Traditional QSAR studies are performed using 2D structural data of molecules. The field was started in the early sixties by Hansch *et al.* [12], Fujita *et al.* [13] and others. Different physicochemical parameters have been used to describe the global properties of the molecule or the contribution from individual substituents. The most commonly used substituent parameters are summarized in Table 18.1. One can also include indicator variables describing the presence or absence of a certain substituent or other structural characteristic, or use molecular fingerprints.

The QSAR equations can be generated using different statistical methods. Multiple linear regression (MLR) is one of the most widely used techniques. There are some important criteria to consider when using multiple linear regression analysis to achieve statistically significant results. There should be sufficient data; it is often considered that at least five compounds are required for each parameter included in the regression analysis. The selected compounds should give a good spread of values of parameters, which should be uncorrelated. Multiple linear regression cannot deal with datasets where the variables are highly correlated and/or the number of variables exceeds the data values. Two methods

Table 18.1 Physicochemical parameters used in traditional QSAR

Hydrophobic parameters	
Partition coefficient	logP
Substituent constant	π
Hydrophobic fragmental constant	f, f′
Distribution coefficient	logD
Apparent partition coefficient	logP′, $logP_{app}$
Solubility parameter	δ
Electronic descriptors	
Hammet constants	$\sigma, \sigma^-, \sigma^+$
Taft's inductive constants	σ^*, σ_I
Swain and Lupton field parameter	F
Swain and Lupton resonance parameter	R
Ionization constant	$pK_a, \Delta pK_a$
Theoretical parameters	
Atomic net charge	q^σ, q^π
Electrostatic potential	V(r)
Energy of highest occupied molecular orbital	E(HOMO)
Energy of lowest unoccupied molecular orbital	E(LUMO)
Steric descriptors	
Taft's steric parameter	E_s
Molar volume	MV
Molecular weight	MW
van der Waals volume	V_W
van der Waals radius	r
Molar refractivity	MR−
Parachor	P_r

are widely used to deal with such situations: principal component regression and partial least square analysis (PLS) [14]. In principal component regression a principal component analysis is first performed then followed by a regression analysis using the first few components. PLS uses latent variables which are constructed from the original independent variables in such a way that they form an orthogonal set.

18.2.3.2 3D QSAR

Traditional 2D QSAR can be applied to congeneric molecules with the same scaffold, however, a problem arises when differences exist in the backbone of the molecules; such a series cannot be described by substituent constants. 3D QSAR methods have been developed to overcome this problem. 3D QSAR models relate the biological activity of ligands to their properties calculated in 3D space. One of the most commonly used methods is Comparative Molecular Field Analysis (CoMFA) [15]. The basic idea behind CoMFA is that the intermolecular interaction energy between the ligand and the receptor correlates with the steric, electrostatic, hydrogen bonding and hydrophobic fields of the ligands. The steric, electrostatic, etc. interaction energies between the ligands and hypothetical probe atoms are calculated at a number of grid points. The fields can be represented as a matrix, in which each row corresponds to a molecule and the columns are the interaction energy values at the grid points, an additional column contains the biological activity. There are usually thousands of field descriptors, which strongly exceed the number of compounds, i.e. the problem is underdetermined. PLS provides an appropriate tool to overcome this problem. The CoMFA procedure is represented schematically in Figure 18.3. CoMFA QSAR equations can be used to predict the affinity of proposed molecules. The CoMFA models can be displayed graphically as coefficient contour maps to aid new compound design.

One of the most difficult aspects of CoMFA and other grid-based 3D QSAR methods is finding the bioactive conformers and the appropriate alignment rules for the ligands. In order to overcome this problem some new methods have recently been developed. WHIM [16], EVA [17] and COMMA (Comparative Molecular Moment Analysis) [18] methods are based on 3D descriptors that are independent of the orientation of the molecules, and do not require the alignment of the molecules. These descriptors, however, are sensitive to the conformation of the molecules, so identification of the bioactive conformers still remains a problem to solve. Another limitation of these methods is that since their descriptors do not consider properties at grid points, rather 3D properties of the molecules as a whole, the interpretation of the resulting models is not as obvious as, e.g., that of CoMFA models (3D graphical display of the models is not possible).

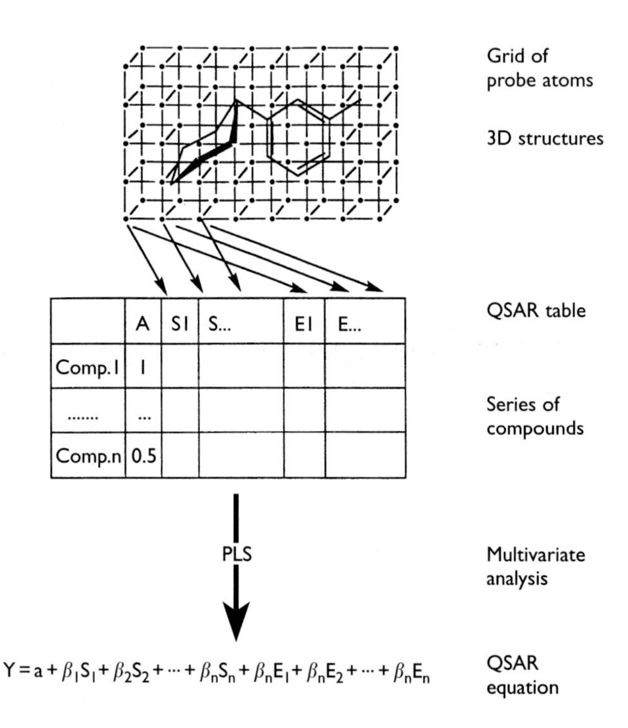

Figure 18.3 Schematic overview of CoMFA method. S_i and E_i denote steric and electrostatic interaction energies with the probe atom at site i.

18.2.4 Case studies

18.2.4.1 Antibacterial agents

The earliest example of a drug designed with the assistance of QSAR is norfloxacin (**1**) [19]. Structural modifications of substituted 1-ethyl-1,4-dihydro-4-oxo-quinoline-3-carboxilic acids (**2**) were guided with the help of 2D QSAR to optimize the antibacterial activity of this series of congeners. One of the compounds predicted to be most potent, norfloxacin has been on the market since 1983 under various brand names including Noroxin.

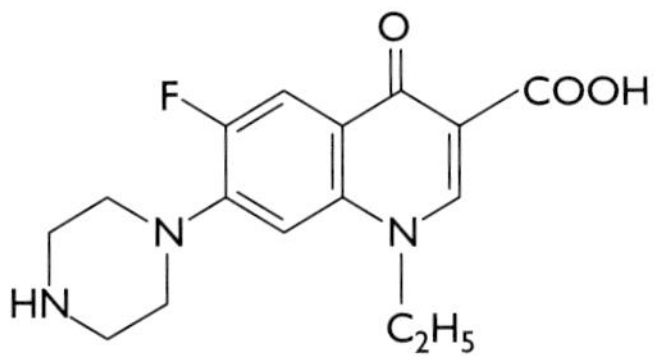

18.2.4.2 Anti-migraine agent

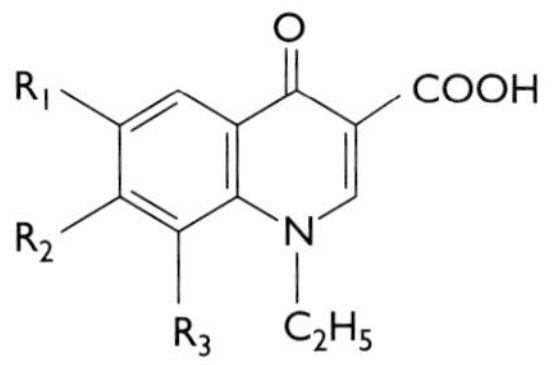

Zolmitriptan (**3**), a 5-HT_{1D}-receptor agonist for the treatment of migraine was discovered with the help of pharmacophore mapping using the active analog approach at Wellcome [20]. It is marketed by Zeneca under the brand name Zomig.

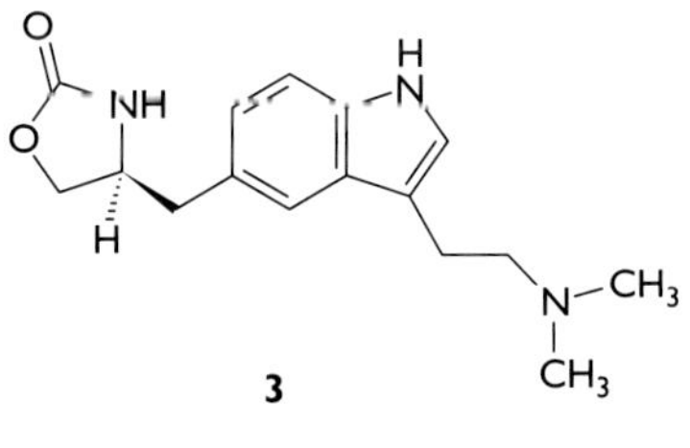

18.2.4.3 HIV integrase inhibitors

Although most of the HIV integrase inhibitors are in preclinical phase we present them as a typical example of 3D database searching. The database of the National Cancer Institute (NCI) was first filtered by an initial pharmacophore derived from the known inhibitor of caffeic acid phenethyl ester (**4**). 267 hits obtained from this search were tested and used to refine and validate the initial pharmacophore model. The refined model was used for the secondary evaluation of the database and yielded ten new, structurally diverse HIV integrase inhibitors from which (**5–8**) were particularly potent [21].

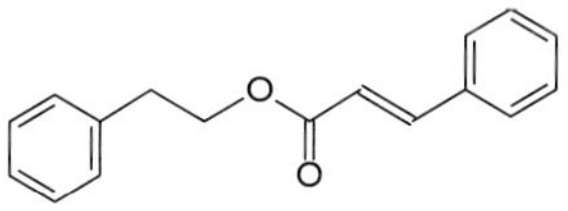

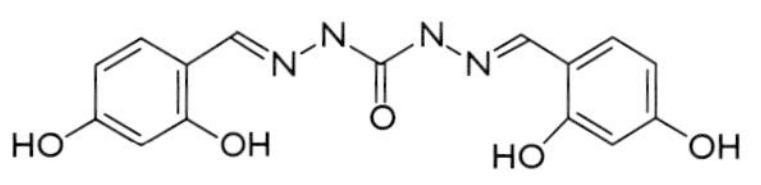

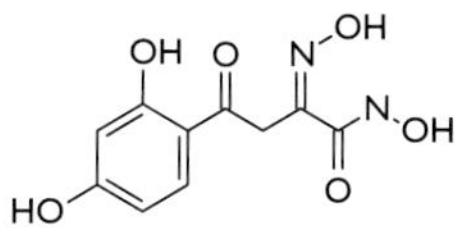

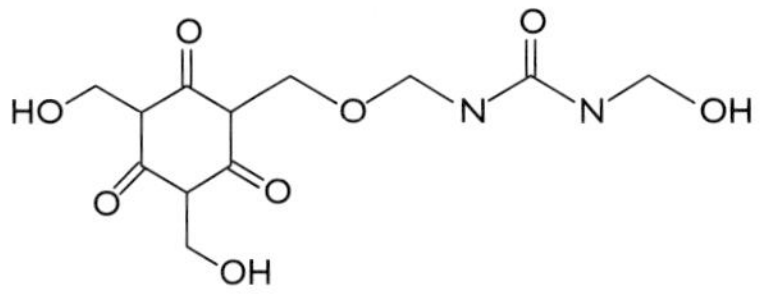

18.2.4.4 Selective 5-HT4 ligands

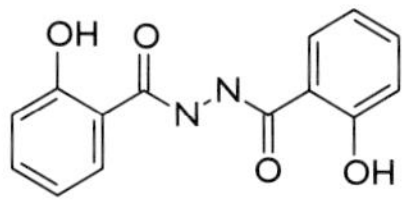

5-HT4 receptors belong to the wide family of G-protein coupled receptors (GPCR) that are very popular targets in the CNS field. The 3D structure of these proteins, including the 5-HT4 receptor, is unknown and therefore GPCRs represent an ideal application domain to ligand-based design. Recently this technique was used to design selective 5-HT4 ligands by using a comparative study of the 5-HT4 and 5-HT3 pharmacophores [22]. Pharmacophore models for these receptors were derived from the alignment of inactive and active ligands of both proteins. Regions of steric and electrostatic tolerance and intolerance were reported as a result of a CoMFA study. The proposed 5-HT4 and 5-HT3 pharmacophores were used to design two new and selective ligands, one for the 5-HT4 (**9**) and one for the 5-HT3 (**10**) target.

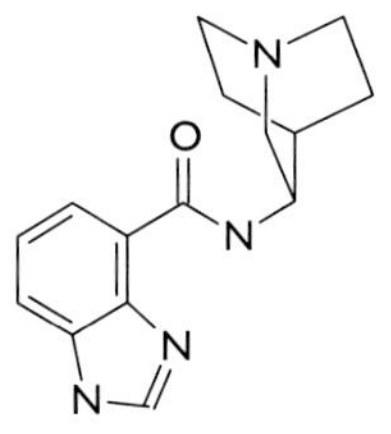

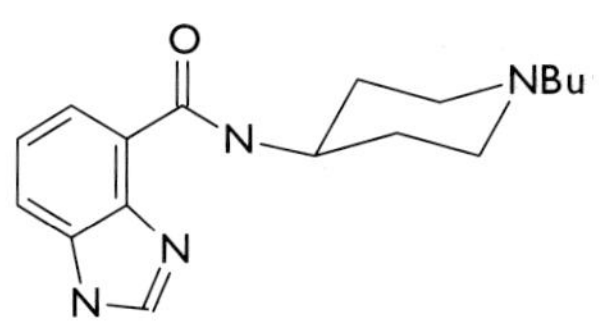

18.3 Structure based design

The conceptual basis of structure-based design was published by Paul Ehrlich and Emil Fischer formulating the famous "lock and key" hypothesis as a rationalization of drug action [1]. Although this concept is now more than 100 years old, its practical application was limited by the lack of structural information on the "locks". Dramatic developments in structural biology and computer techniques enabled the dream to be realized: designing the keys for a particular target. Structure-based design allows us to look for the structural requirements of ligand binding and to optimize favorable interactions to get high affinity substrates. Protein structures can be obtained experimentally from X-ray or NMR studies or theoretically using homology modeling. Experimental structures are deposited in the Protein Data Bank (PDB, http://www.rcsb.org/pdb). Homology modeling is based on sequential similarities found between a target protein with unknown structure and template proteins for which experimental structures are available. Although a high percentage of sequence identity is a condition of an accurate homology model, structural information on sequentially conserved regions around the binding site obtained in this way may help to identify characteristic features responsible for specific binding of ligands. Protein-ligand complex structures are also available from X-ray or NMR investigations and can be obtained by docking calculations. Docking techniques are based on the 3D structure of the macromolecular binding site and used to predict possible low energy binding modes for potential ligands. These methods – in conjunction with algorithms developed for binding free energy calculation – are useful for quantifying the strength of interactions formed between the substrate and its receptor. Docking methodologies can be grouped into three main groups: (i) complementarity-based docking; (ii) molecular dynamics and Monte Carlo-based docking; (iii) fragmental docking.

18.3.1 Complementarity-based docking

One of the earliest ligand-target matching programs is DOCK developed by Kuntz *et al.* [23, 24]. Within the framework of this method, the binding site cavity of the protein is filled with spheres to achieve complementarity to the Connolly surface of the macromolecule. These spheres define a volume that complements the geometry of the binding site and can be used to find matches with potential ligands. Ligands are screened against this volume to identify molecules whose interatomic distances match the distances within the volume. In addition to geometric criteria DOCK considers physicochemical features as well. Ligands are evaluated in a single conformation but different orientations which are scored by an empirical function. Scoring efficiency can be increased by the calculation of interaction energy by molecular mechanics. Although the original version of DOCK assumed rigid conformation of the ligand, a modification developed by Leach can partly consider ligand flexibility [25].

Another description of the protein binding site was suggested by Nussinov *et al.* [26, 27] abstracting critical points from the concave, convex and saddle points of the molecular surface of the protein and using potential binding parameters to find matches between molecules. The geometric hashing technology applied here results in a highly effective technique to match the surface of extended systems. Fitting molecular surfaces can also be performed by least square fit of superimposed pattern of points [28] or by the application of graph theory to find the maximal number of correspondence between points on a receptor and a ligand [29]. The matching problem was solved by the introduction of intermolecular energy function in the GRAMM program which can be smoothed by changing the range of atom–atom potentials [30–33].

Since these methods treat ligands as rigid objects, ligand flexibility can only be considered using multiconformer libraries. The FLOG methodology developed by Miller *et al.* uses DOCK on conformer databases [34]. Another approach reported by Mizutani *et al.* explores binding modes by searching for potential hydrogen bonds [35]. This approach was implemented in GREEN, a grid-based algorithm that includes multiple energy minimization techniques and energy decomposition to facilitate the docking process [36]. Autodock performs the ligand conformational search by a simulated annealing Monte Carlo approach and generates possible ligand conformations which are evaluated by a grid-based scoring protocol [37]. The protein binding site representation is reduced to pharmacore points in ChemProtein developed by Chemical Design. This methodology has also been used by Clark *et al.* in Prometheus [38].

18.3.2 Molecular dynamics and Monte Carlo-based docking

Methods described so far treat receptors as rigid but take ligand flexibility into account by optimization within the binding site or precalculated conformational libraries ("hand-and-glove" analogy). Molecular dynamics simulations allow receptor and ligand flexibility to be considered simultaneously. Molecular dynamics (MD) simulations are widely used for the evaluation of energetically favored binding conformations of protein–substrate complexes. Until recently MD was also the prefered methodology used for the calculation of binding free energies for enzyme–substrate complexes on the basis of the free energy perturbation (FEP) theory [40]. It is well known, however, that a protein becomes trapped in a certain region of conformational space and may never escape – no matter how long the MD simulation runs [40, 41]. Moreover conformations separated by relatively high energy barriers are usually inefficiently sampled in MD. These sampling problems can easily be overridden using MC-based conformational searching algorithms. In addition to this advantage, the statistical treatment applied in MC does not require any extra coupling algorithms which are used in MD simulations [42]. Furthermore, thermal artifacts usually associated with the MD treatment of nonbonded interactions can also be avoided using MC methods. In contrast to conventional Metropolis Monte Carlo (MMC) simulations we introduced a Monte Carlo-based conformational analysis (MCMM) for the evaluation of protein–substrate complexes. The effectiveness of this approach for the calculation of protein–substrate complexes was demonstrated in comparative MC/MD studies [43–45]. One of the most characteristic components of the success was the increased sampling efficiency achieved by the MCMM protocol.

Sampling efficiency of computer simulations also has a characteristic effect on the evaluation of conformer populations. It should be noted that atomic or internal coordinate constraints used for efficient sampling (e.g. SHAKE) can be more readily introduced to MC than to MD. Systems that have multiple conformations separated by high energy barriers represent a special case for adequate sampling. This is the typical situation in docking a small molecular substrate to a macromolecular binding site, when direct interconversion between low energy binding modes is limited. Exploration of the total conformational space of the protein–substrate complex requires adequate sampling of all significantly populated conformers (global sampling) as well as their local conformational space (local sampling). Due to the low probability associated with barrier crossings, conformers separated by high barriers are usually sampled inefficiently in MD. This results in sampling that is limited to the local space of some possible conformations which might even be restricted to the starting conformation. This can be expected when conformers are separated by extremely high barriers. However, torsional barriers, even those not higher than 3–5 kcal/mol, can cause sampling problems in MD simulations [46, 47]. On the other hand, MC methodology has a unique capability to interconvert conformers along high energy conformational pathways. This unique feature of MC was substantiated in a number of highly efficient MC-based conformational search methodologies such as the Monte Carlo Multiple Minimum (MCMM) [48] applied by us, the Systematic Unbounded Multiple Minimum (SUMM) [49] and the Low-Mode (LMOD) [50] conformational searches. These algorithms are all implemented in the MacroModel package [51].

There are characteristic differences between the most widely used MMC method and the MCMM search used by us. Metropolis Monte Carlo algorithm rapidly generates a large number of trial structures whose acceptance is subject to a simple probabilistic rule. A minimum energy conformation X is subjected to random variation of coordinates (e.g. Cartesian or internal coordinates) and the energy of the resulting Y structure is evaluated. If $E_Y < E_X$, structure Y is accepted as a starting structure of energy minimization (or it can simply be accepted without it) to produce a new conformation. If $E_Y > E_X$, structure Y is accepted only if a random number drawn from the uniform distribution in the 0,1 interval is less than the Boltzmann population calculated by the following formula:

$$r = \frac{e^{-E_Y - E_X}}{RT}$$

The Boltzmann probability applied in the Metropolis Monte Carlo method assumes that the system is thermally equilibrated and it can be shown that the Metropolis algorithm, within the limit, explores different regions of the conformational space according to their Boltzmann probability. In conclusion, low energy regions are explored more frequently than high energy regions, however, this is only true after a huge number of MC steps.

The usage directed Monte Carlo conformational search algorithm, however, has proven to be more efficient in focusing the search on the low energy conformers than Metropolis MC and therefore requires much fewer steps. This search is based on the fact that random trial structures generated from low energy conformers usually afford low energy conformations after energy minimization, while minimizations of conformations generated from high energy structures lead to new high energy conformers. Starting structures for this algorithm are chosen from only the low energy regions, random variations to selected internal coordinates are applied, the structure is minimized, and the result is compared with minima found during previous search steps. In the framework of a Monte Carlo conformational search, this sequence is termed a Monte Carlo step and the entire procedure as a Monte Carlo Multiple Minimum (MCMM) search. It should be noted that the conventional MC step – as is usual in MD as well – does not involve energy minimization. The MCSM (Monte Carlo Single Minimum) search developed by Scheraga [52] applies energy minimization for structures generated by MC, but this search could not be restricted to chemically relevant low energy conformations. The MCMM algorithm, however, chooses a range of low energy starting geometries (in our case this range was set to 50 kJ/mol relative to the actual global minimum) and uses such structures uniformly. The method keeps track of the number of times a particular low energy conformation has been found and always uses the one used least frequently to generate the next trial structure. Furthermore, it has been demonstrated that instead of always varying the same number of internal coordinates at each MC step, it is better to choose a random number of such coordinates varying from between one and some maximum number. Selecting starting geometries from the low energy conformations found previously, and using such structures an equal number of times in combination with the random variation of internal coordinates to be changed in a MC step, makes this methodology particularly efficient. One of the main reasons why MCMM is so efficient is that it is not subject to energy barriers. The random perturbation applied is a geometrical operation, it is not affected by the shape of the potential energy surface. The subsequent energy minimization – TNCG with sophisticated line search – can also easily pass high energy barriers and kind of break its way through rough terrain downhill toward some low-energy minimum. Neither Metropolis MC nor MD have this advantage.

The MCMM search, therefore, takes advantage both of energetically favorable torsional variations and of more random jumps into remote regions of the conformational space. Thus MCMM employs a combination of torsionally local search steps and global conformational changes, which allows exploration of virtually all low energy conformations at the potential energy surface. The combined conformational search used in MCMM docking calculations involves variation of the internal degrees of freedom of the substrate and the binding site (rotatable bonds) as well as the external degrees of freedom of the ligand (relative translations and rotations) relative to the binding site. All of the internal and external degrees of freedom of the ligand and a limited set of internal degrees of freedom of the protein are considered.

In summary, Metropolis MC (MMC) is a simulation technique driven by the Metropolis algorithm to surmount energy barriers, whereas MC conformational analysis is a stochastic method to explore conformational space which is not affected by energy barriers at all. The use of rotamer libraries can, in principle, alleviate the low acceptance rate of MMC in a binding site subject to extremely high energy barriers. However, rotamer libraries ignore the specific conformational effect of a particular protein environment. On the other hand, MC conformational search naturally takes the protein environment into account. Conformational studies on protein–substrate complexes were first performed by Guida *et al.* [53, 54] while comparative MC and MD docking calculations were published by our group [43–45]. Results of our comparative MC/MD study revealed that MC conformational analysis applied to proteins can be more effective than conventional MD simulations.

18.3.3 Fragment-based methods

Fragment-based algorithms maximize favorable contacts within the predefined binding site by small molecule building techniques. In contrast to previously described approaches requiring a database of potential ligands, these de novo methods construct new molecules optimized for high binding site affinity. One of the most popular algorithms applied in fragment-based methods is the build-up procedure. Build-up methods place an initial fragment at the binding site and fill the cavity in a sequential manner. The second group of algorithms, called bridging techniques, places the most important functional groups at the binding site and attempts to connect them to yield a new molecule.

18.3.4 Build-up methodology

One of the first attempts using this approach was GROW, originally developed for the design of peptides that fit to a protein binding site [55]. Within the framework of this method, amino acids are gradually added to optimize binding affinity of the resulting peptide. Conformational aspects are considered by rotamer libraries precalculated for amino acid templates.

GroupBuild can be used to design small molecule substrates from common organic templates. Templates are organized for a library and combined, considering potential sites for interaction with the binding site in order to yield molecules with stereoelectronic complementarity [56, 57]. Templates are changed to atoms in LEGEND, a related method published by Nishibata *et al.* [58], where structures are built sequentially from random atoms positioned with random torsions. Pearlman *et al.* [59] also published an atom-based algorithm similar to LEGEND.

In MCDNLG the active site of the protein is filled by carbon atoms, then mutates and eliminates them depending on favored interactions. Gehlhaar *et al.* utilize a Monte Carlo algorithm to decide whether to accept or reject a particular configuration [60]. SMog, another Monte Carlo based approach, joins predefined functional groups within the binding site and scores the resulting ligand with a scoring function directly related to the binding free energy [61].

18.3.5 Bridging methodologies

Bridging techniques utilize the original idea of Böhm, first implemented in LUDI [62–64]. This program joins pre-defined fragments to the hydrophobic, hydrophilic and H-bonding parts of the binding site. Fragments can be connected after positioning or the user can choose a core fragment and LUDI adds suitable functional groups to maximize binding affinity. The synthetic accessibility can also now be considered in the initial phase [64].

BUILDER, another fragment-based algorithm utilizes DOCK to identify irregular lattices of docked molecules that can be searched for pathways that connect molecular fragments. Synthetic feasibility can be considered manually during this fully interactive procedure [65].

In SPROUT connection of pre-docked fragments is based on multiple graph searches connecting prepositioned functional groups with a set of templates [66]. Target sites can be defined by a rule-based approach.

Multiple copy simultaneous search (MCSS) searches for optimal orientation of small organic molecules such as water, methanol or acetonitrile and identifies local minima on the protein surface. Fragments are initially positioned to these minima and are connected by the HOOK algorithm searching for reasonable connections in a scaffold database [67, 68]. New implementation of MCSS allows the automatic generation of diverse virtual libraries of potential ligands [69].

18.3.6 Case studies

18.3.6.1 HIV-1 protease inhibitors

One of the first successful applications of structure-based design leading to a registered drug is the development of an HIV-1 protease inhibitor, indinavir (Crixivan).

Extensive crystallographic analysis revealed that the binding site is located deep inside the target protein. Molecular mechanics-based docking calculations were used to design potent inhibitors and led to a new drug launched in 1996 by Merck [70].

Nelfinavir is also an HIV protease inhibitor developed at Lilly and Aguron using structure-based design [71, 72].

X-ray crystallography in combination with docking calculations played a major role in identifying this compound, marketed as Viracept by Aguron. The X-ray structure of Viracept complexed with HIV-1 protease [72] is shown in Figure 18.4.

18.3.6.2 Acetylcholinesterase inhibitors

An integrated CADD approach utilizing molecular modeling, QSAR and structure-based design was used to discover Donepezil, an acetylcholinesterase inhibitor [73].

Lead optimization was performed by docking calculations based on the high resolution X-ray structure of the target enzyme [74]. Figure 18.5 depicts the binding site of the protein complexed with Donepezil. Esai launched this compound for the treatment of Alzheimer's disease.

18.3.6.3 Influenza virus neuraminidase inhibitors

Since the active site of influenza neuraminidase (NA) was found to be highly conserved in all clinically relevant strains, this enzyme was identified as an attractive target for antiviral treatments. Several random screenings, however, did not result in potent inhibitors of NA [75]. After the structural elucidation of the target protein, an NA structure complexed with sialic acid was published [76]. Rational design was based

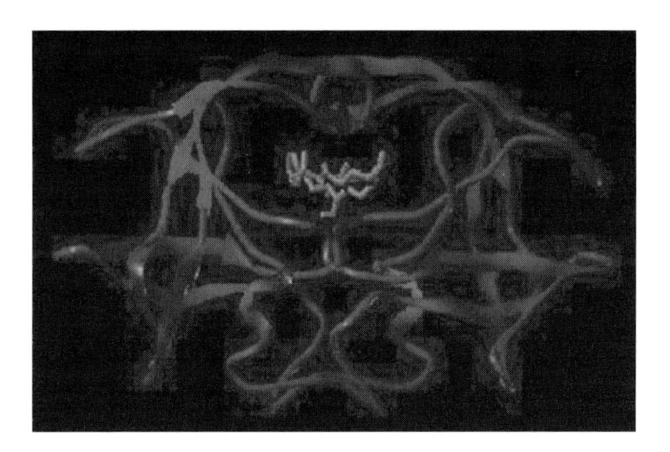

Figure 18.4 (See Colour Plate X.)

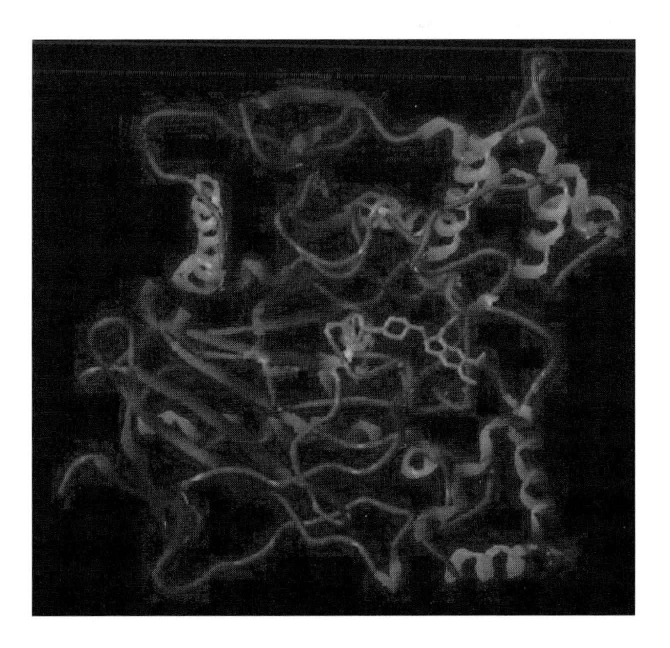

Figure 18.5 (See Colour Plate XI.)

on the high resolution X-ray structure of this complex using small molecule docking. Techniques applied in structure-based design revealed a conserved pocket with negative potential which could be filled by a positively charged nitrogen atom, such as in Zanamivir [77] or a recently published Roche compound, Ro640802 [78].

Zanamivir Ro640802

This proposal was supported by a recently published X-ray structure of the Zanamivir–NA complex [79]. Analysis of the active site environment revealed the existence of the negatively charged pocket responsible for increased affinity of Zanamivir toward NA. Figure 18.6 shows Zanamivir complexed to the NA active site represented as a Connolly surface colored by the calculated electrostatic potential (MEP). Negative regions of MEP (colored blue) are located around the positively charged N-atom of the ligand, maximizing the electrostatic complementarity between the protein and the inhibitor.

18.3.6.4 Dihydrofolate reductase inhibitors

De novo design was successfully applied in the search for selective inhibitors of Pneumocystis carinii dihydrofolate reductase (DHFR), a bacteria that causes opportunistic infections in HIV patients. But clinically these compounds were found to be selective for human DHFR, resulting in several toxic side effects. The search for selective novel DHFR inhibitors was begun by docking compounds from the Fine Chemicals Directory (50,000 compounds) to the binding site of Pneumocystis carinii DHFR yielding 40 compounds [80]. These compounds were checked for anti-DHFR activity and 13 compounds were identified with IC_{50} lower than 150 μM. Selectivity was also evaluated by measuring binding affinities to human DHFR which led to the identification of compound **11**, a 7 μM inhibitor of Pneumocystis carinii DHFR with 25 fold selectivity. This compound can be used as a potential lead for selective DHFR inhibitors.

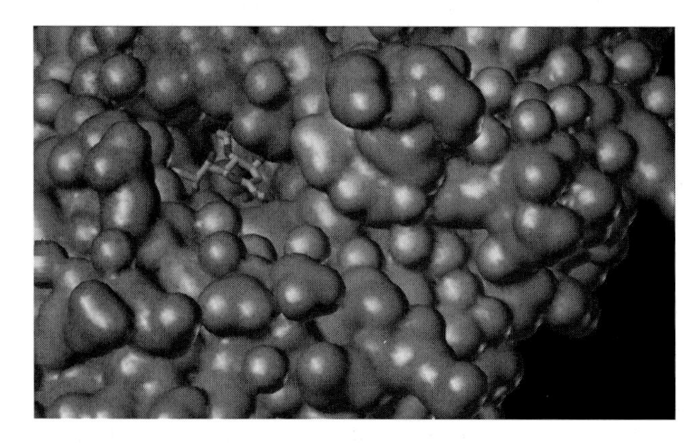

Figure 18.6 (See Colour Plate XII.)

18.4 Summary

Computer assisted drug discovery has been widely applied in most drug discovery programs to facilitate lead discovery and lead optimization. There are several examples both for ligand and structure-based design where the successful application of these methods has helped to find new drugs, several of them already on the market. Although CADD might have a significant impact on speed and effectiveness, this is not a technique that provides a direct route to new drugs. CADD, in combination with medicinal chemistry, combinatorial and HTS technologies, might however, be a key component for success. Examples presented in this chapter clearly demonstrate that cooperative CADD efforts should not be omitted from the game named drug discovery.

Acknowledgement

The authors are grateful to Mr. Zoltán Kovári for helpful discussions. G M K is a Bolyai Fellow of the Hungarian Academy of Sciences.

References

1 Náray-Szabó, G., *J. Mol. Recognition* **6** (1993) 205.
2 Golender, V. E. and Vorpagel E. R., In *3D QSAR in drug design: Theory, methods and applications* (Kubinyi, H. ed.) ESCOM, Leiden, The Netherlands, 1993.
3 Martin, Y. C., Bures, M. G., Danaher, E. A., Delazzer, J., Lico, I. and Pavlik, P. A., *J. Comput. Aid. Mol. Design* **7** (1993) 83.
4 Walters, D. E. and Hinds, R. M., *J. Med. Chem.* **37** (1994) 2527.
5 Allen, F. H., Bellard, S. A., Briece, M. D., Cartwright, B. A., Doubleday, A., Higgs, H., Hummelink, T., Hummelink-Peters, B. G., Kennard, O., Motherwell, W. D. S., Rodgers, J. R., and Watson, D. G., *Acta Crystallogr.* **B35** (1979) 2331.
6 Rusinko III, A., Skell, J. M., Balducci, R., McGarity, C. M., and Pearlman, R. S., CONCORD, University of Texas, Austin TX and Tripos Associates, St. Louis, MO, 1988.
7 Gasteiger, J., Rudolph, C., and Sadowski, J., *Tetrahedron Comput. Methodol.* **3** (1990) 537.
8 Dolata, D. P., Leach, A. R., and Prout, K., *J. Comput. Aided. Mol. Design* **1** (1987) 73.
9 Klebe, G. and Mietzner, T., *J. Comput. Aided. Mol. Design* **8** (1994) 583.
10 UNITY^{TM}, Tripos Associates Inc., St. Louis, MO, USA.

11 Green, J., Kahn, S., Savoj, H., Sprague, P., and Teig, S., *J. Chem. Inf. Comput. Sci.* **34** (1994) 1297.

12 Hansch, C., Maloney, P. P., Fujita, T., and Muir, R. M., *Nature* **194** (1962) 178.

13 Fujita, T., Iwasa, I., and Hansch, C., *J. Am. Chem. Soc.* **86** (1964) 5157.

14 Wold, S., Johansson, E., and Cocchi, M., In *3D QSAR in drug design: Theory, methods and applications* (Kubinyi, H. ed.) ESCOM, Leiden, The Netherlands (1993) pp523–550.

15 Cramer, R.D., Patterson, D.E., and Bunce, J.D., *J. Am. Chem. Soc.* **110** (1988) 5959.

16 Todeschini, R. and Gramatica, P., In *3D QSAR in drug design* (Kubinyi, H., Folkers, G., and Martin, Y. C., eds); Kluwer Academic Publisher, Dodrecht (1998) Vol.2, pp355.

17 Ferguson, A. M., Heritage, T., Jonathon, P., Pack, S. E., Philips, L., Rogan, J., and Snaith, P. J., *J. Comput. Aided Mol. Design* **11** (1997) 143.

18 Silverman, B. D., Platt, D. E., Pitman, M., and Rigoutos, I., In *3D QSAR in drug design* (Kubinyi, H., Folkers, G., and Martin, Y. C., eds.); Kluwer Academic Publisher, Dodrecht (1998) Vol.3, pp183.

19 Koga, H., Itoh, A., Murayama, S., Suzue, S., and Irikura, T. J., *J. Med. Chem.* **23** (1980) 1358.

20 Buckingham, J., Glen, R. C., Hill, A. P., Hyde, R. M., Martin, G. R., Robertson, A. D., Salmon, J. A., and Woollard, P. M., *J. Med. Chem.* **38** (1995) 3566.

21 Hong, H., Neamati, N., Wang, S., Nicklaus, M. C., Mazumder, A., Zhao, H., Burke, T. R. J., Pommier, Y., Milne, G. W. A., *J. Med. Chem.* **40** (1997) 930.

22 Lopez-Rodriguez, M. L., Jose Morcillo, M. J., Benhamu, B., and Rosado, M. L., *J. Comp. Aided Mol. Design* **11** (1997) 589.

23 Kuntz, I. D., *Acc. Chem. Res.* **27** (1994) 117.

24 Gschwend, D. A., Good, A. C., and Kuntz, I. D., *J. Mol. Recognition* **9** (1996) 175.

25 Leach, A. R., and Kuntz, I. D., *J. Comp. Chem.* **13** (1992) 730.

26 Fischer, D., Lin, S. L., Wolfson, H. L., and Nussinov, R., *J. Mol. Biol.* **248** (1995) 459.

27 Sandak, B., Nussinov, R., and Wolfson, H. J., *CABIOS* **11** (1995) 87.

28 Bacon, D. J., and Moult, J., *J. Mol. Biol.* **225** (1992) 849.

29 Kasinos, N., Lilley, G. A., Subbarao, N., and Haneef, I., *Protein Eng.* **5** (1992) 69.

30 Vakser, I. A., *Protein Eng.* **8** (1995) 371.

31 Vakser, I. A., *Protein Eng.* **9** (1996) 37.

32 Vakser, I. A., *Biopolymers* **39** (1996) 455.

33 Vakser, I. A., and Aflalo, C., *Proteins* **20** (1994) 320.

34 Miller, M. D., Kearsley, S. K., Underwood, D. J., and Sheridan, R. P., *J. Comput.-Aided Mol. Design* **8** (1994) 153.

35 Mizutani, M. Y., Tomioka N., and Itai, A., *J. Mol. Biol.* **243** (1994) 310.

36 Tomioka, N. and Itai, A., *J. Comput. Aided Mol. Design* **8** (1994) 347.

37 Morris, G. M., Goodsell, D. S., Huey, R., and Olson, A. J., *J. Comput. Aided Mol. Design* **10** (1996) 293.

38 Clark, D. E., Westhead, D. R., Sykes, R. A. and Murray, C. W., *J. Comput. Aided Mol. Design* **10** (1996) 397.

39 Ajay, Mrucko, M. A., and Stouten, P. F. W. Recent Advances in Prediciton of Binding Free Energies. In *Practical Application of Computer-Aided Drug Design* (Charifson, P. S., ed.) Marcel Dekker, NY. (1997).

40 van Gunsteren, W. F., King, P. M., and Mark, A. E. Q., *Rev. Biophys.* **27** (1994) 435.

41 van Gunsteren, W. F., and Mark, A. E. Q., *Eur. J. Biochem.* **204** (1992) 947.

42 Oda, K., Miyagawa, H., and Kitamura, K., *Mol. Simul.* **16** (1996) 167.

43 Keserű, G. M., Kolossváry, I., and Bertók, B., *J. Am. Chem. Soc.* **119** (1997) 5126.

44 Keserű, G. M., Kolossváry, I., and Székely, I., *Int. J. Quant. Chem.* **73** (1999) 123.

45 Keserű, G. M., Menyhárd, D.K., *Biochemistry* **38** (1999) 6614.

46 Straatsma, T. P., and McCammon, J. A., *J. Chem. Phys.* **90** (1989) 3300.

47 Straatsma, T. P., and McCammon, J. A., *J. Chem. Phys.* **91** (1990) 3631.

48 Chang, G., Guida, W. C., and Still, W. C., *J. Am. Chem. Soc.* **111** (1989) 4379.

49 Goodman, J., and Still, W. C., *J. Comput. Chem.* **12** (1991) 1110.

50 Kolossváry, I., and Guida, W. C., *J. Am. Chem. Soc.* **118** (1996) 5011.

51 Mohamadi, F., Richards, N. G. J., Guida, W. C., Liskamp, R., Lipton, M., Caufield, C., Chang, G., Hendrickson, T., and Still, W. C., *J. Comput. Chem.* **11** (1990) 440.

52 Li, Z., and Scheraga, H. A., *Proc. Natl. Acad. Sci. U.S.A.* **84** (1987) 6611.

53 Guida, W. C., Bohacek, R. S., and Erion, M. D., *J. Comput. Chem.* **13** (1992) 214.

54 Montgomery, J. A., Niwas, S., Rose, J. D., Secrist, J. A., Babu, Y. S., Bugg, C. E., Erion, M. D., Guida, W. C., and Ealick, S. E., *J. Med. Chem.* **36** (1993) 55.

55 Moon, J. B., and Howe, J. W., *Proteins* **11** (1991) 314.

56 Rotstein, S. H., and Murcko, M. A., *J. Med. Chem.* **36** (1993) 1700.

57 Rotstein, S. H., and Murcko, M. A., *J. Comput. Aided Mol. Design* **7** (1993) 23.

58 Nishibata, Y., and Itai, A., *Tetrahedron* **47** (1991) 8985.

59 Pearlman, D. A. and Murcko, M. A., *J. Comput. Chem.* **14** (1993) 1184.

60 Gehlhaar, D. K., Verkhivker, G. M., Rejto, P.A., Sherman, C. J., Fogel, D. B., Fogel, L.J., and Freer, S. T., *Chem. & Biol.* **2** (1995) 317.

61 De Witte, R. S., and Shakhnovich, E. L., *J. Am. Chem. Soc.* **118** (1996) 11733.

62 Böhm, H. J., *J. Comput. Aided Mol. Design* **6** (1992) 61.

63 Böhm, H. J., *J. Comput. Aided Mol. Design* **6** (1992) 593.

64 Böhm H. J., *J. Comput. Aided Mol. Design* **10** (1996) 265.

65 Lewis, R. A., Roe, D. C., Huang, C., Ferrin, T. E., Langridge, R., Kuntz, I. D., *J. Mol. Graphics*, **10** (1992) 66.

66 Gillet, V. J., Newell, W., Mata, P., Myatt, G., Sike, S., Zsoldos, Z., and Johnson, A. P., *J. Chem. Inf. Comput. Sci.* **34** (1994) 207.

67 Miranker, A., and Karplus, M., *Proteins* **11** (1991) 29.

68 Caflish, A., Miranker, A., and Karplus, M., *J. Med. Chem.* **36** (1993) 2142.

69 Caflisch, A., *J. Comput. Aided Mol. Design* **10** (1996) 372.

70 Dorsey, B. D., Levin, R. B., McDaniel, S. L., Vacca, J. P., Guare, J. P., Darke, P. L., Zugay, J. A., Emini, E. A., and Schleif, W. A., *J. Med. Chem.* **37** (1994) 3443.

71 Patick, A. K., Mo, H., Markowitz, M., Appelt, K., Wu, B., Musick, L., Kalish, V., Kaldor, S., and Reich, S., *Antimicrob. Agents Chemother.* 1996, **40**, 292.

72 Kaldor, S. W., Kalish, V. J., Davies, J.F., II, Shetty, B. V., Fritz, J. E., Appelt, K., Burgess, J. A., Campanale, K. M., Chirgadze, N. Y., Clawson, D. K., Dressman, B. A., Hatch, S. D., Khalil, D. A., Kosa, M. B., Lubbehusen, P. P., Muesing, M. A., Patick, A. K., Reich, S. H., Su, K. S., and Tatlock, J. H., *J. Med. Chem.* **40** (1997) 3979.

73 Kawakami, Y., Inoue, A., Kawai, T., Wakita, M., Sugimoto, H., and Hopfinger, A. J., *Bioorg. Med. Chem.* **4** (1996) 1429.

74 Kryger, G., Silman, I., and Sussman, J. L. *Structure Fold. Des.* **7** (1999) 297.

75 Murray, J. F., and Mills, *J. Am. Rev. Respir. Dis.* **141** (1990) 1582.

76 Burmeister, W. P., Henrissat, B., Bosso, C., Cusack, S., and Ruigrok, R. W., *Structure* **1** (1993) 19.

77 von Itzstein, M., Wu, W. Y., Kok, G. B., Pegg, M. S., Dyason, J. C., Jin, B., van Phan, T., Smythe, M. L., White, H. F., Oliver, S. W., Colman, P. M., Varghese, J. N., Ryan, D. M., Woods, J. M., Bethel, R. C., Hotham, V. J., Cameron, J. M., and Penn, C. R., *Nature* **363** (1993) 418.

78 Kim, C. U., Lew, W., Williams, M. A., Liu, H., Zhang, L., Swaminathan, S., Bischeofberger, N., Chem, M. S., Mendel, D. B., Tai, C. Y., Laver, W. G., and Stevens, R. C., *J. Am. Chem. Soc.* **119** (1997) 681.

79 Taylor, N. R., Cleasby, A., Singh, O., Skarzynski, T., Wonacott, A. J., Smith, P. W., Sollis, S. L., Howes, P. D., Cherry, P. C., Bethell, R., Colman, P., and Varghese, J., *J. Med. Chem.* **41** (1998) 798.

80 Gschwend, D. A., Sirawaraporn, W., Santi, D. V., and Kuntz, I. D., *Proteins: Struct. Funct. Genet.* **29** (1997) 59.

Chapter 19

Gene therapy

László Kopper

Contents

In all cases where a disease is caused by a gene defect, the correction (or cure) by gene therapy seems to be an appropriate, and causal, approach. We have just begun to identify the different gene errors and to understand their roles in certain diseases, therefore gene therapy is a relatively "young" modality, with little clinical experience, but with great expectations, and frequently not without fear. Besides the biomedical aspects, numerous social, ethical, legal and economical questions should be answered. Similarly, standardization, and uniform regulations are still hot topics (1–5).

In the US, gene therapy protocols are supervised and licensed by the RAC (Recombinant DNA Advisory Committee) and FDA (Food and Drug Administration). In England, the Gene Therapy Advisory committee (recently Human Genetic Commission) makes these serious decisions. In Europe, the Eurogenethy network tries to harmonize the heterogenous regulations (in many countries such regulations are still missing), emphasizing the need for safety and international interaction. In the US there are currently 357 clinical gene therapy protocols, 218 for cancer patients, 43 for diseases caused by monogene defect, 30 for infectious diseases, 49 for arthritis and other diseases, and 17 under review. In Europe the number of protocols ranges from 120–150, and 70% of them are related to cancer. More than half of these studies are running in England, France and Germany – mainly in Phase I and II. There is still no outstanding clinical result, but we are just at the beginning (Table 19.1).

The concept of gene therapy comes from the observations that certain diseases are caused by functional failure of a gene (monogene diseases), e.g. adenosine deaminase deficiency, Gaucher's disease (glucosylceramide lipoidosis), etc., which, theoretically, can be treated and cured by the introduction of the normal counterpart of the mutant or missing gene into the relevant host cell(s). If the normal copy of the target gene is transfered into a renewing population (e.g. bone marrow stem cells), then only one or a few treatments are necessary during the lifetime of the patient. In other cases, repeated transfer is necessary. As already mentioned, the majority of protocols are related to cancer, therefore – and partly due to the interest of the author – most examples in this chapter are of gene therapy in cancer patients.

Cancer is considered to be a genetic disease (besides the role of many epigenetic changes), which means the crucial importance of the accumulation of gene defects during the development of any tumor. To correct all these defects is probably impossible (at the stage of current

Table 19.1 Efficiency of antitumor gene therapy*

	Protocol	*Patient*	*Evaluable*	*Clinical response*
Cytokine transfection	60	376	237	15
Prodrug therapy	21	104	62	8
Resistance genes	8	?		
Suppressor gene/oncogene	13	78	26	6

* Roth, Cristiano – JNCI, 89, 21, 1996.

knowledge). However, there are many approaches that use gene therapy directly or indirectly against the carcinogenic process (1,6).

The core problem for cancer gene therapy is to correct key faults in the regulation of gene activity, mainly those which *regulate cell proliferation and cell death.* This is usually via inhibition of activated proto-oncogenes (oncogenes) or replacement of missing (mutant or deleted) suppressor genes. Another important area is the *increase in therapeutic efficiency* through the manipulation of immune mechanisms or to support chemo- or radiotherapy, for example, reducing host toxicity thereby allowing the use of higher dose would result in a higher therapeutic index. All these aims can be fulfilled by:

a introduction of the given gene into the given (target) cells – that is gene transfer (often called "gene therapy"); or
b inhibition of unwanted gene function with a complementary nucleotide sequence – this is oligonucleotide or antisense therapy.

19.1 Forms of gene therapy

19.1.1 Gene transfer

Gene transfer aims to substitute a missing function both in normal and cancer cells. In normal cells the target could only be a somatic cell (not the germ cells). The transfered gene, depending on the vector, will be built into the genome or can be expressed episomally. Preferably, gene expression is governed by the appropriate regulation. The best target cells for a long-lasting effect are the stem cells, as bone marrow or peripheral blood stem cells, the existance of tumor stem cells is still unknown and cannot be used. In fact, at least in cancer, all tumor cells should be targeted, which is again hardly possible. Several approaches have been designed to circumvent this problem.

19.1.2 Oligonucleotide/antisense therapy

The DNA or RNA strands carrying the coding nucleotide sequences (for protein synthesis) are known as "sense", and their complementary strands (with complementary nucleotide sequence) are the "antisense" (7–9). When a sequence of the sense strand (or part of it – a gene, or part of a gene) is known, it is easy to synthesize the antisense, which can bind to the sense with high specificity. This binding or complex formation can inhibit the function of the targeted sense sequence – either in the somatic cells, or in viruses, or in any infectious agents.

The antisense nucleotides in the therapy may be oligonucleotides or expressed nucleotides (sequences). The oligonucleotides (oligos) are usually short, single-stranded DNA pieces, built up from 15–25 bases. (This is the length of a given sequence in the genome that could be considered unique, occurs only once, therefore targeting is highly specific.) Oligonucleotides are made artificially, while expressed nucleotide sequences are produced by genes carried by expression vectors. The antisense RNA could have a length from a few dozen to a few thousand bases. For safety and efficiency reasons, the oligonucleotide (or antisense) therapy is mainly used.

Oligonucleotide therapy may have several strategies:

(a) *anti-gene (anti-code, triplex) strategy* – where the oligo binds to DNA inhibiting the transcription of information from DNA to mRNA, possibly due to the inhibition of binding of the activating protein (transcription or transactivation factor) to its relevant sequence. (At the binding site the double helix DNA becomes triplex. Additionally, triplex can be formed when mRNA is sandwiched by two antisense oligos.) The triplex may also block DNA replication. It has been shown, that benzo(e)piridoindol derivatives could enhance the effect of triplexes (probably by interacting with topoisomerase II/DNA complexes).

(b) *Anti-sense strategy* – where the oligo binds to mRNA blocking its "maturation" (splicing), transport to the cytoplasm, or protein synthesis (translation of the code). Besides spatial inhibition caused by complex formation, the efficiency of the antisense oligo depends on its capacity to activate endogenous RNase-H, which digests the RNA component of the complex (mRNA/DNA oligo). Activation, however, could lead to other events, and consequently to unwanted side effects, if the oligo temporarily binds to a non-targeted mRNA – since the enzyme will also be activated in this case, causing the undesired inhibition of that mRNA. One possibility for solving this problem is the use of a chimeric oligo, which has two parts, a RNase-H non-activating methylphosphonate backbone surrounded by a RNase-H activating region with phosphodiester bonds.

(c) *Ribozyme strategy* – where the antisense sequence is placed into an endogenous RNA with catalytic (RNase) activity. This is the ribozyme (10). Therefore, when the ribozyme carrying the antisense sequence forms a complex with the target mRNA, the ribozyme will digest the mRNA. It is possible to design a construction using multiple ribozymes against more than one target, or activating each other in a cascade-like manner. In most cases the target mRNA at the cleavage site should contain a GUC sequence. Ribozymes are very specific, they can "recognize" even one substitution difference. The problem is that at each new target the efficiency of the ribozymes should be proven. When antisense and ribozyme strategies are compared, remarkable differences are found depending on the given biological system and the target gene.

(d) *Sense strategy* – where the opposite of normal regulation is used, i.e. in normal cases the protein (e.g. transcription factor) can bind to DNA, whereas in this strategy the oligo and the sense sequence will inhibit the unwanted protein in the cell (transcription factors, enzymes, or even viral proteins).

19.2 Vectors

One of the most important problems influencing the success of gene therapy is the delivery of the functioning genes or antisense oligos to the required site, specific organ, tissue or cell (if there is such a site). There are different delivery systems (vectors) – indicating that an optimal system is still not available – they can be viral or non-viral. The introduction of a gene or sequence into the cell may occur *in vivo* or *ex vivo*. With *in vivo* application the vector carrying the gene or the sequence may be given in any therapeutic agent, peritumorally, systemically, locally, intracavity, orally, in a depot form, etc. *Ex vivo* means that the transfer takes place outside the organism. The target cells may be autologous, using cells removed previously from the host or allogene, especially useful when the removal or culturing of the patient's tissue (e.g. liver, lung, central nervous system) is difficult. Obviously, *ex vivo* application provides a much higher efficiency for transfer than *in vivo*, moreover, the gene defect can be characterized much better.

19.2.1 Viral vectors

Many observations support the applicability of viral vectors in cancer: gene transfer with viral vectors into cancer cells is more efficient than was expected from experience gained with normal cells; viral vectors can spread well in the extracellular space; transfected cells can induce bystander effects in non-transfected cells; the correction of a single gene defect may cause clinically significant tumor sequences. One problem with viral vectors is that they can be given locally-regionally but not systemically. Furthermore, in cases of repeated application, the success of therapy may be hampered by an antiviral immune response (11–13).

(a) *Retroviruses* – used by most licensed protocols. It seems to be advantageous that the retroviruses can enter dividing cells, and efficiency of *ex vivo* transfection is quite high. This is probably the best vector if we want to integrate foreign DNA into the target cell. The main disadvantages are that its capacity is relatively small (8–12 kB); complement in the sera could inactivate; the titer in the tumor is rather low; large-scale production is not easy (but can be improved by packaging cell lines); with some retroviruses the host choice is limited.

(b) *Adenoviruses* – attention has focused on the Ad5-based viral vector. Adenoviruses contain double-stranded DNA and can be transferred successfully into many cell types independent from their proliferation state. Virus replication requires an intact E1 region. Many data confirm that in most cell types – except hemopoietic cells – few viruses can ensure good transfer and efficient expression. They have been used for HSV-TK gene delivery in brain and liver tumors, and for p53 gene delivery in lung, head and neck, and liver cancer, by direct injection into the tumor. One problem with recombinant virus is that the intact viral genes could lead, at low degree, to virus formation which is toxic (especially neurotoxic). Therefore the duration of gene expression is limited, and the efficacy of repeated treatment is dubious, due to the immune response against viral antigens (immune response was not observed with retroviruses). Several approaches have been developed to eliminate this problem (heat-sensitive mutants, IL-2 treatment to inhibit $T_{H}2$, immunosuppression with cyclosporin, using different serotypes).

(c) *Adeno-associated virus* – contains small, single-stranded DNA part of which could be replaced by the potentially therapeutic genes. This construct induces hardly any immune response. Dividing, non-dividing, and hemopoietic cells can be transfected. The lack of viral genome results in low titer, multiplication could be promoted by helper viruses, e.g. adenovirus.

(d) *Herpes simplex virus* – contains large double-stranded DNA. Attention to HSV has been low because it causes latent infection in the brain. In cancer therapy it is used almost exclusively against brain tumors.

(e) *Vaccinia virus* – the recombinant form can incorporate relatively rather large inserts; it is used mainly to deliver interleukin and co-stimulator genes.

(f) *Poxvirus, baculovirus, lentiviruses* – are also in trials.

(g) *Hybrid vectors* – the aim is to provide vectors which could be integrated into nonmitotic cells (e.g. adenovirus/retrovirus; adenivirus/adeno-associated virus) (14).

19.2.2 Non-viral vectors

This group contains liposomes, molecular conjugates and naked DNA.

(a) *Liposomes* – charged lipid membranes in which the hydrophilic, polyanion nucleic acids – as DNA – remain stable. The liposome/DNA complex fuses with the cell membrane and negatively charged DNA enters the cell. Although the immune response does not develop against liposomes, gene transfer is less efficient than with viral vectors.

(b) *Molecular conjugates* – one part is a protein or synthetic ligand bound to nucleic acids or a DNA-binding part in order to target specific nucleic acids. If the DNA is coupled to the molecular conjugate, the result is a protein : DNA complex. This system can carry nucleic acids of any size, with fairly good specificity and efficiency *in vitro*. *In vivo* application is hampered by the need for non-viral endosomal lysis, or short expression of the gene. However, this system could be the basis for the development of "synthetic viruses" avoiding the use of true viruses.

(c) *Naked DNA* – is the most simple delivery "system", where naked DNA or DNA bound to gold particles is bombarded into the tissue (15). Naked plasmid DNA given intramuscularly has been tried in order to induce antitumor immunity in colorectal cancer and melanoma patients. (Actually, the use of naked DNA – taken up by immune cells, integrated into the host genome, and expressing the required protein – is a means of DNA vaccination). Besides cancer, it has been tested in autoimmune diseases to create immune tolerance against autoantibodies). Obviously, in this case there is no "targeting", the DNA is taken up by cells at the site of the injection.

(d) Additional non-viral methods use *calcium-phosphate* or *electroporation*.

19.3 Challenges of gene therapy

19.3.1 Stability

One of the main problems of antisense therapy is that the oligos could be digested by nucleases – especially 3′ exonucleases, and these enzymes are everywhere in the body – resulting in a very short half-life. The solution is the chemical modification of nucleotide sequences, provided that the modification does not disturb the intracellular uptake, binding to the target sequence, and RNase-H activation. Modifications are generally made on the phosphodiester backbone (e.g. phosphorothioates, methylphosphonates), sometimes the sugars (e.g. morpholino type polymers), and occasionally the bases (mainly 5-substituted pyrimidines). The ends of the oligos can also be involved, e.g. with hydrophobic, fluorescent and catalytic groups. All of these can change the behavior of the oligo, its pharmacokinetic properties, ability to activate RNase-H, and in the case of large substituents stereoisomeres may appear. Recently, nucleic acid-like non-nucleotide structures have been designed (e.g. plastic DNA), or the phosphodiester bonds have been replaced by polyamides (e.g. peptide nucleic acids). Polyanions (e.g. phosphorothioates) use receptor-mediated endocytosis, non-ionic oligos (e.g. methylphosphonates) pinocytosis for intracellular uptake. Efficiency can be enhanced by promoting internalization (e.g. with L-lysin, cholesterin, biotin, ligands which recognize tissue- or organ-specific receptors), or by decreasing the activity of lysosomal enzymes (e.g. using chloroquin or viruses). (Certain viruses, when entering the endosomal system, can break the membrane at low pH, and this endosmolytic activity rescues them, and the bound oligo, from lysosomal digestion.)

19.3.2 Efficiency

In human leukemic cells it was shown a decade ago that, *in vitro*, antisense BCL-2 can decrease the level of intracellular BCL-2, resulting in growth inhibition and apoptosis. Later human B-cell lymphoma with t(14;18) translocation in SCID mice was shown to be inhibited by antisense BCL-2. However, the efficiency was too low. It turned out that one of the potential problems is related to the design of the antisense. When the efficiency of a family of antisenses was checked, many of these showed a more pronounced effect that the original, or that which was made "against" the starting region of translation (with ATG starting codon), a popular site for targeting mRNA. The fact is that intracellular mRNA is not a linear molecule as we imagined at the time of oligo design on paper, but has a three dimensional structure, with loops, allowing intramolecular hybridization. This may explain why the oligo will not bind to the suggested target, instead it forms a complex with sequences distant on paper (or on the computer screen), but which can get very close in space. This variability in the efficiency of the oligo depending on the intramolecular sequence and, similarly, on the length of the oligo, causes difficulties in construction of the antisense. Today, automatic systems attempt to scale down the selection procedure.

In therapeutic experiments there is the question of whether the observed effect is the consequence of antisense action (as was expected), especially if one considers that, e.g., phosphorothioates can interact with proteins to modify their actions – i.e. biological effect is obvious, but not due to the desired action of the antisense. Sometimes the inefficiency is not caused by the failure of the strategy, e.g. the BCL-2 mRNA is blocked effectively, however the biological result is missing since the cell has started to "use" another anti-apoptotic molecule (in this case BCL-X_L) instead of the inhibited BCL-2. It may also be that the antisense alone is unable to efficiently reduce the level of BCL-2 mRNA, but by using chemotherapy in combination, apoptosis could be induced. The most probable explanation is that the antisense increased sensitivity to the agent.

19.3.3 Targeting and specificity

Undoubtedly, one of the most important problems in gene therapy is the specificity of gene expression, which may be influenced by the correct targeting of the vector, by tissue- or cell-specific promoters, by the route used for delivery, and by modulation of immune receptors.

(a) Normally the viral vectors do not "target" specific cells, however their *tissue tropism* results in certain specificity. Adenoviruses usually infect the respiratory epithelium, and are useful for good transfection into the liver as well. Retroviral vectors integrate into dividing cells. Herpes simplex virus has the capacity to infect many cell types, however the neuronal elements show high sensitivity. Viral targeting may be modified by binding specific ligands chemically or with antibody to the viral capsid, or the motif responsible for the virus binding may also be modified. Non-viral targeting uses mainly molecular conjugates and protein/DNA complexes. Many ligands have been tried in order to increase specificity, e.g. folates for folate receptor-expressing cells (ovarian cancer cell), or asialoorosomucoid (ASOR) glycoprotein for liver cells. Malaria proteins have been coupled with ASOR when the expression of ASOR-receptors on the liver cells is low (e.g. cirrhosis, diabetes, liver cancer). Increased EGFR expression enables the use of EGF/DNA complex (lung cancer). In these examples the vector binds to the cell through the receptor followed by receptor-mediated endocytosis. The fusion of endosomes with lysosomes result in the digestion of DNA, provided endosomal lysis is not inhibited (e.g. with viruses such as adenovirus).

(b) Another tool to increase specificity is the application of tissue- or cell-specific *promoters*. According to recent studies one of the strongest promoters is the promoter of the cytomegalovirus, therefore this is advised for cancer therapy. Trials have been undertaken using the promoter of the CEA gene in pancreatic cancer cells, and regulating sequences of human surfactant A gene in small cell lung cancer (obviously in those expressing this protein). Using a 769 base pair part of the tyrosinase gene increased specific expression of β-galactosidase in melanoma cells. Promoters of α1-antitrypsin and albumin (especially in the liver) activated α-foetoprotein promoter in hepatomas, and DF3/MUC1 promoter in breast cancer cells. The promoter of stress-induced grp78 (glucose related protein) reacts to hypoxic conditions, and increased the riporter-gene expression in mouse fibrosarcoma models. To find a universal tumor specific promoter is almost impossible due to the heterogeneity of the tumors.

(c) The *route of delivery* could help the specificity. Direct introduction of retroviral vector-containing cells into the tumor, e.g. by stereotaxis into gliomas or by bronchoscopy into lung cancer, is a frequent application in clinical protocols. The most simple route, would be the systemic, intravenous administration, although it does not promote tissue specificity. In preclinical models the use of peritumoral or intracavitary (e.g. intaperitoneal) treatment produced some effect. In these cases the vectors and gene are introduced into cells previously removed from the patient and injected back after manipulation. This is the usual method for melanomas and hematological tumors.

(d) Tumor targeting could be achieved using cytotoxic *lymphocytes* or TILs (tumor infiltrating lymphocytes) if gene transfer could induce immune receptors. E.g. the T-cell receptor can "identify" MART-1 antigen expressed on Jurkat cells. The expression of tumor-recognizing T-cell receptors would allow the use of TIL or other cytotoxic cells for tumor targeting.

19.3.4 Gene integration and expression

The efficiency of therapy is substantially influenced by the duration of expression of the given gene. Vectors, as retroviruses or adeno-associated viruses incorporated into the host cell genome ensure long-lasting expression, but at the same time they could be genotoxic. In cancer therapy, permanent expression is especially important in antisense sequences. In other cases, such as the introduction of suppressor genes or in prodrug therapy, short expression could cause cell death. Short-term expression is characteristic of adenoviral vectors, because the linear virus remains episomal and therefore is sensitive to digestion. This is relevant for non-viral vectors as well, since the bound nucleic acids have no maintaining sequence.

The most difficult aspect is regulation of gene expression at transcriptional level. For practical reasons it may be easier to focus on inducibility rather than regulation of a promoter (see above). In early growth response, gene promoter sequences were discovered which increased gene expression after radiation. Therefore, radiation-induced therapeutic gene expression caused specific cell death. Another possibility for regulation of gene expression is the use of temperature-sensitive mutants. It has been shown that expression of viral genes is inhibited by introducing temperature-sensitive mutant sequences into the genes.

19.4 Gene therapy applications

19.4.1 Immuntherapy with recombinant DNA expressing cytokines and lymphokines

This approach is based on experiments suggesting that cytokines produced by tumor cells (when cytokine-expressing recombinant DNA was incorporated into tumor cells) have antitumor effects. It is possible that recombinant DNA with tumor vaccines and adoptive transfer of genetically manipulated lymphocytes could increase the antitumor systemic immune response (16).

19.4.2 Prodrug therapy

The essence of this approach is that an enzyme is transferred selectively into tumor cells will convert the systemically administered non-toxic prodrug into a toxic metabolite, in situ, in the tumor cells (10,17) (Table 19.2). The enzyme could be introduced with antibody (ADEPT; antibody directed enzyme prodrug therapy) or with a vector containing the gene of the enzyme (GDEPT, VDEPT; gene (virus) directed enzyme prodrug therapy). The first protocol, which underlined the feasibility of this strategy, was the introduction of a retroviral vector containing the gene of herpes simplex virus thymidine kinase (HSV-TK) into brain tumors, which cleaved systemically administered gancyclovir into gancyclovir triphosphate – but only in the cells expressing HSV-TK. The advantage of this technique is that the vector is taken up selectively by proliferating cells – and in the brain it is likely that only the tumor cells proliferate. Rat glioma models have shown that significant tumor regression occurs

Table 19.2 Prodrug gene therapy

Gene product	*Prodrug*	*Toxic metabolite*
HSV thymidine kinase	Ganciclovir	Ganciclovir triphosphate
Cytosine deaminase	5-fluorocytosine	5-fluorouracil
5′-deoxy-5-fluorouridine	Thymidine phosphorylase	5-fluorouracil
VZV thymidine kinase	6-methoxypurine arabinonucleoside	6-methoxypurine arabinonucleoside monophosphate
Xanthine-guanine phosphoribozyl transferase	6-thioxanthine	6-thioxanthine monophosphate
E. coli purine nucleoside phosphorylase	6-methylpurine-2′ deoxyribonucleoside	6-methylpurine
β-lactamase	Vinca-cephalosporin	Vinca alkaloid
β-glucosidade	Amygdalin	Cyanide
Cytochrome p450 281	Cyclophosphamide	Alkylating metabolite
Nitroreductase	5-(aziridin-l-yl)-2, 4-dinitrobenzamide	5-(aziridin-l-yd-4-hydroxyl-amino-2-nitrobenzamide
Carboxypeptidase	CMDA	Benzoic acid mustard

(Smythe: Current Oncol Rep 2, 17, 2000).

even though only a small percentage of the tumor cells are transfected with HSV-TK genes using a retroviral vector. This phenomenon, that cytotoxicity appears elsewhere besides the transfected cells, is called the bystander effect. The reason for this effect is unknown, but there are suggestions that toxic metabolites or "apoptotic vesicules" can be released from the damaged cells into neighbouring cells, or that an immune response is induced. It has also been shown that the bystander effect can contribute significantly to the antitumor effect, e.g. when 10–20% of the tumor cells were successfully transfected with the enzyme gene, together with the bystander effect – at least in the experimental system – all or almost all tumor cells were killed. Therapeutic response occurred even when only 1–5% of tumor cells were transfected. Another prodrug-cleaving enzyme is cytosine deaminase, missing from mammalian cells, which makes 5-fluorouracil from 5-fluorocytosine. This enables 5-fluorouracil to be given in a much higher dose with lower toxicity. Alkalic phosphatase is used for etoposide phosphate and doxorubicin phosphate. Many other cytostatics already have a prodrug form.

19.4.3 Drug resistance gene

Theoretically the introduction of a gene responsible for drug resistance could decrease drug toxicity – which is an advantage in normal cells, but a disadvantage in tumor cells. Usually the multidrug resistance (MDR1) gene is transfected into hemopoietic stem cells (from bone marrow or peripheral blood), increasing protection of hemopoiesis against the toxic effect of certain cytotoxic agents. In other words, if normal cells are protected, a much higher dose of cytotoxic drugs could be applied. Such protection against nitrosourea toxicity was achieved with the introduction of the gene of O-methylguanidin-DNA-methyltransferase (DNA repair protein), and against methotrexate toxicity with the expression of methotrexate resistant dihydrofolate reductase. Experiments are under way with other compounds as well (aldehyde dehydrogenase, gluthation-S-transferase, superoxide dismutase). Nevertheless, some problems still need to be solved: high-dose therapy is not always accompanied by higher therapeutic efficiency; doses could be limited by non-hematological toxicity; the drug resistence gene could be transferred accidentally into tumor cells hiding in the bone marrow. It is also possible that vectors could be constructed expressing more than one resistance gene.

Alternatively, MDR1 (or other resistence gene) activity can be blocked by antisense strategies. The clinical results obtained so far are promising but not fully convincing.

19.4.4 Inhibition of oncogenes, transfer of suppressor genes

There is hardly any oncogene or suppressor gene that has not been used for gene therapy – at least experimentally. Here are only a few examples.

The expression of unwanted oncogenes may be blocked by antisense oligos. This has been tried in the inhibition of mutant Ki-RAS expression in human lung cancer cells. A triplex strategy was used to inhibit expression of the HER2/neu receptor. Bone marrow purging was achieved by antisense-MYB in both the chronic and acute accelerating phases of CML. Antisense BCR/ABL has also been tested in CML (Table 19.3).

Table 19.3 Antisense therapy

Target gene	*Size*	*Targeted tumor*	*Clinical phase*
Ha-RAS	ps/20-mer	Solid tumors	I
c-RAF	ps/20-mer	Solid tumors	I/II
BCL-2	ps/18-mer	NHL, melanoma, prostate cancer	I/II
p53	ps/20-mer	AML, MDS, (infusion, *ex vivo*)	I
Protein kinase C-α	ps/20-mer	Solid tumors	I/II
Protein kinase A RI-α unit	18mer	Solid tumors	I/II
c-MYB	ps/24-mer	CML (inf, *ex vivo*)	I
BCR/ABL	ps/26-mer	CML (inf, *ex vivo*)	II
VEGF rec:Fltl	Ribozyme	Solid tumors	I

Ps phosphorothioate Kushner, Silverman: Current Oncol Rep 2, 23, 2000.

The most popular gene among suppressors is p53 and many attempts have been made to correct p53 defect either blocking the mutant for or introducing the wild-type p53. The problem is that the biological role of p53 is not entirely understood. There are observations claiming an increase in the effect of cytostatics both in the presence and absence of functioning p53. Adenovirus vectors have been constructed with sequences which can identify and kill only cells with mutant p53, but not with wt-p53. In head and neck cancer the introduction of p53 caused regression in 30% of tumors, however the ratio increased up to 90% when this gene therapy was combined with chemotherapy (cisplatin, 5-fluorouracil). There are data showing, that transfer of p53 could decrease the expression of the MDR1 gene.

19.5 Conclusions

Theoretically it is possible to use gene therapy for prevention, i.e. to prevent the transformation of a premalignant lesion into a malignant one (e.g. introduction of missing gene functions into bronchial dysplasia or Barrett oesophagus). Until now gene therapy in the clinic was performed in patients with a tumor of advanced stage and where all other known therapeutic options have failed. The positive results obtained even in these cases allow us to propose the use of gene therapy in earlier stages of tumor growth. In many tumors the residue or recidiva of the primary tumor is a very tough clinical challenge (e.g. lung cancer, head and neck cancer, pancreatic cancer), gene therapy given locally (intralesionally) or as adjuvant therapy may be recommended.

Although the gene therapy has a very short history (not more than 10 years), some conclusions can already be drawn. Retroviral vectors usually seem to be a safe delivery systems for gene transfer, at least a toxic effect has not been observed yet. To transfer and express a gene in tumor cells *in vivo* viral or non-viral vectors are equally useful. Tumor regression was observed in many clinical trials, but evaluation of the result is not easy. Often the effect was similar to the effect of chemotherapy in advanced stage patients. As a whole it is almost certain that gene therapy will be used together with other therapeutic modalities.

Gene therapy is probably a straightforward approach in a case where only one gene defect is responsible for all clinical problems. However, in cancer, where the disease progresses due to the accumulation of gene defects, it is extremely difficult to estimate the biological role for each of them. The discovery of the gene defects in a given tumor and the use of strategies where multiple gene defects can be targeted simultaneously will surely increase the efficiency of this – causal (!) – therapy.

References

1 Roth, J.A., Cristiano, R.J. (1997) Gene therapy for cancer: what have we done and where are we going? *JNCI 89, 21–39*

2 Scanlon, K.J., Kashani-Sabet, M. (1998) Ribozymes as therapeutic agents: are we getting closer? *JNCI 90, 558–561*

3 Costantini, L.C., Bakowska, J.C., Breakefield, X.O., Isacson, O. (2000) Gene therapy in the CNS. *Gene Therapy 7, 93–109*

4 Asahara, T., Kalka, C., Isner, J.M. (2000) Stem cell therapy and gene transfer for regeneration. *Gene Therapy 7, 451–457*

5 Giannoukakis, N., Thomson, A.W., Robbins, P.D. (1999) Gene therapy in transplantation. *Gene Therapy 6, 1499–1511*

6 Weichselbaum, R.R., Kufe, D. (1997) Gene therapy of cancer. *Lancet 349, (suppl II), 0–12*

7 Kushner, D.M., Silverman, R.H. (2000) Antisense cancer therapy: the state of the science. *Current Oncol Rep 2, 23–30*

8 Persidis, A. (1999) Antisense therapeutics. *Nat Biotechnol 17, 403–404*

9 Flanagan, W.M. (1998) Antisense comes of age. *Cancer and Metastasis Rev 17, 169–176*

10 Smythe, W.R. (2000) Prodrug/drug sensitvity gene therapy: current status. *Current Oncol Rep 2, 17–22*

11 Clackson, T. (2000) Regulation gene expression systems. *Gene Therapy 7, 120–125*

12 Burcin, M.M. (1999) Adenovirus mediated regulable target gene expression *in vivo*. *PNAS 96, 355–360*

13 Schagen, F.H.E., Rademaker, H.J., Fallaux, F.J., Hoeben, R.C. (2000) Insertion vectors for gene therapy. *Gene Therapy 7, 271–272*

14 Recchia, A. (1999) Site-specific integration mediated by a hybrid adenovirus/adeno-associated virus vector. *PNAS 96, 2615–2620*

15 Corr, M. (1996) Gene vaccination with naked plasmid DNA: mechanism of CTL priming. *J Exp Med 184, 1555–1560*

16 Restifo, N.P., Ying, H., Hwang, L., Leitner, W.W. (2000) The promise of nucleic acid vaccines. *Gene therapy 7, 89–92*

17 Ishii-Morita, H., Agrabaria, R., Mullen, C.A. (1997) Mechanism of bystander effect killing in the herpes simplex thymidine kinase gene therapy model of cancer treatment. *Gene Ther 4, 244–251*

Part 5

Molecular pathomechanism of cancer

Chapter 20

Cancer genetics

László Kopper

Contents

20.1 Multiple genetic alterations in cancer cells

A century ago (essentially based on the concept of Virchow) the cell was considered a basic element of many diseases, including cancer. Hundreds of types and subtypes of human tumors are still classified according to the cellular (cytological/histological) characteristics. However, in the past few decades we have moved a dimension further and have begun to understand the molecular machineries operating behind the cell functions as well as some of the molecular changes resulting in different diseases. Although it may sound reductionist, cancer is currently defined as genetic disease of the somatic cells, stating that somatic mutations are what all cancers have in common. Keeping in mind that the precise nature of these mutations and the cell types in which they take place must be understood in each individual type of cancer, our considerations should revolve around three main issues:

1 HOW MANY GENE ERRORS ARE REQUIRED TO PERVERT A NORMAL CELL INTO A CANCER CELL?

Our estimation, deriving from the pioneering work of Knudsen (the two-hit model) and partly from epidemiological studies on age dependence of the incidence of cancer in various populations, is that the number (n) of gene errors required is probably in the range of 2–5. There is no *a priori* reason why n should be the same for all tumors – even those with the same histological type or subtype. It is known that many gene errors identified in various cancers may appear in normal cells without known functional significance. It has also been demonstrated (especially in work from Vogelstein's laboratory) that the morphological changes from a normal to a malignant structure can be – at least in certain cases – characterized by the gradual accumulation of gene errors. The sequence of somatic mutations may reflect the interplay of inherited factors and acquired factors in carcinogenesis. The inheritance of a mutated suppressor gene will decrease the number (n) of somatic mutations required to cause cancer by 1.

2 WHAT ARE THE SPECIFIC GENES IN WHICH THE ERROR(S) OCCURS?

Many different genes may be damaged (mutated) in cancer but they belong to discrete functional sets. All-inclusive classification of cancer genes is difficult, but popular dichotomies are oncogenes versus tumor suppressor genes, gatekeepers versus caretakers, proliferation promoting genes versus pro-apoptotic genes. In general, anything that increases the rate of mutation increases the risk of cancer and anything that eliminates mutant cells decreases the probability. In practice, if we try to classify those genes which, if they mutate, may contribute (according to recent evidence) to the development of cancer, the following list seems to be acceptable: (1) genes encoding growth factors and growth factor receptors; (2) genes participating in signal transduction and otherwise in the cell cycle; (3) transcription controling genes and other genes encoding nuclear proteins; (4) genes involved in DNA repair and in chromosomal replication, mitotic segregation and telomere maintenance; (5) genes responsible for triggering apoptosis of abnormal cells; (6) genes involved in interactions of cells with the extracellular matrix and blood vessels.

3 WHAT IS THE NATURE OF GENE ERRORS THEMSELVES?

Mutation may result in loss or gain of function (to some extent this is in the mind of the beholder). Mutations can be classified according to their molecular nature or their effect on the function of the gene product. Large deletions, frameshift mutations, nonsense mutations and many splice site mutations usually inactivate the gene leading to loss of function, especially in tumor suppressor genes. These types of mutations probably represent the majority of gene errors in many malignancies. Missense mutations influence gene function in only a limited number of cases, e.g. when it modifies an active site or a binding domain. Major gene rearrangements (producing fused or juxtaposed genes) usually result in gain of function.

20.2 Regulation of proliferation and cell death

20.2.1 Protooncogenes/oncogenes; suppressor genes/tumor suppressor genes; cell cycle genes

Tissue homeostasis is maintained by the tissue- or cell-specific balance between cell proliferation and cell death. There are hundreds of regulatory substances to maintain this balance depending on the actual physiological need. Disturbances in the regulation of proliferation and death due to genetic or epigenetic errors may lead to the accumulation of cells which are not required by the host. Alternatively, these cells may gradually become independent from normal regulation (Figure 20.1). Probably the two most important sets of genes which support normal growth control are the protooncogenes and suppressor genes, and their altered forms – oncogenes and tumor suppressor genes – play a crucial role in the development of neoplasia (16).

20.2.1.1 Protooncogenes/oncogenes

"Activated" forms of protooncogenes are oncogenes. (Their similarity with retroviral alternatives helped in their discovery. Viruses stole the cellular prooncogenes, activated them, and when they infected other cells used this altered form to influence the proliferation of the host cell. Continuous proliferation served the interest of the virus maintaining permanent virus formation). Not all oncogenes have a viral counterpart. An oncogene is a particular gene that contributes to tumor formation by producing, usually, endogenous proliferation signals. Oncogenes act dominantly, i.e. the activation of one allele leads to a change in the phenotype (whereas suppressor genes act recessively with the exception of p53). Protooncogenes have different functions; they can be growth factors, membrane or cytoplasmic receptors or other signal-transducing molecules, and transcription factors, consequently, their activation primarily disturb signaling pathways. Many protooncogenes have relatives (forming families, e.g. RAS, MYC), and the activated forms of certain members appear in certain human tumors with differing frequency.

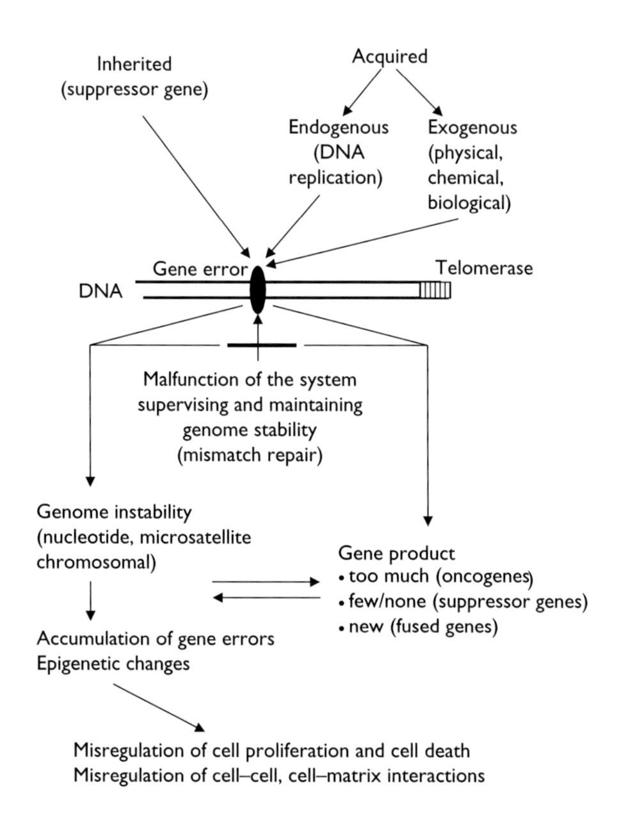

Figure 20.1 Scheme for carcinogenesis.

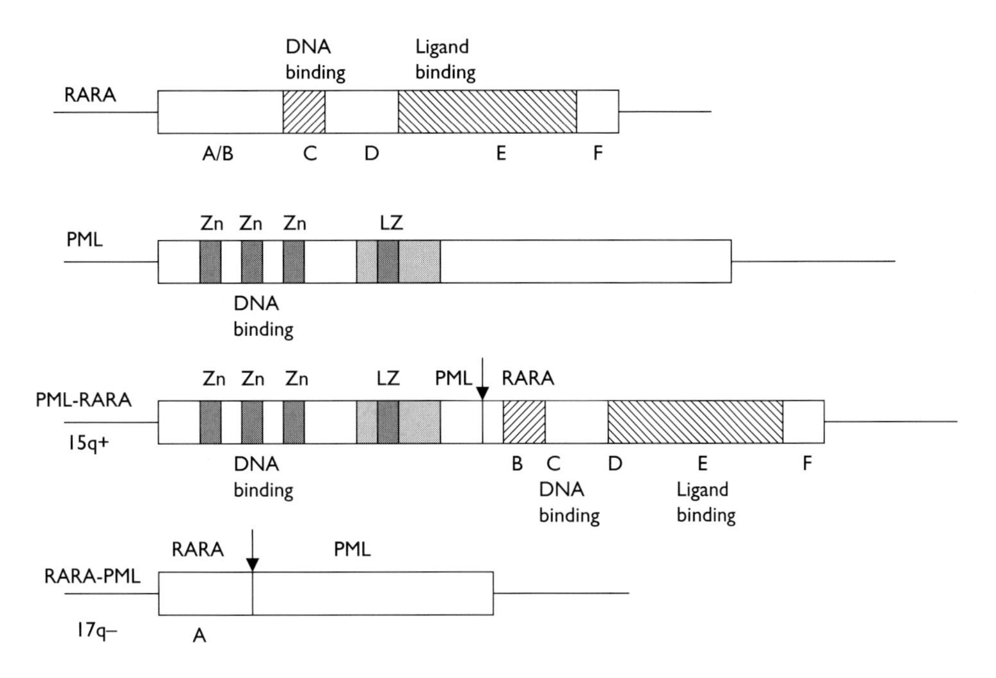

Figure 20.2 Translocation – t(15;17) – between retinoic acid receptor α (RARA) and promyelocytic leukemia gene (PML). RARA has six functional domains (A–F), PML has zink-fingers (Zn) and leucin-zipp (LZ). Translocation may result in two different fusion proteins, from which PML-RARA contains almost all functional domains.

As a result of activation the oncogene will produce either a "new" protein (which therefore could be tumor-specific), or "normal" proteins but in the wrong quantity and at the wrong time. Since activation cause signals stimulating proliferation, these changes could be considered as "gain of function". Some typical consequences of new protein production caused by structural changes are: a) receptors send signals without ligand binding due to mutation or loss of extracellular domain (e.g. ERB-B2, MET, RET); b) signal transducers work continuously (switch off does not operate) (e.g. RAS); c) transcription factors activate genes which they normally do not. The overproduction of normal products may result in: a) amplification (e.g. N-RAS, MYC), or b) translocation (e.g. MYC – where overproduction of MYC can consume all MAX by forming heterodimers, leaving less MAX to form inhibitory heterodimers with MAD or MXL1) (Figure 20.2). It should be mentioned that oncogenes can influence other cell functions besides proliferation (6,7).

20.2.1.2 *Suppressor genes/tumor suppressor genes*

As mentioned before, with the exception of p53, in all suppressor genes both alleles have to be damaged (mainly mutated or deleted) for the manifestation of phenotypic changes (i.e. these gene errors are recessive). The loss of one allele (usually together with a larger chromosome segment) results in the loss of heterozygosity (LOH) (Figure 20.3). Similarly to protooncogenes, suppressor genes also have

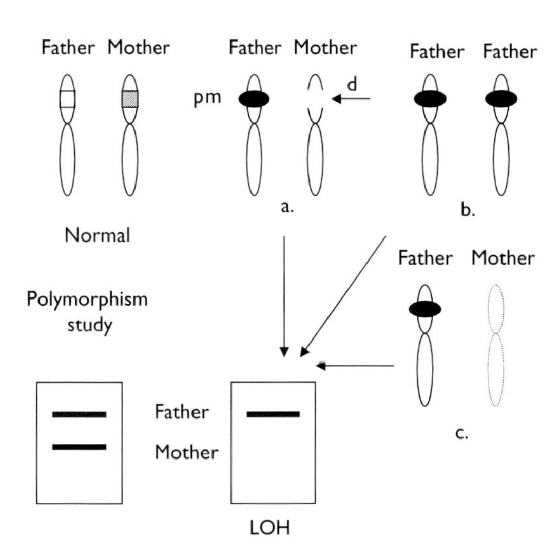

Figure 20.3 Loss of heterozygosity (LOH) in tumors. The first gene error of suppressor genes is almost always point mutation (pm). Loss of heterozygosity (loss of the other allele) may occur by (a) deletion of the allele-containing segment of the chromosome (d – deletion); (b) duplication of the chromosome carrying the mutated allele accompanied by loss of the normal chromosome; (c) loss of the normal chromosome. Chromosomal changes could be detected, e.g., by RFLP (restriction fragment length polymorphism), since the alleles from father and mother are not identical and a restriction enzyme cleaves fragments of different length (heterozygosity). The loss of one allele (only one band remains), or loss of both alleles causes the loss of heterozygosity.

different functions, however they usually represent the other arm of regulation: inhibition of proliferation or induction of apoptosis. They have to be considered in the identification of gene errors, the induction of repair mechanisms, and in the regulation of telomerase or angiogenesis (8,17–20).

The prototype of suppressor genes is the retinoblastoma gene (RB) (its defects are among the most frequent gene errors – besides p53 and p16 – in human malignancies). RB product is pRb (or p105), a typical "pocket-protein" (relatives: p107, p130). In hypo- or non-phosphorylated form, the pRb (and other pocket-proteins) can interact with other proteins, mainly with transcription factors. pRb forms a complex with E2F/DP inhibiting its activating effect on proliferation-supporting genes, i.e. inhibiting the cell in G1 at the R (restriction) point. This is one of the most important check points in the regulation of cell proliferation. When pRb is phosphorylated by cyclinD/CDK4, E2F is released and is able to activate the appropriate genes (the G0-G1 transition is regulated by a family member p130 complexed with E2F4). Target genes for E2F could be the cell cycle genes (cyclinE and A, CDC2, CDC25C), CDK inhibitors (p16, p21), protooncogenes (e.g. MYC, RAS), p107 and E2F1 itself, etc. There are many interactions and regulatory loops at this level, making it rather difficult to estimate the end result of changes in one of the many components. However, it is accepted that the main regulatory pathway of cell proliferation is the MYC/p16/CyclinD/pRb sequence (retinoblastoma pathway) (Figure 20.4).

The loss of pRb activity (due to deletion or mutation in the pocket) essentially cancels the R point, and the cells, even with gene defects, will continue proliferating, doubling the DNA, dividing and transferring the defects to the daughter cells. However, it is more appropriate to say that loss of pRb activity increases the risk of accumulation of gene defects; i.e. the pRb defect alone does not lead to complete misregulation of proliferation (e.g. there are other check points still operating). It seems that the pRb defect is essential ("gatekeeper") in the development of certain tumors (e.g. retinoblastoma, osteosarcoma), while in others (e.g. colon, breast, lung cancer) it is "only" a risk factor. pRb defect is also a prototype for inherited malignancies. In familial retinoblastoma the defect of one pRb allele is inherited (i.e. it is present in germ cells, as well as in somatic cells). To lose pRb function the other allele must be damaged in a somatic (e.g. retinal) cell. Hereditary tumors are generally bilateral and multifocal.

There are many other suppressor genes, e.g. NF1 and NF2, WT1, VHL, BRCA1 and BRCA2, APC, MCC, DCC, PTEN, FHIT, with a wide variety of functions (8). Their defect can occur in different tumors, similarly to oncogenes, however in certain tumors the defect can be considered "specific", e.g. NF1 in neurofibromas, WT1 in Wilms tumor, VHL in renal cell cancer, APC in colon cancer, BRCA1 in breast and ovarian cancer, etc.

20.2.1.3 Cell cycle genes (cyclins, cyclin dependent kinases, cyclin dependent kinase inhibitors)

Cyclins (A, B1–3, C, D1–4, E, F, G, H) operate at different phases of the cell cycle (1). They have a homologous "cyclin box" of about 100 aminoacids to bind to their partners,

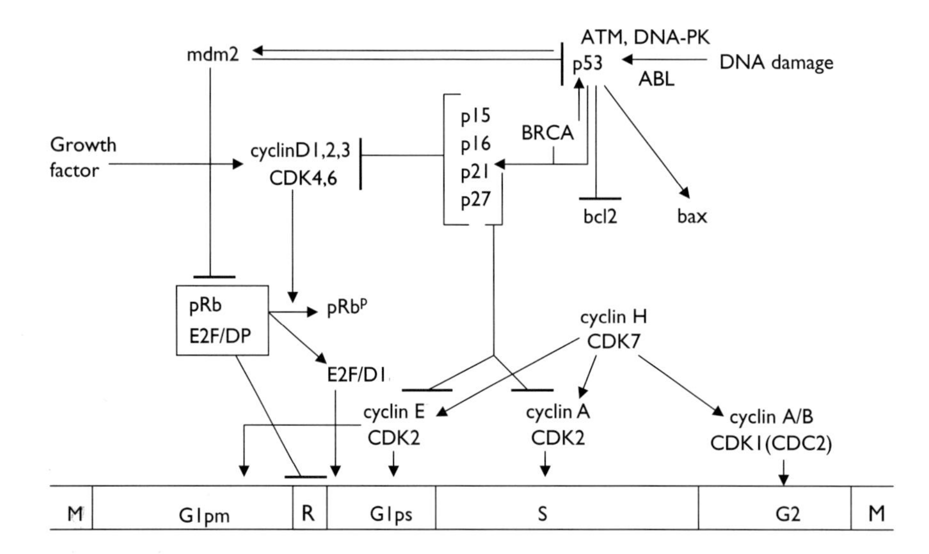

Figure 20.4 Scheme for the p16/cyclinD/CDK4/pRb pathway. In normal cells pRb supervises – mainly – the restriction (R) point. Phosphorylation of pRb by cyclinD/CDK4 releases E2F/DP transcription complex which can activate the genes responsible for cell cycle progression. Errors in this pathway – loss of the function of inhibitors (e.g. pRb, p16, p53) or overproduction of stimulators (e.g. cyclins, CDKs) – lead to insufficiency of the R point, and permanent stimulation of proliferation.

cyclin dependent kinases (CDK). CDKs are activated by complexing with the respective cyclin and by phosphorylation/dephosphorylation. CDKs are inhibited by certain suppressor genes, e.g. $p21^{WAF1/Cip1}$, $p27^{Kip1}$, $p57^{Kip2}$, $p15^{INK4B}$, $p16^{INK4A}$ (most of these can be the target of growth factors, e.g. TGFβ).

The R point is regulated mainly by cyclinD/CDK4 complexes phosphorylating pRb. This action may be inhibited, e.g. by $p21^{WAF1/Cip1}$ activated by p53, or by $p16^{INK4A}$. INK4a (ARF) has an alternatively transcribed reading frame: exons 1β, 2, 3 make $p19^{ARF}$ (human $p14^{ARF}$), while exons 1α, 2, 3 make $p16^{INK4a}$. p16 is active in the retinoblastoma pathway, $p19^{ARF}/p14^{ARF}$ in the p53 pathway, inhibiting MDM2. Further phases of cell cycle fall under the supervision of cyclinE/CDK2, cyclinA/CDK2 – influnced by p21/p53. Entry into mitosis is checked mainly by cyclinB/CDC2.

Defect of p16 is quite common in human malignancies; it is also involved in hereditary tumors such as melanoma and pancreatic cancer. Overexpression of cyclinD1 (PRAD1, BCL1) is relatively frequent in gastrointestinal tumors; translocation involving cyclinD1 (11q13) has been described in lymphomas and parathyroid adenomas (proliferation stimulatory effect of estrogen receptor is often mediated by cyclinD1). Defects of other cell cycle genes or products are hardly known.

20.2.2 *Apoptosis and cancer*

Programmed cell death (apoptosis) is responsible for the removal of unwanted cells, which could be either physiologically useless or damaged but functionally not repaired. Since cancer can be considered as the unlimited accumulation of cells due to the lost balance between cell proliferation and cell death, a defect in the apoptotic program could be a major contributor to cell accumulation (9–11,21).

Loss of apoptotic capacity could influence cancer development and progression, because (1) it produces an "environment" which promotes genetic instability and the increase of gene errors (the cell can cross check-points where the apoptotic program would otherwise be activated); (2) it supports resistance against immunoprotective cells as well as against chemo- or radiotherapy; (3) it helps cell survival in the absence of growth/survival factors and decreases the dependence on oxygen and nutrients.

The apoptotic program is a cascade where a signal activates self-destruction of the cell (Figure 20.5). In most cases these signals (death ligands) bind to their respective receptors (death receptors). The cytoplasmic part of the receptors has a specific sequence (death domain) which transfers the signal through adaptor proteins (e.g. FADD) to one of the initiator procaspases (8 or 10) (Figure 20.5). The initiator (or upstream) caspases will activate the downstream members of the family (caspase 3,6,7) which use their proteolytic capacity to cleave structural components of the cell as well as activate other lytic enzymes (e.g. DNase) for the same purpose.

Besides the death receptor/ligand–caspase pathway, there is an alternative possibility for activation of apoptosis: the mitochondrial–caspase sequence. Under physiological and pathological circumstances certain damage (cytotoxics, radiation, lack of growth factors, ROS) can change the permeability of the mitochondrial membrane which allows the escape of cytochrome C from the intermembranous space. The released cytochrome C forms a complex with Apaf-1 (apoptotic protease activating factor) using dATP and the complex (apoptosome) activates procaspase 9, which transfers the signal to downstream caspases. Permeability of the mitochondrial membrane is regulated essentially by members of the BCL-2 family – comprising pro- and antiapoptotic molecules. The proapoptotics can dimerize with antiapoptotics to inhibit their action (either on membrane permeability or by interfering with other effects, e.g. inhibition of Apaf-1), or they may open the membrane. Certain viral proteins show homology with antiapoptotics preventing cell death, ensuring cell surival and continuous virus production. The connection between the mitochondria and the caspases could be bidirectional: downstream caspases may activate the proapoptotic BID providing the possibility of an amplification loop.

There are many details in programmed cell death which are under intensive study, e.g. different inhibitors of the program, such as decoy receptors, FAP1, FLIP, IAP family, NFκB, etc. Other possibilities also exist for initiation of apoptosis, e.g. via ceramid formation following cell membrane damage, pore formation (by perforin) for granzyme, etc. Recently, a caspase-independent pathway for apoptosis has been described as well.

The relation between apoptosis and cancer is contradictory: essentially any change which works against apoptosis could favour cancer. A classical example is the overproduction of BCL-2 due to the translocation, t(14;18), first described in follicular lymphomas, or inactivation of BAX by mutation (an important target of p53). Lack of nonresponsiveness of death receptors/signals could result in the inactivation of apoptosis. The restoration of apoptosis inducibility in cancer cells is an important strategy in cancer therapy. Such approaches are: (a) inhibition of antiapoptotic members of the BCL-2 family either by antisense oligos or by small binding proteins; (b) inhibition of IAP family members to save caspase activity; (c) inhibition of certain protein kinases (e.g. AKT phosphorylates and inactivates pro-apoptotic BCL-2 proteins) or inhibition of NFκB.

It is important to emphasize that genetic instability, which is a characteristic feature of cancer cells, would support sensitivity towards apoptosis, but this, unfortunately, is overruled by a strong antiapoptotic effect (e.g. a survival signal). The removal of this signal seems to be essential for revitalization of the apoptotic program.

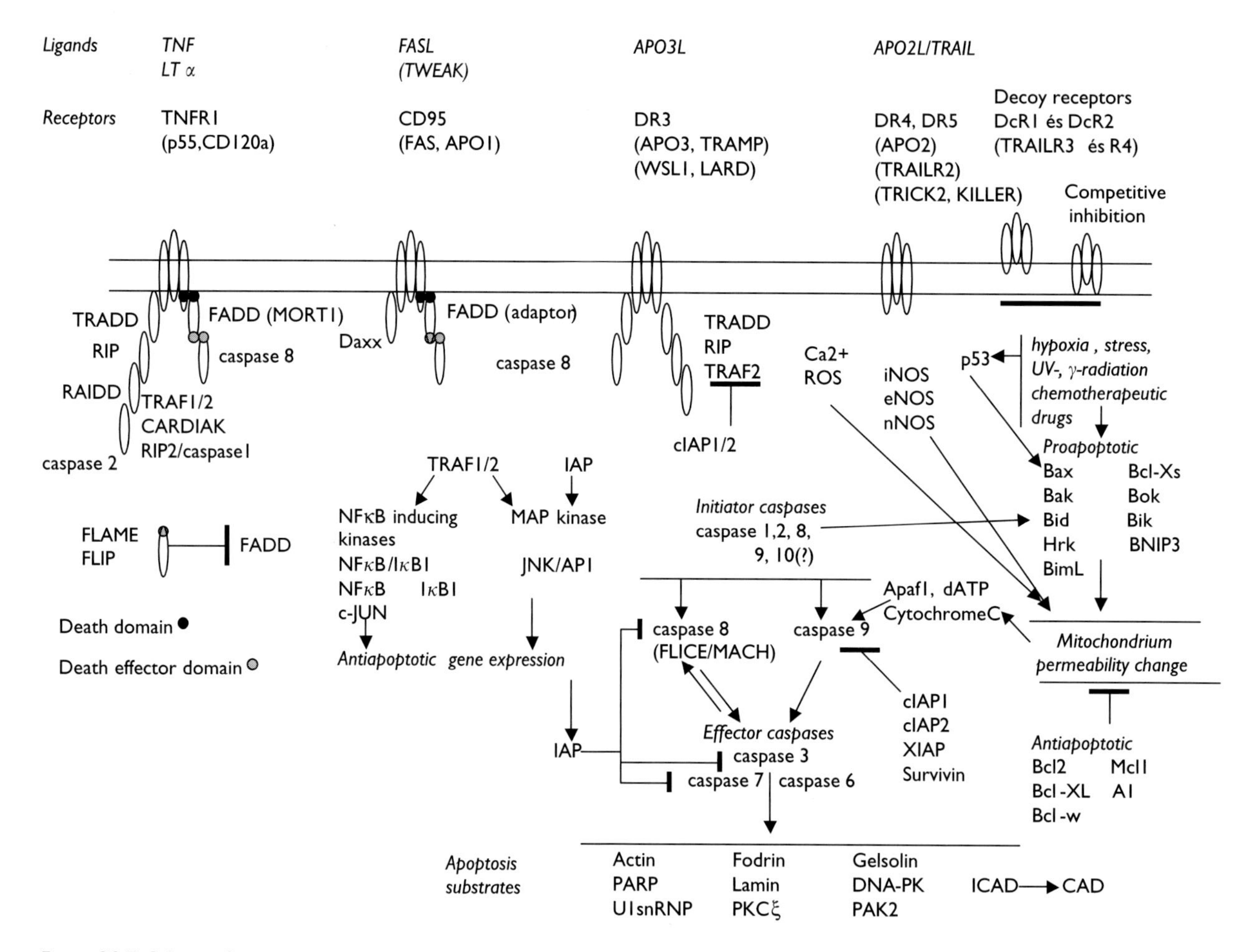

Figure 20.5 Scheme for apoptosis.

20.3 Regulation of gene expression and DNA repair

20.3.1 Methylation

Methylation is a key regulatory step of many chromosomal processes (2). The term methylation refers to the addition of a methyl group to the cyclic carbon 5 of a cytosine nucleotide. This reaction is catalyzed by a conserved family of DNA methyltransferases and supervised by specific transcription repressors. Normally, a methyl-binding protein attaches to methylated CpG units and brings together histone deacetylases and the corepressor protein mSin3A. This complex contributes to the formation of repressive chromatin states and deacetylation of histone-proteins involved in the packaging of DNA in the nucleus. Basically, the methyl group tags a gene so it is turned off, and an unnecessary protein product is not produced in that particular cell. For instance, one of the two X chromosomes or one allele of imprinted genes is inactivated by methylation. Typically 60–65% of cytosine-guanine dinucleotide sequences (CpG islands) are methylated in mammalian adult somatic cells (Figure 20.6).

In cancer, the error in methylation could be a) global hypomethylation (promoting genetic instability and an increase in mutation rate) and b) hypermethylation of otherwise non-methylated CpG islands (CIMP: CpG island methylator phenotype) (Figure 20.6). There are two forms of CpG hypermethylation: a) age-related (A type) (e.g. estrogen receptors in colon cancer, and probably in breast cancer; IGF2, N33, MyoD, versican) – this type is similar to MMR problems (see below), i.e. many loci are involved, but few of them are functionally relevant; b) cancer-related (C type) which involves suppressor genes, repair genes and other genes. For instance:

- Suppressor genes – retinoblastoma gene (only in sporadic retinoblastoma), VHL gene (in clear cell type renal cell cancer, hemangioblastoma), p16 (CDKN2A; one of

Normal cell

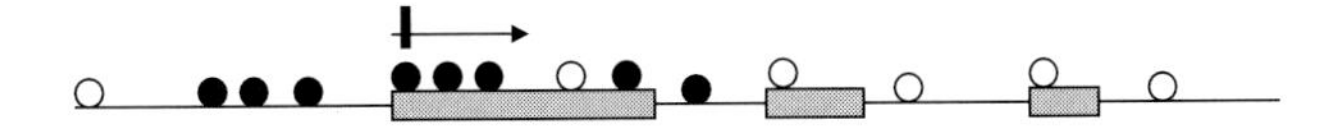

Cancer cell

Figure 20.6 In normal cells a typical gene contains CpG islands, and exons shown by boxes. The open circles represent the unmethylated sites, filled circles the methylated CpG sites. The CpG islands are usually non-methylated (transcription is allowed) while the rare CpG sites outside the island are heavily methylated. In cancer cells loss of methylation in the non-island regions occurs with hypermethylation of CpG sites within CpG island (resulting in blocking of transcription).

the most frequent genetic errors in human malignancies, but rare in SCLC (small cell lung cancer) and in certain leukemias), p15 (CDKN2B; interestingly, it is frequent where p16 methylation is rare, e.g. ALL, AML, Burkitt's lymphoma), BRCA1 (?), p53 related genes (p73, p14ARF);

- DNA repair genes – glutation-S-transferase, π type (prostate cancer, breast cancer, renal cell cancer), O6-methylguanine methyltransferase (lung cancer, colon cancer, lymphomas, brain tumors), MLH1 (colorectal cancer, endometrial cancer, gastric cancer);
- Other genes – E-cadherin (in lobular breast cancer but not in ductal form), steroid receptors (estrogen, androgen, retinoic acid receptors), apoptosis related genes (e.g. γ-interferon mediated pro-apoptotic DNA kinase), endothelin B receptor, thrombospondin, CD44, FHIT, TIMP3, etc.

20.3.2 Genetic instability – DNA repair

Essentially malignancies are genetically unstable, this is the "engine" of heterogeneity and progression. The stability of the genome requires (a) molecular sensors to identify damage, and (b) effective DNA repair.

20.3.2.1 Identification of damage – TP53

DNA, the carrier of genetic information, can suffer innumerable damage either during replications, recombinations, rearrangements, or due to endogenous and exogenous – physical, chemical, biological – genotoxins. The options of the cell to deal with the damage are either to repair it, or, if the repair is incomplete, to induce apoptosis – especially when the damage hits a crucial regulatory gene. (Obviously, the damage is variable – it could be lethal or could improve survival in a given environment). Anything which interferes with the repair mechanisms – both the identification and the actual repair of the damage – may increase the risk of accumulation of gene errors.

There are several genes and gene products which take part in DNA repair. Here only the famous p53 (Molecule of the Year) and the repair enzymes will be introduced.

TP53 (suppressor gene, 17p13) – product, p53 (transcription factor) – is a central player in the maintenance of genomic integrity. P53 action is regulated by the intracellular location, active and inactive conformation, and the intracellular level (influenced by the life-span). Again, there are numerous factors which can influence p53 activity (Figure 20.7). HMDM2 inactivates p53, but p53 activates HMDM2 – resulting in an autoregulatory loop; this interaction is influenced by p14ARF, an alternative product of INK4A, antagonizing HMDM2. Other important members of this orchestra interplaying with p53 are cMYC, ABL, ATM and BRCA1.

p53 behaves either as a molecular sensor alone, as part of a larger damage-recognizing complex (e.g. TFIIH), or collects the messages from other sensors. P53 is activated by phosphorylation at different sites depending on the nature of the damage, which is performed by different kinase families (DNA-PK, casein kinases, MAP kinases, SAP kinases, cyclin dependent kinases). Activation results in stabilization (longer life-span) and increased transactivation capacity. Inactivation is due to dephosphorylation (complex-formation see below), transport to the cytoplasm, and digestion by proteasomes.

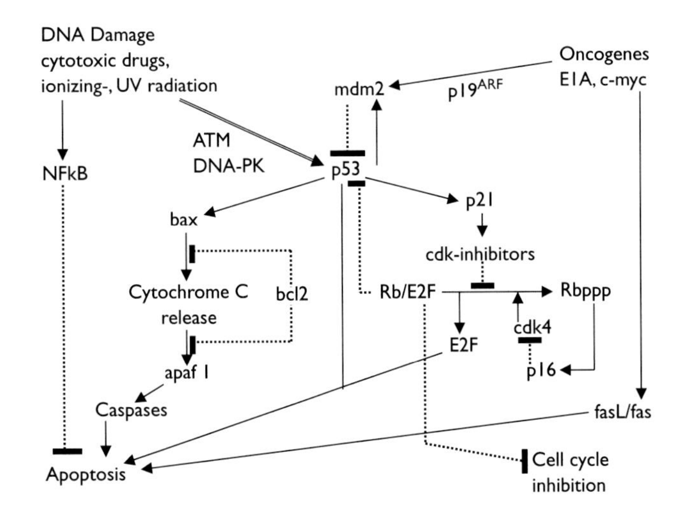

Figure 20.7 The connection of p53 with different cell functions.

The main task of p53 after "sensing" DNA damage is (a) to stop cell proliferation (stop transition of the cell in cell cycle) giving time for repair, (b) in the case of unsuccessful repair to switch on the apoptosis program. In both cases the aim is to avoid the transfer of DNA damage to daughter cells, which could destabilize the genome, allowing accumulation of more damage, and leading to irreversible regulatory chaos. (One must not forget, that although DNA damage can lead to an imbalance between cell proliferation and cell death, the capacity for survival should be untouched in order to progress towards transformed geno- and phenotype). The main effectors of p53 that fulfill these tasks are the WAF1/Cip1 gene (6p21.2, product: $p21^{waf1/cip1}$) for halting proliferation (but $p21^{waf1/cip1}$ can be activated without p53); and BAX for inducing apoptosis (there are many p53 dependent participants of the apoptosis program).

The main reasons for p53 dysfunction are (a) TP53 gene damage, and (b) inactivation of p53. TP53 is mutated in the germ cells in the inherited Li-Fraumeni syndrome (increased risk for cancer formation). In about 50% of human malignancies p53 is either mutated or deleted. Most of the mutations (more than 3,000 are described) are located in the 5–8 exons, between 130–290 codons, and these are mainly missense mutations (Figure 20.8). The mutated p53 can form a complex that inactivates the wild type p53. The normal "condition" for p53 is tetramerization. This is the reason why the effect of p53 damage is dominant in contrast to the damage of other tumor suppressor genes which appears to be recessive. It is important to know that normal p53 can be inhibited by viral proteins, in the same way as pRb.

Recently, relatives of p53 have been described, p73 and p63, which show considerable sequence homology, and share many functions of p53 – except the role in DNA repair.

20.3.2.2 DNA repair

The instability of the genome due to repair problems could be manifest at (a) nucleotide level (base-replacements, deletions, insertions – involving some nucleotides), and (b) at chromosomal level (14,15).

NUCLEOTIDE/MICROSATELLITE INSTABILITY (NIN, MSI)

Genetic instability at nucleotide level involves one or few base pairs. This is rare in human tumors but the consequences could be serious. The reason is the inefficiency of repair systems. Nucleotide excision repair (NER) problems, known as nucleotide instability (NIN), mainly appear in skin tumors (e.g. following xeroderma pigmentosum). The mismatch repair (MMR) problems (identified as MSI, microsatellite instability) were first observed in colon cancer (especially in HNPCC, hereditary non-polyposis colon cancer), in HNPCC-related tumors, such as stomach and endometrial cancer (~10–15%), but in fewer than 2% in other tumors. MMR genes belong to the PMS family (hMLH1, hPMS1, hPMS2), and to the MSH group (hMSH2, hMSH3, hMSH6). A classic example of MMR damage is HNPCC, where MIN is identified in 80% of cases, and germ cell mutations of hMSH2 and hMLH1 in 30–30% of the HNPCC families. Such mutations are relatively rare in sporadic colon cancer, however the frequent inactivation of hMLH1 due to the methylation of its promoter has recently been discovered. The role of alternative transcripts is still in question.

The "instability" of the genome can be tested using microsatellites, highly polymorphic tandem repeats of single nucleotide sequences, located by chance in the genome (usually 10–50 copies of 1–6 base-pairs). The replication of microsatellites is very exact, therefore a proper marker panel

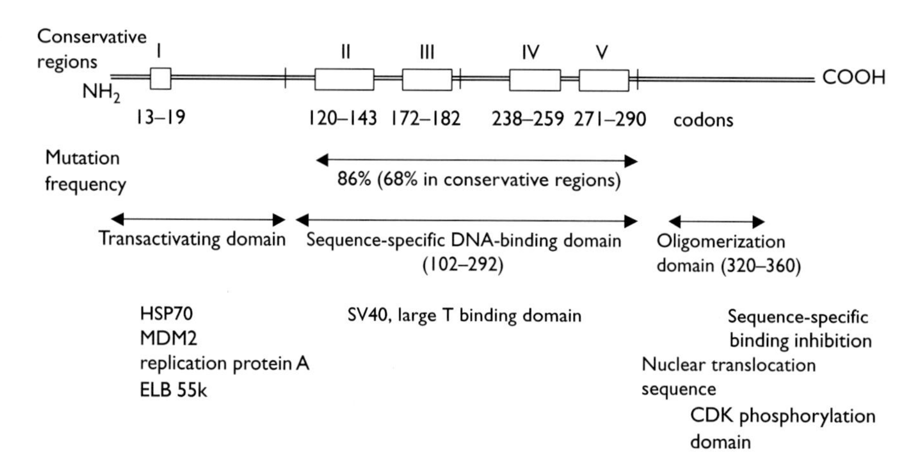

Figure 20.8 Structure and functional elements of the p53 protein.

must be used to judge genomic instability upon the mutation rate of microsatellites. (One has to take into account the spontaneous mutation rate of a given locus). Conventionally the cases can be divided into MSI-H (high) and MSI-L (low) groups, where only H has real significance, the difference between L and MSI negative cases is doubtful. MSI, besides errors in the repair, could be the result of exogenous damaging agents, producing active oxygen radicals as well as lipid adducts.

It was observed, that the MMR error per se does not lead to higher mutation rates. Such errors may occur in normal cells and the change in the microenvironment can "help" to realize this mutator phenotype. It is also important that loss of heterozygosity (LOH) and MSI exclude each other. LOH (mainly related to damage of suppressor genes) is frequent in MSI negative tumors and the opposite is true for MSI positives.

CHROMOSOMAL INSTABILITY (CIN)

It has long been observed that chromosome number (DNA content) may be abnormal (aneuploid) in tumor cells. In colon cancer most of the MSI+ tumors are diploid and MSI- are aneuploid indicating that MSI and CIN probably rule each other out. During progression not only may the number change, but the karyotype also becomes heterogenous, new chromosome variations arising all the time. The reason for CIN is error in the check-points, which otherwise should supervise the location of chromosomes along the mitotic spindle before chromatid separation. Such a problem may occur from errors in genes hMADs and hBUBs. The existence of DNA damage may also increase the risk of mitotic recombination and produce abnormal chromosomal segregation. Translocations are also caused by disturbances in the chromosomes. Complex translocations occur in many tumors, however accidentally, and can differ from the translocation of a similar tumor. The problem is that the cells begin mitosis before the DNA double strand breaks are repaired. The single type of translocation is characteristic of a given tumor type (most of them develop in leukemias and lymphomas). These problems are probably not the consequence of a specific genetic instability, but rather errors of the normal recombination process influenced/regulated by RAG proteins (Figure 20.9).

Amplification is a form of genetic instability and involves only a few genes (while NIN, MSI and CIN involve many genes). In normal cells the amplicon (the multiplied part of the genome) can induce p53-dependent apoptosis, but the cancer cell can survive even with the amplicon.

20.3.3 Telomerase

At birth, the ends of chromosomes (telons, telomeres) contain TTAGGG repeats of about 15,000 base pairs long (4). Telomeres protect the ends of the chromosomes, prevent recombinations, and promote the separation of sister chromosomes during mitosis. At the same time they are the targets of telomere-binding proteins (Figure 20.9).

The gradual shortening of telomeres can behave as a molecular clock. Each cycle of mitosis loses 25–200 base pairs from the end of the telomere, because DNA polymerase is "lazy" and does not synthesize this region. After about one hundred mitotic cycles the cells stop proliferation and reach the stage of senescence. However, reproductive tissues should have a compensatory mechanism that involves the telomerase, a complex containing the template for the telomere and an enzyme, a terminal transferase (telomerase), i.e. ribonucleoprotein reverse transcriptase. This is responsible for synthesizing the sequence missed by DNA polymerase. Continuous telomerase activity would mean that the cell, and its descendants, could live indefinitely (i.e. they would be immortalized), provided nothing else other than the telomere stopped proliferation. Telomerases are active in fetal tissues, reproductive and renewal tissues, inflammatory (clone-forming) cells, and in most if not all malignant cells. The regulation of telomerase activity is still mostly unknown. Activity is probably inhibited by

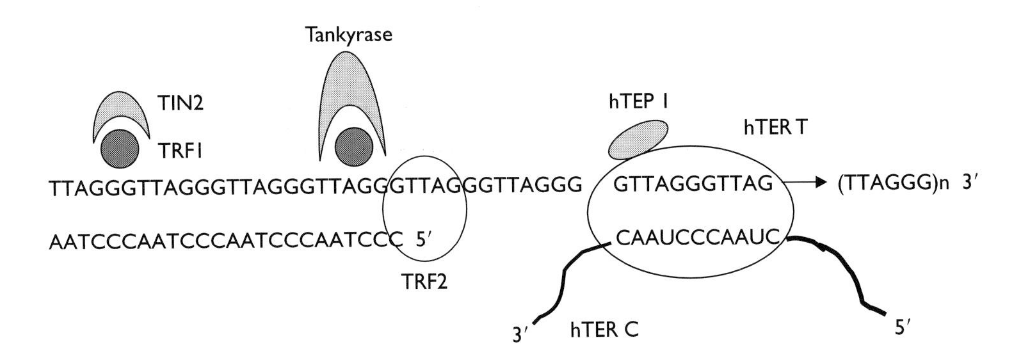

Figure 20.9 The G-rich strand is extended by the telomerase subunits hTERC and hTERT. The C-rich strand is elongated at its 5′ ends by DNA polymerase. Access of telomerase complex to the 3′ end may be regulated by telomere binding proteins.

a product of a gene located in 3p as well as by those proteins involved in the repair of dsDNA breaks (e.g. RAD50, MRE 11, Ku heterodimer).

The role of telomeres/telomerases in senescence and in cancer can be underlined as follows: (a) in somatic cells, including stem cells of renewing tissues, progressive telomere loss is a normal phenomenon; this is the mechanism that regulates cell senescence; (b) cell senescence occurs when one or more telomeres reach a critical length (or shortness); (c) telomerase is inactive (inhibited) in most somatic tissues; the cessation of inhibition results in activation of the enzyme; (d) telomerase or some other mechanism maintains telomere stability in cancer cells – this phenomenon is characteristic of all immortalized tumor cell populations.

However, the link between telomerase activity and cancer is still controversial. Many observations have shown that telomerase activity is lower in premalignant conditions than in the respective invasive cancer which means that telomerase activity could be a warning signal, but does not itself determine the biological behavior. In other studies, telomerase activity was present either in all malignancies, or only in certain cases. Comparing telomerase activity in the washing fluid (desquamated cells) from both normal and cancer patients, the sensitivity proved to be 60–70%, specificity 90–100%. How telomerase activity could be used in prevention, or as a target for therapy is still in question.

20.4 Cancer growth and progression

20.4.1 Tumor growth

In general, the most characteristic feature of neoplasia is the continuous growth in space and in time. Malignancy is further characterized by the progression (invasion and metastasis) of the tumor. (Although these statements are valid in most cases, there are exceptions, e.g. in rare examples of spontaneous regression, growth may stop for a shorter or longer period). In solid tumors the growth rate is best described by *tumor volume doubling time* (Td) which is influenced by three factors: cell cycle time (Tc) of proliferating cells; proportion of the proliferating cells (GF, growth fraction); and cell loss rate (ϕ) of the tumor (13) (Figure 20.10). (Besides these "tumor-related" factors, the change in stromal elements can also modify the volume). In human tumors Td ranges from days to months. In cell cycle the most variable period is G1. The most important feature of the tumor cells (from a clinical point of view) is their capacity to temporarily leave the cycle but maintaining the potential to "step back" and proliferate (serving as a source for early or late, usually therapy-resistant, recidivas). It is important to note that tumors are not the most rapidly dividing tissues in the body (e.g. normal bone marrow may have a higher proliferation rate than marrow-originated leukemia). Chemotherapy is directed mainly against proliferating cells, but this action is not specific therefore side effects are to be expected especially in continuously proliferating normal tissues such as bone

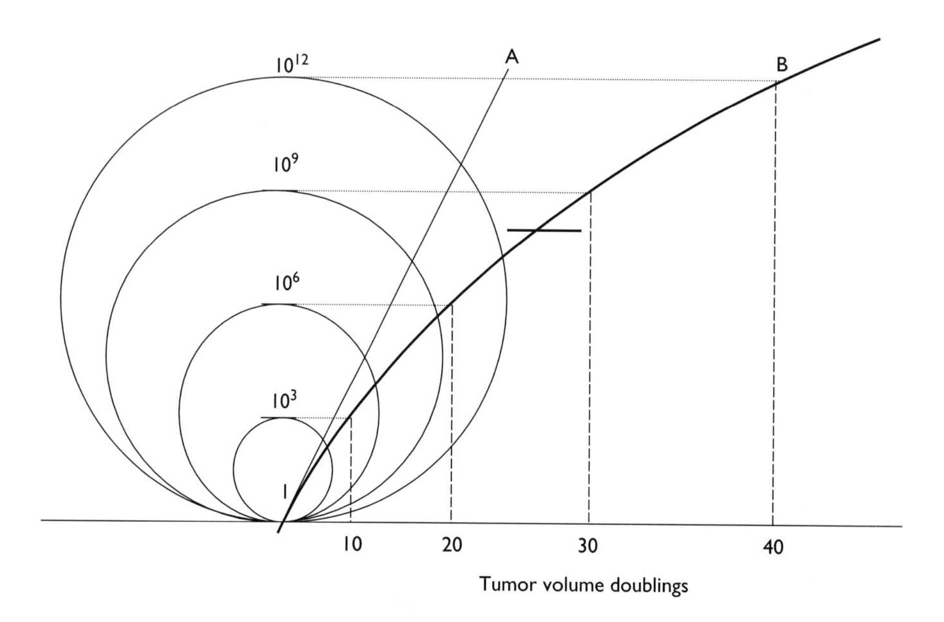

Figure 20.10 Line "A" shows the "ideal" growth curve, "B" represents the curve which is much closer to reality. If ten doublings of cell number results in a three-magnitude increase in cell number, it means that the tumor has already passed a significant part of its life-time at a size (short, thick line ~10^8 cells) which limits clinical discovery.

marrow, epithelium of the intestine or skin, etc., depending on the age of the patient.

Tumors have two components: *parenchyma* (tumor cells) and *stroma* (matrix with vessels, host cells). Stroma is essential for tumor growth (even in hemopoietic malignancies). It provides mechanical support, and conditions for survival and progression. Such conditions include vessel neoformation (angiogenesis) and production of substances which can promote progression (e.g. proteolytic enzymes). In general, the intratumoral host elements – tumor-infiltrating lymphocytes, macrophages, fibroblasts, myofibroblasts, endothelial cells – support the tumor cells instead of acting against them; how tumor cells "manipulate" these normal cells is a key question).

Angiogenesis is probably the most important contribution of the host to the survival of the tumor (5). There are two types of vessels: a) already present, invaded by tumor cells; b) newly formed (i.e. proliferation of endothelial cells under the influence of angiogenic and anti-angiogenic substances, including some protooncogenes and suppressor genes). Tumor cells can also form vascular channels, either by de novo transformation, or after the destruction of the endothelial lining of existing vessels. The structure of intratumoral vessels is more venous-like, and their increased permeability ("leaky" tumor) may lead to accumulation of interstitial fluid. (Due to structural differences the tumor cells may respond differently to certain effects (e.g. heat), than normal vessels; the resorption of increased interstitial fluid is slow due to the lack of lymphatic vessels, making the penetration/extravasation of drugs rather difficult). Angiogenesis can be influenced by extracellular matrix components (especially proteases) and certain integrins. One of the most important factors that can influence angiogenesis, as well as the response to therapy, is hypoxia (e.g. HIF-1 can stabilize VEGF, p53; or can be blocked by pVHL). There are several approaches to blocking angiogenesis, e.g. by inhibiting angiogenesis (anti-VEGF, IL-12, Vitaxin), receptors (SU 5416, SU 101), proteolysis (Marimastat, AG 3340, BAY 12-9566), and other drugs (e.g. thalidomide). It also seems "easier" to specifically kill hypoxic cells rather than make them more oxygenized. Hypoxia-selective cytotoxics are currently undergoing clinical trial, e.g. quinon antibiotics, nitroimidazols, tirapazamine (Figure 20.11).

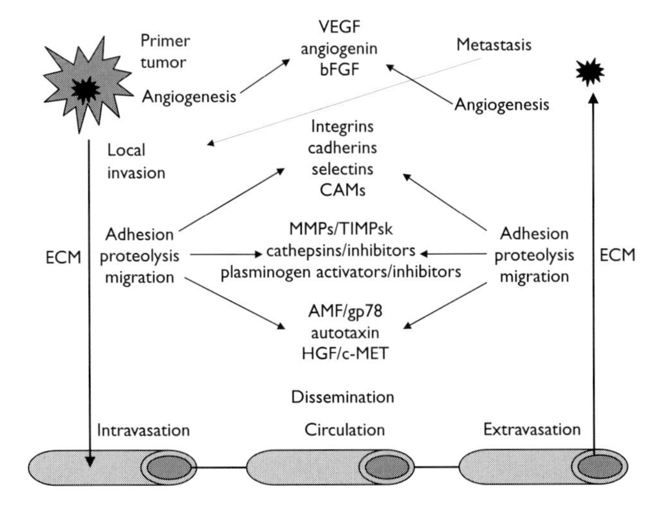

Figure 20.11 Scheme for invasion and metastasis.

20.4.2 Tumor progression (invasion and metastasis)

The biological significance of a tumor is essentially dependent on its capacity to leave the primary tumor, invade the surrounding tissues and make metastases in distant organs (3). This is the true meaning of malignancy.

20.4.2.1 Invasion

Although cells can be passively pressed into the vessels during tumor growth, in most cases invasion is an active process. The cells should be able to "open" the matrix and move further (migrate) using alternative events of adhesion and detachment.

Adhesion is a key step during tumor progression and is influenced by many factors. The homophilic cell-cell contacts are mainly regulated by cadherins, the heterophilic contacts (e.g. tumor cells and endothelial cells) by members of the immunoglobulin superfamily and/or by selectins, and cell-matrix interactions by the integrins.

Cadherins are transmembrane glycoproteins, the cytoplasmic domain binds to catenins (α,β,γ). Loss of E-cadherin or α-catenin function, mainly due to mutation, results in loss of intercellular adhesion between epithelial cells. In the case of mutant APC the level of β-catenin can increase, stimulating proliferation and inhibiting apoptosis through the β-catenin-TCF pathway. (This pathway can be activated by the mutation of β-catenin).

Selectins may contribute, similarly to leukocytes, to the binding of tumor cells to the endothelial surface. In the next step, the interaction between members of the Ig-family, e.g. ICAM-1 and -2, V-CAM, PECAM, may help tumor cells to pass through the vascular wall by binding to integrins on the surface of the tumor cells.

Integrins, adhesion and signaling membrane receptors influence many cellular functions. These capacities are relatively dynamic, regulated by intracellular signals (inside-out) or by extracellular signals (outside-in). Integrins have α and β subunits, forming heterodimers; the extracellular ligand-binding domain is an RGDS sequence. The integrin-mediated interaction of endothelial cells and epithelial cells with the extracellaular matrix is essential for survival. Integrin signals can use FAK as well as ILK kinases, and

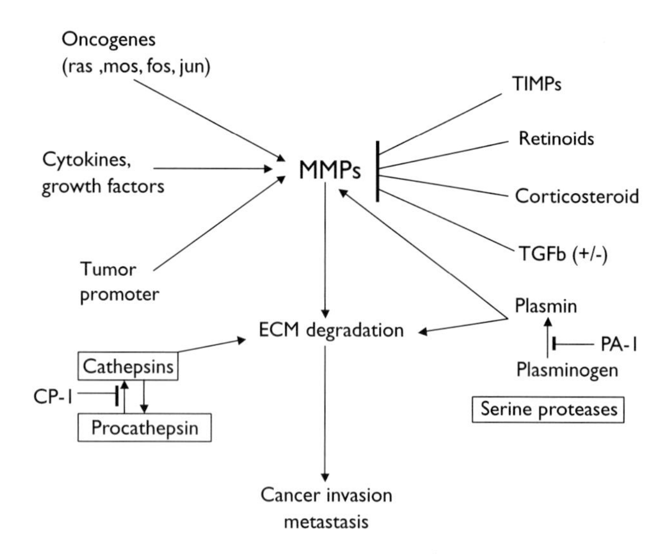

Figure 20.12 Stimulators and inhibitors of matrix metalloproteinases.

are regulated (inhibited) by PTEN (a phosphatase). Changes in the expression of various integrins (increase: $\alpha v\beta 3$; decrease: $\alpha 5\beta 1$) has been described in many human tumors, however the biological significance of these changes is still unknown.

Motility/migration is promoted by the change of cytoskeletal elements as a response to different signals, produced by tumor cells, e.g. autocrine motility factor, paracrine cytokine as hepatocyte growth factor (scatter factor).

Proteolysis, causing decomposition of the matrix (including the basal membrane), can be attributed to three enzyme families: serine proteases (urokinase plasminogen activator, elastase, plasmin, cathepsin G), matrix metalloproteases (gelatinases, interstitial collagenases, stromelysin, matrilysin), cysteine proteases (cathepsin B and L). Regulation of enzyme activity is complex, and is influenced basically by the balance between activating and inhibitory factors (Figure 20.12). For example, matrix metalloproteases (Zn-dependent endopeptidases) are activated from their proform by extracellular proteases, such as plasmin, and inhibited by specific tissue inhibitors (TIMPs). All of these compounds can be produced either by tumor cells, or by intra- or peritumoral normal cells.

20.4.2.2 *Metastasis*

Tumor cells can reach both types of circulation (lymphatic and hematogenic) which have many interconnections. Anatomically the lymphatic spread first reaches the regional lymph nodes (the design of "loco-regional" treatment should take this into account), while the hematogenous spread has different forms depending on the venous or arterial connections, producing mainly hepatic and lung metastases, or systemic, most frequently brain and bone metastasis, respectively. While most tumor cells die in the circulation (and in this respect metastasis is an inefficient process), some can successfully traverse the vessel wall again, and be arrested in a new environment. The next question is, how the tumor cells are supported by that microenvironment (sometimes, unfortunately, it can provide exeptional hospitality, resulting in organ/site specific metastases). The response to these local signals, together with autonomous regulation of the tumor cells, will decide the survival, i.e. the performance of those actions that are characteristic for the primary foci (proliferation, angiogenesis, stroma formation, etc.). Obviously, metastases could be the source of further progression, following the steps of the metastatic cascade.

From the prognosis point of view, the potential development of dormant cells is important (12). These tumor cells can "sit and wait" forming a tiny, but not growing, population of tumor cells, and can be activated months or even years after settlement (e.g. in breast cancer) and can produce late metastases. The identification of residual tumor cells after – seemingly efficient – therapy, has great significance.

It is still an open question whether metastatic capacity is caused by a genetic program or mainly by epigenetic changes. There is probably no genetic lesion which could be responsible for all steps of the metastatic cascade, although some have been suspected (e.g. NM23, KiSS1, KAI1, E-cadherin, MKK4). It is more conceivable that different gene errors can support the completion of certain step(s) of the metastatic cascade, and these errors may be present in the primary tumor or may occur later. Nevertheless, the series of events during tumor progression are accompanied by the continuous selection of cells as a result of interactions between tumor cells and their environment.

20.5 Host-tumor relationship

20.5.1 *Immunological aspects of cancer*

According to the immune surveillance theory one of the basic functions of the immune system is to identify and eliminate transformed cells before they can form a tumor. Although this activity could be very efficient against many potentially tumorous cells, the onset and growth of malignancies indicate the failure of the immune response in such cases.

There is a long list of "strategies" for tumor cells to escape from immune surveillance. For example, a) sneaking through the immune barrier (experimentally proven by repeated injection of small amounts of tumor cells); b) rapid proliferation or growth in a "protected" anatomical site (which immune cells usually cannot reach); c) there is no immunogenic structure (antigen or presenting MHC) on the tumor cells; d) immunogenic structure is hidden by other surface molecules; e) the tumor cell produces soluble factors (ligands and/or receptors – e.g. death receptors, cytokine receptors) which can interfere with immune effectors (or can even stimulate them to support tumor growth – e.g. by producing stroma forming, angiogenic factors).

In order to take advantage of the potential antitumor capacity of the immune system we should know more about: a) how active the immunological defense in a given patient is against the given tumor; b) what the efficiency of the effector cells is and what inhibitory factors are present; c) what factors paralyze the immunogenicity of the tumor cells; d) what biologically active substances (e.g. cytokines and receptors) are produced by the tumor cells and what we can expect in this microenvironment when, e.g., exogenous cytokines are given.

The basic antitumor immune defense is a long-lasting protection offered by a T-cell specific complex effector function involving regulator and helper CD4+ and CD8+ cells. The onset of such defense can be expected only when tumor antigens are presented by professional antigen-presenting cells to the T-cells. The best presenters are the dendritic cells (DC) since they have MHC I-II structures as well as co-stimulators (B7.1, B7.2, CD40, etc.) and can activate the relevant T-cells. It is possible to isolate autologous DC cells from peripheral blood (using CD34+ cells and growth factors) *in vitro* exposed to tumor antigens (tumor lysate) and to use these for vaccination.

An essential precondition of antitumor action is the presence of those phenotypic structures on the tumor cells which could be recognized by the immune system, primarily the cellular immune defense. In most human tumors it has been shown that the main problem is not the lack of tumor epitopes, which would be identified by CD4+ and mainly CD8+ T-cells, but the inability of this recognition to induce a long-lasting immune response to inhibit tumor progression. In other words, an effective immune response would require recognition of a "self" antigen – self, because the tumor epitopes are very similar to host epitopes or slightly modified (e.g. mutated) – and destruction of tumor cells by an "autoimmune" reaction (directed exclusively against the tumor cells).

Identification of antigens serving as epitopes for CD4+ helper cells opened the way for more effective vaccination and immunotherapy. Recent experiments, especially with melanoma patients raise several questions: are melanoma epitopes which can activate CD4+ helper cells individual (self-antigen of a given patient), or tumor specific (here: melanoma specific) or shared by many tumors? Such specific antigens could be, e.g., MART1/Melan-A (gp100) tyrosinase, or the MAGE family in melanomas, or the products of the RAGE gene in renal cell carcinoma.

Before the availability of "tumor-specific" epitopes, immunotherapy used "cell-based" vaccination (complete, lysated or irradiated tumor cells). This approach enabled the introduction of certain genes (primarily cytokine genes, MHC I–II genes, co-stimulator coding genes) to increase immunogenicity of the tumor cells and provoke more effective local T-cell response. Experience with this strategy is rather limited.

Huge efforts were invested into producing monoclonal antibodies against malignancies, however the advantage of this approach is still questionable. Improvement is expected from the use of bifunctional monoclonal antibodies. Similarly, adoptive immunotherapy using mainly peripheral T-cells, killer cells from the patient or TILs (tumor infiltrating lymphocytes) did not produce clinically useful results, although there are specific conditions (especially in leukemias) where adoptive immunotherapy could be useful.

As a whole, we are still just beginning to understand immunological aspects of host-tumor interactions. The application of this knowledge in clinical practice is in the experinmental phase, both in the prevention and therapy of tumors.

References

1 Sherr, C.J. (2000) The Petzcoller Lecture: Cancer cell cycles revisited. *Cancer Res 60, 3689–3695.*

2 Herman, J.G. (1999) Hypermethylation of tumor suppressor genes in cancer. *Seminars in Cancer Biology. 9, 359–368.*

3 Meyer, T., Hart, I.R. (1998) Mechanisms of tumour metastasis. *Europ J Cancer34, 214–221.*

4 Dhaene, K., Van Marck, E., Parwaresch, R. (2000) Telomeres, telomerase and cancer: an update. *Virchows Arch 437, 1–16.*

5 Jones, A., Harris, A.L. (1998) New developments in angiogenesis: a major mechanism for tumor growth and target for therapy. *The Cancer J 4, 209–215.*

6 Schwab, M. (1999) Oncogene amplification in solid tumors. *Seminars Cancer Biology 9, 319–325.*

7 Aman, P. (1999) Fusion genes in solid tumors. *Seminars Cancer Biology 9, 303–318.*

8 Yang, X., Lippman, M.E. (1999) BRCA1 and BRCA2 in breast cancer. *Breast Cancer Res Treat 54, 1–10.*

9 Lowe, S.W., Lin, A.W. (2000) Apoptosis in cancer. *Carcinogenesis 21, 485–495.*

10 Kroemer, G., Dallaporta, B., Resche-Rigon, M. (1998) The mitochondrial death/life regulator in apoptosis and necrosis. *Annu. Rev. Physiol. 60, 619–642.*

11 Bamford, M., Walkinshaw, G., Brown, R. (2000) Therapeutic applications of apoptosis research. *Exper Cell res 256, 1–11.*

12 Hart, I.R. (1999) Perspective: tumour spread – the problems of latency. *J Pathol 187, 91–94.*

13 Steel, G.G. (1997) *Growth kinetics of tumours.* Clarendon Press. Oxford.

14 Lengauer, C., Kinzler, K.W., Vogelstein, B. (1998) Genetic instabilities in human cancers. *Nature 396, 643–649.*

15 Pihan, G.A., Doxsey, S.J. (1999) The mitotic machinery as a source of genetic instability in cancer. *Seminars Cancer Biology 9, 289–302.*

16 Mihich, E., Croce, C.M. (1999) Ninth Annual Pezcoller Symposium: The biology of tumors. *Cancer Res 59, 491–497.*

17 Freedman, D.A., Levine, A.J. (1999) Regulation of the p53 protein by the MDM2 oncoprotein. *Cancer Res 59, 1–7.*
18 Tamura, M., Gu, J., Tran, H., Yamada, K.M. (1999) PTEN gene and integrin signaling in cancer. *JNCI 91, 1820–1828.*
19 Brown, J.M., Wouters, B.G. (1999) Apoptosis, p53 and tumor cell sensitivity to anticancer agents. *Cancer Res. 59, 1391–1399.*
20 Hussain, S.P., Harris, C.C. (1998) Molecular epidemiology of human cancer: contribution of mutationm spectra studies to tumor suppressor genes. *Cancer Res 58, 4023–4037.*
21 Reed, J.C. (1999) Dysregulation of apoptosis in cancer. *J Clin Oncol 17, 2941–2953.*

Chapter 21

Cancer as a communication disorder

György Kéri and Axel Ullrich

Contents

21.1 Overview

Cancer is a very complex disease, however, if we want to make a simple and clear-cut statement about the molecular pathomechanism of cancer we can state that cancer is a communication disorder, the result of an intracellular, intercellular (and to some extent even inter-personal) communication disorder, in which a series of genomic changes can be the cause and the consequence of these communication disorders. Communication in this sense means a message which induces or results in a response. In a healthy organism normal cells do not proliferate without an external message. Cancer cells – for various reasons discussed below – generate a false, mimicked proliferation signal for themselves via oncogenes and other genomic changes. Whether this communication failure is the result of environmental factors and/or external messages (generating changes at the genomic level), or comes from the genetic program, is a question which can be answered only on a case by case basis. However, we have to consider that cells, like human beings, live in a well organized society and in a given ecosystem, which, to a certain extent, determines their receptivity and responsiveness as well as the system-response for the various carcinogenic agents and effects. In other words carcinogenic compounds can be carcinogenic in a given *in vitro* system or in a given organism but the same agent can have different effects in different systems, depending on the circumstances. Clearly, during carcinogenesis changes at the genomic level are critical steps, however the manifestation of these genomic changes and the system response very much depends on the communication state and responsivity of the system (1) (Figure 21.1).

During multistage carcinogenesis the functions of several key genes involved in cell growth control must be damaged. Such genes include not only those involved in cell cycle control of individual cells, but also those involved in the coordination of cell growth throughout a given tissue through cell-cell communication.

For the generation of a malignant tumor, several independent genomic changes are usually required. Functional genomic studies have demonstrated that, at the transcriptional level, several hundred genes' expression can differ in a tumor cell compared to a normal cell from the same tissue. According to functional genomic and differential display studies a major part of these differentially expressed genes are involved in signal transduction, which is an interesting finding

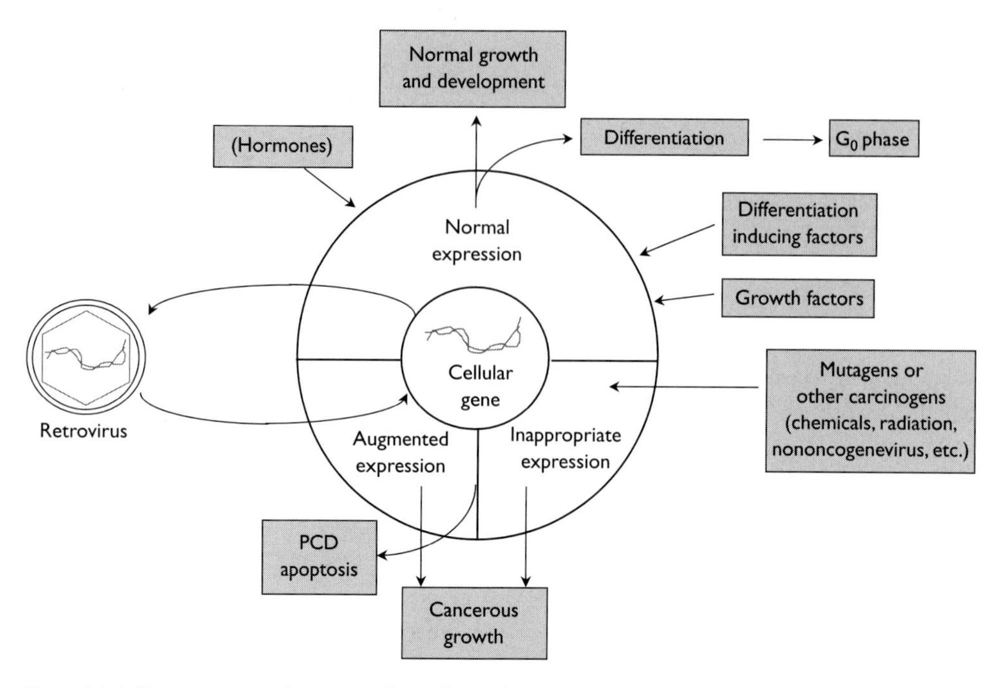

Figure 21.1 Tumorous transformation depends on the communication state and responsiveness of the cell.

if we consider that it is estimated that 3–4% of the entire human genome represent signal transduction genes (2,3).

Carcinogenesis can be conceived as a distortion in intercellular signaling and intracellular signal transduction, which can be originated, for example, from false messages by hormones, cytokines or growth factors or false signal transduction generated by oncogenes. Mammalian tissue development and regeneration takes place within a milieu of regulatory growth factors, cytokines, lymphokines and other intercellular mediators. These affect many parameters of cell development, such as survival, proliferation and differentiation, and certain aspects of cell behavior are influenced by a balance between stimulatory and inhibitory signals. The precise effect of any given factor is determined by the responding cell type, the concentration of the factor, and the presence of other stimuli. Thus some factors may fulfill a variety of functions under different circumstances (4).

The control of cell proliferation and differentiation, as well as the function of differentiated cells can be regulated by the transfer of molecules (e.g. growth factors, cytokines, lymphokines, hormones, neurotransmitters) from one type of cell, over space and distance, to an other cell type (**systemic communication**) or by the transfer of ions and small molecular weight molecules through membrane-bound protein structures, gap junctions, and by direct intercellular contact via the extracellular matrix proteins (**local communication**).

The balance of positive and negative signals to regulate stem, progenitor, and differentiated cells appears to be very different for different types of organs (i.e. closed versus open systems). For example, the kidney or the liver must have a strict regulation of cell proliferation to maintain a steady-state size, while the constant loss of cells from the epithelial layers of skin or gastrointestinal tract must promote a signal for the stem cells to self-renew and to differentiate. It is generally true that highly differentiated cells in a given tissue generate inhibitory signals (systemic or local) which act on the stem cells of the particular tissue, while the loss of the differentiated cells for various reasons, lack of inhibitory signals, or the inability of the stem cells to receive or respond to inhibitory signals would lead to proliferation of the stem cell into another stem cell (renewal) and into a daughter cell destined to differentiate. Once the critical mass of differentiated cells is reached, the inhibitory signals can inhibit the stem cells. Inhibitory signals can come from certain growth inhibitory or differentiation inducing factors, extracellular matrix proteins or via gap junctions. Control of normal embryonic growth and wound healing is a good example of such regulation.

As mentioned earlier, normal cells do not proliferate without an **external signal**, and proliferation is normally closely associated with the activation of growth inhibitory mechanisms. On the other hand, tumor cells generate an overexpressed or mimicked proliferation signal via oncogenes, and also use locally available external signals (growth factors, cytokines, lymphokines – discussed in detail later) to maintain or increase progression, while they have to eliminate growth inhibitory negative signals. In other words they must cut down selectively on intercellular communication, listening only to messages that they want to "hear". The repertoire of cell surface receptor types involved in oncogenesis is quite large, but growth control signal transduction pathways mostly start with specific tyrosine kinases, which eventually stimulate serine kinases (such as protein kinase C) and other signaling mechanism (5,6).

Protein phosphorylation has become widely recognized as a major mechanism for information relay and processing within eukaryotic cells. This ubiquitous form of reversible covalent modification impacts on essentially all cellular processes. Over 1000 different protein kinases catalyze the phosphorylation of hundreds of diverse proteins in the cell. Many of these enzymes form an elaborate network of highly integrated protein kinase cascades. The signaling pathways for oncogenesis include various receptor pathways, like receptor protein tyrosine kinases, the wnt/wingless pathway, the hedgehog pathway, the Notch pathway, the cytokine receptor pathways and some other serpentine receptor-related mitogenic pathways. Several oncoprotein-related intracellular signaling pathways also play crucial roles in oncogenesis: such as the ras pathway, the Rho proteins-related pathways, the MAP kinase (Mitogene activated protein kinase) pathway, the Jak/STAT pathway, etc. (4,7).

A characteristic feature of malignant tumor cells is the lack of contact inhibition and **gap junctional communication** (8). Since gap junctional communication is an essential feature of contact inhibition, cancer cells always have some defect in gap junctional intercellular communication, either because of their inability to adhere, or because of some defects in gap junction structure or its regulation. Interestingly, but not surprisingly, after tumorous transformation, when the basic "lie" of intercellular communication (the mimicked proliferation signals) has been established, from that point on, all the following tumor promoting effects support this "false" communication, or selectively destroy other communication channels. It has been demonstrated that several oncogenes linked to the control of cell proliferation and differentiation affect gap junctional communication as part of their oncogenic function. In addition, several growth factors and chemical tumor promoters inhibit gap junction function and intercellular communication. On the other hand, carcinogen induced cell transformation, *in vitro* led to cells of the transformed foci which were able to perform gap junctional communication between themselves, but not with their normal neighbors, suggesting a selective lack of gap junctional communication between transformed cells and normal cells. Recently, it has been demonstrated that co-transfection of myc and ras oncogenes in the normal rat liver epithelial cell line, WB-F344 results in a loss of functional channels and normal growth regulation: cell-cell communication was significantly decreased and tumorigenicity was induced *in vivo* in adult male rats (9).

A major player in gap junctions is **connexin** the most important gap junctional protein. When connexin genes are transfected into tumor cells, normal cell growth control is often recovered. Certain dominant-negative mutant connexin genes can reverse such tumor suppression. While these results suggest that connexin genes form a family of tumor suppressor genes, so far no connexin gene mutations have been found in human tumors; but two connexin gene mutations have been found in chemically induced rat tumors. On the other hand, recent studies suggest that connexin genes may be inactivated by hypermethylation of their promoter regions, suggesting that epigenetic inactivation of connexin genes may be a mechanism of gap junctional communication disturbance in certain tumors. On the other hand, connexins in many tumor cells are normally expressed but aberrantly localized. The mechanisms of aberrant localization of connexins include lack of an appropriate cell-cell recognition apparatus and aberrant phosphorylation of connexins. These results suggest that gap junctional disorders may occur not only because of aberrant expression of connexin genes themselves, but also as a result of disruption of various control mechanisms of protein functions or phosphorylation cascade.

Cell-cell and cell-matrix **adhesion molecules** and extracellular matrix proteins, beside contributing to gap junctional communication, play a direct regulatory and signaling role in intercellular communication. In cancer cells, as with gap junctions, adhesion molecule mediated intercellular communication is down-regulated and the related inhibitory signals cannot exert their effect. On the other hand, proliferative signals from the stroma and extracellular matrix can reach the tumor cells, providing an additional negative evolutionary pressure for survival. This, again, means that the tumor cells "hear what they want to hear", a strong example of the false communication mentioned earlier for oncogenic signaling. It has to be emphasized though, that a certain level of communication always exists between normal cells and tumor cells (10).

According to the concept of carcinogenesis as a communication disorder, **plasma membrane defects** that produce abnormal electron and proton efflux, and electrical uncoupling of cells through loss of intercellular communication also play important roles in carcinogenesis. These changes can be induced by a wide variety of stimuli, including chemical carcinogens, oncoviruses, inherited and/or acquired genetic defects, and epigenetic abnormalities. The resulting loss of electron/proton homeostasis leads to decreased transmembrane potential, electrical microenvironment alterations, decreased extracellular pH, and increased intracellular pH. Together these effects enhance and sustain the proton/electron efflux and loss of intercellular communication. Low transmembrane potential is functionally related to rapid cell cycling, changes in membrane structure, and malignancy. Intracellular alkalinization affects a variety of pH-sensitive systems including glycolysis, DNA synthesis, DNA transcription and DNA repair, and promotes genetic instability, partly accounting for the accumulation of genetic defects seen in malignancy. Thus, abnormal microenvironment and loss of intercellular communication result in genomic changes and in selective survival and proliferation of malignant cells, providing a good example how communication disorders can be causes and effects of genomic changes and cancer (11).

An additional major intercellular communication disorder of cancer cells is their false **immunological communication**, or a failure of the immune system to communicate properly with tumor cells. Tumor cells survive in a competitive environment because of a lack of recognition by immune cells. In addition, inflammation or other immunological responses related to lymphokines and other mediators can be utilized again by the tumor cells to generate a proliferation or survival signal for accumulating mutations.

21.2 Network signaling via growth factors and other mitogens

Growth factors, cytokines and other mitogenic factors represent a group of extracellular signals that are critically important for influencing a diverse array of cellular responses (the term growth factors here also includes cytokines, chemokines, and other mitogenic factors). In multicellular organisms, the cells' decisions about their survival, growth, proliferation, gene expression, differentiation, and other functions are made on the basis of external stimuli. These stimuli include cell–cell and cell–extracellular matrix (ECM) adhesion, growth factors, hormones, cytokines, chemokines, neuropeptides and other mediators. The ability to integrate information from multiple sources is essential to the cells' ability to respond appropriately to a wide range of conditions and thus accomplish cell fate decisions. Growth of adherent cells *in vitro* is controlled primarily by growth factors and other soluble mitogens and by adhesion to ECM, neither alone is effective. Thus, signals from multiple regulatory factors must converge at the receptor level (12,13) (Figure 21.2).

The effects of growth factors are mediated by high affinity receptors located in the plasma membrane. The interaction of growth factors with their receptors generates primary signals that activate a variety of intracellular signaling enzymes, mainly protein kinases and phosphatases. Changes in protein phosphorylation then lead to the transcription of early response genes that encode transcription factors, which, in turn, induce the transcription of the delayed response genes. The products of the delayed response genes include G1 cyclins and relatives of Cdc2 (cell division cycle proteins) which interact with each other to yield active protein kinases. These kinases are believed to perform two functions: they inactivate the tumor suppressor gene products that restrain the cell cycle engine, and they activate DNA replication and as a consequence cause quiescent somatic cells to leave G_0, progress to G_1 and enter S phase. Following the generation of one or more primary signals, a sequence of metabolic events occurs that includes ionic changes, increased ornithine decarboxylase activity, protein phosphorylation and enhanced glycolytic flux with increased glucose and amino acid uptake, glycolysis, etc. In parallel with and independent of these events, coordinated transcription of hundreds of genes is activated. The immediate early response genes activated within an hour include ornithine decarboxylase, Fos, Jun, Ets, Myc etc., nuclear transcription

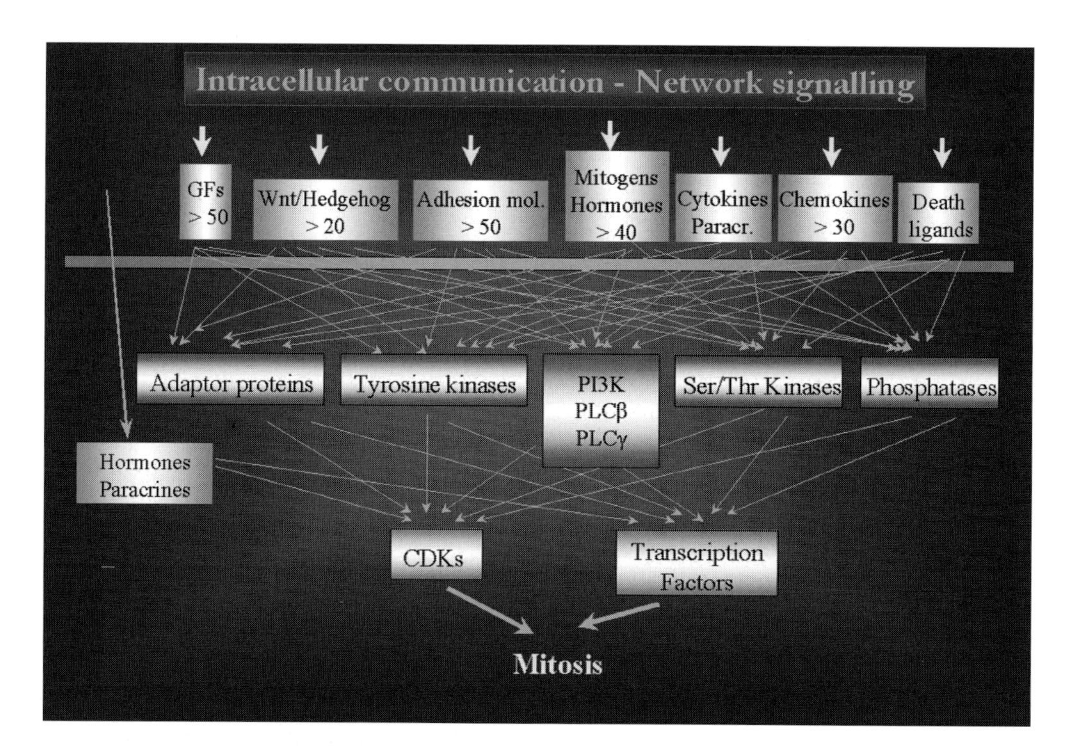

Figure 21.2 (See Colour Plate XIII.)

Table 21.1 Growth factors involved in regulation of cell proliferation

Growth factors acting on receptor tyrosine kinases

(More than 50 RTKs in 19 families estimated number of ligands > 200)
EGFR Family (Epidermal Growth Factor Receptor) Her2, Her3, Her2
PDGFR Family (Platelet-derived Growth Factor Receptor) CSFIR, cKit, Flk2
Insulin and Insulin-like Growth Factor Receptor Family
VEGFRTK/Flk1, Family (Vascular Endothelial Growth Factor Receptor) Flt1, Flt4
FGFRTK Family (Fibroblast Growth Factor Receptor TK)
HGFR/Met Family (Hepatocyte Growth Factor Receptor TK), Ron
Trk Family (Nerve Growth Factor RTK); KLG RTK Family
Axl/ c-Eyk Family; Tie/Tek Family; Eph/Elk/Eck orphan receptor Family; Ryk Family;
DdR/Tkt family; Ros Family
Ltk/Alk family; Ror Family; Torpedo; Torso

Cytokines

On 5 Family of receptors > 40 Cytokines

Class I Cytokine receptors (with associated tyrosine kinase activity):
Most of the Interleukins, Granulocyte Colony-stimulating Factor (G-CSF), GM-CSF,
Erythropoietin (EPO), Leukemia Inhibiting Factor (LIF), Oncostatin M (OSM), Ciliary
Neurotrophic Factor (CNTF), Growth hormone, Prolactin

Class II Cytokine receptors (with associated tyrosine kinase activity):
Interferons, Interleukins
Tumor Necrosis Factors Family
Transforming Growth Factors (TGFβ) Family, Bone Morphogenic Proteins (BMPs)
(acting on Ser/Thr kinase receptor)
Ig Superfamily

Chemokines, known number >30

Chemokines (C-X-C family): IL-8/NAP1 Neutrophil-activating peptide-1, 2; Growth
related peptides (GRO α; β, γ,); Epithelial derived Neutrophil Attractant (ENA-78);
Granulocyte Chemotactic Protein-2 (GCP-2); Platelet factor-4, Platelet basic protein;
Thromboglobulin β; Connective Tissue Activating Protein (CTAP-III.); Stromal cell
Derived Factor (SDF-1α); SDF1-β; Interferon inducible protein (IP-10) etc.
Chemokines (C-C family): Macrophage inflammatory proteins; Macrophage chemotactic
proteins; Eotaxin; TARC etc.
C Chemokines: Lymphotactin (Ltn); ATAC etc.

Mitogens and Hormones – on G-protein coupled receptors

More than 40 ligands on more than 20 receptors
Angiotensin II, Thrombin, Prostaglandins, Adenosine, Lysophosphatydic acid(LPA),
Yeast mating factor, Bradykinin, Bombesin, Endothelin, CCCK and Gastrin etc.
Antiproliferative hormone: Somatostatin

On intracellular receptors

More than 30 hormones, >20 receptors, steroids, small molec. hormones, retinoids etc.

Adhesion molecules and extracellular matrix related growth factors (ECM fragments)

Estimated number >50 (several unidentified and overlapping with GFRTK ligands)
Immunoglobulin gene family such as ICAM, VCAM, CD4
Integrins such as $\alpha 4\beta 1$, $\alpha 6\beta 4$, and cadherins such as E-cadherin and N-cadherin.

Death ligands (overlapping with cytokines and adhesion protein ligands) Fas, Fad

factors, steroid hormone receptors, cytoskeletal and adhesion protein genes (fibronectin, bactin, tropomyosin), steroid receptors, glucose transporter, etc. The coordinated activation of these genes results in synthesis of RNA, proteins and DNA.

Several hundreds of polypeptide growth factors and cytokines regulate cell proliferation in the human body. Aberrant expression of these factors by tumor cells confers growth advantage and competence for metastasis either by autocrine or paracrine mechanisms. Peptide growth factors and their receptors serve as the major molecular signaling system in orchestrating tumor survival, growth, and interactions with the microenvironment. Tumor progression is intimately associated with increasing growth autonomy of the malignant cells, perhaps due to the constitutive production of multiple growth factors. Growth factors not only stimulate the "selected" or "dependent" tumor cells for proliferation, but have autocrine or paracrine roles in other aspects of tumor development and progression (triggering an inflammatory reaction, initiating angiogenesis and fibrous stroma formation, modulating the host immune response, and activating proteolytic enzymes).

A good example of this complex network is signaling via the family of ErbB receptors of epidermal growth factor-like ligands. The ErbB receptor tyrosine kinases, which play a key role in various kinds of tumorous transformations, bind multiple neuregulins and other epidermal growth factor-like molecules, and can be activated with binding proteins either via an autocrine secretory loop or with molecules produced by adjacent tissues. Stromal cells in the vicinity of propagating tumors serve as the prime source for ErbB activating ligands, whereas the extracellular matrix surrounding them represents a reservoir that can increase local ligand concentrations. More than 30 EGF-like ligands bind to ErbB receptors recruited by differential and partially overlapping specificities. Ligand-receptor binding gives rise to 10 possible dimeric complexes, comprising the second level of signaling complexes. The identity of dimer-receptors is dependent on receptor expression and ligand affinity, and results in activation of a variety of downstream pathways. Effector molecules with various protein binding domains serve as adaptors by binding activated receptors and conveying the signals further, activating parallel, interacting, or overlapping cascades, resulting in the activation of various sets of transcription factors. These signaling networks can lead to both differentiation and proliferation, promoting normal or pathological cell growth. This network pattern suggests that a wide variety of growth stimulating signals result in the activation of a multitude of intracellular molecules, only to converge into a few mainstream pathways (14).

Since hundreds of growth regulatory signals can reach the cells, cell fate decisions (growth, proliferation or differentiation) depend on the cooperativity of a given set of signals, while the generated intracellular signals also have to form a coordinated or cooperative network of signals to generate the required output – required from the organism's point of view. To accomplish proper control of cell fate decisions the organism, or system, uses a large series of growth factors, cytokines, lymphokines, hormones and other soluble and membrane-bound mediators.

These growth factors represent very important messengers (communication tools) in intercellular communication, and the large number of such factors means that cell fate decisions are based on a very complex regulatory system. Failures in this communication system (false messages, too much or too low message, overexpression or down-regulation of receptors, increase or loss of sensitivity, etc.) can, in most of cases, be compensated by the redundancy and compensatory function of the network signaling, but cancer as a communication disorder very much depends on the messages and function of these growth factors. From a pathological and therapeutic point of view, the key question is not only why there is false communication (which results in uncontrolled proliferation), but what is the wrong message (lack of message can also be a wrong message), and how can the proper communication or the compensatory threshold function of the network be reestablished (15,16)?

An incomplete list of the most important growth factors and cytokine/lymphokine receptors involved in regulation of cell proliferation is shown in Table 21.1.

21.3 System response: the holographic theory of cancer

The progression of a normal cell to one with malignant features requires the acquisition of multiple and distinct genomic changes which can be both cause and effect of various inter and intracellular communication disorders, resulting in modified growth features that may be unique for specific tumor forms derived from particular tissue-specific tumor cells. The rate at which such changes occur may be contingent upon the accumulation of rare genetic events that enhance the replicative capacity of the progenitor cell, and depends on the given environment and on the interactive or communicative state of the particular cell type.

In a healthy organism cells live in a very well organized society, where interactions are based on cooperativity instead of competition and every cell does its job (according to its "best knowledge") and uses resources only to the extent required for proper functioning. In order to function in the best interest of the organism, cells have to have a very sophisticated and intensive communication system. Intercellular communication has been found to be the critical step in maintaining the dedicated and differentiated functions of cells in a multicellular organism. As discussed earlier, a failure in this communication network can lead to various malfunctions including tumor growth.

Several types of systemic and immunological responses are known to occur in the body with cancer (which is always accompanied by cell injury) at the site of cancer growth, including a primary local inflammatory response, a non-specific systemic response which may be observed physically as rigors, fever, loss of appetite, lassitude and weakness, and an antibody response. These responses are initiated or influenced by cytokines and other mediators (such as prostaglandins), which induce the various metabolic changes of the response to injuries, due to trauma, cancer, infection or autoimmune disease. A great number of mediators have been described, most with a large variety of effects on cells. Most of these mediators can be produced by many different types of cell, so they can interrelate in various ways and their effects sum to form the systemic response to cancer growth. The tumor cell lives in a very competitive environment, and hundreds of supporting or opposing messages reach the tumor cell constantly trying to keep it under control, while several hundred proteins have different levels of expression or function in tumor cells to provide it with selective growth advantages and survival chances.

As we discussed earlier, cancer can be considered as a communication disorder. Whether this communication disorder is a cause or a consequence is hard to define. A cause can always become an effect, which can again become a new cause, while for the generation of a malignant tumor, several causes and effects have to accumulate, interact, and support each other. However, it is clear that in the generation and survival of a malignant tumor the general and functional state, and the communicative capability of the whole organism plays an essential role. The drop is in the sea and the sea is in the drop, says the ancient Chinese saying.

Recently, it has been demonstrated that information is stored in the brain in a holographic manner. This means that a single neural cell can depict the general interrelationship and connection state of the brain during memory storage. In other words the interrelationships and excitation states of a certain brain area are depicted in the connection state of a memory cell. The hologram is formed with interference waves in such a way that in a small portion of the holographic picture the whole picture can be seen, but in much less focus. So, in the drop you can see the whole ocean. If neural cells can "feel" the state of the whole brain, it is reasonable to suppose that the function and state of the various somatic cells can also represent or are strongly influenced by the general state and function of the whole body, while the state of mind or the psyche can also have a strong influence on the functions and communication states of particular somatic cells. This would mean that the communication disorder or the tumorous transformation at cellular level also represents a communication disorder or dysfunction at a higher level of the organism.

This holographic interrelationship of the part and the whole has been described in other aspects of biological and physical sciences, such as quantum physics, protein chemistry, and immunology. Of course, this concept, in a less scientific way, has already been described in holistic medicine which claims that besides addressing a particular disease or tumor locally or directly (with surgery, targeted drugs, radiation, etc.), the whole organism should be also be considered and treated. But the modern holographic picture of cancer gives a deeper insight into interrelationships between the disease or the transformed cells and the rest of the organism, and supports the concept that cancer can be considered as a communication disorder. This is why a primary challenge for molecular medicine and biology is to understand what kind of communication or signal transduction disorders exist between cancer cells and normal cells so that we can try to establish proper communication at various levels (12).

21.4 Signal transduction pathways of tumor cells

There are complex extracellular and endocellular signals controlling the incidence, speed and extent of growth, differentiation, and cell death in mammalian tissues. Hundreds of messages reach the cells continuously, and growth control pathways form a very complex network. The constitutively activated mitogenic signaling pathways play a key role in oncogenesis. In the past years there has been enormous progress in elucidating the details of mitogenic signaling pathways (Figure 21.3).

The most important signaling pathways involved in oncogenesis include:

- receptor protein tyrosine kinase pathways;
- cytokine and chemokine receptor-related pathways;
- seven transmembrane receptor-related pathways;
- Wnt/wingless, Notch and Hadgehog/Patched pathway;
- extracellular matrix/adhesion protein-related pathways;
- intracellular (cytoplasmic and nuclear) oncoprotein-related pathways;
- nuclear receptors and the endocrine/paracrine regulators-related pathways;
- apoptosis related pathways.

These pathways are specified and listed here on the basis of the related oncoproteins, i.e. where the oncogenic signals are generated from.

Of course, various elements of these pathways are common, and they form an elaborated network of signaling cascades. The assembly of enzyme and adaptor protein complexes is emerging as a major mechanism of signal transduction, and may regulate the pleiotropic effects of growth factors and other external signals. Acting somewhat similarly to the subcellular equivalent of a nervous system, this network monitors the environment within and outside the cell, and coordinates the appropriate responses. In the near future a major challenge to biomedical scientists will be the elucidation of the precise connections between kinases and various sensory and effector termini in this network.

We shall briefly discuss all these pathways, pointing out the relevant target molecules for potential therapeutic intervention.

21.4.1 *The role of tyrosine kinases in tumor cells*

Tyrosine kinases, which were the first oncogen products to be identified, are enzymes that regulate signal transduction in cells, leading to a complex series of intracellular reactions, which result in very diverse cellular responses, like proliferation, differentiation, cell motility, changes in cell shape, production of extracellular matrix, and transcription of specific

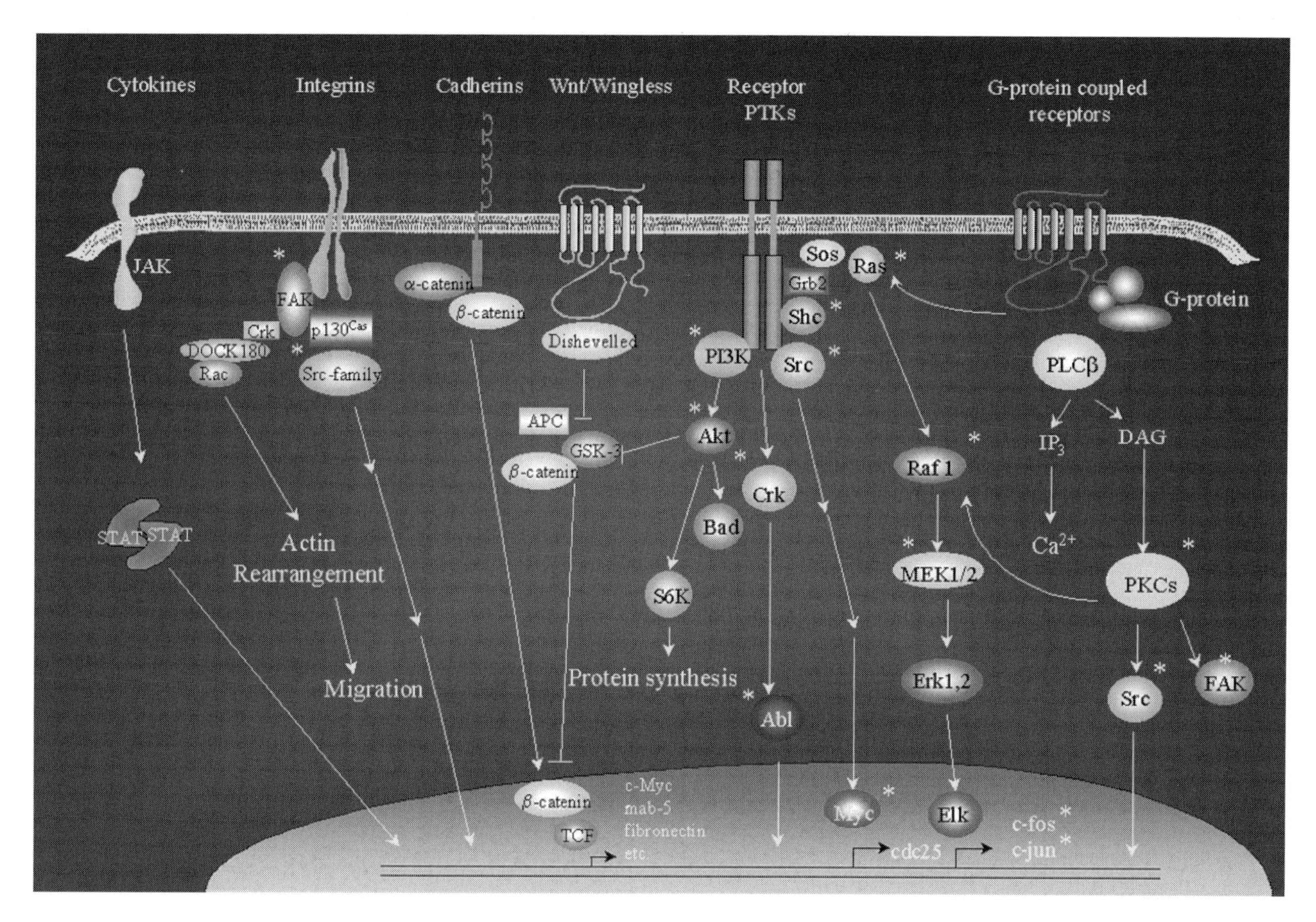

Figure 21.3 The most important signaling pathways involved in oncogenesis. Selected oncoproteins are marked with stars (see Colour Plate XIV).

genes. Aberrant activity of tyrosine kinases, resulting from overactivation or underutilization of the signaling pathways, can lead to several disease states. Many oncogenes found in cancers are derived from tyrosine kinase genes that have been deregulated, leading to constitutive activity. Overexpression or activation of wild type tyrosine kinases can also lead to cancer. Abnormal activation may also result from an autocrine loop, in which a growth factor and its receptor are expressed in the same cell, leading to continuous signaling. Tyrosine kinases exist as two major structural types: transmembrane receptors and cytoplasmic proteins (17).

21.4.1.1 The receptor protein tyrosine kinases

The Receptor Tyrosine Kinases (RTKs) are a family of more than 50 different transmembrane polypeptides with a protein kinase domain in their intracellular portion. RTKs can be divided into at least 19 subfamilies including the epidermal growth factor (EGFR), Insulin RTK, platelet-derived growth factor receptor (PDGFR), vascular endothelial growth factor receptor (VEGFR), fibroblast growth factor receptor (FGFR), hepatocyte growth factor/scatter factor receptor (MET), the neurotrophins (TRKR) as well as a number of orphan receptor subfamilies (18).

Upon binding to their corresponding growth factors, these receptors initiate mitogenesis, differentiation, migration, apoptosis and many other cellular functions. Ligand binding to a receptor tyrosine kinase in most cases induces receptor dimerization. The conformational change, which results from ligand binding and receptor dimerization, leads to interactions between adjacent cytoplasmic domains and activation of the tyrosine kinase. Ligand binding to RTKs induces auto-phosphorylation of the intracellular domains and initiates a cascade of phosphorylation that includes phosphorylation of cytoplasmic substrates and recruitment of various signaling molecules like SH2 domain-containing proteins that bind to specific tyrosine phosphorylated residues. These events result in the activation of other proteins, as well as the production of second messenger molecules, which have regulatory functions. These signals can be transmitted to the nucleus, where expression of specific genes is induced, resulting in a response, such as cell division.

The list of signaling molecules associated with RTKs is very broad and it includes phospholipase Cγ (PLCγ, acting on PtdIns(4,5)P2 hydrolysis), Ras-GTPase activating protein (GAP), phosphatidylinositol 3-kinase (PI3K), tyrosine phosphatases (PTP1 and 2), non-receptor tyrosine kinases (like Src, Fps or Syk), Vav (which is a PH domain containing G-nucleotide exchange factor) and STAT proteins (signal transducer and activator of transcription) as well as adaptor proteins like Grb2, Shc Nck, Crk, etc. The RTK associated signaling molecules form an elaborated network of signaling cascades, and the activation of a particular signaling cascade depends on the actual functional and metabolic state of the cell, the strength of the external signal, the formation of the signaling complexes, the intracellularly available signaling molecules, etc. Signaling molecules can be associated with the RTKs via various protein binding domains, like the SH2 domain (Src homology 2); SH3 domain which binds to a specific proline reach region, the PTB domain, which also binds phophotyrosine within a specific sequence context; the PH domain (plekstrin homology domain) which binds the charged headgroups of specific polyphosphoinositides thus regulating the subcellular targeting of signaling proteins to specific regions of the plasma membrane, or the PDZ domain, which recognizes short specific peptides with a free carboxylate group and are localized to highly specialized submembranous sites, involved in junction formation, receptor or channel clustering and other intracellular signaling events, and the 14-3-3 proteins, which bind to a specific phosphoserine containing region. Most of the signaling molecules contain more than one protein binding domain, thus further increasing interactive and networking possibilities. The receptor tyrosine kinases and receptor associated intracellular tyrosine kinases, associated adaptor molecules and phosphorylation cascades, with Ser/Thr kinases and related phosphatases thus form a very complicated signaling network. Specific receptor-tyrosine phosphorylations are converted to a series of serine/threonine phosphorylations in the cytoplasm. Seven specific transmembrane receptors, adhesion protein or cytokine receptors can also stimulate protein kinase cascades involved in growth regulation. Sequential phosphorylation reactions ultimately converge in the nucleus to phosphorylate and regulate the activity of transcription factors (Figure 21.4).

In the case of the growth-regulatory tyrosine kinases, concurrent growth-inhibitory signals are generated to prevent cellular proliferation from continuing indefinitely. One of the functions of these inhibitory signals is the deactivation of tyrosine kinases via feedback signals and phosphatases. Precise control of these signaling events, both positive and negative, is necessary to maintain normal cellular growth.

Constitutive activation of tyrosine kinases, which can lead to **tumorous transformation**, can occur by several

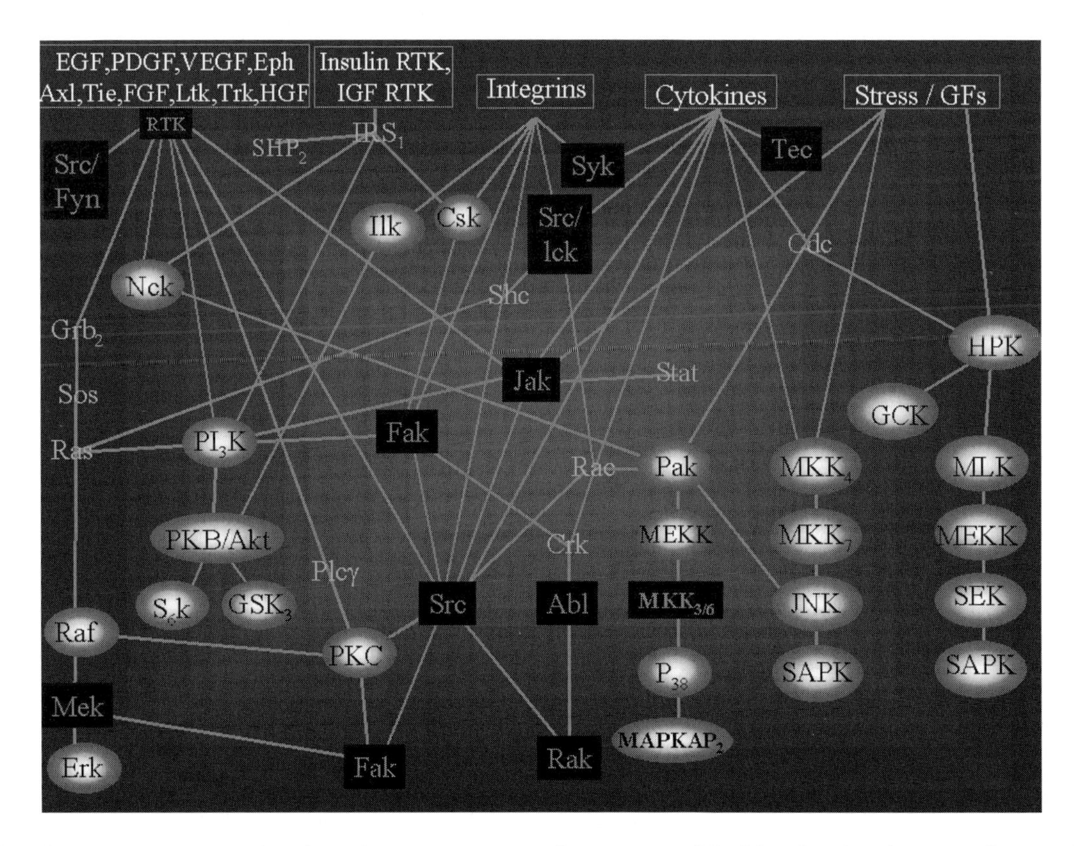

Figure 21.4 Specific receptor tyrosine phosphorrylations are converted to a series of Ser/Thr phosphorylations in the cytoplasm. Tyrosine kinases are shown in boxes, Ser/Thr kinases in ellipses.

mechanisms. The first category comprises various mutations that lead to constitutive kinase activity. Ligand-independent kinase activation is seen in many receptor tyrosine kinase oncogenes isolated from transforming viruses. These alterations include: deletion or mutation within the extracellular domain, alteration of the ATP-binding motif, or alterations of the catalytic portion of the kinase that renders the kinase constitutively activated and transforming. Mutations of the transmembrane domain may lead to ligand-independent kinase activation, as reported for NEU/c-erbB-2. Mutation of the transmembrane domain may also cause oncogenic activation through expanded substrate specificity. Deletions of the intracellular regulatory domains of tyrosine kinases have been found that lead to oncogenic kinase activation. Additionally, mutations that affect negative regulation of a tyrosine kinase, such as a mutation of tyrosine 527 in SRC kinase, or a mutation in a phosphatase binding site, can lead to dysregulated tyrosine kinase activity and oncogenic activation. Several novel amino-terminal sequences are involved in fusion and oncogenic activation of multiple tyrosine kinases. One of the common themes among these amino terminal sequences is their ability to induce oligomerization and activation of the fused tyrosine kinase.

A second category of mechanisms of tyrosine kinase activation involves closure of autocrine growth loops. Autocrine growth loops are possible when aberrant expression or overexpression of an otherwise normal tyrosine kinase occurs in the presence of its associated ligand, or when overexpression of ligand occurs in the presence of its normal receptor. This mechanism of transformation has been described for EGFR, colony stimulating factor-1 receptor (CSF-1R), insulin-like growth factor-I receptor (IGF-IR), hepatocyte growth factor receptor (MET), and platelet-derived growth factor receptor (PDGFR), using *in vitro* assays of tumorigenesis.

The third category includes alterations in regulatory pathways that affect the activity of a tyrosine kinase. For example, mutations which occur in negative regulators in a signaling pathway, such as a kinase or phosphatase, might lead to dysregulated tyrosine kinase activity and oncogenic activation. Other examples are more complicated, but include alterations in substrate availability or constitutive activation of an associated receptor complex.

Permanent activation of receptor tyrosine kinases has been shown in various cases of cancers. Members of the EGFR family and its ligands are often overexpressed or expressed as an autocrine loop (for example in pancreatic cancers, gastric cancers and cell lines, renal carcinoma cell lines and ovarian carcinoma specimens and hepatocellular carcinomas). In a study of primary breast tumors, 59% of samples over-expressed EGF receptors. Her2 is very often overexpressed in breast, ovarian, prostate and colorectal carcinomas and it correlates with tumor grading and poor prognosis. The presence of PDGF and its receptors has been demonstrated, for example in biopsies of human gliomas and prostate tumors as well as in various tumor cell lines, including breast and colon cancers and melanoma. Insulin-like growth factor I and its receptors are also involved in various cancers, including breast cancer, small cell lung cancer, prostate cancer and Wilms' tumor. In a clinical study the TPR-MET fusion protein was detected in more than 50% of gastric cancer biopsy specimens. The role of FGF has been demonstrated with various molecular biological techniques in renal cell carcinoma, certain prostate tumors and esophageal tumors. In the skin of melanoma patients high FGFR-1 expression was seen in invading melanoma cells and stroma. VEGF was found to be the major angiogenic factor in 60% of solid tumors tested. The listed GFRTKs represent very good validated target molecules for therapeutic interventions (17,19,20).

21.4.1.2 Non-receptor tyrosine kinases

Non-receptor tyrosine kinases are cytoplasmic proteins that transduce extracellular signals to downstream intermediates in pathways that regulate cellular growth, activation, and differentiation. Many non-receptor tyrosine kinases are linked to transmembrane receptors, including receptors for peptide hormones, immune recognition, and hematopoietic cytokines. These non-receptor tyrosine kinases are activated by extracellular events, such as binding of ligands to their associated receptors. Other cytoplasmic tyrosine kinases are activated by events such as cellular adhesion, calcium influx, or at particular stages of the cell cycle, without an obvious link to a specific transmembrane receptor. There are presently 8 different classes of non-receptor protein tyrosine kinases, grouped according to structural and functional similarity. These are the Src family, Tec family, Csk family, Fes (Fps) family, Abl family, Syk/Zap70 family, Jak/Tyk family and Ack family (Figure 21.4).

Activation of non-receptor tyrosine kinases is induced in response to the appropriate extracellular signal, and dimerization may or may not be necessary for activation. For example, JAK kinases are activated following ligand-induced dimerization or oligomerization, while members of the SRC family of tyrosine kinases are maintained in an inactive state by phosphorylation of a specific tyrosine residue, tyrosine 527. SRC can be activated by either dephosphorylation of this residue or by phosphorylation at other sites. Similarly to RTKs, activation of non-receptor tyrosine kinases initiates a cascade of phosphorylation that includes autophosphorylation, phosphorylation of cytoplasmic substrates, and recruitment of SH2 domain-containing proteins that bind to specific tyrosine phosphorylated residues. These events result in the activation of other proteins, which have regulatory functions. These signals can be transmitted to the nucleus, where expression of specific genes is induced, resulting in a response such as cell division. In the case of a growth-regulatory tyrosine kinase, concurrent growth-inhibitory signals, mostly via tyrosine phosphatases, are generated to prevent cellular proliferation from continuing indefinitely. One of the functions of these inhibitory signals is deactivation of the tyrosine kinase. Precise control of these signaling events, both positive and negative, is necessary to maintain normal cellular growth. Dysregulation or oncogenic activation of intracellular tyrosine kinases can occur by mutation, aberrant expression or overexpression, or via the failure of related regulatory signals, which would then generate a permanent proliferation signal.

Activation or overexpression of the Abl oncogene, which results in the permanent activation of the Abl cytosolic tyrosine kinase, has been associated with a variety of human leukemias. Chronic myelogenous leukemia (CML) is the most common form of leukemia and it was demonstrated that the 210 kDa form of the Bcr-Abl fusion protein resulting from a (9;22) chromosome translocation is present in 95% of these patients. The transforming function of the Bcr-Abl fusion protein depends on the constitutive and elevated tyrosine kinase activity of the enzyme compared to c-Abl. So the Abl tyrosine kinase has become a very important validated target for signal transduction therapy-based modern antitumor drug research (see the success story of Gleevec). Src family kinases were found to be involved in various types of cancers, especially colon cancers and melanoma, while the involvement of Jak kinase in acute lymphoid leukemia has also been demonstrated (17,19,20).

21.4.2 The cytokine, chemokine and paracrine pathways

21.4.2.1 Lymphokines

Human tumor cells and stromal and immune cells in contact with them constitutively produce a variety of growth factors, cytokines, paracrine mediators, chemoattractants, motility factors, and proteolytic enzymes and their inhibitors. Juxtaposed normal cells, such as fibroblasts, endothelial cells, and monocytes are also potential producers of most of these factors, but, in general, they require specific signals to trigger synthesis and/or to release biologically active factors. Growth stimulation of normal and malignant cells, angiogenesis, and stroma formation within malignant lesions, tissue degradation by invasive tumor cells, cell motility, detachment of tumor cells from lesions and/or attachment to basement membranes involve complex interactions between autocrine, paracrine, and endocrine factors.

Many cytokine receptors lack intrinsic catalytic domains. These receptors consist of a conserved extracellular domain, a transmembrane region and an intracellular domain containing the membrane-proximal so-called "box1" and "box2" motifs. Cytokine receptors couple ligand binding to tyrosine phosphorylations by using non-covalently associated protein tyrosine kinases: the Janus kinases (JAKs). These JAKs bind to box1 and box2 motifs. The known members of the JAK family (JAK1, JAK2, JAK3 and TYK2) have a C-terminal kinase domain, which is immediately preceded by a pseudo-kinase domain; several other homologous regions in the N-terminal sequences have been implicated in interactions with various cytokine receptors. Among the signaling proteins that are recruited to the receptor complex and tyrosine phosphorylated are the cytoplasmic signal transducers and activators of transcription (STATs). Phosphorylated STATs dimerize through reciprocal SH2–PY interactions and translocate to the nucleus, where they bind to specific DNA elements and stimulate transcription. STATs are not only activated by cytokine receptors, but also by RTKs and the so-called 'serpentine' angiotensin II receptor (see below). There is evidence that STATs are at least partly modulated by the Ras–MAPK pathway; conversely, JAK2 has been shown to activate MAPK through the upstream kinase Raf-1.

The TGFβ-family members: TGFβs, activins and bone morphogenic proteins (BMPs) signal through distinct sets of type II serine/threonine kinase receptors, type I receptors and Smad proteins. Five distinct type II receptors and seven type I receptors (also termed activin receptor-like kinases (ALKs) have been identified. ALKs act downstream of type II receptors. Smad proteins (1–8) form hetero-oligomeric complexes that translocate into the nucleus and serve as signal integrators within an extensive intracellular network. The Smad2 protein has recently been reported to act as a common positive effector of both receptor serine/threonine kinases and RTKs; this provides one of the many examples of cross-talk during signal transduction from the cell surface to the nucleus. Smads were found to make functional interactions with a great number of transcription factors and appear to have important roles in the integration of various signals (21).

21.4.2.2 Chemokines

Chemokines are proinflammatory mediators, which have an important role in various pathological conditions, including inhibition or promotion of angiogenesis, inhibition or induction of cellular proliferation and induction of integrin receptors. Chemokines signal by interaction with specific, cell-surface GPCRs. There are more than 50 chemokines reported to date and only 12 types of receptors, so more ligands can bind to the same receptor. The situation is further complicated by the way in which each ligand is presented to the receptor, because chemokines are bound tightly to various extracellular matrix proteins. Thus, chemokine selectivity can be accurately mapped only in the exact physiological environment.

After activation, chemokine receptors have altered sensitivity to repeated stimulation with the appropriate ligand. As with all GPCRs, activated chemokine receptors couple to heterotrimeric G-proteins consisting of subunits. It is the pertussis toxin-sensitive G protein, Gi, that links the receptor to PIP2 hydrolysis through the intermediary of the dissociated subunit, that is implicated for all receptors studied to date. Sequential stimulation of CCR3 by eotaxin, but not any other chemokine, caused homologous desensitization. Thus, receptor regulation represents another control point in chemokine function. Expression rates of receptor and receptor recycling also determine the response to chemokine ligands and might be modified to achieve a therapeutic benefit. It has been demonstrated that chemokines interfere with the serine/threonine kinase pathways and Ras/MAPK pathways, suggesting a complex interrelationship between cytokine, chemokine and other mitogenic pathways (22).

There is now much evidence that mediators other than cytokines and chemokines also have roles in the various changes in intercellular communication and local reactions accompanied by tumor growth. A series of paracrine substances, eicosanoids (prostaglandins, thromboxanes, leukotrienes) and vasoactive substances are involved in local cellular responses, inflammation, platelet aggregation and cell-cell interaction associated with tumor growth and related cell injuries. The role of prostaglandins in inflammation and the role of inflammation in tumorous transformation and growth has long been known. However the direct effect of prostaglandins on tumor growth has been elucidated only recently.

For example, it has been demonstrated that Prostaglandin E2 activates c-fos in human prostate cancer cells via the protein kinase A pathway, and selective COX2 inhibitors reduce colon tumor growth and its proxy markers via inhibition of prostaglandin formation (23,24).

The role of cytokine and chemokine-related signaling in cancer has not been fully elucidated, but it has become clear that they play very important roles in intercellular communication and failures in this communication can result in various pathological states including cancer. Cytokines and chemokines act, as the term implies, by modifying the activity of cells in a variety of ways, including inducing cell division and stimulating the release of other mediators affecting inflammation and cellular metabolism due to cell injury which always accompanies tumorous transformation and growth. Recent results suggest that cell injury induces inflammation, inflammation generates cytokines, chemokines, prostaglandins, leukotrienes and additional mediators, which stimulate proliferation and local immune response and generate a vicious circle for cancer cell stimulation. Thus cytokines and chemokines can contribute to the permanent proliferation signal generated by oncogenes and provide an increased possibility for additional mutation needed for malignant transformation, so inhibiting this vicious circle might become an important therapeutic approach (25).

21.4.3 The Wnt/Wingless and the Hedgehog/Patched pathway

Wnt1 is a secreted factor that plays a role in tumorigenesis. Members of the Wnt gene family are proposed to function in both normal development and differentiation as well as in mammary tumorigenesis. Biochemical characterization of seven Wnt family members has been described: all of them are glycosylated, secreted proteins that are tightly associated with the cell surface or extracellular matrix, suggesting that Wnt family members function in cell to cell signaling in a fashion similar to Wnt-1. The family of Wnt proteins also includes the Drosophila Wingless protein, which plays multiple roles in Drosophila development by influencing cell fate decisions. But how do Wnt proteins generate tumorous transformation? The Frizzled proteins have recently been identified as Wnt receptors. Frizzled family members have seven transmembrane domains and a cysteine reach extracellular domain required for Wingless binding. The signaling pathway of Wnt is not fully understood, but several components of this pathway are implicated in cancer, such as the so called disheveled proteins, glycogen synthase kinase 3 (GSK3) and β-catenin (Figure 21.3). It has been demonstrated that the adenomatous polyposis coli gene product (APC) – which is mutated in most colon cancers – interacts with GSK3 and β-catenin decreasing the amount of free β-catenin. Free β-catenin can signal via the TCF/LEF transcription factors, inducing the expression of genes which block cellular differentiation. Via this signaling cascade the autocrine expression of Wnts could lead to the induced expression of TGFβ family members and the cytokine Hedgehog, as well as to certain transcriptional repressors, thus contributing to tumorous transformation.

Hedgehog is a secreted protein controlling cell fates and growth in many tissues. It binds to the Patched/Smoothened system which is intimately connected with the Wingless/Fizzled system in determining cell fate decisions. Mutations of the human homolog of the Patched gene are responsible for the hereditary nevoid basal cell carcinoma syndrome, characterized by multiple basal cell carcinomas (26).

21.4.3.1 The Notch pathway

Several members of the growing Notch family of receptors have been associated with cancer. Notch mediates messages of various ligands, like Delta/Jagged, related to intercellular communication and cell fate decisions and acts as a suppressor of cellular differentiation. Activated forms of Notch associate with the DNA binding protein CBF1, triggering transcriptional activation of proliferation-related genes through the HES1 promoter. In these cases CBF1, the repressor, becomes an activator. The known Notch-derived oncoproteins (such as Notch1 inducing T-cell leukemias, or Notch2 inducing thymic lymphomas) have deletions in the extracellular domain resulting in permanent activation of the Notch pathway, thus blocking differentiation and causing continuous proliferation (4).

21.4.4 The intracellular (cytoplasmic and nuclear) oncoprotein-related pathways

Cell proliferation in response to growth factors is mediated by specific high affinity receptors. Binding of growth factors, mitogenic factors or specific hormones to their respective receptors activates a network of signaling cascades and stimulates progression through the G1 phase of the cell cycle. Various enzymes and adapter proteins of this signaling cascade can act as oncoproteins (Figure 21.3). These signal cascades ultimately control transcription of both immediate and delayed response genes contributing to the decision to progress through the cell cycle or to arrest in G1 and proceed to differentiation. The signal transduction cascades controlling these responses often utilize sequential protein kinase reactions. In growth factor-related pathways receptor tyrosine phosphorylations are generally converted to a series of serine/threonine phosphorylations in the cytoplasm and the nucleus. Key signaling enzymes are recruited to the plasma membrane through the formation of stable complexes with activated receptors. These interactions are mediated by the conserved, non-catalytic SH2 domains present in the signaling molecules, which bind with high affinity and specificity to tyrosine-phosphorylated sequences on the receptors. Examples of proteins that have SH2 domains that bind to the phosphotyrosine sites on growth factor receptors include phospholipase Cγ, (PLCγ) the p85 subunit of phosphatydolinositol-3 kinase (PI-3K), Shc, Nck, Grb2, GTP-ase activating protein (GAP), Syp and p91. Proteins containing SH2 domain are recruited to specific autophosphorylated phosphotyrosines of the receptor, which provides a signal for the sequential activation of intracellular proteins. Several of the signaling proteins that have an SH2 domain also encode Src homology 3 (SH3) domains, which are involved in specific protein-protein interactions during signal transduction. For example Grb2 is an adaptor protein that binds Sos, which is an exchange catalyst that stimulates GDP dissociation from Ras, allowing GTP to bind, thus forming an activated Ras-GTP complex. Shc and Nck are also adapter proteins with similar functions and oncogenic potential.

Activating mutations in Ras have been found in many kind of human tumors and the frequency of Ras mutation is very high in many human cancers. Recently some Ras-like genes, such as TC21 or R-Ras, have also been shown to mutate in certain human cancers. A number of effectors for Ras proteins have been identified which bind preferentially to Ras in the GTP bound state, e.g. raf1, the P110 PI3 kinase catalytic subunit, Ral GDS, Rin1 and MEKK1. Several of these have been implicated in carcinogenesis. It has been shown that multiple cellular pathways are activated by Ha-Ras and contribute to Ha-Ras-induced mammalian cell transformation. Ras-induced tumorous transformation often involves other members of the Rho family of small G-proteins, like Rho, Rac and Cdc42 proteins. These proteins have important roles in the regulation of the cytoskeleton and plasma membrane topology and were found to be involved in oncogenesis. Among the direct targets of Cdc42 and Rac are several protein kinases. The importance of Rho family small G proteins in transformation is emphasized by the large number of mutant activated GTP exchange factors for this family that have been found to possess fibroblast transforming activity (27).

Adaptor proteins such as Crk, Nck, and Shc, have been implicated in oncogenesis, either through their conversion into viral oncoproteins or because they transform cells when overexpressed. During tumorous transformation Nck may be connected to the actin cytoskeleton and Rho family proteins and Nck-including complexes are translocated to activated growth factor receptor PTKs via the Nck SH2 domain, increasing associated kinase (like Pak1) activity. Pak1 in turn can lead to activation of MAP kinase pathways.

The three known MAPK pathways (which are essential elements of the intracellular signaling pathways) are the ERK, JNK/SAPK and p38 pathways. The JNK/SAPK and the p38 pathways lead to activation of many of the same transcription factors thus contributing to the network signaling. Activated MAPKs are translocated into the nucleus where they activate several transcription factors including the ternary complex factors in the Ets family that stimulate the expression of the c-fos gene and constitutive activation of MAP kinase mediated mitogenic signaling pathways result in transformation. The ERK pathway has been strongly implicated in oncogenesis while the c-Jun protooncoprotein is activated by JNK/SAPK phosphorylation and is required for Ras transformation (28).

The PI3 kinase (PI3K) pathway is another important intracellular signaling pathway which can generate so called "false messages" implicated in carcinogenesis. PI3 kinase generates 3′-phosphoinositides, which activate c-Akt, a proto oncogene product, a protein serine kinase with a phosphoinositides binding domain. In oncogenic signaling, c-Akt is activated either by overexpressed RTKs or by mutationally activated forms of the p110 PI3K catalytic subunit. Akt can phosphorylate several targets including S6kinase and GSK3 and via GSK3 inhibition it provides an intriguing connection to the cell-cell communication-related Wnt pathway, and evidence for network signaling.

Several components of the NFκB pathway have been identified as oncoproteins. NFκB is a transcription factor which plays a very important role in controlling expression of genes involved in immune and inflammatory functions as well as in controlling cell growth and oncogenesis and, with the suppression of apoptotic cascades (such as those induced by tumor necrosis factor (TNF)), expression of oncoproteins and genotoxic stress. Several oncogenic products are known to activate NFκB, including Ras, Raf, Her2/Neu, and Bcr-Abl. It has been shown that the activation of NFκB by growth factors or cytokines suppresses the apoptotic response induced by c-myc expression during growth factor deprivation. NFκB upregulates ICAM-1, a cell adhesion protein known to be upregulated in several cancers. Thus NFκB is activated by various oncogenic signaling cascades and, as a consequence, NFκB can generate false oncogenic communication signals and inhibits proapoptotic genes (29).

Another important cytokine-regulated proliferative signaling pathway is the JAK/STAT pathway. Several cytokines, such as TGFB and various interleukins, activate Janus kinases Jak1 and Jak3 which in turn activate src kinases and, among others, the STAT transcription factors. These proteins were found not to be oncoproteins but various cytokines which generate proliferative signals for tumorous transformation using these pathways.

Induction of STAT1 and STAT5 complexes was shown to occur in cases of B cell acute lymphoblastic leukemia, while failures in the Jak/Stat signaling pathway resulted in failures during lymphoid development.

Since their discovery as key mediators of cytokine signaling, considerable progress has been made in defining the structure-function relationships of Signal Transducers and Activators of Transcription (STATs). In addition to their central roles in normal cell signaling, recent studies have demonstrated that diverse oncoproteins can activate specific STATs (particularly Stat3 and Stat5) and that constitutively-activated STAT signaling directly contributes to oncogenesis. Furthermore, extensive surveys of primary tumors and cell lines derived from tumors indicate that inappropriate activation of specific STATs occurs with surprisingly high frequency in a wide variety of human cancers. Together, these findings provide compelling evidence that aberrant STAT activation associated with oncogenesis is not merely adventitious but instead contributes to the process of malignant transformation. These studies are beginning to reveal the molecular mechanisms leading to STAT activation in the context of oncogenesis, and candidate genes regulated by STATs that may contribute to oncogenesis are being identified. Recent studies suggest that activated STAT signaling participates in oncogenesis by stimulating cell proliferation and preventing apoptosis (25).

Besides false proliferative signals, oncogenic signaling can be generated via failures or lack of inhibitory signals too. The TGFβ/Smad pathway represents a very important growth inhibitory signal and Smad proteins appear to function not only as nuclear effectors for TGFβ family members, but as signal integrators within an extensive intracellular network. Smads mediate growth inhibitory and apoptotic signals and are putative tumor suppressors while other tumor suppressors and oncogene products can bind to Smads and activate and inhibit Smad function respectively.

Key elements of the proliferative or oncogenic signaling cascades are the cell cycle regulatory proteins: the cyclins and cyclin dependent kinases (CDKs). Each phase of the cell cycle is under the control of specific positive regulators (cyclin–CDK complexes) and negative regulators CDK2 interacting protein – CIP, CDK inhibitory proteins – KIP, INK) that either promote or arrest cell cycle progression depending on their relative levels of expression. During cell division cells have to pass the restriction (R) point, as described in the chapter on Cancer Genetics. At this stage CyclinD-CDK2 and CDK4 complexes phosphorylate the retinoblastoma (Rb) protein that normally is found in a hypophosphorylated form bound to transcription factor E2F. Phosphorylation of Rb leads to its dissociation from the transcriptionally active E2F, which than activates transcription of genes necessary for S-phase progression e.g. cyclins A and E.

The CDKs, key regulators of the cell cycle, consist of catalytic subunits that form complexes with proteins known as cyclins. There are at least nine CDKs (CDK1–9). There are at least 15 cyclins (cyclin A through T). Cyclin expression varies during the cell cycle. Cyclins and their cognate CDK catalytic subunits form noncovalent 1:1 complexes to produce CDK holoenzyme. The holoenzyme is activated by phosphorylation of specific residues in the CDK catalytic subunit. This phosphorylation can be catalyzed by Cdk7/cyclinH which is also known as cdk-activating kinase. Specific CDKs – complexed with their respective cyclin partners – operate in distinct phases of the cell cycle. These complexes are, in turn, regulated by a stochiometric combination with small inhibitory proteins called endogenous CDK inhibitors.

The vast majority of human cancers have abnormalities in some components of the cell cycle regulatory pathways, mostly in the Rb pathway. Such abnormalities could result from hyperactivation of CDKs, resulting from the overexpression of positive cofactors (cyclins/CDKs) or a decrease in negative factors (endogenous CDK inhibitors) or Rb gene mutations. Therefore a pharmacological CDK inhibitor that could be used in "mechanism-based therapy" would be of great interest. The loss of endogenous CDK inhibitors suggests poor prognosis. For example loss of p27(Kip1), or p16(Ink4) or p21(waf1) protein predicts a poor outcome in most patients with breast, prostate, lung, colon, gastric carcinoma or melanoma.

From a drug targeting point of view, the most important CDKs are the CDK2/CyclinA complex which is required for the cell's progression through the S phase and a complex of CDK1 (also known as Cdc2) and CyclinB which is required for mitosis, as well as Cdc25 phosphatase which activates Cdk2 and Wee1 kinase which inhibits Cdk2 (30).

The final elements of the oncogenic cascades are nuclear proteins which include the transcription factors discussed in a separate chapter. The number of identified oncoproteins involved in intracellular signaling is close to 100.

21.4.5 The seven transmembrane receptor signaling pathways

Several hormones and paracrines exert mitogenic action via the seven transmembrane receptor associated pathways. Peptide hormones and non-peptides including angiotensin II, thrombin, adenosine, lysophosphatydic acid (LPA), bradykinin, bombesin, endothelin, CCCK and gastrin acting on G-protein coupled receptors (GPCR) can contribute to tumorous transformation with an enhanced proliferative stimulus.

Upon binding of its ligand, a GPCR generally interacts with a heterotrimeric guanine-nucleotide binding protein (G-protein). Heterotrimeric G-proteins consist of three subunits: α, β and γ. The α subunit binds G-nucleotide, contains a GTPase activity and modulates effector enzymes. The β/γ complex enhances receptor interaction with α subunits and also regulates a plethora of effectors directly. Activation of G-proteins leads to the regulation of channels and second messenger-producing enzymes such as adenyl cyclase and phospholipase C.

Active adenyl cyclase (AC) produces the second messenger cyclic AMP from ATP. All nine cloned mammalian ACs can be activated by stimulatory subunits (Gs); several are modulated by inhibitory subunits (Gi) and/or complexes. cAMP can activate the cAMP-dependent protein kinase (PKA), which in turn phosphorylates a wide range of substrates, such as the cAMP responsive element binding protein (CREB). When PKA translocates to the nucleus and phosphorylates CREB, the latter is stimulated to regulate gene transcription.

There are three mammalian phospholipase C (PLC) isoform families: PLC-β, PLC-γ and PLC-δ. Members of the PLC-β family are activated by serpentine receptors and PLC-γ isoforms are stimulated through RTKs. There is cross-talk between AC and PLC pathways: PKA can specifically inhibit Gβ/γ-stimulated PLC-β2. Active PLCs catalyze the hydrolysis of PIP2 to generate the second messengers IP3 and DAG. On the endoplasmatic reticulum, there are IP3-specific receptors, which regulate $Ca2^+$ release into the cytoplasm. Cytosolic $Ca2^+$ can modulate the activity of serine/threonine $Ca2^+$/calmodulin-dependent kinases (CaMKs) via calmodulin. One of the other effects mediated by $Ca2^+$ is regulation of the conventional PKC isoforms. Mitogenic GPCRs can also activate the Ras–MAPK cascade.

At least four G-protein mediated mitogenic signaling pathways have been identified: (1) stimulation of PLC; (2) inhibition of AC; (3) activation of Ras and the downstream Raf–MAPK pathway; and (4) remodeling of the actomyosin cytoskeleton. Activation of the Ras–MAPK cascade involves Gi/subunits, tyrosine kinase activity and recruitment of the adapter protein Grb2 (and therefore Sos, etc.) to the plasma membrane.

Several studies provide compelling evidence that GPCRs use RTKs as signaling intermediates. In a number of cell lines, agonists that activate GPCRs also induce tyrosine phosphorylation of the EGF receptor. These ligands include lysophosphatidic acid (LPA), endothelin, thrombin, bombesin, angiotensin II. Similarly, thrombin increases tyrosine phosphorylation of the insulin-like growth factor 1 receptor. In cells lacking EGF receptor, GPCR can activate the PDGF receptor, although EGF receptor is the preferred RTK. Several studies found that RTK activity was required for GPCR-induced DNA synthesis, activation of mitogen-activated protein kinase (MAPK) and c-fos expression. Src-family kinases may be implicated in activation of RTKs by GPCRs. On the whole, ligand-independent transactivation of RTKs seems to contribute, in a cell-type specific manner, to GPCRs-mediated mitogenic signaling and represents a novel paradigm for cross-talk between signaling pathways (31).

Interestingly, the only endogenous antiproliferative hormone, somatostatin also acts via the seven transmembrane receptor signaling pathways and although there is cross-talk with EGFR here also, this is an inhibitory message, while somatostatin, and especially the recently developed tumor selective somatostatin analog (TT232), inhibits various other signaling events downstream to EGFRTK. Their mechanism of action not only involves inhibition of the EGFRTK pathway, but activation of a MAP kinase cascade as well, resulting in activation of p21 Cip and cell cycle arrest (32) (Figure 21.5).

The role of GPCRs-mediated mitogenic signaling has been well documented in several forms of cancer including gastric, lung, prostate, ovarian cancer and lymphoma. The GPCR mediated mitogenic signals are not usually primary oncogenic signals, but tumor promoting signals which evolve as part of the "false" communication and contribute to the malignant phenotype. Peptide hormones are often secreted by epithelial tumors as part of an autocrine loop.

Recently, for example, it was demonstrated that endothelin 1 (ET-1) is produced in ovarian cancer cell lines and has been shown to act through ET(A) receptors as an autocrine growth factor to promote tumor cell proliferation in vitro. Bradykinin receptors have been demonstrated in several types of tumors and kinins facilitate tumor proliferation and invasion. The observation of PRL-responsive expression of FGF-2 in Nb2 lymphoma cells suggests a previously unrecognized pathway for PRL action in lymphoid cells. Thrombin, a multifunctional protein, has been found to be involved in cellular mitogenesis, tumor growth, and metastasis, in addition to its well known effects on the initiation of platelet aggregation and secretion and the conversion of fibrinogen to fibrin to form blood clots. The role of bombesin in gastrointestinal tumors has been well documented. The hormone bombesin and its mammalian equivalent, gastrin-releasing peptide (GRP), act through specific GRP receptors to affect multiple cellular functions in the gastrointestinal tract. Human gastric cancer SIIA possesses GRP R and rapid induction of c-jun and jun-B gene expression was noted after bombesin treatment; this effect was blocked by specific GRP-R antagonists, indicating that bombesin is acting through the GRP-R. The induction of AP-1, in response to bombesin, is mediated through both PKC- and PTK-dependent signal transduction pathways, which contribute to network signaling in tumors (33).

21.4.6 The extracellular matrix/adhesion protein pathways

Expression of a number of cell-surface molecules that mediate cell-cell interaction is associated with tumor development. These molecules include members of the immunoglobulin gene family such as ICAM (Intracellular Cell Adhesion Molecule-1) VCAM, CD4, Integrins such as $\alpha4\beta1$, $\alpha6\beta4$, and cadherins such as E-cadherin and N-cadherin. In the past decade, there have been major advances in the understanding

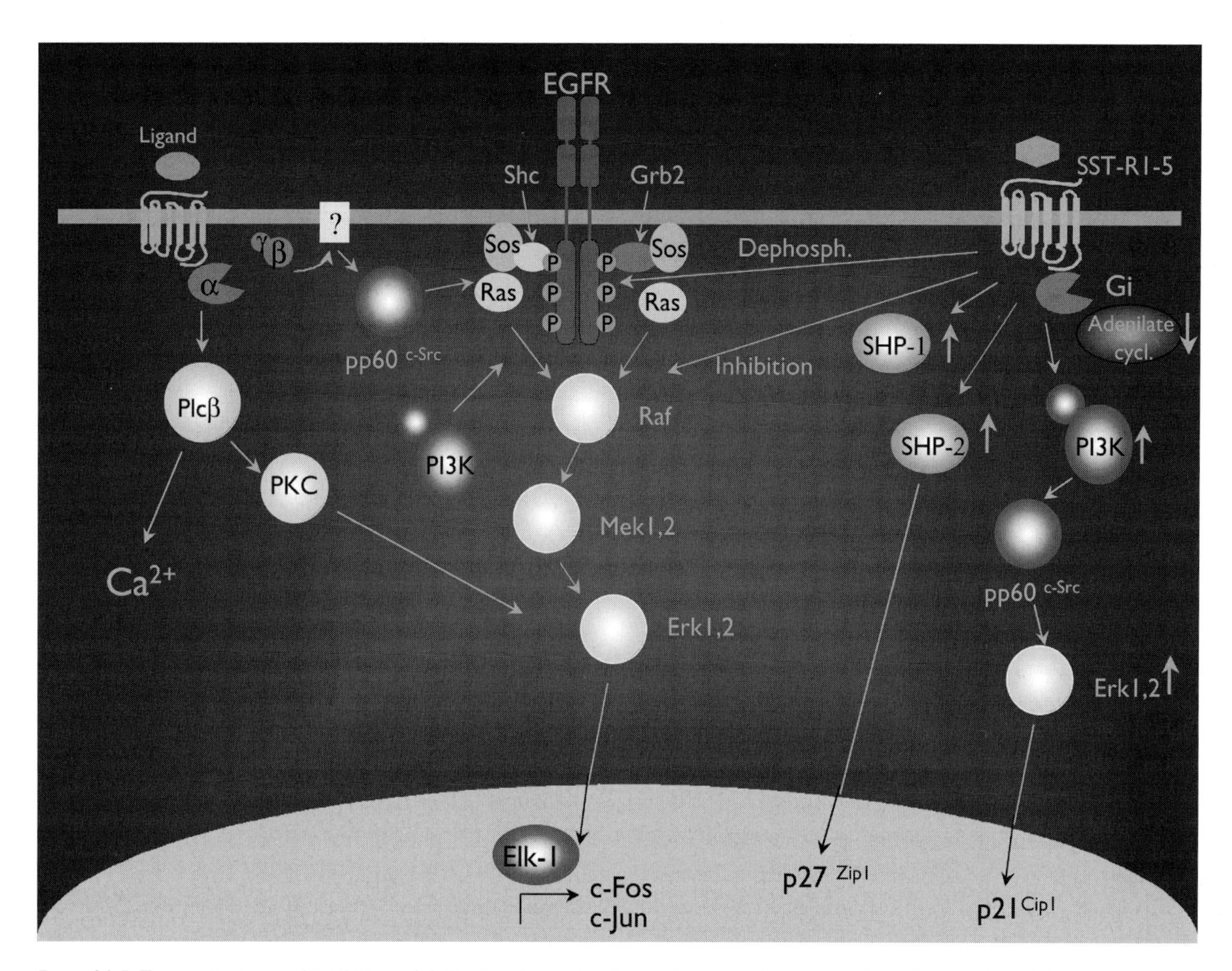

Figure 21.5 Transactivation and inhibition of GFRTK pathways by G-protein-coupled receptors (see Colour Plate XV).

of some of the mechanisms underlying tumor differentiation, invasion, and metastasis, in which cell-cell and cell-matrix adhesion molecules play a critical role. The cadherin/catenin complex and the integrins are the prime mediators of cell adhesion in normal and transformed cells, cadherin/catenin being largely responsible for intercellular adhesion and integrins for cell-extracellular matrix interactions.

Cadherins are Ca-dependent transmembrane proteins which normally help to maintain tissue architecture in adult animals. On the cytoplasmic side cadherin molecules interact with catenins which link cadherin molecules to the cytoskeleton. Through interactions with catenins, cadherins form cell junctions which are crucial for epithelial cells. Many different cadherins exist, the main ones involved in cell growth control are N-cadherin (A-CAM etc.) and E-cadherin (Cell-CAM).

Integrins are a superfamily of transmembrane glycoproteins that mediate cell-cell and cell–matrix interactions (See also Cancer Genetics). Integrin expression appears to be universal, at least one member of the integrin family has been found in every cell/tissue studied. Integrins are heterodimers consisting of noncovalently associated α and β subunits. More than 20 subunits are known from the literature and the various combinations can interact with ligands such as laminin, fibronectin, ICAM, VCAM, vitronectin etc. Integrin ligands include bacterial and viral proteins, coagulation and fibrinolytic factors, complement proteins and cellular counter–receptors. Integrins bind to their counter-receptors only after activation. The conformation of both integrin and its ligand is critical to intergrin-ligand interaction.

Integrins are the major type of cell surface receptors that bind to ligands on adjacent cells or in the extracellular matrix. Integrins are heterodimers of subunits, which consist of a large extracellular, a single transmembrane and a short cytoplasmic domain. Most integrins bind extracellular matrix components like fibronectin, collagen or vitronectin. Upon ligand binding, the integrins cluster. This leads to the formation of focal adhesions, where integrins link to intracellular cytoskeletal complexes. The cytoplasmic domains of integrins do not have intrinsic enzymatic activities and appear to activate intracellular tyrosine kinases by clustering. Integrin engagement elevates focal adhesion kinase (FAK) PY levels and FAK-associated tyrosine kinase activity. The non-receptor tyrosine kinase Src is thought to associated with an auto-phosphorylated form of FAK and to phosphorylate FAK in turn. Both FAK and Src phosphorylate the adapter molecule Shc at multiple sites to create SH2 binding sites for the Grb2 adapter protein; phosphorylation of a FAK tyrosine residue near the C-terminus (by a Src family kinase) also promotes Grb2 binding. This provides a link to the proliferative Ras–MAPK cascade. Certain levels of integrin-dependent signaling to MAPK can also be achieved in the absence of either Src family or FAK activity (Figure 21.3).

Intercellular and cell-matrix adhesion mediated by cadherin/catenin and integrins are likely to play a role in the control of both structural morphology and functional differentiation; hence, any loss of this control mechanism may well facilitate the neoplastic process. Indeed, in cancer cells there is often a co-ordinated down-regulation of both integrins and cadherins, which correlates with tumor dedifferentiation. However, the expression and cellular localization of catenins do not always correlate with cadherin expression, since catenins are rather promiscuous molecules, which interact not only with E-cadherin, but also with growth regulatory and signaling molecules such as epidermal growth factor receptor and the adenomatous polyposis coli gene product.

Receptor-like protein tyrosine phosphatases (RPTPs) are also involved in intercellular communication and cell-cell adhesion. RPTPs contain a variable extracellular domain, a transmembrane domain and one or two intracellular tyrosine phosphatase domains. Depending on the RPTP subtype, the extracellular domain exhibits immunoglobulin- and fibronectin-like regions and other sequence motifs involved in cell–cell adhesion. The subtype I RPTP CD45 is involved in the activation of B- and T-lymphocytes. It stimulates the Src-like kinases Lck and Fyn by dephosphorylating them. The subtype II RPTPs have been shown to modulate cell–cell interaction with a protein complex containing cadherin transmembrane adhesion molecules and catenins (which are linked to the underlying actin cytoskeleton).

The extracellular matrix (ECM) can affect cell behavior either by mobilization of growth factors and growth factor binding proteins using remodeling enzymes as pivotal tools, via direct regulation of receptor-mediated signaling or by modulating the cellular response to growth factors. Most, if not all, transformed cells have abnormal interactions with their extracellular environment. Epithelial cells are sometimes incapable of producing an organized basement membrane, which would normally induce growth arrest, or the malignant cells fail to recognize their ECM because of expression of inappropriate or non-functional integrins. For example, it was found that by correcting cell-ECM interaction malignant breast epithelial cells could be returned to their quiescent differentiated state simply by altering the level of signaling from the appropriate integrins (34,35).

21.4.7 Endocrine/paracrine regulators acting on intracellular receptors

Cell proliferation and differentiation is regulated by a series of endocrine/paracrine regulators which enter cells and exert their effects via intracellular and/or nuclear receptors. Among these types of signaling hormones are the steroid hormones (estrogens, progestins, glucocorticoids, androgens etc) and other small molecular hormones (thyroid hormone, calcitriol (a Vitamine D3 metabolite) and the retinoids). Most of the receptors for these molecules are transcription factors, transcriptional regulators or chromatin regulatory proteins and they can become oncogenic, for example via overstimulation (because of constitutive signals), mutation, chromosomal translocation etc. Some of the oncoproteins generated act as transcriptional repressors. The v-ErbA for example is a repressor of thyroxine receptor (c-ErbA) induced erythroid differentiation genes. The promyelocytic leukemia zinc finger (PLZF) protein, which is fused to the retinoic acid receptor protein as a result of chromosomal translocation, represses expression of a cyclinA promoter reporter gene. Ligand activated intracellular receptors exert their effect by binding directly to chromosomal enhancer sequences termed hormone response elements (HRE) that are located within the regulatory regions of target genes. Steroid hormones regulate the synthesis of stimulatory and inhibitory growth factors, growth factor receptors, and growth factor binding proteins. In turn, growth factor pathways may modulate phosphorylation and function of steroid receptors and potentiate or inhibit their action. For example, mitogenic and inhibitory growth factors and ovarian steroid hormones play important roles as selective modulators of normal mammary development and in the onset and progression of human breast cancer. Ultimately, during the progression of the malignant mammary epithelial cell to hormonal autonomy, overexpression, mutation, or disregulation of key elements of growth factor signal transduction pathways may all play critical roles.

21.5 Antitumor protective mechanisms and therapeutic perspectives

Tumorous transformation and tumor growth is a multicomponent, multistep process and a healthy organism has several defense lines and protective mechanisms to fight this complex phenomenon. During the multistep process of carcinogenesis, environmental factors and effects such as chemicals, radiation, certain viruses and other pathogens induce genomic changes as well as causing disorders in inter- and intracellular communications and they also damage or eliminate existing protective mechanisms and immune surveillance. All the effects that stimulate the replicative capacity of progenitor cells may increase the likelihood of such rare and heritable mutagenic events to occur. In addition, chemicals and effects that damage DNA and favor the reduction of growth restraints may further enhance the likelihood of achieving those genetic changes which, via a "negative evolution" and escape of immune surveillance, support survival of the tumor cells. While the first such events may give only a modest growth advantage and minor changes in cytopathology, the probability that this cell will steadily accumulate genetic changes increases as the proliferative capacity of the cell is enhanced. In addition, DNA repair processes may be overwhelmed or reduced due to genetic alterations, leading to further escalation of DNA damage, which results in further escalation of the failed inter- and intracellular communication. This could result in a further decrease in normal growth controls. Some of these genetic changes may signify successful attempts of the emerging malignant lesion to overcome tissue specific growth or immunologic constraints. This process has been well demonstrated in human colorectal cancer where genetic activation of dominant-acting oncogenes and loss of tumor suppressing genes mark sequential progression to more malignant tumor forms. There are at least seven communication or defense mechanisms which try to correct improper or false communication in the tumor cell which results in permanent proliferation. These levels include endocrine, paracrine and cytokine actions which try to keep the deviant cell under control. The main message from all these communication layers is that the cell should behave according to its destiny, either doing its job or mission (differentiated function) according to system requirements, (or at least not proliferating with a false message), or dying through programmed cell death (Figure 21.6).

As we learn more about the molecular pathomechanism of cancer and its associated communication failures, as well as related regulatory or protective mechanisms, we can optimize, support or use these defense strategies for therapeutic purposes. In other words the new trends in tumor therapy and drug research are not only to kill (or remove) the bad guys, but to do so with less harm to the system and by reestablishing a proper communication network to support differentiated functions. The new trends in therapeutic approaches and drug research in cancer try to accomplish this task.

Modern cancer chemotherapy has been a moderately successful experimental and practical science for about four decades. During this time many useful drugs have been developed, mostly by systematic screening or by molecular modification of existing active drugs. The primary aim for most of these drugs is to kill cancer cells. Since many types of cancer cells are proliferating relatively rapidly, the most effective drugs have been those that are cytotoxic against cells engaged in active growth.

In most cases, cancer chemotherapy does not eradicate all cancer cells and cure the patient completely, but often only prolongs survival or causes tumor shrinkage (a complete or partial response). Individual drugs or combinations of drugs effective in the treatment of one cancer may not be effective for a different type of malignancy. For most chemotherapeutic regimens, there is a relationship between the dose of drug given and the tumor shrinkage achieved (i.e. higher doses give greater tumor cell kill). However, as drug doses are increased, toxicity increases. The effective dose of an antineoplastic agent often causes toxicity to normal tissues, thus causing very serious side effects and dramatic decrease in quality of life. On the other hand, cancer cells are very effective at developing biochemical mechanisms, which allows cellular resistance to a particular antineoplastic agent. In order to overcome the problem of resistance development, clinicians generally use more than one drug (combination chemotherapy). Drugs used in combination therapies generally have different cellular targets to avoid development of drug resistant tumor cell lines. The most effective treatment regimens are combinations of drugs. Cytotoxic combination regimens have been shown to produce higher response rates and longer durations of response and survival than single-agent therapy. A particular drug may produce some tumor shrinkage when used alone, but can produce much better results when used in combination with other antineoplastic drugs.

Since conventional cancer chemotherapy is not effective enough and lacks selectivity between cancer cells and normal cells, great efforts have been directed toward the development of novel selective antitumor drugs which are not toxic for normal tissues. The discovery of novel anticancer agents are based on novel molecular targets through the application of molecular biology methods to cancer biology.

Advances in cancer research have developed from modern technology, allowing the establishment of cancer cell lines and experimental model systems, recombinant DNA technology, robust biological and biochemical functional assays in addition to classical methods such as immunohistochemistry. Further improvements such as DNA array technology will continue to unravel the genetic and epigenetic defects in a far more rapid and comprehensive way that was previously possible. This knowledge will lead to genetic and functional genomic characterization of cancer cells and its specific subtypes, which will translate to new diagnostic and therapeutic strategies. Novel targets include genes and gene products involved in malignant transformation, signal transduction, cancer progression and metastasis. Thus modern anticancer drug discovery is molecular target-based, using genomic and proteomic approaches with bioinformatic tools to identify novel drug discovery targets, and to probe mechanisms of actions and toxicity of potential drug molecules. Medicinal and combinatorial chemistry and high-throughput screening against identified molecular targets are the methods generally used for primary anticancer drug discovery. Computational structure-based

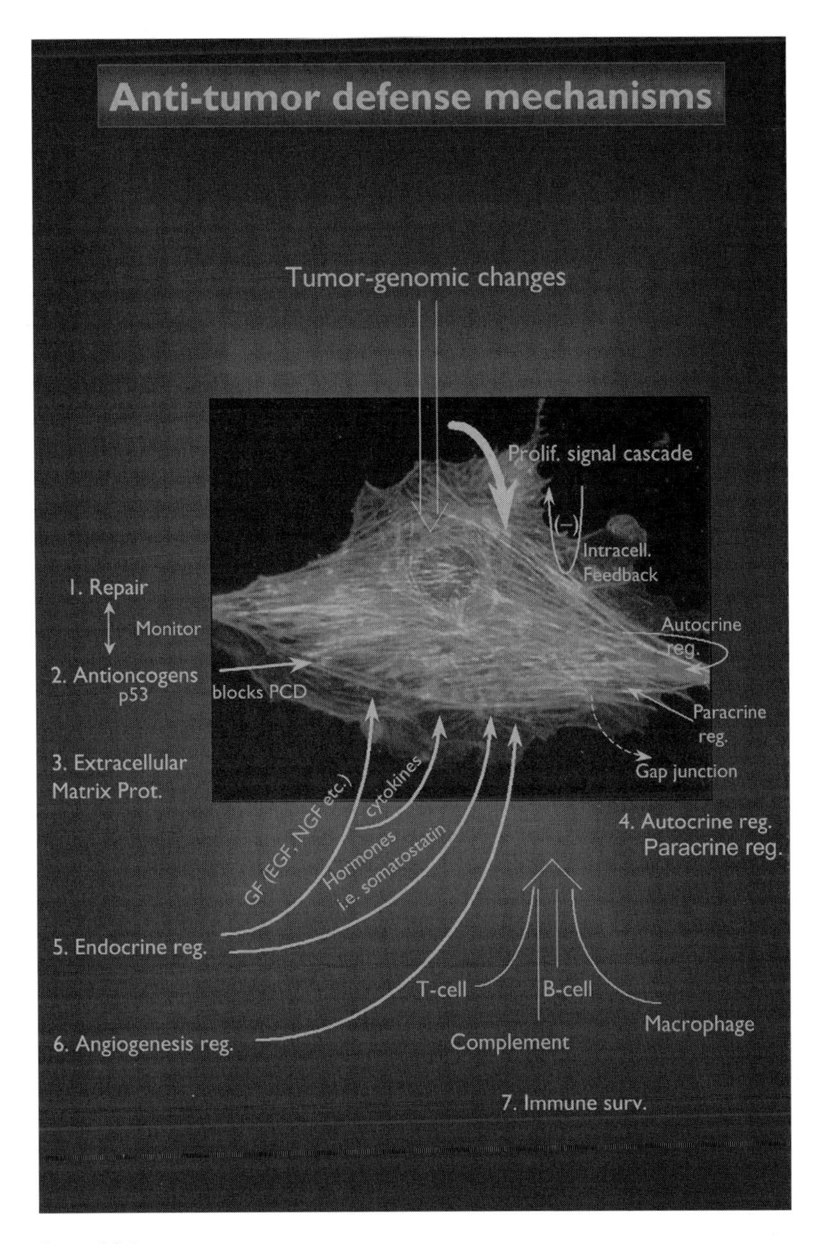

Figure 21.6

drug design, utilizing X-ray crystallographic information, is also becoming rapidly established in cancer drug discovery. One way to cause cell death is to upset the pathways that cells use to control normal growth. All cells have receptors for growth factors, mitogens and hormones, and these systems collectively participate in signal transduction processes that regulate orderly, well-integrated growth and proliferation. Novel molecular targets for anticancer drug discovery include growth factor receptor tyrosine kinases (RTKs) and serine/threonine kinase signal transduction pathway targets; cell cycle targets; apoptosis-related targets; extracellular matrix targets, tumor angiogenesis and metastasis targets.

The current field of anticancer drug discovery and development is based on the concept that, by identifying the genes and proteins responsible for precise molecular abnormalities that create and drive the malignant phenotype, therapies will emerge that are both more effective and better tolerated than traditional cytotoxic agents. Examples of molecularly targeted signal transduction agents include inhibitors of various kinases and oncoproteins. Approaches that target the tumor cell environment include metalloproteinase and integrin inhibitors as well as antiangiogenic strategies. Several specific tyrosine kinase inhibitors against overexpressed HER-2/neu, PDGFR or EGFR are undergiong clinical trials with very promising results. A specific inhibitor of the bcr/abl oncogenic tyrosine kinase (called: Gleevec) has been very successfully launched for clinical practice against chronic myelogenous leukemia (CML), while an EGFRTK inhibitor and a cyclin dependent kinase inhibitor is in Phase III clinical trials. A very potent inhibitor of angiogenesis (specific inhibitor of VEGF-RTK) is also in Phase III clinical trials which promises to become a useful drug for solid tumors. A specific inhibitor of the ras oncogenes, which is activated in many tumors (a Ras farnesyl transferase inhibitor) is also in phase II/III clinical trials. As part of a targeted therapeutic intervention several monoclonal antibodies have recently been used for treatment of patients. Rituximab binds to the CD20 B cell differentiation antigen and is used to treat patients with B cell lymphomas. Another monoclonal antibody Trastuzumab is directed against Her-2, which is overexpressed on metastatic cancers in 25–30% of women with breast carcinoma. A monoclonal antibody against VEGFRTK is in Phase III clinical trials. There are at least eight synthetic matrix metalloproteinase inhibitors (MMPIs) in various phases of clinical trial for the treatment of malignant invasion and metastasis of a number of tumors.

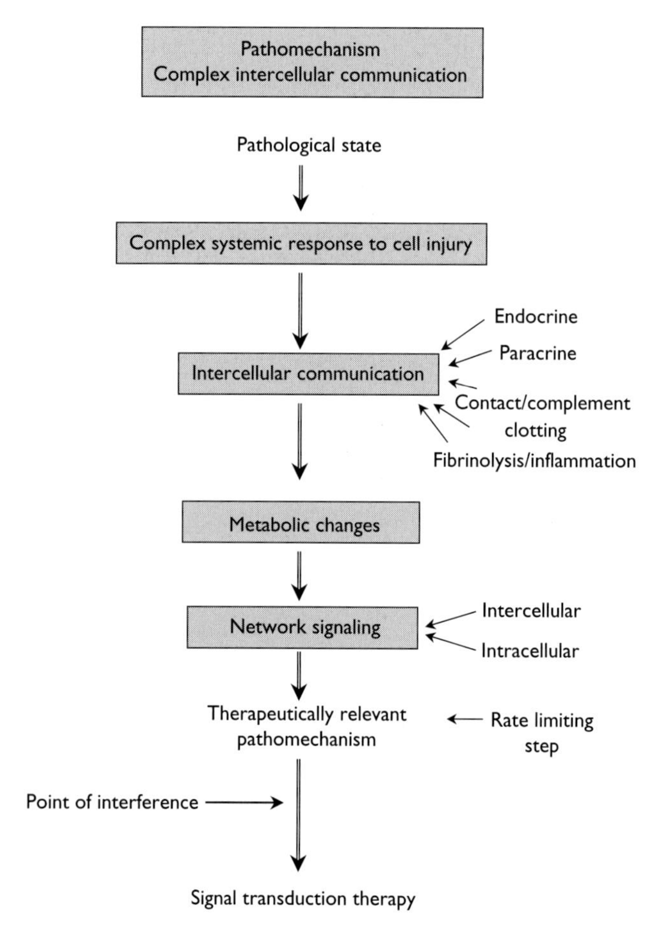

Figure 21.7 Systemic nature and therapeutic approach of the pathological state leading to cancer disease.

Another interesting and creative approach for the specific killing of cancer cells comes from the use of engineered adenoviruses (like ONYX-015, currently in clinical trials) which selectively replicate in, and kill cells that have lost p53 function but are unable to replicate in, and therefore do not affect, cells with normal p53 function.

Immune-based therapy and adoptive cellular therapy as well as gene therapy represent additional promising areas for the management of otherwise unresponsive tumors or as adjuvant therapies. In such approaches lymphokine-activated killer cells, autolymphocyte therapy, and tumor-infiltrating lymphocytes have been used in various clinical protocols with inconsistent benefits.

In summary, we believe that on the basis of our understanding of the systemic nature of the pathological state leading to cancer disease and the inter and intra-cellular communicative disorders involved, there is great potential for the discovery of innovative cancer drugs with improved efficacy and selectivity (Figure 21.7). Novel mechanism-based agents are being discovered by focusing on molecular targets and signal transduction pathways involved in cancer formation, maintenance and progression. On the other hand, mechanism-based targeted drug design has to consider the systemic response and network signaling, remembering the old Chinese saying that "the sea is in the drop and the drop is in the sea" (36,37).

Finally I would like to refer to the Chinese saying which was selected as a motto for this book:

> "Sickness comes from lost integrity
> Sickness comes from lack of inner peace
> The unity of spirit, soul, body and mind
> Will make you happy and shine."

References

1 Cantiey, L.C., Auger, K.R., Carpenter, C., Duckworth, B., Craziani, A., Kapeller, R., Soltoff, S. (1991) Oncogenes and signal transduction. *Cell 64: 281–302.*

2 DeRisi, J., Penland, L., Brown, P.O., Bittner, M.L., Meltzer, P.S., Ray, M., Chen, Y., Su, Y.A., Trent, J.M. (1996) Use of a cDNA microarray to analyse gene expression patterns in human cancer. *Nat Genet 14: 457–460.*

3 Liotta, L. and Petricoin, E. (2000) Molecular profiling of human cancer. *Nature Reviews Genetics (2000) 1: 48–56.*

4 Hunter, T. (1997) Oncoprotein networks *Cell 88: 333–346.*

5 Powis, G., Workman, P. (1994) Signalling targets for the development of cancer drugs. *Anti-Cancer Drug Design 9: 263–277.*

6 Buolamwini, J.K. (1999) Novel anticancer drug discovery. *Current Opinion in Chemical Biology 3: 500–509.*

7 Myers, T.G., Anderson, N.L., Waltham, M., Li, G., Buolamwini, J.K., Scudiero, D.A., Paull, K.D., Sausville, E.A., Weinstein, J.N. (1997) A protein expression database for the molecular pharmacology of cancer. *Electrophoresis 18: 647–653.*

8 Trosko, J.E., Ruch, R.J. (1999) Cell–cell communication in carcinogenesis. *Front Biosci 3: 208–236.*

9 Hayashi, T., Nomata, K., Chang, C.C., Ruch, R.J., Trosko, J.E. (1998) Cooperative effects of v-myc and c-Ha-ras oncogenes on gap junctional intercellular communication and tumorigenicity in rat liver epithelial cells. *Cancer Lett 128(2): 145–54.*
10 Park, C.C., Bissell, M.J. and Barcellos-Hof, M.H. (2000) The influence of the microenvironment on the malignant phenotype. *Molecular Medicine 6, 324–329.*
11 Stern, R.G., Milestone, B.N., Gatenby, R.A. (1999) Carcinogenesis and the plasma membrane. *Med Hypotheses 52(5): 367–72.*
12 Huang, S. (2000) Gene expression profiling, genetic networks, and cellular states: an integrating concept for tumorigenesis and drug discovery. *J. Mol. Med. 469–480.*
13 Guan, K.L. (1994) The mitogen activated protein kinase signal transduction pathway: from cell surface to the nucleus. *Cell Signal 6: 581–589.*
14 Zwick, E., Hackel, P.O., Prenzel, N., Ullrich, A. (1999) The EGF receptor as central transducer of heterologous signalling systems. *Trends Pharmacol Sci 20(10): 408–12.*
15 Schenk, P.W. and Snaar-Jagalska, E. (1999) Signal perception and transduction : the role of protein kinases. *Biochemica et Biophysica Acta 1449: 1–24.*
16 Pawson, T. (1995) Protein modules and signaling networks. *Nature 373: 573–580.*
17 Kolibaba, K.S. and Druker, B.J. (1997) Protein tyrosine kinases and cancer. *Biochem Biophys Acta 1333: F217–248.*
18 Plowman, G.D., Ullrich, A., Shawver, L.K. (1994) Receptor tyrosine kinases as targets for drug intervention. *DN&P 7:334–339.*
19 Levitzki, A., Gazit, A. (1995) Tyrosine kinase inhibition: an approach to drug development. *Science 267:1782–1788.*
20 Strawn, L.M. and Shawver, L.K. (1998) Tyrosine kinases in disease: overview of kinase inhibitors as therapeutic agents and current drugs in clinical trials. *Exp. Oppin. Invest. Drugs 553–573.*
21 Kishimoto, T., Taga, T., Akira, S. (1994) Cytokine signal transduction. *Cell 76: 253–262.*
22 Saunders, J. and Tarby, C.M. (1999) Opportunities for novel therapeutic agents acting on chemokine receptors. *Drug Discovery Today 4: 80–92.*
23 Chen, Y., Hughes-Fulford, M. (2000) Prostaglandin E2 and the protein kinase A pathway mediate arachidonic acid induction of c-fos in human prostate cancer cells. *Br J Cancer 82: 2000–6.*
24 Strockbrugger, R.W. (1999) Nonsteroidal anti-inflammatory drugs (NSAID) in the prevention of colorectal cancer. *Eur J. Cancer Prev Supl 1: S21–5.*
25 Catlett-Falcone, R., Dalton, W.S., Jove, R. (1999) STAT proteins as novel targets for cancer therapy. *Current Oppinion in Oncology 11: 490–496.*
26 Hsieh, J.C., Rattner, A., Smallwood, P.M. and Nathans, J. (1999) Biochemical characterization of Wnt-frizzled interactions using a soluble, biologically active vertebrate Wnt protein. *Proc Natl Acad Sci USA 96: 3546–51.*
27 Campbell, S.L.,Khosravi-Far, R., Rossman, K.L., Clark, G.J., Der, C.J. (1998) Increasing complexiity of Ras signaling. *Oncogene 17: 1395–1413.*
28 Cano, E. and Mahedevan, L.C. (1995) Paralell signal processing among mammalian MAPKs. *Trends Biochem Sci 20: 117–122.*
29 Mayo, M.W. and Baldwin, A.S. (2000) The transcription factor NF-KB : control of oncogenesis and cancer therapy resistance. *Biochem. Biophys Acta 1470: M55–M62.*
30 Senderowicz, A.M. and Sausville, E.A. (2000) Preclinical and Clinical Development of Cyclin-Dependent Kinase Modulators. *Journal of National Cancer Institute 92 376–387.*
31 Prenzel, N., Zwick, E., Daub, H., Leserer, M., Abraham, R., Wallash, C. and Ullrich, A. (1999) EGF receptor transactivation by G-protein-coupled receptors requires metalloproteinase cleavage of proHB-EGF. *Nature 402 884–888.*
32 Keri, G., Erchegyi, J., Horvath, A., Mezo, I., Vantus, T., Bokonyi, G., Seprodi, J., Tepllan, I. Csuka, O., Gaal, D., Szegedi, Z., Szende, B., Roze, C., Kalthoff, H., Ullrich, A. (1996) A tumor selective somatostatin analoog (TT232) with strong *in vitro* and *in vivo* antitumor activity. *Proc Natl Acad Sci USA 93: 12513–518.*
33 Kim, H.J., Ewers, B.M., Litvak, D.A., Hellmich, M.R., Townsend, C.M.Jr. (2000) Signaling mechanism regulating bombeesin mediated AP-1 geneinduction in the human gastric cancer SIIA. *Am. J. Physiol cell Physiol 279: C326–34.*
34 Streuli, C. (1999) Extracellular matrix remodelling and cellular differentiation. *Current Opinion in Cell Biology 11: 634–640.*
35 Boudreau, N. and Bissell, M. (1998) Extracellular matrix signaling: integration of form and function in normal and malignant cells. *Current Opinion in Cell Biology 10: 640–646.*
36 Seymour, L. (1999) Novel anticancer agents in development: exciting prospects and new challenges. *Cancer treatment reviews 25: 301–312.*
37 Gelmon, K.A., Eiseenhauer, E.A., Harris, A.L., Ratain, M.J., Workman, P. (1999) Anticancer agents targeting signaling molecules and cancer cell environment: challenges for drug development. *Journal of National Cancer Institute 91: 1281–1287.*

Part 6

Infectious diseases

Chapter 22

Human immunodeficiency virus

Pathogenesis, replication strategy and the role of antiretroviral drugs

Chris Birch

Contents

22.1 Pathomechanisms of HIV infection

22.1.1 The course of HIV infection

After entering the body through the blood or via a mucosal surface, HIV localizes in regional lymph nodes where it replicates in susceptible cells. The initial infection is facilitated by an interaction between the virus and monocyte-derived dendritic cells found in skin and mucosal surfaces (Reece *et al.*, 1998). The course of primary HIV infection is characterized clinically in up to 70% of individuals by a glandular fever-like syndrome with symptoms including fever, headache, lethargy, sore throat and lymphadenopathy (Pantaleo *et al.*, 1993a). These symptoms generally occur within 3 to 6 weeks of infection, although this period may be longer when the dose of virus transmitted is low (such as in a needle-stick injury).

A high-level viremia, which occurs when symptoms appear and often results in titers of more than one million virions per ml of blood (Piatak *et al.*, 1993), then disseminates virus to other lymphoid organs (lymph nodes, tonsils and adenoids). At this time, a sharp decline in the number of CD4-positive lymphocytes (CD4+ cells) occurs, cytotoxic T-cells are generated, HIV-specific circulating antibodies are produced, and the virus titer in the blood decreases, often by several logs. These events occur in parallel with the entrapment of HIV virions and antigens in the lymphoid organs by follicular dendritic cells, a process which contributes to the induction and maintenance of the immune response (Pantaleo *et al.*, 1993b) but which is insufficient to eliminate the virus from the body.

After these primary events, HIV infection is characterized by a clinically latent, virologically dynamic state, in which the destruction of CD4+ cells as a result of virus replication is fairly closely counterbalanced by their regeneration by the immune system. In fact, there is a slow decline in CD4+ cell numbers approximating 70/μl/year over the course of the infection (Weber, 1998). During this time HIV can be detected and quantified in the blood (although it may be several logs lower than at the time of primary infection) (Piatak *et al.*, 1993), but in the untreated individual its level remains relatively constant, often for years (Ho, 1996). This is the virological "set-point", and its level in terms of quantifiable virus in the blood is predictive of the long-term clinical outcome for the patient (Mellors *et al.*, 1996). Thus, even when the time of infection is not known, the risk of AIDS and death in the untreated patient is directly related to the plasma viral load.

Because the immune system does not have unlimited regenerative capacity, ultimately, during the later stages of infection, a steep decline in CD4+ cell numbers occurs which corresponds with an increase in HIV levels in the blood. In many patients the virus load may reach levels similar to that seen at the time of primary infection (Piatak *et al.*, 1993). This increasing load is associated with a breakdown in the architecture of the lymphoid organs, and a redistribution of HIV to cells outside the lymphoid system, including the brain, lung, colon and liver (Donaldson *et al.*, 1994). In untreated patients who have previously experienced few, if any, symptoms associated with their infection, this decline in CD4+ cells to below 500 per μl of blood signals the possible onset of the opportunistic infections and malignancies associated with AIDS, including pneumocystosis, tuberculosis, cryptosporidiosis, toxoplasmosis, esophageal candidiasis, herpes simplex and cytomegalovirus infections and Kaposi's sarcoma. The timing of acquisition of many of these infections is associated with the degree of decline in immune function as measured by CD4+ cell counts (Crowe *et al.*, 1991).

In untreated patients analyzed as a group, the duration of infection prior to the onset of AIDS to some extent assumes the shape of a bell-shaped curve, with most patients progressing between 11 and 12 years post-infection, but a proportion (20%) progressing rapidly (within 5 years) and about 12% remaining asymptomatic 20 years after infection (Mellors *et al.*, 1996). The availability of antiretroviral drugs to a minority of those individuals infected world-wide with HIV has changed the natural history of HIV infection in these patients. In the context of adherence to these drugs and the subsequent control of virus replication (to below the level of detection in the blood using conventional assays), many treated patients regain normal CD4+ counts (although these may represent an expansion of memory cells rather than naïve cells) and remain free of opportunistic infections. It is unclear at this stage in what proportion of patients, and for how long, this effect of drug therapy will endure, but for many patients it can be measured in years.

The availability of potent antiretroviral drug combinations has raised the prospect that HIV infection might be eliminated if adherence was sustained for sufficient time to overcome the longevity of infected cells. The great majority of such cells have rather short half-lives (Table 22.1). Unfortunately, however, the existence of a small number of long-lived (memory) lymphocytes, perhaps totalling one million within the body, provides a substantial impediment to such a strategy, in that the time required for their complete elimination is measurable in tens of years. Adherence for this length of time to a drug regimen providing complete suppression of virus replication is not a practical reality.

Table 22.1 Peripheral blood cell reservoirs of HIV and the influence of their survival time on the maintenance of HIV infection. The longevity of latently infected memory cells ensures that antiretroviral therapy must be maintained for many years if (theoretical) eradication is to be achieved

Cell type infected	*Estimated half-life*
CD4+ lymphocytes, productively infected	1.1 days
CD4+ lymphocytes, latently infected (HIV DNA not integrated)	8.5 days
CD4+ lymphocytes, containing a defective genome	4–5 months
CD4+ lymphocytes (memory cells), latently infected, with integrated genome	5–6 months
Monocyte-macrophages	14 days

22.2 The HIV replication strategy

22.2.1 Cell surface receptors for HIV

HIV attachment to susceptible cells involves interaction between the viral envelope glycoprotein gp120, the cellular receptor CD4 and members of the chemokine (cytokine) receptor family. While CD4 is the primary receptor for HIV, the so-called secondary cytokine co-receptors most likely represent the primordial receptors for the retroviruses in general (Wyatt & Sodroski, 1998). In the case of HIV, binding of gp120 to CD4 exposes a second domain of the envelope glycoprotein to the chemokine receptor, resulting in a conformational change which leads to fusion mediated by the viral transmembrane protein gp41. At least two co-receptors for HIV-1 have been identified. They normally act as receptors for cytokines that attract leukocytes to areas of inflammation during infection. In concert with CD4, the CCR-5 receptor allows infection of cells of the monocyte lineage by macrophage-tropic strains of HIV that are transmitted sexually, predominate in the early stages of infection and are present throughout infection (Deng *et al.*, 1996; Dragic *et al.*, 1996; Huang *et al.*, 1996). In contrast, the CXCR-4 receptor facilitates infection of T-cell tropic strains that emerge as the infection progresses (Feng *et al.*, 1996).

The essential role of cytokine receptors in the infection process is demonstrated by the markedly reduced susceptibility to infection of individuals homozygous for a 32 base pair deletion (Δ32) in the gene for CCR-5. This results in the production of a truncated protein that is not incorporated into the cell membrane (Liu *et al.*, 1996). Δ32 heterozygous individuals, while still vulnerable to infection, have a slower rate of disease progression than the normal population as evidenced by lower plasma virus load and a slower decline in CD4+ cells following infection (Huang *et al.*, 1996). While accounting for resistance to HIV infection in some multiply-exposed individuals, Δ32 mutations do not explain all cases where HIV infection might have been expected to occur but has not. For example, a cohort of Kenyan prostitutes who are not homozygous for the mutation has remained free of HIV infection despite repeated sexual exposure to the virus (Fowke *et al.*, 1998).

CXCR-4-mediated entry of HIV into susceptible cells has been proposed as a target for antagonists, and one such compound (the bicyclam AMD-3100) has been described, with mechanisms involving inhibition of replication and blocking of apoptosis in uninfected cells (Blanco *et al.*, 2000). It is to be hoped that the requirement of HIV for cytokine receptors in order to gain entry into the cell may present an additional target for drugs, although these will not be able to be classed as antiretroviral agents since they target a cellular protein. Whether such antagonists have a future may depend on the dispensibility of individual chemokine receptors during the virus replication cycle.

22.2.2 *Reverse transcription*

Following entry into the cell, the two copies of RNA carried within the HIV virion are used as a template for the production of double-stranded DNA. This reverse transcription process occurs in the cytoplasm, in a replication complex containing the viral reverse transcriptase (RT), integrase, protease, matrix proteins and cellular histones (Karageorgos *et al.*, 1993). The cell also supplies nucleotides and other factors required for reverse transcription, which initially requires synthesis of negative strand DNA from the input viral RNA template (Peliska and Benkovic, 1992). The primer for this reaction is cellular $tRNA^{lys}$ (Mak and Kleiman, 1997). Completion of negative strand DNA occurs through a "jump" to the second RNA molecule in the same complex (Telesnitsky and Goff, 1993). Virion-associated RNase H activity simultaneously degrades the RNA template. The newly synthesized negative stranded DNA subsequently serves as a template for positive strand synthesis, a process which requires a second "jump" from the growing point at the 5′ end to the 3′ end of the same template molecule (Li *et al.*, 1993). Each of the RNA-dependent DNA polymerase, RNase H and DNA-dependent DNA polymerase functions required for reverse transcription are carried out by the HIV RT, highlighting the importance of this enzyme in the replication of the virus and in potential antiviral strategies.

22.2.3 *DNA transport and integration*

Following its formation, HIV DNA is transported to the nucleus as part of a pre-integration complex, a process regulated by two nuclear localization sequences (NLS) on the p17 matrix protein (Bukrinsky *et al.*, 1993; Haffar *et al.*, 2000). The p17 component is subsequently lost on entry of the complex into the nucleus (Karageorgos *et al.*, 1993). The HIV accessory protein Vpr (also known as protein R) enhances the interaction between the NLS and cellular karyopherin alpha, thereby increasing the karyophilic nature of the pre-integration complex (Bukrinsky and Haffar, 1999). Integration of HIV DNA into the host cell chromosome is mediated by the viral integrase, which is carried within virions and is present in the integrase complex. Although several forms of HIV DNA exist, including relaxed circular and supercoiled circular structures (Jardine *et al.*, 1993), the integration precursor is linear, double-stranded DNA (Farnet and Haseltine, 1990). The integrase removes two terminal 3′ bases from the long terminal repeat (LTR) sequences on each strand of the HIV DNA, forming overlaps at each end and enabling recombination at apparently random sites within the target cellular DNA, which is also cleaved by the integrase (Hahn, 1994). Host DNA repair systems ligate the remaining unjoined DNA ends. In HIV infection, integration only occurs in cells that have been activated through their interaction with factors such as mitogens, cytokines or viral antigens (Stevenson *et al.*, 1992). Although non-activated cells can be infected, HIV DNA does not integrate in them, and is eliminated within hours unless activation occurs. The timing of integration is thus crucial in the overall pathogenesis of HIV infection. Return of memory cells to a non-activated state immediately following integration guarantees longevity to the infection process, since the lifespan of such cells can be measured in months to years (Schrager and D'Souza, 1998) (Table 22.1).

22.2.4 *HIV accessory/regulatory proteins*

Once HIV DNA has been integrated, productive infection is influenced by the interaction between LTR sequences within the U3 region of the genome (Clements and Wong-Staal, 1992) and the HIV regulatory proteins Tat and Rev. These are two of a number of specialized regulatory/accessory proteins that, often in concert with cellular proteins, facilitate the virus replication process (Emerman and Malim, 1998). The other accessory/regulatory proteins are known as Nef, Vif, Vpu and Vpr.

The 14-kD Tat protein acts to enhance transcriptional elongation of all HIV mRNA species through a process involving itself, a cyclin-dependent kinase (Cdk-9) and cyclin T. Binding of this complex to the Tat activation region (TAR), which is present at the 5′ end of all viral mRNAs, results in hyperphosphorylation of the carboxy-terminal domain of cellular RNA polymerase II. As a result, elongation of nascent mRNA occurs, and ensures rapid replication of HIV in the normally short-lived host CD4+ cell (Emerman and Malim, 1998). In the absence of Tat, basal levels of transcription appear to be sufficient to produce the occasional full-length transcript which, following splicing and translation, result in low levels of Tat itself (Cullen, 1993).

Also in the nucleus of infected cells, the 18-kD Rev protein binds in cis to the so-called Rev response element (RRE) which is present on all unspliced (and singly spliced, in the case of *env*) mRNAs. Such mRNAs encode the viral structural proteins, so their efficient transportation to the cytoplasm is essential to the viral replicative strategy. To mediate export of these mRNAs from the nucleus, a nuclear export signal at the carboxy terminus of Rev binds to the cellular factors exportin 1 and Ran guanosine triphosphatase (Ran GTPase), which regulates directionality of nuclear transport (Yoneda, 2000). This complex dissociates in the cytoplasm through Ran GTPase-induced hydrolysis, enabling translation or packaging (into new virions) of the RNA and return of Rev to the nucleus (Emerman and Malim, 1998).

Both Nef and Vpu play important roles in controlling the amount of HIV receptor CD4 expressed at the cell surface. Control of CD4 expression by these accessory proteins occurs at two distinct sites, the endoplasmic reticulum (ER) in the case of Vpu, and the cell membrane for Nef. Vpu is a 16-kD protein that interacts with CD4 in association with the gp160 envelope protein during processing in the ER (Willey *et al.*, 1992). Retardation of the transport of CD4 to the plasma membrane enhances Vpu-CD4 interactions and promotes the subsequent degradation of CD4 through a cellular mechanism involving the ubiquitin-proteosome pathway (Emerman and Malim, 1998; Schubert *et al.*, 1998). The 27-kD Nef protein acts at the cell membrane to remove pre-existing CD4. The mechanism involves the recruitment of CD4 into clathrin-coated pits through an interaction between CD4-bound Nef and AP-2, a cellular adapter complex that recruits transmembrane proteins to the pits (Emerman and Malim, 1998). Subsequent degradation of CD4 occurs within intracellular lysosomes.

Why is Vpu- and Nef-mediated CD4 downregulation important? In the ER, because of the likelihood of envelope protein-CD4 interactions, transport of gp120 via the golgi apparatus to the virion assembly site at the cell membrane may be hindered. Disruption of the interaction between the viral protein and CD4 by a Vpu-induced mechanism may result in the increased efficacy of gp120 incorporation into the cell membrane prior to viral budding. The removal of pre-existing CD4 from the cell surface through the action of Nef may therefore enhance particle formation and virus release. This role for Nef fits well with the observation that individuals infected with *nef* deletion mutants appear to have an altered course of infection towards increased survival time compared to those infected with wild-type virus (Learmont *et al.*, 1992; Deacon *et al.*, 1995). Both Vpu- and Nef-mediated mechanisms therefore contibute to an increased likelihood that infectious virions are ultimately released from the cell.

The 23-kD HIV Vif protein is highly conserved among the lentiviruses (von Schwedler *et al.*, 1993). A small number of its molecules are present in HIV virions (Liu *et al.*, 1995), suggesting a role in the early stages of the replication cycle. This is likely to be preceding or during proviral DNA synthesis, since *vif* mutant viruses are defective in their ability to undergo endogenous reverse transcription (Goncalves *et al.*, 1996). However, evidence also exists for a "late" function for Vif, since *vif*-deficient infection cannot be rescued through the expression of Vif in target cells, and Vif and Gag proteins colocalize in infected cells, suggesting a role in the assembly or budding process (Simon *et al.*, 1997).

22.2.5 *Assembly of structural proteins and virus maturation*

A single genome-length mRNA encodes the 55-kD Gag and 160-kD Gag-Pol proteins, which are synthesized in the ratio of approximately 15 to 1 (Jacks *et al.*, 1988). In the HIV genome, the *gag* and *pol* coding regions overlap in different reading frames and are interrupted by a third gene (coding for the protease). To overcome these blocks to the synthesis of a fusion protein, HIV undergoes a single (−1) ribosomal frameshift in a 241-nucleotide sequence overlap directly between the *gag* and *pol* genes (Jacks *et al.*, 1988). The resulting polyproteins are post-translationally modified by myristoylation at their amino-terminal ends, a process which is required for their intracellular transport and localization to membranes (Park and Morrow, 1991; Faacke *et al.*, 1993).

Cleavage of Gag and Gag-Pol is undertaken by the viral protease. The matrix, capsid, and nucleocapsid proteins are derived from the 55-kD precursor. The protease, RT and integrase enzymes are derived from the 160-kD precursor (Figure 22.1). Because of the absolute requirement for cleavage of HIV structural polyproteins in the process of replication, the viral protease represents a prime target for antiretroviral drugs, and this has been exploited to the extent that at least 6 protease inhibitors have been approved for the treatment of HIV infection (see below). The process of

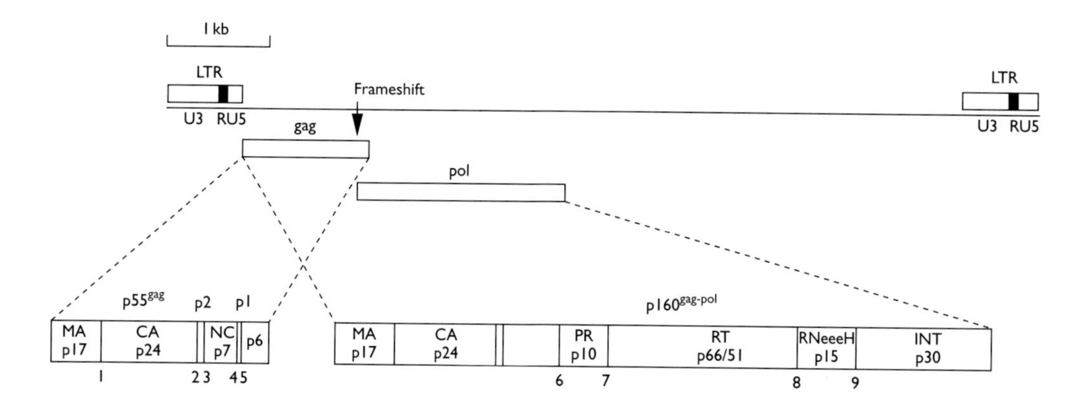

Fig 22.1 HIV protease-catalyzed cleavage of gag and gag-pol polyproteins. Under normal conditions of translation only gag is produced. An occasional frameshifting event produces gag-pol. Up to nine cleavages occur at sites indicated by the numerals. Cleavage sites 4 and 5 are abolished by the frameshift. LTR, long terminal repeat; U3 and U5, unique 3′ and 5′ regions, respectively; MA, matrix; CA, capsid; NC, nucleocapsid; PR, protease; RT, reverse transcriptase; INT, integrase. Molecular weights of proteins (in kD) are indicated.

rational design and synthesis of these inhibitors was facilitated by the relatively early determination of the crystal structure of the enzyme (Wlodawer *et al.*, 1989; Navia *et al.*, 1989).

The primary translation product of *env* mRNA is a 92-kD protein to which oligosaccharide is added in the ER, producing high mannose containing gp160 (Ratner, 1992). While still in the ER, gp160 is modified by α-glucosidases I and II, the latter enzyme being inhibited by the plant derivative castanospermine (Walker *et al.*, 1987). This approach to the inhibition of HIV replication failed clinically, most likely because of the lack of viral specificity of this inhibitor. In contrast to the Gag and Gag-Pol proteins, which are cleaved by the viral protease, cleavage of the HIV envelope glycoprotein precursor (gp160) is undertaken by a cellular, trypsin-like protease within the golgi, where further modification by mannosidases also takes place (Ratner, 1992).

Assembly and release of retrovirus particles can occur in the absence of all gene products except Gag (Hunter, 1994), highlighting the importance of this protein in the process. An association between Gag and Gag-Pol ensures the packaging of viral enzymes. The RNA genome is packaged through interaction between it and sequences on the p7 nucleocapsid protein, specifically, a stem-loop structure on the RNA (Sakaguchi *et al.*, 1993) and cysteine-histidine rich regions on the nucleocapsid (De Roquigny *et al.*, 1992). Packaging and binding of $tRNA^{lys}$ to the primer binding site on RNA has been ascribed as a function of both the RT sequence on Gag-Pol and the nucleocapsid protein (Mak and Kleiman, 1997).

The final step in assembly, marking the start of maturation, involves pinching-off of the immature virus from the cell membrane. The p6 protein, situated at the carboxy terminal end of Gag, plays a role in this process, since its mutation results in particles with an immature morphology which remain attached to the cell membrane by a thin strand (Gottlinger *et al.*, 1991; Wills and Craven, 1991). Processing of Gag and Gag-Pol by the viral protease occurs at the time of, or shortly after, budding (Garnier *et al.*, 1998), although the mechanism by which the protease extricates itself from the fusion protein in order to undertake this function remains unclear. The end result of the complex and coordinated processing and interaction events that make up the assembly and maturation process is an HIV virion containing not only structural proteins, but also the enzymes essential for the establishment of a new round of infection in a susceptible cell.

22.3 Control of HIV infection

22.3.1 *The role of vaccination*

Despite progress made in the treatment of HIV-infected individuals with antiretroviral drugs, the absolute need for strict adherence to a complicated drug regimen, the potential development of drug resistance and cross-resistance, and the adverse effects of long term daily therapy with at least three drugs highlight the complexity of this approach to the control of HIV infection. Also, the infrastructure and finance needed to support the appropriate distribution of antiretroviral drugs is available only to a small minority of the infected population. While strategies aimed at modifying behavioral patterns likely to result in the transmission of HIV will undoubtedly continue to prevent some new infections, ultimately the control of HIV infection, particularly in endemic regions, will only be achieved through vaccination.

Unfortunately, the statement "a vaccine for HIV is at least 10 years away" has been repeated often in the 16 years since the virus was first identified. While attempts to produce a vaccine have been actively pursued by many groups since that time, major obstacles still remain for those attempting this difficult task. Of particular importance is the need for a strong cell-mediated response to the viral component of the vaccine. Such a response is required to eliminate infected cells expressing HIV antigens on their surface in association with MHC class I antigens. This required cytotoxic T lymphocyte (CTL) response does not fit easily with the likely need for the vaccine to be inactivated. While live, attenuated vaccines are more likely to induce CTL (and humoral) responses, the use of a live vaccine raises safety issues, since HIV has the potential to mutate (perhaps regaining virulence) or to undergo recombination with other wild-type retroviruses (to produce non-attenuated pathogenic strains). Also, it is unclear what constitutes true attenuation. Apparently naturally attenuated HIV strains associated with long term non-progression to AIDS in individuals infected through a common source have been described (Learmont *et al.*, 1992). However, recently some of the patients within the cohort have experienced unfavorable changes in virological and immunological markers, suggesting that the virus, which contains *nef* and *nef*-LTR overlap deletions (Deacon *et al.*, 1995), retains a degree of virulence which would make it an unacceptable vaccine candidate.

In the face of difficulties associated with the use of attenuated HIV strains and the need for an appropriate CTL response, what alternatives remain for the successful development of an HIV vaccine? One promising approach involves a schedule of priming with HIV-specific DNA and later boosting with a recombinant fowlpox virus encoding multiple HIV proteins (Kent *et al.*, 1998). In macaques, this results in enhanced HIV-specific CTL and Th responses to Env and Gag proteins contained within the recombinant virus. Animals vaccinated with HIV-1 derived from autologous peripheral blood mononuclear cells using the prime-boost strategy have been subsequently protected against a large intravenous challenge dose of virus (>100 monkey infectious doses). While this HIV-1/macaque system is not considered to be a suitable model for studying the pathogenesis of HIV infection, it is appropriate for evaluating the immune response to an acute infection with the virus. Failure to detect evidence of HIV RNA or DNA in material collected several weeks after challenge is an encouraging sign for the utility of this approach in generating protective immunity.

22.3.2 Antiretroviral drug therapy

At present, despite the number of viral targets presented during HIV replication, only two classes of antiretroviral drug are available for use in the treatment of the HIV infected patient. These are RT inhibitors and protease inhibitors. However, a novel inhibitor of the fusion process described in 22.2.1 (T20, Kilby *et al.*) is in advanced clinical trials and represents an important development in this area.

Those drugs that inhibit the RT can be subdivided into dideoxynucleoside analogue inhibitors (NRTIs) and nonnucleoside analog (NNRTIs) (Table 22.2). The NRTIs require phosphorylation by cellular enzymes before they inhibit RT activity by mechanisms including direct competion with the natural (deoxynucleotide) substrates for the substrate binding site on the enzyme, and chain termination (Furman *et al.*, 1986; Arts and Wainberg, 1996). In the latter case, the absence of a 3′-hydroxyl position on the ribose sugar prevents further addition of nucleotides during reverse transcription. The requirement for phosphorylation by cellular kinases is a disadvantage for the NRTIs, since it reduces their antiviral specificity, in contrast to herpes simplex virus-active nucleoside analogues such as acyclovir which are monophosphorylated by a viral thymidine kinase (Fyfe *et al.*, 1978).

The structurally diverse NNRTIs bind to and inhibit the RT of HIV-1 strains through a non-competitive mechanism with respect to deoxynucleotides, template and primer. Within the 3-dimensional structure of the enzyme, this binding occurs in a region adjacent to but not overlapping the catalytic site, probably reducing the amount of flexibility required during reverse transcription (Arnold *et al.*, 1992).

The crystallographic determination of the HIV protease structure has facilitated the design and synthesis of inhibitors with exceptional potency (Navia *et al.*, 1989; Wlodawer *et al.*, 1989) (Table 22.2). In general, the structure of these inhibitors falls into one of two categories, peptide-based compounds (including the twofold rotational (C2) symmetric structures) and nonpeptide inhibitors. Any chemical modifications made to the basic structures are designed to increase the interaction between the inhibitor and the 3-dimensional pockets (subsites) of the protease. Such inhibitors require

Table 22.2 Antiretroviral drugs currently available, listed by class. The list of recognised mutations is abbreviated for simplicity. Considerable cross-resistance exists between protease inhibitors

Drug class	*Drug*	*Base*	*Recognised resistance mutations*[1]
RT Inhibitor (nucleoside)	Abacavir	Purine (G)	M184V[2]
	Didanosine	Purine (I)[3]	K65R[4], L74V[4]
	Lamivudine	Pyrimidine (C)	M184V
	Stavudine	Pyrimidine (T)	V75T[5]
	Zalcitabine	Pyrimidine (C)	K65R, T69D, L74V, V75T.
	Zidovudine	Pyrimidine (T)	M41L
RT Inhibitor (nonnucleoside)	Delavirdine		K103N[6]
	Efavirenz		K103N[6]
	Nevirapine		K103N[6]
Protease Inhibitor	Amprenavir		I50V[7], I84V
	Indinavir		M46I + V82A/F/T[8]
	Lopinavir[9]		Multiple mutations required
	Nelfinavir		D30N[7], L90M
	Ritonavir		V82A/F, I84V
	Saquinavir		G48V, L90M

1. These mutations may be selected first, and are often associated with phenotypically detectable levels of drug resistance, which may vary from low-level to high-level. Mutations causing resistance to multiple RTIs (Q151M complex (Shafer *et al.*, 1994) and T69S insertions (Winters *et al.*, 1998)) have been described. For detailed summaries of resistance mutations generated by the listed antiretroviral drugs refer to Schinazi *et al.* 2000.
2. Other mutations contribute to phenotypic resistance (K65R, L74V, Y115F).
3. The active form of didanosine (ddI) is dideoxyadenosine triphosphate.
4. Rarely seen clinically.
5. Mutations associated with zidovudine resistance are occasionally selected by stavudine and may contribute to resistance.
6. K103N causes cross-resistance to all NNRTIs. Some other well characterised mutations (eg. Y181C, P236L) cause resistance to one or more NNRTIs.
7. Signature mutation for this drug.
8. These mutations may predispose to indinavir resistance. No single mutation conferring indinavir resistance has been described.
9. Lopinavir is a mixture of ritonavir and ABT-378 (Carillo *et al.*, 1998).

inhibition constants in the nanomolar range before they are considered for preclinical development. Their mechanism of action involves the structural mimicry of a stable tetrahedral intermediate (the transition state) formed during cleavage of the natural substrates by the enzyme (Debouck, 1992).

22.3.3 *The problem of drug resistance*

The process of reverse transcription is not fail-safe, and the error-rate of the HIV RT in terms of its incorporation of incorrect nucleotides has been estimated at 1 in 10,000 (Roberts *et al.*, 1988). This approximates to one error per viral RNA template transcribed. Given that the enzyme does not have exonucleolytic proof-reading ability, and up to 10^{10} HIV virions are produced daily in some individuals (Wei *et al.*, 1995), it is not difficult to conclude that some of these errors will occur in genes whose products can be targeted by antiretroviral drugs, and at bases that directly contribute to a drug resistant phenotype. Such a scenario is supported by the observation that HIV strains containing codon changes known to be associated with phenotypic resistance to protease inhibitors have been shown to exist in patients never treated with this class of antiretroviral drug (Kozal *et al.*, 1996). These strains constitute only a small fraction of the wild-type viruses present (Coffin, 1995) and are likely to be at a replicative disadvantage compared to many of the other quasispecies present. But when cells containing their proviral DNA are exposed to antiretroviral drug pressure, the virions produced from them may gain a selective advantage. Even limited replication in the presence of an antiviral drug will enable the generation of more mutations, some which confer increasing levels of drug resistance, and others which improve virological fitness. Mutations commonly associated with resistence to members of each of the drug classes currently available for therapeutic use are shown in Table 22.2. In the case of nucleoside RT inhibitors, certain mutations, in particular those generated by zidovudine, are able to increase the likelihood of removal of a terminated ddNMP by a process known as excision (Naeger *et al.*, 2002).

22.3.4 *Reducing the risk of antiretroviral drug resistance*

A number of strategies can be used to decrease the risk of drug resistance developing during therapy. Complete suppression of HIV replication following initiation of treatment is of considerable importance, since even a limited amount of replication in the presence of suboptimal drug concentrations may result in the generation of resistance. Maintenance of the virus load to below 50 copies per ml for long periods of time, while not eliminating the virus, does appear to prevent the evolution of strains with new drug resistance mutations (Wong *et al.*, 1997; Finzi *et al.*, 1997). Supporting this, individuals with low-level HIV RNA detected in plasma below 400 copies per ml have an inferior virological response than individuals whose virus is undetectable below this level (Pilcher *et al.*, 1999). However, there is no universal agreement as to when such suppressive therapy should commence. While preventing the generation of potentially drug-resistant strains of HIV at or near the time of primary infection is virologically sound, this needs to be balanced against the difficulties faced by patients trying to adhere to a complex drug regimen on a life-time basis. Long term clinical trials are needed to determine whether a treat-early strategy has advantages or disadvantages over a delayed-start strategy. The latter has the advantage of minimizing the patient's exposure to drugs, thereby avoiding drug-generated resistance and the adverse side effects associated with long term therapy. However, its disadvantages include the possible impairment of immune reconstitution, creation of reservoirs of HIV infection that cannot be accessed by drugs when treatment is initiated, and the establishment of greater numbers of long-lived latently infected cells.

22.4 Conclusions

The HIV replication strategy offers a number of potential targets for antiretroviral drugs. Nevertheless, the accumulation of considerable knowledge about these targets has resulted in the clinical utility of only two drug classes, RT inhibitors and protease inhibitors. Strict adherence to combinations of these drugs is essential to maintain the HIV load at undetectable levels in the blood and other body compartments, and prevent the development of resistance. This is problematic in the context of lifelong administration of these drugs, all of which are associated with some degree of adverse side effects. We have sufficient understanding of the replication of HIV and its pathogenesis to realise that its elimination from the infected individual using antiretroviral drugs is probably not achievable. The future control of HIV lies to some extent in producing effective, non-toxic drugs targeted at other viral proteins. However, such a strategy will only be in reach of a small minority of the globally infected population. To a large extent the die is cast for those who are already infected. A vaccine providing protective immunity for those who are yet to become infected is essential if the spread of HIV is to be successfully controlled.

References

Arnold, E., Jacobo-Molina, A., Nanni, R.G., Williams, R.L., Lu, X., Ding, S. *et al.* (1992). Structure of HIV-1 reverse transcriptase complexed with a double-stranded DNA template-primer: implications for catalysis by and evolution of DNA polymerases. *International Antiviral News 1; 3–4.*

Arts, E.J. and Wainberg, M.A. (1996) Mechanisms of nucleoside analog antiviral activity and resistance during human immunodeficiency virus reverse transcription. *Antimicrobial Agents and Chemotherapy 40: 527–540.*

Blanco, J., Barretina, J., Henson, G., Bridger, G., De Clercq, E., Clotet, B. and Este, J.A. (2000) The CXCR-4 antagonist AMD3100 efficiently inhibits cell-surface-expressed human immunodeficiency virus type 1 envelope-induced apoptosis. *Antimicrobial Agents and Chemotherapy 44: 51–56.*

Bukrinsky, M.I., Haggerty, S., Dempsey, M.P., Sharova, N., Adzhubei, A., Spitz, L., Goldfarb, D., Emerman, M. and Stevenson, M. (1993) A nuclear localization signal within HIV-1 matrix protein that governs infection of non-dividing cells. *Nature 365: 666–669.*

Bukrinsky, M.I. and Haffar, O.K. (1999) HIV-1 nuclear import: in search of a leader. *Frontiers in Bioscience 15: D772–781.*

Carrillo, A., Stewart, K.D., Sham, H.L., Norbeck, D.W., Kohlbrenner, W.E., Leonard, J.M., Kempf, D.J. and Akhteruzzaman, M. (1998) *In vitro* selection and characterisation of human immunodeficiency virus type 1 variants with increased resistance to ABT-378, a novel protease inhibitor. *Journal of Virology 72: 7532–7541.*

Clements, J.E. and Wong-Staal, F. (1992) Molecular biology of lentiviruses. *Seminars in Virology 3: 137–146.*

Coffin, J.M. (1995) HIV population dynamics *in vivo*: implications for genetic variation, pathogenesis, and therapy. *Science 267: 483–489.*

Crowe, S.M., Carlin, J.B., Stewart, K.I., Lucas, C.R. and Hoy, J.F. (1991) Predictive value of CD4 lymphocyte numbers for the development of opportunistic infections and malignancies in HIV-infected persons. *Journal of Acquired Immune Deficiency Syndromes 4: 770–776.*

Cullen, B. (1993) Does HIV-1 Tat induce a change in viral initiation rights? *Cell 73: 417–420.*

De Rocquigny, H., Gabus, C., Vincent, A., Fournie-Zaluski, M.C., Roque, B. and Darlix, J.L. (1992) Viral RNA annealing activities of human immunodeficiency virus type 1 nucleoprotein require only peptide domains outside the zinc fingers. *Proceedings of the National Academy of Science USA 89: 6472–6476.*

Deacon, N.J., Tsykin, A., Solomon, A., Smith, K., Ludford-Menting, M., Hooker, D.J., *et al.* (1995) Genomic structure of an attenuated quasispecies of HIV-1 from a blood transfusion donor and recipient. *Science 270: 988–991.*

Debouck, C. (1992) The HIV-1 protease as a therapeutic target for AIDS. *AIDS Research and Human Retroviruses: 8: 153–164.*

Deng, H., Liu, R., Ellmeier, W., Choe, S., Unutmaz, D., Burkhart, M., *et al.* (1996) Identification of a major co-receptor of primary isolates of HIV-1. *Nature 381: 661–666.*

Dragic, T., Litwin, V., Allaway, G.P., Martin, S.R., Huang, Y., Nagashima, K.A., *et al.* (1996) HIV-1 entry into CD4+ cells is mediated by the chemokine receptor CC-CKR-5. *Nature 381: 667–673.*

Donaldson, Y.K., Bell, J.E., Ironside, J.W., Brettle, R.P., Robertson, J.R., Busuttil, A. and Simmonds, P. (1994) Redistribution of HIV outside the lymphoid system with the onset of AIDS. *Lancet 343: 382–385.*

Emerman, M. and Malin, M.H. (1998) HIV-1 regulatory/accessory genes: keys to unravelling viral and host cell biology. *Science 280: 1880–1884.*

Faacke, M. (1993) A large deletion in the matrix domain of human immunodeficiency gag gene redirects virus particle assembly from the plasma membrane to the endoplasmic reticulum. *Journal of Virology 67: 4972–4980.*

Farnet, C. and Haseltine, W. (1990) Integration of human immunodeficiency virus type 1 DNA *in vitro*. *Proceedings of the National Academy of Science USA 87: 4164–4168.*

Feng, Y., Broder, C.C., Kennedy, P.E. and Berger, E.A. (1996) HIV-1 entry cofactor: functional cDNA cloning of a seven-transmembrane, G protein-coupled receptor. *Science 272: 872–877.*

Finzi, D., Hermankova, M., Pierson, T., Carruth, L.M., Buck, C., Chaisson, R.E., *et al.* (1997) Identification of a reservoir for HIV-1 in patients on highly active antiretroviral therapy. *Science 278: 1295–1300.*

Fowke, K.R., Dong, T., Rowland-Jones, S.L., Oyugi, J., Rutherford, W.J., Kimani, J., *et al.* (1998) HIV type 1 resistance in Kenyan sex workers is not associated with altered cellular susceptibility to HIV type 1 infection or enhanced beta-chemokine production. *AIDS Research and Human Retroviruses 14: 1521–1530.*

Furman, P.A., Fyfe, J.A., St Clair, M.H., Weinhold, K., Rideout, J.L., Freeman, G.A., *et al.* (1986) Phosphorylation of 3′-azido-3′-deoxythymidine and selective interaction of the 5′-triphosphate with human immunodeficiency virus reverse transcriptase. *Proceedings of the National Academy of Science USA 83: 8333–8337.*

Fyfe, J.A., Keller, P.M., Furman, P.A., Miller, R.L. and Elion, G.B. (1978) Thymidine kinase from herpes simplex virus phosphorylates the new antiviral compound 9-(2-hydroxyethoxymethyl) guanine. *Journal of Biological Chemistry 253: 8721–8727.*

Garnier, L., Bowzard, J.B. and Wills, J.W. (1998) Recent advances and remaining problems in HIV assembly. *AIDS 12 (suppl A): S5–S16.*

Goncalves, J., Korin, Y., Zack, J. and Gabuzda, D. (1996) Role of Vif in human immunodeficiency virus type 1 reverse transcription. *Journal of Virology 70: 8701–8709.*

Gottlinger, H.G., Dorfman, T., Sodroski, J.G. and Haseltine, W.A. (1991) Effect of mutations affecting p6 gag protein on human immunodeficiency virus particle release. *Proceedings of the National Academy of Science USA 88: 3195–3199.*

Haffar, O.K., Popov, S., Dubrovsky, L., Agostini, I., Tang, H., Pushkarsky, T., Nadler, S.G. and Bukrinsky, M. (2000) Two nuclear localization signals in the HIV-1 matrix protein regulate nuclear import of the HIV-1 pre-integration complex. *Journal of Molecular Biology 299: 359–368.*

Hahn, B.H. (1994) Viral genes and their products. In: Broder, S., Merigan, T.C. and Bolognesi, D. (eds.) *Textbook of AIDS Medicine.* Williams and Wilkins, Baltimore MD, USA.

Ho, D.D. (1996) Viral counts count in HIV infection. Science 272: 1124–1125.

Huang, Y., Paxton, W.A., Wolinsky, S.M., Neumann, A.U., Zhang, L., He, T., *et al.* (1996) The role of mutant CCR5 allele in HIV-1 transmission and disease progression. *Nature Medicine 2: 1240–1243.*

Hunter, E. (1994) Micromolecular interactions in the assembly of HIV and other retroviruses. *Seminars in Virology 5: 71–83.*

Jacks, T., Power, M.D., Masiarz, F.R., Luciw, P.A., Barr, P.J., Varmus, H.E. (1988) Characterisation of ribosomal frameshifting in HIV gag-pol expression. *Nature (London) 331: 280–283.*

Jardine, D., Tachedjian, G., Locarnini, S. and Birch, C. (1993) Cellular topoisomerase I associated with HIV-1. *AIDS Research and Human Retroviruses 9: 1245–1250.*

Karageorgos, L., Li, P. and Burrell, C. (1993) Characterisation of HIV replication complexes early after cell-cell infection. *AIDS Research and Human Retroviruses 9: 817–823.*

Kent, S.J., Zhao, A., Best, S.J., Chandler, J.D., Boyle, D.B. and Ramshaw, I.A. (1998) Enhanced T-cell immunogenicity and protective efficacy of a human immunodeficiency virus type 1 vaccine regimen consisting of consecutive priming with DNA and boosting with recombinant fowlpox virus. *Journal of Virology 72: 10180–10188.*

Kilby, J.M., Hopkins, S., Venetta, T.M., DiMassimo, B., Cloud, G.A., Lee, J.Y. *et al.* (1998). Potent supression of HIV-1 replication in humans by T-20, a peptide inhibitor of gp 41-mediated virus entry. *Nature Medicine 4: 1302–1307.*

Kozal, M.J., Shah, N., Shen, N., Yang, R., Fucini, R., Merigan, T.C. (1996) Extensive polymorphisms observed in HIV-1 clade B protease gene using high-density oligonucleotide arrays. *Nature Medicine 2: 753–759.*

Learmont, J., Tindall, B., Evans, L., Cunningham, A., Cunningham, P., Wells, J., Penny, R., Kaldor, J. and Cooper, D.A. (1992) Long-term symptomless HIV-1 infection in recipients of blood products from a single donor. *Lancet 340: 863–7.*

Li, P., Stephenson, A.J., Kuiper, L.J. and Burrell, C.J. (1993) Double-stranded strong-stop DNA and the second template switch in human immunodeficiency virus (HIV) DNA synthesis. *Virology 194: 82–88.*

Liu, H., Wu, X., Newman, M., Shaw, G.M., Hahn, B.H. and Kappes, J.C. (1995) The Vif protein of human and simian immunodeficiency viruses is packaged into virions and associates with viral core structures. *Journal of Virology 69: 7630–7638.*

Liu, R., Paxton, W.A., Choe, S., Ceradini, D., Martin, S.R., Horuk, R., *et al.* (1996) Homozygous defect in HIV-1 coreceptor accounts for resistance of some multiply-exposed individuals to HIV-1 infection. *Cell 86: 367–377.*

Mak, J. and Kleiman, L. (1997) Primer tRNAs for Reverse Transcription. *Journal of Virology 71: 8087–8095.*

Mellors, J.W., Rinaldo, C.R., Gupta, P., White, R.M., Todd, J.A. and Kingsley, L.A. (1996) Prognosis in HIV infection predicted by the quantity of virus in plasma. *Science 272: 1167–1170.*

Naeger, L.K., Margot, N.A. and Miller, M./D. (2002). ATP-dependent removal of nucleoside reverse transcriptase inhibitors by human immunodeficiency virus type1 reverse transcriptase. *Antimicrobial Agents and Chemotherapy 46: 2179–2184.*

Navia, M.A., Fitzgerald, P.M., McKeever, B.M., Leu, C.T., Heimbach, J.C., Herber, *et al.* (1989) Three-dimensional structure of aspartyl protease from human immunodeficiency virus. *Nature (London) 337: 615–620.*

Pantaleo, G., Graziosi, C. and Fauci, A.S. (1993) The role of lymphoid organs in the immunopathogenesis of HIV infection. *AIDS 7 (suppl 1): S19–S23.*

Pantaleo, G., Graziosi, C. and Fauci, A.S. (1993) The immunopathogenesis of human immunodeficiency virus infection. *New England Journal of Medicine 328: 327–335.*

Park, J. and Morrow, C.D. (1991) Overexpression of the gag-pol precursor from human immunodeficiency virus type 1 proviral genomes results in efficient proteolytic processing in the absence of virion production. *Journal of Virology 65: 5111–5117.*

Peliska, J.A. and Benkovic, S.J. (1992) Mechanisms of DNA strand transfer catalysed by HIV-1 reverse transcriptase. *Science 258: 1112–1117.*

Piatak, M., Saag, M.S., Yang, L.C., Clark, S.J., Kappes, J.C., Luk, K.C., *et al.* (1993) High levels of HIV-1 in plasma during all stages of infection determined by competitive PCR. *Science 259: 1749–1754.*

Pilcher, C.D., Miller, W.C., Beatty, Z.A., Eron, J.J. (1999) Detectable HIV-1 RNA at levels below quantifiable limits by amplicor HIV-1 monitor is associated with virologic relapse on antiretroviral therapy. *AIDS 13: 1337–1342.*

Ratner, L. (1992) Glucosidase inhibitors for treatment of HIV-1 infection. *AIDS Research and Human Retroviruses 8: 165–173.*

Reece, J.C., Handley, A.J., Anstee, E.J., Morrison, W.A., Crowe, S.M. and Cameron, P.U. (1998) HIV-1 selection by epidermal dendritic cells during transmission across skin. *Journal of Experimental Medicine 187: 1623–1631.*

Roberts, J.D., Bebenbek, K. and Kunkel, T.A. (1988) The accuracy of reverse transcriptase from HIV-1. *Science 241: 1171–1173.*

Sakaguchi, K., Zambrano, N., Baldwin, E.T., Shapiro, B.A., Erickson, J.W., Ominchinski, *et al.* (1993) Identification of a binding site for the human immunodeficiency virus type 1 nucleocapsid protein. *Proceedings of the National Academy of Science USA 90: 5219–5223.*

Schinazi, R.F., Larder, B.A. and Mellors, J.W. (2000) Mutations in retroviral genes associated with drug resistance: 2000–2001 update. *International Antiviral News 8:5, 65–91.*

Schrager, L.K. and D'Souza, P. (1998) Cellular and anatomical reservoirs of HIV-1 in patients receiving potent antiretroviral combination therapy. *Journal of the American Medical Association 280: 67–71.*

Schubert, U., Anton, L.C., Bacik, I., Cox, J.H., Bour, S., Bennink, *et al.* (1998) CD4 glycoprotein degredation induced by human immunodeficiency virus type 1 Vpu protein requires the function of proteasomes and the ubiquitin-conjugating pathway. *Journal of Virology 72: 2280–2288.*

Shafer, R.W., Kozal, M.J., Winters, M.A., Iversen, A.K.N., Katzenstein, D.A., Ragni, M.V., *et al.* (1994) *The Journal of Infectious Diseases 169: 722–729.*

Simon, J.H.M., Fouchier, R.A.M., Southerling, T.E., Guerra, C.B., Grant, C.K. and Malim, M.H. (1997) The Vif and Gag proteins of human immunodeficiency virus type 1 colocalize in infected human T cells. *Journal of Virology 71: 5259–5267.*

Stevenson, M., Bukrinsky, M. and Haggerty, S. (1992) HIV replication and potential targets for intervention. *AIDS Research and Human Retroviruses 81: 107–117.*

Telesnitsky, A. and Goff, S.P. (1993) Strong-stop strand transfer during reverse transcription, p49–84. In: Skalka, A.M. and Goff, S.P. (eds.), *Reverse transcriptase, Vol. 1.* Cold Spring Harbor Laboratory Press, New York, NY.

von Schwedler, U., Song, J., Aitken, C. and Trono, D. (1993) Vif is crucial for human immunodeficiency virus type 1 proviral DNA synthesis in infected cells. *Journal of Virology 67: 4945–4955.*

Walker, B.D., Kowalski, M., Goh, W.C., Kozarsky, K., Krieger, M., Rosen, C., *et al.* (1987) Inhibition of human immunodeficiency virus syncytium formation and virus replication by castanospermine. *Proceedings of the National Academy of Science USA 84: 8120–8124.*

Weber, J. (1998) HIV and sexually transmitted diseases. *British Medical Bulletin 54: 717–729.*

Wei, X., Ghosh, S.K., Taylor, M.E., Johnson, V.A., Emini, E.A., Deutsch, P., *et al.* (1995) Viral dynamics in human immunodeficiency virus type 1 infection. *Nature 373: 117–122.*

Willey, R.L., Maldarelli, F., Martin, M.A. and Strebel, K. (1992) Human immunodeficiency virus type 1 Vpu protein induces rapid degredation of CD4. *Journal of Virology 66: 7193–7200.*

Wills, J.W. and Craven, R.C. (1991) Form, function, and use of retroviral gag proteins. *AIDS 5: 639–654.*

Winters, M.A., Coolley, K.L., Girard, Y.A., Levee, D.J., Hamdan, H., Shafer, R.W., Katzenstein, D.A. and Merigan, T.C. (1998) *Journal of Clinical Investigation 102: 1769–1775.*

Wlodawer, A., Miller, M., Jaskolski, M., Sathyanarayana, B.K., Baldwin, E., Weber, I.T., *et al.* (1989) Conserved folding in retroviral proteases. Crystal structure of a synthetic HIV-1 protease. *Science 245: 616–621.*

Wong, J.K., Hezareh, M., Gunthard, H.F., Havlir, D.V., Ignacio, C.C., Spina, C.A., Richman, D.D. (1997) Recovery of replication-competent HIV despite prolonged suppression of plasma viremia. *Science 278: 1291–1295.*

Wyatt, R. and Sodroski, J. (1998) The HIV-1 envelope glycoproteins: fusogens, antigens, and immunogens. *Science 280: 1884–1888.*

Yoneda, Y. (2000) Nucleocytoplasmic protein traffic and its significance to cell function. *Genes Cells 5: 777–787.*

Chapter 23

Signal transduction in virus infected cells

Implications for therapy

Peter Krajcsi

Contents

23.1 Introduction

Molecular details of viral pathogenesis have been explored at an amazing speed over the past decade. Previously, viruses had been viewed as genetic parasites that exploit cellular metabolism in order to propagate themselves. To usurp the infected cell viruses have evolved various strategies to shut down synthesis of cellular macromolecules and direct the cell machinery to produce viral material. Over millions of years of virus and host coevolution there has been plenty of time for the development of very complex and intimate interactions between the two. Studies on latent and persistent viruses in particular have shown that viruses have developed extremely sophisticated strategies to manipulate the host's cellular regulation in order to transform host cells, replicate, or persist. Cellular homeostasis in multicellular organisms is directed by extracellular signals. Not surprisingly, to exercise total control over host cells, viruses have evolved numerous genes that function as signal interceptors or activators. Some of these genes are of host origin, as shown by homology searches, others may have arisen by convergent evolution (Murphy, 1994; Smith, 1994). In addition to helping us understand viral pathogenesis, these studies provide further insight into cellular regulation.

Cell growth and immunoregulation pathways are the two signaling networks most frequently targeted by viruses (Krajcsi and Wold, 1998). This review focuses on the plasma membrane and cytoplasmic events of virally regulated cell growth and immunoregulation signaling. The paradigm of viral gene products which reprogram cell growth is provided by retroviruses. This material has been extensively reviewed (Kung and Liu, 1997) and is therefore not addressed here.

23.2 Cell death/immunoregulation signaling

A virus must penetrate the host's antiviral immune defenses before it becomes absorbed by a host cell to redirect the host's cell functions to its own replicative advantage. Most of the antiviral defenses directed against infected cells are as radical as possible: complement, cytotoxic T lymphocytes (CTL), natural killer (NK) and phagocytic cells kill the infected cells. In general, viral strategies to counteract cytotoxic immune defenses fall into two groups: (i) prevention of recognition of infected cells and (ii) inhibition of cell death machinery (Krajcsi and Wold, 1998).

A virus must first pass extracellular immune defenses such as the complement cascade and antibody neutralization. Both defenses directly mediate lysis or phagocytosis of free virus and virus-infected cells. A common theme is the highly efficient destruction of the target. Therefore, the most successful viral strategies to evade complement and antibody mediated lysis are those that inhibit association of the virus-infected cell with complement and/or antibody driven phagocytes. Vaccinia virus and herpes viruses have evolved proteins (Table 23.1) that block complement function either by inhibiting complement activation or preventing deposition of the membrane attack complex. Expression of Fc receptors by herpes viruses and varicella zoster virus serves a dual function: it prevents activation of the classical complement pathway and may protect against Fc-facilitated phagocytosis as well.

CTLs have evolved dual strategies to kill virus infected cells. Activated CTLs express membrane-bound Fas-ligand that triggers Fas-mediated cell death. At the same time CTL introduces granzyme B, a serine protease, into the infected cell. Both molecules activate the cascade of apoptotic proteases (caspases). CTL's lytic attack is highly efficient. Therefore, the counterattack by many viruses is based on inhibition of

Table 23.1 Virus factors that regulate cell death/immunoregulation signaling

Virus factor	*Homolog in host*	*Function complement*
Vaccinia VCP	Complement SCR	Binds C4b, blocks complement activation
HVS CCPH	Complement SCR	Binds C4b, blocks complement activation
HVS ORF 15	CD59	Inhibits formation of MAC
HSV C-1		Binds C3b, blocks complement activation
HSV gE–gI		Binds Fc, inhibits complement and phagocytosis
MCMV early gene		Binds Fc, inhibits complement and phagocytosis
VZV gE†		Binds Fc, inhibits complement and phagocytosis
CTL		
Ad12 E1A		Downregulation of MHC-class I at transcriptional level
AdE3 gp19K		Retains MHC class I in ER
MCMV m152		Blocks MHC transport
MCMV early gene		Downregulation of MHC-class I at transcriptional level
HCMV US3		Retains MHC class I in ER
HCMV US2, US11		Dislocates MHC class I from ER to cytosol
HCMV US6		Inhibits TAP-mediated peptide transport
HSV ICP47		Inhibits production and transport of antigenic peptides
HIV-1 Nef		Induces endocytotic degradation of MHC class I
NK cytotoxicity		
HCMV UL18	MHC class I	Binds LIRs and KIRs (?), inhibits NK cytotoxicity
MCMV m144	MHC class I	Binds KIRs (?), inhibits, NK cytotoxicity
TNF/Fas-L/TRAIL		
Shope fibroma virus T2	TNF receptor	Sequesters TNF and blocks TNF function
Cowpox CrmB	TNF receptor	Sequesters TNF and blocks TNF function
Cowpox CrmC	TNF receptor	Sequesters TNF and blocks TNF function
Cowpox CrmA		Inhibits IL-1 convertase, Blocks TNF and Fas L induced cell death
Ad E3-14.7K		Blocks TNF induced arachidonic acid release, inhibits TNF cytotoxicity, interacts with FLICE, blocks Fas-L cytotoxicity
Ad E3-RID		Blocks TNF induced $cPLA_2$ translocation, AA release and TNF cytotoxicity, down-regulates Fas and TRAIL receptors, blocks Fas-L and TRAIL cytotoxicity
Ad E1B-19K	Bcl-2	Sequesters Bax, blocks TNF cytotoxicity, binds FLASH‡, blocks Fas-L cytotoxicity
EBV BHRF1	Bcl-2	Blocks TNF cytotoxicity
Herpes virus, molluscipox virus v-FLIPs IL-1	FADD	Bind FADD, inhibit recruitment of FLICE
EBV gp350		Induces expression of an IL-11R antagonist, blocks IL-1 function
Vaccinia virus B15R	IL-1 receptor	Sequesters IL1-β and blocks IL-1β function
Inflammation		
EBV BCRF1	IL-10	Th2 inducer, suppresses inflammation
ORF gene	IL-10	Th2 inducer, suppresses inflammation
Measles virus		Down-regulates pro Th1 IL-12, suppresses inflammation
IFN		
Rabbit myxoma M-T7	IFN type II receptor	
Pox virus	IFN type II receptor	Sequesters IFN γ and blocks its function
Vaccinia B18R	IFN type I receptor	Sequesters IFN α/β and blocks its function
Ad E1A		Represses STAT 1, binds p300/CBP, suppresses blocks IFN induced JAK/STAT pathway
SV5 V		Targets STAT1 for degradation
HBV terminal protein		Blocks IFN signaling
HCV		Inhibits Jak/STAT signaling

† VZV: varicella-zoster virus; see text for other abbreviations; ‡ FLASH: FLICE-associated huge protein.

major histocompatibility complex class I (MHC class I) mediated presentation of viral peptides on infected cells to avoid recognition. The viral proteins can strike at any point in gene expression and transport of MHC class-I antigen (Table 23.1). This strategy is applied by adenoviruses (Ad), cytomegaloviruses (CMV) and human immunodeficiency virus I (HIV-I). Herpes simplex virus ICP47 protein inhibits production of antigenic peptides or their translocation into ER/cis golgi and in another scenario cytomegalovirus US6 gene product prevents peptide loading of MHC class I by inhibiting peptide translocation into ER acting on the luminal side.

Orthopoxvirus SPI-2 (crmA in cowpox virus) is an exception to the rule since it blocks CTL-induced apoptosis by inhibition of the caspase cascade. In some cell types mutation of the SPI-1 gene in addition to SPI-2 was also necessary to completely abrogate inhibition, thus orthopoxviruses encode two genes that function independently to inhibit antiviral CTL cytotoxicity. The Bcl-2 family viral proteins, Ad E1B-19K, Epstein-Barr virus (EBV) BHRF1, human herpes virus 8 Ksbcl-2, HSV ORF16 and African swine fever virus (ASFV) LMV5-HL (O'Brien, 1998) are also likely CTL signal interceptors since Bcl-2 inhibits both the Fas and the perforin-granzyme branches of CTL cytotoxicity.

NK cells dominate cellular cytotoxicity early after infection. Patients with NK deficiency show increased susceptibility to herpes virus infection (reviewed by Biron, 1997). However, while downregulation of MHC class I will prevent CTL-mediated lysis of infected cells, it renders them susceptible to NK-mediated killing. It has been hypothesized (Kärre & Welsh, 1997) that the MHC class I homolog human cytomegalovirus (HCMV) UL18 and mouse (MCMV) m144 bind inhibitory NK receptors (KIRs) acting as a "viral decoy". Although the proposed interaction of UL18 and m144 with the KIRs themselves has not been demonstrated, it has been shown that both viral proteins inhibit NK cytotoxicity. Instead, interaction studies utilizing UL 18 led to discovery of a new family of proteins, the leukocyte immunoglobulin-like receptors (LIR-1) (Cosman *et al.*, 1997). LIRs are structurally, and functionally, homologous to KIRs. However, LIRs are only expressed on a subset of NK cells. In contrast, they are expressed on most or all monocytes, dendritic and B cells, and a subset of T-cells. UL 18, thus, might inhibit activation of a plethora of immune cells.

Cytokines are an important part of the immune response to many organisms. They contribute to the proliferation and differentiation of lymphocytes and other target cells and also drive inflammation. Most of the infectious organisms seem to target local inflammatory mediators directly and avoid wholescale immunosuppression. In support of this idea, several infectious agents have been identified that produce macromolecules which interfere with the function of the inflammatory cytokine interleukin-1 (IL-1) and tumor necrosis factor (TNF) (Table 23.1). The most direct approach to counter TNF action is taken by the Shope fibroma/myxoma virus and cowpox viruses that encode secreted TNF receptors. Similarly, vaccinia virus expresses a soluble IL-1 receptor that is specific for IL-1β. EBV has evolved a different strategy, the gp350 viral envelope protein induces expression of the IL-1 receptor antagonist that competitively inhibits IL-1 activity. TNF and IL-1 are inducible functions of the immune defense. Expression of IL-1β is blocked by cowpox and vaccinia viruses that inhibit proteolytic maturation of IL-1 through the virally encoded protease inhibitors.

Many viruses encode proteins that can interfere with the intracellular phase of the cytokine signaling pathway. Adenoviruses have evolved a collection of proteins that block TNF cytotoxicity at the signaling phase (Wold, 1993). Specifically, adenoviruses target TNF induced arachidonic acid (AA) release, a signaling event crucial for TNF cytotoxicity. Arachidonic acid metabolites are potent mediators of inflammation. Thus, inhibition of arachidonic acid release suppresses TNF-induced inflammation as well as cell death. In most cells TNF induces a protective response through activation of the nuclear factor kappaB (NF-κB). FIP-3, a cellular protein inhibits TNF-induced NF-κB activation. Ad E3-14.7 K protein rescues the protective response by complex formation with FIP-3 (Li *et al.*, 1999). Thus, E3-14.7 K protein acts through at least two different pathways to protect infected cells against TNF. Another Ad protein, E1B-19K, that can prevent cell death inflicted by TNF, inactivates Bax, Bak, and Bik the proapoptotic cellular proteins. EBV BHRF1, HSV ORF16 and ASFV LMW5-HL protein are also Bcl-2 homolog, so they presumably utilize a similar strategy to inhibit TNF cytotoxicity. E1B-19K and the E3-RID complex also inhibit cell death via the Fas ligand, another member of the TNF family (Imai *et al.*, 1999 and Tollefson *et al.*, 1998). In addition to Fas, RID also down-regulates TRAIL receptors (Wold *et al.*, 1999) making infected cells nonsusceptible to TRAIL-induced cell death. Another recently discovered family of inhibitors of TNF receptor family-mediated apoptosis are the viral flice inhibitory proteins (vFLIPs) that interact with FADD, an adaptor protein that links TNF receptor family proteins to FLICE, and inhibit recruitment of FLICE, the most upstream element of the caspase cascade. VFLIPs have been found in γ-herpes viruses, including HHV-8 and HVS, and tumorigenic human mollusci-poxvirus. Adenovirus 14.7 K takes a shortcut, it directly interacts with FLICE to block TNF induced cell death (Chen *et al.*, 1998).

An indirect route to reduce the inflammatory response is to shift the ratio of T helper 1 (Th1) and T helper 2 (Th2) cells produced during immune responses. The IL-2 and interferon-γ (IFN-γ) producing Th1 cells stimulate cellular cytotoxicity and macrophage mediated inflammatory responses, while Th2 cells boost antibody production. The EBV BCRF1 gene product, the poxvirus orf virus (OV) gene and CMV UL111a are homologs of the Th2 inducer mammalian IL-10: these viral products are likely to be potent suppressors of inflammation. In a different scenario, measles virus infection down-regulates production of pro-Th1 cytokine IL-12 resulting in suppression of inflammatory and CTL responses.

The importance of the IFN system in the antiviral immune response has been demonstrated *in vivo* using interferon receptor knockout mice. Mice lacking either the IFN α/β specific IFN type I receptor or the IFN γ specific type II receptor showed no overt anomalies but were unable to cope with viral infection (Müller *et al.*, 1994). It has also been shown that both IFN systems are essential for antiviral defense (Müller *et al.*, 1994). The anti-interferon strategies bear resemblance to the anti-TNF/IL-1 defense: a soluble type I receptor homolog is encoded by vaccinia virus, functional IFN type II receptors are secreted by cells infected with poxviruses. Signaling through the interferon receptors activates the Janus kinase (JAK)/signal transducers and activators of transcription (STAT) signal transduction pathway. Signal interceptors known to target the JAK/STAT pathway of interferon signaling include simian virus 5 (SV5) V protein and the adenovirus transcription factor E1A. SV5 V protein targets STAT1 for proteosome mediated degradation while Ad E1A inhibits interferon-stimulated genes factor (ISGF) activity at three different levels: (i) by repression of expression of the STAT 1 subunit of ISGF (ii) by binding p300/CBP, a transcriptional adaptor for STAT1 and STAT 2 components of the IFNα and IFNγ signaling apparatus or (iii) by lowering the level of the p48 subunit, an essential component of ISGF3γ. Hepatitis C virus (HCV) has also been shown to inhibit Jak/STAT activity. Although the effect requires viral protein synthesis no single viral protein function has yet been identified (Heim *et al.*, 1999).

The main outcome of interferon response is the blockade of viral protein synthesis. A number of viral gene products, known to inhibit the IFN system, interfere at translational level with the execution phase of the IFN response. This has been reviewed by Smith (1994), and will not be discussed here.

In sum, blockade of antiviral immune surveillance pathways is a major concern for viruses. Viruses that have been shown to cause persistent infections are particularly well-armed against antiviral defenses. Adenoviruses, well-known examples of persistent viruses have evolved 9 gene functions to suppress antiviral immune defense (Table 23.1, reviewed by Wold *et al.*, 1999). Those Ad functions that block antiviral responses at the signaling phase are depicted in Figure 23.1. E3-14.7 K and RID block TNF and Fas-L cytotoxicity. The repertoire of RID complex includes blockage of the TRAIL pathway as well. E3-gp19 K is the anti-CTL arm of the virus, while E1B-19 K is a broad range inhibitor of apoptosis induced by various stimuli.

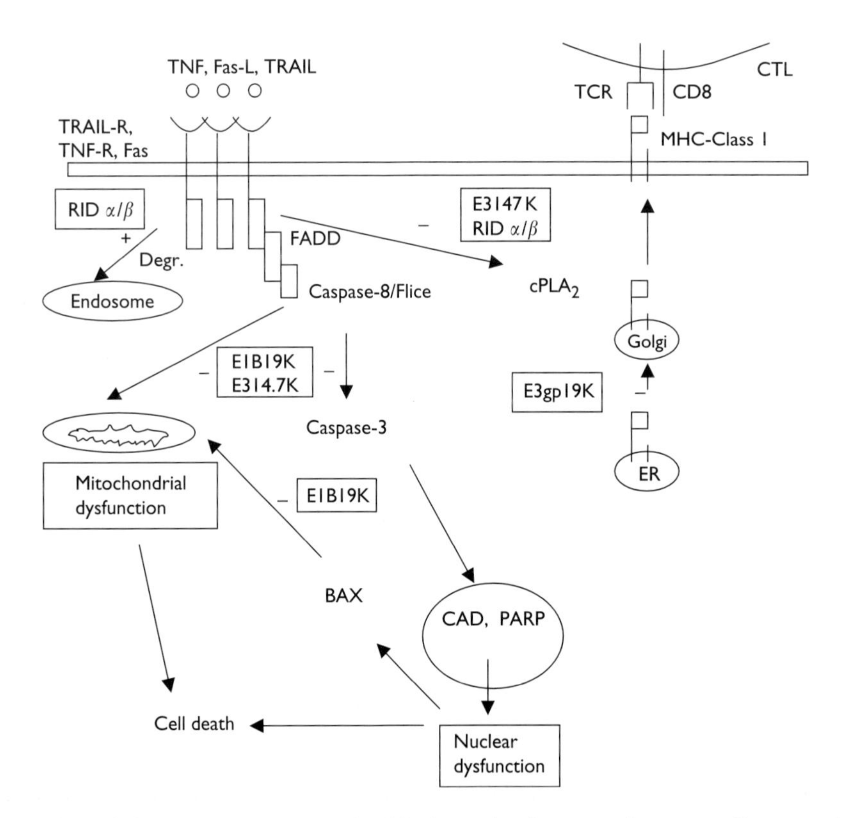

Figure 23.1 Adenovirus proteins that block signaling by antiviral immune effector molecules/cells. See text for abbreviations.

23.3 Cell growth/activation signaling

23.3.1 Signal activators

Subversion of normal growth factor signaling pathways is critical to neoplastic processes. Consistent with this notion, a sizable fraction of oncogenes has been shown to encode components of the cell growth machinery such as growth factors and growth factor receptors. The transforming proteins of many DNA viruses and retroviruses are known to interact with growth factor signaling pathways (Figure 23.2) (Krajcsi and Wold, 1998).

A family of receptor tyrosine kinases that have been described as transmiting growth factor signals are targeted by DNA and retroviruses. There is evidence that bovine papilloma virus (BPV) E5 protein cooperates with colony stimulating factor receptor (CSF-R) and epidermal growth factor receptor (EGF-R) to transform NIH 3T3 cells. The effect of E5 occurs in the absence of receptor stimulation by ligand. It is enhanced by addition of ligand and is associated with inhibition of receptor degradation and persistence of activated receptors at the cell surface. In a similar fashion BPV E5 induces platelet-derived growth factor receptor β dimerization and association with downstream effectors such as p85 subunit of phosphoinositol 3′-kinase (PI3K), phospholipase C-γ (PLC-γ) and Ras-GTPase activating protein (Lai *et al.*, 2000). A different approach is utilized by vaccinia virus. The virally encoded vaccinia virus growth factor (VGF) is homologous to EGF and has been shown to stimulate the protein kinase activity of EGF-R. Although vaccinia virus is not considered a transforming virus it induces hyperplastic responses of cells at the edge of growing lesions during active viral

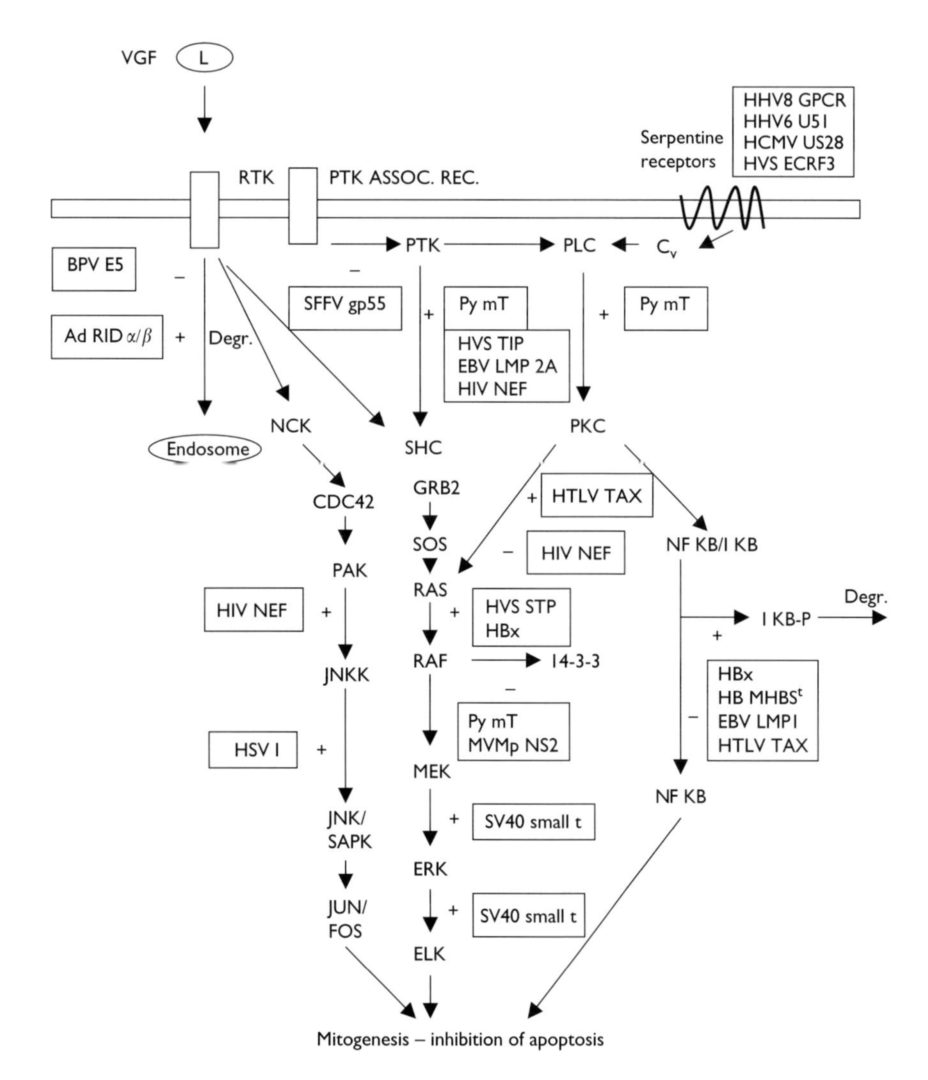

Figure 23.2 Virally regulated cell growth signaling. Degr.: degradation; GRB2: growth factor receptor binding protein 2; G_p: PLC specific G-protein; L: ligand; PTK ASSOC. REC.: protein tyrosine kinase associated receptor; SHC: src homology and collagen; SOS: son of sevenless: −: inhibition; +: activation. See text for other abbreviations.

replication. It has been speculated that the mitogenic signal is beneficial to virus replication (Kim *et al.*, 1995).

Other signal transducing, plasma membrane receptors do not harbor intrinsic tyrosine kinase activity. Instead, they associate with nonreceptor tyrosine kinases that can serve as targets for viral intervention. Middle-T antigen of the oncogenic mouse polyomavirus (Py mT) associates with and activates the cellular tyrosine kinases c-Src, c-Yes and Fyn. Upon complex formation middle T becomes phosphorylated, a prerequisite for binding phosphatidyl inositol-3-kinase (PI3K), the adaptor protein SHC and phospholipase C-γ1 (PLC-γ1). These interactions transduce mitogenic signals (Figure 23.2) normally triggered by growth factors and are essential for polyomavirus-mediated transformation of cells in culture and tumor formation in animals.

Tumorigenic activities of acute transducing retroviruses have been found to reside in the transduced cellular sequences. However, not all retroviruses that induce cell proliferation contain such oncogenic cellular sequences. The acute erythroleukemia-inducing spleen focus forming virus (SFFV) appears to promote cell growth through the viral env gene that encodes a 55 kD glycoprotein (gp55) which interacts with the erythropoietin receptor (EPO-R) in the endoplasmic reticulum and at the plasma membrane. The results strongly argue that SFFV gp55 subverts normal requirements of erythroid cells for EPO by directly binding to and triggering EPO-R.

The major signaling pathway that transduces mitogenic signals from the plasma membrane to the nucleus is the Ras/MAP kinase network. Mitogen activated protein kinases (MAPKs), also known as extracellular signal regulated kinases, are protein serine/threonine kinases that are rapidly activated upon stimulation of a variety of cell surface receptors. They control expression of genes essential for many cellular processes including cell growth and differentiation (Marshall, 1995). These kinases play a central role in mitogenic signaling, as impeding their function prevents cell proliferation in response to a number of growth-stimulating agents. Furthermore, constitutive activation of the MAPK pathway is itself sufficient for tumorigenesis. MAP kinases include extracellular signal-regulated protein kinase (ERK), c-Jun amino terminal kinase/stress activated protein kinase (JNK/SAPK) and p38 subgroups. So far, only the ERK pathway has been implicated in oncogenesis, even though the c-Jun protooncogene is activated by JNK/SAPK phosphorylation and is thought to be required for Ras transformation.

It is plausible that some oncogenic viruses encode proteins that activate the Ras/MAPK pathway. Herpes virus saimiri (HVS) STP-C488 was the first virus-encoded protein shown to associate with cellular Ras in transformed cells. STP-C488 can transform Rat-1 cells, and results in the formation of invasive tumors in nude mice. Expression of STP-C488 activated the ras signaling pathway as evidenced by a 2-fold increase in the ratio of Ras-GTP to Ras-GDP and by the constitutive activation of MAPK. HBx, the transcriptional transactivating protein of hepatitis B virus (HBV), also enhances Ras-GTP formation stimulating both ERK and JNK/SAPK activities. The molecular interactions have not yet been elucidated, but both Ras and Raf-1 are required for activation of c-jun transcriptional activity. The oncogenic potential of HBx has been described as "atypical". The long latency between HBV infection and the development of hepatocellular carcinoma makes it unlikely that Hbx alone directly causes hepatocyte transformation. Since the Ras-Raf pathway has also been implicated in prevention of apoptosis, it is tempting to speculate that HBx might prevent apoptotic cell death of hepatocytes. Simian virus 40 small t antigen utilizes a different strategy to induce cell growth. Small t binds to protein phosphatase A (PP2A) and inhibits its ability to dephosphorylate a variety of phosphoproteins *in vitro*, including extracellular signal regulated kinase (ERK2) and MAPK/ERK kinase (MEK). Expression of small t in SV40 permissive cell line CV 1 leads to an increase in the activity of ERK2 and MEK1 and ultimately causes these cells to proliferate.

The other major pathway known to transmit mitogenic signals is the phosphatidyl inositol/protein kinase C (PKC) system. The phosphoinositidase C family (otherwise known as phosphoinositide specific phospholipase C or PI-PLC) is a receptor-controlled family of enzymes that catalyse the splitting of phosphatidyl-inositol-4,5-bisphosphate (PIP_2) into 1,2-diacyl glycerol (DAG) and inositol-1,4,5-trisphosphate (IP_3). IP_3 induces receptor-mediated release of Ca^{2+} from intracellular stores, and the elevated level of Ca^{2+} in concert with DAG induces translocation and activation of protein kinase C (PKC). Sustained activation of PKC is essential for subsequent responses such as cell proliferation and differentiation. Activation of this pathway by adenoviruses, herpesviruses, HIV-1, HBV and HTLV has been reported. However, the molecular details of these interactions have been investigated in only a few cases. Kaposi's sarcoma-associated herpes virus (human herpes virus 8, HHV 8) encodes a G protein coupled receptor (GPCR) that is a *bona fide* signaling receptor which has constitutive activity in the phosphoinositide-inositolphosphate-protein kinase C pathway, stimulating cellular proliferation and making it a candidate oncogene. Polyomavirus takes an entirely different approach to activate the PI-PKC pathway: the Py mT antigen forms a complex with PLC-γ1, a member of the PI-PLC family. Tyrosine phosphorylation of PLC-γ1 is elevated in cells expressing middle T, suggesting activation of this enzyme. Another enzyme of the phospholipid metabolism that is known to associate with middle T is PI3K. PI3K is a dual specificity kinase and is thought to be important in regulating cell proliferation. The Py mT-PI3K complex formation results in activation of pp70 s6 kinase ($pp70^{S6K}$). The induction of $pp70^{S6K}$ activity most likely plays a role in the emergence of at least some transformation related changes, given the known role of S6 kinase in normal mitogenesis.

The oncogenic retrovirus, human T-cell leukemia virus I (HTLV-I), usurps PKC signaling pathway as well. Tax_I, a viral regulatory protein of HTLV-I, associates with PKC. This interaction stimulates PKC autophosphorylation in the absence of cofactors, leading to the cytosol-to-membrane translocation of the enzyme, suggesting that Tax_I activates PKC.

Two other signal transducing protein complexes targeted by many viruses are the 14-3-3 proteins and the NFκB/IκB complex. The 14-3-3 family comprises at least seven isoforms in mammalian cells. Some of the 14-3-3 proteins were recently reported to interact with cellular products, in particular Raf-1, Bcr and Bcr-Abl, PKC, cdc25 and Bad, and so they are thought to be connected with signal transduction and cell cycle control pathways. Py mT and, more recently, the nonstructural proteins, NS2, of minute virus of mice (MVMp), a parvovirus, have been found to associate with 14-3-3 proteins. Autonomous parvoviruses are characterized by their dependence for growth on cellular factors that are expressed only in proliferating cells. These requirements result in the restriction of productive parvovirus infections to proliferating tissue and are likely to account for the remarkable oncotropism exhibited by some of these viruses. Thus, both of these interactions seem to function in the proliferation or apoptosis inhibitory signaling pathways.

The transcription factor NFκB lies at the crossroads of signaling cascades. It has been implicated in many different cellular regulatory pathways, most notably transformation, inhibition of cell death, immunoregulation and activation of viral gene expression. In resting cells, NFκB/Rel complexes are sequestered in the cytoplasm as latent precursors by physical association with a family of ankyrin motif-rich inhibitory proteins, named IκBα-ε. Cellular activation triggers IκB phosphorylation the subsequent degradation of IκB and the concomitant nuclear translocation of NFκB/rel.

The Tax_I gene product of HTLV-1 is known to induce persistent nuclear expression of various NFκB/Rel factors. Recent studies have demonstrated that Tax induces NFκB activation either by phosphorylation and degradation of IκBα and IκBβ, or by direct interaction with p100/IκBδ via the rel homology domain and with p105/IκBγ through the ankyrin motif. These molecular interactions culminate in induction of IL-2 IL-2Rα expression and establish a positive autostimulatory loop.

Hepatitis B virus encodes two genes to manipulate NF κB signaling. HBx protein utilizes a Ras-dependent and a Ras-independent pathway. The other protein, $MHBs^t$, an ER resident transactivator, acts through a radical-mediated mechanism to activate NFκB signaling. EBV LMP1, in addition to induction of IκB degradation, utilizes a second mechanism to turn on NFκB signaling. LMP1 complexes with tumor necrosis factor receptor associated factor 2 (TRAF2) protein, that has been implicated in TNF-mediated NFκB activation. LMP1 recruits TRAF2 to the membrane and this complex triggers activation of NFκB. Recombinant EBV genetic analysis has indicated that the TRAF binding domain is essential for primary B-lymphocyte transformation. In addition, it may protect the virus from TNF cytotoxicity, since NFκB blocks cytolytic action by TNF. The fact that different oncogenic viruses have evolved three oncogenic proteins to turn on NFκB signaling strongly argues in favor of its role in oncogenesis.

A number of viruses express proteins that regulate nuclear events of mitogenesis (e.g. cell cycle). The mechanism of these interactions has been reviewed elsewhere (Nevins, 1994; Vousden, 1995).

All DNA viruses share a common strategy of programed gene expression and the necessity to induce DNA synthesis in host cells. Early viral proteins are thought to bring about both viral and host DNA replication. Small DNA viruses are especially dependent on cellular enzymes for replication. Blocking cell cycle progression by n-butyrate inhibited replication of the small DNA viruses, polyoma and papilloma, supporting this notion. Although the large T antigen of polyoma virus can by itself induce G_1-S cell cycle transition due to its affinity for retinoblastoma protein (Rb) and p53, it has been shown that the Py mT protein is also essential for efficient virus replication. Thus, mitogenic signaling may serve another function, boosting viral replication by induction of host cell macromolecular synthesis. Large DNA viruses encode a number of replicatory enzymes and so are less dependent on the cellular replication machinery. Nevertheless, a VGF^- mutant of vaccinia virus replicated less efficiently in Swiss 3T3 cells and exhibited an attenuated phenotype in mice and rabbits.

23.3.2 Signal interceptors

Persistent/latent viruses may benefit from interception of signaling pathways transducing growth/activation signals, since induction of cell proliferation/activation pathways may lead to apoptosis of host cells (Gougeon and Montagnier, 1993) or activation of the lytic replication pathway in latently infected cells (Miller *et al.*, 1994).

p56 Lck, a nonreceptor tyrosine kinase, is an ideal target for signal interceptors since it has been found associated with cell surface receptors like CD2, CD4, CD5, CD8, the IL-2 receptor (IL-2-R) and a number of downstream effectors, and thus functions in an array of growth/activation signaling pathways in lymphocytes. On the other hand, Lck has been implicated in T-cell receptor-mediated or HIV-induced T lymphocyte apoptosis. It has been shown that herpes virus saimiri ORF 1 (tip) forms a complex with Lck. HVS tip has two domains homologous to src family kinases, a kinase homology domain and a proline rich src-homology 3 (SH3) binding conserved motif, both of which are required for interaction with Lck. Tip may be responsible for the recruitment and sequestration of Lck. HVS transforms T-cells *in vivo* and *in vitro* to immortalized growth, thus it seems counterintuitive that HVS tip should down-regulate Lck-mediated signal transduction. However, this virus is known to contain another gene, STP, that is capable of functioning as an oncogene in rodent fibroblasts and is required for viral T-cell transformation. Recent results suggest that STP binds and activates Ras. A similar scenario has been found in transformation of B-cells by EBV. LMP2A protein of EBV associates with B-cell tyrosine kinases Lyn and Syk. LMP2A is a constitutive dominant negative modulator of surface immunoglobulin (Ig) signaling through Lyn and Syk. The EBV equivalent of HVS STP is LMP1, a viral oncogene that transforms rodent fibroblasts *in vitro* and is required for viral transformation. These strategies may allow cellular immortalization by HVS and EBV to proceed and at the same time prevent potential adverse effects that may be associated with Lck, Syk or Lyn activation e.g. apoptosis or untimely viral replication.

Inhibition of cell proliferation/activation pathways has been reported for nononcogenic viruses as well. HIV-1 Nef has been shown to interact with cellular signaling apparatus at different points. Nef post-translationally down-regulates CD4 protein via endocytosis, and forms a complex with Lck inhibiting its kinase activity, and protecting virus-infected cells against reinfection and activation through the T-cell receptor (TCR). Moreover, it has been shown that HIV-1 Nef associates with a member of the p21-activated kinase (PAK) family. The recruitment of PAK by membrane associated Nef might facilitate the PAK activation pathways. PAK initiates a cascade leading to the activation of JNK. The significance of the activation of the PAK-JNK pathway for HIV pathogenesis is not known. However, CD28 costimulation, that inhibits TCR-induced apoptosis during a primary T-cell response, is also known to activate JNK, supporting the notion that the primary function of Nef is to inhibit apoptosis of infected cells. Finally, Nef complexes with protein kinase theta (PKCθ) inhibiting the usual translocation of PKCθ from cytosol to the particulate fraction upon phorbol myristate acetate (PMA) or phytohemagglutinine stimulation. Thus, PKCθ is another target for Nef to impair T-cell function. These observations suggest that Nef may act as a positive factor in the HIV-1 life cycle by protecting the cell against virus induced apoptosis and other cell activation processes. Inhibition of apoptosis in HIV-infected cells enhances virus production and facilitates persistent infection.

In another scenario, the adenovirus E3 RID complex stimulates EGF-R internalization and degradation without activating the EGF-R kinase activity (Krajcsi and Wold, 1988). This property initially desensitizes the infected cells to EGF. The biological significance of this effect is unclear. Since EGF is known to induce cytosolic phospholipase A2 activity at both transcriptional and post-transcriptional levels and synergizes with TNF and IL-1 to activate prostaglandin biosynthesis, this receptor down-regulation could be another mechanism by which adenovirus inhibits inflammatory antiviral responses.

23.4 Viral adaptation

How can viruses gain these functions? First alternative: molecular piracy. A majority of the proteins with clear homology to cellular proteins are involved in regulating cellular signaling. The sequence relationship of the host and viral homologues for the RNA viruses is very strong, in most cases about 80% amino acid identity, yet for the DNA viruses it is usually weak, generally less than 40% (Murphy, 1994). The other group of proteins have no sequence similarity, but subserve a similar function. These viruses used a second alternative: convergent evolution (Murphy, 1994). This is a potent selective force behind generation of host defense protein diversity. On the other hand, since the virus mutates much more quickly than the cell, the absolute dependence of the virus on the host may ultimately be a limitation for the redesigning of the viral protein.

Studies on viral proteins regulating cellular signaling have shed more light on the strategies of viruses in the battle for survival against the host. Firstly, viral strategies that affect cell growth and viral replication are mostly intracellular. In contrast, the major battlefield between antiviral immune functions and viral defense is extracellular. In keeping with this, it has been shown that due to coevolutionary pressure the interspecies divergence among proinflammatory and host-defense proteins is three times higher than the average (Murphy, 1994). Even more striking, the divergence of the extracellular domains of receptors for host-defence proteins is usually higher than the divergence of the intracellular domains (Gerard *et al.*, 1998). Secondly, viral genomes are rich sources of host genes: analysis of the HVS genome has led to the discovery of a new cytokine, IL-17 which binds to a novel receptor (Yao *et al.*, 1995). This is neither a surprise nor unprecedented since a number of protooncogenes were explored through the cognate sequences found in transducing retroviruses.

23.5 Implications for therapy

Understanding virus-cell interaction opens new avenues for treatment of viral diseases. Since HIV pathogenesis has been the focus of viral research for the last decade many studies in the HIV field point to therapeutic applications. NF-κB, a major cellular transcription factor, induces activation of proviral DNA. Therefore, agents blocking NF-kB-mediated HIV activation are potential drug candidates for treatment of AIDS. Fasudil hydrochloride, a serine/threonine kinase inhibitor that blocks NF-κB activation has been shown to block replication of HIV-1 (Sato *et al.*, 1998). Members of the MAPK pathway have also been involved in various aspects of HIV pathology. It has been shown that UV-induced activation of provirus is mediated by p38 MAPK, and can be blocked by SB203580, a p38 MAPK specific inhibitor (Taher *et al.*, 2000) while PD98059, an ERK kinase inhibitor blocks HIV induced endothelial permeability (Osinma *et al.*, 2000). PAK1, a p21-activated kinase is known to associate with HIV-1 Nef and this interaction is critical for viral replication. Accordingly, blocking PAK1 by an inhibitory peptide severely impaired HIV replication (Fackler *et al.*, 2000). HIV disarms the host's immune system by killing a large pool of T lymphocytes. The Fas/Fas-L system is involved in apoptosis of T lymphocytes in HIV-infected individuals, and it is known that ceramide generation and mitochondrial dysfunction are major components of the Fas-mediated apoptotic pathway. L-carnitin, a small endogenous molecule, is known to improve mitochondrial function. Accordingly, it has been shown that carnitin treatment slows down T lymphocyte apoptosis in asymptomatic, HIV-1-infected subjects (Moretti *et al.*, 1998).

Recent evidence indicates that the p38 MAPK pathway is important for replication of HCMV. HCMV has evolved two mechanisms to activate the p38 pathway. Initially the HCMV infection inhibits dephosphorylation of p38, later increased p38 activation is mediated by increased activity of upstream kinases (MKK3 and MKK6) (Johnson *et al.*, 2000). Accordingly, the same group has shown that a p38 inhibitor drug inhibited HCMV replication and prevented permissive infection (Johnson *et al.*, 1999).

Cancer gene therapy utilizes the inherent differences between the regulatory apparatus of tumor and normal cells in a reverse fashion, using viruses to cure disease. Human reovirus requires an activated Ras signaling pathway for infection of cultured cells. Reovirus replication-induced cell lysis thus selectively kills cancer cells with active Ras and induces regression in some tumors (Coffey *et al.*, 1998). Adenoviruses are also dependent on the cellular replication apparatus for efficient replication and cell lysis. Adenoviruses, expressing an E1A mutant that is unable to bind the retinoblastoma protein and p300, are unable to drive quiescent cells into cell cycle. These viruses therefore replicate selectively in tumor cells. Indeed, this remarkable tumor specificity has been utilized to make cancer-specific adenovirus vectors (Doronin *et al.*, 2000).

The benefits from studies on molecular aspects of viral pathogenesis have been multiple: (i) providing insights into cellular regulatory pathways (ii) yielding new strategies to fight viral diseases (iii) opening new avenues for using viruses to fight other diseases.

Acknowledgments

This work was supported by OTKA T029622.

References

Biron, C.A. (1997) Activation and function of natural killer cell responses during viral infection. *Curr. Opin. Immunol. 9, 24–34.*

Chen, P., Tian, J., Kovesdi, I., and Bruder, J.T. (1998) Interaction of the adenovirus 14.7 kDa protein with FLICE inhibits Fas ligand-induced apoptosis. *J. Biol. Chem. 273, 5815–5820.*

Coffey, M.C., Strong, J.E., Forsyth, P.A., and Lee, P.W. Reovirus therapy of tumors with activated Ras pathway. *Science, 282, 1332–1334.*

Cosman, D., Fanger, N., Borges, L., Kubin, M., Chin, W., Peterson, L., and Hsu, M-L. (1997) A novel immunoglobulin superfamily receptor for cellular and viral MHC Class I molecules. *Immunity, 7, 273–282.*

Doronin, K., Toth, K., Kuppuswamy, M., Ward P., Tollefson, A.E., and Wold, W.S.M. (2000) Tumor-specific, replication-competent adenovirus vectors overexpressing the adenovirus death protein. *J. Virol., 74, 6147–6155.*

Fackler, O.T., Lu, X., Frost, J.A., Geyer, M., Jiang, B., Luo, B., Abo, A., Alberts, B., and Peterlin, B.M. (2000) p21-activated kinase 1 plays a critical role in cellular activities by Nef. *Mol. Cell Biol. 20, 2619–2627.*

Gerard, C., Bao, L., Orozco, O., Pearson, M., Kunz, D. and Gerard, N.P. (1998) Molecular cloning of the mouse C5a anaphylatoxin receptor. *J. Immunol. 149, 2600–2606.*

Gougeon, M.-L. and Montagnier, L. (1993) Apoptosis in AIDS. *Science 260, 1269–1270.*

Heim, M.H., Moradpour, D., and Blum, H.E. (1999) Expression of hepatitis C virus proteins inhibits signal transduction through the Jak-STAT pathway. *J. Virol. 73, 8469–8475.*

Imai, Y., Kimura, T., Murakami, A., Yasima, N., Sakamaki, K., and Yonehara, S. (1999) The CED-4-homologous protein FLASH is involved in Fas-mediated activation of caspase-8 during apoptosis. *Nature, 398, 777–785.*

Johnson, R.A., Huong, S.M., and Huang, E.S. (2000) Activation of the mitogen-activated protein kinase p38 by human cytomegalovirus infection through two distinct pathways: a novel mechanism for activation of p38. *J. Virol. 74, 1158–1167.*

Johnson, R.A., Huong, S.M., and Huang, E.S (1999) Inhibitory effect of 4-(4-fluorophenyl)-2-(4-hydroxyphenyl)-5-(4-pyridyl) 1H-imidazole on HCMV DNA replication and permissive infection. *Antiviral Res. 41, 101–111.*

Kërre, K. & Welsh, M. (1997) Viral decoy vetoes killer cell. *Nature 386, 446–447.*

Kim, H.S., Lee, Y.H., Min, D.S., Chang J.-S., Ryu, S.H., Ahn, B.-Y. and Suh P.-G. (1995) Tyrosine phosphorylation of phospholipase C-γ1 by vaccinia virus growth factor, *Virology, 214, 21–28.*

Krajcsi, P., and Wold, W.S.M. (1998) Viral proteins that regulate cellular signaling. *J. Gen. Virol., 79, 1323–1335.*

Kung, H.-J. and Liu, J.-L. (1997) Retroviral oncogenesis. N. Nathanson (ed.) *Viral Pathogenesis*, Lippincott-Raven Publishers, Philadelphia, pp 235–266.

Lai, C.C., Henningston, C., and DiMaio, D. (2000) Bovine papillomavirus E5 protein induces the formation of signal transduction complexes containing dimeric activated platelet-derived growth factor beta receptor and associated signaling proteins. *J. Biol. Chem. 275, 9832–9840.*

Li, Y., Kang, J. and Horwitz, M.S. (1997) Interaction of an adenovirus 14.7 K protein, inhibitor of tumor necrosis factor alpha cytolysis with a new member of the GTPase superfamily of signal transducers. *J. Virol. 71, 1576–1582.*

Marshall, C.J., (1995) Specificity of receptor tyrosine kinase signaling: transient versus sustained extracellular-signal regulated kinase activities, *Cell, 70, 179–185.*

Miller, C.L., Lee, J.H., Kieff, E. and Longnecker, R. (1994) An integral membrane protein (LMP2) blocks reactivation of Epstein-Barr virus from latency following surface immunoglobulin crosslinking. *Proc. Natl. Acad. Sci. USA 91, 772–776.*

Moretti, S., Alesse, E., Di Marzio, L., Zazzeroni, F., Ruggeri, B., Marcellini, S., *et al.* (1998) Effect of L-Carnitine on human immunodeficiency virus-1 infection-associated apoptosis: a pilot study. *Blood 91, 3817–3824.*

Murphy, P.M. (1994) Molecular mimicry and the generation of host defense protein diversity. *Cell 72, 823–826.*

Müller, U., Steinhoff, U., Reis, L.F.L., Hemmi, S., Pavlovic, J., Zinkernagel, R.M. and Auget, M. (1994) Functional role of type I and type II interferons in antiviral defense. *Nature 264, 1918–1921.*

Nevins, J.R. (1994) Cell cycle targets of the DNA tumor viruses, *Curr. Opin. Genet. Dev. 4, 130–134.*

O'Brien, V. (1998) Viruses and apoptosis. *J. Gen. Virol. 79, 1833–1845.*

Oshima, T., Flores, S.C., Vaitaitis, G., Coe, L.L., Joh, T., Park, J.H., Zhu, Y., Alexander, B., and Alexander, J.S. (2000) HIV-1 Tat increases endothelial solute permeability through tyrosine kinase and mitogen-activated protein kinase-dependent pathways. *AIDS 14, 475–482.*

Sato, T., Asamitsu, K., Yang, J.P., Takahashi, N., Tetsuka, T., Yoneyama, A., Kanagawa, A., and Okamoto, T. (1998) Inhibition of human immunodeficiency virus type 1 replication by a bioavailable serine/threonine kinase inhibitor, fasudil hydrochloride. *AIDS Res. Hum. Retroviruses, 14, 293–298.*

Smith, G.L. (1994) Virus strategies for evasion of the host response to infection. *Trends Microbol. 8, 81–88.*

Taher, M.M., Hershey, C.M., Oakley, J.D., and Valerie, K. (2000) Role of the p38 and MEK-1/2/p42/44 MAP kinase pathways in the differential activation of human immunodeficiency virus expression by ultraviolet and ionizing radiation. *Photochem. Photobiol. 71, 455–459.*

Tollefson A.E., Hermiston, T.W., Colle, C.E., Tripp, R.A., Dimitrov, T., Toth, K., Wells, C.E., Doherty, P.C., and Wold, W.S.M. (1998) Forced degradation of Fas inhibits apoptosis in adenovirus-infected cells. *Nature, 392, 726–730.*

Vousden, K.H. (1995) Regulation of cell cycle by viral oncoproteins, *Semin. Cancer Biol. 6, 109–116.*

Wold, W.S.M. (1993) Adenovirus genes that modulate the sensitivity of virus-infected cells to TNF. *J. Cell. Biochem. 53, 329–335.*

Wold, W.S.M., Doronin, K., Toth, K., Kuppuswamy, M., Lichtenstein, D.L. and Tollefson, A.E. (1999) Immune responses to adenoviruses: viral evasion mechanisms and their implications for the clinic. *Current Opinion Immunol 11, 380–386.*

Yao, Z., Fanslow, W.C., Seldin, M.F., Rousseau, A.M., Painter, S.L., Comeau, M.R., Cohen, J.I., and Spriggs, M.K., (1995) Herpesvirus Saimiri encodes a new cytokine IL-17, which binds to a novel cytokine receptor. *Immunity, 3, 811–821.*

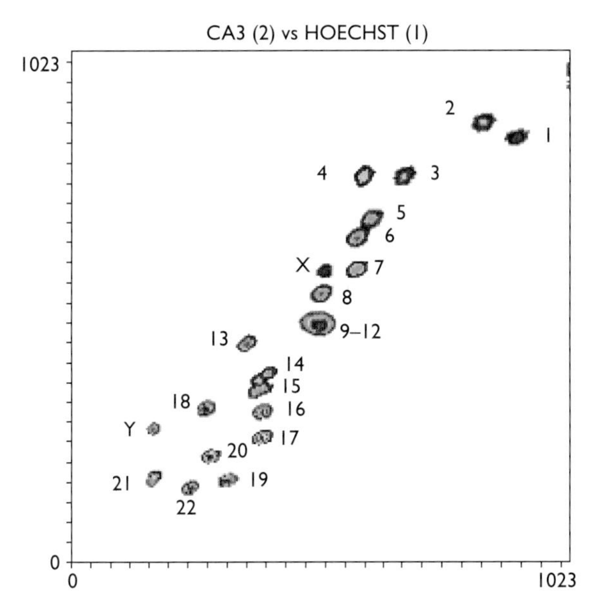

Colour Plate I Separation of human chromosomes on the basis of their relative fluorescence by flow sorting (see Figure 3.3).

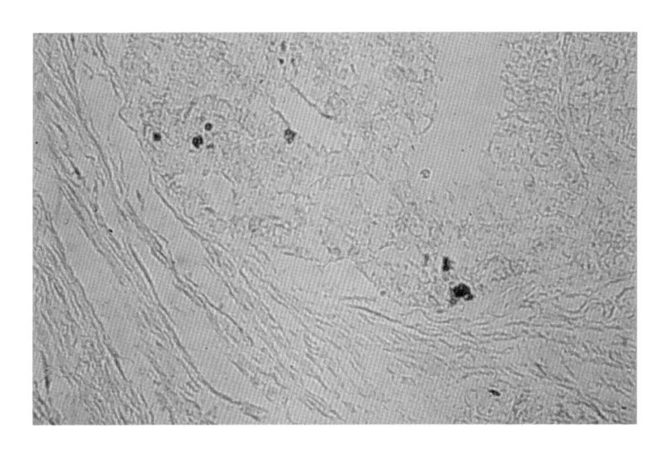

Colour Plate II Apoptotic tumor cells in a poorly differentiated prostate carcinoma, TUNEL reaction, ×400 (see Figure 8.1).

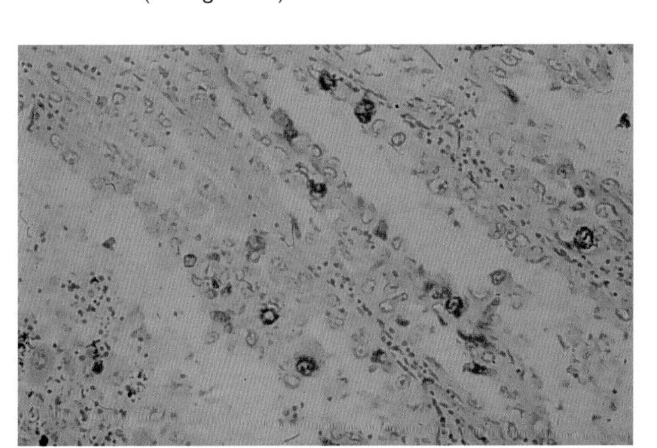

Colour Plate III Adenocarcinoma of the lung. Note numerous apoptotic tumor cells (brown colour), TUNEL reaction, ×400 (see Figure 8.2).

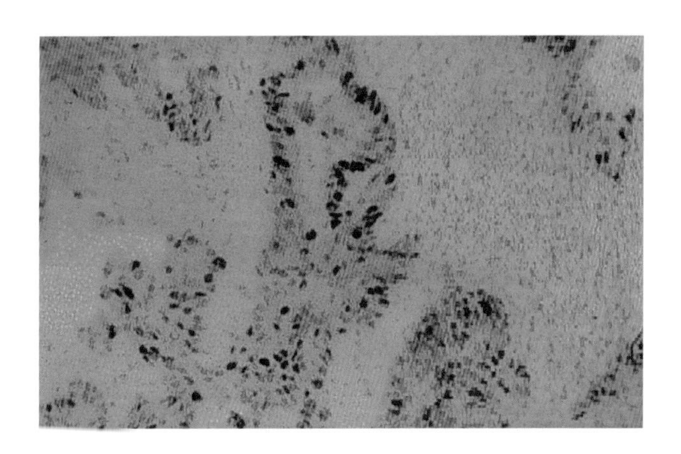

Colour Plate IV Colon biopsy containing both normal and tumor tissue (see Figure 8.3).

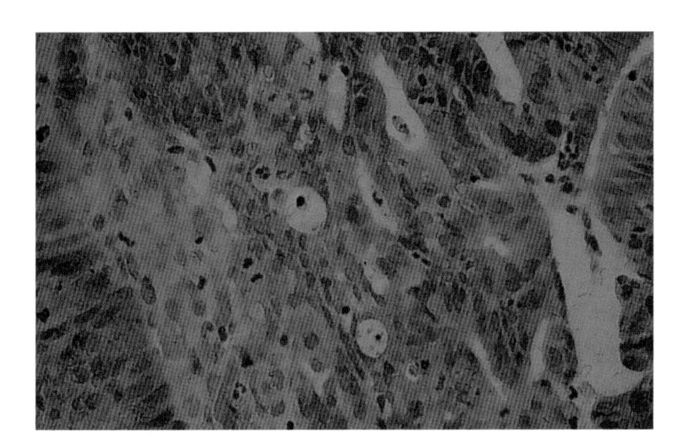

Colour Plate V Apoptotic tumor cells in a colon tumor biopsy, taken from a patient pre-treated with 5-fluorouracyl, H and E, ×400 (see Figure 8.6).

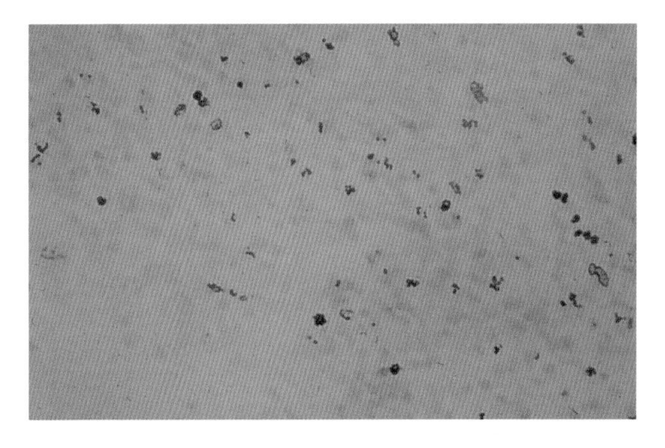

Colour Plate VI Apoptotic cells in periventricular leukomalacia in a pre-term infant, Neurotacs, ×200 (see Figure 8.7).

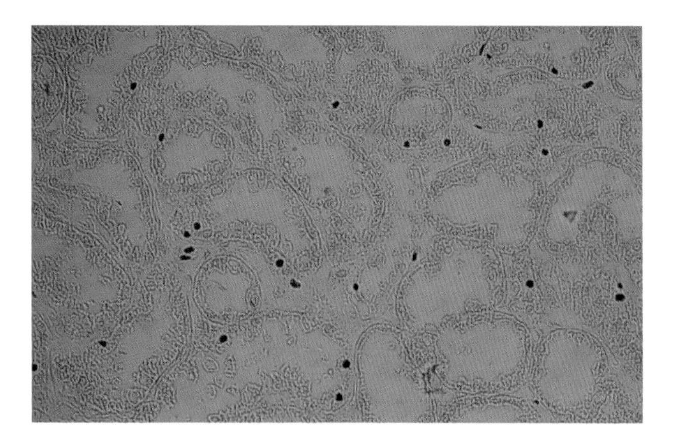

Colour Plate VII Apoptotic cells in a renal biopsy taken from a transplanted kidney 30 minutes after transplantation, TUNEL reaction, ×200 (see Figure 8.8).

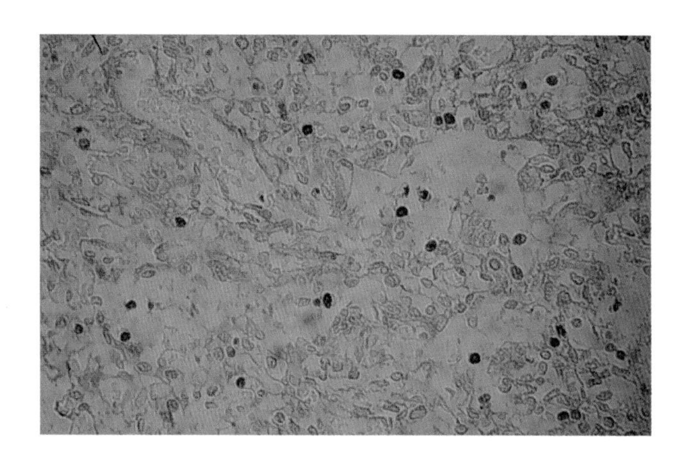

Colour Plate VIII Apoptotic pneumocytes in alveolar walls in a case of bronchopulmonary dysplasia, TUNEL reaction, ×200 (see Figure 8.9).

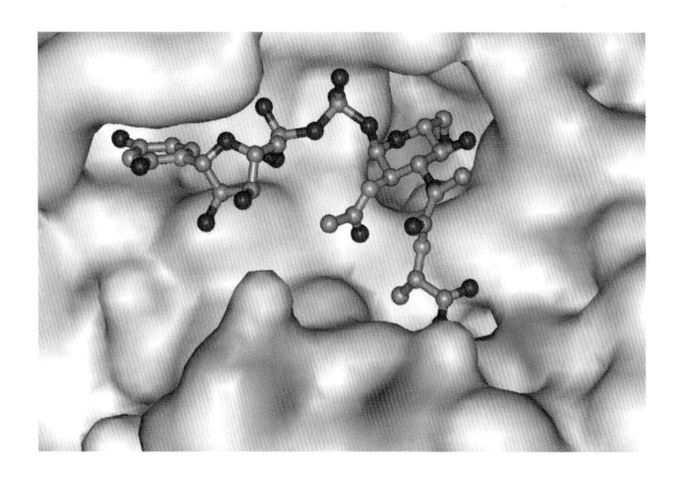

Colour Plate IX (See Figure 17.17.)

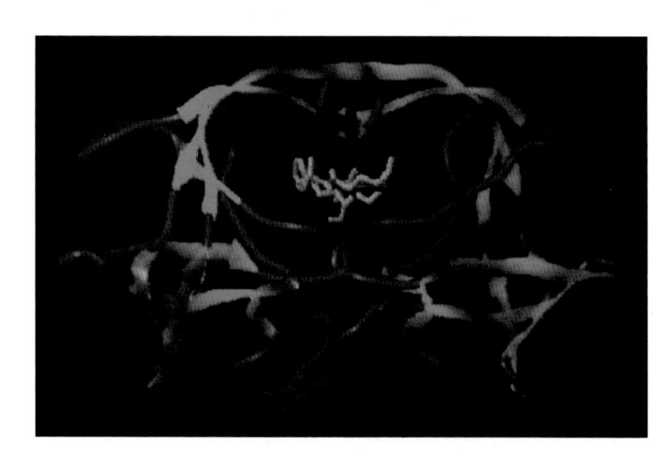

Colour Plate X (See Figure 18.4.)

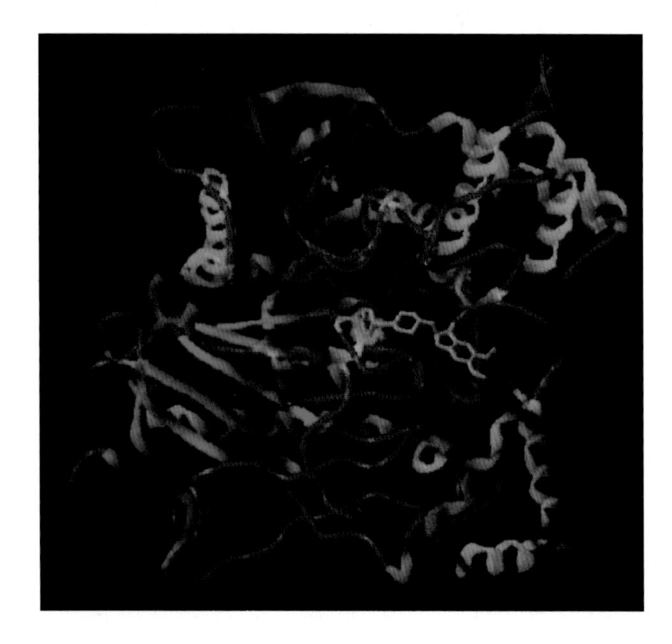

Colour Plate XI (See Figure 18.5.)

Colour Plate XII (See Figure 18.6.)

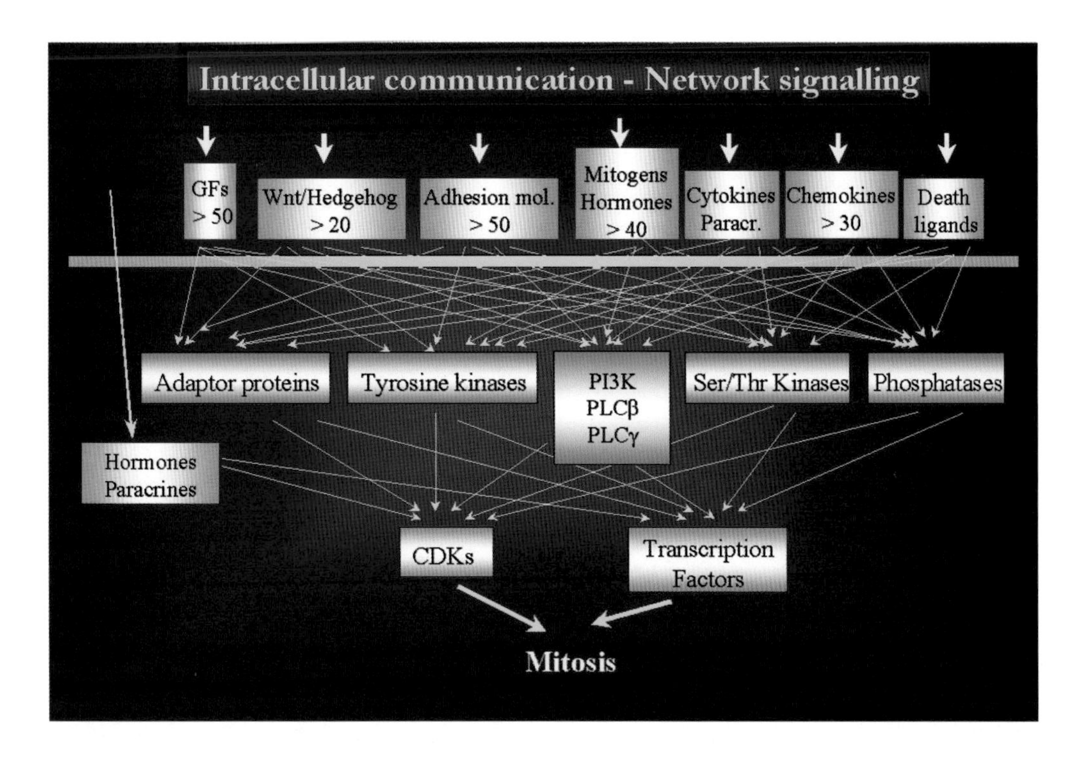

Colour Plate XIII (See Figure 21.2.)

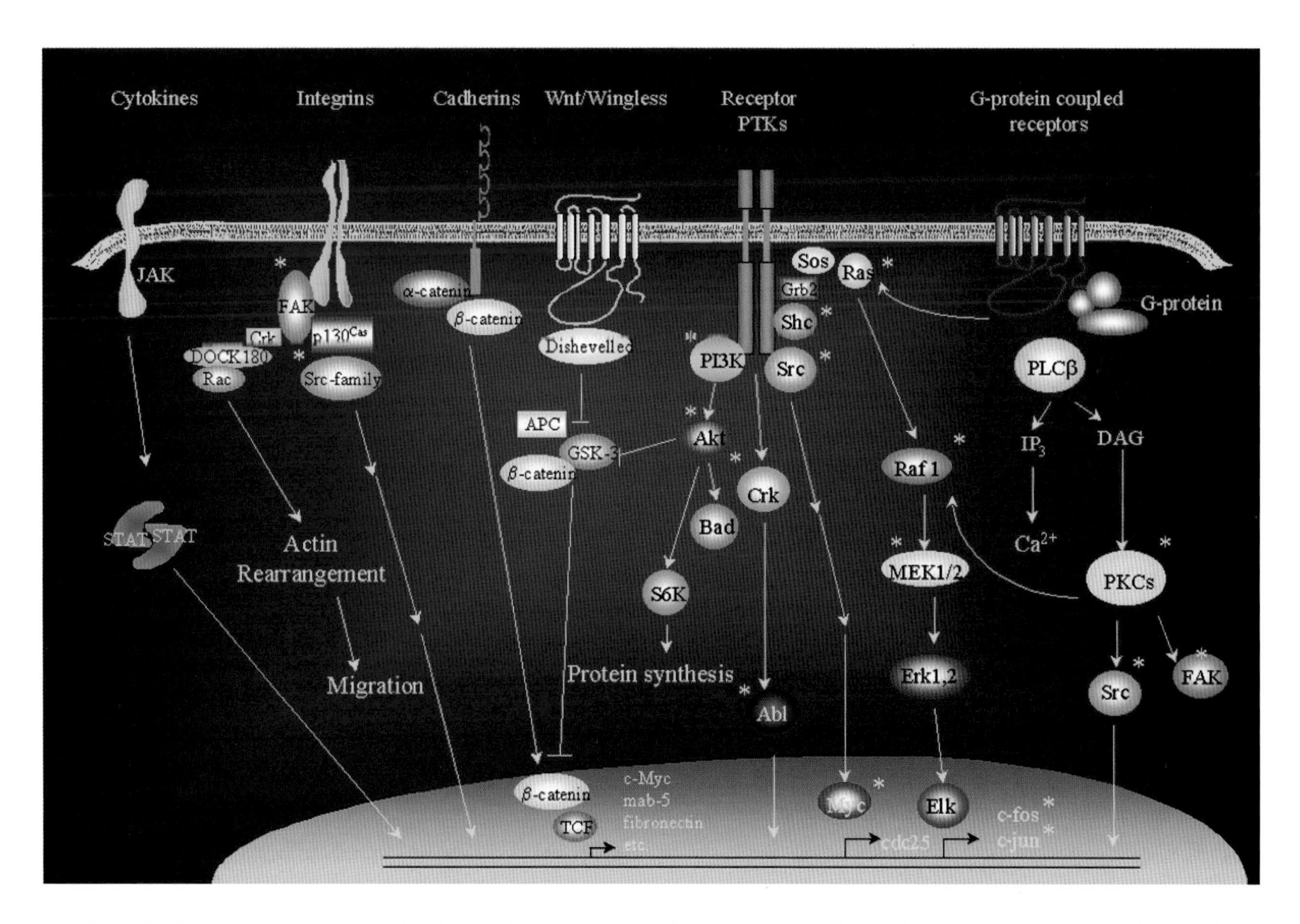

Colour Plate XIV The most important signaling pathways involved in oncogenesis. Selected oncoproteins are marked with stars (see Figure 21.3).

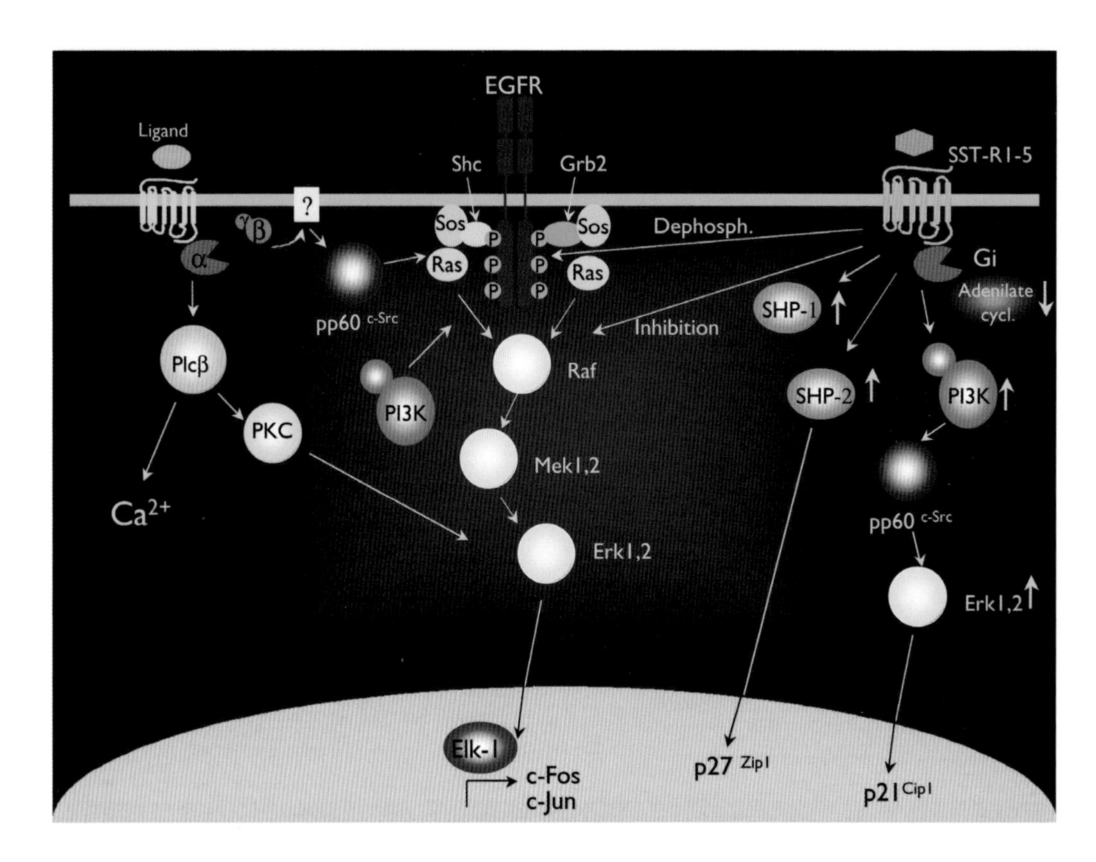

Colour Plate XV Transactivation and inhibition of GFRTK pathways by G-protein-coupled receptors (see Figure 21.5).

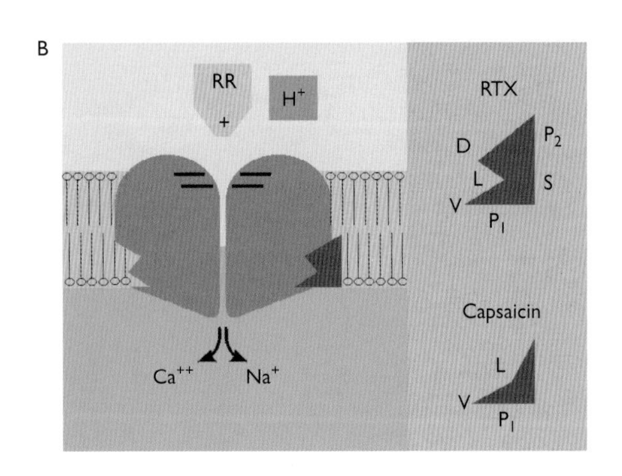

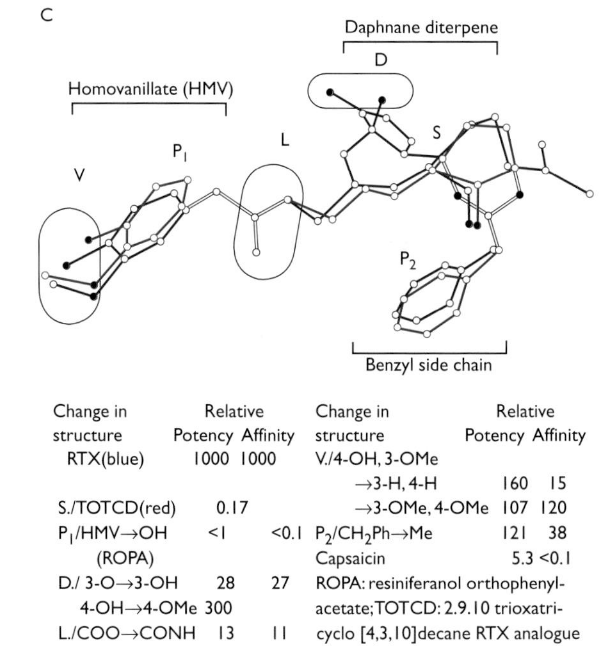

Colour Plate XVI (B) Capsaicin VR1 receptor-cation channel. (C) Polar (D, L, V) and apolar (P1, P2, S) domains of resiniferatoxin (RTX) and capsaicin and their proposed binding to VR1. RR: ruthenium red (see Figure 28.5B & C).

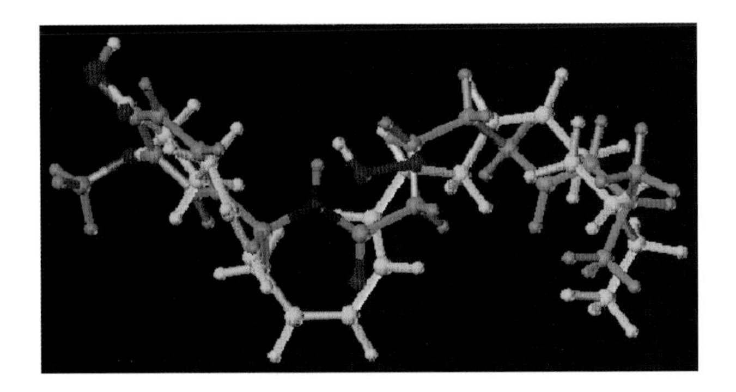

Colour Plate XVII Molecular modeling overlay of minimum-energy conformations of capsaicin (green) and 12-(S)-HPETE (yellow). Oxygen and nitrogen atoms are colored in red and blue, respectively (see Figure 28.4).

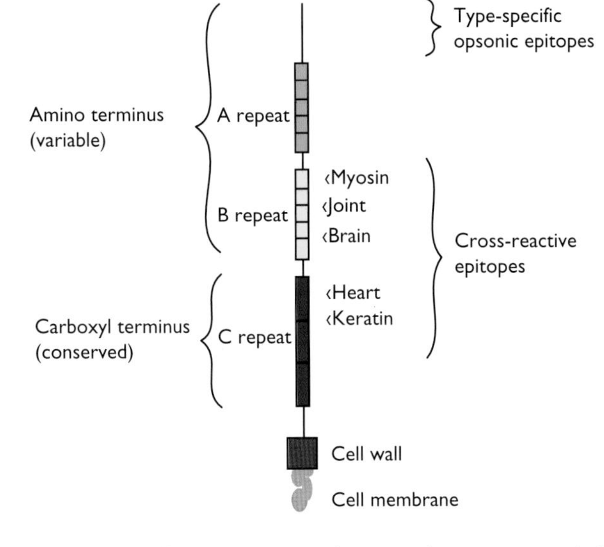

Colour Plate XIX Representation of group A streptococcal M protein (see Figure 24.4).

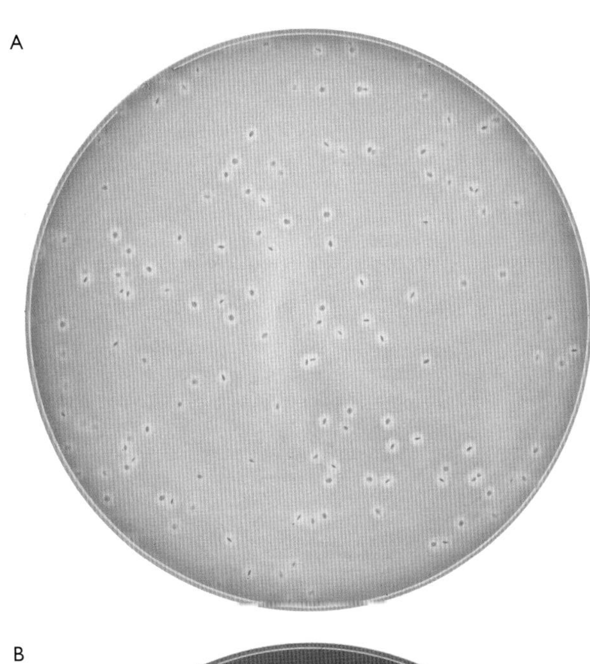

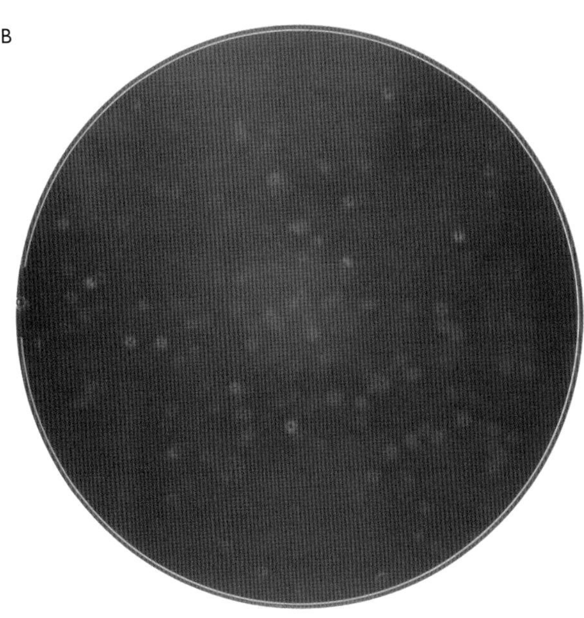

Colour Plate XVIII Growth of group A streptococci on 2% (A) and 10% blood agar plates (B) (see Figure 24.1).

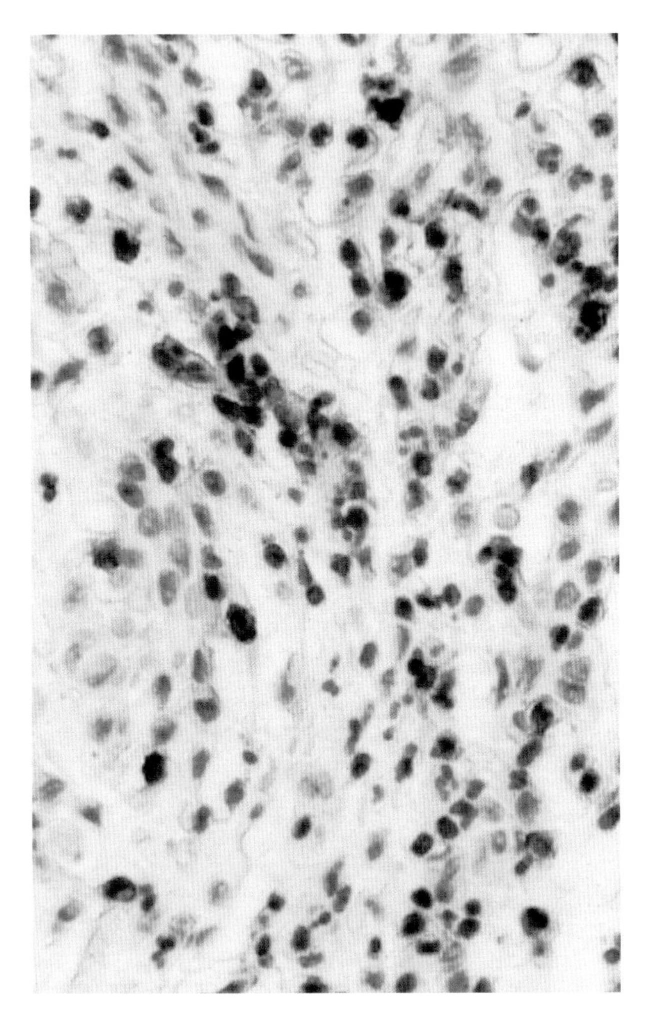

Colour Plate XX Immunohistochemical staining of T-cells in rheumatic heart disease lesions. Activated T-cells were stained with an antibody to CD45RO and developed using the LSAB-2 (Dako, Australia) detection system. T-cells are stained brown (see Figure 24.5).

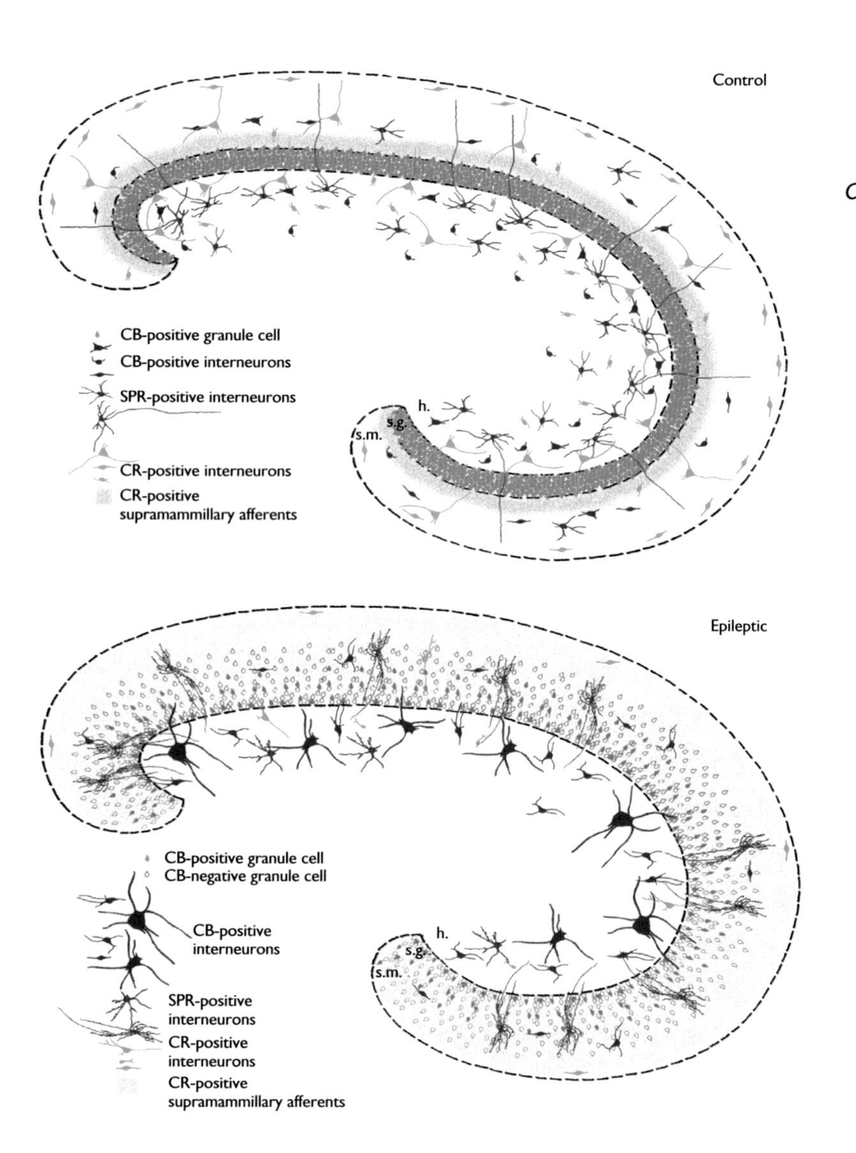

Colour Plate XXI Schematic diagram of the laminar distribution of the examined interneuron subpopulations in control and epileptic dentate gyrus. The most characteristic representatives of each examined cell type were drawn by camera lucida and scanned into the computer. In the epileptic samples, the interneuronal network has changed considerably, even the "resistant" interneurons are modified, and differ from the control. Large hypertrophic CB-containing cells appear in the hilus, their dendritic tree is longer than those in the control (red). The number of CR-positive cells decreases (green), while the CR-positive supramammillary pathway extends to innervate the entire width of the stratum moleculare (green shading). The distribution of SPR-positive cells changes. Most are located in the hilus in the control, whereas the epileptic samples show them to be numerous in the stratum moleculare. In addition, they have a bushy and highly varicose dendritic tree (blue). Granule cells are also shown (grey). Most lose their CB-content (open circle) and migrate up to the stratum moleculare in the epileptic dentate gyrus. All characteristic profound changes have been combined into this diagram, but it should be pointed out that each individual case shows a combination of these abnormalities in cell distribution and morphology at different degrees (see Figure 31.1).

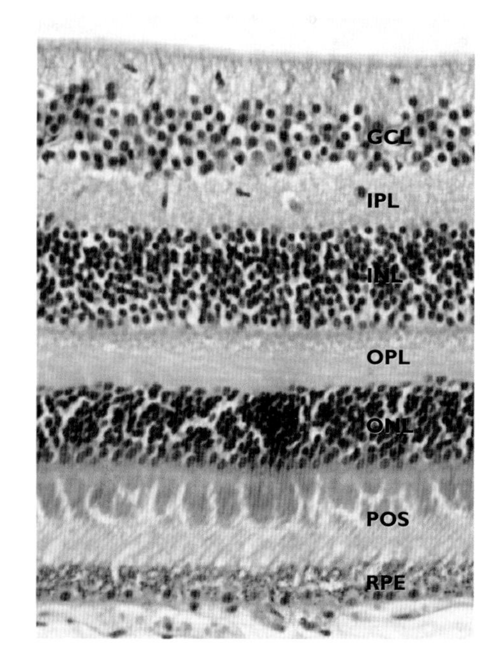

Colour Plate XXII Structure of the retina. Abbreviations: GCL, ganglion cell layer. IPL, inner plexiform layer. INL, inner nuclear layer. OPL, outer plexiform layer. ONL, outer nuclear layer. POS, photoreceptor outer segment. RPE, retinal pigment epithelium (see Figure 34.1).

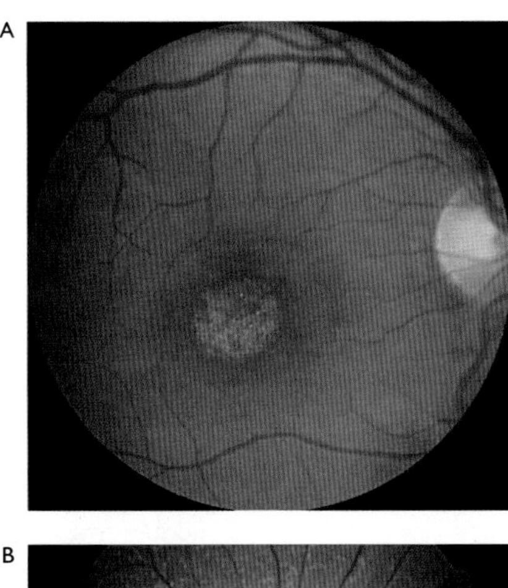

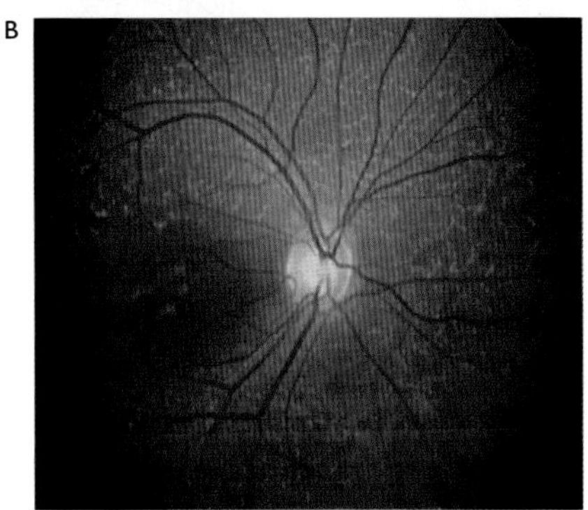

Colour Plate XXIV Fundus images of Stargardt's disease (A) and fundus flavimaculatus (B) (see Figure 34.5).

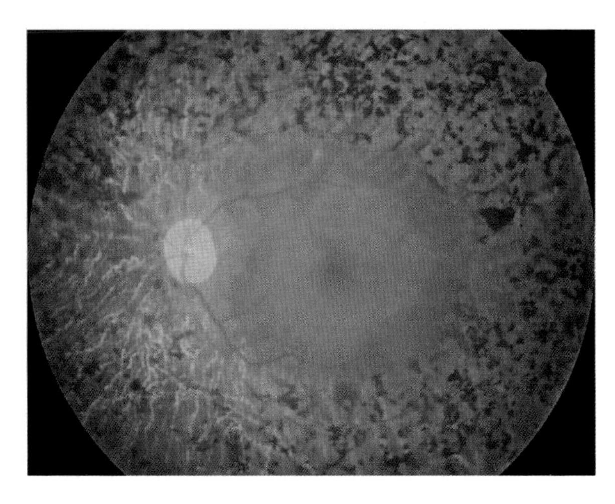

Colour Plate XXIII A fundus image of retinitis pigmentosa (see Figure 34.4).

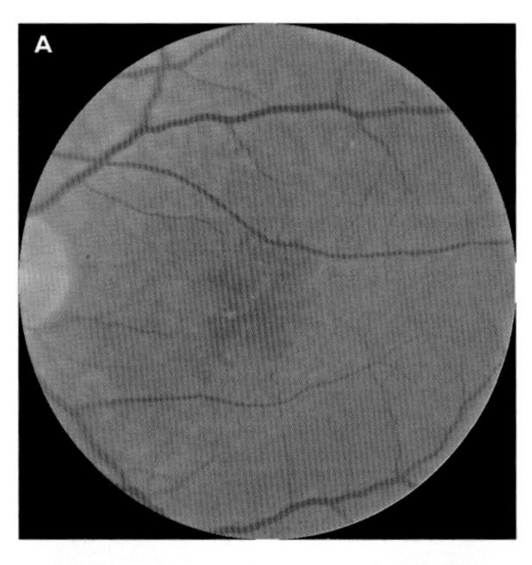

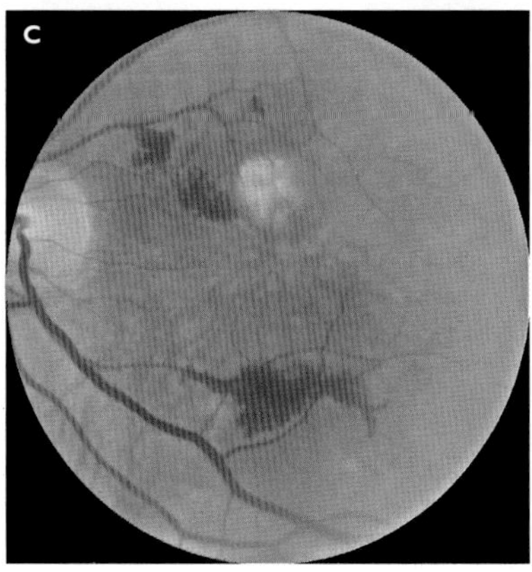

Colour Plate XXV "Dry" (A & B) and "wet" forms (C & D) of age-related macular degeneration (ARMD). Images obtained from fundus photography (A & C) and fluorescein angiography (B & D) (see Figure 34.6A & C).

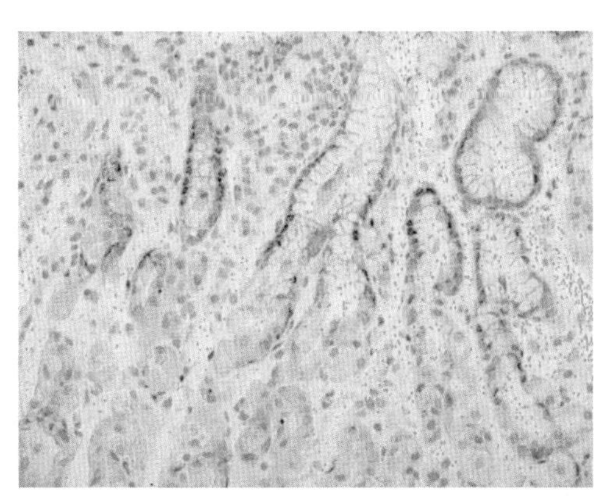

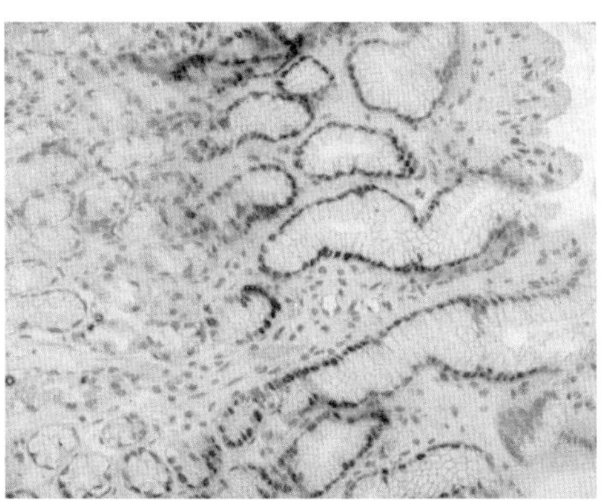

Colour Plate XXVI Proliferation nuclear antigen (PCNA) expression in H. pylori negative and positive gastric mucosa. The positive nuclei are labelled with dark stain. Note that in the case of inflammation the middle third proliferation zone moves to the bottom of the gastric pit (see Figure 40.14).

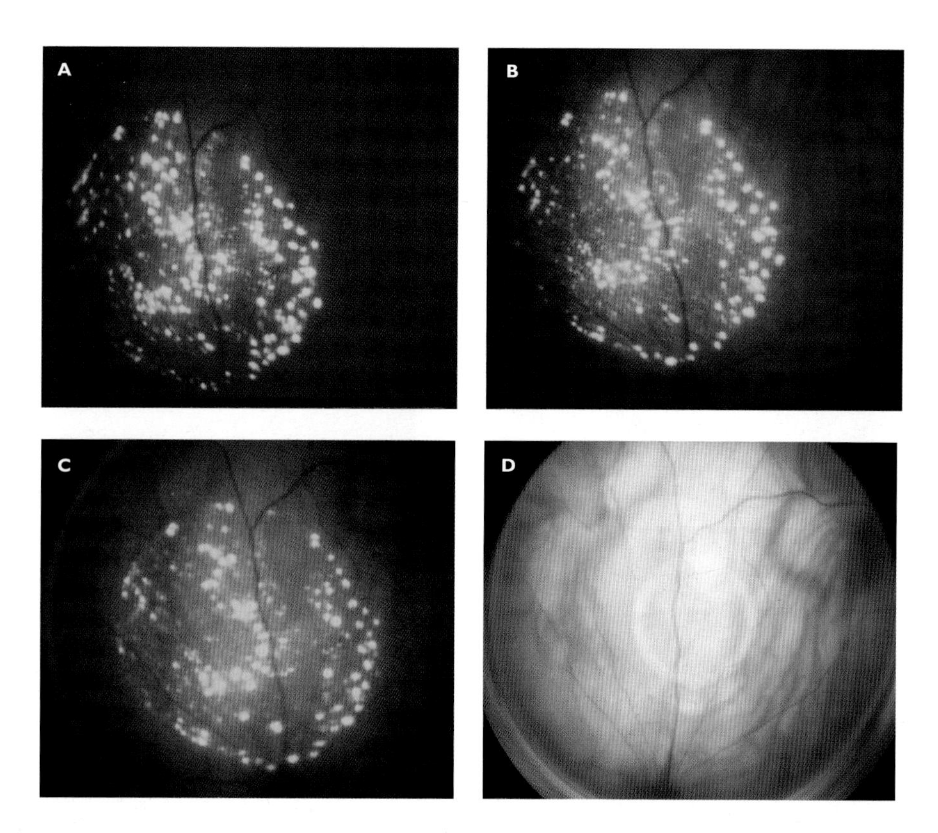

Colour Plate XXVII Long term and stable transgene expression mediated by adeno-associated virus (AAV) following subretinal delivery of a recombinant AAV carrying a enhanced green fluorescent protein (GFP) gene. GFP expression was monitored in real time at 2 (A), 12 (B) and 19 (C) months post injection in the rat. (D) showing the normal retinal image at 19 months post injection (see Figure 34.7).

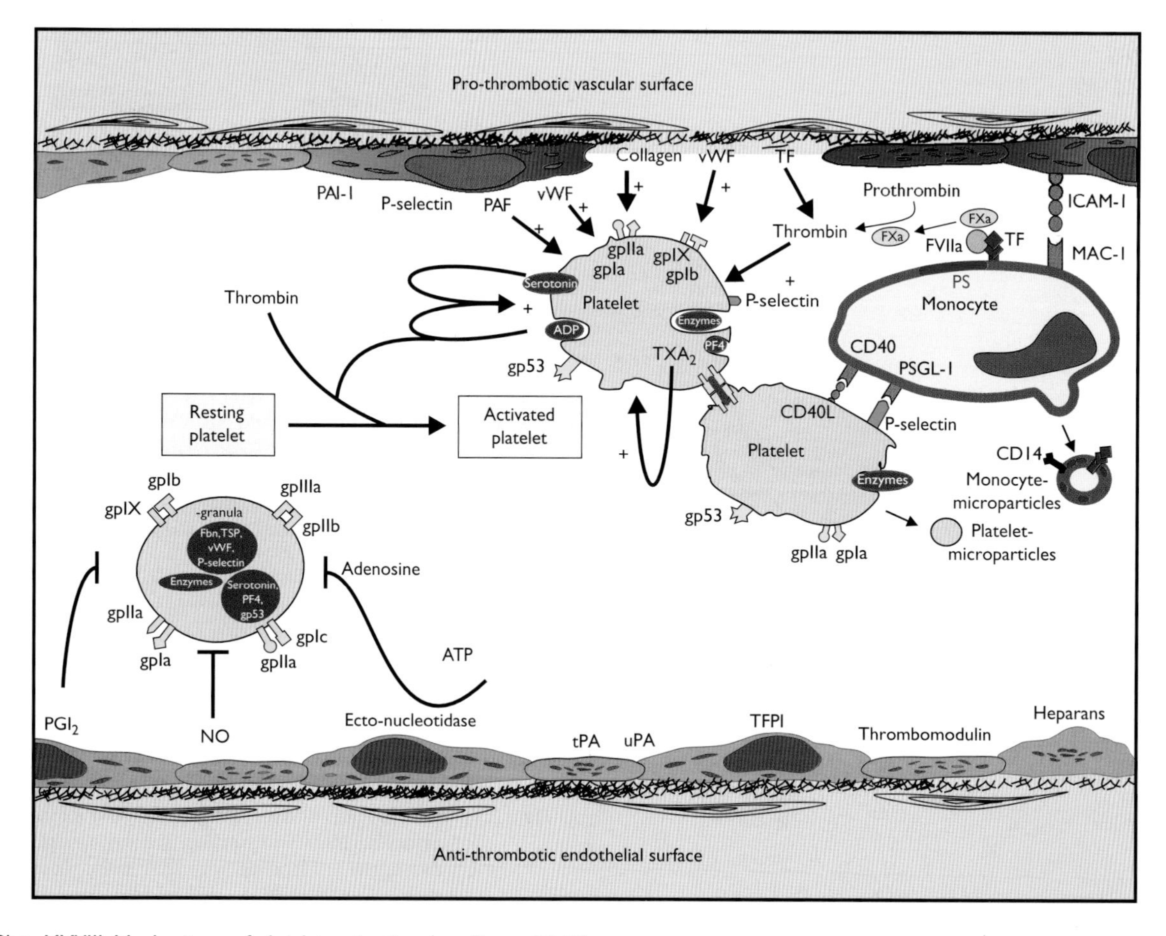

Colour Plate XXVIII Mechanisms of platelet activation (see Figure 35.14).

Chapter 24

The development of vaccines to prevent rheumatic fever

Colleen Olive and Michael Good

Contents

The development of a vaccine against group A streptococcal (GAS) infection would prevent GAS-associated diseases including rheumatic fever (RF). This would have significant public health implications. However, there are two major impediments to the development of a GAS vaccine; the variability in the GAS M protein, which is the target of type-specific protective antibodies, and the potential for the induction of autoimmunity. To develop a safe and effective GAS vaccine, we have therefore taken a multi-epitope approach whereby several non-cross-reactive M protein peptides have been combined into a single immunogen using novel peptide chemistry. This "heteropolymer" vaccine candidate has proved successful in protecting mice from GAS infection and paves the way for future human clinical trials.

24.1 Group A streptococci

There are several types of the organism *Streptococcus* (types A, B, C, D and G), however, the majority of infections in humans are due to the group A streptococcus (*Streptococcus pyogenes*) which can cause a variety of clinical manifestations [1, 2]. The group A streptococci (GAS) are Gram-positive β-hemolytic bacteria as demonstrated by their ability to lyse red cells and produce a zone of clear hemolysis when grown on blood agar (Figure 24.1). Microscopically, the bacteria form cocci each approximately 0.6–1 μm in diameter and produce a characteristic chain pattern of growth as identified by electron microscopy (Figure 24.2).

24.1.1 *Diseases caused by group A streptococci*

A range of illnesses and diseases have been associated with GAS infection [1]. These include the relatively minor pharyngitis ("strep throat"), impetigo (superficial skin infection) and scarlet fever which is characterized by sore throat, fever and rash. More severe invasive infections can be caused by GAS and in particular immunocompromised individuals, cancer patients receiving chemotherapy, and burn victims are at a greater risk [1]. Two such invasive infections are toxic shock syndrome and necrotizing fasciitis both of which may lead to multisystem organ failure. Lastly, the GAS

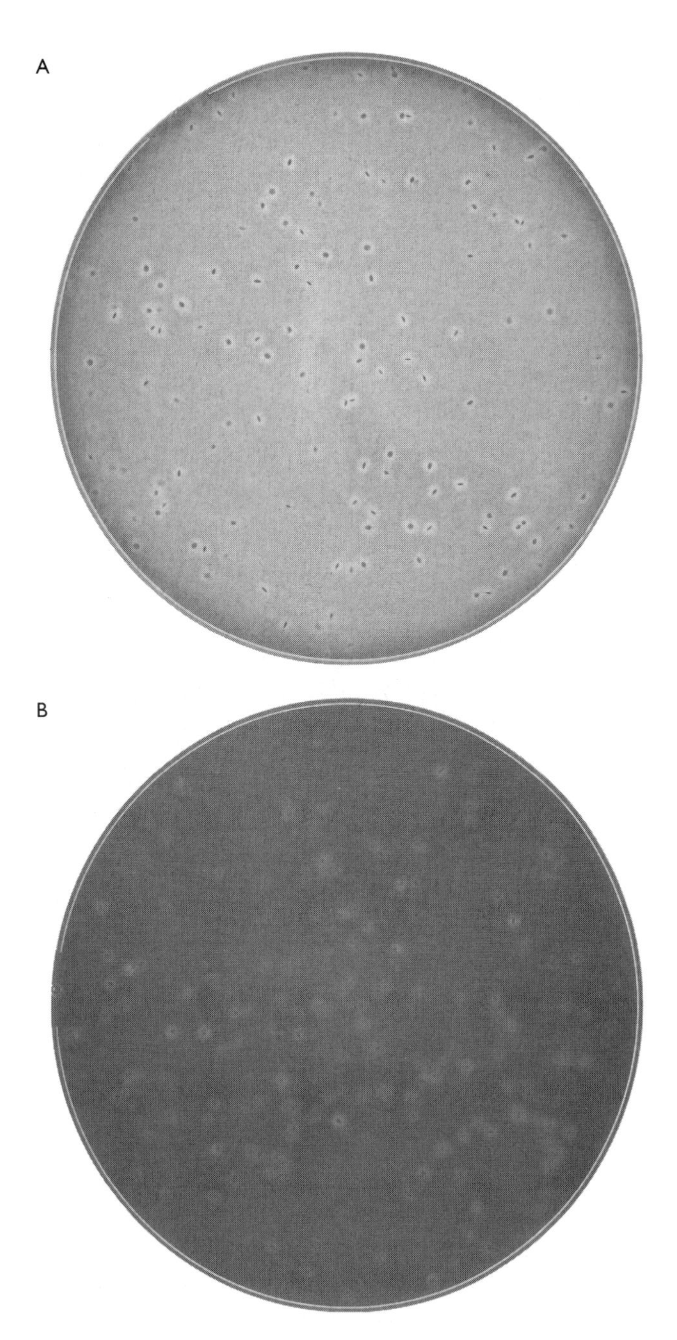

Figure 24.1 Growth of group A streptococci on 2% (A) and 10% blood agar plates (B). The group A streptococci are β hemolytic bacteria which lyse red cells when grown on blood agar plates forming a zone of clear hemolysis surrounding the bacterial colonies (see Colour Plate XVIII).

post-infectious sequelae of rheumatic fever (RF), rheumatic heart disease (RHD) and acute glomerulonephritis (AGN) are a major problem in developing countries and indigenous populations, particularly in Australian Aboriginals [3, 4]. Current treatment for controlling GAS infection is with

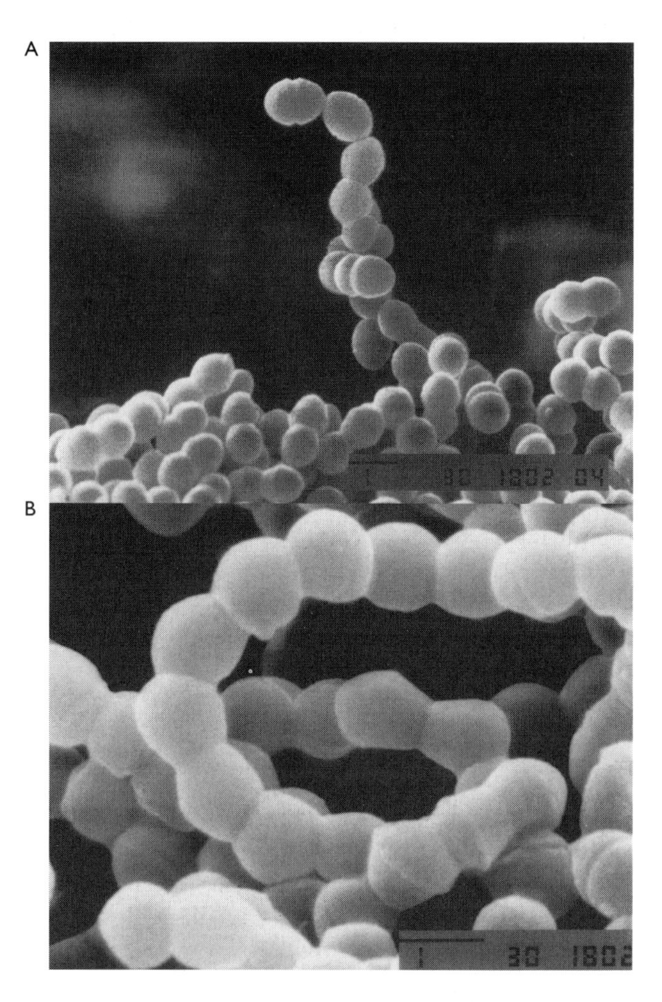

Figure 24.2 Transmission electron micrographs of group A streptococci. ×10,000 (A) and ×19,000 (B) showing the formation of cocci and characteristic chain pattern of growth.

antibiotic therapy [4], however, continual high dose administration of antibiotic is required in cases of repeated episodes of acute RF and RHD and there is poor compliance. This is often associated with the persistence of these GAS-associated diseases.

24.2 Pathogenesis of rheumatic fever and rheumatic heart disease

RF usually follows 2–6 weeks after an untreated acute GAS throat infection with symptoms of polyarthritis, carditis and chorea [3, 4]. It has been reported that up to approximately 3% of untreated GAS throat infections lead to RF [3]. Acute RF can result in rheumatic heart disease (RHD). RHD is a major cardiovascular condition in Australian Aboriginals contributing to high morbidity and mortality in young adult

Aborigines with a mean life expectancy due to RHD of 33 years [5]. A diagnosis of RF is based on several clinical manifestations affecting various target tissues including the heart (pericardium, myocardium and endocardium/valves), joints, brain and skin. Those affecting the heart are unresolvable and can lead to valve damage and heart failure [4]. It is estimated that 12 million people are affected by RF and RHD with 500,000 deaths annually (The World Health Report, 1998). RF and RHD are rare in affluent countries, and confined largely to indigenous communities and developing countries where over-crowding and poor access to health care are contributing factors; Australia's Aboriginal population experience the highest rates of these diseases in the world, even higher than those reported in developing countries [5]. Annual incidence rates of acute RF in Australian Aboriginals have been reported as 651 per 100,000 compared to rates of 5–100 per 100,000 and 0.2–0.5 per 100,000 in developing countries and affluent countries, respectively [5]. The point prevalence of RHD has been reported as 9.6/1000 in Australian Aboriginals compared to 0.6/1000 in affluent countries [5]. Since RF and RHD only ever follow infection with GAS, the current approach for controlling RF and RHD is with long-term penicillin prophylaxis. However, as previously highlighted, this approach is largely inadequate due to poor compliance, highlighting the need for a GAS vaccine to prevent these potentially life-threatening diseases.

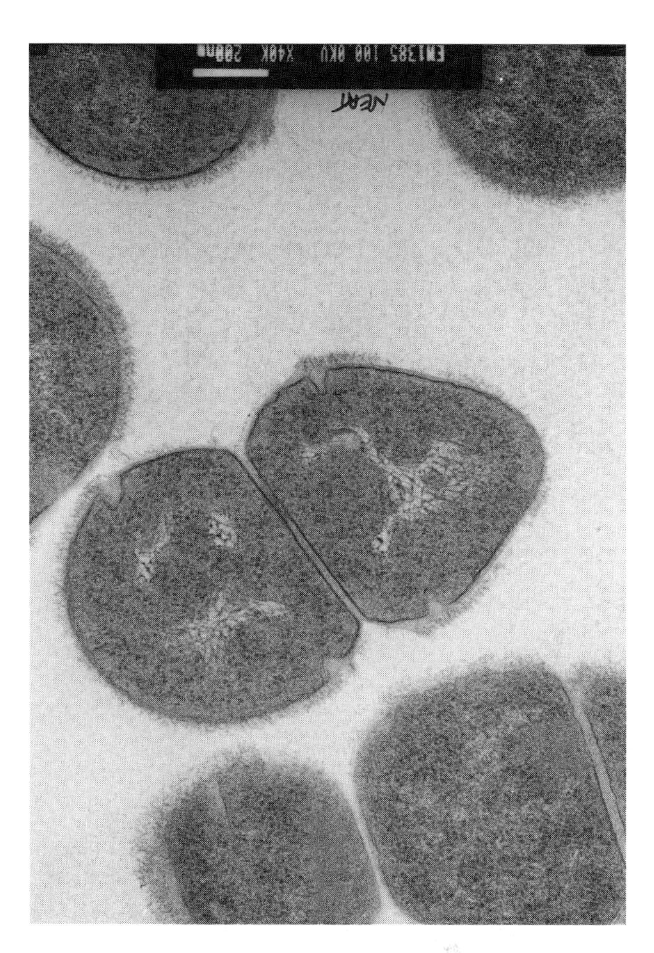

Figure 24.3 Scanning electron micrograph of group A streptococci (×40,000). Surface proteins, including M protein, extend from the bacterial cell wall producing a surface "hairy" appearance.

24.2.1 Biology and diversity of the group A streptococcal M protein

The GAS M protein is the major immunogen and virulence factor in GAS infection, which, by binding to the plasma protein factor H and decreasing the deposition of C3b onto the bacterial surface, allows the bacteria to escape destruction by phagocytosis in the absence of type-specific antibodies [2, 6, 7]. Resistance to phagocytosis may also involve binding of the M protein to immunoglobulins [8]. The M protein is an α helical coiled-coil bacterial surface protein that extends approximately 50 nm from the bacterial surface to form fibrils [7] (Figure 24.3). The coiling of the M protein is generated by a periodicity of heptad sequences with hydrophobic residues each at positions 1 and 4 [2]. Serotype-specific determinants of the M protein are located in the highly variable amino terminus (over 80 serotypes are known) and there are three repetitive domains; A, B and C domains (Figure 24.4); with a fourth D region and proline/glycine-rich region associated with the cell membrane [2, 7]. The amino terminus is highly polymorphic between different GAS strains whereas there is >98% identity in the carboxyl terminus C-repeat region [2]. Previous studies indicate that certain strains of GAS are more commonly associated with infection of the throat, and that these are different to those that cause infection via the skin [3]. Furthermore, RF is believed to follow GAS infection of the respiratory tract whereas AGN is more commonly a consequence of skin infection [3]. Interestingly, in Australian Aboriginals the most prevalent site for streptococcal infection is the skin with pyoderma reported at a higher incidence in Aboriginal children when compared to pharyngitis [3]. Certain strains of GAS have also been associated with RF and termed "rheumatogenic" whereas those strains associated with AGN are "nephritogenic" [3], although this delineation remains somewhat controversial. Additional studies have now identified a molecular basis for the tissue-specific GAS isolates based on genetic distinctions between the *emm* genes, which encode the M proteins [9]. The virulence of GAS strain may also contribute to the severity of disease as indicated by the resurgence of severe GAS infections in the US in the mid-1980s that were associated with the M1 GAS type [10].

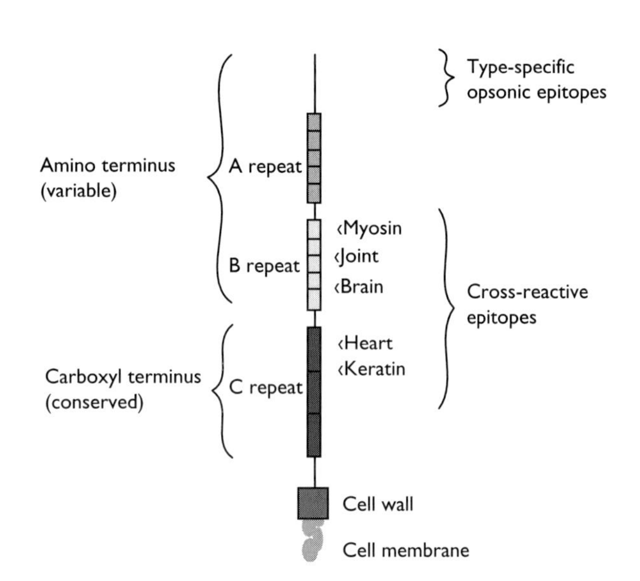

Figure 24.4 Representation of group A streptococcal M protein (see Colour Plate XIX).

24.3 Role of group A streptococcal M protein in rheumatic fever and rheumatic heart disease

M proteins share significant sequence homology with mammalian fibrillar proteins. Of the GAS strains associated with RF, M5, M6 and M24 possess up to 30% overall homology with type 1 keratin, myosin and human α-tropomyosin [6]. B-cell epitopes have been identified within the M protein, especially the B-repeat region [6, 11], that are cross-reactive with human heart tissues, joints, kidney and brain [6, 11–13]. These autoantibody-inducing epitopes have been mapped to the M protein molecule proximal to the type-specific epitopes of the N-terminal region [6, 11] (Figure 24.4). Studies of the specificity of the human T-cell response to the conserved region of the M protein have also identified T-cell epitopes that are shared with peptides representing homologous regions of cardiac and skeletal muscle myosin [14]. The homology observed between GAS M protein and self tissues provides a structural basis to explain the induction of an autoimmune response following GAS infection. The superantigenic nature of the M protein could also contribute to the activation of autoreactive T-cells and trigger the autoimmune process [15]; however, this activity has been questioned due to the possible contamination by pyrogenic toxins which are known mitogens [16].

24.4 Group A streptococcal-specific T-cell responses in rheumatic fever and rheumatic heart disease

Although the exact mechanism for the development of RF and RHD is unknown, evidence indicates an autoimmune basis involving streptococcal-specific antibodies and T-cells that are cross-reactive with human tissue proteins, in particular heart, leading to a local inflammatory response [reviewed in 17]. Subsequent carditis, however, may result from an immune response to other, normally sequestered, cardiac antigens which become exposed following the initial cardiac inflammation. Furthermore, there is evidence that cellular immune responses are critical in the initiating events leading to tissue damage and that the cross-reactivity mediated by antibodies is a secondary immunological response [17]. Strong evidence indicates that T-cell mediated immune responses play a pivotal role in the pathogenesis of RF and RHD [14, 18–21]. Lymphocytes from patients with RF have been shown to be highly reactive to streptococcal antigens from "rheumatogenic" GAS strains when compared with normal subjects [22]. In addition, cytotoxic immune cross-reaction between heart tissue and GAS antigens has been shown [23, 24]. Together, the data suggest that the pathogenesis of RF and RHD may involve a T-cell mediated autoimmune response which is triggered by antigenic mimicry between GAS M protein and heart antigens. It has recently been reported [20] that peripheral blood T-cells from a proportion of RHD patients recognized an extract of myocardial tissue, and that priming of these cells with GAS increased the response to a particular myocardial protein fraction of 50–54 kDa. However, T-cells from control subjects did not recognize this protein, indicating a distinct repertoire of T-cells in patients with RHD.

24.5 The role of heart-infiltrating T-cells in rheumatic heart disease

Previous studies have shown the presence of a mononuclear infiltrate consisting predominantly of T-cells in both acute and chronic rheumatic carditis lesions [19]. The majority of T-cells in acute disease were of the helper phenotype whereas those in chronic disease were more heterogeneous. Immunohistochemical staining of heart valve tissue from RHD patients illustrates the presence mainly of T-cells (Figure 24.5; courtesy of Dr Natkunan Ketheesan, James Cook University, Townsville). Yoshinaga *et al.* [21] established selective T-cell lines by priming peripheral blood and valvular lymphocytes from patients with RHD with GAS membrane antigen. These cell lines were shown to proliferate in response to streptococcal cell wall and cell membrane preparations but no reactivity to M protein was observed. Guilherme *et al.* [18] obtained T-cell clones from heart valves of four RHD patients and examined their reactivity to M5 peptides and cardiac antigens. T-cells were expanded with phytohemagglutinin without prior priming with antigens, and individual clones obtained by limiting dilution. The specificity of peripheral blood T-cell clones was not determined in this study. Of the 107 heart-infiltrating T-cell clones tested, eight clones recognized both a synthetic M

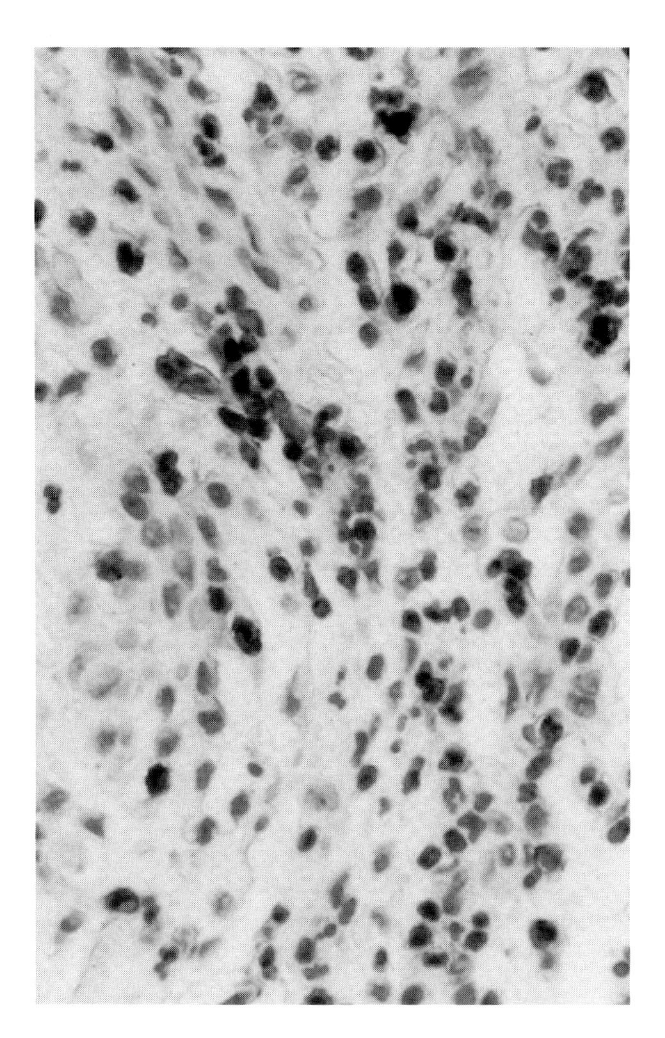

Figure 24.5 Immunohistochemical staining of T-cells in rheumatic heart disease lesions. Activated T-cells were stained with an antibody to CD45RO and developed using the LSAB-2 (Dako, Australia) detection system. T-cells are stained brown (see Colour Plate XX).

peptide and heart and/or aortic valve protein extract supporting the concept of molecular mimicry and autoimmunity in disease pathogenesis.

24.6 Strategies for the development of a group A streptococcal M protein-based vaccine

24.6.1 Recombinant proteins

The development of a broad-based effective vaccine against GAS infection has been impeded by the sequence variability that occurs between different GAS M proteins. A second impediment in the design of a GAS vaccine is the possibility of inducing protective epitopes that are cross-reactive with cardiac and other host tissues. A vaccine candidate based purely on the M protein type-specific determinants is likely, therefore, to provide protection only against a specific GAS strain, whereas there is the possibility that a vaccine based on the M protein B-repeat region will induce cross-reactivity and disease. In addressing the above issues, a recent vaccine strategy has been to develop recombinant multivalent M protein vaccines containing epitopes from the amino terminal region representing multiple M protein serotypes. Dale and colleagues have developed tetravalent and octavelent M protein vaccines [25, 26]. In each case, the vaccines were shown to be immunogenic in rabbits. Furthermore, the antisera produced were shown to be opsonic against the GAS strains represented in the recombinant constructs and there was no cross-reactivity with human tissues, indicating the potential use of these vaccines in inducing broadly protective immune responses. Theoretically, however, these vaccines would provide limited protection against the myriad of GAS infections that occur in human populations. Another approach which has been used for the delivery of recombinant proteins to mucosal sites (the primary portal of entry of GAS into the body) is to express these in commensal bacteria [27]. This approach has been used to demonstrate mucosal and systemic immune responses generated after oral colonization with *Streptococcus gordonii* engineered to express an allergen protein [27], and has the potential for use in the delivery of a GAS vaccine to mucosal surfaces. Protective immunity against mucosal and systemic GAS infection has been demonstrated by oral administration of GAS M protein expressed on attenuated *Salmonella typhimurium* [28]. Secondly, a reduction in nasopharyngeal GAS colonization has been reported in mice immunized intranasally with a vaccinia virus engineered to express the conserved region of the GAS M protein [29]. There are obvious safety concerns, however, associated with using pathogenic organisms for vaccine delivery, in addition to the potential for inducing an autoimmune response when using the entire GAS M protein as a vaccine due to immunological cross-reactivity between GAS antigens and host tissues.

24.6.2 Peptides

We have investigated the potential of the M protein C-repeat region as a GAS vaccine candidate. The C-region is conserved in the majority of the different GAS strains that have been identified [2]. Moreover, using a series of 15 overlapping peptides (p145–p159, Figure 24.6) spanning the entire M protein C-region, a peptide (p145) that is recognized by antibodies in the sera of most adults living in areas of high GAS exposure was identified [30]. The acquisition of these antibodies with age paralleled the acquisition of GAS immunity [31]. Human sera with antibodies to p145 have also been shown to be opsonic against heterologous GAS strains [31, 32]. Similarly, mice immunized with p145 elicited

```
p145 LRRDLDASREAKKQVEKAL
p146           AKKQVEKALEEANSKLAALE
p147                     EANSKLAALEKLNKELEESK
p148                               KLNKELEESKKLTEKEKAEL
p149                                         KLTEKEKAELQAKLEAEAKA

p150 QAKLEAEAKALKEQLAKQAE
p151           LKEQLAKQAEELAKLRAGKA
p152                     ELAKLRAGKASDSQTPDT
p153                               SDSQTPDTKPGNKAVPGKGQ
p154                                          GNKAVPGKGQAPQAGTKPNQ

p155 APQAGTKPNQNKAPMKETKR
p156           NKAPMKETKRQLPSTGETAN
p157                     QLPSTGETANPFFTAAALTV
p158                               PFFTAAALTVMATAGVAAVV
p159                                   MATAGVAAVVKRKEEN
```

Figure 24.6 Overlapping peptides spanning the conserved C-repeat region of GAS M protein.

antibodies that were opsonic and induced protective immunity against GAS infection [33]. However, our studies indicated that p145 contained a T-cell epitope shared with determinants on human heart valve myosin [14], and keratin in mouse [34], and therefore we decided to map the minimal B- and T-cell epitopes within p145 to establish a C-region vaccine that was devoid of any potentially deleterious T-cell autoepitope but which contained a protective B-cell epitope. To accomplish this, a series of conformational peptides was synthesised (J1–J9) that covered the entire p145 sequence [34] (Figure 24.7). To mimic the helical nature of the M protein and maintain the conformational structure of the B-cell epitopes, the p145-specific sequences were embedded within the non-M protein GCN4 sequences to create the necessary periodicity of hydrophobic residues required for α-helical folding. Mouse anti-p145 sera were shown to react with J7, J8 and J9 [34] whereas human anti-p145 sera reacted with J1, J2, J7 and J8 [31]. Moreover, we identified a minimal, non-host cross-reactive peptide (J8, and J14 which spans J7, J8 and J9) from the conserved C-region of the M protein that contained a B-cell epitope only [34]. Following immunization of mice with these peptides (either J8 or J14) in Complete Freund's Adjuvant (CFA), we demonstrated the production of opsonic antibodies and protection following challenge with the homologous GAS strain M6 (unpublished data). Studies by Bessen and Fischetti [35, 36], also indicate the potential use of C-region peptides in reducing the degree of nasopharyngeal GAS colonization and inducing heterologous GAS immunity following intranasal immunization. Together, the data indicate that a vaccine based on the C-region may be an excellent vaccine candidate.

24.6.3 *Multi-epitope approach*

The C-region encoding the p145 epitope was shown to be identical in 50% of the GAS isolates sequenced, with sequence identity ranging from 65 to 95% in the remaining GAS isolates [32]. Using a novel strategy to develop a new and improved GAS vaccine with even broader strain coverage, multiple M protein peptides have been combined from both the amino terminal and conserved regions of the M protein, into a single immunogen, in the design of a multi-epitope "heteropolymer" vaccine which can prevent GAS infection in the mouse model [37]. The GAS vaccine was designed specifically to target strains found in the Australian Aboriginal population. It contains the C-region peptide, J14, in addition to seven amino terminal serotypic M protein peptide sequences found to be common in GAS-endemic communities of the Northern Territory. The rationale for the study was that GAS strains not represented on the heteropolymer by their amino terminal serotypic determinants would potentially be targeted

p145	LRRDLDASREAKKQVEKALE
J1	QLEDKVKQ**LRRDLDASREAK**EELQDKVK
J2	LEDKVKQA**RRDLDASREAKK**ELQDKVKQ
J3	EDKVKQAE**RDLDASREAKKQ**LQDKVKQL
J4	DKVKQAED**DLDASREAKKQV**QDKVKQLE
J5	KVKQAEDK**LDASREAKKQVE**DKVKQLED
J6	VKQAEDKV**DASREAKKQVEK**KVKQLEDK
J7	KQAEDKVK**ASREAKKQVEKA**VKQLEDKV
J8	QAEDKVKQ**SREAKKQVEKAL**KQLEDKVQ
J9	AEDKVKQL**REAKKQVEKALE**QLEDKVQL
J14	KQAEDKVK**ASREAKKQVEKALE**QLEDKVK
Jcon	DKVKQAEDKVKQLEDKVEELQDKVKQLE

Figure 24.7 Chimeric GAS M protein C-region peptides, J1–J9 and J14. These incorporated short peptide-specific sequences of p145 (bold) flanked by the yeast-derived GCN4 sequences which were designed to mimic the α-helical folding of M proteins and maintain the conformational B-cell epitopes. A non-specific sequence was synthesized as a control (Jcon).

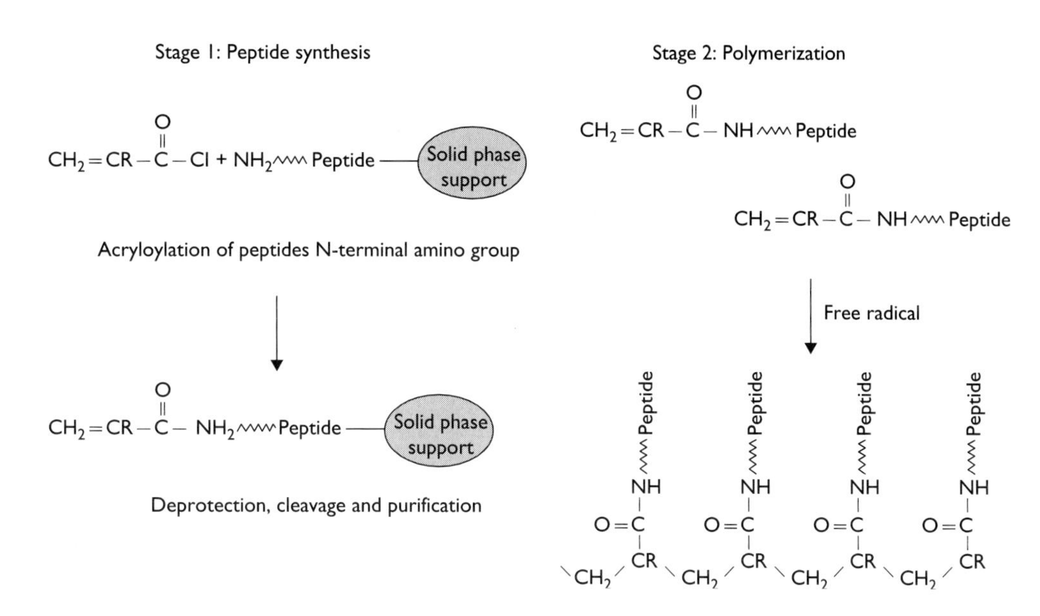

Figure 24.8 Chemical synthesis of the multi-epitope "heteropolymer" GAS vaccine using the acryloylation polymerization protocol.

by antibodies to the C-region. We have also shown that sera from Australian Aboriginals reacted to a number of these peptides in an age-related manner [37] indicating that immunity to these peptides may be important in protection from GAS infection and that protective immunity may potentially be mimicked by vaccination using these peptides. The peptides were linked using novel chemistry [38] that enables the immunogen to display each individual peptide pendant from an alkane backbone (Figure 24.8). Complete protection was observed in mice immunized with heteropolymer and challenged with different GAS strains [37].

24.7 Alternative vaccine approaches to prevent group A streptococcal infection

The GAS M protein is the major virulence factor in GAS infection and the target of type-specific opsonic antibodies [6]. Primary vaccine strategies have therefore focused on the M protein in the induction of protective immunity, however, a number of other non-M protein molecules have been investigated as attractive vaccine candidates. These include surface GAS proteins and extracellular molecules which may also be important GAS virulence factors. A few of these will be discussed.

24.7.1 C5A peptidase

C5a peptidase (SCPA) is a complement component protease that specifically cleaves the human serum chemotaxin C5a and inhibits the migration of polymorphonuclear cells to the site of GAS infection, and hence GAS clearance [39, 40]. Using a mouse air sac model of inflammation, it was shown that GAS that lacked the C5a peptidase were rapidly cleared within four hours after infection when compared to wild-type GAS [39]. C5a peptidase may therefore be an important factor in GAS colonization of the nasopharynx since their initial clearance following respiratory infection may be impeded sufficiently to allow the bacteria to adhere to and invade mucosal epithelial cells. A recent study has investigated the vaccine potential of C5a peptidase against GAS infection [39]. Intranasal immunization of mice with various constructs encoding wild-type or mutant SCPA were assessed for their immunogenicity and affect on GAS colonization of the nasopharyngeal mucosa following GAS challenge. Wild-type SCPA GAS elicited both mucosal IgA and systemic IgG antibody responses, and induced high titre antibodies that neutralized peptidase activity of different GAS strains *in vitro* [39]. Furthermore, SCPA GAS mutants were cleared more rapidly from the throats of mice compared to wild-type GAS leading to reduced colonization.

24.7.2 Hyaluronic acid capsule

There is evidence that the hyaluronic capsule of the GAS may be important in GAS virulence. Highly encapsulated mucoid forms of GAS have been isolated more commonly from patients with severe invasive infections and acute RF when compared to patients with pharyngitis [41]. The hyaluronic capsule may also be important in evading phagocytosis and in resistance to opsonization [42]. In addition, recent data indicate the importance of the hyaluronic capsule in adherence and attachment to epithelial cells possibly by interacting with adhesion molecules, such as CD44 [43].

Antibodies directed against the hyaluronic capsule may therefore have the potential to interfere with GAS colonization.

24.7.3 *Lipoteichoic acid, fibronectin-binding protein and adhesion molecules*

Adherence of GAS to mucosal surfaces precedes GAS colonization and infection. Therefore, a vaccine targeted against adhesion molecules on GAS, such as CD44 ligand [43], may be a useful approach in vaccine design to prevent GAS adherence and subsequent colonization and infection. Another surface GAS molecule, lipoteichoic acid (LTA), is thought to bind with host cell fibronectin and allow other GAS proteins such as M protein to bind to host cells [44]. Moreover, mouse studies have shown that both passive intranasal immunization of LTA can inhibit GAS colonization following intranasal challenge, as well as challenge of unimmunized mice with GAS that has been pretreated with anti-LTA antibodies [45]. Other fibronectin-binding molecules include Sfb protein which mediates GAS adherence to human epithelial cells and plays a key role in the colonization of the upper respiratory tract [46]. Antibodies specific for the fibronectin-binding domain of Sfb protein have been shown to recognize streptococcal fibronectin binding surface proteins in various clinical isolates and were not cross-reactive with heart myosin suggesting the potential of Sfb protein as a GAS vaccine candidate [47].

24.7.4 *Group a streptococcal carbohydrate*

Most human sera contain antibodies to GAS carbohydrate which appears to correlate with age [48]. Sera from rabbits immunized with GAS carbohydrate have been shown to opsonize different GAS strains [48, 49]. However, anti-GAS carbohydrate antibodies from RHD patients have been shown to cross-react with myosin and other α-helical coiled proteins [50] and therefore may be involved in disease pathogenesis in the susceptible host.

24.8 Development and utilization of a group A streptococcal vaccine

One of the major obstacles to overcome in the design of a GAS vaccine is the development of a broad-based vaccine that provides protection against a wide spectrum of GAS strains. Development of a type-specific GAS vaccine based on the M protein would provide protection against a single GAS strain but would offer no protection against other GAS strains, and since there are at least 80 different GAS serotypes this would not be efficacious. Several investigators have therefore explored the potential for using the conserved region of the M protein in vaccine design which would also overcome the obstacle of immunological cross-reactivity. These candidate vaccines have proved extremely promising in the mouse system, in both protection from systemic GAS infection [37] and in the reduction of nasopharyngeal GAS colonization [35, 36], although the antibodies produced are not as opsonic as those elicited by the type-specific amino terminal region. Limited data from humans, however, have shown strong opsonic activity of p145-specific antibodies [31]. Secondly, the scope of protective immunity by a C-region-based GAS vaccine against infection with GAS strains displaying a small percentage of amino acid disparity, remains to be determined. Nevertheless, conserved region epitopes of the GAS M protein represent promising GAS vaccine candidates. By combining multiple highly opsonic M protein amino terminal peptides with a non-host cross-reactive conserved region peptide into a single immunogen it is possible to show complete protection from systemic GAS infection [37]. Since GAS infections occur predominantly in young school age children and present as pharyngitis or skin infections, the former of which, if not treated effectively with antibiotics can lead to the more serious complications of RF and RHD, a heteropolymer GAS vaccine to prevent GAS infection in humans would therefore be most effective when administered to young children. It is likely, however, that the GAS strains currently targeted will differ from those that will predominate in the future, and secondly, this vaccine may not be suitable in the protection from GAS infection in other populations, due to geographic differences in GAS serotype prevalence. A GAS vaccine may have to be re-designed periodically to maintain it's efficacy, and may also need to be designed on a population-specific basis. It is important to emphasize that while RF and RHD are currently rare in developed countries worldwide, there remain potential problems such as the possibility for the emergence of antibiotic-resistant strains of GAS and the re-emergence of more virulent GAS strains that could lead, in the future, to outbreaks of GAS-associated diseases in developed countries, as has previously been observed [10]. The development of a GAS vaccine therefore has global implications regarding public health.

There are several promising GAS vaccine candidates, as reviewed, and the data indicate that a combination of approaches and antigens may be required for controlling and preventing the consequences of GAS infection. New technological strategies and adjuvants for vaccine design and delivery are currently being investigated and will facilitate developments. To date, however, most of the research remains focused on the development of an M protein-based GAS vaccine.

Acknowledgments

We are grateful for funding support from the National Heart Foundation, The Prince Charles Hospital Foundation, the National Health and Medical Research Council, and the Cooperative Research Centre for Vaccine Technology. We also thank Dr Natkunan Ketheesan (James Cook University, Townsville) for his excellent advice and critical comments of this review.

References

1 Bisno, A.L. (1994) Group A streptococcal infections and acute rheumatic fever. *N Eng J Med 325: 783–793*

2 Fischetti, V. (1991) Streptococcal M protein. *Scientific American 264: 32–39*

3 Martin, D.R., Sriprakash, K.S. (1996) Epidemiology of group A streptococcal disease in Australia and New Zealand. *Recent Advances in Microbiology 4: 1–40*

4 Stollerman, G.H. (1997) Rheumatic fever. *Lancet 349: 935–42*

5 Carapetis, J.R., Wolff, D.R., Currie, B.J. (1996) Acute rheumatic fever and rheumatic heart disease in the top end of Australia's Northern Territory. *Medical Journal of Australia 164: 146–49*

6 Robinson, J.H., Kehoe, M.A. (1992) Group A streptococcal M protein: virulence factors and protective antigens. *Immunology Today 13: 362–367*

7 Fischetti, V.A. (1989) Streptococcal M protein: Molecular design and biological behavior. *Clinical Microbiology Reviews 2: 285–314*

8 Raeder, R., Boyle, M.D.P. (1996) Properties of IgG-binding proteins expressed by *Streptococcus pyogenes* isolates are predictive of invasive potential. *J Infect Dis 173: 888–895*

9 Bessen, D.E., Sotir, C.M., Readdy, T.L., Hollingshead, S.K. (1996) Genetic correlates of throat and skin isolates of group A streptococci. *J Infect Dis 173: 896–900*

10 Bronze, M.S., Dale, J.B. (1996) The reemergence of serious group A streptococcal infections and acute rheumatic fever. *The American Journal of Medical Sciences 311; 41–54*

11. Dale, J.B., Beachey, E.H.. (1985) Multiple heart cross-reactive epitopes of streptococcal M proteins. *Journal of Experimental Medicine 161: 113–22*

12 Froude, J., Gibofsky, A., Buskirk, D.R., Khanna, A., Zabriskie, J.B. (1989) Cross-reactivity between streptococcus and human tissue: a model of molecular mimicry and autoimmunity. *Current Topics in Microbiology and Immunology 145: 5–26*

13 Bronze, M.S., Dale, J.B. (1993) Epitopes of streptococcal M proteins that evoke antibodies that cross-react with human brain. *J Immunol 151: 2820–2828*

14 Pruksakorn, S., Currie, B., Brandt, E., Phornphutkul, C., Hunsakunachai, S., Manmontri, *et al.* (1994) Identification of T cell autoepitopes that cross-react with the C-terminal segment of the M protein of group A streptococci. *International Immunology 1235–44*

15 Tomai, M., Schlievert, P.M., Kotb, M. (1992) Distinct T-cell receptor Vβ gene usage by human T lymphocytes stimulated with the streptococcal pyrogenic exotoxins and pep M5 protein. *Infection and Immunity 60: 701–705*

16 Fleischer, B., Schmidt, K.H., Gerlach, D., Kohler, W. (1992) Separation of T cell-stimulating activity from streptococcal M protein. *Infection and Immunity 60: 1767–1770*

17 Carapetis, J.R., Currie, B.J., Good, M.F. (1996) Towards understanding the pathogenesis of rheumatic fever. *Scand J Rheumatol 25: 127–131*

18 Guilherme, L., Cunha-Neto, E., Coelho, V., Snitcowsky, R., Pomerantzeff, P.M.A., Assis, R.V., *et al.* (1995) Human heart-infiltrating T-cell clones from rheumatic heart disease patients recognize both streptococcal and cardiac proteins. *Circulation 92: 415–20*

19 Kemeny, E., Grieve, T., Marcus, R., Sareli, P., Zabriskie, J.B. (1989) Identification of mononuclear cells and T cell subsets in rheumatic valvulitis. *Clinical Immunology and Immunopathology 52: 225–37*

20 El-Demellawy, M., El-Ridi, R., Guirguis, N.I., Alim, M.A., Kotby, A., Kotb, M. (1997) Preferential recognition of human myocardial antigens by T lymphocytes from rheumatic heart disease patients. *Infection and Immunity 65: 2197–2205*

21 Yoshinaga, M., Figueroa, F., Wahid, M.R., Marcus, R.H., Suh, E., Zabriskie, J.B. (1995) Antigenic specificity of lymphocytes isolated from valvular specimans of rheumatic fever patients. *Journal of Autoimmunity 8: 601–13*

22 Read, S.E., Zabriskie, J.B., Fischetti, V.A., Utermohlen, V., Falk, R. (1974) Cellular reactivity studies in patients with streptococcal infections and rheumatic fever. *J Clin Invest 54: 439–442*

23 Yang, L.C., Soprey, P.R., Wittner, M.K., Fox, E.N. (1977) Streptococcal-induced cell mediated immune destruction of cardiac myofibers *in vitro*. *Journal of Experimental Medicine 146: 344–360*

24 Dale, J.B., Beachey, E.H. (1987) Human cytotoxic T lymphocytes evoked by group A streptococcal M proteins. *Journal of Experimental Medicine 166: 1825–1835*

25 Dale, J.B., Chiang, E.Y., Lederer, W. (1993) Recombinant tetravalent group A streptococcal M protein vaccine. *J Immunol 151: 2188–2194*

26 Dale, J.B., Simmons, M., Chiang, E.C., Chiang, E.Y. (1996) Recombinant, octavalent group A streptococcal M protein vaccine. *Vaccine 14: 944–948*

27 Medaglini, D., Pozzi, G., King, T.P., Fischetti, V.A. (1995) Mucosal and systemic immune responses to a recombinant protein expressed on the surface of the oral commensal bacterium *Streptococcus gordonii* after oral colonization. *Proc Natl Acad Sci USA 92: 6868–6872*

28 Poirier, T.P., Kehoe, M., Beachey, E.H. (1988) Protective immunity evoked by oral administration of attenuated *aroA Salmonella typhimurium* expressing cloned streptococcal M protein. *J Exp Med 168: 25–32*

29 Fischetti, V.A., Hodges, W.M., Hruby, D.E. (1989) Protection against streptococcal pharyngeal colonization with a vaccinia: M protein recombinant. *Science 244: 1487–1490*

30 Pruksakorn, S., Currie, B., Brandt, E., Martin, D., Galbraith, A., Phornphutkul, C., Hunsakunachai, S., Manmontri, A., Good, M.F. (1994) Towards a vaccine for rheumatic fever: indentification of a conserved target epitope on M protein of group A streptococci. *Lancet 344: 639–642*

31 Brandt, E.R., Hayman, W.A., Currie, B., Carapetis, J., Wood, Y., Jackson, D.C., *et al.* (1996) Opsonic human antibodies from an endemic population specific for a conserved epitope on the M protein of group A streptococci. *Immunology 89: 331–337*

32 Brandt, E.R., Hayman, W.A., Currie, B., Pruksakorn, S., Good, M.F. (1997) Human antibodies to the conserved region of the M protein: opsonization of heterologous strains of group A streptococci. *Vaccine 15: 1805–1812*

33 Pruksakorn, S., Galbraith, A., Houghten, R.A., Good, M.F. (1992) Conserved T and B cell epitopes on the M protein of group A streptococci: Induction of bactericidal antibodies. *J Immunol 149: 2729–2735*

34 Hayman, W.A., Brandt, E.R., Relf, W.A., Cooper, J., Saul, A., Good, M.F. (1997) Mapping the minimal murine T cell and B cell epitopes within a peptide vaccine candidate from the conserved region of the M protein of group A streptococcus. *Int Immunol 9: 1723–1733*

35 Bessen, D., Fischetti, V.A. (1988) Influence of intranasal immunization with synthetic peptides corresponding to conserved epitopes of M protein on mucosal colonization by group A streptococci. *Infect Immunity 56: 2666–2672*

36 Bessen, D., Fischetti, V.A. (1990) Synthetic peptide vaccine against mucosal colonization by group A streptococci. I. Protection against a heterologous M serotype with shared C repeat region epitopes. *J Immunol 145: 1251–1256*

37 Brandt, E.R., Sriprakash, K.S., Hobb, R.I., Hayman, W.A., Zeng, W., Batzloff, M.R., Jackson, D.C., Good, M.F. (2000) Novel multi-epitope strategy for a group A streptococcal vaccine designed for the Australian Aboriginal population. *Nature Medicine 6: 455–459*

38 Jackson, D.C., O'Brien-Simpson, N.M., Ede, N.J., Brown, L.E. (1997) Free radical induced polymerisation of synthetic peptides into polymeric immunogens. *Vaccine 15: 1697–1705*

39 Ji, Y., Mclandsborough, L., Kondagunta, A., Cleary, P.P. (1996) C5a peptidase alters clearance and trafficking of group A streptococci by infected mice. *Infect Immunity 64: 503–510*

40 Ji, Y., Carlson, B., Kondagunta, A., Cleary, P.P. (1997) Intranasal immunization with C5a peptidase prevents nasopharyngeal colonization of mice by the group A *streptococcus. Infect Immunity 65: 2080–2087*

41 Johnson, D.R., Stevens, D.L., Kaplan, E.L. (1992) Epidemiologic analysis of group A streptococcal serotypes associated with severe systemic infections, rheumatic fever, or uncomplicated pharyngitis. *J Infect Dis 166: 374–382*

42 Dale, J.B., Washburn, R.G., Marques, M.B., Wessels, M.R. (1996) Hyaluronate capsule and surface M protein in resistance to opsonization of group A streptococci. *Infect Immunity 64: 1495–1501*

43 Schrager, H.M., Alberti, S., Cywes, C., Dougherty, G.J., Wessels, M.R. (1998) Hyaluronic acid capsule modulates M protein-mediated adherence and acts as a ligand for attachment of group A streptococcus to CD44 on human keratinocytes. *J Clin Invest 101: 1708–1716*

44 Kehoe, M.A. (1991) New aspects of Streptococcus pyogenes pathogenicity. *Rev Med Microbiology 2: 147–152*

45 Dale, J.B., Baird, R.W., Courtney, H.S., Hasty, D.L., Bronze, M.S. (1994) Passive protection of mice against group A streptococcal pharyngeal infection by lipoteichoic acid. *J Infect Dis 169: 319–323*

46 Talay, S.R., Valentin-Weigand, P., Jerlstrom, P.G., Timmis, K.N., Chhatwal, G.S. (1992) Fibronectin-binding protein of *Streptococcus pyogenes*: sequence of the binding domain involved in adherence of streptococci to epithelial cells. *Infect Immunity 60: 3837–3844*

47 Valentin-Weigand, P., Talay, S.R., Kaufhold, A., Timmis, K.N., Chhatwal, G.S. (1994) The fibronectin binding domain of Streptococcus pyogenes occurs in many group A streptococci and does not cross-react with heart myosin. *Microbial Pathogenesis 17: 111–120*

48 Salvadori, L.G., Blake, M.S., McCarty, M., Tai, J.Y., Zabriskie, J.B. (1995) Group A streptococcus-liposome ELISA antibody titers to group A polysaccharide and opsonophagocytic capabilities of the antibodies. *J Infect Dis 171: 593–600*

49 Dale, J.B., Cleary, P.P., Fischetti, V.A., Kasper, D.L., Musser, J.M., Zabriskie, J.B. (1997) Group A and B streptococcal vaccine development. *Adv Exp Biol 418: 863–869*

50 Adderson, E.E., Shikman, A.R., Ward, K.E., Cunningham, M.W. (1998) Molecular analysis of polyreactive monoclonal antibodies from rheumatic carditis: human anti-*N*-acetylglucosamine/anti-myosin antibody V region genes. *J Immunol 161: 2020–2031*

Chapter 25

Systemic antifungal agents

Ross P. McGeary

Contents

25.1 Introduction

The last 20 years have witnessed a significant increase in the incidence of serious, systemic fungal infections. One major cause of this is the acquired immunodeficiency syndrome (AIDS) epidemic which has left many thousands of individuals with susceptibility to systemic mycoses, leading to a high degree of morbidity and mortality. Another group of immunocompromized patients vulnerable to fungal infection are those receiving treatment for cancer or organ transplantation, and even those undergoing major surgery. Low birth-weight babies in intensive care are also at risk. For recipients of bone marrow transplants, invasive aspergillosis is a leading cause of death, and *Pneumocystic carinii* is a major cause of death in AIDS patients.

While many therapeutic option are available for the treatment of superficial fungal infections, drugs to manage deep-seated mycoses are more limited. The first breakthrough in the treatment of systemic mycoses came with the introduction of nystatin, a polyene antifungal, in 1951. This was soon replaced by the related polyene amphotericin B, in 1956, which still remains the standard by which new antifungal drugs are compared.

Other than the introduction of the nucleoside analog 5-fluorocytosine in 1964, the modern era of antifungal drug has been typified by the introduction of, and improvements made upon, the azole drugs. Miconazole was introduced in 1978, followed by ketoconazole (1981), fluconazole (1989), then itraconazole in 1992. Advances have steadily been made in this class of drug in terms of increasing potency, decreasing toxicity, broader spectrum activity and oral bioavailability.

Six drugs are currently available for use against systemic fungal infections: amphotericin B and its lipid formulations, fluconazole, itraconazole, ketoconazole, miconazole and 5-fluorocytosine. The current worldwide market for antifungals exceeds $2 billion.

25.2 Azoles (miconazole, ketoconazole, itraconazole, fluconazole)

All azole derivatives interfere ultimately with the fungal cell wall membrane. The primary target is lanosterol demethylase (14α-sterol demethylase), a cytochrome P-450 enzyme containing a heme group in its active site. The nitrogens of the azole coordinate to the iron atom in the heme molecule. The enzyme causes 14α-demethylation of lanosterol, and enzyme inhibition leads to a depletion of ergosterol (an essential sterol for cell proliferation) and an accumulation of C14-methyl sterols in the cytoplasmic membrane (Scheme 1). At higher concentrations cell death is caused by gross alteration of the cell membrane, alteration in the function of several membrane-bound enzymes, disruption of the membrane barrier function and increased permeability. Since P-450-dependent enzymes are present in mammalian cells as well as fungi, the clinical value of these agents is presumed to arise from the specificity of their binding. Side effects resulting from administration of the azole may be due to interactions with other P-450 systems.

In general the imidazoles are fungistatic, although miconazole is fungicidal at higher concentrations. Miconazole

(Figure 25.1) is poorly water soluble and poorly absorbed after oral administration and must be solubilized with an amphophilic carrier, such as polyethoxylated castor oil, for intravenous use. It has been used successfully in the treatment of coccidioidomycosis and coccidioidal meningitis. However, significant toxicity associated with its prolonged intravenous administration, its cost, and high rate of relapse following discontinuation of therapy have led to its use now mainly as a topical agent for the treatment of cutaneous and mucocutaneous fungal infections.

Ketoconazole (Figure 25.2) was the first orally administered antifungal azole. It is used for the treatment of systemic and superficial mycoses and it exhibits a similar spectrum of antifungal activity as miconazole. While its *in vitro* activity is lower than miconazole, its *in vivo* activity is superior. It is not effective in the treatment of cryptococcosis or aspergillosis. It is rapidly absorbed from the GI tract, but its oral availability appears to be variable, possibly due to the sensitivity of its absorption on intestinal pH (an increase in pH leads to decreased absorption of the drug), or due to non-linearity of its bioavailability as a function of dose administered. It is not available in an intravenous formulation and it is not appreciably absorbed following topical application. Its serum half life has been reported to be from 1 to 11 hours. It is highly protein bound and does not appreciably cross the blood-brain barrier, although it may be detected in synovial fluid. Ketoconazole has been shown to inhibit the synthesis of testosterone *in vivo*. Relapse rates are fairly high, probably due to its fungistatic action.

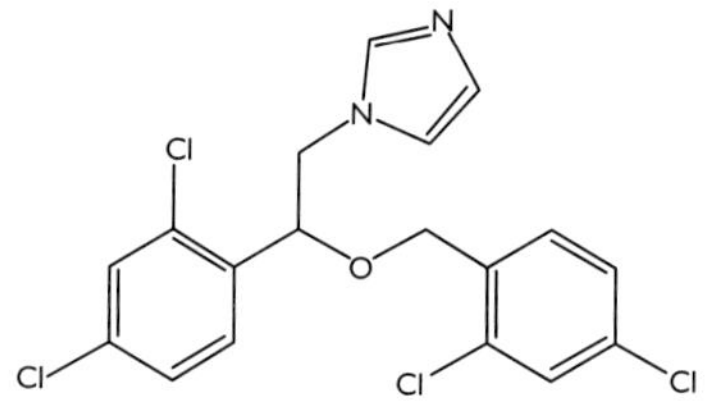

Figure 25.1 Miconazole.

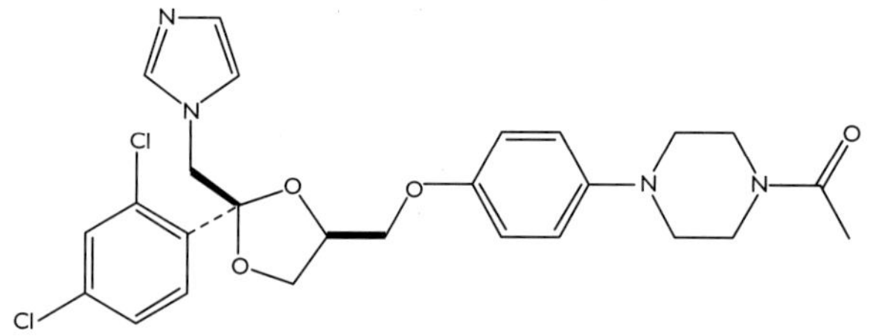

Figure 25.2 Ketoconazole.

Itraconazole (Figure 25.3) is a triazole analog of ketoconazole which is more lipophilic and less soluble. Replacement of the imidazole ring of the azole antifungal agents with a triazole ring gives improved potency, a broader spectrum of activity, greater resistance to metabolic degradation and greater affinity for fungal enzymes than mammalian enzyme targets. Itraconazole is more potent than ketoconazole and less toxic, and is active against *Candida, Cryptococcus* and *Aspergillus* species. It has a long clearance time, 20–30 hours, but its oral absorption varies among individuals and is affected by food and gastric pH. An intravenous formulation has recently become available for clinical use. Itraconazole is almost completely protein bound in the serum. While there is little penetration into the cerebrospinal fluid, it still demonstrates some antifungal activity against cryptococcal and coccidioidal meningitis. Levels of itraconazole in stratum corneum after oral administration are high enough for treatment of tinea corporis and tinea cruris. Lack of parenteral availability and variable oral absorption can make this an unreliable agent, but its more favorable pharmacokinetics profile has made it the drug of choice for some non-life threatening fungal infections.

Of the newer triazoles, fluconazole (Figure 25.4) has a broader spectrum of activity, being highly fungistatic and fungicidal against *Aspergillus* species. Its water solubility is good, and it can be administered either orally or intravenously. Its oral bioavailability is 85–90% and not dependent on gastric pH. It exhibits very low protein binding in the serum (11%), and its half life in the serum is about 30 hours. Unlike other azole antifungals, it penetrates well into the cerebrospinal fluid. Fluconazole's activity towards inhibition of fungal 14-α-demethylase is comparable to that of ketoconazole, but it is much less active against the analogous mammalian enzymes. Fluconazole is effective against many *Candida* species, except *C. glabrata* and *C. krusei*, and it is ineffective against *Aspergillus*. Like itraconazole, it is very effective

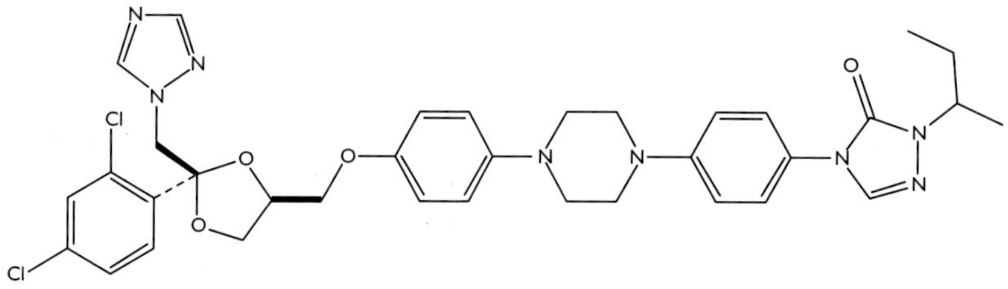

Figure 25.3 Itraconazole.

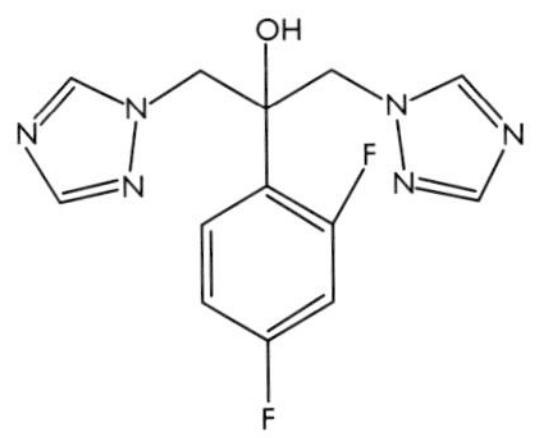

Figure 25.4 Fluconazole.

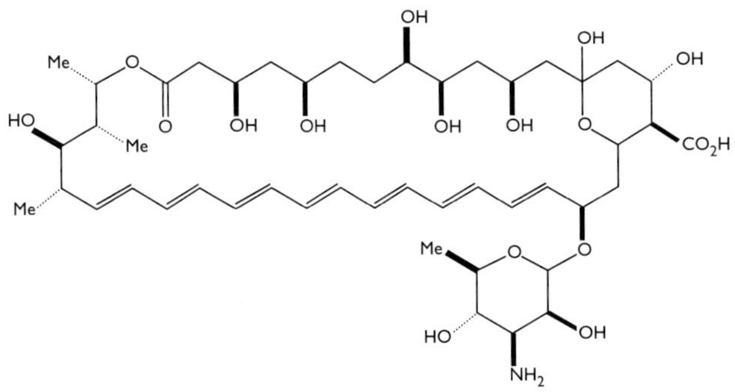

for the treatment of nail and hair infections. It is usually well tolerated and side effects are usually mild, but fetal abnormalities have been noted and its use should be avoided in pregnant women.

25.3 Polyenes (amphotericin B)

Amphotericin B (Figure 25.5) is a polyene macrolide antibiotic that shows a broad spectrum of antifungal activity against most fungal pathogens, including *Aspergillus* and *Candida* species, and is the first choice for the treatment of deeply invasive mycoses. It is fungistatic at low concentrations and fungicidal at higher concentrations. It is isolated from a fermentation process of *Streptomyces nodosus*, and is the only natural product (non-synthetic) antifungal drug in clinical use.

X-ray crystallography has shown that amphotericin B adopts a rigid, rod-shaped conformation with opposing hydrophilic and hydrophobic faces. The molecule is approximately the same length as a phospholipid molecule. Interaction of the hydrophobic faces of the polyene with membrane sterols results in disorganization of the membrane and the production of aqueous pores in which the polyene hydroxyls residues face inward, leading to impairment of barrier function, altered permeability, leakage of vital cytoplasmic components and death of the organism. At low concentrations, however, potassium channel activity is increased, leading only to reversible fungistatic activity. Amphotericin B has a higher affinity for ergosterol, the primary sterol in fungal cell membranes, than for cholesterol, the most common sterol in mammalian cell membranes. It has also been suggested that autooxidation of membrane-bound amphotericin B can lead to the formation of damaging free radicals and may be a significant factor in its fungal toxicity. In addition, amphotericin B may increase the antifungal effects of other drugs by facilitating their penetration though the fungal cell membrane.[1]

Due to its low water solubility, amphotericin B is conventionally solubilized as a micellular formulation with sodium deoxycholate. Amphotericin B is not absorbed from the gut (<5% bioavailability), and so must be administered intravenously. Due to amphotericin B's very poor water solubility, much effort has been devoted to drug delivery technologies.[2]

Figure 25.5 Amphotericin B.

The most widely used of these involves administration of amphotericin B as a liposome complex with two phospholipids, dimyristoylphosphatidylcholine and dimyristoylphosphatidylglycerol. Liposomes are microscopic vesicles consisting of concentrically arranged lipid bilayers. Liposomes are taken up by macrophages and are digested in the secondary lysosomes to release their contents into the vacuole and surrounding cytosol. As well as providing a means of efficient drug delivery, the use of liposomes has resulted in a marked reduction in the toxicity associated with amphotericin B, while retaining its efficacy. This allows for the administration of higher doses of amphotericin B, with associated increased therapeutic response. The lipid components of these liposomes are thought to decrease binding to human cell membranes and so decrease the toxicity of amphotericin B. Among those lipid-complexed formulation of amphotericin now marketed are Amphotericin B colloidal dispersion (ABCD, Abelcet, Amphocil), disc-shaped particles containing amphotericin B and sodium cholesteryl sulfate, ABLC (Amphotec), formulated from amphotericin B and synthetic phospholipids, and amphotericin B liposomal (AmBisome). The high prices of these formulation have, however, limited their use in the clinic to those patients who are unresponsive to, or are unable to tolerate, conventional amphotericin B deoxycholate.

Even after four decades of clinical use, optimal dose and duration of therapy remains unclear. Treatment with amphotericin B frequently leads to severe, acute side effects such as chills and fever, nausea and vomiting. Ongoing treatment often leads to anemia and renal dysfunction which may require limitations to the administered dose and subsequent poor response in the patient. The dose-limiting effect is nephrotoxicity, thought to result from amphotericin B binding to cholesterol in human cells, although the precise mechanism is unclear. Natural or developed resistance is low.

Amphotericin B is highly protein-bound. Its penetration into the cerebrospinal fluid is very low, and its use is limited to those infections where the fungus is exposed to the blood stream. The metabolism and excretion of amphotericin B is poorly understood, and no metabolites have been identified.

The half life of amphotericin B in the body is between 24 and 48 hours.

25.4 Allylamines

The allylamines inhibit ergosterol biosynthesis by acting as inhibitors of the enzyme squalene epoxidase, making the cell unable to synthesize membrane sterols (Scheme 2). In addition, accumulation of squalene in the fungi is thought to be fungicidal due to increased cell permeability, and may be the primary cause of cell death. Terbinafine (Figure 25.6) is highly selective and mammalian squalene epoxidase is several orders of magnitude less sensitive to terbinafine than the fungal enzyme.

The antimycotic activity of terbinafine is good against most species of fungi, except yeasts, specifically *C. albicans*. Terbinafine is highly effective in the treatment of dermatophytes, filamentous fungi and dimorphic fungi. Terbinafine may be administered topically or orally, whereas the related allylamine, naftitine, is applied topically. The oral bioavailability of terbinafine is greater than 70% and the elimination half life is 22 hours. It is strongly bound to plasma proteins and is rapidly distributed to lipophilic tissues, as well as the skin, hair and nail beds. The use of terbinafine is confined almost exclusively to the treatment of onychomycosis, dermatomycosis and vulvovaginal candidiasis. Terbinafine appears to have dramatic *in vitro* synergism when tested in conjunction with fluconazole and itraconazole. It has the potential to be used for the treatment of deep-seated mycoses, but has been little studied for these infections.

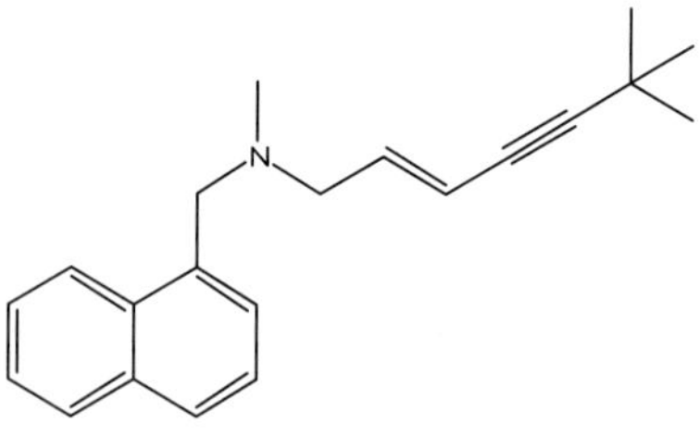

N

Figure 25.6 Terbinafine.

25.5 Nucleic acid inhibitors (5-fluorocytosine)

5-Fluorocytosine (flucytosine) is a water-soluble fluorine analog of the pyrimidine base cytosine. 5-Fluorocytosine is first transported to susceptible fungal cells by the enzyme cytosine permease. Once inside the cell 5-fluorocytosine is converted to 5-fluorouracil by cytosine deaminase, an enzyme found in only some fungi such as *C. albicans*. Only those fungi with this enzyme are sensitive. 5-Fluorouracil competes with uracil, so cellular enzymes then convert 5-fluorouracil into 5-fluorodeoxyuridine monophosphate, which is a potent inhibitor of the enzyme thymidylate synthase, the enzyme which methylates uridine to form thymidine, and an essential enzyme for DNA synthesis (Scheme 3). Fungal DNA synthesis and nuclear division are thus inhibited. 5-Fluorouracil is also incorporated into fungal RNA, causing miscoding and leading to disruption of protein synthesis. 5-Fluorouracil itself is not useful as an antifungal drug since it is poorly taken up by fungi and it is toxic to mammalian cells. Mammalian cells lack the enzyme cytosine deaminase and are therefore unaffected by 5-fluorocytosine.

5-Fluorocytosine is rarely given alone, since fungi rapidly develop resistance, often by down-regulation of the enzyme that coverts 5-fluorouracil into 5-fluorodeoxyuridine monophosphate. It is usually given in combination with fluconazole, or amphotericin B, with which it acts synergistically, allowing for combination therapy leading to lower toxic side effects.

5-Fluorocytosine is used for oral treatment of systemic fungal infections, typically for the treatment of cryptococcal meningitis and for deep-seated candidal infections. It is highly bioavailable and is well absorbed from the intestine. The binding of 5-fluorocytosine to blood proteins is very low, it is widely distributed in body tissues, and it effectively penetrates into the central nervous system. Its half life in blood plasma is about five hours. Tolerance to 5-fluorocytosine is good and large doses can often be given over prolonged periods without significant side effects. In some cases intestinal bacteria may be responsible for converting 5-fluorocytosine to 5-fluorouracil leading to some of the observed toxic effects (chiefly diarrhoea, nausea, vomiting, elevated liver function tests and hepatoxicity). The antifungal

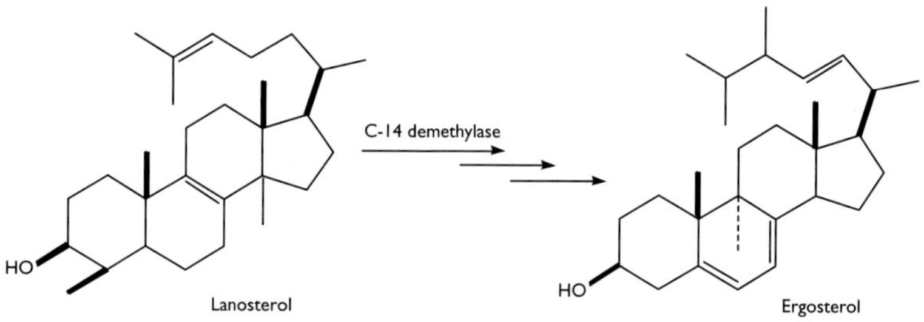

HO
Lanosterol
C-14 demethylase
HO
Ergosterol

Scheme 1

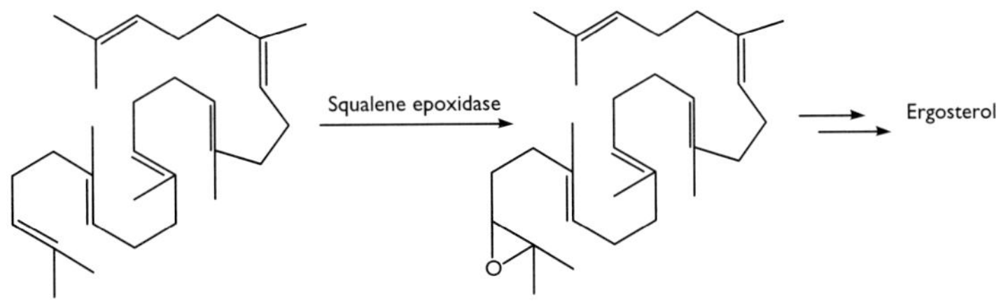

Squalene epoxidase

Ergosterol

Scheme 2

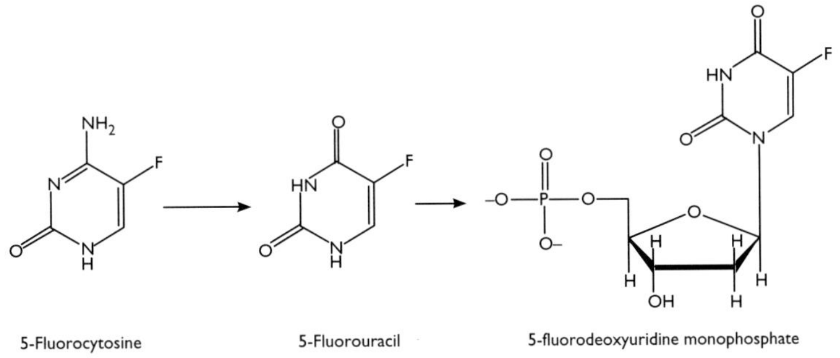

5-Fluorocytosine

5-Fluorouracil

5-fluorodeoxyuridine monophosphate

Scheme 3

spectrum of 5-fluorocytosine is limited to species of *Candida, Cryptococuss, Aspergillus* and dematiaceae.[3] At low doses it is fungistatic, and at higher doses it becomes fungicidal.

25.6 Drugs under clinical investigation

Progress in the development of new antifungal agents has been relatively slow. In part this has been due to the fact that, until recently, the incidence of life-threatening fungal infections has been too low to justify extensive research in this area by pharmaceutical companies. Also slowing progress is that fungi are, like mammalian cells, eukaryotes. This means that any drug which interferes with transcription or DNA synthesis is likely to have potential for human toxicity.[4] With the exception of 5-fluorocytosine, all of the current agents used against systemic mycoses act by interfering with the functional or structural integrity of the fungal cell membrane. This is also the case with those agents that are now in clinical trials. The fungal cell wall consists of three major rigid macromolecular components: chitin, glucan and mannoproteins, and these have emerged as current targets for therapeutic intervention, as outlined below.[5] Several new members of the triazole class of antifungal drugs are currently in clinical development. These have been reviewed by Ablordeppey,[6] Luna[7] and Sheehan.[8]

25.7 Glucan synthesis inhibitors

The major component present in the fungal cell wall is the polymerized carbohydrate (1,3)β-linked glucan. This polymer consists of repeating units of glucose joined by β-glucoside bonds between the C-1 oxygen of one residue and the C-3 oxygen of the next. Some (1,6)β-branching is also present to provide cross-linking and structural rigidity. Its role is largely structural (providing mechanical strength to the wall) although it performs other functions.[9] It is essential for normal fungal growth and development. The enzyme responsible for the polymerization of glucose to (1,3)β-glucan is (1,3)β-D-glucan synthase. This is an enzyme absent in mammals, making it an attractive target for antifungal drugs.

The echinocandins and pneumocandins are a class of naturally occurring cyclic lipo-hexapeptide fungal fermentation metabolites. Their activity derives from specific and non-competitive inhibition of fungal (1,3)β-D-glucan synthase. The resulting loss of glucan in growing cells leads to osmotic shock, cell wall damage and cell lysis.

Echinocandin B (Figure 25.7) is highly active against yeasts, particularly *C. albicans*, and including those strains resistant to fluconazole. Many semisynthetic derivatives have been made by replacing the fatty acid side chain with other acyl moieties. One such derivative, the Eli Lilly compound LY-303366, is currently in phase II trials and has comparable therapeutic efficacy as amphotericin B. LY-303366 is water soluble, fungicidal, and has greater potency and superior oral bioavailability than echinocandin B.

Pneumocandin B is a cyclic hexapeptide closely related to the echinocandins, and several semi-synthetic derivative are also under investigation as possible clinical candidates.[10]

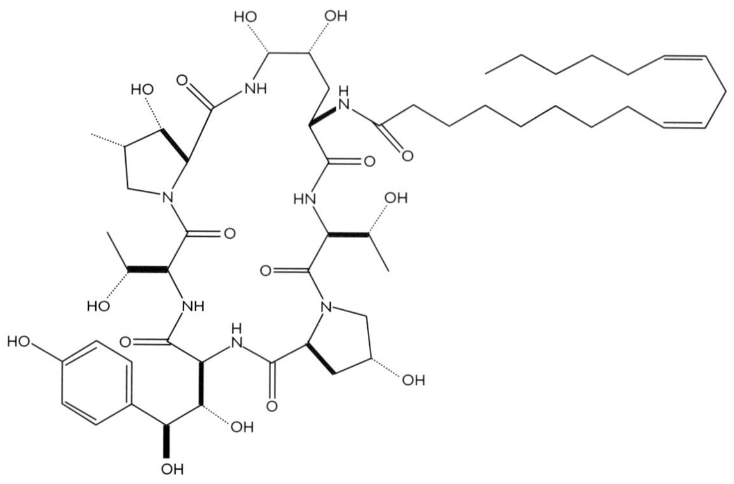

Figure 25.7 Echinocandin B.

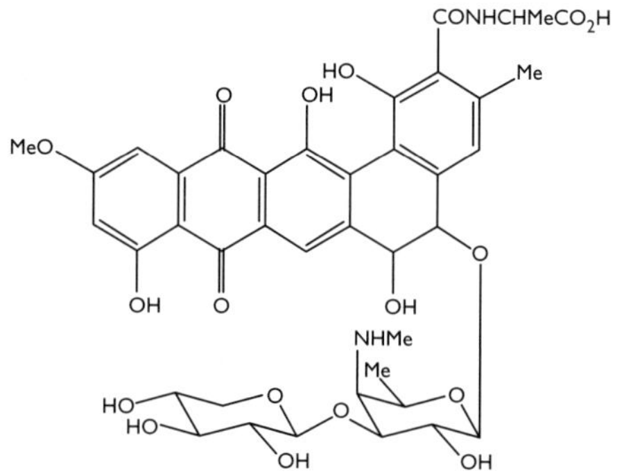

Figure 25.8 Pradimicin A.

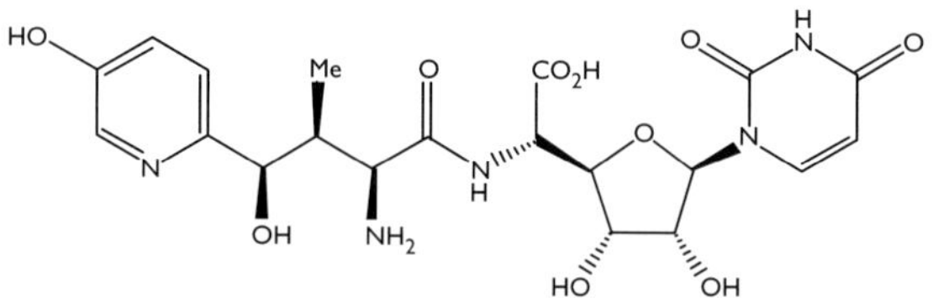

Figure 25.9 Nikkomycin Z.

25.8 Mannoprotein binders

The pradimicins and benanomicins are a class of broad-spectrum fungicidal agents which consist of a disaccharide linked to a dihydrobenzo[α]naphthacenquinone. These compounds are derived form *Actinomycetes* and appear to complex to mannoprotein components of the fungal cell surface. Mannoproteins are branched α-linked mannose polymers attached to proteins via an *N*-acetylglucosamine-*N*-acetylglucosamine moiety. After binding of pradimicins or benanomicins to the mannoproteins, in the presence of calcium ions, the cell membrane is perturbed and leaks its intracellular components, leading to cell death. They do not appear to bind to mammalian cell membranes (Figure 25.8).

25.9 Chitin synthase inhibitors

The nikkomycins and related polyoxins are a family of pyrimidine nucleosides linked to a peptide side-chain. They are produced as fermentation metabolites of *Streptomycetes*. These compounds are competitive inhibitors of fungal chitin synthase enzymes. Chitin, a major constituent of the fungal cell wall, is a linear polymer of (1,4)β-linked *N*-acetylglucosamine, and is prepared by chitin synthase from the precursor UDP-*N*-acetylglucosamine. Chitin is the second most important macromolecule in the fungal cell wall, although its relative amount varies between species, accounting for between 1–10% of the cell wall. Chitin is not present in mammalian cells. It is thought that the nikkomycins and polyoxins are substrate mimics for the chitin precursors. Although its antifungal spectrum is modest, nikkomycin Z has been entered into clinical trials because it shows synergistic effects with other drugs used against several fungal infections, such as glucan synthase inhibitors and the triazoles.

The clinical use of nikkomycin Z (Figure 25.9) may be limited because many fungi have several different isoforms of the chitin synthase enzyme, which may be inhibited to different degrees by these drugs. These inhibitors also require transport into the fungal cell by a peptide permease enzyme which may be a cause of resistance. The drugs may also be susceptible to degradation by endogenous proteases.[4]

25.10 Future prospects

Rational drug development program directed at producing antifungal drugs have focused on identifying fungal molecular targets which, ideally, have no mammalian counterpart, or are significantly differentiated from mammalian analog such that selective targeting is possible. Apart from the previously discussed targets for the fungal cell wall, several other areas are under investigation and may lead to future therapeutics. The peptide cecropin causes lysis of fungal cells and appears to bind selectively to fungal ergosterol. It is relatively inactive against mammalian cells.[11] An aminocatechol has been identified as a selective inhibitor of fungal topoisomerase I.[12] Most potential targets, however, have no know inhibitors. For example, an essential protein factor, EF-3, is known to be required for fungal protein synthesis. EF-3 has no mammalian counterpart and is therefore an attractive target for inhibition.[13] Another protein target is the enzyme responsible for the transfer of mannoproteins to (1,3)β-glucan. This enzyme also has no mammalian counterpart but, again, no inhibitor has yet been identified.[14]

References

1 Kucers, A., Crowe, S.M., Grayson, M.L. and Hoy, J.F. (1997) *The Use of Antibiotics*, Fifth Edition, Butterworth-Heinemann, Oxford.
2 Bekersky, R.M., Fielding, D.B. and Lawrence, I. (1999) Lipid-Based Amphotericin B Formulations: From Animals to Man, *Pharmaceutical Science & Technology Today, 2: 230–236.*
3 Polak, A., Mode of Action Studies, in *Chemotherapy of Fungal Diseases*, Ryley, J.F. (Ed.), Springer-Verlag, Berlin.
4 Groll, A.H, Piscitelli, S.C. and Walsh, T.J. (1998) Clinical Pharmacology of Systemic Antifungal Agents: A comprehensive Review of Agents in Clinical Use, Current Investigational Compounds, and Putative Targets for Antifungal Drug Development, *Advances in Pharmacology, 44: 343–500.*
5 Maertens, J.A. and Boogaerts, M.A. (2000) Fungal Cell Wall Inhibitors: Emphasis on Clinical Aspects, *Current Pharmaceutical Design, 6: 225–239.*
6 Ablordeppey, S.Y., Fan, P., Ablordeppey, J.H. and Mardenborough, L. (1999) Systemic Antifungal Agents Against AIDS-Related Opportunistic Infections: Current Status and Emerging Drugs in Development, *Current Medicinal Chemistry, 6: 1151–1195.*
7 Luna, B., Drew, R.H. and Perfect, J.R. (2000) Agents for Treatment of Invasive Fungal Infections, *Otolaryngologic Clinics of North America, 33: 277–299.*
8 Sheehan, D.J., Hitchcock, C.A. and Sibley, C.M. (1999) Current and Emerging Azole Antifungal Agents, *Clinical Microbiology Reviews, 12: 40–79.*
9 Selitrennikoff, C.P. (1995) *Antifungal Drugs: (1,3)β-Glucan Synthase Inhibitors*, Springer-Verlag, Heidelberg.
10 Klein, L.L. and Li, L. (1999) Design and Preparation of Cyclopeptamine Antifungal Agents, *Current Pharmaceutical Design, 5: 57–71.*
11 DeLucca, A.J., Bland, J.M., Jacks, T.J., Grimm, C., Cleveland, T.E. and Walsh, T.J. (1997) Fungicidal Activity of Cecropin A, *Antimicrobial Agents and Chemotherapy, 41: 481–483.*
12 Fostel, J. and Montgomery, D. (1995) Identification of the Aminocatechol A-3253 as an *In Vitro* Poison of DNA-Topoisomerase I from Candida Albicans, *Antimicrobial Agents and Chemotherapy, 39: 586–592.*
13 Colthurst, D.R., Santos, M., Grant C.M. and Tuite, M.F. (1991) Candida Albicans and Three other Candida Species Contain an Elongation Factor Structurally and Functionally Analogous to Elongation Factor 3, *FEMS Microbial Letters, 80: 45–50.*
14 DeNobel, H. and Lipke, P.N. (1994) Is There a Role for GPI's in Yeast Wall Assembly?, *Trends in Cellular Biology, 4: 42–45.*

Part 7

Diseases of the central and the peripheral nervous system

Chapter 26

Mechanisms of neurodegenerative diseases

Richard Anthony Hughes

Contents

26.1 Introduction

Neurodegenerative diseases are a class of disorders characterized by the selective death of neurons. Neurons in the central nervous system (CNS) are typically *post-mitotic* – that is, they are unable to divide. Thus, once neurons are lost in a neurodegenerative disease, they will not be replaced, as might happen following degenerative disease in other organs (e.g. the liver). As a consequence, neurodegenerative diseases are *chronic* (i.e. long lasting) and *progressive* (i.e. they worsen with time). Also, therapy for neurodegenerative diseases is currently limited to treating the *symptoms* of the disease. A better understanding of the causes and underlying mechanisms of neurodegenerative diseases is likely to lead to treatments which affect the *cause* of the disease, i.e. neuronal death. Given that neurodegenerative diseases are more prevalent in the elderly, and that many countries are showing a shift towards an aging population, finding effective means to treat neurodegenerative diseases has become a major priority for medical researchers, both in academia and industry.

In this chapter, we will first examine characteristics common to neurodegenerative diseases, before looking at five diseases in more detail: Parkinson's disease, Alzheimer's disease, Huntington's disease, amyotrophic lateral sclerosis and transmissible spongiform encephalopathies. Finally, we will consider new approaches to the treatment of neurodegenerative diseases.

26.2 Common features of neurodegenerative diseases

Neurodegenerative diseases can be considered to follow a common pathway (Figure 26.1). This pathway is characterized

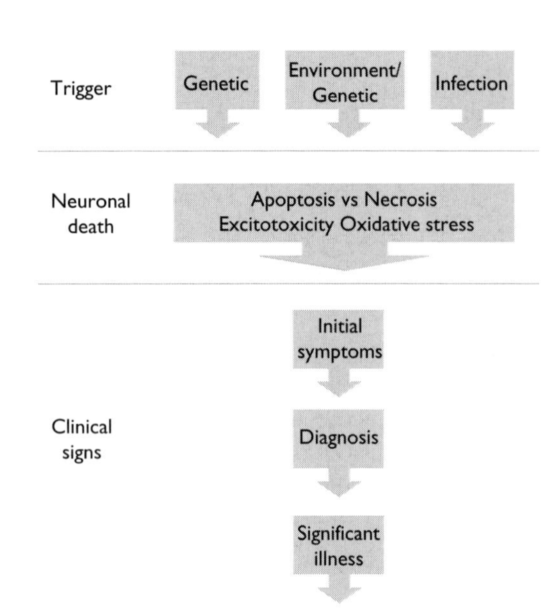

Figure 26.1 The common pathway of neurodegenerative diseases.

by a *trigger* – an event or set of circumstances which initiates the disease. In some neurodegenerative diseases, the trigger is clearly genetic (e.g. Huntington's disease), in others it may have a genetic component which predisposes the patient to another (i.e. environmental) cause (e.g. Parkinson's disease, amyotrophic lateral sclerosis), or it may be an infective agent (e.g. new variant Creutzfeldt-Jakob disease). The trigger leads to the common feature of neurodegenerative diseases: *neuronal death*. The resultant loss of neurons will lead to *clinical signs*. It is a hallmark of neurodegenerative diseases that often a considerable proportion of a population of neurons needs to be lost before symptoms begin to appear. Thus, it is likely that a substantial period of time will have elapsed from the true onset of the disease to the appearance of clinical signs. It may be some time later still before a patient actually presents to a clinician. Thus, neurodegenerative diseases are generally well advanced by the time of diagnosis. It is a major thrust of current research to develop means to diagnose these diseases as early as possible, and so give the best chance for therapies, particularly those that might involve the rescue or repair of *dying* (but not yet dead) neurons. The actual progression of the clinical disease varies from patient to patient, although some diseases generally show slow progression (e.g. Parkinson's disease will progress over decades), while others are more rapid (e.g. patients with amyotrophic lateral sclerosis typically deteriorate rapidly over two to three years).

26.3 Neuronal death: apoptosis and necrosis

Neurons die in neurodegenerative disease by two mechanisms: *apoptosis* and *necrosis*. Apoptosis, or *programmed cell death* is a type of cellular suicide. It is a common feature of development – neurons, for example, undergo a period of naturally-occurring cell death as part of normal embryonic development – as well as in later life. Apoptosis is characterized by:

- condensation of chromatin;
- fragmentation of DNA, leading to the formation of distinctly-sized DNA fragments which can be detected by gel electrophoresis;
- cell "*blebbing*", where the cell membrane pinches itself off to form small buds;
- shrinkage of the cell;
- the *absence of an inflammatory response* (cell debris is removed by macrophages).

A number of events can lead to apoptosis in neurons, including the removal or absence of specific *neurotrophic factors* (proteins which act to prevent apoptosis and therefore promote neuronal survival), *excitoxicity*, and *oxidative stress* (see below). These events lead to altered gene transcription, in particular of a family of highly-specific cysteine aspartate-specific proteases, *caspases*. Caspases cleave a variety of substrate proteins which activates them, resulting in a cascade of events that results in apoptosis. The caspases themselves are synthesized as *zymogens* (enzymes with little or no catalytic activity), which themselves require cleavage for protease activity e.g. by another caspase.

Necrosis, on the other hand, is associated with swelling of a cell, leading to membrane damage (cells "explode"), typically resulting in an inflammatory response. Toxic agents are often able to kill cells directly by necrosis, although exposure to a lower concentration of toxin may cause cell death by apoptosis. Neurons apparently die in neurodegenerative disease by both apoptosis and necrosis. Indeed, excitotoxicity and oxidative stress are probably both able to initiate apoptotic mechanisms in neurons, as well as killing them directly by necrosis.

26.3.1 Excitotoxicity

Excitotoxicity is the name given to the neurotoxic effects of *glutamate*. Glutamate is a widely spread excitatory neurotransmitter. However, it has been known for some decades that even low concentrations of glutamate are able to cause neuronal death in cell culture, and neurodegeneration *in vivo*. The main means by which glutamate causes neuronal death is via intracellular *calcium overload* in the affected neuron:

- glutamate binds to and activates *AMPA*, *NMDA* and *metabotropic* glutamate receptors;
- this leads to an influx of extracellular Ca^{2+} into the neuron via NMDA channels, voltage-operated Ca^{2+} channels and Na^{+}/Ca^{2+} exchange;

- intracellular Ca^{2+} is also released from the endoplasmic reticulum as a result of activation of metabotropic glutamate receptors.

The resultant raised intracellular calcium concentration causes:

- activation of proteases and lipases causing membrane damage;
- generation of *reactive oxygen species* (ROS) which damage membranes, proteins and DNA;
- initiation of apoptosis.

The protective mechanism that neurons normally use against damage caused by glutamate/Ca^{2+} appears to involve the mitochondria and endoplasmic reticulum, which act as *sinks* for Ca^{2+} in cells. Overloading of the mitochondrial stores with Ca^{2+} disrupts mitochondrial function, reducing the amount of energy (in the form of ATP) to power membrane pumps and drive accumulation of Ca^{2+} by the endoplasmic reticulum. It has been suggested that reduced ATP production might be a common factor in rendering neurons susceptible to excitoxicity in neurodegenerative disease.

26.3.2 Reactive oxygen species (ROS) and oxidative stress

ROS, such as *hydrogen peroxide* and the *hydroxyl radical* $HO^{\bullet}$, are produced as part of normal cellular function. Being reactive, they have the capacity to react with cellular components (lipids, proteins and DNA) to produce cellular damage. To prevent this, cells employ a series of antioxidant defense mechanisms, including enzymes such as *superoxide dismutase* (which removes the ROS superoxide anion radical $O_2^{-\bullet}$), and antioxidants such as vitamins E and C. Under certain conditions however, so-called *oxidative stress*, the defense mechanisms cannot cope with the ROS load. This could be due either to deficiencies in the defense system, or increases in the amount of ROS. In this state, ROS may afflict damage upon a cell, of which perhaps the most destructive is membrane damage through *lipoperoxidation*. The state of oxidative stress renders neurons more susceptible to excitotoxic cell destruction. An increased tendency towards states of oxidative stress may therefore contribute towards the susceptibility of an individual to a neurodegenerative disease.

26.4 Parkinson's disease

Parkinson's disease (PD) is a neurodegenerative disease that causes progressive dysfunction of movement. It more commonly affects the elderly, having a prevalence of around 2% in people over 65 years, although it can strike young patients. The three classical clinical signs of PD are:

- *tremor*, typically in the hands, and present primarily at rest;
- *rigidity*, seen as increased resistance to passive movement of limbs;
- *hypokinesia* or *akinesia* (reduced or no voluntary movement), making it difficult for the patient to both initiate and terminate voluntary movement.

In addition, PD is often accompanied by *dementia*, as the neuronal degeneration spreads more widely through the brain.

The classical signs of PD result from the degeneration of neurons in the basal ganglia in the brain, more specifically neurons in the substantia nigra which project to the corpus striatum (Figure 26.2). Because these neurons use the monoamine *dopamine* as a neurotransmitter, their degeneration results in loss of dopamine in the striatum. The major pathological feature of PD is the presence of large inclusions in the cytoplasm of degenerating neurons – so-called Lewy bodies – although their role in the degenerative process is unclear.

26.4.1 Causes of PD

Although the precise cause of neurodegeneration in PD is not known, a number of theories have been proposed. Currently the most popular proposals concern defects in energy metabolism in the mitochondria. The chemical *1-methyl-4-phenyl-1,2,3,6-tetrahydropyridine (MPTP)* is known to

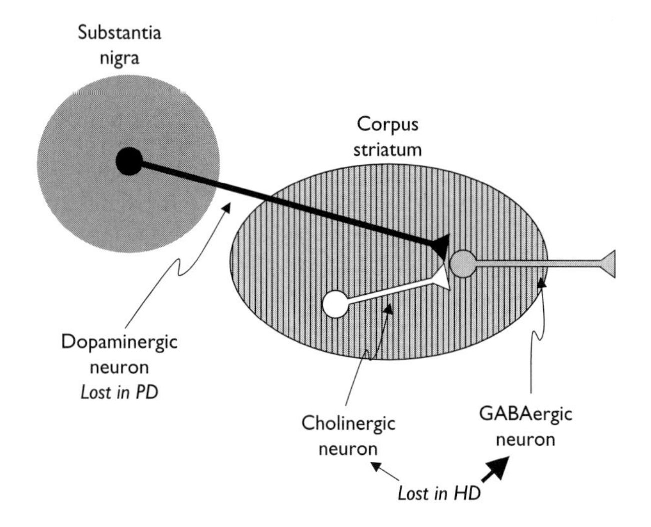

Figure 26.2 Simplified diagram of neuronal connections in the basal ganglia – a collection of nuclei in the brain that are involved in movement. Dopaminergic neurons projecting from the substantia nigra to the corpus striatum are lost in Parkinson's disease (PD). In contrast, neurons in the striatum itself degenerate in Huntington's disease (HD), principally inhibitory GABAergic neurons which project to the globus pallidus (external), as well as some cholinergic neurons within the striatum.

cause a parkinsonian syndrome in humans and other primates, and has been used to develop models of PD for use in research (in the early 1980s there were a series of reports of young drug addicts who had developed a severe form of PD ("frozen addict syndrome"). It turned out that these patients had inadvertently administered MPTP, which was a contaminant in an inexpertly synthesized batch of the opioid agonist meperidine). MPTP is converted to MPP^+ by the enzyme monoamine oxidase B (MAO-B); MPP^+ is taken up selectively by neurons in the substantia nigra, via an active transport mechanism that is normally responsible for the uptake of dopamine. MPP^+ inhibits complex I of the electron transport chain in mitochondria. Indeed, PD patients show a 40% decrease in the activity of complex I activity in the substantia nigra. The resultant dysfunction in energy metabolism kills neurons necrotically, or predisposes them towards apoptosis, by placing them in a state of oxidative stress. These data suggest that PD itself may result from exposure to a particular neurotoxin or neurotoxins, resulting in death (probably by apoptosis) of dopaminergic nigral neurons. The selectivity of such an, as yet, unidentified toxin could be related to the ability of the nigral neurons to concentrate or eliminate the toxin. Glutamate released from neurons in the pars compacta, which are known to be hyperactive in PD, may contribute a further excitotoxic component to PD.

26.4.2 *Current pharmacological treatment of PD*

The mainstay of treatment for PD is the drug *levodopa*. Levodopa is a *prodrug* which is converted to dopamine by the enzyme dopa decarboxylase (Figure 26.3). Once in the brain, dopamine is able to compensate for the lack of dopamine caused by degeneration of nigral neurons, either by facilitating the release of dopamine from remaining nigral neurons, or by flooding the dopaminergic synapses with dopamine. Levodopa is typically coadministered with an inhibitor of dopa decarboxylase which is unable to cross the blood brain barrier, either *carbidopa* or *benserazide*. By reducing the formation of dopamine in the periphery, such a combination reduces the dose of levodopa required by about 10-fold, as well as reducing peripheral side effects. Inhibitors of COMT, such as *entacapone*, are being investigated as a means of prolonging the presence of dopamine following levodopa administration by preventing dopamine metabolism by COMT in the CNS.

Two main side effects are observed with levodopa:

- dyskinesia (involuntary writhing), primarily in the face and limbs;
- on-off effects, where the PD symptoms of hypokinesia and rigidity may suddenly (and quite dramatically) appear for minutes to hours, then just as rapidly disappear. This can cause PD patients being treated with levodopa to "freeze" on the spot.

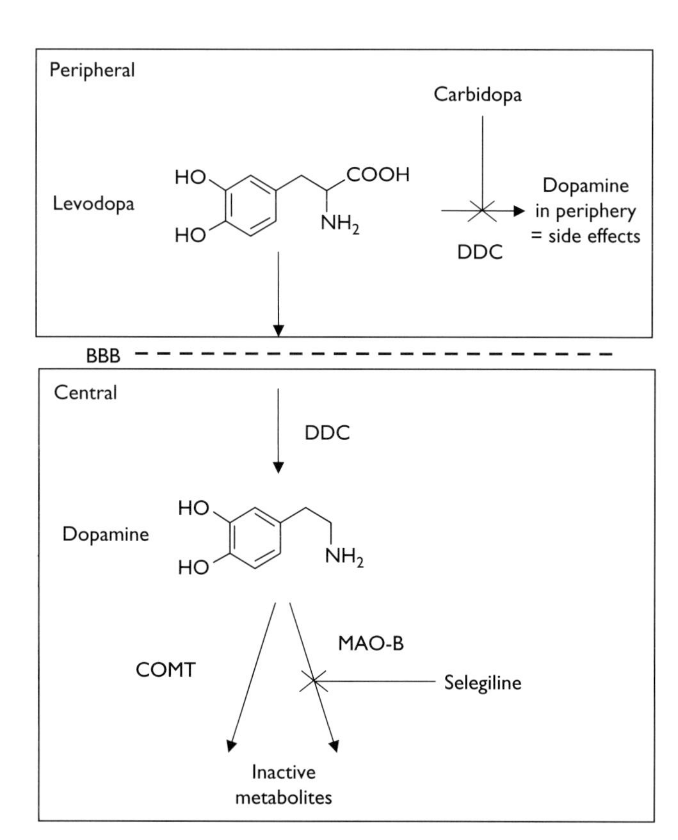

Figure 26.3 Rationale for use of levodopa as a prodrug in the treatment of the symptoms of Parkinson's disease (PD). Levodopa administered orally crosses the blood brain barrier (BBB) and enters the brain, where it is converted into dopamine by the enzyme dopa decarboxylase (DDC). Dopamine acts as a replacement for dopamine lost as a result of the degeneration of dopaminergic neurons in the substantia nigra (see Figure 26.2). The action of dopamine is terminated enzymatically by catechol-O-methyltransferase (COMT) and monoamine oxidase B (MAO-B).The use of a dopa decarboxylase inhibitor (e.g. carbidopa) which is unable to cross the BBB – limiting the formation of dopamine in the periphery – reduces the dose of levodopa required and peripheral side effects. Inhibitors of MAO-B such as selegeline are also useful clinically to treat PD.

Although levodopa is effective in many PD patients to begin with, its effectiveness generally reduces with time. This reduction is likely to be a consequence of the progression of the disease, which is not halted by levodopa.

Other drugs used to treat PD include:

- *selegeline*, an inhibitor of monoamine oxidase B, which probably acts by preventing dopamine breakdown;
- dopamine receptor agonists, such as the D_2 receptor agonist *bromocriptine*, which can act directly at dopaminergic receptors in the substantia nigra.

Unfortunately, none of the drugs used to treat PD are able to halt the progression of the disease. The need for drugs which are targeted towards preventing death of the nigral neurons, or are otherwise able to stem the neurodegenerative process in PD, is great. Such experimental approaches under investigation for PD and other neurodegenerative diseases will be discussed later in this chapter.

26.5 Alzheimer's disease

In 1907, the German psychiatrist Alois Alzheimer first described the pathological signs – *amyloid plaques* and *neurofibrillary tangles* – of the disease that now bears his name. Alzheimer's disease (AD) results from the degeneration of neurons in the *cerebral cortex* and *hippocampus* – areas of the brain involved in memory and higher intellectual function. Many of the neurons that are lost in AD use *acetylcholine (ACh)* as their major neurotransmitter. The death of these cholinergic neurons results in the primary clinical symptom of AD, namely dementia, or loss of intellectual ability.

26.5.1 Amyloid plaques

The so-called amyloid plaques observed in AD are extracellular aggregates of the protein *β-amyloid* (Aβ, see Figure 26.4). Aβ is produced by proteolytic cleavage of a longer membrane-traversing protein, *amyloid precursor protein (APP)*. APP is thought to possess neurotrophic actions. APP is normally cleaved predominantly by the enzyme *secretase α* to release soluble APP, and to a small extent by *secretases β* and *γ* to yield low levels of Aβ-40. However, mutations in APP lead to increased cleavage by secretases β and γ to give large amounts of a slightly larger form of Aβ, *Aβ-42*. Increased expression of *presenilin* genes further enhances the production of Aβ-42, possibly because presenilin gene products act as a cofactor for secretin γ. Aβ-42 readily aggregates into β-sheets, which then give rise to the amyloid plaques. The aggregation of Aβ-42 is enhanced by the expression of a variant form of *apolipoprotein E, ApoE4*. It is not yet understood how β-amyloid plaques might lead to neuronal death; there is evidence for both apoptosis, as well as inflammation indicative of necrosis. However, it is interesting that the APP gene is found on chromosome 21, the chromosome which is duplicated in *Down's syndrome* (Down's syndrome patients also develop dementia associated with increased expression of APP).

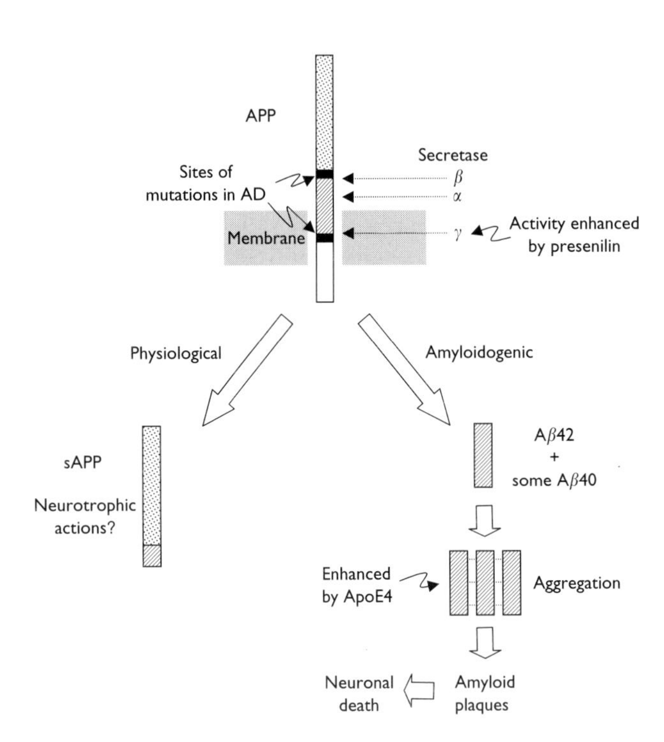

Figure 26.4 Mechanism of formation of amyloid plaques in patients with Alzheimer's disease. Amyloid precursor protein (APP) is a membrane bound protein. It is normally cleaved by the enzyme secretase α to a soluble form (sAPP), which might have neurotrophic actions. Mutations in APP seen in AD increase cleavage by secretases β and γ – whose activity is further enhanced by expression of the gene for presenilin – resulting in the β-amyloid peptide (Aβ, either 42 or 40 amino acid residues). In contrast to sAPP, Aβ42 strongly self-associates into β-sheets, the process being enhanced by apolipoprotein E4 (ApoE4). The sheets further aggregate to give the amyloid plaques which are a feature of AD, and which likely play a key role in the death of neurons.

26.5.2 Neurofibrillary tangles

Neurofibrillary tangles contain the protein *tau*. Tau is a normal component of the *microtubules* of neurons, probably involved in the process of *fast axonal transport*. In AD, tau is abnormally phosphorylated (possibly as a consequence of mutations in tau; amyloid may promote phosphorylation) which causes tau to associate into *paired helical filaments*. Following the death of a neuron, then filaments aggregate extracellularly to form neurofibrillary tangles. As with amyloid plaques, it is not known whether or not neurofibrillary tangles contribute to the death of neurons in AD, although the phosphorylation of tau may disrupt fast axonal transport.

26.5.3 Pharmacological treatment of AD

To date, pharmacological treatment of AD, like that of PD, consists of attempts to replenish then neurotransmitter lost as a consequence of the death of neurons. In contrast to PD, however, where such replacement therapy is able to dramatically improve the quality of life of patients, replacement therapy for the treatment of AD gives modest benefits, at best.

Two drugs are currently approved for clinical use:

- *tacrine*, an inhibitor of *acetylcholine esterase (AChE)*, the enzyme responsible for the breakdown and termination of action of ACh at the neuronal synapse. By inhibiting AChE, tacrine and related compounds would be expected to increase ACh levels. However tacrine also possesses other pharmacological actions, including *inhibition of potassium channels* and *monoamine uptake*, which might contribute to its reported ability to enhance cognition. Because of its short plasma half life (approximately two hours), tacrine needs to be taken four times a day, a problem in patients who may be experiencing short term memory loss. Also, tacrine possesses a number of side effects (e.g. liver toxicity) which further limit its clinical usefulness.

- *donezepil*, also an AChE inhibitor. The *pharmacokinetic profile* of donezepil (longer half life allowing once daily dosing, increased *oral bioavailability*), as well its reduced side effects may make it a more attractive drug than tacrine.

The observation of a reduced prevalence of AD amongst smokers has led to the hypothesis that nicotine, or nicotinic agonists have potential in the treatment of AD. Other potential treatments for AD will be discussed later in the chapter.

26.6 Huntington's disease

The actual triggers behind most neurodegenerative diseases remain elusive. In the case of *Huntington's Disease (HD)*, however, the trigger is known: HD is inherited, caused by a mutation in the gene *HD*, which codes for the protein *huntingtin*. The major signs of HD are *chorea* (jerky, involuntary movements) and dementia, resulting from extensive degeneration of neurons in the striatum and cortex (see Figure 26.2). Symptoms normally appear in adulthood, but patients then deteriorate rapidly before death.

The normal function of huntingtin is unknown, although it is expressed widely both within and outside the CNS. However, the *HD* gene has an *unstable CAG repeat* in the open reading frame of the first exon – CAG is the trinucleotide which codes for the amino acid *glutamine* (Figure 26.5). In normal individuals, the mean number of CAG repeats is 19. Thus, the huntingtin protein that is translated from the gene will typically have 19 Gln residues at its N-terminus. Patients who will develop HD generally have more than 40 CAG repeats (i.e. they will express a huntingtin protein with more than 40 N-terminal Gln residues). It appears that this abnormally elongated form of huntingtin is able to form aggregates with other proteins – including caspases involved in apoptosis – probably after being cleaved into fragments. The enhanced poly-Gln sequence found in HD appears to enhance or otherwise affect some of these protein-protein interactions (Figure 26.5).

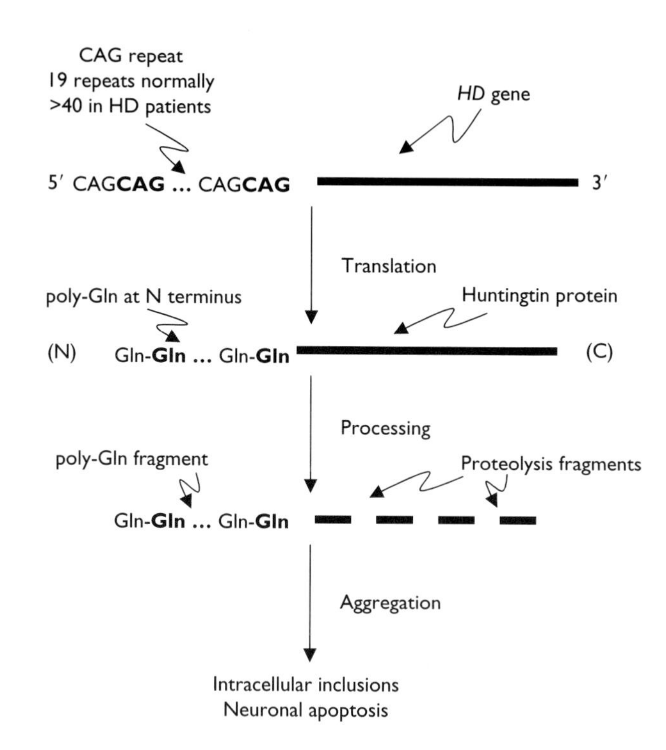

Figure 26.5 Schematic mechanism of neuronal death in patients with Huntington's disease. *HD* gene codes for a protein of unknown function, huntingtin. The 5′ end of the *HD* gene contains a CAG repeat sequence. In normal individuals, there is a median value of 19 CAG repeats; in patients with HD, the number of repeats is greater than 40. This gives rise to expression of a huntingtin protein with an increased number of Gln residues at the N-terminus. This elongated poly-Gln sequence, probably following proteolysis, interacts with a variety of other proteins, leading ultimately to the death of neurons.

26.6.1 Treatment of HD

The loss of GABA-containing neurons in the corpus striatum in patients with HD is believed to give rise to hyperactivity of dopaminergic neurons, projecting from the substantia nigra. Thus, HD is in a sense a mirror image of PD. The symptoms of HD can therefore be relieved to some extent by dopamine antagonists, such as *haloperidol* – dopamine agonists would be expected to exacerbate the symptoms of HD. One would also expect that agonists of GABA receptors, such as *baclofen*, might also be of therapeutic use in HD, analogous to the use of dopamine agonists in PD. However, their effectiveness in the clinic appears to be limited.

Because of the genetic nature of HD, it is now possible to carry out a simple test on an individual to see if they carry the increased number of CAG repeats characteristic of HD. The ethical considerations surrounding a patient being able to learn that they will develop a terminal disease for which there is no cure are enormous. The issues are compounded by the fact that HD might be passed on to children. For this reason, genetic testing for HD is carried out in conjunction with extensive counselling, typically over a period of months. It is interesting that currently only a small proportion of children of patients with HD, who are at considerable risk of developing the disease themselves, actually choose to undergo genetic testing.

26.7 Amyotrophic lateral sclerosis

Amyotrophic lateral sclerosis (ALS, also referred to as *motor neuron disease*) is a disease that results in the degeneration of *upper* (i.e. in the spinal cord) and *lower* (i.e. projecting from the spinal cord to skeletal muscle) *motor neurons*. Motor neurons are responsible for the movement of skeletal muscle. Thus patients with ALS progressively lose the ability to move, even though their intellect may remain intact (a striking example of the specificity of the loss of motor neurons and the sparing of other CNS neurons in ALS is the British physicist Stephen Hawking. Despite being confined to a wheelchair for many years, Hawking is regarded as one of the great physicists of the 20th century. It is curious that he is also one of the longest surviving ALS patients in Great Britain). The majority of cases of ALS (95%) are *sporadic* (i.e. they occur apparently spontaneously); the remainder are inherited (the *autosomal dominant familial* form of ALS). Compared to other neurodegenerative diseases such as PD, ALS disease follows a fairly rapid course, with death resulting from respiratory depression typically some 2–4 years after diagnosis.

26.7.1 Causes of ALS: the role of SOD I mutations

A variety of mechanisms have been proposed for the loss of motor neurons in ALS, including:

- excitotoxicity, as a consequence of abnormalities in glutamate metabolism or transport, or changes in Ca^{2+} binding proteins;
- increased free radical damage, including that resulting from impaired mitochondrial function.

Interest in the possible role of free radicals in ALS was stimulated in 1993, when it was reported that about one-fifth of patients with the inherited form of ALS showed mutations in the enzyme *copper/zinc superoxide dismutase I (SOD I)* gene. SOD I normally catalyzes the conversion of superoxide anion radicals to hydrogen peroxide, and thus plays an important protective role against free radical damage in cells. The mutations in SOD I clearly are critical to the pathogenesis of ALS: transgenic mice which express mutant forms of human SOD I develop a disease which resembles ALS.

It was first thought that the mutations in SOD I would lead to an impairment of catalytic activity of the enzyme. This has since been shown not to be the case, and it is more likely that the mutations in SOD I actually lead to the gain of a new enzymatic function. Proposals for a gain of function of SOD I include:

- participation of SOD I in new free radical-forming reactions, such as the production of destructive hydroxyl radicals from hydrogen peroxide;
- enhanced activity of SOD I, leading to the formation of peroxynitrite from superoxide anion radicals and the neurotransmitter nitric oxide. Peroxynitrite is capable of nitrating tyrosine residues in proteins. Neurofilament light chain, a component of the cytoskeleton of neurons, has a high proportion of tyrosine, and it has been suggested that the nitrosylation of neurofilament may play a role in the pathogenesis of ALS.

26.7.2 Treatment of ALS

Because of:

- its rapid progression;
- the fact that the disease is typically well advanced by the time of clinical diagnosis (around 50% of lower motor neurons will have already been lost by the time a diagnosis of ALS can be made);
- it is not amenable to neurotransmitter replacement therapy.

ALS has been and will probably continue to be, a particularly difficult disease to treat. The only drug to date to be approved for use is the agent *riluzole*. Riluzole is a glutamate antagonist (probably at NMDA receptors), and it can also inhibit the presynaptic release of glutamate. It is tempting to believe that the clinical use of riluzole strongly supports a mechanism for ALS involving glutamate-mediated cytotoxicity. However, the effects of riluzole in retarding the progression of ALS are, at best, very modest.

Several clinical trials have been carried out in patients with ALS using *recombinant neurotrophic factors*. Neurotrophic factors are proteins which promote neuronal survival. They apparently play an important role in regulating neuronal number during embryonic development, and have been shown to prevent neurodegeneration in a variety of experimental situations. *Ciliary neurotrophic factor (CNTF), brain-derived neurotrophic factor (BDNF)* and *insulin-like growth factor I (IGF-I)* have all been administered to ALS patients by daily *subcutaneous* injection in the hope of retarding or arresting the death of motor neurons. Of these factors, only IGF-I was shown to produce any clinical benefit, and was briefly approved for clinical use in the United States. However, it was withdrawn due to lack of efficacy following subsequent monitoring in the clinic. Given that neurotrophic factors are proteins, it is not difficult in hindsight to question the commonsense of administering them subcutaneously – motor neuron cell bodies reside in the CNS, and if the neurotrophic factors are to have any effect on motor neuron survival, they would need to cross the BBB to gain access to them. There is the possibility that they could be carried from the peripheral circulation down motor neuron axons by retrograde axonal transport, but the efficiency and capacity of this process in sick or dying neurons might be limited. Other approaches that might be used to more effectively harness the activity of neurotrophic factors for the treatment of neurodegenerative diseases will be discussed later in this chapter.

26.8 Transmissible spongiform encephalopathies

The emergence of *bovine spongiform encephalopathy* (*BSE*, so-called "mad cow disease") in beef cattle in Britain towards the end of the twentieth century popularized interest in what had been, until then, an obscure collection of diseases, the *transmissible spongiform encephalopathies. Creutzfeldt-Jakob disease (CJD)* is a form of this disease which affects humans, and variants of these diseases are found in many other species (e.g. scrapie in sheep). Most cases of CJD are *sporadic* (i.e. they occur at random without an apparent genetic cause), with the remainder being either familial or *iatrogenic* (i.e. resulting from medical treatment). The latter cases of CJD have occurred when patients have been treated with equipment used on, or tissue derived from, infected patients. However, there have recently emerged a small number of patients with *new variant CJD*. It now appears that these patients have contracted this disease as a result of eating infected beef.

The symptoms of CJD (and similar diseases in other species) typically include rapid, progressive dementia and loss of motor coordination. Death normally ensues within a matter of months. Pathologically, the characteristics of CJD are intracellular vacuoles (giving the brain the "spongy" appearance from which the disease gets its name), amyloid plaques (akin to those seen in AD), and widespread neurodegeneration. The brains of patients can also be stained histologically for a protein called *PrP^Sc^*, which is related to a protein normally present in the brain and other tissues, *PrP^C^*.

26.8.1 Prions

A variety of hypotheses have been proposed as to the nature of the infectious agent in CJD and other transmissible spongiform encephalopathies. However, it now appears that the causative species is a *prion*, a term coined by Stanley Prusiner, who won a Nobel Prize in 1997 for this work. Unlike viruses and bacteria, prions are simply proteins, and contain no genetic material. Normal prion proteins (PrP^C) are expressed widely throughout the body. Although their function is unknown, they appear to adopt a structure which is predominantly *α-helical* (Figure 26.6). Infectious prions (PrP^{Sc}) result from a change in conformation (i.e. a change in shape) of PrP^C, such that it unfolds and adopts a structure which consists mostly of *β-sheet*, with relatively little helix. Although the

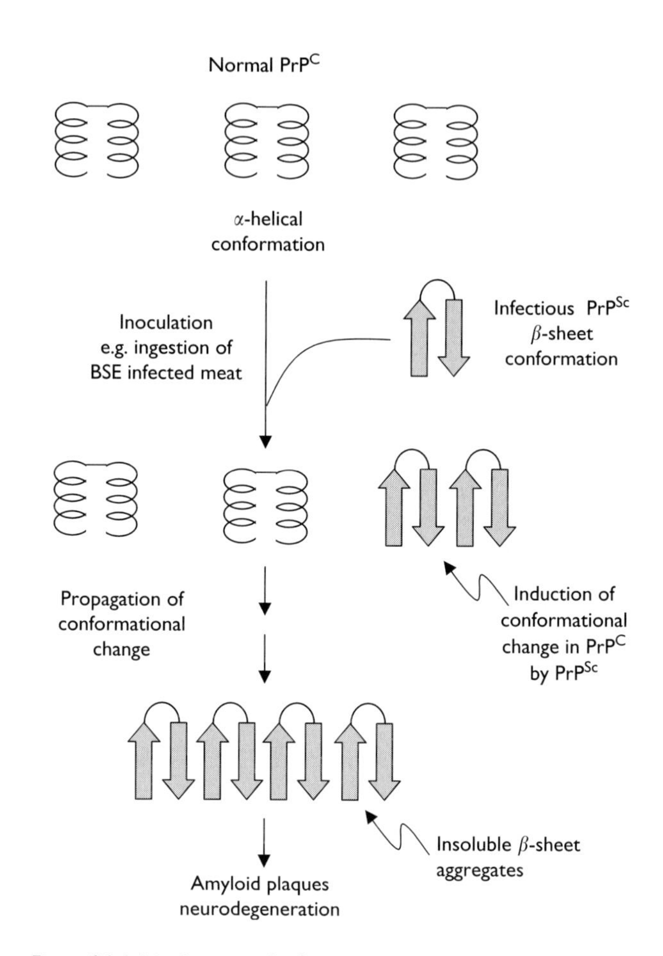

Figure 26.6 Mechanism of infectivity of prions. Normal prion protein (PrP^C) adopts a largely α-helical conformation; infectious prion protein (PrP^{Sc}) exists primarily in a β-sheet conformation. PrP^{Sc} – introduced either by inoculation with infected material (e.g. with BSE-infected meat in the case of new variant Creutzfeldt-Jakob disease), or as spontaneous production as a result of mutation (e.g. in familial Creutzfeldt-Jakob disease) – can initiate a conformation change in normal PrP^C, converting it into PrP^{Sc}. The conformational switch can be further propagated by newly formed PrP^{Sc}. The resultant PrP^{Sc} forms insoluble aggregates, leading to neuronal degeneration.

amino acid sequence of the prion protein is unchanged, the new conformation adopted by PrP^{Sc} is remarkably resistant to proteolysis (this feature of prions is believed to have contributed to the outbreak of BSE. Cattle were being fed a dietary supplement prepared from various carcasses, including sheep – a procedure that had been carried out for decades. The normal process to prepare the supplement included the use of high temperature and treatment with organic solvents to remove fat. In the early 1980s, the organic solvent step was stopped, and continuous heating replaced high temperature batch heating to reduce costs. As a result, it is likely that cattle were inoculated with PrP^{Sc} from sheep with scrapie; as a consequence of this, the cattle developed BSE. The process was further exacerbated by the fact that cattle were fed with supplements prepared from infected cattle. It is hoped that the ban on the feeding of ruminant-derived protein to ruminants in the late 1980s will have brought a halt to the BSE epidemic in Britain). Unlike "native" PrP^C, PrP^{Sc} is in a conformational state which allows it to aggregate. The insoluble collections of PrP^{Sc} form the amyloid deposits which are a feature of the prion diseases.

The change from normal, correctly folded PrP^C to the misfolded, and thus aggregating, amyloid-forming PrP^{Sc} is obviously critical to the development of a prion disease. This change can occur via either:

- the presence of mutations in the PrP gene which favor the formation of the misfolded PrP^{Sc}. Such an event is likely to be responsible for familial cases of CJD (and equivalent diseases in other species);
- infection with PrP^{Sc}, which is able to *induce* the recipient's PrP^C to change conformation to PrP^{Sc}. This could initiate a chain reaction in the recipient, whereby newly-formed PrP^{Sc} from the recipient is able to propagate the conformational change in still more PrP^C.

There is currently no treatment for CJD. Studying the disease has proved problematic, due to the very long *incubation time* for this and other prion diseases (typically in the order of years), and the fact that PrP^{Sc} from one species will not necessarily be infectious to another. The situation is further complicated by the fact that infectivity can only be acquired in some instances by passage through an intermediate species. Although CJD has a very low prevalence, it is difficult to predict the number of people who might develop new variant CJD as a result of eating BSE-infected beef in Britain and elsewhere in Europe. An estimated 170,000 cattle were infected over a ten year period. Although only a small number of patients have subsequently developed new variant CJD (<100), the incubation time in humans is not known. Depending on the efficiency of infection via the gastrointestinal tract with BSE-contaminated meat, the number of patients who will develop new variant CJD could lie anywhere between some hundreds of individuals, to tens of thousands. Nonetheless, the similarity of CJD to AD, both of which are characterized by plaques resulting from deposition of insoluble forms of normally soluble proteins, gives some hope that a common approach may be found to treat these diseases.

26.9 New approaches to treating neurodegenerative diseases

Neurodegenerative diseases have proven to be among the most difficult class of disorders to treat. The pharmacological therapies to date target symptoms only, and with the exception of PD, the effects are modest.

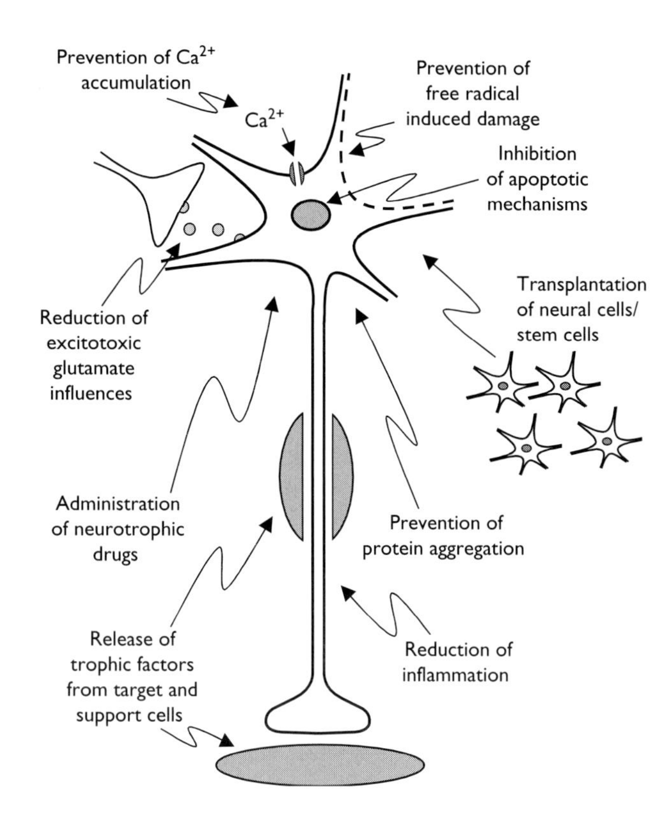

Figure 26.7 Potential targets for the development of novel means to treat neurodegenerative diseases.

As mechanisms behind neurodegenerative diseases have been elucidated, new targets for the development of novel therapeutic strategies have emerged (Figure 26.7). In most cases, these are targeted towards preventing the common feature of neurodegenerative diseases, neuronal death. A selection of strategies under investigation is discussed below.

26.9.1 Neurotrophic factors

Clinical trials of recombinant neurotrophic factors in patients with ALS have so far proved disappointing. This is likely to be largely attributable to the fact that the proteins are unable to cross the BBB and gain access to the CNS. Thus a number of strategies are being investigated, many at clinical trial stage, for getting neurotrophic molecules to their site of action on neurons, and thus effectively harnessing the ability of neurotrophic molecules to promote survival of degenerating neurons:

- *intrathecal administration* (i.e. into the cerebrospinal fluid), either by injection, pump, or cellular implants;
- discovery of *small molecule mimetics of neurotrophic factors*, either by targeted design or random screening. The small molecules would be better suited to crossing the BBB, but would retain neurotrophic actions;

- modification of neurotrophic factor proteins to *facilitate their passage* across the BBB;
- development of compounds which cause the *release* of neurotrophic factors.

Several recombinant neurotrophic factors are currently under investigation for the treatment of *peripheral sensory neuropathies*. These disorders, in which sensory neurons in the *peripheral* nervous system degenerate, can arise as a long term complication of diabetes, or as a side effect of treatment with some antitumour drugs. Because neurotrophic factors will not need to cross the BBB to reach sensory neurons, these trials may prove to be important proofs-of-principle of the use of neurotrophic factors to treat neurodegenerative disease.

26.9.2 Cell transplantation

Progress in the technology of organ transplantation has kept alive hopes that it may be possible to replace neurons lost in neurodegenerative disease. Aside from the potential problems of rejection, survival and correct reconnection of such transplanted neurons, an appropriate source of neurons would need to be found. While adult tissue might be suitable, it is thought that fetal neurons – being more plastic – might better integrate themselves into the brain. This raises obvious ethical issues regarding the use of human fetal tissue for experimentation and therapy. More recently, advances in *stem cell technology* have further sparked interest in this field of research. Stem cells are largely undifferentiated cells with the capacity to give rise to fully differentiated cell types. Scientists are now able to obtain differentiated cells in the laboratory – neurons, hepatocytes etc. – from such primordial cells, suggesting it may be possible to grow neurons *in vitro* suitable for transplantation into a given patient. Ependymal cells – which act as a physiological pool of neural stem cells in the CNS – might also be used as a source of cells for transplantation. It is, however, likely to be some time before the technical and ethical issues in neural cell transplantation can be overcome, and this approach to the treatment of neurodegenerative diseases becomes a reality.

26.9.3 Inhibition of protein aggregation

Several neurodegenerative diseases (e.g. AD, HD and CJD) are characterized by the presence of insoluble protein aggregates, which typically appear in the form of plaques. These aggregates probably play a causative role in the pathogenesis of these diseases. Strategies being examined that might prevent the formation of such aggregates include:

- development of secretase γ inhibitors for the treatment of AD. These compounds would be expected to inhibit the formation of amyloidogenic Aβ42. The pharmaceutical industry has an excellent record in the discovery of protease inhibitors (e.g. inhibitors of angiotensin converting enzyme and HIV protease). If secretase γ does play an important role in Aβ formation, one would expect the chances of obtaining a suitable inhibitor to be very good;
- discovery of compounds able to inhibit the formation of β-sheet structures from Aβ, which would thus prevent subsequent aggregation;
- design of inhibitors of phosphorylation of tau, which might prevent the formation of the neurofibrillary tangles seen in AD.

26.9.4 Inhibition of free radical-induced damage

Given that free radical mechanisms have been reported to play a role in many neurodegenerative diseases, it is tempting to speculate that interfering with these free radical mechanisms in some way might give rise to a general means of preventing or reducing neuronal damage and death. Trials to date with naturally occurring antioxidants such as *α-tocopherol (vitamin E)* have not been successful (α-tocopherol is, however, used clinically to prevent retinal damage that can occur in prematurely born babies as a result of free radical-mediated damage arising from oxygen supplementation). Second generation agents, perhaps with improved radical scavenging or pharmacokinetic properties, may prove to be more successful.

26.9.5 Inhibition of excitotoxicity/ Ca^{2+}-induced damage

Preventing glutamate-mediated excitotoxicity is a major focus for the development of drugs to treat many neurodegenerative diseases. There is a strong school of thought that excitotoxicity is a key process in neurodegeneration. Furthermore, there already exist inhibitors of glutamate release and receptor activation, as well as inhibitors of Ca^{2+} flux, which provide reasonable starting points for the development of new agents. A variety of compounds which target excitotoxicity at various sites are under laboratory and clinical investigation.

26.9.6 Inhibition of apoptosis

As apoptosis is apparently the important mechanism of cell death in neurodegenerative diseases, direct inhibition of this process is an attractive means of therapy. Targets for the prevention of apoptosis include:

- inhibitors of caspases, which play an important initiation role in apoptosis;
- modulators of the *Bcl-2/Bax* family of proteins. Bcl is anti-apoptotic, while Bax promotes apoptosis. Their

expression is regulated by neurotrophic factors, and it is likely, therefore, that they play a role in the survival effect of these factors. Another strategy would be to directly modulate the signaling initiated by these intracellular proteins, perhaps by affecting their dimer/multimer states.

26.9.7 *Antiinflammatory drugs*

Where necrosis, rather than apoptosis, is the means of neuronal death, antiinflammatory drugs may help to alleviate the inflammation and secondary scarring that would result. For example, *non-steroidal antiinflammatory drugs (NSAIDS)* have been reported to show some efficacy in the treatment of AD. The more recently developed inhibitors of *cyclooxygenase 2 (COX-2)*, which will probably show a reduced likelihood of gastrointestinal side effects than non-selective NSAIDs such as aspirin and indomethacin, are attractive agents for examination.

26.10 Conclusions

Neurodegenerative diseases still constitute an unmet medical need. It is hoped that the knowledge gleaned from basic science regarding the mechanisms of these diseases will soon begin to yield rational and effective therapies. With treatment directed against their cause, neurodegenerative diseases may one day cease to be progressive and terminal.

References

1 Dunnett, S.B., Kendall, A.L., Watts, C. and Torres, E.M. (1997) Neuronal cell transplantation in Parkinson's and Huntington's diseases. *British Medical Bulletin 53: 757–776.*
2 Gérard, E., Beyreuther, K., Masters, C.L. and Maloteaux, J.-M. (2000) Prospects for pharmacological intervention in Alzheimer disease. *Archives of Neurology 57: 454–459.*
3 Johnson, R.T. and Gibbs, C.J.Jr. (1998) Medical progress: Creutzfeldt-Jakob disease and related transmissible spongiform encephalopathies. *The New England Journal of Medicine 339: 1994–2004.*
4 Martin, J.B. (1999) Mechanism of disease: Molecular basis of the neurodegenerative diseases. *The New England Journal of Medicine 340: 1970–1980.*
5 Morrison, B. and Morrison, J. (1999) Amyotrophic lateral sclerosis with mutations in superoxide dismutase: a putative mechanism of degeneration. *Brain Research Reviews 29: 121–135.*
6 Skaper, S.D. and Walsh, F.S. (1998) Neurotrophic molecules: Strategies for designing effective therapeutic molecules in neurodegeneration. *Molecular and Cellular Neuroscience 12: 179–193.*

Chapter 27

Neurobiology of pain

Identification of novel targets for new pain management medications

Maree Smith

Contents

27.1 Introduction

27.1.1 Pain: what is it?

Pain is defined by the International Association for the Study of Pain (IASP) as "an unpleasant sensory and emotional experience associated with actual or potential tissue damage, or described in terms of such damage"[1]. Because pain is a personal experience incorporating a patient's emotional response to the nociceptive stimulus, there are no objective measures that allow the severity of one person's pain to be validly compared with that of another. Rather, the many scales that have been devised for quantifying pain and pain relief, are only valid for comparison of changes in the severity of pain within individual patients[2].

27.1.1.1 *Acute pain*

The IASP has defined acute pain as "pain of recent onset and probable limited duration. It usually has an identifiable temporal and causal relationship to injury or disease"[1]. Acute pain, e.g. that which occurs post-operatively or following trauma, is an adaptive process with a physiologically important role. Acute pain generally comprises two phases such that the first phase (lasting seconds) is a mechanism "alerting" the individual to potentially dangerous stimuli. The second subchronic phase (lasting hours to days) can be viewed as a "protective" mechanism characterized by "guarding" of the injured tissue, a process which promotes healing and recuperation.

27.1.1.2 *Chronic pain*

By contrast, chronic pain is defined as "pain lasting for long periods of time. It commonly persists beyond the time of healing of an injury and frequently there may not be any clearly identifiable cause"[1]. Prolonged chronic pain is thus a maladaptive process fulfilling no useful physiological purpose whereby the pain state itself has become the "disease" that requires treatment. Chronic pain may have multiple components including nociceptive pain, inflammatory pain and neuropathic pain which are variously defined as follows[3]:

- *Nociceptive pain*: refers to the normal, acute pain sensation evoked by activation of nociceptors located in non-damaged skin, viscera and other organs in the absence of sensitization;
- *Inflammatory pain*: hypersensitivity that arises in inflamed tissue following sensitization of peripheral nerve terminals;
- *Neuropathic pain*: pain due to a dysfunction of, or damage to, a nerve or group of nerves. Primarily peripheral nerves, although pain due to CNS damage ("central pain") may share these characteristics.

It is this latter type of pain which currently presents the greatest challenge to pain clinicians because drugs which are effective for the relief of moderate to severe nociceptive pain, such as opioids, are often much less effective for the relief of neuropathic pain, particularly when administered by the systemic route.

Although the IASP definition of pain recognizes its multi-dimensional nature incorporating a patient's emotional response, it is only recently that brain imaging techniques have begun to be utilized to investigate this latter aspect in humans[4]. By contrast, a huge research effort has focused on the neurobiology of pain, and so the material covered in this chapter is focused on the sensory-discriminative dimension of pain.

27.1.2 Pain signaling apparatus

27.1.2.1 *How is pain detected?*

In simple terms, pain signals from peripheral tissue exposed to damaging or potentially damaging stimuli are detected by specialized primary afferent nerve fibers (nociceptors) located in the affected tissue. These pain signals are transmitted via the nociceptive primary afferents to the outer layers of the dorsal aspect of the spinal cord (dorsal horn) from where they are relayed to higher centres in the brain which in turn activate the descending inhibitory pain-relieving mechanisms to reduce the severity of the perceived pain to tolerable levels (Figure 27.1)[5,6].

In the next section, the neuronal pathways for the transmission of nociceptive information from peripheral nerve terminals to the dorsal horn (DH) of the spinal cord and then to higher centres in the brain, are outlined. This conceptual framework is then utilized for a consideration of peripheral, spinal and supraspinal (brain) mechanisms involved in the induction of pain by stimulation of peripheral nociceptors, by peripheral nerve injury and by damage to the central nervous system (CNS) itself, highlighting the actual or potential targets for pharmacological intervention. It is only by having an improved understanding of the mechanisms

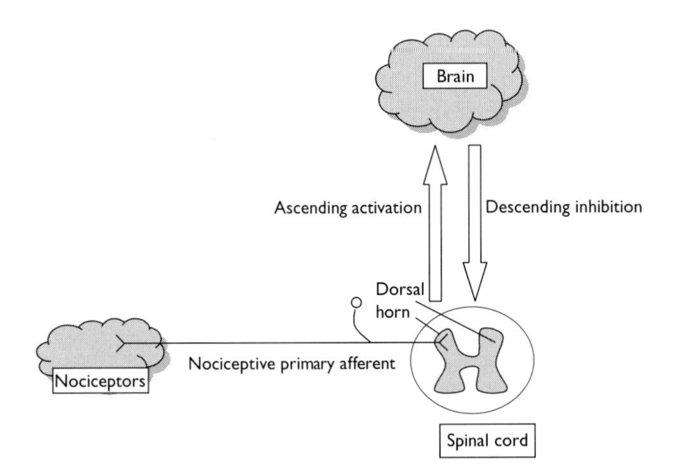

Figure 27.1 Pain signals are detected by specialized primary afferent nerve fibers (nociceptors) in peripheral tissues and then transmitted via nociceptive primary afferents to the dorsal horn in the spinal cord. The pain signals are then relayed to higher centres in the brain thereby activating the descending inhibitory system which serves to reduced the severity of the perceived pain[5,6].

responsible for producing and maintaining pain, that more effective treatments can be identified and/or developed.

27.1.3 Functional characteristics of the pain signaling system

27.1.3.1 Nociceptors

Nociceptors are functionally specialized primary afferent nerve fibers in the skin (cutaneous), muscle, joints, viscera and dura (Figure 27.1), that are responsible for detecting noxious (pain-producing) stimuli that damage or threaten to damage tissue[7]. Nociceptors respond to a broad range of physical (heat, cold, pressure) or chemical (acid, irritants, inflammatory mediators) stimuli but only at stimulus intensities capable of causing tissue damage[5,8]. The axons that relay nociceptive information from the nociceptors to the central nervous system (CNS) are either thin (0.4–1.2 μm), unmyelinated C-fibers with conduction velocities <2.5 m/s or small (2–6 μm) myelinated Aδ-fibers with a conduction velocity in the range 4–30 m/s. By contrast, sensitive cutaneous mechanoreceptors are often supplied by large (>10 μm), fast (30–100 m/s) myelinated Aβ fibers[9]. In the absence of tissue or nerve injury, cutaneous Aβ fibers are responsive only to touch, vibration, pressure and other modes of non-noxious, low intensity mechanical stimuli[9,10]. In the presence of chronic inflammation or nerve injury, Aβ fibers may undergo phenotypic change resulting in their becoming sensitive to noxious stimuli.

Generally, a single discharge of an individual nocisponsive fiber is not perceived as noxious and many nocisponsive units need to be recruited for "pain" to be experienced.

27.1.3.2 Pain characteristics

Activation of cutaneous Aδ nociceptors results in a sensation of pricking or sharp pain, whereas activation of cutaneous C-nociceptors causes burning or dull pain[5]. These are also termed "first" and "second" pain, because pricking pain occurs at a shorter onset latency than does burning pain[11]. By contrast, activation of nociceptors in muscle, bone and joints results in aching or cramping pain[12]. The generally diffuse nature of deep and visceral pain, notably from the gastrointestinal tract, and its frequent referral to other tissues, may be distinguished from the focal quality of cutaneous pain[5].

27.1.4 Treatment of pain in the clinical setting

In the last decade, studies of the neurobiology of pain have revealed that the pain signaling system is highly complex, involving considerably more than the modulation of a single class of ion channels or receptors to produce changes in the excitability of discrete populations of neurons[5,13,14]. Indeed, clinical experience has shown that the most effective analgesics tend to have a wide spectrum of action at a number of sites throughout the nervous system, confirming the view that multiple mechanisms underpin the development and maintenance of pain. However, it is also appreciated that for maximal effectiveness, pain therapy needs to be better targeted at the specific mechanisms underpinning the induction and maintenance of the various pain states, i.e. it needs to be more than symptom management. However, at the present time, rational prescribing on this basis is generally not possible because the underlying pain mechanisms are poorly or incompletely understood.

27.1.5 Major aims of the treatment of clinical pain

1. Normalize responsiveness, i.e. return the pain detection system from a state of hypo- or hypersensitivity to one where the response to defined low- or high-intensity stimuli is correctly perceived as an innocuous or painful sensation respectively[15].
2. Prevent sensitization in the face of stimuli that would otherwise cause peripheral or central sensitization[15].
3. Uncouple the response between a noxious stimulus (e.g. post-surgery or post acute trauma) and the painful sensation it normally elicits, i.e. provide analgesia by increasing inhibition[15].

27.1.5.1 Pharmacological management of pain: current treatment guidelines

The guidelines first developed by the World Health Organization (WHO) in the mid-1980s for the management of chronic cancer pain[16], are generally applicable to the alleviation of pain in both the acute and chronic settings. These guidelines are succinctly summarized by the WHO 3-step Analgesic Ladder (Figure 27.2). For mild pain (Step 1 of Analgesic Ladder), the recommended drugs of choice are the non-opioids such as paracetamol (acetaminophen), the non-steroidal inflammatory drugs (NSAIDs) including aspirin, and the more recently introduced selective COX-2 inhibitors (e.g. celecoxib, rofecoxib). These non-opioid drugs can be combined with one or more adjuvant drugs (tricyclic antidepressants, anticonvulsants, anti-arrhythmics, local anesthetics, etc.), as required.

For patients with moderate pain (Step 2 of the Analgesic Ladder), the WHO guidelines recommend the use of weak opioids such as codeine, dextropropoxyphene and tramadol which are often prescribed in combination with non-opioids and/or adjuvant medications. When moderate pain progresses to severe pain (Step 3 of the Analgesic Ladder), then strong opioids such as morphine, oxycodone, methadone, hydromorphone, fentanyl, etc., are warranted, with morphine being recommended as the drug of choice for the

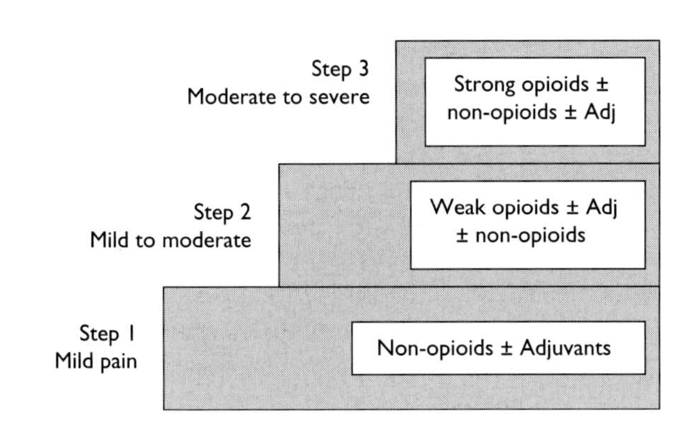

Figure 27.2 WHO Three-Step analgesic ladder[16].

management of moderate to severe cancer pain[16]. These opioid analgesics are often coadministered with a non-opioid analgesic and/or adjuvant medications.

27.1.5.2 *Why are new analgesic drugs needed?*

The currently available analgesics are associated with a plethora of side effects. Furthermore, some pain states, particularly neuropathic pain states, are poorly responsive to many of the above-mentioned analgesic medications. For example, nonsteroidal anti-inflammatory drugs are associated with gastric ulceration and renal impairment particularly in the elderly[17]. Administration of morphine-like drugs in the acute pain setting often results in nausea, vomiting, constipation, itch and sedation, with the ever present possibility of respiratory depression. Following chronic administration of opioid drugs, patients develop tolerance to their pain relieving effects. Although patients may also develop tolerance to some of the opioid-related side effects (sedation, nausea and vomiting), constipation remains prominent[18]. Moreover, following chronic administration, opioids do not provide adequate pain relief for many patients with chronic pain, particularly patients with neuropathic pain or whose nociceptive pain is complicated by a neuropathic component[19]. Although anticonvulsants, tricyclic anti depressants, anti-arrhythmics or local anesthetics may successfully relieve neuropathic pain, this is not guaranteed, and use of these drugs is often dose-limited due to the occurrence of CNS side effects and/or cardiotoxicity[20].

27.1.5.3 *Animal models of pain*

Experience over the last six decades has shown that activity in acute nociceptive tests in rodents (rats, mice) is a good predictor of efficacy for many classes of agents (opioids, α_2-agonists) for the relief of pain in the post-operative or post-trauma settings in humans[6]. Similarly, the efficacy of agents in neuropathic pain models in rodents, e.g. the chronic constriction injury of the sciatic nerve (CCI model) or tight ligation of the L5/L6 spinal nerves, appears to predict efficacy of several classes of agents (Na^+ channel blockers, gabapentin, NMDA antagonists, N-type calcium channel blockers) in human neuropathic pain states[6]. The success of preclinical studies in predicting the efficacy of many classes of agents in relieving multiple human pain states emphasizes the important role of preclinical studies in the development of new pain relieving medications[6].

More recently, preclinical studies in mice with deletions of specific genes, have been successfully utilized in studies of the neurobiology of pain to demonstrate that particular features of the response to acute and persistent injury conditions are differentially influenced by various elements of the pain signaling apparatus[21].

27.1.5.4 *Is it possible to develop better analgesic drugs?*

In the last decade there have been many exciting advances in our understanding of the neurobiology of pain, providing insight into some of the key mechanisms that underpin the development and maintenance of debilitating chronic inflammatory and neuropathic pain states[14]. Most importantly, these recent advances have revealed an unexpectedly high degree of complexity and plasticity in the pain signaling apparatus, characterized by the presence of multiple redundant mechanisms[5,13,14]. These parallel redundancies in the pain signaling system shed light on the considerable difficulty often experienced in the clinical setting in relieving chronic pain with traditional analgesic agents.

Importantly, the frustrations of the 1980s due to the lack of success in developing improved analgesic agents targeting components of the endogenous pain-relieving system (descending inhibitory system), have been replaced with considerable optimism that it will be possible to effectively use the insights gained from preclinical studies of the neurobiology of pain, to develop new pain medicines and/or drug delivery strategies that effectively target the key pathobiochemical mechanisms underpinning the development of difficult to manage persistent pain states such as those associated with nerve injury and chronic inflammation[5,21].

27.2 Neurochemical characteristics of the pain signaling system

27.2.1 *Transmission of nociceptive information from the periphery to the spinal cord*

In addition to their functional characterization, primary afferents may also be classified neurochemically on the basis of the chemicals synthesized in their cell bodies (in the dorsal root ganglia) and released from their terminals. These nociceptive neurotransmitters include glutamate (major excitatory amino acid (EAA) neurotransmitter in the mammalian

CNS) and aspartate, the tachykinin peptides, substance P (SP), neurokinin A (NKA), calcitonin gene related peptide (CGRP), ATP, nitric oxide (NO), prostaglandins (PG's) and neurotrophins (growth factors). Glutamate, SP, NKA and/or CGRP are often co-localized in a subset of capsaicin-sensitive, nocisponsive, C-fibers projecting to the dorsal horn from the skin and other tissues[5,9].

A substantial proportion of C-fibers are sensitive to the vanilloid, capsaicin, the pungent ingredient of hot chilli peppers, and express the VR1 (vanilloid-1) receptor to which capsaicin binds with high affinity[21]. Of the capsaicin-sensitive C-fibers, two major classes can be identified[9,22,23]. The first class contains CGRP and SP, and express the neurotrophin receptor, TrkA, as they are developmentally dependent upon the neurotrophin, NGF[5,21]. The second class do not contain peptides and are defined by the presence of the lectin, IB-4. These C-fibers express the $P2X_3$ subtype of the purinergic receptor[24] and a fluoride-resistant acid phosphotase (FRAP)[25] and are dependent upon glial cell derived nerve growth factor (GDNF)[5]. Each of these C-fiber sub-classes is thought to have a key role in nociceptive transmission in the dorsal horn of the spinal cord. Hence drugs that target the VR-1 or $P2X_3$ receptors represent novel targets for the development of new analgesic medications.

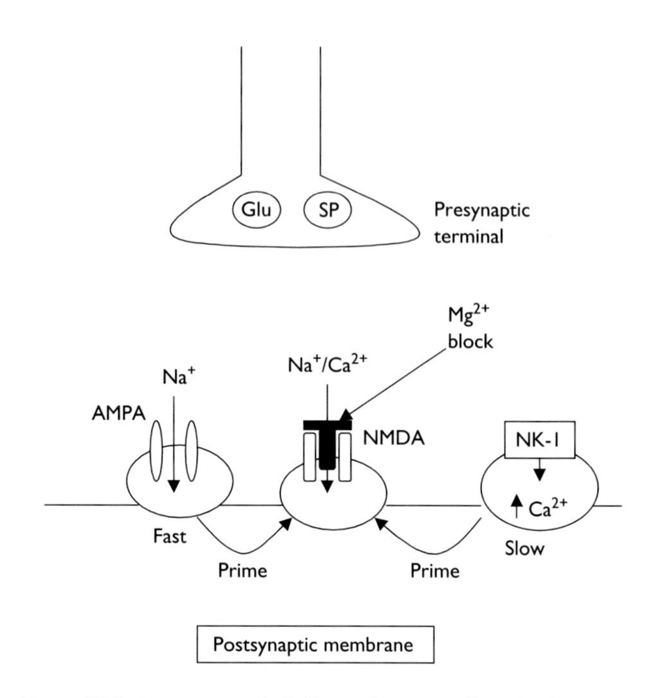

Figure 27.3 Activation of C-fiber afferents often leads to co-release of glutamate and SP. After release, glutamate binds to AMPA receptors causing an initial depolarization which removes the Mg^{2+} block of the NMDA receptor, thereby allowing influx of calcium into the cell. SP binds to NK-1 receptors to produce a long, slow depolarization which further removes the Mg^{2+} block of the NMDA receptor[6].

27.2.2 Nociceptive neurotransmitters and their target receptors

27.2.2.1 Excitatory amino acids and their receptors

As stated above, glutamate and aspartate are the principal EAA neurotransmitters in sensory neurons[27] with glutamate being found in at least 70% of cells in the dorsal root ganglia (DRG). In small diameter primary afferents, glutamate is often co-localized in the same nerve terminals as SP; however glutamate is contained within small open core vesicles and peptides within small dense core vesicles[26,27]. Glutamate acts at both NMDA and non-NMDA receptors. Metabotropic glutamate receptors (mGluRs) are coupled via G-proteins to soluble second messengers whereas ionotropic receptors (NMDA & AMPA) are directly coupled to cation-permeable ion channels (Figure 27.3).

These receptors show a complex pattern of localization on various neuronal classes in the dorsal horn[13,28,29]. Both NMDA and non-NMDA receptors mediate somatosensory inputs such that stimuli of brief duration primarily activate non-NMDA receptors, whereas prolonged stimuli lead to the activation of NMDA receptors[6]. Spinal neurons projecting supraspinally contain glutamate and/or glutamate synthesizing enzymes as well as a range of neuropeptides including SP, dynorphin 1–8, cholecystokinin (CCK), somatostatin (SOM) and vasoactive intestinal peptide (VIP)[6].

NMDA receptors are made up of heteromultimeric subunits across several, distinct classes. These are differentially distributed in the CNS, with the NMDA R1 component predominant in the dorsal horn. NMDA receptors are located in the superficial and deeper laminae of the dorsal horn on terminals of primary afferents as well as on membranes that are post-synaptic to the primary afferent[6]. All NMDA receptors display slow channel kinetics, a certain degree of voltage-dependent Mg^{2+} block and marked permeability to Ca^{2+} following removal of the Mg^{2+} block[13,30–32]. In addition to the glutamate recognition site (NMDA site), there is a modulatory site for glycine, the occupation of which is essential for functioning of the coupled ion channel[13,32,33]. In the dorsal horn, the concentration of glycine is sufficent to saturate these sites[5].

The NMDA receptor complex is not a participant in "normal" synaptic transmission[13]. Rather, spinal NMDA receptors have an important role in the facilitatory process known as "wind-up" whereby repetitive C-fiber (but not A-fiber) input results in a highly augmented response to subsequent C-fiber stimuli[34]. Confirming this, spinally administered NMDA antagonists have been shown to not only block "wind-up" in the spinal cord subsequent to repetitive input from small afferents following induction of tissue injury and inflammation, but to also block the development of thermal hyperalgesia and tactile allodynia[35–38]. Importantly, volatile general anesthetics, e.g. isoflurane, do not prevent this type of activity[13], indicating that pre-emptive analgesia should be considered in the surgical setting to prevent "wind-up".

AMPA receptors are present in the superficial layers (laminae I–II) of the spinal cord and are thought to be located on neurones post-synaptic to the primary afferents and on post-synaptic neurons in the dorsal horn[6]. In contrast to NMDA receptors, AMPA receptors have lower affinity for glutamate, have low voltage-dependence and rapid kinetics and they rapidly desensitize following their selective stimulation[5]. Activation of AMPA receptors leads to potent depolarization of dorsal horn neurons and activation of NMDA receptors resulting in calcium influx into neurons to initiate a cascade of biochemical events that alter responsivity of the cell to subsequent depolarizing stimuli[6]. Blockade of AMPA receptors attenuates synaptic activation of dorsal horn neurons by noxious and non-noxious stimuli[39]. However, because AMPA receptors have widespread roles in CNS function and they lack nociceptive selectivity, AMPA receptors are an unsuitable therapeutic target for the alleviation of pain[13].

Following calcium influx into neurons, NMDA receptors are phosphorylated by protein kinase C (PKC) which counteracts the Mg^{2+} block and allows NMDA receptors to operate under more hyperpolarized potentials[40]. The activity of PKC is synergistically facilitated by increases in intracellular calcium and the generation of diacylglycerol (DAG) following activation of phospholipase C (PLC) by neurokinin 1/2 (NK1/2) receptors and group 1 mGluRs[5].

Eight different metabotropic glutamate receptors (mGluRs) have been cloned and these are classified into 3 groups depending upon the second messenger system to which they are coupled. Group I (clones 1 & 5) mGluRs are positively coupled to PLC and possibly nitric oxide synthase (NOS) whereas both Group II (clones 2 & 3) and Group III (clones 4, 6–8) mGluRs are negatively coupled to adenylate cyclase. In the superficial layers of the dorsal horn, Group I mGluRs predominate and play an important role in the mediation of nociceptive input from C-fibers. Studies involving intrathecal dosing of rodents with mGluR antibodies have shown that mGluR's have a minimal role in acute pain processing, but may have a significant role in the development of hyperalgesia following nerve injury[41,42].

CO-CONTAINMENT OF NEUROTRANSMITTERS IN NERVE TERMINALS

Activation of C-fiber afferents often leads to co-release of glutamate and SP (Figure 27.3), together with several other neuropeptides. After release, glutamate binds to AMPA/KA receptors to evoke an initial depolarization which removes the Mg^{2+} block of the NMDA receptor, thereby allowing influx of calcium into the cell[43]. SP binds to NK-1 receptors and typically produces a long, slow depolarization which leads to local membrane depolarization and furthur removal of Mg^{2+} block at the NMDA receptor[44,45]. Additionally, SP may enhance phosphorylation of the NMDA receptor ionophore, thereby markedly increasing neuronal excitability. Preterminally located NMDA receptors on primary afferent neurons initiate a positive feedback loop driven by the effects of locally released glutamate which, in turn, produces a marked increase in neurotransmitter release (e.g. SP/CGRP) and an increase in excitation[46].

27.2.2.2 Neural plasticity and pain

After inflammation or nerve injury, nociceptors become sensitized, resulting in increased sensitivity and amplified responses so that pain can be produced by low-intensity or normally innocuous stimuli. Thus inflammatory pain occurs as a result of plasticity in the somatosensory system operating at multiple levels including the nociceptors themselves, the dorsal root ganglia (DRG), the dorsal horn (DH) of the spinal cord and supraspinally[47].

27.2.2.3 Peripheral sensitization

Following tissue damage secondary to inflammation, intracellular contents leak into the extracellular fluid, inflammatory cells are recruited and there is upregulated production and release of a broad range of neuroactive compounds (e.g. protons, serotonin (5HT), histamine, adenosine, bradykinin, prostaglandinE_2 (PGE_2), nitric oxide (NO), interleukin-1 (IL-1), tumour necrosis factor alpha (TNFα), interleukin-6 (IL-6), leukemia inhibitory factor (LIF), nerve growth factor (NGF)) by inflammatory and other cells[47]. Exposure of sensory neurons to the contents of this pro-inflammatory "soup" (Figure 27.4), induces alteration in the neurochemical content and function of these sensory neurons. The net result is that the somatosensory system becomes "sensitized" so that it no longer simply mediates normal nociceptive transmission, but innocuous stimuli are detected as painful (allodynia) or there is a heightened response to noxious stimuli (hyperalgesia)[5].

Sensitization of the somatosensory system is secondary to early post-translational changes both in the peripheral terminals of nociceptors which alter transduction sensitivity (peripheral sensitization), and in dorsal horn neurons secondary to activity in C-fibers (central sensitization). Both peripheral and central sensitization alter basal sensitivity to noxious and non-noxious stimuli[47,48]. Additionally, there are longer-lasting transcription-dependent changes in the DRG and in the DH that are due to a complex combination of activity and retrograde transport of specific signal molecules produced as a result of the inflammation[47]. These changes result in a potentiated nociceptive system, and one in which phenotypic changes in low-threshold Aβ-fiber inputs results in the development of tactile hypersensitivity (allodynia)[47]. Both the potentiation of the system and the phenotypic

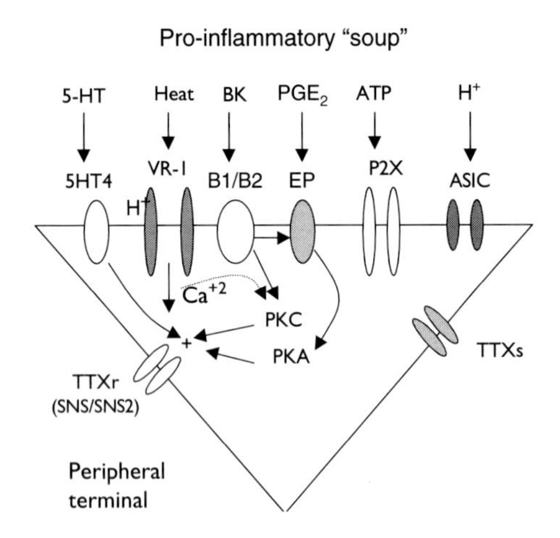

Figure 27.4 After inflammation or nerve injury, nociceptors become sensitized due to exposure to neuroactive compounds in the pro-inflammatory "soup" in the damaged peripheral tissue. These neuroactive compounds are produced by inflammatory and other cells and include protons (H^+), serotonin (5-HT), histamine, adenosine, bradykinin, prostaglandin E_2 (PGE_2), nitric oxide (NO), interleukin-1 (IL-1), tumour necrosis factor alpha (TNFα), interleukin-6 (IL-6), leukemia inhibitory factor (LIF) and nerve growth factor (NGF)[47]. Sensitization of nociceptors results in increased sensitivity and amplified responses so that pain may be produced by normally innocuous stimuli. (Adapted from Woolf and Costigan[47].)

changes manifest as a change in stimulus-evoked rather than basal hypersensitivity[47,49].

POST-TRANSLATIONAL CHANGES

Although nociceptors are defined by their normally high threshold for activation, inflammation or repeated noxious stimuli, can "sensitize" nociceptors such that lower intensity stimuli will initiate activity in the nociceptors. This peripheral sensitization, which can be detected within a very short period, is due to changes either in the transducing receptor molecules themselves or in the Na^+ channels in the nerve terminal[47]. A change in the transducer molecule is best exemplified by VR1, where repeated heat stimuli or exposure to protons progressively augments the inward current through the ion channel[50,51]. Additionally, phosphorylation of membrane-bound receptor/ion channels may occur as many inflammatory mediators activate either protein kinase A or C (Figure 27.4), both of which can phosphorylate receptors. Phosphorylation of the peripherally located tetrodotoxin (neurotoxin from the puffer fish)-resistant (TTXr) sodium channel, SNS/PN3, results in a greater sodium current in

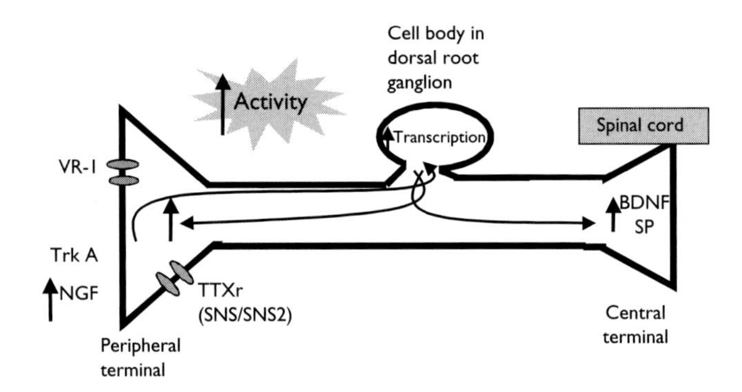

Figure 27.5 Peripheral sensitization occurs as a result of changes in transducer proteins, e.g. VR-1, and as a result of a PKC- and PKA-mediated phosphorylation of TTXr sodium channels. Transcriptional changes in the cell body increase production of transducer molecules (VR-1), ion channels (SNS/SNS2) and synaptic neuromodulators (brain derived neurotrophic factor (BDNF) and substance P (SP), thereby potentiating the system and altering neuronal phenotype[47,52–56]. (Adapted from Woolf and Costigan[47].)

the terminal[52–54]. These sensitizing changes occur locally in the peripheral nerve terminal, independent of transcriptional changes in the neuronal cell bodies in the DRG.

TRANSCRIPTIONAL CHANGES: EFFECTS ON PERIPHERAL SENSITIZATION

Inflammation produces an increase in peripheral levels of nerve growth factor (NGF)[10], a neurotrophin thought to have a key signaling role in inducing concomitant transcriptional changes such as upregulation of VR1-receptors and the Na^+ channels, PN3/SNS and NaN/SNS2 (Figure 27.5)[55,56]. Although peripheral sensitization does not itself require transcription, up-regulated synthesis of components of the pain signaling apparatus will amplify peripheral sensitization. However, after the initiation of inflammation, there will be a time delay of many hours for this up-regulated expression and transport of proteins to take place[47].

27.2.2.4 Central sensitization

Apart from peripheral sensitization which produces an area of increased sensitivity localized to the site of inflammation, any C-fiber input activated at the time of, or during, inflammation will also initiate central sensitization[47]. This input will result in an NMDA receptor-sensitive increase in responsiveness to low- and high-intensity stimuli, both when applied to the site of inflammation (primary hyperalgesia) and in the contiguous non-inflamed area (secondary hyperalgesia).[47] Tactile allodynia and pin-prick hyperalgesia in the zone of secondary hyperalgesia[57] are characteristic NMDA receptor-mediated features of central sensitization[58].

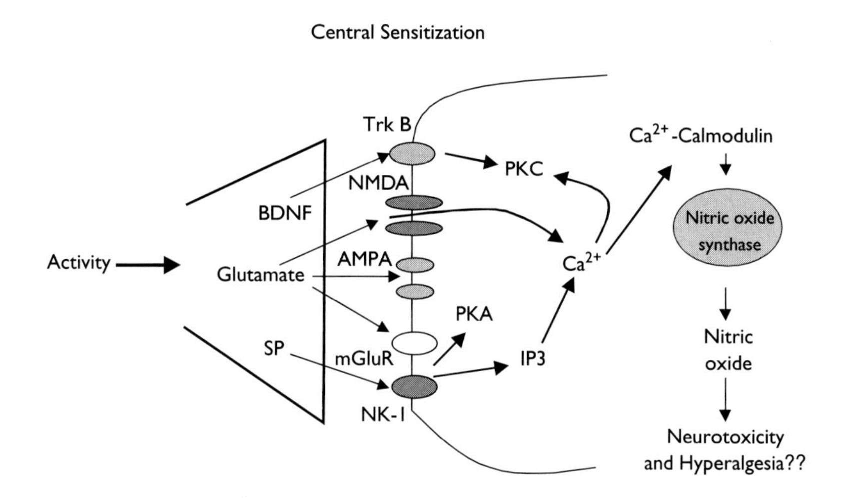

Figure 27.6 Central sensitization is characterized by an NMDA receptor-sensitive increase in responsiveness to low and high intensity stimuli, both when applied to the site of tissue damage (primary hyperalgesia) and in the contiguous non-injured area (secondary hyperalgesia)[47]. Following priming of the NMDA receptor complex, subsequent glutamate release results in activation of the NMDA receptor with calcium influx into the cell. Calcium then acts on a calmodulin-sensitive site to activate nitric oxide synthase (NOS) which converts L-arginine to nitric oxide. It has been proposed that excess nitric oxide production may have a role in hyperalgesia and neurotoxicity[17]. (Adapted from Woolf and Costigan and from Cousins and Power.[17])

Under conditions of peripheral nerve injury, the increase in dorsal horn excitability mediated via NMDA receptors may contribute to the diminished antinociceptive efficacy of μ-opioid agonists. Correspondingly, administration of NMDA antagonists, potentiates the pain-relieving effects of opioids in animal models[5]. However, attempts to replicate these findings in the clinic have been less than spectacular due to either low potency (dextrorphan and dextromethorphan) or prominent side effects (ketamine) of the clinically available NMDA receptor antagonists[17,59]. Attempts to develop more potent NMDA receptor antagonists by the pharmaceutical industry have generally failed due to the inability to separate the neurotoxic side effects from the desired actions[60].

Calcium influx into spinal neurons also activates phospholipase A_2 (PLA_2) to increase both the intracellular concentration of arachidonic acid and the formation of cycooxygenase (COX) and lipoxygenase products[61]. Spinal COX inhibitors suppress thermal hyperalgesia evoked by spinally administered SP or NMDA and the behavioral hyperalgesia resulting from peripheral tissue injury is consistent with the view that prostaglandins have a role in the regulation of spinal nociceptive processing[6]. Moreover, co-administration of opioids and NSAIDs produce antinociceptive synergy[17].

Spinal NMDA receptor-mediated hyperalgesia is blocked by intrathecally administered inhibitors of nitric oxide synthase (NOS), consistent with the proposed role of nitric oxide (NO) as a retrogradely acting messenger on presynaptic terminals[6]. Importantly, COX and NOS inhibitors together with NMDA, non-NMDA and NK-1 antagonists do not alter acutely evoked pain behaviors, consistent with their proposed role as anti-hyperalgesics (normalize pain behavior evoked by non-noxious stimuli) rather than analgesics (reduce response evoked by noxious stimuli)[6]. However, unless NOS inhibitors selective for neuronal NOS can be developed, NOS inhibitors are unsuitable as novel therapeutic agents for the alleviation of chronic pain, because they will also produce severe hypertension due to the loss of the vascular relaxing effects of NO[13].

Presynaptic inhibition of the release of glutamate, tachykinins (SP, NKA) and/or CGRP, or blockade of their postsynaptic actions offer the potential for the interruption of nociceptive transmission in the dorsal horn. However, insights from mice with deletion of the gene required for the synthesis of SP and NKA, have revealed that SP may be a less important pro-nociceptive neurotransmitter in persistent pain states than had been previously thought[21]. This may well explain the disappointing efficacy of SP antagonists in clinical trials[59].

27.2.2.5 *Inhibition of nociceptive neurotransmitter release: N-type calcium channel blockers*

N-type calcium channels modulate the release of the pronociceptive neurotransmitters, glutamate, SP and CGRP in both the central and peripheral nervous systems[5], raising the possibility that it may be possible to develop novel pain-relieving medications by blockade of N-type calcium channels. Indeed at present, the ω-conotoxins, AM336 (also known as CVID) and ziconotide (also known as MVIIA or SNX-111), are in clinical development[62,63] as these

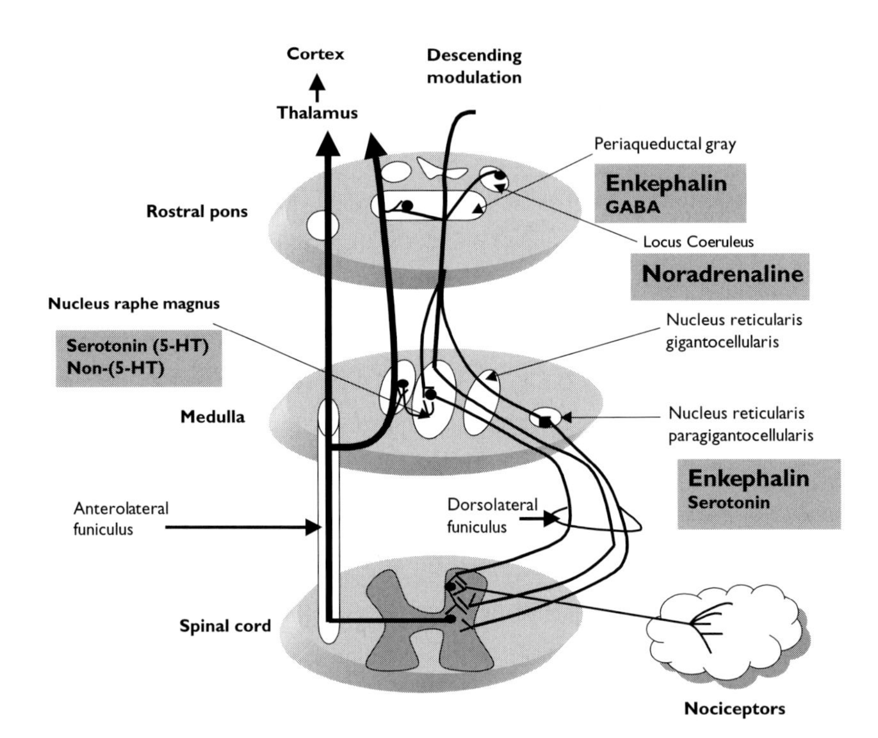

Figure 27.7 Schematic representation of ascending afferent sensory input (left) and descending inhibitory modulatory output (right). Following stimulation of nociceptors in peripheral tissue, an impulse is generated in the primary afferent together with an increase in levels of various endogenous pro-nociceptive agents (e.g. SP, prostaglandins, histamine, 5-HT, bradykinin). Primary afferent nociceptors relay to projection neurons in the dorsal horn, which ascend in the anterolateral funiculus to terminate in the thalamus. Additionally, collaterals of the projection neurons activate multiple higher centres, including the nucleus reticularis gigantocellularis (NRG), from where neurons project to the thalamus, the nucleus raphe magnus (NRM) and the periaqueductal gray (PAG) of the midbrain. Descending fibers from the PAG project to the NRM and reticular formation adjacent to the NRM to activate descending inhibitory neurons that travel via the dorsolateral funiculus to terminate in the dorsal horn of the spinal cord. Descending projections also arise from a number of brainstem sites including the llocus coeruleus (LC)[17]. (Adapted from Cousins and Power[17].)

compounds have been shown to inhibit neuronal N-type calcium channels[63], to inhibit SP release in the rat spinal cord[64] and to relieve pain in both inflammatory[64,65] and neuropathic pain models in rats[62,66].

More recently, multiple splice variants of both central and peripheral N-type calcium channels have been identified but possible differential physiological roles remain to be elucidated[63]. Intriguingly, electrophysiological evidence shows that AM336 and ziconotide have differential potencies for some of these N-type calcium channels subtypes expressed in *Xenopus* oocytes[63]. Moreover, in behavioral studies in rats with chronic FCA-induced inflammatory pain, AM336 given by continuous intrathecal infusion appeared to have a wider therapeutic window than ziconotide (our unpublished results). Whether similar differences between AM336 and ziconotide are found in patients with severe pain that is poorly responsive to other pharmacological interventions, remains to be confirmed. If the ω-conopeptides are found to be efficacious with acceptable levels of side effects in the clinical setting, then orally active, small molecule peptidomimetics that inhibit specific isoforms of N-type calcium channels may be of great benefit for the alleviation of chronic inflammatory and neuropathic pain states in patients.

Interestingly, the ability of the novel anticonvulsant, gabapentin, to reduce mechanical allodynia both in the clinic and in experimental models of neuropathic pain, has been proposed to involve inhibition of neuronal Ca^{2+} channels[5], but this remains to be confirmed experimentally.

TRANSCRIPTIONAL CHANGES: EFFECTS ON CENTRAL SENSITIZATION

One consequence of inflammation-induced transcriptional changes in DRG neurons is that some low-threshold Aβ neurons acquire the neurochemical phenotype typical of C-fibers i.e. they begin to express SP[67]. This change in neurochemical expression together with the inflammation-induced increase in NK-1 receptors in the DH[68] produce not only a potentiated system but one in which the specific type of stimulus that can evoke central sensitization has changed. Stimulus-induced hypersensitivity can thus be mediated by low-intensity Aβ inputs as well as high-intensity C-fiber inputs which manifests as progressive tactile allodynia where light touch produces a progressive increase in excitability of spinal neurons, something they never can elicit in the normal situation[67,69–72].

Although much still remains unknown, the new insights provide opportunities for translation into novel and more targeted approaches to treating inflammatory pain.

SUPRASPINAL CONTRIBUTIONS TO CENTRAL SENSITIZATION

Nociceptor sensitization and central sensitization are considered to underlie the development of primary and secondary hyperalgesia, respectively. Multiple studies confirm the importance of the spinal NMDA receptor in the development and maintenance of central sensitization. However, the significant contribution of descending influences from supraspinal (brain) sites in the development of secondary hyperalgesia must not be overlooked[73]. In animal models of primary and secondary inflammatory hyperalgesia, inactivation of the RVM (rostral ventromedial medulla) blocked the development of secondary but not primary hyperalgesia[73–75]. Additionally, tactile allodynia and thermal hyperalgesia after peripheral nerve injury appears to involve, at least in part, activation of descending facilitatory influences from supraspinal sites, including the RVM, which is important to the development and maintenance of primary but not secondary hyperalgesia[73].

DYSFUNCTION OF CENTRAL INHIBITIONS

It has been proposed that GABA up-regulation may be an intrinsic compensatory mechanism to block NMDA-mediated allodynia in chronic pain states and that failure of these and other inhibitory controls may contribute significantly to the development of pain chronicity[13].

Taken together, central sensitization develops at both spinal and supraspinal levels following inflammation and so optimal treatments may need to act at both spinal and supraspinal targets.

27.2.2.6 *SNS/PN3 Na^+ channels as potential drug targets*

TTXr Na^+ channels have a restricted distribution in the somatosensory system[76–80], being primarily present in capsaicin-sensitive small diameter nociceptive neurons. The first TTXr Na^+ channel cloned is variously referred to as PN3 or SNS[81,82] and is only present in a subpopulation of primary afferent neurons with small diameter cell bodies. A second TTXr Na^+ channel, called NaN[83] or SNS2[55], has also been cloned from sensory neurons and is similarly present in a subpopulation of sensory neurons with a small diameter cell body.

BIOPHYSICAL PROPERTIES

The biophysical properties of the first and most widely studied TTXr Na^+ channels (PN3/SNS) are consistent with their proposed critical role in modulating the excitability of the afferent neurons in which they are located. In particular, PN3/SNS Na^+ channels mediate currents that have high activation thresholds relative to tetrodotoxin sensitive (TTXs) currents[79,84,85] which may explain why nocisponsive afferent neurons have high thresholds for activation. Additionally, PN3/SNS Na^+ channels have high thresholds for steady-state inactivation relative to TTXs currents[79,84,85], consistent with the view that activation of TTXr Na^+ channels contributes to ongoing activity in the presence of a sustained depolarization of primary afferent neurons[85]. Furthermore, PN3/SNS currents recover rapidly after inactivation in contrast to TTXs currents[84,85] which enables TTXr currents to maintain sustained spiking in response to prolonged depolarizations[84,86]. Moreover, the much slower inactivation rates of TTXr Na^+ currents relative to TTXs currents, especially for membrane potentials close to the activation potential, implies that membrane depolarization may be facilitated after the activation of a sustained increase in current carried by TTXr Na^+ channels. The slow inactivation rate of TTXr Na^+ currents also contributes to the broad action potential typically observed in high threshold primary afferent neurons[86].

EFFECTS OF INFLAMMATORY MEDIATORS

Considerable electrophysiological, immunohistochemical and behavioral evidence indicates that alterations in the expression and function of the peripherally selective sodium channel, SNS/PN3, has a key role in the abnormal processing of pain that develops following tissue and peripheral nerve injury[48,87]. Behavioral studies have used selective PN3 oligodeoxynucleotides (ODN's) to "knock down" PN3 expression and this effectively prevented the development of thermal hyperalgesia in a chronic but not acute inflammatory pain model, implying that Na^+ channels do not play a role in normal nociceptive processing[48]. Rather, the PN3/SNS Na^+ channels appear to have a significant role in the development of persistent pain states. Thus development of selective inhibitors of the PN3 Na^+ channel are unlikely to alter normal nociceptive function, although this remains to be confirmed. The behavioral studies using PN3 antisense in rats were also notable for the lack of overt, CNS side effects, consistent with the discrete localization of PN3 to peripheral sensory nerve fibers with an absence of staining in the CNS and cardiac tissue[48]. It also reaffirms the potential that a selective inhibitor of PN3 may result in a clinically useful novel pain-relieving medicine, giving not only an improved therapeutic window over existing therapies, but offering relief from neuropathic pain that is often poorly responsive to current therapies[48].

Directly acting hyperalgesic inflammatory mediators, e.g. PGE_2, 5-HT and adenosine, reduce the activation threshold, increase rates of activation and inactivation and increase the magnitude of TTXr-Na^+ current [54], which could contribute to both the decrease in threshold and the increase in the number of action potentials evoked from a sensitized neuron[89]. The time course of inflammatory mediator-induced modulation of TTXr-Na^+ currents, developing within seconds and attaining a maximum within minutes[54], mirrors the time course for the development of primary hyperalgesia in response to a peripheral injection of directly acting inflammatory mediators[90–92].

SECOND-MESSENGER PATHWAYS

Recent evidence[52] supports the view that PGE_2-induced modulation of TTXr-Na^+ currents in primary afferent neurons involves activation of protein kinase A (PKA) resulting in phosphorylation of SNS/PN3 Na^+ channels[93]. Following phosphorylation, the gating properties of TTXr-Na^+ channels become similar to those induced by inflammatory mediators[93], consistent with the view that inflammatory mediator-induced modulation of TTXr currents reflects a direct phosphorylation of these channels.

27.2.2.7 Role of SNS2/NaN

Porrecca and colleagues[48] have shown that the second cloned TTXr Na^+ channel, NaN/SNS2, is not involved in either the establishment of nociceptive thresholds in control animals or in the maintenance of hyperalgesia and allodynia in a neuropathy model, thereby implying tht a selective inhibitor of NaN/SNS2 is unlikely to be successful for the alleviation of neuropathic pain in humans. However whether NaN/SNS2 Na^+ channels have a role in inflammatory hyperalgesia, remains to be clarified.

From the foregoing, it is clear that targeting of SNS/PN3 TTXr-Na^+ channels may lead to the development of a therapeutic modality for the treatment of hyperalgesia with fewer side effects than currently available modalities[89], due to the restricted distribution of TTXr-Na^+ channels and the observation that a decrease in their expression has little impact on low threshold mechanical transduction.

27.2.2.8 Vanilloid receptors: VR-1 as a potential drug target

As stated earlier, the capsaicin (vanilloid) receptor (VR-1) is expressed by primary afferents in the "pain" pathway. Electrophysiological experiments performed using cloned VR-1 receptors expressed in transfected mammalian cell lines showed that these receptors can be activated by vanilloid compounds, protons or heat (>43 °C), all of which evoke pain in humans or pain-related behaviors in animals[8]. Complementary behavioral studies in VR-1 null mutant mice showed that loss of the VR-1 receptor resulted in marked attenuation of responses to noxious thermal stimuli, indicative of the role of VR-1 in normal thermal nociception. However, the null mutant mice were not insensitive to noxious heat implying that thermal nociception involves multiple transduction mechanisms[8].

Following the induction of chronic inflammation, both wild-type and VR-1 null mutant mice developed mechanical allodynia to a similar extent whereas only the wild-type mice showed thermal hypersensitivity. However, VR-1 does not appear to have a role in sensitization after nerve injury, because both wild-type and null mutant mice exhibited comparable degrees of mechanical and thermal hypersensitivity. Thus, as VR-1 appears to be a critical mediator of the thermal hyperalgesia that occurs following tissue injury, it is a potential therapeutic target for the alleviation of pain following tissue injury[8].

27.2.2.9 Purinergic receptors

$P2X_3$ receptors are coupled to cation-permeable channels on peripheral, nocisponsive C-fiber terminals consistent with their proposed role in nociceptive neurotransmission[5].

27.2.2.10 SP and NKA: insights from null mutant mice

Studies in mice with a deletion of preprotachykinin-A (PPT-A), the gene that encodes SP and NKA[94], have shown that these PPT-A null mutant mice have responses similar to the wild-type mice for mild to moderate acute noxious stimuli, but markedly attenuated responses to high intensity chemical, thermal and mechanical stimuli[21]. These findings indicate that although SP/NKA contribute to the intensity coding of stimuli across modalities, the coding range is limited to one in which the stimulus is very intense. These findings are consistent with previous studies that showed internalization of NK-1 receptors (measure of tachykinin release) occurred only after application of an intense stimulus[95]. More importantly however, there were no significant differences between wild-type and PPT-A null mutant mice in the development of mechanical and thermal allodynia following the development of inflammatory or neuropathic pain states[21], indicating that glutamate has a more important role than SP in the development of persistent pain states. These findings help to explain the lack of success of SP antagonists in the clinic for the relief of chronic pain states[59].

Importantly, preclinical studies using NK2 antagonists which block the central effects of NKA but not SP[96] showed that the hyperexcitability of DH neurons evoked by inflammation was reduced. However, the NK2 receptor is not present in the DH suggesting that another receptor to which NKA binds but is sensitive to NK-2 antagonists, remains to be identified. If confirmed, this may be an important therapeutic target for the development of new analgesic drugs.

27.2.2.11 *Brain-derived neurotropic factor (BDNF)*

BDNF meets many of the criteria required to establish it as a neurotransmitter in small diameter nociceptive neurons[97]. In particular, BDNF is synthesized and packaged into dense core vesicles in nociceptive afferents that terminate primarily in the superficial dorsal horn. Importantly, the post-synaptic cells in this region express full-length trkB receptors (BDNF receptors). Both histological and electrophysiological evidence show that spinal neurons respond to exogenous BDNF to evoke post-synaptic excitability. Additionally, both electrophysiological and behavioral data show that antagonism of BDNF prevents (at least in part) some aspects of central sensitization, implying that BDNF is also necessary for the full expression of this phenomenon[97].

BDNF is constitutively expressed in some nociceptors and it may be important in mature animals for regulating the sensitivity of slowly adapting mechanoreceptors, myelinated fibers required for fine tactile discrimination[98]. BDNF rapidly and specifically induces phosphorylation of postsynaptic NMDA receptors[99] resulting in a 3-fold increase in NMDA receptor open time and a potentiation of NMDA responses. In the spinal cord, NMDA receptor activation plays a central role in the induction and maintenance of central sensitization such that the recruitment of the NMDA receptor is a pivotal event in the increased sensitivity of nociceptive spinal circuits to sensory inputs[100]. Thus it is possible that BDNF may be a key mediator of central sensitization within the spinal cord via an interaction at the NMDA receptor site, possibly via NMDA receptor phosphorylation[97].

In normal animals, the trkB receptor (BDNF receptor) is expressed in low density at the cell surface. However, following the initiation of peripheral inflammation, there is a marked up-regulation of BDNF in trkA-expressing nociceptors in an NGF-dependent fashion and it has also been proposed that inflammation induces translocation of the trkB receptor to the cell surface[97]. In support of this proposal, trkB-IgG (sequesters BDNF) blocked the increased spinal excitability evoked by BDNF in electrophysiological experiments, and reduced behavioral responses following induction of an inflammatory pain insult, indicating that BDNF is released from nociceptive afferent terminals and contributes to the post-synaptic responses in spinal cord neurons[97].

When combined with the rather modest phenotype of SP and NK1 in PPT-A null mutant mice[21,101], these findings suggest that BDNF is an important mediator of the changes that take place in the DH in chronic pain states, and antagonism of BDNF may provide a novel therapeutic target for the development of new agents for the relief of persistent pain.

PROTEIN KINASE Cγ AND PERSISTENT PAIN

Considerable evidence suggests that PKC also contributes to the long-term changes that underlie injury-associated allodynia and hyperalgesia[102–104]. As there are no selective antagonists for the various isoforms of PKC, recent studies have used mice with a deletion of individual PKC isoforms to gain insight into their respective roles in the development and maintenance of persistent pain states.

The distribution of PKCγ is restricted to a subpopulation of interneurones in the inner part of lamina II whereas other isoforms of PKC (e.g. α, $\beta 1$ and $\beta 2$) are found in the DRG, throughout laminae I and II of the dorsal horn, as well as more ventrally[21]. In a series of complementary behavioral, electrophysiological and immunohistochemical studies, PKCγ null mutant mice were found to process "acute" pain messages in a manner indistinguishable from the wild-type mice. By contrast, following peripheral nerve injury, the development of mechanical and thermal allodynia was markedly reduced in PKCγ null mice relative to wild-type mice[25]. Furthermore, nociceptive neurons in the dorsal horn (DH) of PKCγ null mice did not show enhanced responses to non-noxious stimuli unlike wild-type mice, implying that the DH is the locus of the defect. Additionally, there was a marked decrease in SP immunoreactivity and an increase in NK-1 receptor staining in the DH of wild-type mice following transection of the sciatic nerve and NPY immunoreactivity was up-regulated in the DH of wild-type mice with a partial constriction of the sciatic nerve. Importantly, all of these changes were markedly attenuated in the PKCγ null mice, indicative of either defective transmission of pain signals from the site of nerve injury to the DRG and spinal cord, or an altered response by DH neurons to the pain signals sent from the injured peripheral nerve[21]. Basbaum's group support the view that the defect is an abnormal processing of pain signals by PKCγ interneurons of the inner part of lamina II[21]. If future studies reveal that these changes can be prevented or reversed once established, then development of PKCγ selective inhibitors offers the possibility of developing yet another class of novel pain-relieving medications for the treatment of persistent pains in humans[21].

27.2.3 Descending inhibition: endogenous pain relief system

27.2.3.1 *Processing of low intensity stimuli*

The encoding of low threshold mechanical stimuli as innocuous appears to depend completely on the presence of a tonic activation of intrinsic glycine and/or GABAergic neurons within the spinal dorsal horn[5,6] and the presence of high levels of glycine and GABA binding in the dorsal horn[5,6]. Importantly, the GABA-containing terminals are frequently presynaptic to the large central afferent terminal complexes[6].

27.2.3.2 *Processing of high intensity stimuli*

Following activation of high threshold (C-afferent) but not low threshold (A-afferent) inputs, there is increased release of the inhibitory neurotransmitters, enkephalin, serotonin (5HT), noradrenaline (NA), acetylcholine (Ach) and adenosine at one or more levels of the central nervous system[6]. Although these molecules have only a modest effect in the ongoing modulation of nociceptive processing, they have potent effects on stress-induced pain behavior via activation of the endogenous monoamine and opioid receptor systems[105].

OPIOIDS

Three major classes of opioid receptors, μ (MOR), δ (DOR) and κ (KOR), were cloned in the 1990s and these opioid receptors are present at all levels of the neuraxis including multiple supraspinal sites, spinally and on peripheral nerve terminals[13,106]. There are four major classes of endogenous opioid peptides, viz the endomorphins, β-endorphin, the enkephalins and the dynorphins. Endormorphin-1 and endomorphin-2 have 1000-fold selectivity for μ-versus κ- and δ-opioid receptors. Met- and leu-enkephalin have a common precursor and completely overlapping CNS distributions and are released in proximity to both μ- and δ-opioid receptors[106]. The enkephalins have higher binding affinity for δ- than μ-opioid receptors and very low affinity for κ-opioid receptors. β-endorphin has approximately equal affinity for μ- and δ-opioid receptors[106] and dynorphin interacts preferentially with κ-opioid receptors[13].

Opioids interfere with nociceptive processing in the spinal cord and midbrain/brain stem via three major mechanisms[6,13,106], viz:

- Inhibition of pro-nociceptive neurotransmitter release from presynaptic nerve terminals;
- Post-synaptic activation to hyperpolarize output neurons and interneurons;
- Disinhibition in a circuit of two neurons.

Additionally, following the induction of inflammation, considerable immunohistochemical and behavioral evidence in animals shows that peripheral opioid actions are induced which suggests that development of peripherally selective opioids offers the potential for the development of opioid analgesics devoid of CNS side effects[107,108].

Following activation of opioid receptors by the endogenous opioid peptides, excitation evoked by small afferent input is suppressed in the spinal cord[6]. Opioids also act supraspinally to activate descending inhibitory pathways by enhancing the release of 5HT and NA in multiple brain areas[106]. In humans, intrathecal and epidural opioids evoke potent analgesia that is naloxone-reversible in a variety of acute post-operative and chronic pain states[6]. Following systemic morphine administration, morphine activates both spinal and supraspinal opioid receptors resulting in spinal-supraspinal synergistic pain relief[109].

Mu opioid agonists (e.g. morphine, fentanyl, methadone, etc) are the most effective analgesics, whether given systemically or by microinjection into various brain regions or the spinal cord[110]. The lack of effect of morphine in MOR-knockout mice, confirms the widely held view that morphine-like opioid analgesics produce their pain-relieving effects by interacting with μ-opioid receptors[6].

Intrathecally administered μ- or δ-opioid agonists, α_2-adrenoceptor agonists, and neuropeptide Y (NPY) evoke dose-dependent, naloxone-reversible antinociception in the absence of motor dysfunction and this has been shown across a wide range of species including rodents, primates and humans[6]. These effects are thought to be due to concurrent inhibition of nociceptive neurotransmitter release from afferent C-fibers as well as post-synaptic hyperpolarization of wide-dynamic range (WDR) neurons. Thus, even though opioids and α_2-agonists may also act in the ventral (motor) horn, these actions are insignificant at doses that produce profound inhibitory effects on nociceptive processing[6]. Unfortunately, the lack of progress in dissociating the sedative/hypotensive actions of α_2-agonists from their analgesic properties has hampered their clinical development[5].

HETERODIMERIC OPIOID RECEPTORS

More recently, functional heterodimeric opioid receptors comprising δ- and κ-opioid receptors or δ- and μ-opioid receptors, have been reported[111]. The ligand binding and functional properties (decreased cAMP levels) of the new heterodimeric opioid receptor were shown to be distinct from those of either δ- or κ-opioid receptors. Additionally, selective agonists bound cooperatively to the heterodimers and evoked synergistic functional responses (decrease in cAMP levels, potentiation of MAPK phosphorylation)[111]. Given that each of the cloned opioid receptors has multiple splice variants[114], it is interesting to speculate that multiple heterodimeric opioid receptors, comprising different combinations of individual splice variants of μ (MOR), δ (DOR) and κ (KOR) opioid receptors, remain to be characterized in the CNS. Whether the dozen or more different opioid receptor subtypes that have been pharmacologically defined over the last two decades are functional homo/hetero-oligomeric opioid receptor complexes, remains for future investigation. If future studies show that the desired effects of opioids to reduce nociceptive neurotransmission can be selectively activated by targeting specific homo/hetero-oligomeric opioid receptors without the production of CNS side effects, then development of such highly selective novel opioid analgesics would be a major advance.

ADENOSINE

NMDA receptor activation in brain and spinal cord results in an increase in extracellular adenosine concentrations[6], consistent with the proposed inhibitory role of adenosine in normal sensory processing. Intrathecally administered adenosine evokes potent antinociception mediated by adenosine A1 receptors, presumably by decreasing glutamate release[6]. However, in contrast to spinally administered opioid agonists, agonists at adenosine, $GABA_A$ or $GABA_B$ receptors produce only modest antinociception in doses that don't produce motor deficits[6]. This is thought to be because $GABA_B$ and adenosine agonists don't alter C-fiber transmitter release but only the firing of WDR neurons by increasing K^+ conductance, so that at concentrations sufficient to induce hyperpolarization of WDR neurons, there are concurrent direct effects within the motor horn, resulting in a narrow therapeutic index and lack of clinical utility[6].

CANNABINOIDS

In recent years, it has been shown that cannabinoids, including the endogenous cannabinoid anandamide, produce antinociception through multiple mechanisms at both spinal and supraspinal sites in the CNS via activation of CB_1 and/or CB_2 receptors[115]. In common with opioids, cannabinoids not only produce antinociception, but also produce a range of side effects including motor deficits, hypothermia, sedation, hypotension, inhibition of intestinal motility, tolerance and dependence. At present, the currently available cannabinoids are unsuitable for development of novel pain-relieving medications due to the occurrence of motor deficits throughout the antinociceptive dose range, i.e. the therapeutic window is very narrow[115]. Additionally, it has been proposed that cannabinoids may evoke peripheral pain relief via activation of CB_1 and/or CB_2 -like receptors in injured tissue, suggesting that it may be possible to develop novel peripherally active cannabinoid analgesics devoid of centrally mediated side effects, in a manner similar to peripherally selective opioid agonists[6].

NICOTINIC CHOLINERGIC AGONISTS

Microinjection of nicotinic agonists directly into the RVM evokes hotplate and tail flick antinociception in rodents as well as inhibition of pain behaviors in an inflammatory pain model[106]. The pain-modulatory actions of nicotinic agonists in the RVM depend on serotonergic neurons and the nAchR is located predominantly on serotonergic neurons in the RVM[116]. Th nicotinic agonist, metanicotine, has been reported to produce antinociception in mice following exposure to both acute (thermal, mechanical and chemical) and persistent inflammatory stimuli[117]. These findings suggest that metanicotine and related nicotinic agonists may have potential to be developed as a new class of analgesic agents.

ANTI-OPIOID PEPTIDES

CCK-8 Considerable evidence implicates CCK as an endogenous anti-opioid peptide because it has been shown to decrease the antinociceptive effects of morphine and to reverse morphine-induced inhibition of dorsal horn neurons[13,118]. Additionally, CCK_B agonists functionally antagonize the aninociceptive effects of endogenous μ- but not δ-opioid peptides in the RVM (rostral ventral medulla). In a rat model of neuropathic pain (peripheral nerve injury), synthesis of CCK was found to be up-regulated in primary afferent neurons[13] leading to a reduction in potency of intrathecal morphine. Importantly, CCK_B antagonists restored the opioid responsiveness in this pain model[13], suggesting that the CCK_B antagonists in development by the pharmaceutical industry may well enhance/restore morphine analgesia in a variety of chronic pain states[6].

Dynorphin Recently, spinal dynorphin has been implicated as a factor common to both the development of tolerance to the pain-relieving effects of morphine-like drugs as well as the development of pathological pain states[119–121]. For example, spinal administration of dynorphin anti-serum (but not control serum) completely blocked the expression of abnormal pain behaviour and the development of analgesic tolerance in rodents dosed chronically with morphine-like drugs (Vanderah *et al.*, 2001), effectively 'unmasking' the pain-relieving effects of the administered opioid, morphine. Clearly, identification of compounds that can block the unwanted excitatory actions of dynorphin, have the potential to enhance the usefulness of morphine-like drugs for the alleviation of pathological pain states, and to attenuate the development of analgesic tolerance.

27.3 Scope for improved usage of currently available pain-relieving medications

27.3.1 *Drug combinations: additive and synergistic pain relief*

Clearly, complex human clinical conditions, e.g. cancer or trauma-related pain states, involve the activation of multiple pain mechanisms such that efficacious interventions will almost certainly require polypharmacy[6]. The accumulating knowledge of the plasticity in the nociceptive signaling system provides a rational basis for combination therapy[13]. Combination of pain-relieving medications generally falls into two categories, viz:

1. Combination of opioids with other inhibitory agents such as the α_2-adrenoceptor agonists, e.g. spinal opioids (e.g. morphine, fentanyl) plus clonidine;
2. Dual therapy by combining opioids with non-opioids (e.g. CCK and NMDA receptor antagonists, NSAIDs, local anesthetics) in a bid to restore, add to, or potentiate opioid analgesic mechanisms[13].

More recently, combined administration of two opioids that produce their pain-relieving effects through distinctly different populations of opioid receptors, viz oxycodone and morphine, has been shown to evoke markedly synergistic pain relief in preclinical studies[122]. Whether this synergism translates into the clinical setting, remains to be clarified.

27.4 Summary

There is marked plasticity in the pain signaling apparatus following the induction of both nerve and tissue injury. Moreover, the high degree of parallel redundancy in the underlying pathobiochemical mechanisms makes it difficult to successfully alleviate persistent pain states with single therapeutic agents. Additionally, the currently available pain-relieving medications, whether administered alone or in combination, often lack efficacy for persistent pains, and are bedevilled by side effects. However, significant advances in our understanding of the neurobiology of pain over the last decade give cause for great optimism, as a number of potential targets for the development of novel pain-relieving medications have been indentified; including the N-type Ca^{2+} channel blockers, selective PN3/SNS Na^+ channel blockers, VR-1 antagonists, adenosine agonists, SP antagonists, NMDA antagonists, cannabinoid agonists, selective agonists for individual heterodimeric opioid receptors, $P2X_3$ purinergic antagonists, NK-2 antagonists, BDNF antagonists, PKCγ selective inhibitors, nicotinic cholinergic agonists, CCK_B antagonists, dynorphin antagonists, and combination analgesic medications.

References

1 Merskey, H. and Bogduk, N. (eds) (1994) Classification of chronic pain. Descriptions of chronic pain syndromes and definitions of pain terms. 2nd ed. IASP Press, Seattle, Washington.
2 Melzack, R. and Katz, J. (1999) in *Textbook of Pain* 4th ed. Wall, P.D. and Melzack, R. (eds) Churchill Livingstone, London, UK, pp 409–26.
3 Devor, M. and Seltzer, Z. (1999) in *Textbook of Pain* 4th ed. Wall, P.D. and Melzack, R. (eds) Churchill Livingstone, London, UK, pp 129–164.
4 Ingvar, M. and Hsieh, J.-C. (1999) in *Textbook of Pain* 4th ed. Wall, P.D. and Melzack, R. (eds) Churchill Livingstone, London, UK, pp 215–33.
5 Millan, M.J. (1999) Prog in Neurobiol 57: 1–164.
6 Yaksh TL (1999) in: Textbook of Pain Eds Wall PD and Melzack R, 4th ed. p 253–308.
7 Sherrington CS (1906) The integrative action of the nervous system. Scribner, New York.
8 Caterina MJ, Leffler A, Malmberg AM, Martin WH, Trafton J, Petersen-Zeitz KR, Koltzenburg M, Basbaum AI, Julius AI (2000) Science 288 (5464): 306–313.
9 Willis WD and Coggeshall RE (1991) Sensory Mechanisms of the Spinal Cord, Plenum, New York.
10 Woolf CJ, Safieh-Garabedian B, Ma Q-P, Crilly P, Winter J (1994) Neuroscience 62: 327–31.
11 Treede and Mageri W (1995) News Physiol Sci 10: 216–228.
12 Schaible H-G and Grubb BD (1993) Pain 55: 5–54.
13 Dickenson AH (1997) Behav Brain Sci 20: 392–403.
14 Dubner R and Gold M (1999) Proc Natl Acad Sci USA 96: 7627–30.
15 Doubell TP, Mannion RH, Woolf CJ (1999) in "Textbook of Pain" 4th ed. Wall PD and Melzack R (eds) Churchill Livingstone, London, UK, pp 165–181.
16 World Health Organization (1986) Cancer Pain Relief. WHO, Geneva.
17 Cousins M and Power I (1999) in "Textbook of Pain" 4th ed. Wall PD and Melzack R (eds) Churchill Livingstone, London, UK, pp 447–491.
18 Mather LE and Smith MT (1999) in "Opioids in Pain Control – Basic and Clinical Aspects", C. Stein (Ed.), Cambridge University Press, pp 188–211.
19 Mayer DJ, Mao J, Holt J, Price DD (1999) Proc Natl Acad Sci USA 96: 7731–7736.
20 Sindrup SH and Jensen TS (1999) Pain 83: 389–400.
21 Basbaum AI (1999) Proc Natl Acad Sci USA 96: 7739–7743.
22 Averill S, McMahon SB, Clary DO, Reichardt LF, Priestley JV (1995) Eur J Neurosci 7: 1484–1494.
23 McMahon SB, Bennett DLH, Michael GJ, Priestley JV (1997) in Neurotrophic Factors and Pain, eds Jensen TS, Turner JA, Wiesenfeld-Halin Z, IASP Press, Seattle, Washington, pp 353–379.
24 Vulchanova L, Riedl MS, Shuster SJ, Buell G, Surprenant A, North RA and Elde R (1997) Neuropharmacology 36, 1229–1242.
25 Silverman JD and Kruger L (1988) Somatonsens Res 5: 219–246.
26 De Biasi S and Rustioni A (1988) Proc Natl Acad Sci USA 85: 8720–7824.
27 Hokfelt T (1991) Neuron 7: 867–79.
28 Chaplan SR, Malmberg AB, Yaksh TL (1997) J Pharmacol Exp Ther 280: 829–838.
29 Coggeshall RE and Carlton SM (1997) Brain Res Rev 24: 28–66.
30 Conti F, Minelli A, De Biasi S, Melone M (1997) Molec Neurobiol 14: 1–18.
31 Sucher NJ, Awobuluyi M, Choi YB, Lipton SA (1996) Trends Pharmacol Sci 17: 348–355.
32 Hollman M and Heinemann S (1994) Ann Rev Neurosci 17: 31–108.
33 Leeson PD and Iversen LL (1994) J Med Chem 37: 4054–4067.
34 Woolf CJ and King AE (1987) J Neurophysiol 58: 460–79.
35 Coderre TJ and Melzack R (1992) J Neurosci 12: 3665–3670.
36 Dougherty PM and Willis WE (1992) J Neurosci 12: 883–94.
37 Ren K, Williams GM, Hylden JL, Ruda MA, Dubner R (1992) Eur J Pharmacol 219: 235–43.
38 Coderre TJ (1993) Mol Neurobiol 7: 329–46.
39 Dougherty PM, Palecek J, Paleckova V, Sorkin LS, Willis WD (1992) J Neurosci 12: 3025–41.
40 Yu X-M and Salter MW (1999) Proc Natl Acad Sci USA 96: 7697–7704.
41 Fundytus ME, Fisher K, Dray A, Henry JL, Coderre TJ (1998) Neuroreport 9: 731–35.

42 Schoepp DD, Jane DE, Monn JA (1999) Neuropharmacology 38: 1431–1476.
43 Mayer ML and Westbrook GL (1987) Prog Neurobiol 28: 197–276.
44 Duggan AW, Riley RC, Mark MA, MacMillan SJA, Schaible HG (1995) Neurosci 65: 849–58.
45 Marvizon JCG, Martinez V, Grady EF, Bunnett NW, Mayer EA (1997) J Neurosci 17: 8129–8136.
46 Liu H, Mantyh PW, Basbaum AI (1997) Nature 386: 721–24.
47 Woolf CJ and Costigan M (1999) Proc Natl Acad Sci USA 96: 7723–30.
48 Porreca F, Lai J, Biau D, Wegert S, Ossipov MH, Eglen RM, Kassotakis L, Novakovic S, Robert DK, Sangameswaran L, Hunter JC (1999) Proc Natl Acad Sci USA 96: 7640–44.
49 Neumann S, Doubell TP, Leslie TA, Woolf CJ (1996) Nature (London) 384: 360–64.
50 Caterina MJ, Schumacher MA, Tominaga M, Rosen TA, Levine JD, Julius D (1997) Nature (London) 389: 816–24.
51 Tominaga M, Caterina MJ, Malmberg AB, Rosen TA, Gilbert H, Skinner K, Raumann BE, Gold MS, Levine JD, Correa AM (1998) J Neurosci 18: 10345–55.
52 Gold MS, Levine JD, Correa AM (1998) J Neurosci 18: 10345–10355.
53 England S, Bevan S, Docherty RJ (1996) J Physiol (London) 495: 429–40.
54 Gold MS, Reichling DB, Schuster MJ, Levine JD (1996) Proc Natl Acad Sci USA 93: 1108–12.
55 Tate S, Benn s, Hick C, Trezise D, John V, Mannion RJ, Costigan M, Plumpton C, Grose D, Gladwell Z, Kendall G, Dale K, Bountra C, Woolf CJ (1998) Nat Neurosci 1: 653–55.
56 Okuse K, Chaplan SR, McMahon SB, Luo ZD, Calcutt NA, Scott BP, Akopian AN, Wood JN (1997) Mol Cell Biol 10: 196–207.
57 Koltzenburg M, Lundberg LER, Torebjork HE (1992) Pain 51: 207–20.
58 Stubhaug A, Breivik J, Eide PK, Kreunen M, Foss A (1997) Acta Anaesthesiol Scand 41: 1124–32.
59 Munglani R and Hill RG (1999) in "Textbook of Pain" 4th ed. Wall PD and Melzack R (eds) Churchill Livingstone, London, UK, pp 1233–50.
60 Fields HL, Baron R, Rowbotham MC (1999) in "Textbook of Pain" 4th ed. Wall PD and Melzack R (eds) Churchill Livingstone, London, UK, pp 1523–33.
61 Leslie JB and Watkins WD (1985) J Neurosurg 63: 659–668.
62 Bowersox S, Gadbois T, Singh T, Pettus M, Wang Y-X, Luther RR, J. Pharmacol. Exp. Ther., 279 (1996) 1243–1249.
63 Lewis RJ, Nielsen KJ, Craik DJ, Loughnan ML, Adams DA, Sharpe IA, Luchian T, Adams DJ, Bond T, Thomas L, Jones A, Matheson J-L, Drinkwater R, Andrews PR, Alewood PF (2000) J Biol Chem 275 35335–35344.
64 Smith MT, Cabot PJ, Ross FB, Robertson AD, Lewis RJ (2002) Pain 96: 119–27.
65 Malmberg AB and Yaksh TL (1995) Pain 60: 83–90.
66 Scott DA, Wright CE, Angus JA (1999) Proceedings of the Australasian Society for Clinical and Experimental Pharmacologists and Toxicologists 6: 35.
67 Neumann S, Doubell TP, Leslie T, Woolf CJ (1996) Nature 384: 360–364.
68 Krause JE, DiMaggio DA, McCarson KE (1995) Can J Physiol Pharmacol 73: 854–59.
69 Ma Q-P and Woolf CJ (1996) Pain 67: 97–106.
70 Ma Q-P and Woolf CJ (1997) NeuroReport 8, 807–10.
71 Ma Q-P and Woolf CJ (1997) Eur J Pharmacol 322: 165–71.
72 Ma Q-P, Allchorne AJ, Woolf CJ (1998) Pain 77: 49–57.
73 Urban MO, Zahn PK, Gebhart GF (1999) Neuroscience 90: 349–52.
74 Wiertelak EP, Furness LE, Horan R, Martinez J, Maier SF, Watkins, LR (1994) Brain Res 649: 19–26.
75 Wiertelak EP, Roemer B, Maier SF, Watkins LR (1997). Brain Res 748: 143–50.
76 Gold MS and Levine JD (1996) Neurosci Lett 212: 83–86.
77 Arbuckle JB and Docherty RJ (1995) Neurosci lett 185: 70–73.
78 Caffrey JM, Eng, DL, Black JA, Waxman SG, Kocsis JD (1992) Brain Res 592: 283–297.
79 Ogata N and Tatebayashi H (1993) J Physiol (London) 466: 9–37.
80 Roy ML and Narahashi T (1992) J Neurosci 12: 2104–11.
81 Akopian AN, Sivilotti L, Wood JN (1996) Nature (London) 379: 257–62.
82 Sangameswaran L, Delgado SG, Fish LM, Koch BD, Jakeman LB, Steward GR, Sze P, Hunter JC, Eglen RM, Herman RC (1996) J Biol Chem 271: 5953–56.
83 Dib-Hajj SD, Tyrrell L, Black JA, Waxman SG (1998) Proc Natl Acad Sci USA 95: 8963–68.
84 Elliott AA and Elliott JR (1993) J Physiol (London) 463: 39–56.
85 Rush AM, Brau ME, Elliott AA, Elliott JR (1998) J Physiol (London) 511: 771–89.
86 Schild JH and Kunze DL (1997) J Neurophysiol 78: 3198–3209.
87 Novakovic SD, Tzoumaka E, McGivern JG, Haraguchi M, Sangameswaran L, Gogas KR, Eglen RM, Hunter JC (1998) J Neurosci 18: 2174–87.
88 Chabal C, Jacobson L, Russell LC, Burchiel KJ (1992) Pain 49: 9–12.
89 Gold MS (1999) Proc Natl Acad Sci USA 96: 7645–49.
90 England S, Bevan SJ and Docherty RJ (1996) J Physiol 495: 429–440.
91 Taiwo YO, Goetzl EJ, Levine JD (1987) Brain Res 423: 333–37.
92 Cardenas CG, Del Mar LP, Cooper BY, Scroggs RS (1997) J Neurosci 17: 7181–89.
93 Fitzgerald EM, Okuse K, Wood JN, Dolphin AC, Moss SJ (1999) J Physiol (London) 516: 433–446.
94 Cao UQ, Mantyh PW, Carlson EJ, Gillespie A-M, Epstein CH, Basbaum AI (1998) Nature (London) 392: 390–94.
95 Abbadie C, Trafton J, Liu J, Mantyl PW, Basbaum AI (1997) J Neurosci 17: 8049–60.
96 Neugebauer V, Rumenapp P, Schaible HG (1996) Eur J Neurosci 8: 249–260.
97 Thompson SWN, Bennett DLH, Kerr BJ, Bradbury EJ, McMahon SB (1999) Proc Natl Acad Sci USA 96: 7714–18.
98 Carroll P, Lewin GR, Koltzenburg M, Toyka KV, Thoenen H (1998) Nat Neurosci 1, 42–46.
99 Levine ES, Crozier RA, Black IB, Plummer MR (1998) Proc Natl Acad Sci USA 95: 10235–38.

100 Urban L, Thompson SWN, Dray A (1994) Trends Neurosci 17: 432–38.
101 Woolf CJ, Mannion RJ, Neumann S (1998) Neuron 20: 1063–66.
102 Meller ST, Dykstra C, Gehbart GF (1996) Neuroscience 71: 327–35.
103 Mao J, Mayer DJ, Hayes RL, Price DD (1993) J Neurophysiol 70: 470–81.
104 Coderre TJ (1992) Neursci Lett 140: 181–84.
105 Bodnar RJ (1991) Neurobiology 6: 36–49.
106 Fields HL and Basbaum AI (1999) in "Textbook of Pain" 4th ed. Wall PD and Melzack R (eds) Churchill Livingstone, London, UK, pp 309–329.
107 Cabot PJ, Carter L, Gaiddon C, Zhang Q, Schafer M, Loeffler JP, Stein C (1997) J Clin Invest 100: 142–148.
108 Machelska H, Cabot PJ, Mousa SA, Zhang Q, Stein C (1999) Nat Med 4: 1425–1428.
109 Tao R and Auerbach SB (1995) Neuroscience 68: 553–561.
110 Yeung JC and Rudy TA (1980) J Pharmacol Exp Ther 215: 633–642.
111 Jordan BA and Devi LA (1999) Nature 399: 697–700.
112 Gomes I, Jordan BA, Gupta A, Trapaidze N, Nagy V, Devi LA (2000) J Neurosci 20: RC110, 1–5.
113 George SR, Fan T, Xie Z, Tse R, Tam V, Varghese G, O'Dowd BF (2000) J Biol Chem 275: 26128–35.
114 Pan YX, Xu J, Bolan E, Chang A, Mahurter L, Rossi G, Pasternak GW (2000) FEBS Lett 466: 337–40.
115 Manzanares J, Corchero J, Romero J, Fernandez-Ruiz JJ, Ramos JA, Fuentes JA (1999) Trends Pharmacol Sci 20: 287–94.
116 Notcutt W, Price M, Chapman G (1997) Pharm Sci 3: 551–55.
117 Bitner RS, Nikkel AI, Curzon P, Arneric SP, Bannon AW, Decker MW (1998) J Neurosci 18: 5426–5432.
118 Damaj MI, Glassco W, Aceto MD, Martin BR (1999) J Pharmacol Exp Ther 291: 390–8.
119 Malan TP, Ossipov MH, Gardell LR, Ibrahim M, Bian D, Lai J, Porreca F (2000) Pain 86: 185–94.
120 Wang Z, Gardell LR, Ossipov MH, Vanderah TW, Brennan MB, Hochgeschwender U, Hruby VJ, Malan TP Jr, Lai J, Porreca F (2001) J Neurosci 21: 1779–86.
121 Vanderah TW, Ossipov MH, Lai J, Malan TP Jr, Porreca F (2001) Pain 92: 5–9.
122 Ross FB, Wallis SC, Smith MT (2000) Pain 84: 421–428.

Chapter 28

Capsaicin receptors as target molecules on nociceptors for development of novel analgesic agents

János Szolcsányi

Contents

28.1 Introduction

It is a surprising fact that after the turn of the last century target molecules of analgesic drugs commonly used in medical practice are still the same as they were a hundred years ago. Opioids were used before history was recorded and aspirin (acetylsalicylic acid), phenacetin (acetophenetidin) and acetaminophen were introduced at the end of the nineteenth century. Using these compounds, analgesia was achieved by opioid receptor agonism and via inhibition the cyclooxygenase (COX) enzymes, respectively, although certainly these mechanisms were not rationalized for a long period of time. Both groups of target molecules are still the focus of drug development and efforts are in progress to diminish their pronounced side effects, e.g. by developing selective COX-2 inhibitors.

A substantial group of high threshold sensors, the A-delta mechanonociceptors and C-polymodal nociceptors, were described by Burgess and Perl (1967) and Bessou and Perl (1969) in single unit recordings from thin afferent fibers (see Kumazava, Kruger and Mizumara, 1996). Nociceptors are sensory receptors that provide essential information to the central nervous system for differentiation either between noxious and non-noxious stimuli or within the nociceptive or painful stimulation range. Chemical intervention to block the function of these nociceptive primary afferent neurons is a potential means of creating analgesic agents although in neuropathic pain, long term phenotype changes and the role of non-nociceptive low threshold afferents should not be neglected.

Well before the nociceptors were identified by single unit studies, a particular behavioral action of capsaicin, the pungent principle of hot peppers of the *Capsicum* family was discovered in Hungary by N. Jancsó, providing a pharmacological approach (see Szolcsányi, 1993). He observed that rats, mice and guinea-pigs pretreated with high doses of capsaicin became "desensitized" to the nociceptive actions of capsaicin and all types of chemical agents applied to the conjunctiva or other mucosal areas, while the observed protective reflexes to "pain elicited by pinching, pricking, heat or electrical current remained unaltered" (Jancsó, 1960, Szolcsányi, 1993). As his coworker in his last years, I followed this line with his wife, and we obtained further data about the remarkable neuroselective site of action of capsaicin. In our early studies – up to 1975 – the following data suggested that capsaicin might serve as a useful lead molecule for a new type of analgesic drug acting on nociceptors: 1) Capsaicin did not evoke nociceptive protective reflexes in non-mammalian species such as the frog, chicken or pigeon, indicating that irritation evoked by this agent is due to the presence of specific macromolecules in mammalian nociceptive afferents and it cannot be related to some general neurotoxic effect; 2) Local desensitization of the human tongue impaired warmth discrimination, enhanced the noxious heat threshold and the nociceptive threshold to capsaicin, piperine, zingerone and mustard oil, but not to menthol or taste stimuli. These psychophysical assessments indicated that non-nociceptive sensory nerve endings did not become insensitive to chemical stimuli by capsaicin pretreatment. Furthermore, after capsaicin desensitization, cold difference limen and tactile threshold remained unaltered; 3) In rats systemic capsaicin pretreatment caused selective ultrastructural impairment of the small, dark B-type neurons of the trigeminal, nodosal and dorsal root ganglia without causing changes in the large, light, A-type primary afferent neurons or cellular elements of the sympathetic ganglia; 4) Medicinal chemistry with several newly synthesized compounds shed light on some differences in pharmacophores of the capsaicin molecule responsible for the nociceptive agonist effect and on the sensory neuron blocking effect, providing evidence that "desensitization" is not due to an overstimulation of the nociceptors (Szolcsányi 1993, 1996).

These data led to the hypothesis for the existence of a "capsaicin receptor" expressed in mammalian nociceptive primary afferent neurons (Szolcsányi and Jancsó-Gábor, 1975). The neuroselective site of action of capsaicin was supported in further studies by showing that close arterial injection of capsaicin selectively excited the slowest conducting group of C-afferents in the cat saphenous nerve which comprises the axons of C-polymodal nociceptors. Furthermore, it has been proved that capsaicin does not interfere with the cholinergic, adrenergic, purinergic, peptidergic autonomic and intrinsic enteric neurotransmissions in the guinea-pig gut and airways *in vitro* (Szolcsányi 1993, Holzer 1991, Maggi 1995). Subsequently, capsaicin became the limelight of interest outside Hungary, particularly since Tom Jessell, Iversen and Cuello in 1978, and soon after Fred Lembeck's group in Graz and Masanori Otsuka's group in Japan clearly showed that capsaicin selectively releases and depletes substance P from the primary afferent neurons but not from other parts of the nervous system (Holzer 1991, Maggi 1995). Capsaicin in this way was, at that time, the first chemical agent which selectively depleted a sensory neuropeptide.

28.2 Capsaicin-sensitive sensory receptors

Classification of primary afferents with pharmacological action into "capsaicin sensitive" and "capsaicin-insensitive" groups was introduced more than twenty years ago (Szolcsányi 1993) to denote subgroups of chemosensitive afferents (mainly interoceptors) which were difficult to characterize by their adequate stimuli due to the lack of precise knowledge of their natural chemical ligands. Identification of the molecular site of action of capsaicin by cloning the capsaicin (vanilloid) receptor VR-1 from sensory neurons (Caterina *et al.*, 1997) and generation of VR1-null mutant mice (Caterina *et al.*, 2000, Davis *et al.*, 2000) provided an unequivocal molecular basis (see later) for this prediction. The neuroselective site of action of capsaicin among different types of primary afferent neurons was first demonstrated by electrophysiological techniques using sensory physiological criteria.

In the skin, a capsaicin-sensitive subgroup of afferents was identified by single unit recordings from axons of identified and classified sensory receptors (Szolcsányi 1993, 1996). These studies, made by several groups of researchers on cutaneous afferents of different species (rabbit, rat, cat, monkey and humans), clearly showed that capsaicin applied by close arterial or intradermal injections, topically onto the skin or to a skin-nerve preparation *in vitro* selectively excited the C- and Aδ-polymodal nociceptors and some less common types of C-afferents (heat nociceptors, mechano-heat insensitive chemonociceptors and warm receptors). No excitation of the high or low threshold C-mechanoreceptors, C- or Aδ-cold receptors, Aδ-high threshold mechanonociceptor and D-hair, G-hair or field types of rapidly adapting mechanoreceptors, as well as Type I and Type II slowly adapting mechanoreceptors were encountered (Szolcsányi 1993, 1996).

28.2.1 C-Polymodal nociceptors

These sensors are polymodal since they respond to noxious heat, moderate to high mechanical stimuli as well as to pain producing exogenous irritants or endogenous ligands such as bradykinin or protons (low pH), i.e. being integrators of several nociceptive stimuli. They are denoted also as high threshold mechanoheat sensitive C-afferents (CMH) when chemical stimulations are not applied or implication of their role in nociception is not intended to be involved in their name. This group of sensors includes 45–90% of the C-afferents in different species and nerves. Using an electrosearch method to pick up mechano-heat insensitive units in the human skin Törebjörk, Schmelz and Handwerker described 45% of the C-afferents as C-polymodal nociceptors (Belmonte and Cervero, 1996).

28.2.2 Other capsaicin-sensitive sensors

Single unit recordings from cutaneous C-afferents of humans and pigs revealed that C-heat nociceptors and mechano-heat insensitive C-fibers form further groups of high threshold sensors. A large proportion of the latter group proved to be sensory fibers on the ground that they become sensitized to mechanical or heat stimuli after capsaicin or mustard oil application. Although a systemic survey of the proportion of those afferents which are sensitive to capsaicin was not made, the majority of these units seem to belong to the capsaicin-sensitive population (Szolcsányi, 1996).

In rats, few units with polymodal type response characteristics conducted in the A-delta range and were similarly excited by capsaicin given intra-arterially as their C-conducting counterparts. Types of A-delta mechano-heat sensitive nociceptors (AMH) were characterized in detail by Campbell and Meyer. Type I of AMH has a higher threshold above 53 °C and Type II a lower one around 46 °C (Kumazawa *et al.*, 1996, Belmonte and Cervero, 1996). It is tempting to assume that the former group expresses the VRL-1 noxious heat-gated channel while the latter one the capsaicin-sensitive VR1 receptors (see later).

Capsaicin responsive interoceptors with slowly conducting C- or A-delta fibers have also been described. Pulmonary J-receptors, bronchial C-afferents activated by bradykinin or prostaglandins, R2 pelvic chemoceptive renal receptors and U2 type ureteric receptors were characterized as capsaicin-sensitive, being activated by low doses of capsaicin. In the knee joint and skeletal muscle, sensors of slowly conducting fibers forming less defined classes also responded to capsaicin (Szolcsányi, 1996); low threshold mechanoreceptors with fast conducting axons were not excited by the agent.

28.2.3 Desensitization of C-polymodal nociceptors to natural stimuli

The excitability of C-polymodal nociceptors was reduced or abolished after high intra-arterial doses of capsaicin in the rabbit or following systemic pretreatment of the rat. An enhancement of threshold to all types of stimuli (mechanical, noxious heat, capsaicin, bradykinin) was observed without afferent selectivity. Nevertheless, desensitization of a particular unit to one kind of stimulation often was not paralleled by similar changes in responsiveness to other stimuli (Szolcsányi 1993). Topical application of capsaicin to the human skin or blister base or to the eye of animals elicits local desensitization which is characterized in humans by a pronounced increase in noxious heat threshold and impaired magnitude estimation and enhanced difference limen in warmth sensation besides unresponsiveness to pain-producing chemicals as bradykinin or mustard oil. Ultrastructural changes with swollen mitochondria and reduced number of clear vesicles were described in the sensory nerve terminals of the rat cornea (Szolcsányi, 1993).

28.2.4 Long-lasting selective sensory blocking and neurotoxic effects

In rodents a long-term enhancement of noxious heat threshold and a marked loss of responsiveness to various irritant nociceptive chemicals were described after periaxonal, intrathecal application or after systemic pretreatment in the adult and neonate. Central and peripheral terminals of C-afferents degenerate in these animals and destruction of dermal and ureteric terminal axons were clearly documented. The proportion of responsive C-polymodal nociceptors is selectively diminished, and the remaining receptors respond with higher threshold to stimulation (Szolcsányi 1993, 1996). In contrast, treatment of neonatal rats with 50–80 mg/kg subcutaneous (s.c.) capsaicin causes an irreversible chemoanalgesia which is due to substantial loss of B-type and some A-type neurons in the dorsal root ganglia accompanied by a marked indiscriminate loss of C-afferents. It had been claimed that this neurotoxic effect of capsaicin is due to direct action of the agent on chemosensitive neurons, and acute cell death induced by capsaicin takes place within an hour. Recent data in our laboratories (Szolcsányi *et al.*, 1998, Szállási and Blumberg, 1999) however, indicated that after neonatal treatment the loss of neurons develops slowly, i.e. after a week. Furthermore, this process was prevented by treating the animals with NGF starting its s.c. injections one day after capsaicin was given. Thus, it has been concluded that in the neonatal rat, as in the adult, high toxic doses of capsaicin destroy the nerve terminals of the C-polymodal nociceptive fibers and other capsaicin-sensitive afferents. In this unmatured age, however, unlike in adult rats, cell body of B-type neurons of the dorsal root ganglia could not survive the impaired uptake of NGF. In these animals the lack of selectivity in the loss of C-polymodal nociceptive units is probably due to secondary phenotype changes within the population of the remaining sensory neuron population (for further details see Szolcsányi, 1993, 1996 and Szállási and Blumberg, 1999).

It is important to note that after the neurotoxic capsaicin pretreatment, both in the adult rodents and in the neonatal rat treated with NGF, swollen mitochondria with disorganized cristae persist for several weeks in B-type sensory neurons. A-type sensory neurons or neural elements in the sympathetic ganglia showed no ultrastructural changes. The mechanism of this extremely long-lasting or irreversible mitochondrial damage is unclear. In the rat, similar morphological signs in one type of neurons was observed in the preoptic area where central thermodetectors were described (Szolcsányi, 1993, 1996; Holzer, 1991; Szállási and Blumberg, 1999).

28.3 Resiniferatoxin, capsazepine and ruthenium red: new pharmacological tools

28.3.1 Resiniferatoxin

Resiniferatoxin (RTX), the irritant isolated from the latex of a Moroccan cactus-like plant is a resiniferonol ester of *Euphorbia resinifera*. It was introduced into sensory pharmacology by Peter Blumberg and Árpád Szállási. RTX and capsaicin share a common structural moiety of 3-methoxy-4-hydroxybenzyl group, with capsaicin an amide of vanillylamine and RTX an ester of homovanillic acid (Figure 28.1).

RTX had several additional advantages as a lead molecule over capsaicin. In a series of experiments beginning in 1989 the above authors and their associates provided quantitative data showing that RTX is an "ultrapotent" capsaicin analog. The extremely high affinity of RTX was utilized to elaborate a technique to demonstrate specific [^{3}H] RTX binding to rat DRG membranes. Both RTX and capsaicin are highly lipophilic, prone to nonspecific binding which was overcome by the potent RTX but not by capsaicin. Subsequently displacement of RTX binding was used to reveal the binding characteristics of capsaicin to its receptors. Mismatch of biological effects and binding characteristics between RTX and capsaicin was striking. Capsaicin inhibited [^{3}H] RTX binding to DRG neurons by 100,000 times higher molar concentration than RTX. On the other hand, the potency of RTX is only 300–500 times higher than that of capsaicin in respect of inducing $^{45}Ca^{++}$ uptake or single channel activation in DRG neurons (for more details see Szállási and Blumberg, 1999).

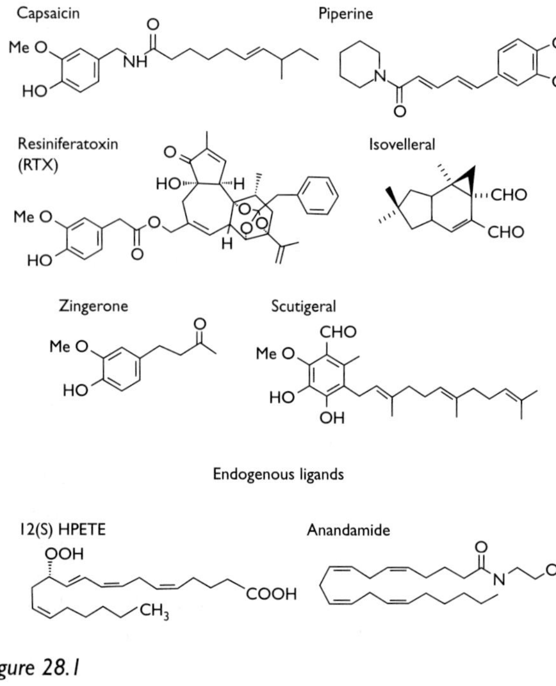

Figure 28.1

28.3.2 Capsazepine and ruthenium red as capsaicin receptor antagonists

Capsazepine and ruthenium red antagonize the effect of capsaicin and resiniferatoxin on the capsaicin receptor. Capsazepine is the first commercially available competitive capsaicin antagonist at the VR1 receptor with moderate potency and selectivity (IC_{50} 100–700 nM in DRG sensory neurons). Capsazepine at and above 1 μM applied in the absence of other drugs inhibited calcium currents (EC_{50} at equilibrium; 1.4 μM or 7.7 μM after 6 min exposure) and also inhibited the function of the nicotinic channels (Docherty *et al.*, 1997; Liu and Simon, 1997; Szállási and Blumberg, 1999).

Ruthenium red, $[(NH_3)_5\,Ru\text{-}O\text{-}Ru\,(NH_3)_4\text{-}ORu(NH_3)_5]Cl_6$, is a polycationic dye which in a concentration range of 50 nM–10 μM antagonizes the actions of capsaicin in different preparations (Amann and Maggi, 1991; Szállási and Blumberg, 1999). The preparation-dependent concentration where its action can be considered selective underlines the usage of appropriate controls. Ruthenium red, due to its large polycationic structure, antagonizes the effect of capsaicin and resiniferatoxin by blocking the cationic channel part of the VR1 receptor. Some examples of the analgesic effect of RTX, capsaicin and ruthenium red *in vivo* are shown in Figure 28.2 (Szolcsányi *et al.*, 1993).

28.4 Various natural putative VR1 agonists

These compounds are less potent than capsaicin in various test objects and their action on VR1 capsaicin receptor is supported but has not been unequivocally substantiated. Nevertheless, owing to their structural diversity they might serve as novel lead molecules to reveal further recognition sites and gating mechanisms on the VR1 protein while zingerone, a structural analog of capsaicin, was used for decades as a prototype of non-desensitizing pungent agents (Szolcsányi, 1993; Holzer, 1991).

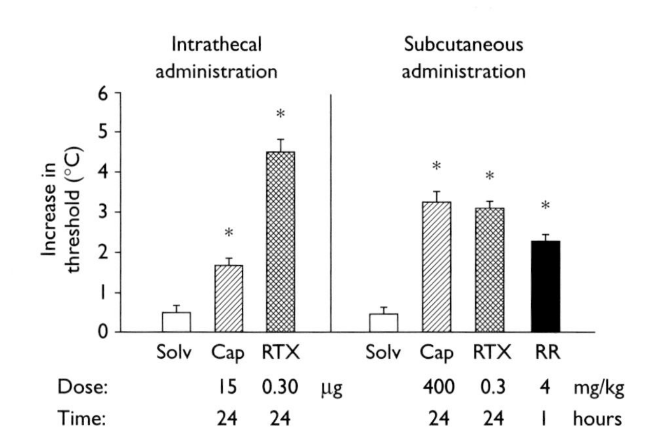

Figure 28.2 Increase in noxious heat threshold after intrathecal or subcutaneous administration of capsaicin (CAP), resiniferatoxin (RTX), ruthenium red (RR) or solvent (Solv). Doses and time intervals between injection and testing are indicated (mean ± SEM n = 5–10 animals; * significant differences from the control values of the same animals $P < 0.05$ or less). Reproduced from Szolcsányi J. *et al.*, 1993. Regul. Peptides 46, 437–439 with permission.

Figure 28.1 shows the chemical structure of some of these natural products. Piperine, is the pungent principle of black pepper (*Piper nigrum*) and isovelleral, the sensory stimulant agent of the mushroom *Lactorius vellereus*, the representative putative VR1 agonist investigated out of several terpenoids containing, α,β-unsaturated 1,4-dialdehydes. Further, less potent members of this group are: warburganal from the bark of the tropical tree Warburgia ugandensis, polygodial of water pepper (*Polygonum hydropiper*), cinnamodial and other sesquiterpenes from *Cinnamosma frograns*. Zingerone and shogaol are the two pungent principles of ginger (*Zingiber officinale*) and scutigeral isolated from a mushroom *Albatrellus ovinus* is a non-pungent triprenyl phenol compound (Holzer, 1991; Szolcsányi, 1993; Szállási and Blumberg, 1999).

Piperine and zingerone have been studied in their irritant and "desensitizing" potencies since the fifties and, using the protective eye-wiping test, zingerone became the prototype for an irritant which elicited reproducible nociception without short term or long term chemoanalgesic effects. Systemic or local application of piperine elicited both effects including cross-tachyphylaxis with capsaicin (Szolcsányi and Jancsó-Gábor, 1975; Szolcsányi, 1993). In cultured rat trigeminal cells, both pungent agents induced inward currents only in cells that could be activated by capsaicin and their responses were inhibited by capsazepine. The EC_{50} of piperine and zingerone was 35.5 μM and 15.5 mM respectively (Liu and Simon, 1996; Szállási and Blumberg, 1999).

Isovelleral was described as a vanilloid receptor agonist because it inhibited [^{3}H]RTX binding by rat trigeminal ganglion or spinal cord, elicited nociceptive protective reflexes when applied into the rat eye, caused cross-tachyphylaxis with capsaicin and induced calcium uptake by DRG neurons in culture which was antagonized by capsazepine (Szállási and Blumberg, 1999). On the VR1 transfected HEK293 cell line, however, a VR1 antagonist effect without agonism was recently reported (Jerman *et al.*, 2000). The reason for this discrepancy is not clear.

Scutigeral also mimicked the capsaicin effect on RTX binding assay (K_i = 19 μM) and on ^{45}Ca uptake in DRG cells (EC_{50} = 1.6 μM) which was antagonized by capsazepine. On the other hand it is non-pungent in human tongue, nor provokes eye-wiping protective reflexes upon instillation into the rat eye (up to 100 μM). Furthermore, it does not elicit inward current in cultured DRG neurons, and reduces the proportion of capsaicin-responsive cells. The mechanism of this unique profile of VR1 action is not known (Szállási and Blumberg, 1999).

28.5 Capsaicin VR1 receptor

The remarkable achievement of cloning the capsaicin receptor of the rat was made by David Julius, Michael Caterina and their associates (Caterina *et al.*, 1997). Capsaicin receptor-encoding 3-kilobase cDNA was identified and transfected to HEK 293 cells and Xenopus oocytes, where it conferred capsaicin or resiniferatoxin sensitivity. These transfected cells responded to capsaicin or resiniferatoxin with inward current and Ca^{++} accumulation with similar characteristics as described earlier for capsaicin-gated cation channels of dorsal root ganglion neuron demonstrated by single patch or whole cell patch clamp recordings (Bevan and Szolcsányi, 1990; Bevan and Docherty, 1993; Vlachova and Vykliczky, 1993; Oh *et al.*, 1996; Nagy and Rang, 1999).

Nomenclature Since the endogenous chemical ligand for the receptor activated by capsaicin was not known (and the issue is still unsettled: see later) the receptor was named for the first time with the aid of the traditionally accepted practice to coin a name referring to the mimicing lead molecule of plant origin (e.g. muscarinic, nicotinic, cannabinoid etc. receptors). Szolcsányi and Jancsó-Gábor, 1975). The name "capsaicin receptor" was also applied to the cloned receptor (Caterina *et al.*, 1997), but at that time a novel label, the "vanilloid receptor" had also been used because of the vanilloid moiety shared in the chemical structures of both capsaicin and resiniferatoxin and because the latter compound was used in binding studies for characterization of the capsaicin (C) type and resiniferatoxin (R) type of receptors (Ács *et al.*, 1997; Szállási and Blumberg, 1999; Szállási *et al.*, 1999). Therefore although the newly cloned receptor for both agonists was still denoted according to its original name (Szolcsányi and Jancsó-Gábor, 1975), it was dubbed "VR1 for vanilloid receptor subtype 1" (Caterina *et al.*, 1997). This term has been considered "somewhat of a misnomer" by the inventors of the vanilloid label (Szállási and Blumberg,

1999), particularly since this structural feature is not common among agonists of the receptor. Furthermore, it is unprecedented and misleading in the name of a receptor to refer to a chemical moiety of some exogenous lead molecules which are missing in the proposed endogenous ligands (Szolcsányi 2000a,b). Thus in this review the name of capsaicin VR1 receptor is used although the VR abreviation does not conform to the "NC-IUPHAR recommendations for nomenclature of receptors" which states that the "name should not include the letter "R" or "r" as an abbreviation for "receptor".

28.5.1 Characteristics of the rVR1 receptor

The capsaicin VR1 receptor of the rat (rVR1) is a protein of 838 amino acids with a relative molecular mass of 95,000 and a predicted membrane topology containing six transmembrane domains (β-sheets) with an additional short hydrophobic stretch (predicted pore region) between transmembrane regions 5 and 6 (Figure 28.3). Its structure resembles TRP-related store-operated calcium channels, the prototypical member of which is the *Drosophila* retinal proteins TRP and TRPL (Caterina *et al.*, 1997) and therefore VR1 and TRPV1 are both in usage to label this protein.

A remarkable feature of this VR1 cation channel is that it is also activated by noxious heat (48 °C) forming by this, "thermal transducer" feature the molecular basis for the selective action of capsaicin on the noxious heat sensitive nociceptors, i.e. the C-polymodal nociceptors as revealed in earlier single unit studies (Szolcsányi and Bevan, 1990; Szolcsányi, 1993, 1996).

Hydrogen ions (pH 6.4) activate this channel (Tominaga *et al.*, 1998) and enhance markedly the effects of both capsaicin (Caterina *et al.*, 1997) and noxious heating (Tominaga *et al.*, 1998). On cultured DRG neuron membrane patches

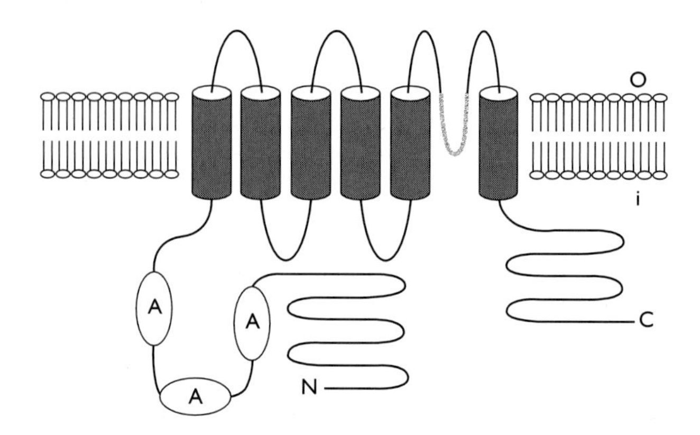

Figure 28.3 Predicted membrane topology and domain structure of VR1. Outer (o) and inner (i) plasma membrane leaflets are indicated. Reprinted by permission from Nature (389, 816–824, 1997) copyright (1997) Macmillan Magazines Ltd.

moderately acidic (pH 6.6) physiological saline greatly increased the probability of capsaicin-gated channels being in open state in outside-out single patch-calmps studies (Baumann *et al.*, 2000). Thus protons express both modulatory and agonist characteristics on native or recombinant VR1 receptors.

Pharmacological characterization of rVR1 expressed in HEK293 cells (Jerman *et al.*, 2000; Alnadaf *et al.*, 2000) and in CHO cells (Szállási *et al.*, 1999; Alnadaf *et al.*, 2000) were made by using the Ca^{++} uptake method (Table 28.1). It turned out that at pH 6.4 the effect of RTX or its structural analog phorbol 12-phenylacetate 13-acetate 20-homovanillate (PPAHV) was not enhanced unlike the response to capsaicin or olvanil. Isovelleral, the most potent irritant of the terpenoid family (Szállási and Blumberg, 1999) induced Ca^{++} response antagonized by capsazepine in rat cultured DRG neurons. On the other hand this compound was a rVR1 antagonist in transfected HEK293 cells like ruthenium red or capsazepine (Jermans *et al.*, 2000).

Voltage- and time-dependent characteristics of the recombinant rVR1 in HEK293 cells were analysed by whole-cell voltage-clamp techniques. Capsaicin elicited inward current (Caterina *et al.*, 1997; Tominaga *et al.*, 1998) with EC_{50} values of 497 nM (Gunthorpe *et al.*, 2000) similar to the Ca^{++} uptake measurements and whole-cell patch-clamp recordings from native DRG cells (Table 28.1). The latter report also showed similar outward rectification (4–6-fold) and reversal close to 0 mV of rVR1 consistent with the gating of non-selective cation channels described earlier for cultured DRG neurons (Bevan and Szolcsányi, 1990; Bevan and Docherty, 1993; Vlachova Vykliczky, 1993; Liu and Simon, 2000). Furthermore, it revealed that, unlike typical ligand-gated channels, rVR1 exhibits non-instantaneous rectification which might lead to enhancement of rVR1-mediated responses *in vivo* and could induce the long duration of action potentials of the capsaicin-sensitive DRG neurons (Gunthorpe *et al.*, 2000).

28.5.2 Localization of capsaicin rVR1 receptors

Immunocytochemical localization of capsaicin rVR1 receptors (Caterina *et al.*, 1997; Tominaga *et al.*, 1998; Guo *et al.*, 1999) and its mRNA (Michael and Priestley, 1999) were in accord with the data of distribution of the receptor binding studies with [^{3}H] RTX (Winter *et al.*, 1993; Szállási and Blumberg, 1999). According to these data, immunoreactivity of rVR1 or in situ hybridization of VR1 mRNA were encountered in the small or medium sized neurons in colocalization with TrkA or IB4 markers in the dorsal root ganglia and recent studies confirmed this in the trigeminal, nodosal and jugular (Helliwell *et al.*, 1998; Szállási and Blumberg, 1999) ganglia also. A small group (1%) of sacral DRG neurons expressed an exceptionally high level of VR1-ir which were not stained by these markers but showed binding to receptor tyrosine kinase (RET) (Michael and Priestley,

Table 28.1 Agonist potency (EC_{50}) and affinity (Kd, Ki) of VR1 capsaicin receptor agonists in VR1 transfected cells lines and cultured rat DRG neurons (nM)

	Transfected cell lines				*Cultured DRG neural membranes*				
	HEK 293 rVR1		*HEK 293 hVR1*	*CHO rVR1*					
	Binding Kd, Ki	*Ca++ resp. EC_{50}*	*Ca++ resp. EC_{50}*	*Binding Kd, Ki*	*Ca++ resp. EC_{50}*	*Binding Kd, Ki*	*Ca++ resp. EC_{50}*	*Inward current EC_{50}*	*Single patch current EC_{50}*
Capsaicin	4000[a]	9[b] 34[c]	6[b] 74[c]	1700[a]	38[a]	600–4900[d]	200–340[adcef]	310[p] 290[m] 680[n]	1100[g]
Resiniferatoxin	0.084[a]	3[b] 6.5[c]	3[b]	0.13[a]	1.4[a]	0.07[a] 0.018–0.046[d]	1–2.5[adfhi]	1–5[h]	
PPAHV		955[c]				3100[k]			
Scutigeral	18000[a]					19000[o]	1600[o]		
Anandamide		2496[b]	2070[b] 1150[c]						11700[l]
12(S) HPETE									8000[l]
15(S) HPETE									8700[l]
5(S) HPETE									9200[l]
LTB_4									11700[l]

Notes
a Szállási *et al.*, 1999; b Alnadaf *et al.*, 2000; c Jerman *et al.*, 2000; d Szállási, Blumberg, 1999; e Wood *et al.*, 1988; f Winter *et al.*, 1993; g Oh *et al.*, 1996; h Winter *et al.*, 1990; i Ács *et al.*, 1997; k Szállási *et al.*, 1996; l Hwang *et al.*, 2000; m Bevan, Docherty, 1993; n Liu, Simon, 1996; o Szállási *et al.*, 1999; p Vlachova, Vykliczky, 1993. For ref. e, h, k, n, o see: d.

1999). Detailed quantitative analysis supported a high proportion of colocalization of VR1-ir with IB4, P_2X_3 (purinoceptor), substance P and CGRP in the L5 dorsal root ganglion (Guo *et al.*, 1999). In the IB_4 positive population colocalization with somatostatin was also described (Michael and Priestley, 1999).

In the periphery, dermal and intraepithelial corneal nerve fibers stained intensively with VR1 although colocalization with neuropeptides was less frequent (Guo *et al.*, 1999). Sciatic and vagal nerves and mucosal and submucosal plexuses of the urinary bladder also showed VR1 positivity (Tominaga *et al.*, 1998).

In the spinal cord intensive immunoreactivity was encountered in the superficial dorsal horn, particularly in the lamina II, while in lamina I either intensive (Tominaga *et al.*, 1998) or sparse (Guo *et al.*, 1999) occurrance was reported. Subcellular distribution of VR1-ir was associated with unmyelinated fibers with clear vesicles localized at the plasma membrane, axonal microtubules and Golgi complex (Tominaga *et al.*, 1998; Guo *et al.*, 1999).

28.5.3 Human capsaicin VR1 receptor

Human VR1 cDNA was identified using the published rat VR1 sequence and hVR1-HEK293 cells has been introduced for large scale screening (Smart *et al.*, 2000; Alnadaf *et al.*, 2000). In these transfected cells capsaicin elicits an inward current and enhances intracellular Ca^{++} concentration (pEC_{50}:7.14). Lowering the pH from 7.4 to 6.4 enhanced the potency of capsaicin. Capsazepine antagonized the capsaicin-induced calcium response (pK_B:7.4). Presumably hVR1 is used in several pharmaceutical companies for high throughput screening (Alnadaf *et al.*, 2000) but few data have appeared in print until now.

28.5.4 VRL-1, the capsaicin-receptor homolog with a high threshold for noxious heat

This cloned protein also contains six transmembrane domain and is structurally related to VR1. This cation channel can be gated by higher temperature than VR1 (50–52 °C vs 44 °C). It is striking, however, that neither capsaicin nor resiniferatoxin evoked currents in Xenopus oocytes or HEK293 cells expressing VRL-1. This capsaicin-insensitive thermotransducer is localized in medium to large diameter neurons of the DRG ganglia with very few costaining with substance P immunoreactivity. It is presumed that VRL-1 is expressed in sensory neurons, the receptor of which has been characterized by single unit studies as type I A-delta mechano-heat-sensitive nociceptors (Caterina *et al.*, 1999).

28.6 Endogenous ligands for the capsaicin VR1 receptor

Acidification which occurs in inflammatory exudates and noxious heat as a physical stimulus can both be considered as natural activators of the VR1 capsaicin receptor. Nevertheless, to reveal an endogenous chemical ligand with possible functional significance for initiation of pain seemed to be an important challenge both scientifically and also for drug research. During the last year, two candidates have been the focus of interest. The first to stir up debate (Szolcsányi, 2000a,b) is the endocannabinoid anandamide (Zygmunt *et al.*, 1999; Smart *et al.*, 2000; Alnadaf *et al.*, 2000) and the most recent is 12-(S)-hydroperoxyeicosatetraenoic acid (12-(S)-HPETE) and some other less potent lipoxygenase products (Hwang *et al.*, 2000).

28.6.1 Anandamide

Anandamide induced an inward current in whole cells and isolated membrane patches in rVR1 expressing Xenopus oocytes and HEK293 cell (Zygmunt *et al.*, 1999) as well as producing an increase in intracellular Ca^{++} in rat cultured trigeminal neurons (Szőke *et al.*, 2000), HEK293 and CHO cell lines transfected with rVR1 or hVR1 (Smart *et al.*, 2000; Alnadaf *et al.*, 2000). In rVR1-transfected HEK293 cells, the pEC_{50} for anandamide was 5.31, (Zygmunt *et al.*, 1999) but it was slightly more potent in hVR1-transfected cells where the Ca^{++} response was measured (pEC_{50} 5.94) (Smart *et al.*, 2000). Both responses were antagonized by low concentrations of capsazepine ($pK_B = 7.40$), but not by the cannabinoid antagonists. Functional relevance for the agonism of anandamide on VR1 was proposed on the basis that anandamide released CGRP from perivascular plexuses of isolated arterial segments of the rat and guinea-pig causing smooth muscle relaxation in 0.1–10 μM concentration range without involvement of the cannabinoid receptors (Zygmunt *et al.*, 1999). As anandamide strongly inhibited the function of the capsaicin-sensitive sensory nerve endings, both in respect of nociception and CGRP-release, at 300 times lower concentration, the functional role of anandamide as a VR1 endogenous ligand was questioned (Szolcsányi, 2000a,b).

28.6.2 Products of lipoxygenases

A series of lipoxygenase (LO) products, anandamide and capsaicin were tested on single channel currents recorded from inside-out membrane patches of HEK293 cells transfected with capsaicin rVR1 receptor and that of cultured neurons from dorsal root and sympathetic ganglia (Hwang *et al.*, 2000). In VR1 transfected HEK cells, capsaicin (100 nM) and 12-(S)-hydroperoxyeicosatetraenoic acid (12-(S)-HPETE) (10 μM) activated identical outward rectifying currents in a capsazepine-sensitive manner, while they were ineffective in control and mock-transfected cells. Cultured sympathetic neurons were not activated. Quantitative analysis with several compounds was made on DRG cell membrane patches and the EC_{50} values of the most potent compounds are seen in Table 28.1. No response was obtained with LTC_4, stearic acid, arachidic acid, PGI_2, PGH_2, PGE_2 or PGD_2. Arachidonic acid activated about 50% of the patches, but after inhibition of 5-LO (REV5901) or 12-LO (baicalein) enzymes or in the presence of nonspecific LO inhibitors (e.g. nordihydroguaiaretic acid), the arachidonic acid-induced currents were markedly inhibited suggesting that "capsaicin receptors are activated preferentially by LO products among lipid messengers" (Hwang *et al.*, 2000). Molecular modeling of minimum-energy conformations of capsaicin and 12(S) HPETE indicates (Figure 28.4) that although the chemical structure of the two molecules differs considerably, S-shaped configuration of their apolar chain with a remarkable overlap of polar key regions favors a common binding site with capsaicin.

28.7 Characteristics of capsaicin receptor VR1 knockout mice

The very recent breakthrough in the field of capsaicin research is the generation of genetically altered mice in which the capsaicin VR1 receptors are removed. Two groups published these elegant results independently and sent their manuscripts within a month to the editors of the journals of Science and Nature and their complementary result led to very similar conclusions (Caterina *et al.*, 2000; Davis *et al.*, 2000). The group of Caterina, Julius, Basbaum and Kolzenburg with their associates deleted an exon encoding part of the fifth and all of the sixth putative transmembrane domains of the channel, together with the intervening pore-loop region, while the British group of John Davis generated a mouse lacking transmembrane domains 2–4 of the mVR1 gene (Figure 28.3). Cellular and behavioral assays on these genetically altered mice seemed to be particularly important, from the perspective of drug research, revealing the significance of capsaicin receptors as target molecules for the discovery of the first putative pain killer with a selective site of action on nociceptors.

Important issues can be answered with the aid of these knockout animals. First of all one could address the question of whether receptors other than VR1 are also involved or not in the biological effects of capsaicin or resiniferatoxin. In earlier studies, multiple vanilloid receptors were proposed (Ács *et al.*, 1997; Szállási and Blumberg, 1999; Liu and Simon, 2000) and the acronym VR1 also refers to possible further receptor subtypes. Furthermore, since the VR1 receptor is gated by noxious heat stimuli (but not below 43 °C) the thermoregulatory heat loss responses induced by capsaicin or resiniferatoxin could not be explained on this ground. Certainly it is the key issue concerning to what extent

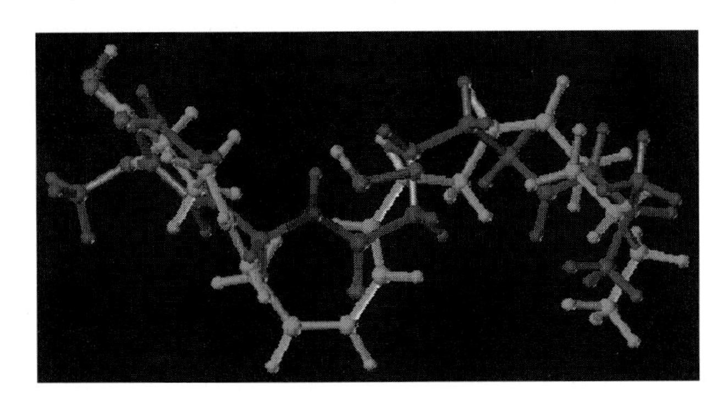

Figure 28.4 Molecular modeling overlay of minimum-energy conformations of capsaicin (green) and 12-(S)-HPETE (yellow). Oxygen and nitrogen atoms are colored in red and blue, respectively. Reprinted by permission from Proc. Natl. Acad. Sci. 97 (2000) 6155–6160, copyright: 2000, National Academy of Sciences, USA (see Colour Plate XVII).

nociception is impaired in the VR1 knockout animals. Finally, side effects of a putative VR1-blocking agents could also appear in these animals.

28.7.1 Abolition of capsaicin and resiniferatoxin actions in VR1 knockout mice

It was somehow surprising that, according to the data of both groups, all the cellular and behavioral effects of capsaicin and resiniferatoxin examined so far were abolished in the VR1 knockout mice.

1 According to both studies none of the cultured neurons prepared from dorsal root ganglia of VR1-null mutant mice responded to 1 μM capsaicin or 300 nM resiniferatoxin with inward current or to 3 μM capsaicin (0/680) or to 300 nM resiniferatoxin (0/261) with an increase in cytosolic free calcium. On the other hand, cells responded normally to GABA, ATP, α_1- β-methylene ATP or KCl and had normal resting membrane potential and voltage-gated sodium currents.
2 None of the C-afferents of VR1-null mutant mice responded to 1 μM capsaicin in the skin-nerve preparation (Caterina *et al.*, 2000).
3 Nociceptive protective reflexes and paw swelling evoked by intraplantar injection of capsaicin (1–5 μg) or resiniferatoxin were absent or almost absent in these animals and aversive drinking of capsaicin solution did not occur (Caterina *et al.*, 2000).
4 Profound falls in body temperature evoked by s.c. injection of 1 mg/kg capsaicin was also completely lacking in VR1-null mutant mice, although basal body temperature did not differ from that of the controls (Caterina *et al.*, 2000).

On the basis of these results, there is no evidence to indicate that, besides the VR1 receptor, some other capsaicin receptor subtypes might be involved in the biological actions of capsaicin or resiniferatoxin in mice. Certainly further experiments are needed, particularly to resolve the apparent contradiction between the differences in gating temperatures required to open the VR1 channel and that needed to regulate body temperature.

28.7.2 Impairment of thermal nociception and hyperalgesia

Before dealing with the nociceptive responsiveness of mice lacking the capsaicin receptor, it is important to note that sensory neurons of the dorsal root ganglia of VR1-null mutant mice showed complete loss of VR1 immunoreactivity without causing changes in the proportion of neurons expressing VRL-1 capsaicin-insensitive but noxious heat responsive ion channel (Caterina *et al.*, 1999). The proportion of neurons with IB_4 binding or containing substance P or N52 200 kD neurofilament protein also remained unaltered (Caterina *et al.*, 2000).

In spite of this clear histological picture the impaired nociception, particularly the responsiveness to noxious heat stimuli, is less conclusive under *in vitro* and *in vivo* conditions.

Neurons from cultured dorsal root ganglia of VR1 gene disrupted mice did not respond (Caterina *et al.*, 2000) or only by slight current (Davis *et al.*, 2000) to heat stimuli below 55 °C. On the other hand, 5 out of 60 cells responded like those in the control VR^{++} mice to temperatures above 55 °C. At this higher temperature range VRL-1 gated channels are opened, which are insensitive to capsaicin.

In the skin-nerve preparation (Caterina *et al.*, 2000) of VR^{--} mice a marked and highly significant drop in the number of heat responsive C-fibers was encountered, but four units had thermal threshold similar to that of the VR^{++} mice and at these units only the heat-evoked discharge was reduced. The apparent contradiction between the data obtained on DRG neurons and on cutaneous sensory receptors remained unresolved.

In behavioral studies, nociceptive thresholds were not determined, instead reflex latencies to different temperatures were measured using the hot plate, tail immersion and radiant paw heating tests. In both studies it was surprising that VR1-null mutant mice exhibited normal behavioral responses in the lower noxious temperature range of 46–48 °C while they exhibited an enhanced reflex latency at 50–58 °C, where intervention of the VRL-1 receptors could trigger nociceptive messages. On the basis of the temperature threshold of 43 °C described for cells transfected with VR1, the opposite behavioral response was expected. Measurement of the nociceptive thresholds instead of the reflex latencies during development might resolve this apparent paradox.

28.7.3 Inflammation-induced thermal hyperalgesia and behavior of VR1 deleted mice

According to both studies, neurogenic inflammation induced by mustard oil, edema evoked by carrageenin and nociceptive responses to mustard oil or mechanical hyperalgesia one day after injection of complete Freund's adjuvant into the hindpaw were similar both in $VR1^{++}$ and $VR1^{--}$ mice. In contrast, hyperalgesia to noxious heat after injection of mustard oil, carrageenin or the Freund's adjuvant were markedly inhibited or eliminated in VR1 knockout mice. Thermal hyperalgesia in the neuropathic pain model of Seltzer was, however, unchanged in mice lacking the VR1 receptors.

Cultured sensory neurons, like polymodal nociceptors and the cloned VR1 receptors are excited by acidification of the tissue or organ bath. The proportion of DRG cells responding to protons (pH5) was remarkably low in VR^{-} mutant mice (9/784 vs 100/735) although several other molecules in the ASICS family expressed by the sensory neurons might be expected to play a more significant role than the VR1 receptors mediating the proton-induced inward current.

Both groups of authors agreed that no gross behavioral or anatomical alterations were observed in the VR1 deleted mice.

28.8 Theory for different recognition sites for capsaicin and resiniferatoxin on VR1 receptor

Similarities and differences in the structure-activity relationships of capsaicin and resiniferatoxin analogues (Walpole *et al.*, 1996) as well as the marked differences in relative potencies and binding characteristics of these two lead molecules (Ács *et al.*, 1997; Szállási and Blumberg, 1999; Szállási *et al.*, 1999; Winter *et al.*, 1993; Wood 1993) have been emphasized for a long time. Distinct potencies for inducing Ca^{++}-uptake and receptor binding of (^{3}HRTX) or replacing [^{3}HRTX] binding led to the logical prediction of the existence of two vanilloid receptors. It was proposed that the R-type receptor, preferentially labeled with RTX, and the C-type, at which capsaicin has a higher potency, form two distinct classes of receptors (Ács *et al.*, 1997; Szállási *et al.*, 1999). Functional studies combined with binding assays on cells expressing the cloned VR1 as well the initial data on VR1 null mutant mice seem to require a new approach to tackle the discrepancies revealed in earlier studies.

Table 28.1 summarizes the EC_{50} values of six compounds for Ca^{++} uptake in both VR1 transfected cells lines and cultured DRG membranes together with the available binding data of K_d for RTX and (^{3}HRTX) binding displacements (K_i) for the other compounds. From these data several conclusions can be deduced: 1) Binding of capsaicin, resiniferatoxin, olvanil and scutigeral occur in parallel on VR1 expressing cells with native DRG cell membranes. Hence the data do not support two classes of receptors and the inventors of this theory themselves (Szállási *et al.*, 1999) considered alternative explanations (see later); 2) Much lower concentrations were needed for capsaicin or olvanil to induce Ca^{++} responses in VR1 transfected cell lines than in native DRG neurons, while similar concentrations were required in the case of resiniferatoxin; 3) Resiniferatoxin and capsaicin had similar EC_{50} values on rVR1 and hVR1 – expressing HEK293 cells.

Detailed pharmacological comparison measuring intracellular Ca^{2+} accumulation of VR1 transfected HEK293 cells with the FLIPR method revealed further discrepancies (Jerman *et al.*, 2000). 1) Potencies of capsaicin and olvanil but not RTX were enhanced at pH 6.4. 2) PPAHV, a resiniferatoxin analog displayed Hill coefficient around 2, indicative of positive cooperativity on cells expressing recombinant rVR1 while its effect on native tissue was non-cooperative. 3) Combined application of capsaicin and RTX is additive while each of them alone display positive cooperativity. 4) Isovelleral is an agonist on native cells and an antagonist on cells expressing recombinant rVR1.

Both groups of authors tentatively explained these discrepancies by supposing "existence of VR1 in multimeric and/or spliced variant form in neurones, leading to distortions in vanilloid potencies in the calcium influx measurements" (Szállási *et al.*, 1999) or by supposing "the presence of additional VR subunits, splice variants, or regulatory proteins" (Jerman *et al.*, 2000) and in a review article the authors of the former group raised several further possibilities to explain the "ultrapotent" VR1 agonist effects of resiniferatoxin (Szállási and Blumberg, 1999).

On the basis of structure-activity relationships (Walpole and Wrigglessworth, 1993; Walpole *et al.*, 1996; Szolcsányi and Jancsó-Gábor, 1975) and further considerations an hypothesis is put forward in which different critical binding domains for capsaicin and RTX are proposed on the same VR1 protein and an important role is attributed to the lipid bilayer around the receptor, fixing these apolar agonists to the vicinity of the key polar binding sites of the protein with hydrophobic forces.

28.8.1 Capsaicin analogs

Structure-activity relationships for capsaicinoids in assessments of their nociceptive and analgesic behavioral responses (Szolcsányi and Jancsó-Gábor, 1975, 1976; Janusz *et al.*, 1993) and in two *in vitro* assays ($^{45}Ca^{2+}$influx into cultured DRG neurons, contraction of guinea-pig ileum) combined with the mouse tail flick analgesic test (Walpole and Wrigglesworth, 1993) resulted in the following main conclusions:

1 Complete removal of the phenolic 4-OH substituent – indicated in Figure 28.5A as a No 1 critical

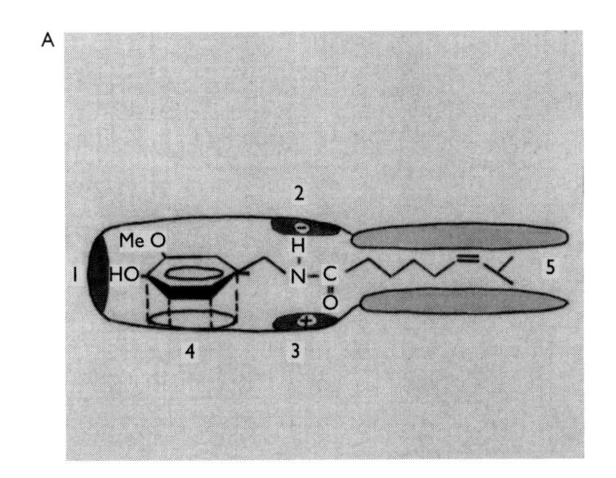

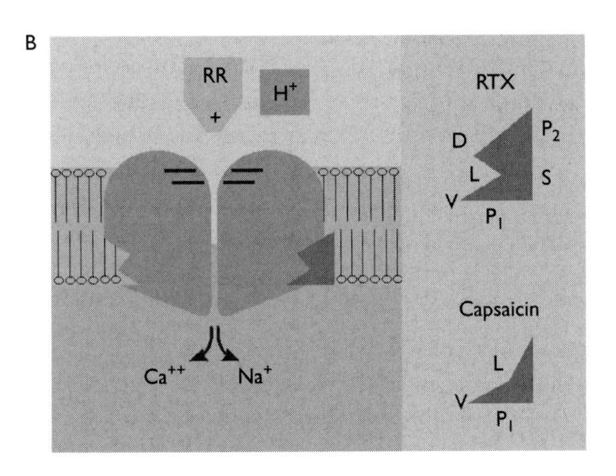

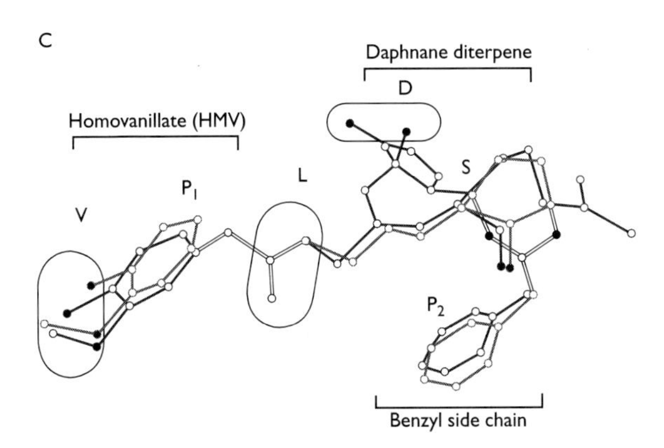

Change in structure	Relative Potency	Affinity	Change in structure	Relative Potency	Affinity
RTX(blue)	1000	1000	V./4-OH, 3-OMe		
			→3-H, 4-H	160	15
S./TOTCD(red)	0.17		→3-OMe, 4-OMe	107	120
P_1/HMV→OH	<1	<0.1	P_2/CH_2Ph→Me	121	38
(ROPA)			Capsaicin	5.3	<0.1
D./ 3-O→3-OH	28	27			
4-OH→4-OMe	300				
L./COO→CONH	13	11			

ROPA: resiniferanol orthophenyl-acetate; TOTCD: 2.9.10 trioxatri-cyclo [4,3,10]decane RTX analogue

Figure 28.5 (A) Schematic representation of the hypothetical capsaicin receptor. (modified from Szolcsányi and Jancsó-Gábor, 1975). Number 1, 2, 3 (blue): H-bonding sites, number 4, 5: apolar areas bound by hydrophobic forces. *(*B) Capsaicin VR1 receptor-cation channel. (C) Polar (D, L, V) and apolar (P1, P2, S) domains of resiniferatoxin (RTX) and capsaicin and their proposed binding to VR1. RR: ruthenium red (for more details see text and Figure 28.5C). Molecular modeling overlay of RTX (blue) and the RTX analogue, TOTCD (red). Overlapping areas (black): ortho ester of P2 ring and the C-20 link (L). Filled symbols are oxygen atoms. For more detail see text (modified from Walpole *et al.*, 1996) (see Colour Plate XVI.)

pharmacophore – leads to a loss of agonist activity, while its substitution with aminoethyl group retains and nitro group reduces its *in vitro* and *in vivo* activity.

2 Replacement of the phenolic ring with pirrole or pyridine ring to form H-bond acceptor sites instead of the 4-OH H-bond donor resulted in inactive compounds similar to more radical changes in this vanilloid region (Walpole and Wrigglesworth, 1993). Nevertheless, if long chained acylamides are formed with 2-amino ethanol or 3-amino propanol (Szolcsányi and Jancsó-Gábor, 1975, 1976) or in anandamide (Zygmunt *et al.*, 1999) the presence of a OH group at a 2–3 carbon distance from the acylamide link but without the benzene ring resulted in agonists for VR1 albeit with low potency.

3 At the link region between the substituted benzyl ring and alkyl chain – this dipolar site is denoted on Figure 28.5A as No 2 and 3; and is seen also as L on Figure 28.5C – amide and reversed amide are equipotent, the ester analogs are less effective and the thiourea moiety shows the highest potency (Walpole *et al.*, 1993). The optimum length between the ring and dipolar moieties was one carbon atom.

4 The optimum chain length for agonist activity was around 8–10 carbon atoms, while for the analgesic action with diminished irritant potency, longer alkyl chain with 12 carbon atoms (Szolcsányi and Jancsó-Gábor, 1976) or oleic acid (olvanil) (Janusz *et al.*, 1993) was more favorable. Aromatic substitutes instead of the flexible alkyl chain were also tolerated. Olvanil and its analog nuvanil have been described in treatments as advantagous for topical agents (Szállási and Blumberg, 1999).

28.8.2 Resiniferatoxin analogs

The elegant medicinal chemistry work of Walpole *et al.* (1996) revealed that although superficial similarities exist – as the "vanilloid" structure – in the molecules of capsaicin and resiniferatoxin (Figure 28.1) the structural criteria for agonism at the VR1 receptor are profoundly different in the two lead molecules. Based on this study, the 3D molecular model of RTX is depicted on Figure 28.5C in a form redrawn to emphasize the chemical components with the polar (encircled) and hydrophobic parts of the molecule besides the steric structural features.

For multiple hydrogen bonding interactions with the receptor, the 4-OH group of the vanilloid region (V), the ester group at the linker area (L) and the 3-CO, 4-OH substituents (D) of the 5-membered ring of the daphnane diterpene are suitable. For hydrophobic forces and steric adjustments, the bulky conformational constraint diterpene skeleton (S) as well as two planar aromatic rings (P_1, P_2) fixed to the molecule with rotating necks are indicated. One of them (P_1) could adjust the V polar region to hydrogen acceptor site of the VR1 receptor.

On Figure 28.5C RTX (blue) and its simplified analog, TOTCD (red), synthesized by Bloomfield *et al.*, 1992 (Walpole *et al.*, 1996) are overlaid in a position where the linker regions (L) and the ortho ester of phenylacetate regions overlap (black lines). Filled circles are the oxygen atoms. It is striking that TOTCD was almost completely inactive although two polar areas (V, L) and the two apolar aromatic rings (P_1, P_2) are in identical positions with RTX.

In the tabular part of the figure, relative potencies are calculated from EC_{50} values of induced $Ca^{2+}{}_i$ accumulation and relative affinities from Kd value of $[^3H]$RTX binding or from Ki values for replacement of RTX measured in cultured dorsal root ganglion cells. The data of Figure 28.5C shows that an inactive congener was also obtained when the vanilloid (HMV) part of the RTX was missing (ROPA), indicating that at least some of the L, V or P_1 regions are still required for action.

Besides these major structural changes, minor substitutions in the molecule resulted in large drops in potency and affinity.

1 Particularly critical is the presence of the 3-keto group (D) in the RTX diterpene moiety. Its replacement by a hydroxyl group decreased markedly both the affinity and potency of the analog, while the presence of 4-OH at D is less important.
2 At the link region (L) a change of ester group to acylamide is not tolerated although in the capsaicinoid series this alteration enhances the potency.
3 At the vanilloid part (V) omission of substitutes of the benzene ring or replacement of the 4-hydroxy with 4-methoxy group retains a significant part of the potency although these changes in the capsaicin molecule led to complete loss of agonism.
4 Omission of the P_2 aromatic ring is also tolerated without a major drop in potency.

In most cases the diminished potency ran parallel with the fall in affinity. Nevertheless, when the substitutes for the aromatic ring at site V were absent or when the P_2 benzene ring was replaced with a methyl group, a more profound decrease in affinity was measured than that in potency. Most striking is, however, the extremely low affinity of capsaicin, with a K_i value more than 50–100 times higher in DRG cells or VR1 transfected HEK293 cells than its EC_{50} (Table 28.1). Taking into consideration that in these binding studies replacement of RTX was measured, in the case of capsaicin, if it uses different critical moieties for binding it might lead to an underestimated value compared to that at its site of action. The following points of the above structure-activity relationships certainly seem to support the assumption that conformational changes of the VR1 molecule to open the channel are triggered, in the case of capsaicin, by polar groups at sites V and L while in the case of RTX the critical moiety for binding and action is the 3-keto group at site D. Furthermore, unlike in the case of capsaicin, the apolar part of P_1 with diminished dipolar hydrogen donor-acceptor L site is sufficient and favorable for the action of RTX. The higher potency of RTX over capsaicin by more than two orders underlines the importance of both the polar D group and the hydrophobic constraint diterpene skeleton (S) in the "ultrapotent" high affinity of this compound (Figure 28.5B). It has already been described that capsaicin opens the VR1 cation channel from its intracellular but not from the extracellular region (Walpole *et al.*, 1996; Jung *et al.*, 1999). It is also likely that RTX, with an even larger apolar backbone, binds to the VR1 protein at the lipid-protein interface and its hydrophobic forces to the lipid bilayer of the plasma membrane play an important role in the precise appositional position of the polar areas in the drug-receptor interaction (Szolcsányi, 2000a). It is tempting to assume or at least to consider that the difference in structure of plasma membrane around the VR1 protein of neurons and transfected cell lines could explain why capsaicin is much more potent in cell lines than in cultured neurons while resiniferatoxin acting on the same VR1 cation channel is similarly potent in both preparations (Table 28.1.)

28.9 Future perspectives

Nociceptive behavioral investigations on transgenic VR1 null mutant mice and on capsaicin pretreated animals as well as pain assessment after topical application of capsaicin or resiniferatoxin on the human skin or mucosal areas clearly showed that the capsaicin-sensitive nociceptors supplied by VR1 receptors are promising targets for development of analgesics with a novel site of action. C-polymodal nociceptors and the VR1 protein are integrators of noxious chemical and heat stimuli of pain signaling particularly in inflammation. Development of new drugs with a site of action on these multifunctional nociceptors is a promising but challenging goal which can open avenues in sensory pharmacology for new types of potent analgesics which are

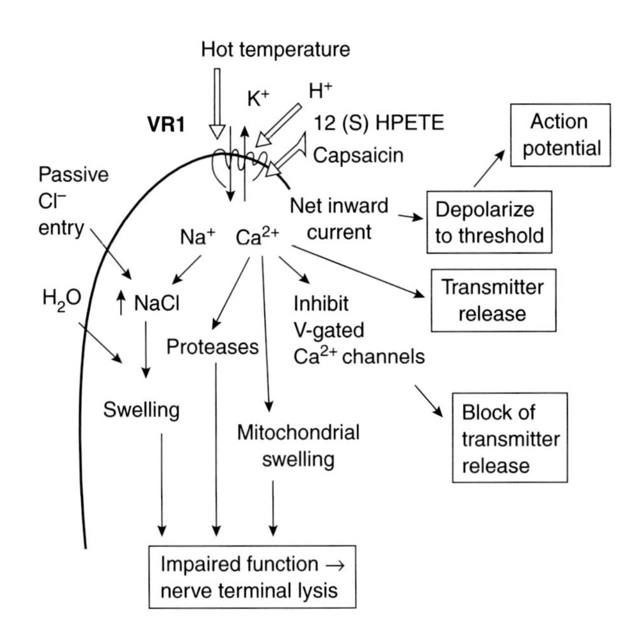

Figure 28.6 Cellular responses to VR1 activation at the capsaicin-sensitive nociceptors.

devoid of the side effects of opioids and COX inhibitors. The means for high throughput screening in this field has already been established and there are several possibilities for intervention in the cellular effects of capsaicin.

Figure 28.6 shows a schematic representation of the chain of events induced by VR1 activation. Hot temperature, endogenous ligands as 12(S)HPETE or exogenous ligands capsaicin, resiniferatoxin and other natural products listed above, open the VR1 cation channel which results in influx of Na^+ and Ca^{2+} and efflux of K^+ (Bevan and Szolcsányi, 1990; Bevan and Docherty, 1993). It is worth noting that the lack of sensory or reflex responses induced by a compound does not necessarily means that it did not open this channel. It could indicate that the induced depolarization of the nerve ending due to its slow time course was insufficient to trigger action potentials in a frequency required to evoke reflexes, nociception or pain. Firing of the C-polymodal nociceptors at frequencies of 0.3 Hz or less is non-painful (Szolcsányi, 1996) and although resiniferatoxin does not elicit the Bezold Jarisch reflex in the rat, it evokes the reflex in the cat with a slower time course than that induced by capsaicin. Furthermore at the VR1-expressing cell lines, both compounds open the same channel but with slow kinetics in the case of RTX (Jerman *et al.*, 2000). Certainly under these conditions the Ca^{2+}-dependent release of sensory neuropeptides as substance P, CGRP and somatostatin indicates VR1 agonism (Maggi, 1995) without pain sensation, but it is important to note that somatostatin released in this way into circulation (Szolcsányi *et al.*, 1998) has antiinflammatory and analgesic effects.

Thus both VR1 channel-blocking agents, slow channel openers, biochemical interventions to enhance VR1 receptor desensitization, endogenous ligand antagonists or their release inhibitors such as the lipoxygenase blocking agents could all produce an analgesic effect. Furthermore, VR1 activation by both capsaicin or resiniferatoxin induces, via Ca^{2a} influx, inhibition of voltage-gated Ca^{2+} channels (Bevan and Docherty, 1993) and mitochondrial swelling with impaired function (Szolcsányi, 1993, 1996, see Figure 28.6), which result in blocking of transmitter release at the peripheral and central terminals of the capsaicin-sensitive nociceptors and inhibition of the responsiveness of these nerve endings to various nociceptive stimuli including the mechanonociceptive ones, respectively. Activation of VR1 with high concentration of capsaicin or RTX for a prolonged period of time also activates Ca^{2+}-dependent intracellular proteases and induces Ca^{2+} overload of mitochondria with high amplitude swelling, which together could cause destruction of the nerve terminals (Szolcsányi and Bevan, 1990; Bevan and Docherty, 1993; Szolcsányi, 1993; Chard *et al.*, 1995). Cell death after systemic pretreatment of the neonatal rat is due to inhibition of NGF uptake from these impaired nerve terminals which develops after several days, contrary to earlier interpretation of direct immediate neurotoxic cell death (Szolcsányi *et al.*, 1998; Szállási and Blumberg, 1999).

Another possible site of action for inhibition of the function of these nociceptive nerve endings is on their pharmacological receptors, activation of which suppresses the VR1-mediated process. Opiates, nociceptin and somatostatin have already been shown to have antinociceptive and antiinflammatory effects in this way (Szolcsányi *et al.*, 1998; Helyes *et al.*, 1997, 2000). All these mechanisms provide examples for drug targets on VR1 expressing capsaicin-sensitive nociceptors by which novel analgesics or antiinflammatory agents suitable for inhibition of the neurogenic inflammatory responses could be developed.

References

Ács, G., Bíró, T., Ács, P., Moderras, S. and Blumberg, P. (1997) Differential activation and desensitization of sensory neurons by resiniferatoxin, *J. Neurosci. 17, 5622–5628.*

Alnadaf, T., Delany, N., Gladwell, Z., Hick, Z., Dale, T., Growther, D., Coward, K., Anand, P., Sanseau, P. and Tate, S. (2000) Functional comparison studies of rat and human vanilloid receptor 1. *Europ. J. Neurosci. 12 Suppl. 11. 381.*

Baumann, T.K. and Martenson, M.E. (2000) Extracellular protons both increase the activity and reduce the conductance of capsaicin-gated channels. *J. Neuroscience 20, RC80, 1–5.*

Belmonte, C. and Cervero, F. (eds) (1996) *Neurobiology of Nociceptors.* Oxford Univ. Press.

Bevan, S., Szolcsányi, J. (1990) Sensory neuron-specific actions of capsaicin: Mechanisms and applications. *Trends Pharmacol. Sci. 11, 330–333.*

Bevan, S.J. and Docherty, R.J. (1993) Cellular mechanisms of the action of capsaicin. In: Wood, J.N. (ed.) *Capsaicin in the Study of Pain*. Academic Press, London, pp. 27–44.

Caterina, M.J. Leffler, A., Malmerg, A.B., Martin, W.J., Trafton, J., Petersen-Zeitz, K.R., Koltzenburg, M., Basbaum, A.J. and Julius, D. (2000) Impaired nociception and pain sensation in mice lacking the capsaicin receptor. *Science 288, 306–313.*

Caterina, M.J., Rosen, T.A., Tominaga, M., Brake, A.J., Julius, D. (1999) A capsaicin-receptor homologue with a high threshold for noxious heat. *Nature 398, 436–441.*

Caterina, M.J., Schumacher, M.A., Tominaga, M, Rosen, T.A., Levine, J.D. and Julius, D. (1997) The capsaicin receptor: a heat-activated ion channel in the pain pathway. *Nature, 389, 816–824.*

Chard, P.S., Bleakman, D., Savidge, J.R. and Miller, R.J. (1995) Capsaicin-induced neurotoxicity in cultured dorsal root ganglion neurons: involvement of calcium-activated proteases. *Neurosci 65, 1099–1108.*

Davis, J.B., Gray, J., Gunthorpe, M.J., Hatcher, J.P., Davey, P.T., Overend, P., Harris, M.H., Latcham, J., Clapham, C., Atkinson, K., Hughes, S.A., Rance, K., Grau, E., Harper, A.J., Pugh, P. L., Rogers, D.C., Bingham, S., Randall, A. and Sheardown, S.A. (2000) Vanilloid receptor-1 is essential for inflammatory thermal hyperalgesia. *Nature 405, 183–187.*

Guo, A., Vulchanova, L., W.J. Li, X. and Elde, R. (1999) Immunocytochemical localization of the vanilloid receptor 1 (VR1): relationship to neuropeptides, the $P2X_3$ purinoceptor and IB4 binding sites. *Europ. J. Neurosci. 11, 946–968.*

Gunthorpe, M.J., Harris, M.H., Prinjha, R.K., Davis, J.B. and Randall, A. (2000) Voltage- and time-dependent properties of the recombinant rat vanilloid receptor (rVR1) *J. Physiol. 525, 747–759.*

Helyes, Zs., Németh, J., Pintér, E. and Szolcsányi, J. (1997) Inhibition of nociceptin of neurogenic inflammation and the release of SP and CGRP from sensory nerve terminals. *Br. J. Pharmacol. 121, 613–615.*

Helyes, Zs., Thán, M., Oroszi, G., Pintér, E., Németh, J., Kéri, Gy. and Szolcsányi, J. (2000) Antinociceptive effect induced somatostatin released from sensory nerve terminals and by somatostatin analogues in the rat. *Neurosci. Lett. 278, 185–188.*

Holzer, P. (1991) Capsaicin: cellular targets, mechanism of action, and selectivity for thin sensory neurons. *Pharmacol. Rev. 43, 143–201.*

Hwang, S.W., Cho, H., Kwak, J., Lee, S-Y., Kang, C-J., Jung, J., Cho, S., Min, K.H., Suh, Y-G., Kim, D. and Oh, U. (2000) Direct activation of capsaicin receptors by products of lipoxygenases: endogenous capsaicin-like substances. *Proc. Natl. Acad, Sci. USA, 97, 6155–6160.*

Janusz, J.M., Buckwalter, B.L., Young, P.A. *et al.* (1993) Vanilloids. 1. Analogs of capsaicin with antinociceptive and antiinflammatory activity. *J. Med. Chem. 36, 2595–2604.*

Jerman, J.C., Brough, S.J., Prinjha, R., Harris, M.H., Davis, J.B. and Smart, D. (2000) Characterization using FLIPR of rat vanilloid receptor (rVR1) pharmacology. *Br. J. Pharmacol. 130, 916–922.*

Jung, J., Hwang, S.W., Kwak, J., Lee, S-Y., Kang, C-J., Kim, W.B., Kim, D. and Oh, U. (1999) Capsaicin binds to the intracellular domain of the capsaicin-activated ion channel. *J. Neurosci. 19, 529–538.*

Kumazawa, T., Kruger, L. and Mizumura, K. (eds) (1996) *The polymodal receptor – a gateway to pathological pain*. Prog. Brain Res. Vol. 113.

Liu, L. and Simon, S.A. (2000) Capsaicin, acid and heat-evoked currents in rat trigeminal ganglion neurons: relationship to functional VR1 receptors. *Physiol. Behav. 69, 363–378.*

Maggi, C.A. (1995) Tachykinins and calcitonin gene-related peptide (CGRP) as cotransmitters released from peripheral endings of sensory nerves. *Prog. Neurobiol. 45, 1–98.*

Michael, G.J. and Priestly, J.V. (1999) Differential expression of the mRNA for the vanilloid receptor subtype 1 in cells of the adult rat dorsal root and nodose ganglia ant its down regulation by axotomy. *J. Neurosci. 19, 1844–1854.*

Nagy, J., Rang, H.P. (1999) Similarities and differences between the responses of rat sensory neurons to noxious heat and capsaicin. *J. Neurosci. 19, 10647–10655.*

Oh, U., Hwang, S.W. and Kim, D. (1996) Capsaicin activates a nonselective cation channel in cultured neonatal rat dorsal root ganglion neurons. *J. Neurosci. 16, 1659–1667.*

Szállási, Á. and Blumberg, P.M. (1999) Vanilloid (capsaicin) receptors and mechanisms. *Pharmacol. Rev. 51, 159–211.*

Szállási, Á., Blumberg, P.M., Annicelli, L.L., Krause, J.E. and Cortright, P.N. (1999) The cloned rat vanilloid receptor VR1 mediates both R-type binding and C-type calcium response in dorsal root ganglion neurons. *Molec. Pharmacol., 56, 581–587.*

Smart, D., Gunthorpe, M.J., Jerman, J.C. Nasir, S., Gray, J., Muir, A.J., Chambers, J.K., Randall, A.D. and Davis, J.B. (2000) The endogenous lipid anandamide is a full agonist at the human vanilloid receptor (hVR1). *Br. J. Pharmacol. 129, 227–230.*

Szolcsányi, J. (1993) Actions of capsaicin on sensory receptors. In: Wood, J.N. (ed) *Capsaicin in the Study of Pain*. Academic Press. London, pp. 1–26.

Szolcsányi, J. (1996) Capsaicin-sensitive sensory nerve terminals with local and systemic efferent functions: Facts and scopes of an unorthodox neuroregulatory mechanisms. *Prog. Brain Res. 113, 343–359.*

Szolcsányi, J. (2000) Are cannabinoids endogenous ligands for the VR1 capsaicin receptor? *Trends Pharmacol. Sci. 21, 41–42.*

Szolcsányi, J. (2000) Anandamide and the question of its functional role for activation of the capsaicin receptor. *Trends Pharmacol. Sci. 21, 203–204.*

Szolcsányi, J., Helyes, Zs., Oroszi, G., Németh, J., Pintér, E. (1998) Release of somatostatin and its role in the mediation of the anti-inflammatory effect induced by antidromic stimulation of sensory fibres of rat sciatic nerve. *Br. J. Pharmacol. 123, 936–942.*

Szolcsányi, J. and Jancsó-Gábor, A. (1975) Sensory effects of capsaicin congeners I. Relationship between chemical structure and pain-producing potency. *Arzneim. Forsch. (Drug Res.) 25, 1877–1881.*

Szolcsányi, J. and Jancsó-Gábor, A. (1976) Sensory effects of capsaicin congeners. II. Importance of chemical structure and pungency in desensitizing activity of capsaicin-type compounds. *Arn. Forsch. (Drug Res.) 26, 33–37.*

Szolcsányi, J., Nagy, J. and Pethő, G. (1993) Effect of CP-96,345 a non-peptide substance P antagonist, capsaicin, resiniferatoxin and ruthenium red on nociception. *Regul. Pept. 46, 437–439.*

Szőke, É., Balla, Zs., Csernoch, L., Czéh, G., and Szolcsányi, J. (2000) Interacting effects of capsaicin and anandamide on intracellular calcium in sensory neurones. *Neuroreport. 11, 1949–1952.*

Tominaga, M, Caterina, M.J., Malmberg, A.B., Rosen, T.A., Gilbert, H., Skinner, K., Raumann, B.E. Basbaum, A.J. and

Julius, D. (1998) The cloned capsaicin receptor integrates multiple pain-producing stimuli. *Neuron 21, 531–543.*

Vlachová, V. and Vykliczky, L. (1993) Capsaicin-induced membrane currents in cultured sensory neurons of the rat. *Physiol. Res. 42, 301–311.*

Walpole, C.S.J., Bevan, S., Bloomfield, G., Breckenridge, R., James, I.F., Ritchie, T., Szállási, Á., Winter, J. and Wrigglesworth, R. (1996) Similarities and differences in the structure-activity relationships of capsaicin and resiniferatoxin analogues. *J. Med. Chem. 39, 2939–2952.*

Walpole, S.J. and Wrigglesworth (1993) Structural requirement for capsaicin agonists and antagonists. In: Wood, J.N. (ed.) *Capsaicin in the Study of Pain*, Academic Press, London, pp. 63–81.

Winter, J., Walpole, C.S.J., Bevan, S. and James, J.F. (1993) Characterization of resiniferatoxin binding sites on sensory neurons: co-regulation of resiniferatoxin binding and capsaicin sensitivity in adult rat dorsal root ganglia. *Neurosci. 57, 747–757.*

Wood, J.N. (ed) *Capsaicin in the Study of Pain*, Academic Press. London (1993).

Zygmunt, P.M., Petersson, J., Andersson, P.A., Chuang, H.-H., Sorgard, M., DiMarzo, V., Julius, D. and Högestätt, E.D. (1999) Vanilloid receptors on sensory nerves mediate the vasodilator action of anandamide. *Nature, 400, 452–457.*

Chapter 29

Migraine

Ilona Jelencsik and Csaba Ertsey

Contents

29.1 Migraine: an introduction

29.1.1 Clinical characteristics and diagnosis

The International Headache Society (IHS) promulgated operational criteria for the diagnosis of headaches in 1988[1]. The classification identifies 12 major categories of headache, which can be divided into two broad groups, the primary and secondary headache disorders. Migraine is the most important form of primary headache.

Migraine is a chronic, paroxysmal disorder, characterized by repetitive attacks of headache, nausea, vomiting, photo- and phonophobia. The frequency, intensity, duration and accompanying symptoms show a great inter-, and intraindividual variability. Therefore, the degree of accompanying debilitation is also variable.

The migraine attack may be preceded by mood disturbances, hunger, thirst, yawning, drowsiness etc. as early as 24 hours before the attack starts. This phase is called the *prodrome.*

In about 30% of the patients the attack is preceded by or associated with neurological disturbances, usually lasting some minutes, not exceeding one hour. The most common form is a variety of visual disturbances ("plus" symptoms, e.g. scintillations, zig-zag lines, fortification spectra, palinopsia and "minus" symptoms, e.g. field defects); less frequently paraesthesia, paresis, speech disturbances and other transient focal neurological signs are experienced. This is the *aura* phase of the attack.

The *headache* phase, consisting of head pain and associated symptoms, is the most characteristic feature of the attack. When the headache gradually subsides, a *resolution* phase signals the termination of the attack (Figure 29.1).

The two main categories of migraine according to the IHS classification are migraine without aura (M0) and

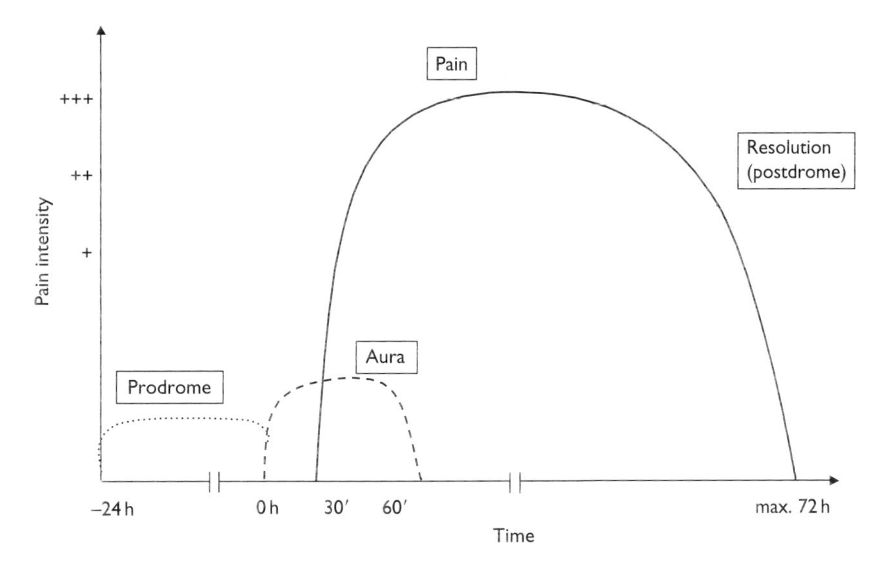

Figure 29.1 Phases of the migraine attack. The two main categories of migraine according to the IHS classification are migraine without aura (M0) and migraine with aura (MA). The most common form is M0.

Table 29.1 The IHS criteria for migraine without aura

1 At least five headache attacks with the following characteristics:
2 The headaches last 4 to 72 hours.
3 The headache must have at least two of the following four characteristics:
  a Unilateral location
  b Throbbing quality
  c Moderate or severe intensity, inhibiting or prohibiting the daily activities
  d Headache is worsened by routine physical activity
4 The headache must have at least one of the following:
  a Nausea and/or vomiting
  b Photophobia and phonophobia
5 Secondary causes of headache are excluded (e.g. neurological examination and/or imaging study)

Modified from reference 1.

migraine with aura (MA). The most common form is M0 (Table 29.1).

In addition to the two main categories, there are some less frequent migraine types: ophthalmoplegic migraine, retinal migraine, childhood periodic syndromes, and complications of migraine. A further category, called "migrainous disorder", denotes any migraine not fulfilling the criteria of other groups.

The diagnosis of migraine is based on a complete headache history from the patient. The localization, quality and severity of the pain are to be clarified. Duration and frequency of the head pain are to be identified, because migraine is a paroxysmal disorder. Some or all of the accompanying symptoms may not be present, however, nausea, phonophobia, and photophobia are highly suggestive of migraine. Before a primary headache such as migraine is diagnosed, secondary headaches and other systemic illnesses should be ruled out. Although not included in the diagnostic criteria, migraine can characteristically be provoked by several factors including hormonal changes (menstruation), stress (emotional and physical), lack of sleep, abrupt changes in weather, dietary factors, etc.

29.1.2 Epidemiology of migraine

In the last 10 years prevalence studies of migraine using the IHS criteria have shown that migraine is a frequent disorder. Rasmussen's study[2] in Denmark yielded a female lifetime prevalence of 25% and male lifetime prevalance of 8%. A Canadian study showed virtually identical prevalence numbers. A French study (Henry *et al.*) showed that 18% of women and 6% of men had migraine. These numbers were similar to those found by Stewart, Lipton, *et al.*[3] in the United States. Thus, migraine prevalence is very similar across industrialized countries, although it has been found to be lower in Asia and Africa. The prevalence of migraine in African and Asian Americans is lower than in whites[3].

29.1.3 Impact of migraine

In Rasmussen's study[2], 15% of migraine patients had 8 to 14 days of migraine per year. In Lipton and Stewart's 1993 study[4], 25% of migraine patients had more than 4 severe migraine attacks per month and 35% had 1 to 3 severe attacks per month. About 10% of migraine patients have more than 52 attacks per year.

Rasmussen found that 32% of men and 49% of women missed work every year due to migraine, while 40% of men and 50% of women had reduced effectiveness at work during attack. According to Lipton and Stewart 84% of migraine patients had at least some disability due to migraine, with 1/3 reporting severe disability. In 1992 Osterhaus estimated a 10 billion USD work loss per year in the US due to migraine. The cost to the international economy has been estimated at 3.5 billion USD per hundred million population.

Migraine imposes an economic burden on society because of the associated costs of medical care and lost productivity. Direct medical cost of migraine includes resources allocated to medical care, such as diagnosis, treatment, medication, and rehabilitation. Osterhaus estimated annual costs for three categories of direct medical service (emergency room visits, clinic visits, and hospitalisation) at 817 USD per patient. Direct medical costs however are only a small proportion of the total economic impact of migraine; lost labour and decreased productivity account for significant indirect costs.

There is "economic comorbidity" with migraine. Stewart *et al.*[3] found that migraine attack frequency goes down as household income increases. The other way to look at this is that migraine causes a downward social drift, as headache related disability disrupts academic and occupational performance. Migraine patients consume 2–5 times more health care than non-migraine patients, thus further compounding the economic costs. Lipton has found[4] that more than a third of patients with migraine never consult physicians for their headache. More than 60% of patients who never consult have severe migraine requiring bedrest. About 40% of those who do consult do not receive a diagnosis of migraine. The individuals with higher incomes are more likely to receive the correct diagnosis.

Quality of life of people with migraine is often worse than that people with other chronic conditions. In a comparison of health scores, the functioning and health perceptions of patients with chronic headache are worse than the health perceptions of patients with diabetes mellitus, chronic diseases, and back problems. The only chronic conditions with similar levels of functional impairment to chronic headache were myocardial infarction and congestive heart failure[5].

Currently, migraine is often underdiagnosed and undertreated. Epidemiological data show that it affects people during their peak productive years (age 25 to 55) and is not without considerable economic impact when considering lost productivity and lost wages. Improving the diagnosis and treatment of migraine might significantly improve the quality of life of migraine patients and, in turn, reduce the economic burdens on society.

29.2 Anatomy

29.2.1 *Pain-producing cranial structures*

The brain has sparse sensory innervation. The stimulation of the cerebral cortex, ependyma, and choroid plexuses does not cause pain. The significant pain-producing intracranial tissues are the blood vessels and the superficial capsule structures comprising the meninges (dura mater)[6]. Stimulation of the meningeal blood vessels can produce throbbing head pain[7]. Furthermore, the manipulation of intracranial blood vessels can activate perivascular sensory nerves and give rise to pain impulses that will be transmitted towards the brain. The stimulation of the intracranial segment of the internal carotid causes pain in the eye, frontal and temporal region. The distension of the middle cerebral artery causes referred pain to the back of the eye, while the vertebral artery causes referred pain to the occiput.

29.2.2 *The trigeminovascular system*

The sensory nerves innervating pain-producing, extracerebral, intracranial conducting arteries, supratentorial meningeal structures and the tentorium arise from the first division of the trigeminal nerve (the ophthalmic branch). The neuronal cell bodies giving rise to these peripheral sensory fibers are to be found within the trigeminal ganglion. Innervation is predominantly ipsilateral and widespread, an anatomical arrangement that could account for the unilateral nature of migraine headache[7]. The infratentorial meninges are innervated by nerve roots C1, C2 and C3 and, in part, cranial nerves IX and X.

Extracranially, the anterior two-thirds of the head are innervated by the branches of the trigeminal nerve. The posterior part is mostly innervated by the nerve roots C2 and C3 with a small contribution from cranial nerves IX and X.

In superior sagittal sinus stimulation studies (an experimental model of migraine headache), fos immunohistochemistry showed the activation of both trigeminal and cervical (C1 and C2 dorsal horn) sensory neurons. Thus, the trigeminal nucleus caudalis and the dorsal horns of the upper cervical region seem to form a functional continuum which has been termed "trigemino-cervical complex"[8].

The central projections of the pseudounipolar trigeminal neurons providing sensory innervation to intracranial blood vessels terminate behind the blood brain barrier within the dorsal root region of the trigemino-cervical nuclear complex that extends from the medulla to the upper cervical spinal cord[8]. These neurons thus provide the pathway to convey pain signals from the intracranial extracerebral vessels into the central nervous system. Because visceral (intracranial) and somatic (extracranial) afferents converge on the same relay neurons in the brainstem, intracranially-originating pain is referred to the superficial territory of the first division of trigeminal nerve and that of the C2 dermatome[7].

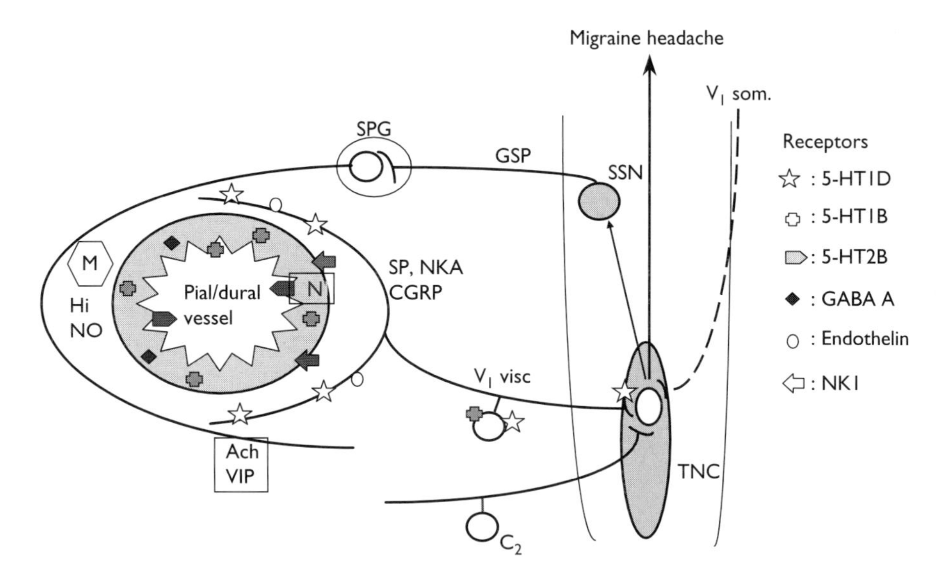

Figure 29.2 The trigeminovascular system. Anatomy: C_2: second cervical ganglion, GSP: greater superior petrosal nerve, SPG: sphenopalatine ganglion, SSN: superior salivatory nucleus, TNC: trigeminal nucleus caudalis, V_1som: somatic axon of trigeminal (ophthalmic branch) neurons, V_1visc: visceral axon of trigeminal (ophthalmic branch) neurons, M: mastocyte Biochemistry: Ach: acetylcholine, CGRP: calcitonin gene-related peptide, Hi: histamine, NKA: neurokinin A, NO: nitric oxide, SP: substance P, VIP: vasoactive intestinal polypeptide.

Perivascular trigeminal sensory nerve fibers contain many vasoactive substances, most notably substance P (SP), neurokinin A (NKA), tachykinin, and calcitonin gene-related peptide (CGRP) (Figure 29.2). SP and NKA cause increased blood vessel permeability and vasodilatation through endothelial receptors and NO release. CGRP does not alter vessel permeability, but is the most potent endogenous vasodilator substance known in humans through its actions directly on vascular smooth muscle and by endothelial NO-like mechanisms[8].

The trigeminal ganglion cells express a variety of serotonin (5-HT) receptors. This topic is dealt with in the chapter on serotonin.

29.2.3 *Pain pathways (central trigeminal connections)*

The second-order sensory neurones in the trigeminocervical complex give rise to the ascending trigeminothalamic tract which transmits pain impulses to the thalamus (ventral posteromedial nucleus and intralaminar nuclei). The thalamus, in turn, relays these impulses to the cortex[8]. Thus signal transduction in the CNS can theoretically be prevented, both in the trigeminocervical complex and at the thalamic level, thereby interrupting the central processing of pain signals.

Collaterals of the trigeminothalamic tract run to the reticular formation, creating the anatomical substrate of the arousal reaction following pain in the cranial region. A connection to the superior salivatory nucleus explains cranial parasympathetic activation in certain types of headache.

29.2.4 *The segmental pain control system*

The central projections of sensory neurons terminating in the dorsal horn of the spinal cord or its analog in the trigeminal nucleus use substance P (SP) as a neurotransmitter. The synaptic transmission is regulated by enkephalin-containing neurons inhibiting presynaptic, and GABAergic neurons inhibiting postsynaptic neurons. These interneurons are, in turn, controlled by descending pathways from the locus ceruleus and the nucleus raphe magnus containing noradrenalin and serotonin, respectively. This array of neurons regulating the transmission of pain impulses is called the segmental pain control system.

29.2.5 *The reticular formation and the brainstem pain control system*

The reticular formation is an important neuronal matrix for the integration of nervous system activities, where inputs converge from almost all somatic and visceral sensory pathways and from the cortex, hypothalamus, corpus striatum, limbic system, and spinal cord. It provides prominent output via the ascending reticular activating system (ARAS) to the same structures. The ARAS plays a key role in alertness, behavior, and affect. Reticular formation neuronal circuits are also involved in the regulation of cardiovascular function, respiration and other visceral responses through influencing cranial nerve nuclei and by descending connections to the autonomic centers in the spinal cord.

Axons of ascending spinothalamic and trigeminothalamic tracts convey pain sensations to the lateral columns of the reticular formation and then to the ARAS. They also activate neurones in the medial column and raphe nuclei. These nuclei are part of the endogenous pain control system[9].

Stimulation of the periaqueductal grey matter (PAG) inhibits the discharge of second-order sensory neurones in the spinal cord and trigeminocervical complex, causing analgesia[9]. This effect is mediated through the serotonergic nucleus raphe magnus. A parallel system, originating in the locus ceruleus, uses noradrenalin as a transmitter. Interestingly, central noradrenergic innervation originating in the locus ceruleus is also essential for maintaining blood brain barrier (BBB) integrity during some pathophysiological states[10]. The consideration that serotonin (5-HT1B/1D) agonist sumatriptan, an effective migraine aborting agent which does not cross the intact BBB, has CNS-related side effects, raises the possibility of BBB disruption during the attack.

Dysfunction in the reticular formation and neighbouring regions could interfere with the pain control mechanisms and disturb neuronal projections that govern the autonomic control of cranial blood vessel caliber and blood flow. On the other hand, alterations in the integrative functions and output of the brainstem may provide an explanation for many of the neurophysiological disturbances and symptoms that precede the migraine attack.

29.2.6 The "migraine generator"

The first direct clinical observation supporting a role of the brainstem in migraine attack generation was seen in patients with no history of migraine who developed migraine-like episodes after surgery to implant electrodes in the periaqueductal gray matter (PAG) and/or the somatosensory area of the thalamus[11].

Recently, PET studies during spontaneous migraine attacks showed the activation of brainstem regions, which persisted after complete relief by a 5-HT1B/1D agonist, from both head pain and accompanying symptoms[12]. These results suggest that these regions may contain an endogenous migraine generator. Key nuclei that are involved include the nucleus raphe magnus and locus ceruleus. The continued activation of this brainstem region suggests that ongoing activity within the migraine generator region could be responsible for headache recurrence which frequently occurs after successful drug therapy.

29.3 Elements of pathophysiology

29.3.1 Genetic basis

Migraine is predominantly an affliction of young people with a strong familial association and early onset of symptoms. These observations suggest that genetic factors are involved in its pathomechanism. Recent epidemiological studies using the IHS criteria on large population samples have underscored this suggestion. In a survey of 4000 persons from the general population[13], the first-degree relatives of probands with migraine without aura (M0) had a 1.9-fold increased risk of M0 while spouses had a 1.5-fold increased risk compared to the general population. This finding indicates that both genetic and environmental factors are important in M0. The first-degree relatives of probands with migraine with aura (MA) had a four-fold increased risk of MA while spouses had no increased risk, indicating that MA is determined largely by genetic factors. The distinct familial patterns suggest that M0 and MA have a different etiology. Complex segregation analysis indicated that both M0 and MA have multifactorial inheritance without generational difference[13].

The interest in the genetics of migraine was further increased when Joutel *et al.* reported that a gene for familiar hemiplegic migraine (FHM), a rare, autosomal dominant variant of migraine with aura, mapped to chromosome 19p13[14]. This locus is close to one identified in Cerebral Autosomal Dominant Arteriopathy with Subcortical Infarcts and Leucoencephalopathy (CADASIL) which is heralded by attacks of migraine with aura in one third of the cases. The defective gene in CADASIL is called Notch3; it encodes a transmembrane receptor protein of, as yet, unknown function.

Further characterization of the FHM locus on 19p13 revealed that it encodes the alpha subunit of a P/Q type brain-specific calcium channel (CACNA1A).[15] CACNA1A is expressed mainly in presynaptic terminals in the cerebellum and brainstem. Its function is modulated by interactions with other receptors, regulatory proteins and neurotransmitters, such as serotonin.

FHM may be regarded as a channelopathy and can be compared to other channelopathies, such as hyperkalemic periodic paralysis, paramyotonia congenita, hypokalemic periodic paralysis, myotonia congenita and episodic ataxia with myokymia. In some of these episodic disorders patients are normal between attacks yet may have profound disability during the attacks. In others, a progressive disability with or without paroxysmal events characterizes the clinical picture. It is of interest in this respect that in approximately 20% of FHM families migraine is associated with a mild permanent cerebellar ataxia.

Different mutations of the same CACNA1A gene may result in different autosomal dominant disease entities. At present, at least 16 different missense mutations causing FHM have been reported: only some, notably T666M and D715E cause cerebellar symptoms.[16] Mutations leading to a truncated protein cause episodic ataxia type 2 (EA-2) whereas expansions of a CAG trinucleotide usually result in spinocerebellar ataxia type 6 (SCA-6). FHM, EA-2 and SCA-6 can be considered as allelic channelopathies, characterized by paroxysmal neurologic symptoms and progressive neurodegeneration.

P-type neuronal Ca2+ channels mediate 5-HT release. Dysfunction of these channels may impair 5-HT release and predispose patients to attacks of migraine or impair their self-aborting mechanism[15].

Another possible role of the CACNA1A is related to the phenomenon of spreading depression (SD) described by Leao[17]. It denotes a wave of depolarization which spreads from the occipital to more rostral cortical areas. Its speed correlates well with the "march" of visual aura symptoms and therefore it is a reasonable assumption that the aura is the manifestation of SD. Ca2+ channels are important in the generation of SD. Moreover, magnesium deficiency has been documented in the cortex of migraine patients[18] and magnesium interacts with Ca2+ channels. Impaired channel function may predispose to more frequent and severe attacks.

CACNA1A mutations account for about 70% of families with FHM. A locus on chromosome 1q21–23, accounting for a further 15% of FHM cases, has been identified by Ducros *et al.* and Gardner *et al.* respectively[19,20]. An unusual presentation of FHM including recurrent episodes of coma lasting 3 to 5 days also shows linkage to this locus. The exact function of the locus is not known; the neuronal small conductance calcium-activated potassium channel gene hKCa3, implicated in schizophrenia, has recently been mapped to the same region[21]. In a large German-American family, linkage to 1q31 and later to 1q21–23 was documented by Gardner *et al.*[20,22] Interestingly, the nonlinkage region between 1q23 and 1q31 contains the CACNL1A6 calcium channel gene, which is a candidate gene in FHM.

Chromosome 19-linked and 1-linked FHM families show somewhat different phenotypes; ataxia and nystagmus are frequent in the first and absent in the latter group, whereas epilepsy and febrile convulsions occur in 1-linked, but not in 19-linked families[23].

The observation that obligate carriers of 19-linked FHM manifest migraine with or without aura but not hemiplegic features has lead to studies concerning the role of this locus on more common forms of migraine. Affected siblings share the marker D19S394, very close in terms of genetic distance to the CACNA1A locus, more frequently than chance alone would allow[24]. In a large family with typical migraine positive linkage lod scores, cosegregation and excessive allele sharing for markers in or around the CACNA1A locus has been documented[25]. Other studies, however, contradict the role of CACNA1A in common forms of migraine[26] and there is no direct evidence of migrainous (non-FHM) families harboring its presently known mutations.

Calcium channels are by no means the only candidate genes in migraine. During the last decade, rapid technological advances in molecular biology have allowed us to analyze some of them. Serotonin, as will be discussed later, plays a key role in migraine. Therefore, the role of genes coding various serotonin receptor subtypes have been investigated. Linkage and association studies of 5-HT2A, 5-HT2C, 5-HT1B and 5-HT1D receptor genes were negative[26]. Therapeutic responses to the 5-HT1B/D agonist sumatriptan were not influenced by the allelic distribution of 5-HT1B and 5-HT1F receptor genes. On the other hand, the study of the serotonin transporter gene (5-HTSERT) polymorphisms showed that migraine with aura patients had an increased frequency of the STin2.12 an STin2.9 alleles, whereas migraine without aura patients had an increased frequency of the STin2.12 alone[27]. This transporter takes up the released serotonin from the synaptic space to the presynaptic terminal, thus modulating serotonin transmission.

Migraine is a feature of MELAS, an acronym for mitochondrial encephalomyopathy, lactic acidosis and stroke-like episodes. As will be discussed later, a dysfunction of the mitochondrial oxidative metabolism could underlie the common migraine types. Although segregation analysis data from migraine patients do not support a maternal cytoplasmic inheritance (which is typical in disorders of the mitochondrial DNA), several papers have reported mitochondrial DNA mutations in people with migraine, patients with migrainous stroke and cyclic vomiting, the latter being considered a migraine equivalent. (*For a review on the subject, see reference 23*).

As suggested by clinical symptoms and efficacy in the acute treatment of antidopaminergic agents acting on the dopamine D2 receptor (DRD2), dopaminergic transmission can play a role in migraine. Migraine with aura patients were found to have a higher frequency DRD2 NcoI C allele than controls or patients with migraine without aura[28]. This finding was denied by Dichgans *et al.*[29] The DRD2 NcoI polymorphism, moreover, involves a silent change (change in nucleotide but not amino acid), which probably means that the DRD2 alleles are markers of genetic association, but cannot explain differences in dopamine receptor function[29]. Other genetic studies examined the endothelial NO synthase gene polymorphism, the role of complement C3, and the HLA system, but no definite conclusions can be drawn at present.

Taken together, the present data indicate that migraine is a multifactorial disease in which genetic (probably polygenic) and environmental factors both play a causative role. A migraine sufferer probably inherits a diathesis or constitution that makes him liable to headache. Triggers could bring on an attack due to the individual's susceptibility which involves the basic neurobiology of migraine.

29.3.2 Serotonin and serotonin receptors

There is a massive body of evidence suggesting that serotonin (5-HT) is involved in migraine. For a review of the first clinical data see *reference 30*.

5-HT is widely distributed throughout the body. Major concentrations occur in the gastrointestinal tract (90%), in platelets (8%), and in the brain. The high amount of platelet-contained 5-HT and its liberation during normal blood clotting may lead to methodological difficulties. Therefore, serum 5-HT levels are not informative in migraine research.

Pharmacological interest has focused on 5-HT, because it was observed that urinary excretion of 5-hydroxyindoleacetic acid (5-HIAA), the main metabolite of serotonin, is increased during the M attack. Clinical observations showed that intravenous 5-HT could stop an acute migraine attack, although the marked side effects prevented its clinical use. Moreover, 5-HT depletion can induce an attack. The platelet 5-HT drops rapidly during the onset of a migraine attack. This decrease, approximately 40%, is to be found in migraine without aura, but not in migraine with aura. A low molecular weight platelet serotonin-releasing factor has been hypothesized but its existence is not fully proved.

Between attacks, migraine (both M0 and MA) patients have lower plasma 5-HT and higher plasma 5-HIAA levels than controls. During the attack, plasma levels are about twice as high as during the attack-free period[31]. It was hypothesized that these changes reflected an increased 5-HT turnover in migraine patients during the attack-free period and an ictal reduction of the 5-HT turnover, presumably due to a fall in the enzymatic degradation rate. The low interictal levels of pharmacologically active circulating 5-HT may affect the cranial vasculature or might reflect the hypoactivity of serotonergic neural pathways.

During the last decade, multiple 5-HT receptor subtypes have been identified. Initially, receptor subtypes were characterized using pharmacological tools only. On the basis of the receptor binding profiles, common secondary messenger coupling and the functional activity of ligands, four main groups of 5-HT receptors, termed 5-HT1, 5-HT2, 5-HT3, 5-HT4, were identified. More recently, molecular biological techniques have confirmed this classification, have unraveled the existence of subgroups in the 5-HT2 group with relatively dissimilar protein structures, and have also led to the identification of novel 5-HT receptors (5-HT1F, 5-HT5, 5-HT6, 5-HT7) enabling them to be cloned, expressed in cultured cell lines and be characterized, both pharmacologically and functionally. Knowledge of 5-HT receptor cDNA sequences has also allowed antibody and antisense techniques to be employed.

In addition to the well known operational criteria (agonist and antagonist rank order, ligand binding), the present classification of 5-HT receptors is based on transductional (effector pathway) and structural (gene, amino acid sequence) data (see Table 29.2). The functional responses mediated by different 5-HT receptors are variable.

The 5-HT1 (negatively) and 5-HT4 (positively) receptors are coupled to adenylyl cyclase to decrease or increase cAMP levels, 5-HT2 receptors to phospholipase C (production of inositol triphosphate and diacylglycerol), while the 5-HT3 receptor forms part of a cation channel[32] (Table 29.2).

All 5-HT receptors have been identified in the brain. The 5-HT1 receptors are most dense in the hippocampus, the dorsal raphe, and the substantia nigra, with lesser concentration in the cortex. 5-HT1 receptors are inhibitory, while the remainder (5-HT2 through 5-HT7) are excitatory. The inhibition of the 5-HT2C subtype is supposed to be responsible for the migraine prophylactic effect of methysergide, pizotifen and cyproheptadine. 5-HT3 has a role in emesis; the 5-HT3 blocker ondansetron is an effective antiemetic. As no role of 5-HT4, 5-HT5, 5-HT6, and 5-HT7 receptors has been confirmed in migraine as yet, they will not be discussed further.

In humans, there are at least five 5-HT1 receptor subtypes, including 5-HT1A, 5-HT1B (formerly 5-HT1Dbeta) 5-HT1D (formerly 5-HTDalfa), 5-HT1E and 5HT1F. All are seven transmembrane, G-protein coupled receptors,

Table 29.2 Classification of 5-HT receptors

5-HT receptor class	*Subtypes*	*Second messenger*	*Antagonist*	*Function*
1	See Table 29.2 for details	Adenylate cyclase (−)		
2	2A 2B 2C	Inositol triphosphate (+)	Pizotifen Methysergide (both at 2C)	Smooth muscle contraction CNS excitation
3		K+	Ondansetron Granisetron	Membrane depolarization
4		Adenylate cyclase (+)	GR113808	Stimulates GI contraction
5	5a 5b		–	
6	Single receptor		–	
7	7a 7b	Adenylate cyclase (+)	–	Circadian rhythm

Modified from references 8 and 33.

Table 29.3 Classification of 5-HT subclass I receptors

5-HT1 receptor subtype	*Agonist*	*Antagonist*	*Function*
A	8-OH-DPAT dihydroergotamine	WAY100165	5-HT release from raphe presynaptic terminals mood satiety
B (previously 1Dβ)	CP-93,129 dihydroergotamine triptans (suma, ele, nara, riza, zolmi)	GR127935	Nerve terminal autoreceptor trigeminal cell receptor cerebral blood vessel receptor
D (previously 1Dα)	Dihydroergotamine triptans (suma, ele, nara, riza, zolmi)	GR127935	trigeminal nerve receptor neurogenic inflammation (?)
E	Rizatriptan	None	unknown
F	LY334370 dihydroergotamine triptans (suma, nara, riza,zolmi)	None	unknown

Modified from references 8 and 34.

Table 29.4 Distribution of 5-HT1B/1D receptors in humans

Site	*5-HT1B*	*5-HT1D*
Extracerebral cranial vessels	+++	±
Cerebral microvessels	−	+
Trigeminal ganglion	++	+++
Coronary vessels	+++	±

Modified from reference 8.

encoded by intronless genes, of between 365 and 422 amino acids with an overall sequence homology of 40%.[33] These receptors are all negatively linked to adenylyl cyclase (Table 29.3, 29.4).

29.3.2.1 5-HT1A

This subtype is widely distributed in the CNS, particularly in the hippocampus, septum and amygdala, areas that are thought to be associated with the control of mood. 5-HT1A autoreceptors are to be found in the raphe nuclei where they inhibit neuronal cell firing and 5-HT release onto postsynaptic sites. Blockade of these autoreceptors enhances the effect of SSRIs[33]. Stimulation of postsynaptic 5-HT1A receptors may cause anxiogenic-like responses[33].

29.3.2.2 5-HT1B

The 5-HT1B receptor is located on human chromosome 6q13 and is concentrated in the basal ganglia, striatum and frontal cortex. It is 93% identical with the similarly localized 5-HT1Dbeta receptor, described in rodents, the difference being attributed to a mutation in the transmembrane spanning region[33]. The differences in the binding affinity of key ligands such as beta-adrenoceptor agonist cyanopindolol are now attributed to this mutation; 5-HT1B and (rat) 5-HT1Dbeta are considered as species homolog of the same receptor. The 5-HT1B receptor subtype is the nerve terminal autoreceptor controling 5-HT release in animals and humans[33]. Nerve cell bodies in the trigeminal ganglion also contain 5-HT1B receptors[35], while they are not present on trigeminal nerve terminals. In the rat, approximately 15% of trigeminal ganglion cells express 5-HT1B receptors. These cells show immunoreactivity for CGRP, whereas only a few of them contain substance P[36]. Thus, 5-HT1B receptors seem to control trigeminal CGRP release and their agonists might reduce neurogenic vasodilation. As many nociceptive trigeminal neurones do not contain 5-HT1B receptors, a direct action of 5-HT1B agonists on primary trigeminal pain transmission seems unlikely. Moreover, the selective 5-HT1B receptor agonist CP-93,129 does not inhibit the firing of trigeminal neurons to noxious electrical stimulation of the middle meningeal artery[37].

Outside the CNS, 5-HT1B receptors are found on cerebral blood vessels. Their stimulation results in closure of arteriovenous shunts in dogs and cats and appears to produce vasoconstriction. Interest in 5-HT1B receptor agonists has been raised by the antimigraine properties of sumatriptan, a non-selective 5-HT1B and 5-HT1D agonist (see below).

29.3.2.3 5-HT1D

This receptor is located on human chromosome 1p36.3-p34.3. 5-HT1D is prefentially neuronal (residing on nerve terminals) and is the most widespread serotonin receptor in the brain. In the rat, 5-HT1D receptor mRNA is found in the caudate, putamen, nucleus accumbens, hippocampus and cortex, and also in the dorsal raphe and locus coeruleus. In the raphe, 5-HT1D receptors function as an autoreceptor modulating serotonin release[33].

Human trigeminal ganglia contain 5-HT1D immunoreactive cells[35] and 5-HT1D receptor-coding mRNA. In the guinea pig, 5-HT1D receptor mRNA was found both in the trigeminal ganglia and in the mesencephalic trigeminal nucleus[38].

It has been proposed that neurogenic inflammation and nociceptive activity within trigeminovascular afferents may be 5-HT1D receptor mediated, due to the presence of 5-HT1D but not 5-HT1B receptor mRNA in guinea pig. A novel selective 5-HT1D agonist, PNU-109291, which blocked plasma extravasation and reduced c-fos expression in the trigeminal nucleus caudalis[39] seems to underline this proposal.

29.3.2.4 5-HT1E

The receptor has been mapped to human chromosome 6q14-q15. Human binding studies and mRNA mapping have reported that 5-HT1E receptors are concentrated in the caudate, putamen, and, at a lower level, in the amygdala, frontal cortex and globus pallidus[33]. The function of this receptor is unknown.

29.3.2.5 5-HT1F

In the rat, mRNA coding for the receptor is concentrated in the dorsal raphe, hippocampus and cortex. 5-HT1F mRNA has also been detected in the trigeminal ganglion, the stimulation of which causes plasma extravasation, a component of neurogenic inflammation. Thus, the 5-HT1F receptor might be a target in antimigraine therapy. A novel selective 5-HT1F receptor agonist, LY334370, has recently been found effective and well tolerated during migraine attacks[40].

29.3.2.6 Triptans

The greatest breakthrough in the acute treatment of migraine in the last 50 years has been the development of a selective 5-HT1B/D receptor agonist family, called triptans. The first compound, sumatriptan was developed by Humphrey and coworkers[41] as the result of a fifteen-year effort. Subsequently, a number of structurally homologous compounds were synthesized, some of which are already being marketed.

The proposed mechanisms of action of the triptans include:

1. Neuronal inhibition blocking the depolarization of trigeminal afferents, which blocks vasoactive peptide release and neurovascular inflammation of the meningeal and dural vasculature[7].
2. Central (brainstem) neural inhibition within the trigeminal nuclei with inhibition of pain transmission[42,43].
3. Vasoconstriction of meningeal, dural and cerebral arteries[44].

Sumatriptan was the first specific antimigraine drug to be tested in a large population using double-blind methods.

EFFICACY OF SUMATRIPTAN

An analysis of 3000 patients' data showed that headache response (a significant reduction of headache severity) was 56% and 58% two hours after the oral administration of 50 or 100 mg sumatriptan, respectively. Thirty-one and 35 per cent of patients became pain-free at two hours post-dose, as compared to 9 per cent of patients on placebo. Sumatriptan also efficiently reduced or abolished the accompanying symptoms. Headache recurrence was a frequent problem, occurring in 30 to 35 per cent of patients. Adverse events were less frequent in the 50 mg group (7% vs. 17% over placebo), with efficacy being nearly the same[45].

The low oral bioavailability of sumatriptan has led to the development of other formulations (subcutaneous autoinjection, nasal spray, suppository). Although sumatriptan is very effective, it has certain drawbacks limiting its use. These include short half-life (T 1/2: 2 hours) resulting in a high attack recurrence rate, low speed of action in the oral form, and frequent side effects. Sumatriptan does not cross the intact BBB[42] which may explain its ineffectivity if administered during the aura phase.

These limitations have encouraged the development of "new triptans", some of which have already been introduced in to clinical practice (Table 29.5).

In short, two triptan groups can be delineated: one with fast relief onset and high potential but also a high recurrence rate (sumatriptan, zolmitriptan, rizatriptan, eletriptan, almotriptan), and a slow onset, lower potency, lower recurrence rate group (naratriptan and frovatriptan). Within these groups, minor differences exist as regards headache response, speed of action, adverse effect frequency and recurrence. At present, the accumulating clinical data are not sufficient to arrive at definitive conclusions about the place of the various triptans in migraine treatment. The use of triptans should probably be "tailored" to the individual migraine patient's needs.

SIDE EFFECTS

The most common untoward effects of triptans include chest pressure/heaviness (up to 40% of patients), flushing, dizziness, paresthesiae, drowsiness and nausea. These adverse

Table 29.5 The pharmacology of triptans

Name	*Oral bioavailability*	T_{max}	$T_{1/2}$
Sumatriptan	14%	2 h	2 h
Zolmitriptan	40%	1 h	3 h
Naratriptan	63–74%	2–3 h	6 h
Rizatriptan	45%	1,6 h	1,8 h
Eletriptan	50%	1 h	4,8 h
Almotriptan	70%	1,5–3 h	
Frovatriptan	24–30%	2–4 h	25 h

From reference 46.

effects are usually mild and of short duration. The more severe but far less frequent side effects of 5-HT1B/1D agonists stem from their vasoconstrictive properties and include coronary vasospasm, myocardial pain and infarction, arrhythmias, as well as rare incidence of stroke and ventricular arrhythmia[47].

CONTRAINDICATIONS

Contraindicatons for triptans include coronary artery disease, Prinzmetal's angina, complicated or hemiplegic migraine, basilar artery migraine, severe uncontrolled hypertension, pregnancy, breastfeeding, diabetes, hepatic disease, severe renal impairment, and hypercholesterolemia. Triptans should be avoided by patients taking MAOIs. Ergots or other vasoconstrictors should not be used with triptans, both concurrently and within 24 hours of drug intake. The use of SSRI is relatively contraindicated because of a potential for the serotonin syndrome. There are no sufficient data concerning the safety of triptans in children and in the elderly (above 65 years).

29.3.3 Pathobiochemistry

29.3.3.1 Energy deficit

A decrease of organic phosphates relative to inorganic phosphates (an index of phosphorylation potential in mitochondria) in migraine patients' brains has been demonstrated using 31-phosphorus magnetic resonance spectroscopy[48]. These findings were confirmed and further extended showing reduced mitochondrial phosphorylation potential and energy reserve in the occipital cortex of M0 and MA patients compared to healthy controls[49,50]. There appears to be a generalized mitochondrial dysfunction in migraine, which has also been demonstrated in platelets and muscle[49,51].

29.3.3.2 Neuroexcitatory amino acids

Elevated plasma levels of glutamate (GLU) and aspartate (ASP) were found in migraine patients between attacks which increased further during the attack[52]. In another study higher levels of platelet GLU were found in MA and higher levels of serum GLU were found in M0 patients (between attacks) compared to healthy adults[53].

Neuroexcitatory amino acids acting on NMDA receptors may be involved in the generation and propagation of spreading depression (SD), as NMDA antagonists block or supress SD and capsaicin induced c-fos expression (a marker of neuronal activation) in the trigeminal nucleus caudalis in animal models[54]. Interestingly, other compounds used either in acute or prophylactic migraine therapy fail to reduce the propagation of SD in animal models[55]. An alternative hypothesis is that activation of the glutamatergic NMDA receptor causes Ca^{2+} influx which may be responsible for the activation of the Ca^{2+} dependent NO synthase (NOS) and, subsequently, NO-stimulated cGMP production[56].

The action of excitatory amino acids is modulated by other neurotransmitters. Serotonin reduces Ca^{2+} and glutamate-induced NO/cGMP release in adult hippocampus slices via the 5-HT1A receptor[57]. Serotonin can also inhibit the potassium-evoked release of glutamate from presynaptic terminals[58] via the 5-HT1D receptor.

29.3.3.3 Magnesium

Magnesium is essential for energy transport of the cell, has major influence on membrane stability and as a consequence, on cortical excitability. Low levels of magnesium might therefore predispose the brain to spontaneous initiation of spreading depression or its activation via trigger factors.

Magnesium has been applied as a prophylactic drug in migraine and has been found to be effective in 3 double-blind placebo controlled trials with a therapeutic gain of 18.4 % over placebo[59], whereas in one trial no significant difference was found compared to placebo[60]. There are some early reports of magnesium-sulfate as an acute migraine treatment. Mauskop *et al.*[61] reported magnesium-sulfate injection to be effective in attacks of migraine and cluster headache and also tension type headache and transformed migraine. The low efficacy rates in therapeutic trials, however, do not indicate a major role for magnesium in attack generation.

29.3.3.4 Opioid peptides

Examination of plasma and CSF beta-endorphin levels in migraine yielded contradictory results. On the other hand, peripheral blood mononuclear cell beta-endorphin concentrations are reduced in migraine patients (both M0 and MA) in the attack-free period. Migraine patients have low plasma and high platelet methionin-enkephalin levels between attacks; both increase considerably during attacks. The significance of these findings is unknown[62].

29.3.3.5 Estrogens

In a subgroup of migraine patients, migraine headache occurs almost exclusively in relation to menstruation (so called menstrual migraine). In others, attacks occur both at menstruation and at other times of the month (menstrually triggered migraine). Menstrual migraine attacks may be precipitated by a sudden drop in estrogen levels and estradiol injections can postpone an attack. The percutaneous application of estradiol, with a more protracted decrease of estrogen levels, can prevent menstrual migraine attacks[63]. Notwithstanding these clinical findings, no consistent differences in ovarian hormone levels were found between menstrual migraine patients and controls[8].

29.3.3.6 Vasoactive neuropeptides

Plasma levels of calcitonin gene-related peptide (CGRP), a marker of trigeminovascular activation, selectively increase during a migraine attack, while other neuropeptide levels, notably substance P, neuropeptide Y and vasoactive intestinal polypeptide (a marker for cranial parasympathetic activation) remain unchanged[64]. Sumatriptan normalizes CGRP levels during migraine attacks[65]. These changes were only observed in blood drawn from the external jugular vein, but not in blood from the anterocubital vein, suggesting trigeminovascular activation during the attacks.

29.3.3.7 Dopamine

Several symptoms accompanying the migraine attack such as nausea, vomiting and blood pressure changes are mediated by dopaminergic activation. The observation that migraine patients are hypersensitive to dopamine (DA) agonists has led to the suggestion that DA might play a major role in migraine pathogenesis. Recently, an increased frequency of the NcoI C dopamine D2 receptor allele was reported in MA patients[28], causing a resurgence of interest in the role of DA in migraine.

The basal nuclei and DAergic structures are involved in the regulation of nociception[66]. Results of clinical studies are contradictory. In humans, dopa agonists (apomorphine, L-dopa) can alleviate severe painful states associated with the thalamic syndrome[67], herpes zoster[68] or Parkinson's disease (PD)[69]. According to other observations L-dopa can induce painful states in PD[70].

The relationship between migraine and PD is also controversial. Observations about male PD patients having a higher prevalence of migraine attacks[71] and migraine patients who noted a reduced frequency and intensity of attacks after developing PD[72] coexist in the literature.

Compared with controls, migraine patients are hypersensitive to dopaminergic agents which produce pronounced vegetative symptoms (nausea, vomiting, orthostatic hypotension). Low doses of apomorphine induce more yawns, the behavioral correlate of dopaminergic transmission, in migraine patients[73]. It is of interest that, in the rat, the activation of D2 receptors is involved in yawning. Neuroendocrine tests also demonstrated the interictal dopaminergic hypersensitivity of migraine patients: L-deprenyl reduced circulating prolactin levels in female migraine patients but not in controls[74]. While these studies suggest a hyperreactive central DA system in migraine, small patient numbers and problems in headache classification suggest that the results should be considered with caution.

Plasma and platelet DA levels[75], cerebrospinal fluid DA and its metabolites[76] have been subject to some controversy, but are usually normal interictally and during attacks.

Some dopamine D2 receptor (DRD2) antagonists, i.e. metoclopramide, chlorpromazine and haloperidol have a limited role in the acute treatment of severe migraine attack[77], while flunarizine, a calcium antagonist with significant DA antagonist properties, and some DA agonists, e.g. bromocriptine and lisuride, may be used as prophylactics. All the above mentioned drugs, however, are also active on a number of other receptors, including muscarinic, histaminic and serotonergic (5-HT2, 5-HT4) receptors.

Taken together, these data suggest an interictal DA receptor hypersensitivity in migraine. As the dopaminergic system is modulated by the serotonergic system[78,79], the hypofunction of the dopaminergic system with subsequent receptor hypersensitivity could be secondary to reduced serotonin turnover.

29.3.3.8 Nitric oxide

Nitric oxide (NO) is a ubiqitous, short-lived non-synaptic messenger implicated in a variety of diseases. NO is formed from L-arginine by a group of enzymes known as NO synthases. The endothelium and neurons express the calcium/calmodulin dependent constitutive NOS (cNOS), responsible for a short-term NO release after receptor stimulation. Inducible NOSs (iNOS) are to be found in macrophages, astrocytes and microglia; endotoxins and cytokines cause long periods of NO release via iNOSs[80].

NO exerts many of its actions by stimulating guanylate cyclase, resulting in the accumulation of cyclic guanosine monophosphate (cGMP). Some of the well-known actions of NO are of particular interest in relation to migraine:

- NO is a potent vasodilator, the most important of the so called endothelium-derived relaxing factors (EDRFs). Moreover, NO deriving from so called nonadrenergic, noncholinergic (NANC) perivascular nerves may be important in the regulation of vascular tone. NANC nerves have been demonstrated around intracranial arteries[81]. Taking into account that the balloon dilatation of intracranial arteries induces headache[82] and large intracranial arteries seem to be dilated during migraine attacks, the vasodilator effect of NO may contribute to migraine pain.
- NO plays an important role in the spinal processing of painful stimuli. Interacting with NMDA receptors[83], NOS is involved in hyperalgesia[84] and the development of tolerance to morphine[85].
- NO donors cause the release of calcitonin gene-related peptide (CGRP) from perivascular nerve endings in animal studies[86]. As discussed earlier, increased levels of CGRP have been found in external jugular vein samples from migraine patients during attacks[64].

NO frequently causes bilateral, short-lived headaches in non-migraine patients, shortly after NO donor administration. A more severe form of this early-onset headache, sometimes fulfilling the IHS criteria for migraine without aura, is also experienced by migraine patients being given an NO donor. However, migraine patients usually also develop a geniune migraine attack some 5.5 hours after NO administration. Thus, migraine patients seem to be hypersensitive to NO donors. According to transcranial Doppler studies, this phenomenon might be associated with an increased middle cerebral artery response to NO[87].

Endogenous NO formation may also play a role in migraine. Histamine, known to provoke attacks in migraine patients, most likely exerts its action by activating endothelial H1 receptors, thereby inducing NO formation[88].

Although the half-life of NO is very short (it decays 50% in about 3 seconds when perfused over a tissue in oxygenated saline[80]), NO can be a causative factor in starting a cascade which eventually leads to a migraine attack.

The data suggesting an important role for NO in the migraine attack raise the possibility of using endothelin antagonists in the acute treatment of migraine. Such an agent, bosentan, however, was found to be uneffective in a double-blind placebo-controlled study[89]. Further studies are needed for the clarification of this issue.

29.3.4 *Neurophysiology*

That a primary dysfunction of the brain might be a cornerstone in migraine pathogenesis is suggested by several clinical observations: behavioral and psychological patterns characteristic of migraine patients, the so called "migraine personality"[90]; comorbidity with depression and anxiety disorders[91]; hypersensitivity to sensory stimuli[92]; and migraine prodromes, i.e. CNS-mediated symptoms preceding the attack by up to 24 hours[93]. The observations of a reduced brain mitochondrial phosphorylation potential[48–50] and abnormal evoked and event-related potentials[94] in migraine sufferers lend further credit to this suggestion.

In migraine patients, increased amplitude of visual evoked potentials (VEP), increased amplitude and reduced habituation of event-related potentials, and contingent negative variation (CNV) have been reported between attacks[95]. These neurophysiological abnormalities might be caused by a common dysfunction of cortical information processing, most probably by lack of habituation of sensory stimuli. Habituation is thought to protect the cortex against sensory overload[94]. According to NMR spectroscopy studies, visual stimulation leads to increased lactate levels in the visual cortex in healthy volunteers. After sustained stimulation, habituation of the visual stimuli coincides with decreased lactate levels[96]. Therefore a habituation deficit may cause biochemical shifts in the CNS.

Migraine can be triggered by a number of factors, but the underlying mechanisms are not clearly understood. Changes in cortical excitability might be a plausible explanation. Indeed, various studies of the visual system have suggested cortical hyperexcitability of migraine patients between attacks[97]. A recent study using transcranial magnetic stimulation for eliciting phosphenes found hyperexcitability of the visual cortex[98]. In the same study, visual stimulation caused an activation and subsequent suppression of activation in contiguous areas of the occipital cortex (functional MRI findings) at a progression speed of 3 to 5 mm/min, similar to the events of cortical spreading depression described by Leao[17]. However, cortical hypoexcitability, instead of hyperexcitability in the interictal period has also been demonstrated[99]. Some of these contradictions can be explained by differences in study design and methodology, selection of migraine patients/controls, and timing of investigation.

29.4 An attempted synthesis of migraine pathogenesis

The previous pages have introduced us to the impressive amount of experimental and clinical data gathered about migraine in the last 30 years. Various models have been proposed to explain these findings in terms of migraine pathomechanism. Yet a "unifying theory" of migraine is still awaited. Rather than proposing a new model, this section is an attempt to offer a synthesis of pertinent data (Figure 29.3).

There appears to be a genetic predisposition to migraine, resulting in abnormal ion channel function or modulation and therefore leading to a baseline alteration in cortical and subcortical neuronal activity. Malfunctioning channels may also cause abnormal calcium entry with mitochondrial Ca^{2+} accumulation and energy failure.

In a genetically predisposed individual, exposure to various triggers may more easily result in altered cortical responses with inappropriate signal transduction from the cortex to the thalamus and hypothalamus. Besides blood pressure, these diencephalic structures regulate respiration and the sleep-wake cycle (hypothalamic "inner clock"), as well as the control of pain. The cortical signals and autonomic activity of the hypothalamus may ultimately activate a "migraine generator" in the brainstem, most likely in the region of the locus ceruleus, dorsal raphe nucleus and periaqueductal gray. The activation of the migraine generator could initiate, via brainstem and hypothalamic centers, many symptoms associated with migraine, such as nausea, vomiting, fatigue and dysthymia. Complex brainstem-cortex connections might be responsible for the generation of spreading depression (SD) and consequent aura symptoms. A reduced mitochondrial energy reserve and a deficit in external stimulus habituation can lead to biochemical shifts which render the cortex more prone to SD. Elevated levels of excitatory amino acids and reduced magnesium levels may favor the same process. The activation of the trigeminovascular system is likely to be responsible for the head pain by causing the dilatation of blood vessels (leading to throbbing pain), plasma extravasation and sterile inflammation. The dilation of blood vessels and the sterile inflammation may be

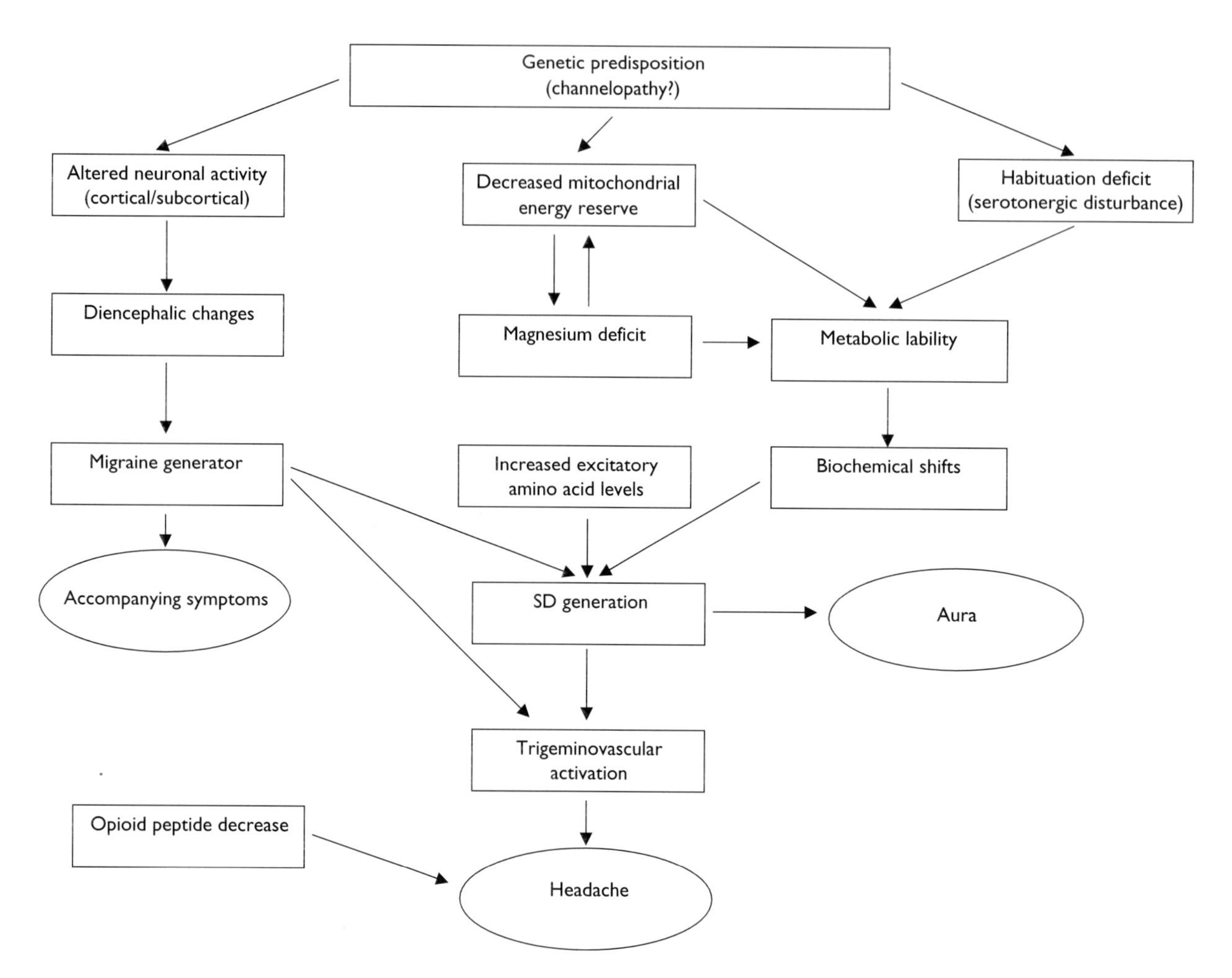

Figure 29.3 A possible explanation of migraine pathogenesis.

coupled to dural mast cell activation and histamine release, which can be triggered by substance P and CGRP release from the trigeminal nerve terminals[100].

29.5 Perspectives on migraine research and therapy

Based on recent data, most current models of migraine pathogenesis postulate a central nervous system dysfunction, while the vascular alterations are regarded as epiphenomena. The trigeminovascular system and the trigeminocervical complex have become therapeutic targets. A better understanding of the trigeminal pain transduction system and its modifiers, and of neural control of the cranial circulation should facilitate the understanding of the complex issue of migraine pathomechanism and management of patients.

The currently used 5-HT1D agonists (triptans), which are the most effective aborting agents, were originally designed as vasoconstrictors of the meningeal vessels. While their exact mode and site of action are not exactly known, an increasing body of evidence suggests that their effects are partly due to central action. Future studies are necessary to further clarify the role of the various serotonin receptor subtypes in migraine. The production of more selective 5-HT1D and 1F receptor agonists could result in more effective attack control, meeting the patients' needs for a fast-acting, efficient and economic oral antimigraine drug with a long-lasting action and fewer side effects. More clinical studies are needed to evaluate the efficacy and tolerability of new triptans and to compare them to sumatriptan. On the other hand, considering our current knowledge of the pathophysiology of the migraine attack, the antimigraine potential of new drugs such as CGRP antagonists, NO inhibitors, spreading depression inhibitors and drugs acting on trigeminal pain signal transduction should be sought.

Future prophylactic strategies should aim at restoring energy reserves, influencing ion channels, and modulating the activity of the migraine generator.

The present models of migraine pathomechanism can explain several features of the attack and also the action of current antimigraine drugs. However, drugs acting at various stages, such as the endothelin antagonist bosentan, a neurokinin 1 receptor antagonist, and two substance P antagonists (RPR100893 and lanepitant) were found ineffective in the acute treatment of migraine. This fact casts some doubt on the present model of migraine pathogenesis, although pharmacological property does not necessarily predict efficacy in aborting migraine attacks. On the other hand, experimental data do not necessarily reflect the alterations that predispose to, ignite and maintain an attack in migraine patients. A more comprehensive model of migraine pathomechanism may answer our present questions.

References

1 Headache Classification Committee of the International Headache Society (1988) Classification and diagnostic criteria for headache disorders, cranial neuralgias and facial pain (Sjaastad O, ed.) *Cephalalgia 8: (Suppl.7): 1–96.*

2 Rasmussen, B.K., Jensen, R., Schroll, M., Olesen, J. (1991) Epidemiology of headache in general population - a prevalence study. *J Clin Epidemiol. 44: 1147–1157.*

3 Stewart, W.F., Lipton, R.B., Celentano, D.D., Reed, M.L. (1992) Prevalence of migraine headache in the United States. Relation to age, income, race, and other sociodemographic factors. *JAMA 267: 64–69.*

4 Lipton, R.B., Stewart, W.F. (1993) Migraine in the United States: a review of epidemiology and health care use. *Neurology 43(suppl 3): S6–S10.*

5 Solomon, G.D., Skobieranda, F.G., Gragg, L.A. (1993) Quality of life and well-being of headache patients: Measurement by the medical outcomes study instrument. *Headache 33(7): 351–358.*

6 Lance, J.W. (1993) The pathophysiology of migraine. In: Dalessio, D.J. and Silberstein, S.D. (eds.) *Wolff's Headache and Other Head Pain,* 6th ed. New York: Oxford University Press, pp. 59–95.

7 Moskowitz, M.A. (1993) The trigeminovascular system. In: Olesen, J., Tfelt-Hansen, P., Welch, K.M.A. *The Headaches.* New York: Raven Press.

8 Silberstein, S.D., Lipton, R.B., Goadsby, P.J. (1998) *Headache in clinical practice.* Oxford: Isis Medical Media, pp. 51.

9 Basbaum, A.I., Fields, H.L. (1978) Endogenous pain control mechanisms: review and hypothesis. *Ann Neurol 4: 451–462.*

10 Harik, S.I., McGunigal, T. (1984) The protective influence of the locus coeruleus on the blood-brain barrier. *Ann Neurol 15: 568–574.*

11 Raskin, N.H., Hosobuchi, Y., Lamb, S. (1987) Headache may arise from the perturbation of brain. *Headache 27: 416–420.*

12 Weiller, C., May, A., Limmroth, V., Juptner, M., Kaube, H., Schayck, R.V., Coenen, H.H., Diener, H.C. (1995) Brainstem activation in spontaneous migraine attacks. *Nature Medicine 1: 658–660.*

13 Russell, M.B.; Iselius, L., Olesen, J. (1996) Migraine without aura and migraine with aura are inherited disorders. *Cephalalgia 16: 305–309.*

14 Joutel, A., Bousser, M.G., Biousse, V., Labauge, P., Chabriat, H. *et al.* (1993) A gene for familial hemiplegic migarine maps to chromosome 19. *Nature Genetics 5: 40–45.*

15 Ophoff, R.A., Terwindt, G.M., Vergouwe, M.N., van Eijk, R., Oefner, P.J. *et al.* (1996) Familial hemiplegic migraine and episodic ataxia type-2 are caused by mutations in the Ca2+ channel gene CACNL1A4, *Cell 87(3): 543–552.*

16 Ducros, A., Denier, C., Joutel, A. *et al.* (1999) Characterization of CACNA1A mutations in familial and sporadic hemiplegic migraine. *Neurology 52(Suppl.2): A273–4.*

17 Leao, A.P. (1994) Spreading depression of activity in cerebral cortex. *J Neurophysiol 7: 359–390.*

18 Ramadan, N.M., Halvorson, H., Vande-Linde, A., Levine, S.R., Helpern, J.A., Welch, K.M.A. (1989) Low brain magnesium in migraine. *Headache 29:416–419.*

19 Ducros, A., Joutel, G.M., Vahedi, K. *et al.* (1997) Familial hemiplegic migraine: mapping of the second gene and evidence for a third locus. *Cephalalgia 17:232.*

20 Gardner, K., Barmada, M., Ptacek, L.J., Hoffman, E.P. (1997) A new locus for hemiplegic maps to chromosome1q31. *Neurology 489:1231–1238.*

21 Austin, C.P., Holder, D.J., Ma, L., Mixson, L.A., Caskey, C.T. (1999) Mapping of hKCa3 to chromosome 1q21 and investigation of linkage of CAG repeat polymorphism to schizophrenia. *Mol Psychiatry 4:261–6.*

22 Gardner, K., Badger, J., Bannada, M.M., O'Connell, J., Hoffmann, E.P. (1998) A two locus model for hemiplegic migraine on Chr1q and candidate gene screening. *Am J Hum Genet (Abstract) 63:154.*

23 Montagna, P. (2000) Molecular genetics of migraine: a review. *Cephalalgia 20:3–14.*

24 May, A., Ophoff, R.A., Terwindt, G.M. *et al.* (1995) Familial hemiplegic migraine locus on chromosome 19p13 is involved in common forms of migraine with and without aura. *Human Genetics 96:604–08.*

25 Nyholt, D.R., Lea, R.A., Goadsby, P.J., Brimage, P.J., Griffiths, L.R. (1998) Familial typical migraine: linkage to chromosome 19p13 and evidence for genetic heterogeneity. *Neurology 50:1428–32.*

26 Monari, L., Mochi, M., Valentino, M.L. *et al.* (1997) Searching for migraine genes: exclusion of 290 cM out of the whole human genome. *Ital J Neurol Sci 1:277–282.*

27 Ogilvie, A.D., Russell, M.B., Dhall, P. *et al.* (1998) Altered allelic distributions of the serotonin transporter gene in migraine without aura and migraine with aura. *Cephalalgia 18:23–26.*

28 Peroutka, S.J., Wilhoit, T., Jones, K. (1997) Clinical susceptibility to migraine with aura is modified by dopamine D2 receptor (DRD2) NcoI alleles. *Neurology 49:201–206.*

29 Dichgans, M., Forderreuther, S., Deiterich, M., Pfaffenrath, V., Gasser, T. (1998) The D2 receptor NcoI allele: absence of allelic association with migraine with aura. *Neurology 51:928.*

30 Anthony, M. (1986) The biochemistry of migraine. In: Vinken, P.J., Bruyn, G.W., Clifford-Rose, F. (eds.) *Handbook of clinical neurology* Vol 4 (48), Headache. Amsterdam, Elsevier Science Publ. 85–105.

31 Ferrari, M., Odink, J., Tapparelli, C., Van Kempen, G.M.J., Pennings, E.J.M., Bruyn, G.W. (1989) Serotonin metabolism in migraine. *Neurology 39:135–146.*

32 Saxena, P.R. (1994) The pathogenesis and pharmacology of migraine. *Rev Contemp Pharmacother 5:259–269.*

33 Hoyer, D., Clarke, D.E., Fozard, J., Hartig, P.R., Martin, G.R. *et al.* (1994) VIIth International union of Pharmacology classification of receptors for 5-hydroxytryptamine (serotonin) *Pharmacol Rev 46:157–204.*

34 Hartig, P.R., Hoyer, D., Humphrey, P.A., Martin, G.R. (1996) Alignment of receptor nomenclature with the human genome: classification of the 5-HT-1B and 5-HT-1D receptor subtypes. *Trends in Pharmacological Sciences 17:103–105.*

35 Smith, D., Shaw, D., Hopkins, R., McAllister, G., Hill, R. *et al.* (1998) Development and characterisation of human 5-HT1B- or 5-HT1D-receptor specific antibodies as unique research tools. *J Neurosci Methods 80:155–161.*

36 Wotherspoon, G., Priestley, J.V. (2000) Expression of the 5-HT1B receptor by subtypes of rat trigeminal ganglion cells. *Neuroscience 95:465–471.*

37 Cumberbatch, M.J., Hill, R.G., Hargreaves, J.R. (1998) The effects of 5-HT1A, 5-HT1B and 5-HT1D receptor agonists on trigeminal nociceptive neurotransmission in anaesthetized rats. *Eur J Pharmacol 362:43–46.*

38 Bonaventure, P., Voorn, P., Luyten, W.H., Jurzak, M., Schotte, A., Leysen, J.E. (1998) Detailed mapping of serotonin 5-HT1B and 5-HT1D receptor messenger RNA nd ligand binding sites in guinea-pig brain and trigeminal ganglion: clues for function. *Neuroscience 82:469–484.*

39 Cutrer, F.M., Yu, X.Y., Ayata, G., Moskowitz, M.A., Waeber, C. (1999) Effects of PNU-109291 a selective 5-HT1D receptor agonist, on electrically induced dural plasma extravasation and capsaicin-evoked c-fos immunoreactivity within the trigeminal nucleus caudalis. *Neuropharmacology 38:1043–1053.*

40 Gossen, D., Granier, L.A., Vandenhende, F. (2000) Oral LY334370 in migraine. Results of Phase 1 studies. *Cephalalgia 20:249.*

41 Humphrey, P.A., Apperley, E., Feniuk, W., Perren, M.J. (1990) A rational approach to identifying a fundamentally new drug for the treatment of migraine. In: Saxena, P.R., Wallis, D.I., Wouters, W., Bevan, P. (eds.) *Cardiovascular pharmacology of 5-hydroxytryptamine: prospective therapeutic applications.* Dordrecht: Kluwer, 1990:416–31.

42 Kaube, H., Hoskin, K.L., Goadsby, P.J. (1993) Inhibition by sumatriptan of central trigeminal neurones only after blood-brain barrier disruption. *Br J Pharmacol 109:788–92.*

43 Goadsby, P.J., Hoskin, K.L. (1996) Inhibition of trigeminal neurons by intravenous administration of the serotonin (5-HT)-1D receptor agonist zolmitriptan (311C90): are brainstem sites a therapeutic target in migraine? *Pain 67:355–359.*

44 Jansen, I., Edvinsson, L., Mortensen, A., Olesen, J. (1992) Sumatriptan is a potent vasoconstrictor of human dural arteries via a 5-HT1-like receptor. *Cephalalgia 12:202–205.*

45 Goadsby, P.J. (1998) A triptan too far? *J Neurol Neurosurg Psych 64:143–47.*

46 Ertsey, C., Áfra, J., Jelencsik, I. (2000) Migraine therapy '2000. *Symposium of the Hungarian Headache Society, March 2000.*

47 Ottervanger, J.P., Valkenburg, H.A., Grobbee, D.E., Stricker, B.H. (1997) Characteristics and determination of sumatriptan-associated chest pain. *Ann Neur 54:1387–1392.*

48 Welch, K.M.A., Levine, S.R., D'Andrea, G., Schultz, L., Helpern, J.A. (1989) Preliminary observations on brain energy metabolism in migraine studied by *in vivo* 31 phosphorus NMR spectroscopy. *Neurology 39:538–541.*

49 Barbiroli, B., Montagna, P., Cortelli, P., Funicello, R., Iotti, S. *et al.* (1992) Abnormal brain and muscle energy metabolism shown by 31P magnetic resonance spectroscopy in patients affected by migraine with aura. *Neurology 42:1209–1214.*

50 Montagna, P., Cortelli, P., Monari, L., Pierangeli, G., Parchi, P. *et al.* (1994) 31P-magnetic resonance spectroscopy in migraine without aura. *Neurology 44:666–668.*

51 Sangiorgi, S., Mochi, M., Riva, R., Cortelli, P., Monari, L., Pierangeli, G. (1994) Abnormal platelet mitochondrial function in patients affected by migraine with and without aura. *Cephalalgia 14: 21–23.*

52 Ferrari, M.D., Odink, J., Bos, K.D., Malessy, M.J.A., Bruyn, G.W. (1990) Neuroexcitatory plasma aminoacids are elevated in migraine. *Neurology 40:1582–1586.*

53 Cananzi, A.R., D'Andrea, G., Perini, F., Zamberlan, F., Welch, K.M.A. (1995) Platelet and plasma levels of glutamate and glutamine in migraine with and without aura. *Cephalalgia 15:132–135.*

54 Mitsikostas, D.D., Sanchez del Rio, M., Waeber, C., Moskowitz, M.A., Cutrer, F.M. (1998) The NMDA receptor antagonist MK-801 reduces capsaicin-induced c-fos expression within rat trigeminal nucleus caudalis. *Pain 76:239–248.*

55 Kaube, H., Goadsby, P.J. (1994) Anti-migraine compaunds fail to modulate the propagation of cortical spreading depression in cat. *Eur Neurol 34:30–35.*

56 Farad, F.M., Breese, K.R. (1993) Nitric oxide mediates vasodilatation in response to activation of N-methyl-D-aspartate receptors in brain. *Circ Res 72:476–80.*

57 Raiteri, M., Maura, G., Bonanno, G., Pittaluga, A. (1986) Differential pharmacology and function of two 5-HT1 receptors modulating transmitter release in rat cerebellum. *J Pharmacol Exp Ther 237:644–648.*

58 Maura, G., Barzizza, S., Folghera, S., Raiteri, M. (1991) Release of endogenous aspartate from rat cerebellum slices and synaptosomes: inhibition mediated by a 5HT2 receptor and by a 5HT1 receptor of a possibly novel subtype. *Naunyn Schmiedelbergs Arch Pharmacol 343:229–36.*

59 Peikert, A., Wilimzig, C., Kohne-Volland, R. (1996) Prophylaxis of migraine with oral magnesium: results from a prospective, multi-center, placebo-controlled and double-blind randomized study. *Cephalalgia 16:257–263.*

60 Pfaffenrath, V., Wessely, P., Meyer, C., Isler, H.R., Evers, S. *et al.* (1996) Magnesium in the prophylaxis of migraine – a double-blind placebo-controlled study. *Cephalalgia 16:436–440.*

61 Mauskop, A., Altura, B.T., Cracco, R., Altura, B.M. (1996) Intravenous magnesium sulfate rapidly alleviates headache of various types. *Headache 36:154–160.*

62 Ferrari, M.D. (1993) Systemic biochemistry (of migraine). In: Olesen, J., Tfelt-Hansen, P., Welch, K.M.A. (eds.) *The Headaches.* New York: Raven Press, 1993.

63 De Lignieres, B., Vincens, M., Mauvais-Jarvis, P., Mas, J.L., Touboul, P.J., Bousser, M.G. (1986) Prevention of menstrual migraine by percutaneous oestradiol. *Br Med J 293:1540.*

64 Goadsby, P.J., Edvinsson, L., Ekman, R. (1990) Vasoactive peptide release in the extracerebral circulation of humans during migraine headache. *Ann Neurol 28:183–187.*

65 Goadsby, P.J., Edvinsson, L. (1991) Sumatriptan reverses the changes in calcitonin gene-realted peptide seen in the headache phase of migraine. *Cephalalgia 11 (Suppl.11):3–4.*

66 Chudler, E.H., Dong, W.K. (1995) The role of the basal ganglia in nociception and pain. *Pain 60:3–38.*

67 Grant, R., Behan, P.O. (1984) Resistant thalamic pain treated by levodopa. *BMJ 289:1272.*

68 Kernbaum, S., Hauchecorne, J. (1981) Administration of levodopa for relif of herpes zoster pain. *J Am Med Assoc 246:132–134.*

69 Nutt, G., Carter, J.H. (1984) Sensory symptoms in Parkinsonism related to central dopaminergic function. *Lancet ii:456–457.*

70 Quinn, N.P., Lang, A.E., Koller, W.C., Marsden, C.D. (1986) Painful Parkinson disease. *Lancet i:1366–1369.*

71 Lorentz, I.T. (1989) A survey of headache in Parkinson's disease. *Cephalalgia 9:83–86.*

72 Van Hilten, J.J. (1992) The migraine-dopamine link: do migraine and Parkinson's disaese coexist? *Clin Neurol Neurosurg Suppl 94:168–170.*

73 Blin, O., Azulay, J.P., Masson, G., Aubrespy, G., Serratrice, G. (1991) Apomorphine-induced yawning in migraine patients: enhanced responsiveness. *Clin Neuropharmacol 14:91–95.*

74 Calabresi, P., Silvestrini, M., Stratta, F., Cupini, L.M., Arigo, G. *et al.* (1993) L-deprenyl test in migraine: neuroendocrinological aspects. *Cephalalgia 13:406–409.*

75 Pradalier, A., Launay, J.M., Dry, J., Dreux, C., Artigou, C. (1985) Migraine commune. Variations des catecholamines conjugées et non conjugées. *Presse Med 14:583–585.*

76 Kovács, K., Bors, L., Tóthfalusi, L., Jelencsik, I., Bozsik, Gy., Kerényi, L. *et al.* (1989) Cerebrospinal fluid (CSF) investigations in migraine. *Cephalalgia 9:53–57.*

77 Cameron, J.D., Lane, P.L., Speechley, M. (1995) Intravenous chloprpomazine vs. metoclopramide in acute migraine headache. *Acad Emerg Med 2:597–602.*

78 Guan, X.M., McBride, W.J. (1989) Serotonin microinfusion into the ventral tegmental area increases accumbens dopamine release. *Brain Res Bull 23:541–547.*

79 Benloucif, S., Keegan, M.J., Galloway, M.P. (1993) Serotonin-facilitated dopamine release *in vivo*: pharmacological characterisation. *J Pharmacol Exp Ther 265:373–377.*

80 Moncada, S., Palmer, R.M.J., Higgs, E.A. (1991) Nitric oxide: physiology, pathophysiology and pharmacology. *Pharmacol Rev 43:109–141.*

81 Nozaki, K., Moskowitz, M.A., Maynard, K.I. *et al.* (1993) Possible origins and distribution of immunoactive nitric oxide synthase-containing nerve fibers in cerebral arteries. *J Cereb Blood Flow Metab 13:70–79.*

82 Nichols, F.T. III., Mawad, M., Mohr, J.P. *et al.* (1990) Focal headache during balloon inflation in the internal carotid and middle cerebral arteries. *Stroke 21:555–559.*

83 Garthwaitem, J., Charlesm, S.L., Chess-Williams, R. (1988) Endothelium-derived relaxing factor release on activation of NMDA receptors suggest a role as intercellular messenger in the brain. *Nature 336:385–388.*

84 Kitto, F., Haley, J.E., Wilcox, G.L. (1992) Involvement of nitric oxide in spinally mediated hyperalgesia in the mouse. *Neuroscience Lett 148:1–5.*

85 Kolesnikov, Y.A., Pick, C.G., Pasternack, G.W. (1992) NG-nitro-l-arginine prevents morphine tolerance. *Eur J Pharmacol 221:399–400.*

86 Wei, E.P., Moskowitz, M.A., Baccalini, P. *et al.* (1992) Calcitonin gene related peptide mediates nitroglycerin and sodium nitroprusside induced vasodilation in feline cerebral arterioles. *Circulat Res 70:1313–1319.*

87 Thomsen, L.L. (1997) Investigations into the role of nitric oxide and the large intracranial arteries in migraine headache. *Cephalalgia 17:873–895.*

88 Toda, N. (1990) Mechanism underlying responses to histamine of isolated monkey and human cerebral arteries. *Am J Physiol 258:H311–317.*

89 May, A., Gijsman, H.J., Wallnoefer, A., Jones, R., Diener, H.C., Ferrari, M.D. (1996) Endothelin antagonist bosentan blocks neurogenic inflammation, but is not effective in aborting migraine attacks. *Pain 67:375–378.*

90 Passchier, J., Andrasik, F. (1993) Migraine: psychological factors. In: Olesen, J., Tfelt-Hansen, P., Welch, K.M.A. (eds.) *The headaches*. New York: Raven Press, pp. 233–240.

91 Merikangas, K.R., Angst, J., Isler, H. (1990) Migraine and psychopathology: results of the Zurich cohort study of young adults. *Arch Gen Psychiatr 47:849–853.*

92 Hay, K.M., Mortimer, M.J., Barker, D.C., Debney, L.M., Good, P.A. (1994) 1044 Women with migraine: the effect of environmental stimuli. *Headache 34:166–168.*

93 Blau, J.N. (1980) Migraine prodromes separated from the aura: complete migraine. *Br Med J 281:658–660.*

94 Schoenen, J. (1994) Pathogenesis of migraine: the biobehavioral and hypoxia theories reconciled. *Acta Neurol Belg 94:79–86.*

95 Schoenen, J., Barkley, G.L. (1993) Neurophysiology (of migraine). In: Olesen, J., Tfelt-Hansen, P., Welch, K.M.A. (eds.) *The Headaches*. New York: Raven Press.

96 Sappey Marinier, D., Galabrese, G., Fein, G., Hugg, J.W., Biggins, C., Weiner, M.W. (1992) Effect of photic stimulation on human visual cortex lactate and phosphates usint 1H and 31P magnetic resonance spectroscopy. *J Cereb Blood Flow Met 12:584–592.*

97 Wray, S.H., Mijovich-Prelec, D., Kosslyn, S.M. (1995) Visual processing in migraineurs. *Brain 118:25–35.*

98 Aurora, S.K., Cao, Y., Bowyer, S.M., Welch, K.M.A. (1999) The occipital cortex is hyperexcitable in migraine: experimental evidence. *Headache 39:469–476.*

99 Afra, J., Mascia, A., Gérard, P., Maertens de Noordhout, A., Schoenen, J. (1998) Interictal cortical excitability in migraine: a study using transcranial magnetic stimulation of motor and visual cortices. *Ann Neurol 44:209–215.*

100 Ottosson, A., Edvinsson, L. (1997) Release of histamine from dural mast cells by substance P and calcitonin gene-related peptide and effect of sumatriptan. In: Olesen, J., Edvinsson, L. (eds.) *Headache Pathogenesis: Monoamines, Neuropeptides, Purines and Nitric Oxide*. Lippincott-Raven, Philadelphia, pp. 167–171.

Chapter 30

Regulatory mechanisms in focal ischemic stroke

Zoltan Nagy, Martin J. Berg and Neville Marks

Contents

30.1 Introduction

Stroke is a major cause of death and disability in Western countries, providing an incentive to study the regulatory mechanisms for the purposes of developing rational therapies. While mechanisms contributing to hallmark hemodynamic and neurochemical alterations are poorly understood, recent studies have established that loss of cerebral blood flow (CBF) rapidly depletes brain regions of their limited energy reserves (glycogen, glucose, oxygen, ATP) leading to an "energy crisis" concomitant to reduction in protein synthesis. In man, a leading cause of stroke is arteriosclerosis, but hallmark morphology can be replicated in experimental animals by *middle cerebral artery occlusion* (MCAO). Focal ischemia is a useful model since blocking an artery induces *ipsilateral* damage comparable to that arising from a brain clot in humans, but in animals can be compared directly to the contralateral hemisphere serving as the control. Severe damage occurs in brain regions proximal to the affected artery forming an ischemic "*core*" with radiating zones of lesser damage or "*penumbra*". Restoration of CBF results in post-ischemic damage arising from a temporary hyperemia followed by a period of hypoperfusion, edema resulting from movement of water and ions (Cl^-, HCO^-) and waves of *cerebral spreading cortical depression* (CSD). Reperfusion results in damage to vessels from changes in blood viscosity, the swelling of astrocytic end-feet that cause occlusions, and by the action of infiltrating monocytes and macrophages. The early goal of reducing infarct size diminishes the risk of premature mortality from edema and transtentorial herniation. The complex spatiotemporal events within the core and penumbra by glia, neurons and endothelial cells currently present challenges for therapeutic intervention.

30.2 Neurochemistry

Several neuronal components alter within minutes-hours-days with overlap including: (1) expression of cIERGs,

Table 30.1 Major hemodynamic changes in ischemia

Decrease in cerebral blood flow (CBF)	Formation of an ischemic core and penumbra Oxygen, glucose, ATP deprivation Decrease of protein synthesis
Post-ischemic damage[1,2]	Reperfusion-Hyperemia-Hypoperfusion- Neovascularization Acidosis (lactate production, sodium influx) Membrane damage (ER, Mt, plasma membrane) Edema (Cl^-, HCO^-) Cerebral Spreading Depression (CSD)
Glial cells	Activation of astrocytes (gliosis) GFAP loss in cores Microglia or macrophage infiltration

Notes

1 Reperfusion results in hyperemia followed by a compensatory hypoperfusion and damage to the microvasculature.

2 Agents that may prevent inflammation include NS-398 (COX-inhibitor: http://us.f93.mail.yahoo.com/ym/Compose? To = CO@-inhibitor) thromboxones or aminoguanidine as an iNOS inhibitor.

For other details see Hossmann, 1994; Kimelberg, 1999; Iadecola, 1999; Phillis *et al.*, 2000).

Table 30.2 Temporal neurochemical changes in focal ischemia

*Time after focal ischemia**	*Category*	*Components*
Minutes to hours	Genetic instability cIERGs	c-*fos*, *krox*-20, 24, *fos*-B, c-*Jun*, *nurr*-1, -77 (TR3)
Hours	Heat shock proteins *(hsp)* or factors Cytokines Inflammatory Adhesion and glial Vasoactive	hsp 70,72, HO-1, HSF, HSE, HIF *TNF-α, Il-1β, -6, MCP-1, CINC* COX-2 ICAM-1, ELAM-1, P-selectin, GFAP FFA, PAF, *IL-6, 10*
	ROS	NOS-SOD, NO, O_2, OH
Days	Apoptosis	p53, bcl-2, proteases (calpains, caspases)
	Necrosis Growth factors	Ca^{2+}, Glu, EAA's BDNF, TGF-β, VEGF, PDGF

The table is not intended to be comprehensive and several alternative designations exist for cellular immediate early response genes (cIERGs).

*Temporal changes are subject to overlap.

(2) Hsps, (3) cytokines and other inflammatory agents, (4) calcium overload, (5) release of reactive oxygen species (ROS), (6) activation of cell-death proteases (caspases, calpain) involved in apoptosis and necrosis, and (7) expression of growth factors. These comprise the major themes of this overview (Table 30.2). Because of space limitations the number of citations are restricted, since earlier work is covered by the monographs cited or can be accessed readily via the Internet using engines such as PubMed, or from summaries in the appended figures. Rather, the purpose of this short account is to focus on the important regulatory mechanisms as targets for therapy.

30.3 Genomic instability (IERGs)

Ischemia induces transient expression of *immediate or early response genes* (cIERGs) of the *fos/jun/krox* multigene families: these act primarily as third messengers for coding proteins that modify neuronal function (Morgan and Curran, 1995; Nowak and Kiessling, 1999). Genes or encoded proteins, despite their short half-lives ($t_{1/2}$), have significant effects on a variety of signaling pathways and other events critical for neuronal function (Figure 30.1). Diversity of expression results from the existence of two interacting families: (1) *leucine zipper* and, (2) *basic zinc finger* genes. The first induces *Fos* and *Jun* proteins in response to ischemia or code for constitutive expression of ATF (*activating transcription factor*) and Maf proteins. The AP-1 (activator protein) family refers to a large family of homo- and heterodimeric transcriptions factors containing *Jun, Fos, ATF* and *CREB* (*c-AMP-response element binding protein*) subunits that bind the DNA consensus sequence TGA G/C TCA, a common regulatory element of

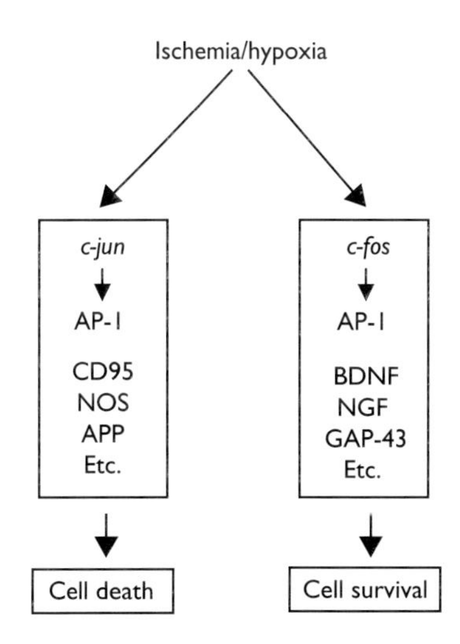

Figure 30.1 Immediate early response genes (IERGs) expression.

many genes. *Fos, Jun and ATF* are 'leucine zipper' components in lieu of 4–5 Leu repeats that form helix-like structures binding to the DNA consensus sequences. *Fos* and *Jun* proteins form dimers and trimers that increase the number of DNA binding motifs and hence functional diversity. The second group of IERGs contains *basic zinc finger* motifs with pairs of cysteine binding metal and includes *KNOX 20, -24* and *NURR*-1, -77 (TR3) (Table 30.2). Components containing AP-1 promoter domains sensitive to ischemia include *amyloid precursor protein* (APP), *glial fibrillary associated protein* (GFAP), iNOS, GAP-43, NGF, prodynorphins, proenkephalins, tyrosine hydroxylase, and vimentin. The increase in IERGs in zones that border the ischemic infarct, or that are present in the penumbra, often denote neurons capable of survival or subject to alteration because of waves of CSD. In the case of CSD induced by Glu activation of NMDA receptors, the IEG response can be suppressed by the antagonist MK-801 (Kiessling *et al.*, 1993; Lindsberg *et al.*, 1993). In focal ischemia, there is peak IEG expression in ipsilateral cortex at 3–8 h with a return to baseline by 2–4 days. Spatiotemporal expression of IEG mRNAs by glia and neurons in the core and penumbra, or the variable sensitivity of AP-1 domains to ROS damage yields unpredictable outcomes in terms of neuronal survival. In the case of selective vulnerability for CA1–CA4 neurons to ischemic insults, there is evidence that this correlates with changes in *growth associated phosphoprotein* - 43, a component with an AP-1 domain (GAP-43) (Tagaya *et al.*, 1995). This provides a plausible explanation along with other neurochemical differences for the earlier morphological findings of Pulsinelli *et al.* (Pulsinelli *et al.*, 1982).

Transgenic or knockout models (*Tg*, *KO*) provide new tools to evaluate neuroprotective agents. A recent example shows protection by estradiol with up-regulation of *estrogen receptor* α: evidence for an essential role comes from lack of protection for ERα but not ERβ in *KO* models. This novel estrogen-mediated protection may lead to altered gene expression downstream, and/or increase in relevant phosphorylation, trophic, or bcl-2 mediated pathways (Dubal *et al.*, 2001).

30.4 Heat shock proteins

These are a family of proteins that respond to heat shock or other forms of stress including unfolding/denaturation (Marks and Berg, 2001). They are highly conserved with regulation by transcription initially by *heat shock transcription factors* (HSF-1 or -2) that form trimers in response to stress which then bind conserved sequences called *heat shock elements* or HSE of relevant genes (Chen and Simon, 1997; Nowak, 1991; Yenari *et al.*, 1998; Wagstaff *et al.*, 1998). Ischemia increases expression of hsp 70 or 72 in glia and endothelial cells as an adaptive response to stress (Rajdev *et al.*, 2000; Hata *et al.*, 2000). Hemeoxygenase (HO-21, hsp32) of glia also responds to ischemic stress but is regulated independently by a *hypoxia inducible factor* (HIF). This factor also forms dimers that bind to other components involved in stress including erythropoietin, tyrosine hydroxylase, iNOS, VEGF, and Glut-1 (Plumier *et al.*, 1997; Guillemin and Krasnow, 1997; Ratcliffe *et al.*, 1998; Flamme *et al.*, 1997; Bergeron *et al.*, 1999). Studies on *Tg* mice show overexpression of CuZn-SOD reduces the ischemic infarct parallel with increase in hsp 72 (Kondo *et al.*, 1997).

30.5 Cytokines

These are polypeptides having protective or cytotoxic properties in response to ischemic stress (Stoll *et al.*, 2000). Cytokines including ~17 kDa TNF α,β, and ~7kD TGF α,β and interleukins (IL) are derived from separate genes by convertase processing of precursors, and play interrelated roles in signaling pathways in response to injury. Variable kinetic responses in their precursor mRNAs are attributed to cell-specific responses by glia, neurons and endothelial cells. In view of cytotoxic properties of some interleukins there is scope to evaluate convertase inhibitors such as those targeting caspase-1 (ICE, interleukin converting enzyme) that may account in part for protective actions of YVAD-fmk described below.

TNFα forms trimers that interact with *fas*, or neurotrophic p55 or p75 receptors for up-regulation of pathways mediated by NF-κB or its IκB subunit following stimulus coupled conversion by proteases, or by ceramides derived from sphingomyelin (Aronowski *et al.*, 2000; Stephenson *et al.*, 2000). Cell-permeable C2-ceramide (D-erythro-N-acetylsphingosine) administered i.c.v. augments bcl-2 mediated events with reduction in infarct volume in hypoxic-ischemic rats (Chen *et al.*, 2001b; Furuya *et al.*, 2001). Interaction with the *fas* receptor recruits pivotal procaspases −8 or 10 via interaction with adaptors that contain cognate death effector domains in hematopoietic or embryological neurons. Mice lacking p55/p75 have higher susceptibility to ischemia probably attributable to down-regulation of NF-κB mediated pathways (Figure 30.5) (Bruce-Keller and Mattson, 1999).

TGFβ interacts with specific receptors I–III and initiates largely neuroprotective pathways in response to ischemia, or *in vitro* to apoptotic or necrotic stimuli (Bruce-Keller and Mattson, 1999). IL-1 of endothelial or microglial cells is cytotoxic (Zhai *et al.*, 1997), but IL-6 and −10 are protective in studies on ischemia, or *in vitro* (Loddick *et al.*, 1998; Wang *et al.*, 1998; Grilli *et al.*, 2000). COX-2 has cytokine-like properties that mediate its protective effects by promoting the synthesis of prostaglandins via the phospholipid-arachidonic pathway (Nagayama *et al.*, 1999).

30.6 Oxygen radicals (NO, ROS)

One consequence of reperfusion is the increase in formation of reactive oxygen species (*ROS*) capable of multiple forms of neuronal damage leading to use of agents to prevent their formation (Figure 30.2). Nitric oxide (*NO*), a vasodilator with multiple neurotransmitter-like properties is formed by conversion of the guanidine nitrogen of L-Arg by action of NO synthase (NOS), or by nonenzymatic reduction of nitrite under acidic conditions (e.g. acidosis by anaerobic Mt reduction of pyruvate to lactate in response to ischemia) (Chan, 1999; Zweier *et al.*, 1995; Sasaki *et al.*, 1999). In ischemic brain, NOS is constitutively expressed (nNOS, −1), or is induced (iNOS, −2), with both forms having potential to damage neurons, although the mechanisms are incompletely understood (Forstermann *et al.*, 1998). A third form eNOS is extracellular and is considered to have protective properties

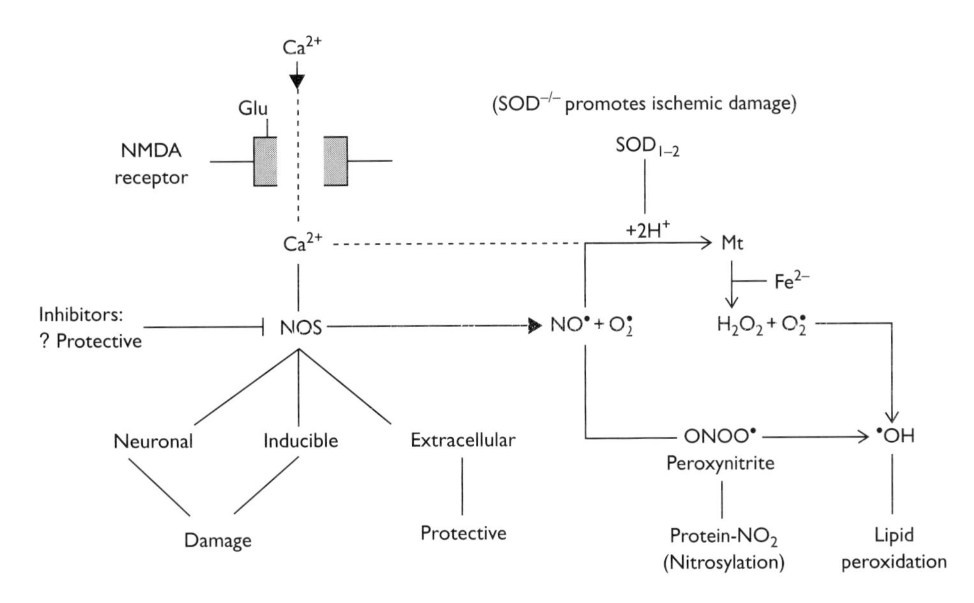

Figure 30.2 Influx of Ca^{2+} via NMDA receptors activates nNOS leading to production of NO• and superoxide O_2• radicals capable of forming ONOO• (peroxynitrite) that degenerates to form hydroxy radicals (•OH) available for lipid peroxidation, protein oxidation and DNA damage, or for nitrosylation. SOD (superoxide dismutase) catalyses formation of peroxide that also forms ••OH hydroxy radicals.

in ischemia. iNOS is induced in response to cytokines, LPS and other agents. Recent studies have identified a novel chaperone PIN that is thought to destabilize NOS dimers that normally are required for its activity. Phosphorylation sites within NOS down-regulate its catalytic properties and provide incentive to synthesize agents such as FK 506 that protect isolated neurons from Glu excitotoxicity (Dawson *et al.*, 1993). Excessive Glu stimulation of NMDA receptors in ischemia with production of NO has led to studies on the use of NOS inhibitors (Dawson *et al.*, 1996), or NOS knockouts as models to examine the different roles for NO in response to ischemia (Huang *et al.*, 1994). NO formation is prevented by i.p. injection of N-(G)nitro-L-arginine (L-NNA) or by the more specific 7-NI (nitroindazole), and in recent studies by the use of antisense probes leading to reduction of ischemic damage (Dawson *et al.*, 1993; Huang *et al.*, 1996; Dirnagl *et al.*, 1999; Parmentier-Batteur *et al.*, 2001). Microelectrode measurement during ischemia shows rapid increase within 1.5–4.0 h from cells within the core and penumbra (Iadecola, 1997).

NO is a short-lived gas that rapidly forms the radical ONOO! (peroxynitrite) capable of nitrosylating heme, Cys, and Tyr moieties resulting in altered neuronal function. ONOO! also has a short $t_{1/2}$ and degenerates to form !OH (hydroxy) via a Fe^{2+} mediated reaction, and leads to lipid peroxidation and DNA damage. NO binds to the heme moiety of guanyl cyclase or cyclooxygenases to increase production of cGMP formation and formation of prostaglandins thereby acting as a regulatory element (Salvemini *et al.*, 1993; Sasaki *et al.*, 1999).

Ischemic release of ROS may overwhelm the restorative capacity of mitochondria or other organelles to remove radicals by action of dismutases (SOD 1–3), glutathione peroxidase, and catalase, or by scavenger roles for antioxidants such as GSH, ascorbic acid and Vitamin E (Chan, 1999). SOD isoforms are present in tissues as CuZn- (−1) or Mn-(−2), or one form is extracellular (−3). Antisense probes to SOD reduce production of NO in PC-12 neural-like cells (Troy *et al.*, 1996), leading to interest in the use of SOD itself as a therapeutic agent, possibly in the form of liposomes to penetrate to the CNS. The potential importance for SOD has led to use of *Tg* or *KO* mice to evaluate the role of this enzyme in ischemia and action of drugs (Chan, 1999; Ginsberg, 1990; Huang *et al.*, 1994; Dawson *et al.*, 1996; Sheng *et al.*, 2000). Overexpression of SOD-1 blocks cytosolic release of cytochrome C in response to focal ischemia, a component that mediates conversion of cell-death proteases or attenuates NF-kB and expression of c-myc, a deleterious protein (Fujimura *et al.*, 2000; Huang *et al.*, 2001).

30.7 Calcium overload

Increase of $[Ca]_I$ in ischemia to neurotoxic levels exceeds the normal compensatory mechanisms of Mt or other organelles for storage, or removal by efflux or exchange, or other mechanisms (Figure 30.3). Binding proteins such as calmodulin, paralbumin, calbindin, calretins and S-100 largely sequester intracellular calcium. Free $[Ca^{2+}]_I$ represents the balance between influx via different *voltage sensitive calcium channels* (VSCC), metabotrophic and ion gated channels, or by

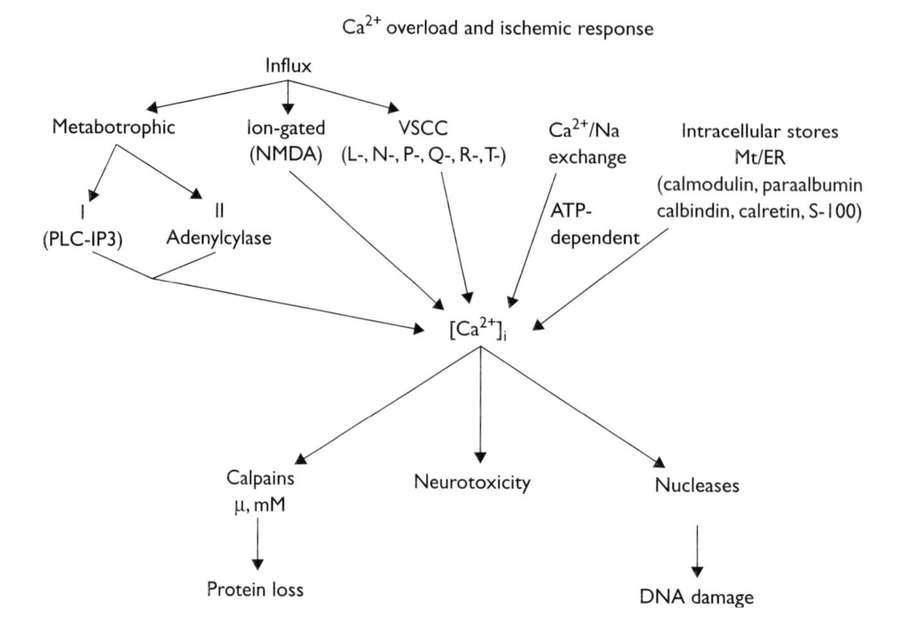

Figure 30.3 Drugs active towards Voltage Sensitive Calcium Channels (VSCC) include phenylalkylamines, benzothiazepines, and dihydropyrines (nifedipine, nimodipene, nitredipene). Specific L- (long-type) VSCC channel agents include PY-108-065, isradipene, flumarizene for N- (neuronal), SNX-111, and for R-, w-conotoxin. MK-801 is a non-competitive inhibitor of NMDA ion-gated channels, dantrolene inhibits ER-Ca^{2+} ATPase, and cyclosporin A inhibits release of Mt-Ca^{2+}.

Na^+/Ca^{2+} exchange (Figure 30.3). Ischemic damage to Mt or other organelles disrupts the homeostasis by release of calcium stores, or changes in efflux/influx or by proton driven Na^+/Ca^{2+} exchange. Damage to Mt leads to translocation of a number of cell death components, changes in Mt potential, increase in formation of PT pores with leakage of components. Among other consequences affecting neuronal survival are release of cytochrome C, AIF (apoptosis inhibitory factor), a caspase co-activator smac/DIABLO, endogenous caspases and Ca^{2+}, and changes in pro and anti-apoptotic bcl-2 proteins. Mt play central roles in neuronal integrity and survival by interaction also with cellular pro-apoptotic p53, proteins kinases, and a potential proapoptotic transcription factor TR3, also known as *Nur77* or *NGFIB* (Table 30.2) (Harrison *et al.*, 2000; Brenner and Kroemer, 2000; Marks and Berg, 2001). Agents suppressing formation of PT pores such as cyclosporin A have neuroprotective properties *in vitro* or *in vivo* lowering loss of calcium or other factors (Li *et al.*, 2000; Halestrap, 1999).

Ca^{2+} activates nucleases and calpains with detrimental effects on neuronal cytoskeletal or regulatory elements (Figures 30.3, 30.4) providing a basis to evaluate agents acting on Ca^{2+} transport including those that target different VSCC channels, or GluR subunits. This may explain the protective effects of the Ca^{2+}/Na^+ channel blocker LY393615, and the VSCC Ca^{2+}/Na^+ blocker NS-7, a phenylpyrimidine analog, both capable of crossing the blood-brain barrier (O'Neill *et al.*, 2001; Tanaka *et al.*, 2000). The NS7-mediated protection led to CREB phosphorylation in the zone bordering the penumbra. On the other hand, dantrolene blocks release of ER-Ca^{2+} dependent ATPases, and protects isolated neurons from ROS, suggesting strategies for the use of this agent or analogs in treatment of ischemia (Mattson, 1997). The effects of Ca^{2+} on the *eukaryotic initiation factor* eIF-2α, and/or on polyribosome aggregation may have significant impact on the rates of protein synthesis (White *et al.*, 2000).

30.8 Apoptosis and necrosis

Programmed cell death or apoptosis represents an orchestrated loss of neurons that is characterized by hallmark blebbing of membranes, cell shrinkage, protein and chromatin fragmentation, DNA degradation, and removal of cellular debris by macrophages. Most morphological changes are caused by a set of cysteine proteases that are activated by apoptotic signals, including ROS, that increase ischemia. Caspases are a superfamily of Asp specific cysteine proteinases that recognize a number of cytoskeletal and regulatory proteins essential for neuronal integrity (Figure 30.4). The conversion of pivotal procaspases containing death effector domains (DED) or caspase associated receptor domains (CARD) initiates sequential processing of downstream members within putative cascades. Distal caspases act as executioners by activating nucleases or inactivating nuclear repair enzymes that account for nicking of single or

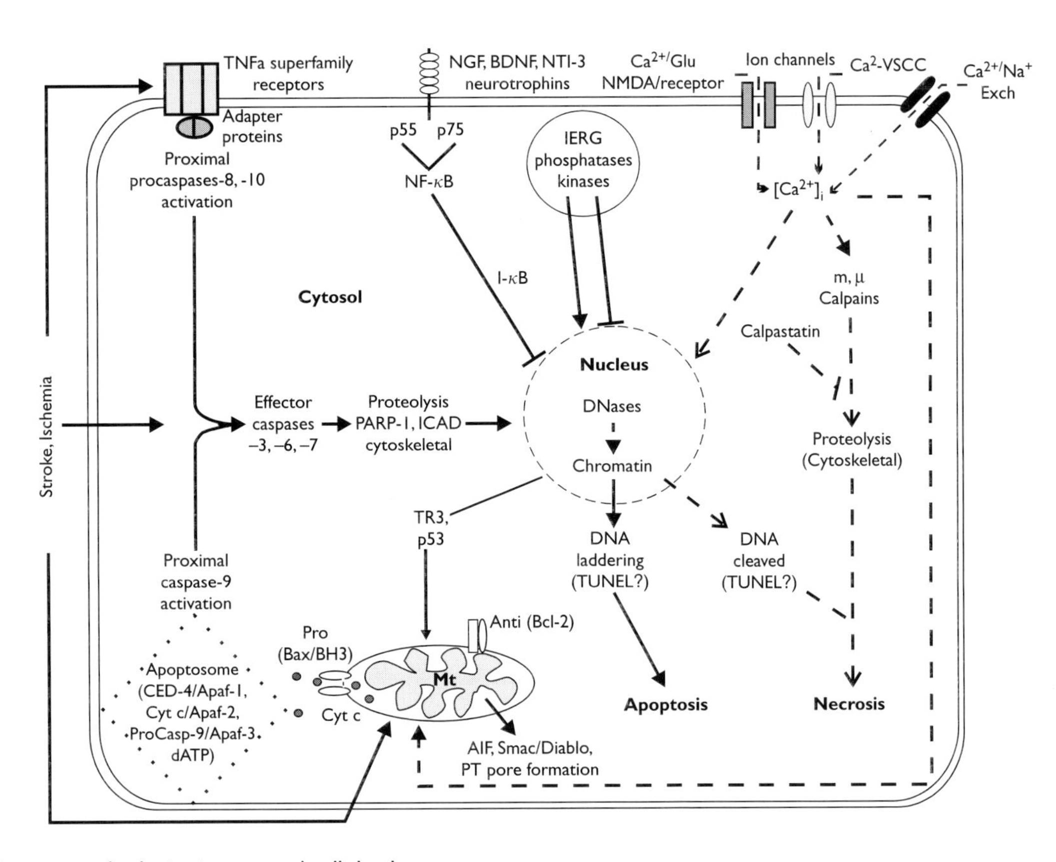

Figure 30.4 Caspases and calpains in neuronal cell death.

double DNA strands (TUNEL staining). However, an increase in TUNEL by itself is not conclusive for apoptosis since it can be confused with other forms of neuronal damage (Charriaut-Marlangue *et al.*, 1996; Zhu *et al.*, 2000). Mutant mice lacking the DNA repair enzyme PARP-1 are subject to fewer lesions following MCAO (Eliasson *et al.*, 1997) that may account for the protective effects of the inhibitor DPQ (Takahashi *et al.*, 1999). DNA damage itself stimulates PARP-1 repair by a reaction consuming NAD^+ and ATP, potentially depriving cells of energy (Ha and Snyder, 2000), which may account for increases in phosphocreatine and NAD^+ following i.c.v. injection of the specific PARP-1 inhibitor 3-aminobenzamide at 10 min prior to MCAO (Plaschke *et al.*, 2000). Studies on MCAO in rat show increase in TUNEL, along with an increase in a spectrum of caspases including −1, −2, −3, −6, and −8 (Krupinski *et al.*, 1997; Asahi *et al.*, 1997; Blomgren *et al.*, 2000a), but fewer in the case of the gerbil (Kinoshita *et al.*, 1997). Intraventricular injection of non-specific caspase inhibitors zDEVD-fmk, zVAD-fmk, and AcYVAD-cmk, reduces infarct volume and improves behavioral scores (Friedlander *et al.*, 1997; Hara *et al.*, 1997b; Hara *et al.*, 1997a; Endres *et al.*, 1998; Rabuffetti *et al.*, 2000). Inactivation of PARP-1 prevents DNA repair contributing to hallmark changes in morphology (Ha and Snyder, 2000).

Several pathways account for recruitment of pivotal procaspases prior to the activation of distal forms (see Figure 30.4). These include interaction of antigens with fas/TNFα receptors followed by binding to adaptors for recruitment of procaspases −8 or −10, or the formation of apoptosomes containing cytochrome C and the CED-4 homolog Apaf-1 in the presence of ATP for conversion of procaspase-9 (Apaf-3) as a pivotal event (Marks and Berg, 1999). The first pathway is down-regulated during neuronal maturation suggesting ischemic effects on mitochondrial cytochrome C may play a role in ischemia of non-mitotic neurons by the activation of proximal procapase-9 followed by a cascade leading to activation of downstream effectors. MCAO in postnatal rats induces marked increase in TUNEL staining within 4 h accompanied by transient increase in caspase-3 (Yu *et al.*, 2000), but the response of this caspase declines with maturation, suggesting recruitment of other cell-death proteases (Hu *et al.*, 2000). Photothrombic ischemia in parietal cortex of rats induces increases in bcl-2 proteins bak, bcl_{XL} and then procaspase-9 prior to −3 in penumbral regions (Isenmann *et al.*, 2000; Schmidt-Kastner *et al.*, 2000). Core neurons may suffer necrotic damage from release of ROS and Ca^{2+} overload leading to swelling, rupture, and loss of components with inflammatory consequences. *Apoptosis* and *necrosis* are operational terms representing extremes and do not cover intermediate forms of cell death before a commitment of neurons to a defined pathway. Other studies suggest ischemia up-regulates caspase-11 in addition to the effector −3 (Martin *et al.*, 1998; Namura *et al.*, 1998; Kang *et al.*, 2000). *In vitro*, a large number of agents or treatments induce apoptosis and prevention by exposure to caspase inhibitors provides evidence that links activation with subsequent neuronal loss (Marks and Berg, 1999). Necrosis *in vivo* is mediated via a caspase-independent pathway that remains to be identified. Existence of multiple proteases (calpains, cathepsins, caspases) indicates cross-talk between enzymes for dismantling key neuronal proteins or nuclear DNA essential for cell integrity and survival (Figure 30.4) (Fujimura *et al.*, 1999a; Fujimura *et al.*, 1999b; Blomgren *et al.*, 2000b; Yamashima, 2000; Puka-Sundvall *et al.*, 2000).

30.9 Neurotrophins

Ischemia increases expression at 2 days or later of various growth factors during the period of neovascularization. These include brain-derived neurotrophic factor BDNF, nerve growth factor NGF, basic fibroblast growth factor (bFGF), platelet derived growth factor (PDGF), transforming growth factor-β (TGF-β) and vascular endothelial growth factor (VEGF) (Koistinaho and Hokfelt, 1997; Kovacs *et al.*, 1996). Injection i.c.v. of BDNF in post natal rat brain protects against ischemic insults by a pathway mediated by the ERK phosphorylation signaling pathway (Han and Holtzman, 2000). Recent studies demonstrate that VEGF, a potential target for HIF gene interaction, contributes to changes in vasculature and permeability following stroke (LaManna *et al.*, 1998; van Bruggen *et al.*, 1999). These include increase of the neurotrophin in the core and penumbra regions, the detection of mRNA in microglia, macrophages and endothelial cells (Levy *et al.*, 1997; Cobbs *et al.*, 1998), and increase of receptors *Flt-1* and *Flk-1* in endothelial cells, and/or glia/neurons parallel to changes in the vascular related transcription factor *Ets-1* (Valter *et al.*, 1999). Interestingly, the *amyloid precursor protein* (APP) associated with production of toxic Aβ fragments in neurodegeneration has trophic properties and is upregulated in ischemia (Kalaria *et al.*, 1993). Animals that overexpress this gene are associated with larger infarct volume following ischemia.

Among properties of neurotrophins of interest is their interaction with tyrosine kinase (TrK) receptors A, -B, and -C. Phosphorylation of receptors in the presence of neurotrophins provides docking sites for phospholipase Cγ, phosphoinositide 3-kinase (PI-3-K) and various adaptors that coordinate neuronal survival (Yuan and Yankner, 2000). These include a new family of PI-3-K-*Akt* (Ser/Thr kinases) and several cell cycle regulators p53, GADD45, MDM2 and 14-3-3. *Akt* targets several key apoptosis regulators such as bcl-2 proteins, several families of transcription factors (Forkhead, CREB, NF-kB), and MAP kinase cascades mediated by ras (raf, MEK, MEKK, NIK) (Figure 30.5). In a study of hippocampal neurons, 15 min of ischemia followed by 36 h reperfusion, analyses showed an increase in caspase-3 and phosphorylation of Akt at Ser^{473} (Ouyang *et al.*, 1999). Since the MAP kinases mediate phosphorylation of bcl-2 proteins or CREB, there is potential divergence of cell-death signaling pathways mediated by NF-κB, ceramides, or other modulators (Chen *et al.*, 2001a) (see Figures 30.4 and 30.5).

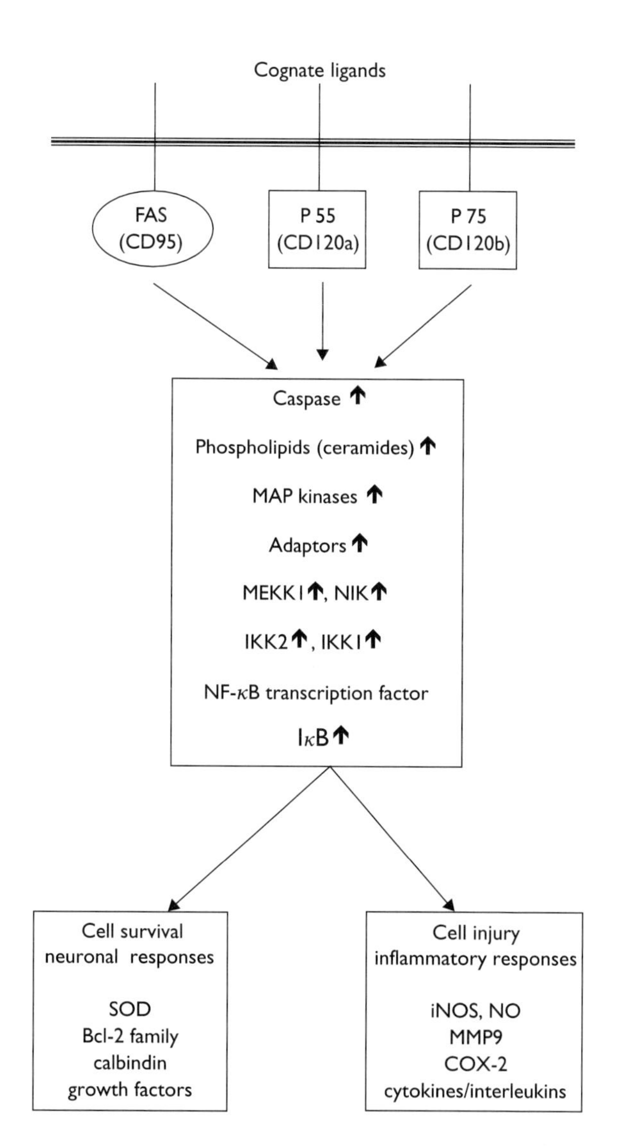

Figure 30.5 NF-κB and signaling pathways in ischemia.

30.10 Concluding comments

This overview attempts to provide a glimpse of the complex temporal hemodynamic and neurochemical consequences of CBF interruption (ischemia, hypoxia, stroke) with emphasis on focal ischemia as a model for evaluating effects of drugs. It is apparent there is little coherence for identifying the primary rate-limiting events for clinical intervention other than rapid restoration of CBF, and the use of anti-clotting and anti-inflammatory agents. Stroke results in changes that are chaotic and markedly heterogeneous for cell types in the core and penumbral zones with different outcomes in terms of neuronal survival. Consequently it is highly unlikely that any single therapy alleviates the complex hemodynamic/biochemical alterations. Effective treatment may require a cocktail of agents to prevent the catastrophic post-ischemic sequelae. Advances can be expected from the use of transgenics, or mutants lacking components identified with ischemia, or genetic methods to replace or enhance components compromised by the trauma, or by use of antisense technology to block translation of relevant damaging components.

References

1 Aronowski, J., Strong, R., Kang, H. S., and Grotta, J. C. (2000) Selective up-regulation of I kappaB-alpha in ischemic penumbra following focal cerebral ischemia. *Neuroreport 11, 1529–1533.*

2 Asahi, M., Hoshimaru, M., Uemura, Y., Tokime, T., Kojima, M., Ohtsuka, T., Matsuura, N., Aoki, T., Shibahara, K., and Kikuchi, H. (1997) Expression of interleukin-1 beta converting enzyme gene family and bcl-2 gene family in the rat brain following permanent occlusion of the middle cerebral artery. *J. Cereb. Blood Flow Metab. 17, 11–18.*

3 Bergeron, M., Yu, A. Y., Solway, K. E., Semenza, G. L., and Sharp, F. R. (1999) Induction of hypoxia-inducible factor-1 (HIF-1) and its target genes following focal ischaemia in rat brain. *Eur. J. Neurosci. 11, 4159–4170.*

4 Blomgren, K., Wang, X. Y., Zhu, C.-L., Karlsson, J.-O., Bahr, B. A., and Hagberg, H. (2000a) Activation of caspase-2, -3, -8, and -9 in neonatal rat hypoxia-ischemia. *Proc. Soc. Neurosci. 26, 245.*

5 Blomgren, K., Zhu, C., Wang, X., Karlsson, J. O., Leverin, A. L., Bhar, B. A., Mallard, C., and Hagberg, H. (2000b) Synergistic Activation of Caspase-3 by m-Calpain after Neonatal Hypoxia-Ischemia – a Mechanism of Pathological Apoptosis. *J. Biol. Chem. 276, 10191–10198.*

6 Brenner, C. and Kroemer, G. (2000) Apoptosis. Mitochondria—the death signal integrators. *Science 289, 1150–1151.*

7 Bruce-Keller, A. J. and Mattson, M. P. (1999) Mechanisms of neuroprotective cytokines: pleiotrophic effects of TNFα and TGFβ on brain injury. *In Cerebral ischemia: molecular and cellular pathopysiology., W.Walz, ed. (Totowa, NJ: Humana Press, Inc.), pp. 125–142.*

8 Chan, P. H. (1999) Oxygen radicals. In Cerebral ischemia: molecular and cellular pathopysiology., W.Walz, ed. (Totowa, NJ: Humana Press, Inc.), pp. 105–123.

9 Charriaut-Marlangue, C., Margaill, I., Represa, A., Popovici, T., Plotkine, M., and Ben-Ari, Y. (1996) Apoptosis and necrosis after reversible focal ischemia: an in situ DNA fragmentation analysis. *J. Cereb. Blood Flow Metab. 16, 186–194.*

10 Chen, J. and Simon, R. (1997) Ischemic tolerance in the brain. *Neurology 48, 306–311.*

11 Chen, Y., Ginis, I., and Hallenbeck, J. M. (2001a) The protective effect of ceramide in immature rat brain hypoxia-ischemia involves up-regulation of bcl-2 and reduction of TUNEL-positive cells. *J. Cereb. Blood Flow Metab 21, 34–40.*

12 Chen, Y., Ginis, I., and Hallenbeck, J. M. (2001b) The protective effect of ceramide in immature rat brain hypoxia-ischemia involves up-regulation of bcl-2 and reduction of TUNEL-positive cells. *J. Cereb. Blood Flow Metab 21, 34–40.*

13 Cobbs, C. S., Chen, J., Greenberg, D. A., and Graham, S. H. (1998) Vascular endothelial growth factor expression in transient focal cerebral ischemia in the rat. *Neurosci. Lett. 249, 79–82.*

14 Dawson, T. M., Steiner, J. P., Dawson, V. L., Dinerman, J. L., Uhl, G. R., and Snyder, S. H. (1993) Immunosuppressant FK506 enhances phosphorylation of nitric oxide synthase and protects against glutamate neurotoxicity. *Proc. Natl. Acad. Sci. U. S. A. 90, 9808–9812.*

15 Dawson, V. L., Kizushi, V. M., Huang, P. L., Snyder, S. H., and Dawson, T. M. (1996) Resistance to neurotoxicity in cortical cultures from neuronal nitric oxide synthase-deficient mice. *J. Neurosci. 16, 2479–2487.*

16 Dirnagl, U., Iadecola, C., and Moskowitz, M. A. (1999) Pathobiology of ischaemic stroke: an integrated view. *Trends Neurosci. 22, 391–397.*

17 Dubal, D. B., Zhu, H., Yu, J., Rau, S. W., Shughrue, P. J., Merchenthaler, I., Kindy, M. S., and Wise, P. M. (2001) Estrogen receptor alpha, not beta, is a critical link in estradiol-mediated protection against brain injury. *Proc. Natl. Acad. Sci. U. S. A 98, 1952–1957.*

18 Eliasson, M. J., Sampei, K., Mandir, A. S., Hurn, P. D., Traystman, R. J., Bao, J., Pieper, A., Wang, Z. Q., Dawson, T. M., Snyder, S. H., and Dawson, V. L. (1997) Poly(ADP-ribose) polymerase gene disruption renders mice resistant to cerebral ischemia. *Nat. Med. 3, 1089–1095.*

19 Endres, M., Namura, S., Shimizu-Sasamata, M., Waeber, C., Zhang, L., Gomez-Isla, T., Hyman, B. T., and Moskowitz, M. A. (1998) Attenuation of delayed neuronal death after mild focal ischemia in mice by inhibition of the caspase family. *J. Cereb. Blood Flow Metab 18, 238–247.*

20 Flamme, I., Frohlich, T., von Reutern, M., Kappel, A., Damert, A., and Risau, W. (1997) HRF, a putative basic helix-loop-helix-PAS-domain transcription factor is closely related to hypoxia-inducible factor-1 alpha and developmentally expressed in blood vessels. *Mech. Dev. 63, 51–60.*

21 Forstermann, U., Boissel, J. P., and Kleinert, H. (1998) Expressional control of the 'constitutive' isoforms of nitric oxide synthase (NOS I and NOS III). *FASEB J. 12, 773–790.*

22 Friedlander, R. M., Gagliardini, V., Hara, H., Fink, K. B., Li, W., MacDonald, G., Fishman, M. C., Greenberg, A. H., Moskowitz, M. A., and Yuan, J. (1997) Expression of a dominant negative mutant of interleukin-1 beta converting enzyme in transgenic mice prevents neuronal cell death induced by trophic factor withdrawal and ischemic brain injury. *J. Exp. Med. 185, 933–940.*

23 Fujimura, M., Gasche, Y., Morita-Fujimura, Y., Massengale, J., Kawase, M., and Chan, P. H. (1999a) Early appearance of activated matrix metalloproteinase-9 and blood- brain barrier disruption in mice after focal cerebral ischemia and reperfusion. *Brain Res. 842, 92–100.*

24 Fujimura, M., Morita-Fujimura, Y., Noshita, N., Sugawara, T., Kawase, M., and Chan, P.H. (2000) The cytosolic antioxidant copper/zinc-superoxide dismutase prevents the early release of mitochondrial cytochrome C in ischemic brain after transient focal cerebral ischemia in mice. *J. Neurosci. 20, 2817–2824.*

25 Fujimura, M., Morita-Fujimura, Y., Sugawara, T., and Chan, P. H. (1999b) Early decrease of XRCC1, a DNA base excision repair protein, may contribute to DNA fragmentation after transient focal cerebral ischemia in mice. *Stroke 30, 2456–2462.*

26 Furuya, K., Ginis, I., Takeda, H., Chen, Y., and Hallenbeck, J. M. (2001) Cell permeable exogenous ceramide reduces infarct size in spontaneously hypertensive rats supporting in vitro studies that have implicated ceramide in induction of tolerance to ischemia. *J. Cereb. Blood Flow Metab 21, 226–232.*

27 Ginsberg, M. C. (1990) Models of cerebral ischemia in the rodent. In Cerebral ischemia and resuscitation, A.Schurr and B.M.Rigor, eds. (Boca Raton, FL: CRC Press), pp. 1–15.

28 Grilli, M., Barbieri, I., Basudev, H., Brusa, R., Casati, C., Lozza, G., and Ongini, E. (2000) Interleukin-10 modulates neuronal threshold of vulnerability to ischaemic damage. *Eur. J. Neurosci. 12, 2265–2272.*

29 Guillemin, K. and Krasnow, M. A. (1997) The hypoxic response: huffing and HIFing. *Cell 89, 9–12.*

30 Ha, H. C. and Snyder, S. H. (2000) Poly(ADP-ribose) polymerase-1 in the nervous system. *Neurobiol. Dis. 7, 225–239.*

31 Halestrap, A. P. (1999) The mitochondrial permeability transition: its molecular mechanism and role in reperfusion injury. *Biochem. Soc. Symp. 66, 181–203.*

32 Han, B. H. and Holtzman, D. M. (2000) BDNF protects the neonatal brain from hypoxic-ischemic injury in vivo via the ERK pathway. *J. Neurosci. 20, 5775–5781.*

33 Hara, H., Fink, K., Endres, M., Friedlander, R. M., Gagliardini, V., Yuan, J., and Moskowitz, M. A. (1997a) Attenuation of transient focal cerebral ischemic injury in transgenic mice expressing a mutant ICE inhibitory protein. *J. Cereb. Blood Flow Metab 17, 370–375.*

34 Hara, H., Friedlander, R. M., Gagliardini, V., Ayata, C., Fink, K., Huang, Z., Shimizu-Sasamata, M., Yuan, J., and Moskowitz, M. A. (1997b) Inhibition of interleukin 1beta converting enzyme family proteases reduces ischemic and excitotoxic neuronal damage. *Proc. Natl. Acad. Sci. U. S. A. 94, 2007–2012.*

35 Harrison, D. C., Roberts, J., Campbell, C. A., Crook, B., Davis, R., Deen, K., Meakin, J., Michalovich, D., Price, J., Stammers, M., and Maycox, P. R. (2000) TR3 death receptor expression in the normal and ischaemic brain. *Neuroscience 96, 147–160.*

36 Hata, R., Maeda, K., Hermann, D., Mies, G., and Hossmann, K. A. (2000) Dynamics of regional brain metabolism and gene expression after middle cerebral artery occlusion in mice. *J. Cereb. Blood Flow Metab 20, 306–315.*

37 Hossmann, K. A. (1994) Viability thresholds and the penumbra of focal ischemia. *Ann. Neurol. 36, 557–565.*

38 Hu, B. R., Liu, C. L., Ouyang, Y., Blomgren, K., and Siesjo, B. K. (2000) Involvement of caspase-3 in cell death after hypoxia-ischemia declines during brain maturation. *J. Cereb. Blood Flow Metab 20, 1294–1300.*

39 Huang, C. Y., Fujimura, M., Noshita, N., Chang, Y. Y., and Chan, P. H. (2001) SOD1 down-regulates NF-kappaB and c-Myc expression in mice after transient focal cerebral ischemia. *J. Cereb. Blood Flow Metab 21, 163–173.*

40 Huang, Z., Huang, P. L., Ma, J., Meng, W., Ayata, C., Fishman, M. C., and Moskowitz, M. A. (1996) Enlarged infarcts in endothelial nitric oxide synthase knockout mice are attenuated by nitro-L-arginine. *J. Cereb. Blood Flow Metab. 16, 981–987.*

41 Huang, Z., Huang, P. L., Panahian, N., Dalkara, T., Fishman, M. C., and Moskowitz, M. A. (1994) Effects of cerebral ischemia in mice deficient in neuronal nitric oxide synthase. *Science 265, 1883–1885.*

42 Iadecola, C. (1997) Bright and dark sides of nitric oxide in ischemic brain injury. *Trends Neurosci. 20, 132–139.*

43 Iadecola, C. (1999) Mechanisms of cerebral ischemic damage. In Cerebral ischemia: molecular and cellular pathopysiology., W.Walz, ed. (Totowa, NJ: Humana Press, Inc.), pp. 3–32.

44 Isenmann, S., Stoll, G., Schroeter, M., Krajewski, S., Reed, J. C., Srinivasan, A., and Bahr, M. (2000) Expression of bak and caspases 3 and 9 following focal cortical ischemia in the rat. *Proc. Soc. Neurosci. 26, 246.*

45 Kalaria, R. N., Bhatti, S. U., Palatinsky, E. A., Pennington, D. H., Shelton, E. R., Chan, H. W., Perry, G., and Lust, W. D. (1993) Accumulation of the beta amyloid precursor protein at sites of ischemic injury in rat brain. *Neuroreport 4, 211–214.*

46 Kang, S. J., Wang, S., Hara, H., Peterson, E. P., Namura, S., Amin-Hanjani, S., Huang, Z., Srinivasan, A., Tomaselli, K. J., Thornberry, N. A., Moskowitz, M. A., and Yuan, J. (2000) Dual role of caspase-11 in mediating activation of caspase-1 and caspase-3 under pathological conditions. *J. Cell Biol. 149, 613–622.*

47 Kiessling, M., Stumm, G., Xie, Y., Herdegen, T., Aguzzi, A., Bravo, R., and Gass, P. (1993) Differential transcription and translation of immediate early genes in the gerbil hippocampus after transient global ischemia. *J. Cereb. Blood Flow Metab 13, 914–924.*

48 Kimelberg, H. K. (1999) Cell swelling in cerebral ischemia. In Cerebral ischemia: molecular and cellular pathophysiology, W.Walz, ed. (Totowa, NJ: Humana Press, Inc.), pp. 45–67.

49 Kinoshita, M., Tomimoto, H., Kinoshita, A., Kumar, S., and Noda, M. (1997) Up-regulation of the Nedd2 gene encoding an ICE/Ced-3-like cysteine protease in the gerbil brain after transient global ischemia. *J. Cereb. Blood Flow Metab. 17, 507–514.*

50 Koistinaho, J. and Hokfelt, T. (1997) Altered gene expression in brain ischemia. *Neuroreport 8, i–viii.*

51 Kondo, T., Reaume, A. G., Huang, T. T., Carlson, E., Murakami, K., Chen, S. F., Hoffman, E. K., Scott, R. W., Epstein, C. J., and Chan, P. H. (1997) Reduction of CuZn-superoxide dismutase activity exacerbates neuronal cell injury and edema formation after transient focal cerebral ischemia. *J. Neurosci. 17, 4180–4189.*

52 Kovacs, Z., Ikezaki, K., Samoto, K., Inamura, T., and Fukui, M. (1996) VEGF and flt. Expression time kinetics in rat brain infarct. *Stroke 27, 1865–1872.*

53 Krupinski, J., Issa, R., Bujny, T., Slevin, M., Kumar, P., Kumar, S., and Kaluza, J. (1997) A putative role for platelet-derived growth factor in angiogenesis and neuroprotection after ischemic stroke in humans. *Stroke 28, 564–573.*

54 LaManna, J. C., Kuo, N. T., and Lust, W. D. (1998) Hypoxia-induced brain angiogenesis. Signals and consequences. *Adv. Exp. Med. Biol. 454, 287–293.*

55 Levy, N. S., Goldberg, M. A., and Levy, A. P. (1997) Sequencing of the human vascular endothelial growth factor (VEGF) 3′ untranslated region (UTR): conservation of five hypoxia-inducible RNA-protein binding sites. *Biochim. Biophys. Acta 1352, 167–173.*

56 Li, P. A., Kristian, T., He, Q. P., and Siesjo, B. K. (2000) Cyclosporin A enhances survival, ameliorates brain damage, and prevents secondary mitochondrial dysfunction after a 30-minute period of transient cerebral ischemia. *Exp. Neurol. 165, 153–163.*

57 Lindsberg, P. J., Frerichs, K. U., Siren, A. L., Hallenbeck, J. M., and Nowak, T. S., Jr. (1996) Heat-shock protein and C-fos expression in focal microvascular brain damage. *J. Cereb. Blood Flow Metab 16, 82–91.*

58 Loddick, S. A., Turnbull, A. V., and Rothwell, N. J. (1998) Cerebral interleukin-6 is neuroprotective during permanent focal cerebral ischemia in the rat. *J. Cereb. Blood Flow Metab 18, 176–179.*

59 Marks, N. and Berg, M. J. (1999) Recent advances in neuronal caspases in development and neurodegeneration. *Neurochem. Int. 35, 195–220.*

60 Marks, N. and Berg, M. J. (2001) Amyloid (TACE, BACE) and presenilin proteases associated with Alzheimer's disease. In The role of proteases in the pathophysiology of neurodegenerative diseases, N.E.Banik and A.Lajtha, eds. (New York, NY: Kluwer Academic/Plenum Publishers), pp. 155–178.

61 Martin, L. J., Al Abdulla, N. A., Brambrink, A. M., Kirsch, J. R., Sieber, F. E., and Portera-Cailliau, C. (1998) Neurodegeneration in excitotoxicity, global cerebral ischemia, and target deprivation: A perspective on the contributions of apoptosis and necrosis. *Brain Res. Bull. 46, 281–309.*

62 Mattson, M. P. (1997) Neuroprotective signal transduction: relevance to stroke. *Neurosci. Biobehav. Rev. 21, 193–206.*

63 Morgan, J. I. and Curran, T. (1995) Immediate-early genes: ten years on. *Trends Neurosci. 18, 66–67.*

64 Nagayama, M., Niwa, K., Nagayama, T., Ross, M. E., and Iadecola, C. (1999) The cyclooxygenase-2 inhibitor NS-398 ameliorates ischemic brain injury in wild-type mice but not in mice with deletion of the inducible nitric oxide synthase gene. *J. Cereb. Blood Flow Metab 19, 1213–1219.*

65 Namura, S., Zhu, J., Fink, K., Endres, M., Srinivasan, A., Tomaselli, K. J., Yuan, J., and Moskowitz, M. A. (1998) Activation and cleavage of caspase-3 in apoptosis induced by experimental cerebral ischemia. *J. Neurosci. 18, 3659–3668.*

66 Nowak, T. S. (1991) Localization of 70 kDa stress protein mRNA induction in gerbil brain after ischemia. *J. Cereb. Blood Flow Metab 11, 432–439.*

67 Nowak, T. S. and Kiessling, M. (1999) Reprogramming of gene expression after ischemia. In Cerebral ischemia: molecular and cellular pathopysiology., W.Walz, ed. (Totowa, NJ: Humana Press, Inc.), pp. 145–215.

68 O'Neill, M. J., Hicks, C. A., Ward, M. A., Osborne, D. J., Wishart, G., Mathews, K. S., McLaughlin, D. P., Stamford, J. A., McCarty, D. R., Patrick, K. E., Roman, C., Fleisch, J. H., Gilmore, J., and Boot, J. R. (2001) LY393615, a novel neuronal Ca(2+) and Na(+) channel blocker with neuroprotective effects in models of in vitro and in vivo cerebral ischemia. *Brain Res. 888, 138–149.*

69 Ouyang, Y. B., Tan, Y., Comb, M., Liu, C. L., Martone, M. E., Siesjo, B. K., and Hu, B. R. (1999) Survival- and death-promoting events after transient cerebral ischemia: phosphorylation of Akt, release of cytochrome C and Activation of caspase-like proteases. J. Cereb. *Blood Flow Metab 19, 1126–1135.*

70 Parmentier-Batteur, S., Bohme, G. A., Lerouet, D., Zhou-Ding, L., Beray, V., Margaill, I., and Plotkine, M. (2001) Antisense oligodeoxynucleotide to inducible nitric oxide synthase protects against transient focal cerebral ischemia-induced brain injury. *J. Cereb. Blood Flow Metab 21, 15–21.*

71 Phillis, J. W., Ren, J., and O'Regan, M. H. (2000) Inhibition of Na(+)/H(+) exchange by 5-(N-ethyl-N-isopropyl)-amiloride reduces free fatty acid efflux from the ischemic reperfused rat cerebral cortex. *Brain Res. 884, 155–162.*

72 Plaschke, K., Kopitz, J., Weigand, M. A., Martin, E., and Bardenheuer, H. J. (2000) The neuroprotective effect of cerebral poly(ADP-ribose)polymerase inhibition in a rat model of global ischemia. *Neurosci. Lett. 284, 109–112.*

73 Plumier, J. C., Armstrong, J. N., Wood, N. I., Babity, J. M., Hamilton, T. C., Hunter, A. J., Robertson, H. A., and Currie, R. W. (1997) Differential expression of c-fos, Hsp70 and Hsp27 after photothrombotic injury in the rat brain. *Brain Res. Mol. Brain Res. 45, 239–246.*

74 Puka-Sundvall, M., Hallin, U., Zhu, C., Wang, X., Karlsson, J. O., Blomgren, K., and Hagberg, H. (2000) NMDA blockade attenuates caspase-3 activation and DNA fragmentation after neonatal hypoxia-ischemia. *Neuroreport 11, 2833–2836.*

75 Pulsinelli, W. A., Brierley, J. B., and Plum, F. (1982) Temporal profile of neuronal damage in a model of transient forebrain ischemia. *Ann. Neurol. 11, 491–498.*

76 Rabuffetti, M., Sciorati, C., Tarozzo, G., Clementi, E., Manfredi, A. A., and Beltramo, M. (2000) Inhibition of caspase-1-like activity by Ac-Tyr-Val-Ala-Asp-chloromethyl ketone induces long-lasting neuroprotection in cerebral ischemia through apoptosis reduction and decrease of proinflammatory cytokines. *J. Neurosci. 20, 4398–4404.*

77 Rajdev, S., Hara, K., Kokubo, Y., Mestril, R., Dillmann, W., Weinstein, P. R., and Sharp, F. R. (2000) Mice overexpressing rat heat shock protein 70 are protected against cerebral infarction. *Ann. Neurol. 47, 782–791.*

78 Ratcliffe, P. J., O'Rourke, J. F., Maxwell, P. H., and Pugh, C. W. (1998) Oxygen sensing, hypoxia-inducible factor-1 and the regulation of mammalian gene expression. *J. Exp. Biol. 201 (Pt 8), 1153–1162.*

79 Salvemini, D., Misko, T. P., Masferrer, J. L., Seibert, K., Currie, M. G., and Needleman, P. (1993) Nitric oxide activates cyclooxygenase enzymes. *Proc. Natl. Acad. Sci. U. S. A. 90, 7240–7244.*

80 Sasaki, M., Dawson, V. L., and Dawson, T. M. (1999) The NO signaling pathway in the brain: neural injury, neurological disorders, and aggression. *In Cerebral signal transduction: from first*

to fourth messengers., M.E.A.Reith, ed. (Totowa, NJ: Humana Press, Inc.), pp. 151–173.

81 Schmidt-Kastner, R., Truettner, J., Zhao, W., Belayev, L., Krieger, C., Busto, R., and Ginsberg, M. D. (2000) Differential changes of bax, caspase-3 and p21 mRNA expression after transient focal brain ischemia in the rat. *Brain Res. Mol. Brain Res. 79, 88–101.*

82 Sheng, H., Kudo, M., Mackensen, G. B., Pearlstein, R. D., Crapo, J. D., and Warner, D. S. (2000) Mice overexpressing extracellular superoxide dismutase have increased resistance to global cerebral ischemia. *Exp. Neurol. 163, 392–398.*

83 Stephenson, D., Yin, T., Smalstig, E. B., Hsu, M. A., Panetta, J., Little, S., and Clemens, J. (2000) Transcription factor nuclear factor-kappa B is activated in neurons after focal cerebral ischemia. *J. Cereb. Blood Flow Metab 20, 592–603.*

84 Stoll, G., Jander, S., and Schroeter, M. (2000) Cytokines in CNS disorders: neurotoxicity versus neuroprotection. *J. Neural Transm. Suppl 59, 81–89.*

85 Tagaya, M., Matsuyama, T., Nakamura, H., Hata, R., Shimizu, S., Kiyama, H., Matsumoto, M., and Sugita, M. (1995) Increased F1/GAP-43 mRNA accumulation in gerbil hippocampus after brain ischemia. *J. Cereb. Blood Flow Metab 15, 1132–1136.*

86 Takahashi, K., Pieper, A. A., Croul, S. E., Zhang, J., Snyder, S. H., and Greenberg, J. H. (1999) Post-treatment with an inhibitor of poly(ADP-ribose) polymerase attenuates cerebral damage in focal ischemia. *Brain Res. 829, 46–54.*

87 Tanaka, K., Nogawa, S., Nagata, E., Suzuki, S., Dembo, T., Kosakai, A., and Fukuuchi, Y. (2000) Effects of blockade of voltage-sensitive Ca(2+)/Na(+) channels by a novel phenylpyrimidine derivative, NS-7, on CREB phosphorylation in focal cerebral ischemia in the rat. *Brain Res. 873, 83–93.*

88 Troy, C. M., Derossi, D., Prochiantz, A., Greene, L. A., and Shelanski, M. L. (1996) Downregulation of Cu/Zn superoxide dismutase leads to cell death via the nitric oxide-peroxynitrite pathway. *J. Neurosci. 16, 253–261.*

89 Valter, M. M., Hugel, A., Huang, H. J., Cavenee, W. K., Wiestler, O. D., Pietsch, T., and Wernert, N. (1999) Expression of the Ets-1 transcription factor in human astrocytomas is associated with Fms-like tyrosine kinase-1 (Flt-1)/vascular endothelial growth factor receptor-1 synthesis and neoangiogenesis. *Cancer Res. 59, 5608–5614.*

90 van Bruggen, N., Thibodeaux, H., Palmer, J. T., Lee, W. P., Fu, L., Cairns, B., Tumas, D., Gerlai, R., Williams, S. P., van Lookeren, C. M., and Ferrara, N. (1999) VEGF antagonism reduces edema formation and tissue damage after ischemia/reperfusion injury in the mouse brain. *J. Clin. Invest 104, 1613–1620.*

91 Wagstaff, M. J., Smith, J., Collaco-Moraes, Y., de Belleroche, J. S., Voellmy, R., Coffin, R. S., and Latchman, D. S. (1998) Delivery of a constitutively active form of the heat shock factor using a virus vector protects neuronal cells from thermal or ischaemic stress but not from apoptosis. *Eur. J. Neurosci. 10, 3343–3350.*

92 Wang, X., Ellison, J. A., Siren, A. L., Lysko, P. G., Yue, T. L., Barone, F. C., Shatzman, A., and Feuerstein, G. Z. (1998) Prolonged expression of interferon-inducible protein-10 in ischemic cortex after permanent occlusion of the middle cerebral artery in rat. *J. Neurochem. 71, 1194–1204.*

93 White, B. C., Sullivan, J. M., DeGracia, D. J., O'Neil, B. J., Neumar, R. W., Grossman, L. I., Rafols, J. A., and Krause, G. S. (2000) Brain ischemia and reperfusion: molecular mechanisms of neuronal injury. *J. Neurol. Sci. 179, 1–33.*

94 Yamashima, T. (2000) Implication of cysteine proteases calpain, cathepsin and caspase in ischemic neuronal death of primates. *Prog. Neurobiol. 62, 273–295.*

95 Yenari, M. A., Fink, S. L., Sun, G. H., Chang, L. K., Patel, M. K., Kunis, D. M., Onley, D., Ho, D. Y., Sapolsky, R. M., and Steinberg, G. K. (1998) Gene therapy with HSP72 is neuroprotective in rat models of stroke and epilepsy. *Ann. Neurol. 44, 584–591.*

96 Yu, S. P., Han, B. H., Holtzman, D. M., and Wei, L. (2000) Focal ischemia induced apoptosis in cerebral cortex of neonatal rats. *Proc. Soc. Neurosci. 26, 246.*

97 Yuan, J. and Yankner, B. A. (2000) Apoptosis in the nervous system. *Nature 407, 802–809.*

98 Zhai, Q. H., Futrell, N., and Chen, F. J. (1997) Gene expression of IL-10 in relationship to TNF-alpha, IL-1beta and IL-2 in the rat brain following middle cerebral artery occlusion. *J. Neurol. Sci. 152, 119–124.*

99 Zhu, C., Wang, X., Hagberg, H., and Blomgren, K. (2000) Correlation between caspase-3 activation and three different markers of DNA damage in neonatal cerebral hypoxia-ischemia. *J. Neurochem. 75, 819–829.*

100 Zweier, J. L., Wang, P., Samouilov, A., and Kuppusamy, P. (1995) Enzyme-independent formation of nitric oxide in biological tissues. *Nat. Med. 1, 804–809.*

Chapter 31

Signal transduction abnormalities in epilepsies

Péter Halász

Contents

Epilepsy is a brain disorder affecting 0.5–1.0% of the population, characterized by recurrent seizures. Seizures are the result of excessive discharges of neo- or archi-cortical neurons firing in abnormal synchrony. Seizure symptoms and seizure consequences are determined by the function of the brain region from which the abnormal discharge originates, by the degree of spread to other structures of the brain, and by the quantity and rate of excitatory and inhibitory neurons participating.

Underlying mechanisms of epileptic hyperexcitability may be classified into two main groups. In the first group epilepsy is initiated by damage producing a lesion, the consequence of which spontaneously recurring seizures develop after a certain latent period. Recent advances indicate that epileptogenic cellular (synaptic) and molecular morphological and functional reorganization is the common mechanism determining this kind of exogenic (symptomatic) epilepsy type. In the second group, mutations of genes encoding ion channels of neurons is the aetiology of epilepsy (Berkovic, 1997). In this group inheritance and not exogenic damage determines the epileptic excitability.

It is traditional to classify epilepsies to "localization related" and "generalized" forms. However it has become clear that the concept of "generalized" seizures and epilepsies was supported mainly by the presence of diffuse bilateral EEG manifestations associated with non-lateralized clinical symptoms. Recently, more sophisticated study of these seemingly bilateral and diffuse EEG symptoms reveals a regional predominance and even focal origin. Coincidentally, in more and more of such cases, neuroimaging studies show localized lesions of different origins and the surgical resection of these lesions can eliminate seizures as well as the "generalized" EEG discharges. Research on the most characteristic generalized EEG pattern, the spike-wave discharge, explored the pathophysiology of the pattern as an epileptic variant of the "closed loop" sleep-like function of the thalamo-cortical system (see below for more details). This could be explained as a system related epileptic disorder linked to epileptic hyperexcitability of the thalamo-cortical system with bilateral widespread cortical representations. Therefore the term "generalized epilepsy" and consequently the classical partial versus generalized dichotomy became more and more meaningless. One solution is offered by the concept of system related classification of epilepsies which is increasingly attractive and useful. The physiological system involved could be a well known macrosystem, such as the limbic system, driven by hippocampal discharges in temporal lobe epilepsy (Engel, 1989; Wieser *et al.*, 1993; Williamson *et al.*, 1993) or a more circumscribed intra-cortical microsystem as in migration disorders (Germano *et al.*, 1996), or an abnormality of gene expression at the receptor protein gating ion – channels of certain nuclei or cortical region as for example the mutation of CHRNA4 receptors of frontal lobe cholinergic network (Sheffer *et al.*, 1994; Oldani *et al.*, 1998) in autosome dominant nocturnal frontal lobe epilepsy (ADNFLE).

In this chapter the results of recent research on the pathomechanism of the two most frequent epileptic syndromes, "temporal lobe epilepsy (TLE)" and "idiopathic generalized epilepsies (IGE)" will be summarized. These two syndromes represent examples of both of the above mentioned (symptomatic versus genetic and localization related versus generalized) dichotomies. An additional important feature of these two syndromes should be emphasized, namely that in both cases epilepsy shares the features of so called "closed loop" type physiologic working mode (Buzsáki, 1996) of thalamocortical (IGE) and the hippocampal (TLE) network. The "closed loop" working mode, both in thalamus and in hippocampus, seems to serve memory processes in the "shadow of the night" during slow wave sleep (Buzsáki, 1996; Steriade, 1999) (Table 31.1).

Synchronous discharges of neuron populations characterizing epileptic hyperexcitability may influence functionally connected structures. Repeated bombardment of target structures by excitatory impulses, transmitting pathological instead of physiological messages promote long lasting changes rendering the target structure "secondary epileptic". Secondary epileptogenesis – a kind of escalating epilepsy in the manner of "electrical metastases" – is a proven mechanism reproducable in laboratory circumstances within the framework of the procedure of "kindling", discovered by Goddard and co-workers in the second half of the sixties (1967, 1969).

Kindling refers to a phenomenon whereby repeated, local application of initially subconvulsive electrical stimulation ultimately results in more and more pronounced afterdischarge and finally in spontaneous intense partial and generalized seizures. Repeated focal injections of chemicals that induce afterdischarge are also capable of producing kindling. The susceptibility for kindling varies and limbic structures have proved to be more sensitive than neocortex. This can be established in every species, including rhesus monkeys and baboons. There are important data showing that the hyperexcitability of the kindled brain is underlied by the formation of novel synaptic circuits (McNamara and Wada, 1997). Kindling was demonstrated to be accompanied by sprouting of the mossy fiber axons of the dentate granule cells of hippocampus in the same way as in other models of temporal lobe epilepsy (TLE, see later) and in human TLE itself (Sutula *et al.*, 1988). Kindling offers an explanation for the so called "secondary epileptogenesis". The existence of secondary epileptogenesis in humans is still questionable, however, there are a lot of data about "mirror focus" development in cases where the evolution of local epileptic disorder due to a tumor induced contralateral changes without any hint of previously existing epileptic disorder in that side (Wilder, 1982 and Morrell, 1979). It was shown that the establishment of a "mirror focus" requires ongoing protein synthesis, as well as intact axoplasmic transport. The same is true for kindling (Morrell *et al.*, 1975). These data, together with the fact that fully developed secondary epileptogenic foci persist after ablation of the primary lesion, suggest that secondary epileptogenesis is likely to be associated with some structural modification at the postsynaptic site (Morrell and Toledo-Morrell, 1999). Certain epileptic mechanisms show obvious evolution by secondary involvement of new structures after long lasting existence of epilepsy. Other epilepsies follow a rather regressive course, or even improvement in an age dependent manner. Kindling and secondary epileptogenesis are clear cut examples that pathological communication between cells may lead to disorders. In our case pathological information (too much excitation) changes the working mode of the cells where the distorted information was communicated realizing that "seizures beget seizures" (Gowers).

Table 31.1 Etiological characteristics and relation to brain physiology in temporal lobe (TLE) and idiopathic generalized epilepsy (IGEP)

	TLE	*IGEP*
Genetic determination (idiopathic)	–	+
Symptomatic (lesional)	+	–
Related system	Hippocampal network	Thalamo-cortical network
Physiological working mode	Close loop bursting mode (like in sleep)	Close loop bursting mode (like in sleep)
Relation to memory processing	++	++

31.1 Epileptic hyperexcitability in the limbic system due to synaptic reorganization in the hippocampal network after injury (model for mesial temporal lobe epilepsy)

31.1.1 *Mesial temporal lobe epilepsy (MTLE) – time sequence of development – animal models of TLE*

TLE is perhaps the most common human epileptic condition in adults and represents an important proportion of childhood epilepsies as well. The common scheme in development of TLE is: a) early epileptogenic damage (frequently complicated febrile seizures or status epilepticus) affecting mainly the sensitive CA1 and CA3 region of the dentate pyramidal cells, b) development of hippocampal sclerosis (detectable on MRI), and a parallel synaptic reorganization, mossy fiber sprouting (MFS) could be observed (Babb *et al.*, 1991), c) the net behavioral result evolves in the form of recurrent complex partial seizures, after some years of long latent period. This sequence of events was modeled in previous years by several animal models either by the technique of electrical stimulation (Repressa *et al.*, 1989) (kindling) or chemical interventions such as kainic acid (Tauck and Nadler, 1985; Repressa *et al.*, 1993) pilocarpin (Turski *et al.*, 1983). The study of these animal models as well as the resected specimens of human TLE treated by surgery made it possible to understand more about the underlying mechanisms.

31.1.2 *Cell loss, synaptic reorganization and hyperexcitability in the hippocampal network (the hen – egg problem)*

Let us take a deeper insight in to the transformation of the hippocampal network in epileptic samples. The main pathway of excitation in the hippocampal network consists of a trisynaptic circuit in which information from the entorhinal cortex (the main afferent gate of the hippocampal structure) is transmitted sequentially, first to dentate granule cells, from there to CA3 pyramidal cells, and finally from CA3 to CA1 pyramidal cells through Schaffer collaterals. The CA3 cells show strong excitatory arborization to the neighboring pyramidal cells and therefore this sector is especially prone to epileptic synchronous discharges, while recurrent excitatory synapses between granule cells are not seen in normal conditions.

Cell death, reactive gliosis and MFS are common features shared by the above mentioned experimental models and human mesial temporal lobe epilepsy (Ben Ari, 1980). The causal chain between these features is still not entirely clear (Chevassus *et al.*, 1997) and a circular causality is probably at work. Damage of the CA4 pyramidal cells may induce MFS to compensate loss of their main targets, on the other hand cell death in CA3 sector may be caused by paroxysmal activity of the afferent granule cells over releasing glutamate (Ben Ari *et al.*, 1980) or, alternatively, the release of Zn^{2+} by mossy fiber synapses could have excitotoxic effects on CA3 postsynaptic cells (Charton *et al.*, 1985).

The timing of events in development of hippocampal epilepsy seems to be another important factor. The susceptibility of hippocampal formation to seizing shows characteristic changes during ontogenetic development. One week old rats are much less prone to seizures than those reaching the second and third weeks of life (Mares, 1973; Michelson *et al.*, 1989; Swann and Moshé, 1997). One factor explaining this could be the time difference in maturation of the cell elements in the hippocampus. Granule cells mature up to postnatal day 15, while pyramidal cells are already mature at birth. In a preparation obtaining "granule cell free" hippocampus by applying postnatal irradiation to selectively destroy dividing cells, epilepsy could not be observed in response to application of epileptogenic agents. The early resistence to epilepsy in the hippocampus is also congruent with the phenomenon of "remodeling" of circuits during early life, corresponding to persistence of only those connections which are functioning and the loss of those which are not functioning (pruning) according to the principle "firing together, wiring together". Therefore the epileptogenicity of hippocampal damage is probably regulated by a "time-window" explaining why early damage is not so "epileptogenic" compared with later damage (Swann and Moshé, 1997). Cell death is present in CA1 and CA4 sectors of the pyramidal neurons and CA2 and partial CA3 is usually preserved. This pattern is similar to the susceptibility to death of the condition of status epilepticus, even without hypoxia, in experimental circumstances (Meldrum *et al.*, 1979).

Recently MFS has become one of the most extensively studied phenomenon in the context of epileptic hippocampal hyperexcitability. It can be detected easily by the Timm stain which reflects the high zinc content of mossy fiber axons in animal as well as in human specimens (Tauck and Nadler, 1985; Represa *et al.*, 1989; Cavazos *et al.*, 1991). The sprouted mossy fibers of granule cells show new collateral branches growing across the granule cell layer of dentate gyrus which reach the inner third of the molecular layer and form recurrent synapses with the granule cell dendrites, providing a dense network for syncronized overexcitation. But newly developed synapses onto GABA-ergic neurons have also been found (Ribak and Peterson, 1991) raising the possibility that inhibition could also take part in the new discharge pattern in which granule cells behave as an epileptic neuron population promoting seizure initiation and propagation through the hippocampal excitatory pathway.

In recent years a large body of evidence has been gathered to support the statement that temporal lobe epilepsy is very poorly treated by traditional antiepileptic drugs developed using methods testing seizure propensity in acute seizure models or even in chronic epileptic animal models. These drugs are effective only in protecting against seizures, they do not prevent the development of epilepsy and they are not able to interfere with deteriorating vicious circles and cascade of events within the framework of synaptic reorganization. Therefore we need a deeper understanding of the whole machinery of cellular plasticity leading to synaptic reorganization and new vantage points are required for therapeutic interventions, probably quite different from the traditional ones. From this point of view the candidates are neuromodulators, growth factors, neuropeptides and related proteins, structural proteins and second messengers.

Neurons utilize neurotransmitters along with one or more neuropeptides. The release of a transmitter or a neuropeptide depends on the frequency of action potentials. Peptides may diffuse for long distances from their site of release to their receptors. Neuropeptides may facilitate or inhibit the action of neurotransmitters influencing transmitter synthesis, or metabolizing enzymes, or receptors (Kito and Miyoshi, 1991).

Metabotrophic receptors are coupled to a G-protein, which has seven membrane-spanning domains which, by changing shape, activates an enzyme on the inside surface of the membrane, generating diffusable messengers such as cyclic adenosine monophosphate (AMP), and cyclic guanosine monophosphate (cGMP). This second messenger will, in turn, act on enzymes such as protein kinases to modify the function of target molecules (receptors, ionic channels, or other enzymes or binding proteins) to change cell excitability or gene expression.

Neuromodulators play a role in adaptation processes during and after seizure expression. Opioid peptides modulate postictal excitability and behavior both in animal experiments and human epileptic events. In epileptic hippocampi, long term changes in opioid peptide receptors and in neuropeptide expression are commonly seen, and NPY is expressed in dentate granule cells, where it is not present normally (Rizzi *et al.*, 1993). Nitric oxide (NO) another important neuromodulator, on one hand helps adaptation to greater oxygen demand during seizures by vasodilatation, and works as an endogenous anticonvulsant since blockers of NO synthesis enhance seizure activity, but when liberated in excessive amounts it can be toxic causing excitotoxic cell death (Wang *et al.*, 1994).

Cyclic AMP, as an ubiquitous second messenger, and membrane phospholipids induce growth factors and immediate early genes, regulate the function of protein kinases and take part in seizure regulation in many ways (Jope *et al.*, 1992). Calcium also plays an important role in transmitter release, long term plasticity, and cell damage. Antiepileptics and neuroprotective agents reduce calcium entry into the neurons by blocking the depolarization of presynaptic terminals.

There are several structural proteins that also have roles in synaptic reorganization in development of epilepsy in temporal lobe models. MFS is, for example, probably favored in the dentate gyrus by the presence of growth promoting substances such as laminin and fibronectin in the granule cell layer (Niquet *et al.*, 1994, 1995). Astrocytes may have a function in this procedure as well. Proliferating astrocytes in the hilus express embryonic Neural Cell Adhesion Molecule (NCAM), fibronectin and vitronectin enhance axonal branching and growth, while reactive astrocytes in CA3 express tenascin antibodies, which is an unfavorable substance for mossy fibers. Another major issue is the involvement of growth factors in sprouting of mossy fibers. An increase of Nerve Growth Factor (NGF) expression in the dentate gyrus in response to kindling has been described (Gall and Isackson, 1989). Seizure dependent changes were observed in levels of fibroblast growth factor (bFGF) and brain derived nerve growth factor (BDNF). Sprouting of mossy fibers was preceded by an important expression of α-tubulin and microtubule associated protein 2 in granule cells (Represa *et al.*, 1993). These data show that mossy fiber collateral branch sprouting originates under the influence of trophic factors. Mossy fiber growth and orientation seems to be guided by astrocytes excreting cell adhesion and substrate molecules.

In the light of these results epilepsy should be considered as a vicious circle where seizures set in motion a cascade of molecular and genomic changes including early genes, trophic factors, cytoskeletal proteins, and adhesion-substrate molecules besides transmitter receptors. These changes contribute to development of long lasting functional and morphological changes summarized as synaptic remodeling, and pave the way for the evolution of epileptic mechanisms.

31.1.3 Reorganization of the interneuronal network in TLE

Synaptic reorganization is not restricted to pyramidal cell loss, glial scar, and sprouting of the mossy fibers. A more widespread reorganization involving the whole interneuronal network develops in the epileptic hippocampus. A large proportion of inhibitory interneurons are preserved in the epileptic human dentate gyrus, but their distribution, morphology and synaptic connections differ from controls. In a study of 20 surgically treated temporal lobe epileptic patients the resected hippocampal tissue was examined (Magloczky *et al.*, 2000). Three types of inhibitory interneurons, containing Ca^{2+} binding proteins (calbindin (CB), calretinin (CR), and substance P (SPR)) were analyzed. The common features of the changes found in the samples include: large hypertrophic CB containing cells appear in the hilus, and their dendritic tree is longer than that of the control; the number of CR-positive cells decreases, while the CR-positive supramammillary pathway extends to innervate the entire width of the stratum moleculare; the distribution of SPR-positive cells changes, most are located in the hilus in the control, whereas the epileptic samples show them to be numerous in the stratum moleculare. In addition, they have a bishy and highly varicose dentritic tree. Most of the granule cells lose their CB content and migrate up to the stratum moleculare in the epileptic tissue. The preserved interneurons show signs of a high rate of metabolic activity. The expansion of the CR-positive supramammillary afferents demonstrates that in addition to MFS, there is a second, extrahippocampal source of excess excitation of dentate granule cells (Figure 31.1). Several morphological features of the changed interneurons seem to be similar to the immature condition of the hippocampus when neurotrophic factors are active, and the increased activity shown in hippocampal epilepsy models could explain the observed morphological changes. Therefore the possibility that neurotrophins have a crucial role in the regulation of cell survival during epilepsy should also be considered, and this would offer new approaches in the development of protective antiepileptic drugs.

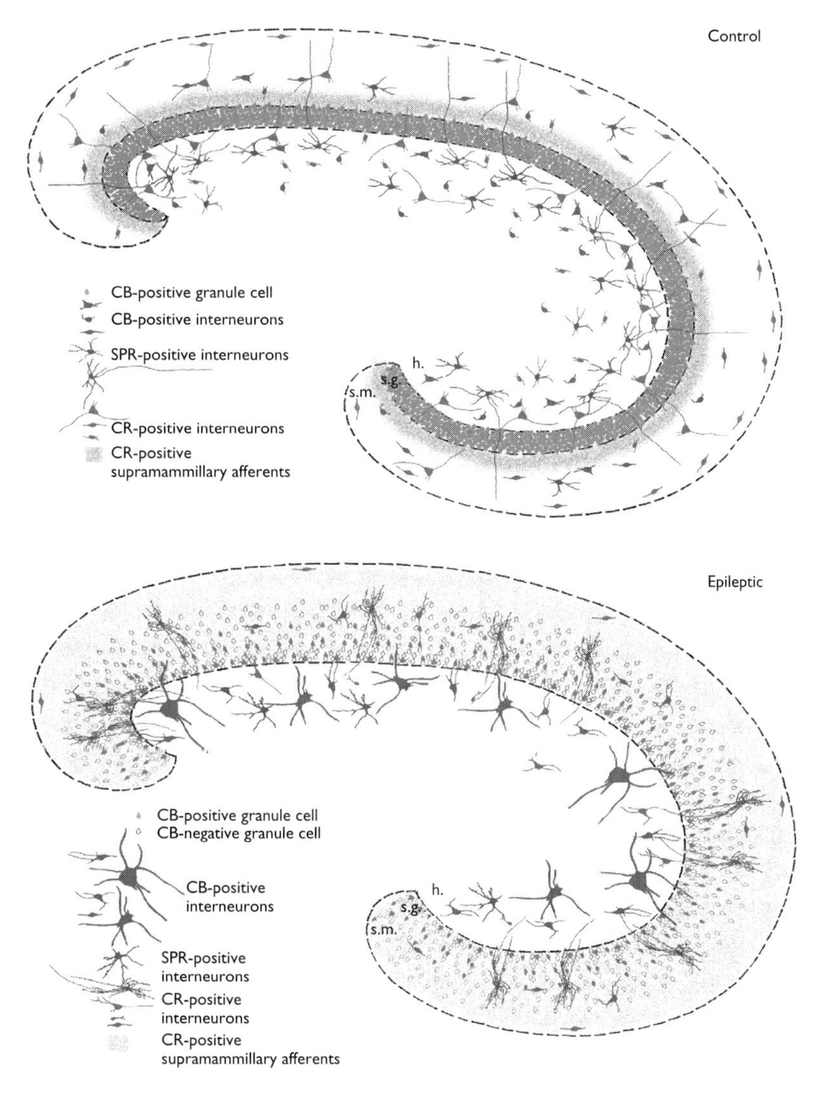

Figure 31.1 Schematic diagram of the laminar distribution of the examined interneuron subpopulations in control and epileptic dentate gyrus. The most characteristic representatives of each examined cell type were drawn by camera lucida and scanned into the computer. In the epileptic samples, the interneuronal network has changed considerably, even the "resistant" interneurons are modified, and differ from the control. Large hypertrophic CB-containing cells appear in the hilus, their dendritic tree is longer than those in the control (red). The number of CR-positive cells decreases (green), while the CR-positive supramammillary pathway extends to innervate the entire width of the stratum moleculare (green shading). The distribution of SPR-positive cells changes. Most are located in the hilus in the control, whereas the epileptic samples show them to be numerous in the stratum moleculare. In addition, they have a bushy and highly varicose dendritic tree (blue). Granule cells are also shown (grey). Most lose their CB-content (open circle) and migrate up to the stratum moleculare in the epileptic dentate gyrus. All characteristic profound changes have been combined into this diagram, but it should be pointed out that each individual case shows a combination of these abnormalities in cell distribution and morphology at different degrees (From Maglóczky *et al.* 2000) (see Colour Plate XXI).

31.1.4 Histological changes in hippocampal sclerosis and their in vivo visual assessment by MRI – clinical studies contributing to the understanding of the role of HS in the pathophysiology of TLE

Histological examination of hippocampal atrophy associated with epilepsy was first described by Sommer in 1880, who documented the selective vulnerability of the sector named after him as "Sommer's-sector". The association of hippocampal atrophy and sclerosis with epilepsy and TLE was first supported by autopsy findings (Bratz, 1899; Stauder, 1936; Meyer *et al.*, 1954; Margerison and Corsellis, 1966). The most prominent changes occur in the Ca1 region of pyramidal cells (Sommer's sector), less severe but important cell loss appears in the Ca3 region and in the hilus of the dentate gyrus (Ca4 or end folium). The Ca2 region and the dentate gyrus shows relative resistance. The neuronal loss can be measured nowadays by three-dimensional counting techniques (Braendgaard and Gundersen, 1986; Gundersen *et al.*, 1988; Oorschot, 1994). There is a characteristic accompanying fibrous astrogliosis proportional to the degree of damage. In the light of recent research, astroglial proliferation seems to be due to the liberation of different plasticity proteins promoting proliferation of glial elements as a part of the reorganization process, not simply due to scar production. It has also been recognized that in a large proportion of cases the granular cell layer is wider than normal and extends into the molecular layer, or may be duplicated (Houser, 1990). The cause of the granule cell dispersion could be a migration disorder, but the role of damage due to seizures cannot be excluded. The most consistent feature of histological changes in hippocampal sclerosis is MFS which may be best demonstrated by Timm's staining and/or dynorphin immunohistochemistry. The recurrent sprouted axons of the granule cells contain zinc-labeled vesicles in their excitatory terminals (Figure 31.2).

Improvement of MR neuroimaging techniques makes it possible to detect HS visually, enables long term follow up, and its quantitative measurement. HS scanning should be performed along the axis of the hippocampus in thin (1–3 mm) sections, and application of inversion recovery and T2-weighted images as well as FLAIR images are necessary. Hippocampal volumetry also proved to be helpful in the hands of several investigators (Jack *et al.*, 1990; Cendes *et al.*, 1993; Adam *et al.*, 1994). Cell loss appeared in the form of "atrophy", reflected by decrease in volume and signal in T1 and inversion recovery images, while astrogliosis appeared as a signal increase in T2-weighted and FLAIR images. MRI-based hippocampal volumes have been shown to correlate with the extent of cell loss. Unilateral HS is found in the 40–60% of patients with intractable TLE, and 60–80% of them have a good outcome after surgery with different degrees of temporal lobe resections, including resection of the head and certain portions of the hippocampal body

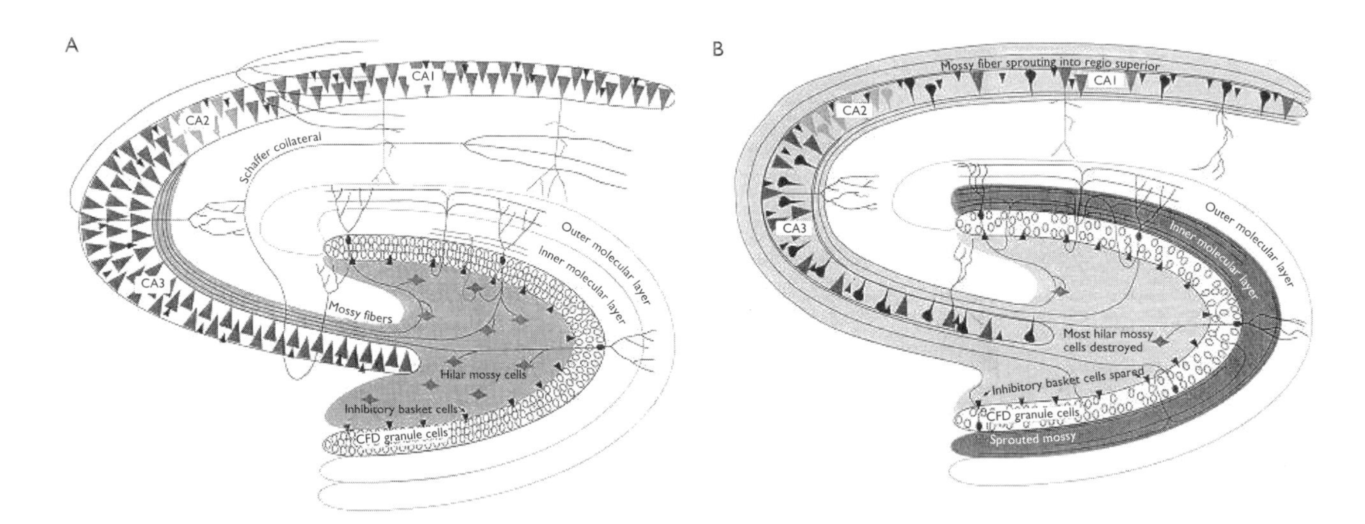

Figure 31.2 Schematic illustration of normal hippocampal anatomy (left) and the probable reactive sproutling changes that occur in hippocampal sclerosis (right). (A) The anatomic areas of the normal fascia dentata and Ammon's horn are labelled. Granule cell dendrites are located in the outer and inner molecular layers. Activation of the granule cell axons, the mossy fibers (light-shaded gray area), can stimulate CA3 pyramids. In addition, hilar mossy cells (rhomboid-shaped) and inhibitory basket cells (solid triangles) are stimulated, thereby regulating granule cells by feedback mechanisms. The axons of mossy cells are in the inner molecular layer. CA3 axons, the Schaffer collaterals, stimulate CA1 pyramids. (B) In hippocampal sclerosis, there is loss of hilar mossy cells, CA3 neurons, and other hippocampal cells. Current evidence indicates that mossy fibers sprout axon collaterals into the inner molecular layer (dark gray shade) and into regio superior (light shaded gray area). Other fibers, such as neuropeptide Y and GABAergic axons, also sprout axon collaterals into the inner molecular layer (dark gray shade). In other words, hippocampal sclerosis is associated with aberrant excitatory (mossy fibers) and inhibitory (GABAergic) axon circuits, which may initiate or contribute to epileptogenesis. (From Matthern *et al.* 1997.)

(Paesschen and Révész, 1997). Bilateral HS has been reported in 50–88% of patients with TLE (Babb, 1991). In the acute seizing stage in TLE, transient swelling of hippocampal formation was demonstrated by serial MRI investigations (Nohria *et al.*, 1994; Jackson *et al.*, 1995).

Besides the classical form, of HS, coexistence of other lesions are frequently demonstrated. This "dual pathology" consists of neoplasms, vascular malformations, post-traumatic scars, cortical dysplasias and microdysgenesis, and inflammatory lesions. Tumors are the second largest pathological group after HS, and among them DNTs, gangliogliomas, oligodendrogiomas and astrocytomas are most frequently found. In these cases febrile convulsions were found less frequently in the patient's history (Raymond, 1994). Therefore the combination of HS and developmental anomalies seems to be an important pathophysiological constellation. Since developmental anomalies are proved to be highly epileptogenic lesions, the possibility of secondary evolution of HS by kindling is one rational scenario. The examples of secondary evolution of HS in extratemporal developmental lesions such as localized periventricular heterotopias (Janszky *et al.*, 1999) provide further support for this kind of origin.

31.2 Epileptic hyperexcitability in the thalamo-cortical system due to a genetic abnormality as the basic mechanism of idiopathic generalized epilepsies

31.2.1 Idiopathic Generalized Epilepsy (IGE)

Idiopathic generalized epilepsies form a well known and important form of epilepsy in childhood and adolescents. Common characteristics are the paroxysms of generalized spike-wave discharges around 3 Hz, lack of neurological and mental deficit symptoms in interictal conditions, and lack of focal features in interictal and ictal EEG and ictal clinical signs. The most frequent and best defined subgroups are absence epilepsy of early school age and juvenile myoclonic form of postpubertal early adulthood period.

31.2.2 IGE and the thalamo-cortical system

The characteristics of the thalamo-cortical system play a key role in the mechanism of idiopathic generalized epilepsies and also in the secondary generalization of localization related epilepsies. The thalamus is the main sensory relay station transmiting sensations from the external world toward the cortical mantle in the awake state. The same structure works in a special closed-loop bursting-mode, shuting down the influence of the external world during slow wave sleep. This working mode is due to oscillations of excitation and inhibition in the intrathalamic and thalamo-cortical network and is reflected by the properties of sleep EEG features such as spindles and slow waves. It is becoming increasingly clear that the physiopathogenesis of idiopathic generalized epilepsy can be explained by the epileptic distorsion of the oscillatory working mode of the system. This is the reason why properties of idiopathic generalized epilepsy are so intimately interconnected with manifestations of sleep, and why it is so sensitive to changes in the sleep-wakefulness continuum.

In the last decade intensive research has explored the characteristics of the "thalamic clock" which has important heuristic aspects both in the pathomechanism of generalized epilepsies and in the physiology and pathology of slow wave sleep. It is likely that further implications in other neurological disorders could be also expected.

31.2.3 Functional properties of the thalamo-cortical system

The basic interconnections in the thalamo-cortical and intrathalamic loops are as follows: neurons of the sensory thalamic nuclei and the cortical pyramidal neurons have a reciprocal, mutually glutaminergic excitatory connection. Both the thalamocortical and corticothalamic projections give an excitatory collateral to the nucleus reticularis thalami (NRT). The latter structure sends inhibitory GABA-ergic innervation to the thalamic neurons and to other reticular nuclei providing a strong inhibitory gating influence which is the main determinator of the rhythmic activity generated by the thalamic network (Steriade *et al.*, 1985; Steriade, 1991) (Figure 31.3). The membrane conductance of thalamic neurons is characterized by the presence of low threshold T-type Ca^{2+} channels, the bursting properties of which have a role in the amplification of thalamic oscillations. In the awake state, T-current is inactivated and does not interfere in the transmission of sensory information relayed by the thalamus. Depolarization of the thalamic neurons is provided in the awake state by the ascending influence of the brain stem activating system. Desinactivation of the low threshold Ca^{2+} current dominates the functional properties of the thalamic neurons when they are hyperpolarized from their waking resting membrane potentials. This occurs in the period of drowsiness and transition from the awake state to slow wave sleep when the activity of the ascending arousal structures decrease. Under this condition the relay neuron will answer to either cortical or afferent sensory stimulation with a rebound Ca^{2+} burst to the inhibitory feedback from the NRT. Since the NRT innervates large thalamic neuron populations, synchronized GABA-A and -B-ergic inhibitory postsynaptic potentials desinactivate low threshold Ca^{2+} current spikes in an important number of thalamic neurons what is reflected by a spindle wave and at the same time initiates rebound excitation. This excitation then reactivates NRT and activates the cortex. Reactivated NRT initiates an inhibitory postsynaptic potential in thalamic neurons again forming a second spindle wave. The hyperpolarization of thalamic neurons reopens the way for the activation of spike burst based on Ca^{2+} current and the whole excitatory-inhibitory cycle involving the NRT and thalamic relay cells is set into motion again. The NRT is the driving force of spindling. Thalamic structures isolated from NRT do not show oscillatory behavior, while the NRT produce spindling even after isolation from the rest of the thalamus.

31.2.4 From "bursting-mode" to spike-wave pattern

It was suspected long ago that thalamic structures have an essential role in the generation of absence epilepsy (Jasper and Drooglever-Fortuyn, 1946; Williams, 1953). After exploration of the phasic inhibitory gating exerted by NRT on the thalamocortical circuit, it became clear that this mechanism could be the basis of the spike-wave pattern in absence epilepsy (Figure 31.4). Gloor showed that the same thalamic volley eliciting spindles on the cortex in normal sleep evoke the spike-wave pattern when cortical excitability is elevated by application of penicillin (Gloor *et al.*, 1990). They also showed that arousal influences – depolarizing the thalamic relay cells – were able to block spike wave paroxysms while shifts toward sleep promoted their appearence. The transition from spindles to spike-wave was hypothesized to be underlied by the increase of strength of cortical recurrent inhibition resulting in a longer lasting slow wave instead of every two spindles. In recent years the possible role of the increased function of low threshold Ca^{2+} current in thalamic (NRT and relay cell) neurons became more and more evident. This would class the disorder among the "channelopathies", and would emphasize the role of the thalamus more than the cortex. However, the thalamocortical circuit may be influenced via several ascending and descending pathways from the brain stem and the cortex. Therefore, it is not surprising that the same distorsion of functions resulting in epileptic spike-wave discharges in the system could stem from influences from different key points in the network. The Montreal School (Gloor, 1988) emphasized the coexistence of three components: increased excitability of the cortex, weakness of the brainstem tonic arousal influence and phasically inhibited thalamo-cortical stream of impulses, now known as "bursting mode" of the thalamocortical system.

31.2.5 Lessons learned from genetics

The recently published single locus genetic absence models show a considerable diversity of underlying mechanisms and heterogeneity of genetic basis resulting in the same absence phenotype. At this moment we have two polygenic rat and five single locus mice absence models (Marescaux *et al.*, 1984;

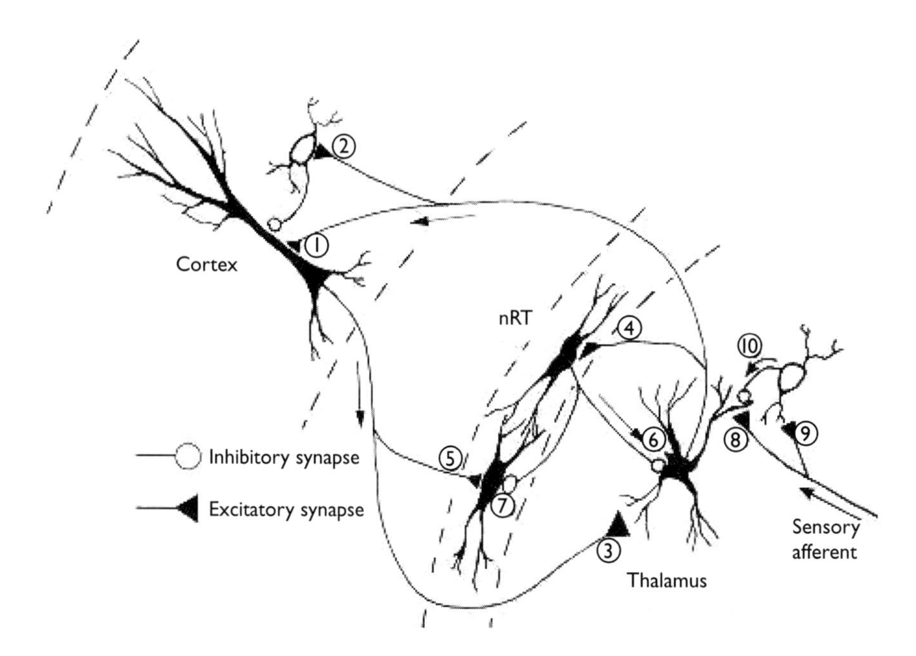

Figure 31.3 The basic thalamocortical circuit. Thalamic neurons in primary sensory relay nuclei project to layers III/IV and V/VI of the cerebral cortex. This projection terminates on both pyramidal neurons (synapse 1) and inhibitory interneurons (synapse 2) in the cortex. Layer VI pyramidal neurons reciprocally innervate the same area of thalamus from which an ascending afferent is received (synapse 3). Both the thalamocortical and the corticothalamic projections send an axon collateral to the nucleus reticularis thalami (NRT; synapses 4 and 5). Nucleus reticularis thalami provides inhibitory GABAergic innervation to the thalamus (synapse 6) and to other NRT neurons (synapse 7). The major sensory afferents to the thalamus synapse, onto the dendrites of both thalamic relay neurons (synapse 8) and inhibitory interneurons (synapse 9). The dendrites of inhibitory interneurons can function as both pre- and postsynaptic elements and can provide inhibitory innervation of thalamic relay neuron dendrites (synapse 10), as well as conventional axonal synaptic connections (synapse 10). (From Coulter 1997.)

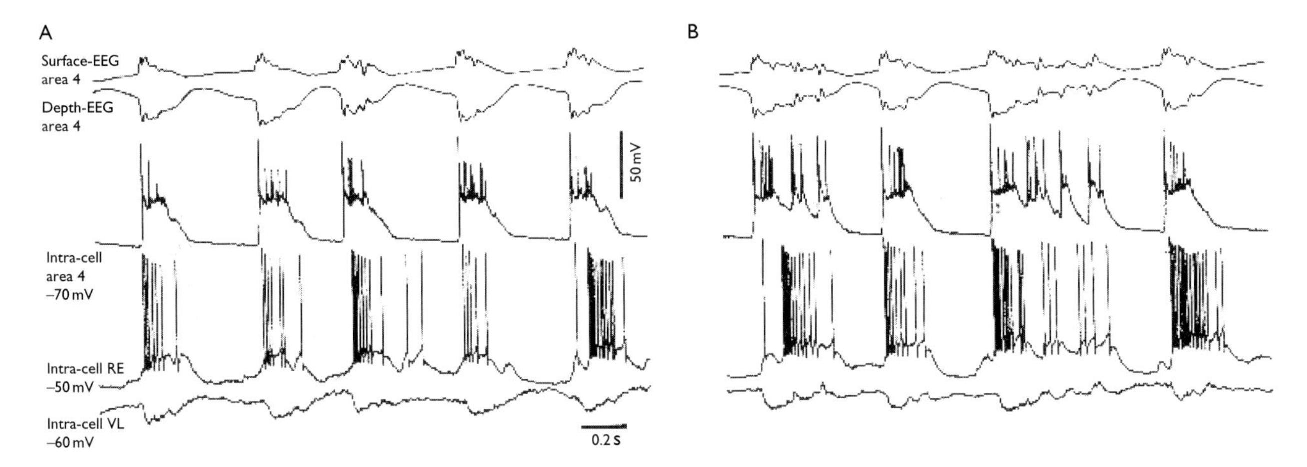

Figure 31.4 Inhibitory processes of thalamocortical cells during cortical seizures (polyspike-wave complexes at ~2 Hz) are due to cortically elicited excitations of GABAergic thalamic reticular neurons. Dual simultaneous intracellular recordings from cortical neuron in area 4 and thalamic ventrolateral (VL) neuron, together with surface- and depth-EEG from area 4, close to the cortical cell. A thalamic reticular (RE) neuron was intracellularly recorded in the same animal, during another spontaneous seizure that displayed EEG features virtually identical to those depicted here. The non-simultaneous RE neuron traces, shown between the cortical and VL neuron traces, demonstrate the temporal similarity of these spontaneous seizures. Note the IPSPs in the VL cell, in close time-relation with paroxysmal depolarizing shifts in cortical cell and spike-bursts in the RE cell. (From Lytton *et al.* 1997.)

Noebels, 1984; Coenen *et al.*, 1992). In the GAERS, polygen rat model increase of the T-type Ca^{2+} current was described in the thalamic reticular nuclei (Avanzini *et al.*, 1987). Among the single locus models in the tottering mouse a gene-linked proliferation of noradrenergic loecus coeruleus axon terminals was shown and the neonatal correction of the inherited hyperinnervation with a selective neurotoxin prevents the expression of epilepsy (Noebels, 1984). In an other mutant, the lethargic mouse, an increase in GABA-B receptors has been reported in thalamocortical neuron populations compared with littermates without epilepsy. The magnitude of the increased number of GABA-B receptors correlated positively with the frequency of absence seizures. The spike-wave pattern was dependent on GABA-B transmission: GABA-B agonists increased while GABA-B antagonists decreased the frequency of absences. Noradrenergic fiber proliferation is not present in this animal. Again, other mutants do not show either of the characteristics of tottering and lethargic mice (Table 31.2).

Due to intensive research in this field and with understanding of the anatomy and physiology of the responsive thalamocortical network, the pharmacological control of spike-wave discharges is becoming increasingly clear.

31.2.6 Pharmacology of spike-wave pattern

A number of mechanisms appear to regulate the ability of the thalamocortical network to undergo oscillatory burstfiring. Circuitry, and the nature of synaptic transmission within the cortico-thalamic network, explains the pharmacological responsivity of idiopathic generalized epilepsies.

RECIPROCAL GLUTAMATE TRANSMISSION BETWEEN THALAMIC RELAY CELLS AND CORTICAL PYRAMID CELLS AND BETWEEN THE PYRAMIDAL CELLS AND NRT (COULTER 1997)

Activation of cortico-thalamic pathways have been shown to generate excitatory postsynaptic potentials in thalamic relay neurons mediated by NMDA receptors. Since diffuse cortical hyperexcitability has been found to be associated with absence-like seizures in the feline penicillin model (Gloor and Fariello, 1988), it is possible that there is an increased excitatory synaptic input to the thalamus from the cortex during the development of spike-wave discharges. However, the role of thalamic NMDA receptors in spike wave paroxysms was not supported in the GHB model (Banerjee and Snead, 1995).

GABA-ERGIC (A AND B) NEUROTRANSMISSION BETWEEN NRT AND THE RELAY CELLS

GABA mimetic agents, either direct or indirect GABA agonists such as muscimol and THIP, exacerbate absence seizures and increase the propensity and synchrony of spike-wave paroxysms in experimental animals, although GABA-A antagonists do not protect against the seizures (Micheletti *et al.*, 1985). Clinical experience with antiepileptics acting on the GABA side seems to support this. Vigabatrin elevating the GABA level in the brain by irreversible inhibition of the GABA transaminase enzyme increases seizure propensity in idiopathic generalized epilepsies and also in the GAERS (Marescaux *et al.*, 1992). The same is true for the GABA reuptake inhibitor drug (tiagabin). Benzodiazepine drugs are exceptions, probably due to the selective increase of

Table 31.2 Genetic characteristics, transmitter/receptor features in single locus and inbred absence models

	Gen	*Network anomaly*	*Receptor anomaly*	*Blocked by ethosuximide*	*Blocked by GABA/B receptor antagonists*
Tottering mouse	Tg, chr 8	Proliferation of nonadrenergic locus coeruleus axon terminals	GABA/A mediated inhibitory defect(?)	++	No
Stargezer mouse	Stg, chr 15	No	?	++	No
Lethargic mouse	Lh, chr 2	No	GABA/B receptor increase	++	++
GAERS	?	No	Increase of T-type Ca++ channels in reticular neurons increase of GHB receptors in thalamus	++	++
Wag/Rij	?	No	?	++	++

GABA-ergic transmission in the cortex having less effect in the thalamus.

GABA-A receptors are not the only group of sites through which GABA mediates its physiological and pathological effects. A second type of receptors are the GABA B receptors which are not coupled to the Cl channels but instead gate Ca^{2+} and/or K^{-} channels. The neuronal response to GABA-B agonist compounds is slow hyperpolarization compared to the faster GABA-A action. GABA-B mediated inhibition seems to be essential for the mechanism of absence seizures. GABA-B receptor activation by Baclofen promotes the generation of burst-firing working mode and consequently the propensity to absence seizures, while GABA-B antagonists (CGP 35348, 36742, 46381) dose-dependently decrease the frequency of absences in the lh/lh mice, GAERS rats, and in GBH and pentylentetracor pharmacological models. Pertussis toxin (PTx) treatment reduces GABA B receptor mediated events. In the GAERS spike-wave discharges were reduced by 70–80% in thalamic nuclei after PTx treatment (Bittiger *et al.*, 1990; Marescaux *et al.*, 1992; Bowery *et al.*, 1995).

INTRINSIC MECHANISMS WHICH ARE RESPONSIBLE FOR DE-ACTIVATING T-TYPE CA^{2+} CHANNELS IN THE RELAY AND NRT CELLS AND INFLUENCE THE ENGAGEMENT OF THE SYSTEM INTO BURST-FIRING

It has long been known that absences could be very well controlled by succinimide which has no efficacy in other epileptic disorders and specifically blocks T-type Ca^{2+} current (Coulter *et al.*, 1989; Coulter *et al.*, 1990).

Each of these mechanisms represent potential targets for existing and newly developing antiepileptic drugs. Extrinsic mechanisms that increase or decrease the propensity of oscillatory working mode may also influence the propensity of spike-wave paroxysms and of absence seizures. Increase of the cholinergic and dopaminergic input attenuates the appearence of absences and spike-wave discharges, and a decrease in their influence has the opposite effect.

As we have seen, the complex nature of the thalamo-cortical network permits several means by which the oscillatory working mode and spike-wave activation could be developed. However, only a few among them have proved to work in animal models or in epileptic patients.

Candidate molecular mechanisms, which may be critically involved in the pathogenesis of absence seizures in selected animal models include the following: 1) altered biophysical properties of T-type Ca^{2+} channels in the GAERS model (Avanzini, 1987); 2) increased numbers of GABA-B receptors in lh/lh mices (Hosford, 1993) and 3) changes in the subunit composition of GABA-A receptors in the GAERS (Hosford, 1995).

31.2.7 Spike-wave pattern and vigilance level changes

There is a very close relationship between vigilance level and expression of spike-wave paroxysms. Spontaneous paroxysms are promoted by transitory decreases of vigilance level of the awake state (Stevens *et al.*, 1971; Passouant *et al.*, 1972; Janz, 1974) after awakening, after lunch, in evening sleepiness, during boring tasks or situations, experimental depression of reticular arousal functions (Gloor and Testa, 1974), and after sleep deprivation. Spontaneous paroxysms are inhibited by a sudden increase in vigilance (Li *et al.*, 1952;

Jung, 1957; Rajna and Lona, 1989), arousals (calling by name), and experimental stimulation of the reticular arousal system (Guerrero-Figoa *et al.*, 1963; Pollen *et al.*, 1963). This relationship stems from the common "bursting" working mode of the thalamo-cortical system sharing by a mechanism that sets into motion shifts toward slow wave sleep and physiopathogenesis of spike-wave pattern. However, the fact that spike-wave activation in the form of absence-like 3 Hz paroxisms occurs selectively in the transitional periods between slow wave sleep and wakefulness, and between slow wave and REM sleep, and that spike-wave pattern is absent in REM sleep both in human (Niedermeyer, 1967, 1972; Passouant, 1971; Halasz *et al.*, 1974; Halasz 1981, 1991; Horita *et al.*, 1991) and animal studies (Shouse *et al.*, 1990; Coenen *et al.*, 1991; Drinkenburg *et al.*, 1991) and is present only in distorted groups during deep slow wave sleep, needs explanation. Studies analyzing this relationship have shown that not only the level of vigilance differs but activation in these transitional periods is closely connected with sudden oscillations of vigilance attached to the, so called, phasic events of sleep. Spontaneous paroxysms (with or without clinical manifestations) have been associated with arousal-dependent phasic events preceded by K-complexes and/or slow waves (Niedermeyer, 1967; Halász 1982). With sensory stimulation these dynamic changes could have been experimentally elicited and studied (Halász, 1982, 1991).

Association of generalized spike-wave pattern in IGEP with arousal instability in SWS can be measured by the cyclic alternating pattern (CAP) phenomenon, the frequency of which is proportional with arousal instability. Sleep EEG analysis of 10 primary generalized patients (Terzano *et al.*, 1989) showed significant prevalence of s-w paroxysms during CAP as compared to NCAP periods (68% vs 32% (p 0.001)), 93% of all the s-w pattern occured in CAP were found in the reactive phase A (p 0.001). In sleep EEG analysis of 10 JME patients (Gigli *et al.*, 1992) spiking rate was significantly higher in CAP A phase (normalized spiking rate = 4.00 ± 0.98) compared to NCAP, and showed strong inhibition in CAP B phase (0.54–0.27).

The link between EEG arousal phenomena and spike-wave paroxysms is in apparent contradiction to the association of spike-wave pattern and sleep-like bursting mode of the thalamo-cortical system. To solve this contradiction we should take into consideration that most of the evoked phasic events during the dominance of the bursting mode show the features of sleep response. They contain clear cut slow wave sleep elements (single or serial K-complexes, slow wave groups) occurring in the same form as in the spontaneously appearing counterparts. Each arousal during slow wave sleep seems to evoke a regulatory rebound shift toward sleep, which seems to be the best activator of the oscillatory mode of thalamo-cortical network and the spike-wave mechanisms as well (Halász, 1982b).

31.3 Signal transduction abnormalities during "closed loop" working mode of hippocampus and thalamo-cortical systems underlying plastic changes as a common mechanism for both forms of epilepsy

Both the hippocampal formation and the thalamo-cortical system have two basic working modes. One is devoted to aroused, preparatory open loop activities and to mediate the inflow from the outer world to the cortex, and the other to nonaroused, consummatory, closed loop activity which serves as a dialog within the brain and mediates plastic changes leading to memory consolidation and learning. The open loop mode usually works during wakefulness while the closed loop mode works in slow wave sleep. In the open loop mode the principal neurons show steady firing whereas in the closed loop mode they work in a bursting or oscillatory manner. In this latter mode, availability of subcortical neurotransmitters is reduced and neuronal synchrony is produced by the intrinsic connectivity of the network. Burst firing results in larger depolarization of its postsynaptic targets and a higher probability of discharge. In this mode calcium entry into the cells elicits a cascade of events exerting long term changes on synaptic connectivity which is the basis for plasticity and gene expression in neurons (Buzsáki, 1966).

In the hippocampus, closed loop activity is characterized by intermittent population burst events, known as "sharp waves". In the thalamo-cortical system, closed loop mode activity is dominated by spindle formation. In the epileptic transformation of functions, interictal manifestations of epilepsy are spike discharges in the hippocampus but spike-wave discharges in the thalamo-cortical system.

Studies of the interictal hippocampal spikes indicate that they are initiated in the Ca2 Ca3 region and the synchronous bursts induce temporally summated "giant" excitatory postsynaptic potentials in the apical dendrites of Ca1 pyramidal cells which may trigger action potentials in their cell bodies. "The same neuronal machinery might be involved in generating physiologically occurring sharp waves and epileptic spikes, since bursting pyramidal cells, collateral excitation, and temporal release from tonic inhibition are the main elements of both neuronal patterns" (Buzsáki, 1989). Pyramidal cells involved in epileptic bursting are not able to provide new sets of initiator cells and interfere with memory process.

Recent research revealed similar conditions in the thalamo-cortical network. The well known "augmenting response" to stimulation of thalamic relay cells is associated with a short-term plasticity process, that is, persistent and progressive increases in depolarizing synaptic responses and decreases in inhibitory responses. Steriade (1999) hypothesized that this sequence of events also seems to serve as memory consolidation in slow wave sleep. Such changes, leading to self sustained oscillations in closed loops, could trigger seizures if not constrained by inhibitory processes. The cortico-thalamic bursting neurons, discharging rhythmic spike bursts, could lead to synaptic changes in target structures which favor alterations required for memory processes. The epileptic spike-wave discharges here also "uses", in exaggerated form, the same mechanisms which serve physiological memory functions and consequently interferes with memory process too.

In terms of signal transduction epilepsy there is an essential communication disorder and also a disorder of learning. We can say epilepsy is a kind of disordered learning in systems in which inbuilt mechanisms provide the possibility of functional and morphological plasticity and which are, in addition, especially prone to synchronization. In these systems long term changes develop either by overexcitation or by rearrangement of the system as compensation for cell loss, shifting the function of cell groups toward synchronous overexcitation*.

The most elementary basis of neuronal memory is the neuronal plasticity phenomenon, which may be modelled by long term potentiation (LTP). The temporal lobe and the hippocampus in particular, are the most robust generators of LTP and also the most susceptible structures for epilepsy.

References

1 Adam, C., Baulac, M., Saint-Hilaire, J.M., Landau, J., Granat, O., Laplane, D. (1994) Value of magnetic resonance imaging-based measurements of hippocampal formations in patients with partial epilepsy. *Arch Neurol, 51, 130–138.*

2 Avanzini, G., Vergnes, M., Spreafico, R., Marescaux, C. (1987) Calcium dependent regulation of genetically determined spike and waves by the reticular thalamic nucleus of rats. *Acta Neurol Scand, 2, 219–223.*

3 Babb, T.L.: Bilateral pathological damage in temporal lobe epilepsy. (1991) *Can. J. Neurol Sci, 18, 645–648.*

4 Babb, T.L., Kupper, W.R., Pretorius, J.K., Crandall, P.H., Levesque, M. (1991) Synaptic reorganization by mossy fibres in human epileptic fascia dentata. *Neuroscience, 4, 351–363.*

5 Banerjee, P.K. and Snead, O.C. (1995) Thalamic NMDA receptors in the γ-hydroxybutyrate model of absence seizures: a cerebral microinjection study in rats. *Neuropharmacology, 34, No1, 43–53.*

6 Ben Ari, Y., Tremblay, E., Otterson, O.P., Maldrum, B.S. (1980) The role of epileptic activity in hippocampus and remote cerebral lesion induced by kainate. *Brain Res, 191, 79–97.*

7 Ben Ari, Y. (1985) Limbic seizures and brain damage produced by kainic acid; mechanism and relevance to human temporal lobe epilepsy. *Neuroscience, 14, 375–403.*

* Epilepsy therefore can be held as a "punishment" for the achievement of neuronal plasticity. Symbolically speaking epilepsy in this sense is a consequence of the original sin "to eat the apple from the tree of knowledge". The burden of epilepsy can be viewed from this angle as suffering from being "expelled from paradise".

8 Berkovic, SF. (1997) Genetics of epilepsy syndromes. In J. Engel and T.A. Pedley (eds.), *Epilepsy: A Comprehensive Textbook*, Lippincott-Raven, Philadelphia, pp. 217–224.

9 Bittiger, H., Froestl, W., Hall, R. *et al.* (1990) Biochemistry, electrophysiology and pharmacology of a new GABAB antagonist: CGP 35348. In Bowery, N.G., Bittiger, H., Olpe, H.R., (eds.), *GABAB receptors in mammalian function*, Wiley, Chichester, pp.47–80.

10 Bowery, N., Richards, D.A., Lemos, T., Whitton, P.S. (1995) GABA transmission in absence epilepsy. In J.S. Duncan, and C.P. Panayiotopoulos, (eds.), *Typical absences and related epileptic sydromes*, Churchill Livingstone Int., New York, pp. 51–58.

11 Braendgaard, H. and Gundersen, H.J.G. (1986) The impact of recent stereological advances on quantitative studies of the nervous system. *J. Neurosci. Meth, 18, 39–78.*

12 Buzsáki, G. (1989) Two-stage model of memory trace formation: a role for "noisy" brain states. *Neuroscience, 31, No 3, 551–570.*

13 Buzsáki, G. (1991) The thalamic clock: emergent network properties. *Neuroscience, Vol. 41, No 2/3, 351–364.*

14 Buzsáki, G. (1996) The hippocampo-neocortical dialogue. *Cerebral Cortex, 6, 81–92.*

15 Cavazos, J.E., Golarai, G., Sutula, T.P. (1991) Mossy fibres reorganization induced by kindling; Time course of development, progression, and permanence. *J. Neurosci, 11, 2795–2803.*

16 Cendes, F., Andermann, F., Gloor, P., Evans, A., Jones-Gotman, M., Watson, C., Melanson, D., Olivier, A., Peters, T., Lopes-Cendes, I., Leroux, G. (1993) MRI volumetric measurement of amygdala and hippocampus in temporal lobe epilepsy. *Neurology, 43, 719–725.*

17 Charton, G., Rovira, C., Ben Ari, Y., Leviel, V. (1985) Spontaneous and evoked release of endogenous Zn^{2+} in the hippocampal MF zone in situ. *Exp Brain Res, 58, 202–205.*

18 Coenen, A.M.L., Drinkenburg, W.H.I.M., Peeters, B.W.M.M., Vossen, J.M.H., Van Luijtelaar, E.L.J.M. (1991) Absence epilepsy and the level of vigilance in rats og the WAG/Rij strain. *Neuroscience & Biobehavioral Reviews, 15, 259–263.*

19 Coenen, A.M.L., Drinkenburg, W.H.I.M., Inoue, M., Van Luijtelaar, E.L.J.M. (1992) Genetic models of absence epilepsy with an emphasis on the WAG/Rij strain. *EpilepsyResearch, 12, 75–86.*

20 Coulter, D.A., Huguenard, J.R., Prince, D.A. (1989) Characterization of ethosuximide reduction of low-threshold calcium current in thalamic neurons. *Annals of Neurology, 25, 582–593.*

21 Coulter, D.A., Huguenard, J.R., Prince, D.A. (1990) Differencial effects of petit mal anticonvulsants and convulsants on thalamic neurons. I. Calcium current reduction. *British Journal of Pharmacology, 100, 800–806.*

22 Coulter, D.A. (1997) Thalamocortical anatomy and physiology. In Engel, J. Jr. and Pedley, T.A. (eds.) *Epilepsy: A Comprehensive Textbook* Lippincott-Raven, Philadelphia, pp. 341–351.

23 Drinkenburg, W.H.I.M., Coenen, A.M.L., Vossen, J.M.H., Van Luijtelaar, E.L.J.M. (1991) Spike-wave discharges and sleep-wake sate in rats with absence epilepsy. *Epilepsy Research*, 92, 218–224.

24 Engel, J. Jr. (1989) Epileptic syndromes. In FA Davis (ed.), *Seizures and Epilepsy*, Philadelphia, pp. 179–220.

25 Gall, C. and Isackson, P.J. (1989) Limbic seizures increase neuronal production of messenger RNA for nerve growth factor. *Science, 245, 758–761.*

26 Germano, I.M., Sperber, E.F., Moshé, S.L. (1996) In Guerrini *et al.* (eds.) R Molecular and experimental aspects of neuronal migration disorders. *Dysplasias of cerebral cortex and epilepsy*, Lippincott-Raven, Philadelphia, pp.27–34.

27 Gloor, P., Testa, G., Guberman, A. (1973) Brain-stem and cortical mechanisms in an animal model of corticoreticular epilepsy. *Trans Am Neurol Assoc, 98, 203–205.*

28 Gloor, P. (1988) Neuropsychological mechanism of generalised spike-and-wave discharge and its implications for understanding absence seizures. In Myslobodsky MS and Mirsky AF (eds.), *Elements of petit mal epilepsy*, Peter Lang, New York, pp.159–209.

29 Gloor, P. and Fariello, R.G. (1988) Generalized epilepsy: Some of its cellular mechanism differ from those of focal epilepsy. *Trends in neuroscience, 11, 63–68.*

30 Gloor, P., Avoli, M., Kostopoulos, G. (1990) Thalamocortical relationships in generalised epilepsy with bilaterally syncronous spike-and-wave discharge. In Avoli, M., Gloor, P., Kostopoulos, G., Naquet, R. (eds.) *Generalized epilepsy: neurobiological approaches*, Birkhauser, Boston, pp.190–212.

31 Guerrero-Figoa, R., Barros, A., De Balbian, V. (1963) Some inhibitory effects of attentive factors on experimental epilepsy. *Epilepsia, 4, 225–240.*

32 Gundersen, H.J.G., Bendtsen, T.F., Korbo, L., Marcussen, N., Moller, A., Nielsen, K., Nyengaard, J.R., Pakkenberg, B., Sorensen, F.B., Vesterby, A., West, M.J. (1988) Some new, simple and efficient stereological methods and their use in pathological research and diagnosis. *APMIS, 96, 379–394.*

33 Halász, P., Dévényi, É. (1974) Petit mal absence in night-sleep with special reference to transitional sleep and REM periods. *Acad. Sci. Hung., 31, 31–45.*

34 Halász, P. (1981) Generalized epilepsy with spike-wave paroxysms as an epileptic disorder of the function of sleep promotion. *Acta Physiol. Acad. Sci. Hung., 57, 51–86.*

35 Halász, P. (1982a) Generalized epilepsy with spike-wave pattern (GESW) and intermediate states of sleep. In Sterman MB, Shouse MN, Passouant P (eds.), *Sleep and epilepsy*, Academic Press, New York, pp.219–237.

36 Halász, P. (1982b) Role of the non-specific phasic activation in sleep regulation and in the pathomechanism of generalized epilepsy with spike-wave pattern. *Doctorial thesis*, Budapest.

37 Halász, P. (1991) Sleep, arousal and electroclinical manifestations of generalized epilepsy with spike wave pattern. *Epilepsy Res. Suppl, 2, 43–48.*

38 Horita, H., Uchida, E., Maekawa, K. (1991) Circadian rhythm of regular spike-wave discharges in childhood absence epilepsy. *Brain Dev, 13(3), 200–202.*

39 Houser, C.R. (1990) Granule cell dispersion in the dentate gyrus of humans with temporal lobe epilepsy. *Brain Res, 535, 195–204.*

40 Jack, C.R.Jr., Sharbrough, F.W., Twomey, C.K., Cascino, G.D., Hirschorn, K.A., Marsh, W.R., Zinsmeister, A.R., Scheithauer, B. (1990) Temporal lobe seizures: Lateralization with MR volume measurements of the hippocampal formation. *Radiology, 175, 423–429.*

41 Janszky, J., Barsi, P., Halász, P., Erőss, L., Rásonyi, Gy. (1999) Temporal lobe epilepsy syndrome with peritrigonal nodular heterotopia. *Clin Neurosci, 52(1–2), 44–50.*

42 Janz, D. (1974) Epilepsy and the sleeping-waking cycle. In: Vincken P.J., Bruyn G.W. (eds.) *Handbook of clinical neurology*, vol 15: The epilepsies. North Holland, Amsterdam, pp.457–490.

43 Jasper, H.H. and Droogleever-Fortuyn, J. (1946) Experimental studies on the functional anatomy of petit mal epilepsy. *Res. Publs. Ass. Res. nerv. ment. Dis, 26, 272–298.*

44 Jope, R.S., Song, L., Kolasa, K. (1992) Inositol trisphosphate, cyclic AMP, and cyclic GMP in rat brain regions after lithium and seizures. *Biol Psychiatry, 31, 505–514.*

45 Jung, R. (1957) Zur Klinik und Elektrophysiologie des "petit mal". 4ème Cong. Internat. d'EEG et Neuropsychologie clinique. *Acta Med. Belg, p. 296.*

46 Kito, S. and Miyosi, R. (1991) Effect of neuropeptides on classic types of neurotransmission in the rat central nervous system. In Kito, S. *et al.* (eds.), *Neuroreceptors Mechanisms in Brain* Plenum Press, New York, pp. 1–11.

47 Li, C.L., Jasper, H.H., Henderson, L. (1952) The effect of arousal mechanisms on various forms of abnormality in the electroencephalogram. *Electroenc Clin Neurophysiol, 4, 513–526.*

48 Au Louis, N.C., Niquet, J., Ben Ari, Y., Represa, A. (1997) Cellular plasticity In J. Engel Jr. and T. A. Pedley (eds.), *Epilepsy: A Comprehensive Textbook*, Lippincott-Raven, Philadelphia, pp. 387–396.

49 Maglóczky, ZS., Wittner, L., Borhegyi, Zs., Halász, P., Vajda, J., Czirják, S., Freund, T. (2000) Changes in the distribution and connectivity of interneurons in the epileptic human dentate gyrus. *Neuroscience, 96, 1,7–25.*

50 Mares, P. (1973) Ontogenetic development of bioelectrical activity of the epileptogenic focus in rat neocortex. *Neuropadiatrie, 4, 434–445.*

51 Marescaux, C., Micheletti, G., Vergnes, M., Depaulis, A., Rumbach, L., Warter, J.M. (1984) A model of chronic spontaneous petit mal-like seizures in the rat: comparison with pentylenetetrazol-induced seizures. *Epilepsia, 25, 326–331.*

52 Marescaux, C., Vergnes, M., Depaulis, A. (1992a) Genetic absence epilepsy in rats from Strasbourg: a review. *Journal of Neural Transmission (Suppl.), 35, 37–69.*

53 Marescaux, C., Vergnes, M., Bernasconi, R. (1992b) GABAB receptor antagonists: potential new anti-absence drugs. *Journal of Neural Transmission (Suppl.), 35, 179–188.*

54 Margerison, J.H. and Corsellis, J.A.N. (1966) Epilepsy and the temporal lobes. A clinical, electroencephalographic and neuropathological study of the brain in epilepsy with particular reference to the temporal lobes. *Brain, 89, 499–530.*

55 McNamara, J.O. and Wada, J.A. (1997) Kindling model. In Engel J. Jr. and Pedley TA (eds.), *Epilepsy: A Comprehensive Textbook*, Lippincott-Raven, Philadelphia, pp. 419–425.

56 Meldrum, B.S., Vigoroux, R.A., Brierley, J.B. (1973) Systemic factors and epileptic brain damage. Prolonged seizures in paralized, artificially ventilated baboons. *Arch. Neurol, 29, 82–87.*

57 Meyer, A., Falconer, M.A., Beck, E. (1954) Pathological findings in temporal lobe epilepsy. *J. Neurol. Neurosurg. Pychiat, 3, 276–285.*

58 Micheletti, G., Marescaux, C., Vergnes, M., Rumbach, L., Warter, J.M. (1985) Effects of GABA-mimetics and GABA antagonists on spontaneous non-convulsive seizures in Wistar rats In Bartholini G., Bossi L., Lloyd KG., Morselli M.L. (eds.), *Epilepsy and GABA receptor agonists*, Raven, New York, pp.129–137.

59 Michelson, H.B., Williamson, J.M., Lothman, E.W. (1989) Ontogeny of kindling: the acquisition of kindled responses at different ages with rapidly recurring hippocampal seizures. *Epilepsia, 30, 672.*

60 Morell, F., Tsuru, N., Hoeppner, T.J., Morgan, D., Harrison, W.H. (1975) Secondary epileptogenesis in frog forebrain: effect of inhibition of protein synthesis. *Can J Neurol Sci, 2, 407–406.*

61 Morell, F. (1979) Human secondary epileptogenic lesions. *Neurology, 29, 558.*

62 Morell, F., Toledo-Morell, L. (1999) From mirror focus to secondary epileptogenesis in man: an historical review. In: Stefan H., Andermann F., Chauvel P., Shorvon S. (eds.), *Advances in Neurology*; vol. 81 Lippincott Williams&Wilkins, Philadelphia, pp.11–23.

63 Niedermeyer, E. (1967) Über auslösende Mechanismen von Kramfpotentialen bei centrenphaler Epilepsie. *Nervenarzt, 38, 72–74.*

64 Niedermeyer, E. (1972) The generalized epilepsies: A clinical electroencephalographic study. (CC Thomas ed.) Springfield, Illinois 247pp.

65 Niquet, J., Ben Ari, Y., Represa, A. (1994) Glial reaction after seizure induced hippocampal lesion: Immunocytochemical characterization of proliferating glial cell. *J. Neurocytol, 23, 641–656.*

66 Niquet, J., Ben Ari, Y., Faissner, A., Represa, A. (1995) Lesion and fibre sprouting in the hippocampus is associed with an increase of tenascin immunoreactivity, an extracellular glycoprotein with repulsive properties. *J. Neurocytol, 24, 611–624.*

67 Noebels, J.L. (1984) A single gene error in noradrenergic axon growth synchronizes central neurons. *Nature, 10, 409–411.*

68 Oldani, A., Zucconi, M., Asselta, A. *et al.* (1998) Autosomal dominant nocturnal frontal lobe epilepsy. A video-polysomnographic and genetic appraisal of 40 patients and delineation of the epileptic syndrome. *Brain, 121, 205–223.*

69 Oorschot, D.E. (1994) Are you using neuronal densities, synaptic densities or neurochemical densities as your definitive data? There is a better way to go. *Prog. Neurobiol, 44, 233–247.*

70 Passouant, P. (1971) Absence or petit mal. Clinical and physiopathological problems. *Rev Esp Otoneuroophtalmol Neurocir, 29, 11–28.*

71 Passouant, P., Cadilhac, J., Ribstein, M. (1972) Sleep privation with eye movements using antidepressive agents. *Rev. Neurol, 127(1), 173–192.*

72 Rajna, P., Lona, C. (1989) Sensory stimulation for inhibition of epileptic seizures. *Epilepsia, 30, 168–174.*

73 Raymond, A.A., Fish, D.R., Stevens, J.M., Cook, M.J., Sisodiya, S.M., Shorvon, S.D. (1994) Association of hippocampal sclerosis with cortical dysgenesis in patients with epilepsy. *Neurology, 44, 1841–1845.*

74 Represa, A., Le Gall Salle, G., Ben Ari, Y. (1989) Hippocampal plasticity in the kindling model of epilepsy in rats. *Neurosci. Lett, 99, 345–350.*

75 Represa, A., Niquet, J., Charriot Marlangue, C., Ben Ari, Y. (1993a) Reactive astrocytes in the kainic acid damaged hippocampus have the phenotype features of type II astrocytes. *J. Neurocytol, 22, 299–310.*

76 Represa, A., Pollard, H., Moreau, J., Ghilini, G., Khrestcharisky, M., Ben Ari, Y. (1993b) Mossy fibres sprouting in epileptic rats is associated with a transient increased expression of tubulin. *Neurosci Lett, 156, 149–152.*

77 Ribak, C.B. and Peterson, G.M. (1991) Intragranular mossy fibres in rats and gerbils taken from synapsus with somatic and proximal dendrites of basket cells in the dentate gyrus. *Hippocampus, 1, 355–364.*

78 Rizzi, M., Monno, A., Samanin, R., Sperk, G., Vezzani, A. (1993) Electrical kindling of the hippocampus is associated with functional activation of neuropeptide Y-containing neurons. *Eur J Neurosci, 5, 1534–1538.*

79 Sheffer, I.E., Bathia, K.P., Lopes-Cendes, I. *et al.* (1994) Autosomal dominant frontal lobe epilepsy misdiagnosed as sleep disorder. *Lancet, 343, 515–517.*

80 Shouse, M.N., King, A., Langer, J. *et al.* (1990) Basic mechanisms underlying seizure-prone and seizure-resistent sleep and awakening states in feline kindled and penicillin epilepsy. In Wada JA (ed.), *Kindling 4*, pp. 313–327.

81 Sommer, W. (1880) Erkrankung des Ammonshorns als aetiologisches Moment der Epilepsie. *Arch. Psychiat. Nervenkr, 10, 631–675.*

82 Stauder, K.H. (1936) Epilepsie und Schlafenlappen. *Arch. Psychiat. Nervenkr, 104, 181–211.*

83 Steriade, M., Deschenes, M., Domich, L., Mulle, C. (1985) Abolition of spindle oscillation in thalamic neurons disconnected from nucleus reticularis thalami. *J. Neurophysiol, 54, 1473–1497.*

84 Steriade, M. (1991) Spindling, incremental thalamocortical responses, and spike and wave epilepsy. In Avoli M, Gloor P, Kostopoulos G, Naquet R (eds.), *Generalized Epilepsy* Birkhauser, Boston.

85 Steriade, M. (1997) Synchronized activities of coupled oscillators in the cerebral cortex and thalamus at different level of vigilance. *Cerebral Cortex, Sep, 7, 583–604.*

86 Steriade, M. (1999) Coherent oscillations and short-term plasticity in corticothalamic networks. *TINS, vol.22, No. 8, 337–345.*

87 Stevens, J.R., Kodama, H., Lonsbury, B., Mills, L. (1971) Ultradian characteristics of spontaneous seizure discharges recorded by radio telemetry in man. *Electroencephalogr Clin Neurophysiol, 31, 313–325.*

88 Sutula, T., He, X.X., Cavazos, J., Scott, G. (1988) Synaptic reorganization in the hippocampus induced by abnormal functional activity. *Science, 239, 1147–1150.*

89 Swann, J. and Moshé, L.S. (1997) Developmental issues in animal models. In: J.Engel Jr. and T.A. Pedley (eds.), *Epilepsy: A Comprehensive Textbook*, Lippincott-Raven, Philadelphia, pp. 467–479.

90 Tauck, D.L. and Nadler, J.V. (1985) Evidence of functional mossy fibre sprouting in hippocampal formation of kainic acid treated rats. *J. Neurosci, 5, 1016–1022.*

91 Terzano, M.G., Parrino, L., Anelli, S., Halász, P. (1989) Modulation of generalized spike-and-wave discharges during sleep by cyclic alternating pattern. *Epilepsia, 30, 772–781.*

92 Turski, W.A., Cavalheiro, E.A., Schwartz, M., Czuczwar, S.J., Kleinrok, Z., Turski, L. (1983) Limbic seizures produced by pilocarpine in rats: behavioural electroencephalographic and neuropathological study. *Behav Brain Res, 9, 315–335.*

93 Wang, Q., Theard, A.M., Pelligrino, D.A., Baughman, V., Hoffmann, W.E., *et al.* (1994) Nitric oxide (NO) is an endogenous anticonvulsant but not a mediator of the cerebral hyperemia accompanying bicuculline induced seizures in rats. *Brain Res, 658, 142–148.*

94 Wieser, H.G., Engel, J. Jr., Williamson, P.D., Babb, T.L., Gloor, P. (1993) Surgically remediable temporal lobe syndromes. In Engel J. Jr.(ed.), *Surgical Treatment of the Epilepsies*, Raven, New York, pp. 49–63.
95 Wilder, B.J. (1982) Experimental studies, models, and phylogenetic aspects of secondary epileptogenesis. In: Mayersdorf and R.P. Schmidt (eds.), *Secondary Epileptogenesis*, Raven, New York, pp. 27–43.
96 Williams, D.A. (1953) A study of thalamic and cortical rhythms in petit mal. *Brain; 76: 50–69.*
97 Williamson, P.D., French, J.A., Thadani, V.M., *et al.* (1993) Characteristics of medial temporal lobe epilepsy II. Interictal and ictal scalp electroencephalography, neuropsychological testing, neuroimaging, surgical results and pathology. *Ann. Neurol, 34: 781–787.*

Chapter 32

Parkinson's disease

Its causes and treatment

Michael S. Starr

Contents

32.1 Introduction

James Parkinson was a medical practitioner in London who is best remembered for his *Essay on the Shaking Palsy*, published in 1817. In it he described how some of his patients experienced great difficulty in moving and performing simple everyday tasks, slowly deteriorating to the point where they could no longer look after themselves. This crippling movement disorder was originally called *paralysis agitans*, after the cardinal symptoms of poverty of movement and tremor, but nowadays we know it as Parkinson's disease.

Freedom of movement is an essential ingredient of daily life that we take for granted. Only when this freedom is impaired in some way, can we begin to appreciate its worth. This is precisely what happens in Parkinson's disease, where the muscles are fully functional while the brain centers that control them are not.

Our brains contain an estimated 100 billion nerve cells. The premature death of a mere few thousand of these is sufficient to cause a breakdown in motor control. The neurons in question belong to one of a group of nuclei known as the basal ganglia. These nuclei form part of the extrapyramidal motor system, whose job is to provide the "gain control" for movement. In Parkinson's disease, it is as though the gain switch is imperceptibly adjusted lower and lower, making muscle movements progressively more difficult.

In the vast majority of cases the genesis of Parkinson's disease is unknown, so it cannot be prevented or cured. The best we can do is to alleviate symptoms in the early stages of the disease with the aid of drugs, but their effectiveness invariably wears off with time. Extensive research and technological innovation have resulted in a number of experimental drug treatments and pioneering surgical procedures, which offer new hope for the future. For the moment, however, the major challenge that faces us is to discover what triggers the neurodegenerative process in the first place, since only then will we be in a position to design effective ways of preventing it.

32.2 Causes of Parkinson's disease

Juvenile or early-onset parkinsonism is not unknown, but as a rule the first diagnosis is made in patients in middle age. Only 10–15% of sufferers show symptoms before the age of 50. The risk of getting Parkinson's disease increases with age, rising from 1 in 500 for the general population to 1 in 60 in the over-50s. It is a worldwide problem that affects men and women equally, with around 500,000 sufferers in Great Britain.

In *idiopathic* Parkinson's disease the cause is unexplained, although biological and environmental factors have to be considered a possibility. Twin studies do not support genetic transmission through an autosomal dominant gene, suggesting instead that several gene loci may be involved. Individuals inherit a susceptibility to the disease and not necessarily the disease itself. There appears to be a close familial link in early-onset Parkinson's disease, but not where the disease develops after the age of 40.

The existence of a neurotoxin that is responsible for causing parkinsonism remains controversial. Whilst there are undoubtedly toxins in our environment that can cause parkinson-like conditions in animals and man, no single toxin or group of toxins has yet been identified that can account for the worldwide occurrence of Parkinson's disease. The finger of suspicion has been pointed at environmental pollutants and agrichemicals, but the epidemiological evidence is not convincing. On the one hand, we do not see clearly defined pockets of Parkinson's disease in highly industrialized regions, where pollution is high, or in communities making extensive use of pesticides, fungicides and herbicides in farming. However, it has been reported there is a higher incidence of the disease in newly industrialized cities in China, and in some rural areas of the United States, which could be seen as supporting the environmental pollutant theory.

Interestingly cigarette smokers are less prone to developing Parkinson's disease than non-smokers, and this is not because smokers are more likely to die of cardiovascular or respiratory complications.

Poisoning with carbon dioxide or manganese salts causes midbrain damage that can lead to parkinsonism, but these account for very few cases. Occasionally midbrain tumors or senile atherosclerosis can have the same effect. Traumatic head injury can also cause parkinsonism. For example, some car crash victims and sportsmen (footballers, boxers) have later succumbed to the disease, the most famous of these being the ex-world heavyweight boxing champion Muhammad Ali.

The socalled "typical" antipsychotic drugs used to treat schizophrenia (such as chlorpromazine, haloperidol), impair the functioning of the basal ganglia and frequently give rise to extrapyramidal side effects in the form of parkinson-like bradykinesia and muscle rigidity. These effects usually disappear when the drug is stopped. Newer "atypical" antipsychotics (such as olanzepine, risperidone) are safer in this respect.

Probably the most celebrated example of Parkinson's disease in the 20th century dates back to a viral epidemic that occurred in the United States in the 1920s. It was known as *encephalitis lethargica*, because in the early stages its victims tended to fall asleep. The virus caused extensive brain damage resulting in a form of Parkinson's disease that started in people at a young age and then slowly worsened over the next 40 years. In his book *Awakenings*, the neurologist Oliver Sacks describes how these postencephalitic parkinsonian patients were scarcely able to move a muscle, and were kept in institutions along with the mentally ill (Jenner, 1990).

32.3 Symptoms

Parkinson's disease is the most frequently encountered disorder of the extrapyramidal motor system of the brain. The primary clinical features are a resting tremor, rigidity of muscles, slowness of movement and impaired postural reflexes (Figure 32.1) Everyday tasks become more difficult, such as dressing oneself, cutting up and chewing and swallowing food, writing a letter, getting out of a low armchair, and so on. Muscles feel stiff and less pliant and movements become labored and slow. Friends will notice a loss of facial expression and blinking, and that speech tends to fluctuate and sometimes loses its volume (Figure 32.2).

- Resting tremor
- Mustle regidity
- Bradykinesia
- Abnormal posture

Figure 32.1 Symptoms of Parkinson's disease.

- Difficulty starting and stopping walking
- Small shuffling steps
- Body stooped
- Increased muscle tone
- Resting tremor of hand
- Small handwriting
- Mask-like face
- Decreased blinking
- Voice weak and expressionless
- Impaired learning and memory
- Depression
- Autonomic disturbances (drooling, sweating)

Figure 32.2 Clinical features of Parkinson's disease.

In addition to losing motor skills, many patients later experience learning and memory difficulties, loss of autonomic control (drooling, sweating and orthostatic hypotension are common) and endocrine disorders, while about a third suffer from depression. Clearly it is not just the motor pathways of the brain that are damaged.

TREMOR

Resting tremor of one limb, which disappears with exercise, is an early diagnostic feature and one which generally precipitates the patient's first visit to the doctor. When it involves an arm the action is often referred to as "pill rolling" or "money counting", because of the rhythmic movements of the fingers and thumb. It is not to be confused with kinetic tremor, an entirely different disorder which appears with movement. Getting a patient to touch a doctor's outstretched finger with their own, or to touch their nose with the affected finger, is a typical diagnostic test. Resting tremor is ameliorated, while kinetic tremor is exacerbated in this simple task. Parkinsonian tremor tends to spread to both limbs on the same side of the body before becoming bilateral.

COGWHEEL RIGIDITY

When the neck or a limb of a parkinsonian patient is passively moved, the examiner feels a ratchet-like or cogwheel jerkiness, because the opposing muscle groups are contracted and oppose relaxation. Additionally, an underlying resistance to passive movement is an indication that there is an increase in muscle tone.

BRADYKINESIA

This is the term used to describe the slow speed with which all voluntary movements are made. It is considered by many to be the most debilitating and frustrating feature of Parkinson's disease. The patient's desire to move is undiminished, but the brain fails to send the appropriate messages to the muscles. This means that starting and stopping physical activity can be problematic and patients will devise mental strategies to overcome their inertia.

POSTURAL DEFECTS

To stop ourselves from falling over we make constant unconscious adjustments to muscles that control balance, by way of postural reflexes. Staying upright in a crowd of people, or changing direction when walking, are more difficult for a parkinsonian patient, who moves with characteristic shuffling steps. At rest the patient assumes a stooped posture. Many patients die as a consequence of the injuries they sustain through falling over. Typical injuries, remembering that we are mainly talking about elderly people, include broken arms, dislocated hips and subdural hematomas from a blow to the head.

32.4 Diagnosis

The diagnosis of Parkinson's disease is not straightforward. The neurologist needs to make a detailed evaluation of the patient's clinical features using a battery of tests. In Great Britain this assessment is made with reference to the Unified Parkinson's Disease Rating Scale (Figure 32.3).

Early	▷ Independent of others ▷ Able to work normally ▷ Lasts 1–5 years ▷ Stable response to levodopa
Moderate	▷ Need some help ▷ Work disability ▷ Lasts 3–5 years ▷ Dose-related levodopa swings
Advanced	▷ Dependent on others ▷ Bound to chair or bed ▷ Lasts 1–5 years ▷ No response or fluctuating response to levodopa

Figure 32.3 Stages of Parkinson's disease.

32.5 Neuropathology

It has long been known that the parkinsonian brain loses pigmented cell bodies in the *substantia nigra*, but only in the 1950s was dopamine discovered to be the neurotransmitter of these cells. Thus Parkinson's disease was the first neurological disorder to be linked with a specific neurotransmitter. The black pigment is neuromelanin, an oxidation product of dopamine, and the loss of pigmentation in the midbrain indicates that the cell bodies of the *nigrostriatal* dopamine neurons have degenerated. The axons of these neurons project forward and innervate the caudate-putamen. We normally expect to lose 8–10% of our dopamine neurons every decade through natural aging, but in Parkinson's disease this process is accelerated. Some of the remaining dopamine neurons contain densely-staining inclusions of cytoskeletal protein, called *Lewy bodies*, which resemble the neurofibrillary tangles of Alzheimer's disease.

The primary neurochemical deficit in Parkinson's disease is therefore a decrease in brain dopamine content. This is more severe in the putamen (>90%) than in the caudate nucleus (60–70%). The reason we do not get Parkinson's disease as the result of losing dopamine cells through natural aging, is that approximately 80% of the nigrostriatal dopamine pathway has to degenerate before visible motor symptoms occur. This is because various compensatory mechanisms are brought into play: (a) remaining dopamine neurons fire more rapidly and release more dopamine; (b) postsynaptic dopamine receptors become supersensitive; and (c) presynaptic dopamine uptake sites are destroyed.

Damage is also evident in the other dopamine pathways of the brain, but to a smaller extent (50–60%).

One of the secondary consequences of losing inhibitory dopaminergic control in the caudate-putamen, is that the cholinergic interneurons there become hyperactive. This tips the balance away from dopamine-mediated inhibition and towards acetylcholine-mediated excitation of cells in the forebrain. The outcome of this excitatory dominance is that the basal ganglia now send instructions back to the motor cortex which prohibit rather than permit movement. One of the aims of antiparkinson drug treatment, therefore, is to normalize this dopamine-acetylcholine imbalance.

Elsewhere in the brain are cholinergic neurones concerned with learning and memory. They arise in the *nucleus basalis of Meynert* in the basal forebrain and project to the cerebral cortex and hippocampus. The 30–80% depletion of these cells in idiopathic (but not postencephalitic) Parkinson's disease, and the deposition of Lewy bodies in the nucleus basalis, does not correlate with age or duration of the illness, but is clearly higher when dementia is additionally present.

For comparison, the equivalent damage to this cholinergic system in demented Alzheimer's patients is 60–90%.

Extensive damage (50% or more) is similarly apparent in noradrenaline- and 5-hydroxytryptamine-containing neurons in the brainstem, which might explain the associated intellectual impairment, changes in mood and loss of smell experienced by many parkinsonian patients. According to research findings noradrenaline protects dopamine neurons from neurotoxic damage, indicating that a reduction of this monoamine could influence the pathogenesis of the disease. If we could find some way of preserving noradrenergic function in the brain, then we might be able to limit the progressive destruction of the nigrostriatal dopamine system.

Deficiencies in GABA and neuropeptides (substance P, cholecystokinin-8 and methionine-enkephalin) have also been detected, these being most evident where dopamine cell loss is severe and not compensated by levodopa administration.

It is important to remember, then, that Parkinson's disease has a complex neuropathology and is not simply a dopamine-deficiency syndrome.

32.6 Brain imaging studies

So far, all of the anatomical and biochemical changes that have been described for the parkinsonian brain have been determined from inspection of brain material after death. Nowadays it is possible to scan the brains of patients and detect abnormalities during life. *Magnetic resonance imaging* (MRI) allows us to pinpoint damage to dopamine neurons, but more specific information can be obtained with *positron emission tomography* (PET). By injecting into the bloodstream a radioactively-labelled drug that selectively binds to dopamine receptors (e.g. ^{11}C-spiperone) or to dopaminergic axon terminals in the brain (e.g. ^{18}F-DOPA), we can tell if these dopaminergic properties are different from normal. PET scans confirm that dopamine receptors are up-regulated and dopaminergic terminals are reduced in the caudate-putamen of patients with chronic Parkinson's disease, but they cannot detect the gradual changes that take place during the early stages of the disease.

A more sensitive technique, called *single photon emission computerised tomography* (SPECT), works on the same principle as PET, but incorporates a more powerful source of radiation. Data from SPECT studies using ^{123}I-cocaine, reveal a progressive loss of dopamine uptake sites, and therefore dopamine terminals, with disease duration. These range from 50% in the early stages to 80% in the advanced stages of parkinsonism. This approach is expensive and is currently reserved for experimental purposes, but it clearly has enormous potential to: (a) diagnose parkinsonism before symptoms appear; (b) follow disease progression; and (c) evaluate the outcome of neuroprotective regimes or corrective surgery.

32.7 Mechanisms of toxicity in Parkinson's disease

Over the years, three main theories have been proposed to explain the early death of dopamine neurones in Parkinson's disease: (a) a neurotoxin, either in the environment or produced endogenously; (b) oxidative stress due to abnormal metabolism of dopamine; and (c) a genetic abnormality.

Evidence can be found to support all three theories, and it is likely there is an element of truth in all of them. For instance, deficiencies in the mitochondrial electron transport chain at Complex 1 have been detected in the substantia nigra of parkinsonian patients. This defect in energy metabolism could be due to an inherited vulnerability, which only manifests itself under conditions of stress. Such a condition might occur if the breakdown of dopamine generated toxic free radicals (oxidative stress), or if the patient was exposed to a toxic agent in the environment.

32.8 MPTP and the case of the frozen addicts

In the late 1970s a number of drug addicts in the United States were admitted to hospital suffering from a mysterious condition, their muscles frozen. It turned out they had all injected MPPP, a home-made designer drug resembling pethidine that was contaminated with 1-methyl-4-phenyl-1,2,3,6-tetrahydropyridine (MPTP) (Figure 32.4). By a cruel twist of fate, MPTP turned out to be one of the most potent neurotoxins for catecholamine-containing neurons known to man, and was almost certainly responsible for their severe parkinson-like state. Some good was to come from this tragic story, however, for the symptomatology and pathology of MPTP-induced illness are so similar to those of idiopathic parkinsonism, that

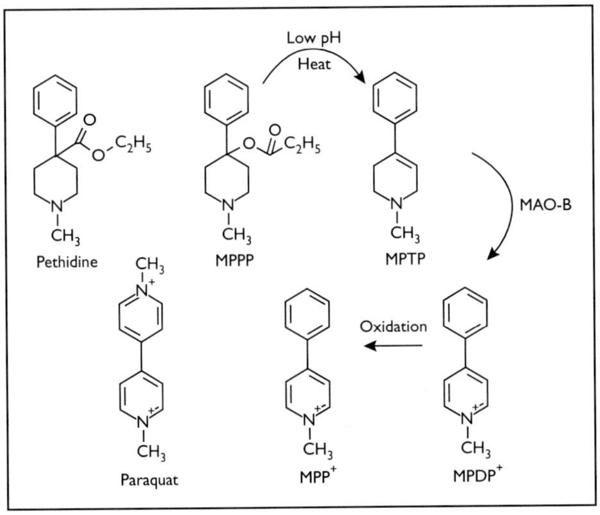

Figure 32.4 MPTP and related structures.

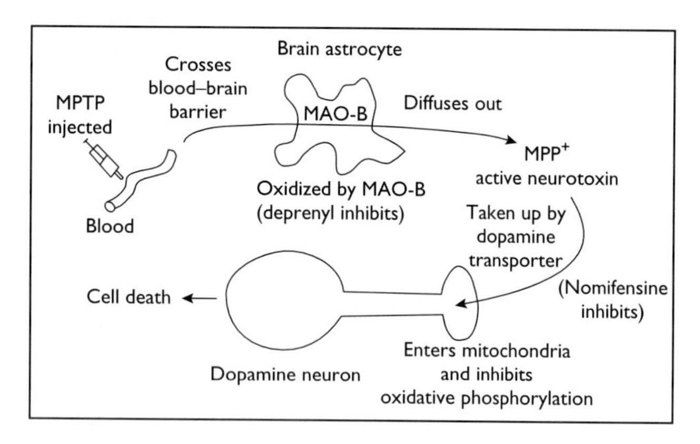

Figure 32.5 Neurotoxicity of MPTP.

MPTP has become an important investigational tool for studying the native disease.

Detailed studies have shown that MPTP readily crosses the blood-brain barrier and enters astrocytes. It is then oxidized by monoamine oxidase type-B (MAO-B) to give MPP^+, the active neurotoxin. MPP^+ accumulates in nigrostriatal dopamine cells via the dopamine transporter and binding to neuromelanin (Figure 32.5). Unbound MPP^+ is actively taken up into mitochondria, where it inhibits NADH coenzyme-Q reductase (Complex 1 of the respiratory chain), thereby blocking oxidative phosphorylation. If ATP formation is prevented, the cell is starved of energy, energy-dependent ion transport fails and cytosolic Ca^{2+} reaches toxic proportions.

Toxic hydroxyl free radicals may also be formed secondary to the cessation of mitochondrial respiration. Although they only exist for a fraction of a second, they are exceedingly reactive and damaging to cells. Such radicals destroy vital enzymes, damage DNA and liberate proteolytic enzymes, as well as disrupting polysaccharides and cell membranes. The net effect of all these changes is that the cell dies (Singer *et al.*, 1987).

32.9 Evidence for an underlying neurotoxic process

Postmortem analyses of parkinsonian brains provide indirect evidence for an ongoing neurotoxic process in the substantia nigra. Levels of the enzyme NADH reductase are significantly lower in the nigra, but not in other brain regions, signifying a lowered capacity of dopamine neurons to carry out oxidative phosphorylation. The parkinsonian nigra also contains 15% fewer polyunsaturated fatty acids and 35% more fatty acid oxidation products than normal, consistent with free radicals damaging membrane lipids and irreversibly altering the structural integrity of the cell. Free radicals are normally detoxified by superoxide dismutase. The activity of this enzyme is significantly elevated (by 30%) in nigral mitochondria of parkinsonian patients, possibly because it has to deal with excessive amounts of free radicals.

MPTP is not implicated in idiopathic Parkinson's disease, and in spite of the passing similarity between MPP^+ and the herbicide paraquat (see Figure 32.4), no firm link with agrichemicals has been established either. Instead, neuroscientists are looking inwards, into the brain itself, in an attempt to identify possible endotoxins.

32.10 Theory of oxidative stress

This theory proposes that the oxidation of dopamine itself generates free radicals, which then destroy the dopamine neurons. Dopamine is oxidized enzymically through MAO (types A and B), or more slowly by spontaneous auto-oxidation, to generate hydrogen peroxide. This is then rendered harmless by glutathione peroxidase in glial cells, using reduced glutathione as cofactor. In Parkinson's disease, levels of glutathione peroxidase and glutathione are abnormally low in the substantia nigra, signaling a reduced capacity to remove hydrogen peroxide (Figure 32.6).

So what happens to this hydrogen peroxide? It can interact with superoxide ions, though this is normally too slow to be significant. The reaction is greatly speeded up in the presence of metal ions (*Fenton* reaction), and it so happens that there are higher than normal quantities of Fe^{3+} in the parkinsonian nigra, as well as lower than normal amounts of the iron-binding protein ferritin. It has been speculated, therefore, that unbound iron catalyses the production of highly reactive hydroxyl free radicals in Parkinson's disease. We cannot tell, however, if altered iron metabolism in the substantia nigra is the cause of parkinsonian neurotoxicity, or is the result of it.

Dopamine, in physiological concentrations, is also capable of initiating programmed cell death (apoptosis) in chick embryo sympathetic neurons. It would be particularly ironic, therefore, if the genetically-regulated program of dopamine cell death was somehow hastened by the cell's own neurotransmitter (Ebadi *et al.*, 1996).

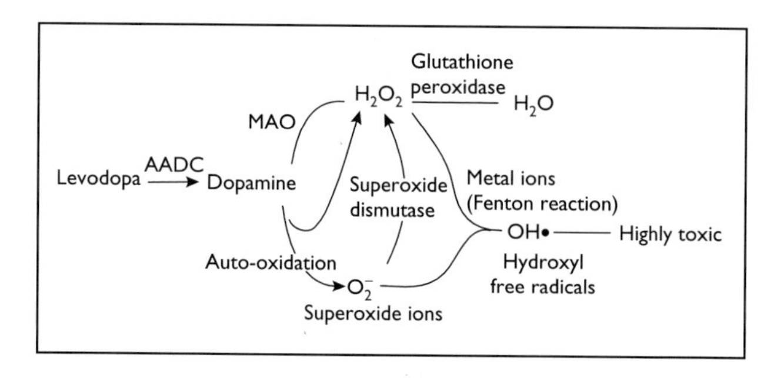

Figure 32.6 Reactive oxygen species formed from dopamine.

32.11 Drug treatments

James Parkinson cautioned against the use of internal medicines, until such time as the nature of parkinsonism was better understood and the efficacy of pharmacotherapy could be demonstrated scientifically. Nevertheless, in his day the recommended treatment consisted of taking laxatives or purgatives, as well as blood-letting. Laxatives would have relieved the constipation, but not the motor incapacity of Parkinson's disease. Throughout the 19th and early 20th centuries the scientific and popular press mentioned a number of "cures", ranging from heavy metals, parathormone, arsenic, heroin, cocaine and the Calabar bean. Non-medical treatments included physical vibrations and electroconvulsive therapy.

Nowadays drugs are used primarily to ameliorate the motor syndrome, although there is a growing emphasis on achieving *neuroprotection* (slowing the progress of the disease) and *neurorescue* (recovery). The choice of treatment depends on a number of factors: (a) which symptoms predominate; (b) the stage of the illness; (c) the age of the patient; and (d) how well an individual responds to a particular drug regimen (Figure 32.7). Achieving good therapeutic results with as few adverse effects as possible is a skillful procedure that requires careful monitoring of the patient, and regular adjustments to the timing and dosages of drugs (Kopin, 1993).

ANTIMUSCARINICS

The Belladonna alkaloid atropine was given originally to curb drooling in patients who had lost autonomic control, when it was noticed that tremor showed a marked improvement. We now know this is because atropine restores the dopamine-acetylcholine balance in the caudate-putamen. A wide variety of modern synthetic antimuscarinic agents are currently available, but only modest clinical improvement can be achieved. Tremor responds particularly well and muscle rigidity can be partially alleviated, whilst bradykinesia is not improved. There is no clinical advantage to using mixtures of different antimuscarinics, whereas combining an antimuscarinic with a dopamine agonist is the preferred method of use. Antimuscarinics are now second-line treatment for Parkinson's disease, although there is something to be said for starting a patient on these drugs where possible, in order to save levodopa for when it is most needed.

A major problem with all antimuscarinics is that they cause troublesome side effects by blocking the muscarinic actions of acetylcholine in the periphery. Patients should contact their doctor if they experience rapid heart rate, severe eye pain or urinary retention. They should also be made aware that antimuscarinic drugs slow gastric emptying, and will therefore delay the intestinal absorption of levodopa when the two are administered together. Antimuscarinics should be avoided in older patients over the age of 65 years, and in patients with a history of psychiatric problems, since blockade of acetylcholine in cortical regions poses a serious risk of dementia-like symptoms, with mental deterioration and memory loss. Prolonged blockade of acetylcholine receptors causes receptor up-regulation, inducing a state of cholinergic hypersensitivity. Abrupt withdrawal of treatment is therefore dangerous, as it can cause a rebound worsening of parkinsonian symptoms.

Old	▷ Antimuscarinics (benzhexol, benztropine, procyclidine) ▷ Peripheral AADC inhibitors (carbidopa, benserazide) ▷ Levodopa ▷ Dopamine agonists (bromocriptine, lisuride, pergolide, apomorphine) ▷ Peripheral dopamine antagonists (domperidone) ▷ Amantadine ▷ Atypical antipsychotics (clozapine)
New	▷ Dopamine D2 agonists (ropinirole, pramipexole, cabergoline) ▷ Atypical antipsychotics (olanzepine, sertindole, risperidone) ▷ MAO-B inhibitors (deprenyl) ▷ Peripheral COMT inhibitors (entacapone) ▷ Central COMT inhibitors (tolcapone)
Experimental	▷ Dopamine D1 agonists (dihydrexidine) ▷ Glutamate antagonists (dextromethorphan) ▷ Dopamine uptake blockers (nomifensine) ▷ Antioxidants (vitamin E) ▷ Neurotrophins (GDNF)

Figure 32.7 Drug treatments for Parkinson's disease.

LEVODOPA AND AADC INHIBITORS

Dopamine replacement with levodopa has been the mainstay of antiparkinson treatment for the past 40 years, and it is true to say that all parkinsonian patients receive levodopa at some time in their lives. Levodopa is the biological precursor of dopamine and the transformation is catalysed by the enzyme aromatic L-amino acid decarboxylase (AADC). Levodopa is always formulated with a peripherally-active AADC inhibitor (*benserazide, carbidopa*), since this enables the dose and cost of levodopa to be greatly reduced and eliminates unwanted effects of dopamine outside the brain (nausea, orthostatic hypotension, cardiac arrhythmias). AADC is found in many places in the brain, which means that levodopa is not just converted to dopamine in the remaining dopamine neurons, but in brain capillaries, 5-HT neurons and glia as well.

PROBLEMS ENCOUNTERED WITH LEVODOPA

Clinical studies have shown that symptomatic relief with levodopa is related directly to plasma levels of the drug, which in turn are governed by its uptake from the bowel. Levodopa should therefore not be taken with a protein-rich meal, as this generates other amino acids that compete for uptake and delay levodopa's absorption across the intestinal wall. Slow-release forms of levodopa are better in this

Central	▷ Loss of efficacy ("wearing off") ▷ Dyskinesias ▷ Motor fluctuations ▷ "on-off" phenomenon ▷ Psychiatric disturbances ▷ Insomnia
Peripheral	▷ Nausea and vomiting ▷ Cardiac arrhythmias ▷ Orthostatic hypotension

Figure 32.8 Problems with levodopa.

respect, as they achieve more stable plasma levels for a longer period of time.

After about 5 years of regular use, the overall clinical improvement obtained with levodopa declines and serious complications arise. Firstly the clinical effect is shortened, with a delayed response (*start-up* period) and a shorter duration of effect (*wearing-off phenomenon*) (Figure 32.8). There may be severe akinesia on waking (*morning akinesia*) before the first dose of medication. Raising the levodopa dose will overcome these problems, but this can cause abnormal involuntary movements (*dyskinesias*) to appear. These may be *choreic* and consist of brief, jerky and unpredictable actions that affect different parts of the body in a random fashion. Or they may be *dystonic*, characterized by twisting repetitive movements and abnormal postures. Patients are powerless to control these dyskinetic activities, but given a choice generally prefer this limited form of mobility to no mobility at all.

With chronic levodopa therapy the patient's sensitivity to levodopa may become erratic and fluctuate from day to day. Psychiatric complications can arise, with patients complaining of visual hallucinations, voices in the head, paranoid delusions, altered sleep patterns and nightmares. These are due to excessive amounts of dopamine being released by neurons of the mesocorticolimbic dopamine pathway.

The most perplexing complication is the *on-off* phenomenon, in which all of the usual parkinsonian symptoms can suddenly and unpredictably reappear, in spite of adequate brain dopamine levels. Minutes or hours later they can just as suddenly disappear again and mobility is restored. The *on-off* phenomenon requires a serious re-evaluation of the patient's treatment. The following may be beneficial: (a) initiating a low-protein diet; (b) carefully titrating the levodopa dose; (c) giving more frequent, smaller doses or a slow-release formulation of levodopa; (d) giving levodopa with a dopamine receptor agonist (e.g. *bromocriptine*) or MAO-B inhibitor (e.g. *deprenyl*). As a last resort a *drug holiday* may be warranted, in which all drug treatment is withdrawn for up to a week, in an attempt to buy a few extra months of responsiveness. Withdrawal must be done gradually and under strict medical supervision.

Because the antiparkinsonian efficacy of levodopa is so precious, many physicians consider it best to reserve its use until such time as the patient's symptoms are not adequately controlled by other therapeutic measures. The patient is therefore best started on an antimuscarinic, amantadine or deprenyl, or a combination of these agents. When levodopa is eventually introduced, the dose should be kept low from the outset (typically 100 mg 2–3 times daily), should be tailored to the patient's needs, and should always be given with an adequate amount of benserazide or carbidopa.

COMT INHIBITORS

Levodopa is also a substrate for catechol-O-methyl transferase (COMT) and the resulting metabolite, 3-O-methyldopa, competes with levodopa for intestinal absorption and brain uptake. Parkinsonian patients who have high plasma levels of 3-O-methyldopa and respond poorly to conventional levodopa treatment, can be given a peripherally-acting (*entacapone*) COMT inhibitor (Figure 32.9). *Tolcapone* can also be used to inhibit COMT centrally and improve the level of levodopa in the brain.

DOPAMINE RECEPTOR AGONISTS

The effects of dopamine on motor behavior are mediated through two different classes of dopamine receptor, called D_1 and D_2. As far as the relief of parkinsonian symptoms is concerned, D_2 agonists are used clinically whilst D_1 agonists are used experimentally.

Semisynthetic derivatives of the ergolines, found in the plant parasite Ergot, are potent agonists at dopamine D_2

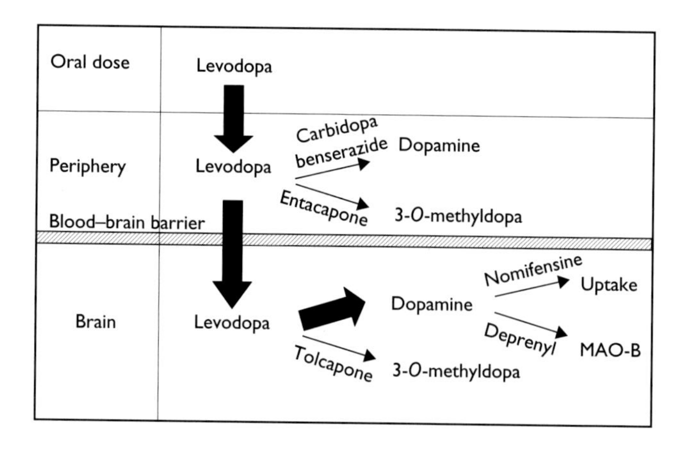

Figure 32.9 Fate of levodopa.

receptors. They were introduced into antiparkinsonian therapy in the mid-1970s. *Bromocriptine* was the first to be used, followed by the longer-acting and more potent drugs *lisuride* and *pergolide*. Given at bedtime they can relieve morning akinesia and dystonia. As a general rule they are inferior to levodopa when given alone, and their use is restricted by prominent side effects (postural hypotension, nausea, vomiting, psychiatric disturbances). Patients are therefore counselled to take drugs with a meal and to rise slowly from a sitting or a lying position. They are less prone to inducing involuntary movements and so can be used to stabilize levodopa-induced motor fluctuations, leading to increased *on* time and improved motor scores.

Apomorphine is a potent D_1/D_2 agonist which is inactive orally, but elicits a pronounced, short-lived antiparkinsonian effect when injected subcutaneously. Because it is a strong emetic its use in parkinsonism has been largely ignored, but the advent of *domperidone* has caused clinicians to re-evaluate this drug in patients experiencing difficulties with levodopa. Occasional and unpredictable *on/off* fluctuations can be treated with a pen self-injection system. Where the oscillations to levodopa are more frequent, it is possible to infuse apomorphine continuously with the aid of a minipump. Dramatic reductions in the time spent in the *off* state have been achieved with this treatment regimen.

Medical opinion is divided on whether it is better to introduce dopamine agonists early or late in combination therapy with levodopa. It has been customary to use dopamine agonists to help stabilize patients with erratic levodopa responses, after several years of levodopa treatment. However, fewer long-term complications are said to be experienced in patients who receive a dopamine agonist/levodopa mixture from the outset, but this observation requires confirmation in carefully controlled trials.

AMANTADINE

This drug was first employed as an antiviral agent in the treatment of influenza A and only discovered by chance to alleviate parkinsonian motor symptoms. Its actions are long-lasting, though weak, and it works in about 50% of patients. Amantadine can be used alone, or together with an antimuscarinic or levodopa. It is a safe drug with adverse effects being reported by less than 10% of patients (anticholinergic and psychiatric effects, orthostatic hypotension, edema of the ankles). Insomnia can be avoided by taking amantadine early in the day. Occasionally patients develop a harmless purplish mottling of the skin (*livedo reticularis*) which disappears when amantadine is discontinued.

The dopaminomimetic and antimuscarinic properties of amantadine are too weak to account for its antiparkinsonian activity. Instead, this is now believed to be due to glutamate antagonism.

DEPRENYL

Deprenyl was originally introduced as an antidepressant, since it inhibits MAO-B in small doses (10 mg or less) and MAO-A at higher doses (see Figure 32.9). The selectivity for MAO-B over MAO-A avoids the "cheese effect", in which tyramine-rich foods (cheese, red wine, smoked or aged meats) cause a hypertensive crisis in patients receiving a MAO-A inhibitor. Levodopa is contraindicated with MAO-A inhibitors for the same reason. Deprenyl only weakly attenuates parkinsonian symptoms by itself, but allows the dose of levodopa to be reduced by 20–30%. Patients whose response to levodopa fluctuates from dose to dose are improved by deprenyl, and the wearing-off effect to levodopa can be delayed by 6–24 months.

Deprenyl protects against MPTP-induced toxicity in the laboratory, but attempts at rescuing dopamine neurons in idiopathic Parkinson's disease have proved to be controversial (see Figure 32.5). Results of the 1987 *Deprenyl and Tocopherol Antioxidative Therapy of Parkinsonism* (DATATOP) clinical study, indicated deprenyl delayed the need for levodopa by 6–24 months and increased life expectancy in previously untreated patients. A later trial concluded patients were actually made worse after 5 years treatment with deprenyl, and so the matter remains unresolved.

Adverse reactions to deprenyl are infrequent and are similar to those of levodopa. Patients should be cautioned against increasing the dose, when side effects associated with MAO-A inhibition may appear.

DOPAMINE RECEPTOR ANTAGONISTS

An obvious drawback to using dopamine receptor agonists is their unwanted activation of dopamine D_2 receptors outside the brain, and also in regions of the brain that do not possess a blood-brain barrier (*area postrema* or "vomiting centre"). At therapeutic doses *domperidone* selectively blocks peripheral D_2 receptors, eliminating the cardiovascular and gastrointestinal effects of apomorphine and the ergots. There is no need to use domperidone with levodopa, as peripheral dopamine formation has already been prevented with benserazide or carbidopa.

In the brain levodopa also stimulates dopamine synthesis in the mesocorticolimbic dopamine neurons, leading to psychiatric disturbances. These should be viewed seriously as they can cause confusion, especially in elderly patients. The standard treatment for this condition is *clozapine*, an atypical antipsychotic drug. Clozapine does not disturb motor regulation, as it blocks dopamine receptors in limbic and not motor pathways. Open-label trials with clozapine have shown it may have a beneficial action of its own in Parkinson's disease, by ameliorating tremor and levodopa-induced dyskinesias. A major drawback with clozapine, however, is that it causes agranulocytosis, requiring careful monitoring of patients.

Antimuscarinics, amantadine and directly acting dopamine agonists can all exacerbate neuropsychiatric toxicity. Stopping or reducing the doses of these agents often allows the patient to continue on therapeutically effective doses of levodopa, removing the necessity for adjunctive treatment with a dopamine receptor antagonist.

32.12 New and experimental treatments

DOPAMINE D_1 RECEPTOR AGONISTS

The attraction of agonists which selectively stimulate dopamine D_1 receptors (e.g. SKF 38393), is that they display potent antiparkinsonian activity in rodent models, but with fewer side effects. Unfortunately this turns out to be a species-specific effect, and attempts to relieve parkinsonian akinesia in MPTP-treated monkeys and human volunteers with SKF 38393, met with failure. If anything, SKF 38393 made their condition worse. Newer, more potent and *full* D_1 receptor agonists have since been synthesized (e.g. A-86929, dihydrexidine), which look more promising in primate tests. However, the acute production of epileptic seizures, and the appearance of dyskinesias with chronic use, suggest such drugs will have a limited application in man.

SECOND GENERATION D_2 AGONISTS

A multitude of selective dopamine D_2 receptor agonists have been synthesized and tested for antiparkinsonian activity (e.g. lergotrile, mesulergine, terguride, PHNO), but have been abandoned because of animal toxicity or human adverse effects. *Cabergoline* is a new ergot-based D_2 agonist with an exceptionally long half life (~65 hours), while *ropinirole* and *pramipexole* are novel agonists with a stronger affinity for D_3 over D_2 receptors. All three compounds are said to be safe and well tolerated by man. They increase the *on* time of chronic Parkinson's disease patients, allowing the dose of levodopa to be reduced.

Some clinicians advocate starting patients on D_2 agonists, but up till now the success of this approach has been limited by the low efficacy of drugs such as bromocriptine and lisuride. Preliminary trials with cabergoline and ropinirole show these are equivalent in effect to levodopa in new patients, and their propensity for inducing dyskinesias is also low. These second generation D_2 agonists therefore represent a significant advance in antiparkinsonian therapy.

SECOND GENERATION ATYPICAL ANTIPSYCHOTICS

The bone marrow toxicity of clozapine has stimulated the search for safer antipsychotics which retain clozapine's atypical profile. *Risperidone*, *sertindole* and *olanzepine* meet these criteria, and their efficacy in treating psychosis in Parkinson's disease has already been demonstrated in brief open-label clinical trials (Hagan *et al.*, 1997).

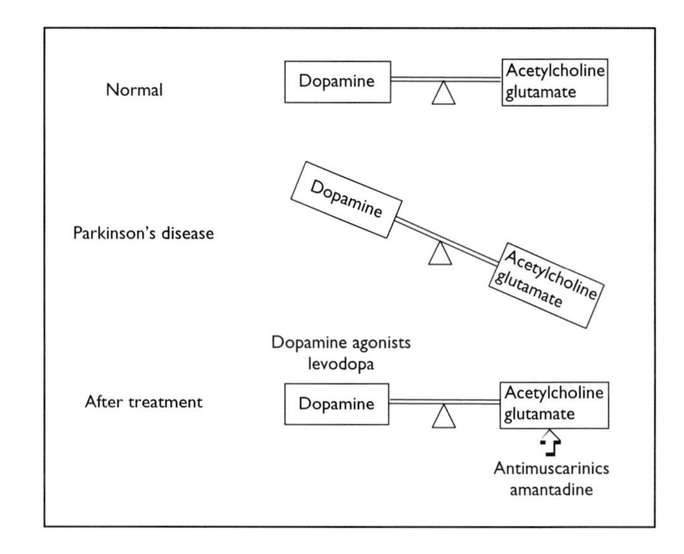

Figure 32.10 Dopamine/acetylcholine/glutamate balance.

GLUTAMATE ANTAGONISTS

Some of the brain circuits which regulate dopamine's effects on movement are excitatory and use glutamate as their transmitter. In parkinsonism these circuits become overactive, and so glutamate antagonists have been tested to see if they relieve parkinsonian symptoms (Figure 32.10). There are many new and selective NMDA and AMPA receptor antagonists to choose from, and several of these combat parkinsonian rigidity and akinesia in animal tests. Unfortunately, the abundance of glutamate receptors across the brain means that antagonists exhibit many unwanted effects. These include motor abnormalities, psychostimulation and neurotoxicity, as well as learning and memory difficulties. Glutamate is the focus of research into new treatments for epilepsy and ischemia, and it is hoped that advances in these fields will identify safer antagonists that can be used to good effect in Parkinson's disease. In addition to ameliorating parkinsonian symptoms, such drugs could have the added benefit of being neuroprotective, and slow or even stop the progression of the disease.

The same result should theoretically be possible with drugs that suppress the synaptic release of glutamate. To date only two such drugs have been evaluated in this way, the antiepileptic *lamotrigine* and the κ opioid agonist *enadoline*, with mixed results. A double-blind trial with *lamotrigine* in Parkinson's disease patients failed to show any symptomatic improvement, whilst *enadoline* enabled parkinsonian monkeys to function normally.

It is clearly too early to predict the outcome of this line of research, but one interesting fact stands out, which is that a number of clinically useful antiparkinsonian drugs (*amantadine, memantine, budipine*) are all non-competitive, use-dependent antagonists of the NMDA receptor-ionophore. All three compounds potentiate levodopa in the clinic, which research suggests is due to an increase in the activity of the enzyme that decarboxylates levodopa into dopamine.

DOPAMINE UPTAKE BLOCKERS

Synaptic dopamine is taken back into the dopamine neuron by a high affinity transporter protein. Inhibiting this uptake process could benefit the parkinsonian patient in two ways. Firstly the concentration of dopamine in the synapse would be increased, allowing dopamine to act for longer at postsynaptic dopamine receptors and so potentiating the clinical response to a given dose of levodopa. Secondly, a number of neurotoxins (e.g. MPTP) are rendered harmless by dopamine uptake inhibitors (e.g. *nomifensine, mazindol*), which are therefore neuroprotective. Future trials will tell us if these drugs are effective or not in man.

ANTIOXIDANTS

α-Tocopherol (vitamin E) is a powerful antioxidant and protects unsaturated cell lipids from oxidative attack by free radicals. It should therefore increase the resistance of parkinsonian patients to any free radicals generated by oxidative stress. Preliminary reports indicate that vitamin E supplements may lessen the severity of the disease, presumably by replacing the action of the protective enzymes that are deficient in parkinsonism. Only patients diagnosed early and treated immediately will get the full benefit of antioxidant therapy.

NEUROTROPHIC FACTORS

Neurotrophins are essential not only for the development and differentiation of neurons, but also for their survival. One of these neurotrophic factors derived from glial cells (GDNF), specifically promotes the survival of dopamine neurons. In experimental studies, GDNF has been found to rescue dopamine cells from the damage caused by neurotoxins. There is consequently much excitement that GDNF may be used to protect surviving dopamine neurons against further damage in idiopathic Parkinson's disease.

NEURONAL TRANSPLANTATION

The brain has no way of replacing dopamine neurons once they die, so surgeons have resorted to transplantation. The first attempts to replace dopamine-releasing neurons into the brain of a parkinsonian patient began in the 1980s, when one of the patient's own adrenal medullas was removed and pieces of tissue were placed into the caudate-putamen. This early work involved a fairly complex operation, which was somewhat hit-and-miss and the grafts did not survive. Nowadays operations have been greatly streamlined and employ multiple deposits of nigral dopamine cells taken from aborted fetuses. Suspensions of the neurons are injected bilaterally into the caudate-putamen through small holes drilled in the front of the skull. This can be done under local anesthesia and the whole procedure takes about an hour.

Several hundred patients worldwide have now received dopaminergic grafts. PET scans show that grafts can survive and liberate dopamine for several years, in spite of continued degeneration of the patient's own dopamine cells. Successful transplants do not mean the patient is cured, simply that responsiveness to levodopa is regained. Many grafts do not work at all, and there is still much uncertainty as to whether the benefits of the operation outweigh the risks.

To eliminate graft rejection and ethical issues regarding the use of fetuses, the way forward lies with genetic engineering, in which non-neuronal cells taken from the patient (e.g. skin fibroblasts, muscle cells) are transfected with genes that express dopamine synthesis and release. These modifed cells are cultured and then immortalized, to stop further division, before being implanted into the patient's brain. It can only be a matter of time before the early successes this technique has had in animals, are repeated in man (Freed *et al.*, 1991).

NEUROSURGICAL APPROACHES

The last decade has seen a rekindling of interest in stereotaxic neurosurgery for the treatment of parkinsonism. The principle is that overactive nervous pathways are cut in order to normalize motor control by the basal ganglia. Lesions of the thalamus *(thalamotomy)* have long been known to tame muscle tremor. Similarly, lesions of the globus pallidus (*pallidotomy*) are now being resurrected to reduce parkinsonian tremor, akinesia and muscle rigidity, although the real benefits are a restored levodopa response. Improved surgical precision has overcome the earlier complications of pallidotomy and avoids damage to healthy neurons.

Instead of lesioning overactive neurons, a safer alternative is to shut them down with high frequency stimulation via tiny stimulus electrodes embedded in the brain. Wires from the electrodes are tucked away under the skin and connected to a stimulator buried in the chest, which the patient is able to control with a magnet. Electrical stimulation of the *subthalamic nucleus* in the basal ganglia brings about an immediate and dramatic relief from parkinsonian akinesia and rigidity. The advantage of this procedure is that the intensity of the stimulus can be individually tailored to the patient's needs (Limousin *et al.*, 1995).

References

Singer, T.P., Trevor, A.J. and Castagnoli, N. (1987) Biochemistry of the neurotoxic action of MPTP: or how a faulty batch of "designer drug" led to parkinsonism in drug abusers. *Trends in Biochemical Sciences 12: 266–270* (historical)

Ebadi, M., Srinivasan, S.K. and Baxi, M.D. (1996) Oxidative stress and antioxidant therapy in Parkinson's disease. *Progress in Neurobiology 481: 1–19* (toxicity)

Kopin, I.P. (1993) The pharmacology of Parkinson's disease therapy: an update. *Annual Review of Pharmacology and Toxicology 32:467–495* (therapy and toxicity)

Hagan, J.J., Middlemiss, D.N., Sharpe P.C., and Poste, G.H. (1997) Parkinson's disease: prospects for improved drug therapy. *Trends in Pharmacological Sciences 18:156–163* (new D_2 agonists)

Klockgether, T. Löschmann, P.A. and Wüllner, U. (1994) New medical and surgical treatments for Parkinson's disease. *Current Opinion in Neurology 7:346–352* (recent advances)

Jenner, P. (1990) Parkinson's disease: clues to the cause of cell death in the substantia nigra. *Seminars in the Neurosciences 2:117–126* (toxicity)

Greenamayre, J.T. and O'Brien, C.F. (1991) N-Methyl-D-aspartate antagonists in the treatment of Parkinson's disease. *Archives of Neurology 48:977–991* (therapy)

Limousin, P., Pollak, P., Benazzouz, A., Hoffmann, D., Le Bas, J.F., Brousolle, E., Perret, J.E. and Benabid, A.L. (1995) Effect on parkinsonian signs and symptoms of bilateral subthalamic nucleus stimulation. *Lancet 345:91–95* (new surgical technique)

Freed, W.J., Poltorak, M., Takashima, H., LaMarca, M.E. and Ginns, E.I. (1991) Brain grafts and Parkinson's disease. *Journal of Cellular Biochemistry 45:261–267* (neuronal transplantation)

Fagerström, K.O., Pomerlau, O., Giordani, B. and Stelson, F. (1994) Nicotine may relieve symptoms of Parkinson's disease. *Psychopharmacology 116:117–119* (therapy)

Olson, L. (1997) The coming of age of the GDNF family and its receptors: gene delivery in a rat Parkinson model may have clinical implications. *Trends in Neurosciences 20:277–279* (therapy)

Chapter 33

Physiological role of amyloid precursor protein (APP) and *β*-amyloid peptides

B. Penke and T. Harkány

Contents

33.1 Introduction

Alzheimer's disease (AD) is the most common form of age-related neurodegenerative disorders. The pathogenesis of these and other neurodegenerative diseases have remained unclear, and effective treatments are currently lacking. However, recent studies from three diverse disciplines, neuropathology, genetics, and biophysics, have begun to converge on novel targets for therapeutic attack.

Alzheimer's disease counts for about 2/3 of the cases of dementia. It was first described by the German physician Alois Alzheimer (1907). This is a progressive dementia associated with an almost total extirpation of the neurons in some regions of the brain. At the onset of clinical symptoms, typically at the age of 65–70 years, the brain has already suffered irredeemable damage; the disease progresses unceasingly to total mental breakdown and death in 3–10 years. It is estimated that in the USA 5 million people suffer from AD and need health and social assistance. As the prevalence of AD rises exponentially with age, this number will increase with increasing average age of the population.

These problems have elicited intense research on AD and related fields. Progress in basic neuroscience and application of molecular biological research methods has brought a detailed understanding of many of the mechanisms underlying neuronal death and development of AD. However, we still do not have a complete picture of the molecular pathomechanism of AD, which would be of major importance for designing prevention and treatment strategies.

33.1.1 Morphopathological presentation of AD

The most characteristic feature of the brain in AD is the presence of extracellular starch-like deposits (senile plaques, amyloid plaques). Their main component is the β-amyloid (Aβ) peptide (Glenner *et al.*, 1984). The initial, amorphous form is called a "diffuse plaque". Diffuse plaques are often present in healthy aged persons. In the next step of maturation, the plaque is called a "neuritic plaque". It can be observed as a denser deposit of fibrillar amyloid with β-sheet structure. This kind of plaque is often surrounded by degenerating neurites, and also by reactive astrocytes and microglia (Kalaria, 1993).

A second characteristic of AD is the presence of intracellular filamentous deposits in the degenerating neurites and cell bodies, called neurofibrillary tangles (NFT) (Morris, 1995). Their main component is the abnormally (hyper-)phosphorylated tau protein, (a microtubule-associated protein) (Lippa, 1993). NFTs are critical lesions in AD, but their presence is not limited to AD, as they can be observed in a variety of other neurological diseases. This shows that this cytoskeletal alteration can develop as a secondary – albeit important – response to a variety of cerebral insults.

Another marker of AD is amyloid angiopathy, the presence of Aβ peptide in the walls of leptomeningeal and small cortical vessels. Parenchymal and vascular amyloid depositions are not related to each other (Lippa, 1993, Terry *et al.*, 1994). The distortion of brain capillaries is a general event in AD.

Extensive neuronal and synaptic losses are also characteristic for AD: 30–60% for neurons and 25–50% for synapses as compared with age-matched controls. Their degrees correlate powerfully with dementia severity (Terry *et al.*, 1994). The loss is most severe in the hippocampus, magnocellular basal nucleus, entorhinal cortex and neocortex. The different cell populations are selectively affected: cholinergic neurons are the most vulnerable; glutamatergic, adrenergic, serotoninergic neurons are also severely affected; while certain populations of GABAergic neurons with intracellular Ca^{2+} (Ca_i^{2+}) binding proteins are relatively well preserved. Finally, mitochondria of the neurons are also severely damaged in AD brain.

33.1.2 Epidemiology of AD

The epidemiological study of AD with special emphasis on genetic causes was among the most fruitful areas of research in the past few years. Risk factors were found as follows:

- Age is the most important risk factor for dementias of all kinds, including AD. The prevalence of AD shows an exponential rise as a function of age, with rates doubling every 5 years in the population aged over 65 years (Katzman and Kawas, 1994).
- In case of a family history of AD (in first degree relatives), the risk of developing the disease increases approximately 4-fold. This points to genetic causes of AD, which constitute the second most important risk factor for the disease.
  - There are several well established genes associated with AD (Selkoe, 1999).
- The existence of certain situations is often found in the prehistory of AD patients and can be viewed as environmental, precipitating factors of the disease. The most important are the following factors:
  - repeated hypoxia (by divers and mountain climbers) causing brain damage
  - head injury (defined as one sufficient to produce unconsciousness) up to 30 years prior to the onset of AD, either accidental or repeated (e.g. in the case of the prize fighters)
  - some illnesses, including depression (occurring at 10 or fewer years before AD onset), atherosclerosis, coronary disease, diabetes, and hypothyroidism, were also shown to increase the prevalence of AD (Katzman, 1994).

In the popular press, exposure to aluminum has been the most feared risk factor for AD. However, none of the extensive European and American case-control studies has found a relationship of AD to aluminum antacid intake or use of aluminum cookware (Katzman, 1994).

33.1.3 Molecular pathomechanism of AD

A model of the molecular pathomechanism of AD, shared by many scientists, can be shortly expressed in one sentence: an abnormally elevated concentration of Aβ peptide, generated by enzymatic cleavage of the β-amyloid precursor protein (βAPP), induces a cascade of pathological events in the neurons (increase of Ca_i^{2+} concentration, hyperphosphorylation of tau-protein, NFT deposition etc.), leading to cell death and onset of the clinical symptoms of AD. This is the "amyloid cascade" model (Hardy and Allsop, 1991). This view, which attributes a central role to Aβ peptide in AD pathogenesis, emerged on the basis of a series of crucial discoveries, mainly in 1990–1991, showing that the Aβ peptide (deposited in the brain in AD) has direct neurotoxic effect (Yankner *et al.*, 1990), and that mutations of its precursor (βAPP) in proximity to the Aβ-linked enzymatic cleavage sites (Figure 33.1) cause familial AD or another amyloidotic disease in an autosomal dominant manner. As βAPP and (in a low concentration) β-amyloid peptides (Aβ) are normal components of the living organism, it is important to investigate their physiological role.

33.1.4 Physiological role and metabolism of βAPP

The β-amyloid precursor protein isoforms comprise a family of neuronal proteins with synaptic localization, whose physiological roles in brain have been poorly understood. βAPP is an ubiquitous transmembrane protein whose heterogenity arises from both alternative splicing and post-translational processing. Non-neuronal cells express 751- and 770-residue splice forms throughout the body. Neurons express a more abundant 695-residue isoform. The difference between the 751/770 and 695-residue forms is the presence of a 56-amino acid motif in the longer forms which is homologous to the Kunitz-type serine protease inhibitors (reviewed by Selkoe, 1994). All of them are rapidly induced proteins in the injured cortex, the induction and secretion of βAPP is a response to subcortical lesions of cholinergic innervation.

One of the most important physiological roles of βAPP is cell surface adhesion and membrane matrix stabilization (Beer *et al.*, 1995) βAPP interacts with selected integrins and shares similar cellular machinery to promote cell adhesion. βAPP may be glycosylated as a proteoglycan (PG-βAPP) and this population of βAPP may act as a neuritic outgrowth and branching promoter. Appican, the chondritin sulfate proteoglycan form of βAPP, seems to be a very strong cell adhesion molecule (Pangalos *et al.*, 1996).

The normal physiological role of βAPP in the brain was evaluated in a cross-sectional analysis of mice deficient in βAPP. These "βAPP-null mice" developed age-dependent deficits in cognitive function and also had impairments in long-term potentiation. In addition, the brains of βAPP-null mice had marked reactive gliosis in many areas, especially in the cortex and hippocampus. Normal βAPP may serve an essential role in the maintenance of synaptic function during aging (Dawson *et al.*, 1999). *In vivo* and *in vitro* studies suggest that βAPP is an important factor for viability, growth, morphological and functional plasticity of nerve cells. Furthermore, βAPP and its derivatives have an important role in learning and memory processes. Altered expression of the βAPP gene in aged animals or in genetically modified animals leads to memory deficits. By contrast, secreted forms of βAPP have recently been shown to facilitate learning and memory processes in mice (Dodart *et al.*, 2000). βAPP shows a preferential localization at synaptic sites of human brain and at neuromuscular junctions. βAPP is further found on vesicular elements of neuronal perikarya, dendrites and axons. The synaptic localization of βAPP implies an important role of βAPP in physiological synaptic activity.

βAPP enhances neuron viability and modulates neuronal polarity. A line of transgenic knockout mice selectively deficient in βAPP survive and breed, but exibit motor dysfunction and brain gliosis, consistent with a physiological role for βAPP in neuron development. Experimental data demonstrate that cell-associated βAPP contributes to neuron viability, axogenesis, and arborization, it modulates axon growth, dendrite branching, and dendrite numbers (Perez *et al.*, 1997).

βAPP proved to be a potentiating agent for the neurotrophic activity of NGF. βAPP is a rapidly induced protein in the injured cortex and may render compromized neurons more responsive to low levels of NGF and other neurotrophins (Wallace *et al.*, 1997). βAPP might trigger neurite degeneration that is followed by caspase-dependent apoptosis in neural cultures. It may contribute to the profound cell death associated with AD.

βAPP has an about 23-residue hydrophobic region near its C-terminal region, this serves as an anchor, binding the protein into internal membranes (e.g. endoplasmic reticulum, Golgi and trans-Golgi network, endosome) and into the plasmalemma. βAPP molecules undergo specific endoproteolytic cleavage during and after their transport through the secretory pathway to the cell surface by specific proteases, the so called secretases (Figure 33.1).

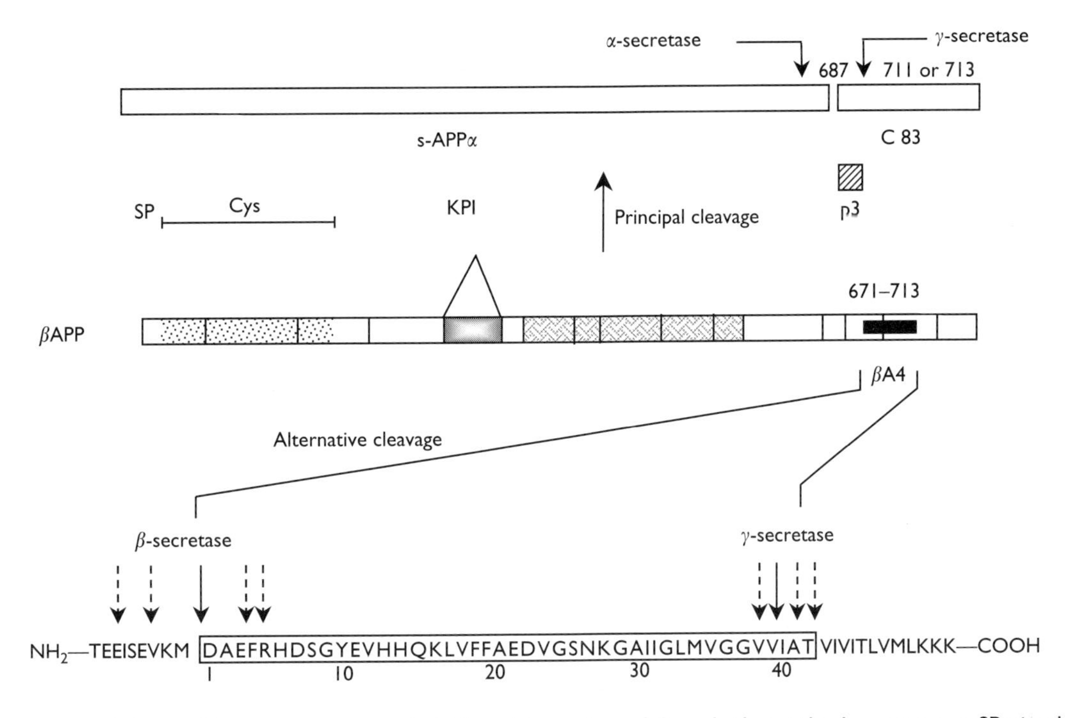

Figure 33.1 Processing of β-amyloid precursor protein (βAPP) and formation of β-amyloid peptides by secretases. SP: signal peptide; Cys: cysteine-rich region; KPI: Kunitz-type serine-protease inhibitor; βA4: 4 kD β-amyloid peptides. s-APPα: soluble amyloid -protein; C83: C-terminal fragment of 83 amino acid residues; p3: 24 or 26 amino acid short peptides.

10
Asp – Ala – Glu – Phe – Arg – His – Asp – Ser – Gly – Tyr –

20
–Glu – Val – His – His – Gln – Lys – Leu – Val – Phe – Phe –

30
–Ala – Glu – Asp –Val – Gly – Ser – Asn – Lys – Gly – Ala –

39 ↓ 40 ↓ 41 42
–Ile – Ile – Gly – Leu – Met – Val – Gly – Gly – Val – Val – Ile – Ala – Thr

Figure 33.2 The amino acid sequence of β-amyloid peptides ($A\beta_{1-39}$, $A\beta_{1-40}$, $A\beta_{1-42}$ and $A\beta_{1-43}$).

The main endoproteolytic cleavage is the scission between amino acids 16 and 17 of the Aβ-region by protease(s) called α-secretase(s). This cleavage results in formation of a large, soluble protein fragment (sAPP-α) and a membrane-retained fragment containing 83 amino acids (C83). The enzyme(s) show little sequence specificity. These enzyme(s) are not clearly identified, they are probably membrane-anchored metalloproteases (Sisodia, 1992). An alternative cleavage mode by the β- and γ-secretases lead to Aβ formation (Figure 33.1). These enzymes also show a little sequence specificity, therefore Aβ comprises a series of peptides containing 39, 40, 42 and 43 amino acid residues (Figure 33.2).

Aβ peptides are secreted by normal cells and detected as circulating peptides in the plasma and cerebrospinal fluid in healthy humans. It seems that under normal conditions a much higher portion of the total cellular βAPP undergoes cleavage by α-secretase than by β-secretase. After α-secretase cleavage, the C83 fragment can be further metabolized by the enzyme γ-secretase resulting in a short Aβ peptide called p3 (having a molar mass of about 3 kD).

β-Secretase has been identified by Vassar *et al.* (1999) as a transmembrane aspartic protease. An intensive search was made for γ-secretase (Hardy and Israel, 2000) and now it has been identified as a special transmembrane aspartyl protease which cleaves the βAPP peptide chain within the membrane (Wolfe *et al.*, 1999). More recent investigation might prove that γ-secretase is identical with presenilin 1 (Kimberley *et al.*, 2000).

33.2 Physiological role of Aβ peptides and their catabolism

Exposure of nerve, glial and endothelial cells to $A\beta$ results in the modification of a wide variety of cellular functions which, may differ significantly depending on the stereochemical characteristics, aggregation state, and concentration of the $A\beta$ fragment applied. Initially, Yankner *et al.* (1990) demonstrated that $A\beta$ exerts bi-phasic actions on neurons in culture. While pM-nM $A\beta$ concentrations promote neuronal differentiation and increase neuronal viability in a neurotrophin-like manner, $A\beta$ concentrations in the μM-range are toxic to neurons. Beside its neurotrophin-like activity, a direct neuromodulatory role, affecting cholinergic neurotransmission in the brain, was proposed for low concentrations of $A\beta$ (Auld, 1998). This hypothesis was based on the observation that pM to nM $A\beta$ concentrations inhibited the release of acetylcholine (ACh), decreased $[ACh]_i$ and choline-acetyltransferase (ChAT) activity without concurrent signs of apparent neurotoxicity.

Receptor binding experiments of Wang *et al.* (2000) have shown that $A\beta$ 1–42 binds with high affinity to $\alpha 7$ nicotinic acetylcholine receptor (α7nAchR). This receptor is a neuronal pentameric cation channel and modulates calcium homeostasis and release of the neurotransmitter acetylcholine. Wang's work suggests that the physiologically relevant neuronal receptor is just the α7nAChR.

Micromolar ranges of $A\beta$ concentration are toxic to neurons inducing necrotic and apoptotic cell death. In this concentration $A\beta$ peptides aggregate to toxic supramolecular systems, which are resistent against proteolytic enzymes. Recent studies show that the steady-state amount of $A\beta$ peptides in the brain is determined by a balance between their production and their degradation (Iwata, 2000). In this respect, not only the activity of β- and γ-secretases seems to be important, but also their degradation and clearance. Investigation of the mechanism of catabolism has shown that $A\beta_{1-42}$ peptide underwent full degradation through limited proteolysis conducted by neutral endopeptidase (NEP). Up-regulation of this enzyme could reduce the risk of developing Alzheimer's disease by preventing $A\beta$ accumulation.

In summary, the release of minute amounts of $A\beta$ during normal metabolism, and the proposed neuromodulatory action of $A\beta$ on cholinergic neurons hint at physiological functions for $A\beta$ which in turn may yield pathological dimensions during noxious conditions that suppress the metabolic control of $A\beta$ production. Excessive $A\beta$ release super-stimulates $A\beta$ signaling pathways, perturbes the intracellular ion homeostasis and ultimately leads to cell death.

33.3 Aggregation of Aβ-peptides, toxic folding, cell signaling and the toxic effect of Aβ-peptides: the neurotoxic cascade

The central role of $A\beta$ in the pathogenesis of Alzheimer's disease provide the rationale for summarizing recent knowledge on aggregation of $A\beta$ peptides, the ligand-like interaction of $A\beta$-aggregates with putative neuronal and glial receptors as well as signaling pathways critically involved in cellular $A\beta$ toxicity.

33.3.1 *β-Amyloid aggregation, toxic folding*

Above a threshold concentration ($>10^{-6}$ M) $A\beta$ undergoes highly ordered protein aggregation in aqueous solutions resulting in the formation of senile plaques in AD. Following β- and γ-secretase-mediated catabolism of APP, Aβ, released in a monomeric form in the extracellular space, begins a self-aggregation process, which, through the transient formation of several discrete intermediate Aβ species, leads to Aβ condensation as extremely insoluble protein fibrils with characteristic morphological features (80- to 150 Å fibrils) (Figure 33.3).

Particular steps of this aggregation process, the influence of mutation-related amino acid substitutions both in the APP and Aβ sequences, and the distinct morphological features of Aβ fibrils have recently been studied extensively.

Toxic folding of different polypeptides and proteins are well known from the literature and some of them are summarized in Table 33.1.

Aβ exists predominantly as a 40 or 42 amino acid residue-containing peptide in AD. The biological activity of Aβ_{1-40} and Aβ_{1-42} is shared by the undecapapetide Aβ_{31-35} (Pákáski *et al.*, 1998), which was therefore designated as the active center of Aβ. The presence of the C-terminal dipeptide sequence in Aβ_{1-42}, and the concomitant augmentation of hydrophobicity, results in a several-fold increase in the dynamics of peptide assembly into filaments, as compared with Aβ_{1-40}, and to the formation of unusually stable fibrils even by nonapeptides (Aβ_{34-42}). Aβ fibril formation is a nucleation-dependent process, resembling crystallization, as the presence of pre-formed fibrils and negatively charged surfaces (Figure 33.4) elicit immediate 'seeding' of metastable Aβ solutions.

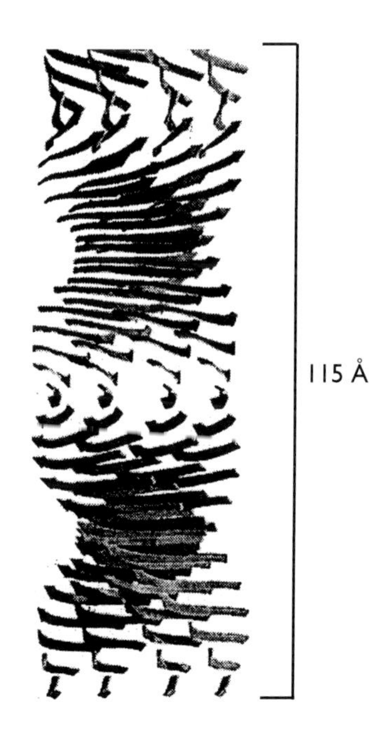

Figure 33.3 Aggregation of βA-monomers to amyloid fibril.

Table 33.1 Toxic folding of proteins

1 Scrapie Creutzfeld-Jacob	Prion protein	Aggregation in synapses
2 Alzheimer's disease	β-Amyloid	Aggregation, plaques
3 Amyloidosis	Transthyretin lysosim	Aggregation
4 Cataract	Crystallin	Aggregation

Shortly after its release to the extracellular space in pM–nM concentrations, Aβ may exist as monomers. However, during the highly ordered process of Aβ fibrillogenesis, discrete but transient Aβ dimers, tetramers and intermediate oligomer Aβ species, also termed protofibrils, exist between the two stable states of Aβ, the monomeric protein and the fibril (Harper and Lansbury, 1997). Aβ monomers, and dimers, also termed low molecular weight Aβ, do not alter the normal metabolic activity of neurons in culture. In solution, monomeric or dimeric Aβ forms are in equilibrium with protofibrils (Walsh *et al.*, 1999). According to Tjernberg *et al.* (1999), formation of Aβ protofibrils occurs by the sequential assembly of anti-parallel dimers and/or tetramers in a β-sheet conformation. The β-structure (β-sheet and β-turn) content of protofibrils ranges from 45 to 50%. Protofibrils reach approximately 40% of the height of the product fibril, generally do not exceed a length of 400 nm, and appear as beaded chains when examined electron microscopically. As Walsh *et al.* (1999) suggested, the small globular assembly, constituted by as few as five or six Aβ molecules, may represent the structural unit for protofibril formation. Aβ_{1-40} and Aβ_{1-42} assemble *via* analogous steps, although Aβ_{1-42} protofibrils are assembled more rapidly. Recent data from Lambert and colleagues (1998) demonstrate that soluble, ligand-like Aβ oligomers (between 17 to 42 kDa) – presumably protofibrils – that are non-fibrillar and readily diffusible act as neurotoxins on pyramidal and granule cells in hippocampal slices at nanomolar concentrations. These and other studies (Hartley *et al.*, 1999) provided evidence against the widely appreciated principle that fibrillar Aβ is the only directly toxic Aβ species, whereas soluble Aβ peptides lack direct toxicity. *In vitro* aggregation of Aβ peptides results in the formation of fibrils identical to those found in AD, as Aβ fibrils of both origins exhibit structural, conformational and staining similarities. While the β[16–20] sequence is essential to Aβ-Aβ binding, stabilization of the fibrillar state of Aβ is confined partly to a β–turn of β[26–29] tetrapeptide region and to hydrophobic interactions in the 29–42 domain.

In summary, the above experimental data indicate that distinct molecular forms of Aβ affect cellular viability in a differential manner. While low molecular weight Aβ species do not compromise physiological neuronal functions, Aβ protofibrils and mature fibrils act as potent neurotoxins in the central nervous system.

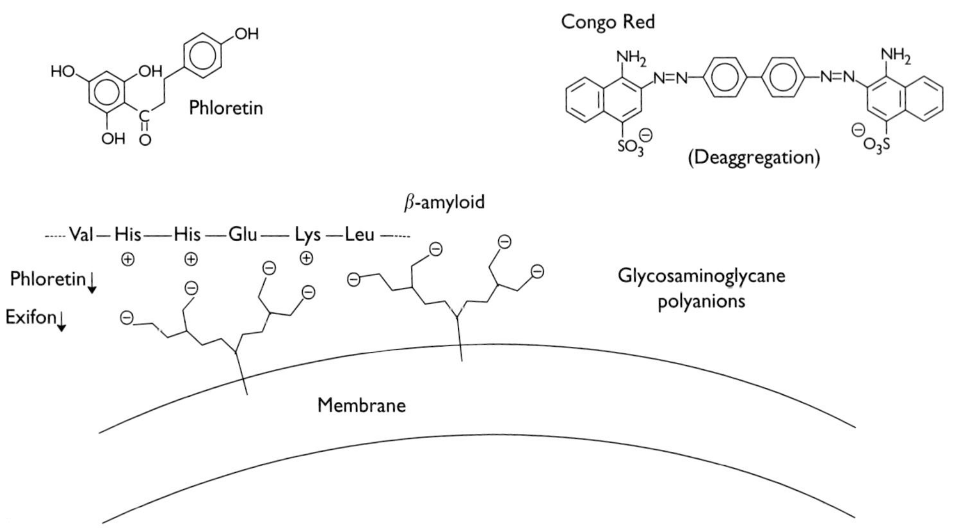

Figure 33.4 Amyloid aggregation on the cell membrane over physiological concentration on glycosaminoglycane polyanions. Phloretin and exifon prevent binding and aggregation on the cell surface. Congo Red also prevents aggregation by binding to the basic center HHQK of Aβ peptides.

33.3.2 Interaction of Aβ peptides with putative neuronal and glial receptors and membrane structures

The diversity of conceivable physiological (neurotrophic, modulatory) and pathological (neurotoxic) actions of Aβ on virtually all cells of the central nervous system led to the assumption that Aβ afflicts cellular functions by its selective interaction with putative cell-surface receptors. On the other hand, the pronounced lipophilicity of Aβ accounts for its profound interaction with biological membranes. Aβ peptides are known to bind specifically to the scavenger receptors expressed in microglia and macrophages (El Khoury *et al.*, 1996) and the ubiquitous receptor for advanced glycation end products (RAGE, Yan *et al.*, 1996). Aβ binding to scavenger receptors may induce cell death by generating free radicals, whereas binding to RAGE may promote Aβ clearance. However, the respective receptor function and distribution pattern for scavenger receptors and RAGE cannot fully explain the Aβ-induced pathophysiology of Alzheimer's disease (cognitive and memory impairment).

Aβ destabilizes the intracellular Ca^{2+} homeostasis (Mattson *et al.*, 1992), and augments the generation of reactive oxygen intermediates (for review see Behl, 1999). For explanation of these effects, a number of receptors were proposed as molecular targets for Aβ. Generation of overt inward Ca^{2+} fluxes via glutamate receptors, and a subsequent rise of the intracellular free Ca^{2+} concentration ($[Ca^{2+}]_i$) are substantial components of excitotoxic signaling, which were postulated to mediate a final common pathway for neurological disorders. Excessive activation of the N-methyl-D-aspartate (NMDA) subclass of glutamate receptors was reported to lead to excitotoxic neuronal death under pathological conditions. Therefore, extensive research was devoted to the identification of the role of NMDA receptors in Aβ toxicity. Cowburn *et al.* (1994) demonstrated that Aβ exerts differential effects at the glycine and, presumably, also at the glutamate recognition sites of the NMDA receptor, and modulates NMDA receptor activity. Whether Aβ directly affects NMDA receptor function by its binding to modulatory sites of the receptor remains to be elucidated.

Cholinergic neurons of the magnocellular nucleus basalis (MBN) are endowed with neurotrophin receptors, particularly of the non-selective low-affinity neurotrophin receptor ($p75^{NTR}$). Since the expression of $p75^{NTR}$ correlates with cellular Aβ sensitivity, $p75^{NTR}$ has been regarded as candidate receptor for Aβ (Yaar *et al.*, 1997).

An intriguing question remains as to whether additional subclasses of putative neuronal receptors also exhibit high-affinity binding for distinct molecular forms of Aβ. Lambert *et al.* (1998) reported that oligomer Aβs act as potent neurotoxins via their binding to trypsine-sensitive receptor-like binding sites. Although functional biochemical assays support the existence of the above receptors, their molecular structures are yet to be identified.

Aβ peptides interact with cell membranes and also with membranes of subcellular organelles (mitochondria, Golgi complex and endoplasmic reticulum). Several experimental data suggest that Aβ peptides can form cation-selective pores in lipid bilayers (Kim *et al.*, 1999). However, short aggregated, Aβ fragments ($A\beta_{31-35}$, $A\beta_{34-39}$, etc,) (Pákáski, 1998) are unable to form such channels and are still neurotoxic. We assume that Kanfer *et al.* (1999) are right by stating that the

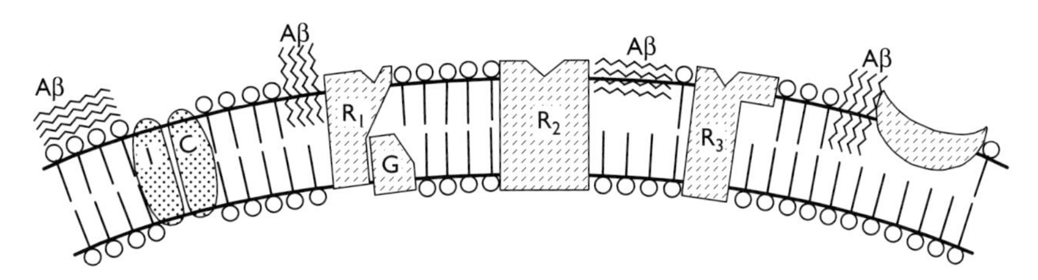

Figure 33.5 Interaction of aggregated Aβ peptides with cell membranes, multitude of receptor sites and ion channels. IC: ion channel; R_1, R_2 and R_3: receptors; G: G-protein; Aβ: aggregated Aβ peptides.

membrane perturbation caused by Aβ peptides is the basis for their biological effects. These peptides have amphiphil character and form a highly aggregated structure which interacts at the cell membrane with a multitude of receptor sites and also physically inserts into the membrane matrix (Figure 33.5). This membrane insertion affects the membrane fluidity and potentially influences the function of resident membrane proteins, including some receptors.

Hertel *et al.* (1997) proved that Aβ peptides have strong electrostatic interaction with cell membranes. Inhibition of the interaction between Aβ and membrane prevents β-amyloid induced toxicity.

The newest results of Lorenzo *et al.* (2000) have shown that fibrillar Aβ binds to a subset of detergent-soluble membrane proteins and also to βAPP. βAPP may be involved in the mechanism of Aβ neurotoxicity: APP-null neurons are less vulnerable to Aβ toxicity. Thus Aβ neurotoxicity may be mediated by the interaction of fibrillar Aβ with neuronal membrane proteins, notably βAPP.

33.4 Intracellular consequences of β-amyloid toxicity: the inflammatory and neurotoxic cascades, apoptosis and necrosis

Chronically elevated concentrations of Aβ in the brain leads gradually to oligomerization and, eventually, fibrillization of the peptide, then deposition as diffuse and, later, as mature plaques. It is hypothesized, that Aβ accummulation and diffuse plaque formation is associated with local microglial activation, cytokine release, reactive astrocytosis and a multiprotein inflammatory response (Rogers *et al.*, 1996), including complement activation (C1q component) by Aβ and triggering the classical cascade (Selkoe, 1999). Long-lasting anti-inflammatory medication reduces the prevalence of Alzheimer's disease. It supports the hypothesis that glial activation as a consequence of inflammatory mechanisms exacerbate neurotoxic cascades leading to late-life dementia.

It has been proposed that a glial inflammatory process and the direct neurotoxic effect of aggregated Aβ peptides on neurons (sustained binding of Aβ to cell surface proteins, lipophilic interaction of Aβ with membranous structures) activate signal transduction cascades. Two major molecular cascades, namely *excitotoxicity* and *free radical-induced oxidative stress*, have been implicated in Aβ cytotoxicity. Although these molecular pathways utilize substantially different second messengers, and may activate separate sets of genes executing cell death; they converge at numerous intracellular signaling events, share intracellular target molecules, and thereby may effectively trigger neuronal death.

33.4.1 Excitotoxic events

The postulate that excitotoxic mechanisms play cardinal roles in the pathogenesis of AD was based on extensive neuroanatomical and biochemical evidence indicating both pre- and post-synaptic disruption of excitatory aminoacidergic pathways, pathological alterations of brain Ca^{2+} signaling, and the ability of Aβ itself, or in synergy with glutamate to dysregulate cellular Ca^{2+} homeostasis.

Exposure of membrane bilayers to Aβ results in the formation of cation-permeable transmembrane Aβ pores with high conductances for Ca^{2+}. Additionally, Aβ probably regulates the activity of NMDA receptor channels, and thereby may generate overt inward Ca^{2+} fluxes. Besides such *direct* actions of Aβ on Ca^{2+} mobilization and translocation through the plasma membrane, a complex glutamate-mediated excitotoxic cascade with *indirect* involvement of Aβ also emerged. The Aβ-evoked increased neurotransmitter release is selective for aspartate and glutamate. Taking the increasing susceptibility of nerve cells to glutamate excitotoxicity during aging into account, it can be assumed that the Aβ-induced glutamate release may yield pathological dimensions in AD. The above observations support the postulate of Geula and his colleagues (1998) that "aging renders the brain vulnerable to Aβ toxicity".

Whereas the particular molecular interaction of Aβ with glutamate receptors is still obscure, excessive Ca^{2+} entry via NMDA (and AMPA) receptors is undoubtedly a key signal for sustained membrane depolarization and for the generation of uncontrolled intracellular Ca^{2+} bursts executing excitotoxic cell death. The critical involvement of membrane depolarization in Aβ toxicity is supported by the beneficial actions of K^+ channel openers that protect nerve cells against $A\beta$ toxicity by their ability to hyperpolarize the plasma membrane and decrease Ca^{2+} influx (Harkány *et al.*, 2000). $A\beta$ induces intracellular changes characteristic for excitotoxicity. Exposure to $A\beta$ elicits sustained activation of Ca^{2+}-permeable receptor channels, which results in a pathological enhancement of inward Ca^{2+} currents and a substantial increase in the $[Ca^{2+}]_i$. Elevation of $[Ca^{2+}]_i$ may progress further via the activation of intracellular Ca^{2+}-mobilizing second messengers, such as inositol(1,4,5)trisphosphate (IP_3) and inositol(1,3,4,5)tetrakisphosphate (IP_4), and their action on respective receptors inducing Ca^{2+}-mediated Ca^{2+}-release from intracellular Ca^{2+} stores. As Ca^{2+} is a universal signal not only for life but also death, and a prolonged increase in $[Ca^{2+}]_i$ can be lethal to neurons, the potential of $A\beta$ to elevate $[Ca^{2+}]_i$ can directly predispose cells to neurodegeneration.

Glial cells contribute significantly to the physiological integrity of the glutamatergic system by the regulation of extracellular glutamate concentrations via high-affinity Na^+-dependent active uptake mechanisms mediated by glutamate transporters. Arrest of extracellular glutamate uptake by glial cells leads to overt stimulation of neuronal glutamate receptors, sustained membrane depolarization, generation of uncontrolled firing (prolonged bursts), and subsequent excitotoxic cell death. Altered βAPP expression and $A\beta$ generation may contribute to the decline of glutamate uptake in AD.

In conclusion, a pathophysiological model of $A\beta$ excitotoxicity implicates the $A\beta$-induced dysfunctions of both neurons and glial cells. In response to several stressors, neural cells increase their βAPP expression, which, if accompanied by β-secretase βAPP processing, yields intact $A\beta$. Subsequent accumulation of potentially neurotoxic $A\beta$ species in the extracellular space activates microglia that releases numerous neurotoxic inflammatory cytokines and glutamate, as well as hampers astroglial glutamate uptake. The rise of extracellular $A\beta$ and glutamate concentrations leads to excitotoxic damage to neurons. While $A\beta$ potentially forms Ca^{2+}-permeable channels in the plasma membrane and acts on the NMDA receptor channel, glutamate activates AMPA and NMDA receptors and elicits sustained membrane depolarization. Finally, $[Ca^{2+}]_i$ reaches pathological concentrations and induces cell death, predominantly through apoptotic mechanisms. Characteristic features of oxidative cellular damage therefore involve enhanced disorganization of the plasma membrane and that of intracellular organelles, DNA fragmentation, altered catabolic pathways of peptides, destabilization of intracellular ion homeostasis, decrement of repair mechanisms and decline in the expression of antioxidant enzymes. Since aging is associated with the temporal enhancement of oxidative damage to neurons, and accelerated oxidative neuronal damage is evident in AD, an oxidative stress hypothesis has been proposed to describe cellular events leading to neuronal loss in AD (Behl, 1997). $A\beta$ induces oxidative stress to neurons (i) by acting as a free radical following distortions of its electronic structure, (ii) by interacting with membranous cellular components by its lipophilic nature, and (iii) by initiating receptor-mediated pro-inflammatory signaling pathways.

As reported by Hensley *et al.* (1994), $A\beta$ may generate free radicals in aqueous solution and they proposed that "radicalization" of $A\beta$ may be a critical component of its neurotoxicity. High affinity binding of $A\beta$ on membranes, (both on the cell surface and intracellularly) may alter the turnover of lipids and peptides by modifying the local micro-homeostasis in a manner allowing abnormal enzymatic degradation of membrane components.

Another means for $A\beta$ toxicity to induce oxidative stress is its binding to RAGE receptors on virtually all types of cells of the central nervous system. Such a receptor-ligand-like interaction may have several consequences: $A\beta$ translocates into the cytosol, and internalized $A\beta$ acts on, as yet unidentified, recognition sites of intracellular organelles, such as the mitochondria and endoplasmic reticuli. Damage to mitochondrial membranes results in a significant decrease in the mitochondrial transmembrane potential and impairs the terminal oxidation chain. As a consequence of energy depletion, cytotoxic reactive oxygen species (ROS) are produced, and released into the cytosol. Furthermore, Ca^{2+} uptake and Ca^{2+} regulation of both mitochondria and endoplasmic reticuli apparently become compromized, which lead to the subsequent elevation of $[Ca^{2+}]_i$.

Behl and colleagues (1997) demonstrated that $A\beta$ toxicity is associated with the intracellular accumulation of hydrogen peroxide (H_2O_2), and designated H_2O_2 as a potential mediator of $A\beta$ toxicity.

33.4.2 A common fatal cascade leading to cell death

Events of excitotoxicity and oxidative stress do not occur solely in the course of cytotoxicity. They trigger each other by common intracellular effectors, and in a complementary fashion contribute to the acceleration and execution of cell death. In contrast with several other agents, which apparently exhibit either primary excitotoxic (e.g. glutamate) or oxidative mechanism, amyloids may act in *both* fashions. Interaction of $A\beta$ with cell-surface receptors, moderate amounts of $A\beta$ induce excitotoxicity by the elevation of $[Ca^{2+}]_i$. Secondly, sustained lipophilic interaction of $A\beta$ aggregating on the cell surface may augment membrane lipid peroxidation, generate ROS and compromise nerve cells primarily in an oxidative stress pathway. It can be assumed that these toxic pathways converge at the level of mitochondria and other intracellular Ca^{2+} stores. A pathologically elevated $[Ca^{2+}]_i$ exhausts the buffering capacity for intracellular Ca^{2+} pools, particularly that of the mitochondria and endoplasmatic reticuli, and orchestrates Ca^{2+}-mediated

Ca^{2+} release from intracellular stores. Subsequently, damage to the mitochondria results in the enhanced production of ROS, and in the translocation of "death" factors involved in apoptotic cell death into the cytosol. Similarly, intracellular accumulation of Aβ may increase the oxidative breakdown (peroxidation) of intracellular membranous organelles – such as the mitochondria – impair the cellular energy supply, and release free Ca^{2+} into the cytosol. Accordingly, elevated $[ROS]_i$ and $[Ca^{2+}]_i$ predisposes cells to degeneration. The amyloid cascade is summarized in Figure 33.6.

The pathways summarized above ultimately converge to a common and fatal cascade resulting in the death of neurons. Degeneration of neurons may follow two scenarios: necrosis or apoptosis, the latter also termed programmed cell death. While necrosis occurs as a sudden consequence of intense deleterious stimuli, milder neurotoxic stimulation leads to apoptotic cell death that occurs over a relatively long period of time, ranging from hours to several days. During necrosis, nerve cells exhibit extensive rapid disintegration and loss of neurites, swelling of the somata, and immediate damage of membranous cytoplasmic organelles, such as the Golgi apparatus, mitochondria and endoplasmic reticuli, followed by breakdown of the plasma membrane. DNA degradation occurs fairly late in the process, possibly as the consequence of cell death. Apoptotic cell death is characterized by transient blebbing of the plasma membrane, followed by the condensation of nuclear chromatin and the fragmentation of nuclear DNA into oligonucleosomal-sized fragments. Subsequently, cytoplasmic organelles start to degenerate, and both the nucleus and the cytoplasm become compartmentalized into membrane-bound apoptotic bodies.

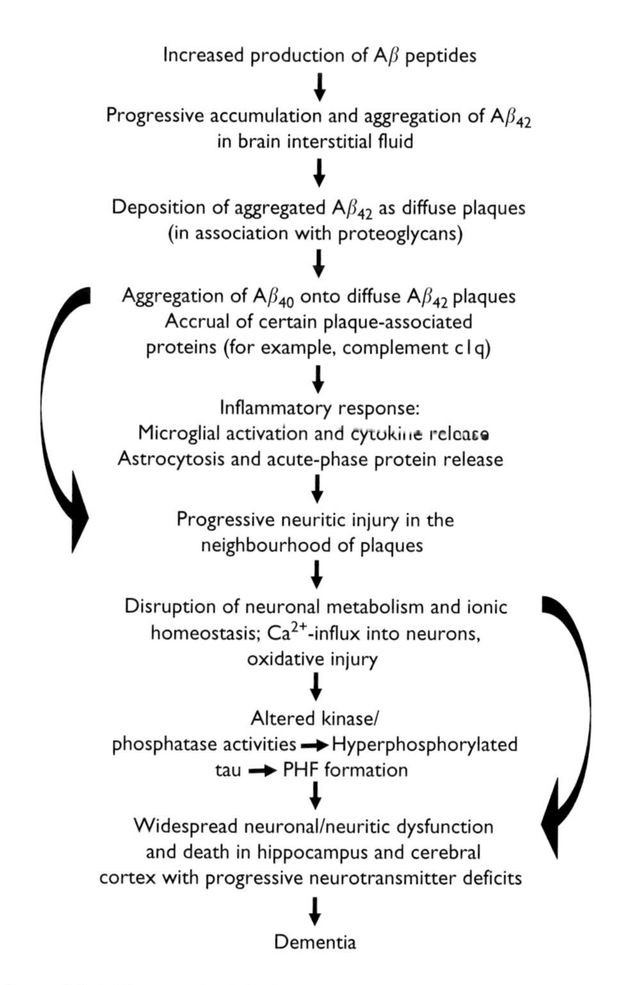

Figure 33.6 The amyloid (inflammatory and neurotoxic) cascade leading to neuron death and dementia (Selkoe 1999).

Many studies have demonstrated apoptotic cell death elicited by Aβ and related peptides. The relevance of Aβ-induced apoptosis to neurodegenerative changes in AD is supported by the increased numbers of neurons and glial cells exhibiting DNA fragmentation, and increased expression of genes associated with apoptosis in human AD tissue. While programmed cell death of neurons is due to the activation of complex signaling cascades and increased expression of a wide spectrum of apoptosis-related death genes, here we will address the emerging role of caspases as mediators of Aβ toxicity. Caspases are cysteine proteases and recognize four amino acid-long substrates as their targets and cleave C-terminal to an obligatory aspartic acid residue. Their relevance to amyloidosis and to AD derives from the fact that caspases were reported to cleave APP and generate potentially amyloidogenic fragments, and become activated in cells exposed to Aβ.

Both excitotoxic injury and oxidative stress may result in caspase activation through substantially distinct molecular mechanisms. Ca^{2+} influx via NMDA receptors was suggested to lead to successive activation of caspases, followed by the formation of ROS and lipid peroxidation. In contrast, uncoupling of the mitochondrial terminal oxidation chain as a consequence of metabolic compromise elicits not only the abrupt generation of ROS but also increases the translocation of cytochrome C (Cyt-c) and other apoptotic proteases from the mitochondria into the cytosol. Thereafter, multimers of Cyt-c, apoptotic protease activating factor 1, and pro-caspases interact as a caspase-activating complex, also termed the aposome (Cain *et al.*, 1999). When activated, the aposome executes cleavage of pro-caspases to effector caspases and boosts a fatal programmed cell death pathway.

A "terminal effector" role for caspases in Aβ-induced apoptosis derives from recent observations indicating that activation of caspases occurs hand-in-hand with the progression of apoptotic cell death. Caspases become activated in a sequential manner, following cleavage from their respective pro-caspase forms, which might be a phase-specific event during apoptosis. Pharmacological blockade of any steps of the activation cascade may yield considerable neuroprotection, as it abrogates further downstream events of caspase activation.

Activation of death pathways following exposure to Aβ may ultimately lead to the altered expression of Aβ-responsive genes. Santiard-Baron and colleagues (1999) identified a subset of genes, which become activated after exposure of NT2 cells to $A\beta_{25-35}$. $A\beta_{25-35}$ induced, among others, the expression of the growth arrest and DNA damage-inducible gene (gadd45), the mitogen-activated protein kinase/Erk-kinase kinase-1 (MEKK1) gene, a growth factor-inducible immediate-early gene (CYR61), immediate-early genes, such as c-fos and c-jun, and the basic fibroblast growth factor gene (bFGF). The induction of numerous genes, involving c-jun, c-fos, fosB, ngfi-B, and iκB was demonstrated, in response to Aβ in both c-Jun deficient and wild-type neurons. Interestingly, the products of these genes are involved in DNA repair, stress signaling, and synthesis of neurotrophic (growth) factors that are known to rescue neurons from Aβ toxicity. Further attempts to explore expressional changes of inducible genes in response to Aβ toxicity may be of great value in the characterization of possible pharmacological targets and/or novel neuroprotective mechanisms.

33.5 The role of lipids and vascular factors in Alzheimer's disease

Vascular factors and lipids play an important role in the pathogenesis of AD. Several longitudinal studies established that hypertension, diabetes, atrial fibrillation and smoking are risk factors for dementia, including AD. Preliminary data suggest that treating these risk factors may prevent or delay dementia. It is also becoming evident that strokes and ischemic processes not only co-exist but may precipitate or potentiate AD.

There is a growing body of evidence for membrane lipid defect in AD (Ginsberg *et al.*, 1993). An inherent tendency towards the destabilization of cellular membranes in AD-brain has been reported. In general, all fractions from AD-brain show significantly lower levels of ethanolamine glycerophospholipids and significantly higher levels of serine glycerophospholipids than the control brain. Membrane destabilization may facilitate Aβ deposition. As docosahexaenoic acid (DHA) represents a very important membrane stabilization factor, low serum DHA level is a significant risk factor for Alzheimer's dementia (Kyle *et al.*, 1999).

Apolipoprotein-E (apoE), a protein long known for its role in cholesterol transport and plasma lipoprotein metabolism, has recently emerged as a major genetic risk factor for AD. One of the 3 common alleles of apoE, the apoE4 is over represented in AD subjects compared with age- and sex-matched controls. The genetic and epidemiologic evidence suggests that apoE is a major susceptibility gene for AD (Weisgraber and Mahley, 1996). Its role in development is unknown, however, its role in cholesterol and lipid transport provides important insight into the mechanism of action.

33.6 Diagnosis of Alzheimer's disease

Early diagnosis of Alzheimer's disease is of utmost importance in the selection of subjects for preventive treatment, particularly when coupled with assessment of genetic risk factors and family history of AD. It would be highly desirable to measure a substance or substances in blood or urine samples or cerebrospinal fluid (CSF) that would lead to a positive diagnosis of Alzheimer's disease without the need for specialized dementia clinics. Miller *et al.* (1993) performed pioneering work in the diagnosis of AD using noninvasive proton MR spectroscopy: Alzheimer's patients showed a 22% increase in myo-inositol compared with healthy subjects. More recently, Zubenko and coworkers (1999) have found an increased fluidity of platelet membranes in AD patients; these investigations were performed with fluorescence spectroscopy and electron spin resonance spectroscopy. The most recent result in the field of AD-diagnosis is the direct visualization of neuritic Aβ plaques using radiolabelled Aβ, administered intravenously in mice (Wengenack, 2000). The labelled peptide crosses the blood-brain barrier and labels at least a subset of the hallwork neuritic plaques in the brain. Altered CSF levels of both $A\beta_{1-42}$ and tau protein have already been reported by Motter *et al.* (1995) to have useful sensitivity and specificity in confirming diagnosis of AD. Imaging the amounts of Aβ deposits in hippocampus and temporal cortex, where AD-type dementia begins, could be a very useful tool for selection of patients for trials, and monitoring of drug efficacy during and after those trials. We except to soon enter the era of meaningful risk assessment for AD. Plasma $A\beta_{1-42}$ and tau will be measured, documented and followed over time. Sensitive psychometric screening of cognitive dysfunction will be coupled with functional MR imaging of the brain. Direct semiquantitative or quantitative imaging of Aβ deposits by radioisotope methods will help to determine whether an agressive antiamyloid therapy should be started.

33.7 Prevention, treatment and pharmacotherapy of Alzheimer's disease

Despite the incomplete understanding of Alzheimer's disease, sufficient progress has been achieved for prevention and treatment of the disease (see the review of Cacabelos *et al.*, 2000). Figure 33.7 shows the possibility of prevention and rational drug design.

Antioxidants, free radical scavengers, could protect neurons from the downstream effect of the accumulation of Aβ. Compelling experimental data indicate that blockade of either superoxide or hydroxyl radical formation attenuates Aβ toxicity. Vitamin E and ascorbate (vitamin C) have been demonstrated to protect against Aβ-induced oxidative stress under a wide variety of experimental conditions both *in vitro* and *in vivo*.

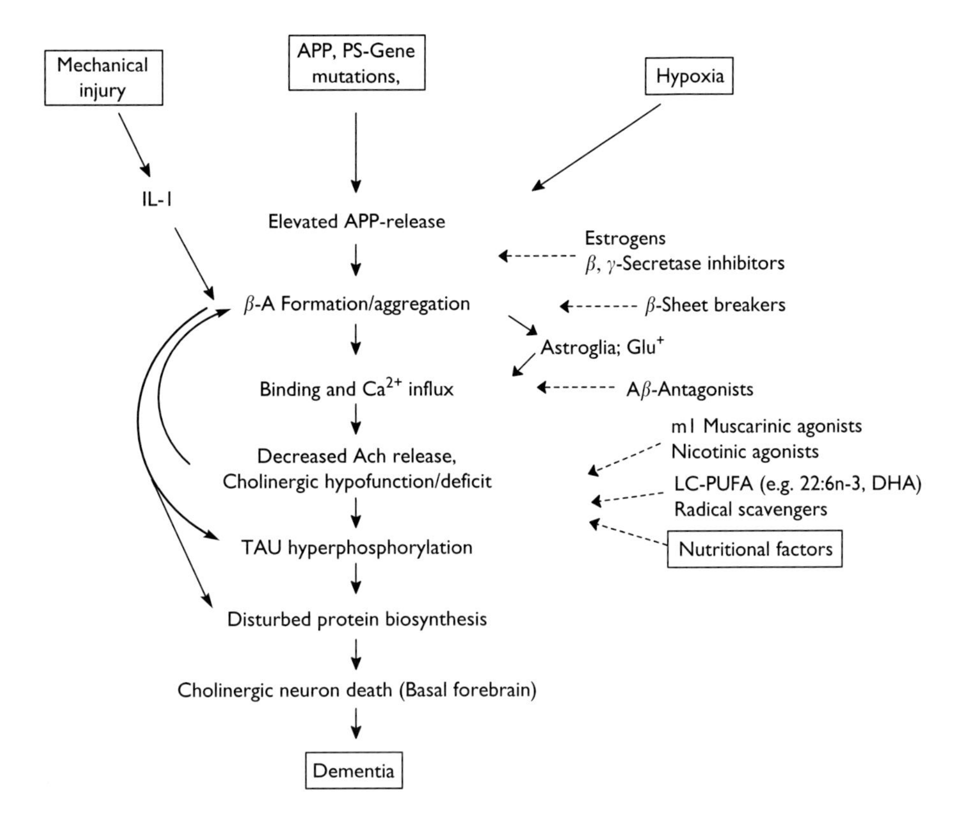

Figure 33.7 Hereditary and other factors (mechanical injury, hypoxia, nutritional factors) lead to Aβ overproduction and aggregation triggering the "amyloid cascade". Possibilities for prevention of AD and for rational drug design are given on the right side. PS: presenilins; LC-PUFA: long-chain poly-unsaturated fatty acids; DHA: docosahexaenoic acid, Glu +: glutamate increase.

Calcium channel blockers may also protect neurons from Aβ toxicity. However, caution should be taken because these blockers are associated with an increased risk of cancer in aged populations and can induce hypotension with episodes of brain hypoperfusion and cognitive deterioration.

Neurorestorative factors, e.g. neurotrophins and estrogens, could conceivably rescue synapses and cell bodies from injuries. Comprehensive clinical studies demonstrate the efficacy of estrogen replacement therapy in reducing the risk, and delaying the onset, of AD in post-menopausal women. 17β-estradiol exerts its beneficial effects by a significant reduction in Aβ-generation just by the selective stimulation of βAPP by α-secretase. Estrogen has been shown to be a trophic factor for cholinergic neurons and preserves cholinergic function during aging and in AD.

Antibody against β-amyloid was found to clear senile plaques by Schenk *et al.* (1999) and Morgan (2000). Mouse experiments were successful in clearing plaques and preventing memory loss. Anti-Aβ-antibodies penetrate into the brain through the blood-brain barrier. The pertinent question (whether such a treatment is associated with behavioral improvement, or is just a way to clear a by-product) has a positive answer: Janus (2000) found that Aβ-peptide immunization reduces behavioral impairment simultaneously with clearing of plaques. Clinical trials are already in progress.

Use of *long-chain polyunsaturated fatty acids* (LC-PUFA) as food components is a relatively new method for prevention of AD, although this method is already protected with a lot of patents. LC-PUFAs (first of all, eicosapentaenoic and docosahexaenoic acid) exert their beneficial effects by significant reduction of Aβ formation by selective inhibition of γ-secretase. Processing of βAPP occurs in membrane rafts and the activity of β- and γ-secretase is highly dependent on membrane composition. If the raft contains a high amount of cholesterol, β-secretase has high activity and large amounts of βA peptides will be generated (Frears *et al.*, 1999). Lovastatin and similar drugs decrease cholesterol biosynthesis and decrease formation of Aβ by inhibition of secretase activity. If LC-PUFAs are built into the membrane, a highly ordered membrane structure forms and the microenvironment of the membrane-bound γ-secretase changes dramatically, also causing a decrease in γ-secretase activity.

Anti-inflammatory drugs interfere with aspects of microglial and astrocytic responses in the brain and are therefore neuroprotective. Cyclooxygenase-2 inhibitors (e.g. aspirin™) have proved to be effective in the prevention or postponement of the beginning of Alzheimer's disease, but with well known side effects (e.g. gastric ulcer).

Substitutive treatment tries to maintain or restore the original neurotransmitter (initially acetylcholine) concentration in the brain. The first generation of anti-dementia drugs began with cholinesterase inhibitors when tacrine™ was approved in 1993. Other cholinergic enhancers activate muscarinic or nicotinic acetylcholine receptors at the postsynaptic level. Compounds belonging to monoaminergic enhancers tend to potentiate noradrenergic and dopaminergic transmission, some MAO-B inhibitors (selegiline™, delagil™) proved to be useful in treatment of AD. Aminoacidergic regulators, e.g. the excitotoxic inhibitor NMDA antagonist (memantine™), act as neuroprotectants blockading Ca^{2+} entry in palliative strategies.

Figure 33.8 summarizes the most recent possibilities for rational drug design by preventing Aβ formation or aggregation to neurotoxic Aβ fibrils: use of β- and γ-secretase inhibitors, β-sheet breakers and functional antagonists.

β-and γ-secretase inhibitors. While the processing of βAPP by β- and γ-secretases represents a minor, alternative proteolytic pathway, these enzymic cleavages are very important for Aβ generation. Specific inhibition of these two enzymes represents a rational drug design for AD therapy. Olson and Thompson (2000) reviewed the secretase inhibitors and their use as therapeutics. The tripeptide aldehyde Cbz-Val-Leu-Leu-H proved to be a β-secretase inhibitor decreasing the formation of both $A\beta_{1-40}$ and $A\beta_{1-42}$ with approximately equal efficacy. More recently peptidomimetics of complex structure have come under investigation as β-secretase blockers. Also potent γ-secretase inhibitors have been identified having peptide aldehyde or peptidomimetic character. Reports that γ-secretase inhibitors are approaching clinical evaluation indicates that lowering of Aβ production will soon be tested in humans.

Inhibitors of Aβ oligomerization and fibrillization. The idea of inhibiting or considerably reducing the aggregation of Aβ peptides by chemically disrupting the formation of β-pleated polymers, termed "*β-sheet breaking*", dates back to the early 90s, when it was demonstrated that the aggregation properties and amyloidogenicity of Aβ are dependent on the integrity of a hydrophobic domain localized to the β[17–21] region, and postulated that site-directed conformational modification of this pentamer "core" may lead to the development of anti-Aβ agents.

In 1996, Tjernberg *et al.* reported systematic construction of peptidic compounds that inhibit Aβ aggregation either via binding to Aβ fibrils and subsequent β-sheet breaking, or modifying the aggregation kinetics and higher order structure of fibrillar Aβ, and paved the way to the generation of an extensive family of β-sheet breakers (BSBs). Soto *et al.* (1998) provided experimental evidence on the activity of the Leu-Pro-Phe-Phe-Asp pentapeptide in an acute Aβ lesion model in rats, as this BSB inhibited amyloidogenesis and formation of Aβ deposits, and disassembled pre-formed Aβ fibrils. The preventative effect of BSBs on amyloid aggregation is shown in Figure 33.9.

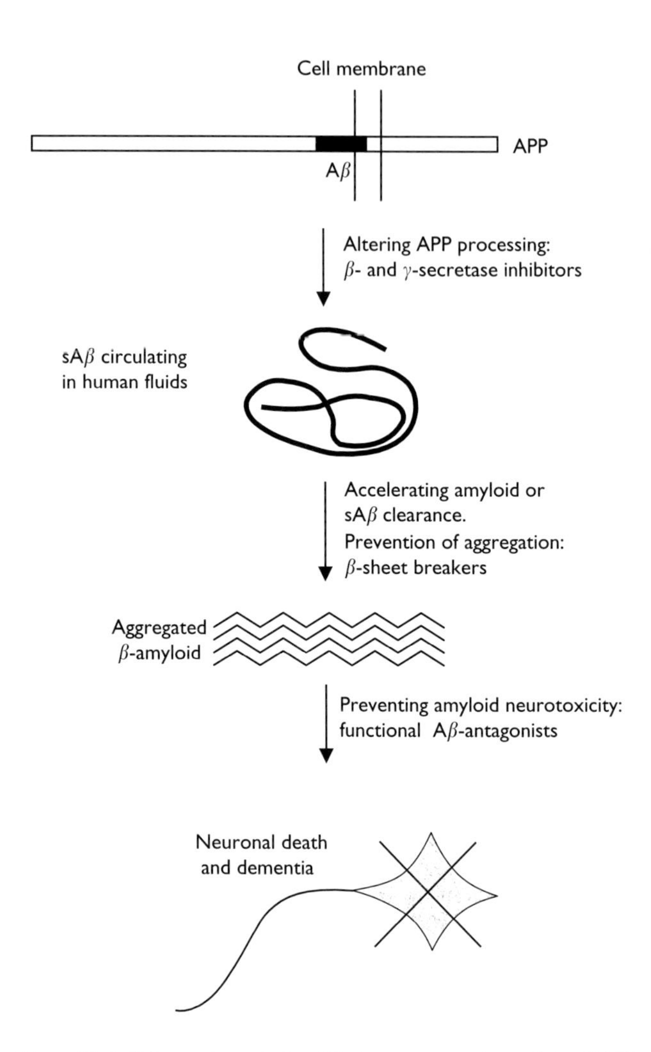

Figure 33.8 Schematic and simplified representation of the βAPP processing and Aβ aggregation to neurotoxic fibrils and some possibilities for rational drug design.

Several, unrelated small molecules have been shown to prevent Aβ fibrillogenesis or inhibit Aβ toxicity. Sulfonated dyes, such as Congo red, small sulfonated anions and benzofuran-based-compounds have been described as inhibitors of amyloid formation. Intensive research work in this field is in progress.

Functional Aβ antagonists are peptidic compounds counteracting Aβ toxicity by ligand-like displacement of Aβ from presumed cell surface recognition sites (receptors) and ameliorating pathological intracellular signaling cascades. Laskay *et al.* (1997) found that a tetrapeptide derivate of Aβ, propionyl-Ile-Ile-Gly-Leu amide, is a functional antagonist of Aβ

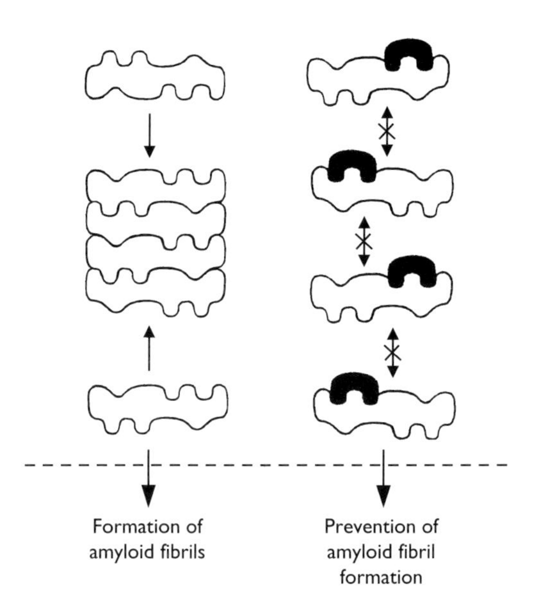

Figure 33.9 Specifically bound small molecules (β-sheet breakers) prevent amyloid fibril formation.

peptide preventing toxic effects *in vitro* and also *in vivo*. Further studies show that this type of peptide has a very important interaction with cell membranes. According to Kanfer (1999), membrane perturbation triggered by Aβ is the basis for its biological effects and involves a series of membrane proteins. Functional Aβ antagonists may prevent just the membrane perturbing effect of Aβ peptides.

References

Auld, D.S., Kar, S. and Quirion, R. (1998) β-Amyloid peptides as direct cholinergic neuromodulators: a missing link? *Trends Neurosci. 21, 43–49.*

Beer, J., Masters, C.L. and Beyreuther, K. (1995) Cells from peripheral tissues that exhibit high APP expression are characterized by their high membrane fusion activity. *Neurodegeneration 4, 51–59.*

Behl, C. (1997) Amyloid β-protein toxicity and oxidative stress in Alzheimer's disease. *Cell Tissue Res. 290, 471–480.*

Behl, C. (1999) Alzheimer's disease and oxidative stress: implications for novel therapeutic approaches. *Prog. Neurobiol. 57, 301–323.*

Behl, C., Davis, J.B., Lesley, R., Schubert, D. (1994) Hydrogen peroxide mediates amyloid β protein toxicity. *Cell 77, 817–822.*

Cacobelos *et al.* (2000) Pharmacological treatment of Alzheimer's disease: from psychotropic drugs and cholinesterase inhibitors to pharmacogenomics. *Drugs of Today 36, 415–499.*

Cain, K., Brown, D.G., Langlais, C. and Cohen, G.M. (1999) Caspase activation involves the formation of the aposome, a large (~700 kDa) caspase-activating complex. *J. Biol. Chem. 274, 22686–22692.*

Cowburn, R.F., Messamore, E., Li-Li, M., Winblad, B. and Sundström, E. (1994) β-Amyloid related peptides exert differential effects on [^{3}H]MK-801 binding to rat cortical membranes. *Neuroreport 5, 405–408.*

Dawson, G.R. *et al.* (1999) Age-related cognitive deficits, impaired long-term potentiation and reduction in synaptic marker density in mice lacking the β-amyloid precursor protein. *Neuroscience 90, 1–13.*

Dodart, J.C., Mathis, C. and Ungerer, A. (2000) The β-amyloid precursor protein and its derivatives: from biology to learning and memory process. *Rev. Neurosci. 11, 75–93.*

El Khoury, J., Hickman, S.E., Thomas, C.A., Cao, L., Silverstein, S.C. and Loike, J.D. (1996) Scavenger receptor-mediated adhesion of microglia to β-amyloid. *Nature 382, 716–719.*

Frears, E.R., Stephens, D.J., Walters, C.E., Davies, H. and Austen, B.M. (1999) The role of cholesterol in the biosynthesis of β-amyloid. *Neuroreport 10, 1699–1705.*

Geula, C., Wu, C-K., Saroff, D., Lorenzo, A., Yuan, M. and Yankner, B.A. (1998) Aging renders the brain vulnerable to amyloid β-protein neurotoxicity. *Nature. Med. 4, 827–831.*

Ginsberg, L. *et al.* (1993) Evidence for a membrane lipid defect in Alzheimer disease. *Mol. Chem. Neuropathol. 19, 37–46.*

Glenner, G.G., Wong, C.W. (1984) Alzheimer's disease: initial report of the purification and characterization of a novel cerebrovascular amyloid protein. *Biochem. Biophys. Res. Commun. 120, 885–890.*

Hardy, J. and Allsop, D. (1991) Amyloid deposition as the central event in the aetiology of Alzheimer's disease. *Trends Pharmacol. Sci. 12, 383–388.*

Hardy, J. and Israēl, A. (2000) In search of γ-secretase. *Nature 398, 466–467.*

Harkány, T., Ábrahám, I., Kónya, C., Nyakas, C., Zarándi, M., Penke, B. and Luiten, P.G.M. (2000) Mechanism of β-amyloid neurotoxicity perspectives of pharmacotherapy. *Review Neurosci. 1, 329–382.*

Harper, J.D. and Lansbury, P.T. Jr. (1997) Models of amyloid seeding in Alzheimer's disease and scrapie: mechanistic truths and physiological consequences of the time-dependent solubility of amyloid proteins. *Annu. Rev. Neurosci. 66, 385–407.*

Hartley, D.M., Walsh, D.M., Ye, C.P., Diehl, T., Vasquez, S., Vassilev, P.M., Teplow, D.B. and Selkoe, D.J. (1999) Protofibrillar intermediates of amyloid β-protein induce acute electrophysiological changes and progressive neurotoxicity in cortical neurons. *J. Neurosci. 1, 8876–8884.*

Hensley, K., Carney, J.M., Mattson, M.P., Aksenova, M., Harris, M., Wu, J.F., Floyd, R.A. and Butterfield, D.A. (1994) A model for β-amyloid aggregation and neurotoxicity based on free radical generation by the peptide: relevance to Alzheimer's disease. *Proc. Natl. Acad. Sci. USA 91, 3270–3274.*

Hertel, C., Terzi, E., Hauser, N., Jacob, R.R., Seelig, J. and Kemp, J.A. (1997) Inhibition of the electrostatic interaction between β-amyloid peptide and membranes prevents β-amyloid-induced toxicity. *Proc. Natl. Acad. Sci. USA 94, 9412–9416.*

Iwata, N. *et al.* (2000) Identification of the major $A\beta_{1-42}$ –degrading catabolic pathway in brain parenchyma: supression leads to biochemical and pathological deposition. *Nature Med. 6, 143–150.*

Janus, C., *et al.* (2000) Aβ peptide immunization reduces behavioural impairment and plaques in a model of Alzheimer's disease. *Nature 408, 979–982.*

Kalaria, R.N. (1993) The immunopathology of Alzheimer's disease and some related disorders. *Brain Pathol 3, 333–347.*

Kanfer, J.N., Sorrentino, G. and Siter, D.S. (1999) Amyloid β-peptide membrane perturbation is the basis for its biological effects. *Neurochem. Res. 24, 1621–1630.*

Katzman, R., Kawas, C.H. (1994) The epidemiology of dementia and Alzheimer disease. In: R.D. Terry, R. Katzman and K.L. Bick (eds), *Alzheimer Disease*, Raven Press, New York, pp.105–122.

Kim, H.J., Suh, Y.H., Lee, M.H. and Ryu, P.D. (1999) Cation selective channels formed by a C-terminal fragment of β-amyloid precursor protein. *Neuroreport 10, 1427–1431.*

Kimberley, W.T., Xia, W., Rahmati, T., Wolfe, M.S. and Selkoe, D.J. (2000) The transmembrane aspartates in presenilin 1 and 2 are obligatory for γ-secretase activity and amyloid β-protein generation. *J. Biol. Chem. 275, 3173–3178.*

Kyle, D., Schaefer, E., Patton, G. and Beiser, A. (1999) Low serum docosahexaenoic acid is a significant risk faktor for Alzheimer's dementia. *Lipids 34, S245b.*

Lambert, M.P., Barlow, A.K., Chromy, B.A., Edwards, C., Freed, R., Liosatos, M., Morgan, T.E., Rozovsky, I., Trommer, B., Viola, K.L., Wals, P., Zhang, C., Finch, C.E., Krafft, G.A. and Klein, W.L. (1998) Diffusible, nonfibrillar ligand derived from $A\beta_{1-42}$ are potent central nervous system neurotoxins. *Proc. Natl. Acad. Sci. USA. 95, 6448–6453.*

Laskay, G., Zarándi, M., Varga, J., Jost, K., Fónagy, A., Torday, C., Latzkovits, L. and Penke, B. (1997) A putative tetrapeptide antagonist prevents β-amyloid-induced long-term elevation of $[Ca^{2+}]_i$ *Biochem. Biophys. Res. Commun. 235, 479–481.*

Lippa, C.F., Hamos, J.E., Smith, T.W., Pulaski-Salo, D. and Drachman, D.A. (1993) Vascular amyloid deposition in Alzheimer's disease. Neither necessary nor sufficient for the local formation of plaques or tangles. *Arch. Neurol. 50, 1088–1092.*

Lorenzo, A. *et al.* (2000) Amyloid β interacts with the amyloid precursor protein: a potential toxic mechanism in Alzheimer's disease. *Nature Neuroscience 3, 460–464.*

Mattson, M.P., Cheng, B., Davis, D., Bryant, K., Lieberburg, I. and Rydel, R.E. (1992) β-Amyloid peptides destabilize calcium homeostasis and render human cortical neurons vulnerable to excitotoxicity. *J. Neurosci. 12, 376–389.*

Miller, B.L., Moats, R.A., Shonk, T., Ernst, T., Woolley, S. and Ross, B.D. (1993) Alzheimer's disease: Depiction of increased cerebral myo-inositol with proton MR spectroscopy. *Radiology 187, 433–437.*

Morgan, D., *et al.* (2000) $A\beta$ peptide vaccination prevents memory loss in an animal model of Alzheimer's disease. *Nature 408, 982–985.*

Morris, J.C. (1995) Relationship of plaques and tangles to Alzheimer's disease phenotype. In: A.M. Goate, and F. Ashall (eds), *Pathobiology of Alzheimer's disease*, Academic Press, London, pp. 193–223.

Motter, R., Vigo-Pelfreg, C. and Kholodenko, D. (1999) Reduction of βA1–42 in the CSF of patients with AD. *Ann. Neurol. 38, 634–648.*

Olson, R.E. and Thompson, L.A. (2000) Secretase inhibitors as therapeutics for Alzheimer's disease. *Annual Reports in Med. Chem. 35, 31–40.*

Pakaski, M., Farkas, Z., Kasa, P. Jr., Forgon, M, Papp, H., Zarandi, M., Penke, B. and Kasa, P. Sr. (1998) Vulnerability of small GABAergic neurons to human β-amyloid pentapeptide. *Brain Res. 796, 239–246.*

Pangalos, M.N., Shioi, J., Efthimiopoulos, S., Wu, A. and Robakis, N.K. (1996) Characterization of appican, the chondroitin sulfate proteoglycan form of the Alzheimer amyloid precursor protein. *Neurodegeneration 5, 445–451.*

Perez, R.G., Zheng, H., Van der Ploeg, L.H. and Koo, E.H. (1997) The β-amyloid precursor protein of Alzheimer's disease enhances neuron viability and modulates neuronal polarity. *J. Neurosci. 17, 9407–9414.*

Rogers, J. *et al.* (1996) Inflammation and Alzheimer's disease pathogenesis. *Neurobiol. Aging 17, 681–686.*

Santiard-Baron, D., Gosset, P., Nicole, A., Sinet, P.M., Christen, Y. and Ceballos-Picot, I. (1999) Identification of β-amyloid-responsive genes by RNA differential display: early induction of a DNA damage-inducible gene, gadd45. *Exp. Neurol. 158, 206–213.*

Schenk, D. *et al.* (1999) Immunization with $A\beta$ attenuates Alzheimer's disease-like pathology in the PDAPP mouse. *Nature 400, 173–177.*

Selkoe, D.J. (1994) Cell biology of the amyloid β-protein precursor and the mechanism of Alzheimer's disease. *Annu. Rev. Cell. Biol., 10, 373–403.*

Selkoe, D.J. (1999) Translating cell biology into therapeutic advances in Alzheimer's disease. *Nature Suppl.* 399, A 23–A 31.

Sisodia, S.S. (1992) β-Amyloid precursor protein cleavage by a membrane-bound protease. *Proc. Natl. Acad. Sci. USA 89, 6075–6079.*

Soto, C., Kindy, M.S., Baumann, M. and Frangione, B. (1996) Inhibition of Alzheimer's amyloidosis by peptides that prevent β-sheet conformation. *Biochem. Biophys. Res. Commun. 226, 672–680.*

Terry, R.D., Masliah, E. and Hansen, L.A. (1994) Structural basis of the cognitive alterations. In.: R.D. Terry, R. Katzman and K.L. Bick (eds), *Alzheimer Disease.* Raven Press, New York, pp. 179–196.

Tjernberg, L.O., Naslund, J., Lindquist, F., Johansson, J., Karlstrom, A.R., Thyberg, J., Terenius, L. and Nordstedt, C. (1996) Arrest of beta-amyloid fibril formation by a pentapeptide ligand. *J. Biol. Chem. 271, 8545–8548.*

Tjernberg, L.O., Callaway, D.J.E., Tjernberg, A., Hahne, S., Lilliehöök, C., Terenius, L. J. and Nordstedt, C. (1999) A molecular model of Alzheimer amyloid β-peptide fibril formation. *J. Biol. Chem. 274, 12619–12625.*

Vasser, R. *et al.* (1999) β-Secretase cleavage of Alzheimer's amyloid precursor protein by the transmembrane aspartic protease BACE. *Science 285, 735–741.*

Wallace, W.C., Akar, C.A. and Lyons, W.E. (1997) Amyloid precursor protein potentiates the neurotrophic activity of NGF. *Brain. Res., Mol. Brain. Res. 52, 201–212.*

Walsh, D.M., Hartley, D.M., Kusumoto, Y., Fezoui, Y., Condron, M.M., Lomakin, A,. Benedek, G.B., Selkoe, D.J. and Teplow, D.B. (1999) Amyloid β-protein fibrillogenesis. Structure and biological activity of protofibrillar intermediates. *J. Biol. Chem. 274, 25945–25952.*

Wang, H-Y. Lee, D.H.S., D'Andrea, M.R., Peterson, P.A., Shank, R.P. and Reitz, A.B. (2000) β-Amyloid 1–42 binds to α7 nicotinic acetylcholine receptor with high affinity. *J. Biol. Chem. 275, 5626–5632.*

Weisgraber, K.H. and Mahley, R.W. (1996) Human apolipoprotein E: the Alzheimer's disease connection. *FASEB J. 10, 1985–1994.*

Wengenack, T.M., Curran, G.L. and Poduslo, J.F. (2000) Targeting Alzheimer amyloid plaques *in vivo. Nature Biotechnology 18, 868–872.*

Wolfe, M.S., De Los Angeles, J., Miller, D.D., Xia, W. and Selkoe, D.J. (1999) Are presenilins intramembrane-cleaving proteases? Implications for the molecular mechanism of Alzheimer's disease. *Biochemistry 38, 11223–11230.*

Yaar, M., Zhai, S., Pilch, P.F., Doyle, S.M., Eisenhauer, P.B., Fine, R.E. and Gilchrest, B.A. (1997) Binding of β-amyloid to the p75 neurotrophin receptor induces apoptosis. *J. Clin. Invest. 100, 2333–2340.*

Yan, S.D., Chen, X., Fu, J., Chen, M., Zhu, H., Roher, A., Slattery, T., Zhao, L., Nagashima, M., Morser, J., Migheli, A., Naworth, P., Stern, D. and Schmidt, A.M. (1996) RAGE and amyloid β-peptide neurotoxicity in Alzheimer's disease. *Nature 382, 685–691.*

Yankner, B.A., Duffy, L.K., Kirschner, D.A. (1990) Neurotrophic and neurotoxic effects of amyloid β-protein: reversal by tachykinin neuropeptides. *Science 250, 279–282.*

Zubenko, G.S., Kopp, U., Seto, T. and Firestone, L.L. (1999) Platelet membrane fluidity individuals at risk for Alzheimer's disease: a comparison of results from fluorescence spectroscopy and electron spin resonance spectroscopy. *Psychopharmacology 145, 175–180.*

Chapter 34

Eye diseases

P. Elizabeth Rakoczy, Wei-Yong Shen and Chris J. Barry

Contents

34.1 Introduction

Until recently, diseases of the eye have been diagnosed by clinical observations of effected patients. These observations then lead to the formulation of a hypothesis regarding the origin or pathomechanism of the disease. During the last twenty years, molecular genetics have provided an unparalleled insight of the pathogenesis of diseases including the eye. It was Mendel who first demonstrated that phenotype variation among members of a species was caused by differences in specified elements later termed genes. The concept of genetic linkage analysis was developed during the first half of the twentieth century and laboratory analysis of genes has become viable during the last twenty years. Genetic technologies have changed the diagnosis of diseases with Mendelian inheritance forever. With the availability of the human gene sequence, identification of new genes and linkage of mutations to diseases could be complete in the forseeable future.

This chapter will demonstrate the advancement of molecular genetics on diseases of the retina. We selected inherited retinal diseases as they are probably the most untreatable disorders in the eye. Thus, revolutionary diagnosis and treatment techniques are expected to have a significant impact on understanding of these diseases. We begin with a short description of the structure and function of the retina to assist understanding and help with the subsequent diagnostic and therapeutic discussions.

34.2 The structure and function of the retina

The retina covers the back of the eye and its main function is to convert light into an electrical signal that is then delivered to the visual cortex where our perception of vision occurs. The retina is made up from a complex array of cells that are organized in a well-defined structure (Figure 34.1). The outermost layer is the retinal pigment epithelium (RPE), a tightly packed mono cell layer. The RPE has several metabolic-biochemical, transport, optical and physical functions. Of these, phagocytosis of photoreceptor outer segments and participation in the visual cycle are the most important. The RPE is responsible for the phagocytosis and digestion of 10% of the adjacent photoreceptor outer segments per day. This process, which continues throughout life, makes the RPE cells the most phagocytic cell type in the body[1].

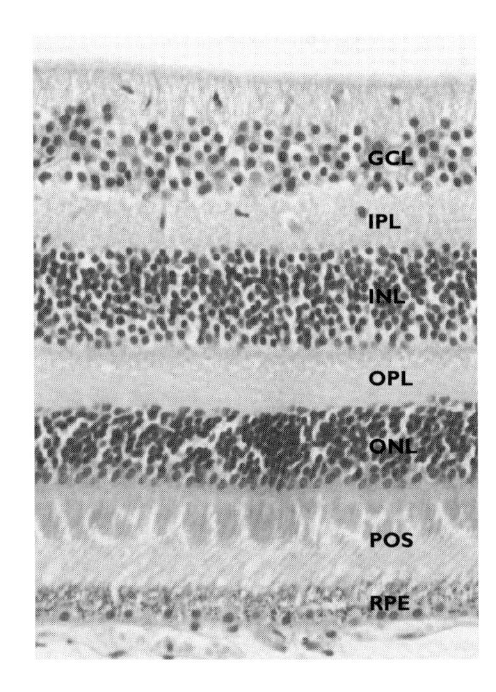

Figure 34.1 Structure of the retina. Abbreviations: GCL, ganglion cell layer. IPL, inner plexiform layer. INL, inner nuclear layer. OPL, outer plexiform layer. ONL, outer nuclear layer. POS, photoreceptor outer segment. RPE, retinal pigment epithelium (see Colour Plate XXII).

A

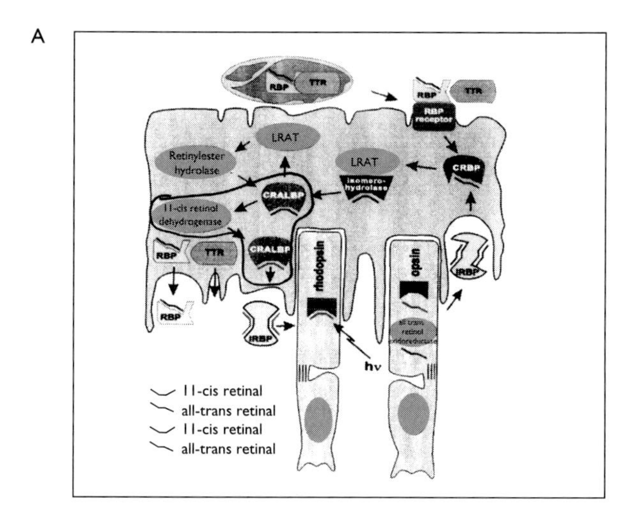

B

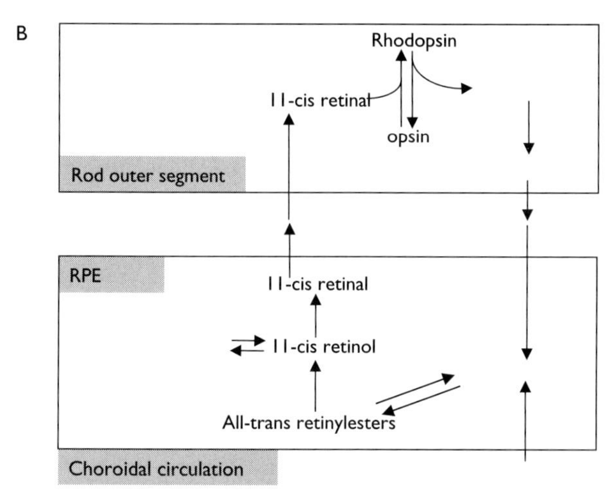

Figure 34.2 Enzymes (A) and productions (B) in visual cycle. (A) is adapted from Dr. Dean Bok, J Cell Sci 1993 (Suppl 17): 189–195. (B) is adapted from reference 2.

To perform these tasks the RPE contains large amounts of lysosomal enzymes that are responsible for the enzymatic digestion of the phagocytosed photoreceptor outer segments. These include: acid phosphatases, cathepsins, lipases, sulphatises and glucurunidases[1]. The visual cycle (Figure 34.2) is the process that converts all-trans retinol (vitamin A derivative) into visual chromophore, 11-cis retinal. There is a complicated enzymic system present, containing oxidative and isomeric enzymes such as retinyl ester hydrolase and 11-cis-retinol dehydrogenase isomero hydrolase (Figure 34.2A). These convert all-trans retinol recycled from the photoreceptors and all-trans retinol absorbed from the blood by the RPE into 11-cis retinal. One of the first processes is the isomeric change of all-trans retinol into 11-cis retinol. 11-cis retinol is an unstable compound and there are several proteins such as Cellular Retinaldehyde Binding Protein (CRALBP) and Cellular Retinal Binding protein (CRB) present to ensure its stability throughout the oxidation process.

Finally, 11-cis retinal is transported out of the RPE cells into the photoreceptors (Figure 34.2B). It has been estimated that the RPE cells process approximately 1 million chromophores per second[2].

The outer segments of 10–12 layers of photoreceptors, are adjacent to the RPE layer. The nuclei of photoreceptors are called the outer nuclear layer (ONL). Photoreceptors are the cells where photoactivation and phototransduction take place. Phototransduction is the process that converts light into an electrical signal. The excitation phase of phototransduction begins with photoactivation of a visual pigment, which induces a conformational change that catalyses the activation of a photoreceptor G protein, transducin. Each activated α subunit of transducin, now bound to GTP, displaces the inhibitory subunit of a cGMP phosphodiesterase (cGMP PDE), which then catalyses the hydrolysis of cGMP to GMP, leading to a cascade that results in membrane hyperpolarization and a graded attenuation in neurotransmitter release at the photoreceptor synapse[2].

The nuclei of the second inner layer of cells are called the inner nuclear layer (INL). The INL contains bipolar cells, onto which the photoreceptors synapse, amacrine cells, processing visual information in the retina, and Müller and horizontal cells. The innermost cell layer of the retina contains the ganglion cells (GCL), the output units of the retina. Ganglion cell axons track along the inner surface of the retina coming together at the optic disc to form the optic nerve.

34.3 Genetic linkage and mutation analysis

The majority of genetic disease cannot be studied with "traditional" molecular biological techniques such as cloning and sequencing. For most diseases, information regarding the function of the causative gene or biochemical knowledge is limited. Such difficulties can be overcome by genetic linkage analysis. Genetic linkage analysis enables us to identify the chromosomal location of the disease gene without any prior genetic knowledge or hypothesis of the pathogenesis. It is based on the observation that an allele of two genes located very close to each other on the same chromosome tend to be inherited together or cosegregate. As the distance between a genetic marker and the disease gene increases, homologous recombination becomes more likely. The frequency of this cross over is called the recombination factor. Linkage analysis is fundamentally a calculation of probability based on an observed association between the inheritance of a known chromosomal marker allele and the presence of a disease phenotype and it is measured by the lod score, representing the probability of linkage versus no linkage. By convention, a positive lod score of 3 or more is accepted as proof of linkage (1000 : 1 odds favoring linkage versus no linkage) whereas a negative lod score of -2 or less is evidence against linkage (100 : 1 odds of no linkage versus true linkage). If a marker generates a lod score of between -2 and $+3$, it yields no useful information and is considered to be "uninformative". Usually, successful linkage analysis requires several generations. Linkage analysis has been successfully used to identify genes responsible for autosomal dominant, recessive and X-linked retinitis pigmentosa, corneal dystrophy, North Carolina macular dystrophy, Best's disease, autosomal dominant and recessive Stargardt's disease[3–7].

Once gene linkage is obtained, the interval for the gene can be narrowed by maximizing information from affected individuals who are recombinants. "Fine mapping" of the gene may be performed by isolating additional markers. In the absence of any cytogenetic abnormalities, two methods are used to identify the disease gene: positional cloning and the candidate gene approach[7]. Both of these methods are dependent on the technique of genetic linkage analysis, since their first step is to localize a gene to a particular chromosomal region. In contrast to linkage analysis, positional cloning and the candidate gene approach have a higher degree of resolution and can actually lead to the identification and characterization of the mutation of a gene causing the disease. It may be possible to localize the disease gene to a region near a polymorphic gene marker known to cosegregate with the disease phenotype, but this region may still contain several million base pairs. Therefore, one must create a library of smaller, overlapping DNA fragments which cover the entire region that putatively contains the disease gene. When the DNA interval is less than one million base pairs, physical mapping can begin using yeast artificial chromosomes, cosmids and plasmid artificial chromosomes. Molecular characterization of these fragments can identify a potential gene that may be the disease gene. These genes can be found via different methods. One can compare the DNA sequence in each fragment to that of known genes that have been conserved in other species during evaluation. Alternately, one could search for exon/intron boundaries or hypomethylated sites that would indicate the presence of a coding sequence. Since irrelevant genes may be found during this process, it is necessary to determine which potential gene exhibits the tissue expression pattern consistent with the disease phenotype through molecular techniques such as Northern blot analysis. Transcripts of these genes are more likely to be from the mutant gene in question. Finally, one must identify a mutation in the sequence of a gene that is present in individuals with the disease phenotype but not in normal individuals. Mutations can be identified using techniques such as heteroduplex cleavage or by detection of the conformation change produced by mutations in DNA molecules. Ultimately, the gene must be sequenced to determine the exact nature of the mutation. Successful genetic linkage is dependent upon a close collaboration between ophthalmologists, molecular biologists and genetic epidemiologists. The sequence of events is outlined in Figure 34.3.

Gene mapping of autosomal dominant retinitis pigmentosa (ADRP) may be a good example to explain the process of genetic linkage analysis and recognition of point mutation of the disease gene in patients with retinitis pigmentosa. Retinitis pigmentosa is a group of hereditary retinal degenerations that affect about 1 in 3500 at birth[8]. This type of retinal degeneration is characterized by the occurrence of night blindness during adolescence and loss of vision in the mid-peripheral field followed by the far peripheral field in adulthood. Most patients are legally blind by the age of 40. Findings on ophthalmic examination include attenuated retinal vessels and intra-retinal pigment around the mid-periphery in the zone where rods are normally in maximal concentration (Figure 34.4). Retinitis pigmentosa is genetically heterogeneous, being transmitted as an autosomal dominant trait in some families and as an autosomal recessive trait or an X-linked trait in other families[8]. One of the first successful gene linkages was the discovery of a restriction-fragment-length polymorphism (RFLP) from the long arm of chromosome 3, named CRI-C17, which was tightly linked to the disease trait in a large Irish family with ADRP[9]. Since rhodopsin, the visual pigment expressed in rods, is also within this region of the chromosome, Dryja *et al.*[8] detected mutations in the rhodopsin gene from DNA of the patients with ADRP using polymerase chain reaction and sequencing techniques. They identified a point mutation of the rhodopsin gene in condon

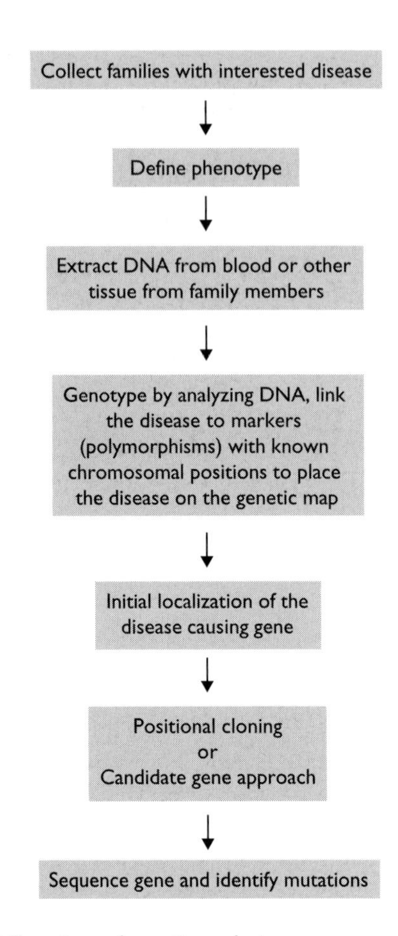

Figure 34.3 Paradigm of genetic analysis.

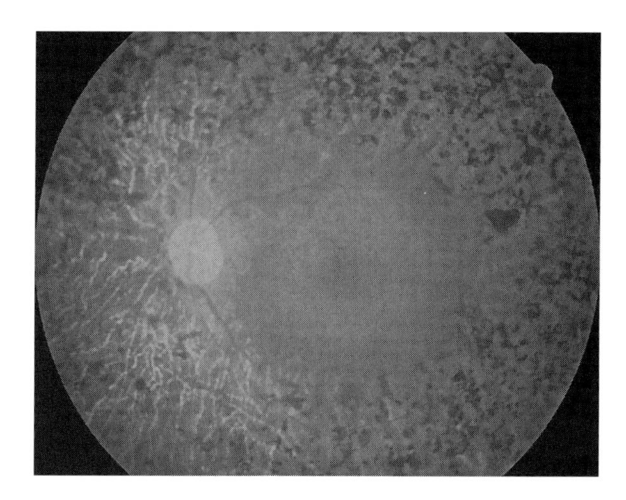

Figure 34.4 A fundus image of retinitis pigmentosa (see Colour Plate XXIII).

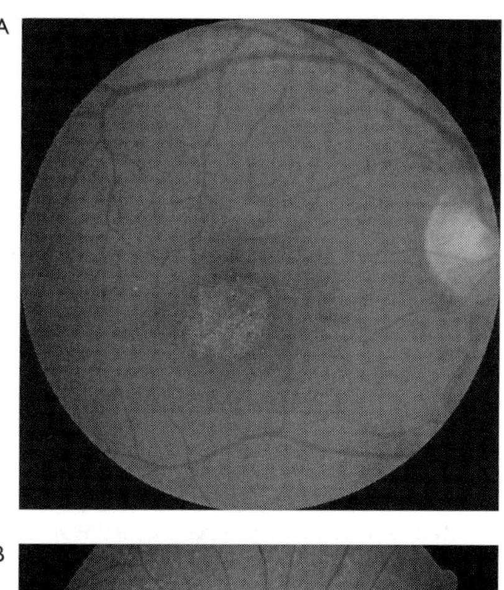

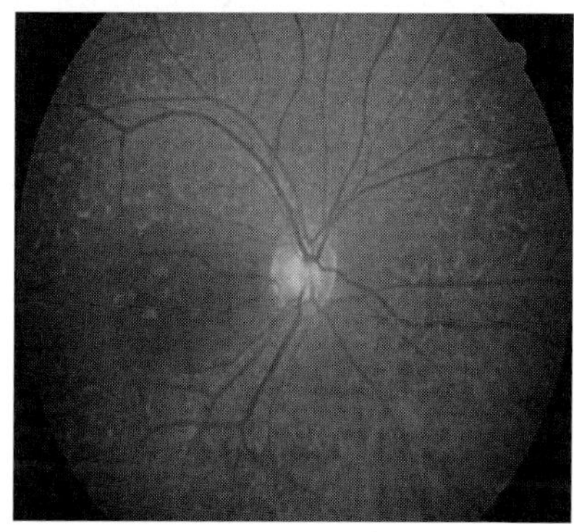

Figure 34.5 Fundus images of Stargardt's disease (A) and fundus flavimaculatus (B) (see Colour Plate XXIV).

23, resulting in a nucleotide base change (cytosine→adenine). DNA replication is normally a remarkably accurate process. A single nucleotide substitution or point mutation in a DNA sequence of the patients with ADRP alters the code in a triplet of bases and causes the replacement of one amino acid by another in the gene product, which corresponds to a proline to histidine substitution in the 23rd amino acid of the rhodopsin. Since proline is highly conserved among the normal population, this suggested that this mutation could be responsible for ADRP. This mutation was found in about 12% of the patients with ADRP and in none of 102 unaffected individuals[10]. Other point mutations, involving codons 347 and 58 have also been found in an additional 6 percent of patients with ADRP[8]. Generally speaking, approximately 30% of cases of ADRP are caused by rhodopsin gene mutations, most of which produce single amino acid substitution[8].

Among inherited macular dystrophies such as North Carolina macular dystrophy, Best's disease and Stargardt's disease; Stargardt's disease is the most common hereditary recessive macular dystrophy[7,11,12]. It is estimated that 1:10,000 people are affected by this form of macular dystrophy and it accounts for 7% of all retinal dystrophies[11,12]. Stargardt's disease is characterized by juvenile to young adult onset, central visual impairment, progressive bilateral atrophy of the macular RPE and neuroepithelium with the frequent appearance of orange-yellow flecks distributed around the macula and/or the midretinal periphery (Figure 34.5A). A clinically similar retinal disorder, fundus flavimaculatus, often displays later age of onset and slower progression (Figure 34.5B). From genetic linkage analysis, it has been concluded that these two macular dystrophies are most likely allelic autosomal recessive disorders with slightly different clinical manifestations. Recently, Allikmets *et al.*[11] have mapped a gene encoding an ATP-binding cassette (ABC) transporter to the 2-cM (centiMorgan) interval at 1p13–p21 where previously it was shown to harbor the Stargardt's disease gene by genetic linkage analysis. This gene, named photoreceptor cell specific ATP-binding transport gene (ABCR), is expressed exclusively and at high levels in rod but not cone photoreceptors in the retina of patients with Stargardt's disease. Mutational analysis of ABCR in Stargardt's disease families revealed a total of 19 different mutations including homozygous mutations in two families with consanguinous parentage, indicating that ABCR is the causal gene of Stargardt's disease/fundus flavimaculatus[11]. Autosomal dominant forms of Stargardt's disease have been mapped to chromosome 13q34 and to the long arm of chromosome 6 between loci D6S313 (6q13–q16.2) and D6S252 (6q14–q16.2)[13].

Age-related macular degeneration (ARMD) is the most common cause of acquired visual impairment in the elderly over 65 years old in the Western world[14]. Clinically, ARMD is divided into two subtypes, "dry" and "wet", in which 80% of patients have the "dry" form of ARMD, characterized by one or more of the following hallmarks: the presence of cellular debris (drusen) in or under the RPE, irregularities in the pigmentation of the RPE, or geographic atrophy (Figures 34.6A,B). About 20% of patients have the exudative or "wet"

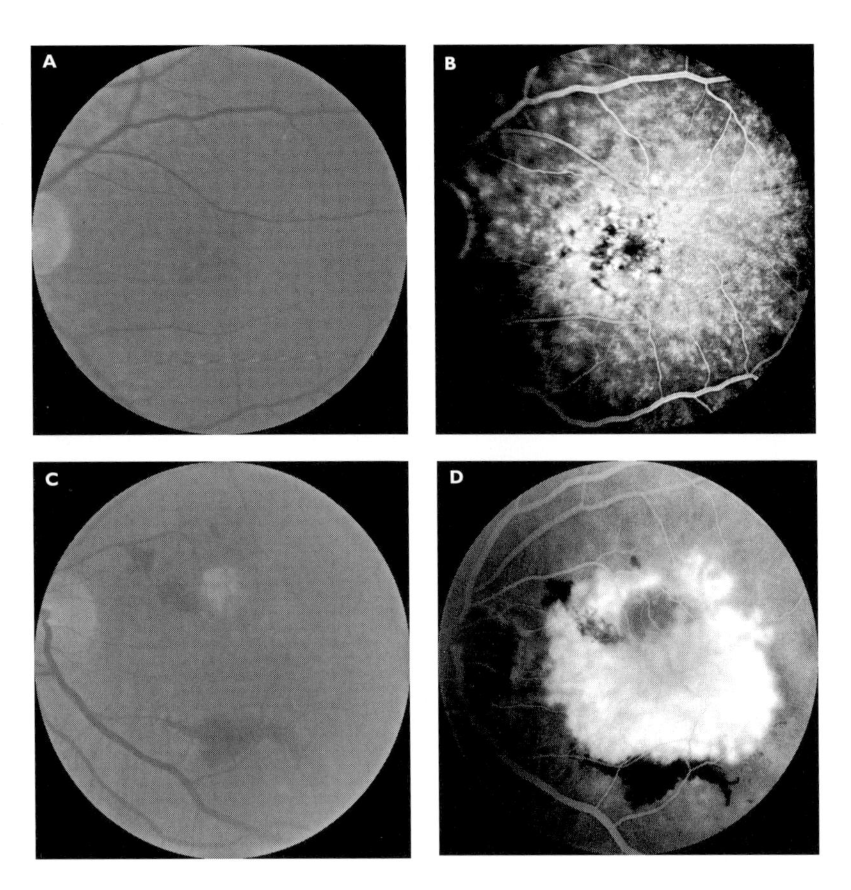

Figure 34.6 "Dry" (A & B) and "wet" forms (C & D) of age-related macular degeneration (ARMD). Images obtained from fundus photography (A & C) and fluorescein angiography (B & D). (D) showing a choroidal neovascular membrane, the severe complication at the late stage of "wet" form of ARMD (see Colour Plate XXV).

form of ARMD, characterized by serous detachment of the RPE or choroidal neovascularization (CNV) or both (Figures 34.6C,D). ARMD is a multifactoral disorder that is associated with environmental risk factors such as cigarette smoking, diet, cholesterol level and oxidative stress[15]. The recently identified photoreceptor-specific gene, ABCR, also known to be defective in Stargardt's disease, has been found to be associated with ARMD[16–18]. However, this genetic study lacked a control group. Data from recent investigations have shown controversy on the relationship between allelic variation in the ABCR gene and the prevalence of ARMD, suggesting that some variants in the ABCR gene may increase susceptibility to ARMD but are not a major genetic risk factor for ARMD[19,20]. Additional genetic studies are needed to more fully evaluate the role of ABCR in ARMD.

Table 34.1 lists selected examples of eye diseases where the underlying affected genes have been identified by genetic analysis. They represent only a minority of affected inherited eye diseases. To date several hundred mutant genes have been identified. Knowledge of these mutations not only enabled physicians to set up genetic consulting services but advances in biochemistry, molecular biology and functional research have opened up the possibility for developing animal models and subsequent treatment strategies. With the availability of the human genome sequence and new techniques such as Microarray, the search and success of linking genetic mutations to disease phenotypes will speed up and genetic testing will become part of a routine diagnosis.

34.4 Animal models for retinal degenerations and dystrophies

One of the most important components of developing a new therapy is the existence of animal models for human diseases. Until recently, all models of retinal diseases were hereditary animal models. During the past decades these animal models were essential to assist our understanding of human conditions. However, they are not always appropriate to model a specific human disease. It was in 1981 when Wagner performed the first transgenesis by transplanting a rabbit β-globin gene into a mouse embryo[21]. But it was only in 1992 that the first transgenic mouse of autosomal

Table 34.1 Selected examples of genes responsible for inherited eye diseases

Gene	*Disease*
Rhodopsin	Dominant retinitis pigmentosa (RP), recessive RP Congenital stationary night blindness (CSNB)
Transducin	CSNB
Arrestin and rhodopsin kinase	Oguchi disease (a variant of CSNB)
cGMP phosphodiesterase (cGMP PDE)	Recessive RP Dominant CSNB
Cyclic nucleotide gated channel	Recessive RP
Guanylate cyclase and	Laber congenital amaurosis
Guanylate cyclase activating protein	Dominant cone-rod dystrophy
Peripherin/rds	Dominant RP Digenic RP Dominant macular dystrophies Cone-rod degenerations
Rod outer segment membrane protein-1	Digenic RP
Rab escort protein-1	Choroideremia
Cellular retinaldehyde binding protein (CRALBP)	Several recessive inherited retinal degenerations
RPE65	Several forms of recessive early onset retinal degenerations
Retinal specific ATP-binding cassette transporter (ABCR)	Recessive Stargardt's disease Age-related macular degeneration?
Bestrophin	Best disease
Tissue inhibitor of metalloproteinase-3 (TIMP-3)	Sorsby fundus dystrophy
Ornithine amino transferase	Gyrate atrophy
Mitochodrial genome	Leber hereditary optic neuropathy

dominant retinitis pigmentosa with a rhodopsin mutation (Pro23His) was produced and characterized[22]. In the past few years, progress in genetic engineering has led to the rapid growth of transgenic animal models for retinal degenerations where transgenic animals carry constructs that lead to disruption or overexpression of candidate genes[23]. A well structured review has been presented by Hafezi *et al.*[23] on recently developed transgenic animal models as well as on models based on spontaneous mutations and induced retinal degenerations. This paper summarized the models of retinal degeneration in several species of animals, including mouse, rat, cat, chicken and larger animals such as dog and pig. Most of the genes involved are implicated in phototransduction such as rhodopsin, peripherin, and β-phosphodiesterase genes. In addition, there are also well known inducible animal models available for retinal degeneration. Werner Noëll developed the first inducible animal model for retinal degeneration using bright light damage in 1966[24]. With inducible animal models, the animals are mature and the stimulus provided is flexible and adjustable. Therefore, more or less severe photoreceptor damage can be obtained. These new models constitute a powerful and suitable tool to investigate the role of specific genes in inherited retinal degenerations and extend the spectrum of animal models suitable for the newly emerging field of retinal gene therapy.

34.5 Gene therapy for inherited retinal degeneration

Advances in molecular biology have had a profound impact on basic science research in vision and ophthalmology. The molecular revolution in medicine brings with it the prospect of gene therapy, aiming to introduce a functional copy of the defective or target gene into human somatic cells to correct inherited and acquired eye diseases. Ideally, four basic prerequisites should be met for any genetic therapy for an ocular disease. First, the gene delivery system must be efficient and nontoxic. Second, the genetics of the target disease should be well characterized so that an appropriate therapeutic approach can be selected. Third, expression of the therapeutic gene needs to be properly controlled. Finally, an experimental animal model of the target disease should be available for proof-of-principle and preclinical testing for the therapy. Modified recombinant viruses, which are replication deficient as a result of deletions in essential viral genes, are currently the most commonly used vectors for *in vivo* gene transfer in the eye. Viral vectors that have been used for delivering transgenes into ocular cells include retrovirus, herpes simplex virus (HSV), adenovirus (Ad) and adeno-associated virus (AAV)[25]. Each vector system has its advantages and disadvantages (Table 34.2). Retrovirus primarily targets dividing cells and this has limited its utility in the eye as the majority of ocular cell types are terminally differentiated and non-replicating. The other DNA viruses, HSV, Ad and AAV, are all efficient at transducing non-neuronal cells in the eye, in particular the RPE cells. Transduction of the photoreceptors by these vectors varies according to different investigators[26–28]. It may be a major challenge for ocular gene therapy to further improve the efficiency of transgene delivery on neuronal cell and, in particular, photoreceptor cells. An important issue needing to be addressed for retinal gene therapy is the safety following intraocular gene delivery of viral vectors. As part of the central nervous system, any damage to the retinal neural cells induced by viral vectors would be irreversible and permanent. HSV and Ad vectors currently suffer from a number of drawbacks. With most of the currently available HSV and Ad vectors, *in vivo* transduction has been limited to a matter of days and weeks. In addition, recent reports have more concerns for safety following HSV and Ad subretinal injection[29, 30]. AAV vectors have generated considerable interest in recent years. They are potentially attractive since a wild type of AAV is not associated with any pathology in humans. In addition, recombinant AAV vectors are deleted for all virally encoded proteins, therefore reducing their immunogenicity. Recent results from different research groups, including our own data (Figure 34.7), have demonstrated that AAV mediated transgene expression in the RPE and photoreceptor cells remains stable for up to 2 years in the primate and more than 3 years in rats without any obvious retinal toxicity in any species[27,31]. Currently, AAV is a very promising vector for ocular gene therapy targeting inherited retinal diseases.

The greatest stride in progress involving ocular gene therapy is the development of gene-based strategies for inherited retinal degeneration such as retinitis pigmentosa (RP). This is because the genetics of many forms of RP are relatively well understood. More than twenty different genes responsible for RP have been identified, of which 18 have been cloned. In addition, a number of transgenic animal models also exist in which *rd* and *rds* mice are two of the extensively characterized mouse strains for recessive and dominant models of RP[23]. The photoreceptor cell degeneration and loss in *rd* mouse are due to a defect of a recessive null mutation in the β subunit of rod phosphodiesterase (β-PDE) gene. Photoreceptor loss begins within 1 week of birth, with 50% loss by 2 weeks. Bennett *et al.*[32] have slowed the photoreceptor degeneration in the *rd* mouse by subretinal injection of a recombinant adenovirus carrying the β-PDE gene in 5-day-old *rd* pups. In this work, the rescue possibly resulted from the therapeutic gene (β-PDE) directly delivered to the photoreceptor cells, and may also be due to the effective transduction of RPE cells with the secretion of β-PDE gene product. However, the short term transgene expression and potential retinal toxicity have limited adenovirus as a vector

Table 34.2 Properties and current practices of commonly used viral vectors for ocular gene therapy

Vectors	*HSV*	*Ad*	*AAV*
Size of viral genome (kb)	152	35	4.7
Insert capacity (kb)	15~30	7	4.5
Diameter (nm)	200	80	20
Expression cassette	10~30	1	1
Existence	Episomal	Episomal	Integrated, episomal
Optimal titers	10^{10}pfu/ml	10^{12}pfu/ml	10^{10}tu/ml
Defective/recombinant	Defective	Recombinant	Defective
Possible contamination	Yes/no	Minimal	Minimal
Prominent host cells transduced-			
AC	TM, CE, IPE, GC, MC, RPE	TM, CE, IPE, GC	TM, CE
IV	TM, CE, IPE, GC, MC, RPE	TM, CE, IPE, GC, MC	TM, CE, GC, MC
SR	GC, MC, PR, RPE	PR, RPE	PR, RPE
Onset of transgene expression	Fast < 2 days	Fast < 2 days	Delayed, mostly > 2 weeks
Longevity	Days	Days ~ weeks	years
Immunogenicity	High	High	Low
Suitability for retinal gene therapy	Not current generation	Not current generation	Promising
Application in the eye	Proof-of-principle	Proof-of-principle	Gene therapy

Abbreviations: HSV, herpes simplex virus. Ad, adenovirus. AAV, adeno-associated virus. pfu, plaque forming unit. tu, transfection unit. AC, anterior chamber injection. IV, intravitreal injection. SR, subretinal injection. TM, trabecular meshwork. CE, corneal endothelium. IPE, iris pigment epithelium. GC, ganglion cell. MC, Müller cell. RPE, retinal pigment epithelium. PR, photoreceptor.

for therapeutic purposes. Currently the recombinant adenoviral vector is more commonly accepted as a vector for proof-of-principal rather than practical therapeutic applications. A good example for AAV-mediated gene therapy in the eye is the recent report that the photoreceptor structure and function in the *rds* mouse have been restored following subretinal injection of a recombinant AAV encoding the Prph2 gene[33]. The gene Prph2 encodes a photoreceptor-specific membrane glycoprotein, peripherin-2, also known as peripherin/rds. Prph2 is inserted into the rim of photoreceptor outer segment discs in a complex with rod outer segment membrane protein-1. The complex is necessary for the stabilization of the discs, which are renewed constantly throughout life[2]. Mutations in Prph2 have been shown to result in a variety of photoreceptor dystrophies, including ADRP and macular dystrophy[2]. A common feature of these diseases in humans is the loss of photoreceptor function, as seen in the *rds* mouse, which is homozygous for a null mutation in Prph2[2, 23]. Employing a recombinant AAV encoding Prph2 gene, subretinal gene delivery resulted in stable generation of outer segment structures and formation of new stacks of discs containing both Prph2 and rhodopsin, which are morphologically similar to normal outer segments. In addition, the re-establishment of the outer segment structures also resulted in electrophysiological correction, demonstrated as a significant b-wave response in dark-adapted electroretinography (ERG)[33]. In another recent study, AAV mediated ribozyme rescue of photoreceptor cells has been achieved in P23H transgenic rat. The AAV-delivered ribozyme targeted to the P23H mutation in rhodopsin markedly slowed the rate of photoreceptor degeneration for at least 8 months and also resulted in significantly greater ERG amplitudes for at least 180 days post-natal[34].

In summary, advances in molecular biology have opened up new avenues in understanding the etiology of inherited eye diseases and the application of molecular techniques already contributes to their diagnosis and classification. The easy accessibility of the eye favors the development of ocular gene therapy. The techniques of viral vector construction are not only useful in proof-of-principle, but also enable further delivery of a therapeutic gene to correct eye diseases, suggesting the feasibility of AAV-based gene therapy for inherited retinal diseases. The AAV vector has emerged as a very promising vector for ocular gene therapy and it is currently best poised for initial clinical trials for the treatment of inherited eye diseases.

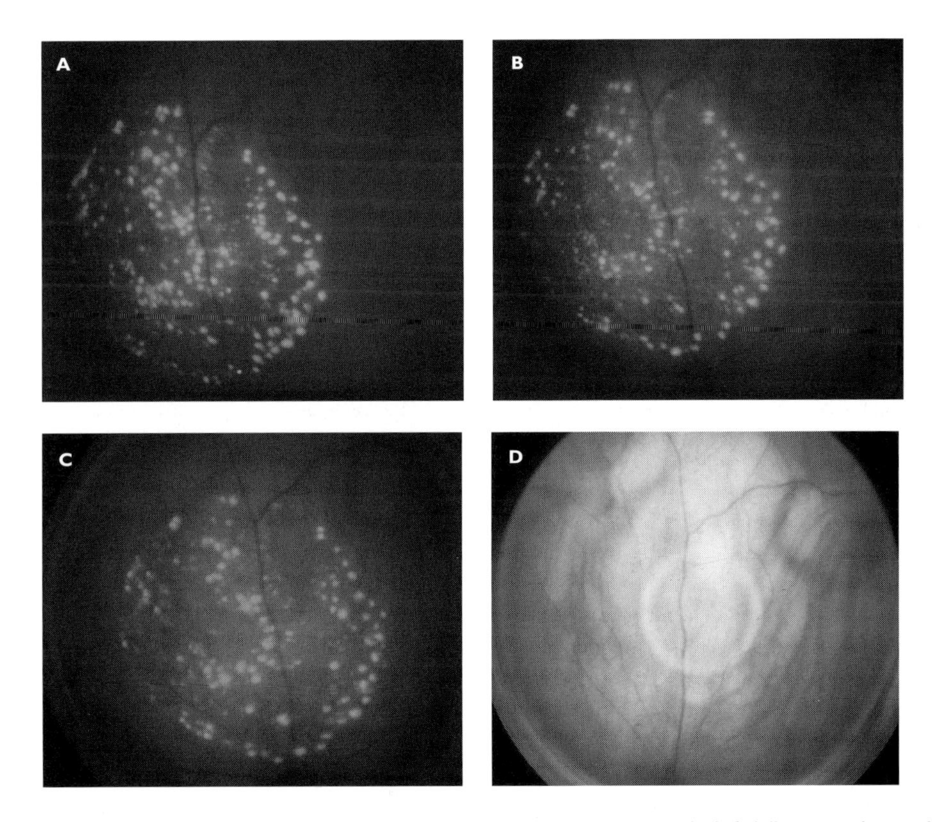

Figure 34.7 Long term and stable transgene expression mediated by adeno-associated virus (AAV) following subretinal delivery of a recombinant AAV carrying a enhanced green fluorescent protein (GFP) gene. GFP expression was monitored in real time at 2 (A), 12 (B) and 19 (C) months post injection in the rat. (D) showing the normal retinal image at 19 months post injection (see Colour Plate XXVII).

References

1 Kennedy, C.J., Rakoczy, P.E., Constable, I.J. (1995) Lipofuscin of the retinal pigment epithelium – A review. *Eye. 9:763–771.*
2 Rattner, A., Sun, H., Nathans, J. (1999) Molecular genetics of human retinal disease. *Annu Rev Genet. 33:89–131.*
3 Sullivan, L.S., Heckenlively, J.R., Bowne, S.J. *et al.* (1999) Mutations in a novel retina-specific gene cause autosomal dominant retinitis pigmentosa. *Nat Genet. 22:255–259.*
4 Banerjee, P., Lewis, C.A., Kleyn, P.W. *et al.* (1998) Homozygosity and physical mapping of the autosomal recessive retinitis pigmentosa locus (rp14) on chromosome 6P21.3. *Genomics. 48:171–177.*
5 Shastry, B.S. (1994) Retinitis pigmentosa and related disorders – phenotypes of rhodopsin and peripherin rds mutations. *Am J Med Genet. 52:467–474.*
6 Callaghan, M., Hand, C.K., Kennedy, S.M., FitzSimon, J.S., Collum, L.M.T., Parfrey, N.A. *et al.* (1999) Homozygosity mapping and linkage analysis demonstrate that autosomal recessive congenital hereditary endothelial dystrophy (CHED) and autosomal dominant CHED are genetically distinct. *Br J Ophthalmol. 83:115–119.*
7 Zhang, K., Nguyen, T.H.E., Crandall, A., Donoso, L.A. (1995) Genetic and molecular studies of macular dystrophies – recent developments. *Surv Ophthalmol 40:51–61.*
8 Dryja, T.P., McGee, T.L., Hahn, L.B. *et al.* (1990) Mutations within the rhodopsin gene in patients with autosomal dominant retinitis pigmentosa. *N Engl J Med. 323:1302–7.*
9 McWilliam, P., Farrar, G.J., Kenna, P. *et al.* (1989) Autosomal dominant retinitis pigmentosa (ADRP): localization of an ADRP gene to the long arm of chromosome 3. *Genomics. 5:619–22.*
10 Dryja, T.P., McGee, T.L., Reichel, E. *et al.* (1990) A point mutation of the rhodopsin gene in one form of retinitis pigmentosa. *Nature. 343:364–366.*
11 Allikmets, R., Singh, N., Sun, H. *et al.* (1997) A photoreceptor cell-specific ATP-binding transporter gene (ABCR) is mutated in recessive Stargardt macular dystrophy. *Nat Genet. 15:236–46.*
12 Maugeri, A., van Driel, M.A., van de Pol, D.J. *et al.* (1999) The 2588G→C mutation in the ABCR gene is a mild frequent founder mutation in the Western European population and allows the classification of ABCR mutations in patients with Stargardt disease. *Am J Hum Genet. 64:1024–35.*
13 Zhang, K., Bither, P.P., Park, R., Donoso, L.A., Seidman, J.G., Seidman, C.E. (1994) A dominant Stargardt's macular dystrophy locus maps to chromosome 13q34. *Arch Ophthalmol. 112:759–764.*
14 Bressler, N.M., Bressler, S.B., Fine, S.L. (1988) Age-related macular degeneration. *Surv Ophthalmol. 32:375–413.*
15 Hirvela, H., Luukinen, H., Laara, E.Sc.L, Laatikainen, L. (1996) Risk factors of age-related maculopathy in a population 70 years of age or older. *Ophthalmology. 103:871–877.*
16 Allikmets, R., Shroyer, N.F., Singh, N. *et al.* (1997) Mutation of the Stargardt disease gene (ABCR) in age-related macular degeneration. *Science. 277:1805–1807.*
17 Allikmets, R. (2000) Further evidence for an association of ABCR alleles with age-related macular degeneration. The International ABCR Screening Consortium. *Am J Hum Genet. 67:487–491.*
18 Souied, E.H., Ducroq, D., Rozet, J.M. *et al.* (2000) ABCR gene analysis in familial exudative age-related macular degeneration. *Invest Ophthalmol Vis Sci. 41:244–247.*
19 De La Paz, M.A., Guy, V.K., Abou-Donia, S. *et al.* (1999) Analysis of the Stargardt disease gene (ABCR) in age-related macular degeneration. *Ophthalmology. 106:1531–1536.*
20 Stone, E.M., Webster, A.R., Vandenburgh, K. *et al.* (1998) Allelic variation in ABCR associated with Stargardt disease but not age-related macular degeneration. *Nat Genet. 20:328–329.*
21 Wagner, T.E., Hoppe, P.C., Jollick, J.D., Scholl, D.R., Hodinka, R.L., Gault, J.B. (1981) Microinjection of a rabbit beta-globin gene into zygotes and its subsequent expression in adult mice and their offspring. *Proc Nat Acad Sci U S A. 78:6376–6380.*
22 Olsson, J.E., Gordon, J.W., Pawlyk, B.S. *et al.* (1992) Transgenic mice with a rhodopsin mutation (Pro23His): a mouse model of autosomal dominant retinitis pigmentosa. *Neuron. 9:815–830.*
23 Hafezi, F., Grimm, C., Simmen, B.C., Wenzel, A., Reme, C.E. (2000) Molecular ophthalmology: an update on animal models for retinal degenerations and dystrophies. *Br J Ophthalmol. 84:922–927.*
24 Noëll, W.K. (1965) Aspects of experimental and hereditary retinal degeneration. In: Graymore, C.N. (ed.) *Biochemistry of the retina*. London: Academic Press, pp. 51–72.
25 Walther, W., Stein, U. (2000) Viral vectors for gene transfer – A review of their use in the treatment of human diseases. *Drugs 60:249–271.*
26 Bennett, J., Wilson, J., Sun, D.X., Forbes, B., Maguire, A. (1994) Adenovirus vector-mediated *in vivo* gene transfer into adult murine retina. *Invest Ophthalmol Vis Sci. 35:2535–2542.*
27 Bennett, J., Maguire, A.M., Cideciyan, A.V. *et al.* (1999) Stable transgene expression in rod photoreceptors after recombinant adeno-associated virus-mediated gene transfer to monkey retina. *Proc Nat Acad Sci U S A. 96:9920–9925.*
28 Rakoczy, P.E., Shen, W.Y., Lai, M., Rolling, F., Constable, I.J. (1999) Development of gene therapy-based strategies for the treatment of eye diseases. *Drug Dev. Res. 46:277–285.*
29 Ali, R.R., Reichel, M.B., Hunt, D.M., Bhattacharya, S.S. (1997) Gene therapy for inherited retinal degeneration. *Br J Ophthalmol. 81:795–801.*
30 Reichel, M.B., Ali, R.R., Thrasher, A.J., Hunt, D.M., Bhattacharya, S.S., Baker, D. (1998) Immune responses limit adenovirally mediated gene expression in the adult mouse eye. *Gene Ther. 5:1038–1046.*
31 Hauswirth, W.W., Beaufrere, L. (2000) Ocular gene therapy: Quo vadis? *Invest Ophthalmol Vis Sci. 41:2821–2826.*
32 Bennett, J., Tanabe, T., Sun, D.X. *et al.* (1996) Photoreceptor cell rescue in retinal degeneration (rd) mice by *in vivo* gene therapy. *Nat Med. 2:649–654.*
33 Ali, R.R., Sarra, G.M., Stephens, C. *et al.* (2000) Restoration of photoreceptor ultrastructure and function in retinal degeneration slow mice by gene therapy. *Nat Genet. 25:306–310.*
34 LaVail, M.M., Yasumura, D., Matthes, M.T. *et al.* (2000) Ribozyme rescue of photoreceptor cells in P23H transgenic rats: Long-term survival and late-stage therapy. *Proc Natl Acad Sci U S A. 97:11488–11493.*

Part 8

Cardiovascular diseases

Chapter 35

Atherosclerosis

Gerd Schmitz, Michael Torzewski, Stefan Barlage and Wolfgang Drobnik

Contents

35.1 Atherosclerosis: an inflammatory disease

Atherosclerosis is the primary cause of ischaemic heart disease (IHD) and stroke and the leading cause of death and long-term disabilities in westernized societies. Several important environmental and genetic risk factors are associated with atherosclerosis. Over the past decade, it has become evident that atherosclerosis is not simply an inevitable degenerative consequence of aging, but rather a chronic inflammatory condition connecting altered cholesterol metabolism and other risk factors to the development and progression of the atherosclerotic lesion and its sequelae like plaque rupture and thrombosis[220].

Eight histologically characteristic types of lesion were identified, and the Roman numerals with which they were designated give the usual sequence of lesion progression (Table 35.1, Figure 35.1). Lesion types I to IV are formed primarily through the accumulation of lipid in the intima. In types I and II, accumulation is mainly intracellular, in macrophages and smooth muscle cells (SMCs). Type III lesions contain, in addition, at least as much lipid extracellulary. Type IV is the stage at which so much extracellular lipid has accumulated that the SMCs at the core of the intima are displaced, atrophic, or dead. Clumps of mineral are now found in dead cells and extracellulary in the lipid core. Although the arterial lumen may not be obstructed that much at this stage, the disruption and change of the intima may precipitate events (fissure, hematoma, thrombosis) that will suddenly reduce the lumen. Lesions that have developed such complications are designated as type VI.

In response to changes in the intima caused by the accumulated lipid alone, and more so if thrombotic deposits are present, intimal SMCs generate additional matrix, particularly collagen (type V lesions). Type VII is a lesion that is largely mineralized. The mineral (calcium) replaces earlier accumulation of lipid and dead cells. Similar lesions can be produced in animals by drastically reducing high serum cholesterol. Type VIII lesions consist mainly of layers of collagen, but they lack lipid. Such lesions could also be the end result of lipid regression, or they could be the consequence of a thrombus that had formed in a part of the artery lacking lipid accumulation[340].

Cellular processes in atherogenesis are, in principle, not different from those in chronic inflammatory-fibroproliferative diseases such as liver cirrhosis, rheumatoid arthritis, glomerulosclerosis, pulmonary fibrosis, or chronic pancreatitis. Atherosclerotic lesions are the result of a series of highly specific cellular and molecular responses to various endogenous

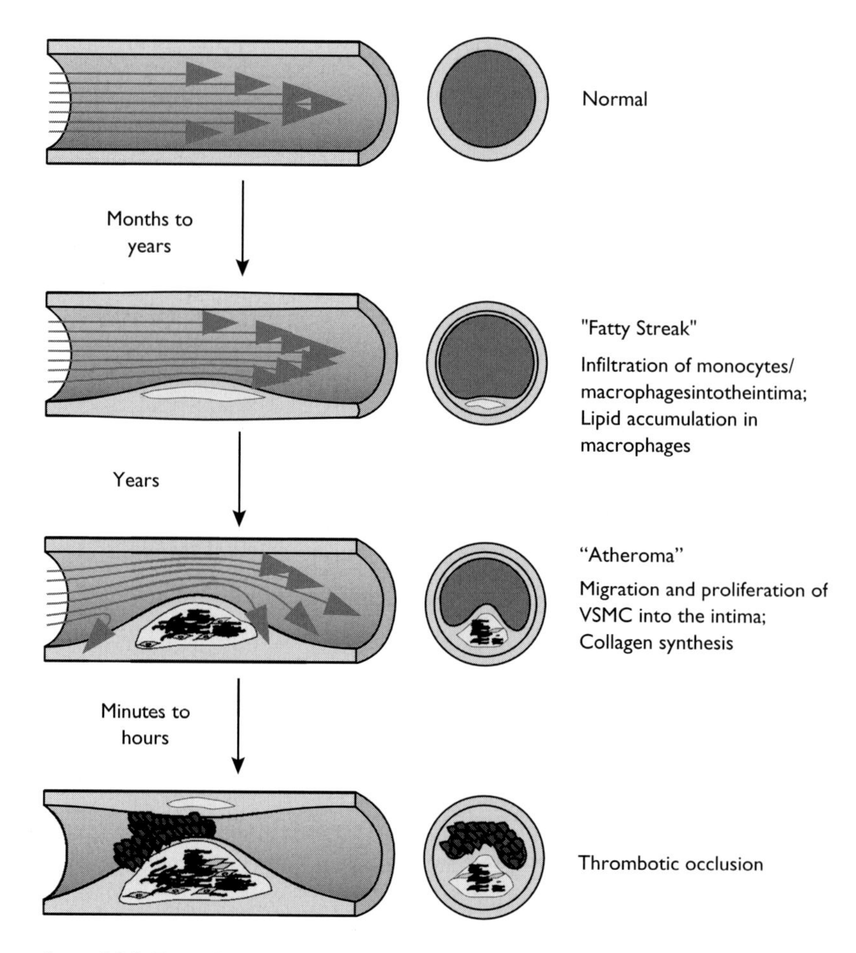

Figure 35.1 Example sequence of atherosclerotic lesion development.

Table 35.1 Histological classification and terms used for atherosclerotic lesions

Type	*Synonym*	*Effect on wall thickness and arterial lumen*	*Clinical correlation*
Type I lesion	Initial lesions		
Type IIa lesion	Progression-prone type II	No or minimal increase in wall thickness; no reduction in lumen	Clinically silent
Type IIb lesion	Progression-resistent type II		
Type III lesion	Preatheroma		
Type IV lesion	Atheroma	Increased wall thickness but often only minimal lumen reduction	
Type V lesion	Fibroatheroma	Additional increases in wall thickness; progressive reduction of the lumen	Clinically overt or silent
Type VI lesion	Lesion with surface defect, hematoma-hemorrhage, and/or thrombotic deposit	Thrombotic deposits increase lesion thickness; lumen often completely blocked	
Type VII lesion	Calcific lesion	Wall thickened, lumen reduced	
Type VIII lesion	Fibrotic lesion		

risk factors and potential exogenous antigens. These responses are mediated by endothelial cells, monocyte-derived macrophages, SMCs and specific subtypes of T lymphocytes. Activation of these cells leads to the release of a wide spectrum of inflammatory hydrolases, cytokines, chemokines, and growth factors followed by cellular lipid accumulation and proliferation of smooth muscle cells as well as formation of fibrous tissue. The modified response-to-injury hypothesis of atherosclerosis that emphasizes endothelial dysfunction rather than denudation as the first step in atherosclerosis was recently extended suggesting that the key initiating event in early atherosclerosis is the subendothelial retention of cholesterol-rich, atherogenic lipoproteins bound to arterial proteoglycans (response-to-retention hypothesis)[388]. Following adherence to endothelial cells, defined subpopulations of circulating monocytes that express the LPS receptor CD14 and the FcγRIII/CD16 are supposed to extravasate into the subendothelial space[318]. Within the vessel wall, phagocytic monocytes rapidly transform to foam cells characterized by the excessive uptake of atherogenic lipoproteins (Table 35.2) by receptor-mediated endocytosis. Cellular uptake of these lipids and lipoproteins is mediated by charge and motif receptors (scavenger receptors) directly recognizing non-opsonized ligands[405]. Alternatively, modified lipids and lipoproteins may be opsonized by either innate (complement components, C-reactive protein (CRP), serum amyloid P (SAP), serum amyloid A_2 (SAA_2)) and/or specific opsonins (immunoglobulins) prior to cellular uptake mediated by different opsonin receptors including complement receptors, pentraxin family receptors and/or Fcγ receptors[405]. Continuous exposure to modified lipoproteins is supposed to trigger a chronic inflammatory process within the lesion. At the monocyte/macrophage level, this leads to the differentiation of either a phagocytic or an antigen-presenting phenotype with enhanced expression of procoagulant and proinflammatory genes as well as genes associated with lipid metabolism.

At the T lymphocyte level, T lymphocytes, of both helper and suppressor phenotypes, have been identified within human atherosclerotic lesions[250,387] displaying a shift during lesion progression. In early atherosclerotic lesions, T lymphocytes of suppressor phenotype appeared to predominate over helper cells. However, initial stages in the development of atherosclerosis also involve the infiltration of the arterial intima with Th1 cells reactive to heat shock protein (hsp) 65/60, expressed by endothelial cells in areas that are subject to increased hemodynamic stress[387] and Th1 cells responding to ox-LDL by proliferation and cytokine secretion[345]. In advanced atherosclerotic lesions, the cytokine expression demonstrates the presence of a predominantly pro-inflammatory, Th1-type T-cell response[112]. Animal models, however, indicate that in addition to the inflammatory cascade, a specific immune response mediated by T-cells may play a role in lesion progression, rather than lesion development[81,94,316]. How T-cells acquire antigens in atherogenesis is still uncertain. Recent findings that dendritic cells (DCs) are present in the arterial wall and that their numbers increase in atherosclerotic lesions suggest that DCs could play a crucial role in T-cell activation in atherosclerosis[41].

Despite the predominance of T-cells, occasional B- and NK cells were observed in advanced human atherosclerotic lesions[325,403]. Both early fatty streaks and advanced atherosclerotic plaques of hypercholesterolemic apolipoprotein E knockout mice contained $CD22^+$ B-cells suggesting that B-cells participate in the local immune response in this experimental model[69]. In addition, the detection of antibodies as specific opsonins against different epitopes of lipids and lipoproteins in human plasma stresses the importance of a humoral immune response in atherogenesis (see above). In aortic tissue, some of the infiltrating cells include NK cells

Table 35.2 Atherogenic lipids and lipoproteins

Lipids/lipoproteins	*Atherogenic molecular component*	*Reference*
Oxidized LDL (ox-LDL)	Hydroperoxy and hydroxy fatty acids, secondary aldehydic lipid peroxidation products, 7 b-hydroperoxycholesterol	Steinberg *et al.*, 1989 Steinberg, 1997 Colles *et al.*, 1996
Acetylated LDL (ac-LDL)	Acetylated lipids	Goldstein *et al.*, 1979
Enzymatically degraded LDL (E-LDL)	Free cholesterol, lysophospholipids	Bhakdi *et al.*, 1995
LDL-proteoglycan complexes	Basic amino acid residues 3359–3369 in site B of apo-B100	Boren *et al.*, 1997
Lipoproteins aggregated by PLA2, PLC	Lysophosphatidylcholine, phosphatidic acid, lysophosphatidic acid	Frank and Fogelman, 1989 Aviram *et al.*, 1995
LDL hydrolyzed by sphingomyelinase	Ceramide, phosphocholine	Schissel *et al.*, 1998
LDL hydrolyzed by phospholipase C	Lysophosphatidylcholine, phosphatidic acid, lysophosphatidic acid	Suits *et al.*, 1989
LDL modified by advanced glycation end products (AGEs)	67 amino acid domain located 1791 residues N-terminal to the LDL receptor binding site of apo-B	Bucala *et al.*, 1995
	modification of residues adjacent to the putative LDL receptor binding site	Wang *et al.*, 1998
Gangliosides (sialic acid containing glyco-sphingolipids)	sialic acid	Wen *et al.*, 1999

and it was hence suggested that these cells may play a critical role in vascular cell injury by releasing perforin and thus causing atherosclerotic aneurysms[325].

At the SMC level, complement activation by e.g. CRP, SAA_2 or hydrolase modified LDL takes place even in the early atherosclerotic lesion. Sublytic complement attack on SMCs leads to the local release of monocyte chemotactic protein 1 (MCP-1), a specific monocyte chemoattractant[405].

35.2 Risk factors and their relation to atherogenicity

Many potential "new" risk factors may predispose to atherosclerotic diseases. These include fibrinogen, C-reactive protein and other inflammatory markers; hyperhomocysteinemia; low vitamin E intake; insulin resistance; lipoprotein(a); and fetal malnutrition. However, clinical trial proof of a causal role of these risk factors and CHD may be difficult to obtain unless selective interventions are developed to lower them. In contrast, elevated plasma levels of atherogenic lipoproteins such as low-density lipoprotein (LDL) and remnant lipoproteins and low levels of HDL cholesterol, are well established as major risk factors for the development of atherosclerosis and reduction of LDL plasma levels will predictably represent the most important prophylactic measure against development and progression of atherosclerosis.

Disorders of lipid metabolism and its constituents are regarded as a prerequisite for the development of atherosclerotic lesions and its sequelae. The relative abundance of the different plasma lipoproteins appears to be of primary importance, as raised levels of atherogenic lipoproteins are a prerequisite for most forms of the disease. The principal transport vehicle of cholesterol in the circulation, low density lipoprotein (LDL), must still be regarded as the most atherogenic lipoprotein species. LDL plays a central role in atherogenesis and every diminution, genetic or acquired, of the LDL-receptor mediated LDL catabolism, causes acceleration of the formation of atheromatous lesions. Association of elevated levels of LDL to coronary heart disease (CHD) was demonstrated in epidemiological studies and supported by studies of genetic disorders and animal models. Studies have now revealed that LDL is structurally heterogeneous, based on its size and density. Multiple subclasses have been characterized in several populations of normal and hyperlipidemic subjects. Some of these subclasses are linked with an atherogenic potential.

Familial hypercholesterolemia (FH) is characterized by high plasma LDL-cholesterol levels due to genetic LDL receptor defects. The low-density particles are the end-product of intraplasmatic lipolysis of triglyceride-rich VLDL and chylomicrons (Figure 35.2) that are metabolized by lipoprotein lipase (LPL) into intermediate density lipoproteins and further converted to LDL by hepatic lipase (HL). Finally LDL particles are taken up in cholesterol requiring cells via the LDL receptor (Figure 35.2, top). A large body of evidence (including experimental, clinical, epidemiological data as well as the results of large trials with lipid lowering drugs) has accumulated to establish that these atherogenic particles are one of the major causative factors of atherosclerosis and its complications. Besides mutations in the LDL

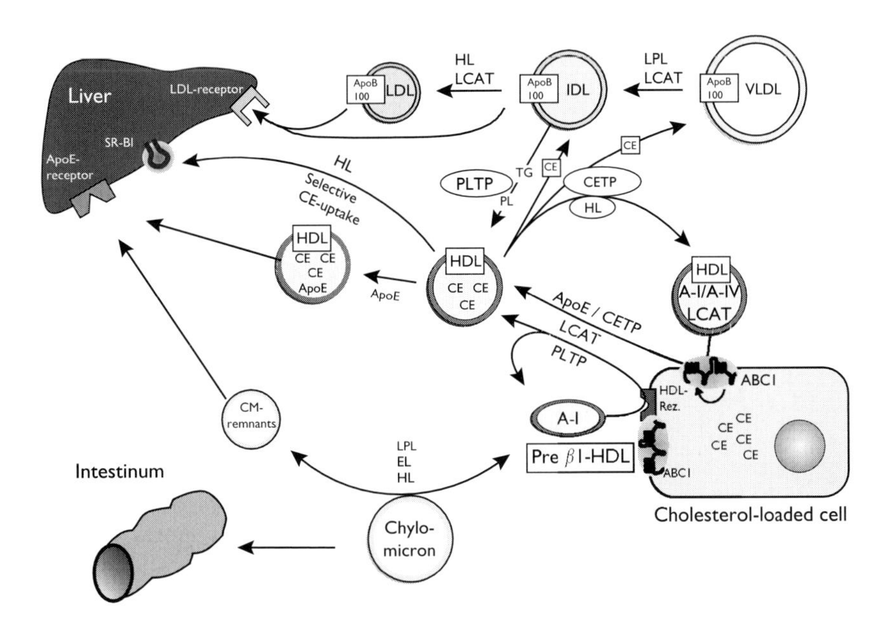

Figure 35.2 LDL metabolism and reverse cholesterol transport.

receptor in FH, defects in the LDL ligand protein apolipoprotein B100 are the underlying cause of familial defective apolipoprotein B-100 (FDB). The LDL cholesterol elevation is less pronounced in FDB than in FH and the age-specific prevalence of atherosclerotic cardiovascular disease (CVD) is lower in FDB than in FH[242].

Specific therapeutic interventions should be initiated to decrease the risk for developing cardiovascular disease, mainly coronary heart disease. The therapeutic intervention includes both a diet low in saturated fatty acids and cholesterol and statins which are now the first line therapy. Fibrates are proposed to those who have mixed hyperlipidemia with dominating triglyceride and low HDL cholesterol, and to those who do not tolerate statins, and where LDL-apheresis is associated with statins in the rare cases of homozygous familial hypercholesterolemia[165].

Various clinical studies have identified lipoprotein(a) [Lp(a)] as an independent risk factor for CHD. In particular, the severity of this pathological condition correlates well with the molecular weight (MW) of the apolipoprotein(a) [apo(a)] isoforms supporting Lp(a) (low MW = high risk; high MW = low risk), which are characterized by a high degree of individual variability. Since apo(a) isoforms are genetically determined, a genetic predisposition for CHD has been identified, i.e. the apo(a) phenotype is considered a marker of genetic susceptibility for cardiovascular risk (Table 35.2)[151].

Patients with familial combined hyperlipidemia (FCHL) exhibit a lipid profile – the so called atherogenic lipoprotein phenotype – that is associated with elevated triglyceride levels, low levels of high density lipoprotein and a preponderance of atherogenic, small, dense LDL particles (Table 35.2). Such individuals are at an increased risk of CHD events, regardless of their total circulating LDL mass. Evidence suggests that when plasma triglycerides exceed a critical threshold, this favors the formation of small, dense LDL from larger, less dense species. Lipid-lowering agents that are capable of lowering triglyceride levels below this threshold value will cause a shift to a less dense and, therefore, less atherogenic LDL profile[24].

Apolipoprotein E plays a key protective role in atherosclerosis. Its capacity to safeguard against this disease can be attributed to at least three distinct functions. Plasma apolipoprotein E maintains overall plasma cholesterol homeostasis by facilitating efficient hepatic uptake of lipoprotein remnants. Apolipoprotein E released from lesional macrophages in concert with apolipoprotein A-I facilitate cellular cholesterol efflux from macrophage foam cells within the intima of the lesion. Lesion apolipoprotein E directly modifies both macrophage- and T lymphocyte-mediated immune responses that contribute to this chronic inflammatory disease. Type III hyperlipoproteinemia (HLP) is a genetic disorder characterized by accumulation of remnant lipoproteins in the plasma and development of premature atherosclerosis. Although receptor binding-defective forms of apolipoprotein (apo) E are the common denominator in this disorder, a number of apparent paradoxes concerning its pathogenesis still exist. However, studies in transgenic animals are resolving the mechanisms underlying this disorder[225]. Finally, apolipoprotein E plays a critical role in neurobiology. The apolipoprotein E4 allele is the major susceptibility gene related to the occurrence and early age of onset of Alzheimer's disease. It is probable that one of the major functions of apolipoprotein E in the central nervous system is to mediate neuronal repair, remodeling, and protection, with apolipoprotein E4 being less effective than the E3 and E2 alleles. The isoform-specific effects of apolipoprotein E are currently being unraveled through detailed structure and function studies[224].

It is believed that HDL exerts its anti-atherogenic effects through the process of delivering cholesterol from peripheral tissues back to the liver for removal from the body (i.e. reverse cholesterol transport) (Figure 35.2). The metabolic life cycle of HDL lipid and apolipoproteins during reverse cholesterol transport involves both its modification in plasma by lipid transfer proteins and the clearance from plasma of HDL lipid and protein mediated by hepatic cell surface proteins. Association of reduced levels of HDL has also been demonstrated by numerous epidemiological studies and supported by studies of genetic diseases and animal models. In the homozygous state, null mutations of apoAI (Apo AI deficiency) result in the virtual absence of HDL and early CHD. In 1999, Tangier disease, a rare recessive disorder characterized by very low levels of circulating HDL, was shown to be due to mutations in the gene for the ATP-binding-cassette (ABC) transporter 1, providing an excellent candidate cause for more common forms of HDL deficiency[317]. Other common genetic variations contributing to CHD and its risk factors by low HDL levels, are a promoter polymorphism of hepatic lipase (HL), multiple polymorphisms in the ApoAI-CIII-AIV cluster, common null mutations (Japanese) and missense polymorphisms of cholesteryl ester transfer protein (CETP) and missense mutations of lipoprotein lipase[220].

Effective drugs for lowering cholesterol have been developed. In particular, the statins lower levels of atherogenic lipoproteins and dramatically decrease clinical events and mortality from atherosclerosis. Over the past decade, a number of further promising new targets have been identified, as discussed above. However, agents that block disease at the level of the vessel wall or that raise anti-atherogenic HDL are required.

35.3 Current therapeutic approaches reducing the pool size of atherogenic lipoproteins or enhancing the protective effects of HDL against CHD

35.3.1 *Reducing the atherogenicity of LDL*

The therapeutic benefits of lowering LDL cholesterol were first shown in the Lipid Research Clinics Coronary Primary Prevention Trial (LRCCPPT) where almost 2000 men with moderate hypercholesterolemia received the non-resorbable, sterol binding anion-exchanger cholestyramine (24 g/d) for more than 7 years[1]. During this period, their mean plasma LDL-cholesterol was 12.6% lower than that of placebo-treated controls and, overall, cholestyramine-treated subjects sustained a 19% reduction in CHD death and non-fatal myocardial infarction ($p < 0.05$) compared to the placebo group, but no decrease in mortality. This trial proved for the first time the benefit of lowering LDL cholesterol and was the forerunner of the statin trials, which achieved much greater reductions in LDL and provided evidence for a reduction in cardiovascular events during statin therapy.

Three secondary prevention trials using simvastatin in the Scandinavian Simvastatin Survival Study (4S)[3] and pravastatin (Cholesterol and Recurrent Events[310] and Long-term Intervention with Pravastatin in Ischemic Disease (LIPID)[9]) and two primary prevention trials involving pravastatin (West of Scotland Coronary Prevention Study)[327] and lovastatin (Air Force/Texas Coronary Atherosclerosis Prevention Study (AFCAPS/TexCAPS))[88] reduced the risk of CHD by 31% and total mortality by 21% with a mean decrease in LDL cholesterol of 28%. These benefits were equally evident in men and women at all ages[203].

Moreover, meta-analysis of these and other trials using simvastatin, lovastatin or pravastatin, involving almost 10,000 patients, also showed a 27% decrease in the risk of stroke[78].

In the Post-Coronary Artery Bypass Graft Trial, patients on high-dose lovastatin had a lower incidence of saphenous vein graft disease and re-do coronary artery bypass grafts than patients on low-dose therapy[7]. In the Atorvastatin versus Revascularisation Therapy trial, patients with stable angina treated with atorvastatin had fewer ischemic events than those who underwent angioplasty[280].

Prevention of CHD is a major challenge in the overall management of patients with type 2 diabetes (NIDDM). The usefulness of statins for treating diabetic dyslipidemia is the subject of ongoing clinical trials in specific diabetic populations. Vascular disease accounts for approximately 70% of death in this high-risk population[290]. CHD is increased 2–4 fold even at time of diagnosis[39,198] and the relative risk increase for women is greater and the cardioprotection conferred by the premenopausal state is lost[34]. Diabetic patients are more likely to die with their first myocardial infarction, early and late mortality rates are increased[138,241] and outcome after revascularization is worse[6].

Several interventions have been shown to reduce morbidity and mortality rates in NIDDM with established CHD and there is good evidence for the use of ACE inhibitors[11], aspirin[2], β-adrenergic blocking agents[181] and lipid lowering drugs[9,119,137].

The pathogenesis of CHD in NIDDM is multifactorial, but diabetic dyslipidemia is likely to be an important risk factor and is a current focus for therapeutic primary and secondary intervention[129,136,373]. The Heart Protection Study (HPS) includes a total of 20,536 individuals and contains a large number of NIDDM patients with and without established vascular disease[10]. This is a 5 year, double-blind, randomized comparison of simvastatin versus placebo and of anti-oxidant vitamins versus placebo and the results are expected to emerge in 2001. It will provide the first large-scale randomized evidence of the effects of lipid lowering in a wide range of men and women with diabetes. The Lipids in Diabetes Study (LDS), coordinated by the Diabetes Trials Unit (University of Oxford, UK), is the first large-scale randomized study to examine the combination of statin and a fibrate and the results are expected in 2005. The Collaborative Atorvastatin Diabetes Study (CARDS), coordinated by the Department of Epidemiology and Public Health at the University College London, UK, is a multicentre, randomized, placebo-controlled, double-blind clinical trial of atorvastatin in NIDDM patients with normal or moderately elevated LDL cholesterol levels with no previous CHD or other macrovascular disease. The study is expected to report in 2004. Atorvastatin as Prevention of CHD in patients with NIDDM (ASPEN) is an ongoing study that involves approximately 2250 patients in 70 centres in the USA, Europe and Australia. ASPEN is due to complete in 2003.

From the overall analysis of all statin trials it is obvious that statins have pleiotropic CHD-reducing properties over and above that attributable to their effect in lowering LDL cholesterol.

35.3.2 Pleiotropic effects of statins

Aside from their lipid-lowering activity, statins seem to possess relevant additional properties that target platelet and endothelial dysfunction, leukocyte adhesion, macrophage stimulation, inflammatory and procoagulant responses and SMC transformation, as summarized in Table 35.3, to impair progression of atherosclerosis and to stabilize the atherosclerotic plaque[215,255].

Besides cholesterol synthesis, the mevalonate pathway (Figure 35.3) leads to the formation of dolichols, which have an essential role in lipoprotein synthesis, ubichinone, involved in electron transport, and isoprenoids, which take part in the farnesylation or geranylation of many proteins, including those needed for cell proliferation[73,72]. Since virtually all cells possess the mevalonate pathway, it is not surprising that the influence of statins is not limited to the cardiovascular system. The inhibition of osteoclast formation by statins has the potential to inhibit the progression of osteoporosis and the mechanism involves the mevalonate pathway by impaired prenylation of G-proteins that participate in the activation of osteoclasts, including the reorganization of the cytoskeleton, vesicular fusion and apoptosis[104,135]. Apoptosis as well as the antimitogenic

Table 35.3 In vitro pleiotropic effects of statins

Target	*Stimulation*	*Inhibition*	*References*
Endothelium	ENOS tPA uPA	PAI-1 Tissue factor Endothelin	Bellosta *et al.*, 1998; Brandes *et al.*, 1999; Colli *et al.*, 1997; Emdres *et al.*, 1998; Essig *et al.*, 1998; Hernandez-Perera *et al.*, 1998; Laufs *et al.*, 1998
Platelets		Adhesion Aggregation	Farnier *et al.*, 1998
Macrophages		MMP-9 secretion Chemotaxis Tissue factor activity LDL oxidation LFA-1 interaction with VCAM-1	Bellosta *et al.*, 1998; Colli *et al.*, 1997; Giroux *et al.*, 1993; Kallen *et al.*, 1999; Kreuzer *et al.* 1991
NK cells		Cytotoxicity	Katznelson *et al.*, 1998; Muldoon *et al.*, 1997; Rudich *et al.*, 1998
T-cells		Cytotoxicity	Katznelson *et al.*, 1998; Muldoon *et al.*, 1997; Rudich *et al.*, 1998
B-cells		Activation	Rudich *et al.*, 1998
SMCs	Apoptosis TGF-1	Proliferation Migration c-fos, c-jun PDGF ras and rho proteins ACE activity	Bellosta *et al.*, 1998; Corsini *et al.*, 1996; Essig *et al.*, 1998; Muller *et al.*, 1999; Vrtovsnik *et al.*, 1997
Osteoclasts		Proliferation	Fisher *et al.*, 1999
Adipogenesis		Proliferation	Gujarro *et al.*, 1998
Fibroblasts, various tumor cells	Apoptosis G1-arrest PI2-kinase inhibitor	S-phase progression	Crick *et al.*, 1994; Dimitroulakos *et al.*, 1999; Gosh *et al.*, 1997; Hunakova *et al.*, 1997; Lee *et al.*, 1998; Marcelli *et al.*, 1998; Muller *et al.*, 1998; Rao *et al.*, 1998; Rubins *et al.*, 1998; Shipman *et al.*, 1998; Tan *et al.*, 1999

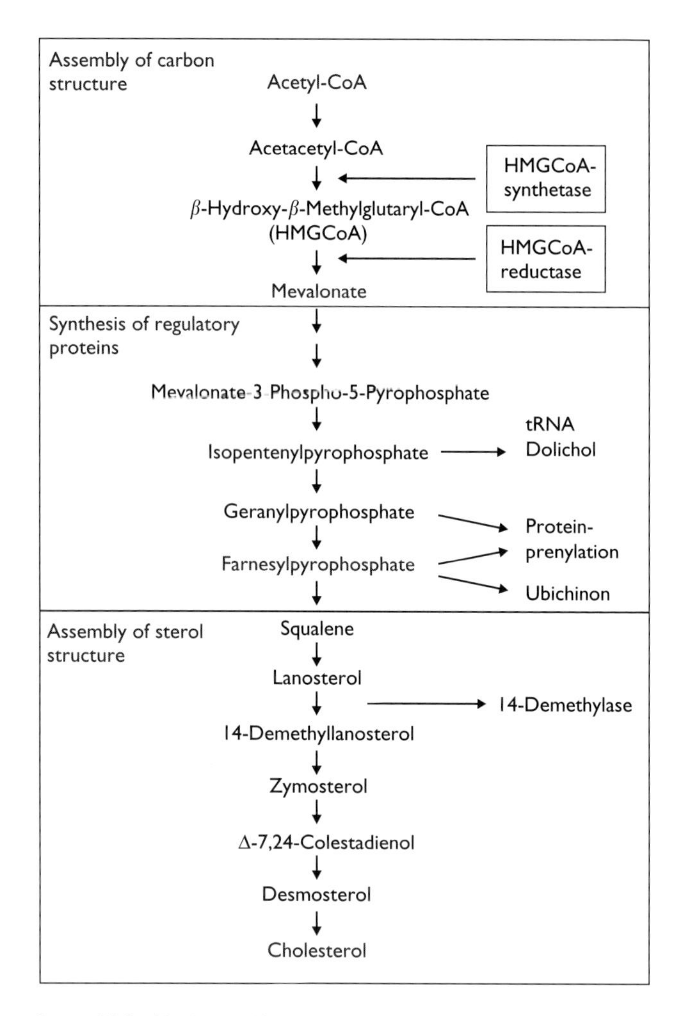

Figure 35.3 Cholesterol/mevalonate pathway.

activity of statins[240,334] also seem to be linked to the mevalonate pathway.

The reduction of plasma fibrinogen and of plasma viscosity, an improved deformability of red blood cells[189,327,372], a stimulation of tissue plasminogen activator (tPA) activity, a reduced plasminogen activator inhibitor (PAI-1) activity[98] and a reduced platelet activation[99] are all significant indicators for a shift of the hemostatic balance towards anticoagulation.

Despite there being no relationship between cholesterol and stroke incidence[4] in the 4 S study[3], in the Cholesterol and Recurrent Events Study (CARE)[310] and in the LIPID study[222], stroke events were significantly reduced by 28%, 31%, and 19%, respectively. Recently, the Carotid Atherosclerosis Italian Ultrasound Study (CAIUS) indicated that 6 month statin treatment produced a significant reduction in the progression of intima-media thickness.

Thus, statins cannot be regarded as pure lipid-lowering agents, because they trigger numerous cellular responses as inhibitors of the mevalonate pathway.

35.3.3 Therapeutic modulation of hypertriglyceridemia and the particle size of apolipoprotein B-containing lipoproteins

35.3.3.1 *Atherogenicity of triglyceride-rich lipoproteins and small dense LDL*

It has long been debated whether hypertriglyceridemia can predict risk of cardiovascular disease statistically independent from HDL[152,159,169]. However, given the complex, multifactorial nature of the pathophysiology involved, the concept of statistical "independence" of the effects of triglycerides on risk makes limited sense[362]. Triglycerides are carried in lipoproteins of different atherogenicity and different genetic entities are involved, such as familial hypertriglyceridemia (FH) and familial combined hyperlipidemia (FCH) in which there is hypercholesterolemia as well as hypertriglyceridemia[27,51]. FCH is associated with a higher risk of CHD[51]. The particle size of triglyceride-rich lipoproteins[169,261], however, is obviously a critical determinant for the atherogenicity of triglyceride-rich lipoproteins leading to the paradoxical situation that the highest CHD risk is associated with moderate rather than severe hypertriglyceridemia where large particles (chylomicrons) predominate that may be too large to enter the arterial wall[261].

Mild hypertriglyceridemia, low HDL cholesterol and a smaller and denser species of LDL constitute a common atherogenic lipoprotein phenotype[25] that is associated with other risk factors for atherosclerosis such as insulin resistance and procoagulant conditions including hyperfibrinogenemia, impaired fibrinolysis and raised PAI-1, hypertension and hyperuricemia.

Impaired VLDL lipolysis favors small dense LDL formation[269], which contains less cholesteryl ester, free cholesterol and phospholipids (Figure 35.2). Small dense LDL contain fewer polyunsaturated fatty acids and less abundant peroxidation products of C18 and C20 fatty acids[60] as a result of a loss of protection against oxidation. Platelet activating factor acetyl hydrolase (PAFAH) as well as paraoxonase may confer resistance to the oxidation of small dense LDL. PAFAH resides on LDL either as a tightly adherent form that predominates on larger LDL or one that is more readily displaced which predominates on small LDL[233]. In diabetic women, small dense LDL were found to be more susceptible to oxidation[132]. Higher free fatty acid and PAI-1 concentrations were also found associated with small dense LDL[101,284].

Evidence for the inheritance of small dense LDL is derived from genome-wide search studies[26,111,295]. Major quantitative trait loci for small dense LDL were identified on chromosome 3 (including Apo D) and chromosome 4 (including microsomal transfer protein (MTP)) with another possible locus on chromosome 6. Larger LDL particles were possibly linked to genes on chromosome 19, which may involve the LDL receptor. Small dense LDL is also associated with the LPL gene[153], the Apo E4 genotype[266] and a promoter variation in CETP[355]. In families with FCH, LDL size was also linked to the hepatic lipase gene[16].

In diabetic individuals progressive albuminuria was found to be associated with increases in small dense LDL pool size[330] and increased hepatic lipase activity[127], the latter of which promotes small dense LDL accumulation[399]. Physical exercise and loss of body mass reduce small dense LDL and increase HDL_2[398]. Increases in LDL size and decreases in small dense LDL are also found during consumption of n-3 polyunsaturated fatty acid preparations[54,366].

35.3.3.2 Fibrates resembling peroxisomal proliferation activator receptors (PPAR) agonist as targets for treatment of hypertriglyceridemia and small dense LDL

Fibrates lower plasma triglycerides and raise HDL efficiently and with few side effects. Fibrates may also have a hypoglycemic and thus anti-diabetic effect, as a consequence of their hypolipidemic action. Fibrates increase LDL size through multiple effects mediated by the activation of the PPAR nuclear receptors (Figure 35.4). PPARs are a group of lipid-activated nuclear receptors which regulate genes involved in lipid and glucose metabolism[182,206]. The PPARs were first cloned as the nuclear receptors that mediate the effects of synthetic compounds called peroxisome proliferators on gene transcription. PPARs are ligand-dependent transcription factors: activation of target gene transcription depends on the binding of the ligand to the receptor. First, they bind a specific element in the promoter region of target genes. PPAR and some other nuclear hormone receptors bind the promoter only as a heterodimer with the receptor for 9-*cis* retinoic acid, RXR (retinoid X receptor). Second, they activate transcription in response to binding of the hormone (ligand) (Figure 35.4). For the PPAR:RXR heterodimer, binding of the ligand of either receptor can activate the complex, but binding of both ligands simultaneously is more potent.

Three PPAR isotypes have been identified: α, β (also called δ and NUC1) and γ. Since their discovery in the early 1990s, it has become clear that PPARs are crucial in the genetic regulation of complex pathways of mammalian metabolism, including fatty acid oxidation and lipogenesis. Whereas PPAR-α promotes fatty acid oxidation under conditions of lipid catabolism such as fasting, PPAR-γ acts at the level of the adipose tissue and promotes lipogenesis under anabolic conditions. PPAR-α is highly expressed in liver,

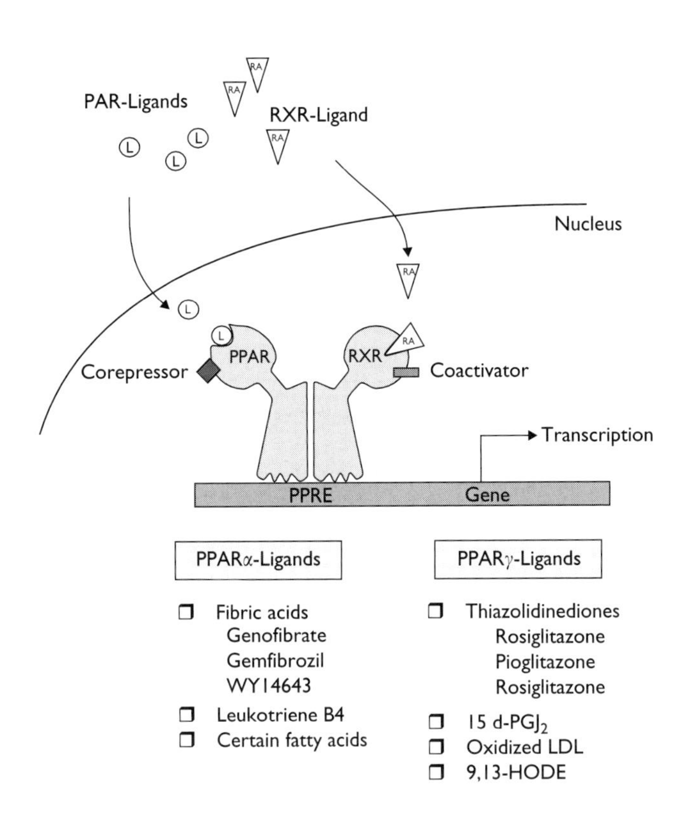

Figure 35.4 Transcriptional regulation by PPARs and their ligands.

heart, muscle and kidney, as well as in cells of the arterial wall, while PPAR-γ is expressed at high levels in white adipose tissue, where it activates adipocyte differentiation.

PPAR-γ is also highly expressed in foam cells of macrophage and SMC origin raising many questions about its roles and its ligands in cardiovascular disease.

The scavenger receptors SR-A and CD36, expressed on the surface of macrophages, bind to and internalize chemically modified LDL, causing the conversion of macrophages to pro-atherogenic foam cells. Oxidized lipids and antidiabetic thiazolidinediones (TZD) activate the nuclear receptor PPAR-γ, leading to increased expression of CD36. By itself, this effect increases foam cell lipids and might lead to atherosclerosis. However, it was recently demonstrated that this might be counter-regulated by reduced expression of SR-A and increased expression of another nuclear receptor, LXR, which induces the ATP cassette transporter ABCA1 that is involved in reverse cholesterol transport (Figure 35.2; Figure 35.6). The net effect of activation of PPAR-γ is thus assumed to be anti-atherogenic in foam cells.

PPAR-γ has also been implicated in inflammation and in monocytes/macrophages, PPAR-γ has been shown to reduce cytokine production of TNFα, interleukin-1β and interleukin-6 by inhibiting the activity of pro-inflammatory transcription factors such as AP-1, STAT and NF-κB. This anti-inflammatory effect of PPAR-γ could be beneficial in the treatment of atherosclerosis. In addition, PPAR-γ may reduce expression of metalloproteinases such as MMP-9, which are implicated in plaque destabilization.

These findings warrant an in-depth investigation of the gene regulatory mechanisms of PPAR ligands. Synthetic agonists for both PPAR-α (e.g. fibrates) and PPAR-γ (e.g. thiazolidinediones) might be useful (Figure 35.4). Fibrates, which include gemfibrozil, bezafibrate and fenofibrate, bind PPAR-α with high affinity and it is believed that most of their effects on lipid and lipoprotein metabolism are mediated by PPAR-α to increase the activity of lipoprotein lipase as well as Apo AI and Apo AII secretion.

High affinity PPAR-γ ligands known as thiazolidinediones (TZDs) have been postulated to be a potential new class of anti-atherosclerotic and anti-diabetic drugs. In diabetic individuals, the PPAR-γ agonist Rosiglitazone besides improving diabetes control also increased LDL particle size[349]. As for the anti-atherosclerotic mechanism, TZDs have been shown to inhibit the secretion of inflammatory cytokines. However, it was observed that TZD's effect on cytokine gene expression and secretion was not altered in macrophages taken from PPAR-γ-null mice. It was suggested that another PPAR-related receptor, such as PPAR-α, mediates the anti-inflammatory effects of TZDs in the PPAR-γ null macrophages. Indeed, PPAR-α expression is activated by high concentrations of TZDs and its ability to regulate transcription of some of the same macrophage genes as PPAR-γ was demonstrated. PPAR-γ ligands could also signal through a different nuclear receptor or by a completely distinct pathway.

However, the effect of TZDs and 15-deoxy-$D^{12,14}$-prostaglandin J_2, which was originally thought to be a selective PPAR-γ ligand, on atherosclerosis and inflammation do not always correspond. This suggests that at least one type of ligand, probably 15-deoxy-$D^{12,14}$-prostoaglandin J_2, also acts independently of PPAR-γ, although PPAR-γ-independent effects of TZDs have been reported as well. Use of antisense and/or gene-targeting technology should clarify the role of PPAR-γ in inflammation and atherosclerosis. Without this information and without clear data from clinical trials, the link between PPAR-γ and atherosclerosis, as well as the therapeutic benefit of PPAR ligands, remains speculative.

The question arises, however, whether the fascinating molecular mechanisms of PPAR modulation are associated with clinical benefit as defined by reductions in rates of death by CHD. Up to 1996, 12 trials of therapy with fibrates have been published so far including more than 21,000 patients[52]. Overall, these trials reported no benefit in terms of reduction in risk of coronary death. Since 1996 four additional fibrate-trials with more than 6000 patients were published[12,93,110,307]. Two of them were major trials[169,307]. The Veterans Affairs Cooperation Studies Program High-density lipoprotein Cholesterol Intervention Trial (VA-HIT) was very encouraging, because treatment with gemfibrozil produced a significant reduction in the combined incidence of fatal and non-fatal coronary events. There was no significant reduction in coronary death, however.

The results of the Bezafibrate Infarction Prevention study (BIP) were disappointing, because they demonstrated no significant effect of treatment with bezafibrate on either the primary end-point of the trial or on rates of coronary death.

On the basis of the evidence now available, fibrates may have a fairly limited indication. It is possible that modulation of LDL size in the hypertriglyceridemic[227] and probably insulin-resistant subsct[361] of the Helsinki Heart Study and the VA-HIT conferred cardiovascular benefit.

35.3.3.3 The effect of niacin on lipid and lipoprotein metabolism

In pharmacologic doses, niacin reduces total plasma cholesterol, triglycerides, VLDL, LDL and increases HDL and Apo AI levels[177,197]. It also decreases small dense LDL and Lp(a) (Figure 35.5)[400].

Various clinical trials including the Coronary Drug Project, the Cholesterol Lowering Atherosclerosis Study and the Familial Atherosclerosis Treatment Study indicated that treatment with niacin significantly reduced mortality, coronary events, and retarded the progression and induced regression of coronary atherosclerosis[177,197].

Niacin significantly inhibits triglyceride production at two synthetic sites (Figure 35.5)[173]: a) fatty acid synthesis from acetate, and b) esterification of fatty acids to form triglycerides. However, niacin does not alter the synthesis of cholesterol, cholesteryl esters or phospholipids in hepatocytes. Niacin may be acting selectively on fatty acid synthase and diacylglycerol acyl-transferase.

The niacin-mediated inhibition of triglyceride synthesis and availability may limit Apo B lipidation leading to the delayed translocation of Apo B across the ER membrane, which in turn may increase intracellular Apo B degradation[173] and decrease VLDL production and Apo B secretion[130]. Niacin, however, does not affect Apo B mRNA expression, LDL uptake and MTP activity.

Niacin-mediated inhibition of triglyceride synthesis may inhibit the generation of large TG-rich VLDL particles which in turn leads to a reduction of small dense LDL particles[347] and also decreases fatty acid release from adipocytes.

Plasma levels of HDL and Apo AI are regulated by a balance between synthetic and catabolic processes. Niacin is the most potent clinically used agent that increases circulating HDL cholesterol and Apo AI by decreasing the fractional catabolic rate of apo AI[40,328].

Niacin selectively decreases the hepatic removal of HDL Apo AI, but not HDL cholesteryl esters and does not influence Apo AI synthesis[172]. Niacin obviously inhibits the removal of HDL Apo AI at the level of the putative

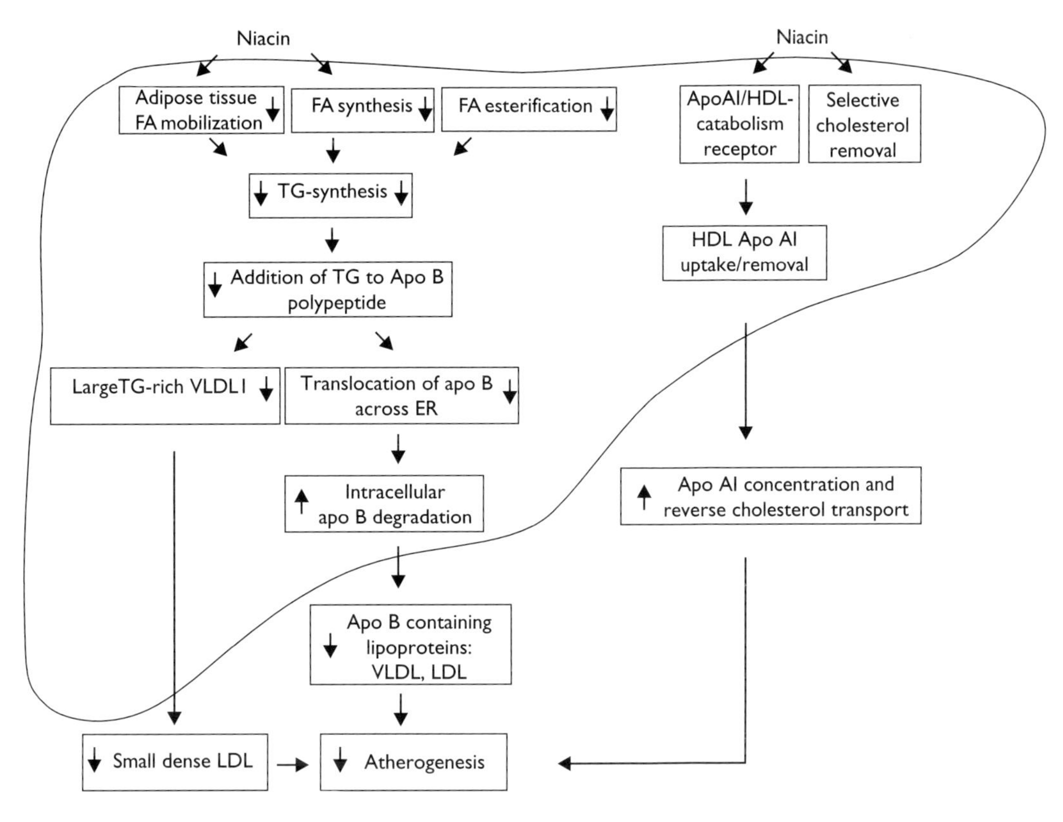

Figure 35.5 Effects of niacin on hepatic lipoprotein metabolism.

Apo AI/HDL catabolism receptor pathway, but not SR-BI mediated processes, which are selective for HDL cholesteryl ester uptake[301,376].

Despite the fact that niacin has been proven a valuable drug, the side effect profile (e.g. flush) significantly reduces the compliance rate, and the magnitude of the reduction in cholesterol is not as convincing as for the statins. The statins, on the other hand, fail to reduce plasma TG and small dense LDL which provides space for new targets such as MTP inhibitors and thyromimetics.

35.3.4 Raising HDL cholesterol

High density lipoprotein cholesterol (HDL cholesterol) is inversely associated with CHD risk[123] consistent with several antiatherogenic activities of HDL. HDL is involved in reverse cholesterol transport from peripheral cells back to the liver for the synthesis of bile acids and to steroidogenic organs for steroid hormone synthesis, and mediates the efflux of cholesterol from lipid-loaded cells[117,294,304,341]. Beyond its function in reverse cholesterol transport, HDL has other antiatherosclerotic properties related to anti-inflammatory, anticoagulant and antioxidants effects (Table 35.4). HDL inhibits the oxidation of LDL, the formation of methionine sulphoxides in Apo AI and Apo AII and promotes the enzymatic degradation of chemically modified lipids through the action of paraoxonase (arylesterase), PAF acylhydrolase and LCAT[28,115,221,341].

HDL particle constituents also modulate platelet activation and the expression of endothelial cell adhesion molecules and E-selectin, and thereby inhibit endothelial adhesion and transmigration of monocytes into the vessel wall[22,67,211,251,260,396].

In order to understand better the anti-atherogenicity in terms of the search for potential drug targets, HDL metabolism can be divided into three steps: first, the synthesis and secretion of HDL precursors from the mucosa and hepatocytes or the formation of HDL precursors during intraplasmatic lipolysis of triglyceride-rich lipoproteins; second, the conversion of HDL precursors to mature HDL; and third, catabolism of HDL lipid and protein components.

35.3.4.1 HDL precursor formation

Cholesterol-poor HDL precursors (Figure 35.2) are either secreted as nascent HDL from the intestinal mucosa and hepatocytes or are released as "surface remnants" from chylomicrons and VLDL during lipoprotein lipase (LPL)-mediated hydrolysis of triglycerides for fatty acid delivery to peripheral cells and "core remnant" formation[309].

Table 35.4 Antiatherogenic effects of HDL

Mechanism	*Site of synthesis*	*Comments*
Reverse cholesterol transport		
Apo A-I	Intestine, liver	Promotes cholesterol efflux, LACT activator
Apo A-IV		Transports lipids, LCAT activator
Apo A-II		Activator of LPL and HL, modulates LCAT activity
LCAT	Liver	Mediates cholesterol esterification within the plasma compartment
CETP	Liver, intestine, spleen, adrenal glands	Facilitates lipid transfer between lipoproteins and lipoprotein interconversation
Apo E	Liver, brain, spleen, lung, adrenal glands, ovary, kidney, muscle, may be ubiquitous	Mediates cholesterol efflux from peripheral cells and the removal of remnant lipoproteins by the liver
Apo C-I	Liver	Activator of LCAT and LPL
Apo C-II		Activator of LCAT and LPL
Apo C-III		Inhibitor of LPL and the uptake of TG-rich lipoproteins
Apo D	Liver, intestine, brain, placenta, adrenal glands, pancreas, spleen	Bilirubin binding, potential radical scavenger
Inhibition of oxidation		
Paraoxonase	Liver, kidney	Promotor polymorphisms are associated with serum levels
LCAT	Liver	Antioxidant activity prevents the formation of oxidized lipids
Inhibition of platelet activation/aggregation		
?		HDL reduces platelet activation and subsequent aggregation
Inhibition of endothelial adhesion molecule expression		
?		HDL reduces VCAM, ICAM-1 and E-selectin expression on endothelial cells, effects probably depend on the lipid composition of the HDL particle
Clearance of bacterial lipopolysaccharide		
LPS-binding protein (LBP)	Liver	Mediates LPS-binding and detoxification of LPS through HDL
PLTP		Mediates LPS-binding and detoxification of LPS through HDL
Regulation of complement activation and hemostasis		
Apo J (clusterin)	Brain, testis, ovary, liver, heart, lung, spleen	Inhibits cytolytic reaction of complement C5b-9 complex
TFPI	Liver	Mixed disulfide linkages with Apo AII, inhibitor of Tissue factor and factor Xa

Alternatively, HDL precursors are generated during interconversion of HDL_2 and HDL_3 by CETP, phospholipid transfer protein (PLTP) and hepatic lipase[309].

The synthesis rate of the main HDL-protein component Apo AI is a critical determinant for HDL pool size and upregulation of Apo AI expression apparently protects against atherosclerosis[36,96,175,357]. The formation of HDL particles can also be enhanced by overexpression of Apo AII or Apo AIV. Atherosclerosis was found to be decreased in mice by overexpressing human Apo AII but increased by overexpression of murine Apo-AI[97,175,351]. Transfection of Apo AIV on the background of Apo E deficient mice inhibited the progression of atherosclerosis without changing HDL cholesterol[69,91]. Sera from Apo AI as well as Apo AIV transgenic mice have increased cholesterol efflux capacity[108,342] and in men, a case-control study recently unravelled an independent association of impaired Apo AIV levels with the presence of CHD[195].

The association of genetic Apo AI deficiency is inconsistent. Lack of Apo AI is not associated with increased susceptibility for atherosclerosis in mice[213]. Total lack of Apo AI results in HDL deficiency with plane xanthoma[199] which is associated with corneal clouding but not enhanced risk of CHD. The combined deficiency of Apo AI-CIII-AIV[275], which resembles an interactive coregulated gene cluster, however, is associated with early onset and severe CHD.

Cubilin was recently identified as a high affinity receptor for endocytosis of HDL and lipid-poor Apo AI[140,191,245]. Cubilin may play an important role in the renal clearance of filterable Apo AI/HDL and in the maternal/fetal transport of cholesterol in the placenta. Cubilin is a 460 kD membrane protein expressed in the apical membrane of various absorptive epithelia and was previously known as the receptor for vitamin B_{12} uptake in the intestine[192,246]. The coexpression, coassociation and putative coregulation of cubilin and megalin[140,246,343] has led to the assumption of a functional relationship of the two receptors. Megalin might also be involved in HDL metabolism by direct binding to HDL apolipoproteins and Apo E, Apo J, Apo H (β_2 glycoprotein 1) are megalin ligands[118,247,389] that are associated with various HDL subsets. Cubilin seems to be an important receptor for lipid-poor Apo AI/HDL particle catabolism in the kidney to eliminate either excess HDL precursor molecules or recycled Apo AI particles[245].

Fibrates and estrogens are the only drugs known to increase Apo AI production, although with relatively low efficacy and without proven antiatherogenicity[49,113].

35.3.4.2 *Conversion of HDL precursors to mature HDL*

During conversion of lipid-poor HDL precursors to mature cholesterylester-rich HDL, the initial particles acquire phospholipids and unesterified cholesterol from cells and Apo B-containing lipoproteins (Figure 35.2). The major players in the plasma compartment are lecithin-cholesteryl-acyltransferase (LCAT)[312], CETP[102,353] and PLTP[160] and at the cellular side the ATP cassette transporter ABCA1[317] that facilitates phospholipid/cholesterol co-transport from cholesterol and sphingolipid-enriched raft-domains at the cell membrane to HDL particles that are either rich in Apo AI, Apo AIV or Apo E. These lipoproteins are important sorting signals for later recognition of the mature HDL particles by hepatic receptors (Apo-AI degradation receptor) or in peripheral cells such as adipocytes (Apo E), the blood-brain barrier (Apo AIV) or LPS receptors (CD14) on circulating monocytes/macrophages (LBP).

Human ABCA1 is a full-size ABC transporter and the prototype of ABC proteins of the A subfamily[188,202,317] that is expressed in a variety of organs with highest expression levels in placenta, liver, lung, adrenal glands and macrophages. Genetic deficiency of ABCA1 causes familial HDL deficiency syndromes in man[44,50,205,298,308] and mice[267]. The ABCA1-deficient mice show an impaired absorption of the lipophilic vitamins (A, E, K) indicative for a cotransport-function of ABCA1[267] and also an impaired cholesterol absorption[406]. Intriguingly, analysis of all, as yet, known mutations in ABCA1 revealed that their specific localization appears to determine a tropism of either the RES, as seen in the classical Tangier-disease phenotype (yellow tonsils, splenomegaly), or the vessel wall, as in cases of HDL deficiency associated with atherosclerosis[317]. The available data thus support the notion that ABCA1 functions as a cholesterol-responsive exporter of cholesterol and choline-backbone phospholipids, which serves a gate-keeper function in HDL maturation, acting as a mobile molecule that may shuttle between the plasma membrane and the Golgi complex as a constituent of a vesicular transport route (Figure 35.6). Moreover, ABCA1 is critically involved in the maturation of monocytes to macrophages and the targeting of these cells to either the vessel wall or the RES, thus playing an important role for the macrophage response in CHD.

The promoter of the ABCA1 gene contains binding motifs for several transcription factors (M56) including sterol regulatory binding protein (SREBP), liver-X-receptor/retinoid-X-receptor (LXR/RXR), the repressor zinc finger transcription factor ZNF-202[407] and an E-Box/SP1/SP3 complex[408] that is responsible for cAMP/ATP-dependent upregulation of ABCA1. The promoter responsive elements reflect the functional observations that ABCA1 expression is upregulated by cholesterol, oxysterols and cAMP stimulation[205,267]. Interestingly, the Th-1 response cytokine IFN-γ[271] and systemic inflammatory responses downregulate ABCA1 expression and the formation of cholesterylester-rich mature Apo AI-HDL responsive for reverse cholesterol transport[409]. Instead phospholipid and triglyceride-rich apo E-HDL are formed to deliver phospholipid precursors and fatty acid precursors to peripheral cells to maintain membrane integrity and energy generation. Upregulation of ABCA1 may become an interesting target for antiatherosclerotic drug therapy to increase reverse cholesterol transport as well as inhibition of monocyte/macrophage maturation and extravasation

The role of LCAT in modulating CHD can be derived from genetic models. Patients with classical LCAT deficiency[312] suffer from HDL deficiency, progressive kidney insufficiency and corneal clouding. However, neither classical LCAT deficiency and fish eye disease nor LCAT deficiency in mice, despite almost total HDL deficiency, suffer from premature atherosclerosis[312]. Overexpression of LCAT increased HDL-cholesterol in both mice and rabbits but was anti-atherogenic in rabbits and pro-atherogenic in mice[312] which may be partially explained by a lack of the CETP gene in mice. Thus transgenic coexpression of human CETP and LCAT in mice reduced both HDL cholesterol and atherosclerosis[107]. This is consistent with data derived from family studies with CETP gene mutations and markedly elevated HDL cholesterol[148,149,353,397,402].

Moderate overexpression of PLTP increases HDL cholesterol associated with increased production of pre-β-HDL

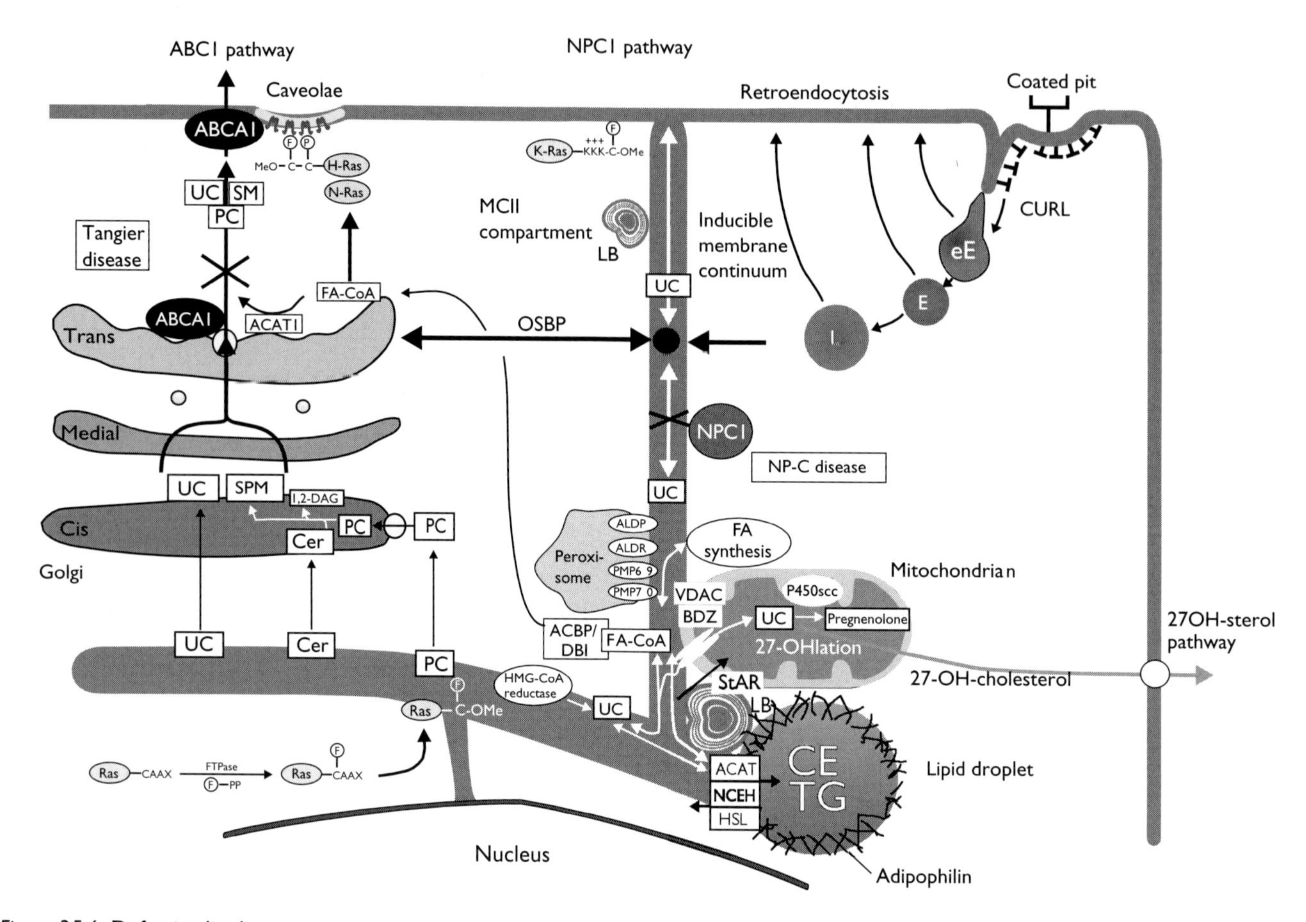

Figure 35.6 Defective lipid transport routes in Niemann-Pick C disease and Tangier disease and their putative compensatory pathways to and from the medial Golgi are indicated (thick arrows).

and increased cholesterol efflux[375] whereas enhanced overexpression of PLTP and PLTP deficiency are associated with impaired HDL cholesterol in mice[170,171,375]. Upregulation of LCAT and PLTP and inhibition of CETP[260] might be interesting targets that have to be evaluated in the foreseeable future.

35.3.4.3 *Catabolism of HDL lipid and protein components*

Mature HDL and their lipid and protein components are removed from the circulation by direct receptor-dependent pathways which include the scavenger receptor BI (SR-BI)[194,371], apo E receptors[80] and may be Apo AI receptors and also by indirect mechanisms that involve the action of CETP[354], hepatic lipase[68,363] and endothelial lipase (Figure 35.2)[293]. The remodelling of HDL_2 into HDL_3 by HL and PLTP, like the conversion of HDL_3 into HDL_2 by PLTP, regenerate apo AI rich, lipid-poor pre β-HDL precursor particles[170,309,354,363,375]. In hepatocytes and steroidogenic tissues, SR-BI binds HDL and mediates selective uptake of cholesteryl esters without internalizing HDL proteins[194,371].

Overexpression of SR-BI in mice decreased HDL cholesterol and atherosclerosis and led to the disappearance of atherogenic lipoproteins[20,190] whereas SR-BI deficiency increases HDL cholesterol and atherosclerosis[370]. SR-BI is down-regulated by cholesterol and estradiol and up-regulated by polyunsaturated fatty acids in hepatocytes and corticotropin and gonadotropins in various steroidogenic tissues[194,371].

SR-BI mediated removal of HDL lipids in the liver may involve both Apo E and HL as ligands or coreceptors[19,200] and familial HL deficiency in men is associated with elevated HDL cholesterol and remnant lipoproteins as well as premature CHD[68,363]. Low HL activity in the general population is also associated with the presence of atherosclerosis. The beneficial role of HL overexpression is still under debate[68,363] and, as yet, it is uncertain whether therapeutic modulation of HL will provide a clinical benefit.

The differential effects of estrogens and testosterone on HL are considered important for the regulation of HDL cholelsterol by these steroid hormones[49,356] but with the controversial data from the HL mice models the relevance of HDL elevating effects by estrogens and the HDL reducing effects by testosterone for CHD prevention remains open.

Endothelial lipase (EL) generates free fatty acids for cellular uptake preferentially from HDL phospholipids and not from triglycerides as shown for HL and LPL[150,168,293]. EL overexpression lowers HDL cholesterol[168]. It has to be proven, however, whether this promotes pro- or anti-atherogenic effects.

Apo E deficient mice develop severe atherosclerosis[80], whereas macrophage-specific overexpression of Apo E prevents atherosclerosis without changes in the plasma concentration of atherogenic remnant lipoproteins. Transplantation of Apo E deficient bone marrow into Apo E wild-type mice leads to the occurence of atherosclerosis in the absence of atherogenic lipoproteins[80,219]. These data indicate that local apo E synthesis in macrophages within the vessel wall may be more important than hepatic Apo E induction which elevates Apo E systematically and promotes hypertriglyceridemia[157].

The anti-atherogenic effects of Apo E may include the stimulation of cholesterol efflux from macrophages[201,217,346] and the inhibition of VCAM-1 expression on endothelial cells as well as SMC proliferation[166].

35.3.4.4 *Efficacy of therapeutic HDL modulation*

Several intervention trials with statins and fibrates indicated that patients with low HDL cholesterol experienced greater benefit from hypolipidemic therapy than individuals with normal or high HDL cholesterol levels[31,125,278]. These results indicate that HDL cholesterol levels may not reflect the underlying effects of these drugs on HDL metabolism which rather modify HDL formation rates or catabolism.

For the fibrates, only the Helsinki Heart Study and the VA-HIT study showed a significant association between modulation of HDL cholesterol and a reduction in CHD incidence[227,307].

Statins increase HDL cholesterol by 5–10% and it appears that those statins which are less efficient in reducing LDL cholesterol are more potent in increasing HDL cholesterol[79,82,174] and atorvastatin was found to increase HDL cholesterol at low dosages but decreased HDL at high dosages.

Pravastatin was found to increase both the synthesis and catabolism of Apo AI[314]. However, since the Apo AI promoter does not contain a sterol regulatory element, statins may have rather complex interactions with HDL metabolism.

SR-BI upregulation by statins may enhance HDL catabolism by the selective uptake of HDL cholesterol by the liver[194,371] which was found to be associated with anti-atherosclerotic effects in two genetic mice models[20,190]. Downregulation of ABCA1[202] and CETP activity[133] upon treatment with statins would also decrease HDL cholesterol.

Fibrates, via their PPARα effects increase HDL precursor formation, maturation and catabolism of HDL and increase HDL cholesterol by approximately 10%. The contradictory data from the various intervention studies, however, may indicate that differences in the etiology of low HDL cholesterol determines the efficacy of fibrate treatments.

Nicotinic acid increases HDL cholesterol by more than 20% due to a decrease in the fractional catabolic rate of Apo AI independent from the selective uptake of HDL cholesterol mediated by SR-BI[378] and long-term treatment reduced death rates from CHD[56].

Probucol is a non-polar antioxidant that substantially lowers plasma cholesterol[235,253]. Besides LDL cholesterol probucol also reduces HDL cholesterol[144,183]. Probucol treatment increases the selective uptake of HDL cholesteryl esters by modifying HDL in a way that causes enhanced interaction with SR-BI[302] which might be partly responsible for the beneficial effects of probucol on atherosclerosis and restenosis[358].

Hormonal replacement therapy with estrogens increases HDL cholesterol by increasing Apo AI synthesis and decreasing HDL catabolism through inhibition of HL and SR-BI[49,106].

In estrogen/progestin combinations, derivatives of 19-nortestosterone (e.g. levonorgestrel) and derivatives of hydroxyprogesterone (e.g. medoxyprogesterone acetate) diminish the HDL-raising effect of estrogens.

In the Heart Estrogen/Progestin Replacement Study (HERS) and the Women's Health Initiative[105,158], therapy with estrogens and medroxyprogesterone acetate did not reduce coronary event rates in post-menopausal women with CHD despite increasing HDL cholesterol by 8% but increased the rate of both coronary and venous thrombosis in the first year of treatment. It appears that the hypertriglyceridemic and procoagulant effects of estrogens offset the potential beneficial effects on HDL cholesterol, Lp(a) and vascular function.

To prove the efficacy of HDL-modifiying drugs in preventing CHD, it will be necessary to include new surrogate markers, other than HDL cholesterol, which better reflect the anti-atherogenic properties of defined lipid and protein components and the role of individual HDL particle populations.

35.3.5 New pharmacologic targets for lipid-lowering therapies

The statins as inhibitors of cholesterol synthesis (Figure 35.3) and isoprenoids have been proven as potent cholesterol-lowering agents and for the reduction of the risk of both cardiovascular and total mortality.

For the reduction of atherogenic triglyceride rich lipoproteins, small dense LDL, Lp(a) and the modulation of the anti-atherogenicity of HDL there is still room for new pharmacologic targets.

35.3.5.1 *Inhibition of bile acid transporters*

From the experience with bile acid resins (e.g. cholestyramine), their poor compliance and their relatively low efficacy in the reduction of total plasma cholesterol due to feedback up-regulation of endogenous cholesterol biosynthesis,[322] two strategies emerged, the development of more specific biopolymer-based resins that bind more bile acid per gram of resin[208] and the development of inhibitors for the apical sodium-dependent bile acid transporter (ASBT) or ileal bile acid transporter (Figure 35.4). The potency of this target is obvious from bile acid metabolism because >95% of the bile acids released into the small intestine are recycled back to the liver, at least 60% of which are selectively reabsorbed by ASBT (Krag) that is highly expressed in the distal half of the ileum and kidney[77]. The topology of the protein includes 8–9 transmembrane segments[139] and ASBT is able to efficiently transport bile acids and Na^+ at a 1:2 stoichiometry into the mucosa cell[381]. These bile acids appear in the hepatic portal circulation where they are re-extracted from the plasma via another transport protein, the Na^+-taurocholate co-transporter peptide[236]. The total bile acid pool flows through this pathway 8–10 times/day yielding a total circulation of 40–50 g bile acids/day.

ASBT (Figure 35.7) offers an attractive target for a new class of cholesterol-lowering drugs that block the transporter but without the need to be absorbed, thus reducing the potential for adverse systemic effects.

In the meantime several compounds listed in Figure 35.8 have been synthesized that significantly reduce plasma cholesterol[383] and Na+ taurocholate transport in brush border membranes of the intestinal mucosa[303].

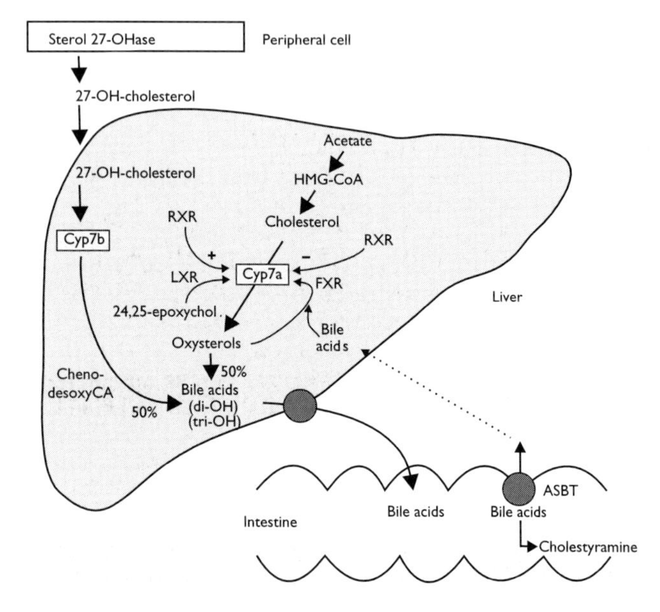

Figure 35.7 Regulation of sterol and bile acid metabolism.

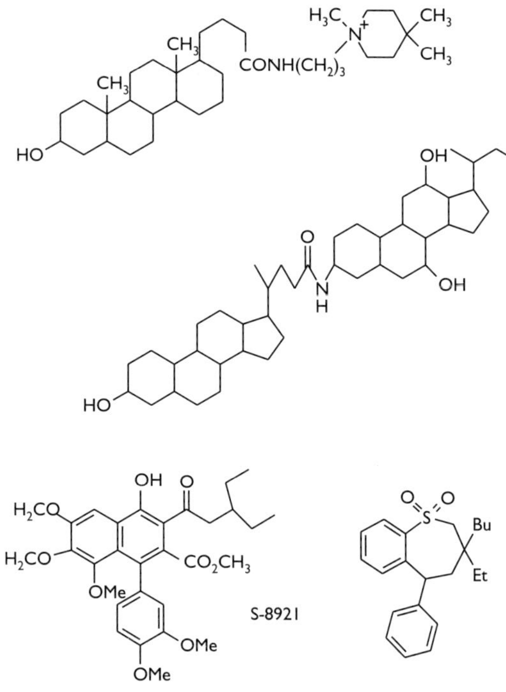

Figure 35.8 Inhibition of the ileal bile acid transporter.

Twice daily administration of 2164U90 to lipid-fed rats caused a 67% reduction in serum VLDL and LDL and a small increase in HDL. S-8921 inhibits both ileal bile acid transport and cholesterol absorption in the intestine[162] and in rats fed a lipid-rich diet this compound decreased elevations in plasma cholesterol by 70% after only 3 days of oral administration. Several heterozygous mutations have been identified in ASBT that render the transporter incapable of efficiently transporting bile acids[392]. Plasma cholesterol levels in the affected patients, however, were found to be similar to normal individuals. This observation indicates that pharmacologic blockade of ASBT must be high since other compensatory bile acid transport mechanisms might exist. Such drugs will significantly increase the luminal concentration of bile acids with expected adverse effects already known, in part, from bile acid resin therapy.

Moreover, bile acids, specifically deoxycholic acid, promote cell proliferation and may have tumor-promoting capacity (e.g. colon cancer) especially in the presence of other carcinogens[249]. This tumor-promoting effect can be reversed by a variety of compounds, especially butyrate, a short chain fatty acid produced by fermentation of fibers in the gut[35] and ursodeoxycholic acid, a secondary bile acid[92].

35.3.5.2 *Inhibition of cholesterol absorption*

Similar to bile acids, blockade of the absorption of dietary cholesterol might have a significant impact on plasma cholesterol levels. The precise mechanism by which cholesterol is absorbed was unclear[84] until recently when the genes for β-sitosterolemia were identified as the two half-size ATP casette transporters ABCG5 and ABCG8 that are located in close proximity and opposite orientation on chromosome 2p21 with a bidirectional promoter and common regulatory elements[37,163,209,281]. In β-sitosterolemia, a rare autosomal recessive disorder, affected patients hyper-absorb not only cholesterol but also a variety of other sterols, including plant and shellfish sterols from the intestine, and the major plant sterol species is sitosterol which provided the name for this disorder. Patients with β-sitosterolemia have very high levels of plant sterols in their plasma and develop tendon and tuberous xanthomas, accelerated atherosclerosis and premature coronary artery disease. These data suggest that ABCG5 and ABCG8 normally cooperate to limit intestinal sterol absorption and to promote biliary excretion of sterols and that mutated forms of these transporters disrupt the mucosal absorption barrier and predispose to sterol accumulation and atherosclerosis. A number of small molecules, typically plant-derived sterols, or analogs thereof, interfere with cholesterol absorption. β-sitosterol and its 5α-saturated form, sitostanol, are poorly absorbed and decrease intestinal cholesterol absorption by approximately 15% which can be increased to 37% by administering sitostanol as a lecithin micelle[146,268]. Sitostanol obviously interferes with the ABCG5 and ABCG8 dependent cholesterol absorption.

By analogy, two plant saponins, tisqueside and pamaquestide, inhibit cholesterol absorption in rabbits by 50% with a higher potency of pamaquestide over tisqueside. Nonsterol molecules such as (3R, 4S)-1,4, bis-(4-methoxyphenyl)-3-(3-phenylpropyl)-2-azetidinone (SCH 48461) potently inhibited cholesterol absorption in hamsters and monkeys, by 68% and 95% respectively. SCH 48461 might act directly on the transporter or may affect transporter expression.

Similarly, recent studies have shown that retinoid acid X receptor (RXR) agonists such as LGN268 are potent inhibitors of cholesterol absorption in the mouse[299]. With the recent identification of ABCG5 and ABCG8, modulation of cholesterol absorption through this target might offer new therapeutic applications. In addition to ABCG5 and ABCG8 the lipoprotein scavenger receptor SR-BI was also presented as a candidate gene involved in mucosal cholesterol transport. SR-BI was found on the apical side in the brush border of mucosa cells along the luminal surface of the small intestine to bind cholesterol containing micelles. Whether inhibition of SR-BI with a non-resorbable compound is a feasible target needs further experimental proof.

Once cholesterol enters the intestinal enterocyte it is transported to the ER lumen or the Trans Golgi Network (TGN) where cholesterol is esterified intraluminally (Figure 35.6) upon packaging into chylomicrons by the enzyme acyl-CoA: cholesterol O-acyltransferase 2 (ACAT 2) which is expressed only in the intestine and the liver[17,57,263]. It appears that this ACAT 2-mediated esterification prevents the back-diffusion of cholesterol rather than participating directly in the absorptive process (Figure 35.9).

In contrast, the second ACAT-1 gene is expressed in many tissues with the highest levels in macrophages, adrenal glands, dermal sebaceous glands and preputial glands[237,244] and is also found in atherosclerotic lesions. Selective ACAT-1 deficiency has been generated in two mouse models of

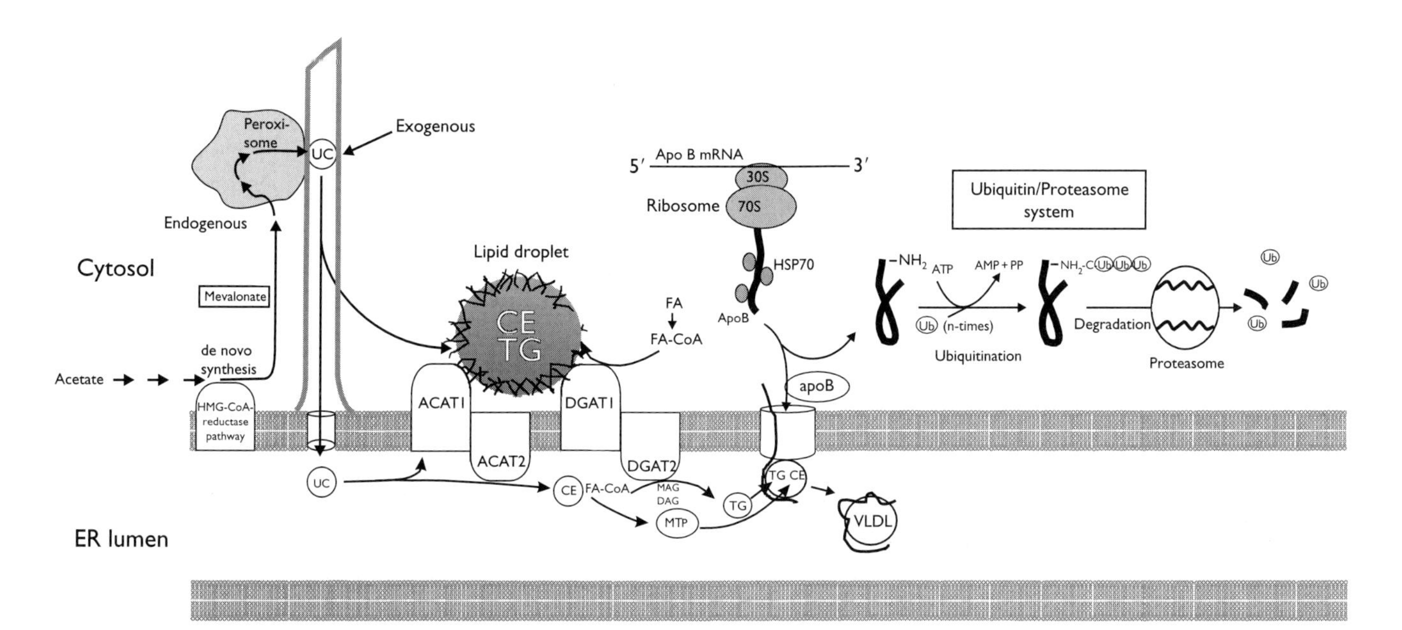

Figure 35.9 Assembly of triglyceride rich lipoproteins.

atherogenesis including Apo E deficient mice and LDL receptor deficient mice[13]. ACAT-1 (−/−) mice have greatly reduced cholesteryl ester levels in the adrenal cortex and peritoneal macrophages but hepatic cholesterol esterification, intestinal cholesterol absorption and plasma cholesterol levels are normal. There is also extensive deposition of unesterified cholesterol in the skin and brain. Bone marrow transplantation experiments revealed that ACAT-1 deficiency in macrophages is sufficient to cause dermal xanthomas in LDL receptor-deficient mice. ACAT-1 deficiency did not prevent the development of atherosclerotic lesions in Apo E deficient and LDL receptor deficient mice, despite causing lower serum cholesterol levels. The lesions in ACAT-1 deficient mice are atypical in composition with reduced amounts of neutral lipids and a paucity of macrophages in advanced lesions.

The marked alteration in cholesterol homeostasis indicates that selectively inhibiting ACAT-1 under conditions of severe hyperlipidemia may have detrimental consequences. Current ACAT inhibitors were developed before the recent findings that two mammalian ACAT genes exist and there are only a few compounds that exerted hypolipidemic effects and reduced atherosclerosis[207,232,257,395]. Inhibition of ACAT by Avasimibe decreases both VLDL and LDL Apo B production in miniature pigs[53]. Cholesterol balance studies indicated that some ACAT-inhibitors may change cholesterol metabolism by reducing lipid secretion and increasing the supply of free cholesterol as a substrate and inducer of cholesterol 7α-hydroxylase resulting in enhanced production of bile acids[283].

However, this research field awaits re-examination of the various ACAT-inhibitors[283] to understand the specificity towards ACAT-2 and the cross-reactivity with ACAT-1. Moreover, an ACAT-2 (−/−) mouse model is urgently required.

Similar to cholesterylester formation by ACAT-1 and ACAT-2, triglyceride synthesis has been assumed to occur primarily through acyl-CoA:diacylglycerolacyltransferase (DGAT), a microsomal enzyme that catalyzes the final step in the glycerol-phosphate pathway[58]. Recently, DGAT deficient mice were presented[333]. These mice are lean and resistant to diet-induced obesity that involves increased energy expenditure and increased activity. DGAT deficiency alters triglyceride metabolism in various tissues including the mammary gland, where lactation is defective. In the duodenum, however, white adipose tissue, as well as other cell types, has residual triglyceride synthesis capacity indicative that either multiple DGATs exist (similar to ACAT) or that alternatively fatty acyl groups may be transferred directly or from diacylglycerol or phospholipid to a second diacylglycerol acceptor through a transacylase activity that does not involve a fatty acyl CoA intermediate[210]. DGAT inhibition might be an interesting target to disrupt triglyceride synthesis in diet-induced obesity and different forms of hypertriglyceridemia.

35.3.5.3 Stimulation of bile acid synthesis and secretion

Quantitatively, the bile acid synthesis pathways are the most important routes for elimination of cholesterol carbon from the body (Figure 35.7)[323].

Bile acids are synthesized by two biochemical pathways. The acidic or classical pathway in the liver starts with the hydroxylation of cholesterol by cholesterol 7α-hydroxylase (Cyp 7a). 7α-OH-cholesterol is then hydroxylated by C27-hydroxylase leading to the 7–27-dihydroxy primary bile acid chenodeoxycholic acid (Figure 35.7).

Alternatively, hydroxylation of 7α-OH-cholesterol at C12 and C27, via a branch point in the pathway, leads to the 7–12–27-trihydroxy bile acid, cholic acid.

A different pathway begins with 27-hydroxylation (sterol 27-hydroxylase) which may occur preferentially in extrahepatic tissues such as spleen, macrophages and endothelial cells[379] but accounts as for as much as 50% of bile acid synthesis. 27-OH cholesterol is taken into the liver and further hydroxylated by oxysterol 7α-hydroxylase (Cyp 7b). Additional biotransformation produces chenodeoxycholic acid. As a drug target, most interest has focused on cholesterol 7α-hydroxylase (Cyp 7a), the rate-limiting enzyme of the classical pathway. Cyp 7a is regulated by numerous hormones and by plasma cholesterol levels and bile acids. Cholesterol-rich diet increases Cyp 7a transcription while bile acids recycled to the liver from the intestine via ASBT are potent inhibitors of Cyp 7a[379]. Increasing Cyp 7a expression might lead to reductions in plasma cholesterol[282,336,337]. Two nuclear receptors play opposing roles in regulating Cyp 7a expression. Liver X receptor (LXR), a member of the steroid receptor superfamily, is highly expressed in the liver. LXR exists in two isoforms and the α-isoform is responsible for regulating Cyp 7a (Figure 35.7)[279].

24,25-epoxy cholesterol and other oxysterols are potent LXRα agonists and in the presence of these ligands LXRα forms a heterodimer with RXR and then binds to a response element in the Cyp 7a promoter, thus increasing transcription of Cyp 7a and enhanced bile acid synthesis and secretion.

The antagonist of this pathway is another member of the nuclear receptor family, the farnesyl X receptor (FXR), which also forms a heterodimer with RXR[401]. The ligand for FXR is a bile acid, most potently chenodeoxycholic acid[272]. Ligand binding to FXR reduces transcriptional activity of the Cyp 7a promoter in concert with other transcription factor interactions. Thus either LXR agonists or FXR antagonists would be expected to increase Cyp 7a expression and enhance bile acid secretion. In the human Cyp 7a promoter sequence, however, LXR binding sites are not present[61] and therefore the rodent data must be interpreted carefully. Moreover, any drug that increases bile acid synthesis and secretion must overcome the recycling transport capacity of

ASBT in the ileum and the effect of chronically elevated bile acid concentration in the distal portions of the intestine needs to be considered (Figure 35.7).

35.3.5.4 Inhibitors of microsomal transfer protein (MTP)

Assembly and secretion of VLDL requires the presence of apolipoprotein B (apoB), phosphatidylcholine, and neutral lipids (Figure 35.9)[311]. Available evidence suggests that the rate of VLDL secretion is not regulated by the rate of apoB production[288]. ApoB is thus constitutively overproduced, and excess apoB that is unable to associate with lipid is degraded in the endoplasmic reticulum (ER)[311]. Consequently, it may be the supply of lipid into the ER that governs hepatic VLDL secretion. The discovery of a neutral lipid transfer activity catalysed by the ER microsomal triglyceride transfer proteine (MTP) in the ER has given an insight into how triacylglycerol (TAG) is incorporated into the nascent VLDL particle. These findings demonstrated that MTP is required for the assembly of VLDL in the liver and chylomicrons in the intestine (Figure 35.9)[300].

The MTP is a heterodimer with a unique structure composed of the multifunctional enzyme, protein disulfide-isomerase (PDI, 58 kDa), and a unique large subunit (88 kDa)[385]. Studies have shown that the large subunit confers lipid transfer activity to the MTP complex[384]. MTP is found as a soluble protein within the lumen of the ER of liver and intestine and is required for the assembly of VLDL and chilomicrons[48]. Mutations in MTP which result in an absence of MTP function have been shown to cause abetalipoproteinemia.

Lipid-binding sudies and kinetic analysis of MTP-catalyzed transport of lipid between membranes indicated that MTP binds and shuttles lipid molecules between membranes[23]. The results indicated that there are two distinct classes of lipid molecule binding sites present on MTP and that the sites differentially bind and transport phospholipid and neutral lipid. One class is responsible for catalyzing lipid transfer and can be occupied by triglycerides, cholesteryl esters, or phospholipids[23]. On the basis of the lipid transport specifity of MTP[386], this sites appear to selectively bind and transport triglycerides and cholesteryl esters compared to phospholipids. Neutral lipid transfer is disproportionally fast when compared to its membrane concentration[23]. A second class of lipid binding sites on MTP appears to be specific for phospholipids. This class transports lipid slowly and is distinct from the sites which rapidly transport lipid molecules between membranes. Although the role of this stable binding site is unclear, there is evidence for similar sites in the literature. For example, phosholipase A_2 contains an activator phospholipid binding site which is necessary for enzyme activity[14]. The differential binding of phospholipids and neutral lipids to MTP not only indicates the presence of more than one class of lipid binding site but also suggests they have distinct functions[23].

Compounds that inhibit MTP's lipid transfer activity (Figure 35.10) have been identified which decrease triglyceride secretion and lower plasma lipid levels even at low levels of LDL receptor expression (Watanabe heritable hyperlipidemic rabbits) in various animal species. In addition to its lipid transfer activity, Apo B and MTP bind with high affinity and these interactions are affected by the length and degree of lipidation of Apo B. Lysine and arginine residues in the N-terminal region of Apo B are critical for its binding to MTP and within this region Apo B and MTP can interact at multiple sites. AG/-S17 has been developed as a novel antagonist that inhibits Apo B-MTP binding without affecting MTP's lipid transfer activity but decreasing efficiently Apo B secretion[30].

Apo B binding may represent MTP's chaperone activity that assists in the transfer from the ER membrane to the lumen and in the net lipidation of nascent Apo B and may be essential for lipoprotein assembly and secretion.

Another MTP-inhibitor (CP-10447) was shown to accumulate newly synthesized Apo B in the translocation channel across the ER exerting a selective and negative effect on the synthesis of Apo B at the stage of elongation. A caveat comes from the observation that the first generation MTP inhibitors may have a risk for the development of fatty liver due to triglyceride storage in lipid droplets[167] similar to the genetic deficiency of MTP as the underlying cause of the human disease abetalipoproteinemia.

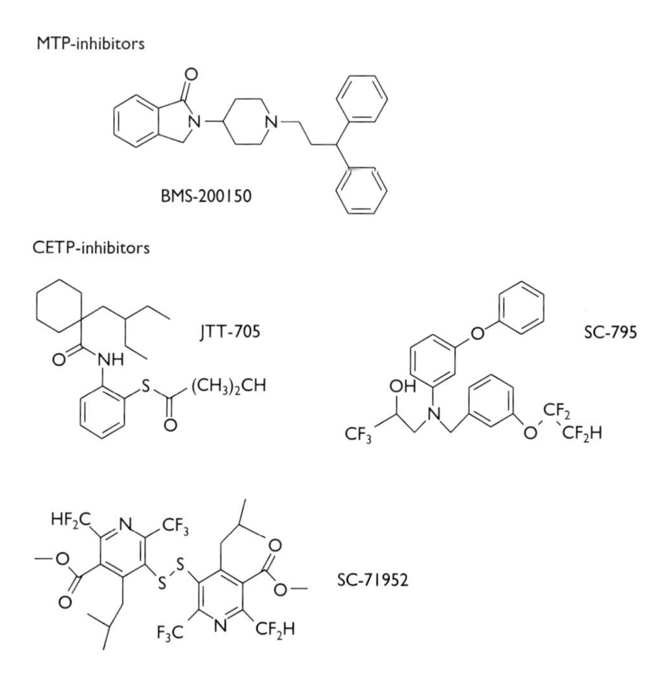

Figure 35.10 Inhibitors of microsomal transfer protein (MTP) and cholesteryl ester transfer protein (CETP).

35.3.5.5 *Inhibitors of cholesteryl ester transfer protein (CETP)*

CETP is structurally and functionally similar to MTP in that it has sequence homologies to MTP and catalyzes the transfer of neutral lipids and phospholipids from HDL to the VLDL/LDL pathway (Figure 35.2). CETP binds cholesteryl esters in a 1 : 1 molar ratio and phospholipids in about a 10 : 1 molar ratio[348]. CETP has a higher affinity for membranes than MTP[23] and, on the basis of its amino acid composition, an overall more hydrophobic character than MTP[89,326], which may explain the lipid binding specificity. CETP is capable to transfer triglycerides, cholesteryl esters, phospholipids, and possibly other hydrophobic compounds found in lipoproteins, however under normal condition in humans, CETP mediates a net movement of cholesteryl esters from HDL to VLDL, in exchange for triglyceride[291].

A number of synthetic CETP inhibitors have been developed in the meantime (Figure 35.10). JTT-705 achieved marked inhibition of CETP in human plasma and caused significant elevation of HDL cholesterol in rabbits[329]. (R)SC-795 is a member of another recently discovered class of CETP inhibitors. (R)SC-795 binds approximately 5000-fold more efficiently to CETP than the natural ligand cholesteryl ester. SC71958, a substituted analog of dithiobisnicotinic acid dimethyl ester, was also identified as a potent inhibitor that, with time, irreversibly inactivates CETP, presumably by reacting with one of the three cysteines of CETP[154].

A peptide inhibitor of CETP was recently detected by screening a bacteriophage display library with the sequence xaa-Arg-Met-Arg-Tyr-xaa that may be useful in studying CETP/lipoprotein interactions[45].

CETP inhibitors may lead to a compensatory up-regulation of the Apo E/HDL pathway with an enhanced formation of particles that are potentially targeted to peripheral tissues such as the adipocyte compartment. A first paper provides evidence, however, that the CETP inhibitor JTT-705 attenuates atherosclerosis in rabbits[264].

35.3.5.6 *Effects of thyromimetics on lipoprotein metabolism*

Although L-triiodothyronine (L-T_3) lowers plasma cholesterol and triglycerides, this hormone is not used as a lipid lowering agent because of its cardiotoxic effects. The cholesterol lowering effect of thyroid hormones and thyromimetics arises from their interactions with the nuclear receptors for L-T_3. The liganded receptors regulate the expression of several hepatic genes involved in cholesterol metabolism leading to increases in expression of LDL-receptors and several lipolytic enzymes.

Novel synthetic thyromimetics (Figure 35.11) are designed specifically to eliminate the cardiac side effects and target the liver as one of the major sites of cholesterol metabolism.

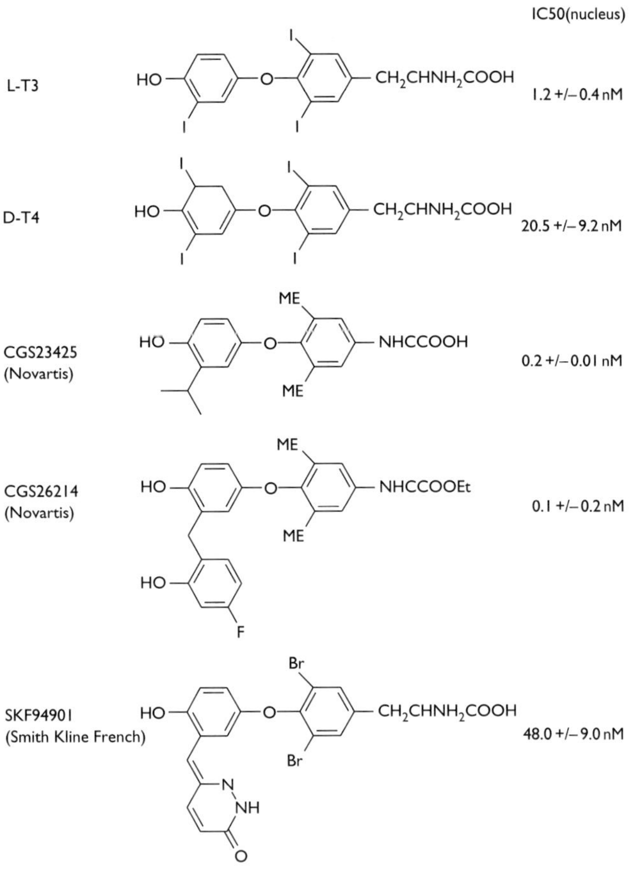

Figure 35.11 Thyromimetics.

Selective actions of thyromimetics on hepatocytes may arise from differences in cytoplasmic binding, active transport at the plasma membrane, the activities of a putative stereospecific cytoplasm to nucleus transport system and differential affinities to the two major isoforms of nuclear T3 receptors in hepatocytes compared to other cell types.

Several novel thyromimetics have been developed that reduce LDL cholesterol and increase plasma Apo AI concentrations by enhancing transcriptional activity of the Apo AI gene[359].

Principally, thyromimetics may be useful for the prevention and reversal of atherosclerosis with the caveat that the adverse cardiotoxic and neurotoxic effects need to be eliminated.

35.3.5.7 *Putative anti-atherosclerotic effects of antipsychotic drugs*

The peripheral benzodiazepine receptor (PBR, syn: mitochondrial benzodiazepine receptor, peripheral benzodiazepine binding or acceptor sites, p-sites, mitochondrial diazepam binding inhibitor (DBI) receptor (Figure 35.6))[273], is present with high density in virtually all peripheral tissues like heart, several glandular and secretory tissues (adrenal and salivary glands, epithelium, testes, ovary, kidney) as well as in brain[85,228,320]. Microradiography and immunohistochemistry showed a high level of PBR expression in astrocytes and microglia after injury indicative that the PBR is involved in the inflammatory and gliosis response following brain injury[196] and a chemotactic influence of the PBR on monocytes was also reported. In blood cells, a low expression was observed in platelets and erythrocytes, a higher one was found on peripheral blood mononucleated cells (lymphocytes $<$ monocytes $=$ polymorphonuclear cells)[55,196,223].

Subcellular studies have shown that the high affinity receptor in hepatocytes was localized predominantly in mitochondria (Figure 35.6), whereas in extra-hepatic cells such as epithelial cells, testes and hematopoietic cells, PBR was localized preferentially in the plasma membrane[393,394] and other cellular organelles such as the nucleus and peroxisomes[142]. The lack of a distinct mitochondrial target sequence and the largely hydrophobic nature of PBR make it feasible that this protein could associate with various cellular membranes. Differential localization of PBRs may also be derived from the existence of chaperone proteins and PBR-associated proteins that may modulate membrane association and function. The PBR gene was located in the 22q13.3 band of the human genome and has five membrane-spanning domains, with amino teminus within the mitochondrial periplasm and the carboxyl terminus externally[116]. The PBR is in a complex with three other components:

- The 18 kDa component on the outer mitochondrial membrane appears to contain the benzodiazepine binding site and is hence often termed the PBR.
- The 35 kDa protein, called porin, or voltage-gated anion channel (VDAC) which is located between outer and inner mitochondrial membrane (junction). This pore exhibits some anion selectivity and voltage-dependence, it seems to allow passage of all molecules whose molecular mass is less than 6 kDa.
- The third part of this complex is the adenosine nucleotide carrier (ADC) located on the inner mitochondrial membrane.

Although the physiological role of the PBR is still unclear, it has been shown to be involved in a number of cellular and biochemical events including hormone-stimulated steroid biosynthesis. Available data suggest that the PBR promotes differentiation and inhibits cell proliferation and it may have a growth controling function. Although the role of cholesterol in tumorigenesis is not clear, it is known that steroids are important factors in human carcinogenesis. Numerous studies implicate a role for cholesterol in the mechanisms underlying cell proliferation, such as the first step of cancer progression. The PBR was found in aggressive metastatic human breast tumor biopsies, primarily in and around the nucleus[142], and a strong correlation between the expression of PBR ligand activity and the invasive and chemotactic potential, as well as expression of the breast tumor marker CD44 was described in this study. Functionally, they found that the PBR is responsible for increased cholesterol transport into the nuclei of the highly metastatic cells. At nanomolar concentration the PBR-ligands, such as PK1195, Ro5-4864, stimulated cell growth.

The accumulation of cholesterol in the nuclear membrane was suggested to inhibit the activation of nuclear membrane sterol response element binding properties (SREBPs). Conversely, it was shown that PBR-ligands reduce the presence of cholesterol in the nuclear membrane in tumor cells[142]. The PBR protective effects relevant for the field of atherosclerosis may be due to cholesterol transport function across cellular membranes into the organelle lumen. The physiological significance of PBR has been linked in part to the cholesterol delivery process. The mitochondrial benzodiazepine receptor appears to be a key factor in the flow of cholesterol into mitochondria to permit the initiation of steroid hormone synthesis. Studies of PBR mRNA expression have demonstrated that the number of mRNA copies per cell correlated with the densities of PBR found in immune cells[55].

The functional importance of diazepam binding inhibitor (DBI) is its participation in promoting stereogenesis via a PBR mediated increase in mitochondrial cholesterol targeting the cytochrome P450 side-chain cleveage enzyme. The second functional impact of this polyprotein on lipid dynamics follows from its identity with the intracellular medium-long chain acyl-CoA-ester-binding protein (ACBP), an acyl-CoA carrier *in vivo* (Figure 35.6)[193]. ACBP can efficiently extract acyl-CoA esters from phospholipid membranes and donate them to suspended mitochondria or microsomes for beta-oxidation. Acyl-CoA is an essential intermediate in the absorption and transport of lipids, because the incorporation of free fatty acids into mono- and diacylglycerols requires their conversion to the corresponding CoA derivates. The diacylglycerol ester is denoted in water at extremely low solubility and the disturbance influences the structure of biomembranes, which could be a possibile explanation for the protective effect of these elements. Furthermore, peptides with sequences identical to DBI have been shown to function as paracrine or autocrine modulators of cell proliferation and cell function. Specifically, high levels of DBI mRNA were observed in liver, adrenal gland, testes, heart and kidney.

DBI participates in the regulation of lipid homeostasis and stimulated under conditions that promote lipogenesis (androgens, insulin), and is directly controlled by SREPs. DBI acts as an endogenous protein ligand, that has the ability to cooperatively displace benzodiazepine analog, such as diazepam, from its allosteric place in gamma-aminobutyric acid type A ($GABA_A$) receptor in the mammalian central neuronal system. This negative allosteric modulator of $GABA_A$ receptor can be co-released with gamma-aminobutyric acid (GABA from primary cultures of cortical neurons after treatment with veratridine) or high potassium-induced depolarization.

Benzodiazepine drugs have been reported to have anti-atherosclerotic effects in rabbits and roosters and to alter the pattern of circulating lipoproteins in humans, acting as inhibitors of cholesterol esterification by acetyl-CoA: cholesterol transferase (ACAT) in atheromatous rabbit aortas, and inhibitors of lecithin: cholesterol acyltransferase (LCAT) in the plasma compartment. Diazepam was shown to be a potent inhibitor of arterial ACAT with 50% inhibition occurring at approximately 0.06 mM concentration and 90% inhibition occurring at about 0.25 mM concentration. The hypocholesterolemic effect may also be due to impaired absorption of cholesterol via inhibition of intestinal ACAT. The possibility that diazepam affects arterial metabolism directly through inhibition of ACAT may, in part, account for the apparent anti-atherosclerotic effect of diazepam in animal models. In hepatocytes, an inhibitory effect of diazepam was observed on sterologenesis. A similar effect was exerted by other membranophilic drugs, such as local anesthetic and antiarrythmic lidocaine and neuroleptic chlorpromazine. In the LRC Program Prevalence Study, benzodiazepine users were found to have significantly higher levels of plasma VLDL-cholesterol and triglyceride. Repeated administration of diazepam is known to accumulate N-desmethyldiazepam as a pharmacological active metabolite that may affect LCAT activity. Table 35.5 summarizes the effects of antipsychotic drugs on lipid and lipoprotein metabolism that may exert anti-atherogenic properties.

35.4 Dysregulation of endothelial cell integrity in atherosclerosis

35.4.1 Endothelial cell activation in atherosclerosis

The endothelium acts as a highly specialized, metabolically active barrier between blood and the underlying tissues, maintaining vascular tone, thrombosis resistance and a selective permeability to cells and proteins (Figure 35.12)[277]. In contrast to the physiological state, endothelial perturbation leads to a preponderance of vasoconstrictory mechanisms and a pro-coagulant and pro-inflammatory character of the endothelial cell layer (Figure 35.12). This altered state, also known as endothelial cell activation, can be triggered and propagated by various atherogenic factors such as smoking, hypercholesterolemia or diabetes, and many of the mechanisms involved share a common intracellular mechanism, which involves activation of the I-κB/NF-κB transcriptional control system.

35.4.1.1 Altered control of vascular tone

Vascular tone is the result of a delicate balance between various stimuli triggering either vasoconstriction or vasodilation

Table 35.5 Effects of antipsychotic drugs on lipid and lipoprotein metabolism

Neuroleptics	*Selected drugs*	*Influence on lipid metabolism*
Phenothiazine	Chlorpramosine, trifluorperazine	Reduction of modified LDL catabolism, inhibition of LDL- and non-LDL-dependent esterification, inhibition of FCH abd CHE accumulation, lowering of HDL-C /HDL-C3/, apoA-I and A-II level, activation of CEH, direct inhibition of ACAT activity, decrease of sphingomyelinase activity, potent inhibition of: calmodulin (calmodulin antagonist), PKC, PLA2, inhibition of Ba^{2+} current permeating through Ca^{2+} channels.
	Levomepromazine	Lowering of HDL-C /HDL-C3/, apoA-I and A-II level.
Dibenzodiazepine	Clozapine	Increase of serum TG level.
Anticonvulsives		
Hydantoin derivates	Phenytoin	Increase of HDL-C, TG, TCH, gamma-glutamyltransferase activity. Decrease level of LDL and apoB.
Dibenzazepine	Carbamazepine	Increase in HDL2 -cholesterol, LDL/females, apoA-I, TCH, phospholipid.
	Oxcarbamazepine	Influence on the voltage-dependent Na^+ and K^+ channels.
Valproic acid	Valproamid	Decrease level of LDL-cholesterol, TCH and apoA-I.
Barbituric acid	Phenobarbital	Increase in TCH/females, HDL2, 3-cholesterol/ HDL3-phospholipid, decrease the plasma concentration of LDL, VLDL, activition of microsomal enzyme system.
Hypnotics		
Benzodiazepine	Diazepam	Lowering TL and TCH, increase basal aldosterone secretion, Inhibitory effect on voltage-dependent Ca^{2+} channel.
	Lorazepam	Reduction of serum TL, TCH, TG, elevation of blood glucose

(Figure 35.12, Figure 35.13). Atherogenic risk factors such as hypercholesterolemia, smoking or diabetes impair the endothelium-dependent vasodilation and chronic endothelial dysfunction is one of the earliest signs of atherosclerotic vessel wall changes, occuring much earlier than obvious alterations of the structural integrity of the endothelial cell layer[178,296]. Loss of the vasodilatatory capacity increases the sensitivity of the vessel to vasoconstrictors such as norepinephrine, which may result in paradoxical coronary vasoconstriction, e.g. upon exercise, and acute episodes of angina associated with daily life activities.

Several substances mediate vasoconstriction (Figures 35.12 and 35.13) and, although most clinical studies in this field have been focused on the modulation of vasoconstrictor mechanisms to treat hypertensive patients, a great deal of data indicates that these mediators may also act as pro-atherogenic factors. Angiotensin II plays a central role in the regulation of systemic arterial pressure through its systemic synthesis via the renin-angiotensin-aldosterone cascade[114]. The major effects are mediated through angiotensin II binding to the high-affinity AT_1-receptor, which is a seven transmembrane domain receptor, coupled to G-proteins and various effector proteins such as adenylate cyclase, phospholipases C, A, and D and also a Ca^{2+} channel. Angiotensin II induces vasoconstriction via its action on vascular smooth muscle cells and may act pro-atherogenically through alteration of the redox-state, and the expression of redox-sensitive genes such as MCP-1 and VCAM will facilitate monocyte adhesion and migration toward the vessel wall[62,180,268]. Antagonism of the binding of angiotensin II to the high-affinity AT_1-receptor can be achieved by inhibition of the angiotensin converting enzyme or direct receptor blockade. Therefore, angiotensin II is an excellent target for pharmacological blockade, as it not only plays a pivotal role in the regulation of blood pressure but also acts as an important modulator of structural and functional vessel wall changes in atherosclerosis[243,365]. Additional binding sites for angiotensin II have been found, revealing completely opposing effects, when compared to the AT_1-receptor. Although only the AT_2-receptor has been characterized in detail, it may be suggested that blockade of the AT_1-receptor could be beneficial, by shifting angiotensin II binding to the AT_2-receptor.

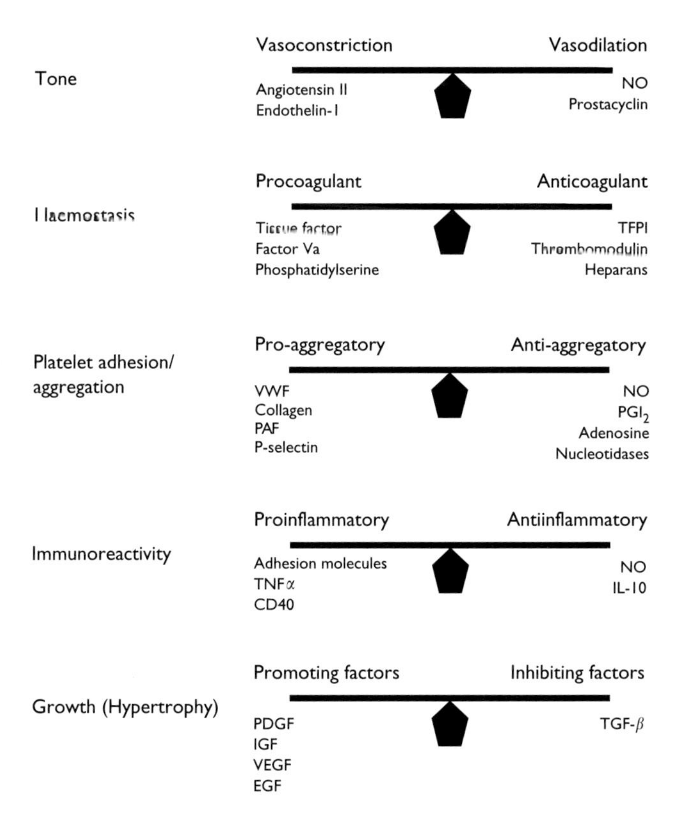

Figure 35.12 Endothelial cell and vascular pathology.

Endothelin-1 acts as a potent vasoconstrictor and pro-mitogenic peptide produced by endothelial cells (Figure 35.12)[243]. Several isoforms (ET-1, ET-2, ET-3, ET-4) and two receptor subtypes (A and B) have been identified: The ET-A receptor, a heptahelical G-protein coupled receptor, mediates smooth muscle cell constriction, whereas the ET-B receptor has partially antagonistic actions, including vasodilation via NO-release from endothelial cells. ET-1 receptors have been found to be upregulated in human atherosclerotic lesions[391]. Plasma ET-1 concentrations are significantly elevated in hypercholesterolemic subjects and LDL enhances ET-1 secretion from endothelial cells *in vitro*[226,155]. Selective ET-A receptor antagonists as well as non-selective ET-A/B antagonists have been developed, with only a few clinical studies completed up to now, focusing on the treatment of congestive heart failure, systemic hypertension and cerebral vasospasm. However, animal studies in apo E-knock out mice suggest that administration of endothelin receptor antagonists results in a significant reduction of atherosclerosis[29,66].

Endothelial cells are also responsible for the regulated synthesis of two potent, short-lived vasodilator substances, prostacyclin and nitric oxide (NO), which both provide accurate control of vascular tone (Figure 35.13)[248,277]. NO, which is a small reactive inorganic molecule, is one of the end-products of a reaction in which L-arginine is metabolized to L-citrulline by the endothelial isoform of nitric oxide synthetase (eNOS)[297]. eNOS is expressed constitutively and its activity is enhanced by an increase in intracellular calcium with shear stress as the primary regulator or cellular stimulation via agonists such as acetylcholine, bradykinin, vasopressin, and adenosin. The underlying mechanisms by which risk factors such as hypercholesterolemia impair endothelium-dependent relaxation still remain to be elucidated in detail. However, considerable evidence now indicates that NO bioactivity is particularly sensitive to oxidative stress[90,179]. Hypercholesterolemia, diabetes and hypertension are known to be associated with an increase in superoxide within the vessel wall, which will result in a reduction of NO bioactivity[126]. Lipid peroxidation products and oxidized LDL (oxLDL) may

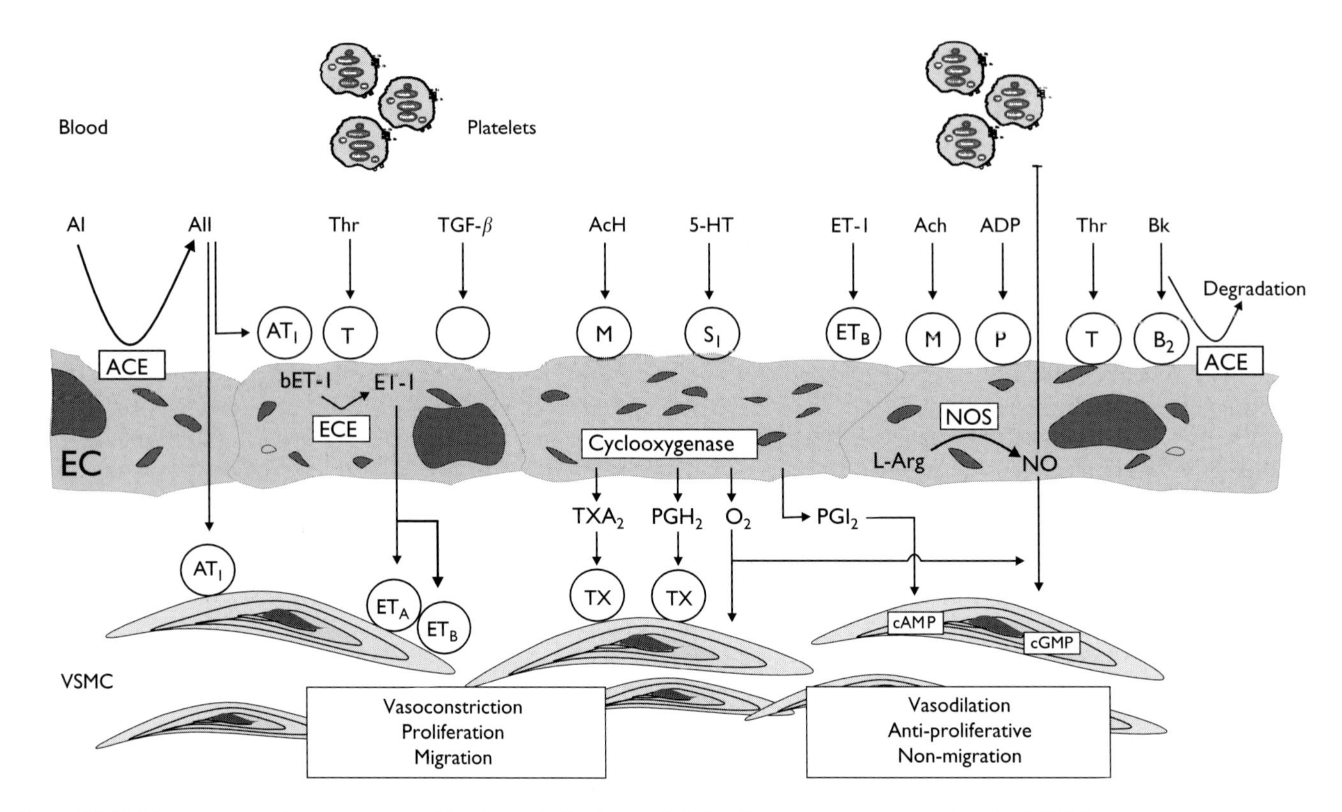

Figure 35.13 Vasoactive mediators generated by the endothelium and their effects on smooth muscle cells (SMCs).

either directly inactivate NO or reduce eNOS in endothelial cells[63,214]. Strategies to reduce plasma lipid and vascular lipid peroxidation, therefore, improve NO bioactivity. Treatment of patients with HMG-CoA-reductase inhibitors lowers plasma lipid levels and restores endothelial cell function, probably also through up-regulation of eNOS synthesis[95]. The water soluble antioxidant vitamin C improves endothelium-dependent vasodilation, whereas the data reported for vitamin E, which is thought to protect LDL from oxidation, are still conflicting[350]. Furthermore, supplementation of the eNOS cofactor tetrahydrobiopterin improves endothelium dependent vasodilation in patients with hypercholesterolemia and the effects of estrogen replacement therapy in post-menopausal women and of angiotensin concerting enzyme inhibitors also seem to be substantially mediated by an increase in vascular NO bioactivity[134,147,186].

35.4.1.2 *Expression of leukocyte adhesion molecules*

Resting endothelial cells in most blood vessels do not constitutively express molecules that can attract leukocytes or platelets from flowing blood[122]. Moreover, the production of NO, adenosine or IL-10 tend to inhibit adhesion and activation of leukocytes[186,332]. However, stimulation of endothelial cells causes a rapid and transient translocation of various adhesion molecules to the cell surface, such as P-selectin, ICAM-1 and VCAM, which trigger leukocyte adhesion and diapedesis[59]. In the context of atherosclerosis, it is important to note that the expression of VCAM is sensitive to oxidant stress, and both hyperlipidemia and hyperglycemia lead to the generation of pro-oxidant molecules that can up-regulate endothelial synthesis of VCAM and the monocyte-selective chemokine MCP-1, providing a plausible link to the early accumulation of monocytes in the vessel wall during development of the early atherosclerotic lesion[87,230,352]. Angiotensin II stimulates VCAM-1 expression via NF-κB activation induced by intracellular oxidative stress, whereas estrogens or PPARα-activators may inhibit VCAM-1 expression via distinct mechanisms, indicating that anti-adhesive drug therapy may complement current therapeutic strategies[231,270,287,331].

35.4.1.3 *Loss of the antithrombotic endothelial surface*

The endothelial cell layer maintains blood flow by balancing hemostatic and fibrinolytic functions of the plasmatic coagulation system via several mechanisms (Figure 35.14). Antithrombin, bound to the endothelial surface via surface glycoaminoglycans, binds thrombin, and the inactivated complex becomes released and rapidly cleared by the liver. Thrombomodulin on the endothelial cell surface binds thrombin and changes its conformation such that it is far more efficient at cleaving and activating circulating protein C, which again inactivates factors Va and VIIIa. Two main regulators of fibrinolysis are continuously released into the blood stream by the endothelial cell, tissue plasminogen activator (tPA) and its inhibitor, plasminogen activator inhibitor-1 (PAI-1), with PAI-1 normally in excess. Whereas the expression of the main initiator of the coagulation system, tissue factor, has not been conclusively demonstrated for endothelial cells *in vivo*, endothelial cells represent a major source for its main inhibitor, the tissue factor pathway inhibitor (TFPI)[128,276].

At sites of activated endothelium, the capacity to control local thrombin generation gets increasingly lost, due to a down-regulation of thrombomodulin expression, an increased activity of PAI-1, and, probably, as suggested from *in vitro* experiments with pro-inflammatory cytokines, the expression of tissue factor (Figure 35.14). This, finally, could result in a substantial production of thrombin, even at sites where integrity of the endothelial cell layer still prevents contact between blood and the subendothelial matrix. The formation of thrombin at the endothelial surface and within the vessel wall is a major prothrombotic and pro-inflammatory event. Thrombin must be regarded as a pleiotropic agent, since, besides cleaving fibrinogen and supporting positive feedback to the formation of prothrombinase, this protease activates membrane receptors on a variety of cells, such as endothelial cells, VSMCs, platelets and monocytes[75].

35.4.2 Platelet activation

Experimental, clinical and epidemiological studies have demonstrated a unique role for platelets in the pathogenesis of atherothrombotic diseases, which have made obvious the need to develope specific and effective drugs for the modulation of platelet function. Platelets circulate in the blood as non-adhesive, non-nucleated, disk shaped cells, which, upon activation, undergo a series of discrete transformations. Platelet activation (Figure 35.14), as a receptor-mediated response of resting platelets to specific stimuli, may be induced by subendothelial matrix proteins (e.g. von Willebrand factor (vWF), collagen) or specific soluble mediators such as thrombin, ADP,

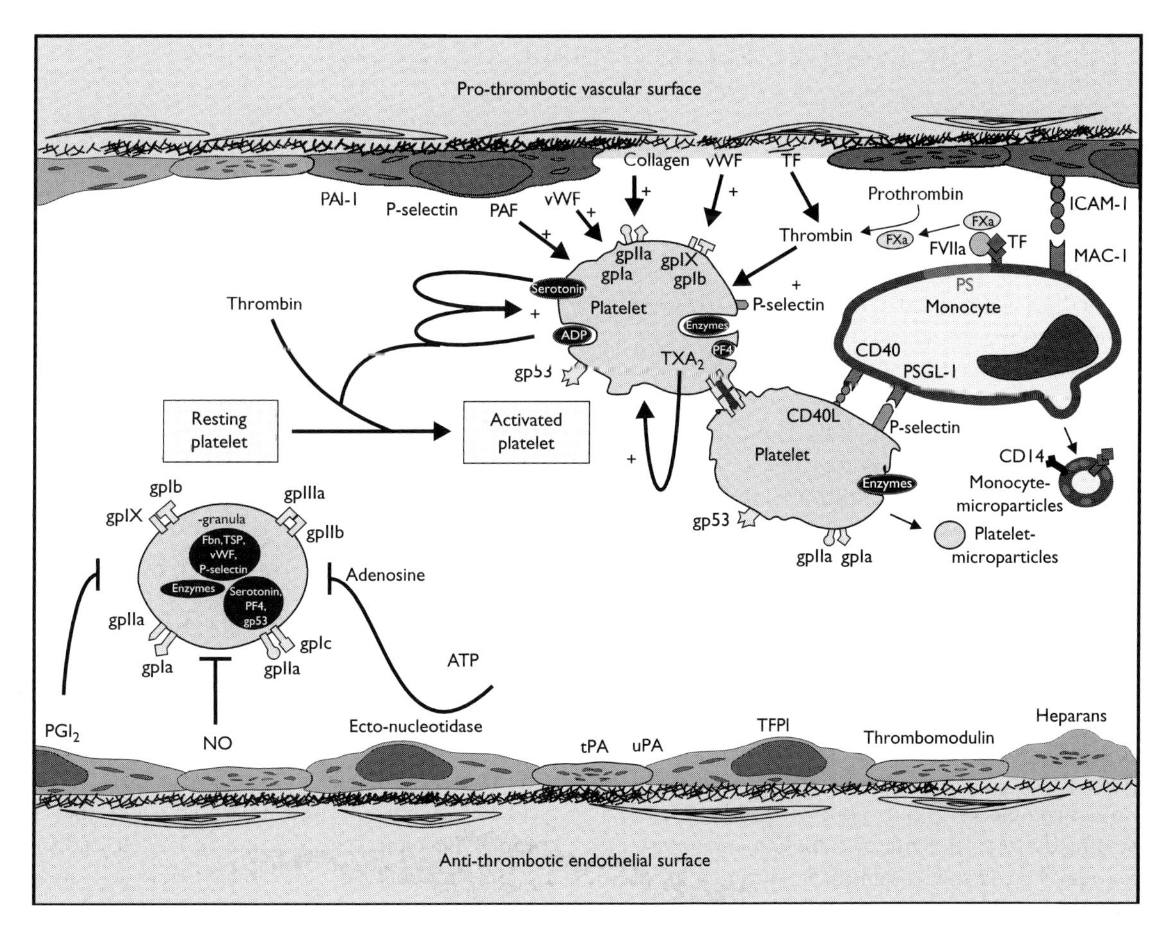

Figure 35.14 Mechanisms of platelet activation (see Colour Plate XXVIII).

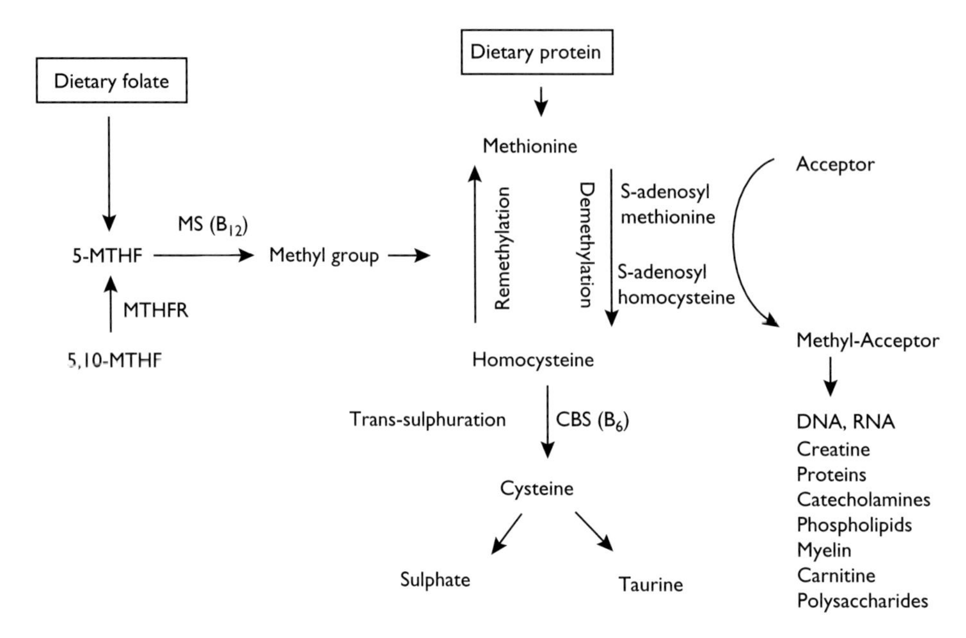

Figure 35.15 Folate, methionine and homocysteine metabolism.

epinephrine or PAF. These receptor-mediated signaling processes trigger the reorganization of the cytoskeleton and platelet shape change, as well as a redistribution of glycoproteins, the expression of a procoagulant surface, and activation of the fibrinogen receptor GPIIb/IIIa[319]. Moreover, a degranulation of platelets results in the release of mediators such as ADP or serotonin, which further stimulate cellular activation, and growth factors such as PDGF, which may contribute to cellular proliferation, e.g. of VSMC. Platelet activation is a prerequisite for platelet aggregation, which finally results in the formation of a hemostatic plug as a natural response to injury. On the other hand, life-threatening occlusion of, e.g. coronary arteries upon plaque rupture causes mortality from atherosclerotic disease. Therefore, certain strategies have been pursued to modulate platelet function (Table 35.6), which involve the inhibition of specific platelet agonists and modulation of the intracellular cAMP/cGMP-system, antagonism of platelet aggregation via inhibition of fibrinogen binding and inhibition of platelet thromboxane A_2 production (Figure 35.14).

35.4.2.1 *Inhibition of specific platelet agonists*

Thrombin, which is the most important soluble platelet agonist, acts through a G-protein coupled receptor on the platelet surface. The thrombin receptor (PAR-1, PAR-4), as a seven transmembrane domain receptor, becomes activated through thrombin mediated cleavage of the extracellular N-terminus and formation of an internal receptor ligand. This activates PLCβ via G_i-proteins, thereby leading to phosphoinositide hydrolysis and formation of DAG and IP_3, which in turn activate PKC and recruit Ca^{2+} from intracellular stores[74,76]. Ca^{2+} plays a central role in all platelet functional responses. Up to now, pharmacological inhibition of thrombin action at the receptor level is not feasible, but systemic inhibition of thrombin action through thrombin inhibitors such as hirudin, bivalrudin or argatroban also reduces platelet activation. In contrast to heparin, direct thrombin inhibitors can inactivate fibrin-bound thrombin, which remains enzymatically active and causes thrombus growth, as well as free thrombin[382]. In the OASIS-2 (Organisation to Assess Strategies for Ischaemic Syndromes-2) trial, hirudin was more effective than heparin in diminishing rates of death, myocardial infarction, and angina at both 72 hours and 7 days after unstable coronary artery disease events[109].

Three types of ADP receptors contribute to cellular signal transduction in platelets. The P2X-receptor is coupled to an ion-channel, mediating a rapid Ca^{2+} influx, whereas the $P2Y_1$-receptor mediates inositol phosphate hydrolysis. The $P2Y_T$-receptor is a G-protein coupled receptor and mediates the inhibition of adenylate cyclase[289]. Ticlopidine and clodipogrel are two ADP-receptor antagonists which cause irreversible inhibition of ADP effects[292]. In the CAPRIE (Clopidogrel versus Aspirin in Patients at Risk of Ischemic Events) study, the annual risk of ischemic stroke, myocardial infarction and death from vascular disease was significantly lower during treatment with clopidogrel 75 mg/day than aspirin 325 mg/day after a mean follow-up of 1.9 years. Clopidogrel provided even greater reductions in the risk of recurrent ischemic events than aspirin in patients with a history of coronary artery bypass surgery, diabetes mellitus and

Table 35.6 Platelet antagonistic drugs

Type of antagonism	*Selected drugs*	*Comments*
Thrombin inhibitors	Hirudin	Systemic inhibition of plasma thrombin activity
ADP-receptor antagonists	Ticlopidin	Irreversible inhibition of ADP-receptor activity
	Clodipogrel	Irreversible inhibition of ADP-receptor activity
Thromboxan receptor antagonists	Daltroban, Vapiprost	Currently of no clinical value
Thromboxan synthetase inhibitors	Dazixiben, Dazmegrel	Currently of no clinical value
Cyclooxygenase inhibitors	Aspirin	Most common antiplatelet drug
Stimulators of cAMP/cGMP	Prostacyclin, Iloprost, PGE_1	Limited clinical value due to systemic side effects
Phosphodiesterase inhibitors	Cilostazol	Inhibits degradation of cAMP
GPIIb/IIIa receptor antagonists	Abciximab	Monoclonal Fab fragment, first clincally relevant GPIIb/IIIa receptor antagonist
	Eptifibatide	Cyclic heptapeptide, RGD-type competitive antagonist
	Tirofiban	Non-peptide GPIIb/IIIa receptor antagonist
Miscellaneous drugs	Molsidomin	Release of NO from endothelial cells

in those receiving concomitant lipid-lowering therapy. Moreover there was a significant reduction in the incidence of hospitalization in patients treated with clopidogrel[5,292].

Drugs which increase the intracellular level of cAMP or cGMP, e.g. through activation of adenylat-/guanylatecyclase or inhibition of phosphodiesterase (PDE), inhibit platelet aggregation and secretion initiated by various agonists[143]. Prostacyclin or stable analog such as PGE_1 or iloprost, represent strong antagonists of platelet function, although side effects on hemodynamics and their pharmacokinetic characteristics limit their clinical use for platelet antagonism[252]. PDE-inhibitors (e.g. Cilostazol) block the degradation of intracellular cAMP or cGMP[164].

35.4.2.2 Inhibition of platelet aggregation

A key event in platelet activation is the conformational activation of the fibrinogen receptor (GPIIb/IIIa). The change of the spatial structure of the receptor complex allows fibrinogen binding to the receptor, which represents the key mechanism of platelet aggregation with surface bound fibrinogen serving as a bridging molecule between platelets. Platelet aggregation at sites of vascular injury represents an effective mechanism to prevent blood loss in cases of vascular damage, but may also be harmful in cases of plaque rupture at sites of atherosclerotic vascular lesions or upon therapeutic vessel wall angioplasty (PTCA)[390]. Nowadays, inhibition of platelet aggregation can be achieved via inhibition of fibrinogen binding to its receptor using antibody fragments targeted to the fibrinogen receptor or other competitive, low-molecular weight receptor antagonists (Table 35.6)[313]. The first receptor antagonist introduced into clinical practice was a Fab fragment of the chimeric 7E3 antibody (Abciximab, ReoPro), which binds with high affinity to the platelet fibrinogen receptor, and also to other receptors such as the vitronectin receptor[70,71]. Low molecular weight antagonists such as the cyclic heptapeptide eptifibatide or the non-peptidic antagonist tirofiban are competitive antagonists with shorter half-life and duration of action and no cross-reactivity toward the fibronectin- or vitronectin-receptor[360]. In the CAPTURE (c7E3 Fab Anti Platelet Therapy in Unstable Refractory Angina) trial, patients with refractory unstable angina were treated with abciximab or placebo, in addition to standard treatment from 16 to 24 hours preceding coronary intervention through 1 hour after intervention. Treatment with abciximab was associated with a reduction of frequent ischemia and a reduction of total ischemic burden in patients with refractory unstable angina[187]. In the PRISM (Platelet Receptor Inhibition in Ischemic Syndrome Management in Patients Limited by Unstable Signs and Symptoms) study, when administered together with heparin and aspirin, the platelet glycoprotein IIb/IIIa receptor inhibitor tirofiban was associated with a lower incidence of ischemic events in patients with acute coronary syndromes than in patients who received only heparin and aspirin[8].

Antagonists of the GPIIb/IIIa receptor mainly prevent platelet aggregation, whereas platelet adhesion to activated or damaged areas of the endothelial cell layer and subsequent degranulation of the platelet is not prevented. Platelet adhesion may be initiated by components of the subendothelial matrix such as collagens or von Willebrand factor (vWF) immobilized to collagens. Whereas the interaction of vWF and its receptor, the GPIb/V/IX complex, mainly mediates platelet adhesion and activation at higher shear rates, collagens promote platelet adhesion only at low shear stress[18]. Several receptors for collagens have been described, which differ with regard to their particular impact on adhesion and signaling[15]. The integrin Ia-IIa ($\alpha_2\beta_1$) has been proposed to mainly mediate ligand binding, whereas the GPVI receptor protein has been demonstrated to mediate collagen induced signal transduction. Thus, inhibition of the integrin Ia-IIa collagen interaction might primarily prevent platelet adhesion, whereas inhibition of GPVI might modulate platelet activation and thereby the recruitment of additional platelets and the size of the platelet plug[65]. As for other non-platelet specific integrins, inhibition of integrin receptors may also affect, e.g., leukocyte function and, therefore, additional effects, either beneficial or harmful, may arise from these drugs (Table 35.6).

35.4.2.3 *Autocrine stimulation in platelets*

Platelet activation results in a number of substances released from intracellular stores or generated by the platelet metabolism, which are able to induce an autocrine reinforcement of the activation process as well as stimulation of endothelial cells or leukocytes. Thromboxan A_2 is, next to ADP, one of the most important release products of the platelets. Upon activation of the platelet, phospholipase A_2 releases arachidonic acid (AA) from phospholipids and free AA will rapidly be further metabolized, in the presence of cylooxygenase activity, toward the platelet agonist thromboxane A_2. Aspirin, which is the most commonly used drug for the prevention, treatment and prophylaxis of ischemic heart disease, blocks cyclooxygenase by irreversible acetylation of the enzyme[274]. Although this inhibition of cyclooxygenase persists for the rest of the platelet's circulating lifespan, aspirin has a limited effect on overall platelet function, causing only a modest inhibition of platelet aggregation and only a slight decline in hemostatic function at a clinical level. Thus, inhibition of the TXA_2 pathway represents a highly selective but limited antithrombotic effect[321].

35.4.3 *Homocysteine*

The observation by McCully in 1969 that vascular pathology of homocysteinemia was common regardless of the source of the defect suggested that homocysteine may be responsible for premature atherosclerosis and venous thromboem-bolism observed in patients with hyperhomocysteinemia[234]. Mechanisms suggested for these effects include endothelial dysfunction, promotion of oxidation of low-density lipoprotein cholesterol, vascular SMC proliferation and coagulation abnormalities[348]. Epidemiologic data have accumulated to suggest that individuals with even moderately elevated levels of homocysteine (e.g. fasting blood levels exceeding approximately 16 μmol/l) may have small to moderate increased risk of CHD[89]. Homocysteine levels are inversely related to both intake and plasma levels of folate (Figure 35.15). Less strong evidence indicates an inverse relationship between folate intake and coronary heart disease risk. It is likely that current estimates of dietary folate requirements are lower than optimal. Folic acid supplementation reliably reduces homocysteine levels, and may also modifiy endothelial function independent from this effect on homocysteine. Such treatment is cheap and appears to be essentially free of risk. However, until present randomized control trials are complete, it will not be known definitively whether or not increasing folate intake reduces cardiovascular risk.

Taken together, these findings have led to a call for screening of homocysteine levels as a regular component of CHD prevention. In contrast to cross-sectional and case-control studies, however, which tend to report strong association between plasma homocysteine and vascular risk, many prospective epidemiologic studies have not consistently supported an association. The few prospective studies that have reported a positive association often included patients with pre-existing vascular disease, suggesting alternative, non-causal explanations for the association[64]. Thus, while homocysteine may represent a causal factor in atherothrombosis, it is also possible that homocysteine is a marker of preclinical disease, or a consequence of other factors more closely linked to risk. Randomized trials are required to test reliably whether lowering homocysteine levels will decrease the risk of atherosclerotic vascular disease. Current guidelines from the American College of Cardiology and the American Heart Association do not support population-based homocysteine screening to determine cardiovascular risk.

Although some controversy remains, the available evidence supporting an association between moderately elevated levels of homocysteine and stroke is fairly strong. Studies are in progress to determine whether treatment with folate in combination with vitamins B_6 and B_{12} will reduce the risk of stroke in patients with increased serum homocysteine[120].

35.5 Monocytes/macrophages and dendritic cells in atherosclerosis

Cholesterol-loaded macrophages are critical components of atherosclerotic lesions and recent *in vivo* studies have demonstrated the importance of macrophages in lesion progression[47,124,131,332]. Vessel wall macrophages and antigen presenting cells (APCs) form different lineages that originate from CD34+ progenitor cells in the bone marrow (Figure 35.16)[318].

In addition to phagocytic macrophages the recent finding that increased numbers of dendritic cells are found in atherosclerotic lesions[42,43] suggests that DCs as the major classes of APCs could play a crucial role in T-cell activation in atherosclerosis. Their current use in cancer immunotherapy, autoimmunity and transplantation combined with cellular vaccination[33,185,335,338,344,364,380] has made it possible to produce *in vitro* DCs with desirable properties therapeutically instrumental in the treatment of atherosclerosis.

In contrast to DCs, phagocytic macrophages express high numbers of scavenger receptors and Fcγ receptors (CD11b), and are CD14/LPS receptor positive. Based on current knowledge, the initially more restricted definition of scavenger

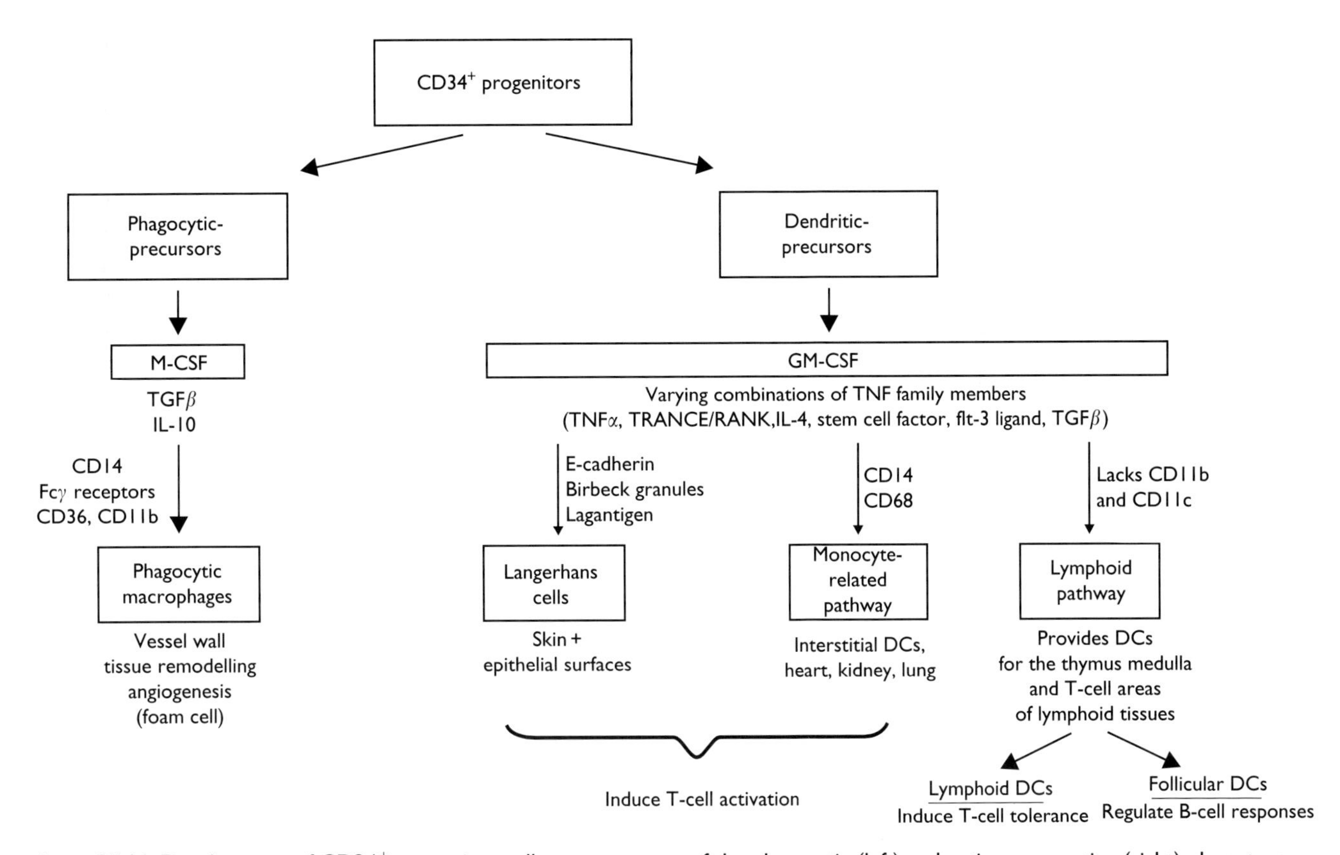

Figure 35.16 Development of CD34$^+$ progenitor cells to monocytes of the phagocytic (left) and antigen presenting (right) phenotype.

receptors (SRs), i.e. as receptors for modified lipoproteins on phagocytic cells, may have to be extended to a more generalized view on receptors involved in scavenger function in host defense, inflammation and tissue remodeling[405]. This notion would include a wide spectrum of pattern recognition receptors that facilitate the clearance of chemically aberrant molecules and opsonized structures (Figure 35.17) and, concomitantly, are involved in the modulation of adhesion, migration, differentiation and also host response processes[318].

Considerable evidence has accumulated demonstrating that integrins are in functional cooperation with these membrane receptors to coordinate monocyte adhesion, migration and differentiation to phagocytically active scavenger cells, which involves integrins as common transducers for a panel of integrin-linked specific receptors. This paradigm includes GPI-linked receptors (uPAR, LPS receptor CD14, FcγRIII/CD16) as well as transmembrane receptors (LRP, SRAs, CD36) and it is likely that future studies will identify additional cooperative interactions[318]. As a first breakthrough, we were able to demonstrate that both LPS and ceramide induced co-association of CD14, complement receptor 3 (CD11b/CD18), CD36 and decay accelerating factor (CD55)[410]. Clustering of signaling competent receptors may therefore provide an interesting mechanism by which different ligands induce distinct cellular processes in sepsis and cardiovascular disease. In particular CD14, which is expressed in elevated density on monocytes during inflammatory processes following myocardial ischemia[238,239], might be an interesting therapeutic target. It is tempting to speculate that disintegrins[285,313] and LPS inhibitors[410] leading to disruption of signaling competent receptor clusters represent promising therapeutic tools in atherosclerosis.

Moreover, lineage and differentiation dependent subsets of circulating blood monocytes and dendritic cell precursors express defined integrin patterns that parallel specific functional properties towards phagocytic capabilities or functions in antigen presentation, processes involved in foam cell formation and inflammatory reaction important for lesion progression in atherosclerosis. In normal subjects, five different subpopulations of blood monocytes were discriminated by cell expression densities of various antigens involved in extravasation, uptake of atherogenic lipoproteins, differentiation and inflammation. Thereby, the Fcγ receptors, in particular CD16a/FcγRIII, together with the LPS receptor CD14, turned out to be key players for defining monocyte subpopulations. Their heterogeneous expression suggested a different capacity of IgG-dependent phagocytosis[305]. The pool size of CD14^{dim}CD16$^+$ monocytes correlates to

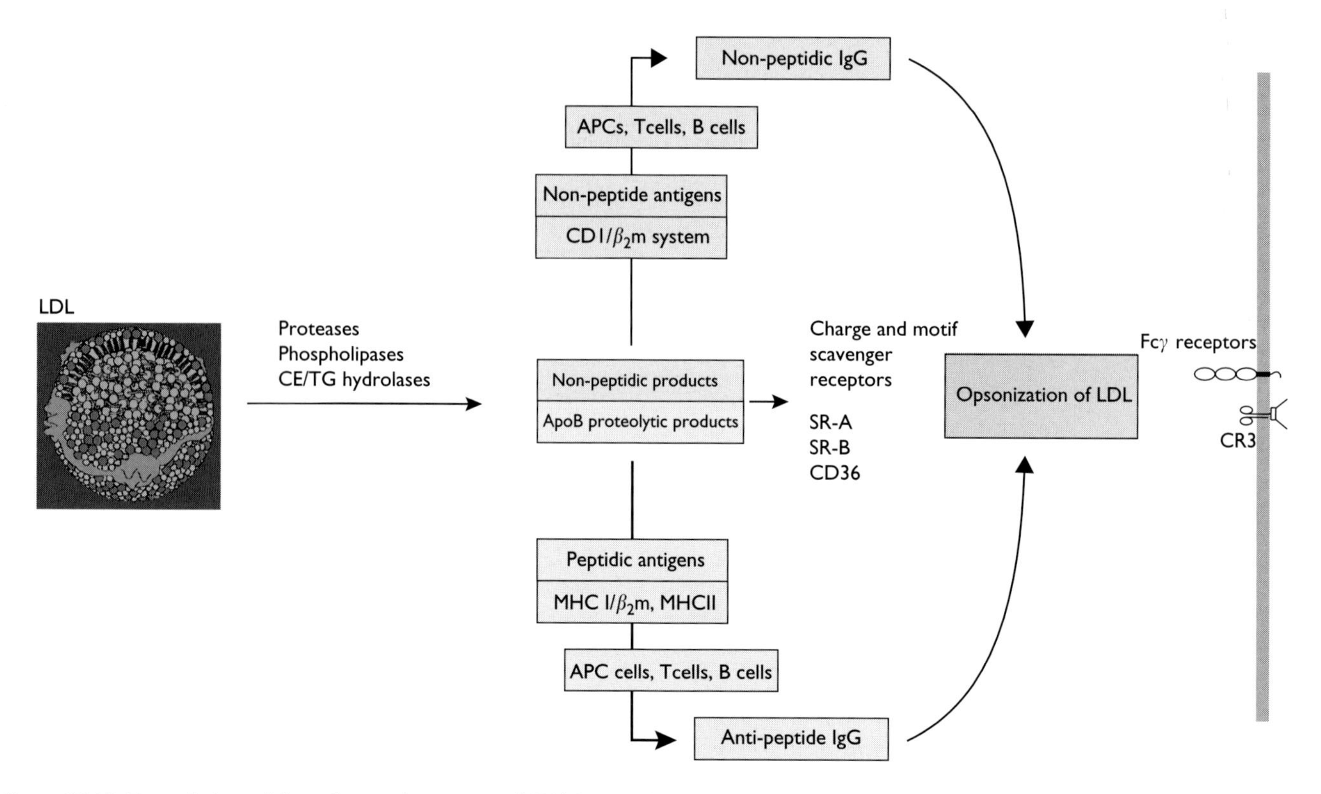

Figure 35.17 Degradation of low density lipoprotein (LDL) by the hydrolytic host defense machinery and cellular uptake of opsonized/non-opsonized modified lipoproteins.

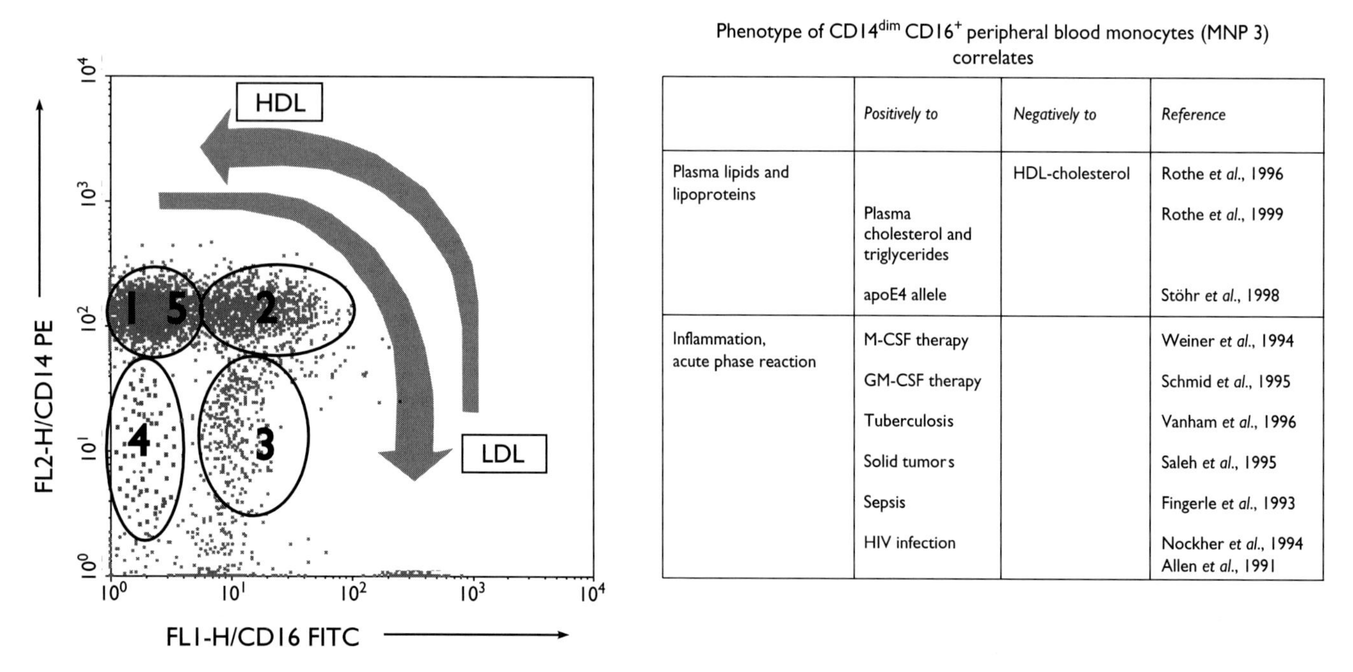

Phenotype of CD14^{dim} CD16^{+} peripheral blood monocytes (MNP 3) correlates

	Positively to	*Negatively to*	*Reference*
Plasma lipids and lipoproteins		HDL-cholesterol	Rothe *et al.*, 1996
	Plasma cholesterol and triglycerides		Rothe *et al.*, 1999
	apoE4 allele		Stöhr *et al.*, 1998
Inflammation, acute phase reaction	M-CSF therapy		Weiner *et al.*, 1994
	GM-CSF therapy		Schmid *et al.*, 1995
	Tuberculosis		Vanham *et al.*, 1996
	Solid tumors		Saleh *et al.*, 1995
	Sepsis		Fingerle *et al.*, 1993
	HIV infection		Nockher *et al.*, 1994 Allen *et al.*, 1991

Figure 35.18 Correlation of the pool size of CD14^{dim}CD16^{+} monocytes (MNP 3) to plasma lipids and lipoprotein metabolism as well as inflammation and the acute phase reaction. Subpopulations of mononuclear phagocytes (MNP 1–5) were discriminated by flow-cytometry (left) due to their expression pattern of CD16a and CD14.

plasma lipids and lipoprotein metabolism as well as inflammation and the acute phase reaction, indicating a link between peripheral blood monocyte heterogeneity and cardiovascular risk factors (Figure 35.18)[306].

The mechanisms discussed here may lead to the identification of new drug targets that affect the functional cooperation of integrins with scavenging and signaling events at the level of circulating monocytes and precursors of dendritic cells[41]. However, because macrophages represent highly heterogeneous cells that exhibit major phenotypic and functional differences and also implicate specific patterns of integrin expression and activity, future research has to consider differences in the functional cooperation of various β-integrins with other membrane proteins in specific subsets of monocytes and macrophages.

35.6 Smooth muscle cells as therapeutic targets

Intimal hyperplasia is the process by which the cell population increases within the innermost layer of the arterial wall. It occurs pathologically in atherosclerosis and after angioplasty. The underlying causes of intimal hyperplasia are migration and proliferation of smooth muscle cells (SMCs) provoked by injury, inflammation, and stretch[256]. Migration and proliferation of SMC are regulated by a diversity of synergistic and opposing growth factors such as platelet derived growth factor (PDGF), epidermal growth factor (EGF), basic fibroblast growth factor (bFGF), insulin-like growth factor (IGF), transforming growth factor-β (TGF-β), and various other cytokines[254].

Among the possible pathophysiological triggers leading to SMC migration and proliferation, activation of the complement system is emerging as a key modulator[38,367] and provides a promising target for therapeutic strategies to prevent atherosclerotic lesion development (see below). Complement activation contributes significantly to atherosclerotic lesion development[38,367]. Cholesterol-induced atherosclerotic lesion formation is reduced in complement-deficient animals[315]. The presence of the macromolecular membrane attack complex (MAC) on cell surfaces of SMCs in early atherosclerotic lesions of human coronary arteries suggests that complement activation occurs even at a very early stage[369]. Induction of complement activation and assembly of C5b-9 may induce cell injury and lysis. However, because of expression of complement regulatory molecules or other defense mechanisms, MAC does not appear to cause substantial lysis of nucleated cells. Instead, sublytic assembly of C5b-9 on SMC induce their activation with release of chemokines[368] and proliferation through complex signal transduction pathways. Recent data show that C5b-9 is mitogenic for human aortic SMC in culture[258].

So far, the underlying upstream molecular mechanisms mediating these effects are unknown. Very recently, it was demonstrated that C5b-9-attack causes activation of PDGF β-R as determined by immunoprecipitation and western blotting. Under identical conditions no such effects of C5b-9 on IGF-1-receptor-activation were observed. Inhibition of PDGF β-R-activation by the selective PDGF-R tyrosine kinase blocker AG1295 reduces C5b-9-induced SMC-proliferation to baseline levels and inhibits C5b-9-induced phosphorylation of PDGF β-R (Torzewski J *et al.*, submitted). These data suggest that C5b-9 may stimulate SMC proliferation during atherogenesis by activating the PDGF β-R. Taken together, the above mentioned data establish a link between the concept of growth factor activation and the novel hypothesis of the role of complement in atherogenesis.

Due to the above mentioned results, PDGF and its receptor seem to be valuable targets for signal transduction therapy in atherosclerosis. Signal transduction therapy to intercept intimal hyperplasia can, in principle, be achieved by novel chemotherapeutic agents.

Another target for signal transduction therapy may be the modulation of protein-tyrosine kinases (PTKs). PTKs regulate cell proliferation, cell differentiation, and signaling processes. Uncontrolled signaling from receptor tyrosine kinases and intracellular tyrosine kinases can lead to inflammatory responses and to diseases such as cancer, atherosclerosis, and psoriasis[212]. PTK can be classified as: (a) receptor tyrosine kinases (RTKs), which receive signals directly through the extracellular domain, which binds a growth factor; and (b) cellular tyrosine kinases (CTKs), which are signal transducers. In the case of RTKs major pathways of signal transduction have been unravelled recently and are therefore targets for signal transduction interception. RTK signaling can, in principle, be inhibited by blocking ligand binding, receptor dimerization, RTK activity or recruitment of signaling molecules. The *in vivo* effect of dominant negative growth factors and their receptors actually validates RTK as targets for drugs. Since RTK must dimerize as an initial step for signal transduction, an inactive receptor, when expressed, can generate inactive dimers thus producing a dominant negative effect[176]. Dominant negative PDGF or PDGF-R inhibits the growth of a tumor which expresses a PDGF/PDGF-R autocrine loop[377]. A truncated receptor can inactivate wild-type receptor function by forming ligand-dependent receptor complexes (probably heterodimers) that are incapable of mediating the early steps of signal transduction like receptor autophosphorylation, association of phosphatidylinositol-3 kinase with the receptor, and calcium mobilization[374].

Inhibition of growth factors by growth factor antagonists is a possible approach to interruption of mitogenic signaling. The validity of this approach is exemplified by suramin. In a mouse model of venous bypass graft atherosclerosis, suramin inhibited SMC migration and proliferation *in vivo* and *in vitro* by blocking PDGF-initiated PDGF receptor and

MAPK-AP-1 signaling[156]. In the rabbit aorta, suramin has an inhibitory effect on neointimal thickening and intimal SMC proliferation after intimal injury, but has no effect on platelet function[21].

Trapidil (triazolopyrimidine), another PDGF antagonist, was useful in preventing intimal proliferation after PTCA, especially in patients with more than 30% residual stenosis after PTCA[265]. Moreover, the Studio Trapidil versus Aspirin nella restenosi Coronarica (STARC) demonstrated that trapidil reduces restenosis after PTCA at a dosage of 100 mg TID and favorably influences the clinical outcome thereafter[229]. Trapidil exerts its antimitogenic effects on SMC by direct activation of protein kinase A (PKA). Thus, PKA-medited inhibition of the Raf-1/MAP kinase pathway may be involved in the antimitogenic actions of the compound[46].

Upon ligand binding, RTKs are activated by autophosphorylation and by actively exposing the kinase site for phosphorylation and activation of downstream signal transducers. Thus, inhibition of receptor autophosphorylation, and inhibition of the ability to phosphorylate exogenous substrate, are valid targets. Tyrphostins, a large family of PTK blockers, block a variety of RTKs. Early *in vitro* studies demonstrated that tyrphostins are potent reversible inhibitors of PDGF-induced mitogenesis in SMCs which act by inhibiting the tyrosine kinase activity of the PDGF-R and the subsequent signaling cascade. In an experimental animal model, local short-term treatment with tyrphostin AG-51 produced a 49% reduction in intimal hyperplasia in experimental vein grafts[161]. In the rat injury model, it was demonstrated that perivascular controlled release delivery of the tyrphostin AG-17 inhibits neointimal formation[121]. Among the tyrphostins, AG1295 has the most striking selectivity for PDGF-R. AG1295 and its derivatives block PDGF-R autophosphorylation with an IC_{50} value of 0.5 μM in intact Swiss 3T3 cells and do not block EGFR even at 50 μM. It exerts a marked inhibitory effect on the activation, migration, and proliferation of porcine and human SMCs *in vitro* and an approximately 50% inhibitory effect on neointimal formation after balloon injury in porcines and rats[32,103].

The mitogen-activated protein kinases (MAPKs) have been shown to play an important role in transducing extracellular signals into cellular responses[83,324]. It is likely that MAPK triggers phosphorylation of transcription factors either directly or by activating other kinases. Kinase activation is followed by an increase in c-fos and c-jun gene expression and enhanced transcription factor AP-1 DNA-binding activity. Current evidence suggests that mammalian cells express at least three groups of MAPKs: extracellular signal-regulating kinases (ERK), $p38^{MAPK}$s, and c-Jun N-terminal kinases[86,141,216]. Recently, it was shown that SMC migration is regulated by activation of p38 (MAPK) resulting in the phosphorylation of heat shock protein 27, which may modulate F-actin polymerization[145]. Acute hypertension and angioplasty rapidly induced MAPK activation in the arterial wall[404]. Inhibition of MAPK activation therefore provides another basis for therapeutic intervention on vascular diseases. Novel insulin sensitizing agents, thiazolidinediones, inhibit tyrosine kinase-dependent growth factor action in SMC, are potent inhibitors of SMC proliferation through MAPK inhibition and down-regulation of c-fos expression and may therefore be a useful agent for prevention of atherosclerosis[184,204,259]. The availability of a safe, oral agent which inhibits the actions of numerous growth factors on SMCs holds promise as an anti-restenotic agent, although larger studies in animal models are required.

Antibodies to growth factors may be another approach to prevention of intimal hyperplasia. The development of an intimal lesion in the carotid artery of athymic nude rats, induced by intra-arterial balloon catheter de-endothelialization was inhibited by a polyclonal antibody to PDGF[100]. In adult rat carotid arteries, systemic injection of a neutralizing antibody against bFGF prior to balloon catheterization significantly decreased the induced SMC proliferation by approximately 80%[218]. However, so far no antibodies are available for clinical use and therefore it is not possible to assess the therapeutic potential of anti-growth factor antibodies.

Gene therapy represents alternative and, most likely, extremely promising approaches[262]. For example, the proliferation of SMCs can be strongly inhibited by antisense oligonucleotides complementary to the mRNA of proliferating cell nuclear antigen[339].

References

1 Anonymous (1984) The Lipid Research Clinics Coronary Primary Prevention Trial results. I. Reduction in incidence of coronary heart disease. *JAMA 251: 351–364.*

2 Anonymous (1994) Collaborative overview of randomised trials of antiplatelet therapy–I: Prevention of death, myocardial infarction, and stroke by prolonged antiplatelet therapy in various categories of patients. Antiplatelet Trialists' Collaboration. *BMJ 308: 81–106.*

3 Anonymous (1994) Randomised trial of cholesterol lowering in 4444 patients with coronary heart disease: the Scandinavian Simvastatin Survival Study (4S). *Lancet 344: 1383–1389.*

4 Anonymous (1995) Cholesterol, diastolic blood pressure, and stroke: 13,000 strokes in 450,000 people in 45 prospective cohorts. Prospective studies collaboration. *Lancet 346: 1647–1653.*

5 Anonymous (1996) A randomised, blinded, trial of clopidogrel versus aspirin in patients at risk of ischaemic events (CAPRIE). CAPRIE Steering Committee. *Lancet 348: 1329–1339.*

6 Anonymous (1996) Comparison of coronary bypass surgery with angioplasty in patients with multivessel disease. The Bypass Angioplasty Revascularization Investigation (BARI) Investigators. *N Engl J Med 335: 217–225.*

7 Anonymous (1997) The effect of aggressive lowering of low-density lipoprotein cholesterol levels and low-dose anticoagulation on obstructive changes in saphenous-vein coronary-artery bypass grafts. The Post Coronary Artery Bypass Graft Trial Investigators. *N Engl J Med 336: 153–162.*

8 Anonymous (1998) Inhibition of the platelet glycoprotein IIb/IIIa receptor with tirofiban in unstable angina and non-Q-wave myocardial infarction. Platelet Receptor Inhibition in Ischemic Syndrome Management in Patients Limited by Unstable Signs and Symptoms (PRISM-PLUS) Study Investigators. *N Engl J Med 338: 1488–1497.*

9 Anonymous (1998) Prevention of cardiovascular events and death with pravastatin in patients with coronary heart disease and a broad range of initial cholesterol levels. The Long-Term Intervention with Pravastatin in Ischaemic Disease (LIPID) Study Group. *N Engl J Med 339: 1349–1357.*

10 Anonymous (1999) MRC/BHF Heart Protection Study of cholesterol-lowering therapy and of antioxidant vitamin supplementation in a wide range of patients at increased risk of coronary heart disease death: early safety and efficacy experience. *Eur Heart J 20: 725–741.*

11 Anonymous (2000) Effects of ramipril on cardiovascular and microvascular outcomes in people with diabetes mellitus: results of the HOPE study and MICRO-HOPE substudy. Heart Outcomes Prevention Evaluation Study Investigators. *Lancet 355: 253–259.*

12 Anonymous (2000) Secondary prevention by raising HDL cholesterol and reducing triglycerides in patients with coronary artery disease: the Bezafibrate Infarction Prevention (BIP) study. *Circulation 102: 21–27.*

13 Accad, M., Smith, S.J., Newland, D.L. *et al.* (2000) Massive xanthomatosis and altered composition of atherosclerotic lesions in hyperlipidemic mice lacking acyl CoA:cholesterol acyltransferase 1. *J Clin Invest 105: 711–719.*

14 Adamich, M., Roberts, M.F., Dennis, E.A. (1979) Phospholipid activation of cobra venom phospholipase A2. 2. Characterization of the phospholipid–enzyme interaction. *Biochemistry 18: 3308–3314.*

15 Alberio, L., Dale, G.L. (1999) Review article: platelet collagen interactions: membrane receptors and intracellular signalling pathways. *Eur J Clin Invest 29: 1066–1076.*

16 Allayee, H., Dominguez, K.M., Aouizerat, B.E. *et al.* (2000) Contribution of the hepatic lipase gene to the atherogenic lipoprotein phenotype in familial combined hyperlipidemia. *J Lipid Res 41: 245–252.*

17 Anderson, R.A., Joyce, C., Davis, M. *et al.* (1998) Identification of a form of acyl-CoA:cholesterol acyltransferase specific to liver and intestine in nonhuman primates. *J Biol Chem 273: 26747–26754.*

18 Andrews, R.K., Lopez, J.A., Berndt, M.C. (1997) Molecular mechanisms of platelet adhesion and activation. *Int J Biochem Cell Biol 29: 91–105.*

19 Arai, T., Rinninger, F., Varban, L. *et al.* (1999) Decreased selective uptake of high density lipoprotein cholesteryl esters in apolipoprotein E knock-out mice. *Proc Natl Acad Sci USA 96: 12050–12055.*

20 Arai, T., Wang, N., Bezouevski, M., Welch, C., Tall, A.R. (1999) Decreased atherosclerosis in heterozygous low density lipoprotein receptor-deficient mice expressing the scavenger receptor BI transgene. *J Biol Chem 274: 2366–2371.*

21 Asada, Y., Tsuneyoshi, A., Marutsuka, K., Sumiyoshi, A. (1994) Suramin inhibits intimal thickening following intimal injury in the rabbit aorta *in vivo*. *Cardiovasc Res 28: 1166–1169.*

22 Ashby, D.T., Rye, K.A., Clay, M.A., Vadas, M.A., Gamble, J.R., Barter, P.J. (1998) Factors influencing the ability of HDL to inhibit expression of vascular cell adhesion molecule-1 in endothelial cells. *Arterioscler Thromb Vasc Biol 18: 1450–1455.*

23 Atzel, A., Wetterau, J.R. (1993) Mechanism of microsomal triglyceride transfer protein catalyzed lipid transport. *Biochemistry 32: 10444–10450.*

24 Austin, M.A. (2000) Triglyceride, Small, Dense Low-density Lipoprotein, and the Atherogenic Lipoprotein Phenotype. *Curr Atheroscler Rep 2: 200–207.*

25 Austin, M.A., King, M.C., Vranizan, K.M., Krauss, R.M. (1990) Atherogenic lipoprotein phenotype. A proposed genetic marker for coronary heart disease risk [see comments]. *Circulation 82: 495–506.*

26 Austin, M.A., Krauss, R.M. (1986) Genetic control of low-density-lipoprotein subclasses. *Lancet 2: 592–595.*

27 Austin, M.A., McKnight, B., Edwards, K.L. *et al.* (2000) Cardiovascular disease mortality in familial forms of hypertriglyceridemia: A 20-year prospective study. *Circulation 101: 2777–2782.*

28 Aviram, M. (1999) Does paraoxonase play a role in susceptibility to cardiovascular disease? *Mol Med Today 5: 381–386.*

29 Babaei, S., Picard, P., Ravandi, A. *et al.* (2000) Blockade of endothelin receptors markedly reduces atherosclerosis in LDL receptor deficient mice: role of endothelin in macrophage foam cell formation. *Cardiovasc Res 48: 158–167.*

30 Bakillah, A., Nayak, N., Saxena, U., Medford, R.M., Hussain, M.M. (2000) Decreased secretion of ApoB follows inhibition of ApoB-MTP binding by a novel antagonist. *Biochemistry 39: 4892–4899.*

31 Ballantyne, C.M., Herd, J.A., Ferlic, L.L. *et al.* (1999) Influence of low HDL on progression of coronary artery disease and response to fluvastatin therapy. *Circulation 99: 736–743.*

32 Banai, S., Wolf, Y., Golomb, G. *et al.* (1998) PDGF-receptor tyrosine kinase blocker AG1295 selectively attenuates smooth muscle cell growth *in vitro* and reduces neointimal formation after balloon angioplasty in swine. *Circulation 97: 1960–1969.*

33 Banchereau, J., Steinman, R.M. (1998) Dendritic cells and the control of immunity. *Nature 392: 245–252.*

34 Barrett-Connor, E.L., Cohn, B.A., Wingard, D.L., Edelstein, S.L. (1991) Why is diabetes mellitus a stronger risk factor for fatal ischemic heart disease in women than in men? The Rancho Bernardo Study. *JAMA 265: 627–631.*

35 Bartram, H.P., Scheppach, W., Englert, S. *et al.* (1995) Effects of deoxycholic acid and butyrate on mucosal prostaglandin E2 release and cell proliferation in the human sigmoid colon. *JPEN J Parenter Enteral Nutr 19: 182–186.*

36 Benoit, P., Emmanuel, F., Caillaud, J.M. *et al.* (1999) Somatic gene transfer of human ApoA-I inhibits atherosclerosis progression in mouse models. *Circulation 99: 105–110.*

37 Berge, K.E., Tian, H., Graf, G.A. *et al.* (2000) Accumulation of dietary cholesterol in sitosterolemia caused by mutations in adjacent ABC transporters. *Science 290: 1771–1775.*

38 Bhakdi, S. (1998) Complement and atherogenesis: the unknown connection. *Ann Med 30: 503–507.*

39 Bierman, E.L. (1992) George Lyman Duff Memorial Lecture. Atherogenesis in diabetes. *Arterioscler Thromb 12: 647–656.*

40 Blum, C.B., Levy, R.I., Eisenberg, S., Hall, M., Goebel, R.H., Berman, M. (1977) High density lipoprotein metabolism in man. *J Clin Invest 60: 795–807.*

41 Bobryshev, Y.V. (2000) Dendritic cells and their involvement in atherosclerosis. *Curr Opin Lipidol 11: 511–517.*

42 Bobryshev, Y.V., Lord, R.S. (1995) Ultrastructural recognition of cells with dendritic cell morphology in human aortic intima. Contacting interactions of Vascular Dendritic Cells in athero-resistant and athero-prone areas of the normal aorta. *Arch Histol Cytol 58: 307–322.*

43 Bobryshev, Y.V., Lord, R.S. (1998) Mapping of vascular dendritic cells in atherosclerotic arteries suggests their involvement in local immune-inflammatory reactions. *Cardiovasc Res 37: 799–810.*

44 Bodzioch, M., Orso, E., Klucken, J. *et al.* (1999) The gene encoding ATP-binding cassette transporter 1 is mutated in Tangier disease. *Nat Genet 22: 347–351.*

45 Bonin, P.D., Bannow, C.A., Smith, C.W., Fischer, H.D., Erickson, L.A. (1998) A peptide inhibitor of cholesteryl ester transfer protein identified by screening a bacteriophage display library. *J Pept Res 51: 216–225.*

46 Bonisch, D., Weber, A.A., Wittpoth, M., Osinski, M., Schror, K. (1998) Antimitogenic effects of trapidil in coronary artery smooth muscle cells by direct activation of protein kinase A. *Mol Pharmacol 54: 241–248.*

47 Boring, L., Gosling, J., Cleary, M., Charo, I.F. (1998) Decreased lesion formation in CCR2-/- mice reveals a role for chemokines in the initiation of atherosclerosis. *Nature 394: 894–897.*

48 Brett, D.J., Pease, R.J., Scott, J., Gibbons, G.F. (1995) Microsomal triglyceride transfer protein activity remains unchanged in rat livers under conditions of altered very-low-density lipoprotein secretion. *Biochem J 310 (Pt 1): 11–14.*

49 Brinton, E.A. (1996) Oral estrogen replacement therapy in postmenopausal women selectively raises levels and production rates of lipoprotein A-I and lowers hepatic lipase activity without lowering the fractional catabolic rate. *Arterioscler Thromb Vasc Biol 16: 431–440.*

50 Brooks-Wilson, A., Marcil, M., Clee, S.M. *et al.* (1999) Mutations in ABC1 in Tangier disease and familial high-density lipoprotein deficiency. *Nat Genet 22: 336–345.*

51 Brunzell, J.D., Schrott, H.G., Motulsky, A.G., Bierman, E.L. (1976) Myocardial infarction in the familial forms of hypertriglyceridemia. *Metabolism 25: 313–320.*

52 Bucher, H.C., Griffith, L.E., Guyatt, G.H. (1999) Systematic review on the risk and benefit of different cholesterol-lowering interventions. *Arterioscler Thromb Vasc Biol 19: 187–195.*

53 Burnett, J.R., Wilcox, L.J., Telford, D.E. *et al.* (1999) Inhibitionof ACAT by avasimibe decreases both VLDL and LDL apolipoprotein B production in miniature pigs. *J Lipid Res 40: 1317–1327.*

54 Calabresi, L., Donati, D., Pazzucconi, F., Sirtori, C.R., Franceschini, G. (2000) Omacor in familial combined hyperlipidemia: effects on lipids and low density lipoprotein subclasses. *Atherosclerosis 148: 387–396.*

55 Canat, X., Carayon, P., Bouaboula, M. *et al.* (1993) Distribution profile and properties of peripheral-type benzodiazepine receptors on human hemopoietic cells. *Life Sci 52: 107–118.*

56 Canner, P.L., Berge, K.G., Wenger, N.K. *et al.* (1986) Fifteen year mortality in Coronary Drug Project patients: long-term benefit with niacin. *J Am Coll Cardiol 8: 1245–1255.*

57 Cases, S., Novak, S., Zheng, Y.W. *et al.* (1998) ACAT-2, a second mammalian acyl-CoA:cholesterol acyltransferase. Its cloning, expression, and characterization. *J Biol Chem 273: 26755–26764.*

58 Cases, S., Smith, S.J., Zheng, Y.W. *et al.* (1998) Identification of a gene encoding an acyl CoA:diacylglycerol acyltransferase, a key enzyme in triacylglycerol synthesis. *Proc Natl Acad Sci USA 95: 13018–13023.*

59 Celi, A., Lorenzet, R., Furie, B., Furie, B.C. (1997) Platelet-leukocyte-endothelial cell interaction on the blood vessel wall. *Semin Hematol 34: 327–335.*

60 Chancharme, L., Therond, P., Nigon, F., Lepage, S., Couturier, M., Chapman, M.J. (1999) Cholesteryl ester hydroperoxide lability is a key feature of the oxidative susceptibility of small, dense LDL. *Arterioscler Thromb Vasc Biol 19: 810–820.*

61 Chen, J., Cooper, A.D., Levy-Wilson, B. (1999) Hepatocyte nuclear factor 1 binds to and transactivates the human but not the rat CYP7A1 promoter. *Biochem Biophys Res Commun 260: 829–834.*

62 Chen, X.L., Tummala, P.E., Olbrych, M.T., Alexander, R.W., Medford, R.M. (1998) Angiotensin II induces monocyte chemoattractant protein-1 gene expression in rat vascular smooth muscle cells. *Circ Res 83: 952–959.*

63 Chin, J.H., Azhar, S., Hoffman, B.B. (1992) Inactivation of endothelial derived relaxing factor by oxidized lipoproteins. *J Clin Invest 89: 10–18.*

64 Christen, W.G., Ridker, P.M. (2000) Blood Levels of Homocysteine and Atherosclerotic Vascular Disease. *Curr Atheroscler Rep 2: 194–199.*

65 Clemetson, K.J. (1999) Platelet collagen receptors: a new target for inhibition? *Haemostasis 29: 16–26.*

66 Clozel, M. (2000) Endothelin receptor antagonists: current status and perspectives. *J Cardiovasc Pharmacol 35: S65–S68.*

67 Cockerill, G.W., Saklatvala, J., Ridley, S.H. *et al.* (1999) High-density lipoproteins differentially modulate cytokine-induced expression of E-selectin and cyclooxygenase-2. *Arterioscler Thromb Vasc Biol 19: 910–917.*

68 Cohen, J.C., Vega, G.L., Grundy, S.M. (1999) Hepatic lipase: new insights from genetic and metabolic studies. *Curr Opin Lipidol 10: 259–267.*

69 Cohen, R.D., Castellani, L.W., Qiao, J.H., Van Lenten, B.J., Lusis, A.J., Reue, K. (1997) Reduced aortic lesions and elevated high density lipoprotein levels in transgenic mice overexpressing mouse apolipoprotein A-IV. *J Clin Invest 99: 1906–1916.*

70 Coller, B.S. (1999) Binding of abciximab to alpha V beta 3 and activated alpha M beta 2 receptors: with a review of platelet-leukocyte interactions. *Thromb Haemost 82: 326–336.*

71 Coller, B.S., Anderson, K., Weisman, H.F. (1995) New antiplatelet agents: platelet GPIIb/IIIa antagonists. *Thromb Haemost 74: 302–308.*

72 Corsini, A., Bellosta, S., Baetta, R., Fumagalli, R., Paoletti, R., Bernini, F. (1999) New insights into the pharmacodynamic and pharmacokinetic properties of statins. *Pharmacol Ther 84: 413–428.*

73 Corsini, A., Bernini, F., Quarato, P. *et al.* (1996) Non-lipid-related effects of 3-hydroxy-3-methylglutaryl coenzyme A reductase inhibitors. *Cardiology 87: 458–468.*

74 Coughlin, S.R. (1999) How the protease thrombin talks to cells. *Proc Natl Acad Sci USA 96: 11023–11027.*

75 Coughlin, S.R. (2000): Thrombin signalling and protease-activated receptors. *Nature 407: 258–264.*

76 Coughlin, S.R. (2000) Thrombin signalling and protease-activated receptors. *Nature 407: 258–264.*

77 Craddock, A.L., Love, M.W., Daniel, R.W. *et al.* (1998) Expression and transport properties of the human ileal and renal sodium-dependent bile acid transporter. *Am J Physiol 274: G157–G169.*

78 Crouse, J.R., Byington, R.P., Hoen, H.M., Furberg, C.D. (1997) Reductase inhibitor monotherapy and stroke prevention. *Arch Intern Med 157: 1305–1310.*

79 Crouse, J.R., Frohlich, J., Ose, L., Mercuri, M., Tobert, J.A. (1999) Effects of high doses of simvastatin and atorvastatin on high-density lipoprotein cholesterol and apolipoprotein A-I. *Am J Cardiol 83: 1476–7, A7.*

80 Curtiss, L.K., Boisvert, W.A. (2000) Apolipoprotein E and atherosclerosis. *Curr Opin Lipidol 11: 243–251.*

81 Dansky, H.M., Charlton, S.A., Harper, M.M., Smith, J.D. (1997) T and B lymphocytes play a minor role in atherosclerotic plaque formation in the apolipoprotein E-deficient mouse. *Proc Natl Acad Sci USA 94: 4642–4646.*

82 Davignon, J., Hanefeld, M., Nakaya, N., Hunninghake, D.B., Insull, W.J., Ose, L. (1998) Clinical efficacy and safety of cerivastatin: summary of pivotal phase IIb/III studies. *Am J Cardiol 82: 32J–39J.*

83 Davis, R.J. (1995) Transcriptional regulation by MAP kinases. *Mol Reprod Dev 42: 459–467.*

84 Dawson, P.A., Rudel, L.L. (1999) Intestinal cholesterol absorption. *Curr Opin Lipidol 10: 315–320.*

85 De Souza, E.B., Anholt, R.R., Murphy, K.M., Snyder, S.H., Kuhar, M.J. (1985) Peripheral-type benzodiazepine receptors in endocrine organs: autoradiographic localization in rat pituitary, adrenal, and testis. *Endocrinology 116: 567–573.*

86 Derijard, B., Raingeaud, J., Barrett, T. *et al.* (1995) Independent human MAP-kinase signal transduction pathways defined by MEK and MKK isoforms. *Science 267: 682–685.*

87 Doi, H., Kugiyama, K., Oka, H. *et al.* (2000) Remnant lipoproteins induce proatherothrombogenic molecules in endothelial cells through a redox-sensitive mechanism. *Circulation 102: 670–676.*

88 Downs, J.R., Clearfield, M., Weis, S. *et al.* (1998): Primary prevention of acute coronary events with lovastatin in men and women with average cholesterol levels: results of AFCAPS/TexCAPS. Air Force/Texas Coronary Atherosclerosis Prevention Study. *JAMA 279: 1615–1622.*

89 Drayna, D., Jarnagin, A.S., McLean, J. *et al.* (1987) Cloning and sequencing of human cholesteryl ester transfer protein cDNA. *Nature 327: 632–634.*

90 Drexler, H. (1999) Nitric oxide and coronary endothelial dysfunction in humans. *Cardiovasc Res 43: 572–579.*

91 Duverger, N., Tremp, G., Caillaud, J.M. *et al.* (1996) Protection against atherogenesis in mice mediated by human apolipoprotein A-IV. *Science 273: 966–968.*

92 Earnest, D.L., Holubec, H., Wali, R.K. *et al.* (1994) Chemoprevention of azoxymethane-induced colonic carcinogenesis by supplemental dietary ursodeoxycholic acid. *Cancer Res 54: 5071–5074.*

93 Elkeles, R.S., Diamond, J.R., Poulter, C. *et al.* (1998) Cardiovascular outcomes in type 2 diabetes. A double-blind placebo-controlled study of bezafibrate: the St. Mary's, Ealing, Northwick Park Diabetes Cardiovascular Disease Prevention (SENDCAP) Study. *Diabetes Care 21: 641–648.*

94 Emeson, E.E., Shen, M.L., Bell, C.G., Qureshi, A. (1996) Inhibition of atherosclerosis in CD4 T-cell-ablated and nude (nu/nu) C57BL/6 hyperlipidemic mice. *Am J Pathol 149: 675–685.*

95 Endres, M., Laufs, U., Huang, Z. *et al.* (1998) Stroke protection by 3-hydroxy-3-methylglutaryl (HMG)-CoA reductase inhibitors mediated by endothelial nitric oxide synthase. *Proc Natl Acad Sci USA 95: 8880–8885.*

96 Eriksson, M., Carlson, L.A., Miettinen, T.A., Angelin, B. (1999) Stimulation of fecal steroid excretion after infusion of recombinant proapolipoprotein A-I. Potential reverse cholesterol transport in humans. *Circulation 100: 594–598.*

97 Escola-Gil, J.C., Marzal-Casacuberta, A., Julve-Gil, J. *et al.* (1998) Human apolipoprotein A-II is a pro-atherogenic molecule when it is expressed in transgenic mice at a level similar to that in humans: evidence of a potentially relevant species-specific interaction with diet. *J Lipid Res 39: 457–462.*

98 Essig, M., Nguyen, G., Prie, D., Escoubet, B., Sraer, J.D., Friedlander, G. (1998) 3-Hydroxy-3-methylglutaryl coenzyme A reductase inhibitors increase fibrinolytic activity in rat aortic endothelial cells. Role of geranylgeranylation and Rho proteins. *Circ Res 83: 683–690.*

99 Farnier, M., Davignon, J. (1998) Current and future treatment of hyperlipidemia: the role of statins. *Am J Cardiol 82: 3J–10J.*

100 Ferns, G.A., Raines, E.W., Sprugel, K.H., Motani, A.S., Reidy, M.A., Ross, R. (1991) Inhibition of neointimal smooth muscle accumulation after angioplasty by an antibody to PDGF. *Science 253: 1129–1132.*

101 Festa, A., D'Agostino, R.J., Mykkanen, L., Tracy, R., Howard, B.V., Haffner, S.M. (1999) Low-density lipoprotein particle size is inversely related to plasminogen activator inhibitor-1 levels. The Insulin Resistance Atherosclerosis Study. *Arterioscler Thromb Vasc Biol 19: 605–610.*

102 Fielding, C.J., Havel, R.J. (1996) Cholesteryl ester transfer protein: friend or foe? *J Clin Invest 97: 2687–2688.*

103 Fishbein, I., Waltenberger, J., Banai, S. *et al.* (2000) Local delivery of platelet-derived growth factor receptor-specific tyrphostin inhibits neointimal formation in rats. *Arterioscler Thromb Vasc Biol 20: 667–676.*

104 Fisher, J.E., Rogers, M.J., Halasy, J.M. *et al.* (1999) Alendronate mechanism of action: geranylgeraniol, an intermediate in the mevalonate pathway, prevents inhibition of osteoclast formation, bone resorption, and kinase activation *in vitro. Proc Natl Acad Sci USA 96: 133–138.*

105 Fitzpatrick, L.A., Litin, S.C., Bell, M.R. (2000) The Women's Health Initiative: a heart-to-HRT conversation. *Mayo Clin Proc 75: 559–561.*

106 Fluiter, K., van der Westhuijzen, D.R., van Berkel, T.J. (1998) *In vivo* regulation of scavenger receptor BI and the selective uptake of high density lipoprotein cholesteryl esters in rat liver parenchymal and Kupffer cells. *J Biol Chem 273: 8434–8438.*

107 Foger, B., Chase, M., Amar, M.J. *et al.* (1999) Cholesteryl ester transfer protein corrects dysfunctional high density lipoproteins and reduces aortic atherosclerosis in lecithin cholesterol acyltransferase transgenic mice. *J Biol Chem 274: 36912–36920.*

108 Fournier, N., Atger, V., Paul, J.L. *et al.* (2000) Human ApoA-IV overexpression in transgenic mice induces cAMP-stimulated cholesterol efflux from J774 macrophages to whole serum. *Arterioscler Thromb Vasc Biol 20: 1283–1292.*

109 Fox, K.A. (1999) Implications of the Organization to Assess Strategies for Ischemic Syndromes-2 (OASIS-2) study and the results in the context of other trials. *Am J Cardiol 84: 26M–31M.*

110 Frick, M.H., Syvanne, M., Nieminen, M.S. *et al.* (1997) Prevention of the angiographic progression of coronary and vein-graft atherosclerosis by gemfibrozil after coronary bypass surgery in men with low levels of HDL cholesterol. Lopid Coronary Angiography Trial (LOCAT) Study Group. *Circulation 96: 2137–2143.*

111 Friedlander, Y., Kark, J.D., Sinnreich, R., Edwards, K.L., Austin, M.A. (1999) Inheritance of LDL peak particle diameter: results from a segregation analysis in Israeli families. *Genet Epidemiol 16: 382–396.*

112 Frostegard, J., Ulfgren, A.K., Nyberg, P. *et al.* (1999) Cytokine expression in advanced human atherosclerotic plaques: dominance of pro-inflammatory (Th1) and macrophage-stimulating cytokines. *Atherosclerosis 145: 33–43.*

113 Fruchart, J.C., Duriez, P., Staels, B. (1999) Peroxisome proliferator-activated receptor-alpha activators regulate genes governing lipoprotein metabolism, vascular inflammation and atherosclerosis. *Curr Opin Lipidol 10: 245–257.*

114 Garbers, D.L., Dubois, S.K. (1999) The molecular basis of hypertension. *Annu Rev Biochem 68: 127–155.*

115 Garner, B., Waldeck, A.R., Witting, P.K., Rye, K.A., Stocker, R. (1998) Oxidation of high density lipoproteins. II. Evidence for direct reduction of lipid hydroperoxides by methionine residues of apolipoproteins AI and AII. *J Biol Chem 273: 6088–6095.*

116 Gavish, M., Bachman, I., Shoukrun, R. *et al.* (1999) Enigma of the peripheral benzodiazepine receptor. *Pharmacol Rev 51: 629–650.*

117 Genest, J.J., Marcil, M., Denis, M., Yu, L. (1999) High density lipoproteins in health and in disease. *J Investig Med 47: 31–42.*

118 Gliemann, J. (1998) Receptors of the low density lipoprotein (LDL) receptor family in man. Multiple functions of the large family members via interaction with complex ligands. *Biol Chem 379: 951–964.*

119 Goldberg, R.B., Mellies, M.J., Sacks, F.M. *et al.* (1998) Cardiovascular events and their reduction with pravastatin in diabetic and glucose-intolerant myocardial infarction survivors with average cholesterol levels: subgroup analyses in the cholesterol and recurrent events (CARE) trial. The Care Investigators. *Circulation 98: 2513–2519.*

120 Goldstein, L.B. (2000) Novel Risk Factors for Stroke: Homocysteine, Inflammation, and Infection. *Curr Atheroscler Rep 2: 110–114.*

121 Golomb, G., Fishbein, I., Banai, S. *et al.* (1996) Controlled delivery of a tyrphostin inhibits intimal hyperplasia in a rat carotid artery injury model. *Atherosclerosis 125: 171–182.*

122 Gonzalez-Amaro, R., Sanchez-Madrid, F. (1999) Cell adhesion molecules: selectins and integrins. *Crit Rev Immunol 19: 389–429.*

123 Gordon, D.J., Rifkind, B.M. (1989) High-density lipoprotein–the clinical implications of recent studies. *N Engl J Med 321: 1311–1316.*

124 Gosling, J., Slaymaker, S., Gu, L. *et al.* (1999) MCP-1 deficiency reduces susceptibility to atherosclerosis in mice that overexpress human apolipoprotein B. *J Clin Invest 103: 773–778.*

125 Gotto, A.M.J., Whitney, E., Stein, E.A. *et al.* (2000) Relation between baseline and on-treatment lipid parameters and first acute major coronary events in the Air Force/Texas Coronary Atherosclerosis Prevention Study (AFCAPS/TexCAPS). *Circulation 101: 477–484.*

126 Griendling, K.K., Alexander, R.W. (1997) Oxidative stress and cardiovascular disease. *Circulation 96: 3264–3265.*

127 Groop, P.H., Elliott, T., Ekstrand, A. *et al.* (1996) Multiple lipoprotein abnormalities in type I diabetic patients with renal disease. *Diabetes 45: 974–979.*

128 Gross, P.L., Aird, W.C. (2000) The endothelium and thrombosis. *Semin Thromb Hemost 26: 463–478.*

129 Grundy, S.M., Benjamin, I.J., Burke, G.L. *et al.* (1999) Diabetes and cardiovascular disease: a statement for healthcare professionals from the American Heart Association. *Circulation 100: 1134–1146.*

130 Grundy, S.M., Mok, H.Y., Zech, L., Berman, M. (1981) Influence of nicotinic acid on metabolism of cholesterol and triglycerides in man. *J Lipid Res 22: 24–36.*

131 Gu, L., Okada, Y., Clinton, S.K. *et al.* (1998) Absence of monocyte chemoattractant protein-1 reduces atherosclerosis in low density lipoprotein receptor-deficient mice. *Mol Cell 2: 275–281.*

132 Guerci, B., Antebi, H., Meyer, L. *et al.* (1999) Increased ability of LDL from normolipidemic type 2 diabetic women to generate peroxides. *Clin Chem 45: 1439–1448.*

133 Guerin, M., Lassel, T.S., Le Goff, W., Farnier, M., Chapman, M.J. (2000) Action of atorvastatin in combined hyperlipidemia: preferential reduction of cholesteryl ester transfer from HDL to VLDL1 particles. *Arterioscler Thromb Vasc Biol 20: 189–197.*

134 Guetta, V., Quyyumi, A.A., Prasad, A., Panza, J.A., Waclawiw, M., Cannon, R.O. (1997) The role of nitric oxide in coronary vascular effects of estrogen in postmenopausal women. *Circulation 96: 2795–2801.*

135 Guijarro, C., Blanco-Colio, L.M., Ortego, M. *et al.* (1998) 3-Hydroxy-3-methylglutaryl coenzyme a reductase and isoprenylation inhibitors induce apoptosis of vascular smooth muscle cells in culture. *Circ Res 83: 490–500.*

136 Haffner, S.M. (1999) Patients with type 2 diabetes: the case for primary prevention. *Am J Med 107: 43S–45S.*

137 Haffner, S.M., Alexander, C.M., Cook, T.J. *et al.* (1999) Reduced coronary events in simvastatin-treated patients with

coronary heart disease and diabetes or impaired fasting glucose levels: subgroup analyses in the Scandinavian Simvastatin Survival Study. *Arch Intern Med 159: 2661–2667.*

138 Haffner, S.M., Lehto, S., Ronnemaa, T., Pyorala, K., Laakso, M. (1998) Mortality from coronary heart disease in subjects with type 2 diabetes and in nondiabetic subjects with and without prior myocardial infarction. *N Engl J Med 339: 229–234.*

139 Hallen, S., Branden, M., Dawson, P.A., Sachs, G. (1999) Membrane insertion scanning of the human ileal sodium/bile acid co-transporter. *Biochemistry 38: 11379–11388.*

140 Hammad, S.M., Stefansson, S., Twal, W.O. *et al.* (1999) Cubilin, the endocytic receptor for intrinsic factor-vitamin B(12) complex, mediates high-density lipoprotein holoparticle endocytosis. *Proc Natl Acad Sci USA 96: 10158–10163.*

141 Han, J., Lee, J.D., Bibbs, L., Ulevitch, R.J. (1994) A MAP kinase targeted by endotoxin and hyperosmolarity in mammalian cells. *Science 265: 808–811.*

142 Hardwick, M., Fertikh, D., Culty, M., Li, H., Vidic, B., Papadopoulos, V. (1999) Peripheral-type benzodiazepine receptor (PBR) in human breast cancer: correlation of breast cancer cell aggressive phenotype with PBR expression, nuclear localization, and PBR-mediated cell proliferation and nuclear transport of cholesterol. *Cancer Res 59: 831–842.*

143 Haslam, R.J., Dickinson, N.T., Jang, E.K. (1999) Cyclic nucleotides and phosphodiesterases in platelets. *Thromb Haemost 82: 412–423.*

144 Hayek, T., Chajek-Shaul, T., Walsh, A., Azrolan, N., Breslow, J.L. (1991) Probucol decreases apolipoprotein A-I transport rate and increases high density lipoprotein cholesteryl ester fractional catabolic rate in control and human apolipoprotein A-I transgenic mice. *Arterioscler Thromb 11: 1295–1302.*

145 Hedges, J.C., Dechert, M.A., Yamboliev, I.A. *et al.* (1999) A role for p38(MAPK)/HSP27 pathway in smooth muscle cell migration. *J Biol Chem 274: 24211–24219.*

146 Heinemann, T., Axtmann, G., von Bergmann, K. (1993) Comparison of intestinal absorption of cholesterol with different plant sterols in man. *Eur J Clin Invest 23: 827–831.*

147 Heitzer, T., Brockhoff, C., Mayer, B. *et al.* (2000) Tetrahydrobiopterin improves endothelium-dependent vasodilation in chronic smokers: evidence for a dysfunctional nitric oxide synthase. *Circ Res 86: E36–E41.*

148 Hirano, K., Yamashita, S., Kuga, Y. *et al.* (1995) Atherosclerotic disease in marked hyperalphalipoproteinemia. Combined reduction of cholesteryl ester transfer protein and hepatic triglyceride lipase. *Arterioscler Thromb Vasc Biol 15: 1849–1856.*

149 Hirano, K., Yamashita, S., Nakajima, N. *et al.* (1997) Genetic cholesteryl ester transfer protein deficiency is extremely frequent in the Omagari area of Japan. Marked hyperalphalipoproteinemia caused by CETP gene mutation is not associated with longevity. *Arterioscler Thromb Vasc Biol 17: 1053–1059.*

150 Hirata, K., Dichek, H.L., Cioffi, J.A. *et al.* (1999) Cloning of a unique lipase from endothelial cells extends the lipase gene family. *J Biol Chem 274: 14170–14175.*

151 Hobbs, H.H., White, A.L. (1999) Lipoprotein(a): intrigues and insights. *Curr Opin Lipidol 10: 225–236.*

152 Hokanson, J.E., Austin, M.A. (1996) Plasma triglyceride level is a risk factor for cardiovascular disease independent of high-density lipoprotein cholesterol level: a meta-analysis of population-based prospective studies. *J Cardiovasc Risk 3: 213–219.*

153 Hokanson, J.E., Brunzell, J.D., Jarvik, G.P., Wijsman, E.M., Austin, M.A. (1999) Linkage of low-density lipoprotein size to the lipoprotein lipase gene in heterozygous lipoprotein lipase deficiency. *Am J Hum Genet 64: 608–618.*

154 Hope, H.R., Heuvelman, D., Duffin, K. *et al.* (2000) Inhibition of cholesteryl ester transfer protein by substituted dithiobisnicotinic acid dimethyl ester. Involvement Of a critical cysteine [In Process Citation]. *J Lipid Res 41: 1604–1614.*

155 Horio, T., Kohno, M., Yasunari, K. *et al.* (1993) Stimulation of endothelin-1 release by low density and very low density lipoproteins in cultured human endothelial cells. *Atherosclerosis 101: 185–190.*

156 Hu, Y., Zou, Y., Dietrich, H., Wick, G., Xu, Q. (1999) Inhibition of neointima hyperplasia of mouse vein grafts by locally applied suramin. *Circulation 100: 861–868.*

157 Huang, Y., Liu, X.Q., Rall, S.C.J. *et al.* (1998) Overexpression and accumulation of apolipoprotein E as a cause of hypertriglyceridemia. *J Biol Chem 273: 26388–26393.*

158 Hulley, S., Grady, D., Bush, T. *et al.* (1998) Randomized trial of estrogen plus progestin for secondary prevention of coronary heart disease in postmenopausal women. Heart and Estrogen/progestin Replacement Study (HERS) Research Group. *JAMA 280: 605–613.*

159 Hulley, S.B., Rosenman, R.H., Bawol, R.D., Brand, R.J. (1980) Epidemiology as a guide to clinical decisions. The association between triglyceride and coronary heart disease. *N Engl J Med 302: 1383–1389.*

160 Huuskonen, J., Ehnholm, C. (2000) Phospholipid transfer protein in lipid metabolism. *Curr Opin Lipidol 11: 285–289.*

161 Huynh, T.T., Davies, M.G., Barber, L., Svendsen, E., Hagen, P.O. (1998) Local inhibition of tyrosine kinase activity markedly attenuates the development of intimal hyperplasia in experimental vein grafts. *J Surg Res 77: 104–111.*

162 Ichihashi, T., Izawa, M., Miyata, K., Mizui, T., Hirano, K., Takagishi, Y. (1998) Mechanism of hypocholesterolemic action of S-8921 in rats: S-8921 inhibits ileal bile acid absorption. *J Pharmacol Exp Ther 284: 43–50.*

163 Ikeda, S., Mochizuki, A., Sarker, A.H., Seki, S. (2000) Identification of functional elements in the bidirectional promoter of the mouse Nthl1 and Tsc2 genes. *Biochem Biophys Res Commun 273: 1063–1068.*

164 Ikeda, Y. (1999) Antiplatelet therapy using cilostazol, a specific PDE3 inhibitor. *Thromb Haemost 82: 435–438.*

165 Illingworth, D.R. (2000) Management of hypercholesterolemia. *Med Clin North Am 84: 23–42.*

166 Ishigami, M., Swertfeger, D.K., Hui, M.S., Granholm, N.A., Hui, D.Y. (2000) Apolipoprotein E inhibition of vascular smooth muscle cell proliferation but not the inhibition of migration is mediated through activation of inducible nitric oxide synthase. *Arterioscler Thromb Vasc Biol 20: 1020–1026.*

167 Jamil, H., Gordon, D.A., Eustice, D.C. *et al.* (1996) An inhibitor of the microsomal triglyceride transfer protein inhibits apoB secretion from HepG2 cells. *Proc Natl Acad Sci USA 93: 11991–11995.*

168 Jaye, M., Lynch, K.J., Krawiec, J. *et al.* (1999) A novel endothelial-derived lipase that modulates HDL metabolism. *Nat Genet 21: 424–428.*

169 Jeppesen, J., Hein, H.O., Suadicani, P., Gyntelberg, F. (1998) Triglyceride concentration and ischemic heart disease: an eight-year follow-up in the Copenhagen Male Study. *Circulation 97: 1029–1036.*

170 Jiang, X., Francone, O.L., Bruce, C., *et al.* (1996) Increased prebeta-high density lipoprotein, apolipoprotein AI, and phospholipid in mice expressing the human phospholipid transfer protein and human apolipoprotein AI transgenes. *J Clin Invest 98: 2373–2380.*

171 Jiang, X.C., Bruce, C., Mar, J., *et al.* (1999) Targeted mutation of plasma phospholipid transfer protein gene markedly reduces high-density lipoprotein levels. *J Clin Invest 103: 907–914.*

172 Jin, F.Y., Kamanna, V.S., Kashyap, M.L. (1997) Niacin decreases removal of high-density lipoprotein apolipoprotein A-I but not cholesterol ester by Hep G2 cells. Implication for reverse cholesterol transport. *Arterioscler Thromb Vasc Biol 17: 2020–2028.*

173 Jin, F.Y., Kamanna, V.S., Kashyap, M.L. (1999) Niacin accelerates intracellular ApoB degradation by inhibiting triacylglycerol synthesis in human hepatoblastoma (HepG2) cells. *Arterioscler Thromb Vasc Biol 19: 1051–1059.*

174 Jones, P., Kafonek, S., Laurora, I., Hunninghake, D. (1998) Comparative dose efficacy study of atorvastatin versus simvastatin, pravastatin, lovastatin, and fluvastatin in patients with hypercholesterolemia (the CURVES study). *Am J Cardiol 81: 582–587.*

175 Kalopissis, A.D., Chambaz, J. (2000) Transgenic animals with altered high-density lipoprotein composition and functions. *Curr Opin Lipidol 11: 149–153.*

176 Kashles, O., Yarden, Y., Fischer, R., Ullrich, A., Schlessinger, J. (1991) A dominant negative mutation suppresses the function of normal epidermal growth factor receptors by heterodimerization. *Mol Cell Biol 11: 1454–1463.*

177 Kashyap, M.L. (1998) Mechanistic studies of high-density lipoproteins. *Am J Cardiol 82: 42U–48U.*

178 Keaney, J.F. (2000) Atherosclerosis: from lesion formation to plaque activation and endothelial dysfunction [In Process Citation]. *Mol Aspects Med 21: 99–166.*

179 Keaney, J.F.J., Xu, A., Cunningham, D., Jackson, T., Frei, B., Vita, J.A. (1995) Dietary probucol preserves endothelial function in cholesterol-fed rabbits by limiting vascular oxidative stress and superoxide generation. *J Clin Invest 95: 2520–2529.*

180 Keidar, S. (1998) Angiotensin, LDL peroxidation and atherosclerosis. *Life Sci 63: 1–11.*

181 Kendall, M.J., Lynch, K.P., Hjalmarson, A., Kjekshus, J. (1995) Beta-blockers and sudden cardiac death. *Ann Intern Med 123: 358–367.*

182 Kersten, S., Desvergne, B., Wahli, W. (2000) Roles of PPARs in health and disease. *Nature 405: 421–424.*

183 Kesaniemi, Y.A., Grundy, S.M. (1984) Influence of probucol on cholesterol and lipoprotein metabolism in man. *J Lipid Res 25: 780–790.*

184 Kihara, S., Ouchi, N., Funahashi, T. *et al.* (1998) Troglitazone enhances glucose uptake and inhibits mitogen-activated protein kinase in human aortic smooth muscle cells. *Atherosclerosis 136: 163–168.*

185 Kirk, C.J., Mule, J.J. (2000) Gene-modified dendritic cells for use in tumor vaccines. *Hum Gene Ther 11: 797–806.*

186 Kitakaze, M., Node, K., Takashima, S., Minamino, T., Kuzuya, T., Hori, M. (2000) Cellular mechanisms of cardioprotection afforded by inhibitors of angiotensin converting enzyme in ischemic hearts: role of bradykinin and nitric oxide. *Hypertens Res 23: 253–259.*

187 Klootwijk, P., Meij, S., Melkert, R., Lenderink, T., Simoons, M.L. (1998) Reduction of recurrent ischemia with abciximab during continuous ECG-ischemia monitoring in patients with unstable angina refractory to standard treatment (CAPTURE). *Circulation 98: 1358–1364.*

188 Klucken, J., Chler, C. *et al.* (2000) ABCG1 (ABC8), the human homolog of the Drosophila white gene, is a regulator of macrophage cholesterol and phospholipid transport. *Proc Natl Acad Sci USA 97: 817–822.*

189 Kohno, M., Murakawa, K., Yasunari, K. *et al.* (1997) Improvement of erythrocyte deformability by cholesterol-lowering therapy with pravastatin in hypercholesterolemic patients. *Metabolism 46: 287–291.*

190 Kozarsky, K.F., Donahee, M.H., Glick, J.M., Krieger, M., Rader, D.J. (2000) Gene transfer and hepatic overexpression of the HDL receptor SR-BI reduces atherosclerosis in the cholesterol-fed LDL receptor-deficient mouse. *Arterioscler Thromb Vasc Biol 20: 721–727.*

191 Kozyraki, R., Fyfe, J., Kristiansen, M. *et al.* (1999) The intrinsic factor-vitamin B12 receptor, cubilin, is a high-affinity apolipoprotein A-I receptor facilitating endocytosis of high-density lipoprotein. *Nat Med 5: 656–661.*

192 Kozyraki, R., Kristiansen, M., Silahtaroglu, A. *et al.* (1998) The human intrinsic factor-vitamin B12 receptor, cubilin: molecular characterization and chromosomal mapping of the gene to 10p within the autosomal recessive megaloblastic anemia (MGA1) region. *Blood 91: 3593–3600.*

193 Kragelund, B.B., Knudsen, J., Poulsen, F.M. (1999) Acyl-coenzyme A binding protein (ACBP). *Biochim Biophys Acta 1441: 150–161.*

194 Krieger, M. (1999) Charting the fate of the "good cholesterol": identification and characterization of the high-density lipoprotein receptor SR-BI. *Annu Rev Biochem 68: 523–558.*

195 Kronenberg, F., Stuhlinger, M., Trenkwalder, E. *et al.* (2000) Low apolipoprotein A-IV plasma concentrations in men with coronary artery disease. *J Am Coll Cardiol 36: 751–757.*

196 Kuhlmann, A.C., Guilarte, T.R. (2000) Cellular and subcellular localization of peripheral benzodiazepine receptors after trimethyltin neurotoxicity. *J Neurochem 74: 1694–1704.*

197 Kwiterovich, P.O.J. (1998) State-of-the-art update and review: clinical trials of lipid-lowering agents. *Am J Cardiol 82: 3U–17U.*

198 Laakso, M., Lehto, S. (1998) Epidemiology of risk factors for cardiovascular disease in diabetes and impaired glucose tolerance. *Atherosclerosis 137 Suppl: S65–S73*

199 Lackner, K.J., Dieplinger, H., Nowicka, G., Schmitz, G. (1993) High density lipoprotein deficiency with xanthomas. A defect in reverse cholesterol transport caused by a point mutation in the apolipoprotein A-I gene. *J Clin Invest 92: 2262–2273.*

200 Lambert, G., Chase, M.B., Dugi, K., Bensadoun, A., Brewer, H.B.J., Santamarina-Fojo, S. (1999) Hepatic lipase promotes the selective uptake of high density lipoprotein-cholesteryl esters via the scavenger receptor B1. *J Lipid Res 40: 1294–1303.*

201 Langer, C., Huang, Y., Cullen, P. *et al.* (2000) Endogenous apolipoprotein E modulates cholesterol efflux and cholesteryl ester hydrolysis mediated by high-density lipoprotein-3 and lipid-free apolipoproteins in mouse peritoneal macrophages. *J Mol Med 78: 217–227.*

202 Langmann, T., Klucken, J., Reil, M. *et al.* (1999) Molecular cloning of the human ATP-binding cassette transporter 1 (hABC1): evidence for sterol-dependent regulation in macrophages. *Biochem Biophys Res Commun 257: 29–33.*

203 LaRosa, J.C., He, J., Vupputuri, S. (1999) Effect of statins on risk of coronary disease: a meta-analysis of randomized controlled trials. *JAMA 282: 2340–2346.*

204 Law, R.E., Meehan, W.P., Xi, X.P. *et al.* (1996) Troglitazone inhibits vascular smooth muscle cell growth and intimal hyperplasia. *J Clin Invest 98: 1897–1905.*

205 Lawn, R.M., Wade, D.P., Garvin, M.R. *et al.* (1999) The Tangier disease gene product ABC1 controls the cellular apolipoprotein-mediated lipid removal pathway [see comments]. *J Clin Invest 104: R25–R31.*

206 Lazar, M.A. (2001) Progress in cardiovascular biology: PPAR for the course. *Nat Med 7: 23–24.*

207 Lee, H.T., Sliskovic, D.R., Picard, J.A. *et al.* (1996) Inhibitors of acyl-CoA: cholesterol O-acyl transferase (ACAT) as hypocholesterolemic agents. CI-1011: an acyl sulfamate with unique cholesterol-lowering activity in animals fed noncholesterol-supplemented diets. *J Med Chem 39: 5031–5034.*

208 Lee, J.K., Kim, S.U., Kim, J.H. (1999) Modification of chitosan to improve its hypocholesterolemic capacity. *Biosci Biotechnol Biochem 63: 833–839.*

209 Lee, M.H., Lu, K., Hazard, S. *et al.* (2001) Identification of a gene, ABCG5, important in the regulation of dietary cholesterol absorption. *Nat Genet 27: 79–83.*

210 Lehner, R., Kuksis, A. (1993) Triacylglycerol synthesis by an sn-1,2(2,3)-diacylglycerol transacylase from rat intestinal microsomes. *J Biol Chem 268: 8781–8786.*

211 Lerch, P.G., Spycher, M.O., Doran, J.E. (1998) Reconstituted high density lipoprotein (rHDL) modulates platelet activity *in vitro* and *ex vivo*. *Thromb Haemost 80: 316–320.*

212 Levitzki, A., Gazit, A. (1995) Tyrosine kinase inhibition: an approach to drug development. *Science 267: 1782–1788.*

213 Li, H., Reddick, R.L., Maeda, N. (1993) Lack of apoA-I is not associated with increased susceptibility to atherosclerosis in mice. *Arterioscler Thromb 13: 1814–1821.*

214 Liao, J.K., Shin, W.S., Lee, W.Y., Clark, S.L. (1995) Oxidized low-density lipoprotein decreases the expression of endothelial nitric oxide synthase. *J Biol Chem 270: 319–324.*

215 Libby, P., Aikawa, M. (1998) New insights into plaque stabilisation by lipid lowering. *Drugs 56 Suppl 1: 9–13.*

216 Lin, A., Minden, A., Martinetto, H. *et al.* (1995) Identification of a dual specificity kinase that activates the Jun kinases and p38-Mpk2. *Science 268: 286–290.*

217 Lin, C.Y., Duan, H., Mazzone, T. (1999) Apolipoprotein E-dependent cholesterol efflux from macrophages: kinetic study and divergent mechanisms for endogenous versus exogenous apolipoprotein E. *J Lipid Res 40: 1618–1627.*

218 Lindner, V., Reidy, M.A. (1991) Proliferation of smooth muscle cells after vascular injury is inhibited by an antibody against basic fibroblast growth factor. *Proc Natl Acad Sci USA 88: 3739–3743.*

219 Linton, M.F., Fazio, S. (1999) Macrophages, lipoprotein metabolism, and atherosclerosis: insights from murine bone marrow transplantation studies. *Curr Opin Lipidol 10: 97–105.*

220 Lusis, A.J. (2000) Atherosclerosis. *Nature 407: 233–241.*

221 Mackness, M.I., Mackness, B., Durrington, P.N. *et al.* (1998) Paraoxonase and coronary heart disease. *Curr Opin Lipidol 9: 319–324.*

222 MacMahon, S., Sharpe, N., Gamble, G. *et al.* (1998) Effects of lowering average of below-average cholesterol levels on the progression of carotid atherosclerosis: results of the LIPID Atherosclerosis Substudy. LIPID Trial Research Group. *Circulation 97: 1784–1790.*

223 Maeda, S., Miyawaki, T., Nakanishi, T., Takigawa, M., Shimada, M. (1998) Peripheral type benzodiazepine receptor in T lymphocyte rich preparation. *Life Sci 63: 1423–1430.*

224 Mahley, R.W., Huang, Y. (1999) Apolipoprotein E: from atherosclerosis to Alzheimer's disease and beyond. *Curr Opin Lipidol 10: 207–217.*

225 Mahley, R.W., Huang, Y., Rall, S.C.J. (1999) Pathogenesis of type III hyperlipoproteinemia (dysbetalipoproteinemia). Questions, quandaries, and paradoxes. *J Lipid Res 40: 1933–1949.*

226 Mangiafico, R.A., Malatino, L.S., Santonocito, M., Spada, R.S., Polizzi, G., Tamburino, G. (1996) Raised plasma endothelin-1 concentrations in patients with primary hypercholesterolemia without evidence of atherosclerosis. *Int Angiol 15: 240–244.*

227 Manninen, V., Tenkanen, L., Koskinen, P. *et al.* (1992) Joint effects of serum triglyceride and LDL cholesterol and HDL cholesterol concentrations on coronary heart disease risk in the Helsinki Heart Study. Implications for treatment. *Circulation 85: 37–45.*

228 Marangos, P.J., Patel, J., Boulenger, J.P., Clark-Rosenberg, R. (1982) Characterization of peripheral-type benzodiazepine binding sites in brain using [3H]Ro 5-4864. *Mol Pharmacol 22: 26–32.*

229 Maresta, A., Balducelli, M., Cantini, L. *et al.* (1994) Trapidil (triazolopyrimidine), a platelet-derived growth factor antagonist, reduces restenosis after percutaneous transluminal coronary angioplasty. Results of the randomized, double-blind STARC study. Studio Trapidil versus Aspirin nella Restenosi Coronarica. *Circulation 90: 2710–2715.*

230 Marui, N., Offermann, M.K., Swerlick, R. *et al.* (1993) Vascular cell adhesion molecule-1 (VCAM-1) gene transcription and expression are regulated through an antioxidant-sensitive mechanism in human vascular endothelial cells. *J Clin Invest 92: 1866–1874.*

231 Marx, N., Sukhova, G.K., Collins, T., Libby, P., Plutzky, J. (1999) PPARalpha activators inhibit cytokine-induced vascular cell adhesion molecule-1 expression in human endothelial cells. *Circulation 99: 3125–3131.*

232 Matsuda, K. (1994) ACAT inhibitors as antiatherosclerotic agents: compounds and mechanisms. *Med Res Rev 14: 271–305.*

233 McCall, M.R., La Belle, M., Forte, T.M., Krauss, R.M., Takanami, Y., Tribble, D.L. (1999) Dissociable and nondissociable forms of platelet-activating factor acetylhydrolase in human plasma LDL: implications for LDL oxidative susceptibility. *Biochim Biophys Acta 1437: 23–36.*
234 McCully, K.S. (1969) Vascular pathology of homocysteinemia: implications for the pathogenesis of arteriosclerosis. *Am J Pathol 56: 111–128.*
235 McLean, L.R., Thomas, C.E., Weintraub, B., Hagaman, K.A. (1992) Modulation of the physical state of cellular cholesteryl esters by 4,4′-(isopropylidenedithio)bis(2,6-di-t-butylphenol) (probucol). *J Biol Chem 267: 12291–12298.*
236 Meier, P.J., Eckhardt, U., Schroeder, A., Hagenbuch, B., Stieger, B. (1997) Substrate specificity of sinusoidal bile acid and organic anion uptake systems in rat and human liver. *Hepatology 26: 1667–1677.*
237 Meiner, V., Tam, C., Gunn, M.D. *et al.* (1997) Tissue expression studies on the mouse acyl-CoA: cholesterol acyltransferase gene (Acact): findings supporting the existence of multiple cholesterol esterification enzymes in mice. *J Lipid Res 38: 1928–1933.*
238 Meisel, S.R., Pauzner, H., Shechter, M., Zeidan, Z., David, D. (1998) Peripheral monocytosis following acute myocardial infarction: incidence and its possible role as a bedside marker of the extent of cardiac injury. *Cardiology 90: 52–57.*
239 Meisel, S.R., Shapiro, H., Radnay, J. *et al.* (1998) Increased expression of neutrophil and monocyte adhesion molecules LFA-1 and Mac-1 and their ligand ICAM-1 and VLA-4 throughout the acute phase of myocardial infarction: possible implications for leukocyte aggregation and microvascular plugging. *J Am Coll Cardiol 31: 120–125.*
240 Melichar, B., Ferrandina, G., Verschraegen, C.F., Loercher, A., Abbruzzese, J.L., Freedman, R.S. (1998) Growth inhibitory effects of aromatic fatty acids on ovarian tumor cell lines. *Clin Cancer Res 4: 3069–3076.*
241 Miettinen, H., Lehto, S., Salomaa, V. *et al.* (1998) Impact of diabetes on mortality after the first myocardial infarction. The FINMONICA Myocardial Infarction Register Study Group. *Diabetes Care 21: 69–75.*
242 Miserez, A.R., Muller, P.Y. (2000) Familial defective apolipoprotein B-100: a mutation emerged in the mesolithic ancestors of Celtic peoples? *Atherosclerosis 148: 433–436.*
243 Miyauchi, T., Masaki, T. (1999) Pathophysiology of endothelin in the cardiovascular system. *Annu Rev Physiol 61: 391–415.*
244 Miyazaki, A., Sakashita, N., Lee, O. *et al.* (1998) Expression of ACAT-1 protein in human atherosclerotic lesions and cultured human monocytes-macrophages. *Arterioscler Thromb Vasc Biol 18: 1568–1574.*
245 Moestrup, S.K., Kozyraki, R. (2000) Cubilin, a high-density lipoprotein receptor. *Curr Opin Lipidol 11: 133–140.*
246 Moestrup, S.K., Kozyraki, R., Kristiansen, M. *et al.* (1998) The intrinsic factor-vitamin B12 receptor and target of teratogenic antibodies is a megalin-binding peripheral membrane protein with homology to developmental proteins. *J Biol Chem 273: 5235–5242.*
247 Moestrup, S.K., Schousboe, I., Jacobsen, C., Leheste, J.R., Christensen, E.I., Willnow, T.E. (1998) beta2-glycoprotein-I (apolipoprotein H) and beta2-glycoprotein-I-phospholipid complex harbor a recognition site for the endocytic receptor megalin. *J Clin Invest 102: 902–909.*
248 Mombouli, J.V., Vanhoutte, P.M. (1999) Endothelial dysfunction: from physiology to therapy. *J Mol Cell Cardiol 31: 61–74.*
249 Morvay, K., Szentleleki, K., Torok, G., Pinter, A., Borzsonyi, M., Nawroth, R. (1989) Effect of change of fecal bile acid excretion achieved by operative procedures on 1,2-dimethylhydrazine-induced colon cancer in rats. *Dis Colon Rectum 32: 860–863.*
250 Munro, J.M., van der Walt, J.D., Munro, C.S., Chalmers, J.A., Cox, E.L. (1987) An immunohistochemical analysis of human aortic fatty streaks. *Hum Pathol 18: 375–380.*
251 Naqvi, T.Z., Shah, P.K., Ivey, P.A. *et al.* (1999) Evidence that high-density lipoprotein cholesterol is an independent predictor of acute platelet-dependent thrombus formation. *Am J Cardiol 84: 1011–1017.*
252 Negrescu, E.V., Grunberg, B., Kratzer, M.A., Lorenz, R., Siess, W. (1995) Interaction of antiplatelet drugs *in vitro*: aspirin, iloprost, and the nitric oxide donors SIN-1 and sodium nitroprusside. *Cardiovasc Drugs Ther 9: 619–629.*
253 Nestel, P.J., Billington, T. (1981) Effects of probucol on low density lipoprotein removal and high density lipoprotein synthesis. *Atherosclerosis 38: 203–209.*
254 Newby, A.C., George, S.J. (1993) Proposed roles for growth factors in mediating smooth muscle proliferation in vascular pathologies. *Cardiovasc Res 27: 1173–1183.*
255 Newby, A.C., Libby, P., van der Wal, A.C. (1999): Plaque instability–the real challenge for atherosclerosis research in the next decade? *Cardiovasc Res 41: 321–322.*
256 Newby, A.C., Zaltsman, A.B. (2000) Molecular mechanisms in intimal hyperplasia. *J Pathol 190: 300–309.*
257 Nicolosi, R.J., Wilson, T.A., Krause, B.R. (1998) The ACAT inhibitor, CI-1011 is effective in the prevention and regression of aortic fatty streak area in hamsters. *Atherosclerosis 137: 77–85.*
258 Niculescu, F., Badea, T., Rus, H. (1999) Sublytic C5b-9 induces proliferation of human aortic smooth muscle cells: role of mitogen activated protein kinase and phosphatidylinositol 3-kinase. *Atherosclerosis 142: 47–56.*
259 Nishio, E., Watanabe, Y. (1999) Troglitazone inhibits alpha1-adrenoceptor-induced DNA synthesis in vascular smooth muscle cells. *Eur J Pharmacol 374: 127–135.*
260 Nofer, J.R., Walter, M., Kehrel, B. *et al.* (1998) HDL3-mediated inhibition of thrombin-induced platelet aggregation and fibrinogen binding occurs via decreased production of phosphoinositide-derived second messengers 1,2-diacylglycerol and inositol 1,4,5-tris-phosphate. *Arterioscler Thromb Vasc Biol 18: 861–869.*
261 Nordestgaard, B.G. (1996) The vascular endothelial barrier–selective retention of lipoproteins. *Curr Opin Lipidol 7: 269–273.*
262 O'Brien, T., Simari, R.D. (2000) Gene therapy for atherosclerotic cardiovascular disease: a time for optimism and caution. *Mayo Clin Proc 75: 831–834.*
263 Oelkers, P., Behari, A., Cromley, D., Billheimer, J.T., Sturley, S.L. (1998) Characterization of two human genes encoding acyl coenzyme A:cholesterol acyltransferase-related enzymes. *J Biol Chem 273: 26765–26771.*

264 Okamoto, H., Yonemori, F., Wakitani, K., Minowa, T., Maeda, K., Shinkai, H. (2000) A cholesteryl ester transfer protein inhibitor attenuates atherosclerosis in rabbits. *Nature 406: 203–207.*

265 Okamoto, S., Inden, M., Setsuda, M., Konishi, T., Nakano, T. (1992) Effects of trapidil (triazolopyrimidine), a platelet-derived growth factor antagonist, in preventing restenosis after percutaneous transluminal coronary angioplasty. *Am Heart J 123: 1439–1444.*

266 Okumura, K., Matsui, H., Kawakami, K. *et al.* (1999) Relationship between the apolipoprotein E and angiotensin-converting enzyme genotypes and LDL particle size in Japanese subjects. *Clin Chim Acta 285: 91–103.*

267 Orso, E., Broccardo, C., Kaminski, W.E. *et al.* (2000) Transport of lipids from golgi to plasma membrane is defective in tangier disease patients and Abc1-deficient mice. *Nat Genet 24: 192–196.*

268 Ostlund, R.E.J., Spilburg, C.A., Stenson, W.F. (1999) Sitostanol administered in lecithin micelles potently reduces cholesterol absorption in humans. *Am J Clin Nutr 70: 826–831.*

269 Packard, C.J., Demant, T., Stewart, J.P. *et al.* (2000) Apolipoprotein B metabolism and the distribution of VLDL and LDL subfractions. *J Lipid Res 41: 305–318.*

270 Panes, J., Perry, M., Granger, D.N. (1999) Leukocyte-endothelial cell adhesion: avenues for therapeutic intervention. *Br J Pharmacol 126: 537–550.*

271 Panousis, C.G., Zuckerman, S.H. (2000) Interferon-gamma induces downregulation of Tangier disease gene (ATP-binding-cassette transporter 1) in macrophage-derived foam cells. *Arterioscler Thromb Vasc Biol 20: 1565–1571.*

272 Parks, D.J., Blanchard, S.G., Bledsoe, R.K. *et al.* (1999) Bile acids: natural ligands for an orphan nuclear receptor. *Science 284: 1365–1368.*

273 Parola, A.L., Yamamura, H.I., Laird, H.E. (1993) Peripheral-type benzodiazepine receptors. *Life Sci 52: 1329–1342.*

274 Patrono, C., Ciabattoni, G., Patrignani, P. *et al.* (1985) Clinical pharmacology of platelet cyclooxygenase inhibition. *Circulation 72: 1177–1184.*

275 Patsch, W., Sharrett, A.R., Chen, I.Y. *et al.* (1994) Associations of allelic differences at the A-I/C-III/A-IV gene cluster with carotid artery intima-media thickness and plasma lipid transport in hypercholesterolemic-hypertriglyceridemic humans. *Arterioscler Thromb 14: 874–883.*

276 Pearson, J.D. (1999) Endothelial cell function and thrombosis. *Baillieres Best Pract Res Clin Haematol 12: 329–341.*

277 Pearson, J.D. (2000) Normal endothelial cell function. *Lupus 9: 183–188.*

278 Pedersen, T.R., Olsson, A.G., Faergeman, O. *et al.* (1998) Lipoprotein changes and reduction in the incidence of major coronary heart disease events in the Scandinavian Simvastatin Survival Study (4S). *Circulation 97: 1453–1460.*

279 Peet, D.J., Turley, S.D., Ma, W. *et al.* (1998) Cholesterol and bile acid metabolism are impaired in mice lacking the nuclear oxysterol receptor LXR alpha. *Cell 93: 693–704.*

280 Pitt, B., Waters, D., Brown, W.V. *et al.* (1999) Aggressive lipid-lowering therapy compared with angioplasty in stable coronary artery disease. Atorvastatin versus Revascularization Treatment Investigators. *N Engl J Med 341: 70–76.*

281 Pollner, R., Schmidt, C., Fischer, G., Kuhn, K., Poschl, E. (1997) Cooperative and competitive interactions of regulatory elements are involved in the control of divergent transcription of human Col4A1 and Col4A2 genes. *FEBS Lett 405: 31–36.*

282 Poorman, J.A., Buck, R.A., Smith, S.A., Overturf, M.L., Loose-Mitchell, D.S. (1993) Bile acid excretion and cholesterol 7 alpha-hydroxylase expression in hypercholesterolemia-resistant rabbits. *J Lipid Res 34: 1675–1685.*

283 Post, S.M., Zoeteweij, J.P., Bos, M.H. *et al.* (1999) Acyl-coenzyme A:cholesterol acyltransferase inhibitor, avasimibe, stimulates bile acid synthesis and cholesterol 7alpha-hydroxylase in cultured rat hepatocytes and *in vivo* in the rat. *Hepatology 30: 491–500.*

284 Prescott, J., Owens, D., Collins, P., Johnson, A., Tomkin, G.H. (1999) The fatty acid distribution in low density lipoprotein in diabetes. *Biochim Biophys Acta 1439: 110–116.*

285 Price, D.T., Loscalzo, J. (1999) Cellular adhesion molecules and atherogenesis. *Am J Med 107: 85–97.*

286 Pueyo, M.E., Gonzalez, W., Nicoletti, A., Savoie, F., Arnal, J.F., Michel, J.B. (2000) Angiotensin II stimulates endothelial vascular cell adhesion molecule-1 via nuclear factor-kappaB activation induced by intracellular oxidative stress. *Arterioscler Thromb Vasc Biol 20: 645–651.*

287 Pueyo, M.E., Gonzalez, W., Nicoletti, A., Savoie, F., Arnal, J.F., Michel, J.B. (2000) Angiotensin II stimulates endothelial vascular cell adhesion molecule-1 via nuclear factor-kappaB activation induced by intracellular oxidative stress. *Arterioscler Thromb Vasc Biol 20: 645–651.*

288 Pullinger, C.R., North, J.D., Teng, B.B., Rifici, V.A., Ronhild, D.B.A., Scott, J. (1989) The apolipoprotein B gene is constitutively expressed in HepG2 cells: regulation of secretion by oleic acid, albumin, and insulin, and measurement of the mRNA half-life. *J Lipid Res 30: 1065–1077.*

289 Puri, R.N., Colman, R.W. (1997) ADP-induced platelet activation. *Crit Rev Biochem Mol Biol 32: 437–502.*

290 Pyorala, K., Laakso, M., Uusitupa, M. (1987) Diabetes and atherosclerosis: an epidemiologic view. *Diabetes Metab Rev 3: 463–524.*

291 Quig, D.W., Zilversmit, D.B. (1990) Plasma lipid transfer activities. *Annu Rev Nutr 10: 169–193.*

292 Quinn, M.J., Fitzgerald, D.J. (1999): Ticlopidine and clopidogrel. *Circulation 100: 1667–1672.*

293 Rader, D.J., Jaye, M. (2000) Endothelial lipase: a new member of the triglyceride lipase gene family. *Curr Opin Lipidol 11: 141–147.*

294 Rader, D.J., Maugeais, C. (2000) Genes influencing HDL metabolism: new perspectives and implications for atherosclerosis prevention. *Mol Med Today 6: 170–175.*

295 Rainwater, D.L., Almasy, L., Blangero, J. *et al.* (1999) A genome search identifies major quantitative trait loci on human chromosomes 3 and 4 that influence cholesterol concentrations in small LDL particles. *Arterioscler Thromb Vasc Biol 19: 777–783.*

296 Reddy, K.G., Nair, R.N., Sheehan, H.M., Hodgson, J.M. (1994) Evidence that selective endothelial dysfunction may occur in the absence of angiographic or ultrasound atherosclerosis in patients with risk factors for atherosclerosis. *J Am Coll Cardiol 23: 833–843.*

297 Rees, D.D., Palmer, R.M., Moncada, S. (1989) Role of endothelium-derived nitric oxide in the regulation of blood pressure. *Proc Natl Acad Sci USA 86: 3375–3378.*

298 Remaley, A.T., Rust, S., Rosier, M. *et al.* (1999) Human ATP-binding cassette transporter 1 (ABC1): genomic organization and identification of the genetic defect in the original Tangier disease kindred. *Proc Natl Acad Sci USA 96: 12685–12690.*

299 Repa, J.J., Turley, S.D., Lobaccaro, J.A. *et al.* (2000) Regulation of absorption and ABC1-mediated efflux of cholesterol by RXR heterodimers. *Science 289: 1524–1529.*

300 Ricci, B., Sharp, D., O'Rourke, E. *et al.* (1995) A 30-amino acid truncation of the microsomal triglyceride transfer protein large subunit disrupts its interaction with protein disulfide-isomerase and causes abetalipoproteinemia. *J Biol Chem 270: 14281–14285.*

301 Rigotti, A., Trigatti, B.L., Penman, M., Rayburn, H., Herz, J., Krieger, M. (1997) A targeted mutation in the murine gene encoding the high density lipoprotein (HDL) receptor scavenger receptor class B type I reveals its key role in HDL metabolism. *Proc Natl Acad Sci USA 94: 12610–12615.*

302 Rinninger, F., Wang, N., Ramakrishnan, R., Jiang, X.C., Tall, A.R. (1999) Probucol enhances selective uptake of HDL-associated cholesteryl esters *in vitro* by a scavenger receptor B-I-dependent mechanism. *Arterioscler Thromb Vasc Biol 19: 1325–1332.*

303 Root, C., Smith, C.D., Winegar, D.A., Brieaddy, L.E., Lewis, M.C. (1995) Inhibition of ileal sodium-dependent bile acid transport by 2164U90. *J Lipid Res 36: 1106–1115.*

304 Rothblat, G.H., de la Llera-Moya, M., Atger, V., Kellner-Weibel, G., Williams, D.L., Phillips, M.C. (1999) Cell cholesterol efflux: integration of old and new observations provides new insights. *J Lipid Res 40: 781–796.*

305 Rothe, G., Gabriel, H., Kovacs, E. *et al.* (1996) Peripheral blood mononuclear phagocyte subpopulations as cellular markers in hypercholesterolemia. *Arterioscler Thromb Vasc Biol 16: 1437–1447.*

306 Rothe, G., Herr, A.S., Stohr, J., Abletshauser, C., Weidinger, G., Schmitz, G. (1999) A more mature phenotype of blood mononuclear phagocytes is induced by fluvastatin treatment in hypercholesterolemic patients with coronary heart disease. *Atherosclerosis 144: 251–261.*

307 Rubins, H.B., Robins, S.J., Collins, D. *et al.* (1999) Gemfibrozil for the secondary prevention of coronary heart disease in men with low levels of high-density lipoprotein cholesterol. Veterans Affairs High-Density Lipoprotein Cholesterol Intervention Trial Study Group. *N Engl J Med 341: 410–418.*

308 Rust, S., Rosier, M., Funke, H. *et al.* (1999) Tangier disease is caused by mutations in the gene encoding ATP-binding cassette transporter 1. *Nat Genet 22: 352–355.*

309 Rye, K.A., Clay, M.A., Barter, P.J. (1999) Remodelling of high density lipoproteins by plasma factors. *Atherosclerosis 145: 227–238.*

310 Sacks, F.M., Pfeffer, M.A., Moye, L.A. *et al.* (1996) The effect of pravastatin on coronary events after myocardial infarction in patients with average cholesterol levels. Cholesterol and Recurrent Events Trial investigators. *N Engl J Med 335: 1001–1009.*

311 Salisbury, B.G., Davis, H.R., Burrier, R.E. *et al.* (1995) Hypocholesterolemic activity of a novel inhibitor of cholesterol absorption, SCH 48461. *Atherosclerosis 115: 45–63.*

312 Santamarina-Fojo, S., Lambert, G., Hoeg, J.M., Brewer, H.B.J. (2000) Lecithin-cholesterol acyltransferase: role in lipoprotein metabolism, reverse cholesterol transport and atherosclerosis. *Curr Opin Lipidol 11: 267–275.*

313 Scarborough, R.M., Kleiman, N.S., Phillips, D.R. (1999) Platelet glycoprotein IIb/IIIa antagonists. What are the relevant issues concerning their pharmacology and clinical use? *Circulation 100: 437–444.*

314 Schaefer, J.R., Schweer, H., Ikewaki, K. *et al.* (1999) Metabolic basis of high density lipoproteins and apolipoprotein A-I increase by HMG-CoA reductase inhibition in healthy subjects and a patient with coronary artery disease. *Atherosclerosis 144: 177–184.*

315 Schmiedt, W., Kinscherf, R., Deigner, H.P. *et al.* (1998) Complement C6 deficiency protects against diet-induced atherosclerosis in rabbits. *Arterioscler Thromb Vasc Biol 18: 1790–1795.*

316 Schmitz, G., Herr, A.S., Rothe, G. (1998) T-lymphocytes and monocytes in atherogenesis. *Herz 23: 168–177.*

317 Schmitz, G., Kaminski, W.E., Orso, E. (2000) ABC transporters in cellular lipid trafficking. *Curr Opin Lipidol 11: 493–501.*

318 Schmitz, G., Orso, E., Rothe, G., Klucken, J. (1997) Scavenging, signalling and adhesion coupling in macrophages: implications for atherogenesis. *Curr Opin Lipidol 8: 287–300.*

319 Schmitz, G., Rothe, G., Ruf, A. *et al.* (1998) European Working Group on Clinical Cell Analysis: Consensus protocol for the flow cytometric characterisation of platelet function. *Thromb Haemost 79: 885–896.*

320 Schoemaker, H., Boles, R.G., Horst, W.D., Yamamura, H.I. (1983) Specific high-affinity binding sites for [3H]Ro 5-4864 in rat brain and kidney. *J Pharmacol Exp Ther 225: 61–69.*

321 Schror, K. (1997) Aspirin and platelets: the antiplatelet action of aspirin and its role in thrombosis treatment and prophylaxis. *Semin Thromb Hemost 23: 349–356.*

322 Schulman, K.A., Kinosian, B., Jacobson, T.A. *et al.* (1990) Reducing high blood cholesterol level with drugs. Cost-effectiveness of pharmacologic management. *JAMA 264: 3025–3033.*

323 Schwarz, M., Lund, E.G., Russell, D.W. (1998) Two 7 alpha-hydroxylase enzymes in bile acid biosynthesis. *Curr Opin Lipidol 9: 113–118.*

324 Seger, R., Krebs, E.G. (1995) The MAPK signaling cascade. *FASEB J 9: 726–735.*

325 Seko, Y., Sato, O., Takagi, A. *et al.* (1997) Perforin-secreting killer cell infiltration in the aortic tissue of patients with atherosclerotic aortic aneurysm. *Jpn Circ J 61: 965–970.*

326 Sharp, D., Blinderman, L., Combs, K.A. *et al.* (1993) Cloning and gene defects in microsomal triglyceride transfer protein associated with abetalipoproteinaemia. *Nature 365: 65–69.*

327 Shepherd, J., Cobbe, S.M., Ford, I. *et al.* (1995) Prevention of coronary heart disease with pravastatin in men with hypercholesterolemia. West of Scotland Coronary Prevention Study Group. *N Engl J Med 333: 1301–1307.*

328 Shepherd, J., Packard, C.J., Patsch, J.R., Gotto, A.M.J., Taunton, O.D. (1979) Effects of nicotinic acid therapy on

plasma high density lipoprotein subfraction distribution and composition and on apolipoprotein A metabolism. *J Clin Invest 63: 858–867.*

329 Shinkai, H., Maeda, K., Yamasaki, T., Okamoto, H., Uchida, I. (2000) bis(2-(Acylamino)phenyl) disulfides, 2-(acylamino)benzenethiols, and S-(2-(acylamino)phenyl) alkanethioates as novel inhibitors of cholesteryl ester transfer protein. *J Med Chem 43: 3566–3572.*

330 Sibley, S.D., Hokanson, J.E., Steffes, M.W. *et al.* (1999) Increased small dense LDL and intermediate-density lipoprotein with albuminuria in type 1 diabetes. *Diabetes Care 22: 1165–1170.*

331 Simoncini, T., Maffei, S., Basta, G. *et al.* (2000) Estrogens and glucocorticoids inhibit endothelial vascular cell adhesion molecule-1 expression by different transcriptional mechanisms. *Circ Res 87: 19–25.*

332 Smith, J.D., Trogan, E., Ginsberg, M., Grigaux, C., Tian, J., Miyata, M. (1995) Decreased atherosclerosis in mice deficient in both macrophage colony-stimulating factor (op) and apolipoprotein E. *Proc Natl Acad Sci USA 92: 8264–8268.*

333 Smith, S.J., Cases, S., Jensen, D.R. *et al.* (2000) Obesity resistance and multiple mechanisms of triglyceride synthesis in mice lacking Dgat. *Nat Genet 25: 87–90.*

334 Soma, M.R., Baetta, R., De Renzis, M.R. *et al.* (1995) *In vivo* enhanced antitumor activity of carmustine [N,N'-bis(2-chloroethyl)-N-nitrosourea] by simvastatin. *Cancer Res 55: 597–602.*

335 Song, W., Kong, H.L., Carpenter, H. *et al.* (1997) Dendritic cells genetically modified with an adenovirus vector encoding the cDNA for a model antigen induce protective and therapeutic antitumor immunity. *J Exp Med 186: 1247–1256.*

336 Spady, D.K., Cuthbert, J.A., Willard, M.N., Meidell, R.S. (1995) Adenovirus-mediated transfer of a gene encoding cholesterol 7 alpha-hydroxylase into hamsters increases hepatic enzyme activity and reduces plasma total and low density lipoprotein cholesterol. *J Clin Invest 96: 700–709.*

337 Spady, D.K., Cuthbert, J.A., Willard, M.N., Meidell, R.S. (1998) Overexpression of cholesterol 7alpha-hydroxylase (CYP7A) in mice lacking the low density lipoprotein (LDL) receptor gene. LDL transport and plasma LDL concentrations are reduced. *J Biol Chem 273: 126–132.*

338 Specht, J.M., Wang, G., Do, M.T. *et al.* (1997) Dendritic cells retrovirally transduced with a model antigen gene are therapeutically effective against established pulmonary metastases. *J Exp Med 186: 1213–1221.*

339 Speir, E., Epstein, S.E. (1992) Inhibition of smooth muscle cell proliferation by an antisense oligodeoxynucleotide targeting the messenger RNA encoding proliferating cell nuclear antigen. *Circulation 86: 538–547.*

340 Stary, H.C. (2000) Natural history and histological classification of atherosclerotic lesions: an update. *Arterioscler Thromb Vasc Biol 20: 1177–1178.*

341 Stein, O., Stein, Y. (1999) Atheroprotective mechanisms of HDL. *Atherosclerosis 144: 285–301.*

342 Stein, Y., Stein, O., Duverger, N. *et al.* (2000) Clearance of cationized LDL cholesterol from a muscle depot is not enhanced in human apolipoprotein A-IV transgenic mice. *Arterioscler Thromb Vasc Biol 20: 179–184.*

343 Steinberg, D. (1996) A docking receptor for HDL cholesterol esters. *Science 271: 460–461.*

344 Steinbrink, K., Wolfl, M., Jonuleit, H., Knop, J., Enk, A.H. (1997) Induction of tolerance by IL-10-treated dendritic cells. *J Immunol 159: 4772–4780.*

345 Stemme, S., Faber, B., Holm, J., Wiklund, O., Witztum, J.L., Hansson, G.K. (1995) T lymphocytes from human atherosclerotic plaques recognize oxidized low density lipoprotein. *Proc Natl Acad Sci USA 92: 3893–3897.*

346 Stohr, J., Schindler, G., Rothe, G., Schmitz, G. (1998) Enhanced upregulation of the Fc gamma receptor IIIa (CD16a) during *in vitro* differentiation of ApoE4/4 monocytes. *Arterioscler Thromb Vasc Biol 18: 1424–1432.*

347 Superko, H.R., Krauss, R.M. (1992) Differential effects of nicotinic acid in subjects with different LDL subclass patterns. *Atherosclerosis 95: 69–76.*

348 Swenson, T.L., Brocia, R.W., Tall, A.R. (1988) Plasma cholesteryl ester transfer protein has binding sites for neutral lipids and phospholipids. *J Biol Chem 263: 5150–5157.*

349 Tack, C.J., Smits, P., Demacker, P.N., Stalenhoef, A.F. (1998) Troglitazone decreases the proportion of small, dense LDL and increases the resistance of LDL to oxidation in obese subjects. *Diabetes Care 21: 796–799.*

350 Taddei, S., Virdis, A., Ghiadoni, L., Magagna, A., Salvetti, A. (1998) Vitamin C improves endothelium-dependent vasodilation by restoring nitric oxide activity in essential hypertension. *Circulation 97: 2222–2229.*

351 Tailleux, A., Bouly, M., Luc, G. *et al.* (2000) Decreased susceptibility to diet-induced atherosclerosis in human apolipoprotein A-II transgenic mice. *Arterioscler Thromb Vasc Biol 20: 2453–2458.*

352 Takahara, N., Kashiwagi, A., Nishio, Y. *et al.* (1997) Oxidized lipoproteins found in patients with NIDDM stimulate radical-induced monocyte chemoattractant protein-1 mRNA expression in cultured human endothelial cells. *Diabetologia 40: 662–670.*

353 Tall, A.R. (1995) Plasma cholesteryl ester transfer protein and high-density lipoproteins: new insights from molecular genetic studies. *J Intern Med 237: 5–12.*

354 Tall, A.R., Jiang, X., Luo, Y., Silver, D. (2000) 1999 George Lyman Duff memorial lecture: lipid transfer proteins, HDL metabolism, and atherogenesis. *Arterioscler Thromb Vasc Biol 20: 1185–1188.*

355 Talmud, P.J., Edwards, K.L., Turner, C.M. *et al.* (2000) Linkage of the cholesteryl ester transfer protein (CETP) gene to LDL particle size: use of a novel tetranucleotide repeat within the CETP promoter. *Circulation 101: 2461–2466.*

356 Tan, K.C., Shiu, S.W., Pang, R.W., Kung, A.W. (1998) Effects of testosterone replacement on HDL subfractions and apolipoprotein A-I containing lipoproteins. *Clin Endocrinol (Oxf) 48: 187–194.*

357 Tangirala, R.K., Tsukamoto, K., Chun, S.H., Usher, D., Pure, E., Rader, D.J. (1999) Regression of atherosclerosis induced by liver-directed gene transfer of apolipoprotein A-I in mice. *Circulation 100: 1816–1822.*

358 Tardif, J.C., Cote, G., Lesperance, J. *et al.* (1997) Probucol and multivitamins in the prevention of restenosis after coronary angioplasty. Multivitamins and Probucol Study Group. *N Engl J Med 337: 365–372.*

359 Taylor, A.H., Stephan, Z.F., Steele, R.E., Wong, N.C. (1997) Beneficial effects of a novel thyromimetic on lipoprotein metabolism. *Mol Pharmacol 52: 542–547.*

360 Tcheng, J.E. (1999) Differences among the parenteral platelet glycoprotein IIb/IIIa inhibitors and implications for treatment. *Am J Cardiol 83: 7E–11E.*

361 Tenkanen, L., Manttari, M., Manninen, V. (1995) Some coronary risk factors related to the insulin resistance syndrome and treatment with gemfibrozil. Experience from the Helsinki Heart Study. *Circulation 92: 1779–1785.*

362 Tenkanen, L., Pietila, K., Manninen, V., Manttari, M. (1994) The triglyceride issue revisited. Findings from the Helsinki Heart Study. *Arch Intern Med 154: 2714–2720.*

363 Thuren, T. (2000) Hepatic lipase and HDL metabolism. *Curr Opin Lipidol 11: 277–283.*

364 Timmerman, J.M., Levy, R. (1999) Dendritic cell vaccines for cancer immunotherapy. *Annu Rev Med 50: 507–529.*

365 Timmermans, P.B. (1999) Angiotensin II receptor antagonists: an emerging new class of cardiovascular therapeutics. *Hypertens Res 22: 147–153.*

366 Tinker, L.F., Parks, E.J., Behr, S.R., Schneeman, B.O., Davis, P.A. (1999) (n-3) fatty acid supplementation in moderately hypertriglyceridemic adults changes postprandial lipid and apolipoprotein B responses to a standardized test meal. *J Nutr 129: 1126–1134.*

367 Torzewski, J., Bowyer, D.E., Waltenberger, J., Fitzsimmons, C. (1997) Processes in atherogenesis: complement activation. *Atherosclerosis 132: 131–138.*

368 Torzewski, J., Oldroyd, R., Lachmann, P., Fitzsimmons, C., Proudfoot, D., Bowyer, D. (1996) Complement-induced release of monocyte chemotactic protein-1 from human smooth muscle cells. A possible initiating event in atherosclerotic lesion formation. *Arterioscler Thromb Vasc Biol 16: 673–677.*

369 Torzewski, M., Torzewski, J., Bowyer, D.E. *et al.* (1997) Immunohistochemical colocalization of the terminal complex of human complement and smooth muscle cell alpha-actin in early atherosclerotic lesions. *Arterioscler Thromb Vasc Biol 17: 2448–2452.*

370 Trigatti, B., Rayburn, H., Vinals, M. *et al.* (1999) Influence of the high density lipoprotein receptor SR-BI on reproductive and cardiovascular pathophysiology. *Proc Natl Acad Sci USA 96: 9322–9327.*

371 Trigatti, B., Rigotti, A., Krieger, M. (2000) The role of the high-density lipoprotein receptor SR-BI in cholesterol metabolism. *Curr Opin Lipidol 11: 123–131.*

372 Tsuda, Y., Satoh, K., Kitadai, M., Takahashi, T., Izumi, Y., Hosomi, N. (1996) Effects of pravastatin sodium and simvastatin on plasma fibrinogen level and blood rheology in type II hyperlipoproteinemia. *Atherosclerosis 122: 225–233.*

373 Turner, R.C., Millns, H., Neil, H.A. *et al.* (1998) Risk factors for coronary artery disease in non-insulin dependent diabetes mellitus: United Kingdom Prospective Diabetes Study (UKPDS: 23). *BMJ 316: 823–828.*

374 Ueno, H., Colbert, H., Escobedo, J.A., Williams, L.T. (1991) Inhibition of PDGF beta receptor signal transduction by coexpression of a truncated receptor. *Science 252: 844–848.*

375 van Haperen, R., van Tol, A., Vermeulen, P. *et al.* (2000) Human plasma phospholipid transfer protein increases the antiatherogenic potential of high density lipoproteins in transgenic mice. *Arterioscler Thromb Vasc Biol 20: 1082–1088.*

376 Varban, M.L., Rinninger, F., Wang, N. *et al.* (1998) Targeted mutation reveals a central role for SR-BI in hepatic selective uptake of high density lipoprotein cholesterol. *Proc Natl Acad Sci USA 95: 4619–4624.*

377 Vassbotn, F.S., Andersson, M., Westermark, B., Heldin, C.H., Ostman, A. (1993) Reversion of autocrine transformation by a dominant negative platelet- derived growth factor mutant. *Mol Cell Biol 13: 4066–4076.*

378 Vega, G.L., Grundy, S.M. (1994) Lipoprotein responses to treatment with lovastatin, gemfibrozil, and nicotinic acid in normolipidemic patients with hypoalphalipoproteinemia. *Arch Intern Med 154: 73–82.*

379 Vlahcevic, Z.R., Stravitz, R.T., Heuman, D.M., Hylemon, P.B., Pandak, W.M. (1997) Quantitative estimations of the contribution of different bile acid pathways to total bile acid synthesis in the rat. *Gastroenterology 113: 1949–1957.*

380 Wan, Y., Bramson, J., Carter, R., Graham, F., Gauldie, J. (1997) Dendritic cells transduced with an adenoviral vector encoding a model tumor-associated antigen for tumor vaccination. *Hum Gene Ther 8: 1355–1363.*

381 Weinman, S.A., Carruth, M.W., Dawson, P.A. (1998) Bile acid uptake via the human apical sodium-bile acid cotransporter is electrogenic. *J Biol Chem 273: 34691–34695.*

382 Weitz, J.I., Leslie, B., Hudoba, M. (1998) Thrombin binds to soluble fibrin degradation products where it is protected from inhibition by heparin-antithrombin but susceptible to inactivation by antithrombin-independent inhibitors. *Circulation 97: 544–552.*

383 Wess, G., Kramer, W., Enhsen, A. *et al.* (1994) Specific inhibitors of ileal bile acid transport. *J Med Chem 37: 873–875.*

384 Wetterau, J.R., Aggerbeck, L.P., Laplaud, P.M., McLean, L.R. (1991) Structural properties of the microsomal triglyceride-transfer protein complex. *Biochemistry 30: 4406–4412.*

385 Wetterau, J.R., Combs, K.A., Spinner, S.N., Joiner, B.J. (1990) Protein disulfide isomerase is a component of the microsomal triglyceride transfer protein complex. *J Biol Chem 265: 9801–9807.*

386 Wetterau, J.R., Zilversmit, D.B. (1985) Purification and characterization of microsomal triglyceride and cholesteryl ester transfer protein from bovine liver microsomes. *Chem Phys Lipids 38: 205–222.*

387 Wick, G., Kleindienst, R., Schett, G., Amberger, A., Xu, Q. (1995) Role of heat shock protein 65/60 in the pathogenesis of atherosclerosis. *Int Arch Allergy Immunol 107: 130–131.*

388 Williams, K.J., Tabas, I. (1998) WThe response-to-retention hypothesis of atherogenesis reinforced. *Curr Opin Lipidol 9: 471–474.*

389 Willnow, T.E., Goldstein, J.L., Orth, K., Brown, M.S., Herz, J. (1992) Low density lipoprotein receptor-related protein and gp330 bind similar ligands, including plasminogen activator-inhibitor complexes and lactoferrin, an inhibitor of chylomicron remnant clearance. *J Biol Chem 267: 26172–26180.*

390 Wilson, J.M., Ferguson, J.J. (1999) Platelet-endothelial interactions in atherothrombotic disease: therapeutic implications. *Clin Cardiol 22: 687–698.*

391 Winkles, J.A., Alberts, G.F., Brogi, E., Libby, P. (1993) Endothelin-1 and endothelin receptor mRNA expression in

normal and atherosclerotic human arteries. *Biochem Biophys Res Commun 191: 1081–1088.*

392 Wong, M.H., Oelkers, P., Dawson, P.A. (1995) Identification of a mutation in the ileal sodium-dependent bile acid transporter gene that abolishes transport activity. *J Biol Chem 270: 27228–27234.*

393 Woods, M.J., Williams, D.C. (1996) Multiple forms and locations for the peripheral-type benzodiazepine receptor. *Biochem Pharmacol 52: 1805–1814.*

394 Woods, M.J., Zisterer, D.M., Williams, D.C. (1996) Two cellular and subcellular locations for the peripheral-type benzodiazepine receptor in rat liver. *Biochem Pharmacol 51: 1283–1292.*

395 Wrenn, S.M.J., Parks, J.S., Immermann, F.W., Rudel, L.L. (1995) ACAT inhibitors CL 283,546 and CL 283,796 reduce LDL cholesterol without affecting cholesterol absorption in African green monkeys. *J Lipid Res 36: 1199–1210.*

396 Xia, P., Vadas, M.A., Rye, K.A., Barter, P.J., Gamble, J.R. (1999) High density lipoproteins (HDL) interrupt the sphingosine kinase signaling pathway. A possible mechanism for protection against atherosclerosis by HDL. *J Biol Chem 274: 33143–33147.*

397 Yamashita, S., Sakai, N., Hirano, K. *et al.* (1997) Molecular genetics of plasma cholesteryl ester transfer protein. *Curr Opin Lipidol 8: 101–110.*

398 Yu, H.H., Ginsburg, G.S., O'Toole, M.L., Otvos, J.D., Douglas, P.S., Rifai, N. (1999) Acute changes in serum lipids and lipoprotein subclasses in triathletes as assessed by proton nuclear magnetic resonance spectroscopy. *Arterioscler Thromb Vasc Biol 19: 1945–1949.*

399 Zambon, A., Austin, M.A., Brown, B.G., Hokanson, J.E., Brunzell, J.D. (1993) Effect of hepatic lipase on LDL in normal men and those with coronary artery disease. *Arterioscler Thromb 13: 147–153.*

400 Zambon, A., Hokanson, J.E., Brown, B.G., Brunzell, J.D. (1999) Evidence for a new pathophysiological mechanism for coronary artery disease regression: hepatic lipase-mediated changes in LDL density. *Circulation 99: 1959–1964.*

401 Zavacki, A.M., Lehmann, J.M., Seol, W., Willson, T.M., Kliewer, S.A., Moore, D.D. (1997) Activation of the orphan receptor RIP14 by retinoids. *Proc Natl Acad Sci USA 94: 7909–7914.*

402 Zhong, S., Sharp, D.S., Grove, J.S. *et al.* (1996) Increased coronary heart disease in Japanese-American men with mutation in the cholesteryl ester transfer protein gene despite increased HDL levels. *J Clin Invest 97: 2917–2923.*

403 Zhou, X., Hansson, G.K. (1999) Detection of B cells and proinflammatory cytokines in atherosclerotic plaques of hypercholesterolaemic apolipoprotein E knockout mice. *Scand J Immunol 50: 25–30.*

404 Zou, Y., Hu, Y., Metzler, B., Xu, Q. (1998) Signal transduction in arteriosclerosis: mechanical stress-activated MAP kinases in vascular smooth muscle cells (review). *Int J Mol Med 1: 827–834.*

405 Schmitz, G., Torzewski, M. (2001) Atherosclerosis: an inflammatory disease. In: *Inflammatory and Infectious Basis of Atherosclerosis*, Editor Jay L. Metha, Birkhäuser Verlag Basel, Switzerland.

406 Drobnik, W., Lindenthal, B., Lieser, B. *et al.* (2001) ATP-binding cassette transporter A1 (ABCA1) *affects total body* sterol metabolism. *Gastroenterology 120: 1203–1211.*

407 Porsch-Ozcurumez, M., Langmann, T., Heimerl, S. *et al.* (2001) The zinc finger protein 202 (ZNF202) is a transcriptional repressor of ATP binding cassette transporter A1 (ABCA1) and ABCG1 gene expression and a modulator of cellular lipid efflux. *J Biol Chem 276: 12427–12433.*

408 Langmann, T., Porsch-Ozcurumez, M., Heimerl, S. *et al.* (2002) Identification of sterol-independent regulatory elements in the human ATP-binding cassette transporter A1 promoter: role of Sp1/3, E-box binding factors, and an oncostatin M-responsive element. *J Biol Chem 277: 14443–14450.*

409 Barlage S., Frohlich, D., Bottcher, A. *et al.* (2001) ApoE-containing high density lipoproteins and phospholipid transfer protein activity increase in patients with a systemic inflammatory response. *J Lipid Res 42: 281–290.*

410 Pfeiffer, A., Bottcher, A., Orso, E. *et al.* (2001) Lipopolysaccharide and ceramide docking to CD14 provokes ligand-specific receptor clustering in rafts. *Eur J Immunol 2001 31; 3153–3164.*

Chapter 36

Disorders of the blood coagulation-fibrinolytic system

Raymund Machovich

Contents

Most human cells have no direct contact with the external environment and their living conditions are ensured by blood. An essential criterion for normal blood circulation is the ability of the blood to arrest bleeding at sites of tissue injury and to prevent thrombus formation inside the blood vessels under normal conditions. The balance between these two extreme states is rigorously controlled by cellular and molecular components such as the endothelial cells, hepatocytes, platelets and the blood coagulation-fibrinolytic system, and is designated as normal hemostasis. Any disorganization in hemostasis may lead to hemorrhagic or/and thrombotic disease; the latter is the most serious medical problem of our time in developed countries.

The purpose of this chapter is to survey the role of the blood coagulation-fibrinolytic system in hemostasis, its pathological alterations and treatment of the disorders.

36.1 Blood coagulation

An important feature of circulating blood is the ability to lose its fluidity. A molecular component, fibrinogen, and a cellular element, the platelet, play crucial roles in this process.

36.1.1 Conversion of fibrinogen to fibrin

Fibrinogen, a water phase soluble molecule (340 kDa) synthesized and secreted by the liver, is present in the blood circulation at a concentration of approximately 10 μM. When its two short N-terminal portions (fibrinopeptide-A and -B) are released by thrombin, the remaining part of the molecule (fibrin monomer) loses its solubility and several fibrin monomers bind together through noncovalent interactions (denoted as fibrin polymerization). In a subsequent step, crosslinks (isopeptide bonds) are formed among polymerized fibrin monomers by the action of a transglutaminase (activated Factor XIII; F XIIIa), whereby fibrin is stabilized.

The fibrin formation is initiated by thrombin, which, exists in the blood in a proenzyme form (prothrombin). Thus, the conversion of fibrinogen to fibrin is determined by the amount and activity of thrombin formed during prothrombin activation. The amount of thrombin depends on the rate of prothrombin activation, whereas its activity is limited by its inhibitor system (mainly by antithrombin; a blood plasma protein also synthesized by the liver).

36.1.2 Prothrombin activation

Prothrombin (a 70 kDa protein, synthesized by the liver) is activated by a serine protease, Factor Xa. The rate of activation is slow. When other components, such as Ca^{2+}, Factor Va (a protein without any enzyme activity) and the phospholipid surface of activated platelets are also present, there is an approximately 300000-fold rate-enhancement. In this complex (prothrombinase complex) both Factor Xa and prothrombin are bound to the phospholipid structure via Gla-domains (Figure 36.1). Inhibition of Gla formation in the proteins inhibits the rate of conversion of prothrombin to thrombin; inhibition of the vitamin K cofactor function with coumarin derivatives (Dicumarol, Warfarin, Syncumar, etc.) is frequently used in clinical practice for prevention of thrombus formation and growth (Figure 36.1, inset).

Factor Xa, however, also exists in a proenzyme form (Factor X), which is activated by either Factor VIIa or Factor IXa (Figure 36.2). The reactions occur in complexes, thus, an extremely high amount of thrombin is formed in a short

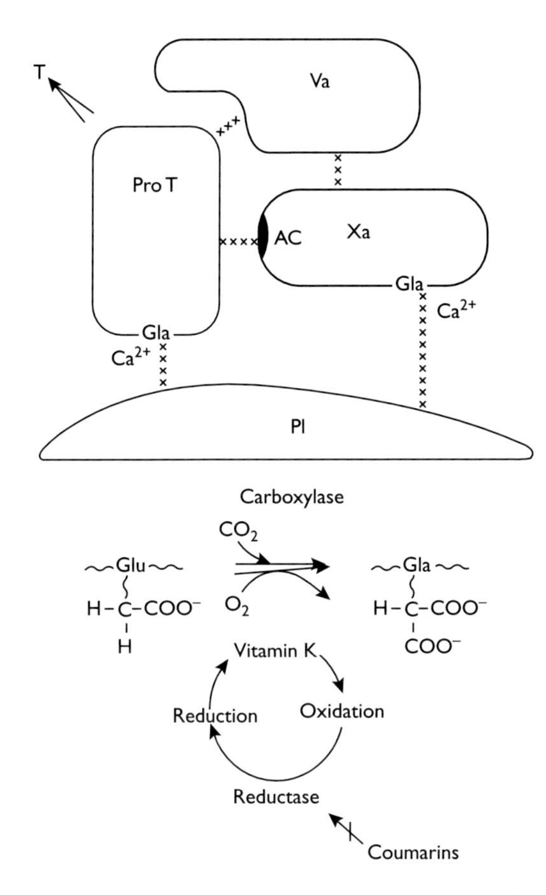

Figure 36.1 The prothrombinase complex. Prothrombin (ProT) is converted to thrombin (T) by activated Factor X (Xa) in the presence of activated Factor V (Va) and phospholipid membrane of activated platelet (PL). Ca^{2+} and Gla-domains of the proteins are prerequisites for the binding of prothrombin and Factor Xa to phospholipid. Gla (γ-carboxyglutamic acid) is formed from CO_2 and Glu (glutamate) in the protein by a carboxylase enzyme using the reduced form (hydroquinone) of vitamin K as a cofactor. During the reaction, vitamin K is oxidized (epoxide form) and thereafter is reduced in two steps (not shown here) by NADH dependent reductase and dehydrogenase enzymes. Coumarin derivatives (vitamin K analogs), inhibit the enzymes of the vitamin K cycle. Wide arrow denotes conversion, narrow arrow action, crossed arrow inhibition, throughout figures in this chapter.

period after initiation of the blood coagulation cascade. It is useful in external tissue damage to prevent blood loss but is dangerous inside the blood circulation provoking thrombosis.

The key to initiation of blood coagulation is a protein, the tissue factor, present on the surface of all cells which are not in direct contact with blood, while cellular elements of blood circulation (endothelial cells, platelets, red blood cells, neutrophils, etc.) do not express tissue factors under normal conditions. Under some pathological circumstances, however,

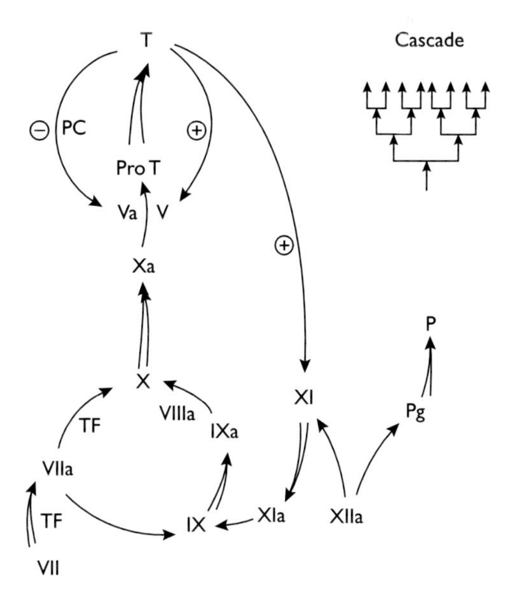

Figure 36.2 Initiation, amplification and control of blood coagulation. Blood coagulation is initiated by tissue factor (TF), a protein present on the surface of cells which are not in direct contact with blood. When Factor VII (VII) in blood plasma meets tissue factor, it is activated and the active form (VIIa) efficiently activates Factor X (X) directly and/or indirectly via factor IX (IX) activation. Both reactions are accelerated by TF. Finally Factor Xa (Xa) activates prothrombin (ProT) in the prothrombinase complex (see also Figure 36.1). All these reactions occur in complexes, where Ca^{2+}, phospholipids and cofactor proteins (tissue factor, Factor V and Factor VIII) are also present. In the complexes extremely high rate enhancements occur and the enzymes formed are protected against the endogenous inhibitor system (except one; not detailed here). Since a single enzyme catalyzes the activation of several zymogens, and the generated enzymes act on further zymogens, the system serves as amplifier of the blood coagulation, called a cascade. The cascade is controlled by positive (⊕) and negative (⊖) feedback mechanisms; thrombin may convert Factor XI (XI) to activated Factor XI (XIa), which in turn, activates Factor IX. Furthermore, thrombin directly activates Factor V (V) and Factor VIII (VIII), and inactivates activated Factor V (Va) and activated Factor VIII (VIIIa) via the Protein C-thrombomodulin system (PC, see also Figure 36.3). Tissue factor may be expressed on the monocyte and endothelial cell surface during Gram-negative infection (the bacterial endotoxin, a lipopolysaccharide induces tissue factor mRNA synthesis) and when TF appears in the blood a generalized thrombosis is induced (Disseminated Intravascular Coagulopathy; DIC). Blood coagulation may be initiated by activated Factor XII (XIIa), which activates Factor XI directly, this reaction, however, does not seem to play a role *in vivo*, at least, for thrombus formation; rather it is fibrinolytic via initiation of plasminogen activation. Patients deficient in Factor XII do not bleed.

tissue factor expression may occur on endothelial cells and/or monocytes leading to generalized thrombosis and, as a consequence, fibrinolysis. Factor V and Factor VIII are proteins without any known biological function present in blood circulation, however, when one peptide bond in the molecule is hydrolyzed by thrombin, they become a cofactor (Factor Va and Factor VIIIa, without enzyme activity) accelerating activation rates in the complexes (see Figure 36.1 for Factor Va in prothrombinase complex; Factor VIIIa with Factor X and Factor IXa complex is not shown). After additional degradation by activated Protein C (aPC), these cofactor functions of Factor Va and Factor VIIIa are diminished (Figure 36.3). Activated Protein C, however, exists in the blood in a proenzyme form (Protein C), which is activated by thrombin. The rate of activation of Protein C is enhanced by several orders of magnitudes when thrombin is bound to thrombomodulin, present on the endothelial cell surface (Figure 36.3).

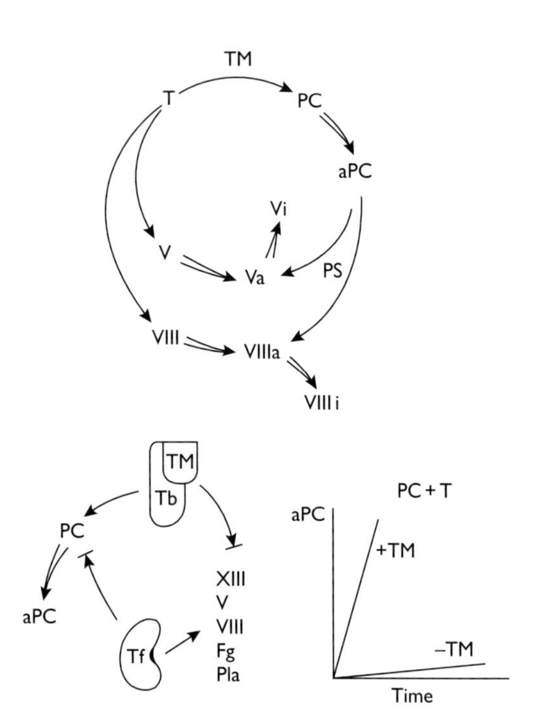

Figure 36.3 The Protein C-thrombomodulin system. In the blood coagulation cascade (see also Figure 36.2) thrombin (T) converts both Factor V (V) and Factor VIII (VIII) to an active conformation (Va and VIIIa), thus these reactions mean a positive feedback control. Thrombin is able to activate Protein C (PC) and its active form (aPC) inactivates Factor Va and Factor VIIIa (Vi and VIIIi) resulting in negative feedback regulation (inhibition of thrombin formation). Activation of Protein C is a slow reaction, but in the presence of thrombomodulin (TM) on the surface of endothelial cells, there is a several orders of magnitude rate enhancement (inset). At the same time, thrombomodulin bound thrombin (Tb) loses its activity on fibrinogen (Fg), platelets (Pla), Factor XIII (XIII), Factor V (V) and Factor VIII (VIII), while free thrombin (Tf) acts very efficiently on these substrates. Any defect in the Protein C-thrombomodulin system (decreased Protein C synthesis, inactive protein formation, damage in thrombomodulin, etc.) causes thrombophilia. A special defect of Factor V (a mutation at the bond sensitive to aPC) prevents Factor Va inactivation; this disorder, known as a PC-resistance, occurs in 2–7% of the general population and is the cause of approximately 60% of the cases of thrombosis during pregnancy and of recurrent thrombosis.

36.1.3 Thrombin inactivation

A crucial step in the blood coagulation cascade is its termination. Several endogenous protease inhibitors exist in the blood circulation, among them antithrombin (antithrombin III) seems to be the most important, at least from a medical viewpoint. Antithrombin forms an equimolar complex with thrombin, in which all its enzyme activities are blocked. The rate of inactivation is accelerated by heparin (Figure 36.4).

36.1.4 Platelet activation

As mentioned earlier, blood coagulation takes place in transiently formed compartments (in complexes), where activated platelet membrane surface (a phosphatidyl-serine rich surface, formerly called platelet factor 3) plays an essential role. Under normal conditions, native platelets (enucleated

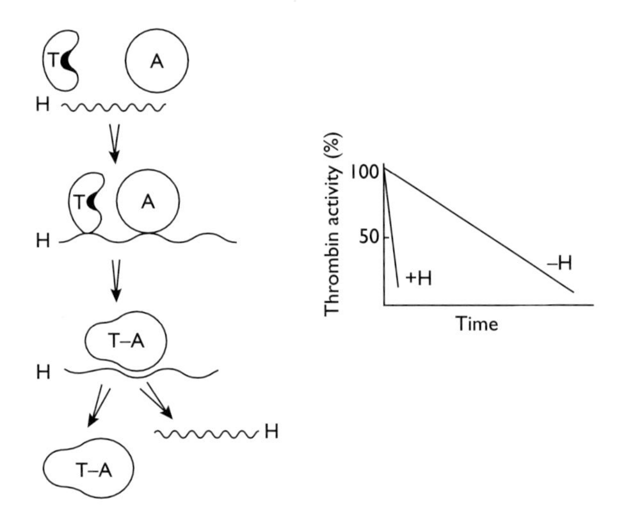

Figure 36.4 Role of heparin in thrombin inactivation by antithrombin. Thrombin (T) forms an equimolar complex with antithrombin (A), where all of its enzyme activities are lost. When heparin (H) is present, both components (enzyme and inhibitor) bind to the polysaccharide and the reaction is extremely accelerated (inset). The affinity of heparin to the thrombin-antithrombin complex (T-A) decreases and the released heparin may participate in another thrombin inactivation reaction. Antithrombin defects, both quantitative and functional, predispose to thrombosis (thrombophilia).

cell fragments of megakaryocytes with a volume of approximately 7 μm^3 and blood count of about 300,000 per microliter) are distributed in a homogeneous suspension. When activated by thrombin, collagen, platelet activating factor (PAF), ADP or thromboxane A_2 (TXA_2), disk-shaped platelets (3 μm in diameter) develop filopodia extensions, release granulum contents and translocate phosphatidyl-serine to the outer layer of the cell membrane. Activated platelets may adhere (platelet adhesion) to a foreign surface (e.g. glass, damaged blood vessel wall) if von Willebrand factor (synthesized and secreted by endothelial cells) is present and/or they may associate with each other (platelet aggregation) in the presence of Ca^{2+} and fibrinogen. Platelet aggregation contributes to plug formation, preventing blood loss or causing stasis of blood circulation (e.g. myocardial infarction, stroke). Activation of platelet is initiated by thrombin via thrombin receptors; when the N-terminal portion of the protein is released by the enzyme, a conformational change of the remnant molecule coupled to a G-protein activates phospholipase C resulting in inositol triphosphate and cytosolic Ca^{2+} increase. Thereafter Ca^{2+} activates both the platelet contractile system and phospholipase A_2; as a result, the cytoskeleton of the platelet is rearranged (filopodia expression), granule content is secreted, phosphatidyl-serine is translocated, and the activated phospholipase A_2 induces thromboxane A_2 (TXA_2) synthesis from arachidonic acid. Both ADP (released from granules) and TXA_2 (secreted) activate neighboring platelets, serving as an amplification of platelet aggregation. ADP, binding to its platelet receptor coupled to G_i-protein, decreases the cAMP synthesis via inhibition of adenylate cyclase. As a result, the platelet becomes more sensitive to activation. In contrast, prostacycline (PGI_2) binds to G_s-protein and activates adenylate cyclase, and the elevated cAMP level protects platelets against activation.

Prostacycline, however, is synthesized and secreted by endothelial cells. The signal transduction mechanism is similar to platelet thromboxane A_2 synthesis; when phospholipase A_2 is activated in endothelial cells (e.g. via endothelial thrombin receptor), prostacycline is synthesized from arachidonic acid and released. In turn, PGI_2 inhibits platelet aggregation.

36.2 Fibrinolysis

Almost parallel with the insoluble fibrin formation, it is solubilized by fibrinolytic enzymes under normal conditions. Although several proteases degrade fibrin, plasmin is considered as the main fibrinolytic enzyme.

36.2.1 Plasminogen activation

Plasmin exists in a zymogen form in the blood, as plasminogen. Thus, a prerequisite of plasmin action is its formation, the activation of plasminogen.

Plasminogen (92 kDa) is synthesized by hepatocytes and secreted into the blood circulation, where its concentration is fairly stable, approximately 2 μM. It is converted to plasmin by plasminogen activators when its conformation is changed (e.g. it is bound to fibrin or its N-terminal portion is hydrolyzed by plasmin or elastase). The activation is a simple reaction; a peptide bond between Arg_{561} and Val_{562} is hydrolyzed by plasminogen activators (Figure 36.5). There are endogenous plasminogen activators, e.g. urokinase-type (uPA) and tissue-type (tPA) and exogenous, e.g. streptokinase. All of them are used in clinical practice. While tPA and uPA, synthesized by endothelial cells and tumor cells, act directly on plasminogen, streptokinase, a product of Streptococcus haemolyticus, binds to human plasminogen, and in the complex plasminogen becomes a plasminogen activator and activates free plasminogen. The catalytic efficiency of uPA is approximately an order of magnitude higher than that of tPA, but the latter is extremely stimulated in the presence of cofactors, such as fibrin, endothelial cell membrane, actin, myosin, some extracellular matrix components and denatured proteins (Figure 36.6). Since endothelial cells synthesize plasminogen activator inhibitors as well, it is an open question at present how the levels of activators and inhibitors are controlled.

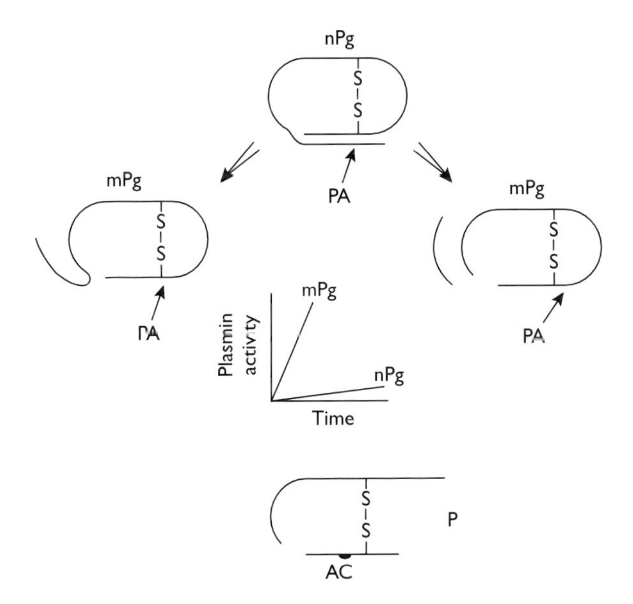

Figure 36.5 Plasminogen activation. Native plasminogen (nPg) is not a susceptible substrate to plasminogen activators (PA); modification of its N-terminal portion either conformationally or proteolytically (mPg) makes peptide bond Arg_{561}-Val_{562} sensitive to PA, and the rate of plasmin formation is enhanced (inset). After hydrolysis of a single peptide bond a two-chain protein (held together with S-S bridges) is formed and the active center (AC) of the serine protease plasmin (P) is expressed on the small chain. Plasmin, among several proteins, digests fibrin(ogen).

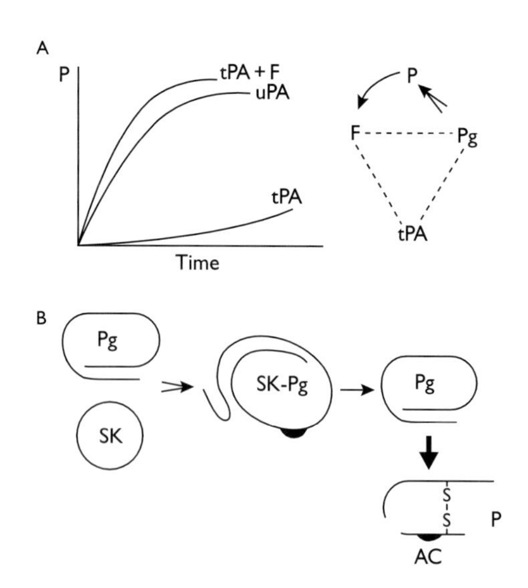

Figure 36.6 Endogenous and exogenous plasminogen activators. (A) Among the endogenous activators, urokinase type plasminogen activator (uPA) is a more efficient enzyme than tissue-type plasminogen activator (tPA), in the presence of fibrin (F), however, tPA becomes as efficient as uPA (fibrin does not affect activation by uPA; not shown). In a ternary complex of plasminogen (Pg), tPA and fibrin, the rate of plasminogen activation is accelerated and the plasmin (P) formed degrades fibrin, thus the cofactor function of fibrin diminishes. (B) Streptokinase (SK), an exogenous protein without any enzyme activity (synthesized by Streptococcus haemolyticus), forms a complex with human plasminogen (SK-Pg), and in the complex an active site of plasminogen develops, which acts like an endogenous activator, converting free plasminogen to plasmin.

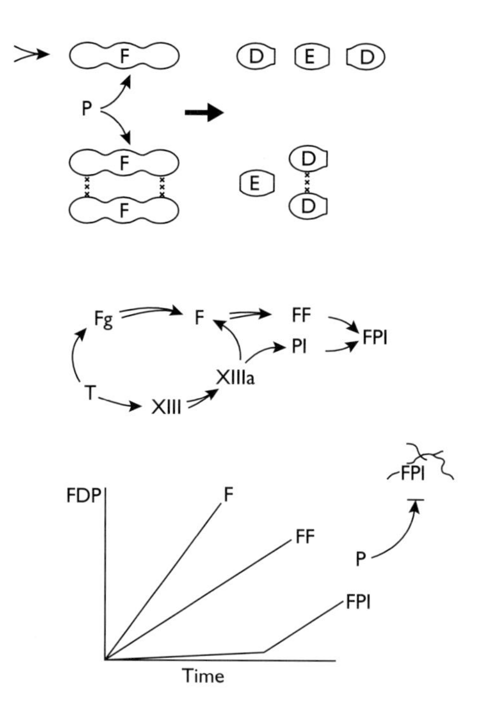

Figurere 36.7 Fibrinogen conversion to fibrin and degradation of the various fibrin structures. When thrombin (T) removes fibrinopeptides (not shown) from fibrinogen (Fg), the fibrin monomers (F) formed lose their solubility and polymerize. At the same time, thrombin converts Factor XIII (XIII) to activated Factor XIII (XIIIa), which, in turn, crosslinks fibrin monomers within polymerized fibrin (FF) and crosslinks plasmin inhibitor (PI) to fibrin (FPI). Thus, during fibrinolysis, plasmin (P) encounters at least three different fibrin structures. Degradation of fibrin monomers by plasmin results in two fragments D (D) and one fragment E (E). When fibrin is crosslinked, the degradation products (FDP) are fragment E and D-dimer (DxxxD), which are also soluble. The rate of degradation of the various fibrin structures, however, is different. While non-crosslinked fibrin is easily digested, fibrin crosslinked with plasmin inhibitor is not digested until the amount of plasmin is not higher than the inhibitor. Thereafter, the rate of degradation is equivalent to that of the crosslinked fibrin. Accordingly, "old" fibrin crosslinked with plasmin inhibitor is "protected" against plasmin digestion.

When formed, plasmin may digest fibrin and several other proteins. Its activity on fibrin degradation is determined by the rate of plasminogen activation and by the rate of its inactivation. There are several types of protease inhibitors present in the blood, among them plasmin inhibitor (formerly named α_2-antiplasmin) is one of the most efficient inhibitors in the mammalian system (second-order rate constant for inactivation of plasmin in solution is $4 \times 10^6 M^{-1}s^{-1}$). It is remarkable that this inhibitor can be bound covalently by Factor XIIIa to fibrin, where it retains its function (see later).

36.2.2 *Fibrin degradation*

While polymerized fibrin monomers are easily digested by plasmin, the rate of degradation of crosslinked fibrin crosslinked with plasmin inhibitor by Factor XIIIa is almost zero (Figure 36.7). Thus, "old" fibrin (crosslinked) is relatively resistant to plasmin digestion.

Plasmin, on the other hand, is rapidly inactivated in the blood plasma by plasma protease inhibitors, primarily by plasmin inhibitor. If plasminogen activation occurs on the fibrin surface, the plasmin formed binds to its substrate and the rate of its inactivation by plasmin inhibitor slows down; plasmin is "protected" against the inhibitor system (Figure 36.8).

36.3 Pathological changes in the blood coagulation-fibrinolytic system

Diseases of hemostasis may be hereditary or acquired with clinical symptoms of bleeding or thrombosis or both. In this review only defects are discussed, the treatment of which is clear and explicit on the basis of the mechanisms described in this chapter.

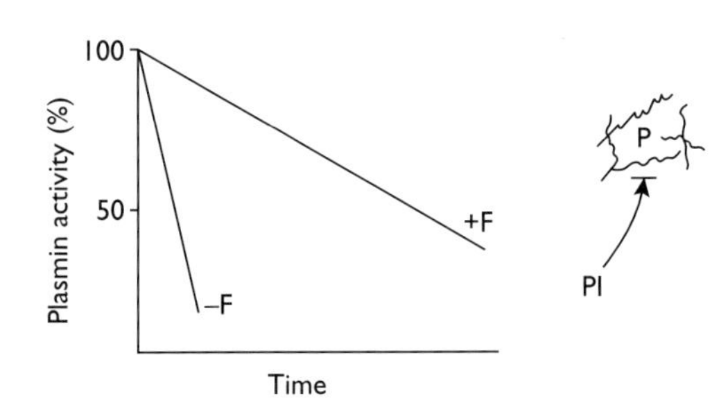

Figure 36.8 Inactivation of plasmin by plasmin inhibitor: effect of fibrin. Free plasmin (P) is inactivated very rapidly ($k'' \approx 430 \times 10^4 M^{-1}s^{-1}$) with plasmin inhibitor (PI), whereas plasmin bound to fibrin (F) is inactivated slowly ($k'' \approx 1 \times 10^4 M^{-1}s^{-1}$). That is, plasmin digesting fibrin is "protected" against inactivation by plasmin inhibitor (inset).

Defects of blood coagulation factors (except their inhibitors and the Protein C system) and of inhibitors of the fibrinolytic enzymes generally cause bleeding problems. Blood transfusion or isolated factor administration is used for treatment (not detailed in this survey).

When the blood coagulation proenzymes are activated (especially when tissue factor expression is facilitated either locally or in a generalized form), or the inhibitor system of blood coagulation proteases, the Protein C system or enzymes of the fibrinolytic system are deficient, thrombophilia occurs. In many cases, however, the cause of thrombosis is not known, only clinical symptoms indicate increased thrombus formation (e.g. myocardial infarction, stroke, deep vein thrombosis, pulmonary embolism). Most frequently endothelial damage underlies the problems, but it remains hidden until serious clinical symptoms appear.

36.3.1 Treatment of thrombosis

Thrombosis treatment generally means prevention; inhibition of thrombus growth and thrombus formation using vitamin K antagonists, heparin, hirudin and/or inhibitors of platelet activation. A real therapy is the initiation of fibrinolysis with tPA, uPA or streptokinase to dissolve the thrombus. The most frequent thrombotic diseases and their treatments are summarized in Table 36.1.

Vitamin K antagonists, coumarins, inhibit thrombin formation via decreasing rate-enhancements in the blood coagulation complexes (Figure 36.1). It is important to keep in mind that coumarins inhibit amino acid modification in proteins, therefore efficient treatment is achieved with a delay in the range of days and in the case of overdose bleeding may develop depending on the synthesis rate of blood coagulation factors containing Gla-domain. Thus, monitoring vitamin K-dependent anticoagulation is mandatory (not detailed here).

Table 36.1 The main therapeutic interventions in the most frequent thrombotic diseases. Thrombosis is generally initiated by endothelial damage, at that time, tissue factor induces blood coagulation and/or collagen platelet activation. Many other pathological conditions (e.g. cancer, inflammation, etc.) may also provoke thrombosis (not detailed here)

	Coumarins	*Heparin*	*Platelet activation inhibitors*	*Fibrinolysis initiation*
Acute myocardial infarction	+	+	+	+
Deep vein thrombosis	+	+	−	+
Pulmonary embolism	+	+	?	+
Stroke	?	?	+	?

Notes: The drugs mentioned here may be administered alone (e.g. heparin for prevention of venous thromboembolism in surgery), but are generally used in combinations. Inhibitors of platelet activation are useful primarily in arterial thrombosis where platelet aggregation is the main mechanism in thrombus formation.

Positive and negative symbols indicate useful and not requested approach, respectively. Question mark stands for consideration of the treatment by medical experts.

Clinical use of heparin is based on its accelerating effect on the inactivation of thrombin by antithrombin (Figure 36.4). It may be administered alone or in combination with coumarins or fibrinolytic agents in the treatment of acute myocardial infarction (AMI), deep vein thrombosis (DVT), pulmonary embolism (PE) and stroke. A mini-dose heparin administered subcutaneously is used during surgery to prevent AMI and DVT. Heparin action is very rapid, its half-life in blood circulation is short, and in addition, when overdosed, protamin, as antidote, neutralizes it. In the future, perhaps hirudin will be used because it inactivates thrombin directly, it is as efficient as heparin, there are no immunological complications and recombinant forms of the protein are available. It seems to be promising especially in antithrombin deficiency.

Therapeutically, inhibition of platelet activation is a frequently used tool in the management of thrombosis. The most popular drug is aspirin, which inhibits cyclooxygenase, thereby preventing thromboxane A_2 synthesis and the amplification of platelet aggregation. A new trend is to create compounds which interfere with platelet receptors participating in platelet adhesion and aggregation (e.g. thrombin receptor, fibrinogen receptor, ADP receptor, etc.).

36.3.2 Fibrinolytic therapy

When tPA, uPA or streptokinase is administered to a patient, plasminogen present in the blood is activated (Figure 36.6), but plasmin forms randomly and is inactivated immediately by plasmin inhibitor present in high concentration in the blood plasma. Plasmin formed on fibrin networks, however, binds to its substrate and escapes rapid inactivation (Figure 36.8). An apparent advantage of tPA is that it also binds to fibrin, whereby plasminogen activation occurs on the fibrin surface where plasmin is protected against the inhibitor. This mechanism is logical, but in practice tPA has no essential advantage in efficiency over other activators. Streptokinase, although efficient, may cause immunological problems, anaphylactic reactions, etc. In the future, perhaps a fibrin-dependent form of streptokinase will be used. If the N-terminal residues (59 amino acids) are removed, the remaining portion of streptokinase in complex with plasminogen becomes fibrin-dependent plasminogen activator. It seems to be an efficient, and not expensive, drug. Fibrinolytic therapy is generally efficient with close to 50% of plasminogen activated (concentration of plasminogen is 2 μM, whereas the plasmin inhibitor is 1 μM in blood), that is, the consumption of the fibrinolytic inhibitor is a prerequisite for thrombus dissolution. An additional clinical aspect of fibrinolysis is that cross-linked fibrin and especially fibrin cross-linked with plasmin inhibitor are dissolved at a slower rate than polymerized fibrin (Figure 36.7), indicating the need for urgent (within a few hours) initiation of the fibrinolytic therapy.

Acknowledgments

This work was supported by Grants OTKA T-031891, ETT 529/96 and FKFP 006/97. I am grateful to Dr. Krasimir Kolev and Dr. István Léránt for helpful criticism.

References

Bachmann, F. (1994) The plasminogen-plasmin enzym system. In: Colman, R.W., Hirsh, J., Marder, V.J., Salzman, E.W. (eds.) *Hemostasis and Thrombosis*, J.B. Lippincott Comp., Philadelphia, pp. 1592–1622.

Bates, S.M. and Hirsh, J. (1999) Treatment of venous thromboembolism. *Thromb. Haemost. 82, 870–877.*

Collen, D. (1999) The plasminogen (fibrinolytic) system. *Thromb. Haemost. 82, 259–270.*

Hirsh, J., Ginsberg, J.S. and Marder, V.J. (1994) Anticoagulant therapy with coumarin agents. In: Colman, R.W., Hirsh, J., Marder, V.J., Salzman, E.W. (eds.) *Hemostasis and Thrombosis*, J.B. Lippincott Comp., Philadelphia, pp. 1567–1583.

Kolev, K, Léránt, I, Tenekejiev, K. and Machovich, R. (1994) Regulation of fibrinolytic activity of neutrophil leukocyte elastase, plasmin and miniplasmin by plasma protease inhibitors. *J. Biol. Chem. 269, 17030–17034.*

Machovich, R. (1984) Inactivation of thrombin by plasma protease inhibitors. In: Machovich, R. (ed.) *The Thrombin*. CRC Press, Inc. Boca Raton, Florida, Vol. II. pp. 1–24.

Machovich, R. (1988) Hemostasis. In: Machovich, R. (ed.) *Blood Vessel Wall and Thrombosis*. CRC Press, Inc., Boca Raton, Florida, Vol. I. pp. 1–80.

Machovich, R. and Owen, W.G. (1990) The elastase-mediated pathway of fibrinolysis. *Blood Coagulation and Fibrinolysis 1, 79–90.*

Mann, K.G. (1999) Biochemistry and Physiology of Blood Coagulation. *Thromb. Haemost. 82, 165–174.*

Moran, N. and FitzGerald, G.A., (1994) Mechanism of action of antiplatelet drugs. In: Colman, R.W., Hirsh, J., Marder, V.J., Salzman, E.W. (eds.) *Hemostasis and Thrombosis*, J.B. Lippincott Comp., Philadelphia, pp. 1623–1637.

Nemerson, Y. (1994) The tissue factor pathway of blood coagulation. In: Colman, R.W., Hirsh, J., Marder, V.J., Salzman, E.W. (eds.) *Hemostasis and Thrombosis*, J.B. Lippincott Comp., Philadelphia, pp. 81–93.

Reed, G.L., Houng, A.K., Liu, L., Parhami-Seren, B., Matsueda, L.H., Wang, S. and Hedstrom, L. (1999) A catalitic switch and the conversion of streptokinase to a fibrin-targeted plasminogen activator. *Proc. Natl. Acad. Sci. USA 96: 8879–8883.*

Salzman, E.W., Hirsh, J. and Marder, V.J. (1994) Clinical use of heparin. In: Colman, R.W., Hirsh, J., Marder, V.J., Salzman, E.W. (eds.) *Hemostasis and Thrombosis*, J.B. Lippincott Comp., Philadelphia, pp. 1584–1591.

Chapter 37

Angiogenesis

József Tímár, Balázs Döme and Sándor Paku

Contents

37.1 Physiological mechanisms and regulation of angiogenesis

37.1.1 *Angiogenic factors and their receptors*

37.1.1.1 *Positive regulators of angiogenesis (Table 37.1)*

At the site of vessel injury, adhered platelets secrete both positive and negative regulators of angiogenesis, mainly from internal α-granules. These positive regulators include: vascular endothelial growth factor-A (VEGF-A), VEGF-C, basic fibroblast growth factor (bFGF), hepatocyte growth factor (HGF), angiopoietin-1 (ANG1) insulin-like growth factor-1 and -2 (IGF-1,2), epidermal growth factor (EGF), platelet-derived growth factor (PDGF), and sphingosine 1-phosphate. One of the common themes is that the majority of the receptors for these factors belong to the tyrosine kinase receptor family (VEGFR flt1, KDR, PDGFR, HGFR, c-met, EGFR, c-erb1). One of the positive angiogenic regulators, hepatocyte growth factor, affects both stimulation and (via the generation of cryptic fragments) suppression of angiogenesis. In the bone marrow, the endothelial stem cells are regulated by G-CSF and their differentiation toward the endothelial linage is driven solely by this factor. A unique mitogen for endothelial cells, platelet-derived endothelial growth factor is actually a surface enzyme, thymidine phosphorylase.

Table 37.1 Physiological pro-angiogenic factors

Factor	*Receptor*
Growth factors	
VEGF165	FLT1
VEGF121	FLK1/KDR
VEGF189	
VEGF206	
PLGF-152/131	Neuropilin
Ang-1	TIE2
FGF1-9	FGFR
HGF	c-met
IGF-1,2	IGFR
PDGF	PDGFR
EGF	EGF
G-CSF	GCSFR
PD-ECGF (thymidine phosphorylase)	
Cytokines	
TNFα	TNFR1
IL-1β	IL-1R
IL-6	IL-6R
IL-8	IL-8R
Chemokine	
PBSF/SDF1	CXCR4
Hormones	
Estrogen	ER-β
androgen	AR
leptin	OB-Rb
Bioactive lipids	
PAF	PAFR
PGE1,2	PGR
TXA2	TXR
12-HETE	HETE-R (?)
Matrix proteins	
Thrombin	THRR
Fibrin	$\alpha v\beta 3$,V-cadherin
CYR61, CTGF	$\alpha v\beta 3$

Note: EGF: epidermal growth factor, FGF: fibroblast growth factor, G-CSF: granulocyte-colony-stimulating factor, HGF: hepatocyte growth factor, IGF: insulin-like growth factor, PAF: platelet-activating factor, PDGF: platelet-derived growth factor, PD-ECGF: platelet derived endothelial growth factor, PGE1,2: prostaglandin E1,2, PLGF: placental growth factor, TXA2: thromboxaneA2, VEGF: vascular endothelial growth factor.

Activated platelets, as well as inflammatory effector cells, can produce bioactive lipids upon activation. These lipids are oxygenase metabolites of arachidonic acid and among them are angiogenic cyclooxygenase products (PAF, PGE1,2) as well as lipoxygenase products such as TXA2 or 12-S-HETE (Nie *et al.*, 2000a,b).

In the cytokine family several members can serve as endothelial mitogens, most of them are inflammatory cytokines, such as TNFα, IL-1β, IL-6, IL-8 and even a chemokine, PBSF. Several matrix proteins which are involved in hemostasis and coagulation are endothelial mitogens themselves, like fibrin and thrombin. The recent discovery that endothelial cells express estrogen receptor-β (ERβ) provide further support for those theories that sex hormones (estrogen, androgen or even leptin) may have angiogenic effects, at least in sex organs (Folkman, 1995; Bussolino *et al.*, 1996; Browder *et al.*, 2000; Jain *et al.*, 2000; Thompson *et al.*, 2000).

37.1.1.2 *Negative regulators of angiogenesis (Table 37.2)*

ENDOSTATIN – COLLAGEN XVIII

The proteolytic product of the carboxy-terminal globular domain of collagen XVIII is a highly potent inhibitor of endothelial cell proliferation, migration as well as angiogenesis, called endostatin. The inhibitory molecule is a 180 amino acid-long residue and is present in serum, urine and in various tissues mostly bound to elastic fibres (O'Reilly *et al.*, 1997).

HGF

In contrast to the multiple pro-angiogenic activities of HGF, alternative processing of the HGF α-chain mRNA generates anti-angiogenic HGF fragments consisting of either the first kringle domain (NK1) or the first two kringle domains (NK2). These first two kringles contain the HGF binding site for its receptor, c-met. NK1 and NK2 suppress HGF-induced endothelial cell migration and abrogate HGF-induced angiogenesis in the rat cornea. These observations led to

Table 37.2 Physiological anti-angiogenic factors

Factor	*Receptor*
Growth factors	
NK1,2,4 (HGF)	c-met
TGFβ	TGFR-I-II
ANG-2	TIE-2
Cytokines	
IFN-α,β,γ	IFN-R
PR-39	P47phox, p130cas
PF-4	HSPG
MMP-inhibitors	
TIMP1,2,3	
Matrix proteins	
TSP-1	$\alpha v\beta 3$, HSPG, CD36
HmwKallikrein-d5	
Prothrombin- F1-2	THR
Cleaved AT-III	
Angiostatin (plasmin fragment)	uPAR, ATP-synthase
Endostatin (collagen XVIII fragment)	$\alpha 1\beta 1, \alpha 2\beta 1, \alpha 3\beta 1, \alpha v\beta 3$
Serpins	
PAI-1	uPA, $\alpha v\beta 3$, uPAR
PEDF	
Maspin	
Heparinase-I, III	
Heparinase-I, III	HSPG

Note: HSPG: heparan sulphate proteoglycan, IFN: interferon, PEDF: pigment epithelium-derived factor, TGF: transforming growth factor, THR: thrombin receptor, TIMP: tissue inhibitor of metalloprotease TSP: thrombospondin, uPA: urokinase-type plasminogen activator.

recombinant construction of NK4, which contains all four kringle domains of HGF. HGF/NK4 is a more potent antagonist of c-met activation by HGF. HGF/NK4 potently inhibits tumor growth *in vivo* by increasing tumor cell apoptosis without affecting the proliferation rate of tumor cells. A similar pattern of tumor inhibition occurs through angiogenesis inhibition. Taken together, the anti-tumor activity of HGF/NK4 *in vivo* is at least partly mediated through anti-angiogenic activity. Thus, expression of NK1 or NK2 cryptic cleavage of HGF into NK1, NK2, or NK4 could counterbalance HGF-induced angiogenesis *in vivo*.

PLATELET FACTOR 4 (PF4)

Unique to platelets, PF4 binds surface heparin-like glycosaminoglycans on endothelial cells, thereby quenching the anti-thrombotic activity of antithrombin III (AT-III) and allowing a clot to form. Nearly two decades ago, PF4 was the first hemostatic protein demonstrated to be an inhibitor of angiogenesis *in vivo*. One mechanism for the initial endothelial cell inhibition following platelet secretion is that PF4 blocks heparin-like glycosaminoglycans that function as critical, low affinity receptors for heparin-binding endothelial growth factors on the surface of endothelial cells. PF4 also directly neutralizes the heparin-binding region of growth factors. Further, a heparin-independent pathway of PF4 inhibition of endothelial cell growth exists. The endothelial cell stimulatory activity of epidermal growth factor and VEGF-A^{121}, endothelial mitogens that lack heparin affinity, is susceptible to PF4 inhibition. Moreover, an analog of PF4 that lacks heparin affinity (rPF4–241) inhibits angiogenesis.

THROMBOSPONDIN (TSP-1)

TSP-1 is the most abundant constituent of platelet α-granules and participates in efficient platelet aggregation. TSP-1 is a large (450 kDa), modular glycoprotein complexed with active transforming growth factor-β1 (TGF-β1) in α-granules and, upon release, can activate latent TGF-β1 secreted by endothelial cells. TSP-1 binds fibrin, fibronectin, plasminogen, surface heparin-like glycosaminoglycans, CD36 and $\alpha_v\beta_3$ integrins on activated endothelials cells, and $\alpha_{IIb}\beta_3$ integrins on activated platelets. TSP-1 may re-adjust growth factor and integrin signaling pathways between endothelial cells and the fibrin clot and prevent endothelial cell motility induced by fibrin, TSP-1 stimulates endothelial cell adhesion and spreading but blocks the chemokinetic response of endothelial cells to bFGF (Jimenez *et al.*, 2000).

TGF-β1

Platelet α-granules are a rich source for active TGF-β1. TGF-β1 promotes the formation of quiescent capillary tubules *in vivo* and mediates potent inhibition of endothelial cell proliferation and migration. TGF-β1 blocks the proliferation of endothelial cells even to supramaximal concentrations of bFGF. *In vivo*, however, TGF-β1 induces angiogenesis that is thought to reflect recruitment of macrophages, which secrete endothelial cell growth factors.

PLASMINOGEN ACTIVATOR INHIBITOR TYPE 1 (PAI-1), α_2-ANTIPASMIN, AND α_2-MACROGLOBULIN

Regulation of plasminogen activation is critical to the sequence of stable fibrin clot formation followed by controlled fibrin digestion. PAI-1 is maintained in an active conformation in complexes with vitronectin within platelet α-granules. Platelet-derived PAI-1 prevents initial fibrinolysis of platelet-rich thrombi but is less effective in the inhibition of the endothelial cell membrane-associated plasminogen activator (uPA) activity that is generated by endothelial sprouts. By limited plasmin generation within the clot structure, PAI-1 can suppress angiogenesis. By scavenging plasmin, platelet-derived and fibrin-bound α_2-macroglobulin may also negatively regulate angiogenesis.

It is an interesting issue that some newly discovered cytokines, such as ANG2, use the same receptor as the proangiogenic factors, ANG1 and TIE2, however, its biological effect is inhibition of the mitogenic signal. This cytokine acts as an extracellular signal-transduction modulator which cooperates with VEGF and ANG1. Other cytokines such as IFNs have strong but non-specific antiangiogenic effect, probably by down-regulation of FGF expression in proliferating endothelial cells. Recent studies on human retina have found that it produces and contains a potent anti-angiogenic factor, produced by the pigment epithelium, PEDF, which prevents vascularization of various tissues in the eye (Folkman, 1995; Bussolino *et al.*, 1996; Dawson *et al.*, 1999; Browder *et al.*, 2000; Jain *et al.*, 2000; Thompson *et al.*, 2000).

37.1.1.3 Negative regulators of angiogenesis within the coagulation cascade

HIGH MOLECULAR WEIGHT KININOGEN (HMWK) DOMAIN 5

HMWk circulates in plasma bound to pre-kallikrein. Contact activation of this complex can begin the coagulation cascade. Kallikrein-cleaved HMWK (Hka) and vitronectin compete for binding to the endothelial cell urokinase receptor. Cryptic generation of Domain 5 (Lys^{420}-Ser^{513}) from Hka inhibits the migration of endothelial cells to vitronectin and fibronectin, both components of the fibrin clot. Domain 5 of HMWK also inhibits endothelial cell proliferation and is anti-angiogenic on the chicken chorioallantoic membrance.

FRAGMENT-1 AND -2 OF PROTHROMBIN

Activated coagulation factor Xa cleaves factor II (prothrombin) to yield thrombin and a two-kringle amino-terminal domain (fragment 1–2). Thrombin then cleaves fragment 1–2 of prothrombin into single-kringle fragment-1 and fragment-2. Thrombin induces angiogenesis *in vivo* via cleavage of the tethered ligand of the thrombin receptor on endothelial cells without the requirement for fibrin formation. Simultaneously, these two amino-terminal kringle domains of prothrombin are released. Fragment-1 and fragment-2 of prothrombin inhibit the proliferation of endothelial cells *in vitro* and angiogenesis *in vivo*. Thus, the stimulatory effects of thrombin on endothelial cells would be antagonized by kringle by-products released upon activation of prothrombin.

AT-III

In the presence of heparin, AT-III avidly inhibits the activated form of factors II (thrombin) and X in plasma. This inactivation is very inefficient when coagulation factors are bound to the anionic phospholipid surface of activated platelets and endothelial cells. AT-III thus serves an important physiologic role in limiting the extent of an evolving clot to the area of vascular injury. Thrombin and neutrophil elastase can cleave the thrombin-binding site of AT-III. Once generated, cleaved AT-III (anti-angiogenic AT-III) becomes a potent inhibitor of endothelial cell proliferation *in vitro* and angiogenesis *in vivo*. Thus, the angiogenic activity of thrombin generated at the site of clotting may be balanced not only by fragment-1 and -2 of prothrombin but also through production of anti-angiogenic AT-III (Folkman, 1995; Bussolino *et al.*, 1996; Browder *et al.*, 2000; Jain *et al.*, 2000; Thompson *et al.*, 2000).

37.1.1.4 Negative regulators of angiogenesis within the fibrinolytic system

Plasminogen is bound to the clot structure and is initially prevented from activation. Angiostatin, kringles 1–4 of plasminogen, is a circulating inhibitor of angiogenesis originally discovered by its ability to prevent the growth of cancer metastases (O'Reilly *et al.*, 1994). Angiostatin potently and specifically inhibits endothelial cell proliferation *in vitro* and angiogenesis *in vivo*. Further, portions of all five kringle domains of plasminogen/plasmin possess anti-angiogenic activity. Angiostatin binds to the $\alpha\beta$-subunits of ATP synthase on the surface of endothelial cells, potentially inducing H^+ cytoplasmic influx into endothelial cells and cytolysis. Several mechanisms have been demonstrated to generate biologically active angiostatin. These include: (1) cleavage by active matrix metalloproteinase (MMP)-2, MMP-3, MMP-7, and MMP-9; (2) cleavage by a tumor cell-derived plasmin thiolreductase; and (3) cleavage of plasminogen on the surface of macrophages by granulocyte-macrophage colony-stimulating factor-induced metalloelastase (MMP-12). Angiostatin also governs the rate of plasminogen activation through non-competitive inhibition of tissue-type plasminogen activator. Thus, generation of angiostatin may regulate the speed of endothelial cell migration and proliferation into the clot both directly and through feedback inhibition of plasminogen activation (Folkman, 1995; Bussolino *et al.*, 1996; Browder *et al.*, 2000; Jain *et al.*, 2000; Thompson *et al.*, 2000).

37.1.1.5 ECM-matrix interactions of endothelial cells (Tables 37.1, 37.2, 37.3)

The extracellular matrix (collagen I–III, basement membrane components, vitronectin and fibronectin) is required for endothelial cell function, their survival and proliferation signaling through the appropriate integrin receptors mediated by the consensus RGD peptide domain. Furthermore, remodeling of blood vessels and the reorganization of the cytoskeleton also involve integrin function. The predominant matrix receptors for quiescent endothelial cells are $\alpha 2\beta 1$ (collagen) and $\alpha v\beta 1$, $\alpha 5\beta 1$ (fibronectin) receptors. On the other hand, proliferating endothelial cells swich integrin

Table 37.3 Endothelial cell extracellular matrix interactions

Receptor	*Ligand*
Integrins	
$\alpha5\beta1$	FN
$\alpha v\beta3$	VN, FN, vWF, FBG, OSP, TSP, den-coll, perlecan
$\alpha v\beta5$	VN, RGD-peptides, HIV-TAT
$\alpha v\beta6$	TNC, FN
$\alpha1\beta1,\alpha2\beta1,\alpha3\beta1$	Collagen
Proteoglycans	
perlecan	VN
CD44s	HA

Note: den-coll: denatured collagen, FBG: fibrinogen, FN: fibronectin, HA: hyaluronic acid, OSP, osteopotin, TNC: tenascin, TSP: thrombospondin, vWF: von Willebrand factor (factor VIII), VN: vitronectin.

expression toward the $\beta3$ family of vitronectin receptors, $\alpha v\beta5$ and $\alpha v\beta3$. $\alpha v\beta5$ recognizes vitronectin not only by the RGD domain but also by a heparin-binding one as well. Matrix-bound vitronectin is cleared by endothelial cells by a cell surface heparan sulphate proteoglycan (HSPG). It is interesting, that a HSPG component of the basement membrane, perlecan, is also a ligand for $\alpha v\beta3$ trough its RGD domain. The native collagen is not a ligand of $\alpha v\beta3$, but its MMP-degraded fragments expose available RGD sites. Similarly, although fibrinogen is not a ligand for $\alpha v\beta3$, a plasminolytic fragment of it expressing a carboxyl-terminal dodecapeptide is an appropriate ligand. $\alpha v\beta3$ signals through FAK and PI3K as well as PKC, inducing differentiation of proliferating endothelial cells forming tubes and primordial capillaries. $\alpha v\beta3$ plays a critical role in endothelial cell spread and migration as well as in leukocyte and platelet interactions and survival, the latter signal involving p53. Ligation of $\alpha v\beta3$ integrin initiates a signaling cascade involing PY-phosphorylation of FAK and paxillin, activation of PKC and Ca^{++} influx. Furthermore, matrix ligand engagement promotes signaling of TNFα, LPA or bombesin as well. The multifunctional matrix receptor, CD44s, is an activation marker of endothelial cells, which express it following bFGF or VEGF activation and use them for hyaluronic acid recognition.

Proteolysis has a major role in the regulation of angiogenesis. Interestingly, though proliferating and migrating endothelial cells are characterized by MMP-2,-9 and uPA activity, the various degradation products of the matrix proteins (such as plasmin, collagen XVIII, prothrombin) are strong inhibitors of angiogenesis. Natural inhibitors of MMPs and TIMPs, are also angiogenesis inhibitors, similar to the serpin-type uPA inhibitors. Finally, since the majority of angiogenic cytokines are heparin-binding proteins bound to heparan sulphate proteoglycan(s) at the surface of endothelial cells, heparanase I–III is also considered to be a potent angiogenesis inhibitor (Sasisekharan *et al.*, 1994; Griffioen *et al.*, 1997; Elicieri *et al.*, 1999; Lau *et al.*, 1999; Zhang *et al.*, 2000).

37.1.2 Mechanism of angiogenesis, vasculogenesis and arteriogenesis

Blood vessels in the embryo form through vasculogenesis; that is, through *in situ* differentiation of undifferentiated precursor cells (angioblasts) to endothelial cells that assemble into a vascular labyrinth. Historically, the term angiogenesis was first used to described the growth of endothelial sprouts from pre-existing postcapillary venules. More recently, this term has been used to generally denote the growth and remodeling process of the primitive network into a complex network. This involves the enlargement of venules, which sprout or become divided by pillars of periendothelial cells (intussusception) or by transendothelial cell bridges, which then split into individual capillaries. New vessels in the adult arise mainly through angiogenesis, although vasculogenesis also may occur.

37.1.2.1 Vasculogenesis: the formation of a primitive network

Endothelial and hematopoietic cells share a common progenitor (the hemangioblast). In the yolk sac, hemangioblasts form aggregates in which the inner cells develop into hematopoietic precursors and the outer population into endothelial cells. Angioblasts may migrate extensively before *in situ* differentiation and plexus formation. Vascular endothelial growth factor (VEGF), VEGF receptor (VEGFR) 2 and basic fibroblast growth factor (bFGF) influence angioblast differentiation, whereas VEGFR1 suppresses hemangioblast commitment. The molecular mechanisms of how transforming growth factor (TGF)-$\beta1$ and TGF-β receptor II affect vasculogenesis remain mostly undetermined. Molecules mediating interactions between endothelial cells and matrix macromolecules, fibronectin or matrix receptors (α_5 integrin), also affect vasculogenesis.

Little is known about the mechanisms governing endothelial cell fate: *Ets-1, Hex, Vezf1, Hox* and *GATA* family members, basic helix-loop-helix factor and their inhibitors of differentiation may be involved. Such molecules may be of therapeutic value, as they could determine the "decision" of endothelial cell's to become angiogenic during pathological conditions (called "angiogenic switch"). The fate of endothelial cells to become integrated into arteries or veins is mediated by the bHLH transcription factor gridlock at the angioblast stage, and, subsequently, by members of the ephrin family, signals that are also involved in guidance of axons and repulsion of neurons. It was once believed that endothelial precursors only existed during embryonic life. However, endothelial precursor cells have been identified in bone marrow and in peripheral blood in adults. VEGF, granulocyte-monocyte colony-stimulating factor (GM-CSF), bFGF and insulin-like growth factor (IGF)-1 stimulate their differentiation and mobilization. Such precursors colonize angiogenic sites and vascular prostheses in the adult and may hold promise for future therapy.

37.1.2.2 *Angiogenesis: sprouting and remodeling*

Angiogenic sprouting is one, but not the only, mechanism of blood vessel formation in the adult; however, it has been studied most extensively. The molecular basis of angiogenesis in the embryo seems to differ from that of pathological angiogenesis in the adult. Several steps have been determined.

VASODILATION, ENDOTHELIAL PERMEABILITY AND PERIENDOTHELIAL SUPPORT

Angiogenesis starts, with vasodilation, a process involving nitric oxide. Vascular permeability increases in response to VEGF, thereby allowing extravasation of plasma proteins that lay down a provisional scaffold for migrating endothelial cells. This increase in permeability is mediated by the formation of fenestrations, vesiculo-vacuolar organelles and the redistribution of platelet endothelial cell adhesion molecule (PECAM)-1/CD31 and vascular endothelial (VE)-cadherin, and involves Src kinases. Although permeability is good for angiogenesis, excessive vascular leakage can be bad, leading to circulatory collapse, intracranial hypertension, formation of adhesion, metastasis, premenstrual discomfort or blindness. Angiopoietin (ANG) 1, a ligand of the endothelial TIE2 receptor, is a natural inhibitor of vascular permeability, tightening pre-existing vessels. When acutely administered to adult vessels, ANG1 protects against plasma leakage without profoundly affecting vascular morphology.

For endothelial cells to emigrate from their resident site, they need to loosen interendothelial cell contacts and to relieve periendothelial cell support; that is, mature vessels become destabilized. ANG2, an inhibitor of TIE2 signaling, may be involved in detaching smooth muscle cells and loosening the matrix. Proteinases of the plasminogen activator, matrix metalloproteinase (MMP), chymase or heparanase families influence angiogenesis by degrading matrix molecules and by activating or liberating growth factor (bFGF, VEGF and IGF-1), sequestered within the extracellular matrix. Urokinase-type plasminogen activator (u-PA) is essential for revascularization of myocardial infarcts, whereas antagonists of the u-PA receptor inhibit tumor angiogenesis. MMP-3, MMP-7 and MMP-9 affect angiogenesis in neonatal bones and tumors, whereas tissue inhibitors of MMPs 1, 3 or PEX, the naturally occurring fragment of MMP-2, by preventing binding of MMP-2 to $\alpha_v\beta_3$ inhibit tumor angiogenesis.

ENDOTHELIAL CELL PROLIFERATION AND MIGRATION

Once the path has been cleared, proliferating endothelial cells migrate to distant sites. VEGF, placental growth factor (PLGF), VEGF-B VEGF-C, VEGF-D and their receptors VEGFR2, VEGFR3 and neuropilin-1 (a co-receptor of VEGFR2) have specific functions: VEGF and its receptor VEGFR2 affect embryonic, neonatal and pathological angiogenesis and are therapeutic targets, although much remains to be learned about the involvement of the distinct VEGF isoforms or of the heterodimers of VEGF family members. $VEGF_{120}$ alone initiates but does not complete angiogenesis. VEGFR3 is involved in embryonic angiogenesis and is expressed in pathological angiogenesis, whereas VEGF-C (a ligand of VEGFR3) is angiogenic in adult pathology. The angiogenic or lymphoangiogenic activity of VEGF-C depends on its processing. Truncation of VEGFR1 at the tyrosine kinase domain does not impair embryonic angiogenesis, but the involvement of VEGFR-1 signaling during pathological angiogenesis remains undetermined. Indeed, the loss of PLGF specifically impairs pathological but not physiological angiogenesis, by increasing the responsiveness of VEGFR2 to VEGF through increased VEGFR2 tyrosine phosphorylation. Loss of VEGF-B affects coronary function after coronary occlusion. Ang1 phosphorylates tyrosine in Tie2 and is chemotactic for endothelial cells, induces sprouting and potentiates VEGF, but fails to induce endothelial proliferation. In contrast to VEGF, Ang1 itself does not initiate endothelial network organization, but stabilizes networks initiated by VEGF, presumably by stimulating the interaction between endothelial and periendothelial cells. This indicates that Ang1 may act at later stages than VEGF. ANG2, at least in the presence of VEGF, is also angiogenic. Low levels of phosphorylated TIE2 have been detected in the adult quiescent vasculature, indicating involvement of TIE2 in vascular maintenance.

Members of the fibroblast growth factor and platelet-derived growth factor (PDGF) family are redundant during normal development, but they affect angiogenesis when administered, probably by recruiting mesenchymal or inflammatory cells. TGF-β1 and tumor necrosis factor (TNF)-α can either stimulate or inhibit endothelial growth. Molecules involved in cell-cell or cell-matrix interactions, such as the $\alpha_v\beta_3$ integrin, which localized MMP-2 at the endothelial cell surface, mediate endothelial spreading, explaining why $\alpha_v\beta_3$ antagonists inhibit angiogenesis. PECAM-1 and EphrinB2 may also be involved in pathological angiogenesis. Nitric oxide, a downstream effector of VEGF, TGFβ-1 and other angiogenic factors, is not essential for embryonic vascular development, but affects pathological angiogenesis and improves the re-endothelialization of denuded vessels.

Angiogenic sprouting is controlled by a balance of activators and inhibitors. Angiogenesis inhibitors suppressing the proliferation or migration of endothelial cells include angiostatin (an internal fragment of plasminogen), endostatin (a fragment of collagen XVIII), antithrombin III, interferon-β, leukemia inhibitory factor and platelet factor 4.

LUMEN FORMATION

Endothelial cells often assemble as solid cords that subsequently acquire a lumen. Intercalation or thinning of endothelial cells and fusion of pre-existing vessels allow vessels to increase their diameter and length. $VEGF_{189}$ decreases luminal diameter, whereas $VEGF_{121}$, $VEGF_{165}$ and their receptors increase lumen formation, in addition to increasing vessel length. In certain tissues (such as psoriatic skin), VEGF mainly exerts morphogenetic activity by enlarging existing vessels. ANG1 in combination with VEGF also increases luminal diameter. Other molecules affecting lumen formation are integrins ($\alpha_v\beta_3$ or α_5) and the myocyte enhancer binding factor 2C (MEF2C) transcription factor. Excessive proteolysis may lead to cystic assembly of endothelial cells and prevent tube formation. Thrombospondin (TSP)-1 is an endogenous inhibitor of lumen formation.

ENDOTHELIAL SURVIVAL

Once assembled in new vessels, endothelial cells become quiescent and survive for years. The importance of endothelial survival is demonstrated by finding that reduced survival causes vascular regression in the embryo. Endothelial apoptosis is a natural mechanism of vessel regression in the retina and ovary after birth and a frequent effect of (therapeutic) inhibitors of angiogenesis. Endothelial apoptosis is induced through deprivation of nutrients or survival signals when the lumen is obstructed by spasms, thrombi or the shedding of dead endothelial cells, or when a change in the angiogenic gene profile occurs. For example, exposure of premature babies to hyperoxia reduces VEGF levels and causes vessel regression in the retina. The survival function of VEGF depends on an interaction between VEGFR2 β-catenin and vascular endothelial (VE)-cadherin. ANG1 also promotes, whereas ANG2 suppresses, endothelial survival, at least in the absence of angiogenic stimuli. Disruption of the interaction with matrix macromolecules, using $\alpha_v\beta_3$ antagonists or the desintegrin accutin, also results in endothelial apoptosis, but, as $\alpha_v\beta_3$ is only expressed in proliferative cells, pre-existing quiescent blood vessels remain unaffected. Different vascular beds may have specific survival mechanisms, such as brain-derived neurotrophic factor for coronary endothelial cells. Hemodynamic forces are essential for vascular maintenance, as physiological shear stress reduces endothelial turnover and abrogates TNF-α mediated endothelial apoptosis. Endothelial apoptosis can be also induced by nitric oxide, reactive oxygen species, angiostatin, TSP-1, the metallospondin METH-1, interferon-γ, tissue factor pathway inhibitor and vascular endothelial growth inhibitor (VEGI).

Several endothelial survival factors (VEGF, ANG1 and $\alpha_v\beta_3$) suppress p53, p21, p16, p27 and *Bax*, whereas they variably activate the survival PI3-kinase/Akt, p42/44 mitogen-activated protein kinase, Bcl-2, A1 and survivin pathways. The mechanism of action remains unknown for many other angiogenesis inhibitors, including prothrombin kringle-1 and kringle-2, TSP-2, PECAM-1 antagonists, interleukin 4 and 12, interferon-α, cyclooxygenase-2 (Cox2)-inhibitors, 1,15-dihydroxyvitamin-D_3 and the N-terminal fragment of prolactin.

ENDOTHELIAL DIFFERENTIATION

To accommodate local physiological requirements, endothelial cells acquire specialized characteristics that are determined in part by the host tissue. For example, an interaction of astroglial cells expressing glial fibrillary acidic protein, pericytes and normal angiotensinogen levels is essential for development of the blood-brain barrier. In contrast, endothelial cells in endocrine glands, involved in the exchange of particles, become discontinuous and fenestrated; this is possibly mediated by interactions between VEGF and the extracellular matrix.

REMODELING

So far, very little is known about the local guiding of endothelial cells into correct patterns and three-dimensional networks, a goal for therapeutic angiogenesis. Maturation of the endothelial network involves remodeling and "pruning" capillary-like vessels with uniform size, and irregular organization into a structured network of branching vessels. Intussusception, resulting in replacement of vessels by matrix, underlies "pruning" and branching. Gene inactivation studies indicate a morphogenetic function for the distinct VEGF isoforms and VEGFR3, the endothelial "orphan" receptor Tie1, the T-cell-leukemia protein stem cell leukemia factor/tal-1, *Tel* (a member of the Ets family of transcription factors), the GTP-binding protein $G\alpha_{13}$ Jagged, chemokine receptor 4, vascular cell adhesion molecule 1, α_4 integrin and fibronectin (Isner *et al.*, 1999; Carmeliet, 2000).

37.1.2.3 Hemostasis

Angiogenesis is the process of sprouting and configuring new blood vessels from pre-existing blood vessels, whereas the hemostatic system maintains the liquid flow of blood by regulating platelet adherence and fibrin deposition. Both systems normally appear quiescent, yet both systems remain poised for repair of injury. With vessel injury, a rapid sequence of reactions must occur to occlude the vessel wall, detect and prevent hemorrhage. Activated platelets link the margins of the defect and form a provisional barrier that is quickly enmeshed with polymerized fibrin. This clot structure initially requires immobilized vascular endothelial cells to anchor the clot and prevent further bleeding. Thereafter, endothelial cells at the clot margins become mobile, dismantling and invading the cross-linked fibrin structure to rebuild a new vessel wall.

Although the positive and negative regulators that control the delicate balance of platelet reactivity and fibrin deposition have been elucidated over the past four decades, analogous proteins that control endothelial cell growth and inhibition have only been discovered within the past decade. Hemostasis and angiogenesis are becoming increasingly interrelated. Proteins generated by the hemostatic system coordinate the spatial localization and temporal sequence of clot/endothelial cell stabilization followed by endothelial cell growth and repair of a damaged blood vessel. During the first day of the clot formation and stabilization, any initiation of angiogenesis prompted by activated platelets such as thrombin and fibrin as well as HGF have to be inhibited. Afterwards, a tightly controlled repair of the vessel wall and/or angiogenesis will start to avoid re-bleeding. This complex regulation is based on cryptic fragments of platelet-secreted hemostatic proteins such as prothrombin fragment 1 and 2 as well as AT-III. Platelets immediately start the secretion of such proteins (PF4, TSP-1 and TGFβ1) following endothelial injury which support endothelial survival rather than mitogenesis and migration. The fibrinolysis around endothelial cells is initiated by the expression of uPA/uPAR system and also by MT1-MMP. Even in that phase angiostatin, variants of HGF, NK1,2,4 or kallikrein-fragments are produced in significant amount to prevent overt angiogenesis activated by proangiogenic factors released from the activated platelets (Broder *et al.*, 2000).

37.1.2.4 Wound healing

Cutaneous wound formation induces neoangiogenesis within 2–4 days accompanied by epithelial damage, followed by regression for 7–10 days. Resting endothelial cells of the pre-existing vessels are stimulated by serum factors, primarily by fibrinogen/fibrin and vitronectin resulting in the increased expression of $\alpha v\beta 3$, $\alpha v\beta 3$ and $\alpha v\beta 6$ integrins. In the meantime, inflammatory cells produce a wide range of angiogenic cytokines including VEGF, bFGF, TNFα, IL-6 etc. The appearance of $\alpha v\beta 3$ (as well as $\alpha v\beta 5$ or $\alpha v\beta 6$) on the surface of endothelial cells provide a powerful integrator for accumulation of stromal cell-produced MMP-2 at the surface facilitating focal degradation of the subendothelial basement membrane as the initial step of the formation of new vessels. It is important to note that in various organs, various angiogenic mitogens induce different endothelial αv-integrins: bFGF and TNFα relay on $\alpha v\beta 3$ while VEGF and TGFβ use $\alpha v\beta 5$ in activated endothelial cells (Christofidou-Salamidou *et al.*, 1997).

37.1.2.5 Sex organs (endometrium, ovary, placenta)

In the endometrium the proliferation pattern of endothelial cells does not show menstrual cycle specificity. However, during the proliferation stage significant vascular growth takes place. Histological analysis of blood vessels in the endometrium indicate that endothelial cells show highly different mitogenic acitivies in respect of the functional versus basal layer of the endometrium where proliferating cells can only be detected in the former zone inside pre-existing vessels. This is supported by the fact that $\alpha v\beta 3$ integrin is expressed only on the surface of endothelial cells in existing vessels of the functional zone. Furthermore, during menstruation, when the steroid levels are lowest, stromal and epithelial cells produce maximum amounts of VEGF under control of hypoxia. On the other hand, the expression of KDR and flt-1 exhibit strict dependence on the cycle phase: in the early proliferation phase and at the late stage of the secretory phase, endothelial cells express KDR, while in the secretory phase they express predominantly flt1. The former seems to be responsible for angiogenesis while the latter is responsible for increased vascular permeability.

In the ovarian corpus luteum a sequential hormone-dependent angiogenesis, vascular maturation and regression take place. In the ovary VEGF/VEGFRs are constitutively expressed independent from the cycle phase. ANG1 and ANG2 are also expressed throughout the cycle, however their ratio is significantly different: there is a predominance of ANG2 over ANG1 in the regression phase of the corpus luteum.

In the placenta, trophoblasts produce PLGF while mesenchymal cells and macrophages are responsible for VEGF production. In addition, trophoblasts express ANG1 and TIE2 in the first trimester, while ANG2 appears at term.

Although these data do not suggest a strong indication for a direct hormonal regulation of the expression of angiogenic factors in the female genital organs, recent data indicate that the promoter region of VEGF contains steroid responsive elements (SRE), a clear indication for the potential for sex hormone regulation. Furthermore, it was also demonstrated that estrogen induces VEGFR2 expression in endothelial cells of the retina, suggesting that some elements of the angiogenic cascade are directly regulated by sex hormones (Rogers *et al.*, 1998; Oehler *et al.*, 2000).

37.1.3 Molecular mechanisms of the production of angiogenic factors (Figure 37.1)

Hypoxia induces a unique transcription factor, HIF1α which serves as a nuclear oxygen sensor. When produced, it activates the expression of several genes involved in the defense mechanism against hypoxia and repairs the damage induced. These include VEGF, VEGFR1, phosphofructokinase, glucose transport protein-1, erythropoietin etc. All of

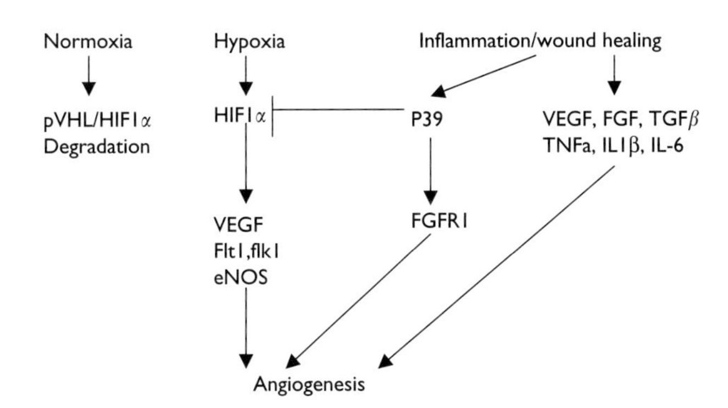

Figure 37.1 Molecular pathways of angiogenesis. HIF: hypoxia-inducible factor, FGF: fibroblast growth factor, FGFR: FGF-receptor, NOS: nitric oxide synthase, TGF: transforming growth factor, TNF: tumor necrosis factor, VEGF: vascular endothelial growth factor, pVHL: von Hippel-Lindau protein.

these genes have the hypoxia responsive element in their promoter regions (HRE) which are binding sites for HIF1. HIF1 has two forms, α and β. HIF1α is inducible by hypoxia, while HIF1β is constitutive and corresponds to ARNT aryl-hydrocarbon receptor nuclear translocator. Under normal oxygen levels, HIF1α is quickly degraded by a ubiquitin-proteasome pathway. Under hypoxia HIF1α degradation is inhibited by a protein which is the product of the pVHL gene. Therefore, hypoxia does not induce increased gene transcription of HIF1α but stabilizes the protein (Brown, 2000).

During inflammation, nucleated cells migrate into the area of interest, become activated and start producing cytokines which are angiogenic including VEGF, bFGF, TNFα, IL-1b, IL-6. Recently it was found that macrophages, one of the major cellular components of the inflammatory exudate, produce a peptide, PR39, which can also inhibit the ubiquitin-proteasome pathway and therefore stabilizes HIF1α, similar to pVHL. PR39-induced HIF1α stabilization results in the increased expression of VERGF, VEGFRs, FGFR1 and NOS3. Accordingly, inflammatory angiogenesis depends on VEGF/VEGRs as well as on the FGF/FGFR system (Li *et al.*, 2000).

37.2 Angiogenesis-dependent diseases

37.2.1 Genetic disorders, vascular malformations

In a rare, autosomal recessive genetic defect, multiple hemangiomas develop in the cerebellum, brain stem and the retina, called von Hippel-Lindau disease (VHL). This highly vascularized tumor is composed of dilated capillaries frequently forming cavernous vessels, where the endothelial cells are not fully matured and rarely produce vWF. The genetic disorder is a mutation of the VHL gene which produces pVHL which binds to HIF1α. The role of pVHL is to bind HIF1α, and promote ubiquitination and degradation. However, the mutated pVHL is unable to bind HIF1α, accordingly it accumulates resulting in the transcriptional activation of genes carrying HRE elements such as VEGF, VEGFR, EPO etc. As a result hemangiomas develop in the brain of these patients. Interestingly, the serum level of EPO is also elevated. Due to its function, the VHL gene is now considered a tumor suppressor gene in endothelial tumors (Brown, 2000).

Hereditary hemorrhagic teleangiectasia (HHT) is an autosomal dominant genetic disorder (Osler-Weber-Rendu disease) resulting in vessel proliferation, but more specifically arterio-venous dilatations in various organs and tissues. It is a disease of the vessel enlargement phase. The gene responsible for those alterations is endoglin, a signaling element of the TGFβ receptor system. TGFβ signals through its receptor, TGFRI and II. Type II phosphorylates type I upon binding TGFβ, activin of BMP followed by intracytoplasmic phosphorylation of SMADs. Two receptors cooperate with TGFRs, betaglycan and endoglin providing the appropriate conformation of TGFβ for the signaling receptor (HHT1). Furthermore, it is also suggested that endoglin cooperates with the VEGFRs in their signaling. ALK-1 is an accessory molecule for both TGF/TGFR-II as well as activin/activin II receptors and is preferentially expressed by vascular endothelium. Missense, and nonsense mutations, deletions, insertions in ALK-1, another TGFβ receptor is responsible for the HHT2 (Jacobson, 2000).

37.2.2 Arteriosclerosis

The formation of new blood vessels (vasculogenesis and angiogenesis) is essential in diseases such as diabetes. Conversely, the deficiency of functional blood vessels in cardio- and cerebrovascular syndromes contributes to a variety of ischemic symptoms, including angina, intermittent claudication and loss of mental function in transient ischemic attacks. Poor perfusion in diabetes is also the leading cause of diabetic limb amputation.

A pathological type of arteriogenesis is the 20-fold enlargment of pre-existing collateral arterioles after occlusion of a supply artery. As a result of the increased collateral flow, endothelial cells express monokines (monocyte chemotactic protein 1) and monocyte adhesion molecules (such as intracellular adhesion molecule 1). The recruited monocytes infiltrate the vessel wall and destroy the media, using proteinases and death factors (TNF-α). Activated endothelial cells then up-regulate bFGF, PDGF-B and TGF-β1, thereby inducing the re-growth of smooth muscle cells and vessel enlargement.

37.2.2.1 Remodeling

The large thoracic vessels undergo considerable remodeling during development. Genetic analysis has shown that loss of MFH-1, dHand or Msx1, Pax-3, Prx1, retinoid acid receptors, the neurofibromatosis type-1 gene product, Wnt-1 connexin 43 or endothelial-1 induce aortic arch malformations. Prostaglandins mediate closure of the neonatal ductus arteriosus. Signals involved in neuronal patterning also seem to be involved in vascular patterning. In the avian heart, there is a close spatial juxtaposition between coronary arteries and Purkinje cells of the myocardial conduction system. Endothelial-1, locally generated in the coronary artery, is an instructive cue for the differentiation of cardiomyocytes into Purkinje cells. Loss of semaphorin-3C or of neurophilin-1, a receptor for neurorepulsive semaphorins, induces abnormal patterning of the large thoracic vessels. Arterial rarefication also occurs during pulmonary or systemic hypertension. An imbalance between endothelial-1 and nitric oxide initially induces vasospasms, but when sustained, this progresses to irreversible vascular loss. Loss of PLGF or u-PA protects against pulmonary vascular remodeling.

37.2.2.2 Modulation of vascular growth

Hypoxia-inducible transcription factors (HIF1β, HIF1α and HIF2α) trigger a coordinated response of angiogenesis and arteriogenesis by inducing expression of VEGF, VEGFR1, VEGFR2, neurophilin-1, Ang2, nitric oxide synthase, TGFβ-1, PDGF-BB, endothelin-1, interleukin-8, IGF-II, TIEl, cyclooxygenase-2 and so on. The von Hippel-Lindau tumor suppressor gene product inhibits the expression of hypoxia-inducible target genes during normoxia by promoting HIF1α degradation. Gene inactivation studies have shown that angiogenesis, not vasculogenesis, is regulated by hypoxia. Tumors lacking HIF1β or HIF1α fail to develop vascularization and lack hypoxic induction of VEGF expression, whereas stabilization of HIF1α by peptide regulator 39 induces angiogenesis in the severly hypoxic myocardium. Hypoxia-inducible factors and hypoxia-response elements are now being tested for angiogenic (gene) therapy of tissue ischemia. Metabolic stimuli, including hypoglycemia and low pH, also stimulate vessel growth, but their mechanisms remain to be determined.

37.2.2.3 Involvement of mechanical factors

Vasculogenesis occurs mostly independently, whereas angiogenesis coincides with the onset of flow and may be influenced considerably by flow. As a result of the higher blood pressure in the capillaries proximal to the aorta, coronary arteries become covered by smooth muscle cells earlier than veins. Remodeling of the developing thoracic arteries or collateral vessels after arterial occlusion also depends on flow. Gene inactivation studies have shown that shear stress-induced vascular remodeling is affected by nitric oxide and P-selectin, that the response of resistance arteries to flow is determined by vimentin, and that vascular tone is affected by bFGF. Mechanical forces affect vascular function through shear stress-responsive gene transcription (Carmeliet, 2000).

37.2.3 Retinopathy (Figure 37.2)

The retina of the eye is supported by retinal and choroidal vessels. The cornea and vitreous fluid contain a unique angiogenesis inhibitor, PEDGF, which is considered to be responsible for the angiogenic inactivity of the adult retina. Damage to these vessels result in ischemic lesion of the retina as manifest in retinopathies of the newborn, diabetic retinopathies and retinal venous occlusions. Neoangiogenesis in the retina is initiated by ischemic injury or hyperglycaemia-induced PKC activation. All of these factors lead to the initiation of VEGF expression in retinal pigment epithelium. The developing new vessels cross the entire retina and start to growth on the vitreous interface. Cytokines, initially produced by the injured pigment epithelia, followed by the activated endothelial cells attract inflammatory and mesenchymal cells which destroy existing anatomic borders as well as producing a new pathologic extracellular matrix. These alterations ultimately lead to detachment of the retina and blindness. In diabetic retinopathies the moderate expression of VEGF primarily induces increased permeability and only initiates endothelial proliferation in a later stage. Glial cells in the hypoxic retina not only produce VEGF, but IL-8

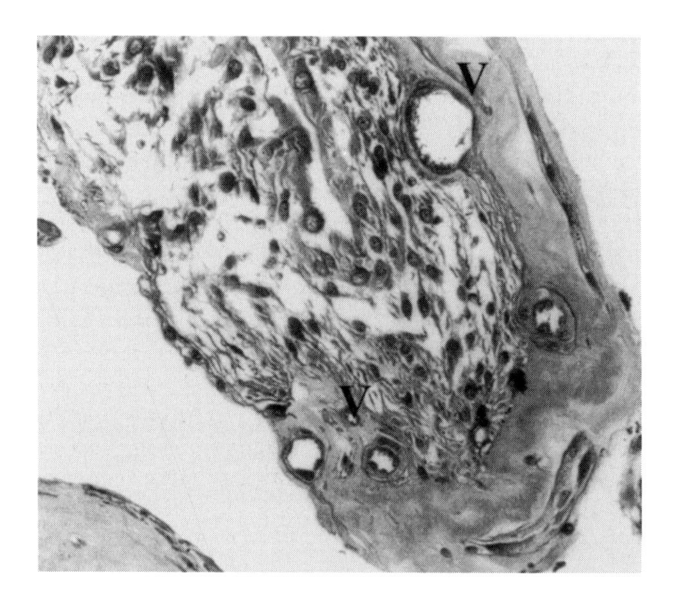

Figure 37.2 Light microscopy of proliferative retinopathy. Toluidine blue staining. The matrix-rich connective tissue contains new vessels of various level of differentiation (V). High power view.

and HGF as well, and pericytes start to express bFGF, all responsible for neoangiogenesis (Lee *et al.*, 1998).

37.2.4 AIDS/Kaposi's sarcoma, HIV-TAT

It was recognized early in the AIDS epidemic, that chronic infection with HIV results in unique vascular disorders and later development of vascular tumor, i.e. Kaposi's sarcoma. However, the pathomechanism of these complications of HIV infection were not known for a long time. HIV-TAT is the transactivating protein of the AIDS virus responsible for viral transcription. Meanwhile this protein has several biological activities based on the multimodular nature of the protein: RGD, heparin-binding domain and chemokine-like sequences. Therefore, TAT is able to induce proliferation and migration of endothelial as well as transformed endothelial cells (Kaposi), induces uPA expression but only in concert with other inflammatory cytokines such as IL-1, TNFα, IFNγ. Interestingly, heparin and heparin-like carbohydrates enhance the angiogenic potential of TAT. Accordingly, one can consider TAT as a viral heparin-binding growth factor which is further supported by the observation that TAT binds the VEGFR, KDR. Meanwhile TAT is not a viral oncogen, since it is unable to transform endothelial cells, which is achieved by the transforming effect of HS8 virus in Kaposi's sarcoma (Bussolino *et al.*, 1996).

37.2.5 Cancer

37.2.5.1 Molecular mechanism of tumor-induced angiogenesis

Tumor-induced neoangiogenesis is the hallmark of malignant tumors, without which solid tissue cannot grow beyond 1–2 mm^2. The genetic background of the "angiogenic switch" during tumor progression is not fully understood, but recent discoveries of the main angiogenic factors, VEGF, bFGF, PDGF, suggest that the switch is able to turn on the expression of the genes of these factors in tumors. Later studies identified several anti-angiogenic factors as well, suggesting that the "angiogenic switch" might also control the expression of these factors. Tumor-induced neoangiogenesis, therefore, means the predominance of pro-angiogenic over anti-angiogenic machinery in cancer. It is now accepted that cancer cells may use physiological pathways to turn on the pro-angiogenic genotype. In this case hypoxic tumor cells in the growing tumor tissue, which express wt-HIF1 gene, activate the expression of HRE-containing angiogenic factor genes including the main angiogenic mitogen, VEGF. Amplification of several oncogens in cancers also leads to the overexpression of pro-angiogenic factors. Loss of wtP53 expression is a characteristic genetic alteration in a considerable proportion of human cancers. An angiogenic consequence of this fact is that such cells would not be able to express thrombospondin, one of the most significant physiological angiogenesis inhibitors. The results of these genetic alterations is the imbalance in the gene expression of pro- and anti-angiogenic factors facilitating the development of the angiogenic switch in cancer.

37.2.5.2 Mechanisms of neoangiogenesis and blood supply in tumors

The first step of tumor-induced angiogenesis is the alteration of the basement membrane over the entire circumference of venules near to the tumor tissue, characterized by loss of electron density and known as gel-sol transition, probably mediated by matrix metalloproteinases or plasminogen activators, secreted by the tumor cells. This alteration must be partly responsible for the initiation of endothelial cell division and migration. On the other hand, angiogenesis factors such as bFGf or VEGF, produced by the tumor cells or host cells, can be liberated from the endothelial basement membrane during this process. Loosening of interendothelial cellular contacts does not occur, suggesting that loss of contact inhibition is not responsible for initiation of cell division and migration (Figure 37.3A). Only the tips of the emigrating endothelial cells are free of the basement membrane (Figure 37.3B). During migration, endothelial cells are arranged in parallel maintaining their polarity (basal-luminal), consequently a slit-like lumen is formed between the endothelial cells (Figure 37.4). This lumen is continuous with the lumen of the original vessel and is sealed by intact inter-endothelial junctions. According to this model no extra stimulus is necessary to induce lumen formation and the retained polarity of endothelial cells allows the continuous deposition of the basement membrane.

Numerous data suggest that the growth of the new capillary sprouts is not oriented towards the tumor, instead the process yields a high density anastomosing network of capillaries at the periphery of tumor cell islands, a process leading to the observed phenomenon that vessel density is much higher around the tumor than inside. A possible explanation is that the continuously growing tumor island incorporates the vessels at the periphery of tumor nests, thereby thinning out the network, whereas the network continuously develops at the advancing tumor-connective tissue interface (Figure 37.5).

The intussusceptive-type growth of vessels yields a high number of large caliber vessels situated outside the area of the active capillary growth – this probably does not contribute to nutrition of the tumor significantly, but rather, provides more sites for sprouting (Paku, 1998).

Recently, a new form of tumor vascularization was suggested based on the findings that tumor cells themselves are able to form channels *in vitro* and the existence of such tumor sinuses was suggested *in vivo* in various human cancers. Furthermore, the genetic background of such channel

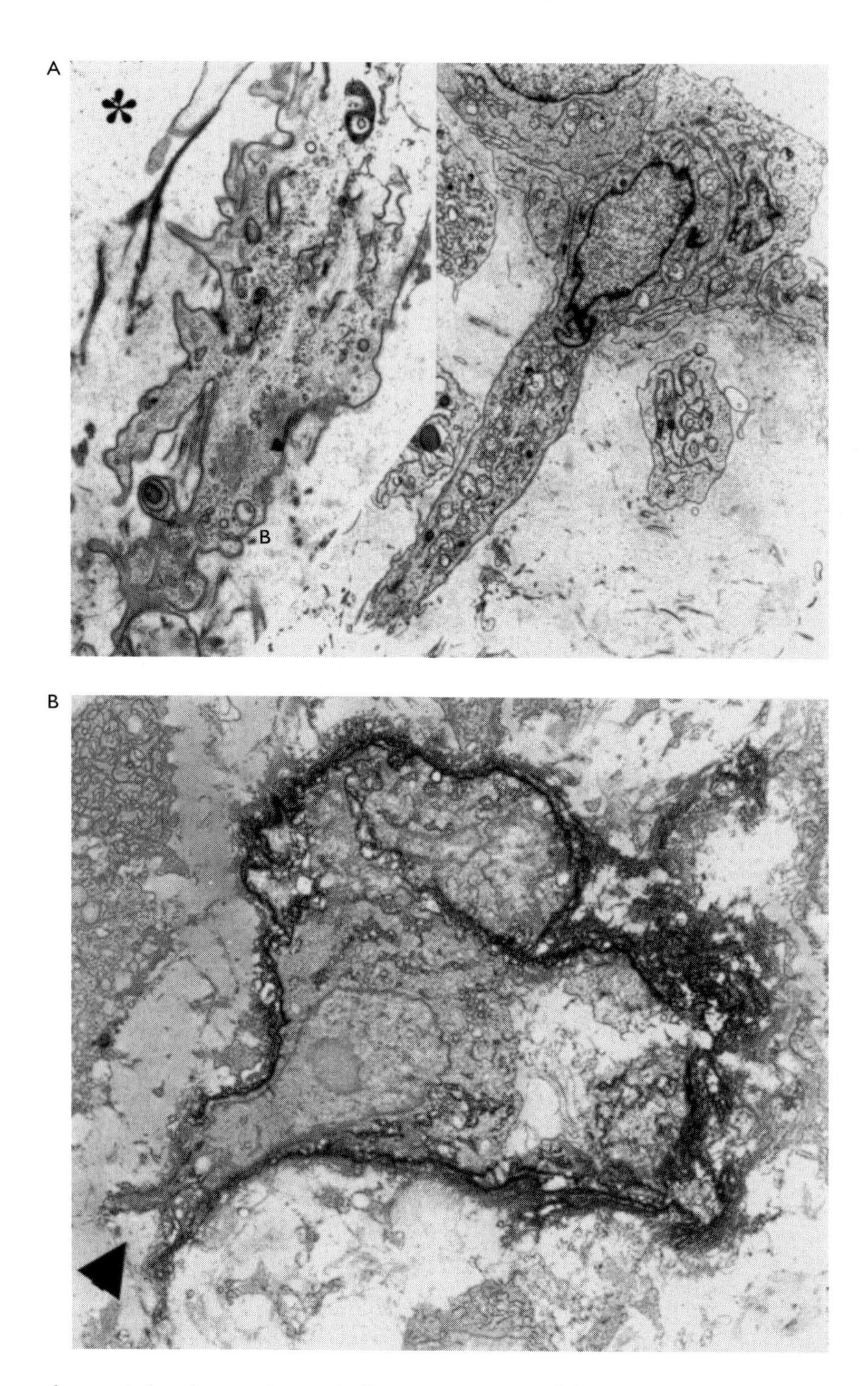

Figure 37.3 Early events of tumor-induced neoangiogenesis. Electron microscopy. (A) Interendothelial cell contacts are maintained when endothelial cell migrates out from the vessel. Insert (*): basement membrane (B) is continous during migration. (B) Tip of the new vessel sprouts lack basement membrane component, laminin (arrow).

formation was also discovered: tumor cells can ectopically express endothelial-specific genes such as TIE1, uPA, HGF/c-met, thereby recapitulating an embryonic geno- and phenotype. This mechanism suggests that tumor cells can form channels, which are connected accidentally to existing intratumoral microvessels in an undiscovered manner (Figure 37.6C). However, there are other theoretical pathways which may lead to the formation of such sinuses (Tímár *et al.*, 2000).

An alternative mechanism for the blood supply of malignant tumors is when the growing tumor tissue incorporates

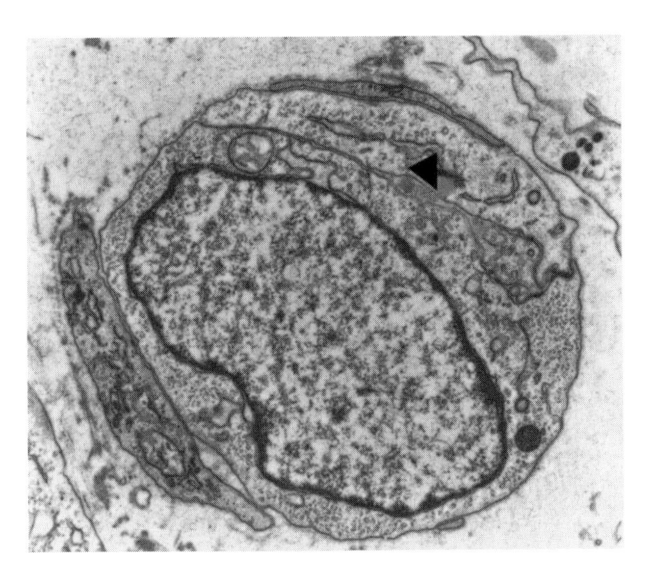

Figure 37.4 Ultrastructural appearance of a young vessel. Note the slit-like lumen (arrow).

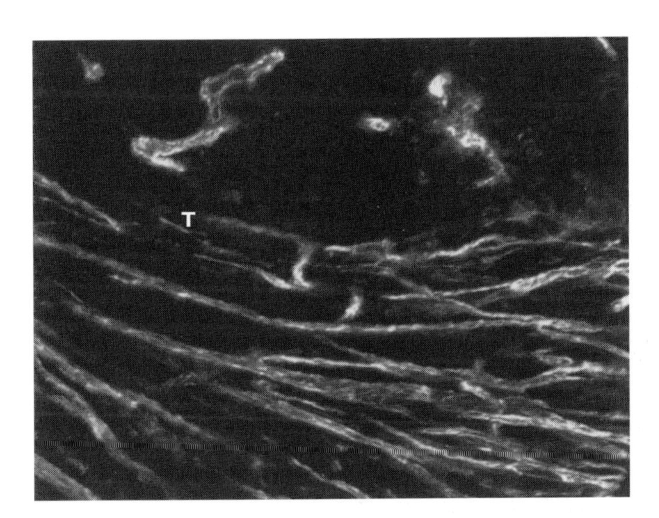

Figure 37.5 Identification of peritumoral vessel network around human melanoma. Composed image of serial confocal images. Vessels are identified by CD31 marker and fluorescent immunohistochemistry. Note the absence of tumor-penetrating vessels. T = tumor.

the pre-existing host vessels, called "cooption". It was demonstrated that such tumors produce pro- and anti-angiogenic factors, the balance of which fundamentally influence both the angiogenetic process as well as remodeling of the coopted vasculature. It was observed that the predominance of anti-angiogenic factors (such as ANG2) in the center of the tumors results in apoptotic death and degeneration of endothelial cells of the incorporated vessels (Figure 37.6B) which can lead to the formation of sinuses lined by tumor cells and still containing red blood cells (Figure 37.6C). Theoretically these channels might maintain their connections to the "surviving" intratumoral microvessels (Figure 37.6A) by the surviving subendothelial matrix network. If such a connection exists, tumor cells lining such sinuses can gain easy access to the microcirculation (the primary event of the hematogenous spread) without the struggle of the complex process of intravasation (Carmeliet, 2000; Tímár *et al.*, 2000).

ANGIOGENESIS IN CANCER METASTASES

The models of angiogenesis above described are based on observations made in tissues that contain high amounts of connective tissue fibers, and can also be valid in tumors such as melanomas, skin, breast and colon cancer. The connective tissue has two important roles in angiogenesis: it allows the build-up of a gradient of the angiogenic factor and provides space for vessel sprouting. The question arises, what kind of angiogenesis takes place in highly vascularized organs such as liver, lungs or adrenals, which are the most frequent sites of metastasis in humans?

A new process of angiogenesis was described in liver metastases of experimental cancers, showing remarkable differences to the models described above. Neither dilatation, nor increase in the number of the vessels, nor enhanced proliferation of endothelial cells could be observed around the metastases, suggesting the absence of sprouting activity in this region. This indicates that the angiogenic stimulus could not reach the endothelial cells around the tumor, which can be explained by the high vascularity of the liver being responsible for diluting or flushing out the factor. The sparse connective tissue cannot provide sufficient space for diffusion of the angiogenic factor, and also impedes the migration of endothelial cells, and thereby the development of new capillaries. Initiation of endothelial cell proliferation is caused by direct contact between tumor cells and endothelial cells. Invading tumor cells migrate along the basement membrane of sinusoids and larger vessels, detaching the endothelial cells from their own basement membrane (Figure 37.7A). Interestingly the basement membrane is not degraded and remains on the surface of hepatocytes (Figure 37.7B), even when these cells became enclosed by the tumor. The observation that proliferative activity is restricted to endothelial cells situated inside the metastases can be explained by results demonstrating that mRNAs of VEGF receptors (KDR, flt-1) are more strongly expressed in metastases of colorectal tumors than in the surrounding liver tissue. The proliferating endothelial cells form large convoluted vessels penetrating deep into the metastases (Figure 37.8A), lacking an immunohistochemically detectable basement membrane (Figure 37.8B) and are continuous with the

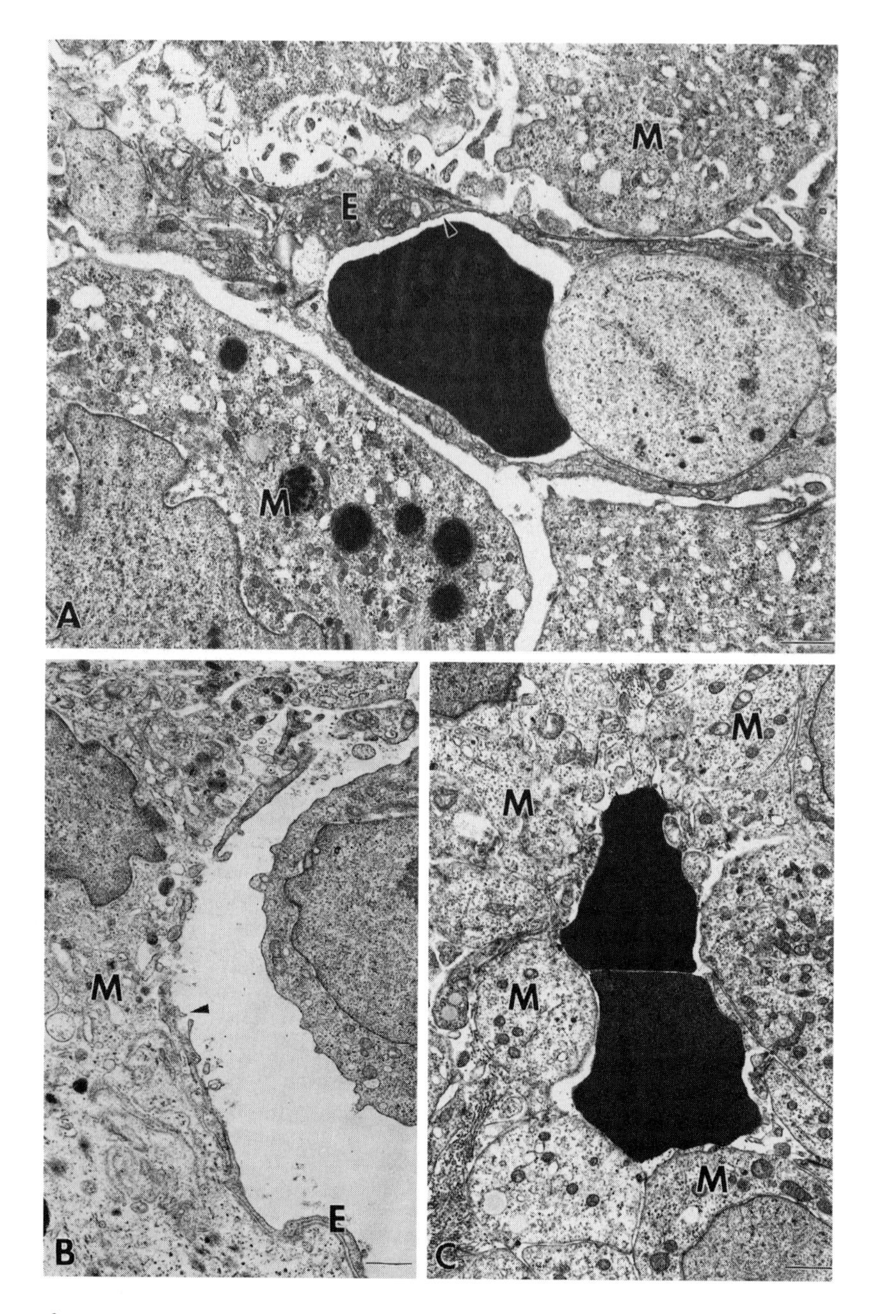

Figure 37.6 Ultrastructure of various intratumoral vessel-types in human melanoma. (A) Normal vessel in human melanoma. (B) Degenerating vessel. Note the fragmentation of the endothelium (E). (C) Tumor-cell lined sinus in melanoma containing red blood cell. M = melanoma cell.

sinusoidal system of the liver, consequently supplied mainly by the portal vein (Paku, 1998).

PROGNOSTIC CONSIDERATIONS

Tumors (benign as well as malignant) cannot grow beyond 1–2 mm^3 in size without vascularization, which is achieved by secreting potent angiogenic factors. However, only malignant and metastatic tumor cells can use vessels for hematogenous dissemination, therefore, tumor induced angiogenesis (TIA) is a necessary but not sufficient prerequisite for metastasis. In the past decade, systematic studies have provided ample evidence that the extent of tumor progression and metastasis correlate with TIA in the case of breast, urogenital, lung and GI tract cancers. Accordingly, vascularization of malignant tumors now serves as a clinicopathological prognostic factor. However, the role of angiogenesis in tumor progression further supports the clinical importance of tumor angiogenesis

A

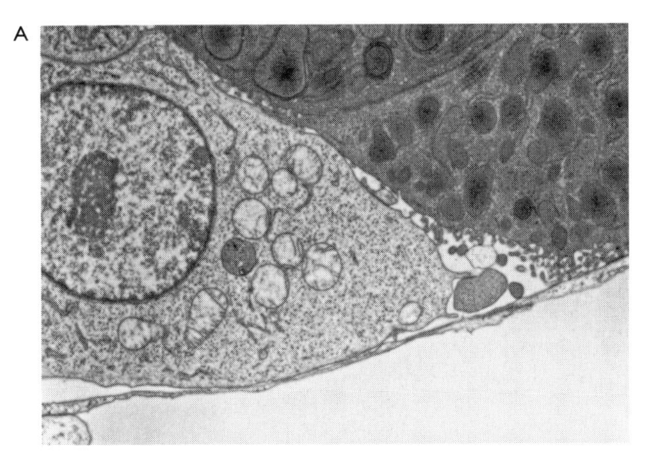

B

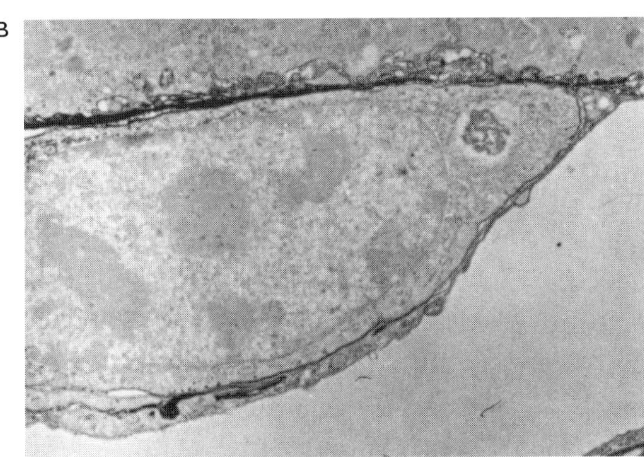

Figure 37.7 Early stage of tumor-induced neoangiogenesis in liver metastasis. Electron microscopy. (A) Migrating tumor cell under liver sinusoid. Note the intact endothelial cover. (B) The sinusoidal basement membrane (laminin) is intact during perisinusoidal tumor cell migration.

A

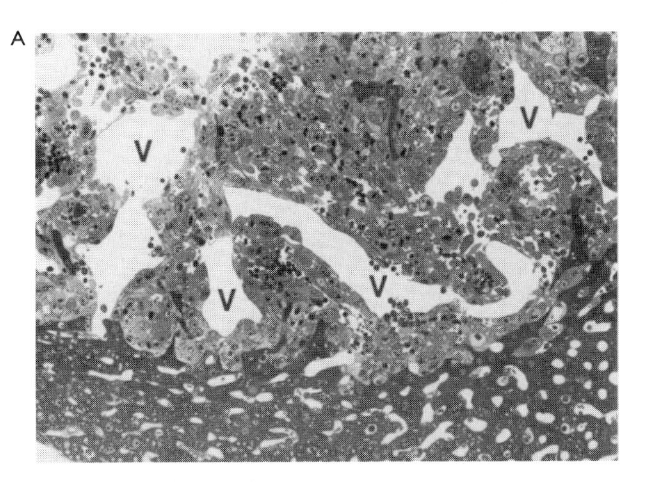

B

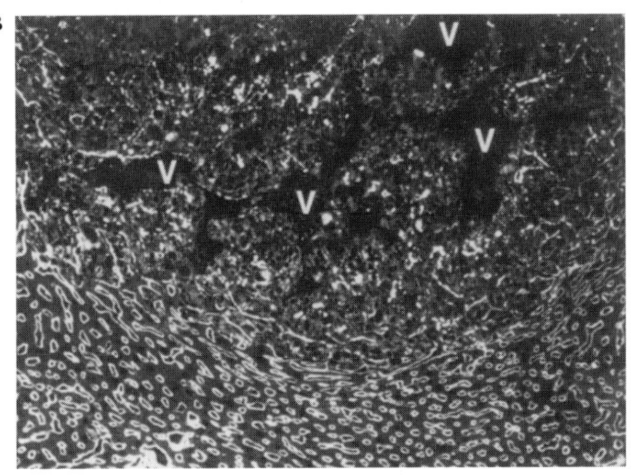

Figure 37.8 Morphology of vessels in liver metastasis. (A) Light microscopy. Large convoluted vessels (V) penetrating liver metastasis. (B) Fluorescent immunohistochemistry of laminin. Note the absence of basement membrane component from the intratumoral vessels (V).

as a necessary target to treat not only primary tumors but the metastatic lesion as well (Folkman, 1995; Kerbel, 2000).

THERAPEUTIC CONSIDERATIONS

The above results strongly suggest that the type of vasculature and blood supply can be dependent on the localization of metastases in organs with both arterial and venous blood supply. Accordingly, these and other observations support the possibility that the tissue architecture of organs, in which angiogenesis is taking place, has an important impact on the process of angiogenesis and consequently on the structure of the developing neovasculature. On the other hand, the properties of tumor cells can also determine the outcome of the angiogenic response. For example, the previously described process of angiogenesis in liver metastases can also be valid in the case of anaplastic tumors, which have high invasive ability and poor stroma-inducing potential. On the contrary, well differentiated tumors showing the same tissue architecture in primary tumors as in metastases, probably elicit a different type of angiogenesis. Accordingly, the same "anti-angiogenic intervention" effective in primary tumors may not be effective in their metastasis or that of one particular cancer may not be effective in a different type of cancer.

The methods used widely in the therapy of primary and secondary liver tumors (hepatic artery infusion, hepatic artery ligation and chemoembolization) are based on the notion that most liver metastases have an arterial blood supply. On the other hand, it is generally accepted that during angiogenesis, new vessels originate mainly from postcapillary venules. Some data suggest that a considerable portion of human and experimental liver metastases have a portal blood supply. Accordingly, in the future more attention should be paid to the portal blood supply during liver metastasis therapy (Paku, 1998).

Table 37.4 Therapeutic angiogenesis trials

Rationale	*Therapy*	*Indication*
Angiogenic growth factor		
Recombinant protein	Rh-PDGF-BB	Diabetic ulcer
	RhFGF-1	Myocardial ischemia
Gene therapy	VEGF	Limb & heart ischemia
Recombinant protein	G-CSF	Limb ischemia
synthetic peptide	bFGF&VEGF	Venous ulcer
FGF-concentration	Cu-peptide	Hair implant
transmyocardial revascularization	Laser therapy	Ischemic heart disease
Oxygenization	Hyperbaric oxygen	Ischemic wound

Notes: bFGF: basic fibroblast growth factor, Cu: copper, G-CSF: granulocyte-colony-stimulating factor, PDGF: platelet-derived growth factor, Rh: recombinant human, VEGF: vascular endothelial growth factor.

37.3 Angiogenesis therapy

37.3.1 *Therapeutic angiogenesis (Table 37.4)*

Therapeutic angiogenesis can be defined as the use of biological agents, bioactive materials or environmental conditions to stimulate growth of new vessels, to restore or augment circulatory perfusion of tissues, to reverse ischemia or to accelerate healing. This approach is now used clinically in cases of heart, cerebrovascular diseases, in critical limb ischemia, delayed wound healing and in peptic ulcer disease. Despite the variety of existing angiogenic mechanisms, most ischemic tissues are generally unable to produce an appropriate response to reverse the disease process including heart, brain, limb and skin.

37.3.1.1 *Angiogenic growth factors*

Among the angiogenic growth factors (Table 37.1), six are now undergoing clinical trials including the most potent and universal factors such as VEGF, a/bFGF, PDGF and G-CSF. These studies employ human recombinant proteins as well as exploiting gene technology for delivering naked DNA or using adenoviral vectors to transfect endothelial cells. Clinical application of G-CSF is based on the observation that, contrary to expectation, embryonic angiogenesis (mobilization of bone marrow endothelial stem cells to the site of neoangiogenesis) is possible in adults too, since they are continously present, though in low number, in the circulation (Isner *et al.*, 1999; Carmeliet, 2000).

37.3.1.2 *Bioactive synthetic materials*

In major wounds and skin loss such as burns, biocompatible materials containing matrices and angiogenesis modulators can be used to promote the healing process. A wound dressing containing a synthetic copper peptide which can accumulate FGF can be used efficiently to promote healing. An alternative to this is the use of a tissue-engineered biological coat containing cells which produce large quantities of angiogenic factors such as FGF or VEGF (Rubanyi, 2000).

37.3.1.3 *Laser therapy*

In the case of heart ischemia, transmyocardial revascularization has been created by employing a laser beam. Following photoacoustic injury, myocytes, inflammatory cells and platelets provide a rich source of angiogenic factors. The device has now been designed for intraoperative use (bypass surgery) as well as a transcutaneous device (Rubanyi, 2000).

37.3.1.4 *Oxygen therapy*

Hyperbaric oxygenation is primarily designed to treat surface wounds. Patients are placed into a chamber, where the oxygen pressure is gradually increased to a much higher level than normal physiologically. At 2.0ATA blood plasma begins to carry an increased amount of oxygen providing a systemic oxygen burst. It has been shown that high oxygen tension stimulates proliferation of endothelial cells and induce angiogenesis in ischemic tissues (Rubanyi, 2000).

37.3.2 *Pharmaceutical inhibition of angiogenesis*

The recognition of the angiogenesis dependence of diseases such as malignant tumor growth and progression as well

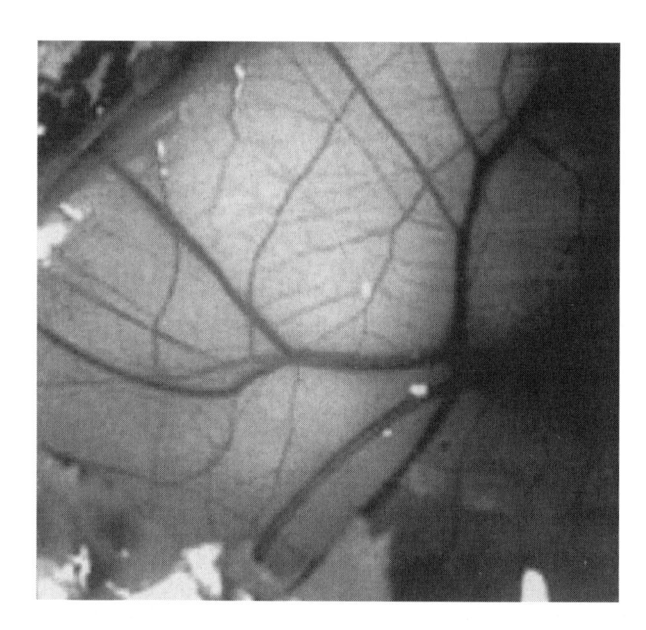

Figure 37.9 Chicken chorioallantoic membrane in a 7 day old egg. Note the well developed vessel network. Macrophotography.

Table 37.5 Classical therapeutic agents with anti-angiogenic potential.

Drug	*Target*	*Primary indication*	*Antiangiogenic effect*
Vinblastine	MT depolimerisation	Anticancer agent	Endothelial permebility
Taxol	MT stabilization	Anticancer agent	Endothelial proliferation
Dolestatin	MT stabilization	Anticancer agent	Endothelial proliferation
Combrestatin A	MT stabilization	Anticancer agent	Endothelial proliferation
Flavone acetic acid		Anticancer agent	Angiogenesis
Irsogladine		Anti-gastric ulcer agent	Angiogenesis
Radicicol		Antimicrobial agent	Angiogenesis
Cyclosporine	Signal transduction	Immunosuppressant	Angiogenesis
Captopril	ACE blocker	Hypertension	Vascular development

ACE: acethylcoline estherase, MT: microtubule.

as retinopathies stimulated extreme translational research activity over the past decade. Much data has now accumulated on the major molecular pathways controling angiogenesis in adults and in these diseases. It has become evident, that successful interference with pathological neoangiogenesis can be achieved in several ways. However, major attention has been paid to tumor progression, where antiangiogenic approaches may fundamentally change the clinical management. It has been necessary to develop new *in vitro* and *in vivo* models where the pro- and anti-angiogenic agents can be tested safely. These are now available as human endothelial cell cultures (veins and arteries), migration assays, *in vitro* tube formation in collagen or basement membrane gels, *in vivo* assays of chicken chorioallantois membrane for angiogenesis (Figure 37.9), rabbit corneal packet assay, availability of human and rodent endothelial markers, morphometric and functional tests.

An interesting development in the field is that conventional therapeutic agents designed to treat angiogenesis-related or unrelated diseases have been found to have "side effects" of antiangiogenic potential. These include mostly anticancer drugs but others are also became members of this "club" (Table 37.5).

On the other hand, molecular technologies have also been successfully applied to promote pharmaceutical research in this field. As a result, there is now a continuously growing battery of natural and synthetic compounds which have been shown to be antiangiogenic (Table 37.6). Furthermore, several members of this family have now entered clinical trials for the treatment of advenced cancer as well as some ocular diseases (Schiller, 1999).

37.3.2.1 Endothelial growth factor ligands as pharmaceutical targets

Since research over the past few years identified the major endothelial mitogenic growth factors (VEGF, FGF, PDGF, HGF), these ligands and their molecular interactions with their signaling or accesory receptors have become a major pharmacological target. Recognition of the common chemical nature of these endothelial growth factors, i.e. their heparin-binding potential, provides an easy though non-specific anti-angiogenic target. The earliest inhibitors were sulphated polysaccharides (D54152) and cationic peptides (protamine) or molecules (Suramin). Identification of the peptide sequences of the major angiogenic mitogens facilitated the identification of heparin-binding cationic peptides of VEGF or HGF. These usually small molecular inhibitors act as competitive antagonists for the heparin-binding ligand preventing endothelial receptor activation. Although their concentration has to be much higher *in vivo* than the physiological ligand, they can be safely administered since they are relatively non-toxic compounds. Meanwhile the most specific approach here is to target the major and the most specific endothelial mitogen, VEGF165. The recent discovery of the secondary and tertiary structure of VEGF165 provides new information for the design of ligands, monoclonal antibodies or peptidomimetics which can eliminate the effect of VEGF. The humanized anti-VEGF165 monoclonal antiobodies are now entering clinical trials following their success in various experimental angiogenesis models including tumor progression models. An old, clinically introduced cytokine, interferon (α/β) is now used clinically to

Table 37.6 Pharmaceutical angiogenesis inhibitors

Agent	*Target*	*Clinical trial*
2-Metoxy-estradiol	Hormone	
	ERβ	+
	Growth factor ligand	
D54152-sulphated polysaccharide	FGF	
Protamine (cationic protein)	Heparin-binding factors	
VEGF-cationic peptide	VEGF	
HGF-cationic peptide	HGF	
Suramin (polysulphonated naphtyl urea)	FGF, VEGF, HGF	+
Anti-VEGF Ab	VEGF165	+
DistamycinA	bFGF, PDGF-B	
Interferon α and β	bFGF	+
	Growth factor receptor	
PF-4	FGFR	+
Anti-VEGFR2 Ab	VEGFR2	+
	Growth factor signaling	
Isoflavonoids	Non-spec. PTK-inhibitors	+
Staurosporine derivatives	Non-spec. PTK-inhibitors	
SU5416 (adenin-mimetic)	VEGFR2-PTK-inhibitor	+
Tyrphostins	VEGFR2-PTK-inhibitor	
Isoquinolines	PDGFR-PTK-inhibitor	+
Retinoic acid	PKC-inhibitor	+
Carboxyamidotriazole	Ca2+channel-blocker	+
	Transcription	
Thalidomide	αv and β3 integrin genes	+
	Matrix ligand	
Rh-Angiostatin	uPAR, ATP-synthase	+
Rh-Endostatin	Collagen receptors	+
Cyclic-RGD peptide	αvβ3	
Accutin-disintegrin	αvβ3, αvβ5	
benzodiazepines	αvβ3	
	Matrix receptor	
Humanized anti-αvβ3 Ab	αvβ3	+
	Protease	
Rh-PAI-2	uPA	
Amiloride	uPA	
P-amidobenzamidine	uPA	
Anti-uPA Ab	uPA	
Anti-uPAR Ab	uPAR	
L-phanylalanin-N-methylamides (Batimastat, Marimastat)	MMP-2, MMP-9	+
AG3340	MMPs	+
Minocycline	MMPs	

Ab: antibody, ER: estrogen receotir, PKC: protein kinase C, PTK: protein tyrosine kinase, MMP: matrix metalloproteinase, Rh: human recombinant, uPA: urokinase-plasminogen activator, uPAR: uPA-receptor.

down-regulate FGF expression both in endothelial as well as in stromal and tumor cells (Malonne *et al.*, 1999; Schlaepi *et al.*, 1999; Schiller, 1999; Bae *et al.*, 2000; Kerbel, 2000; Thompson *et al.*, 2000).

37.3.2.2 Endothelial mitogen receptors as pharmaceutical angiogenesis targets

Recognition and identification of the major endothelial growth factors made it possible to identify their receptors as well (flt1, KDR). The VEGF165 receptor (KDR) is almost exclusively expressed in endothelial cells providing a highly specific endothelial target. Monoclonal antibodies are now available which recognize various epitopes on the extracellular domain of KDR. Some antibodies, such as DC101, are able to interfere with the signaling of KDR similar to some anti-c-erb2 monoclonals. Although it is not a very promising single agent for anti-angiogenic therapy of tumors, combination with low dose conventional cytostatic agent (such as vinblastin) make it possible to reduce unwanted side effects of the cytostatic drug maintaining the antitumoral effect. PF-4 is a natural inhibitor of angiogenesis targeting endothelial cells and the recombinant variant can now be used to inhibit ligantion of the FGF-receptor (Kerbel, 2000).

37.3.2.3 *Angiogenic signaling as a pharmaceutical target*

Identification of the receptors of the major angiogenic mitogens revealed that the majority of these receptors fall into the category of tyrosine kinase (TK) growth factor receptors (flt1, KDR, c-met, c-kit etc). Accordingly, the tyrosine kinase activity of proliferating endothelial cells is a natural drug target. In the early phase of such investigations relatively non-specific kinase inhibitors, such as staurosporine, were used successfully to cope with the signaling activity of endothelial cells. Later, more specific TK-inhibitors became available which could specifically inhibit signaling of the KDR-(SU5416) or the PDGF-receptor (isoquinolines). These TK-inhibitors are highly specific antiangiogenic agents which have now entered clinical trials to treat advanced cancer. Some downstream steps of angiogenic signaling, such as PKC-activation or Ca-signaling, provide powerful though less specific anti-angiogenic targets (retinoic acid or other PKC inhibitors and carboxyamidotriazole, respectively). In tumor progression, since tumor cells as well as endothelial cells are both targets for therapeutic interventions due to the involvement of various mitogenic signaling pathways, the use of such broad spectrum TK inhibitors may provide a rational alternative for antitumor therapy (Luzzi *et al.*, 1998; Malonne *et al.*, 1999; Schlaepi *et al.*, 1999; Kerbel, 2000).

37.3.2.4 *Endothelial cell extracellular matrix interaction as a pharmaceutical target*

Neoangiogenesis is characterized by changes in the endothelial cell-ECM interactions. Mitogenic stimuli introduced by angiogenic factors (bFGF, VEGF, HGF, PDGF) induce expression of new adhesion molecules such as $\alpha v\beta 3$ or $\alpha v\beta 5$ integrins, characteristic of proliferating and migrating endothelial cells (Elicieri *et al.*, 1999). On the other hand, the permeable vessel wall leaks out soluble matrix ligands such as vitronectin, fibrinogen/fibrin or tenascin which are key matrix components of the remodeling vessel (Browder *et al.*, 2000). Mitotically activated endothelial cells have to degrade, at least partially, their subendothelial basement membrane, therefore the mitogens induce protease secretion as well (uPA and MMP-2, MMP-9). All of these events are critical for the initiation of angiogenesis and therefore may serve as drug targets.

Ligand competition for the $\alpha v\beta 3$ integrin is enough to immobilize and detach endothelial cells from their matrix, inducing their programmed cell death (Elicieri *et al.*, 1999). This can be achieved in several ways: using cyclic RGD-peptides (DeNardo *et al.*, 2000), chemical RGD-mimetics (Nicolau *et al.*, 1998) or snake venom ligands (disintegrins with $\alpha v\beta 3$ specificities, Yeh *et al.*, 1998). However, the most powerful matrix ligands to prevent or inhibit endothelial cell proliferation, migration and neonagiogenesis are degradation products of the perivascular matrix proteins, plasmin and collagen XVIII, angiostatin (O'Reilly *et al.*, 1994) and endostatin (O'Reilly *et al.*, 1997) respectively. These proteins are highly effective *in vitro* and *in vivo* at coping with both the tumor induced angiogenesis as well as other types of neo-angiogenesis such as retinopathies (Malonne *et al.*, 1999; Schiller, 1999; Kerbel, 2000). Their therapeutic window is extremely wide, they can be administered for a very long period without significant resistance developing in endothelial cells. Since recombinant human forms are now available, both of these agents have entered clinical trials to treat advanced and/or primary cancer.

The expression of $\alpha v\beta 3$ integrin is a hallmark of proliferating endothelial cells, therefore this integrin itself provides a feasible drug target. Some anti-$\alpha v\beta 3$ monoclonal antibodies have been proven to be anti-angiogenic, partially because the ligation induces apoptosis of the endothelial cells, partially because of the block of integrin signaling. Development of humanized anti-$\alpha v\beta 3$ monoclonal antibodies (Vitaxin and RheoPro) made it possible to use such agents clinically, similar again to the anti-erbB2 antibody treatment of advanced breast cancer (Malonne *et al.*, 1999). The $\alpha v\beta 3$ integrin on the surface of endothelial cells not only serves as matrix ligand but also provides a surface site for protease binding and accumulation (MMP-2 and uPA), critical in the initial step of endothelial cell release from the pre-existing vessel (Stetler-Stevenson, 1999). Therefore, the dramatic *in vivo* and clinical effects of anti-$\alpha v\beta 3$ antibody treatment of angiogenesis might be partially due to these effects.

Migration of endothelial cells is impossible without secretion and function of the key matrix degrading proteases, MMP2 and uPA, therefore recognition of this fact has also made these enzymes targets for therapeutic interventions. Interestingly, although there are several different forms of therapeutic uPA inhibitors now available, (chemical inhibitors, anti-uPA antibodies, or anti-uPA-receptor antibodies), none of these have entered clinical trials (Malonne *et al.*, 1999). On the other hand, metalloprotease inhibitors are the first anti-angiogenic agents that have entered clinical trials (Schiller, 1999). There are now both broad-spectrum MMP inhibitors as well as MMP-2 and MMP-9-specific inhibitors available – the enzymes are used primarily by migrating endothelial cells and invading tumor cells. This dual target theory is most probably behind competition between companies to develop more and more MMP-inhibitor variants.

One of the most interesting (and promising) anti-angiogenic agent to date is thalidomide, which was introduced a couple of decades ago as a sedative agent but was withdrawn due to its teratogenicity. Later it was found that the teratogenic effect of thalidomide is due to its powerful angiogenesis inhibitory potential. Although it is still not completely understood, it seems that thalidomide inhibits expression of genes having SP1 promoter sites including αv and β3 integrins, the key adhesion molecules of proliferating and migrating endothelial cells. Recently, thalidomide has become the first successful anti-angiogenic agent applied clinically to treat retinal neovascularization as well as a broad range of cancers including prostate, breast, Kaposi's sarcoma, glioma and melanoma (Schiller, 1999; Stephens *et al.*, 2000).

37.4 Conclusion

Recognition of the physiological and pathological significance of angiogenesis in various human diseases has initiated studies to understand the underlying molecular mechanisms. Identification of those pathways of angiogenesis clearly underlined the potential molecular therapeutic targets for treatment. Discovery of a broad range of new drugs has not only created a new subdiscipline, angiogenesis pharmacology, but has begun to provide clinical treatments for previously untreatable diseases such as cancer or blindness.

References

1 Bae, D.G., Gho, Y.S., Yoon, W.H., and Chae, C.B. (2000) Arginic-rich anti-vascular endothelial growth factor peptides inhibit tumor growth and metastasis by blocking angiogenesis. *J.Biol.Chem., 275, 13588–13596.*

2 Bowder, T., Folkman, J., and Pirie-Shepherd, S. (2000) The hemostatic system as a regulator of angiogenesis. *J. Biol. Chem., 275, 1521–1524.*

3 Brown, J.M. (2000) Exploiting the hypoxic cancer cell: mechanisms and therapeutic strategies. *Molec. Med. Today, 6, 157–162.*

4 Bussolino, F., Albini, A., Camussi, G., Presta, M., Viglietto, G., Ziche, M., and Persico, G. (1996) Role of soluble mediators in angiogenesis. *Eur. J. Cancer, 32A, 2401–2412.*

5 Carmeliet, P. (2000) Mechanisms of anigogenesis and arteriogenesis. *Nat. Med., 6, 389–395.*

6 Christofidou-Solomidou, M., Bridges, M., Murphy, G.F., Albelda, S.M., and DeLisser, H.M. (1997) Expression and function of endothelial cell αv integrin receptors in wound-induced human angiogenesis in human skin/SCID mice chimeras. *Am. J. Path., 151, 975–983.*

7 Dawson, D.W., Volpert, O.V., Gillis, P., Crawford, S.E., Xu, H-J., Benedict, W., and Bouck, N.P. (1999) Pigment epithelium-derived factor: a potent inhibitor of angiogenesis. *Science, 285, 245–251.*

8 DeNardo, S.J., Burke, P.a., Leigh, B.R., O'Donell, R.T., Miers, L.A., Krogen, L.A., Goodman, S.L., Matzku, S., Jonczyk, A., Lamborn, K.R., and DeNardo, G.L. (2000) Neovascular targeting with cycle RGD peptide (cRGDf-ACHA) to enhance delivery of radioimmunotherapy. *Cancer Biothe. Radiopharmaceut., 15, 71–79.*

9 Elicieri, B.P., and Cheresh, D.A. (1999) The role of αv integrins during angiogenesis: insights into potential mechanisms of action and clinical development. *J. Clin. Invest., 103, 1227–1230.*

10 Folkman, J. (1995) Angiogenesis in cancer, vascular, rheumatoid and other disease. *Nat. Med., 1, 27–31.*

11 Griffionen, A.W., Coenenm, M.J.H., Damen, C.A., Hellwing, S.M.M., Vooys, W., Blijham, G.H., and Groenewegen, G. (1997) CD44 is involved in tumor angiogenesis; an activation antigen on human endothelial cells. *Blood, 90, 1150–1159.*

12 Isner, J.M., and Asahara, T. (1999) Angiogenesis and vasculogenesis as therapeutic strategies for postnatal neovascularization. *J. Clin. Invest., 103, 1231–1235.*

13 Jacobson, B.S. (2000) Hereditary hemorrhagic telangiectasia a model for blood vessel growth and enlargement. *Am. J. Path, 156, 737–742.*

14 Jain, R.K., and Munn, L.L. (2000) Leaky vessels? Call Ang1! *Nat. Med., 6, 131–132.*

15 Jiménez, B., Volpert, O.V., Crawford, S.E., Febbraio, M., Silverstein, R.L., and Bouck, N. (2000) Signals leading to apoptosis-dependent inhibition of neovascularization by thrombospondin-1. *Nat. Med., 6, 41–48.*

16 Kerbel, R.S. (2000) Tumor angiogenesis: past, present and the near future. *Carcinogensis, 21, 505–515.*

17 Lau, L.F., and Lam, S.C-T. (1999) The CCN family of angiogenic regulators: the integrin connection. *Exp. Cell Res., 248, 44–57.*

18 Lee, P., Wang, C.C., and Adamis, A.P. (1998) Ocular neovascularization: an epidemiologic review. Ocular neovascularization: an epidemiologic review. *Surv. Ophthalmol., 43, 245–269.*

19 Li, J., Post, M., Volk, R., Gao, Y., Li, M., Metais, C., Sato, K., Tsai, J., Aird, W., Rosenberg, R.D., Hampton, Th.G., Li, J., Sellke, F., Carmeliet, P., and Simon, M. (2000) PR39. A peptide regulator of angiogenesis. *Nat. Med., 6, 49–55.*

20 Li, W.W., Jaffe, M., Li, V.W., and Tsakayannis, D. (2000) Lessons to be learned from clinical trials of angiogenesis modulators in ischemic diseases. In G.M. Rubanyi, (eds.), *Clinical trials of angiogenesis modulators*, Marcel Dekker, Inc. New York, Basel, pp. 519–536.

21 Luzzi, K.J., Varghese, H.J., MacDonald, I.C., Schmidt, E.E., Koh, E.C., Morris, V.L., Marshall, K.E., Chambers, A.F., and Groom, A.C. (1998/1999) Arginie-rich anti-vascular endothelial growth factor peptides inhibit tumor growth and metastasis by blocking angiogenesis. *J. Biol. Chem., 4, 373–379.*

22 Malonne, H., Langer, I., Kiss, R., and Atassi, G. (1999). Mechanisms of tumor angiogenesis and therapeutic implications: angiogenesis inhibitors. *Clin. exp. Metast., 17, 1–14.*

23 Murohara, T., Ikeda, H., Duan, J., Shintani, S., Sasaki, Ki, Eguchi, H., Onitsuka, I., Matsui, K., and Imaizumi, T. (2000) Transplanted cord blood-derived endothelial precursor cells

augment postnatal neovascularization. *J. Clin. Invest., 105, 1527–1536.*

24 Nicolaou, K.C., Trujilo, J.I., Jandeleit, B., Chibale, K., Rosenfeld, M., Diefenbach, B., Cheresh, D.A., and Goodman, S.L. (1998) Design, synthesis and biological evaluation of non-peptide integrin antagonists. *Bioorg. Med. Chem. 6, 1185–1208.*

25 Nie, D., Lamberti, M., Zacharek, A., Li, L., Szekeres, K., Tang, K., Chen, Y., and Honn, K.V. (2000a) Thromboxane A(2) regulation of endothelial cell migration, angiogenesis, and tumor metastasis. *Biochem. Biophys. Res. Commun., 267, 245–251.*

26 Nie, D., Tang, K., Diglio, C., and Honn, KV. (2000b) Eicosanoid regulation of angiogenesis: role of endothelial arachidonate 12-lipoxygenase. *Blood, 95, 2304–2311.*

27 Oechler, M.K., Rees, M.C., and Bicknell, R. (2000) Steroids and the endometrium. *Curr. Med. Chem., 7, 543–560.*

28 O'Reilly, M.S., Boehm, T., Shing, Y., Fukai, N., Vasios, G., Lane, W.S., Flynn, E., Birkhead, J.R., Olsen, B.R., and Folkman, J. (1997) Endostatin: an endogenous inhibitor of angiogenesis and tumor growth. *Cell, 88, 277–285.*

29 O'Reilly, M.S., Holmgren, L., Shing, Y., Chen, C., Rosenthal, R.A., Moses, M., Lane, W.S., Cao, Y., Sage, E.H., and Folkman, J. (1994) Anigostatin: a novel angiogenesis inhibitor that mediates the suppression of metastases by a Lewis lung carcinoma. *Cell, 79, 315–328.*

30 Paku, S. (1998) Current concepts of tumor-induced angiogenesis. *Pathol. Oncol. Res., 4, 62–75.*

31 Rogers, P.A., Lederman, F., and Taylor, N. (1998). Endmetrial microvascular growth in normal and dysfunctional states. *Hum. Reprod. Update, 4, 503–580.*

32 Sasisekharan, R., Moses, M.A., Nugent, M.A., Cooney, Ch.L., and Langer, R. (1994) Heparinase inhibits neovascularization. *Proc. Nat. Acad. Sci USA, 91, 1524–1528.*

33 Schiller, J. (1999). Early trials of antiangiogenic agents. *Anticancer in Oncology, 15, 3–10.*

34 Schlaeppi, J-M., and Wood, J.M. (1999) Targeting vascular endothelial growth factor (VEGF) for anti-tumor therapy, by anti-VEGF neutralizing monoclonal antibodies or by VEGF receptor tyrosine-kinase inhibitors. *Cancer Metastasis Rev., 18, 473–381.*

35 Stephens, T.D., and Fillmore, B.J. (2000) Hypothesis: Thalidomine embryopathy-proposed mechanism of action. *Teratology 61, 189–195.*

36 Stetler-Stevenson, W.G. (1999) Matrix metalloproteinases in angiogenesis: a moving target for therapeutic intervention. *J. Clin. Invest.,103, 1237–1241.*

37 Thompson, W.D., Li, W.W., and Maragoudakis, M. (2000) The clinical manipulation of angiogenesis: pathology, side-effects, surprises, and opportunities with novel human therapies. *J. Pathol., 190, 330–337.*

38 Tímár, J., Tóth, J. (2000) Tumor sinuses – vascular channels. *Pathol. Oncol. Res., 6, 83–86.*

39 Yeh, Ch.H., Peng, H-Ch., and Huang, T-F. (1998) Accutin, a new desintegrin, inhibits angiogenesis *in vitro* and *in vivo* by acting as integrin $\alpha_v\beta_3$ antagonist and inducing apoptosis. *Blood 92, 3268–3276.*

40 Zhang, M., Volpert, O., Shi, Y.H., and Bouck, N. (2000) Maspin is an angiogenesis inhibitor. *Nat. Med., 6, 196–199.*

Part 9

Endocrinal and gastrointestinal disorders

Chapter 38

Diabetes and the heart

Lindsay Brown

Contents

38.1 Cardiovascular disease and diabetes: the extent of the problem

Cardiovascular disease remains the major cause of death in developed countries causing 40–45% of all deaths[1]. The incidence of cardiovascular disease markedly increases with age with symptoms present in more than 60% of people over 75 years of age yet in less than 9% of those under 35 years of age. Many factors increase the risk of cardiovascular disease, in particular hypertension, hypercholesterolemia, obesity, diabetes, insufficient physical activity, high alcohol use and tobacco smoking. Several of these risk factors tend to cluster; this is referred to as the Metabolic Syndrome – glucose intolerance leading to Type 2 or non-insulin-dependent diabetes together with hypertension, hyperlipidemia and central obesity[2,3].

Obesity and diabetes are major health issues reaching epidemic proportions in developed countries[3]. Diabetes is estimated to affect about 6% of the population in the USA or 12–15 million people, with up to half being undiagnosed[4,5]. The prevalence of diabetes increases with age up to 25–30% with another 10–25% of adults having impaired glucose tolerance[6]. These rates are greater in groups who have recently changed from a traditional to modern lifestyle such as Mexican-Americans, African-Americans, Micronesians and Australian Aborigines; as an example, the prevalence of Type 2 (non-insulin-dependent) diabetes among Pima Indians increased to 10 times that of the general US population over a period of 30 years to 1967[7]. Further, the prevalence increased 6-fold among Pima Indian adolescents from 1967–76 to 1987–96 to 50.9 per 1000, 15 to 19-year olds, about 20-fold higher than in other comparable American Indian groups[8]. For many patients, diabetes is diagnosed or aggressively treated only when they develop one of the characteristic diabetic complications such as retinopathy which leads to blindness, impaired kidney function leading to end-stage renal disease, painful diabetic neuropathy which may lead to ulcers or amputations, or heart disease and stroke. Ultimately, up to 80% of Type 2 (non-insulin-dependent) diabetics die from cardiovascular disease[5,9,10].

Many of the aging population are thus likely to have both diabetes and cardiovascular disease. In humans, diabetes is associated with long term cardiovascular damage (especially endothelial disease, fibrosis and cardiomyopathy) with a much higher risk of coronary disease, heart failure, myocardial infarction and death[11]. This chapter will consider the mechanisms for the development of diabetic complications in the cardiovascular system and discuss actual and potential therapeutic avenues to reduce this adverse risk.

38.2 Diabetes mellitus: a chronic metabolic disorder

In patients with diabetes mellitus, an increased blood glucose concentration (hyperglycemia) causes an increased thirst, hunger and urine volume. Glucose is the major energy source for all cells. The control of blood glucose is necessary since energy must be supplied to all cells at all times despite intermittent food intake and variable demands, such as the level of exercise. The major regulatory hormone for intermediary metabolism is insulin (Figure 38.1), produced and secreted by the β-cells of the islets of Langerhans of the pancreas. The single most important discovery in the history of diabetes was that of insulin by Frederick Banting,

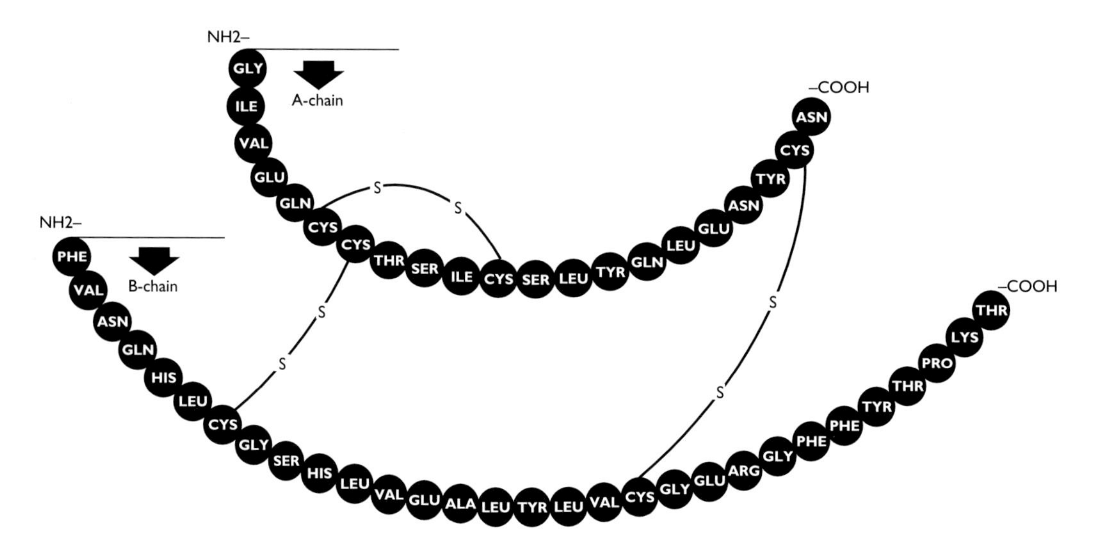

Figure 38.1 The structure of human insulin.

John Macleod, James Collip and Charles Best in Toronto in 1921/22 which won them the Nobel prize in 1923[12]. Prior to this, Type 1 (insulin-dependent) diabetes was a universally fatal disease; the Greek physician Areteus had described the terrible deaths of these patients following "a melting down of the flesh and limbs into the urine" in the second century AD[12]. Insulin was the first protein whose amino acid sequence was determined[13] leading to the Nobel Prize for Frederick Sanger in 1958, and also the first hormone for which a radioimmunoassay was developed[14].

The two major immediate causes of diabetes are either a failure of insulin production (insulin-dependent or Type 1 diabetes) or a failure of glucose-dependent tissues to respond to insulin (non-insulin-dependent or Type 2 diabetes). The WHO report[2] recommends that the terms "Type 1" and "Type 2" be used; the terms "insulin-dependent" or "non-insulin-dependent" are considered to be confusing and led to patients being classified on treatment rather than pathogenesis. In Europe, North America and Australia, Type 1 diabetes accounts for 5–15% of all diabetics. Type 1 diabetics usually have an earlier onset of disease, many around the age of puberty, are usually thin and are ketoacidosis-prone. In contrast, the typical Type 2 diabetic has a much later onset of symptoms, often after 50 years of age, is usually obese and is unlikely to exhibit ketoacidosis. Diabetes has no distinct etiology, pathogenesis or definitive and curative therapy. The complete clinical syndrome involves hyperglycemia, large-vessel disease especially atherosclerosis, small-vessel disease especially to the retina and kidney, and neuropathy[15].

38.3 Animal models of diabetes

Animals, especially rats, mice and dogs, are essential in understanding the changes in cardiovascular structure and function in diabetes. Alloxan became the first identified diabetogenic chemical in 1943[16]. It shows selective inhibitory and cytotoxic effects on pancreatic β-cells but the precise mechanism is unknown. As an example, alloxan-diabetic rats were used recently to determine the effects of a β_3-adrenoceptor agonist on lipid metabolism[17].

Diabetes can be induced by selective destruction of the insulin-producing β-cells of the pancreas with a single, rapid injection of streptozotocin, a glucose moiety with a very reactive nitrosourea group from the mould *Streptomyces griseus*. This procedure, first introduced in 1963, has since been used in over 7600 PubMed citations probably making this the second most used animal model for disease after the spontaneously hypertensive rat (SHR); cardiovascular changes following streptozotocin have been reviewed[18]. Streptozotocin induces the characteristic cardiovascular and renal features of humans with uncontrolled Type 1 diabetes[19]. Streptozotocin doses of 50–65 mg/kg lead to hyperglycemia (20–30 mM) but severe ketosis does not develop and the rats may survive for at least 24 weeks without exogenous insulin. Higher doses (75 mg/kg and above) result in spontaneous ketosis and death within days if insulin is not given. Streptozotocin-diabetic rats are usually normotensive and do not develop atherosclerosis, unlike most human diabetics. Streptozotocin administration to spontaneously hypertensive rats (SHR) produces a model of Type 1 diabetic hypertensive humans; these rats show progressive cardiac deterioration[20] and have been used to measure responses of antihypertensive drugs on hypertrophy and vascular permeability[21]. A different rat model, the Cohen-Rosenthal diabetic hypertensive rat, was developed by cross-breeding the Cohen diabetic rat with SHR and continued feeding of a copper-poor sucrose diet. Chronic treatment with enalapril (ACE inhibitor) or losartan (angiotensin receptor antagonist) lowers blood pressure but only enalapril improves diabetic control, indicating that blood glucose reduction in this model is dependent on bradykinin accumulation[22].

The spontaneously diabetic BioBreed (BB/DP) rat discovered in 1974 may be a closer mimic of human Type 1 diabetes than chemically-induced diabetes as these rats depend on insulin for survival. In this model, cell-mediated autoimmune destruction of the pancreatic β-cells leads to insulin deficiency, hyperglycemia, increased plasma free fatty acids and glycosuria. Isolated myocytes from BB/DP rats show altered excitation-contraction coupling partly because of abnormalities in the myocyte similar to those in human diabetes[23]. Inhibition of the Na^+/H^+-exchanger protects isolated hearts from these spontaneously diabetic rats from ischemic injury[24]. The non-obese spontaneously diabetic NOD mouse resembles the BB/DP rat but overt diabetes develops in 80% of females and only 20% of males[25]. Studies with NOD and ApoE-deficient mice suggest that changes in circulatory pro-inflammatory cytokines are active participants in early atherosclerotic lesion development[26].

Models of Type 2 diabetes, preferably with moderate hypertension and obesity, are very relevant since at least 85% of human diabetics are Type 2 diabetics and many of these are hypertensive and obese. The genetically-determined obese Zucker rat fulfils these criteria: Type 2 diabetes with moderately elevated blood pressure, marked obesity, progressive renal damage and hypercholesterolemia with leptin resistance and moderate insulin resistance[27,28].

The Otsuka Long Evans Tokushima fatty (OLETF) rat has many similarities to human Type 2 diabetes[29]. These CCK-A receptor-deficient rats[30] develop mild obesity, postprandial hyperglycemia, hyperinsulinemia and prolonged deceleration and low peak velocity of transmitral inflow with left ventricular fibrosis[31] and impairment of endothelium-dependent relaxation in the isolated aorta and renal artery[32]. Overexpression of the PDGF β receptor and fibronectin in aortic smooth muscle cells may play an important role in the pre-diabetic accelerated intimal thickening in these rats[33].

The male JCR:LA-cp rat[34] possesses the major elements of the metabolic syndrome, showing obesity, insulin resistance, hyperinsulinemia and hypertriglyceridemia with euglycemia when homozygous for the autosomal recessive *cp* gene. The mutation of the *cp* gene creates a stop codon in the extracellular domain of the leptin receptor[35], leading to the absence of membrane-bound receptors and no insulin-mediated glucose uptake. These rats are atherosclerosis-prone, developing advanced intimal lesions, ischemic lesions of the heart and vasculopathy with hyperproliferative and hyperplastic vascular smooth muscle cells. Treatment with the new insulin sensitiser, S15261, increased insulin sensitivity, decreased insulin levels and reduced the vasculopathy of the JCR:LA-cp rat, indicating possibly effective treatment of the cardiovascular complications of the metabolic syndrome in human diabetics[36].

Hyperglycemia with hyperinsulinemia, hypertriglyceridemia, hypertension and renal impairment can be established by a fructose-enriched diet for 4–10 weeks[37,38]. The endothelin antagonist, bosentan, normalized blood pressure and protected against end-organ damage in this model[38].

Nutrition-evoked diabetes and moderate obesity can be induced by a high-energy diet in the gerbil *Psammomys obesus*, nicknamed the sand rat, a diurnal rodent living in the arid habitat of the North African and Eastern Mediterranean deserts[39]. β-Cell glucotoxicity results from the inability of proinsulin biosynthesis to keep pace with chronic insulin hypersecretion resulting in depletion of insulin stores[39].

Although transgenic mice are widely used as models of human disease, there are relatively few studies in the area of diabetes research. Male LDL receptor-deficient ($LDLR^{-/-}$) mice fed a high-fat diet become hyperglycemic and hyperinsulinemic with substantial hypercholesterolemia and macrophage-laden lesions which do not normally progress to mature atherosclerotic plaques[40]. Transgenic ablation of brown adipose tissue leads to obesity, hyperglycemia, insulin resistance, hypertension, left ventricular hypertrophy with eccentric remodelling and fibrosis[41,42]. Transgenic models of autoimmune diabetes have been developed where dendritic cell-mediated cell transport initiates autoimmune responses[43]. Such models may be useful to delineate basic pathogenetic mechanisms and evaluate therapeutic strategies.

38.4 Cardiovascular disease in diabetes

Insulin regulates the balance of metabolic fuels received by the heart and also regulates energy metabolism in the heart[44]. Abnormalities in myocardial energy metabolism, especially decreased myocardial glucose and lactate uptake and increased use of free fatty acids and ketone bodies, probably contribute to the greater mortality in diabetics[45]. In diabetic humans, it has been difficult to demonstrate direct effects of an increased glucose concentration on myocytes leading to a selective cardiomyopathy independent of vascular disease following endothelial dysfunction. In chronically diabetic rats and rabbits, abnormalities in contractile proteins and regulatory proteins, especially myosin and troponin, can be clearly demonstrated[46]. Regulation of cardiac calcium fluxes in diabetes is controversial. Both a disturbed active calcium transport and depressed passive calcium buffering may lead to depressed myocardial performance and predispose to cardiac injury[47]. The renin-angiotensin system may be involved in the cardiac changes: plasma angiotensinogen decreases in streptozotocin-diabetic rats while the density of cardiac angiotensin (AT1) receptors increases[48]. Fibrosis (excessive collagen deposition) is also a characteristic of the streptozotocin-diabetic rat heart; this could be reversed by administration of either pirfenidone or the aldosterone antagonist, spironolactone[49].

Coronary artery disease and its clinical consequences of angina, myocardial infarction and sudden cardiac death are markedly increased in patients with Type 1 or Type 2 diabetes[11]. The key accessory to vascular disease in diabetics is a damaged endothelium. Endothelial dysfunction, for example in conduit arteries[50], precedes early diabetic microangiopathy in diabetic humans. The biochemical basis for endothelial damage in the presence of high glucose concentrations is controversial[51]. Many factors have been indicated as causative for endothelial dysfunction from animal and *in vitro* studies including hyperglycemic pseudohypoxia, activation of protein kinase C, increased expression of TGFβ (transforming growth factor-β) and vascular endothelial growth factor, non-enzymatic glycation, oxidative stress, activation of the coagulation cascade, increased expression of TNFα (tumour necrosis factor-α) and high levels of insulin and its precursors[51]. It is very difficult to determine the relative contributions of each mechanism in Type 2 diabetics since vascular damage occurs very early in the diabetic process when few, if any, patients are diagnosed. Vascular damage, in particular endoneurial microangiopathy, may underlie the markedly increased prevalence of cardiovascular autonomic neuropathy in Type 1 diabetics[52]; this complication is associated with a coronary risk profile indicating a high risk of future coronary artery disease.

Endothelium impairment and loss of endothelial production of vasodilators such as NO may contribute to the hypertension, dyslipidaemia and insulin resistance which characterize diabetes. Further, elevation in oxidative stress markers in a pro-oxidant environment may inactivate any NO which is produced. Augmentation of antioxidant defenses with vitamins C and E normalizes oxidant stress and improves endothelium-dependent vasodilatation and insulin sensitivity but clinical efficacy in primary or secondary preventative trials has been difficult to demonstrate[53]. A more effective approach may be to prevent free radical generation. Diabetic endothelium produces increases in both superoxide radicals and hydrogen peroxide leading to enhanced intracellular production of hydroxyl radicals[54]. Inhibiting free radical generation in Type 2 diabetics with the xanthine oxidase inhibitor, allopurinol, improves endothelial dysfunction in these patients but not in matched control subjects[55].

38.5 Therapeutic approaches to reduce hyperglycemia

The treatment of Type 1 diabetes with insulin remains one of the major medical triumphs of the last century. The refinements since this breakthrough by Banting and Best in 1921–22 have included the introduction of longer-acting insulin preparations in the 50's, the use of purified monocomponent insulins and the changeover to pork insulin in the 60's and 70's, the introduction of human insulin from recombinant technology in the 80's and now the introduction of synthetic insulins with varying durations of action. The pharmacokinetics of insulin administration are well-covered in standard pharmacology textbooks[56,57]. Insulin treatment is obligatory in Type 1 diabetics and optional in Type 2 diabetics; orally active hypoglycemic drugs are only used in Type 2 diabetics.

Non-pharmacological measures such as education, diet, exercise and weight loss are the foundations of every treatment plan for the management of patients with Type 2 diabetes. These procedures are active at many levels; for example, physical activity enhances insulin sensitivity[58] while restriction of total and saturated fat intake with increased complex carbohydrates and dietary fibre enhances the action of insulin[59].

There are three major abnormalities associated with Type 2 diabetes: impaired insulin secretion, excessive hepatic glucose output and insulin resistance in skeletal muscle, liver and adipose tissue. The orally active drugs available for treatment act primarily by increasing insulin availability (sulphonylureas), delaying gastrointestinal glucose absorption (α-glucosidase inhibitors), suppressing excessive hepatic glucose output (biguanides) and reducing insulin resistance at target tissues (thiazolidinediones or glitazones) (Figure 38.2). Future possibilities include gluconeogenesis inhibitors to decrease hepatic glucose production, lipolysis inhibitors to decrease free fatty acid production, carnitine palmitoyltransferase inhibitors to increase fat oxidation and β_3-adrenoceptor agonists to stimulate energy expenditure and increase non-oxidative glucose disposal.

The sulphonylureas such as chlorpropamide and tolbutamide bind to and inhibit the ATP-activated K^+-channels on the β-cells of the pancreas. This permits intracellular calcium build-up and the resultant stimulus-secretion coupling leads to insulin release. Thus, the sulphonylureas mimic the actions of increased ATP following a meal in non-diabetic patients. Logically, an increased release of insulin would seem to be of little benefit in Type 2 diabetics with hyperinsulinemia and insulin resistance. This may explain why chronic treatment with sulphonylureas is satisfactory in only 20–30% of these patients. ATP-activated K^+-channels are also present in the heart where they may induce cardioprotection[60,61] and a lower incidence of ventricular arrhythmias[62]. Diabetes and acute hyperglycemia attenuate reductions in infarct size produced in dogs by diazoxide, a selective mitochondrial K_{ATP} channel agonist, suggesting that glucose may be a critical modulator of these channels during ischemia[63]. Antiarrhythmic drugs may also act on the same channel but at a different site to induce insulin release[64]. The major adverse effects of the sulphonylureas are hypoglycemia, weight gain and hyperinsulinemia. A possible excess in cardiovascular mortality with the sulphonylureas has been controversial for over 20 years but has not been reported in recent major trials. The United

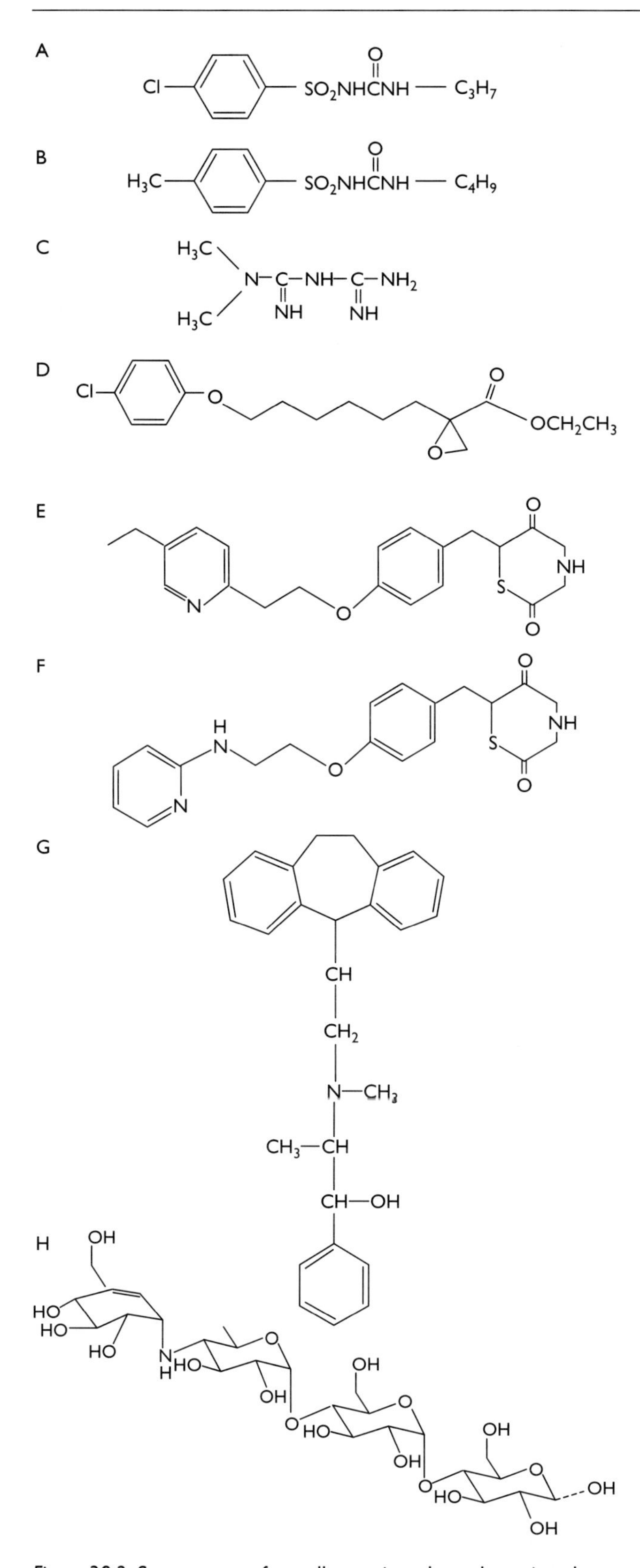

Figure 38.2 Structures of orally active hypoglycemic drugs. (A) chlorpropamide, (B) tolbutamide, (C) metformin, (D) etomoxir, (E) pioglitazone, (F) rosiglitazone, (G) trecadrine, (H) acarbose.

Kingdom Prospective Diabetes Study (UKPDS) is the largest randomized study comparing the effects of diet alone, sulphonylureas, insulin or metformin on glycemic control and long-term diabetic complications. This study provides conclusive proof that control of blood glucose concentrations and hemoglobin A_{1c} together with tight blood pressure control markedly decreases the risk of diabetic complications, including myocardial infarction and microvascular complications[65,66].

α-Glucosidases, the intestinal enzymes that break down starches into monosaccharides before absorption, can be reversibly inhibited by acarbose, thus delaying the digestion and absorption of carbohydrates. Acarbose reduced postprandial glucose and hemoglobin A_{1c} levels alone or in combination therapy with tolbutamide[67]. Acarbose is well-tolerated but may cause carbohydrate malabsorption at higher doses.

Metformin, the only biguanide now available, appears to reduce insulin resistance through myriad effects on carbohydrate, lipid and lipoprotein metabolism to suppress excessive glucose output from the liver and on fibrinolysis[68]. Metformin treatment improves both insulin resistance and endothelial function measured as acetylcholine-stimulated forearm blood flows[69]. The most dreaded adverse effect is lactic acidosis with an incidence of about 3 cases per 100,000 patient-years and a mortality rate similar to that from sulphonylurea-induced hypoglycemia[68].

The major recent advance in the treatment of Type 2 diabetes has been the introduction of the glitazones or thiazolidinediones, referred to as the insulin sensitizers[4,70]. These compounds are highly selective and potent agonists for the nuclear peroxisome proliferator-activated receptor-γ (PPAR-γ) present in key target tissues for insulin action such as adipose tissue, skeletal muscle, liver and blood vessels. The glitazones regulate the genes controlling various metabolic pathways, for example an up-regulation of p85-α phosphatidylinositol 3-kinase (p85-α PI3K), stimulation of glucose transporter gene expression and reduction of TNF-α and hepatic glucokinase expression. The PPAR-γ agonists control proliferation and migration of vascular cells, which are fundamental responses to injury in arteries, thereby preventing intimal lesion development[71]. PPAR-γ activation by troglitazone inhibited early atherosclerotic lesion development in diabetic lipoprotein receptor-deficient mice[40]. Research into gene regulatory mechanisms of PPAR ligands is very active as this promises to lead to pharmacological targets for the treatment of atherosclerosis and diabetes[72]. Troglitazone was withdrawn from the US market in March 2000 when the FDA determined that the risk of idiosyncratic hepatotoxicity outweighed its potential benefits. So far, the newer derivatives, rosiglitazone and pioglitazone, appear free of this adverse effect.

Inhibition of fatty acid oxidation is a potential mechanism for controlling blood glucose concentrations by increasing glucose utilisation in Type 2 diabetes. Further, increased free fatty acids inhibit myocardial function by increasing oxygen requirements and allowing accumulation of potentially toxic intermediates of these fatty acids. The carnitine palmitoyltransferase (CPT) system is a potential target as this three-component system is necessary to transfer long-chain fatty acids into the mitochondrial matrix. The irreversible CPT 1 inhibitor, etomoxir, ameliorates myocardial dysfunction in streptozotocin-diabetic rats possibly by increasing myocardial 1,2-diacylglycerol[73]. Reversible inhibition of CPT 1 by SDZ-CPI-975 lowers blood glucose without increasing myocardial glucose uptake or inducing cardiac hypertrophy in contrast to etomoxir[74]. In rats with aortic constriction, etomoxir prevents the transition from compensated to failing hypertrophy by increasing sarcoplasmic reticulum calcium uptake proportionally to the cardiac hypertrophy[75]. Another key step in carbohydrate and fat metabolism is AMP-activated protein kinase; activation of this system by exercise or future pharmacological manipulation may partially correct the metabolic changes in Type 2 diabetes[76]. This metabolic correction may also be achieved by lipolysis inhibitors, for example adenosine A1 receptor agonists[77].

Agonists at β_3-adrenoceptors have shown potent lipolytic effects in brown and white adipose tissue and thus have potential anti-obesity and anti-diabetic applications. As an example, a new agonist, trecadrine, reduces hepatic glucose output in alloxan-diabetic rats and stimulates glucose uptake in skeletal muscle possibly by increasing the ratios of the GLUT1/GLUT4 transporters[78].

38.6 Therapeutic approaches to reduce cardiovascular symptoms

Since most diabetics will die from cardiovascular disease, and this will include patients with moderate to good long term blood glucose control, it is essential to specifically treat cardiovascular disease in addition to therapy aiming to control blood glucose concentrations to minimize end-organ damage. The UKPDS results clearly show that tight control of blood pressure combined with reduction of blood glucose concentrations is necessary to reduce the incidence of cardiovascular disease in diabetics[65,66]. What is the optimal treatment of hypertension in diabetics? The Appropriate Blood Pressure Control in Diabetes (ABCD) Trial results suggest that ACE inhibitors such as enalapril should be the initial antihypertensive treatment[79]; these findings were confirmed by a meta-analysis of four clinical trials that directly compared ACE inhibitors to alternative antihypertensive medications in hypertensive Type 2 diabetics[80]. A separate analysis of two of these trials and the Hypertension Optimal Treatment (HOT) trial concludes that the combination of a calcium antagonist with an ACE inhibitor is a rational therapeutic choice in patients with co-existing hypertension and diabetes[81]. The conclusion of a study of 719 elderly diabetics in the Swedish Trial in Old Patients with Hypertension-2 (STOP Hypertension-2) is that adequate treatment of hypertension is the important objective, whether with diuretics, beta-adrenoceptor antagonists, ACE inhibitors or calcium antagonists[82]. This is an ongoing issue with several large trials in progress specifically aiming to determine the optimal treatment of hypertension in diabetes. Hypertension is the most important risk factor for heart failure. No trials have determined the optimal treatment of heart failure in diabetics; the assumption is that treatment for all high-risk cardiac patients includes ACE inhibitors or angiotensin receptor antagonists, diuretics, beta-adrenoceptor antagonists and possibly digoxin[11,83]. Treatment is essential as diabetic patients with heart failure following acute myocardial infarction have a worse prognosis than non-diabetic patients with heart failure[84].

The symptoms of coronary artery disease, angina, myocardial infarction and sudden death, are all increased in diabetics, especially women[11,85]. In elderly diabetics with acute myocardial infarction, insulin-treated diabetics have the highest risk of mortality, followed by diabetics treated with oral hypoglycemic drugs and followed by diet-controlled diabetics[86]. Patients with diabetes have more diffuse coronary atherosclerosis, a greater prevalence of mild, moderate and severe stenoses and a two-fold higher occlusion rate compared to non-diabetic patients[87]. Diabetic patients undergoing coronary revascularization with coronary artery bypass grafting surgery have a better survival than patients with percutaneous coronary intervention[88]. In diabetic patients with myocardial infarction treated with either revascularization procedure, the most effective cardiac rehabilitation program focuses on aggressive diabetic control and enhancement of exercise capacity to reduce mortality[89]. Mortality in diabetic patients undergoing percutaneous coronary intervention is markedly reduced by platelet inhibition with abciximab, a potent inhibitor of glycoprotein IIb/IIIa receptors[90]. Blockade of these receptors by tirofiban improves the effectiveness of heparin and aspirin in preventing major ischemic events and death in diabetic patients with acute coronary syndromes[91]. Diabetic patients with myocardial infarctions clearly benefit from treatment with thrombolytics, β-adrenoceptor antagonists and antiplatelet drugs[92]. ACE inhibitors given early following a myocardial infarction significantly improve mortality, possibly more in diabetic than non-diabetic patients[93], although not all trials have shown this effect[11].

Drug treatment can only reduce morbidity and mortality if patients receive treatment and are compliant with therapy. Lipid-lowering therapy is remarkably effective in decreasing cardiovascular mortality, in both diabetics and non-diabetics[11,94]. However, in a large group of patients with myocardial infarction, including diabetics, lipid-lowering medications were part of the discharge regime in only 31.7% of patients[95]. Similarly, the evidence for the benefit of continued low aspirin dosage in patients with cardiovascular disease is very strong yet only 20% of adults with diabetes took aspirin regularly in 1988–1994[96]. Thus, delivering the benefits to diabetic patients is a major public health issue; addressing this problem could significantly improve the quality and quantity of life for diabetic patients with cardiovascular symptoms.

38.7 Summary

The presence of cardiovascular disease in diabetics is a key determinant of patient survival. Diabetic patients have a much higher risk of coronary disease, heart failure, myocardial infarction and death. There have been remarkable advances in understanding the pathophysiological processes underlying the development of these symptoms as well as in treatment modalities for these conditions. However, much more needs to be done, especially in the modification of the risk factors for type 2 diabetes and in delivering benefits to patients but also in improved treatment options. This is necessary to ensure that the current epidemic of diabetes does not reverse the gains in cardiovascular health achieved over the last 30 years within the next 30 years.

References

1 Anonymous (2001) http://www.aihw.gov.au/publications/health/hsvd.html
2 Definition, diagnosis and classification of diabetes mellitus and its complications (1999) World Health Organisation, Department of Noncommunicable Disease Surveillance, Geneva.
3 Zimmet, P., Boyko, E.J., Collier, G.R., de Courten, M. (1999) Etiology of the metabolic syndrome: potential role of insulin resistance, leptin resistance, and other players. *Ann N Y Acad Sci 892:25–44.*
4 Dagogo-Jack, S. (1997) Pathophysiology of type 2 diabetes and modes of action of therapeutic interventions. *Arch Int Med 157:1802–1817.*
5 Anonymous (2001) The dangerous toll of diabetes. http://www.diabetes.org/ada/facts.asp
6 Valle, T., Tuomilehto, J., Eriksson J. (1997) Epidemiology of NIDDM in Europids. In: Alberti, K.G.M.M., Zimmet, P., DeFronzo, R.A., Keen, H. (eds): *International textbook of diabetes mellitus*, 2nd edition, London: John Wiley & Sons pp.125–142.
7 Trevisan, R., Vedovato, M. Tiengo, A. (1998) The epidemiology of diabetes mellitus. *Nephrol Dial Transplant 13(suppl 8):2–5.*
8 Fagot-Campagna, A., Pettitt, D.J., Engelgau, M.M., Burrows, N.R., Geiss, L.S., Valdez, R., Beckles, G.L., Saaddine, J., Gregg, E.W., Williamson, D.F., Narayan, K.M. (2000) Type 2 diabetes among North American children and adolescents: an epidemiologic review and a public health perspective. *J Pediatr 136:664–672.*
9 Harris, M.I., Flegal, K.M., Cowie, C.C., Eberhardt, M.S., Goldstein, D.E., Little, R.R., Wiedmeyer, H.M., Byrd-Holt, D.D. (1998) Prevalence of diabetes, impaired fasting glucose and impaired glucose tolerance in US adults. *Diabetes Care 21: 518–524.*
10 Zimmet, P., McCarty, D. (1995) The NIDDM epidemic: global estimates and projection: a look into the crystal ball. *IDF Bull 40(3):8–16.*
11 Butler, R., Macdonald, T.M., Struthers, A.D., Morris, A.D. (1998) The clinical implications of diabetic heart disease. *Eur Heart J 19:1617–1627.*
12 Strakosch, C. (2000) The discovery of insulin. Ramsay Health Care, Brisbane, Australia.
13 Brown, H., Sanger, F., Kitai, R. (1955) The structure of pig and sheep insulin. *Biochemical Journal 60:356–365.*
14 Yalow, R.S. (1978) Radioimmunoassay: a probe for the fine structure of biological systems. *Science 200:1236–1245.*
15 Pickup, J.C., Williams, J. (1997) Textbook of diabetes, 2nd ed, Blackwell Science, Oxford.
16 Dunn, J.S., Mcletchie, N.G.B. (1943) Experimental alloxan diabetes in the rat. *Lancet ii:384–387.*
17 Milagro, F.I., Martinez, J.A. (2000) Effects of the oral administration of a beta3-adrenergic agonist on lipid metabolism in alloxan-diabetic rats. *J Pharm Pharmacol 52:851–856.*
18 Tomlinson, K.C., Gardiner, S.M., Hebden, R.A., Bennett, T. (1992) Functional consequences of streptozotocin-induced diabetes mellitus, with particular reference to the cardiovascular system. *Pharmacol Rev 44:103–150.*
19 National Diabetes Data Group (1979) Classification and diagnosis of diabetes mellitus and other categories of glucose intolerance. *Diabetes 28:1039–1057.*
20 Van Zwieten, P.A., Kam, K.L., Pijl, A.J., Hendriks, M.G.C., Beenen, O.H.M., Pfaffendorf, M. (1996) Hypertensive diabetic rats in pharmacological studies. *Pharmacol Res 33:95–105.*
21 Hulthén, U.L., Cao, Z., Rumble, J.R., Cooper, M.E., Johnston, C.I. (1996) Vascular hypertrophy and albumin permeability in a rat model combining hypertension and diabetes mellitus. Effects of calcium antagonism, angiotensin converting enzyme inhibition, and angiotensin II-AT_1-receptor blockade. *Am J Hypertens 9:895–901.*
22 Rosenthal, T., Erlich, Y., Rosenmann, E., Cohen, A. (1997) Effects of enalapril, losartan, and verapamil on blood glucose and glucose metabolism in the Cohen-Rosenthal diabetic hypertensive rat. *Hypertension 29:1260–1264.*
23 Ren, J., Bode, A.M. (2000) Altered excitation-contraction coupling in ventricular myocytes from spontaneously diabetic BB rats. *Am J Physiol 279:H238–244.*
24 Ramasamy, R., Schaefer, S. (1999) Inhibition of Na^+ -H^+ exchanger protects diabetic and non-diabetic hearts from ischemic injury: insight into altered susceptibility of diabetic hearts to ischemic injury. *J Mol Cell Cardiol 31:785.*
25 Makino, S., Kunimoto, K., Muraoka, Y., Mizushima, Y., Katagiri, K., Tochino, Y. (1980) Breeding of a non-obese, diabetic strain of mice. *Exp Anim 29:1–13.*
26 Huber, S.A., Sakkinen, P., Conze, D., Hardin, N., Tracy, R. (1999) Interleukin-6 exacerbates early atherosclerosis in mice. *Arterioscler Thromb Vasc Biol 19:2364–2367.*
27 Pederson, R.A., Campos, R.V., Buchan, A.M.J., Chisholm, C.B., Russell, J.C., Brown, J.C. (1991) Comparison of the enteroinsular axis in two strains of obese rat: the fatty Zucker and JCR:LA-corpulent. *Int J Obes 15:461–470.*
28 Kasiske, B.L., O'Donnell, M.P., Keane, W.F. (1992) The Zucker model of obesity, insulin resistance, hyperlipidemia, and renal injury. *Hypertension 19(suppl 1):I110–I115.*

29 Kawano, K., Nirashima, T., Mori, S., Saitoh, Y., Kurosumi, M., Natori, T. (1992) Spontaneous long term hyperglycaemic rat with diabetic complications. Otsuka Long-Evans Tokushima (OLETF) strain. *Diabetes 4:1422–1428.*

30 Nakamura, H., Kihara, Y., Tashiro, M., Kanagawa, K., Shirohara, H., Yamamoto, M., Yosikawa, H., Fukumitsu, K.-I., Hirohata, Y., Otsuki, M. (1998) Expression and the CCK-A receptor-mediated biological functions in Otsuka Long-Evans Tokushima (OLETF) rats. *J Gastroenterol 33:702–709.*

31 Mizushige, K., Yao, L., Noma, T., Kiyomoto, H., Yu, Y., Hosomi, N., Ohmori, K., Matsuo, H. (2000) Alteration in left ventricular diastolic filling and accumulation of myocardial collagen at insulin-resistant prediabetic stage of a type II diabetic rat model. *Circulation 101:899–907.*

32 Kagota, S., Yamaguchi, Y., Nakamura, K., Kunitomo, M. (2000) Altered endothelium-dependent responsiveness in the aortas and renal arteries of Otsuka Long-Evans Tokushima Fatty (OLETF) rats, a model of non-insulin-dependent diabetes mellitus. *Gen Pharmacol 34:201–209.*

33 Tamura, K., Kanzaki, T., Tashiro, J., Yokote, K., Mori, S., Ueda, S., Saito, Y., Morisaki, N. (2000) Increased atherogenesis in Otsuka Long-Evans Tokushima fatty rats before the onset of diabetes mellitus: association with overexpression of PDGF β-receptors in aortic smooth muscle cells. *Atherosclerosis 149: 351–358.*

34 Russell, J.C., Bar-Tana, J., Shillabeer, G., Lau, D.C.W., Richardson, M., Wenzel, L.M., Graham, S.E., Dolphin, P.J. (1998) Development of insulin resistance in the JCR:LA-cp rat: role of triacylglycerols and effects of MEDICA. *Diabetes 47:770–778.*

35 Wu-Peng, X.S., Chua, S.C. Jr., Okada, N., Liu, S.-M., Nicolson, M., Leibel, R.L. (1997) Phenotype of the obese Koletsky (*f*) rat due to Tyr763Stop mutation in the extracellular domain of the leptin receptor. *Diabetes 46:513–518.*

36 Russell, J.C., Ravel, D., Pegorier, J.-P., Delrat, P., Jochemsen, R., O'Brien, S.F., Kelly, S.E., Davidge, S.T., Brindley, D.N. (2000) Beneficial insulin-sensitizing and vascular effects of S15261 in the insulin-resistant JCR:LA-cp rat. *J Pharmacol Exp Ther 295: 753–760.*

37 Shimoni, Y., Ewart, H.S., Severson, D. (1998) Type I and II models of diabetes produce different modifications of K^+ currents in rat heart: role of insulin. *J Physiol 507:485–496.*

38 Cosenzi, A., Bernobich, E., Plazotta, N., Seculin, P., Bellini, G. (1999) Bosentan reduces blood pressure and the target-organ damage induced by a high-fructose diet in rats. *J Hypertens 17:1843–1848.*

39 Leibowitz, G., Yuli, M., Donath, M.Y., Nesher, R., Melloul, D., Cerasi, E., Bross, D.J., Kaiser, N. (2001) β-Cell glucotoxicity in the *Psammomys obesus* model of type 2 diabetes. *Diabetes 50(suppl 1): S113–S117.*

40 Collins, A.R., Meehan, W.P., Kintscher, U., Jackson, S., Wakino, S., Noh, G., Palinski, W., Hsueh, W.A., Law, R.E. (2001) Troglitazone inhibits formation of early atherosclerotic lesions in diabetic and nondiabetic low density lipoprotein receptor-deficient mice. *Arterioscler Thromb Vasc Biol 21:365–371.*

41 Cittadini, A., Mantzoros, C.S., Hampton, T.G., Travers, K.E., Katz, S.E., Morgan, J.P., Flier, J.S., Douglas, P.S. (1999) Cardiovascular abnormalities in transgenic mice with reduced brown fat. An animal model of human obesity. *Circulation 100:2177–2183.*

42 Hamann, A., Benecke, H., Le Marchand-Brustel, Y., Susulic, V.S., Lowell, B.B., Flier, J.S. (1995) Characterization of insulin resistance and NIDDM in transgenic mice with reduced brown fat. *Diabetes 44:1266–1273.*

43 Ludewig, B., Zinkernagel, R.M., Hengartner, H. (2000) Transgenic animal models for virus-induced autoimmune diseases. *Exp Physiol 85:653–659.*

44 Brownsey, R.W., Boone, A.N., Allard, M.F. (1997) Actions of insulin on the mammalian heart: metabolism, pathology and biochemical mechanisms. *Cardiovasc Res 34:3–24.*

45 Stanley, W.C., Lopaschuk, G.D., McCormack, J.G. (1997) Regulation of energy substrate metabolism in the diabetic heart. *Cardiovasc Res 34:25–33.*

46 Malhotra, A., Singh, V. (1997) Regulation of contractile proteins in diabetic heart. *Cardiovasc Res 34:34–40.*

47 Pierce, G.N., Russell, J.C. (1997) Regulation of intracellular Ca^{2+} in the heart during diabetes. *Cardiovasc Res 34:41–47.*

48 Brown, L., Wall, D., Marchant, C., Sernia, C. (1997) Tissue-specific changes in angiotensin II receptors in streptozotocin-diabetic rats. *J Endocrinol 154:355–362.*

49 Miric, G., Dallemagne, C., Endre, Z., Margolin, S., Taylor, S.M., Brown, L. (2001) Reversal of cardiac and renal fibrosisby pirfenidone and spironolactone in streptozotocin-diabetic rats. *Br J Pharmacol, 133:687–694.*

50 Lekakis, J., Papamichael, C., Anastasiou, H., Alevizaki, M., Desses, N., Souvatzoglou, A., Stamatelopoulos, S., Koutras, D.A. (1997) Endothelial dysfunction of conduit arteries in insulin-dependent diabetes mellitus without microalbuminuria. *Cardiovasc Res 34:164–168.*

51 Stehouwwer, C.D.A., Lambert, J., Donker, A.J.M., van Hinsbergh, V.W.M. (1997) Endothelial dysfunction and pathogenesis of diabetic angiopathy. *Cardiovasc Res 34:55–68.*

52 May, O., Arildsen, H., Damsgaard, E.M., Mickley, H. (2000) Cardiovascular autonomic neuropathy in insulin-dependent diabetes mellitus: prevalence and estimated risk of coronary heart disease in the general population. *J Intern Med 248: 483–491.*

53 Laight, D.W., Carrier, M.J., Änggård, E.E. (2000) Antioxidants, diabetes and endothelial dysfunction. *Cardiovasc Res 47:457–464.*

54 Pieper, G.M., Langenstroer, P., Siebeneich, W. (1997) Diabetic-induced endothelial dysfunction in rat aorta: role of hydroxyl radicals. *Cardiovasc Res 34:145–156.*

55 Butler, R., Morris, A.D., Belch, J.J.F., Hill, A., Struthers, A.D. (2000) Allopurinol normalizes endothelial dysfunction in Type 2 diabetics with mild hypertension. *Hypertension 35: 746–751.*

56 David, S.N., Granner, K. (1996) Insulin, oral hypoglycaemic agents, and the pharmacology of the endocrine pancreas. In: Hardman, J.G., Gilman, A.G., Limbird, L.E.: *Goodman and Gilman's The pharmacological basis of therapeutics*, 9th ed, McGraw Hill, New York, pp 1487–1517.

57 Rang, H.P., Dale, M.M., Ritter, J.M. (1999) The endocrine pancreas and the control of blood glucose. In *Pharmacology*, 4th ed, Churchill Livingstone, Edinburgh, pp. 385–398.

58 Horton, E.S. (1986) Exercise and physical training: effects on insulin sensitivity and glucose metabolism. *Diabetes Metab Rev 2:1–17.*

59 Grundy, S.M. (1991) Dietary therapy in diabetes mellitus: is there a single best diet? *Diabetes care 14:796–801.*

60 Cohen, M.V., Baines, C.P., Downey, J.M. (2000) Ischemic preconditioning: from adenosine receptor to K_{ATP} channel. *Ann Rev Physiol 62:79–109.*

61 Pain, T., Yang, X.-M., Critz, S.D., Yue, Y., Nakano, A., Liu, G.S., Heusch, G., Cohen, M.V., Downey, J.M. (2000) Opening of mitochondrial K_{ATP} channels triggers the preconditioned state by generating free radicals. *Circ Res 87: 460–466.*

62 Schotburgh, C.E., Wilde, A.A.M. (1997) Sulfonylurea derivatives in cardiovascular research and in cardiovascular patients. *Cardiovasc Res 34:73–80.*

63 Kersten, J.R., Montgomery, M.W., Ghassemi, T., Gross, E.R., Toller, W.G., Pagel, P.S., Warltier, D.C. (2001) Diabetes and hyperglycemia impair activation of mitochondrial K_{ATP} channels. *Am J Physiol 280:H1744–H1750.*

64 Horie, M., Ishida-Takahashi, A., Ai, T., Nishimoto, T., Tsuura, Y., Ishida, H., Seino, Y., Sasayama, S. (1997) Insulin secretion and its modulation by antiarrhythmic and sulfonylurea drugs. *Cardiovasc Res 34:69–72.*

65 Stratton, I.M., Adler, A.I., Neil, A.W., Matthews, D.R., Manley, S.E., Cull, C.A., Hadden, D., Turner, R.C., Holman, R.R., on behalf of the UK Prospective Diabetes Study Group (2000) Association of glycaemia with macrovascular and microvascular complications of type 2 diabetes (UKPDS 35): prospective observational study. *BMJ 321: 405–412.*

66 Adler, A.I., Stratton, I.M., Neil, A.W., Yudkin, J.S., Matthews, D.R., Cull, C.A., Wright, A.D., Turner, R.C., Holman, R.R., for the UK Prospective Diabetes Study Group (2000) Association of systolic blood pressure with macrovascular and microvascular complications of type 2 diabetes (UKPDS 36): prospective observational study. *BMJ 321:412–419.*

67 Conniff, R.F., Shapiro, J.A., Seaton, T.B., Gray, C.A. (1995) Multicenter, placebo-controlled trial comparing acarbose (Bay b 5421) with placebo, tolbutamide, and tolbutamide-plus-acarbose in non insulin-dependent diabetes mellitus. *Am J Med 98:443–451.*

68 Bailey, C.J., Turner, R.C. (1996) Metformin. *N Engl J Med 334: 574–579.*

69 Mather, K.J., Verma, S., Anderson, T.J. (2001) Improved endothelial function with metformin in Type 2 diabetes mellitus. *J Am Coll Cardiol 37:1344.1350.*

70 Mudaliar, S., Henry, R.R. (2001) New oral therapies for type 2 diabetes mellitus: the glitazones or insulin sensitisers. *Ann Rev Med 52:239–257.*

71 Hsueh, W.A., Jackson, S., Law, R.E. (2001) Control of vascular cell proliferation and migration by PPAR-(gamma): a new approach to the macrovascular complications of diabetes. *Diabetes Care 24:392–397.*

72 Lazar, M.A. (2001) Progress in cardiovascular biology: PPAR for the course. *Nature med 7:23–24.*

73 Hayashi, K., Okumura, K., Matsui, H., Murase, K., Kamiya, H., Saburi, Y., Numaguchi, Y., Toki, Y., Hayakawa, T. (2001) Involvement of 1,2-diacylglycerol in improvement of heart function by etomoxir in diabetic rats. *Life Sciences 68: 1515–1526.*

74 Deems, R.O., Anderson, R.C., Foley, J.E. (1998) Hypoglycemic effects of a fatty acid oxidation inhibitor in rats and monkeys. *Am J Physiol 274:R524–R528.*

75 Rupp, H., Vetter, R. (2000) Sarcoplasmic reticulum function and carnitine palmitoyltransferase-1 inhibition during progression of heart failure. *Br J Pharmacol 131:1748–1756.*

76 Winder, W.W., Hardie, D.G. (1999) AMP-activated protein kinase, a metabolic master switch: possible roles in Type 2 diabetes. *Am J Physiol 277:E1–E10.*

77 Van Schaick, E.A., Zuideveld, K.P., Tukker, H.E., Langemeijer, M.W.E., Ijzerman, A.P., Danhof, M. (1998) Metabolic and cardiovascular effects of the adenosine A_1 receptor agonist N^6-(*p*-sulfophenyl)adenosine in diabetic Zucker rats: influence of the disease on the selectivity of action. *J Pharmacol Exp Ther 287: 21–30.*

78 Milagro, F.I., Gómez-Ambrosi, J., Forga, L., Martínez, J.A. (1999) A β_3-adrenergic agonist increases muscle GLUT1/GLUT4 ratio, and regulates liver glucose utilization in diabetic rats. *Diab Obes Metab 1:97–104.*

79 Estacio, R.O., Schrier, R.W. (1998) Antihypertensive therapy in Type 2 diabetes: implications of the Appropriate Blood Pressure Control in Diabetes (ABCD) Trial. *Am J Cardiol 82: 9R–14R.*

80 Pahor, M., Psaty, B.M., Alderman, M.H., Applegate, W.B., Williamson, J.D., Furberg, C.D. (2000) Therapeutic benefit of ACE inhibitors and other antihypertensive drugs in patients with type 2 diabetes. *Diabetes care 23:888–892.*

81 Parving, H.-H. (1998) Calcium antagonists and cardiovascular risk in diabetes. *Am J Cardiol 82:42R–44R.*

82 Lindholm, L.H., Hansson, L., Ekbom, T., Dahlof, B., Lanke, J., Linjer, E., Schersten, B., Wester, P.O., Hedner, T., de Faire, U. (2000) Comparison of antihypertensive treatments in preventing cardiovascular events in elderly diabetic patients: results from the Swedish Trial in Old Patients with Hypertension-2. STOP Hypertension-2 Study Group. *J Hypertens 18:1671–1675.*

83 Erdmann, E. (2000) The management of heart failure – an overview. *Bas Res Cardiol 95(suppl 1): I3–I7.*

84 Melchior, T., Rask-Madsen, C., Torp-Pedersen, C., Hildebrandt, P., Kober, L., Jensen, G. (2001) The impact of heart failure on prognosis of diabetic and non-diabetic patients with myocardial infarction: a 15-year follow-up study. *Eur J Heart Failure 3:83–90.*

85 Malmberg, K., Yusuf, S., Gerstein, H.C., Brown, J., Zhao, F., Hunt, D., Piegas, L., Calvin, J., Keltai, M., Budaj, A., for the OASIS Registry Investigators (2000) Impact of diabetes on long-term prognosis in patients with unstable angina and non-Q-wave myocardial infarction. Results of the OASIS (Organization to Assess Strategies for Ischemic Syndromes) Registry. *Circulation 102:1014–1019.*

86 Berger, A.K., Breall, J.A., Gersch, B.J., Johnson, A.E., Oetgen, W.J., Marciniak, T.A., Schulman, K.A. (2001) Effect of diabetes mellitus and insulin use on survival after acute myocardial infarction in the elderly (The Cooperative Cardiovascular Project). *Am J Cardiol 87:272–277.*

87 Ledru, F., Ducimetrière, P., Battaglia, S., Courbon, D., Beverilli, F., Guize, L., Guermonprez, J.-L., Diébold, B. (2001) New diagnostic criteria for diabetes and coronary artery disease: insights from an angiographic study. *J Am Coll Cardiol 37:1543–1550.*

88 Niles, N.W., McGrath, P.D., Malenka, D., Quinton, H., Wennberg, D., Shudbrokks, S.J., Tryzelaar, J.F., Clough, R., Hearne, M.J., Hernandez, F. Jr., Watkins, M.W., O'Connor, G.T.,

for the Northern New England Cardiovascular Disease Study Group (2001) Survival of patients with diabetes and multivessel coronary artery disease after surgical or percutaneous coronary revascularization: results of a large regional prospective study. *J Am Coll Cardiol 37:1008–1015.*

89 Yu, C.-M., Lau, C.-P., Cheung, B.M.-Y., Fong, Y.-M., Ho, Y.-Y., Lam, K.-B., Li, L.S.-W. (2000) Clinical predictors of morbidity and mortality in patients with myocardial infarction or revascularization who underwent cardiac rehabilitation, and importance of diabetes mellitus and exercise capacity. *Am J Cardiol 85:344–349.*

90 Theroux, P., Alexander, J. Jr., Pharand, C., Barr, E., Snapinn, S., Ghannam, A.F., Sax, F.L. (2000) Glycoprotein IIb/IIIa receptor blockade improves outcomes in diabetic patients with unstable angina/non-ST-elevation myocardial infarction; results from the Platelet Receptor Inhibition in Ischemic Syndrome Management in Patients Limited by Unstable Signs and Symptoms (PRISM-PLUS) study. *Circulation 102: 2466–2472.*

91 Bhatt, D.L., Marso, S.P., Lincoff, A.M., Wolski, K.E., Ellis, S.G., Topol, E.J. (2000) Abciximab reduces mortality in diabetics following percutaneous coronary intervention. *J Am Coll Cardiol 35:922–928.*

92 Löwel, H., Koenig, W., Engel, S., Hörmann, A., Keil, U. (2000) The impact of diabetes on survival after myocardial infarction: can it be modified by drug treatment? Results of a population-based myocardial infarction register follow-up study. *Diabetologia 43:218–226.*

93 Nesto, R.W., Zarich, S. (1998) Acute myocardial infarction in diabetes mellitus. Lessons learned from ACE inhibition. *Circulation 97:12–15.*

94 Pyörälä, K., Pedersen, T.-R., Kjekshus, J., Faergeman, O., Olsson, A., Throgeirsson, G. (1997) for the Scandinavian Simvastatin Survival Study group: Lowering cholesterol levels improved the prognosis of diabetic patients with coronary heart disease. *Diabetes Care 20:614–620.*

95 Fonarow, G.C., French, W.J., Parsons, L.S., Sun, H., Malmgren, J.A., for the National Registry of Myocardial Infarction 3 participants (2001) Use of lipid-lowering medications at discharge in patients with acute myocardial infarction. Data from the National Registry of Myocardial Infarction 3. *Circulation 103:38–44.*

96 Rolka, D.B., Fagot-Campagna, A., Narayan, K.M. (2001) Aspirin use among adults with diabetes: estimates from the Third National Health and Nutrition Examination Study. *Diabetes Care 24:197–201.*

Chapter 39

The leptin hormonal system as a target for new drugs designed to control food intake and body weight

Andrew Constanti

Contents

Obesity is an increasing health problem in industrialized Western societies. In 1994, the adipose tissue-specific *obese (ob)* gene and its peptide product leptin were discovered. The new discovery of leptin has prompted a multitude of studies on its important function in regulating body weight, and has opened a whole new field of research. Although the anorectic action of leptin on the body has been reviewed extensively, our knowledge of its physiological role(s) is still incomplete and is an ever-growing subject. There is now much interest being generated in the interaction of leptin with other peripheral and neuroendocrine mechanisms to control growth, reproduction, adrenal and thyroid function, and the immunological response. The possibility of the leptin hormonal system as a target for the basis of new slimming drugs is also a factor under consideration and is generating a lot of current interest. This review will attempt to give a summary of the main physiological actions of leptin in the body and to explore the possibility of the leptin feedback system being used as a basis for the development of new drugs to combat obesity.

39.1 Introduction

Obesity (defined as an excess of body fat, or body mass index [BMI] $\geq 30\,kg/m^2$) remains an ever-increasing health problem in late twentieth-century Western societies, and is associated with many common diseases and health problems including non-insulin dependent diabetes mellitus (Wannamethee and Shaper, 1999), hypertension (Huang and Reddy, 1999; Kokot *et al.*, 1999), coronary artery and gall bladder disease (Frayn and Coppack, 1992; Rosengren *et al.*, 1999; Bray *et al.*, 1999), osteoarthritis (Felson and Chaisson, 1997), obstructive sleep apnoea (Rollheim *et al.*, 1997) and even certain types of cancer (Ford, 1999; Lagergren *et al.*, 1999). It is claimed that obesity affects up to 30% of the population in industrialized countries, and its incidence is still on the increase (Rosembaum *et al.*, 1997; Bjorntorp, 1997); it is also considered to be one of the most significant preventable causes of ill health in the UK today, and presents a formidable financial burden on the health services (McIntyre, 1998). Understandably, there has been much research effort over the years devoted to the study of mechanisms controling appetite and body weight, and also to the development of effective therapeutic interventions that could improve the quality of life of the patient, as well as reduce the risk of developing cardiovascular disease and other chronic medical conditions; the discovery of a "miracle" weight-loss drug however, has remained elusive. The recent expansion of knowledge about the hormonal control of metabolism and body composition now offers the hope that such new agents will become available over the next decade.

It is now becoming clear that human obesity is a complex, chronic, multifactorial condition that requires lifelong treatment, and that genetic, developmental and environmental factors are likely to be involved in its pathogenesis (Bouchard *et al.*, 1993; Leibel *et al.*, 1993; Green *et al.*, 1995; Beales and Kopelman, 1996; Schalling *et al.*, 1999). The former notion that obesity can be fully explained by inappropriate eating, coupled with a general lack of willpower is therefore unsatisfactory (Friedman and Halaas, 1998). In recent years, a number of new genes have been discovered, that when mutated, have a profound effect on body weight in rodents (Zhang *et al.*, 1994; Tartaglia *et al.*, 1995; Chen *et al.*, 1996) and also (in some instances) in man (Clement *et al.*, 1996, 1998; Reed *et al.*, 1996; Montague, 1997; Schalling *et al.*, 1999). Although it is it unwise to extrapolate directly from findings in animals, these findings do provide important clues for elucidating the genetics of human obesity. Additionally, knowledge of these genes may ultimately lead to the development of novel drugs targeted at novel biochemical pathways involved in the regulation of body weight.

39.2 Physiological background: the mouse *ob* and *db* mutations

In 1953, Kennedy originally proposed a "lipostatic" theory of weight regulation which suggested that body weight was controlled by a circulating hormonal signal, most probably produced by adipose tissue in proportion to the amount of body fat stores, which further controlled appetite and energy expenditure. His model suggested that when body fat stores were altered by changes in energy balance, the circulating factor exerted a "feedback" effect on the brain to alter appetite, so that ultimately, food intake and energy expenditure were matched. Subsequent work by Hervey (1959) indicated that the site of action of this factor in the brain might be the ventromedial hypothalamus.

Twenty years after Kennedy's observations, Coleman (1973) showed in a classic series of "parabiotic" (cross-circulation) experiments, that such a circulating "satiety factor" controlling food intake, actually existed. These parabiotic experiments involved surgically joining the circulatory systems of two strains of severely obese mice (*ob/ob* and *db/db*) with those of normal (wild-type) animals and to each other. It was shown that in mice possessing the *obese* gene (*ob*) mutation, the production of the circulating anorexic factor was absent, whereas mice with the *diabetes* gene (*db*) mutation failed to respond to circulating levels of this agent. It was proposed that the *ob* gene was normally responsible for encoding the putative satiety factor itself, whereas the *db* gene encoded the cellular receptor that sensed this signal (Coleman, 1973, 1978, 1982).

The two autosomal recessive mouse strains *ob/ob* and *db/db* have proved to be highly useful models for the molecular genetic analysis of obesity in humans. The obesity mutation *ob* originally appeared spontaneously in a colony of laboratory mice and was first characterized phenotypically by Ingalls *et al.*, in 1950; mice homozygous for this mutation (*ob/ob*) developed a profound obesity (approximately three times normal mouse body weight) that resembled the human morbid obesity syndrome (Coleman, 1978; Friedman and Leibel, 1992). Interestingly, *ob/ob* mice also exhibited a reduced body temperature (torpor), infertility, diabetes, hypercortisolemia, abnormal thyroid function, delayed growth and impaired cellular immunity, suggesting that the missing satiety factor might normally be capable of exerting a wide range of physiological and endocrine effects in the body. The *diabetes* mutation *db* was originally reported by Hummel *et al.* (1996), and led to an identical obese phenotype as seen in the *ob* mouse, but in addition, animals developed a severe spontaneous hyperglycemia, hyperinsulinemia and insulin resistance syndrome resembling Type 2 diabetes in humans.

39.3 The *ob* gene and leptin

In 1994, Friedman and his colleagues working at the Howard Hughes Medical Institute in New York, first reported the cloning of the *ob* gene in mice (Zhang *et al.*, 1994). Further characterization of this *ob* gene identified that it encoded for a novel 167-amino acid protein termed *leptin* (from the Greek word *leptos* meaning "thin") which was synthesized primarily by white adipose tissue, and secreted into the blood stream as part of a central negative feedback pathway, to control the size of body fat stores; mutations in the *ob* leptin gene were therefore responsible for the observed deficiency in circulating leptin levels and development of gross obesity in the *ob/ob* mice. It also became clear that the mutant diabetic *db/db* mice developed obesity because they were *resistant* to the action of leptin; this was shown to be caused by a mutation (abnormal splicing) of the high affinity leptin *receptor* molecule (mouse OB-R; Chen *et al.*, 1996; Lee *et al.*, 1996).

The human homolog gene *OB* (nowadays termed *LEP*, and located on chromosome 7q31.3) was shown to possess an 84% nucleotide sequence homology to the mouse gene (now termed *Lep* and located on chromosome 6) (Green *et al.*, 1995); leptin was also found to be expressed at lower concentrations in non-adipose tissue e.g. heart, placenta (Green *et al.*, 1995; Masuzaki *et al.*, 1997; Ashworth *et al.*, 2000) or gastric epithelium (Bado *et al.*, 1998), suggesting a multifunctional physiological role for this hormone in the body.

The discovery of leptin and the leptin receptor genes immediately generated much interest and excitement among workers in the fields of nutrition, endocrinology and metabolism, and raised the question of whether some forms of human obesity could simply be explained by a deficiency in circulating leptin, or by defects in the leptin signaling system. Pharmaceutical companies also quickly realised the potential of the leptin system as the basis of a new generation of "intelligent" slimming drugs. Initial studies in rodents indeed demonstrated that exogenous administration of recombinant leptin to obese *ob/ob* mice led to a reduced food intake, increased locomotor activity and dramatic weight loss, as well as a significant improvement in their reproductive, endocrine and immune status; in addition, administration of recombinant leptin depleted fat reserves even in normal wild-type (lean) animals (Pelleymounter, 1995; for review see Houseknecht *et al.*, 1998). Leptin treatment however, had no corrective effect when injected into obese *db/db* mice (Halaas *et al.*, 1995), nor did it exert any effect on *in vitro* lipolysis of adipose cells (Frűhbeck *et al.*, 1998a).

Initial screening of several obese populations, surprisingly failed to reveal any mutations in the human *OB* gene (Considine *et al.*, 1995, 1996b; Maffei *et al.*, 1996; Niki *et al.*, 1996); moreover, obesity appeared to be associated with an *increase* rather than a decrease in serum leptin levels (Considine *et al.*, 1996a; Caro *et al.*, 1996; Hassink *et al.*, 1996). A strong positive correlation was found between the BMI and serum leptin concentrations, suggesting that the obese condition in most humans represents a leptin-*resistant* as opposed to a leptin-*deficient* state; this might then reflect abnormalities in the leptin receptor signaling cascade in such individuals. By comparison, only a small proportion of obese patients showed relatively normal leptin levels (Maffei *et al.*, 1995; Considine, 1996a). Not surprisingly, the administration of recombinant human leptin to obese patients with normal or elevated leptin levels, has shown only limited effects on body weight (see commentaries by Strosberg and Issad, 1999; Van Gaal *et al.*, 1999). In a recent clinical trial, subcutaneous injections of recombinant leptin produced only a moderate dose-dependent weight loss in some obese subjects (Heymsfield *et al.*, 1999). These findings suggest that mutations in the coding regions of the *OB* gene do not constitute a common cause of increased body weight in humans, and raise serious doubts as to the clinical usefulness of recombinant leptin as a *general* treatment for obesity; however, giving leptin alongside a low calorie balanced diet to obese patients with low or relatively normal plasma leptin levels might have a beneficial effect.

The studies of Clement *et al.* (1996) and Reed *et al.* (1996) indicated that the human *OB* gene (or other nearby genes) may be a candidate for genetic predisposition to gross obesity in some families, and the report of Montague *et al.* (1997) later provided the first clear genetic evidence that leptin was an important regulator of energy balance in humans, by describing severe early-onset obesity in two children (associated with a very low serum leptin level) caused by a frame-shift mutation of the leptin gene. More recently, Strobel *et al.* (1998) reported an obese patient with a low leptin level, that possessed a missense mutation in the *OB* gene, analogous to that observed in the mutant *ob/ob* mice. Also Hager *et al.*, 1998 provided evidence from an obese population study, that a single DNA base defect within the untranslated exon 1 region of the *OB* gene might be involved in obesity, by affecting leptin concentrations.

Thus, leptin replacement therapy using daily injections of recombinant leptin could still prove of positive benefit for some rare cases of extreme obesity, where a direct association between a genetic mutation and low leptin levels can be identified. Alternatively, leptin gene therapy using adenovirus-mediated delivery of the human leptin gene could be used to provide a continuous chronic secretion of leptin, as found to be effective in the obese (*ob/ob*) mouse model (Morsy *et al.*, 1998).

39.4 Leptin biology/physiology

The biology/physiology of leptin and its involvement in the pathogenesis of obesity and other disease states has been extensively reviewed in recent years; considerable information is also available on the structural characteristics of the leptin molecule, leptin receptor subtypes and signaling pathways, leptin binding proteins, regulation of secretion patterns, clearance mechanisms and functional effects (reviewed by Frűhbeck *et al.*, 1998b; Friedman and Halaas, 1998; Mantzoros and Moschos, 1998; Dallongeville *et al.*, 1998; Elmquist *et al.*, 1998; Stephens and Caro, 1998; Lonnqvvist *et al.*, 1999; Trayhurn *et al.*,1999; Marti *et al.*, 1999). The involvement of leptin in the pathophysiology of conditions such as anorexia nervosa, polycystic ovary syndrome, cancer, nephropathy, diabetes mellitus, thyroid disease, Cushing's syndrome and growth hormone deficiency is also well documented (e.g. see Haluzik *et al.*, 1999; Sir-Petermann *et al.*, 1999; O'brien *et al.*, 1999; Mantzoros, 1999; McNeely *et al.*, 1999; Pinkney *et al.*, 1998; Weise *et al.*, 1999; Jorgensen *et al.*, 1998). In addition, leptin appears to be necessary for the normal function of T lymphocytes involved in the immune response (reviewed by Matarese, 2000).

It thus seems clear that the hormonal influence of leptin in the body is quite complex, and extends far beyond its primary role in regulating energy balance; such complexity could have an important bearing on any potential usefulness of novel anti-obesity drugs designed to utilize the leptin system as a target.

39.4.1 Leptin structure

The tertiary structure of the leptin protein was originally reported by Zhang *et al.* (1997). The human leptin molecule is thought to have similar structural features to members of the long-chain helical cytokine family that includes interleukin (IL)-6 and growth hormone (Madej *et al.*, 1995). The four-alpha-helix bundle structure consists of 146 amino acids, with one disulphide bond located at the C-terminal end between cysteine residues 96 and 146; this bond appears to be crucial for maintaining protein stability and for biological activity (Zhang *et al.*, 1997; Rock *et al.*, 1996). Interestingly, short fragment peptides derived from the C-terminal region of the leptin protein (between amino acid residues 106–140) still show a leptin-like action when administered in *ob/ob* mice (Grasso *et al.*, 1997) or adult rats (Frűhbeck *et al.*, 1998c); this raises the possibility of developing active peptide analog of leptin as research tools or for potential therapeutic use in humans.

39.4.1.1 Leptin receptors and signal transduction

The leptin receptor (OB-R) was first cloned from mouse choroid plexus cDNA by Tartaglia *et al.* (1995). OB-R mRNA was found to be expressed not only in choroid plexus, but also in a wide range of peripheral tissues e.g. liver, heart, skeletal muscle, pancreas, ovaries, testes, adipose tissue, spleen, as well as the hypothalamus (the brain centre responsible for satiety); in fact, OB-mRNA was alternatively spliced to encode at least six (one long and five short) receptor isoforms (designated OB-R a to f) all of which share the same extracellular domain, although only the full-length OB-Rb isoform (with a long cytoplasmic region) is considered to be the functional, high affinity leptin receptor involved in leptin signaling in the brain (Friedman and Halaas, 1998); the short leptin receptor isoforms localized in non-neuronal cells in the meninges, choroid plexus, and brain capillary endothelium may be important for mediating the saturable transport of leptin across the blood-brain barrier and into the cerebrospinal fluid (CSF) and also for plasma binding and clearance of leptin from the blood respectively (Bjorbaek *et al.*, 1997; Uotani *et al.*, 1999).

Abnormal spicing of the hypothalamic OB-R in C57BL/Ks *db/db* mice, results in a mutant protein, with a missing cytoplasmic region, unable to activate the signaling cascade (Lee *et al.*, 1996; Baumann *et al.*, 1966). OB-R mutations have also been demonstrated in so-called Zucker 'fatty' (*fa/fa*) rats (homologous to *db/db* mice: Iida *et al.*, 1996) and genetically hypertensive Koletsky (*fak/fak*) rats (Takaya *et al.*, 1996), both defects resulting in hyperphagia and obesity in these animals. Early studies failed to report OB-R mutations in human obese subjects (Rolland *et al.*, 1998; Takaya *et al.*, 1998); however, more recently, leptin receptor defects have been described in some patients with severe obesity (Chen and Garg, 1999). Mutations in the leptin receptor gene have also been associated with an autosomal recessive form of human hypogonadotropic hypogonadism (Layman, 1999).

OB-Rb is a single membrane-spanning protein (found particularly in the hypothalamus) similar to the gp130 class I cytokine signal-transducing component of the IL-6 receptor, and linked to the novel JAK-STAT signal transduction system (*janus* kinase (JAK)-signal transducer and activator of transcription (STAT); Bjorbaek *et al.*, 1997; for recent reviews see Cattaneo *et al.*, 1999; Heim, 1999). Following leptin binding, the receptor is believed to undergo a dimerization which activates the associated JAK-2 protein tyrosine kinase to autophosphorylate internal tyrosine residues on the receptor protein; this creates temporary docking sites for a subset of cytosolic STAT proteins (Ihle, 1996), which subsequently dissociate, dimerize, and translocate to the nucleus, where they bind to specific response elements (in the promoter of target genes) on the nuclear DNA to affect gene transcription (see reviews by Hoey and Schindler, 1998; Dallongeville *et al.*, 1998); current evidence suggests that the STAT3 transcription factor protein may in fact, mediate the important feedback action of leptin in the hypothalamus (Hakansson and Meister, 1998).

Detailed knowledge of the JAK-STAT signaling pathway has led to the development of knockout mice in which coding for various STAT proteins has been disrupted by mutation or targeted deletion (reviewed by Heim, 1999). Such knockout mice (depending on gene deleted) show defects in a range of cytokine-dependent processes, e.g. interferon, interleukin or prolactin/growth hormone signaling; however, to date, no specific STAT-3 knockout animals have been developed to test for possible deficiencies in leptin signaling. Accordingly, dysregulation of the JAK-STAT pathway in humans has been associated with certain acute leukemias, lymphomas and inherited immumodeficiency syndromes (Gouilleux-Gruart *et al.*, 1977; Heim, 1999), but not obesity. It may be noted that internal regulation of the JAK-STAT system occurs via two new families of negative regulatory molecules, SOCS (suppressors of cytokine signaling) and PIAS (protein inhibitors of activated STATs), which act to suppress the signal transduction pathway (Starr and Hilton, 1998, 1999; Hilton, 1999). It is therefore tempting to speculate that disease states leading to *over*expression of these regulator proteins could give rise to *hypo*responsiveness of the leptin transduction pathway with consequent development of hyperphagia and obesity. Interestingly, on the basis of experiments carried out on obese mouse models, the leptin-inducible inhibitor protein SOCS-3 has been suggested to be a potential mediator of leptin resistance in obesity (Bjorbaek *et al.*, 1998).

To date, specific drugs which directly modify SOCS (or STAT) protein functions are not available; however, the future development of such agents could prove potentially useful for probing the patency of the leptin signaling system in obese populations.

39.4.1.2 Leptin uptake into the brain: a rate limiting step for leptin action?

In rodents, the short isoforms of the leptin receptor are abundantly present in the leptomeninges and the choroid plexus of the fourth ventricle, which is the site of cerbral-spinal fluid (CSF) production (Guan *et al.*, 1997; Mercer *et al.*, 1998) and may be involved in the transport of leptin into the brain by a saturable (insulin-independent) uptake mechanism (Banks *et al.*, 1996; Caro *et al.*, 1996); how leptin is cleared from the CSF however, is not currently understood. Interestingly, the leptin CSF/serum ratio measured in obese people is abnormally low, relative to lean individuals, indicating that the capacity of the leptin transport system may be limited in the face of high peripheral leptin levels (Caro *et al.*, 1996; Schwartz *et al.*, 1996b; see also Banks *et al.*, 2000). Since leptin must get across the capillary endothelial blood-brain barrier to exert its important feedback effects on the hypothalamus, any limitation (or defect) in uptake could confer an apparent "leptin resistance" as generally observed with obese subjects. Experiments in both the Zucker and Koletsky obese rat models have also indicated a deficiency in the plasma-to-CSF transport of leptin, although it was suggested that leptin could enter the CSF through a non-leptin receptor-mediated mechanism, which may be saturated at normal physiological plasma leptin levels (Wu-Peng *et al.*, 1997).

The existence of such a limiting (or dysfunctional) barrier for leptin entry into the brain clearly limits the usefulness of any therapeutic approach for obesity, based on peripherally administered leptin; however, future advances in systems for peptide drug delivery into the central nervous system could eventually overcome this problem (Prokai, 1998) and provide an exciting new range of anti-obesity pharmaceuticals.

39.4.1.3 Leptin binding proteins

Many hormones circulate in the blood, bound to serum proteins; this binding may ultimately modulate the bioactivity and bioavailability of the hormone on target tissues as well as clearance from the body. Leptin is bound in the circulation by at least two high-affinity binding proteins, one of which may be the soluble form of the leptin receptor (OB-Re), which lacks the hydrophobic transmembrane and intracellular domains (Houseknecht *et al.*, 1996; Sinha *et al.*, 1996b). Interestingly, Sinha *et al.* (1996b) reported that in obese (leptin resistant) individuals, most of the circulating leptin existed in a free (unbound) form, whereas in lean subjects, the majority of circulating leptin was in the bound state. The significance of this finding is currently unclear; however, it could indicate that leptin (like growth hormone and some cytokines), becomes more active *in vivo* when associated with binding proteins (Turyn *et al.*, 1997; Heaney and Golde, 1993).

It seems that the kidney is the most important site for the clearance of endogenous leptin from the circulation, followed by metabolic degradation in the renal tubules (Cumin *et al.*, 1997; Meyer *et al.*, 1997); consequently, leptin has a relatively short half-life in the bloodstream (~25 min) which does not appear to be related to tissue adiposity. The raised plasma leptin concentration observed in obese humans cannot therefore be explained simply in terms of a decreased leptin clearance (Klein *et al.*, 1996). Optimization of the pharmacokinetics of recombinant leptin would seem to be a valid strategy for improving its therapeutic usefulness.

39.4.1.4 Control of leptin secretion

Leptin is considered to be a "satiety factor" which regulates the appetite by informing the hypothalamus of body fat concentrations; leptin levels thus appear to be related to the total body fat mass, the white adipocyte size *per se* and the relative adipose tissue distribution in the body. Normal mean (basal) plasma levels of leptin in lean individuals ranges between ~7–10 ng/ml (16 ng/ml ≡ 1 nM) (Pinkney *et al.*, 1998). As the amount of fat tissue increases, blood plasma concentration of leptin rises; this increased level of leptin is then sensed by the *arcuate nucleus* of the hypothalamus (adjacent to the third ventricle), which in turn regulates food intake and energy expenditure accordingly. Different fat stores, however, show differences in leptin production: thus, the expression of leptin mRNA is higher in subcutaneous than in visceral (omental) fat depots, possibly due to a difference in adrenergic (sympathetic) lipolytic responsiveness of the adipocytes (Hube *et al.*, 1996). Also, women tend to show higher (~40%) plasma leptin levels than men, most likely due to their greater percentage of peripheral body fat (for more detailed review see Mantzoros and Moschos, 1998). Unlike insulin, plasma leptin concentrations only appear to respond to diet after chronic extremes of fasting or overfeeding (decline or rise respectively) (Kolaczynsky *et al.*, 1996); the lowered concentrations rapidly recover to normal levels upon re-feeding.

Leptin production by white fat cells can be influenced by certain hormones; these include: insulin, glucocorticoids, estrogens and progesterone (which are stimulatory), and androgens, catecholamines (operating through β_3-adrenoceptors) and certain cytokines (e.g. tumor necrosis factor-α [TNF-α]) which are inhibitory (for review see Mantzoros and Moschos, 1998); the relevance of these hormonal modulatory effects in the pathogenesis of various human obesities or other disease states however, remains to be determined.

39.5 Diurnal and pulsatile variation in plasma leptin levels

It has been shown that like some other hormones (e.g. prolactin or TSH), plasma leptin levels in humans show a circadian variation (24 hour periodicity), being highest between midnight and early morning, and lowest in the late morning/early afternoon (Sinha *et al.*, 1996a); this could have the effect of suppressing the appetite during sleeping hours. By comparison, the opposite is observed for cortisol concentrations in humans, where the highest levels occur during the early morning. In addition, leptin secretion shows pulsatility (1–7 oscillations/24 hours) which is maintained (although more frequent) in obese subjects (Sinha *et al.*, 1996c); the mechanisms underlying this leptin pulsatility and its physiological significance remain unclear.

39.5.1 Effects of leptin on the endocrine system

It appears that the characteristic fall in circulating leptin level that occurs during fasting or chronic undernutrition can result in several neuroendocrine hormonal changes that are normally associated with starvation; these include a suppression of gonadal and thyroid activity, an increase in glucocorticoid ("stress" hormone) secretion via the adrenal axis (Ahima *et al.*, 1996), and a decrease in growth hormone release (Vuagnat BA *et al.*, 1998, although see also Ho *et al.*, 1998). These changes are believed to form part of an important adaptive mechanism that operates to conserve energy when food availability is limited (particularly evident in rodents; Himms-Hagen, 1999), and can be reversed by exogenous leptin administration. Normal leptin secretion thus appears to be essential for normal neuroendocrine function.

39.6 Leptin and the hypothalamic-pituitary-gonadal (HPG) axis

There is strong evidence to suggest that leptin can affect gonadal function indirectly via the hypothalamic-pituitary-gonadal (HPG) axis. Leptin is thus believed to be the long-elusive signal that triggers the onset of puberty in children, when a "critical" body fat mass is attained; this could be achieved by stimulation of hypothalamic GnRH neurons (and also of pituitary gonadotroph cells) by leptin, with a consequent increase in release of pituitary gonadotrophins, LH and FSH (see review by Strosberg and Issad, 1999). In keeping with this hypothesis, obese patients with congenital mutations in leptin or leptin receptor genes, show hypogonadism and a failure to enter into puberty (Strobel *et al.*, 1998; Clement *et al.*, 1998). A delay in puberty onset as a consequence of low dietary intake (and *hypo*leptinemia) is also well recognized in female juveniles suffering from anorexia nervosa or in some female "elite" juvenile athletes (anorexia athletica) (Matejek *et al.*, 1999). Girls apparently show a higher serum leptin concentration before, during, and after puberty than boys, even after adjusting for the development of greater female adiposity (Demerath *et al.*, 1999); this sexual dimorphism could be partly due to the opposing effects of estrogens and testosterone on leptin production in females and males as mentioned above.

The long form of the leptin receptor (OB-R) exists in large quantities in both the gonads as well as the hypothalamus (Karlsson *et al.*, 1997; Caprio *et al.*, 1999); leptin therefore appears to be capable of affecting gonadal function directly. However, recent experiments in rodents indicate that leptin *inhibits* testosterone production in testicular Leydig cells *in vitro* (Caprio *et al.*, 1999). In addition, leptin has been shown to *inhibit* steroid production by bovine ovarian thecal cells (Spicer and Francisco, 1998); these results suggest that leptin (perhaps at high plasma concentrations) may also act as a peripheral *inhibitory* metabolic signal to the human reproductive system via *direct* actions at the testicular and ovarian level.

39.6.1 Leptin and the hypothalamic-pituitary-thyroid (HPT) axis

Prolonged fasting in experimental animals is generally associated with a profound depression of the hypothalamic-pituitary-thyroid (HPT) axis, resulting in low plasma levels of thyroid hormone (T_3/T_4) as well as (paradoxically) thyroid releasing hormone (TRH) and thyroid stimulating hormone (TSH) (reviewed by Orban *et al.*, 1998); this down-regulation of the HPT axis is believed to be partly due to a suppression of proTRH gene expression in neurons of the hypothalamic paraventricular nucleus (PVN), with a consequent reduction in hypothalamic TRH release, and can be reversed by systemic administration of leptin (Legradi *et al.*, 1998). There is also some evidence that the central hypothalamic melanocortin system (see below) may also participate in the mechanism by which leptin influences the HPT axis (Kim *et al.*, 2000).

Since hyper- or hypothyroidism is known to influence energy expenditure and body composition in humans, it was of interest to determine if pathological thyroid status could modulate plasma leptin concentration independently of BMI. However, despite conflicting data, the currently emerging view is that circulating levels of leptin are not significantly altered by thyroid dysfunction (Sesmilo *et al.*, 1998; Dallongeville *et al.*, 1998).

39.7 Leptin and hypothalamic-pituitary-growth hormone

Spontaneous pulsatile secretion of growth hormone (GH) appears to be highly sensitive to acute perturbations in nutritional status; however, data obtained from animal or human studies on this aspect are conflicting. Whereas food deprivation in rats (associated with decreased secretion of leptin) eliminates GH pulsatility (Vuagnat *et al.*, 1998), fasting in humans is associated with a paradoxical *increase* in secretion of GH (Ho *et al.*, 1988); the reasons for this apparent species difference in nutritional regulation of GH release remain unclear.

GH secretion, either basal or evoked by pharmacological stimuli acting via the hypothalamus (e.g. insulin-induced hypoglycaemia) or directly on the pituitary (exogenous growth hormone releasing hormone [GHRH]) is markedly reduced in human obesity (reviewed by Scacchi *et al.*, 1999). Leptin-deficient *ob/ob* mice or leptin-insensitive *db/db* mice are also known to be growth hormone deficient (Larson *et al.*, 1976; Sinha *et al.*, 1979), and it has been observed that plasma growth hormone levels in obese *ob/ob* mice (or normal lean rats) are increased by exogenous leptin administration (Stephens and Caro, 1998). Leptin may thus normally be important in regulating GH secretion (and consequent production of the insulin-like growth factor -I [IGF-I]) via a complex influence on the hypothalamic-pituitary axis, i.e. on hypothalamic growth hormone releasing hormone (GHRH) and somatostatin release (Carro *et al.*, 1999). Conversely, GH itself would not appear to affect leptin secretion independently of its effects on the adipose tissue mass (Considine, 1997).

39.8 Leptin and the hypothalamic-pituitary-adrenal (HPA) axis

There is now considerable evidence to suggest that leptin is an important modulator of HPA axis function in both animals and humans. Thus, leptin can *inhibit* the HPA axis both at the hypothalamic (by inhibiting corticotrophin releasing hormone [CRH] release) and adrenal cortical levels (Pralong *et al.*, 1998). Indeed, the characteristic increase in glucocorticoid secretion produced in response to stress stimuli can be blocked by exogenous leptin administration (Heiman *et al.*, 1997). By contrast, the decreased levels of plasma leptin that occur during fasting are associated with activation of the HPA axis and a *rise* in glucocorticoid release; (obese *ob/ob* mice that lack leptin, also show a chronic activation of the HPA axis that is attenuated by exogenous leptin administration; Ahima *et al.*, 1996).

The abdominal obesity associated with chronic glucocorticoid oversecretion in patients with Cushing's syndrome is well recognized; it is also generally accepted that an increase in glucocorticoid production or hyperresponsiveness of the hypothalamic-pituitary-adrenal axis accompanies and contributes to the development of many other human obesity phenotypes (reviewed by Chalew *et al.*, 1995). Like most obese individuals, plasma leptin levels in Cushing's subjects are elevated compared to controls with comparable amounts of body fat (Weise *et al.*, 1999); this suggests that under such abnormal metabolic conditions, the usual inhibitory feedback interaction of leptin with the HPA axis (as well as its effects on appetite suppression) may be impaired (i.e. glucocorticoid-induced leptin resistance) (Ur *et al.*, 1996; Solano and Jacobson, 1999).

High (pharmacological) doses of exogenously-administered glucocorticoids can certainly increase circulating leptin levels in healthy humans (Miell *et al.*, 1996), most likely by promoting leptin synthesis; however, elevations of plasma cortisol in the physiological range are claimed not to affect leptin secretion in the short term (Nye *et al.*, 2000). A glucocorticoid-induced increase in leptin secretion during conditions of stress might be envisaged to contribute to the anorexia and weight loss seen in certain chronic disease states (e.g. major depression) or moderate surgical trauma; however, evidence for such a firm link has yet to be established (Deuschle *et al.*, 1996; Stratton *et al.*, 1997).

39.9 Leptin and insulin

Although much interest has focused on leptin as the major "satiety" factor in the body, the peripheral secretion of insulin also occurs at levels proportional to the increased body adiposity and, like leptin, exerts a direct negative feedback effect on the hypothalamic arcuate nucleus (see below) to regulate appetite and energy homeostasis; the "adiposity signal" relayed by insulin may, however, influence food intake by acting through non-peptide neurotransmitter systems (Baskin *et al.*, 1999). The question arises whether leptin affects insulin secretion and *vice versa*, and whether these two peptides can act synergistically to control adiposity and energy balance.

It is known that the injection of recombinant leptin into obese (*ob/ob*) mice not only reduces their body weight, but also corrects their hyperinsulemia (Pelleymounter *et al.*,1995), suggesting that leptin may reduce insulin release *in vivo*. Indeed, *ob/ob* pancreatic β cells express the long isoform of the leptin receptor (OB-R), and respond to applied leptin with a suppression of insulin secretion, through activation of ATP-sensitive K^+ channels (Kieffer *et al.*, 1997); a similar direct inhibition of insulin release (and proinsulin gene expression) by leptin in *human* isolated pancreatic islets has also been reported (Seufert *et al.*, 1999). In contrast, the production of leptin in humans is *stimulated* by insulin, most likely through a long term trophic effect on adipocyte cells (Kolaczynski *et al.*, 1996). In view of these findings, it has been proposed that an *adipoinsular axis* involving insulin and leptin may be important in maintaining nutrient balance in the body (Kieffer and Habener, 2000); according to this model, an increase in adiposity would tend to increase plasma leptin, thereby curtailing insulin production and any further increase in fat mass (due to *adipogenic* effects of insulin; i.e. stimulation of terminal differentiation of new adipocytes from adipocyte precursor cells.). Dysregulation of the adipoinsular axis due to defective leptin reception by β-cells, might then be envisaged to contribute to the abdominal obesity and development of chronic hyperinsulinemia associated with type II diabetes (it is noteworthy that obese Type 2 (non-insulin-dependent) diabetics show similar plasma leptin levels to those of nondiabetic humans with the same BMI; Tasaka *et al.*, 1997).

In rodents, leptin (acting centrally via the ventromedial hypothalamus) and peripheral insulin have been shown to have a synergistic role in maintaining energy balance in certain peripheral tissues (e.g. heart, brown adipose tissue and skeletal muscles) by augmenting glucose uptake through a sympathetic β3-adrenergic mechanism (Haque *et al.*, 1999); indeed, the sympathetic nervous system (acting through peripheral β-receptors) is now considered an essential element of the leptin feeding control system, and disturbances of either leptin or β3-adrenergic receptor signaling are likely to predispose to obesity and Type 2 diabetes (Nonogaki, 2000).

39.9.1 Leptin, hypothalamic neuropeptides and feeding behavior

The mechanism by which leptin exerts its satiety effects in the brain has understandably attracted considerable research attention, and is likely to be quite complex. It has been known for some time that the direct central administration of leptin into *ob/ob* or diet-induced obese mice could reduce food intake and body weight, indicating an effect on neuronal networks controling feeding and energy balance (Campfield *et al.*, 1995). It is also well recognized that intracerebroventricular infusion of the central neuropeptide Y (NPY) (possibly acting via NPY-Y5 receptors; Criscione *et al.*, 1998; Duhault *et al.*, 2000) has opposing *orexigenic* effects on appetite and feeding behavior, implying that leptin's actions might somehow be mediated through an interference with NPY neuronal signaling (Stephens *et al.*, 1995). In keeping with these findings, chronic systemic administration of leptin into *ob/ob* mice has been shown to decrease the observed overexpression of hypothalamic NPY mRNA (Schwartz *et al.*, 1996a); also, intracerebroventricular leptin functionally inhibits the feeding behavior induced in these animals by NPY when both peptides are co-applied (Smith *et al.*, 1996). The most recent evidence suggests that the nitric oxide system may mediate these opposing effects of NPY and leptin on food intake (Morley *et al.*, 1999).

Rather surprisingly, NPY-deficient knockout mice (apart from being susceptible to seizures) show normal food intake and body weight, and are still responsive to leptin, although to a greater extent than wild-type controls (Erickson *et al.*, 1996); this could be explained if some compensation (by other orexigenic peptides?) for the lack of NPY during development had occurred. By contrast, mice heterozygous for the *anx* mutation (*anx/anx*) that exhibit anomalous processing of both NPY and pro-opiomelanocortin (POMC) in the arcuate nucleus (see below), show a lethal anorexic syndrome (Broberger *et al.*, 1999). Since NPY is abundant in the human hypothalamus (Elias *et al.*, 1998), it seems likely that a dysfunction of the NPY hypothalamic projection could result in a clinical obesity syndrome; the NPY Y5 receptor/transduction system might therefore prove to be another useful target for the development of new anti-obesity drugs (Dumont *et al.*, 2000).

Within the hypothalamus, the paraventricular nucleus (PVN) and the lateral hypothalamic/perifornical areas (LHA/PFA) are believed to be primary sites involved in regulating food intake and body weight. Distinct subsets of leptin-sensitive neurons in the hypothalamic arcuate nucleus (ARC) have been shown to innervate the PVN and the LHA/PFA regions and to contain either the endogenous *orexigenic* peptides – neuropeptide Y (NPY) co-localized with AGRP (agouti-related protein) or the *anorexigeneic* peptides – pro-opiomelanocortin (POMC), (the precursor of melanocyte stimulating hormone, α-MSH [melanocortin]) co-localized with the cocaine- and amphetamine-regulated transcript (CART) (reviewed by Schwartz *et al.*, 2000). According to the current model, leptin is thought to differentially modulate the activity of NPY/AGRP and POMC/CART neurons that project to the PVN and LHA/PFA thereby providing a link between circulating leptin and feeding behavior; thus, a rise in leptin secretion following an increase in adiposity, *stimulates* POMC/CART neurons, and *inhibits* NPY/AGRP neurons, leading to a *decrease* in appetite and food intake.

Interestingly, AGRP is a potent competitive antagonist at hypothalamic melanocortin MC4 receptors, through which α-MSH exerts its central *anorexigenic* effects (Ollmann *et al.*, 1997); these peptides therefore transmit opposite regulatory signals through a single receptor. Indeed, the importance of the melanocortin system is becoming increasingly apparent in understanding how circulating leptin levels signal a change in peripheral adiposity to the brain. Knockout mice lacking the MC4 receptor gene, have been shown to develop a maturity onset obesity syndrome characterized by hyperphagia, hyperinsulinemia, and hyperglycemia (Huszar *et al.*, 1997); moreover, mutations in the human MC4R gene are associated with a dominantly inherited form of obesity (Yeo *et al.*, 1998). Recent studies have also highlighted two rare cases of severe early onset obesity resulting from null alleles of the human POMC gene (Krude *et al.*, 1998). The development of specific MC4R agonists that mimic the hypothalamic melanocortinergic pathway may thus provide an exciting new mode of therapy for human obesity in the future (Benoit *et al.*, 2000).

The possibility of some other neuropeptides, e.g. melanin concentrating hormone (MCH), orexins A/B and galanin [all *orexigenic*] or corticotropin-releasing hormone [CRH], urocortin (a CRH-like peptide), thyrotrophin releasing hormone (TRH), glucagon-like peptide-1 (GLP-1) and neurotensin [all *anorexigenic*], being involved in the central control of food intake has also been advanced (for recent reviews see Frühbeck *et al.*, 1998b; Inui, 1999; Schwartz *et al.*, 2000), indicating further potential target sites for the design of novel antiobesity agents (reviewed by Bray and Tartaglia, 2000).

39.9.1.1 Direct effects of leptin on hypothalamic neurons

Electrophysiologically, both leptin and NPY, have been shown to reduce evoked (glutamatergic) excitatory transmission in the rat arcuate nucleus in hypothalamic brain slices *in vitro*. Leptin also decreased the input membrane resistance of arcuate neurons, induced an outward (hyperpolarizing) membrane current at positive potentials, and modulated Ca^{2+} signals in acutely isolated cells (Glaum *et al.*, 1996); subsequently, Spanswick *et al.* (1997) showed that this inhibitory hyperpolarizing effect of leptin on the arcuate neurons was due to direct activation of ATP-sensitive potassium (K_{ATP}) channels in the cell membrane, that was blocked by applying the specific sulphonylurea antagonist tolbutamide. As these channels are distributed fairly ubiquitously in the brain (Dunn-Meynell *et al.*, 1998), as well as in numerous peripheral tissues (e.g. smooth muscle, pancreatic

β-cells), it seems unlikely that selective K_{ATP} channel *openers* (KATPCOs: e.g. cromakalim, diazoxide) could ever prove useful as anti-obesity agents, unless compounds exhibiting an effective tissue selectivity can eventually be designed (Lawson, 2000).

39.10 Concluding remarks

The discovery of the *ob* gene and leptin has clearly provided exciting new insights into the mechanisms controlling body weight and composition, and has fuelled an exponential surge in research activity in this area. Between the period of June 1995 to July 2000, an *Advanced PubMed Medline* search using the keyword "*leptin*" retrieved 2832 published papers on the subject, and the field will undoubtedly continue to grow rapidly. Leptin research has also revealed several interesting new physiological targets for leptin action in the body (e.g. the endocrine and immune systems) which may lead to novel therapeutic uses for leptin agonists and antagonists in treating a variety of other human disease states.

In view of the underlying complexity that has now been revealed in the mechanisms controling food intake and energy homeostasis in the body, it is becoming increasingly clear that the effective treatment of obesity may eventually involve the use of a combination of drugs acting at different loci within the weight-regulatory system (together with a program of calorie control and exercise), rather than a single "miracle cure" approach using leptin alone. Based on current knowledge of this regulatory system, two basic novel approaches to obesity treatment can be suggested as a basis for future novel drug design:

- *Enhancing* hypothalamic inhibitory effects of *anorexigenic* agents like leptin itself or "downstream" peptides like α-MSH or CART, or exploiting the properties of other anorectic peptides, e.g. GLP-1, CRH/urocortin, or the cytokine ciliary neurotrophic factor (CNTF) (and its derivative *Axokine*).
- *Inhibiting* the actions of *orexigenic* factors like NPY, AGRP, MCH or hypocretins/orexins.

Since obesity is a chronic disorder, treatment is also likely to be chronic, which raises important issues of tolerability and potential toxicity of any newly developed drug. Also, with any such novel anorexigenic therapies, the possibility that following long term receptor blockade, counter-regulatory mechanisms may be induced to restore appetite cannot be discounted. Certainly, the need to develop a new range of appetite suppressant drugs to help obese patients lose weight, would seem ever more important since the recently announced withdrawal of conventional centrally-acting anorectic agents such as fenfluramine, dexfenfluramine, phentermine and amfepramone (diethylpropion) in the European Community, due to adverse reports of primary pulmonary hypertension and heart valve disorders. The currently approved newer anti-obesity drugs orlistat and sibutramine appear to be better tolerated.

Thus, despite the many major advances in obesity research that have been made in recent years, much work still needs to be carried out to better understand the complexities of body weight regulation in humans, and the many important central and peripheral functions of the leptin hormonal system and its various effector pathways. Whether all this research effort will lead to an effective and safe treatment of obesity in the near future remains to be seen.

References

1 Ahima, R.S., Prabakaran, D., Mantzoros, C., Qu, D., Lowell, B., Maratos-Flier, E., Flier, J.S. (1996) Role of leptin in the neuroendocrine response to fasting. *Nature, 382, 250–252.*

2 Ashworth, C.J., Hoggard, N., Thomas, L., Mercer, J.G., Wallace, J.M., Lea, R.G. (2000) Placental leptin. *Rev. Reprod., 5, 8–24.*

3 Bado, A., Levasseur, S., Attoub, S., Kermorgant, S., Laigneau, J.P., Bortoluzzi, M.N., Moizo, L., Lehy, T., Guerre-Millo, M., Le Marchand-Brustel, Y., Lewin, M.J. (1998) The stomach is a source of leptin. *Nature, 394, 790–793.*

4 Banks, W.A., Clever, C.M., Farrell, C.L. (2000) Partial saturation and regional variation in the blood-to-brain transport of leptinin normal weight mice. *Am J Physiol Endocrinol Metab., 278, E1158–E1165.*

5 Banks, W.A., Kastin, A.J., Huang, W., Japan, J.B., Maness, L.M. (1996). Leptin enters the brain by a saturable system independent of insulin. *Peptides, 17, 305–311.*

6 Baskin, D.G., Figlewicz-Lattemann, D., Seeley, R.J., Woods, S.C., Porte, D.Jr., Schwartz, M.W. (1999) Insulin and leptin: dual adiposity signals to the brain for the regulation of food intake and body weight. *Brain Res., 848, 114–123.*

7 Baumann, H., Morella, K.K., White, D.W., Dembski, M., Bailon, P.S., Kim, H., Lai, C.F., Tartaglia, L.A. (1996) The full-length leptin receptor has signalling capabilities of interleukin 6-type cytokine receptors. *Proc. Natl. Acad. Sci U.S.A. 93, 8374–8378.*

8 Beales, P.L., Kopelman, P.G. (1996) Obesity genes. *Clin Endocrinol., 45, 373–378.*

9 Benoit, S.C., Schwartz, M.W., Lachey, J.L., Hagan, M.M., Rushing, P.A., Blake, K.A., Yagaloff, K.A., Kurylko, G., Franco, L., Danhoo, W., Seeley, R.J. (2000) A novel selective melanocortin-4 receptor agonist reduces food intake in rats and mice without producing aversive consequences. *J Neurosci., 20, 3442–3448.*

10 Bjorbaek, C., Elmquist, J.K., Frantz, J.D., Shoelson, S.E., Flier, J.S. (1998) Identification of SOCS-3 as a potential mediator of central leptin resistance. *Mol Cell, 1, 619–625.*

11 Bjorbaek, C., Uotani, S., da Silva, B., Flier, J.S. (1997) Divergent signaling capacities of the long and short isoforms of the leptin receptor. *J Biol Chem., 272, 32686–32695.*

12 Bjorntorp, P. (1997) Obesity, *Lancet, 350, 423–426.*

13 Bouchard, C., Despres, J.P., Mauriege, P. (1993) Genetic and nongenetic determinants of regional fat distribution. *Endocr Rev., 14, 72–93.*

14 Bray, G.A. (1999) Uses and misuses of the new pharmacotherapy of obesity. *Ann Med., 31, 1–3.*

15 Bray, G.A., Tartaglia, L.A. (2000) Medicinal strategies in the treatment of obesity. *Nature, 404, 672–677.*

16 Broberger, C., Johansen, J., Brismar, H., Johansson, C., Schalling, M., Hokfelt, T. (1999) Changes in neuropeptide Y receptors and pro-opiomelanocortin in the anorexia (anx/anx) mouse hypothalamus. *J Neurosci., 19, 7130–7139.*
17 Campfield, L.A., Smith, F.J., Guisez, Y., Devos, R., Burn, P. (1995) Recombinant mouse OB protein: evidence for a peripheral signal linking adiposity and central neural networks. *Science, 269, 546–549.*
18 Caprio, M., Isidori, A.M., Carta, A.R., Moretti, C., Dufau, M.L., Fabbri, A. (1999) Expression of functional leptin receptors in rodent Leydig cells. *Endocrinology, 140, 4939–4947.*
19 Caro, J.F., Kolaczynski, J.W., Nyce, M.R., Ohannesian, J.P., Opentanova, I., Goldman, W.H., Lynn, R.B., Zhang, P.L., Sinha, M.K., Considine, R.V. (1996) Decreased cerebrospinal-fluid/serum leptin ratio in obesity: a possible mechanism for leptin resistance. *Lancet, 348, 159–161.*
20 Carro, E., Senaris, R.M., Seoane, L.M., Frohman, L.A., Arimura, A., Casanueva, F.F., Dieguez, C. (1999) Role of growth hormone (GH)-releasing hormone and somatostatin on leptin-induced GH secretion. *Neuroendocrinology, 69, 3–10.*
21 Cattaneo, E., Conti, L., De-Fraja, C. (1999) Signalling through the JAK-STAT pathway in the developing brain. *Trends Neurosci., 22, 365–369.*
22 Chalew, S., Nagel, H., Shore, S. (1995) The hypothalamic-pituitary-adrenal axis in obesity. *Obes Res., 3, 3713–82.*
23 Chen, H., Charlat, O., Tartaglia, L.A., Woolf, E.A., Weng, X., Ellis, S.J., Lakey, N.D., Culpepper, J., Moore, K.J., Breitbart, R.E., Duyk, G.M., Tepper, R.I., Morgenstern, J.P. (1996) Evidence that the diabetes gene encodes the leptin receptor: identification of a mutation in the leptin receptor gene in db/db mice. *Cell, 84, 491–495.*
24 Chen, D., Garg, A. (1999) Monogenic disorders of obesity and body fat distribution. *J. Lipid Res., 40, 1735–1746.*
25 Clement, K., Garner, C., Hager, J., Philippi, A., LeDuc, C., Carey, A., Harris, T.J., Jury, C., Cardon, L.R., Basdevant, A., Demenais, F., Guy-Grand, B., North, M., Froguel, P. (1996) Indication for linkage of the human OB gene region with extreme obesity. *Diabetes, 45, 687–690.*
26 Clement, K., Vaisse, C., Lahlou, N., Cabrol, S., Pelloux, V., Cassuto, D., Gourmelen, M., Dina, C., Chambaz, J., Lacorte, J.M., Basdevant, A., Bougneres, P., Lebouc, Y.,Froguel, P., Guy-Grand, B. (1998) A mutation in the human leptin receptor gene causes obesity and pituitary dysfunction. *Nature, 392, 398–401.*
27 Coleman, D.L. (1973) Effects of parabiosis of obese with diabetes and normal mice. *Diabetologia, 9, 294–298.*
28 Coleman, D.L. (1978) Obese and diabetes: two mutant genes causing diabetes-obesity syndrome in mice. *Diabetologia, 14, 141–148.*
29 Coleman, D.L. (1982) Diabetes-obesity syndromes in mice. *Diabetes, 31(Suppl. 1 Pt 2), 1–6.*
30 Considine, R.V. (1997) Weight regulation, leptin and growth hormone. *Hormone Res., 48 Suppl. 5:116–121.*
31 Considine, R.V., Considine, E.L., Williams, C.J., Nyce, M.R., Magosin, S.A., Bauer, T.L., Rosato, E.L., Colberg, J., Caro, J.F. (1995) Evidence against either a premature stop codon or the absence of obese gene mRNA in human obesity. *J Clin Invest., 95, 2986–2988.*
32 Considine, R.V., Sinha, M.K., Heiman, M.L., Kriauciunas, A., Stephens, T.W., Nyce, M.R., Ohannesian, J.P., Marco, C.C., McKee, L.J., Bauer, T.L. *et al.* (1996a) Serum immunoreactive-leptin concentrations in normal-weight and obese humans. *N Engl J Med., 334, 292–295.*
33 Considine, R.V., Considine, E.L., Williams, C.J., Nyce, M.R., Zhang, P., Opentanova, I., Ohannesian, J.P., Kolaczynski, J.W., Bauer, T.L., Moore, J.H., Caro, J.F. (1996b) Mutation screening and identification of a sequence variation in the human ob gene coding region. *Biochem Biophys Res Commun., 220, 735–739.*
34 Criscione, L., Rigollier, P., Batzl-Hartmann, C., Rueger, H., Stricker-Krongrad, A., Wyss, P., Brunner, L., Whitebread, S., Yamaguchi, Y., Gerald, C., Heurich, R.O., Walker, M.W., Chiesi, M., Schilling, W., Hofbauer, K.G., Levens, N. (1998) Food intake in free-feeding and energy-deprived lean rats is mediated by the neuropeptide Y5 receptor. *J. Clin. Invest., 102, 2136–2145.*
35 Cumin, F., Baum, H.P., Levens, N. (1997) Mechanism of leptin removal from the circulation by the kidney. *J Endocrinol.,155, 577–585.*
36 Dallongeville, J., Fruchart, J.C., Auwerx, J. (1998) Leptin, a pleiotropic hormone: physiology, pharmacology, and strategies for discovery of leptin modulators. *J Med Chem., 41, 5337–5352.*
37 Demerath, E.W., Towne, B., Wisemandle, W., Blangero, J., Chumlea, W.C., Siervogel, R.M. (1999) Serum leptin concentration, body composition, and gonadal hormones during puberty. *Int J Obes Relat Metab Disord., 23,678–685.*
38 Deuschle, M. Blum, W.F., Englaro, P., Schweiger, U., Weber, B., Pflaum, C.D., Heuser, I. (1996) Plasma leptin in depressed patients and healthy controls. *Horm Metab Res., 28, 714–717.*
39 Duhault, J., Boulanger, M., Chamorro, S., Boutin, J.A., Della Zuana, O., Douillet, E., Fauchere, J.L., Feletou, M., Germain, M., Husson, B., Vega, A.M., Renard, P., Tisserand, F. (2000) Food intake regulation in rodents: Y5 or Y1 NPY receptors or both? *Can. J. Physiol. Pharmacol., 78,173–185.*
40 Dumont, Y., Cadieux, A., Doods, H., Fournier, A., Quirion, R. (2000) Potent and selective tools to investigate neuropeptide Y receptors in the central and peripheral nervous systems: BIB03304 (Y1) and CGP71683A(Y5). *Can. J. Physiol. Pharmacol.,78,116–125.*
41 Dunn-Meynell, A.A., Rawson, N.E., Levin, B.E. (1998) Distribution and phenotype of neurons containing the ATP-sensitive K+ channel in rat brain. *Brain Res., 814, 41–54.*
42 Elias, C.F., Saper, C.B., Maratos-Flier, E., Tritos, N.A., Lee, C., Kelly, J., Tatro, J.B., Hoffman, G.E., Ollmann, M.M., Barsh, G.S., Sakurai, T., Yanagisawa, M., Elmquist, J.K. (1998) Chemically defined projections linking the mediobasal hypothalamus and the lateral hypothalamic area. *J Comp Neurol., 402, 442–459.*
43 Elmquist, J.K., Maratos-Flier, E., Saper, C.B., Flier, J.S. (1998) Unraveling the central nervous system pathways underlying responses to leptin. *Nat Neurosci., 1, 445–450.*
44 Erickson, J.C., Clegg, K.E., Palmiter, R.D. (1996) Sensitivity to leptin and susceptibility to seizures of mice lacking neuropeptide Y. *Nature, 381(6581): 415–421.*
45 Felson, D.T., Chaisson, C.E. (1997) Understanding the relationship between body weight and osteoarthritis. *Baillieres Clin Rheumatol., 11, 671–681.*
46 Ford, E.S. (1999) Body mass index and colon cancer in a national sample of adult US men and women. *Am J Epidemiol., 150, 390–398.*

47 Frayn, K.N., Coppack, S.W. (1992) Insulin resistance, adipose tissue and coronary heart disease. *Clin Sci (Colch)., 82, 1–8.*

48 Friedman, J.M., Halaas, J.L. (1998) Leptin and the regulation of body weight in mammals. *Nature, 395, 763–769.*

49 Friedman, J.M., Liebel, R.L. (1992) Tackling a weighty problem. *Cell, 69, 217–220.*

50 Frühbeck, G., Aguado, M., Gomez-Ambrosi, J., Martinez, J.A. (1998a) Lipolytic effect of *in vivo* leptin administration on adipocytes of lean and ob/ob mice, but not db/db mice. *Biochem Biophys Res Commun., 250, 99–102.*

51 Frühbeck, G., Jebb, S.A., Prentice, A.M. (1998b) Leptin: physiology and pathophysiology. *Clin Physiol., 18, 399–419.*

52 Frühbeck, G., Garcia-Granero, M., Martinez, J.A. (1998c) Age-related differences in the thermogenic and ponderal effects following the administration of fragment peptides from the rat ob protein. *Regul Peptides, 73, 83–87.*

53 Glaum, S.R., Hara, M., Bindokas, V.P., Lee, C.C., Polonsky, K.S., Bell, G.I., Miller, R.J. (1996) Leptin, the obese gene product, rapidly modulates synaptic transmission in the hypothalamus *Mol Pharmacol., 50, 230–235.*

54 Gouilleux-Gruart, V., Debierre-Grockiego, F., Gouilleux, F., Capiod, J.C., Claisse, J.F., Delobel, J., Prin, L. (1997) Activated Stat related transcription factors in acute leukemia. *Leuk Lymphoma, 28, 83–88.*

55 Grasso, P., Leinung, M.C., Ingher, S.P., Lee, D.W. (1997) *In vivo* effects of leptin-related synthetic peptides on body weight and food intake in female ob/ob mice: localization of leptin activity to domains between amino acid residues 106-140. *Endocrinology, 138,1413–1418.*

56 Green, E.D., Maffei, M., Braden, V.V., Proenca, R., DeSilva, U., Zhang, Y., Chua, S.C.Jr., Leibel, R.L., Weissenbach, J., Friedman, J.M. (1995) The human obese (OB) gene: RNA expression pattern and mapping on the physical, cytogenetic, and genetic maps of chromosome 7. *Genome Res., 5, 5–12.*

57 Guan, X.M., Hess, J.F., Yu, H., Hey, P.J., van der Ploeg, L.H. (1997) Differential expression of mRNA for leptin receptor isoforms in the rat brain. *Mol Cell Endocrinol., 133, 1–7.*

58 Hager, J., Clement, K., Francke, S., Dina, C., Raison, J., Lahlou, N., Rich, N., Pelloux, V., Basdevant, A., Guy-Grand, B., North, M., Froguel, P. (1998) A polymorphism in the 5' untranslated region of the human ob gene is associated with low leptin levels. *Int J Obes Relat Metab Disord., 22, 200–205.*

59 Hakansson, M.L., Meister, B. (1998) Transcription factor STAT3 in leptin target neurons of the rat hypothalamus *Neuroendocrinology, 68, 420–437.*

60 Halaas, J.L., Gajiwala, K.S., Maffei, M., Cohen, S.L., Chait, B.T., Rabinowitz, D., Lallone, R.L., Burley, S.K., Friedman, J.M. (1995) Weight-reducing effects of the plasma protein encoded by the obese gene. *Science, 269, 543–546.*

61 Haluzik, M., Papezova, M., Nedvidkova, J., Kabrt, J. (1999) Serum leptin levels in patients with anorexia nervosa before and after partial refeeding, relationships to serum lipids and biochemical nutritional parameters. *Physiol Res., 48, 197–202.*

62 Haque, M.S., Minokoshi, Y., Hamai, M., Iwai, M., Horiuchi, M., Shimazu, T. (1999) Role of the sympathetic nervous system and insulin in enhancing glucose uptake in peripheral tissues after intrahypothalamic injection of leptin in rats. *Diabetes, 48, 1706–1712.*

63 Hassink, S.G., Sheslow, D.V., de Lancey, E., Opentanova, I., Considine, R.V., Caro, J.F. (1996) Serum leptin in children with obesity: relationship to gender and development. *Pediatrics, 98, 201–203.*

64 Heaney, M.L., Golde, D.W. (1993) Soluble hormone receptors. *Blood, 82, 1945–1948.*

65 Heim, M.H. (1999) The Jak-STAT pathway: cytokine signalling from the receptor to the nucleus. *J Recept Signal Transduct Res., 19, 75–120.*

66 Heiman, M.L., Ahima, R.S., Craft, L.S., Schoner, B., Stephens, T.W., Flier, J.S. (1997) Leptin inhibition of the hypothalamic-pituitary-adrenal axis in response to stress. *Endocrinology, 138, 3859–3863.*

67 Hervey, G.R. (1959) The effects of lesions in the hypothalamus in parabiotic rats. *J. Physiol., 145, 336–352.*

68 Heymsfield, S.B., Greenberg, A.S., Fujioka, K., Dixon, R.M., Kushner, R., Hunt, T., Lubina, J.A., Patane, J., Self, B., Hunt, P., McCamish, M. (1999) Recombinant leptin for weight loss in obese and lean adults: a randomized, controlled, dose-escalation trial. *JAMA, 282, 1568–1575.*

69 Hilton, D.J. (1999) Negative regulators of cytokine signal transduction. *Cell Mol Life Sci., 55, 1568–1577.*

70 Himms-Hagen, J. (1999) Physiological roles of the leptin endocrine system: differences between mice and humans. *Crit Rev Clin Lab Sci., 36, 575–655.*

71 Ho, K.Y., Veldhuis, J.D., Johnson, M.L., Furlanetto, R., Evans, W.S., Alberti, K.G., Thorner, M.O. (1988) Fasting enhances growth hormone secretion and amplifies the complex rhythms of growth hormone secretion in man. *J Clin Invest., 81, 968–975.*

72 Hoey, T., Schindler, U. (1998) STAT structure and function in signaling. *Curr Opin Genet Dev., 8, 582–587.*

73 Houseknecht, K.L., Baile, C.A., Matteri, R.L., Spurlock, M.E. (1998) The biology of leptin: a review. *J. Anim. Sci., 76, 1405–1420.*

74 Houseknecht, K.L., Mantzoros, C.S., Kuliawat, R., Hadro, E., Flier, J.S., Kahn, B.B. (1996) Evidence for leptin binding to proteins in serum of rodents and humans: modulation with obesity. *Diabetes, 45, 1638 1643.*

75 Huang, Z., Reddy, A. (1999) Weight change, ideal weight and hypertension. *Curr Opin Nephrol Hypertens., 8, 343–346.*

76 Hube, F., Lietz, U., Igel, M., Jensen, P.B., Tornqvist, H., Joost, H.G., Hauner, H. (1996) Difference in leptin mRNA levels between omental and subcutaneous abdominal adipose tissue from obese humans. *Horm Metab Res., 28, 690–693.*

77 Hummel, K.P., Dickie, M.M., Coleman, D.L. (1966) Diabetes, a new mutation in the mouse. *Science, 153,1127–1128.*

78 Huszar, D., Lynch, C.A., Fairchild-Huntress, V., Dunmore, J.H., Fang, Q., Berkemeier, L.R., Gu, W., Kesterson, R.A., Boston, B.A., Cone, R.D., Smith, F.J., Campfield, L.A., Burn, P., Lee, F. (1997) Targeted disruption of the melanocortin-4 receptor results in obesity in mice. *Cell, 88, 131–141.*

79 Ihle, J.N. (1996) STATs: signal transducers and activators of transcription. *Cell, 84, 331–334.*

80 Iida, M., Murakami, T., Ishida, K., Mizuno, A., Kuwajima, M., Shima, K. (1996) Substitution at codon 269 (glutamine → proline) of the leptin receptor (OB-R) cDNA is the only mutation found in the Zucker fatty (fa/fa) rat. *Biochem Biophys Res Commun., 224, 597–604.*

81 Ingalls, A., Dickie, M., Snell, G. (1950) Obese, a new mutation in the house mouse. *J. Hered. 41, 317–318.*

82 Inui, A. (1999) Feeding and body-weight regulation by hypothalamic neuropeptides-mediation of the actions of leptin. *Trends Neurosci, 22, 62–67.*

83 Jorgensen, J.O., Vahl, N., Dall, R., Christiansen, J.S. (1998) Resting metabolic rate in healthy adults: relation to growth hormone status and leptin levels. *Metabolism, 47,1134–1139.*

84 Karlsson, C., Lindell, K., Svensson, E., Bergh, C., Lind, P., Billig, H., Carlsson, L.M., Carlsson, B. (1997) Expression of functional leptin receptors in the human ovary *J Clin Endocrinol Metab., 82, 4144–4148.*

85 Kennedy, G.C. (1953) The role of depot fat in the hypothalamic control of food intake in the rat. *Proc. R. Soc. Series B, 140, 578–592.*

86 Kieffer, T.J., Habener, J.F. (2000) The adipoinsular axis: effects of leptin on pancreatic beta-cells. *Am J Physiol Endocrinol Metab., 278, E1–E14.*

87 Kieffer, T.J., Heller, R.S., Leech, C.A., Holz, G.G., Habener, J.F. (1997) Leptin suppression of insulin secretion by the activation of ATP-sensitive K+ channels in pancreatic beta-cells. *Diabetes, 46, 1087–1093.*

88 Kim, M.S., Small, C.J., Stanley, S.A., Morgan, D.G., Seal, L.J., Kong, W.M., Edwards, C.M., Abusnana, S., Sunter, D., Ghatei, M.A., Bloom, S.R. (2000) The central melanocortin system affects the hypothalamo-pituitary thyroid axis and may mediate the effect of leptin. *J Clin Invest., 105,1005–1011.*

89 Klein, S., Coppack, S.W., Mohamed-Ali, V., Landt, M. (1996) Adipose tissue leptin production and plasma leptin kinetics in humans. *Diabetes, 45, 984–987.*

90 Kokot, F., Adamczak, M., Wiecek, A., Cieplok, J. (1999) Does leptin play a role in the pathogenesis of essential hypertension? *Kidney Blood Press Res., 22, 154–160.*

91 Kolaczynski, J.W., Nyce, M.R., Considine, R.V., Boden, G., Nolan, J.J., Henry, R., Mudaliar, S.R., Olefsky, J., Caro, J.F. (1996a) Acute and chronic effects of insulin on leptin production in humans: Studies *in vivo* and *in vitro. Diabetes, 45, 699–701.*

92 Kolaczynski, J.W., Ohannesian, J.P., Considine, R.V., Marco, C.C., Caro, J.F. (1996b) Response of leptin to short-term and prolonged overfeeding in humans. *J Clin Endocrinol Metab., 81, 4162–4165.*

93 Krude, H., Biebermann, H., Luck, W., Horn, R., Brabant, G., Gruters, A. (1998) Severe early-onset obesity, adrenal insufficiency and red hair pigmentation caused by POMC mutations in humans. *Nat. Genet., 19, 155–157.*

94 Lagergren, J., Bergstrom, R., Nyren, O. (1999) Association between body mass and adenocarcinoma of the esophagus and gastric cardia. *Ann Intern Med., 130, 883–890.*

95 Larson, B.A., Sinha, Y.N., Vanderlaan, W.P. (1976) Serum growth hormone and prolactin during and after the development of the obese-hyperglycemic syndrome in mice. *Endocrinology, 98, 139–145.*

96 Lawson, K. (2000) Potassium channel openers as potential therapeutic weapons in ion channel disease. *Kidney Int., 57, 838–845.*

97 Layman, L.C. (1999) The molecular basis of human hypogonadotropic hypogonadism. *Mol. Genet. Metab., 68, 191–199.*

98 Lee, G.H., Proenca, R., Montez, J.M., Carroll, K.M., Darvishzadeh, J.G., Lee, J.I., Friedman, J.M. (1996) Abnormal splicing of the leptin receptor in diabetic mice. *Nature, 379, 632–635.*

99 Legradi, G., Emerson, C.H., Ahima, R.S., Rand, W.M., Flier, J.S., Lechan, R.M. (1998) Arcuate nucleus ablation prevents fasting-induced suppression of ProTRH mRNA in the hypothalamic paraventricular nucleus. *Neuroendocrinology, 68, 89–97.*

100 Leibel, R.L., Bahary, N., Friedman, J.M. (1993) Strategies for the molecular genetic analysis of obesity in humans. *Crit Rev Food Sci Nutr., 33, 351–358.*

101 Lonnqvist, F., Nordfors, L., Schalling, M. (1999) Leptin and its potential role in human obesity. *J Intern Med., 245, 643–652.*

102 Madej, T., Boguski, M.S., Bryant, S.H. (1995) Threading analysis suggests that the obese gene product may be a helical cytokine. *FEBS Lett., 373, 13–18.*

103 Maffei, M., Halaas, J., Ravussin, E., Pratley, R.E., Lee, G.H., Zhang, Y., Fei, H., Kim, S., Lallone, R., Ranganathan, S. *et al.*(1995) Leptin levels in human and rodent: measurement of plasma leptin and ob RNA in obese and weight-reduced subjects. *Nat Med Nov; 1(11): 1155–61.*

104 Maffei, M., Stoffel, M., Barone, M., Moon, B., Dammerman, M., Ravussin, E., Bogardus, C., Ludwig, D.S., Flier, J.S., Talley, M., *et al.* (1996) Absence of mutations in the human OB gene in obese/diabetic subjects. *Diabetes, 45, 679–682.*

105 Mantzoros, C.S. (1999) Leptin in renal failure. *J Ren Nutr., 9, 122–125.*

106 Mantzoros, C.S., Moschos, S.J. (1998) Leptin: in search of role(s) in human physiology and pathophysiology. *Clin Endocrinol., 49, 551–567.*

107 Marti, A., Berraondo, B., Martinez, J.A. (1999) Leptin: physiological actions. *J Physiol Biochem., 55, 43–49.*

108 Masuzaki, H., Ogawa, Y., Sagawa, N., Hosoda, K., Matsumoto, T., Mise, H., Nishimura, H., Yoshimasa, Y., Tanaka, I., Mori, T., Nakao, K. (1997) Nonadipose tissue production of leptin: leptin as a novel placenta-derived hormone in humans. *Nat Med., 3, 1029–1033.*

109 Matarese, G. (2000) Leptin and the immune system: how nutri-tional status influences the immune response. *Eur Cytokine Netw., 11, 7–14.*

110 Matejek, N., Weimann, E., Witzel, C., Molenkamp, G., Schwidergall, S., Bohles, H. (1999) Hypoleptinaemia in patients with anorexia nervosa and in elite gymnasts with anorexia athletica. *Int J Sports Med., 20, 451–456.*

111 McIntyre, A.M. (1998) Burden of illness review of obesity: are the true costs realised? *J R Soc Health., 118, 76–84.*

112 McNeely, M.J., Boyko, E.J., Weigle, D.S., Shofer, J.B., Chessler, S.D., Leonnetti, D.L., Fujimoto, W.Y. (1999) Association between baseline plasma leptin levels and subsequent development of diabetes in Japanese Americans. *Diabetes Care, 22, 65–70.*

113 Mercer, J.G., Moar, K.M., Hoggard, N. (1998) Localization of leptin receptor (Ob-R) messenger ribonucleic acid in the rodent hindbrain. *Endocrinology, 139, 29–34.*

114 Meyer, C., Robson, D., Rackovsky, N., Nadkarni, V., Gerich, J. (1997) Role of the kidney in human leptin metabolism. *Am J Physiol., 273, E903–907.*

115 Miell, J.P., Englaro, P., Blum, W.F. (1996) Dexamethasone induces an acute and sustained rise in circulating leptin levels in normal human subjects. *Horm Metab Res., 28, 704–707.*

116 Montague, C.T., Farooqi, I.S., Whitehead, J.P., Soos, M.A., Rau, H., Wareham, N.J., Sewter, C.P., Digby, J.E., Mohammed, S.N., Hurst, J.A., Cheetham, C.H., Earley, A.R., Barnett, A.H., Prins, J.B., O'Rahilly, S. (1997) Congenital leptin deficiency is associated with severe early-onset obesity in humans. *Nature, 387, 903–908.*

117 Morley, J.E., Alshaher, M.M., Farr, S.A., Flood, J.F., Kumar, V.B. (1999) Leptin and neuropeptide Y (NPY) modulate nitric

oxide synthase: further evidence for a role of nitric oxide in feeding. *Peptides, 20, 595–600.*

118 Morsy, M.A., Gu, M.C., Zhao, J.Z., Holder, D.J., Rogers, I.T., Pouch, W.J., Motzel, S.L., Klein, H.J., Gupta, S.K., Liang, X., Tota, M.R., Rosenblum, C.I., Caskey, C.T. (1998) Leptin gene therapy and daily protein administration: a comparative study in the ob/ob mouse. *Gene Ther., 5, 8–18.*

119 Niki, T., Mori, H., Tamori, Y., Kishimoto-Hashirmoto, M., Ueno, H., Araki, S., Masugi, J., Sawant, N., Majithia, H.R., Rais, N., *et al.* (1996) Human obese gene: molecular screening in Japanese and Asian Indian NIDDM patients associated with obesity. *Diabetes, 45, 675–678.*

120 Nonogaki, K. (2000) New insights into sympathetic regulation of glucose and fat metabolism. *Diabetologia, 43, 533–549.*

121 Nye, E.J., Bornstein, S.R., Grice, J.E., Tauchnitz, R., Hockings, G.I., Strakosch, C.R., Jackson, R.V., Torpy, D.J. (2000) Interactions between the stimulated hypothalamic-pituitary-adrenal axis and leptin in humans. *J Neuroendocrinol., 12, 141–145.*

122 O'brien, S.N., Welter, B.H., Price, T.M. (1999) Presence of leptin in breast cell lines and breast tumors. *Biochem Biophys Res Commun., 259, 695–698.*

123 Ollmann, M.M., Wilson, B.D., Yang, Y.K., Kerns, J.A., Chen, Y., Gantz, I., Barsh, G.S. (1997) Antagonism of central melanocortin receptors *in vitro* and *in vivo* by agouti-related protein. *Science, 278, 135–138.*

124 Orban, Z., Bornstein, S.R., Chrousos, G.P. (1998) The interaction between leptin and the hypothalamic-pituitary-thyroid axis. *Horm. Metab. Res., 30, 231–235.*

125 Pelleymounter, M.A., Cullen, M.J., Baker, M.B., Hecht, R., Winters, D., Boone, T., Collins, F. (1995) Effects of the obese gene product on body weight regulation in ob/ob mice. *Science, 269, 540–543.*

126 Pinkney, J.H., Goodrick, S.J., Katz, J., Johnson, A.B., Lightman, S.L., Coppack, S.W., Mohamed-Ali, V. (1998) Leptin and the pituitary-thyroid axis: a comparative study in lean, obese, hypothyroid and hyperthyroid subjects. *Clin Endocrinol., 1998, 49, 583–588.*

127 Pralong, F.P., Roduit, R., Waeber, G., Castillo, E., Mosimann, F., Thorens, B., Gaillard, R.C. (1998) Leptin inhibits directly glucocorticoid secretion by normal human and rat adrenal gland. *Endocrinology, 139, 4264–4268.*

128 Prokai, L. (1998) Peptide drug delivery into the central nervous system. *Prog Drug Res., 51, 95–131.*

129 Reed, D.R., Ding, Y., Xu, W., Cather, C., Green, E.D., Price, R.A. (1996) Extreme obesity may be linked to markers flanking the human OB gene. *Diabetes, 45, 691–694.*

130 Rock, F.L., Altmann, S.W., van Heek, M., Kastelein, R.A., Bazan, J.F. (1996) The leptin haemopoietic cytokine fold is stabilized by an intrachain disulfide bond. *Horm Metab Res., 28, 649–652.*

131 Rolland, V., Clement, K., Dugail, I., Guy-Grand, B., Basdevant, A., Froguel, P., Lavau, M. (1998) Leptin receptor gene in a large cohort of massively obese subjects: no indication of the fa/fa rat mutation. Detection of an intronic variant with no association with obesity. *Obes Res., 6, 122–127.*

132 Rollheim, J., Osnes, T., Miljeteig, H. (1997) The relationship between obstructive sleep apnoea and body mass index. *Clin Otolaryngol, 22, 419–422.*

133 Rosenbaum, M., Leibel, R.L., Hirsch, J. (1997) Obesity. *N Engl J Med., 337, 396–407.*

134 Rosengren, A., Wedel, H., Wilhelmsen, L. (1999) Body weight and weight gain during adult life in men in relation to coronary heart disease and mortality. A prospective population study. *Eur Heart J., 20, 269–277.*

135 Scacchi, M., Pincelli, A.I., Cavagnini, F. (1999) Growth hormone in obesity. *Int. J. Obes. Relat. Metab. Disord., 23, 260–271.*

136 Schalling, M., Johansen, J., Nordfors, L., Lonnqvist, F. (1999) Genes involved in animal models of obesity and anorexia. *J Intern Med., 245, 613–619.*

137 Schevitz, R.W. (1997) Crystal structure of the obese protein leptin-E100. *Nature, 387, 206–209.*

138 Schwartz, M.W., Baskin, D.G., Bukowski, T.R., Kuijper, J.L., Foster, D., Lasser, G., Prunkard, D.E., Porte, D.Jr., Woods, S.C., Seeley, R.J., Weigle, D.S. (1996a) Specificity of leptin action on elevated blood glucose levels and hypothalamic neuropeptide Y gene expression in ob/ob mice. *Diabetes, 45, 531–535.*

139 Schwartz, M.W., Peskind, E., Raskind, M., Boyko, E.J., Porte, D.Jr.(1996b) Cerebrospinal fluid leptin levels: relationship to plasma levels and to adiposity in humans. *Nat Med., 2, 589–593.*

140 Schwartz, M.W., Woods, S.C., Porte, D.Jr., Seeley, R.J., Baskin, D.G. (2000) Central nervous system control of food intake. *Nature, 404, 661–671.*

141 Sesmilo, G., Casamitjana, R., Halperin, I., Gomis, R., Vilardell, E. (1998) Role of thyroid hormones on serum leptin levels. *Eur J Endocrinol., 139, 428–430.*

142 Seufert, J., Kieffer, T.J., Leech, C.A., Holz, G.G., Moritz, W., Ricordi, C., Habener, J.F. (1999) Leptin suppression of insulin secretion and gene expression in human pancreatic islets: implications for the development of adipogenic diabetes mellitus. *J Clin Endocrinol Metab., 84, 670–676.*

143 Sinha, M.K., Ohannesian, J.P., Heiman, M.L., Kriauciunas, A., Stephens, T.W., Magosin, S., Marco, C., Caro, J.F. (1996a) Nocturnal rise of leptin in lean, obese, and non-insulin-dependent diabetes mellitus subjects. *J Clin Invest., 97, 1344–1347.*

144 Sinha, M.K., Opentanova, I., Ohannesian, J.P., Kolaczynski, J.W., Heiman, M.L., Hale, J., Becker, G.W., Bowsher, R.R., Stephens, T.W., Caro, J.F. (1996b) Evidence of free and bound leptin in human circulation. Studies in lean and obese subjects and during short-term fasting. *J Clin Invest., 98, 1277–1282.*

145 Sinha, M.K., Sturis, J., Ohannesian, J., Magosin, S., Stephens, T., Heiman, M.L., Polonsky, K.S., Caro, J.F. (1996c) Ultradian oscillations of leptin secretion in humans. *Biochem Biophys Res Commun., 228, 733–738.*

146 Sinha, Y.N., Baxter, S.R., Larson, B.A., Vanderlaan, W.P. (1979) Levels of prolactin, growth hormone and insulin in genetically diabetic (db/db) mice. *Proc. Soc. Exp. Biol. Med., 161, 78–81.*

147 Sir-Petermann, T., Piwonka, V., Perez, F., Maliqueo, M., Recabarren, S.E., Wildt, L. (1999) Are circulating leptin and luteinizing hormone synchronized in patients with polycystic ovary syndrome? *Hum Reprod., 14, 1435–1439.*

148 Smith, F.J., Campfield, L.A., Moschera, J.A., Bailon, P.S., Burn, P. (1996) Feeding inhibition by neuropeptide Y. *Nature, 382, 307.*

149 Solano, J.M., Jacobson, L. (1999) Glucocorticoids reverse leptin effects on food intake and body fat in mice without increasing NPY mRNA *Am J Physiol., 277, E708–E716.*

150 Spanswick, D., Smith, M.A., Groppi, V.E., Logan, S.D., Ashford, M.L. (1997). Leptin inhibits hypothalamic neurons by activation of ATP-sensitive potassium channels. *Nature, 390, 521–525.*

151 Spicer, L.J., Francisco, C.C. (1998) Adipose obese gene product, leptin, inhibits bovine ovarian thecal cell steroidogenesis. *Biol Reprod., 58, 207–212.*

152 Starr, R., Hilton, D.J. (1998) SOCS: suppressors of cytokine signalling. *Int J Biochem Cell Biol., 30, 1081–1085.*

153 Starr, R., Hilton, D.J. (1999) Negative regulation of the JAK/STAT pathway. *Bioessays, 21,47–52.*

154 Stephens, T.W., Basinski, M., Bristow, P.K., Bue-Valleskey, J.M., Burgett, S.G., Craft, L., Hale, J., Hoffmann, J., Hsiung, H.M., Kriauciunas, A. *et al.* (1995) The role of neuropeptide Y in the antiobesity action of the obese gene product. *Nature, 377, 530–534.*

155 Stephens, T.W. Caro, J.F. (1998) To be lean or not to be lean. Is leptin the answer? *Exp Clin Endocrinol Diabetes 106, 1–15.*

156 Stratton, R.J., Dewit, O., Crowe, E., Jennings, G., Villar, R.N., Elia, M. (1997) Plasma leptin, energy intake and hunger following total hip replacement surgery. *Clin Sci., 93, 113–117.*

157 Strobel, A., Issad, T., Camoin, L., Ozata, M., Strosberg, A.D. (1998) A leptin missense mutation associated with hypogonadism and morbid obesity. *Nat Genet., 18, 213–215.*

158 Strosberg, A.D., Issad, T. (1999) The involvement of leptin in humans revealed by mutations in leptin and leptin receptor genes. *Trends Pharmacol Sci., 20, 227–230.*

159 Takaya, K., Ogawa, Y., Hiraoka, J., Hosoda, K., Yamori, Y., Nakao, K., Koletsky, R.J. (1996) Nonsense mutation of leptin receptor in the obese spontaneously hypertensive Koletsky rat. *Nat Genet., 14, 130–131.*

160 Takaya, K., Ogawa, Y., Masuzaki, H., Matsuoka, N., Hosoda, K., Nakao, K. (1998) Leptin Receptor. *Nippon Rinsho, 56, 1925–1930. [Article in Japanese]*

161 Tartaglia, L.A., Dembski, M., Weng, X., Deng, N., Culpepper, J., Devos, R., Richards, G.J., Campfield, L.A., Clark, F.T., Deeds, J., *et al.* (1995) Identification and expression cloning of a leptin receptor, OB-R. *Cell, 83, 1263–1271.*

162 Tasaka, Y., Yanagisawa, K., Iwamoto, Y. (1997) Human plasma leptin in obese subjects and diabetics. *Endocr J., 44, 671–676.*

163 Trayhurn, P., Hoggard, N., Mercer, J.G., Rayner, D.V. (1999) Leptin: fundamental aspects *Int J Obes Relat Metab Disord., 23, Suppl 1, 22–28.*

164 Turyn, D., Dominici, F.P., Sotelo, A.I., Bartke, A. (1997) Growth hormone-binding protein enhances growth hormone activity *in vivo. Am J Physiol., 273, E549–556.*

165 Uotani, S., Bjorbaek, C., Tornoe, J., Flier, J.S. (1999) Functional properties of leptin receptor isoforms: internalization and degradation of leptin and ligand-induced receptor downregulation. *Diabetes, 48, 279–286.*

166 Úr, E., Grossman, A., Despres, J.P. (1996) Obesity results as a consequence of glucocorticoid induced leptin resistance. *Horm Metab Res., 28, 744–747.*

167 Van Gaal, L.F., Wauters, M.A., Mertens, I.L., Considine, R.V., De Leeuw, I.H. (1999) Clinical endocrinology of human leptin. *Int J Obes Relat Metab Disord., 23, Suppl 1, 29–36.*

168 Vuagnat, B.A., Pierroz, D.D., Lalaoui, M., Englaro, P., Pralong, F.P., Blum, W.F., Aubert, M.L. (1998) Evidence for a leptin-neuropeptide Y axis for the regulation of growth hormone secretion in the rat. *Neuroendocrinology, 67, 291–300.*

169 Wannamethee, S.G., Shaper, A.G. (1999) Weight change and duration of overweight and obesity in the incidence of type 2 diabetes. *Diabetes Care 22, 1266–1272.*

170 Weise, M., Abad, V., Considine, R.V., Nieman, L., Rother, K.I. (1999) Leptin secretion in Cushing's syndrome: preservation of diurnal rhythm and absent response to corticotropin-releasing hormone. *J Clin Endocrinol Metab., 84, 2075–2079.*

171 Wu-Peng, X.S., Chua, S.C.Jr., Okada, N., Liu, S.M., Nicolson, M., Leibel, R.L. (1997) Phenotype of the obese Koletsky (f) rat due to Tyr763Stop mutation in the extracellular domain of the leptin receptor (Lepr): evidence for deficient plasma-to-CSF transport of leptin in both the Zucker and Koletsky obese rat. *Diabetes, 46, 513–518.*

172 Yeo, G.S., Farooqi, I.S., Aminian, S., Halsall, D.J., Stanhope, R.G., O'Rahilly, S. (1998) A frameshift mutation in MC4R associated with dominantly inherited human obesity. *Nat. Genet., 20, 111–112.*

173 Zhang, F., Basinski, M.B., Beals, J.M., Briggs, S.L., Churgay, L.M., Clawson, D.K., Dimarchi, R.D., Furman, T.C., Hale, J.E., Hsiung, H.M., Schoner, B.E., Smith, D.P., Zhang, X.Y., Wery, J.P., Schevitz, R.W. (1997). Crystal structure of the obese protein leptin-E100. *Nature, 387, 206–209.*

174 Zhang, Y., Proenca, R., Maffei, M., Barone, M., Leopold, L., Friedman, J.M. (1994). Positional cloning of the obese gene and its human homologue. *Nature, 372, 425–432.*

Note added in proof

Since this manuscript was submitted, two novel papers have appeared relating to the effects of leptin in the CNS. Shanley *et al.* (2001) found that leptin (50 nM) facilitated induction of long term potentiation (LTP), (a form of synaptic plasticity that may be involved in learning and memory) via a mechanism involving enhancement of postsynaptic NMDA receptor-mediated Ca^{2+} influx. These results raise the interesting possibility that defects in leptin production or receptor function in the brain could be associated with cognitive dysfunction. More recently, Shanley *et al.* (2002) showed that low concentrations (10 nM) of leptin were able to hyperpolarize and inhibit hippocampal neurons by directly activating BK Ca^{2+}-activated K^+ channels, a mechanism different from that shown to operate on hypothalamic arcuate neurons by Spanswick *et al.* (1997) and suggesting a more widespread modulatory effect of this hormone on central neuronal excitability.

References

Shanley, L.J., Irving, A.J., Harvey, J. (2001) Leptin enhances NMDA receptor function and modulates hippocampal synaptic plasticity. *J. Neurosci., 21, RC186 (1-6).*

Shanley, L.J., Irving, A.J., Rae, M.G., Ashford, M.L., Harvey, J. (2002) Leptin inhibits rat hippocampal neurons via activation of large conductance calcium-activated K^+ channels. *Nature Neurosci., 5, 299–300.*

Chapter 40

Regulation of gastric secretion, Helicobacter pylori and inflammation of the gastric epithelium

László Prónai, Béla Molnár and Zsolt Tulassay

Contents

Regulation of gastric secretion, and its interaction with a bacterium called Helicobacter pylori, is a rather complex phenomenon which, in certain circumstances, leads to the development of peptic ulceration. There are several molecules, such as hydrochloric acid, pepsinogen, regulatory peptides, and other chemical messengers secreted by the stomach. In addition, the gastric parietal cell is a complex structure that is controlled by a variety of hormones and neurotransmitters. These agents interact with specific surface receptors and initiate cascades of information to the cell nucleus as highly organized and complex signal transduction pathways. The bacterium H. pylori attaches to the gastric epithelial cell surface and stimulates inflammatory inter- and intracellular signal transduction cascades. It also binds to EGF receptors of gastric epithelial cells and thus interacts with cell growth and repair. This chapter focuses on cellular and intercellular regulation with particular emphasis on acid secretion and H. pylori binding.

40.1 History

Although at present it seems evident that the stomach secrets hydrochloric acid (HCl), this issue was settled only in 1825 by William Beaumont, when he collected the secreted gastric juice of Alexis St. Martin, a patient with post traumatic gastrocutaneous fistula and sent it to three chemists, all of whom demonstrated it to be acidic (Figure 40.1).[1] Pavlov showed that the sight and smell of food are powerful stimuli of gastric acid secretion, and one of his colleagues, Bechterew, reported in 1911 that electrical stimulation of the frontal cortex had a similar effect, and could be inhibited by vagotomy[2]. The presence of a chemical stimulator, called "gastrin", was first demonstrated by Edkins in 1905[3], but it was purified only in 1964 and synthesized in 1965 by Gregory[4]. Histamine, another powerful stimulus of acid secretion, was discovered by Lim in 1922[5]. The secretion of pepsinogen, the main digestive enzyme secreted by the gastric mucosa was described by Langley in 1886[6]. Hormonally mediated inhibition of acid secretion was reported by Feng in 1929[7], but until the late 1970s, considerable controversy surrounded the relative roles of gastrin, acetylcholine and histamine in mediating acid secretion. Soll and colleagues later confirmed that there are individual receptors for each substance (gastrin, histamine, acetylcholine) on isolated canine cells[8]. Gastrin receptors on parietal cells were isolated by Soll in 1984 and somatostatin receptors in 1987 by Park *et al.*[9].

Figure 40.1 A reproduction of Beaumont and his patient St. Martin (c. 1822) in the room of his home at Fort Mackinac. (Adapted from Modlin IM. From Prout to the Proton pump. 1st ed. 1995 Schnetztor-Verlag GmbH Konstanz.)

The existence of spiral organisms in the gastric epithelium was known from the beginning of the 20th century. In 1938 Dounges described these organisms in 43% of 242 stomach autopsy specimens. Warren and Marshall performed a landmark study in 1983, proving the correlation between H. pylori and peptic ulcer disease[10].

40.2 Anatomy of gastric mucosa

The epithelial surface of the stomach consists of thick vascular folds, called rugae, invaginated with microscopic gastric pits. Each pit opens into four to five gastric glands. The epithelial cells lining the gastric glands are highly specialized and different from the surface epithelial cells. Glands from the cardiac region of the stomach provide the transition from the esophageal squamous epithelium to gastric columnar epithelium. They contain mucous and endocrine cells and comprise less than 5% of the gastric gland area. Most gastric glands occur in the oxyntic area (75%) and are responsible for acid secretion (Figure 40.2). They include parietal, chief, mucous neck, endocrine, and enterochromaffin cells. The pyloric glands cover the gastric antrum and pylorus and contain gastrin cells (G cells), mucous cells and other endocrine cells (Figure 40.3). Each of these cell types has evolved into a highly specialized secretory cell that contributes to gastric secretion.

40.3 Hormone specific insights

Hormones with more or less specific effect on gastric secretion can be subdivided into six families; (1) Gastric-Cholecystokinin Family, (2) Pancreatic Polypeptide Family, (3) Tachykinin Family, (4) Somatostatin Family, (5) Secretin Family, and (6) Tyrosine Kinase Receptor Family (Table 40.1).

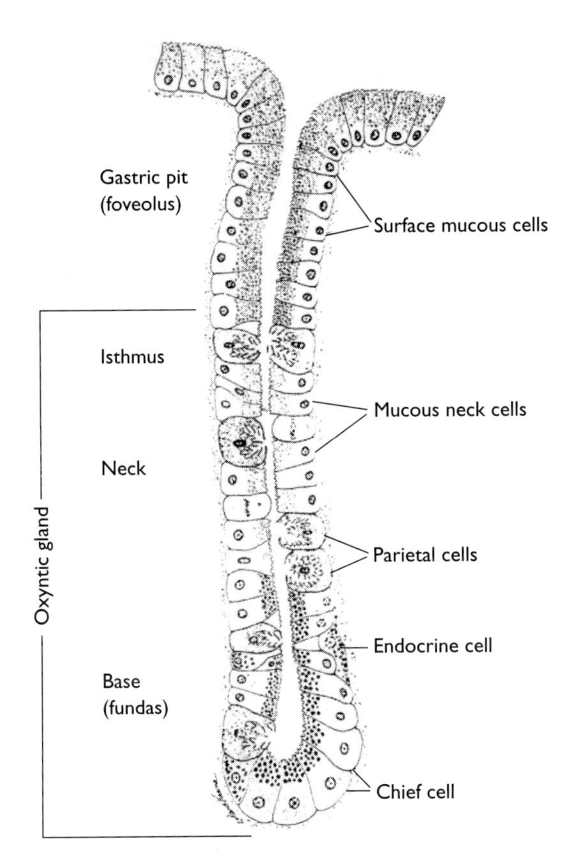

Figure 40.2 Oxyntic gastric gland. (Adapted from Ito S. Winchester RJ. The final structure of the gastric mucosa in the cat. J Cell Biol 1963; 16: 541.)

40.3.1 Gastrin

Gastrin (and somatostatin) belong to the gastrin-cholecystokinin family of peptides which share the same carboxyl-terminal pentapeptide-amide (Figure 40.4). The major biologically active forms of gastrin are 17- and 34-amino acid peptides that have a tyrosine residue six residue form of the carboxyl terminus, which may (gastrin-17-II, gastrin-34-II) or may not (gastrin-17-I, gastrin-34-I) be sulfated. This post-translational modification of gastrin appears to have no functional significance, all forms being equally potent and efficacious at traditional gastrin receptors. The half life of gastrin-17 is 7 min and that for gastrin-34 is 30 min. Thus, the majority of gastrin in the circulation during fasting is gastrin-34, whereas the major form released after a meal is gastrin-17. The major site of gastrin expression is open endocrine cells (G cells) in the gastric antrum. Low levels of gastrin are present in the vagus nerve and in pituitary corticotrophs, and transient expression is observed in fetal pancreatic islets and colon. This hormone is found in a number of neoplasms, the highest in islet cell tumors, called gastrinoma (Zollinger-Ellison syndrome)[12]. Gastrin secretion from antral G cells is regulated by luminal, paracrine, endocrine, and neural stimuli (Figure 40.3). Ingestion of a meal is a strong stimulant of gastrin secretion. Acid secretion results in lowered intragastric pH and reduced G cell secretion of gastrin. This is mediated by the paracrin effect of somatostatin. The neuroendocrine mediator, GRP, has stimulant effects on secretion by the G cell. Sympathetic and parasympathetic innervation have more complex effects.

The classical target of gastrin is the parietal cell, which secrets hydrochloric acid (HCl). This effect is mediated by

Table 40.1 Hormone families in the gastro-intestinal (GI) tract

Family (receptor family)	*Hormone*	*Main target action/organ*
Gastrin-Cholecystokinin (Group 3)	Gastrin	stomach parietal cell HCl ↑
	Cholecystokinin (CCK)	pancreatic acinar cell ↑
Pancreatic Polypeptide (Group 3)	Pancreatic Polypeptide	? (islet cell tumor marker?)
	PYY	pancreas, stomach ↓
	NPY	vasoconstriction?
Tachykinin (Group 3)	Substance P (Neurokinin A)	motility of GI tract
	Gastrin Releasing Peptide (GRP)	neurons of the GI tract
Somatostatin (Group 3)	Somatostatin	major inhibitor of secretion
Secretin (Group 3)	Secretin	bicarbonate secretion ↑
	Vasoactive Intestinal Peptide (VIP)	neurotransmitter, relaxation
	PACAP	relaxation ?
	Gastric Inhibitory Peptide (GIP)	insulin production ↑
	Glucagon	insulin production ↑
Tyrosine Kinase Receptor (Group 1)	Epithelial Growth Factors (EGF)	secretion ↓, cell growth ↑
	Fibroblast Growth Factors (FGFs)	angio-, tumorigenesis
	Insulin-Like Growth Factors	differentiation of tissues

Notes: ↑, stimulation; ↓, inhibition; Group 1, single transmembrane receptors; Group 2, receptors with subunits; Group 3, G protein-coupled receptors.

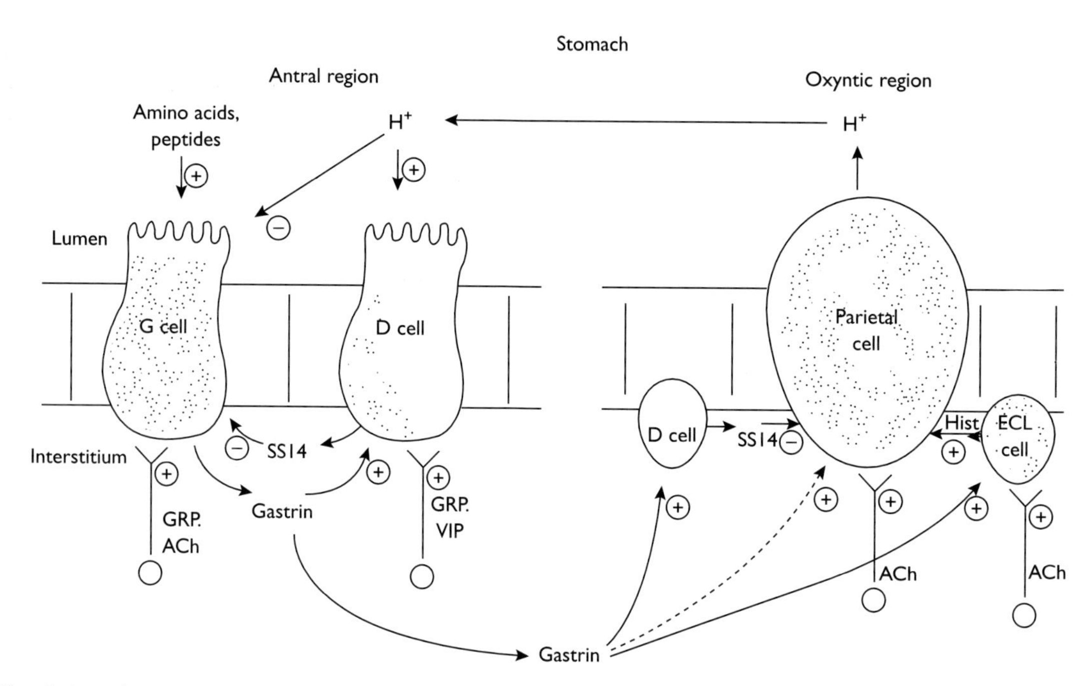

Figure 40.3 Regulation of gastric acid secretion by gastrin. Shown are some of the key components of this cycle: +, stimultory events, acetylcholine (Ach) stimulates the ECL cell, the parietal cell, and the gastrin-producing G cell. Gastrin-releasing peptide (GRP) stimulates both the G cell and D cell. Gastrin stimulates the D cell, parietal cell, and ECL cell, which then secretes histamine to stimulate the parietal cell. Somatostatin (SS14) secreted by the D cell inhibits both the G cell and parietal cell. Acid (H+) secreted by the parietal cell has regulatory effects as well. Ecl, enterochromaffin-like; VIP, vasoactive intestinal peptide; His, histamine.

a G protein-coupled receptor, which is identical to the most abundant CCK receptor (type B CCK receptor) present in the central nervous system. This receptor is in the rhodopsin/ß-adrenergic receptor family and is most closely related to the type A CKK receptor. The structural specificity of ligand binding to these two receptors is quite distinct (Table 40.2). Type B CCK/gastrin receptor recognizes all forms of gastrin and CCK almost equally. In contrast, type A CCK receptor recognizes gastrin poorly. Type B CCK receptor is expressed on enterochromaffin-like cells (ECL) of the gastric mucosa, on smooth muscle cells at various levels along the gastrointestinal tract, and on diffuse domains of the brain. Gastrin secretion can be inhibited by antagonists acting at the receptor (L-365, 200). Of interest, isolated parietal cells respond weakly to gastrin alone. This response is augmented in the presence of histamine and acetylcholine. Gastrin has also been shown to stimulate histidine decarboxylase, a key enzyme in histamine synthesis.

In addition to its classical acute effects in stimulating acid and histamine secretion, gastrin also acts as a growth factor to stimulate mucosal proliferation. Its role as growth factor may be unique, with different tissue-specific expression during development and adult life. During the fetal period, gastrin stimulates islet cell growth in the pancreas. The trophic effect of gastrin on oxyntic epithelium has been suggested by the hyperplasia associated with Zollinger-Ellison syndrome of hypergastrinemia, and by the atrophy associated with low levels of gastrin postantrectomy. Chronic high-level gastrin stimulation has also been correlated with ECL hyperplasia.

Table 40.2 Specificity of cholecystokinin (CCK) receptors

Approximate ligand affinities (nM)		
Ligand	*Type A CCK Receptor*	*Type B CCK Receptor*
CCK-8	1	2
CCK-8-desulfate	500	10
CCK-4	10.000	10
Gastrin-17	1000	10
L-364, 718	1	100
L-356, 260	300	2

In humans, more than 95% of gastrin peptides normally secreted are fully processed and amidated, with a small amount of Gly-extended forms. Increased synthesis of Gly-extended gastrin is secreted in achlorhydria, gastrinomas and colon carcinomas. Gly-extended gastrin may act as growth factor but it has a less potent effect on acid secretion. Gly-extended gastrin has been reported in pancreatic tumour cell line AR42J and in Swiss 3T3 cells[13]. Both reports support actions independent of the classical CCK-B receptor. The postulated relationship between gastrin and colon carcinoma is still unclear.

40.3.2 *Somatostatin*

There are two molecular forms of somatostatin that are 28 and 14 amino acids in length. Both forms contain a critical disulfide bond, giving them a cyclic structure. Somatostatin is synthesized and secreted from both neurons of the neuromuscular layer (10%) and enteroendocrine cells of the mucose (90%), with the latter (D cells) present either open or closed in the gastric mucosa. Somatostatin acts as a hormone, paracrine transmitter, neurotransmitter, and neuromodulator. SST-14 with a half-life of 1–3 min is found predominantly in the stomach, pancreas and neuromuscular layer of the intestine, whereas SST-28 with a half-life of approximately 15 min is found within the mucosa of the bowel. Most of the biologic effects of somatostatin are inhibitory. It inhibits gastric, pancreatic, biliary and even salivary secretion, and blocks glucagon, gastrin, and insulin release. It also inhibits motor events along the gut, and decreases splanchnic and portal blood flow.

40.3.3 *Epithelial growth factors*

Epithelial growth factor (EGF) belongs to the single transmembrane group of enzymes. It is a 53-amino acid single-chain polypeptide that contains three intramolecular disulfide bonds. There are four members of the EGF receptor subfamily, including the EGF receptor encoded by ErbB and proteins encoded by ErbB-2, ErbB-3, and ErbB-4. All are single transmembrane glycoproteins that incorporate two distinct cysteine-rich regions in the ectodomain and tyrosine kinase catalytic domain and typical sites of potential phosphorylation in cytoplasmic regions. Receptor dimerization or oligomerization is a typical initial step in signaling that is induced by agonist binding. This is followed by cross-phosphorylation on tyrosine residues of the receptor and the recruitment of kinase substrates and other signal proteins. Many of these substrates have Src-homology-2 (SH2) domains. A large number of potential signaling cascades have been described, including the Ras, Raf, and MAP kinase pathways and the JAK/STAT pathways. EGF signaling also includes activation of phosphatidylinositol pathways, with activation of isoenzymes of protein kinase C and increases in intracellular Ca^{2+}.

Many gastrointestinal functions can be affected by this family of hormones and receptors. Included among the most prominent effects are inhibition of gastric acid secretion, protection against injury, and stimulation of intestinal cell growth.

40.4 Receptors

Several receptors that are critical for gastric secretory function have been cloned. These include the histamine H_2, somatostatin, M_3-muscaridin, and gastrin or CCK-B receptors. Analyses of the amino acid sequences of these cloned receptors indicate that they belong to the family of G protein-coupled receptors (see Figure 40.4 and Table 40.1). Receptor regulation (e.g. uncoupling, sequestration, down-regulation) appears to involve kinase-mediated phosphorylation events targeted at serine and threonine residues found within the carboxyl terminal proportion of these receptors. Detailed structure-function analyses of the receptors important in gastric secretory function are in progress.

Initial studies showed that histamine, carbachol, and gastrin increased canine parietal cell oxygen uptake. Later studies, using oligonucleotid probes, contradicted the idea that acetylcholine and gastrin could directly activate parietal cells, since expression of receptors for these substances could only be verified on immunocytes in the gastric lamina

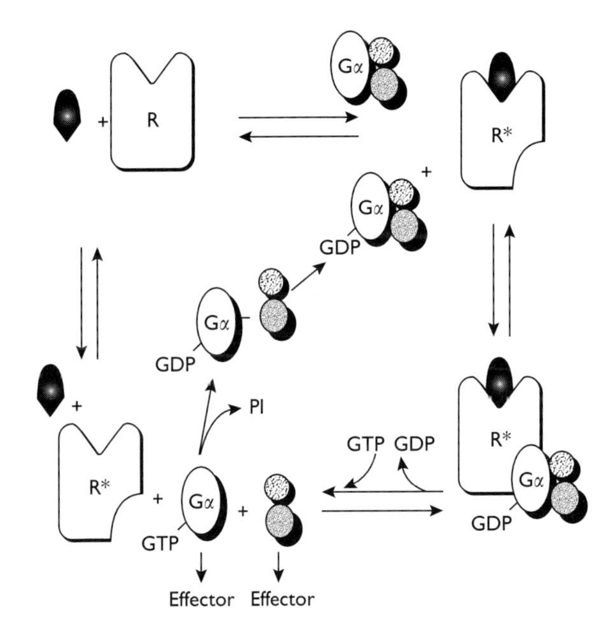

Figure 40.4 Cycles of G protein-coupled receptor activation and proximal signaling events occurring at the level of the plasma membrane. The earliest step is an induced confrontational change in the receptor that facilitates its association with the G protein. At that point, the G protein is in its GDP bound dimeric state. The receptor association facilitates GTP exchange for the GDP, which in turns results in the dissociation of the ternary complex and dissociation of the GTP-bound α subunit from the dimer. Both of these can move on interaction with effectors in other compartments within the cell. The α subunit of the G protein has intrinsic GTPase activity that results in the elimination of one of the phosphatases and reassociation of the trimeric form of the G protein.

propria but not on parietal cells. Recently Diaz and colleagues re-examined this controversy by using in situ hybridization, histochemistry and autoradiography with a highly selective H_2-receptor radioligand (^{125}I-aminopotentidine)[15]. A specific signal for H_2-specific gene transcripts was detected only within parietal cells of the gastric mucosa, thus confirming the initial hypothesis that histamine, gastrin and acetylcholine act directly on specific receptors of parietal cells.

40.4.1 Histamine H_2 receptor

Histamine H_2 receptor is a member of the trimeric G coupled protein receptor family (Figure 40.5). Although the critical role of the H_2 receptor in regulating gastric acid secretion had been well established, the structural components of this receptor that determine H_2 selectivity were unknown. The success of Gantz and colleagues in cloning the H_2-histamine receptor provided essential tools required for studying this question further[16]. The current concept is that aspartic acid residue (Asp98) in the third transmembrane domain is essential for histamine binding and action, and aspartic acid (Asp186) in the fifth transmembrane defines H_2 selectivity. A threonine (Thr190) in the fifth transmembrane domain is important in establishing the kinetics of histamine binding but is not essential for H_2 selectivity. Binding of histamine results in activation of adenylate cyclase by interaction with G_s trimeric G protein and also a small Ca^{2+} signal due to weak interaction with G_q trimeric G protein. The reason for the presence of both signals may be that the cAMP cascade is necessary for pump activation and that elevation of intracellular calcium is necessary for redistribution of the pump to microvilli of the secretory canaliculi.

40.4.2 Muscarinic receptor

The presence of a specific muscarinic receptor on canine parietal cells is supported by specific blockade of carbachol's biologic effect with atropine. Pharmacological studies indicate that parietal muscarinic receptor is of the M_3 subtype, which has been confirmed by molecular cloning of this receptor. Rat parietal cells have also been shown to have M_3 muscrinic receptors that stimulate aminopyrine uptake, increase inositol phospholipid turnover, and bind [N-methyl-^{3}H]scopolamine.

40.4.3 Gastrin receptor

The cDNA clone encoding the gastrin receptor found on parietal cells have been isolated by Kopin and colleagues[17]. They determined that it is a member of the G protein-coupled receptor family, and when expressed, it has a molecular weight similar to that reported in cross-linking studies. Expression of cloned receptor confirmed that it is coupled to membrane inositol phospholipid turnover and mobilization of intracellular Ca^{2+}. Song and colleagues cloned the gene encoding the human gastrin or CKK-B receptor and localized it to a region of chromosome 11 (11p15.4)[18]. It appears that the gene produces two different receptor proteins as a result of alternative RNA splicing. The functional

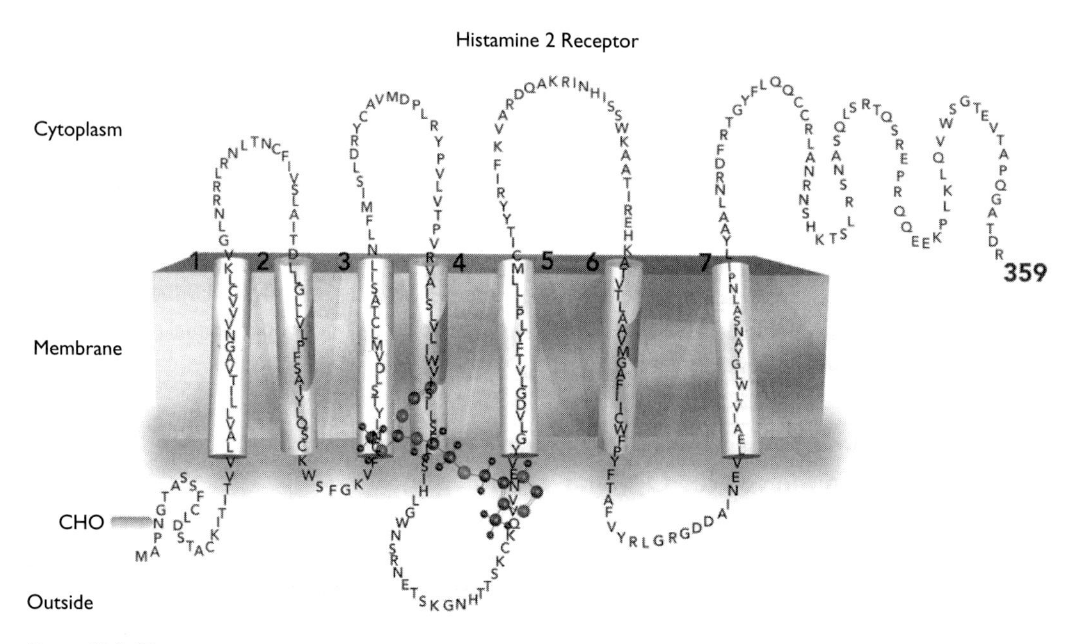

Figure 40.5 The amino acid sequence of the human histamine H_2-receptor.

difference between the two receptor isoforms is not yet elucidated.

As with many peptides, gastrin is synthesized as a precursor molecule that undergoes post-translational processing to become a C-terminus amidated (glycine extended form, G-Gly) product which is presumed to be the sole biologically active form. Seva and colleagues demonstrated that G-Gly exerts growth-promoting effects through activation of non CCK-B receptors[19]. The exact role of G-Gly is now under extensive investigation. It increased dose-dependently both the expression and transcription of the gene encoding the α-subunit of H^+/K^+-ATPase, suggesting that G-Gly may have functional role in potentiating gastric acid secretion by enhancing the expression of the gene encoding the protein responsible for hydrogen ion generation.

40.4.4 *Somatostatin receptor*

Somatostatin receptors are in the rhodopsin/ß-adrenergic receptor family of the G protein-coupled superfamily of receptors. There are five recognized receptor subtypes (SSTR1-5) cloned. These receptors are most closely related to the opioid receptors, with which they share about 40% homology. These receptors couple adenylate cyclase and may activate protein tyrosinase phosphatase. SSTR5 is unique among the receptor subtypes because it has a higher affinity for somatostatin-28 than for somatostatin-14. SSTR2 is found in the stomach and intestine. The development of highly specific and selective agonist has led to more precise functional characterization of these five different receptors. In particular, SSTR2 appears to mediate inhibition of gastric acid secretion. Similarly, Prinz and colleagues demonstrated that SSTR2 is the predominant somatostatin receptor subtype expressed on rat ECL cells where it mediates inhibition of histamine release[20].

40.4.5 *Other receptors*

The presence of other receptors on parietal cells, such as prostaglandins E_2 (PGE_2), and I_2 (PGI_2) has been suggested but awaits confirmation. Ding and colleagues have documented gene expression of the prostaglandin receptor subtypes EP_3 and EP_4 on rat gastric parietal cells[21].

Although it is fairly clear that H3 receptors are expressed on ECL cells, serving as autocrine regulation of histamine release, their presence on parietal cells remains controversial.

Epidermal growth factor (EGF) and transforming growth factor-α (TGFα) have been implicated in the inhibitory modulation of gastric acid secretion. TGFα shares structural homology with EGF and is expressed within parietal cells. Both of these growth factors inhibit parietal cell function in similar fashion.

The presence of secretin, glucagon, and opioid receptors on parietal cells has been suggested but requires confirmation.

40.5 Cellular basis of acid secretion

40.5.1 *Supracellular regulation of gastric acid secretion*

An overview of the interactions of ligands and receptors involved in acid secretion is summarized in Figure 40.3. The parietal cell has stimulatory receptors for gastrin, acetylcholine, and histamine. Several lines of evidence suggest that parietal cells in humans may be exposed continuously to basal levels of acetylcholine and histamine. Gastrin can account for most of the postprandial increase in gastric acid secretion. Although the postprandial response is partly the result of a direct effect of gastrin on parietal cells, a significant component may be the result of gastrin's ability to stimulate histamine release from ECL cells. The parietal cell also has inhibitory somatostatin receptors that counteract the secretory effects. Vagal nerve fibers may enhance their acid stimulatory effect through stimulatory muscarinic receptors on G cells and inhibitory muscarinic receptors on D cells. Adrenergic fibers appear to stimulate D-cell secretion and inhibit release by histamine-containing cells, counterbalancing the vagal effects. Gastrin and CKK stimulate parietal cells and D cells. Gastrin-releasing peptide acts as neurotransmitter that stimulates G cells, and somatostatin may function as a paracrine G cell inhibitor.

40.5.2 *Intracellular signal transduction*

Numerous signal transduction pathways have been studied and characterized in gastric parietal cells.

40.5.2.1 *Classic signal transduction pathways*

Receptors linked to adenylate cyclase influence intracellular levels of cAMP. Such receptors are coupled to inhibitory (G_i) or stimulatory (G_s) GTP-binding proteins (Figure 40.6). G_i attenuates adenylate cyclase activity, decreasing cAMP levels.

The second major intracellular signal transduction cascade involves the turnover of membrane phospholipids, specifically the inositol phospholipids. Receptors are linked to this pathway through a G protein and the initial events that are activated by occupancy of their receptor leads to phospholipase C-induced hydrolysis of phosphatidylinositol bisphosphate (PIP_2) to diacylglycerol (DAG) and inositol triphosphate (IP_3). IP_3 causes the release of Ca^{2+} from intracellular stores, and DGA promotes the translocation of Ca^{2+}-phospholipid-dependent protein kinase from the cytoplasm to its active site on the cell membrane. Increases in $[Ca^{2+}]_i$ activates various calcium-dependent enzyme systems, such as the calmodulin kinases, and promotes the translocation and activation of PKC.

Histamine's acid stimulatory action appears to be mediated primarily through cAMP production in parietal cells.

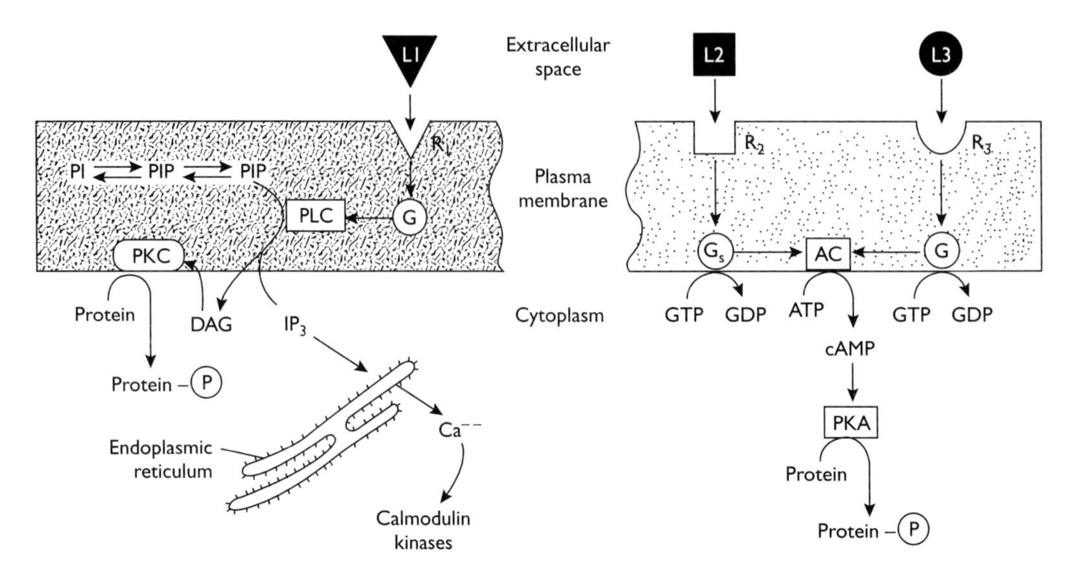

Figure 40.6 Signal transduction pathways in parietal cells. Ligands (L) interact with receptors (R) to initiate the target cell response. The left panel shows the membrane inositol phospholipid pathway activated by gastrin and acetylcholine (L_I). A guanine nucleotide-binding protein (G) activates phospholipase C (PLC), which hydrolyzes phosphotidylinositol 4,5-bisphosphate $(PIP)_2$ to diacylglycerol (DAG) and inositol 1,4,5-triphosphate (IP3). DAG activates protein kinase C (PKC), which phosphorylates (P) target proteins, while IP_3 induces intracellular calcium mobilization, which activates kinases. The right panel depicts the cAMP pathway used by histamine (L_2) and somatostatin (L_3) in parietal cells. The histamine receptor (R_2) acts through a stimulatory protein (G_s), which activates adenylate cyclase (AC) and generates cAMP. Increased levels of cytoplasmic cAMP activate protein kinase A (PKA) and result in phosphorylation of parietal effector proteins. The somatostatin receptor (R_3) activates an inhibitory G protein G_i, which inhibits adynylate cyclase and decreases cAMP generation.

Gastrin and carbachol appear to stimulate acid secretion by Ca^{2+}-dependent pathways. Both gastrin and carbachol cause a time-dependent decrease in PIP_2 and increase in IP_3.

Intracellular increase of Ca^{2+} in parietal cells leads to the activation of the Ca^{2+}/calmodulin-dependent protein kinase II (CaMKII).

The action of PKC in parietal cells is complex. Direct activation of PKC with phorbol esters results in enhanced acid secretion. The regulation is, however, rather complex and requires further studies.

40.5.2.2 New signal transduction pathways

Recent studies indicate that, in addition to the well established signal transduction pathways, mammalian parietal cells contain multiple members of the mitogen activated protein kinases or extracellular signal-regulated protein kinases (MAPKs/ERKs) (Figures 40.7 and 40.8).

The activation of ERKs was initially described as a response to cell stimulation with epidermal growth factor (EGF). Activation of the EGF receptor by ligand binding is known to induce receptor dimerization and autophosphorylation on tyrosine residues and lead to the assembly of a multiprotein complex. Some of these protein-protein interactions occur through modular binding domains known as SH2 and SH3. The phosphotyrosine residues present on the activated EGF receptor allow the binding of the SH2 domain of Grb2

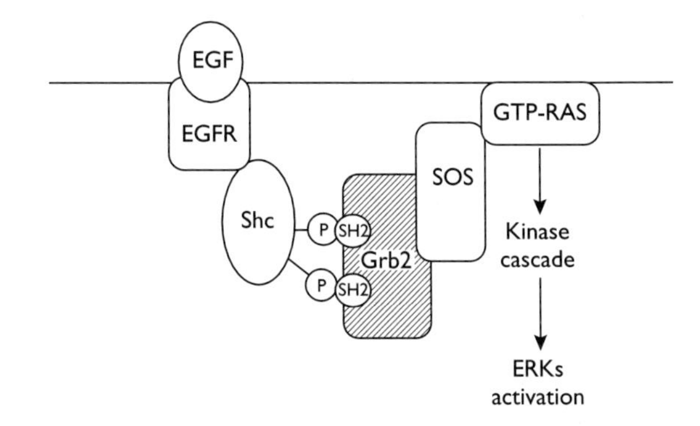

Figure 40.7 Schematic presentation of extracellular signal-regulated protein kinase (ERKs) induction by growth factors. Activation of the epidermal growth factor receptor (EGFR) by ligand binding induces receptor autophosphorylation on tyrosine residues. This event allows the binding of the SH2 domain of Grb2 to the adapter protein Shc and the recruitment of the guanine nucleotide exchanger factor Sos to the membrane in proximity to the small guanosine triphosphate (GTP)-binding protein Ras. Sos promotes the association of Ras with GTP and the GTP-bound activated form of Ras induces a kinase cascade that leads to the activation of ERKs.

(growth factor receptor bound protein-2) to the adaptor protein Shc. This leads to the recruitment of guanine nucleotide exchange factors with proline-rich SH3 domain-binding sites to the membrane in proximity to the isoprenylated small

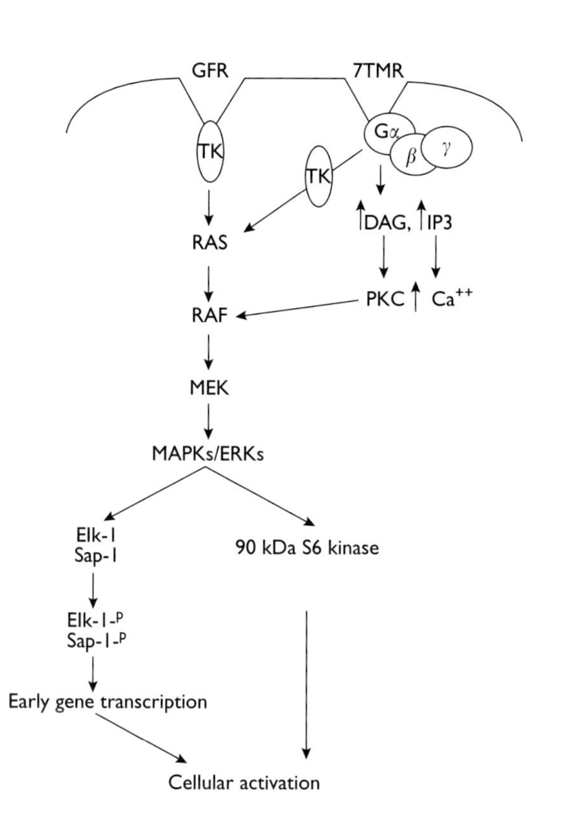

Figure 40.8 Intracellular targets of mitogen-activated protein kinases/extracellular signal-regulated protein kinase (MAPKs/ERKs) induction. Ligand binding to both growth factors and seven transmembrane receptors induces Ras activation. This leads to the activation of a kinase cascade that includes Raf, MAP kinase kinase, or MAP/ERK kinase (MEK) and MAPK. The MAP kinases or ERKs phosphorylate numerous cellular proteins, including downstream protein kinases, such as the 90-kd S6 kinase RSK and transcription factors such as Elk-1, or Sap-1 that regulate transcription of the early response gene c-fos. PKC might activate c-Raf directly, suggesting that an alternative pathway for MAPK activation might require a signaling cascade involving PKC, Raf, and MEK, leading finally to MAPK induction.

GTP-binding protein Ras. Exchange factors such as Sos promote the association of Ras with GTP, and the GTP-bound activated form of Ras binds members of the Raf family of protein kinases. The functional relevance of ERK in the stomach has been the focus of continued investigation. Under acute conditions, EGF had an inhibitory effect on gastric acid secretion, whereas prolonged administration of EGF increase both basal and maximal acid output *in vivo* and acid production in isolated cells *in vitro*.

The MAPKs are proline-directed protein kinases that phosphorylate numerous cellular proteins and transcription factors and thus regulate transcription of the early response gene *c-fos*. The proto-oncogen *c-fos* is a member of a well-characterized family of transcription factors that includes nuclear proteins such as c-myc and c-jun, which are known to be expressed within minutes of cell activation.

40.5.2.3 *Inhibitory effect of somatostatin*

Somatostatin inhibits histamine-induced parietal cells secretion through to inhibition cAMP generation. In addition, somatostatin is also able to inhibit acid secretion induced by dibutyryl cAMP. The action of somatostatin is distal to the activation of intracellular signal transduction cascades. The mechanism is yet to be determined, but somatostatin inhibits the induction of the early response gene c-*fos* in canine parietal cells.

40.5.3 *Acid secretion of the parietal cell*

The gastric epithelium secrets fluid almost isotonic HCl through an active transport process. The parietal cell alone is responsible for producing acid at a pH of 0.8. Thus the H^+ ion gradient increases 2.5 million-fold. Significant amounts of mitochondrial energy are required for this activity. It has been suggested that the family of isoenzymes, kinases/phosphocreatine (CK/CDr), which catalyze the exchange of high-energy phosphate groups between phosphocreatine and ADP, are involved. Isoform BB-CK (brain) has been identified in parietal cells.

The stimulation of parietal cell leads to morphologic transformation, rapid changes in enzyme location and activity, and opening of ion channels. The resting parietal cell contains a collapsed canalicular system and cytoplasmic tubulovesicles containing the proton pump, H^+/K^+-ATPase. The stimulated cell rapidly develops a richly interdigitating intracellular canalicular system bulging with microvilli and loss of cytoplasmic tubulovesicles. This increases the cell membrane surface area 5–10-fold. The fusion of tubulovesicles with the canalicular system translocates H^+/K^+-ATPase from vesicular membranes to the canalicular membrane, where it actively pumps H^+ ions in exchange for K^+. For each proton secreted, an intracellular OH^- ion is generated. This alkaline challenge is handled by carbonic anhydrase II (CAII)-mediated conversion of OH^- to $HCO3^-$, which is exchanged for Cl^- at the basolateral membrane (Figure 40.9).

Induction of specific gene transcription has been shown for both H^+/K^+-ATPase and CAII. Maximal transcription for both CAII and H^+/K^+-ATPase was found 60 min after gastrin stimulation, and 20 min for carbachol and histamine stimulation. The induction of actin gene expression may serve as a particularly useful marker for acid secretion instead of using H^+ generation.

The apical membrane of the parietal cell contains the H^+/K^+-ATPase pump and K^+ and Cl^- conductances. The basolateral membrane also has K^+ conductance, Cl^-/HCO_3^-

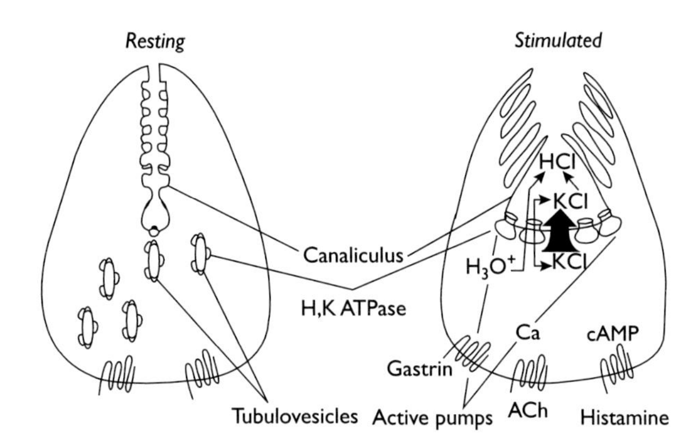

Figure 40.9 Functional transformation of a secreting parietal cell. This model illustrates a resting parietal cell (left) with collapsed secretory canaliculi and cytoplasmic tubulovesicles expressing H^+/K^+-ATPase and stimulated parietal cell (right) with formed secretory canaliculi expressing active H^+/K^+-ATPase pumps (Adapted from ref. 21).

exchangers, Na^+/H^+ exchangers, and H^+/K^+-ATPases to maintain cellular homeostasis during resting stages.

40.5.3.1 The gastric proton pump

The gastric H^+/K^+-ATPase or H^+ ion pump belongs to the type P ATP-ase family[23]. Significant progress has been made towards understanding the structural characteristics of H^+/K^+-ATPase (Figure 40.10). There are two isoforms, both are α-β heterodimers. The gene corresponding to the human α subunit has been cloned and sequenced. It contains 22 exons and has a corresponding deduced protein length of 1035 amino acids. The α subunit contains the phosphorylation consensus sequence and ATP binding domain. It consists of 10 transmembrane segments with a large cytoplasmic domain. Labeling studies suggest that the proton pump inhibitors omeprazole, lansoprazole, pantoprazole and rabenprazole, bind to the cysteine residues located at positions 813 and 822. These residues lie near a predicted extracytoplasmic loop between transmembrane domains 5 and 6, and may be essential for proton pump inhibition.

The cDNA of the β subunit of H^+/K^+-ATPase has been isolated and sequenced. The average length of this subunit ranges from 290–294 amino acids and hydropathy plots suggest a single transmembrane domain. Although the role of the ß subunit has not been established, it is required for assembly and stability of the gastric pump. H^+/K^+-ATPase can be completely inactivated by the so called proton pump inhibitors. These compounds become cyclic in the presence of acid and react with available sulfhydryl groups to form a covalent interaction that irreversibly inactivates the enzyme. The blockade of proton pump transport is virtually complete and has been used clinically.

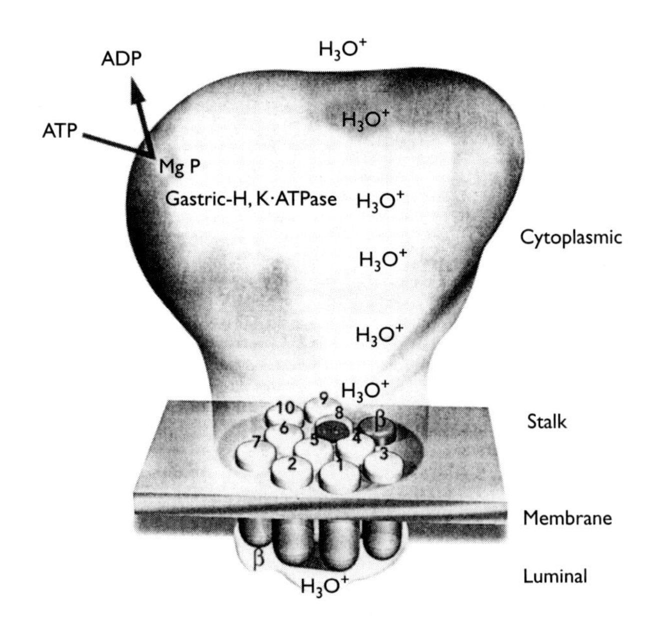

Figure 40.10 Three dimensional schematic model of the gastric H^+/K^+-ATPase. The model shows the 10 membrane segments of the α subunit and the single transmembrane domain of the β subunit.

40.5.4 Other gastric secretory products

Histamine plays a critical role in regulating gastric acid secretion through activation of the parietal cell H_2 receptor. Histamine is generated in ECL cells through decarboxylation of histidine by histidine decarboxylase (HDC). HDC is transcriptionally activated by both gastrin and acetylcholine through the PKC and MAPK-dependent pathway.

Pepsinogen is secreted by chief cells of the stomach. It belongs to the group of aspartic proteases, its role is to digest collagen, a major component of meat. It has 7 isoenzymes, fraction 1–2 migrates toward the anode on electrophoresis and is immunologically similar (pepsinogen I), while fraction 6–7 migrates slower (pepsinogen II). The presence of different isozymogens could be the result of multiple gene loci, multiple alleles at a single gene locus, post-transcriptional processing of RNA, or post-translational modification. Genetic analysis has revealed that the two groups have different gene loci. A human pepsinogen I gene has been isolated, sequenced, and localized to chromosome 11q12–13. Pepsinogen II is a single PGC gene located on chromosome 6.

Secretin, isoproterenol, and VIP stimulate pepsinogen through cell surface receptors linked to the activation of cAMP formation. Cholinergic agonists and CKK stimulate pepsinogen release by activating the IP_3 and $[Ca^{2+}]_i$ signaling pathway.

Several peptide hormones including somatostatin, neuropeptide Y, and PYY have been found to inhibit pepsinogen release through a pertussin toxin-sensitive mechanism.

Mucus gel secreted by the gastric epithelium serves as a barrier, protecting the gastric mucosa from acid. The barrier consists of an unstirred layer of mucus, bicarbonate, sulfate phospholipids, and water. The prominent pH gradient extending from the lumen (pH2) to the epithelial cell surface (pH7) is maintained by this gel.

Mucin is a high molecular weight glycoprotein secreted by surface mucus cells, mucous neck cells and glandular mucus. It consists of serine, threonine, and proline residues. Partial and complete cDNA clones for mucin have been isolated, and deduced peptide sequences confirm this amino acid preponderance. The gene for mucus production is localized on chromosome 11 (11p15.4–11p15.5). The major portion of mucin is heavily glycosylated.

Intrinsic factor (IF) is a 45-kd glycoprotein present in gastric secretions, essential for cobalamin absorption (vitamin B_{12}). Only a single human IF exists, located on chromosome 11.

40.6 Histamine H_2-receptor antagonists

The first series of H_2-receptor antagonists was synthesized in Welwyn Garden City by Smith, Klein & French (SK&F) and introduced in 1970. Black and his colleagues generated the first H_2-receptor antagonists, burimamide and then cimetidine. The latter reached the market in 1977. Accepting that the imidazole ring is essential for H_2 antagonism, a group from Glasgow bypassed the SK&F patent, and announced ranitidine to the world. Currently four H_2-receptor antagonists are available on the market worldwide (Table 40.3).

H_2-receptor antagonists exhibit reversible competitive inhibition of histamine-induced acid secretion *in vivo*, but famotidine has slower onset of dissociation from the receptor, suggesting some degree of noncompetitive inhibition. As much as 90% inhibition of vagal- and gastrin-induced acid secretion occurs with H_2-receptor antagonists, reflecting importance of histamine potentiating or mediating effects of cholinergic- and gastrin-stimulated acid secretion. Near complete basal and nocturnal acid secretion can be achieved by H_2-antagonists, and they are widely used in clinical practice for the treatment of acid related diseases.

40.7 Proton pump inhibitors

The development of the first series of drugs was due to a combination of serendipity, mechanism and conviction that ATPase was the best target to control acid secretion (Figure 40.10). Following a long series of chemical modification from an antiviral agent, pylidine 2-acetamide (imidazole structure), a thyreotoxic substance, timoprazole, was produced and finally a potent proton pump inhibitor (PPI), omeprazole was produced in 1979. It was launched on to the market in 1988. Currently 4 proton pumps are marketed around the world (Table 40.3).

Table 40.3

H2-receptor antagonists (daily eqv. doses)	*Proton pump inhibitors (daily eqv. doses)*
Cimetidine (4×200 mg)	Omeprazole (2×20 mg)
Ranitidine (2×150 mg)	Lansoprazole (2×30 mg)
Nizatidine (2×150 mg)	Pantoprazole (2×40 mg)
Famotidine (2×20 mg)	Rabenprazole (2×20 mg)

Since PPIs are acid labile, they must be formulated so as to pass through the stomach without exposure to acid. In the case of omperazole this was achieved by micro-encapsulation. When the PPI is absorbed in the gut, it reaches the parietal cell through the circulation. PPIs diffuse through the canalicular membrane of the parietal cell. They accumulate in acidic conditions, are protonated, converse the sulfenamide and react with one or more cysteines in the catalytic subunit of the H^+/K^+-ATPase. The PPIs inhibit ATPase activity or acid transport at different rates. The order of reactivity is rabenprazole>omeprazole=lansoprazole>pantoprazole. The location of the site of binding of PPIs on the pump identifies the cysteine critical for inhibition of enzyme activity. The key reaction of inhibition of H^+/K^+-ATPase is binding to cys 822 which is followed by covalent derivatization of cys 813. The reaction is irreversible, therefore acid secretion depends solely on pump synthesis. The half-life of a proton pump is 48 hours, therefore effective inhibition of acid secretion can be maintained clinically if the PPIs have a half-life of 48 hours.

40.8 Helicobacter pylori

40.8.1 Microbiological characteristics of helicobacter pylori

Helicobacter pylori (H. pylori) is a shaped gram-negative rod, approximately 0.5 by 3 μm in size, containing four to seven sheathed flagella at one pole. The spiral form and the flagellas are essential factors in the penetration of the bacteria into the mucus and epithelium of the stomach wall. This bacteria is very sensitive to environmental changes. For culture, microaerofil media and growth factors are required. It is known that culture of H. pylori leads to a coccoid form which is suggested to be a dormant form for the survival of the bacterium in unfavourable environments.

H. pylori produces some important factors, which supports its survival and virulence in the stomach's acidic milieau. One of the most important characteristics is the urease enzyme. The urease enzyme generates a local ammonia cloud around the bacteria protecting it from the acid produced. The urease builds up from the urea ammoniac and bicarbonate. This is an essential factor in producing ammonia-bicarbonate. The bicarbonate group is important in the neutralization of acid produced by the stomach. However, urease also plays an important role in disrupting the tight junctions between epithelial cells.

As the bacteria enters the gastrointestinal tract, it penetrates into the mucus layer. This position supports the survival of the bacteria. H. pylori in the mucus is protected from both the acidic gastric content and from the immune response.

Further known features of the bacteria include the production of protease, lipase, catalase, superoxide dismutase, and platelet activating factors. Lipase can have an important role in the deconstruction of the phospholipid layer on the epithelial cell surface. In this way the epithelial cells can be attacked by the gastric acid products leading to apoptosis and enhancing carcinogenesis[39, 40].

40.8.2 Antigens of H. pylori

H. pylori is a heterogenous bacteria expressing different virulence factors. The factors include the vacuolating cytotoxin (VacA), cytotoxin associated gene protein (Cag family, A,B,C,D,E,F,G,H,I,L,M,N,O,P,Q,R,T,S), the cytokine production inducing PicB factor (cagE), and a new factor called IceA (induced by contact with the epithelium)[25].

The vacA cytotoxin is activated by gastric acid. This cytotoxin causes vacuolated alterations in the epithelia. The activated cytotoxin penetrates the cytoplasm and causes the vacuolization. The vacA gene shows mosaicism. It contains one from four different signal sequences (s1a,s1b,s1c,s2) and one of two middle regions (m1, m2). The m2 subtype is further divided by m2a and m2b. The s1/m1 and s1/m2 type H. pylori strains show high and low degree cytotoxin production, respectively. The s2/m2 strains show minor or no cytotoxin production. The m1 positivity correlates with severe cytotoxicity.

There is close correlation between the vacA s1, m1 and cagA genetic markers. The vacA s1 strains were found to be significantly correlated with peptic ulcer disease. The importance of the m region could not be verified.

The specific role of Vac in development of gastric ulcer was shown by Pie and Tarnawski[26]. They found that H. pylori culture supernatant containing soluble Vac products inhibits the binding of epithelial growth factor (EGF) to its receptor (EGF-R). EGF was discovered in 1962. It inhibits gastric acid secretion, mediates mucosal adaptation, and accelerates healing of experimental and clinical gastric ulcers. In this way reparative epithelial cell proliferation is inhibited on the ulcer margin. On the other hand, the EGF-initiated intracellular signal transduction pathways are reduced.

These cag factors are coded by the so called pathogenicity island. This pathogenic island is segmented into two parts: right (cagI) and left (cagII) separated by the insertion segment (IS605). This segment contains the 41 bp long right (tnpB) and 33 bp long left (tnpA) unit. This insertion segment can be found twice in the pathogenic island (Figure 40.11). The cagA(120–140 kDa) is expressed by almost 60% of the bacteria. The close correlation between cagA and IL-8 production has been proven by several studies. Therefore the cagA positive status is the sign of increased virulence of the bacteria. Mutation in cagE, cagG, cagH, cagI, cagL and cagM decreases IL-8 production.

Recently, contradictory data appeared for the direct correlation between the protein expression of the pathogenic island (PAI), ulcer, and gastric cancer. First results reported that CagA expression represents increased risk for pathologic alteration. Later studies could not support this conclusion.

The picB cytotoxin is responsible for the destruction of the epithelial cell membrane. In the presence of picB the permeability of the epithelial cells increase and so called leaky cells appear. The picB is identical with cagE. The picA is probably identical with cagC and cagD.

The iceA cytotoxin (induced by contact with epithelium) was shown to be correlated with peptic ulcer disease as well. The ice gene has two allele variants, iceA1 and iceA2. The iceA allele is independent from the vacA and cagA status, but significant correlation can be found with ulceration.

40.8.3 Adhesion of H. pylori to the gastric epithelial cell

It was found in cell culture studies using H. pylori and AGS and KATO III cell lines that the attachment occurs very quickly, within minutes. The attachment happened in every case at the aflagellated end of the bacterium[27, 28] (Figure 40.12). Adhesion also occurs in the coccoid form of the bacteria. Shortly after attachment, internalization of the bacteria can be detected[29]. In spite of the fact that this finding was shown both *in vivo* and *in vitro*, its importance is not yet clear.

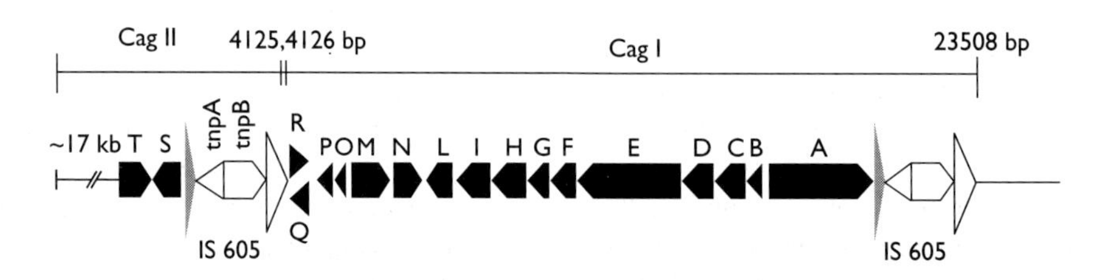

Figure 40.11 The structure of the pathogenic island (PAI) in Helicobacter pylori.

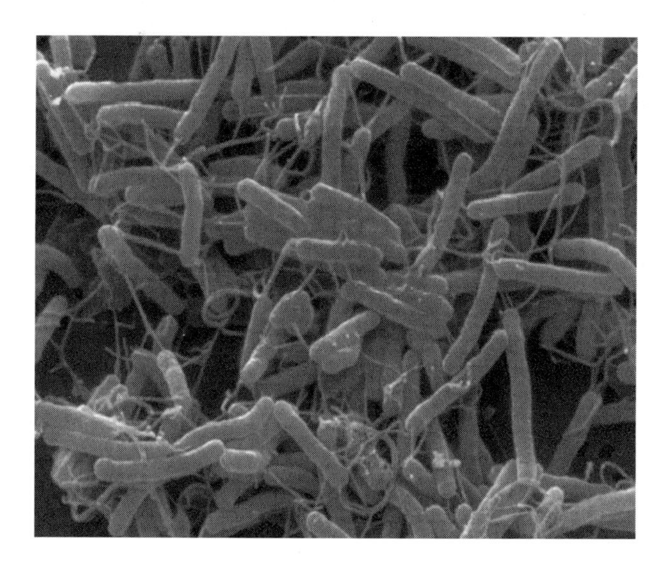

Figure 40.12 H. pylori on the surface of gastric epithelial cells.

The adhesins of the bacteria are also missing, however, several candidates are known; the N-acetyl-neuraminillactose-binding fibrillar hemagglutinin, phosphatidyl-lethanolamine, and gangliotetraosyl ceramid binding protein. The recently defined adhesin molecule (BabA) is a ligand for epithelial cell attachment. Another recently defined adhesin molecule is the heat-shock like protein on the bacterial surface.

The lipopolysaccharide layer of the bacteria (LPS) is also variable. The bacterial alfa-3-fucosyl-transferase genes are responsible for changes in the LPS.

The receptor for attachment on the epithelial cell side is not well defined. One of the targets proven to play an important role is the Class II major histocompatibility complex[30]. This particular protein is used not only for adhesion but also for induction of apoptosis. In this process T helper cells also co-operate producing IFN-γ. In the case of cross-linking, the MHC II complex with H. pylori bacterial super-antigens leads to apoptosis. The presence of both factors increases apoptosis by several magnitudes. Another target of adhesion is the Lewis B blood group antigen. The adhesion is pH dependent.

40.8.4 Action on cell membrane and intracellular signaling pathways

Immediately after the attachment, localized loss of microvilli can be detected using electron microscope studies[31]. Formation of cup and pedestal is also reported around the attached bacteria. With the attachment of H. pylori to gastric epithelial cells, several alterations can be found in the cytoskeleton. Actin, α-actinin and talin proteins are reorganized[32]. The complexes composed of actin and actin-binding proteins (α-actinin, talin, vinculin, paxillin and tensin) are in close cooperation with membrane receptor families. They act as a network for transmembrane signaling pathway between the extracellular and the cytoplasmic environments. Tyrosine phosphorylation occurs immediately. In PAI positive H. pylori strains this phosphorylation is done on a 145 kD protein with unknown function. This alteration was not found in PAI negative or cagE, cagF, cagG, cagH, cagI, CagG mutant strains.

In kinase inhibition studies, the following pathways were evaluated: CaM kinase, Protein kinase A, protein kinase C, protein kinase G, MEK (Mitogen activated protein/EGFR kinase), cAMP dependent and cGMP dependent kinases.

40.9 Inflammation in the gastric epithelium

First we summarize the major steps in the inflammation process from the defensive side:

1 After the appearance and detection of the damaging factor, circulating leukocytes have to enter from the blood vessels into the extravasal sphere towards the infection to control the damage (trafficking).
2 Leukocytes have to be able to locate the place of infection in the extravasal field (chemotaxis).
3 The production of free radicals is necessary to complete the infection, however, an over production may cause damage to host cells (priming and induction).

This model of inflammation is in close correlation with the immune response:

a The inflammatory response can be the effector mechanism of the immune reaction, where neutrophils can phagocytose the antibody-coated bacteria. The inflammatory response is an additional defence of the organism in case immunoprotection is unsatisfactory.
b Inflammation is also a component of wound healing. The migration of inflammatory cells and repair of the epithelium is regulated by the same cytokines.

The cellular compartment of the inflammatory response includes the following cell types:

1 Neutrophils, which differentiate in the bone marrow before entering the peripheral blood. They enter the gastrointestinal tissue by binding to adhesion molecules in the postcapillary venules. They can be found in the gastrointestinal tract for up to two days then they pass into the lumen. Neutrophils cannot proliferate. Neutrophils can produce reactive oxygen species for destroying the damaging factor.
2 Macrophages are derived from circulating monocytes. They can proliferate in the gastrointestinal tissue, however,

migration is still important. As monocytes differentiate, they develop the capacity for phagocytosis, proliferation and bacterial killing. The duration that macrophages remain in the gastrointestinal tissue is not known.

3 Lymphocytes are responsible for antibody production against specific antigens and are also able to fight cellular targets by direct cell destruction.

40.9.1 Leukocyte adhesion and its molecular basis

The first, and one of the most important factors in the trafficking of leukocytes to the gastrointestinal extravasal is the adhesion of leukocytes to the postcapillary venule wall. The adhesion is a multistep process, where the interplay of selectin, integrin and the immuno-globulin superfamily of adhesion molecules and chemokine factors are required (Table 40.1).

In inflammation selectins are expressed on the surface of leukocytes and endothelial cells[33]. The three major types are L-selectin (leukocyte), E-selectin (endothelial cells) and P-selectin (platelets, endothelial cells)[34]. L-selectin expressing leukocytes can adhere to E- or P-selectin expressing endothelial cells. Selectin expression is activated by IL-1, TNF-α, leukotriene B4 (LTB4), and lipopolysaccharide (LPS). Selectin bonds are responsible for leukocyte adhesion to and rolling on the endothelial cells. These bonds are weak. These weak bonds are replaced by stronger ones, formed between β2-integrins expressed on the surface of leukocytes and intercellular adhesive molecules ICAM 1 and 2.

Integrins form a larger group of adhesion molecules. Integrins are divided by their β subunits. β1 is involved in the lymphocyte and β2 in the leukocyte adhesion procedures. Some integrins (CD11a/CD18) are also expressed in the basal state of leukocytes. Cytokines or inflammatory processes do not enhance this expression. Expression of other integrins (CD11b/CD18) are influenced by bacterial products (formyl-methionin-leucyl-phenylalanine, FMLP), TNF-α, IL-1, Il-8, LTB4.

The 3rd group of important binding molecules is the immunoglobulin superfamily of adhesion molecules. The most important members of this family are ICAM-1 and ICAM-2. Unlike ICAM-1, expression of ICAM-2 is not influenced by cytokine production. ICAM-1 binds to CD11a/CD18 and CD11b/CD18, whereas ICAM-2 binds only CD11a/CD18.

The basic role of selectin molecules is the adhesion of leukocytes to the endothelium. Some important intracellular pathway alterations have been described recently.

Following the attachment by adhesion molecules, activation of leukocytes is induced by protein tyrosine kinases (PKC), G proteins or Ras-like GTP binding proteins. The activation of phospholipase C (PLC) generates phosphatidyl inositol 4,5-biphosphonate (PI2P 1,2-dyacilglycerol (DAG) and inositol 1,4,5-triphosphate (IP3). IP3 mediates the release of intracellular Ca and DAG activates PKC.

Another signaling pathway is also activated, the mitogen-activated protein kinase (MAP or ERK). The MAP kinases can be activated through upstream activators like the MAP kinase kinase. Recent studies have shown that L-selectin is closely correlated with the MAP kinase pathway. L-selectin activates the intracellular respiratory burst of neutrophils and can be inhibited by protein tyrosine kinase inhibitors.

40.9.2 Leukocyte chemotaxis and activation of respiratory burst

The process of migration out of the endothelial wall to the site of the injury or bacterial inflammation is called chemotaxis.

Table 40.4 Adhesion molecules involved in the gastric inflammatory process

Adhesion molecule	*Family*	*Basal expression*	*Stimulated expression*	*Counter-receptor*
Neutrophil L-Selectin (LAM-1)	Selectin	+	+	E-selectin and syaleted molecules
CD11a/CD18	β2 integrin	+	−	ICAM-1, ICAM-2
CD11b/CD18	β2 integrin	+	+	ICAM-1
Endothelial				
E-selectin	Selectin	−	+	L-selectin and syaleted molecules
P-selectin	Selectin	+	+	L-selectin and syaleted molecules
ICAM-1	Ig-super gene	−	+	CD11a/CD18, CD11b/CD18
ICAM-2	Ig super gene	−	+	CD11a/CD18

During chemotaxis the leukocyte must be prepared to produce reactive oxygen species in maximal amounts. This procedure is called priming. The stimulation of oxygen production is called activation.

Chemoattractant is a molecule that stimulates the migration of neutrophils, monocytes, eosinophils and other cells. Chemotaxis is the migration of a cell in the direction of the higher concentration of a chemoattractant. The following agents can induce chemotaxis of neutrophils: FMLP, IL-8, complement component C5a, PAF, and LTB4. Receptors are distributed over the cell surface and migration occurs in the direction of the receptor where ligand binding has occurred.

Circulating neutrophils are not capable of producing reactive oxygen species in maximal amounts. After entering the tissue, neutrophils are primed by IL-1, TNF-α, GM-CSF or LPS. Neutrophils can also be primed by exposure to the extracellular matrix. In this process superoxide, singlet oxygen and hydroxyl radicals are produced. Neutrophils kill phagocytosed bacteria with reactive oxygen species, generated intracellularly within the phagosome. A chemotactic agent in neutrophils was also detected in H. pylori infection.

40.9.3 Inflammatory cytokines

Cytokines are glycosylated proteins with molecular weights of 6 to 60 kD. Cytokines express their effect through specific receptors at very low concentrations. A general grouping of cytokines includes three groups:

a Hematopoietic cytokines, controling cell proliferation and differentiation in the bone marrow (they include IL-3, IL-7, IL-11, GM-CSF, and G-CSF).
b Regulatory cytokines (IL-2, 4,5,9,10,12,13, IFN γ, and TGF-β). This groups acts primarily on lymphocytes.
c Pro-inflammatory cytokines promote, prolong and amplify the inflammatory response. IL-1, IL-6, TNF-α, IL-8 are the major components of this group (Figure 40.13).

40.9.3.1 IL-1

IL-1 is a prototypical pro-inflammatory cytokine. The α group is membrane-bound and affects the T-cells; the β group is secreted and binds to receptors on neutrophils, bone marrow cells, B-cells, and endothelial cells. The production of IL-1 is induced by LPS, phagocytosis, and cytokines. Epithelial cells can also express IL-1. The feedback mechanism for diminishing IL-1 production includes the glucocorticoids, PGE2, and IL-1 receptor antagonists. IL-1 promotes the infiltration of neutrophils and monocytes into the gastrointestinal tissue by inducing the expression of endothelial adhesion molecules. IL-1β is an efficient acid production inhibitor, as well.

40.9.3.2 IL-6

IL-6 shows similar activity to IL-1 and TNF-α. This is produced by macrophages, T-cells, B-cells fibroblasts and endothelial cells. LPS, viral infection, and other cytokines stimulate the production of IL-6. Circulating IL-6 is probably the most important stimulus for the production of acute phase proteins. IL-6 promotes the differentiation of B-cells and plasma cell development.

40.9.3.3 IL-8

IL-8 is a prototype of the pro-inflammatory protein family. These chemokines are also divided into two subfamilies, based on their amino acid sequence. Macrophages, endothelial cells, epithelial cells and fibroblasts produce IL-8. IL-8 is a chemoattractant for monocytes and neutrophils.

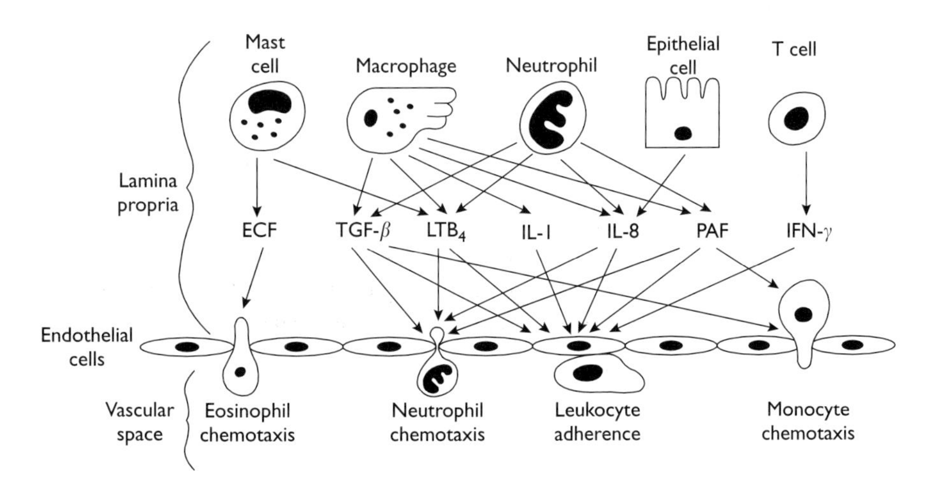

Figure 40.13 Interaction between epithelial cells, cytokines and inflammatory cells. Role of cytokines in the chemotaxis of circulating immune response cells. Note that epithelial cells can produce IL-8 after adhesion and induction by H. pylori.

40.9.3.4 TNF-α

TNF-α is a major pro-inflammatory cytokine. The major source for TNF-α production is the macrophage, in response to LPS, IL-1, and phagocytosis. Epithelial cells also produce this cytokine after bacterial infection.

TNF-α promotes chemotaxis and enhances the production of reactive oxygen species and nitric oxide. The non-immunologic effect of TNF-α includes weight loss, fever and malaise, shock and vascular collapse. TNF-α regulates the expression of ICAM-1 on vascular endothelial cells. Induction of the expression of MHC class II molecules on intestinal epithelial cells can contribute to Helicobacter pylori-induced apoptosis.

40.9.3.5 TGF-β

TGF-β is another regulatory cytokine with proinflammatory properties. TGF-β has three isoforms. The TGF-β cytokine family includes the peptide growth factors, like the insulin like growth factor, fibroblasts growth factor and trefoil factor. TGF-β is chemotactic for T-cells, neutrophils and monocytes. It enhances epithelial migration, allowing surviving epithelial cells to cover defects in the epithelial monolayer. In wounded epithelium TGF-β increases, in healthy epithelium it decreases cell proliferation.

40.9.3.6 IFN-γ

IFN-γ is both a regulatory and pro inflammatory cytokine. IFN-γ is made by T-cells and K-cells. It upregulates MHC class I and II antigens, ICAM-1 and FC receptors. IFN-γ potentiates TNF-α production in LPS-stimulated macrophages augmenting the production of reactive oxygen species[36].

40.9.3.7 Lipid mediators

Prostaglandins (PG), leukotrines (LTB) and platelet activating factor (PAF) are included in the lipid mediators of inflammation. They enhance vascular permeability and activate neutrophils. In prostaglandin synthesis two distinct cyclooxygenases are involved, COX-1 is expressed continuously by the gastrointestinal lamina propria, muscle, epithelial and vascular endothelial cells. The expression of COX-2 is induced by IL-1, TNF-α and other pro-inflammatory cytokines[36]. PGE2 promotes epithelial cell injury. Administration of COX-2 inhibitors inhibits spontaneous healing.

40.9.3.8 Nitric oxide

Nitric oxide (NO) is a small, biologically active compound formed by nitric oxide synthase (NOS) which oxidises the guanidine nitrogen of arginine[37]. The biologic effect of NO includes vasodilatation, neurotransmission and inflammatory response. Three different isoforms of NOS exist.

NOS-1 is present in neurons. NO produced in excess by NOS-1 may contribute to the disorders of bowel motility including ileus.

NOS-3 is found in vascular endothelial cells, causing vascular smooth muscle relaxation and dilatation.

NOS-2 is inducible by pro-inflammatory cytokines (IL-1), LPS. In macrophages the large quantity of NO produced is used for killing bacteria and tumor cells. NOS-2 can be expressed in gastrointestinal epithelia in inflammatory states.

40.9.3.9 Epithelial cells, H. pylori and gastric inflammation

During infection, the host produces antibodies against H. pylori antigens. Some of the antibodies target surface molecules, preventing adhesion. There are antibodies against HSP and Lewis molecule structures.

The HP-MP 1 antigen is a 16 kDa protein, which induces IL-1α, IL-8, and TNF-α production in leukocytes. Gastric biopsy samples of H. pylori infected patients showed increased IL-1B, IL-6, IL-7, IL-8, IL-10 and TNF-α production *in vivo*[37, 38]. Similar findings were shown using KATO and AGS cell lines *in vitro*. It is not clear whether IL-8 production (IL-8 is a potent chemotactic and activation factor for neutrophils!) is a primary result of H. pylori infection or a secondary response to IL-1 or TNF-α induction.

Another recent study investigated the heterogeneity and polymorphism of the IL-1B gene in humans. Depending on the genetic code, differing amounts of IL-1B is produced which influences the degree of inflammation. Furthermore, during inflammation, both iNOS and COX-2 are expressed in gastric and intestinal epithelial cells.

Knowledge about the intracellular control of inflammation-related proteins is limited. Some of the above factors have an NFκB site in their promoters. NFκB is a transcription factor with importance in the regulation of synthesis of several inflammatory related proteins. In unstimulated cells, NFκB is bound to IκB in the cytoplasm. During activation, IκB is phosphorylated and degraded, releasing NFκB. This way NFκB enters the nucleus, and binds to its promoter regions of the target genes. Another important transcription factor is NF-IL6.

40.10 H. pylori-caused epithelial cell damage and ulceration

40.10.1 Biologic organization of the gastric epithelia

The epithelial cells of the gastric epithelium are assembled like a multicellular sheet. The requirement for this higher organization includes the following: cytoskeletal stabilization of single cells, cell-cell adhesion, formation of cell-cell junctions and provision of a foundation by the basement membrane.

40.10.1.1 The epithelial sheet

The gastric cavity is lined by the epithelial sheet. The gastric epithelia functions as a separating wall between the external environment and the organism. This environment is quite hostile for the entire organism and the epithelia itself. Therefore, the gastric epithelia is prepared to handle injury and damage, that is the gastric epithelia is able to regenerate. The steady state between cell loss, due to physiologic or pathologic cell loss, and cell proliferation maintains the integrity of the epithelium. During uncompensated cell loss, erosion and ulceration can be detected. Increased and uncontrolled cell proliferation can be found in malignant alterations as well.

40.10.1.2 Basement membrane

The gastric epithelia resides on a basement membrane of 20 to 40 nm. The basement membrane of the alimentary tract seems to be similar to the basement membrane found in other organs. It contains laminin, heparan sulfate proteoglycan and type IV collagen. These components can have a significant effect on the epithelia such as enhancing proliferation, migration and differentiation.

40.10.1.3 Adhesive recognition of neighboring cells

The adhesive cell-cell interactions trigger the establishment of long term cooperative interaction between cells. E-cadherin is a crucial molecule for the initial cell-cell recognition and adhesion. Defects in adhesion mediated by components of cadherin and cadherin binding proteins (including catenins and abnormal APC gene) may mediate the loss-of-contact inhibition that characterizes carcinogenesis[39].

40.10.1.4 Intercellular junctions

Gastric epithelial cells have a continuous, circumferential tight junction. These intercellular junctions seal the paracellular space and stabilize the epithelial monolayer. Directly below a tight junction lies a zone in which the lateral membrane of adjacent cells adhere by means of e-cadherin based interactions. The gastric surface cells have a very important junction type, the so called "gap junction". The gap junctions consist of aggregates of membrane particles called connexins which behave like conduits to allow direct passage of ions and small molecules between adjacent cells.

40.10.2 Gastric epithelial cell growth and turnover

Gastric epithelial turnover occurs, on average, twice in a week by a process of proliferation, migration, senescence and sloughing. For example, gastric surface cells arise from the proliferative zone of the gastric pit, migrate to the surface, where they become differentiated and ultimately slough into the lumen. The life-time of gastric epithelial cells is influenced by physiologic agents of the stomach.

The duration of mitosis in the antrum is between 6 and 7 hours. The newly generated cells migrate from the proliferating zone in the direction of the lumen. The stem cells of the gastric gland are probably the mucus neck cells. They are located the middle third of the gastric pit. They can differentiate into specialized gastric, parietal, chief, G, and zymogen cells.

^{3}H-thymidine labeling, Ki67 and PCNA immunohistochemistry are the suggested methods, beside quantitative DNA analysis by flow- and TV image cytometry, for the determination of proliferation (Figure 40.14).

40.10.3 Epithelial response to disease and injury

Epithelial injury can lead either to the complete loss of the epithelial cell or break down of the membrane. Membrane injury can be either partial or complete. In the case of partial damage, the integral membrane proteins and the lipid compartment are damaged. In the complete form, which can be caused artificially by trypsin, rapid repair is seen in animal models. This indicates that, in addition to cell proliferation, an intracellular repair mechanism acts against partial external gastrointestinal cell injury.

For external or internal injury, epithelial cells respond with enhanced proliferation in an effort to replenish cells in the eroded area. An increased fraction of progenitor cells is stimulated to move through the S, G_2 and M phases of the cell cycle. The S phase of the cell cycle takes 6 to 12 hours, which is not rapid enough to cover the prompt loss of epithelial integrity. Therefore, other responsive elements of the epithelial barriers are activated as well.

Epithelial cells undergo a process called restitution, characterized by rapid spreading of epithelial cells shouldering the wound. The conversion from the tall columnar form to a spread and flattened cell phenotype is also essential.

40.10.3.1 Role of EGF-R, TGF-α and other growth factors

Signals for stimulation of restitution are currently under evaluation. However, it is known that EGF, fibroblasts growth factor (FGF), and hepatocyte growth factor (c-met) enhance the number of motile spreading cells around the ulceration.

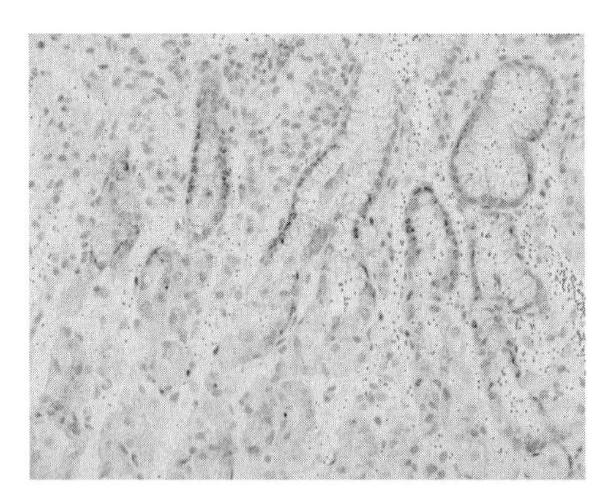

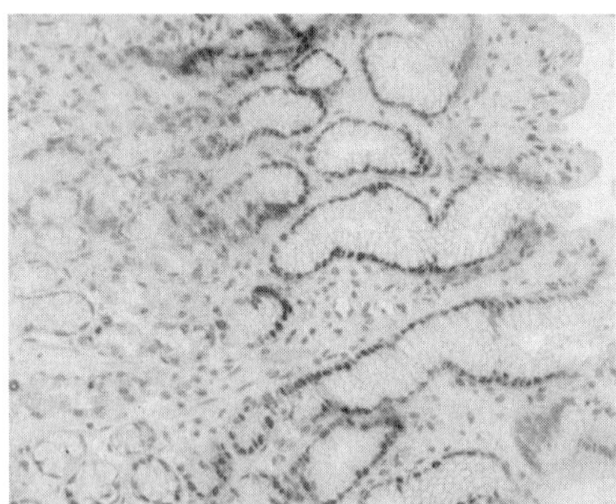

Figure 40.14 Proliferation nuclear antigen (PCNA) expression in H. pylori negative and positive gastric mucosa. The positive nuclei are labelled with dark stain. Note that in the case of inflammation the middle third proliferation zone moves to the bottom of the gastric pit (see Colour Plate XXVI.)

Both TGF-β1 and TGF-α1 have essential roles in this process. Hence they are expressed in the epithelial cells intracellularly, the loss of integrity of epithelial cell membrane, with consecutive extracellular presence of these factors directly induces the gastric epithelial repair mechanisms. The epithelial restitution process undoubtedly includes cytoskeletal events and cell-matrix interactions, as well.

Mitosis of epithelial cells are induced by EGF, TGF-α, platelet derived growth factor (PDGF), FGF, and trefoil growth peptides. Of these growth factors, EGF and TGF-α are essential. EGF is the primary growth factor for gastric epithelial cell repair, however, its receptor, the EGFR-R is also activated by TGF-α. As mentioned previously, injury of gastric epithelial cells liberates TGF-α, this way activating cell proliferation of neighbor intact cells.

40.10.3.2 Direct and indirect effects of helicobacter on epithelial cell death and proliferation

In the presence of H. pylori and inflammation, epithelial cell turnover is increased. There are contradictory results in the literature about the timing and extent of apoptosis and proliferation.

Gastritis itself – as an inflammatory process – increases epithelial cell proliferation and apoptosis. Apoptosis is increased significantly when H. pylori is also present (Table 40.5).

In gastritis with intestinal metaplasia, however, cell turnover has different characteristics. The turnover in this group is significantly slower, as compared to gastritis without intestinal metaplasia. The presence of H. pylori, on the other hand, increases cell proliferation. Successful H. pylori eradication seems to reconstitute a healthy cell turnover. In gastritis with intestinal metaplasia the lower acid production may be one of the factors responsible for slower epithelial cell turnover (Table 40.6).

Based on *in vitro* studies, it is not yet clear which of the H. pylori cytotoxins induce increased cell death. The role of urea, Cag, Vac, Ice and Pic should be considered as primary direct cytolytic factors.

Table 40.5 Proliferation and apoptosis in gastric biopsy specimen

	PCNA LI	*AgNOR*	*TUNNEL*
Normal epithelium	46.4 ± 10.7	230.1 ± 94.3	15 ± 1[+]
Gastritis without intestinal metaplasia			
Hp positive	54.7 ± 19.1	291.4 ± 44.3	29 ± 2*[+]
Hp negative	53.2 ± 20.7	277.8 ± 57.54	13 ± 1*
Gastritis with intestinal metaplasia			
Hp positive	50.2±13.7	306.4±51.4	13±4
Hp negative	44.7±20.6	308.2±34.1	11±4

Source: Unger Z. Pronai, L. Molnar B, Tulassay Z. own data.

Notes
PCNA LI: % of PCNA positive cells.
AgNOR: number of dots in 100 cells.
Tunnel: % of Tunnel positive apoptotic cells.

Table 40.6 Proliferation and apoptosis in gastric biopsy specimen

	PCNA	*AgNOR*	*TUNNEL*
Normal epithelium	46.46 ± 10.72	230.12 ± 94.30	15 ± 14
Gastritis without intestinal metaplasia			
Hp − before eradication	44.7 ± 20.7	308.4 ± 34.3	11 ± 3
Hp + before eradication	52.0 ± 18.9	294.5 ± 58.9	13 ± 4
Hp − after eradication	45.8 ± 11.07	266.333 ± 15.95	16 ± 3
Hp+after eradication	53.26 ± 22.19	310.2 ± 24.6	15 ± 52

Note: The differences in the intestinal metaplasia group were not significant.

Another important mechanism is the binding of H. pylori to EGF receptors. EGF-receptor expression is induced in the marginal zone of the ulcer. However, the works of Tarnawski *et al.*[40] demonstrated that vacA toxin of H. pylori can bind to EGF-R receptor, and it thus competes with EGF. This way, vacA has the following pathophysiologic effect on reconstitutive epithelial proliferation:

1. healing is delayed in experimental gastric ulcers in rat;
2. epithelial cell proliferation is inhibited in the ulcer margin;
3. binding of EGF to its receptor is significantly reduced;
4. EGF-induced signal transduction pathways are not triggered in the presence of vacA.

40.10.4 Complex model of gastric ulceration

At present we have four major factors which influence and play a role in gastric inflammation and ulceration:

1. *The production of acid.*
   The phrase "No acid no ulcer" has been known for a century. Acid overproduction leads to ulceration.
2. *The presence of H. pylori and its quantitative and qualitative characteristics.*
   It is not yet clear whether the amount of H. pylori plays a role in ulceration. The bacterial cytolytic factors causing ulcerations are also under re-evaluation.
3. *The inadequate immune response produces not only antibacterial but also cytolytic agents.*
   Epithelial cell damage can be caused, not only by external aggressive factors, but also by the host's immune response. The mechanism of epithelial ulceration by uncontrolled immune response has been partially determined.
4. *Insufficient host regenerative resources.*
   There is limited data available about the role of the host regeneration capabilities. The power of stem cells to replicate changes with age. However, this replication may also be influenced by direct induction of growth factors (amount of produced EGF, TGF-α), which needs further clarification.

At present we do not have methods and markers to evaluate all of these factors. However, the latest technological developments (e.g. biochips) forecast a new era for multiparameter genetic and functional testing.

References

1 Beaumont, W. (1883) Experiments and observations on the gastric juice and the physiology of digestion Cambridge, MA: Harvard University Press, 1929.
2 Bechterew, W. (1911) Die Functionen der Nervencentra. Drittes heft: Hemispheren des Grosohirus. Jena: Gustav Fischer.
3 Edkins, J.S. (1905) On the chemical mechanism of gastric secretion. *Proc R Soc 756: 376.*
4 Gregory, R.A. (1967) Isolation and chemistry of gastrin. In: Code, C.F. (ed.) *Handbook of physiology: alimentary canal.* Washington, DC: American Physiological Society, pp. 827.
5 Lim, R.K.S. (1922) The question of gastric hormone. *Q J Exp Physiol 13: 79.*
6 Langley, J.N., Edkins, J.S. (1886) Pepsinogen and pepsin. *J Physiol (Lond) 7: 371.*
7 Feng, T-P., Hou, H-C., Lim, R.S.K. (1929) On the mechanism of inhibition of gastric secretion by fat. *Clin J Physiol 3: 371.*
8 Soll, A.H. (1980) Secretagogue stimulation of [^{14}C]aminopyrine accumulation by isolated canine parietal cells. *Am J Physiol 238: 366.*
9 Park, J., Chiba, T., Yamada, T. (1987) Mechanisms for direct inhibition of canine gastric parietal cells by somatostatin. *J Biol Chem 262: 14190.*
10 Marshall, B.J., Warren, J.R. (1984) Unidentified curved bacilli in the stomach of patients with gastritis and peptic ulceration. *Lancet, 8390:1311–1315.*
11 Miller, L.J. (1999) Gastrointestinal Hormones and Receptors, In: Yamada, T. (ed) *Textbook of Gastroenterology*, 3rd Edition, Lipincott Williams & Wilkins, Philadelphia, pp: 35–66.
12 Zollinger, R.M., Ellison, E.H. (1955) Primary peptic ulcerations of the jejunum associated with islet cell tumors of the pancreas. *Ann Surg 142: 709.*
13 Seva, C., Dickinson, C.J., Yamada, T. (1994) Growth-promoting effects of glycine-extended progastrin. *Science 265: 410.*
14 Del Valle, J., Todisco, A. (1999) Gastric Secretion, In: Yamada, T. (ed.) *Textbook of Gastroenterology*, 3rd Edition, Lipincott Williams & Wilkins, Philadelphia, pp: 278–319.
15 Diaz, J., Vizuete, M.L., Traiffort, E. *et al.* (1994) Localization of the histamine H_2 receptor and gene transcripts in rat stomach: back to parietal cells. *Biochem Biophys Res Commun 3: 1195.*
16 Gantz, I., Schäffer, M., Del Valle, J. *et al.* (1991) Molecular cloning of a gene encoding histamine H_2 receptor. *Proc Natl Acad Sci USA 88: 429.*
17 Copin, A.S., Lee, Y.M., McBridge, E.W. *et al.* (1992) Expression cloning and characterization of the canine parietal cell gastrin receptor. *Proc Natl Acad Sci USA 89: 3605.*
18 Song, I., Brown, D.R., Wiltshire, R.N. *et al.* (1993) The human gastrin/cholecystokinin type B receptor gene: alternative splice donor site in exon 4 generates two variant mRNAs. *Proc Natl Acad Sci USA 90: 9085.*
19 Seva, C., Dickinson, C.J., Yamada, T. (1994) Growth-promoting effects of glycine-extended progastrin. *Science 265: 410.*
20 Prinz, C., Sachs, G., Walsh, J.H. *et al.* (1994) The somatostatin receptor subtype on rat enterochromaffinlike cells. *Gastroenterology 107: 1067.*
21 Ding, M., Kinoschite, Y., Kishi, K., *et al.* (1997) Distribution of prostaglandin E receptors in the rat gastrointestinal tract. *Prostaglandins 53: 199.*
22 Hershey, S.J., Sachs, G. (1995) Gastric acid secretion. *Phys Rev 75: 155.*
23 Modlin, I.M., Sachs, G. (1998) *Acid Related Diseases*, Schnetztor-Verlag GmbH, Konstanz, pp: 18–31.
24 Modlin, I.M., Sachs, G. (1998) Inhibition of the histamine 2 receptor. In: *Acid related diseases*, Schnetztor-Verlag GmbH, Konstanz, pp: 121–126.
25 Censini, S., Lange, C., Ziang, Z. *et al.* (1996) Cag a pathogenicity island of Helicobacter pylori, encodes type-I specific and

disease-associated virulence factors. *Proc. Natl. Acad. Sci USA. 93: 14648–14653.*

26 Pie, R., Tarnawski, A. (1998) Signal transduction cascades triggered by EGF receptor activation. Relevance to gastric injury repair and ulcer healing. *Dig. Dis. Sci. 43: 14–22.*

27 Segal, E.D. (1997) Consequences of attachment of Helicobacter pylori to gastric cells. *Biomed Pharmacother, 51: 5–12.*

28 Wyle, F.A., Tarnawski, A., Dabros, W. (1991) Campilobacter pylori interactions with gastric cell tissue culture. *J. Clin. Gastroenterol. 12: 99–103.*

29 Nilius, M., Bode, G., Buchler, M. (1994) Adhesion of Helicobacter pylori and Echerichia coli to human and bovine surface mucus cells *in vitro. Eur. J. Clin invest 24: 454–463.*

30 Fan, X., Crowe, S.E., Behar, S. *et al.* (1998) The effect of Class II Major histocompatibilty complex expression on adherence of Helicobacter pylori and induction of apoptosis in gastric epithelial cells: A mechanism for T Helper Cell Type1-mediated damage. *J. Exp. Med. 187: 1659–1669.*

31 Hemalatha, S.G., Drumm, B., Shermann, P.M. *et al.* (1995) Expression of interleukin-8 and CD54 by human gastric epithelium after H. pylori infection *in vitro. Gastroenterology, 108: 65–74.*

32 Segal, E.D., Lanmge, C., Covacci, A., Tompkins, L.S., Falkow, S. (1997) Induction of host signal transduction pathways by Helicobacter pylori. *Proc. Natl. Acad. Sci., USA, 94: 7595–7599.*

33 Crockett-Torabi, E. (1998) Selectins and mechanisms of signal transduction. *J. Leuc. Biol. 63: 1–14.*

34 Eppiheimer, M.J., Wolitzky, B., Anderson, D.C. (1996) Heterogeneity of expression of E and P-selectins *in vivo. Circ. Res. 79: 560–568.*

35 Furfaro, S., Berman, J.S. (1992) The relation between cell migration and activation in inflammation: beyond adherence. *Am. J. Respir. Cell. Mol. Biol 7: 248–254.*

36 Stenson, W.F. (1997) Cyclooxygenase 2 and wound healing in the stomach. *Gastroenterology 112: 645–655.*

37 Salzmann, A.L., Eaves-Pyles, T., Linn, S.C. (1998) Bacterial induction of inducible nitric oxide synthase in cultured human intestinal epithelial cells. *Gastroenterology 114: 93–99.*

38 Noach, L.A., Bosma, N.B., Jansen, J. (1994) Mucosal tumor necrosis factor-alpha, interleukin-1 beta and interleukin-8 production in patients with Helicobacter pylori infection. *Scand. J. Gastroenterol. 29: 425–434.*

39 Blaser, M.J. (2000) Linking Helicobacter to gastric cancer. *Nature 6: 376–377.*

40 Ramai, P., Tarnawski, A. (1998) Signal Transduction cascades triggered by EGF receptor activation. Relevance to gastric injury repair and ulcer healing. *Dig. Dis. Sci. 43: 14–22.*

Chapter 41

Osteoporosis

György Mészáros, Tamás Kardon and Gyula Poór

Contents

Osteoporosis is one of the most important diseases facing the aging population because of the high prevalence and incidence of fractures, the enormous costs in health care funds, and the substantial effect in terms of suffering and disability.

By the definition of the World Health Organization, an estimated 30% of postmenopausal white women in the USA might have osteoporosis of the hip, spine or distal forearm (Melton L.J.III., 1995). More extensive data indicate that 19% of postmenopausal white women in the USA have bone density of the proximal femur more than 2.5 SD below the young normal mean (Looker *et al.*, 1995). Comparable data are not yet available for men in this survey, but a study in the UK suggested that about 6% of men had hip BMD more than 2.5 SD below the young normal mean for women compared to 22% of British women (Kanis, 1994).

Hip fracture is the most serious complication of osteoporosis and is associated with considerable morbidity and mortality. The incidence of hip fractures increases exponentially with age in both sexes resulting from a combination of declining BMD and an increased incidence of falls with advancing age. It is estimated that about 1.7 million hip fractures occured worldwide in 1990, about 1.2 million in women and 500,000 in men. Due to increasing life expectancy, the number of hip fractures could rise to 6.26 million by 2050 (Cooper *et al.*, 1992). The overall excess mortality associated with hip fracture has been estimated at 12 to 20 per cent with most of the excess deaths occuring in the first 6 months (Poór *et al.*, 1994). Half of survivors are unable to walk unassisted and are confined to long term care in nursing homes (Poór *et al.*, 1995). In contrast with hip fractures the prevalence of vertebral deformities proved to be equal among women and men in the large European Vertebral Osteoporosis Study (O'Neill *et al.*, 1996). Vertebral fractures are also associated with an increased mortality at 5 years, however, the increase in mortality is gradual over this period, compared with hip fractures (Cooper, 1993). Distal forearm fractures are not associated with an increased mortality, but are not free of long term disability either.

The financial costs of osteoporosis are difficult to estimate accurately, because they include acute hospital care, long term residential care and more indirect costs, including loss of working days and pharmacological preventive strategies. In the USA, the total cost of fractures is estimated at $20 billion p.a. with over $8 billion attributed to hip fractures (Praemer *et al.*, 1992).

The recent awareness that osteoporosis is a detectable, treatable and partly preventable disease of the elderly comes at a time of rapid progress in better understanding of the physiology and pathophysiology of bone cell function, in measurement of bone mass and in drug research.

41.1 Bone physiology

Bones are composed of two types of bone tissue: compact bone which forms the outer cortex of bones and trabecular or cancellous bone found mainly in vertebral bodies, the pelvis and the ends of long bones. Despite its hardness, bone remains a complex and dynamic, metabolically active tissue, which undergoes continuous remodeling throughout life. It is constantly resorbed and accreted by bone cells, the activity of which can be modified by many factors. Bone formation (osteoblastic activity) and bone resorption (osteclastic activity) constitute bone remodeling or turnover, a process which takes place on bone surfaces (Figure 41.1). Under physiological circumstances osteoblastic and osteoclastic activity is "coupled" and constitutes the basic multicellular unit (BMU) of bone. Should this process become imbalanced – "uncoupling" – excessive osteoclastic resorption or defective osteoblastic function will cause net loss of bone tissue. Prerequisites for normal mineralization are Vitamin D3 ($1,25(OH)_2D_3$), normal levels of calcium, phosphorous and alkaline phosphatase, and physiological pH.

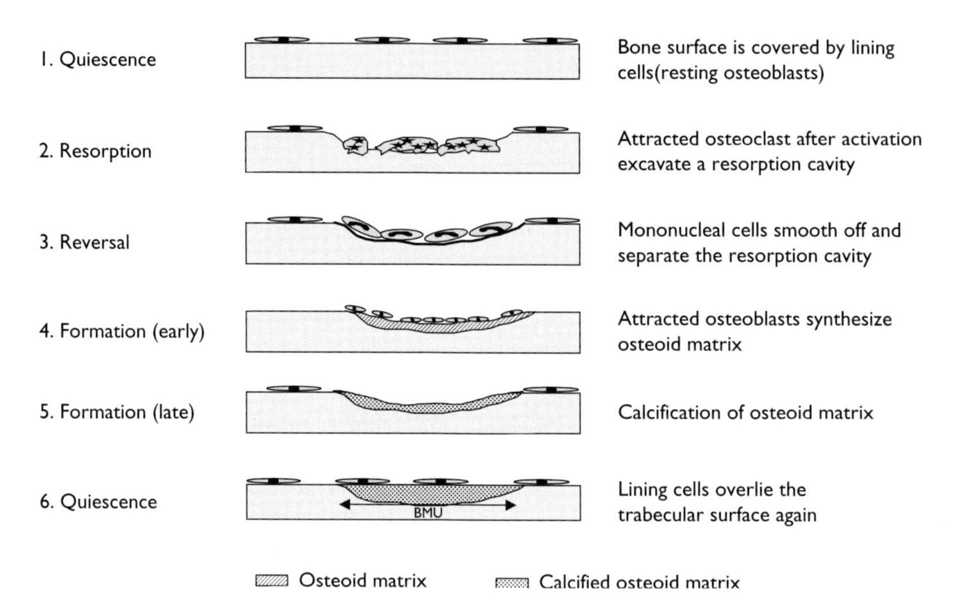

Figure 41.1 Sequence of trabecular bone remodeling.

Table 41.1 Hormonal and local regulators of bone remodeling

- Hormones
  - Polypeptide hormones
    - Parathyroid hormone
    - Calcitonin
    - Insulin
    - Growth hormone
  - Steroid hormones
    - 1,25-Dihydroxyvitamin D_3
    - Glucocorticoids
    - Sex steroids
  - Thyroid hormones
- Local factors
  - "Classic" polypeptide growth factors
    - Insulin-like growth factors
    - Transforming growth factor-β family of peptides
    - Fibroblast growth factors of heparin binding growth factors
    - Platelet-derived growth factor
  - Cytokines from the immune and hematological systems
    - Interleukins
    - Tumor necrosis factor
    - Colony-stimulating factors
    - γ-Interferon
  - Other factors
    - Prostaglandins

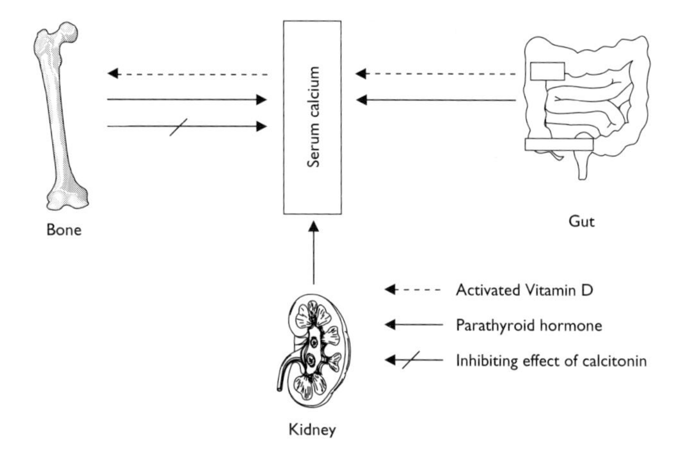

Figure 41.2 The action of the three major hormones on calcium metabolism.

Bone metabolism is regulated by a variety of hormonal and local factors that act on bone forming and bone resorbing cells (Table 41.1). The action of the three major hormones – parathyroid hormone (PTH), calcitonin and activated Vitamin D3 – is demonstrated in Figure 41.2. PTH results in calcium release from the bone by osteoclasts raising the ionic plasma calcium level. The PTH receptor is coupled with a heterodimeric G-protein, and activates adenylate cyclase. Calcitonin acts against PTH, and results in a decrease in the plasma Ca^{2+} level. Calcitonin receptors are also connected with two different types of heterotrimeric G-proteins, and the net result of the receptor-ligand interaction is an intracellular elevation of Ca^{2+}. This hormone activates adenylate cyclase and phospholipase-Cβ, respectively. Osteoclasts have no receptors for these hormones, so they act by releasing cytokines and other mediators from the osteoblasts. Vitamin D3 acts through an intracellular receptor, and affects the transcription of several proteins which, e.g., influence the transport of Ca^{2+}. Vitamin D3 synthesis can be up-regulated by PTH, so the effect on bone homeostasis could be enhanced in this way.

Various growth factors, which can stimulate the proliferation and activity of bone cells and bone formation, have been identified in human bone. They play a major role in segmental bone defects and in fracture repair as well. Insulin-like growth factors (IGFs), transforming growth factors (TGFs) and bone morphogenic proteins (BMPs) are important regulators of bone metabolism. The two IGF genes are expressed by skeletal cells, but IGF-I is more potent than IGF-II. They have modest mitogenic activity for cells of osteoblastic lineage, and enhance the differentiated function of osteoblasts. IGF-I has an inhibitory effect on stimulated bone resorption, suggesting that IGF-I inhibits existing osteoclasts, and alternatively that IGF-I interferes with osteoblast-derived factors that stimulate existing osteoclasts. TGFβ is one of the most abundantly occuring growth factors of the bone matrix, most of which is in latent form and needs to be activated to exert its biologic activity. Estrogen deficiency results in decreased production of TGFβ from bone cells and can play a role in the pathogenesis of bone loss (Kundu *et al.*, 1999). There are also other hormones/peptides which have a greater or lesser effect on bone homeostasis. These are osteogenic growth peptide, substance P, calcitonin gene related peptide, amylin, parathyroid hormone related protein and some neuropeptides.

Osteoclasts and osteoblasts derive from two separate cell lines. Mesenchymal stem cells are the origin of osteoblasts, and hemopoietic monocyte-macrophage precursor cells are responsible for osteoclasts. Recently, the discovery that some tumor necrosis factor (TNF) family members can influence the differentiation of these cell lines, raises the possibility of cross-talk between osteoblast and osteoclast development. The first step was the identification of osteoprotegerin (OPG), its older name is osteoclastogenesis inhibitory factor. This is a naturally ocurring glycoprotein which is a soluble member of the TNF receptor family. OPG is not only expressed in bone forming cells. In adult human tissues high levels of OPG mRNA have been detected in the heart, lung, kidney, placenta, liver and thyroid gland. The study of this protein led to the identification and isolation of a RANKL (receptor activator of nuclear factor – κB ligand or osteoclast differentiation factor). This is present on osteoblast/stromal cells, and it binds a transmembrane receptor on hemopoietic osteoclast precursor cells to RANK (receptor activator of NF-κB). The interaction of RANK and RANKL initiates a signaling and gene expression cascade. This process results in differentiation and maturation of osteoclast precursor cells to active osteoclasts. OPG binds to RANKL, inhibits the interaction with RANK and this leads to the blocking of osteoclast development (Figure 41.3). Osteoclasts are multinucleated giant cells that form from the fusion of mononuclear precursors – the number of nuclei can reach twenty. The regulation of this fusion is altered by OPG and RANKL, where OPG inhibits differentiation, fusion and activation while RANKL has the opposite effect. Osteoclasts must undergo apoptosis in order to be inactivated and this is also activated by OPG and inhibited by RANKL. The relationship between osteoblasts and osteoclast differentiation, because of the lack of receptors on the latter, can be discussed in terms of the effect of such hormones on osteoclasts, which act only on

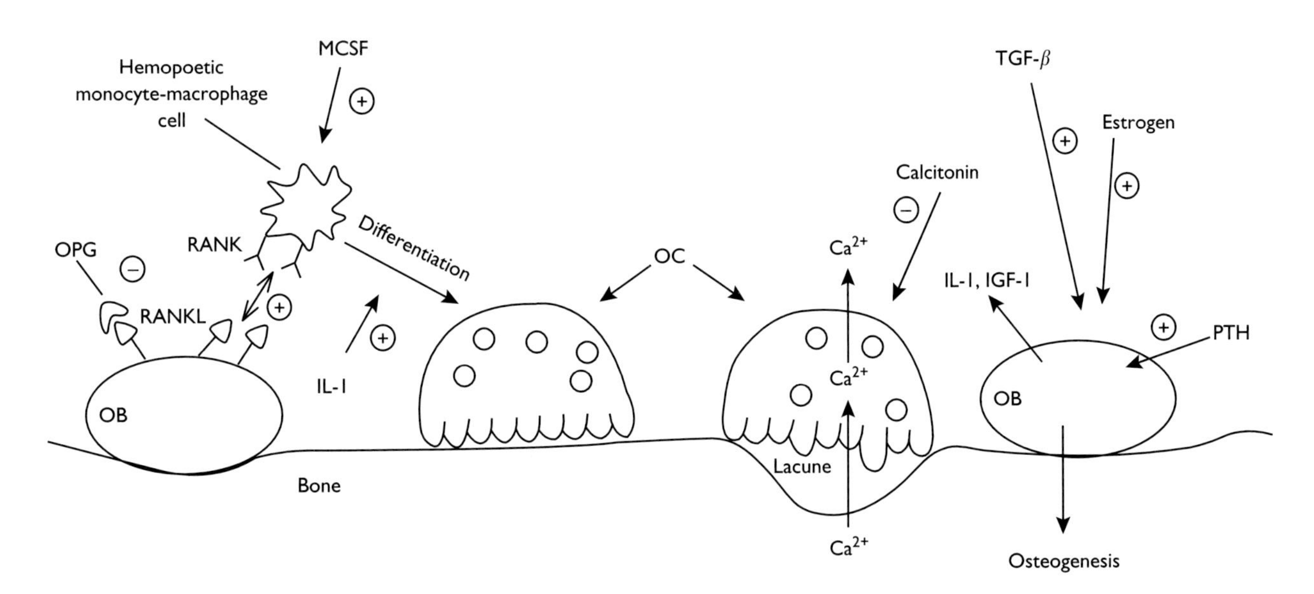

Figure 41.3 Role of some mediators, which may influence bone forming/resorbing cells. OB: osteoblast; OC: osteoclast; MCSF: macrophage colony stimulating factor; IL-1: interleukin 1; OPG: osteoproteregin; RANK: Receptor activator of NF-κB; RANKL: Receptor activator of NF-κB ligand; IGF-1: insulin like gowth factor-1; PTH: parathyroid hormone; TGF-β: Transforming growth factor β.

Table 41.2 Regulation of RANKL and OPG by systemic factors

	RANKL	*OPG*
Stimulation	Dexamethasone	17-β-estradiol
	Vitamin D3 ($1,25(OH)_2D_3$])	BMP-2
		TGF-β
	PTH	
	PGE_2	IL-1 α and -β
	IL-11	TNF-α and β
	TNF-α	
Inhibition		Hydrocortisone
	17-β-estradiol	Dexamethasone
	TGF-β	Vitamin D3 ($1,25(OH)_2D_3$])
		PTH
		PGE_2

Notes: PTH: parathyroid hormone; PGE_2: Prostaglandin-E2; BMP: bone morphogenic protein; TGF: tumor growth hormone; IL: interleukin; TNF: tumor necrosis factor (Modified from Aubin *et al.* 2000, and Hofbauer 1999.)

osteoblasts such as PTH or ($1,25(OH)_2D_3$) (Table 41.2). The interaction between the two cell lineages is required for osteoclast formation, e.g. the production of osteoclast differentiation factor (ODF) by osteoblast/stromal cells as a response to the above mentioned hormones (Aubin *et al.*, 2000).

What biological effects does OPG have? Overexpression of OPG in transgenic mice shows a marked decrease in the number of trabecular osteoclasts and the species develop progressive osteoporosis. It should be mentioned that OPG −/− mice develop a very fast calcification of the aorta and renal arteries, which suggest that OPG may be involved in the prevention of vascular calcification. Hofbauer (1999) wrote a recent review, where these findings are clearly shown in detail. *In vivo* effects of exogenous administration of OPG resulted in a marked increase in trabecular bone mass, which may also be a step forward to human clinical administration.

Osteogenic proteins are those compounds which can induce bone formation or bone regeneration. Many similar agents related to the transforming growth factor-β (TGF-β) superfamily belong to this group. Up to now more than 30 related members of this family have been identified, including growth and differentiation factors (GDFs), inhibins, activins, bone morphogenic proteins (BMPs) and TGFβs. In 1999, Geesing *et al.* showed that two osteogenic proteins, BMP-2 and BMP-7 (also named osteogenic protein-1, OP-1), were effectively taking part in bone regeneration. The observation of induction of ectopic bone formation by demineralized bone matrix led to intensive research on the BMPs. In order to be an active inducer, BMPs must undergo dimerization. The induction can occur both in a homo-, or in a heterodimer form. These dimers are interconnected with covalent bindings, namely seven disulphide bonds, which are prerequisite for bone induction (Wozney *et al.*, 1988). Table 41.3 shows the currently identified BMPs.

Table 41.3 Bone morphogenetic proteins (BMP) and their role in osteoinduction

Bone morphogenetic proteins	*Bone induction*
BMP-2	Bone induction in subcutaneous pockets in rats
BMP-3	Bone induction in subcutaneous pockets in rats
BMP-4	Bone induction in subcutaneous pockets in rats
BMP-5	NR; Role in the initiation of formation of particular skeletal elements
BMP-6	Bone induction in subcutaneous pockets in nude mice
BMP-7 (OP-1)	Bone induction in subcutaneous pockets in rats
BMP-8 (OP-2)	NR
BMP-12	NR; Inhibition of terminal myoblast differentiation
BMP-13	NR; Inhibition of terminal myoblast differentiation; Ectopic induction of tendon and ligament in rats
BMP-14	NR; Ectopic induction of tendon and ligament in rats

OP: osteogenic protein; NR: no report on osteoinductivity. (Modified from Groeneweld *et al.* 2000.)

Our knowledge of the signal transduction pathway of the BMPs is not well established yet. The receptors for BMPs are composed of two different types of membrane-bound serine/threonine kinases: BMPR-1A and BMPR-1B for type I and type II BMP receptors, respectively. Ligand binding activates the type II receptor and phosphorylates the type I receptor. The receptor I activated in this way phosphorylates a member of the Smad family of intracellular proteins. Smads are a group of mediators and regulators of the TGFβ/BMP family. They can be divided into classes I-III. Some Smads from group I form heterodimeric complexes with a class II Smad, and these are able to regulate molecular transcriptional responses directly, while the Smads belonging to group III inhibit signaling from TGFβ(/BMP (Padgett *et al.*, 1998). The use of some BMPs (BMP-2, BMP-7) showed that they have osteoregenerative capacities (Stevenson *et al.*, 1994; Ripamonti *et al.*, 1996) in healing bones, but no observation in connection with osteoporosis has been described so far. The clinical use of osteogenic proteins in the treatment of osteoporosis will be studied intensively in the upcoming years, and it will be of great importance in human care.

41.2 Genes determining bone mass

Bone mass at a certain age depends on the maximal bone mass achieved during growth and on the rate of bone loss in later adult life (Figure 41.4). Major factors determining bone mass include genetic background, aging, menopause and environmental factors such as calcium uptake, physical activity, alcohol, smoking or drugs.

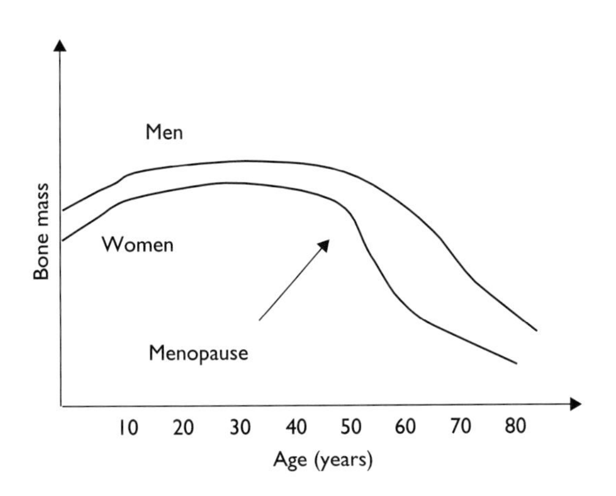

Figure 41.4 Pattern of change in bone mass with age in males and females.

Though a genetic influence on bone mass is well established, the number of genes involved, their chromosomal location, the magnitude of their effects, and their interaction with each other and with other risk factors are not properly defined. Since bone mass gain during the period of childhood, postmenopausal bone loss, and further losses in the elderly are determined by different sets of endogenous and exogenous factors, the relative influence of specific genes on the risk of osteoporosis may vary greatly with age.

Asian women have a 40–50% lower risk of hip fracture compared to Caucasian women. Asian women paradoxically have lower, and African-American women have higher bone density than Caucasian women (Pollitzer *et al.*, 1989). This data mean some genetic components, which may influence bone mass, but don't exclude other (environmental or lifestyle) factors which can also affect the appearance of osteoporosis. Family studies, twin studies and linkage analysis studies were used to find the genetic factors showing a higher risk of osteoporosis. Studies on heritability showed no direct single gene mutations which follows Mendelian inheritance. Although some bone diseases can be caused by mutations in single genes (Cockerill *et al.*, 1997), e.g. Vitamin D3 resistant rickets and the Vitamin D3 receptor gene, osteogenesis imperfecta and genes for type I pro-collagen (Kuivaniemi *et al.*, 1991), these mutations do not occur in the majority of people with osteoporotic fractures. In family studies the purpose was to see whether a trait runs in a family. Surprisingly, the BMD relationship was stronger between premenopausal mothers and daughters than between postmenopausal mothers and daughters (Lutz *et al.*, 1990) which also suggests the influence of other factors that can modify the genetic influence on the skeleton. Some studies revealed that daughters with mothers having a history of low bone density or osteoporosis also have lower bone density (Seeman *et al.*, 1989), however the intestinal calcium absorption was also found to be correlated in mothers and daughters, with the relationship being strongest in pairs where the mother had high bone density (Danielson *et al.*, 1997). Applied segregation analysis for determining the degree of familial resemblence under varying genetic or enviromental factors among families in France found no evidence of a major gene influence on BMD but supported a polygenic effect exerting its greatest influence on peak bone mass (Gueguen *et al.*, 1995). Twin studies take advantage of the fact that monozygotic twins are genetically identical clones and will be the same for genetically determined traits. The data from twin studies consistently show that bone density has a strong genetic component (50–90% of variability is determined by a genetic component, depending on the measured site) at a number of sites (Pocock *et al.*, 1987; Dequeker *et al.*, 1987). Bone from appendicular skeletal sites (femur neck) is comprised primarily of cortical bone and may be less influenced by genetic factors than trabecular bone-rich axial skeletal sites (lumbar spine).

The rate of bone loss over time seems to be genetically controlled only at axial sites, whereas at skeletal sites, which are rich in cortical bone, genetic factors do not have a significant effect (Kelly *et al.*, 1993). The effect of genetic influences on bone loss is reflected in the strong genetic effect on serum markers for bone formation (osteocalcin, bone specific alkaline phosphatase) and to a lesser extent on urinary bone resorption markers (Kelly *et al.*, 1991). Linkage analysis studies have not been commonly used for the identification of genes associated with bone density. In 22 members of a family with a phenotype of high spinal bone density, evidence was found for linkage of this trait to chromosome 11, near chromosomal marker D11S987 (Johnson *et al.*, 1997). Other polymorphisms in the Vitamin D3 receptor (VDR) and collagen type I-alpha-I genes have been associated with low bone density – none of them resides on the chromosome mentioned above. Genetic markers are often based on restriction fragment length polymorphisms (RFLPs) caused by random mutations in DNA that lead to variations in specific endonuclease cut sites. A search (Ferrari *et al.*, 1999) of polymorphic gene markers for osteoporosis led to the identification of several candidates for the genetic determination of peak bone mass and rate of bone loss, such as the VDR-3′ and 5′ genotypes, the estrogen receptor (ER) genotypes and the collagen type I-alpha-1 gene (COL1A1) polymorphisms (Table 41.4).

Vitamin D3 ($1,25(OH)_2D_3$) is an important hormonal regulator of bone and mineral metabolism, in which the VDR mediates biological actions. RFLPs of VDR are common (Cooper *et al.*, 1996). If these polymorphisms could influence the level or function of the VDR, they could have an important effect on mineral metabolism and BMD. A study on *Bsm*I VDR polymorphism was carried out in Australia on 250 Caucasian twins, aged 17–70 years (Morrison *et al.*, 1994). They found that postmenopausal women with the *BB* genotype (where *B* represents the absence of the *Bsm*I restriction

Table 41.4 Frequency of some allelic polymorphisms in candidate genes associated with osteoporotic risk

Chromosomal location	*Gene symbol*	*Markers*	*Genotype frequency*	
			White	*Asian*
12q13–q14	VDR-3′ end	*BSMI*	BB *18%*, Bb *46%* bb *36%*	BB 2%, Bb *21%*, bb *77%*
		TaqI	TT *36%*, Tt *48%*, tt *16%*	TT *78%*, Tt *21%*, tt *1%*
17q21.3–q22	COL1A1	*MscI*	SS *67%*, Ss *30%*, ss *3%*	*Absent (Korean)*
7q21.3–q22	COL1A2	*PvuI*	PP *14%*, Pp *47%*, pp *40%*	*ND*
6q25.1	ER	*XbaI*	XX *10%*, Xx *48%*, xx *42%*	XX *5%*, Xx *31%*, xx *64%*

Polymorphic markers are designated as endonuclease restriction sites. ND – not determined.

site and *b* represents the presence of the cut site) would reach the BMD "fracture threshold" 10 years earlier than the *bb* genotypes. This greater decline in BMD in the *BB* genotype group could significantly increase their risk of bone fracture. Recent linkage analyses found no evidence to support a relationship between VDR genotype and peak bone mass at the spine or femoral neck in either Caucasians or African-Americans (Hustmyer *et al*, 1994; Peacock, 1995; Uitterlingen, 1996).

The phenomenon that estrogen deficiency in postmenopausal women is associated with increased bone turnover and acceleration of bone loss leading to an increased susceptibility to bone fractures is well known (Wood *et al.*, 1998). Estrogen receptors are represented in osteoclasts, and osteoblasts, which raises its possible direct effect on bone formation. Mutations in the estrogen receptor in mice as well in human beings are associated with bone loss (Sano *et al.*, 1995).

Collagen type I is the major protein of bone encoded by COL1A1 and COL1A2 genes. G-to-T polymorphism was reported by Grant *et al.*, (1996) in the promoter region of COL1A1 at a recognition site for the transcription factor Spl, which is related to bone mass and osteoporotic fracture. In two populations of British women, G/T heterozygotes at the polymorphic Spl (*Ss*) site had significantly lower BMD than did G/G homozygotes (*SS*), and BMD was still lower in T/T homozygotes (*ss*). For a marker of prediction of the risk of developing osteoporosis it is noteable that the *Ss* and *ss* genotypes were over-represented in patients with severe osteoporosis and vertebral fractures, compared with controls.

Other candidate genes, which associate with osteoporotic risk are the α_2-HS-glycoprotein (AHSG) gene, interleukin-6 (IL-6) gene, and apolipoprotein E(4) (Apo E) gene. AHSG is a noncollagenous protein of bone matrix, which may take part in the early stages of bone development. IL-6 influences bone metabolism by regulating osteoclast and osteoblast function, and may mediate the effects of sex hormone deficiency on bone.

Shiraki *et al.* (1997) examined the relationship between phenotypes of Apo E and BMD in postmenopausal Japanese women. In their study they found a significant gene-dose effect from the Apo E4 allele on BMD of the lumbar spine and total body. The group representing the negative homozygotes for Apo E allele (Apo E4 -/-) had the highest BMD. Further research is needed to find the real relationship between genetic components and bone mass.

41.3 Diagnosis of osteoporosis

Osteoporosis is a disease characterized by low bone mass and microarchitectural deterioration of bone tissue, leading to enhanced bone fragility and a consequent increase in fracture risk. The operational definition of osteoporosis considers the disease as a value for bone mineral density (BMD) or bone mineral content (BMC) 2.5 SD or more below the young adult mean (WHO Study Group, 1994). BMD or BMC can easily and accurately be evaluated by different types of bone densitometry.

Low BMD, however, is not identical with the diagnosis of osteoporosis. Further comprehensive investigations are needed to clarify the nature of the underlying process. In the case of decreased BMD the aim of the medical evaluation is to exclude other metabolic bone diseases, like osteomalacia, hyperparathyroidism or Paget's disease of bone, that can mimic osteoporosis.

The exact diagnosis can usually be made by clinical history-taking, by routine (Table 41.5) and specific (PTH, 25 OHD, etc.) biochemical tests, by X-rays and sometimes by histology.

The next steps are to separate primary and secondary forms (Table 41.6), to separate postmenopausal and senile types of primary osteoporosis and to provide baseline measurements to evaluate the effectiveness of subsequent treatment (Poór *et al.*, 1999).

41.4 Drug treatment of osteoporosis

A number of therapeutic agents have actions on bone and are used in the treatment of osteoporosis. They are classified into agents which predominantly reduce bone resorption and bone loss via blockage of osteoclasts, and those which stimulate bone formation through osteoblasts thereby increasing bone mass (WHO Study Group, 2000). Treatments for osteoporosis need to be taken for long periods of time in order to have a significant effect on bone mass.

Estrogens have profound antiresorptive action on bone cells via interleukin-6 and other local factors, and estrogen

Table 41.5 Routine laboratory characteristics of metabolic bone diseases

Metabolic bone diseases	*Serum calcium*	*Serum phosphate*	*Urinary calcium*	*Urinary phosphate*	*Alkaline phosphatase*	*TRP (%)*
Osteoporosis	N	N	N	N	N	N
Osteomalacia	N/↓	↓	↓	↓/N	↑	N
Primary hyperparathyroidism	↑	↓	↑	↑	↑	↓
Paget's disease of bone	N	N	N	N	↑	N

TRP = tubular reabsorption of phosphate; N = normal; ↑ = increased; ↓ = decreased.

Table 41.6 Common conditions leading to osteoporosis

- Medical disorders
  - Hyperthyroidism
  - Rheumatoid arthritis
  - Malabsorption due to gastrointestinal and hepatic diseases
  - Chronic renal failure
  - Chronic neurologic disorders
  - Chronic obstructive lung diseases
  - Type I diabetes mellitus
  - Primary hypogonadism
  - Prolactinoma
  - Hypercortisolism
  - Post-transplantation
  - Mastocystosis
  - Anorexia nervosa
- Drugs
  - Prolonged glucocorticoid therapy
  - Excessive thyroid therapy
  - Chemotherapy
  - Anticoagulants
  - Anticonvulsants
  - Gonadotropin-releasing hormone agonists or antagonists

Table 41.7 Molecules that can be used in the present and future treatment of osteoporosis

Inhibitors of bone resorption	*Stimulators of bone formation*
Estrogens	Monofluorophosphate and other organic fluorides
Raloxifen and other selective estrogen receptor modulators	Parathyroid peptide hormone
Tibolone	Vitamin D and active forms
Ipriflavone	Androgens and anabolic steroids
Alendronate, risedronate and other bisphosphonates	Selective Vitamin D receptor modulators
Calcitonins	Bone morphogenic proteins and TGFβ family
Calcium	Statins
Osteoprotegerin and other local factors	

replacement therapy (ERT) is one of the major methods in preventing and treating osteoporosis at the time of menopause. Observational studies have shown an antifracture effect of ERT as well. It can help reduce menopausal symptoms and the risk of colon cancer, however the effect on coronary heart disease and Alzheimer's disease remains controversial. Estrogens may increase the risk of endometrial and breast cancers altrough results are inconsistent on this matter. Recent comprehensive data proved that due to the increased risk of breast cancer as well as cardio- and cerebrovascular events overall health risk exceed benefits from long-term use of combined estrogen plus progestin therapy (Writing Group for the Women's Health Initiative Investigators, 2002). Controlled clinical trials demonstrated that raloxifene, a selective estrogen-receptor modulator (SERM), prevents bone loss and vertebral fractures, and does not appear to have any stimulatory effect on the breast or on the endometrium. Tibolone is a synthetic tissue specific compound with the effect of estrogen, progestogen and androgen, while ipriflavone belongs to the phytoestrogens.

Bisphosphonates are the most effective agents for both prevention and therapy of osteoporosis. Like estrogens, it inhibits bone resorption through decrease of activity and survival of osteoclasts. Inhibition of the mevalonate pathway and protein prenylation represent the main underlying mechanism. Well designed and executed multicenter, randomized, controlled trials have proven the efficacy of alendronate and risedronate in the prevention of hip and vertebral fractures. Alendronate and other bisphosphonates are poorly absorbed and upper gastrointestinal side effects have been reported. The antiosteoclastic and bone mass increasing effect of calcitonins have been used in the therapy of osteoporosis for many years, however, the fracture-prevention effectiveness of the drug has not been well documented.

The administration of low dose, intermittent parathyroid hormone, organic fluoride, vitamin D compounds or androgens/anabolic steroids deserve special mention because these drugs stimulate bone formation rather than inhibiting bone loss. Adequate calcium intake and exercise are important adjuncts to other therapies, but alone do not prevent osteoporosis.

It is reasonable to anticipate five to ten new therapies for osteoporosis becoming available over the next ten years. A number of new bisphosphonates and selective estrogen-receptor modulators, the administration of different local factors, like osteoprotegerin or bone morphogenic proteins, vitamin D receptor modulators, statins and other new approaches may offer alternatives to the drugs currently used in the treatment of osteoporosis.

References

1 Aubin, J.E. and Bonnelye, E. (2000) Osteoprotegerin and its ligand: a new paradigm for regulation of osteoclastogenesis and bone resorption. *Medscape Women's Health 5,(2):4.*

2 Cockerill, F.J, Hawa, N.S., Yousaf, N., Hewison, M., O'Riordan, J.L. and Farrow, S.W. (1997) Mutations in the vitamin D receptor gene in three kindreds associated with hereditary vitamin D resistant rickets. *J. Clin. Endokrinol. Metab. 82, 3156–3160.*

3 Cooper, C., Atkinson, E.J., Jacobsen, S.J., O'Fallon, W.M., Melton, L.J.III. (1993) Population-based study of survival following osteoporotic fractures. *Am J Epidemiol 137, 1001–1005.*

4 Cooper, C., Campion, G., Melton, L.J.III. (1992) Hip fractures in the elderly: a worldwide projection. *Osteoporosis Int 2, 285–289.*

5 Cooper, G.S. and Umbach, D.M. (1996) Are vitamin D receptor polymorphisms associated with bone mineral density? A meta-analysis. *J. Bone. Miner. Res. 11, 1841–1849.*

6 Danielson, M.E., Cauley, J.A., Zmuda, J.M., Ngo, D., Charron, M. and Heaney, R.P. (1977) Fractional calcium absorption is correlated in mothers and daughters. *J. Bone. Miner. Res. 12,S491.*

7 Dequeker, J., Nijs, J., Verstraeten, A., Geusens, P. and Gevers, G. (1987) Genetic determinants of bone mineral content at the spine and radius: a twin study. *Bone 8,207–209.*

8 Eisman, J.A. (1999) Genetics of osteoporosis. *Endocr. Rev. 20, 788–804.*

9 Ferrari, S., Rizzoli, R. and Bonjour, J.P. (1999) Genetic aspects of osteoporosis. *Curr. Opin. Rheumatol. 11, 294–300.*

10 Geesing, R.G.T., Hoefnagels, N.H.M. and Bulstra, S.K. (1999) Osteogenic activity of OP-1, bone morphogenic protein-7 (BMP-7), in a human fibular defect model. *J. Bone Joint Surg. Br. 81, 710–718.*

11 Grant, S.F., Reid, D.M. Blake, G., Herd, R., Fogelman, I. and Ralston, S.H. (1996) Reduced bone density and osteoporosis associated with a polymorphic SpI binding site in the collagen type I alpha 1 gene. *Nat. Genet. 14, 203–205.*

12 Groeneweld, E.H.J and Burger, E.H. (2000) Bone morphogenetic proteins in human bone regeneration. *Eur. J. Endocrinol. 142, 9–21.*

13 Gueguen, R., Jouanny, P., Guillemin, F., Kuntz, C., Pourel, J. and Siest, G. (1995) Segregation analysis and variance components analysis of bone mineral density in healthy families. *J. Bone. Miner. Res. 10, 2017–2022.*

14 Hofbauer, L.C. (1999) Osteoprotegerin ligand and osteoprotegerin: novel implications for osteoclast biology and bone metabolism. *Eur. J. Endocrinol. 141, 195–210.*

15 Hustmyer, F.G., Peacock, M., Hui, S., Johnston, C.C., Christians, J (1994) Bone mineral density in relation to polymorphism at the Vitamin D receptor gene locus. *J. Clin Invest 94, 2130–2134.*

16 Johnson, M.L., Gong, G., Kimberling, W., Recker, S.M., Kimmel, D.B. and Recker R.B. (1997) Linkage of a gene causing high bone mass to human chromosome 11. *Am. J. Hum. Genet. 60, 1326–1332.*

17 Kanis, J.A., Melton, L.J.III., Christiansen, C., Johnston, C.C., Khaltaev, N. (1994) Perspective: the diagnosis of osteoporosis. *J Bone Min Res 9, 1137–1141.*

18 Kelly, P.J., Hopper, J.L. Macaskill, G.T., Pocock, N.A. Sambrook, P.N. and Eisman, J. (1991) Genetic factors in bone turnover. *J. Clin. Endocrinol. Metab. 72, 808–813.*

19 Kelly, P.J., Nguyen, T., Hopper, J.L., Pocock, N.A., Sambrook, P.N. and Eisman, J. (1993) Changes in axial bone density with age: a twin study. *J. Bone. Miner. Res. 8, 11–17.*

20 Kuivaniemi, H., Tromp, G. and Prockop, D.J. (1991) Mutations in collagen genes: causes of rare and some common diseases in humans. *FASEB J. 5, 2052–2060.*

21 Kundu, B., Khare, S.K. and Singh, G. (1999) Role of polypeptides in the treatment and diagnosis of osteoporosis. *Peptides 20, 523–537.*

22 Looker, A.C., Johnston,C.C.Jr., Wahner, H.W., Dunn, W.L., Calvo, M.S., Harris, T.B., Heyse, S.P., Lindsay, R.L. (1995) Prevalence of low femoral bone density on older US women from NHANES III. *J Bone Min Res 10, 796–802.*

23 Lutz, J. and Tesar, R. (1990) Mother-aughter pairs: spinal and femoral bone density and dietary intakes. *Am. J. Clin. Nutr. 52, 872–877.*

24 Meier, C.A. (1998) Role of novel antiresorptive agents for the prevention and treatment of osteoporosis. *Eur. J. Endocrinol. 139, 18–19.*

25 Melton, L.J. III., (1995) Perspective: how many women have osteoporosis now? *J Bone Min Res 10, 175–177.*

26 Morrison, N.A., Qi, J.C., Tokita, A., Kelly, P., Crofts, L., Nguyen, T., Sambrook, P.N. and Eisman, J.A. (1994) Prediction of bone density from vitamin D receptor alleles. *Nature 367, 284–287.*

27 O'Neill, T.W., Felsenberg, D., Varlow, J. *et al.* (1996) The prevalence of vertebral deformity in European men and women: The European vertebral osteoporosis study. *J Bone Miner Res 11, 1010–1018.*

28 Padgett, R.W., Cho, S.H. and Evangelista, C. (1998) Smads are the central component in transforming growth factor-beta signaling. *Pharmacol. Ther. 78, 47–52.*

29 Peacock, M. (1995) Vitamin D receptor gene alleles and osteoporosis: A contrasting view. *J Bone Miner Res 10, 1294–1297.*

30 Pocock, N.A., Eisman, J.A., Hopper, J.L., Yeates, M.G., Sambrook, P.N. and Ebert, S. (1987) Genetic determinants of bone mass in adults. A twin study. *J Clin Invest 80, 706–10.*

31 Pollitzer, W.S. and Anderson, J.J.B. (1989) Ethnic and genetic differences in bone mass: a review with a hereditary vs environmental perspective. *Am J Clin Nutr 50, 1244–1259.*

32 Poór, Gy., Jacobsen, S.J., Melton, L.J.III (1994) Mortality following hip fracture. In: Vellas, B.J., Albarade, J.L., Garry, P.I. (eds) *Epidemiology and aging. Facts and Research in Gerontology*. Serdi Publisher, Paris, pp. 91–109.

33 Poór, Gy., Atkinson., E.J., Lewallen, D.G., O'Fallon, W.M., Melton, L.J.III. (1995) Age-related hip fractures in men: clinical spectrum and short-term outcomes. *Osteoporosis Int 5, 419–426.*

34 Poór, Gy., Szathmári, M. (1999) Managing the patient with osteoporosis. *Rheumatology in Europe 28, 102–105.*

35 Praemer, A., Furner, S., Rice, D.P. (1992) Musculoskeletal conditions in the United States. In: Rosemont, I.L. (ed) *American Academy of Orthopaedic Surgeons*, 145–170.

36 Ripamonti, U., Van Den Heever, B., Sampath, T.K., Tucker, M.M., Rueger, D.C. and Reddi, A.H. (1996) Complete regeneration of bone in the baboon by recombinant human osteogenic protein-1 (hOP-1, bone morphogenetic protein-7). *Growth Factors. 13, 273–89.*

37 Sano, M., Inoue, S., Hosoi, T., Ouchi, Y., Emi, M.,Shiraki, M. and Orimo, H. (1995) Association of estrogen receptor dinucleotide repeat polymorphism with osteoporosis. *Biochem. Biophys. Res. Commun. 217, 378–383.*

38 Seeman, E., Hopper, J.L., Bach, L.A., Cooper, M.E., Parkinson, E., McKay, J. and Jerums, G. (1989) Reduced bone mass in daughters of women with osteoporosis. *N. Engl. J. Med. 320, 554–558.*

39 Shiraki, M., Shiraki, Y., Aoki, C., Hosoi, T., Inoue, S., Kaneki, M. and Ouchi, Y. (1997) Association of bone mineral density with apolipoprotein E phanotype. *J. Bone Miner. Res. 12, 1438–1445.*

40 Stevenson, S., Cunningham, N., Toth, J., Daavy, D. and Reddi, A.H. (1994) The effect of osteogenin (a bone morphogenic protein) on the formation of bone in orthotopic segmental defects in rats. *J. Bone Joint Surg. Am. 76, 1676–1687.*

41 Uitterlingen, A.G., Pols, H.A.P., Burger, H. (1996) A large scale population-based study of the association of Vitamin D receptro gene polymorphismus with bone mineral density. *J Bone Miner Res, 11, 1241–1248.*

42 WHO Study Group (1994) Assessment of fracture risk and iths application to screening for postmenopausal osteoporosis. WHO Technical Report Series, 843, Geneva

43 Wood, R.J. and Fleet, J.C. (1998) The genetics of osteoporosis: vitamin D receptor polymorphisms. *Annu. Rev. Nutr. 18, 233–58.*

44 Wozney, J.M., Rosen, V., Celeste, A.J., Mitsock, L.M., Whitters, M.J., Kriz, R.W., Hewick, R.M. and Wang, E.A. (1988) Novel regulators of bone formation: molecular clones and activities. *Science 242, 1528–1534.*

45 WHO Study Group (1994) Assessment of fracture risk and its application to screening for postmenopausal osteoporosis. WHO Technical Report Series. 843. Geneva.

46 WHO Study Group (in press) *WHO strategy for the prevention and management of osteoporosis.* WHO Technical Report Series, Geneva.

47 Writing Group for the Women's Health Initiative Investigators (2002) Risk and benefits of estrogen plus progestin in healthy postmenopausal women. *Jama 288, 321–323.*

Part 10

Drug applications

Chapter 42

Modulation of drug effects by ABC transporters

Gergely Szakács and Balázs Sarkadi

Contents

42.1 Introduction

Cancer is the second leading cause of death in developed countries, exceeded only by heart disease. About half a million people die of cancer every year in the USA alone, that is more than 1,500 people a day. Chemotherapy is a major weapon in our fight against malignant diseases. Still, in numerous cases even combination chemotherapy is ineffective: the tumor cells may first respond but in a relapse become resistant to multiple agents. There are several forms of cancer drug resistance, based on the activation of specific drug metabolizing enzymes or defense mechanisms, protecting the cellular targets of anticancer drug action. These include, e.g., decreased levels and/or mutations in the topoisomerase protein, enhanced DNA repair, increased activity of dehydrofolate reductase, reduced methotrexate influx, or even a combination of mechanisms, preventing drug-induced apoptosis of the tumor cells.

42.1.1 *Multidrug resistance*

The phenomenon called multidrug resistance (MDR) is based on an ATP-dependent active extrusion of multiple chemotherapeutic agents from tumor cells. This chapter deals exclusively with the function of a group of membrane proteins which cause multidrug resistance in cancer by keeping the intracellular concentration of cyototoxic agents below a cell-killing threshold. These multidrug transporter proteins belong to the superfamily of ATP Binding Cassette (ABC) transporters, present from bacteria to man. ABC proteins may cause cancer drug resistance, however, they also play a major role in general pharmacology, that is in modulating the absorption, distribution and excretion of numerous pharmacological compounds. In this chapter we discuss the role of ABC transporters in cancer drug resistance, in the protection of our body against xenobiotics, and in modulating a variety of pharmacological responses.

42.2 Identification of membrane proteins that maintain an ATP-dependent efflux of cytologic compounds: the multidrug resistance (MDR) proteins

Upon the introduction of several effective cytotoxic agents into cancer chemotherapy in the 1960s and 70s, initial enthusiasm was severely reduced by the recognition of the clinical MDR phenomenon. Unresponsiveness of various tumors, even to combination chemotherapy, prompted major research effort in this regard. The first successful attempts to identify the mechanisms and molecules responsible for cancer MDR made use of *in vitro* model systems. In the early 1970s, when tumor cells were cultured in increasing concentrations of cytotoxic drugs, these cells became resistant not only to the selecting agent, but also to other structurally and functionally unrelated drugs.

42.2.1 P-glycoprotein – MDR1

Cancer cell cross resistance to drugs, in some cases reaching extreme levels, followed the pattern of clinical drug resistance: it included anthracyclines, vinca-alkaloids and epipodophyllotoxins, i.e. the majority of compounds applied in chemotherapy. Biochemical examination of the resistant cells showed that MDR was correlated with increased expression of a large, 170 kDa membrane-bound glycoprotein. The first hints regarding the function of the resistance glycoprotein came from studies which showed that the expression of this protein was associated with reduced cellular drug accumulation. It was also shown that cellular ATP depletion abolished this phenotype, and based on these observations, the 170 kDa membrane protein was named P- (permeability)glycoprotein (P-gp, MDR1). A major finding was that the addition of unrelated hydrophobic compounds, e.g. verapamil, a calcium channel blocker, could reverse tumor cell drug resistance in P-gp/MDR1 expressing cells both *in vitro* and *in vivo*. Cloning of P-gp/MDR1 was achieved in the mid 1980s, and conclusive evidence for the role of this protein in multiple drug resistance came from experiments demonstrating that its cDNA conferred resistance upon transfection to drug-sensitive cells. Sequence comparisons indicated that P-gp/MDR1 shares homology with certain components of bacterial active transport systems. All these observations strongly indicated that an energy-dependent active transport mechanism was responsible for decreased drug accumulation in the cancer cells, and this efflux is maintained by P-glycoprotein. Studies in the early 1990s on heterologously expressed, as well as on purified and reconstituted MDR1, showed that P-gp/MDR1 is indeed an ATP-driven pump, characterized by drug-stimulated ATPase activity, and this protein is sufficient to promote the efficient export of drugs.

The understanding of the basic molecular mechanism of MDR prompted research attempting to predict the success or failure of therapy, as well as to develop chemosensitizing agents for MDR reversal. Analysis of clinical samples by using monoclonal antibodies and other assays supported the assumption that the clinically refractory state of several tumors were indeed associated with increased expression of P-glycoprotein. At the same time, numerous negative results in this field strongly questioned the exclusive role of P-gp/MDR1 in transport-related cancer multidrug resistance.

42.2.2 Cancer MDR associated with other ABC transporter proteins

After cloning and characterization of P-gp/MDR1 it soon became evident that other efflux-pumps may also play significant roles in transport-associated drug resistance. At the time of writing this chapter, non-P-gp/MDR1 mediated clinical resistance is mainly attributed to MRPs (Multidrug Resistance-associated Proteins) and BCRP (see Figure 42.1). MRP1 was identified in 1992 in a drug-resistant human lung cancer cell line, which did not overexpress P-glycoprotein. We now have a good deal of evidence indicating the presence and function of MRP1 in several cancer cell types, including leukemia and solid tumors, which have the enigmatic MDR phenotype without a detectable P-gp/MDR1 expression. Moreover, combined MRP1 and P-gp/MDR1 expression has been revealed in several cancer types. In addition to MRP1, several MRP homologs have been cloned. Overexpression of MRP2 was definitely shown to confer cancer MDR, while the role of the other homologs in this regard remains to be clarified. It was only recently revealed

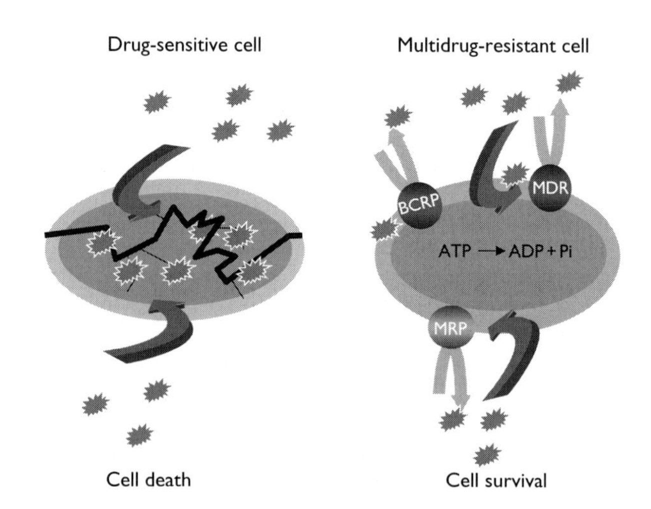

Figure 42.1 MDR proteins confer resistance by an energy-dependent extrusion of cytotoxic drugs from cancer cells.

that another ABC transporter may be a significant player in cancer drug resistance. This protein, the BCRP or Mitoxantrone Resistance Protein (MXR) was found in anthracycline-resistant breast cancer cells that were selected with anthracyclines in the presence of P-gp/MDR1 inhibitors, e.g. verapamil. It has been demonstrated that overexpression of BCRP/MXR is present in clinically drug-resistant tumor samples.

42.3 MDR proteins belong to the superfamily of ABC transporters

The primary structure of MDR proteins shows that they belong to the superfamily of ATP binding Cassette (ABC) proteins. ABC transporters have been identified in prokaryotic as well as eukaryotic organisms. In bacteria, they constitute the largest protein family, in some species encoding from 20 to more than 75 such proteins per genome (ABC transporters account for at least 5% of all proteins in *Escherichia coli*). LmrA, an ABC transporter providing antibiotic resistance in *Lactococcus lactis*, was expressed in lung fibroblast cells and was shown to be able to confer the same pattern of resistance as MDR1. Thus, human MDR proteins can be considered close relatives of ancient proteins which promote the transport of solutes through a lipid bilayer against a concentration gradient. Due to the progress of the Human Genome Project and other mass sequencing efforts, several human ABC proteins have been discovered over the past years. The estimated total number of the human ABC transporters is about 50 (see http://nutrigene.4t.com/humanabc.htm).

42.3.1 *Basic features of human ABC transporters*

Selected members of the human ABC transporter family, with established medical importance, are summarized in Table 42.1. Some of these proteins either directly form (e.g. CFTR) or regulate (e.g. SUR1) plasma membrane ion channels, and are modulated by ATP binding and cleavage. Others, as detailed below, work as ATP-dependent active transporters of various substrates.

Among the human ABC transporters, mutations in the ABCA1 gene result in Tangier disease, a rare genetic disorder characterized by an abnormal lipoprotein profile and the accumulation of cholesterol esters in various tissues. According to recent studies, ABCA1 is probably responsible for cholesterol transport through the cell membrane, thus providing protection for the artery walls against unwanted lipid deposition. ABCR is expressed in the retinal rod photoreceptors, and was found to be mutated in Stargardt disease (STGD) and in several other forms of macular dystrophy. The human multidrug transporter P-gp/MDR1 and its close relatives, MDR3 and sisterPgP, are ATP-dependent transporters of hydrophobic drugs, phospholipids, and bile acids, respectively. The TAP1–TAP2 ABC transporter complex is involved in endoplasmic reticulum peptide transport, required for MHC-I-dependent antigen presentation. The MRP subfamily currently consists of 12 members. The two relatively well characterized members of this group, MRP1 and MRP2, have been shown to actively transport both hydrophobic drugs and organic anions, e.g. glutathione and glucuronide conjugates. The mutations of CFTR (Cystic Fibrosis Transmembrane Conductance Regulator) is causative in cystic fibrosis disease, and the SulfonylUrea Receptor protein, SUR1, serves as an ATP sensitive regulator of a potassium channel in pancreatic B-cells as a critical element in the regulation of glucose-induced insulin secretion. ALD-protein related adrenoleukodystrophy is a severe X-linked neurodegenerative disorder, which is characterized by the accumulation of unbranched saturated fatty acids in the central nervous system.

Table 42.1 Medically important human ABC transporters

Subfamily	*Protein*	*Medical relevance*
ABCA	ABCA1(ABC1)	Tangier disease – lipid metabolism
	ABCA4(ABCR)	Stargard macular distrophy
ABCB	P-gp/MDR1	Multidrug resistance in cancer
	TAP1	Antigen presentation (MHCI)
	TAP2	Antigen presentation (MHCI)
	MDR2/3	Progressive familial intrahepatic cholestasis 3
	ABCB7	X-linked sideroblastic anemia with ataxia
	Sister P-gp, BSEP	Progressive familial intrahepatic cholestasis 2
ABCC	MRP1	Multidrug resistance in cancer
	MRP2	Dubin-Johnson syndrome, Multidrug resistance in cancer
	CFTR	Cystic fibrosis
	SUR1	Familial persistent hyperinsulinemic hypoglycemia of infancy
ABCD	ALD	Adrenoleuko-dystrophy
ABCG	BCRP	Multidrug resistance in cancer

42.3.2 *Building blocks of human ABC transporters*

Despite of their divergent functions, ABC transporters show a high level of sequence homology, with conserved motifs. The common molecular architecture consists of a four-domain arrangement, with two transmembrane domains (TMD) and two ATP binding cassettes (ABCs).

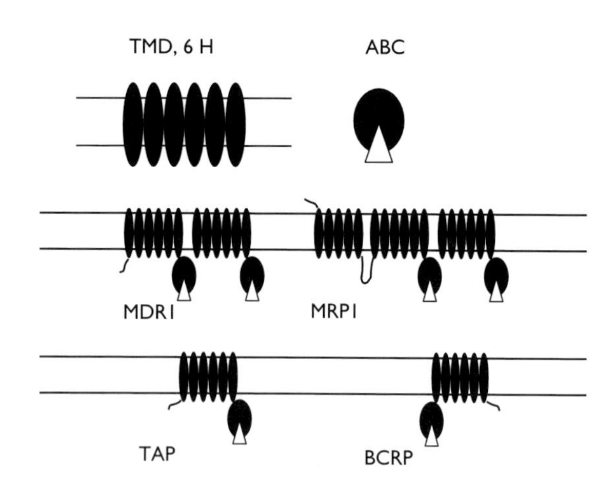

Figure 42.2 Building blocks of human ABC transporters.

In the ABC transporter proteins the characteristic membrane-embedded TMDs are usually composed of six transmembrane (TM) helices. The ABC unit, a 200–250 amino acid "mini" protein, harbors two short, conserved peptide motifs (Walker A and Walker B). A third conserved sequence, called the "ABC signature" motif, diagnostic to the entire superfamily, is located between the Walker A and B sequences.

The minimal structural requirement for an active ABC transporter seems to be the combination of two TMD and two ABC units. These may be present within one polypeptide chain (ABCA1, MDR1, MRP1) or half-ABC transporters may form dimers. A single BCRP chain contains only half of the MDR1-like structure, moreover, the order of the TMD and ABC domains is reversed. This protein probably works as a homodimer, while other half-ABC transporters (e.g. the TAP1/TAP2 complex) are believed to function as heterodimers. In MRP1, the core structure is extended with an N-terminal, 5-helix transmembrane domain, and a connecting cytoplasmic loop.

42.3.3 *Molecular mechanism of human ABC transporters*

It is generally accepted that ABC transporters, sharing homologous domain arrangements, have common functional features. The exact molecular mechanism of the transport process is currently unknown. Still, we have some important basic information concerning the mechanism of ATP binding and cleavage in these proteins. In the case of the active transporters the specificity of recognizing the transported substrates is presumably encoded in the TM domains, whereas the ABC domains fuel the actual transport. In the entire protein ATPase activity is coupled to substrate movement, and substrates significantly enhance the ATPase activity through a tight molecular "coupling" of the TMDs to the ABCs. The two ABC domains show a positive cooperativity, and during ATP hydrolysis they form a transition state intermediate. This intermediate does not involve covalent phosphorylation of the enzyme (as in the P-type ATPases), but is characterized by an occluded form of the ADP moiety. The formation of the transition state intermediate is also greatly stimulated by the transported drug substrates, and cooperativity between the two ABC domains is a requirement for this partial reaction as well.

At present, there is no high-resolution three-dimensional structure available for any of the mammalian ABC transporters, nor for their separate ABC domains. As low resolution structural analysis yields only limited information, models of the function of MDR proteins are based on biochemical approaches. The use of labeled drug and ATP analogs have resulted in several models describing how the drugs are recognized and passed through the lipid bilayer. The technique of site-directed mutagenesis, where amino acids are altered individually, has revealed residues that are important in substrate recognition, ATP binding or intramolecular signaling.

Perhaps the most intriguing aspect of the MDR transporters is their unique ability to transport a wide variety of substrates. Known substrates for the MDR transporters represent structurally and functionally diverse compounds, many of which are of natural origin (see below). In the case of MDR1, the main determinant of specificity seems to be the ability of transported drugs to enter the lipid phase. Recognition of substrates can take place within the membrane or in the cytosolic compartment. In the former case, drugs may be transported directly from the inner to the outer leaflet of the membrane (flippase model), or, in the "hydrophobic vacuum cleaner" model, MDR1 may remove drugs from the lipid bilayer to the extracellular compartment. Alternatively, drugs may pass through a pore (presumably formed by the TMDs) from the cytosolic compartment (see Figure 42.3).

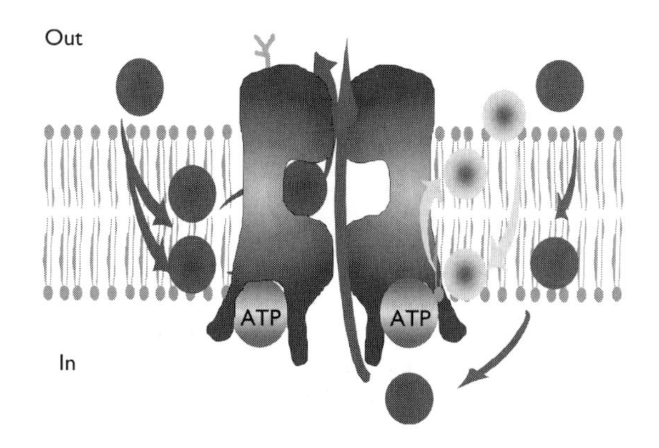

Figure 42.3 Suggested transport modes of the MDR1 protein.

42.4 Cancer therapy and the MDR proteins: development of resistance modifiers

Due to their "promiscuity", MDR proteins transport a large number of cytotoxic compounds (with different cellular targets and mechanisms of action), that are frequently applied in cancer chemotherapy. Both MDR1 and MRP1 are localized mainly in the plasma membrane, and they have overlapping, but distinct transport features. MDR1 carries out the extrusion of large, hydrophobic, uncharged or slightly positively charged molecules. In contrast, MRP1 also transports negatively charged compounds, especially glutathione- or glucuronate conjugates, and may sequester some of its substrates into intracellular vesicles (see Table 42.2).

According to preliminary results, cells overexpressing BCRP show a unique pattern: they are resistant to anthracyclines and mitoxantrone, but remain sensitive to *vinca* alkaloids and cisplatin.

42.4.1 Methods for preventing clinical multidrug resistance

Circumvention of clinical MDR should significantly improve therapeutic response in a large number of tumor patients. One way to achieve this goal would be to develop anticancer agents which do not interact with any of the multidrug transporters. However, as cytotoxic drugs have to penetrate the cell membrane and the MDR proteins have an extremely wide recognition pattern, this seems to be a remote possibility. There are several other suggested methods to prevent the expression or function of multidrug transporters (see below), but pharmacological modulation seems to be the first choice at present.

Agents which inhibit the function of P-gp/MDR1 and/or MRP1, either competitively or non-competitively, are good candidates for such pharmacological modulation. These compounds are expected to increase the cytotoxic action of MDR-related drugs by preventing the extrusion of anticancer drugs from the target cells. In fact, as mentioned earlier, the first demonstration of the possibility of reversal of P-glycoproetin multidrug resistance was achieved by cancer-unrelated hydrophobic pharmacological agents, such as verapamil.

Table 42.2 Major chemotherapeutic compounds transported by the P-gp/MDR1 and the MRP1 proteins

Pgp/MDR1		*MRP1*
Taxol	Anthracyclines (e.g. doxorubicin, daunorubicin, idarubicin, mitoxantrone), vinca-alkaloids (vincristine, vinblastine), epipodophyllotoxins (e.g. etoposide and teniposide), taxanes (e.g. taxol and taxotere), actinomycin D	Cisplatin and metal complexes, alkylating agents (e.g. chlorambucil), methotrexate

42.4.2 Multidrug resistance modifying agents

The co-application of a non-toxic "MDR-modulating" or "MDR-reversing" compound with combination chemotherapy may significantly improve cancer cure rates. Indeed, we can witness a growing interest in the use of such clinically active agents. Biochemical investigations led to the identification of several compounds, as diverse in structure as the transported drugs. The first generation consisted of drugs that were already in clinical use. Calcium channel blockers (Verapamil, Diltiazem, Azidopine), quinine derivatives, calmodulin inhibitors (Trifluoroperazine, Chlorpromazine) or the immunosuppressive agent, Cyclosporin A, were all shown to interact with MDR transporters *in vitro* and *in vivo*. Unfortunately, it soon became evident that the oncological application requires too high a dosage of these compounds, which prevented further clinical testing.

The second generation of drug resistance modulators consisted of derivatives of the above compounds, which had less pronounced effect on their original target, but retained their inhibition of the MDR pumps. Two prominent examples of this group are R-verapamil and PSC-833, the latter being a cyclosporin analog without immunosuppressive effect. Several such compounds, including PSC-833 are currently being tested in major clinical trials. However, we are already aware of the possible differences in the substrate-specificities and inhibitor-sensitivities of the drug resistance protein(s) expressed in different tumor cells, thus therapeutic intervention requires proper diagnosis and more protein-specific modulator agents. As described later, the physiological and pharmacological function of MDR proteins further emphasizes the need for selective blocking agents in a clinical setting.

The third generation of MDR modifiers would be molecules specifically devised to interact with specific MDR transporters. We do not know much about such agents, as they are mostly in the early development phase in various research laboratories, but one example may be the group of hydrophobic peptide derivatives (Reversins), developed in our laboratory. These small hydrophobic molecules interact with P-gp/MDR1 with high affinity and selectivity. Reversins were shown to have a strong inhibitory effect on P-gp/MDR1-mediated drug transport and to eliminate this type of drug resistance without any toxic effect in control cells.

42.4.3 Additional methods to prevent multidrug resistance in cancer

A distinct mode of MDR reversal includes the use of monoclonal antibodies. Several antibodies were reported to inhibit *in vitro* P-gp/MDR1 mediated drug efflux. Since immunoglobulins cannot enter the cells, antibodies with intracellular epitopes are not suitable for *in vivo* inhibition. On the other hand, antibodies that bind to extracellular epitopes of the transporter may provide a supplement to chemical agents for the reversal of MDR in clinical cancer.

Another potential method to eliminate MDR is the use of macromolecular carriers. Conjugation of drugs to various drug carriers has a wide range of application (induction of immune response, production of epitope-specific antibodies, etc). Antitumor agents have been efficiently bound to such carriers. The resulting pro-drug complexes have novel pharmacological properties: they enter cells by specific uptake mechanisms (e.g. endocytosis), which avoid MDR mediated extrusion. The active, toxic form is released intracellularly, following the cleavage of pH sensitive bonds in the lysosomes.

Finally, retrovirus-mediated gene transfer of MDR1 cDNA allows protection of specific target cells. Chemotherapy is often limited by the general toxicity of the applied drugs. As normal bone marrow cells have little or no expression of P-gp/MDR1, they are particularly susceptible. Early results from clinical trials suggest that the hematopoietic cells can be protected. Stem cells from patients with advanced cancer were transduced with an MDR cDNA-containing retrovirus. The patients were undergoing autologous bone marrow transplantation as part of a high-dose chemotherapy treatment. The transduced cells were reinfused, followed by high dose chemotherapy. Although the continuous high-level expression of P-glycoprotein in bone marrow cells remains inefficient, this strategy may prove useful in the protection of sensitive tissues against unwanted toxicity.

42.5 Physiological roles of MDR proteins – effects in general pharmacology

42.5.1 *MDR proteins are present in normal tissues*

Although medical attention focuses primarily on the MDR proteins in cancer cells, their ubiquitous presence in normal tissues suggests a broader impact. High levels of expression of P-gp/MDR1 can be found in the epithelial cells of the lower gastro-intestinal tract, liver, kidney, as well as the capillary endothelia of the blood-brain barrier. Other tissues expressing P-gp/MDR1 include the placenta, early hematopoietic cells and certain immunocompetent cells. MRP1 is found in most cell types and was shown to act as a high affinity, primary active transporter of the cysteinyl leukotriene, LTC4 in mast cells.

The physiological functions of MDR1 and MRP1 were studied most efficiently in knock-out mouse models, disrupting one or more of the MDR genes. Surprisingly, P-gp/MDR1, and even MRP1 knockout mice were viable and

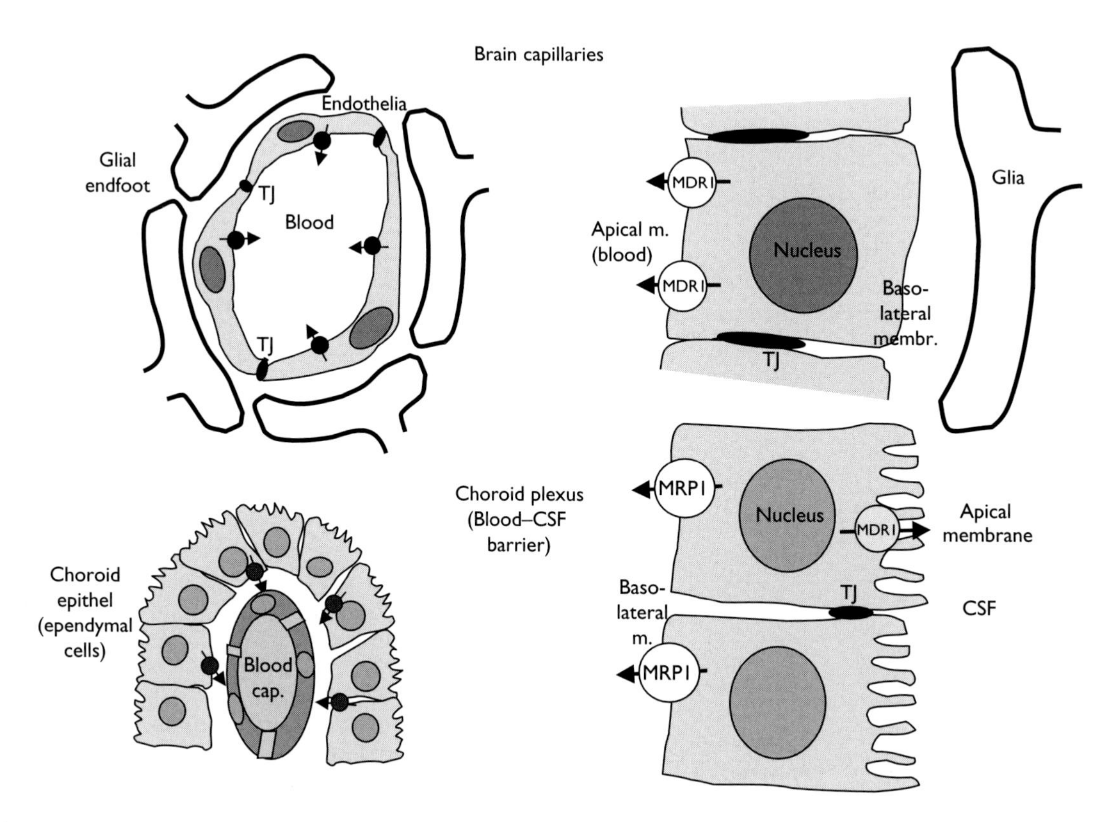

Figure 42.4 Multidrug transporters in the blood-brain barrier.

fertile with no obvious physiological abnormality. These experiments indicated that the lack of both P-gp/MDR1 and MRP1 is compatible with life, supporting the feasibility of pharmacological circumvention of multidrug resistance. However, careful analysis of the knock-out mice showed increased drug absorption, decreased elimination, and an abnormal sensitivity with increased concentration of certain drugs in the brain compartments. These findings are in agreement with the high epithelial-specific expression of Pgp/MDR1 and MRP1 in several tissues, including the blood-brain barrier.

42.5.2 Multidrug transporters in the blood-brain barrier

The blood-brain barrier and the blood-cerebrospinal fluid (CSF) barrier function together in order to isolate the brain from circulating drugs, toxins, and xenobiotics. P-gp/MDR1, expressed in the luminal surface of cerebral capillary endothelial cells, prevents the entry of toxic agents into the brain area. The blood-CSF drug-permeability barrier is primarily localized to the epithelium of the choroid plexus. The barrier function of the choroid plexus is based on the concerted action of both MDR proteins. Here, P-gp/MDR1 localizes subapically, conferring an apical-to-basal trans-epithelial permeation barrier. Conversely, MRP1 localizes basolaterally, conferring an opposing basal-to-apical drug-permeation barrier. Together, these transporters coordinate secretion and reabsorption of natural product substrates and therapeutic drugs (see Figure 42.4).

42.5.3 Multidrug transporters in secretory epithelia

Secretory epithelia in the liver and kidney are predominant areas for the expression of MDR proteins (see Figure 42.5). ABC transporters in the liver have a central role in both the hydrophobic and hydrophilic elimination of endogenous or exogenous compounds. The canalicular membrane of the hepatocytes contains P-gp/MDR1, which mediates the excretion of hydrophobic, mostly cationic metabolites into the bile, while MDR2 in the same membrane compartment is involved in phosphatidylcholine (PC) secretion. MDR2 was shown to work as a phospholipid flippase: it moves PC actively from the inner layer of the canalicular membrane to its outer layer, where a currently unknown extraction mechanism removes this lipid into the bile fluid. The P-gp/MDR1-homolog protein sP-gp is exclusively expressed in the liver epithelial cell apical membrane and is responsible for bile acid secretion.

MRP2, located in the canalicular membrane is responsible for the excretion of conjugated bile salts and other anionic conjugates into the bile – deficiency in MRP2 function causes Dubin-Johnson syndrome. The basolateral membrane of the liver cells contains MRP1, which can extrude glutathione and other anionic conjugates into the blood, probably protecting the liver cells in case of an overproduction of these metabolites. The functions of MRP3–6 (residing mostly in the basolateral membranes) and BCRP (in the apical membranes) in the liver are currently largely unknown.

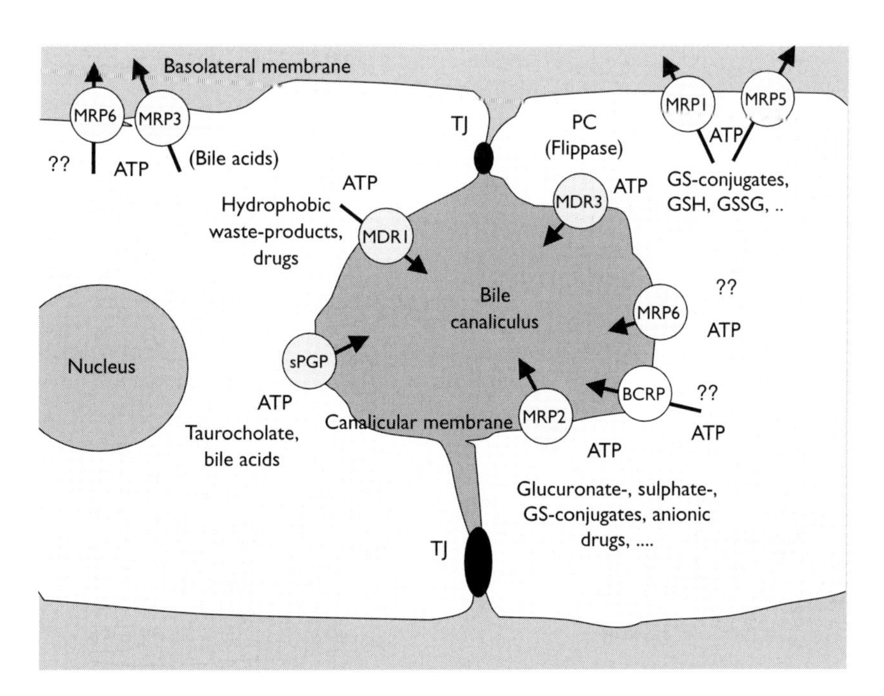

Figure 42.5 Function of the MDR transporters in hepatocytes.

42.5.4 *Additional transport functions of multidrug transporters*

In addition to the widely used chemotherapeutic agents, MDR1 was shown to extrude several other less frequently used drugs, such as protein synthesis inhibitors (puromycin, emetine), antibiotics (valinomycin, gramicidine) and HIV protease inhibitors (Ritonavir, Indinavir, Saquinavir). The pattern of expression in the body suggests that MDR transporters have a major impact on the pharmacokinetic properties of a large number of pharmacological substrates. The presence of P-gp/MDR1 in the gastrointestinal tract influences oral bioavailability and absorption, whereas MRP1 and MRP2 seem to play an important role in the secretion of drug–conjugates and anionic drugs in the liver and kidney. The protection of cells or whole compartments, the penetration of drugs into the brain and fetus are also partially governed by these ABC transporters.

42.6 *Summary*

In the human body the combined action of various ATP-dependent ABC transporters is a major determinant of the desired therapeutic as well as the unfavorable effects of a large number of pharmacological agents. Cancer chemotherapy is untowardly affected by expression of the multidrug resistance transporters (P-gp/MDR1, MRP1–2, and BCRP) in tumor cells, as these membrane proteins actively extrude a wide variety of anticancer drugs. However, both the above proteins and other members of the ABC transporter group, especially by acting in the bowl, liver, kidney, and the blood-brain barrier, have a major role in regulating the absorption, metabolism, and secretion of various organic compounds. In fact, the predominant physiological function of the multidrug transporters is most probably to provide xenobiotic resistance by the elimination of environmental toxins. "Pharmacogenomics", that is the individual adjustment of clinical therapy according to a personal genetic background, will have to be based on a profound knowledge of ABC transporters involved in drug resistance and tissue-dependent drug transport.

Key references

1 Ambudkar, S.V., Dey, S., Hrycyna, C.A., Ramachandra, M., Pastan, I., Gottesman, M.M. (1999) Biochemical, cellular, and pharmacological aspects of the multidrug transporter. *Annu Rev Pharmacol Toxicol, 39: 361–98.*

2 Borst, P. (ed.) (1997) Multidrug Resistant Proteins. *Seminars in Cancer Biology, Vol 8, 3.*

3 Clynes, M. (ed.) (1998) Multiple Resistance in Cancer 2, *Kluwer Acad. Pub.*

4 Higgins, C.F. (1992) ABC transporters-from microorganisms to man. *Annu Rev Cell Biol 8: 67–113.*

5 Marie, J.P., Legrand, O. (1999) MDR1/P-GP expression as a prognostic factor in acute leukemias *Adv Exp Med Biol,457:1–9.*

6 Sarkadi, B., Kuchler, K., Szakács, G. (eds.) (2000) *Structure and Function of ABC Transporters.* BBA Biomembranes, Special Issue, Vol. 1461 No. 2.

7 Schinkel, A.H., Smit, J.J.M., van Tellingen, O., Beijnen, J.H., Wagenaar, E., van Deemter, L., Mol, C.A.A.M., van der Valk, M.A., Robanus-Maandag, E.C., te Riele, H.P.J., Berns, A.J.M., Borst, P. (1994) Disruption of the mouse mdr1a P-glycoprotein gene leads to a deficiency in the blood-brain barrier and to increased sensitivity to drugs. *Cell, 77: 491–502.*

8 Vallabhaneni, V.R., Dahlheimer, J.L., Bardgett, M.E., Snyder, A.Z., Finch, R.A., Sartorelli, A.C. and Piwnica-Worms, D. (1999) Choroid plexus epithelial expression of MDR1 P glycoprotein and multidrug resistance-associated protein contribute to the blood-cerebrospinal-fluid drug-permeability barrier. *PNAS, 96: 3900–3905.*

Chapter 43

Peptide or protein based drug delivery and targeting systems

Ferenc Hudecz, György Kóczán and Judit Reményi

Contents

43.1 Introduction

Biological molecules have realised and potential use as targeting carriers or delivery systems of drugs, toxins, radionuclides, enzymes, oligonucleotides and genes. Such conjugates have been developed to alter the biodistribution and/or bioavailability of the attached bioactive molecule, which might result in decreased toxic side effects, pyrogenicity and/or modified immunogenicity. These constructs are carefully designed to combine the beneficial characteristics of the carrier and of the coupled moiety. Regarding the latter, the following objectives may be achieved:

a altered solubility,
b improved stability against normal biodegradation,
c increased circulation half-life, delayed plasma and/or tissue clearance,
d improved localization at desired sites and
e altered immunogenicity.

The properties of the carrier largely influence the fate of the conjugate, and accordingly will differ from those of free drugs given by the same route.

Peptide/protein carriers can be classified as molecules with or without recognition units (e.g. antibodies, lectins, hormones, transferrin), which enable specific binding to cell surface receptor (Figure 43.1). Delivery with homing device equipped carriers are suitable for "active targeting" of entities attached to well defined acceptor/receptor structures at the

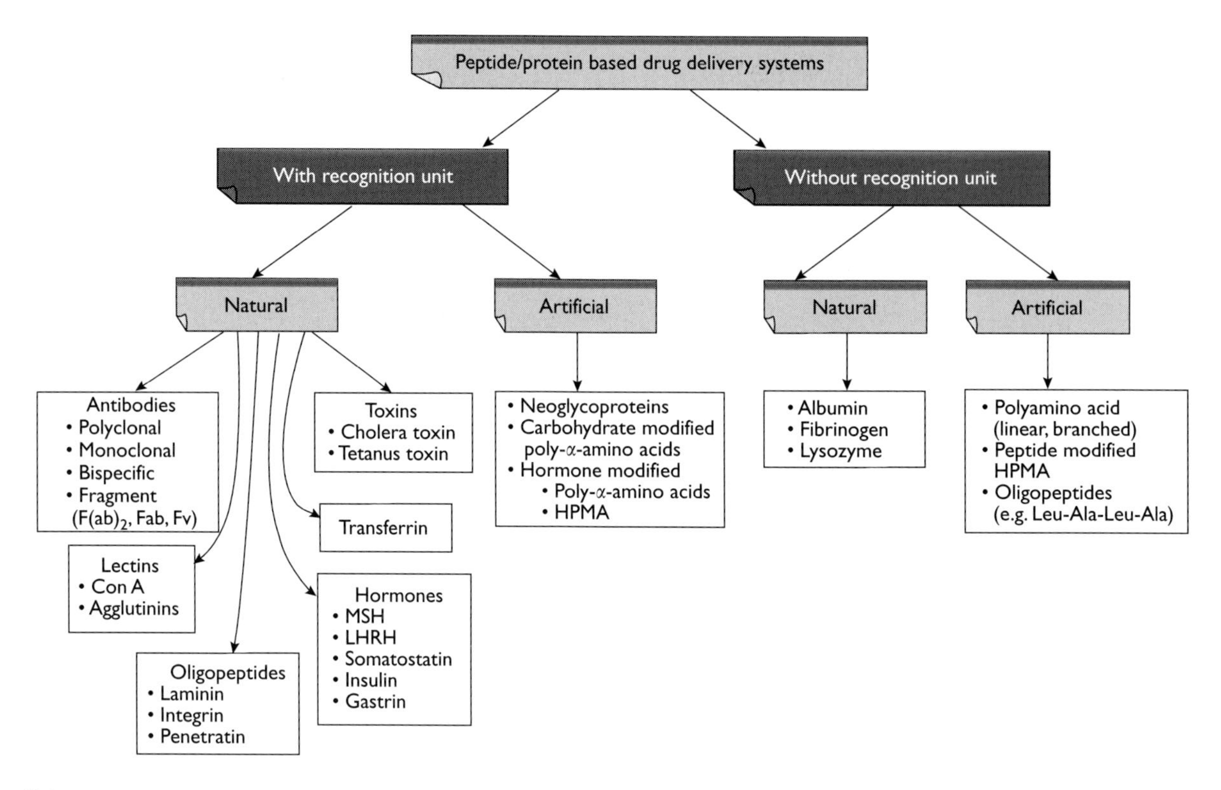

Figure 43.1

cell membrane or in the cytosol. Synthetic macromolecules such as linear or branched chain poly-α-amino acids, polydepsipeptides and peptide-spacer containing non-peptidic polymers (N-(2-hydroxypropyl)methacrylamide, HPMA) have also been used as soluble drug conjugates for "passive targeting" resulting in altered pharmacokinetics (e.g. accumulation in tumor), decreased non-specific toxicity and immunogenicity. Linkage between bioactive compounds and carrier molecules could combine these components directly or by insertion of a "spacer" unit to facilitate enzymatic release and/or an "intermediate carrier" to increase the number of molecules coupled to the carrier without influencing recognition (Figure 43.2).

Peptide/protein carriers can also be grouped according to their origin (native vs. artificial), size (e.g. oligopeptides vs. macromolecules) and means of production (isolation vs. chemical or biochemical synthesis).

It should also be noted that protein/polypeptide carriers (e.g. keyhole limpet hemocyanin, bovine serum albumin (BSA), synthetic branched polypeptides) are frequently applied in immunology to induce specific immune responses against the attached hapten. The function of the carrier in these conjugates could be considered to be targeting of epitopes to immunocompetent cells for production of epitope specific antibody or T-cell responses (Ben-Yedidia and Arnon, 1997). Due to the complexity of this issue and the limited space available, immunological carrier related delivery/

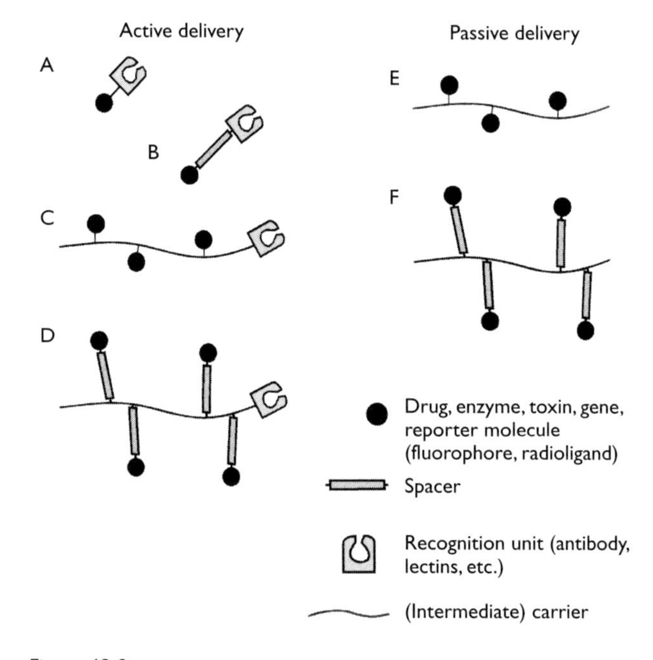

Figure 43.2

targeting will not be comprehensively reviewed, but mentioned only sporadically here.

A wide variety of compounds including natural (nucleic acids, carbohydrates, hormones or proteins) and synthetic products (e.g. HPMA, poly-amino acids) have been used as small-molecular-mass or macromolecular carriers. Bioactive molecules have been combined with carriers by covalent bonding or non-covalent complex formation. There is an increasing interest in the field of micro- or nanotechnology produced delivery/targeting (e.g. liposomes, microcapsules, micro- and nanospheres, particles, adhesive polymers). These supermolecular structures frequently apply covalent and complex linkages. However, this chapter deals only with peptide/protein-based soluble conjugates in which a covalent bond has been formed between the targeting/delivering carrier moiety and the attached entity.

First, an outline will be provided on peptide/protein carriers and their developments for delivery, followed by an example outlining practical chemical approaches to improve therapeutic efficacy of a drug, methotrexate (MTX), by targeting.

43.2 Protein/peptide carriers with recognition unit

A large number of carrier–conjugate systems are made up of proteins such as antibodies (Abs), lectins, neoglycoproteins, toxins, and transferrin. The most outstanding example of carriers with a specific recognition unit is the use of target cell (mainly tumor cell) specific Abs, their engineered versions and fragments. The first reported application of tumor specific γ-globulin raised in hamster and conjugated with MTX to treat mice injected with L1210 murine leukemia was by Mathe *et al.* (1958). Subsequently, a variety of antitumor therapeutic agents have been conjugated and studied including adriamycin, chlorambucil, radioisotopes (e.g. ^{131}I) and toxins (e.g. diphtheria).

43.2.1 Antibodies, antibody-related fragments and chimeras

By the late 1970s it was clear that some benefits could be obtained by the use of immunoglobulin conjugates in the field of tumor targeting. After the demonstration of the existence of tumor specific antibodies in 1965 there was great interest in devising more effective combinations. Reviews are available for inspection of early results of the use of poly- and monoclonal antibodies (MoAbs) for targeting of drugs, toxins, enzymes and radionuclides (Moolten *et al.*, 1982; Edwards, 1983; Hnatowich *et al.*, 1983; Poznansky, 1983; Vitetta *et al.*, 1983; Poznansky and Juliano, 1984). From the 1980s new strategies using bispecific MoAbs, antibody fragments and, later, chimera proteins have emerged. At the same time, more sophisticated chemical methods of conjugation were developed to provide less heterogeneous products. The previously applied coupling strategies based on homobifunctional reagents (e.g. glutaraldehyde, diazocompounds) were replaced by heterobifunctional reagents (e.g. SPDP, carbodiimide) or by periodate oxidation.

Here we outline some of the crucial findings of the past 30 years, which led to the present application of immunoglobulins, mainly MoAbs for site specific delivery of a wide range of bioactive or biomedically important entities.

43.2.1.1 Conjugates with direct attachment

Antibodies have been utilised for the preparation of conjugates with chlorambucil (Ghose and Nigam, 1972), trenimon (Linford *et al.*, 1974), adriamycin/daunomycin (Hurwitz *et al.*, 1975), methotrexate (Mathe *et al.*, 1958) and toxins from bacterium *Corynebacterium diphtheriae* (Moolten and Cooperband, 1970), from plant *Ricinis communis* and from other sources (see reviews by Thorpe and Ross, 1982; Kreitman, 1999), and enzymes (Poznansky, 1983) with antitumor or antibacterial activity (Knowles *et al.*, 1973).

Evaluation of antibodies as carriers of attached entities has been greatly enhanced by the availability of MoAbs produced by the revolutionary hybridoma technique of Köhler and Milstein (1975). This made the use of structurally homogeneous immunglobulin preparations possible (Kosmas *et al.*, 1993). Monoclonal antibody conjugates with cytotoxic agents (anthracyclines, neocarzinostatin, mitomycin C), melanocyte stimulating hormone (MSH) and radioisotopes have been prepared for targeting. Various immunotoxins (A chain from diphtheria toxin, ricin, abrin toxin, T-2 toxin, gelonin and a pore-forming CytA δ-endotoxin have been produced and tested for experimental (Wawrzynczak, 1992) and clinical application. For improved antitumor effect, chlorambucil-diphtheria toxin has been conjugated to Abs.

The *cis*-aconityl-daunomycin derivative-based method was developed for linking daunomycin (Dau) to the MoAbs with different isoelectric points, and a carbodiimide reagent used to produce a peptide bond (Hudecz *et al.*, 1990). Four to five molecules of Dau bound to an Ab with minimal loss of antibody activity showed selective cytotoxicity against human gastric cancer cells *in vitro*. The conjugate recognized the tumor as efficiently as the unconjugated Ab and suppressed the growth of human gastric carcinoma in nude mice significantly (Li *et al.*, 1992).

Daunomycin has been attached to MoAb through a succinylated tetrapeptide arm. The conjugate inhibited binding of the ^{3}H labelled antibody to MCF7 cells as efficiently as unconjugated Ab. After 24 hr incubation at pH 4.8, in the presence of rat liver lysosomal enzymes, 60% of Dau was released from the conjugates (Aboud-Pirak *et al.*, 1989).

It has been found that not all antibodies are suitable for creating an immunotoxin, and a large number of Abs may need to be screened. For a faster and more economical means of identifying potential targeting antibodies is to use a second immunotoxin, an anti-IgG antibody that is coupled to the toxin and test the immune complex for its ability to internalize *in vitro* (Kohls and Lappi, 2000).

From 1972, elemental boron or boron derivatives (Barth *et al.*, 1992 and 1999) have also been incorporated into antibodies to be utilised for boron neutron capture therapy (BNCT). Such studies performed before the advent of MoAbs showed poor specificity and strong immune response. Now mouse MoAbs are widely available and readily prepared, while humanized monoclonal antibodies are being developed. Reviews on both animal model studies and clinical trials have been published by Barth *et al.* (1992 and 1999).

Authors have demonstrated the possibility of using antibodies to direct isotopes to tumor cells for either diagnosis or therapy (Buchsbaum *et al.*, 1999). Poly- or monoclonal Abs (IgG or IgM) were modified by direct radiolabeling with ^{125}I, ^{131}I (Boonkitticharoen and Laohathai, 1992) and ^{188}Re (Seitz *et al.*, 1999) or by incorporation of chelating moieties for ^{111}In, ^{99m}Tc, ^{90}Y (Li *et al.*, 1995; Arano *et al.*, 1996; Virgolini *et al.*, 1998) or of a synthetic heptapeptide substrate for protein kinase mediated phosphorylation by ^{32}P-γ-ATP (Foxwell *et al.*, 1988).

43.2.1.2 Conjugates with intermediate carriers

The early observation of the reverse relationship between the average degree of substitution and Ab recognition of the target antigen led to the concept of the "intermediate carrier" in 1975 (Bernstein *et al.*, 1978). This allowed the conjugation of more drug molecules per immunoglobulin without significant change in Ab binding to cells. The concept, together with other antibody related delivery approaches, is depicted in Figure 43.3.

Dextran (Bernstein *et al.*, 1978), amino-dextran (Shih *et al.*, 1991) or poly-α-amino acid of Glu or Lys were used for coupling p-phenylenediamine, MTX or daunomycin/adriamycin (Bernstein *et al.*, 1978) to antibodies for tumor cell targeting. EDTA- or DTPA-chelate-linked high molecular weight polylysine modification of MoAb resulted in major increases in polyvalent cation binding without loss of antigen binding (Torchilin *et al.*, 1987). Branched chain polypeptides were used as intermediate carriers for borono-Phe with maintained antigen recognition (Mező *et al.*, 1996*b*).

In a similar approach α-glucosidase enzyme (Poznansky, 1983) or MTX (Garnett *et al.*, 1983) attached to albumin was targeted at rat liver hepatocytes by polyclonal or at osteosarcoma cells by monoclonal antibodies.

A photoactive drug, chlorin-e6, attached to HPMA (Krinick *et al.*, 1990) or dextran (Rakestraw *et al.*, 1990) carrier has been conjugated with MoAbs and the photodynamic effect was studied *in vitro*. Similar constructs containing adriamycin or Dau, HPMA and monoclonal antibody or its $F(ab)_2$ fragment specific for neoplastic T-cells caused a significant retardation of tumor growth and an extension of the life span of treated mice (Rihová *et al.*, 1988 and 2000).

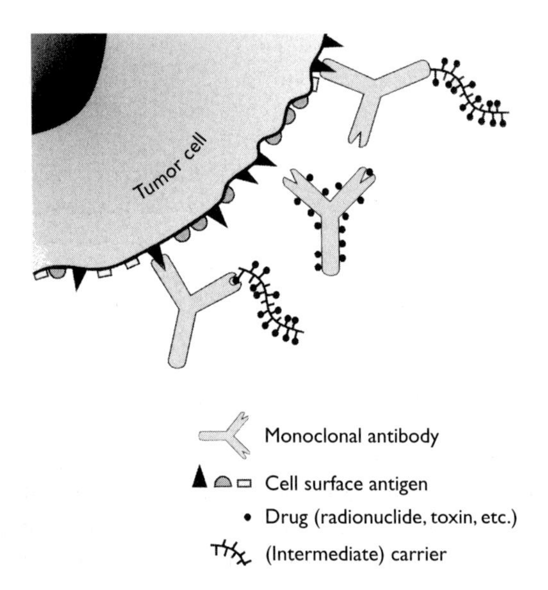

Figure 43.3

43.2.1.3 Conjugates with bispecific antibodies

Using a bispecific MoAb recognizing both carcinoembryonic antigen (CEA) and the cytostatic vinca alkaloids, specific tumor localization of drugs was reported (Corvalan *et al.*, 1987). A bispecific MoAb, reactive with MTX and a tumor-associated antigen (gp72) has been produced. A particular feature of this antibody is that it reacts with albumin conjugated MTX; this would allow *in vivo* targeting of conjugates, increasing many fold the number of molecules of drug carried or localizing to pre-targeted antibody. Purified hybrid Ab specifically enhanced the *in vitro* cytotoxicity of MTX-HSA for gp72 positive tumor cells (Pimm *et al.*, 1990).

Hybrid antibodies with dual specificity have been proposed to deliver ricin to immunoglobulin-bearing cells (Raso and Griffin 1982), while CEA-expressing cells were highly effectively targeted by MoAb bearing ricin A chain and CEA specificity. Kinetics for inhibition of cellular protein synthesis for native diphtheria toxin was compared with that of bispecific Ab recognizing acid-sensitive epitopes on diphtheria toxin and transferrin receptors on target cells. Identical patterns were observed indicating the release of a toxin subunit from the Ab-combining site (Raso *et al.*, 1997). Bispecific $F(ab)_2$ Ab fragment has also been utilised for toxin delivery. The ribosome-inactivating protein, saporin was targeted at human B-cell lymphoma through CD22 and this construct was highly cytotoxic *in vitro* (Bonardi *et al.*, 1993).

In BNCT for tumor-localizing boron compounds, including the sulfhydryl-containing polyhedral borane, sodium borocaptate ($Na_2B_{12}H_{11}SH$), boronated starburst dendrimer and boronophenylalanine, bispecific MoAbs have been applied to brain tumors and melanoma (Liu *et al.*, 1996).

A bispecific Ab binding to substituted adenines with high affinity and to surface antigen on canine hematopoietic target cell antigen was used for targeting adenine conjugated radionuclides or Dau. Radiolabeling performed by complexation of ^{111}In or ^{90}Y with adenine attached DOTA or DTPA and MoAb-mediated, dose-dependent delivery of ^{111}In-labelled adenine-chelate derivatives to myeloid cells has been demonstrated (Krüger *et al.*, 1998). Bispecific Abs have been used to target ^{125}I-, ^{111}In- and ^{99m}Tc-labelled haptens to cell subsets and increased clinical performance was observed (Le Doussal *et al.*, 1992). In clinical studies, a bifunctional $F(ab)_2$ fragment obtained from two different Fab fragments by thioether linkage was injected and allowed to localize at a tumor site. This construct had specificity to CEA and another specificity to EDTA, tagged with ^{111}In. High tumor to normal tissue ratios were achieved (Stickney *et al.*, 1991).

43.2.1.4 Conjugates with antibody fragments

The feasibility of using Ab fragments like $F(ab)_2$ for target-specific delivery were first demonstrated in the late 1970s. In this study Dau was attached to a fragment of polyclonal IgG. Soon after, $F(ab)_2$ fragment-toxin hybrids were tailored with ricin A chain (Raso and Griffin, 1980) and diphtheria toxin (Ross *et al.*, 1980) and applied with some success.

$F(ab)_2$ fragment of a toxic, T-cell activating, anti-CD3 MoAb was prepared and applied as a targeting molecule for the cytotoxic idarubicin. This immunoconjugate was more effective against both acute and chronic rejection than other conjugates or whole antibody (Han *et al.*, 1999*b*). Anti-CD3 MoAb $F(ab)_2$ fragments were conjugated to deglycosylated A chain of ricin toxin for targeting T-cells in established graft-versus-host disease. The immunotoxin was potent and selective in inhibiting T-cell mitogenesis *in vitro* (Vallera *et al.*, 1995).

Site-specific introduction of metal-chelating groups into $F(ab)_2$ fragments of an anti-lymphoma Ab possessing a natural Asn-linked light chain carbohydrate and an anti-CEA antibody grafted with a light chain carbohydrate site was reported. After introduction and labeling with ^{90}Y or ^{111}In of the chelating agents, retained antigen-binding ability *in vitro* and tumor targeting *in vivo* was demonstrated (Govindan *et al.*, 1995).

In comparative studies MoAbs, $F(ab)_2$ and Fab fragments were studied. Monoclonal antibody and its $F(ab)_2$ fragment, radiolabelled with ^{131}I, were tested for tumor localization and imaging in nude mice bearing a squamous cell carcinoma xenograft. Findings suggest that the superior tumor to non-tumor ratios render the $F(ab)_2$ fragment more qualified for specific targeting of radioisotopes to tumor xenografts. In a different system the conjugates of ^{125}I-labelled $F(ab)_2$ and Fab fragments with neocarzinostatin were injected intravenously into mice with pancreatic carcinoma xenografts. The localization of ^{125}I-labelled Fab fragments was markedly greater. Kinne *et al.*, (1999) have analyzed the *in vivo* metabolic profile of ^{99m}Tc-anti-CD4 monoclonal antibodies and their Fab fragments in rats. Fab showed higher urine excretion and lower plasma levels than the complete MoAb, but still more favorable tissue-to-blood ratios following scintigraphy of the arthritic joints.

Cytotoxic derivatives of an even smaller structural unit of the human IgG, fragment Fab with A chain of diphtheria toxin or of ricin (Raso and Griffin, 1980) have been reported against human colorectal carcinoma and L1210 leukemia, respectively. ^{99m}Tc labelled Fab fragments of a mouse/human chimera Ab were produced by Narula *et al.*, (1997) and their potential for non-invasive imaging was explored. Unequivocal visualization of atherosclerotic lesions was possible in rabbits indicating that ^{99m}Tc-based immunoimaging of the vascular lesions may be feasible using smaller Ab fragments. It has also been demonstrated that Fab fragment of an anti-CEA IgG could be labelled not only with ^{99m}Tc, but also with single-photon-emitting ^{94m}Tc suitable for positron emission tomography (Griffiths *et al.*, 1999).

Genetic engineering can produce novel Ab fragments with improved properties for applications such as tumor targeting *in vivo*. Due to their ease of isolation from phage display libraries and their ability to recognize conserved antigens, single-chain Fv (scFv) molecules are rapidly becoming commonplace. However, the monovalent nature of the scFv molecule along with their rapid elimination from circulation, has limited the utility of scFv for applications in the fields of cancer imaging and therapy. Recently, a number of strategies have been evaluated for improving the *in vivo* efficacy of scFv molecules (Adams and Schier, 1999).

A scFv-*Pseudomonas* exotoxin A fusion protein was reported with pronounced antitumor activity and selectivity towards epidermal growth factor (EGF) receptor-overexpressing tumor cells (Schmidt *et al.*, 1997). A similar construct was engineered for specific elimination of B-cells responsible for an autoimmune blistering disease (*Pemphigus vulgaris*). *In vitro* studies revealed that the chimera toxins showed dose-dependent toxic activity (Proby *et al.*, 2000). In another chimera, bovine seminal ribonuclease, an unusual member of the ribonuclease superfamily, because of its remarkable anti-tumor and immunosuppressive properties, has been used as toxin. Immunotoxins were prepared with a single-chain Fv directed against the oncofetal antigen placental alkaline phosphatase with or without C-terminal "KDEL" endoplasmic reticulum retention signal peptide (Deonarain and Epenetos, 1998).

A scFv version of a murine Ab used for human melanoma radioimmunoimaging has been produced with conserved binding specificity and appended C-terminal cysteine residue for thiol-specific modification (Neri *et al.*, 1996). A hybrid humanized and thermally stable anti-epithelial glycoprotein-2 (EGP-2) scFv was engineered and radiolabelled with ^{99m}Tc. This chimera showed favorable blood clearance and efficient enrichment at solid lung tumor xenografts applicable for the diagnostics and therapy of EGP-2-positive tumors in patients (Willuda *et al.*, 1999). Pietersz *et al.*, (1998) have produced a recombinant metallothioneine containing scFv of anti-CEA antibody labelled site-specifically with ^{99m}Tc for use as a diagnostic imaging agent in colorectal cancer. Despite a rapid serum clearance, adequate localization to tumor was demonstrated enabling early gamma camera imaging in xenografted mice.

Monomeric (27 kDa) and dimeric (55 kDa) forms of a scFv were constructed with either a 28 or 14 amino acid connecting peptide, expressed in *E. coli* and compared for imaging potential (Wu *et al.*, 1996). The shorter linker favored formation of dimers. ^{125}I-scFv monomers and dimers exhibited rapid clearance kinetics and targeted very well to xenografts in tumor-bearing mice. Non-covalent dimers of scFv (also known as diabodies) are stable, easy to produce and show excellent tumor localization compared to monomeric scFv, probably due to their increased mass and valency.

A novel engineered Ab fragment (V_L-V_H-C_H3, or "minibody") with bivalent binding to CEA was produced by fusion of a murine anti-CEA scFv to the human IgG_1 C_H3 domain. In minibodies either a two-amino acid linker or human IgG_1 hinge plus an additional 10 residues were used to join V_H and C_H3. *In vivo* tumor targeting studies demonstrated rapid, high tumor uptake in mice bearing LS174T human colon carcinoma xenografts (Hu *et al.*, 1996). A similar approach was developed by Uherek *et al.*, (1998) for cell type-specific uptake and intracellular delivery of DNA. Modular fusion protein comprising distinct functions was designed. The N terminal DNA-binding domain of the yeast transcription factor Gal4 was connected to a C-terminal antibody fragment specific for the tumor-associated ErbB2 antigen *via* an internal diphtheria toxin translocation domain as an endosome escape activity.

Even a small synthetic pentadecapeptide derived from the third heavy-chain complementarity-determining region (CDR-3H) of a tumor-associated MoAb was shown to retain its epitope specificity. The peptide was radiolabelled with ^{99m}Tc and injected intravenously to image malignant lesions in 26 women with primary, recurrent, or metastatic breast cancer. Visualization of breast tumors and their metastases was obtained shortly after administration indicating new opportunities for breast cancer imaging and possibly therapy (Sivolapenko *et al.*, 1995).

43.2.2 Lectins

Carbohydrate binding proteins (also referred to as agglutinines or lectins) (Inbar and Chet, 1997) recognize specific mono- or oligosaccharide moieties, therefore it is tempting to apply this feature to targeting (Wadhwa and Rice, 1995). Some are glycoproteins, such as soybean agglutinin, and others are not, e.g. concanavalin A (Con A).

Con A, a mannose-binding lectin, was first utilised for delivery of Dau, MTX and chlorambucil, enzymes, like dextranase, trypsin and L-asparaginase (reviewed by Shier, 1979). *Cis*-aconityl-linked doxorubicin-wheat germ agglutinin was investigated for targeting colon carcinoma cells *in vitro* (Wirth *et al.*, 1998). The antiproliferative activity of the conjugate was observed. The lectin-delivered prodrug yielded 160% of the cytostatic activity of free doxorubicin on human colonic Caco-2, but not on human lymphoblastic Molt-4 cells. Using the same lectin, Monsigny *et al.* reported rapid internalization and subsequent release of daunomycin from a conjugate which contained a Leu or Asp amino acid spacer (Monsigny *et al.*, 1980).

Using peanut (*Arachis hypogea*) agglutinin or wheat germ agglutinin as recognition units, conjugates were prepared for the treatment of colon diseases. In this conjugate cyclosporin A containing aromatic azo-bond was attached to HPMA copolymer as a non-biodegradable intermediate carrier. The incubation of the conjugate with rat cecal contents resulted in the cleavage of the azo bond by azoreductase and release of the cyclosporin derivative (Lu *et al.*, 2000).

Chain A of diphtheria toxin and of ricin were conjugated by disulphide bridges with Con A. MTX and chlorambucil were also coupled to various types of lectins including ricin and abrin. *In vivo* experiments showed prolonged survival time of mice with sarcoma 180. A subunit of *Wisteria fluoribunda* lectin showed remarkable cytotoxicity when conjugated with diphtheria toxin fragment (Uchida *et al.*, 1978).

Galaptin, an endogenous galactoside-binding lectin and α-L-fucosidase were coupled to target enzyme-deficient lymphoid cells from a fucosidosis patient. Conjugates were effectively bound and internalized in a lactose inhibitable manner indicating that galaptin-α-L-fucosidase conjugates may be useful for enzyme replacement therapy of fucosidosis and also that galaptin as a transport vehicle may be applied to delivery to cells bearing galaptin acceptors (Allen *et al.*, 1990).

Recently it has been demonstrated that lectins might be used for direct delivery of genes to target cells. For this, histone modified with various lectins (Con A, *Maackia amurensis* agglutinin, *Sambucus nigra* agglutinin, or wheat germ agglutinin) and polylysine conjugate was employed and complexed with pCMVlacZ gene. The results indicate that these lectin-polypeptide/protein conjugates were capable of transfering genes as vectors to human airway epithelial cells possessing appropriate receptors (Yin and Cheng, 1994).

Lectin mediated bioadhesive drug delivery systems are also considered to offer a number of possibilities for gastrointestinal drug targeting systems in the future. However, it has been pointed out that before advanced drug delivery systems using lectins can be realised, rigorous evaluation of their toxicity and immunogenicity will be required (Woodley, 2000).

43.2.3 Hormones

As a result of their receptor specificity, attempts have been made at site specific targeting of bioactive molecules by hormones. Until the 1980s hormones investigated included human placental lactogen, human chorionic gonadotropin, epidermal growth factor (EGF) and α-melanotropin (MSH). Interest was renewed when it was found that certain cell surface receptors are overexpressed in tumor tissue and could serve as targets for anticancer drugs as radionuclides attached to receptor ligands.

Human placental lactogen (hPL) has been used for targeting diphtheria toxin fragment A to the lactogenic plasma membrane receptor of the mammary gland. The conjugate retained binding activities of its separate subunits, but failed to inhibit protein synthesis in target cells. This indicates that the hormone receptor cannot mediate the entry of toxin A chain-hPL from membrane-bound conjugate into the cytosol. Human chorionic gonadotropin (hCG) has been cross-linked with the toxic A chain of ricin for surface receptor targeting. Cells possessing receptor for the β-subunit of hCG were killed, but the conjugate was less efficient than free ricin. No effect was observed in control cells bearing no hormone receptor. In a third study, ricin or diphtheria toxin fragments were attached to EGF. There were no marked differences between the two conjugates in receptor binding, however ricin A-EGF was toxic, while its counterpart with diphtheria fragment A was virtually non-toxic at the expected concentration range. EGF was also used to target tumor cells expressing increased number EGF-receptors. Photocytotoxic compound Sn-(IV)chlorin-e6 monoethylene-diamine [SnCe6(ED)] was conjugated to EGF through an intermediate carrier, such as dextran or human serum albumin (HSA). The affinity of EGF for its receptor was substantially impaired when conjugated in EGF-dextrane-SnCe6(ED), in contrast to EGF-HSA-SnCe6(ED) conjugate, which displayed high photocytotoxicity on MDA-MB-468 cells (Gijsens *et al.*, 2000). Rigaudy *et al.*, (1989) reported the effect of D-Ala^2-D-Leu^5-enkephalin coupled to ellipticine or 9-hydroxyellipticine. *In vitro* the conjugates retained both opioid receptors and DNA affinities and were proposed for the treatment of human colorectal and pulmonary carcinomas possessing high levels of opioid receptors. Dau was attached to thyrotropin directly and indirectly with lactalbumin intermediate carrier. Both conjugates inhibited ^{3}H Leu incorporation and cell proliferation of rabbit thyroid cells and were more effective than free Dau. The inhibition was cell specific (Kaneko, 1981).

New groups of hormone-attached cytotoxic compounds with antitumor activity have been developed and studied in detail by Schally and Nagy (1999). These constructs contain MTX or doxorubicin derivatives and analogs of hormonal peptides such as LHRH, bombesin, and somatostatin that can be targeted at certain tumors possessing receptors for those peptides.

43.2.3.1 α-Melanotropin (MSH)

MSH was selected as a vector for site-specific delivery of daunomycin to receptor on melanoma cells. Dau was linked to MSH via the ε-amino groups of Lys and its terminal amino group. The conjugate as well as the free drug was internalized, taken up by the nuclei of melanoma cells and was toxic *in vitro* (Varga *et al.*, 1977). A chimera toxin containing genetically coupled diphtheria toxin with hormone MSH displayed selective toxicity to those cells carrying MSH specific surface receptors. Murphy *et al.*, (1987) favor the genetic engineering approach for the assembly of chimera toxins.

MSH has also been utilized as a recognition unit in HPMA copolymer based conjugates containing doxorubicin bound via the lysosomally degradable spacer Gly-Phe-Leu-Gly for site-specific delivery to malignant melanoma (O'Hare *et al.*, 1993). Polymer-bound MSH, like free MSH, was able to stimulate tyrosinase activity in B16F10 cells *in vitro*, confirming the ability of conjugated hormone to interact with the MSH receptor. Conjugates were cytotoxic to a mouse melanoma cell line *in vitro*. When administered *i.p.* or *i.v.* to C57BL/6J mice bearing B16F10 tumor, HPMA copolymers containing spacer bound doxorubicin (with or without MSH) significantly increased animal survival. In mice with established subcutaneous B16F10 tumors, conjugates containing MSH were significantly more effective than those without MSH.

A chelating derivative of MSH has been synthesized for the development of a melanoma imaging agent. Two MSH molecules were cross-linked by diethylene-triamine pentaacetic acid (DTPA). This compound, labelled with ^{111}In, was equipotent with MSH *in vitro* and *in vivo* in DBA/2 mice with Cloudman S91 melanoma offering an alternative to antibody targeting in the imaging of malignant melanoma (Bard *et al.*, 1990). Other radiolabelled (^{99m}Tc, ^{188}Re or ^{125}I), linear, short chain analogs of MSH were also examined (Chen *et al.*, 1999). In mice with melanoma tumor uptake highlights the potential of the ^{99m}Tc labelled [Nle^4,D-Phe^7]MSH containing Ac-Cys-Gly-Cys-Gly chelation moiety (Bard *et al.*, 1990).

Biotinylated derivative of ACTH$^{1-17}$ (Bagutti and Eberle, 1993) as well as photoaffinity-, fluorophore-labelled or biotinylated MSH analog ([Nle^4,D-Phe^7]MSH) (Erskine-Grout *et al.*, 1996) were also developed for targeting receptors. All probes containing the core motif His-Phe-Arg-Trp bind the receptor on melanoma cells with similar intrinsic efficacies to the endogenous peptide.

For increased sensitivity of fluorescence detection of specific melanotropin receptors, a class of multivalent MSH-poly(vinyl alcohol) conjugates were prepared and applied successfully to all of the melanoma cell lines, both mouse and human, melanotic as well as amelanotic (Sharma *et al.*, 1994). The conjugates consisted of multiple copies of both [Nle^4,D-Phe^7]MSH analog and FITC-fluorophore, both arranged in a pendent fashion on the biologically inert intermediate carrier macromolecule. This methodology was also proposed for receptor specific targeting of (chemo)therapeutic agents.

43.2.3.2 Luteinizing hormone-releasing hormone (LHRH)

MTX selectively coupled through its Glu γ-carboxyl group to the free amino group of two analogs of LHRH [Glp-His-Trp-Ser-Tyr-D-Lys(MTX)-Leu-Arg-Pro-Gly-NH_2] and [Ac-Ser-Tyr-D-Lys(MTX)-Leu-Arg-Pro-NH-Et] (Nagy *et al.*, 1993). Receptor binding of these conjugates as well as an anthraquinone containing variant was investigated. The hormonal activity after *in vivo* administration was retained. Chronic treatment of mice with estrogen-independent MXT mammary tumors with agonist [D-Lys^6]LHRH linked to MTX produced significant down-regulation of membrane receptors for LHRH and inhibition of tumor growth.

New cytotoxic peptide conjugates were also prepared containing 14-O-glutaryl esters of doxorubicin or 2-pyrrolino-doxorubicin and [D-Lys^6]LHRH analog (Rékási *et al.*, 1993). These compounds show high-affinity binding and are much less toxic and more effective *in vivo* than their respective free cytotoxic drug in inhibiting tumor growth in LHRH receptor-positive models of human ovarian, mammary, or prostatic cancer (Nagy *et al.*, 2000*a*). Further improvement was achieved by the suppression of the activity of serum carboxylesterase enzymes which could hydrolyse conjugates to release the cytotoxic component, before targeting is complete (Nagy *et al.*, 2000*a*). It has been indicated that the cytotoxic effect of the [D-Lys^6] LHRH linked 2-pyrrolinodoxorubicin could be exerted through induction of apoptosis.

43.2.3.3 Somatostatin (SST)

Since SST receptors are also found in various human neoplasms and the receptor subtypes to which octapeptide analogs bind with high affinity have been identified, Schally's group have synthesized several cytotoxic SST analogs containing MTX (Nagy *et al.*, 1993) or doxorubicin derivatives (Nagy *et al.*, 1998). MTX was selectively coupled through its Glu γ-carboxyl group to the free amino group of a SST analog (MTX-D-Phe-Cys-Tyr-D-Trp-Lys-Val-Cys-Thr-NH_2) (Nagy *et al.*, 1993). Hormone analog conjugates of anthracyclines were prepared either with doxorubicin or its superactive 2-pyrrolino-derivative (Nagy *et al.*, 1998; Nagy *et al.*, 2000*a*). *In vitro* tests on various human cancer cell lines expressing receptors for SST (e.g. MKN-45 gastric cancer, MDA-MB-231 breast cancer, PC-3 prostate cancer, and MIA PaCa-2 pancreatic cancer) demonstrated antiproliferative activity of conjugates (Nagy *et al.*, 1998). *In vivo* experiments with nude mice with xenografts revealed significant inhibition of tumor growth of estrogen-independent MDA-MB-231, MX-1 and MCF-7-MIII 1 week after injection of a single dose of 2-pyrrolino-derivative (Kahán *et al.*, 1999).

For localization of metastatic gastroenteropancreatic and neuroendocrine tumors, lymphomas and adenocarcinomas by SST receptor scintigraphy radioligand was targeted with octreotide analogs containing tyrosine or phenylalanine for ^{123}I labeling or chelating moiety (DTPA (Hammond *et al.*, 1994), DOTA (1,4,7,10-tetraazacyclo-dodecane-N′,N″,N‴,N⁗-tetraacetic acid) (Albert *et al.*, 1998)) for ^{111}In incorporation.

43.2.3.4 Gastrin/bombesin

Five peptide fragments, based on the C-terminal sequence of bombesin (6–14) or (7–14), were selected as carriers for doxorubicin and 2-pyrrolino-doxorubicin. The peptide component of the conjugate contained reduced peptide bond (CH_2-NH or CH_2-N) and/or substitution by non-native amino acid at various positions. Cytotoxic conjugates of these analogs displayed specific receptor binding and exerted inhibitory effects on the *in vitro* growth of various human cell lines (e.g. pancreatic, lung, prostate and gastric) possessing appropriate receptors (Nagy *et al.*, 1997). Recent studies showed that one of these conjugates with Gln-Trp-Ala-Val-Gly-His-Leu-Ψ(CH_2-NH)-Leu-NH_2 and 2-pyrrolino-doxorubicin was capable of inhibiting the growth of H-69 SCLC tumors in xenograft nude mice after *i.v.* injection. This indicates that the cytotoxic bombesin analog could be considered for targeted therapy of tumors, such as SCLC, that express bombesin receptors (Kiaris *et al.*, 1999).

Czerwinski *et al.*, (1998) have recently studied conjugates containing the gastrin/cholecystokinin type B receptor specific heptagastrin, the C-terminal heptapeptide of gastrin and potently cytotoxic ellipticine derivative, 1-[3-[N-(3-aminopropyl)-N-methylamino]propyl]amino-9-methoxy-5,11-dimethyl-6H-pyrido[4,3-b]carbazole. *In vitro* data show that the conjugate is transported to lysosomes, whereas the receptor is recycled to the plasma membrane. The lysosomal processing depends on the linker between the heptapeptide and the drug. Heptagastrin linked to ellipticine via a succinoyl-substituted pentapeptide, Ala-Leu-Ala-Leu-Ala, is more toxic to cholecystokinin type B receptor-positive cells than to isogenic cells lacking the receptor. The conjugated drug eradicated all receptor-positive tumor cells *in vivo* without producing any general toxicity. Cholecystokinin analogs with specific and high-affinity for receptor were modified by chelating DTPA- and DOTA-moieties capable of incorporating ^{111}In for imaging rat and human tissues (Reubi *et al.*, 1998).

43.2.3.5 Insulin

The possibility of using insulin, which is transported into the brain by receptor-mediated transcytosis, as a peptide carrier for delivery of enzyme across the blood-brain barrier was investigated. Horseradish peroxidase conjugated with insulin, or its fragments, was tested and one fragment obtained by trypsin digestion showed high affinity and scarcely any hypoglycemic activity in mice (Fukuta *et al.*, 1994). Insulin was proposed to direct enzyme-albumin conjugates too (Poznansky *et al.*, 1984).

43.2.4 Neoglycoproteins, glycopeptides

Glycoproteins do not generally show site selectivity. But after the removal of terminal sialic acid moieties from the carbohydrate branches, certain liver cells recognize asialoglycoproteins. Galactosyl-terminated glycoproteins like fetuin (asialofetuin, AF) has been investigated as hepatocyte-specific drug carrier. *Ectromelia virus* infected liver cells were treated with trifluorothymidine or adenine-9-arabinofuranoside (ara-A) containing AF conjugate and the inhibition of DNA synthesis was documented (Fiume *et al.*, 1979).

It is interesting to note that fetuin glycopeptide conjugate of various proteins proved to be capable of increasing hepatic uptake. Asialo-human-α-1 acid glycoproteins were derivatized with low molecular weight polylysine and complexed with an antisense DNA (67-mer). The results of biodistribution studies showed that the accumulation of the asialoglycoprotein anti-sense complex in the liver was rapid and greatly exceeded the accumulation of the sialoglycoprotein antisense analog or antisense alone (Lu *et al.*, 1994).

Neoglycoproteins could also be produced by introduction of covalently linked carbohydrate moieties to proteins. Thioglycosides of D-galactose, D-glucose, N-acetyl-D-glucosamine, and D-mannose were covalently attached to *Aspergillus oryzae* α-amylase, hen's eggs lysozyme, and BSA by amidination, diazo coupling, and amide formation (Krantz *et al.*, 1976).

Lactosamine modified serum albumin is also capable of binding galactosyl-terminated glycoprotein-specific receptors on hepatocytes. Its ara-A derivative proved to be efficient in targeting, and was devoid of immunogenicity. Anti-hepatitis virus activity of adenine arabinoside monophosphate (ara-AMP) or acyclovir monophosphate, either free or conjugated with lactosaminated HSA was tested. The results indicate that the dosages of these antiviral agents required to inhibit woodchuck hepatitis virus growth can be sharply reduced by coupling the drugs to lactosaminated HSA (Ponzetto *et al.*, 1991). Ara-AMP-lactosaminated HSA conjugate administered to hepatitis B virus-infected patients for 28 days, exerted an antiviral activity to the same extent as the free drug without producing any clinical side effects (Fiume *et al.*, 1997). By similar concept, maleoyl derivatives of alkylated vinblastine or A chain of ricin or diphtheria toxin was coupled to the same protein and proved to be effective on hepatocytes and experimental tumors *in vitro*.

Enhanced hepatocyte uptake and liver targeting of MTX were achieved by galactosylated albumin (Han *et al.*, 1999*a*). A novel, highly efficient DNA delivery system was composed of galactosylated histones (H1, H2a, H2b, H3, and H4), and covalently conjugated DNA. This construct was used to transfect HepG2 cells, which displayed the asialoglycoprotein receptor (Chen *et al.*, 1994)

The antileishmanial potency of doxorubicin (Sett *et al.*, 1993) and MTX (Chakraborty *et al.*, 1990) conjugated to mannose-HSA was demonstrated in peritoneal macrophages infected with *Leishmania donovani*. Human macrophages bind and internalize mannosylated and 6-phosphomannosylated BSA-attached antiviral drugs, which inhibits the multiplication of herpes virus more efficiently than that of the free drug. It has been reported that the hepatic uptake of the mannose 6-phosphate modified HSA conjugate is elevated with increasing density of mannose 6-phosphate/insulin-like growth factor II receptor, expressed, in particular, by hepatic stellate cells during liver fibrosis (Beljaars *et al.*, 1999). Mannose 6-phosphate modified serum albumin was also coupled with antisense oligonucleotide and more efficient internalization was achieved (Bonfils *et al.*, 1992). Small synthetic glycopeptides with D-mannose residues covalently linked through a spacer arm to the α- and ε-amino groups of lysine, dilysine, and oligolysine have been proposed to target pharmacologic agents to macrophages through mannose receptor uptake. Doebber *et al.*, (1982) have enhanced macrophage uptake of glucocerebrosidase, probably through a mannose receptor, by conjugating the enzyme with trimannosyldilysine, a synthetic glycopeptide.

The pharmacokinetic analysis of mannose-BSA using ^{111}In-labeling has demonstrated that the uptake process of ^{111}In-Man-BSA takes place at fewer binding sites and with a greater internalization rate in comparison with other liver-specific carriers such as galactosylated, succinylated and cationized BSAs (Ogawara *et al.*, 1999).

In accord with the above findings it has been found that depending on the type (galactose, mannose or lactosamine) and number of carbohydrate residues in modified HSA, naproxen could be selectively delivered to different cell types of the liver. Similar conclusions were reported on binding and endocytosis of glycosylated albumin, with or without MTX, to L1210 cells. The highest uptakes were obtained with fucosylated serum albumin containing 25 ± 5 residues of fucose.

Neoglycoproteins conjugated to cytotoxic drugs (etoposide, *cis*-diaminedichloroplatinum II or MTX) were capable of inhibiting DNA synthesis of human embryonal carcinoma cells. A greater than 10-fold increase in cytotoxic capacity to the nonglycosylated carrier-drug conjugate, nearly reaching the level of toxicity of the freely diffusible drug, was observed. These results together with others on the preparation of nonimmunogenic, neutral glycosylated polymers suggest a potential use for custom-made glycosylated carriers in the targeting of therapeutic agent (Gabius *et al.*, 1987).

43.2.5 Toxins

The nontoxic B subunit of *E. coli* heat-labile enterotoxin has been combined with a peptide from the small subunit of ribonucleotide reductase of herpes simplex virus (HSV) for a receptor mediated delivery. This nonapeptide (YAGAVVNDL) corresponds to the C-terminal region and specifically inhibits the viral enzyme *in vitro*. The recombinant protein produced specifically inhibits virus replication and ribonucleotide reductase activity in quiescent HSV type 1 infected Vero cells indicating the usefulness of toxin subunit carrier-mediated delivery (Marcello *et al.*, 1994).

Atoxic fragment C of tetanus toxin has been shown to promote uptake of covalently coupled enzymes such as β-N-acetylhexosaminidase A, superoxide dismutase, horseradish peroxidase and β-galactosidase into neuronal cells *in vitro* and *in vivo*. Recombinant C fragment covalently bound to polylysine was capable of binding to DNA and transfecting cells (Knight *et al.*, 1999).

Toxins (non-toxic form of anthrax toxin (Ballard *et al.*, 1998), including cholera toxin (Drew *et al.*, 1992), have been applied as immunogenic carriers to deliver cytotoxic T-cell epitope from ovalbumin (Ballard *et al.*, 1998) or from glycoprotein D [gD(1–23)] of Herpes simplex virus type 2 (Drew *et al.*, 1992). Increased epitope-specific responses were recorded.

43.2.6 Transferrin

Transferrin is the major iron-binding protein in the plasma of vertebrates and an essential growth factor for proliferating malignant cells. The specific receptor binds diferric transferrin and is endocytosed into the cell. Dau attached to transferrin was less toxic than the free drug under *in vitro* circumstances, but markedly more active upon malignant cells compared to normal cells *in vivo* (Lemieux *et al.*, 1992).

One novel approach for the biological delivery of peptide drugs is to incorporate the sequence of the peptide into the structure of a natural transport protein, such as human serum transferrin. A peptide sequence cleavable by the HIV type 1 protease (VSQNYPIVL) was inserted into various regions of transferrin and tested for function. Analysis of the mutant proteins provided a number of important findings:

a they retained native human serum transferrin function,
b the inserted peptide sequence was surface exposed, and most importantly,
c two of these mutants could be cleaved by HIV-1 protease.

In conclusion, this investigation has validated the use of human serum transferrin as a carrier protein for functional peptide domains introduced into its structure using protein engineering (Ali *et al.*, 1999).

43.2.7 Oligopeptides

The potential of small peptides as delivery molecules lies largely in their ability to act as a carrier or targeting unit (Ringrose 1983). As carriers they are capable of releasing bioactive molecules (drug, enzyme, oligonucleotides etc.) at or inside the target cell. As targeting agents these molecules recognize target cells because of specific surface receptors. In this case the presence of attached bioactive components (e.g. reporter molecules, enzymes, toxins) at or inside the target cells is due to the specific interaction between the homing moiety and cellular receptor.

43.2.7.1 Carriers without receptor recognition unit

The possibility of using amino acids and peptides as non-toxic carriers or "pro-drugs" of cytotoxic agents has been extensively researched since the 1950s. In a dipeptide variant of an antibacterial drug, alafosfalin (L-Ala-L-1-aminoethylphosphonic acid), the N-terminal L-Ala residue serves as a carrier for the 1-aminoethylphosphonate "warhead" moiety, which after internalization is cleaved off and inhibits the alanine racemase enzyme in bacteria (Atherton *et al.*, 1980).

Based on increased production of hydrolytic enzymes of tumor cells, some of which are, in part, selective for certain tumor (e.g. γ-glutamyltransferase, plasminogen activator, catepsin B), a number of drug-modified substrates have been produced (Ringrose, 1983). Bioactive primary and secondary amines, when acylated with the benzyloxycarbonyl-Gly-Phe dipeptide, are transported into pinocytic cells (e.g. macrophages, tumor cells), much faster than the parent compounds. The N-γ-glutamyl derivatives of aminophenols, and p-fluoro-L-phenylalanine or sulfamethoxazole were readily cleaved by γ-glutamyl transferase and proposed as targeting units in certain pathological conditions, where the enzyme level is increased. Peptidyl and aminoacyl derivatives of Dopa (L-3,4-dihydroxyphenylalanine) and dopamine have also been tried as a means of targeting the drug to the kidney. Increased Dopa concentration was observed, which was explained by the action of γ-glutamyl transpeptidase.

Masquelier *et al.*, (1980) have prepared amino acid and dipeptide derivatives of Dau and have used them *in vivo* on L1210 leukemia cells. It was concluded that Leu-Dau and Ala-Leu-Dau could be activated inside, or in the close vicinity of, tumor cells which display a high aminopeptidase activity. Amino acid conjugates of vinblastine, vincristine, vindesine and their derivatives linked by an amide bond were prepared and some showed significant antitumor activity in mice (Collard *et al.*, 1989).

To target cytotoxic therapy specifically to metastatic prostate cancer sites, an inactive prodrug was synthesized by coupling the primary amine of doxorubicin to the COOH-terminal carboxyl of a seven-amino acid peptide carrier (i.e., His-Ser-Ser-Lys-Leu-Gln-Leu). The peptide was documented to be hydrolyzable specifically by the serine protease prostate-specific enzyme secreted at high level by human prostate cancer cells to liberate the active cytotoxin L-leucyl-doxorubicin. Specific cytotoxic response on human prostate cancer cells was demonstrated (Denmeade *et al.*, 1998).

Oligopeptides as bridging spacer moieties (e.g. Gly-Gly, Gly-Phe-Leu-Gly and Gly-Phe-Phe-Leu) were also introduced to link covalently non-biodegradable HPMA copolymers and doxorubicin, daunomycin or puromycin, providing enzymatic drug-release in the lysosomes (Rihova *et al.*, 1988; Duncan *et al.*, 1987; Seymour *et al.*, 1991). Such HPMA copolymer-based conjugates modified further with galactose (Seymour *et al.*, 1991) or galactoseamine (Duncan *et al.*, 1986) are in phase I/II clinical trial as polymer therapeutics (Vasey *et al.*,1999). Similar tetrapeptides (Gly-Gly-Phe-Gly and Gly-Phe-Gly-Gly) have been used with doxorubicin and carboxymethylpullulan and the resulting conjugates were found to be more potent than doxorubicin after a single *i.v.* injection in rats with Walker 256 carcinoma (Nogusa *et al.*, 1997).

43.2.7.2 Carriers with receptor recognition unit

Increasing number of oligopeptides are being considered for targeting bioactive molecules. Much emphasis has been placed on localizing the recognition unit of various proteins of interest. Frequently, sequences responsible for binding to receptor and/or acceptor structure oligopeptides are synthesized and applied for targeting. Here we summarize those compounds that have been published for delivery purposes.

The laminin peptide fragments GYIGSR-NH_2 and CDPGYIGSR-NH_2 are known to bind to a laminin receptor expressed at higher than normal levels in malignant tumor cells, particularly those of breast and colon carcinomas. Oligopeptides with chelating DTPA moiety were prepared for targeting ^{111}In for specific tumor localization (Mokotoff *et al.*, 1997). Peptide YIGSR was also used as a targeting unit attached to the non-bidegradable polyvinyl-pyrrolidone and its inhibitory effect on experimental lung metastasis of B16-BL6 melanoma cells was examined. The antimetastatic effect of polymer-bound YIGSR was more than 100-fold greater than that of native peptide *in vitro*. After *i.v.* injection the conjugate showed more than a 15-fold longer plasma half-life relative to the pentapeptide (Mu *et al.*, 1999).

Recently we prepared several DTPA-peptide derivatives containing the KDEL signal motif for targeting of non-radioactive $^{115}In^{3+}$ to intracellular receptors in the Golgi apparatus with the potential for visualization by electronmicroscopy (Nagy *et al.*, 2000*c*).

The D-Tyr^4 analog of a selective alpha(v)beta3 integrin antagonist, cyclo(-Arg-Gly-Asp-D-Phe-Val-) was labelled with ^{125}I. This conjugate with preserved selectivity exhibits high affinity for the alpha(v)beta3 integrin *in vitro* and *in vivo* and, as a first reagent of this kind proved to be suitable for the investigation of angiogenesis and metastasis *in vivo* (Haubner *et al.*, 1999). A recombinant, multifunctional protein has been designed for optimized, cell-targeted DNA delivery to integrin receptor-bearing cells. This hybrid construct comprises a viral peptide ligand for integrin alpha(v)beta3 as the recognition unit, polylysine domain for complexing DNA, and a β-galactosidase protein as a purification tag and DNA-shielding agent (Aris *et al.*, 2000).

Arg-Gly-Asp (RGD) sequence has been incorporated or added to porcine pancreatic phospholipase A_2 to induce specific disruption of activated platelets. It has been concluded that the enzyme can be targeted at activated platelets by introducing RGD in a C-terminal extension with a minimum distance (42 amino acids) between RGD and the enzyme. However, more hydrolytic activity is required to eliminate activated platelets (Bekkers *et al.*, 1995).

For new antifibrotic therapies, a specific drug carrier to hepatic stellate cells was developed to inhibit cell activities. Cyclic peptide moieties, with an RGD motif and disulfide bond, recognizing collagen type VI receptors (CGRGDSPC) and HSA as a carrier were conjugated. *In vivo* experiments showed preferential distribution of the conjugate to both fibrotic and normal rat livers. Immunohistochemical analysis demonstrated that the conjugate was predominantly bound to targeted cells in fibrotic livers (Beljaars *et al.*, 2000).

Several organ and tumor-homing peptides selected *in vivo* from a phage peptide library have shown specific binding to different receptors expressed on the vasculature of the target tissue. Targeted delivery of doxorubicin to angiogenic vasculature using these peptides with the Arg-Gly-Asp or Asn-Gly-Arg motif in animal models decreased toxicity and increased the therapeutic efficacy of the drug (Pasqualini, 1999).

Peptidyl derivatives of cytosine-9-arabinofuranoside (ara-C) containing peptide T (a 23-residue peptide derived from the V3 loop of gp120 of HIV-1) recognized CD4+ cells. The conjugates were able to activate CD4 receptor mediated chemotaxis in monocytes and their *in vitro* antiproliferative activity was demonstrated on a panel of leukemia, lymphoma and carcinoma cell lines derived from human tumors (Manfredini *et al.*, 2000)

For membrane targeting, a positively charged peptide (PIP_2-binding peptide, PBP) interacting specifically with a naturally occurring negatively charged phospholipid, phosphatidylinositol 4,5-bisphosphate (PIP_2) has been co-linearly produced with the fusogenic carboxy-terminal part of the β-amyloid peptide (Aβ, amino acids 29–40), involved in Alzheimer's disease. The specific properties of both the Aβ(29–40) and the PBP peptide are conserved in the chimera peptide (Decout *et al.*, 1998).

The internalization of homeodomains and of homeopeptides derived from the third helix of the homeodomain of Antennapedia, a *Drosophila* transcription factor, is used by some investigators to target exogenous hydrophilic compounds into live cells (Prochiantz, 1999). Adriamycin was coupled covalently to two peptides, D-penetratin and SynB1 and its ability to cross the blood-brain barrier was studied. With the conjugated forms increased uptake was found suggesting that the vectorized drug bypasses P-glycoprotein. *In vivo* experiments in mice showed elevated drug concentration in the brain compared with free drug and a significantly decreased adriamycin concentration in the heart (Rousselle *et al.*, 2000).

Peptides from Tat protein of HIV-1 were conjugated to β-galactosidase, horseradish peroxidase, RNase A, and domain III of *Pseudomonas* exotoxin A to assess their transport capability. Tat peptide (residues 1–72 or 37–72) chimera was effective on all cell types tested, showing cellular uptake. In mice, treatment with Tat-β-galactosidase chimera resulted in delivery to several tissues, with high levels in heart, liver, and spleen (Fawell *et al.*, 1994).

A conjugate consisting of a signal import peptide derived from Kaposi fibroblast growth factor and a polycationic polylysine was also produced and complexed with 5′ fluorescently-labelled oligonucleotide. The peptide-polylysine carrier was effective in promoting the cellular uptake of oligonucleotide and has the potential to overcome the problem of oligonucleotide endosomal entrapment and degradation (Dokka *et al.*, 1997).

Phosphodiester antisense oligonucleotides were coupled to a fusogenic peptide, derived from the Influenza hemagglutinin envelope protein by a disulfide or thioether bond. Improved anti-HIV potency of antisense oligodeoxynucleotides (anti-TAT) and oligophosphorothioates was observed (Bongartz *et al.*, 1994). Similar findings were reported with peptide-antisense conjugates containing phosphorothioate oligonucleotide for inhibition of P-glycoprotein expression.

Antitumor imidazotetrazines mitozolomide and temozolomide were introduced to DNA minor and major groove-binding peptidic motifs by solid phase peptide synthesis. When mitozolomide was conjugated to the DNA minor groove-binding peptide $(SPKK)_2$-NH_2 a strong preference for binding with $[dA–dT]_2$ sequences was observed by circular dichroism studies, consistent with the construct making non-covalent interactions within the minor groove. This conjugate showed a >100-fold DNA alkylating activity compared with the free imidazotetrazine as measured by a Taq polymerase assay. Unexpectedly, alkylation was restricted to guanine sites in the major groove of DNA irrespective of the targeting property of the peptidic ligand (Arrowsmith *et al.*, 1999).

A small peptide of repetitive acidic amino acid was proposed as a carrier for selective drug delivery to bone by its binding to hydroxyapatite. Hexapeptide $(Asp)_6$ conjugate of fluorescein isothiocyanate (FITC) was evaluated. The binding of $(Asp)_6$-FITC, but not FITC itself was observed. Targeting to bones after *i.v.* injection was demonstrated *in vivo* (Kasugai *et al.*, 2000).

43.3 Protein/peptide carriers without recognition unit

43.3.1 *Albumin*

Albumin-based drug carrier systems have been developed in the field of chemotherapy to improve passive tumor targeting (Figure 43.1) properties of anti-cancer drugs and the uptake of antiviral or trypanocidal drugs. The rationale of the use of albumin-drug conjugates was the assumption that the conjugates, due to their size, would be localized to cells with a high protein intake. It has been also proposed that the interaction of a macromolecule-attached drug at the site of action is improved relative to that of the free drug. Such a concept was experimentally applied to antiviral chemotherapy using 5-fluoro-deoxyuridine or ara-C with rabbit serum albumin. These conjugates were then used with success on *Ectromeli virus* infected Kupfer cells of the liver *in vitro* and *in vivo*.

In cancer research, antitumour drugs like MTX (Stehle *et al.*, 1997), actinomycin D (Szekerke and Driscoll, 1977), ara-C (Szekerke and Driscoll, 1977), phosphoramide dichloride (see review by Poznansky, 1983) and Dau (Lagueux *et al.*, 1984) were coupled with albumin. Higher, more prolonged serum concentration, decreased excretion rate and elevated drug levels in certain tumors relative to those of free drug were detected. The effect was essentially not dependent on the origin of albumin (e.g. mouse, bovine, human) (Stehle *et al.*, 1999).

Vinblastine-23-oyl amino acid derivatives and the analogous deoxy vinblastine derivatives were synthesized by linking amino acid carbocyclic esters to the vinca-23-oyl moiety, through an amide linkage. Their chemotherapeutic activities on P388, L1210 leukemias and 6C3HED lymphosarcoma in mice were evaluated in comparison to those of the parent alkaloids vinblastine, vincristine, and vindesine. This conjugate was found to be stable in blood and serum up to 48 hours. Lysosomal hydrolases liberate about 50 per cent of the tritiated drug after 48 hours.

Daunomycin has been conjugated with various proteins/polypeptides. The coupling strategies are outlined in Figure 43.4. In succinylated serum albumin conjugates of Dau, an amide bond was formed between the amino group of the drug and a carboxyl side chain of the protein, either directly or with the incorporation of a peptide spacer arm varying from one to four amino acids (Trouet *et al.*, 1982). During *in vitro* incubation with lysosomal hydrolases, intact drug could be released extensively only from conjugates prepared with a tri- or tetrapeptide spacer arm. About 60% of bound drug was released from albumin-Leu-Ala-Leu-Dau and 75% from albumin-Leu-Ala-Leu-Ala-Dau after 10 hr of incubation. A pH optimum of 5.5 was observed for enzymatic release. These conjugates remained very stable in the presence of serum. When tested *in vivo* against the *i.p.* form of L1210 leukemia, the conjugates in which Dau was linked to serum albumin directly or via one amino acid were completely inactive. However, the conjugate with a tetrapeptide spacer arm was more active than free drug inducing a high percentage of long term survivors.

Dau coupled with maleylated BSA was effective against J774A.1 tumor cells in BALB/c mice in lower dosage than

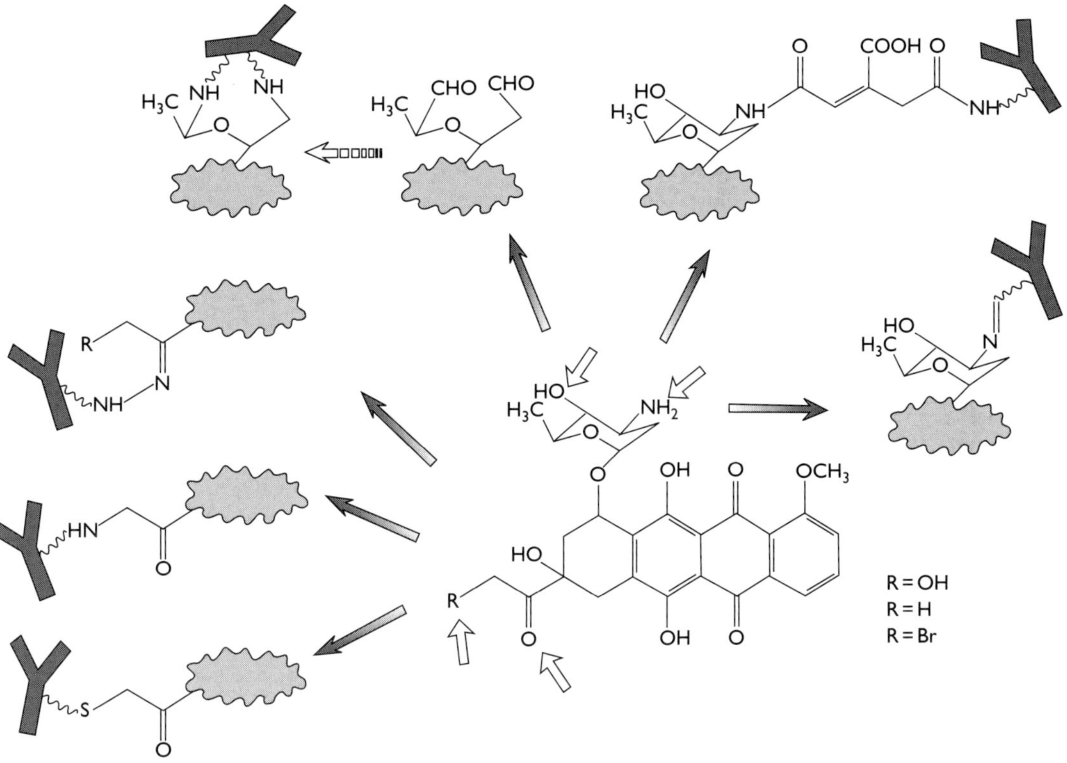

Figure 43.4

the free drug. Binding and internalization of conjugate by macrophage cells were observed *in vitro*. When injected into mice, the drug conjugate was cleared rapidly from the circulation and accumulated in the macrophage-rich tissues, liver, lung and spleen (Mukhopadhyay *et al.*, 1992).

Dau conjugates with BSA (or with ferritin) were used against *Trypanosoma brucei rhodesience*. Drug attached to BSA by a labile glutaraldehyde linkage was effective in *in vitro* and *in vivo* circumstances, while the conjugate comprising a stable succinyl linkage did not show any trypanocidal activity (Golightly *et al.*, 1988).

Several enzymes (e.g. uricase, L-asparaginase, α-glucosidase, superoxide dismutase) have been modified by covalent introduction of albumin and studied for enzyme replacement therapy (Poznansky, 1983 and 1988). In these constructs increased stability, improved blood clearance and antitumor effect (Poznansky, 1983) and decreased immunogenicity (Remy and Poznansky, 1978) were observed.

Albumin has not only been used for delivery of antitumor agents. As a carrier it is coupled with a number of pharmacologically active amines (e.g. histamine, dopamine, norepinephrine), anti-inflammatory agent naproxen, anti-viral azidothymidine, ionophore monensin and immunomodulatory muramyl dipeptide. Prolonged hormonal effects of estrogen-albumin conjugate have been explained by the slow hydrolytic release of the steroid *in vivo*. Cytotoxic peptides, linear amanitin, and cyclic phalloidin containing conjugates showed preferential uptake by hepatocytes and efficient killing of macrophages compared to free peptides.

Cationized albumin (isoelectric point value greater than 8) enters cerebrospinal fluid rapidly from blood. Therefore the directed delivery of peptides through the blood-brain barrier was examined by coupling ^{3}H-β-endorphin to unlabelled cationized albumin (pI = 8.5–9) using the bifunctional reagent, N-succinimidyl-3-(2-pyridyldithio)proprionate (SPDP). The conjugate was rapidly bound and endocytosed by isolated bovine brain capillaries (Kumagai *et al.*, 1987).

43.3.2 *Fibrinogen, lysozyme*

Several authors have examined the potential use of plasma fibrinogen as a carrier of antitumour drugs like methotrexate, N,N-(*bis*-2-chloroethyl)-phenylene diamine and phosphoramide dichloride. Lysozyme has been tested as a low molecular weight protein drug carrier that release drugs site-specifically in the kidney (Franssen *et al.*, 1992). Different drugs (e.g. naproxen) were attached to lysozyme both directly (amide bond) and via different spacers: oligopeptides (amide bond), (poly)α-hydroxy acids (ester bond), and a pH sensitive *cis*-aconityl spacer (amide bond). The capability of the kidney to release the parent drug from conjugates bearing suitable spacer was demonstrated (Haas *et al.*, 1997)

43.3.3 Synthetic polypeptides

Synthetic polymeric polypeptides are utilised as macromolecular carriers for "passive" delivery of bioactive molecules (Figure 43.1). These conjugates exhibit improved solubility, beneficial pharmacological properties (extended blood survival, specific tissue accumulation), diminished cytotoxicity and increased therapeutic efficacy as compared to free drugs and radionuclides (Duncan, 1992; Takakura and Hashida, 1995). Such polypeptides are also applied as intermediate carriers in target specific "active" delivery of drugs, radionuclides etc. by coupling to homing devices like antibodies, hormones, or receptor specific sugar residues (Figure 43.2).

43.3.3.1 Linear polypeptides

Among synthetic linear poly-α-amino acids, polyanionic poly(L-glutamic acid), poly(L-aspartic acid) and polycationic poly(L-lysine) have been used for delivery of antitumor agents, enzymes or albumin.

Polyglutamic acid was conjugated with p-phenylenediamine, mitomycin C or adriamycin. Compared with free drug, polyaspartic acid derivative of Dau was less toxic to HeLa cells *in vitro*, but proved to be more effective against murine tumors tested *in vivo* (Zunino *et al.*, 1984). No such effect was observed with mitocycin C conjugate (Roos *et al.*, 1984). Recently, dexamethasone as a model drug coupled with poly(L-Asp) using an ester bond was investigated to selectively release the drug in the large intestine (Leopold and Friend, 1995). In *in vivo* experiments, increased efficacy and reduced toxicity were reported. Poly(ethyleneglycol)-poly (L-Asp) copolymer attached adriamycin was reported to be active against P388 tumor by forming a micelle (Yokoyama *et al.*, 1990).

Poly(L-lysine) or poly(D-lysine) as lysosomotropic agent was coupled with various antitumor drugs (Arnold, 1985) such as the nicotinamide antimetabolite, 6-aminonicotinamide, or MTX (Ryser and Shen, 1978), adriamycin, Dau (Shen and Ryser, 1981; Zunino *et al.*, 1984), mitomycin C (Roos *et al.*, 1984) and proteins (HSA, horse radish peroxidase). Poly(D-Lysine) conjugate of MTX with disulphide bond was applied to the analysis of reductive processes during endocytosis (Shen *et al.*, 1985). Enhanced cellular uptake of protein conjugates and of MTX-poly(L-Lys) by the drug resistant cell was observed *in vitro*. Poly(L-Lys) conjugate of Dau, compared to the poly(L-Asp)-Dau conjugate, produced no enhancement of drug activity *in vivo* experiments against P388 leukemia (Zunino *et al.*, 1984).

The effect of various spacers (succinic acid, glutaric acid, suberic acid, dithiodiacetic acid, dithiodipropionic acid, oxalic acid, phtalic acid) between Dau and poly(L-Lys) were studied. Under *in vitro* circumstances the activity of all conjugates was only 10–15% that of the free drug against HeLa cells using ^{3}H thymidine uptake assay (Arnold, 1985).

One of the earliest attempts to target polypeptide-drug conjugate with the aid of antibodies was made in 1975. Poly(L-Glu) as an intermediate carrier coupled with p-phenylenediamine was attached to mouse lymphoma specific polyclonal immunoglobulins. This preparation showed greater antitumor activity than did components or mixtures of the conjugate.

Neoglycoprotein-like variants of synthetic poly-α-amino acids were invented by coupling of carbohydrate residues to poly(L-Lys). Galactose containing poly(L-Lys) coupled ara-AMP and acyclovir are active. Preclinical studies are now underway with conjugates obtained using lactosaminated poly(L-Lys) as the hepatotropic carrier and ara-AMP, ribavirin or azidothymidine (Di Stefano *et al.*, 1995). These conjugates have some advantages over those prepared with lactosamine modified HSA: they can be administered by the intramuscular route and can have a heavy drug load.

For the targeting of macrophage mannose receptors and the internalization of the norfloxacin antibiotic, which is active against some intracellular bacteria, poly(L-Lys citramide imide) was conjugated with the antibiotic and mannosyl moieties. This carrier, which derived from two metabolites, citric acid and L-Lys, is known to be biocompatible and slowly degradable under slightly acidic conditions. The mannosyl groups were incorporated through a biodegradable glycolic spacer arm. Competition studies between the conjugate and glucose oxidase, a mannosyl-bearing non-human protein indicated specific binding of the drug-macromolecule to the mannosyl receptor-bearing membranes of macrophages (Gac *et al.*, 2000).

43.3.3.2 Branched chain polypeptides

Very few systematic studies have been reported on structural and functional factors required for an optimal polypeptide carrier. To this end we have initiated systematic structure-function type studies for rational selection of polypeptides as synthetic macromolecular carrier. For this we have prepared new groups of branched chain polypeptides with the general formula of poly[Lys(X_i-DL-Ala_m)] (XAK) (Figure 43.5), poly[Lys(X_i-DL-Ser_m), (XSK), poly[Lys(X_i)] (X_iK) or poly[Lys(DL-Ala_m-X_i)] (AXK) where i~1, m~3, and X represent an additional optically active amino acid residue (Hudecz *et al.*, 1985, Mező *et al.*, 1997; Hudecz, 1995a; Hudecz *et al.*, 1999). These polymeric polypeptides were characterized by their size, chemical (primary structure, polydispersity, solution conformation) and biological (*in vitro* cytotoxicity, pyrogenicity, biodegradation, immunoreactivity and biodistribution) properties (Hudecz *et al.*, 1985; Clegg *et al.*, 1990; Hudecz, 1995*b* and 1999; Mező *et al.*, 1997). It has also been shown that the composition of the polymeric

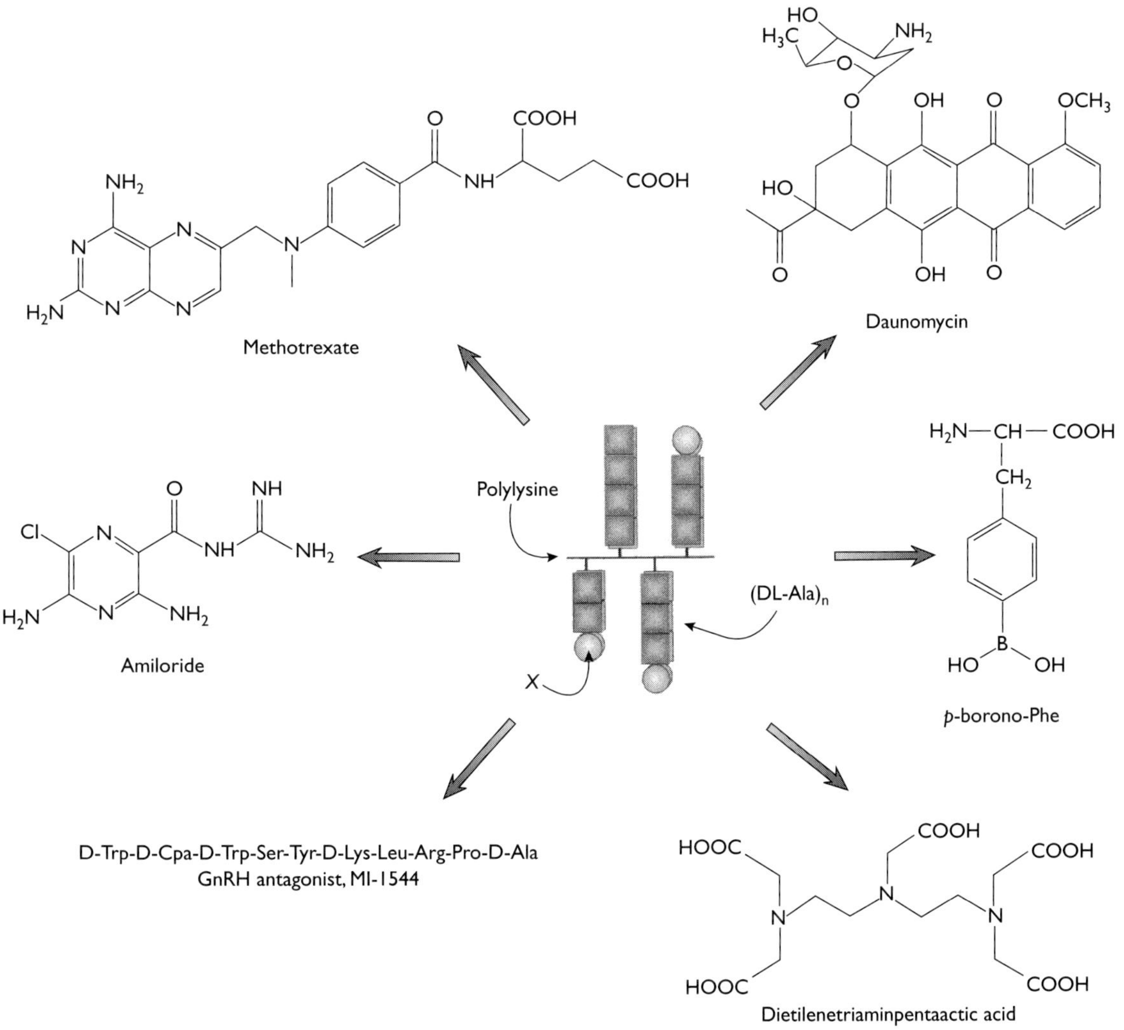

Figure 43.5

component has a marked influence on their interaction with phospholipid mono- or bilayers (Nagy *et al.*, 2000*b*).

Recently we have prepared several conjugates in which various antitumor agents have been covalently attached to selected structurally-related branched polymeric polypeptides (Figure 43.5). Conjugates of Dau (Hudecz *et al.*, 1992*a*; Reményi *et al.*, 2000), MTX (Hudecz *et al.*, 1993; Kóczán and Hudecz, 2000), GnRH antagonist (Mező *et al.*, 1996*a*; Vincze *et al.*, 1994), amiloride (Pató *et al.*, 1999), boron compounds (Mező *et al.*, 1996*b*) and several radionuclides (Pimm *et al.*, 1995; Perkins *et al.*, 1998) have been synthesized. The solution conformation, *in vitro* cytotoxicity, biodistribution and antitumor properties were analyzed.

Antitumor activity of Ac-[D-Trp$^{1-3}$, D-Cpa^{2}, D-Lys^{6}, D-Ala^{10}]-GnRH-poly[Lys(Ac-Glu$_i$-DL-Ala$_m$)] (Ac-EAK) conjugate has been clearly documented (Mező *et al.*, 1996*a*; Vincze *et al.*, 1994). The coupling of the acid labile derivative of daunomycin to poly[Lys(Glu$_i$-DL-Ala$_m$)] (EAK) resulted in compensation of the immunosuppressive effect of the drug and this polymeric conjugate *in vivo* was very effective against L1210 leukemia producing 66–100% long term survivors (>60 days) in mice (Gaál and Hudecz, 1998). Carrier dependent anti-*Leishmania donovani* effect of MTX branched polypeptide conjugates were observed *in vitro* and *in vivo* (Kóczán *et al.*, 1998 and 2000).

43.4 The chemical approach of methotrexate targeting: an example

Methotrexate (or L-Amethopterine hydrate, L-4-amino-N^{10}-methylpteroyl-glutamic-acid), a folic acid antagonist, has proved to be of high therapeutic value in various clinical conditions. Since its introduction in 1948 for the treatment of acute lymphoblastic leukemia it became widely used in cancer chemotherapy as an anti-neoplastic and immunosuppressive agent. Recent studies indicate its potential activity against parasitic diseases. As an antimetabolite MTX exerts severe side effects, such as bone marrow depression, hepatotoxicity, nephrotoxicity, pulmonary complications, and neurotoxicity.

Utilising the results of the numerous structure-activity relationship studies it seemed to be straightforward to try to increase the clinical activity and selectivity, or decrease the toxic side effects of the drug by conjugating it to different carriers.

43.4.1 Synthesis

Methotrexate (Figure 43.6A) is supplied by more companies (e.g. Aldrich, Lederle) at high purity >98%). It is a yellow crystalline powder that is not soluble in distilled water, slightly acidic buffers or even in human plasma. However, it is readily soluble at basic pH (e.g. in 0.1 M $NaHCO_3$ or PBS, pH 7.4) and in organic solvents like DMF, DMSO, methanol or dioxane. The crystalline compound contains water molecules, which can interfere with certain coupling reactions (e.g. the symmetrical anhydride method), however the carbodiimide or active ester methods are not so sensitive for water. Prolonged drying over P_2O_5 could result in an anhydrous compound (Kulkarni *et al.*, 1981). MTX is stable in its solid form and in neutral or acidic solutions, but at higher pH (>11) it decomposes to its keto-derivative.

43.4.1.1 Coupling strategies

The methods for the conjugation of methotrexate to carriers predominantly involve the glutamic acid moiety. Activation of α- and γ-carboxyl groups has been achieved by

a N,N′-dicyclohexyl-carbodiimide in DMF (Kulkarni *et al.*, 1981) or *in situ* in the presence of carrier using water soluble carbodiimide like EDC (Kulkarni *et al.*, 1981; Sanzgiri *et al.*, 1990; Shih *et al.*, 1988) and CMC (Przybylski *et al.*, 1978),
b ester formation with N-hydroxysuccinimide (Ryser and Shen, 1978; Shih *et al.*, 1988), or
c synthesis of mixed anhydride (Kulkarni *et al.*, 1981).

MTX hydrazide was also examined to introduce a hydrazone bond with the carbohydrate moieties in IgG (Kralovec *et al.*, 1989). Since MTX has two reactive carboxylic groups, during conjugation the mixture of two isomers (α and γ) can be formed. Rosowsky and Yu (1978) studied the relative reactivity of the two groups and found that the formation of γ isomer is favored. In a different study the yield of the γ isomer was found to be 2.3 times higher than of the α isomer. With the total synthesis of some MTX derivatives with only one carboxylic group, conjugates with unambiguous structure can be produced (Heath *et al.*, 1986). It should be noted that none of the above reported methods have been used for selective derivatization of α- or γ-carboxyl groups of MTX for coupling to carrier proteins.

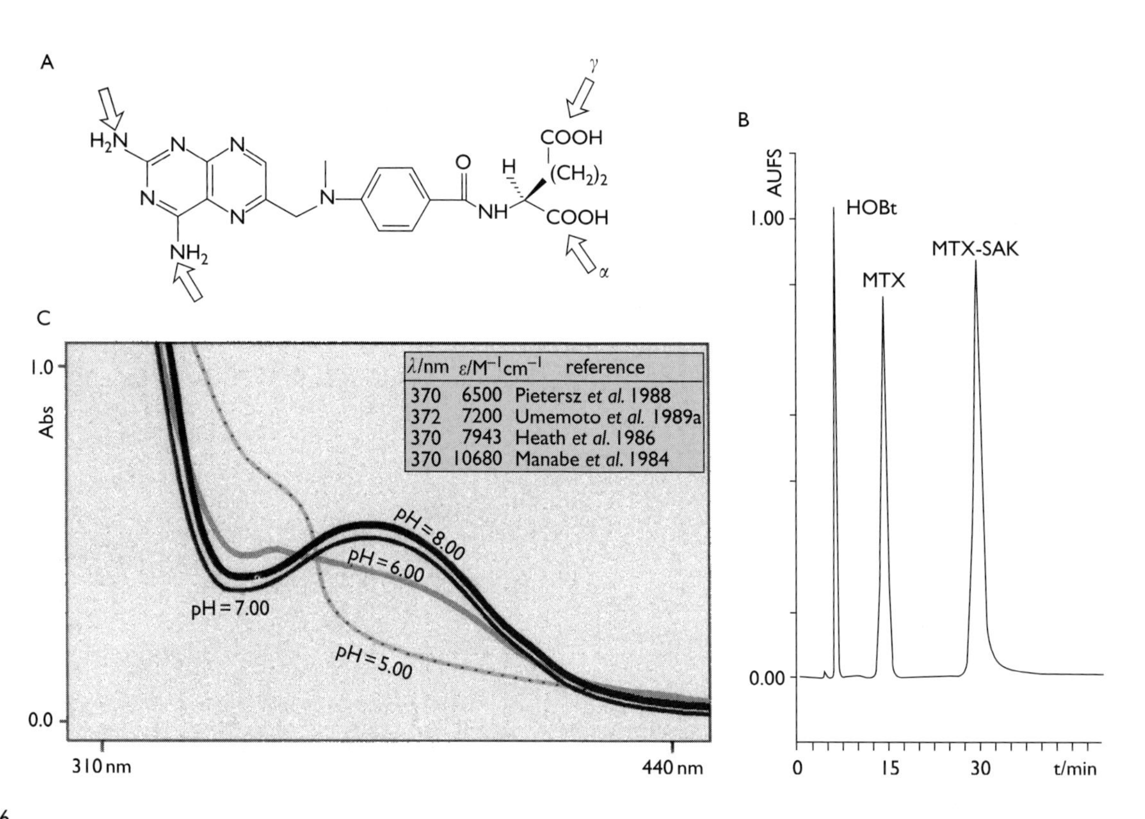

Figure 43.6

The two amino groups at the pteroic moiety at positions 2 and 4 are much less reactive than those of common aliphatic amines, e.g. prolonged incubation of MTX with a coupling reagent (e.g. DCC in the presence of tertier amine) does not result in the formation of MTX-oligomers. Also no dimer formation was observed in DMF solution of the N-hydroxy-succinimide active ester of MTX, which is stable for months at 4 °C (Kulkarni *et al.*, 1981). The diazonium derivative of the carrier, that reacts spontaneously with the amino groups of MTX and form a benzidine type compound was also applied. A more frequently used method is applied for the conjugation of polysaccharides or glycoproteins. In this case the carbohydrate moiety is oxidized with periodate to form aldehyde groups which also react with the aromatic amino groups of MTX and the resulting Schiff base can be reduced to a stable secondary amine with $NaBH_4$ (Manabe *et al.*, 1984). However, the literature is ambiguous about the stability of MTX in the presence of $NaBH_4$. Since the amino group at position 4 was reported to be necessary for the biological effect of MTX, this method can produce totally ineffective derivatives. Aminodextran conjugate of MTX has been linked as described above (Manabe *et al.*, 1984).

Conjugates have also been synthesized by incorporation of cysteine or 2-mercapto-ethylamine (Umemoto *et al.*, 1989*a*), tetrapeptide (Leu-Ala-Leu-Ala) (Umemoto *et al.*, 1989*b*) and of $(Glu)_n$, where n = 2–5 oligopeptides (Abraham *et al.*, 1990) as spacers.

43.4.1.2 *Coupling methods in practice*

By far the most popular coupling technique is the utilization of a water-soluble carbodiimide (for a summary see Table 43.1). The presence of HOBt could suppress the possible side reactions (e.g. dehydration). A slight excess of MTX

Table 43.1 Summary of different coupling methods, analytical procedures reported in literature*

Coupling method/ solvent	*Analytical/purification methods*	*Carrier/composition of conjugate*	*Spacer*	*Reference*
Carbodiimide method				
EDC	GF	BSA-MTX_{22}	no	Bore *et al.* 1984
EDC	DIA	methylated-BSA-MTX	no	Hendel *et al.* 1976
EDC, pH = 6.0	UF, UV, Lowry, R, HPLC	Gal_{24}-BSA-MTX_{10-12}	no	Han *et al.* 1999*a*
EDC, 0.01 M Tris/HCl, pH = 8.5	GF, AF, UV, R, IE, PAGE	Con-A-$MTX_{5.3}$	no	Tsuruo *et al.* 1980
EDC, 0.01 M phosphate, pH = 7.5	UV, R, GF	IgG-$MTX_{2.5}$	no	Hansen *et al.* 1987
EDC		IgG_1-HSA-MTX_{62}	-CO-C_2H_4-S-CH_2CO-	Li *et al.* 1990
EDC, PBS	GF, UV, R, Stab.	IgG_{2b}-$(HSA\text{-}MTX_{32})_{1\text{-}3}$	-CO-C_2H_4-S-CH_2CO-	Garnett *et al.* 1983
EDC, 0.04 M $NaHCO_3$, pH = 7.4	GF, UV, Stab.	IgG_1-Dextran-MTX_{30-50}	amino-dextran	Shih *et al.* 1988
EDC, PBS, pH = 6.4–6.5	GF, HPLC, UV, R	IgG_{2b}-HSA-MTX	-CO-C_2H_4-S-CH_2CO-	Garnett and Baldwin 1986
EDC, PBS, pH = 7.2	GF, DIA, UV, HPLC	poly(L-Lys)-MTX	$(Glu)_n$	Abraham *et al.* 1990
EDC, H_2O, pH = 7.0	GF, TLC, AAA	poly(D-Lys)-MTX	$(Gly)_3$	Shen *et al.* 1985
CMC, H_2O/dimethyl-sulphoxide, pH = 3	DIA, UV, IR	poly(iminoethylene)-MTX	no	Przybylski *et al.* 1978
DCC, *multistep synthesis*	UV, DIA, HPLC	IgG_{2a}-MTX	-*LALA*-N_2H_2CO-CH_2-	Umemoto *et al.* 1989*b*
Active ester method				
DCC+NHS, PBS, pH = 7.3	GF, DIA, RIA, Stab.	IgG_{2b}-$MTX_{1.6-1.9}$	no	Pimm *et al.* 1988
DCC+NHS, PBS, pH = 7.0	GF, UV	$F(ab)_2$-MTX	no	Kulkarni *et al.* 1985
DCC+NHS, PBS	GF, UV, PAGE, R	IgG-HSA-MTX_{23}	HSA with complex linker	Pietersz *et al.* 1988
Anhydride method				
^{i}Bu-O-CO-Cl, $NaHCO_3$	GF	β-D-galactosidase-MTX	no	Al-Bassam *et al.* 1979
Ac_2O		γ-globulin-MTX	no	Latif *et al.* 1980
Other methods				
BOP, *total synthesis, selective γ-coupling*	HPLC, MS, NMR	LHRH–MTX	no	Nagy *et al.* 1993
hydrazon, PBS	GF, UV, Lowry, NMR	IgG-MTX_{4-7}	no	Kralovec *et al.* 1989*b*
azide, PBS	GF, UV, Lowry, NMR	IgG-MTX_{4-7}	no	Kralovec *et al.* 1989*b*
Schiff-base/$NaBH_4$, PBS	GF, UV	IgG_1-Dextran-$MTX_{4.0-9.3}$	dextran	Manabe *et al.* 1984
SPDP, PBS	GF, TLC, AAA	Poly-(D-Lys)-MTX	-NH-C_2H_4-S-S-C_2H_4-CO-	Shen *et al.* 1985
SPDP, PBS, pH = 7.5	IR, UV, DIA	IgG_{2a}-MTX	-NH-C_2H_4-S-S-C_2H_4-CO-	Umemoto *et al.* 1989*a*

*Abbreviations used
AAA: amino acid analysis, AF: affinity methods, DIA: dialysis, GF: gel filtration, HPLC: high performance liquid chromatography, IE: isoelectric focusing, IR: infra red spectroscopy, Lowry: protein-content measurement, MS: mass spectrometry, NMR: magnetic resonance spectrometry, PAGE: polyacrylamide gelelectrophoresis, R: radioactive methods, RIA: radioimmune assay, Stab.: stability studies, TLC: thin layer chromatography, UF: ultrafiltration, UV: UV spectrometry. BOP: benzotriazole-1-yl-oxy-tris(dimethylamino)-phosphoniumhexafluorophosphate, CMC: 1-cyclohexyl-3-(-2-morpholinoethyl)carbodiimide metho-p-toluenesulphonate, DCC: NN′-dicyclohexylcarbodiimide, EDC: 1-ethyl-3-(-3-dimethylaminopropyl)-carbodiimide, NHS:N-hydroxysuccinimide, PBS: phosphate buffered saline, SPDP: 3-(2-pyridyldithio)-propionic acid N-hydroxysuccinimide ester.

and preactivation (10 min) with the carbodiimide reagent can overcome dimer formation between the carrier molecules. For the preactivation step DMF can be used as a solvent. The activated MTX-derivative is than added drop-wise to the water or buffer solution of the carrier protein. There is an optimum pH for coupling; the elevation of pH will increase the number of free, deprotonated amino groups of protein, but the hydrolysis rate of the active MTX will increase at the same time. Controling the pH in the range of 6–8 in general results in acceptable coupling efficiency. Although methanol is widely used as a solvent for preactivation it should be avoided since MTX methyl esters can be formed as stable side products. Examples of the asymmetrical anhydride method have also been reported. In this case the MTX solution in DMF is incubated with isobuthyl-chloroformate at 0–10 °C for 10–30 minutes (Al-Bassam *et al.*, 1979).

As an alternative to the carbodiimide method, active ester derivative of MTX can be prepared (N-hydroxy-succineimide+DCC in DMF) and isolated. This compound can be stored for a longer period of time (Kulkarni *et al.*, 1981). Recently a novel coupling method based on reagents BOP or HBTU was proposed. In this case the use of HOBt as an additive is highly encouraged to avoid the possible nitrile formation from the glutamine or asparagine residues of the protein/polypeptide.

All these methods utilise the carboxylic groups of the MTX to form amide bonds with the free amino groups of the carrier. This bond is stable at physiological conditions; there is no report of leakage of the drug from the conjugate. However, neither of the aforesaid methods has enough selectivity to avoid the reaction with Ser/Thr hydroxyl and Cys thiol groups, such as the formation of ester or thioester bound MTX. These compounds have much lower stability thus their formation should be avoided. Treatment of conjugates with NH_2OH followed by gel filtration could be of help in overcoming this problem (Hudecz *et al.*, 1992*b*).

43.4.1.3 *Coupling efficiency*

The coupling efficiency depends on numerous factors (e.g. isoelectric point of the protein/polypeptide, number of reactive groups, pH, concentration, solvent). Therefore the prediction of the composition of the conjugate is difficult, however Table 43.1 could give some hints. It is good practise to make some pilot experiments with different input stoichiometry to obtain the stoichiometry-degree of substitution relationship that can be used to adjust the composition of the end-product.

43.4.1.4 *Spacers and carriers*

The reported spacers and macromolecular carriers are summarized in Table 43.1. In general there are two objectives to using linkers in MTX conjugates. One is to obtain a high MTX content without diminishing the target-recognition unit of the carrier (e.g. antibody). In this case a polymer (e.g. HSA, dextran) is conjugated in a 1:1 molar ratio to the carrier. The polymer can carry several MTX residues (Shih *et al.*, 1988). The other reason for the incorporation of a spacer is to have a specific cleavage site for the "programmed" release of the drug (e.g. the Leu-Ala-Leu-Ala sequence (Umemoto *et al.*, 1989*b*) to be cleaved by lysosomal enzymes).

A great number of diverse macromolecules have been used in the synthesis of MTX conjugates. The conjugation reaction should be optimized with respect to the properties of the selected carrier. The biological activity (e.g. enzymatic activity or antigen recognition) should be preserved. One possibility is to keep the number of the conjugated species as low as possible. In this case highly potent drugs (e.g. toxin) should be used. Alternatively, the homing site of the molecule should be protected during the conjugation procedure by providing the substrate during the reaction (e.g. conjugating MTX to Con-A in the presence of α-methyl-D-mannoside (Tsuruo *et al.*, 1980), or using the antibody-antigen complex during conjugation of MTX to the antibody portion).

43.4.2 *Purification*

Purification methods such as reversible precipitation, dialysis, gel filtration, preparative HPLC, affinity chromatography are used extensively. The UV absorption of MTX at 370 nm makes detection easy. Details are listed in Table 43.1. The homogenity of the conjugate is commonly checked by SDS-PAGE or gel exclusion chromatography. With these methods the high molecular weight cross-linked side-products can be detected. Low affinity impurities can be detected with affinity chromatography. Recently HPLC became the method of choice for the analysis of MTX conjugates (Kóczán and Hudecz, 2000, Figure 43.6B). Time required for analysis is short and the sensitivity is high. High pore sized reversed phase (C4–C18) packings are optimal.

Sterilization of the end-product can be achieved with filtration, however sometimes the solutions of conjugates are rather viscous and the recovery during filtration is low. In this case working in a sterile environment and starting from sterile compounds can help.

43.4.3 *Determination of the composition*

Since the structure of the conjugates are rather complex, analytical methods of high efficiency are needed to determine the composition and the average degree of substitution. Different approaches are used (e.g. fluorimetry (Kinkade *et al.*, 1974), radioactive technics, RIA (Aherne *et al.*, 1977), ELISA (Al-Bassam *et al.*, 1979)), however the most commonly applied method for determining the MTX content is based on UV absorption spectroscopy. Since MTX contains ionisable groups the UV spectra is highly pH dependent. Due to the different conditions under which UV measurements are performed a great diversity of ε-values have been collected in the literature. Selected ε values are depicted in Figure 43.6C. These ε values can be used only in those cases where the aromatic pteroic ring of MTX is not affected by the conjugation reaction.

Reliable values for MTX content can be gained after the removal of free and ester bound MTX from the conjugate sample. For this purpose HPLC methods were developed by us. Incubating the conjugate with NH_2-OH releases the ester bound MTX as a hidroxamic-acid derivative (Hudecz *et al.*, 1992*b*), and 3 different components can be detected and measured in the chromatogram.

The determination of the carrier quantity in the conjugate may be more complex than that of the MTX moiety. Although MTX absorbs at 280 nm, proteins can be measured by UV absorption at this wavelength (Shih *et al.*, 1988). In this case the absorbance value should be corrected with the absorbance of MTX. However because of the propagation of the error of the absorbance measurements the standard error of this method is always high.

Biological methods applied frequently (e.g. immunological, enzyme activity) are less reliable for measuring the protein or MTX content, since the conjugation can alter the antigen binding or enzyme activity. In our laboratory it was found that the protein content that is measured using the Lowry method is also highly dependent on the MTX content of the conjugate.

Amino acid analysis after treatment with 6 M HCl at 120°C provides some guidance. Under this condition not only the protein/peptide part, but MTX is also hydrolyzed and stoichiometrically releases glutamic acid. A promising alternative is to use mass spectrometry for the determination of conjugate composition

Acknowledgment

Experimental work summarized in this paper was supported by grants from the Hungarian-Spanish (5/1998) and Hungarian-Indian Intergovernmental Programme (4/1996), from the Hungarian Research Fund (OTKA No. T-3024, T-4217, T-014964, T-03838) and from the Hungarian Ministry of Welfare (ETT No. T405, 017/1993, 115/1996, 12/2000). The composition of this review chapter was supported by grant from the Hungarian Ministry of Education (1/047 Medichem).

References

1 Aboud-Pirak, E., Lesur, B., Rao, K.S.P., Baurain, R., Trouet, A. and Schneider, Y.J. (1989) Cytotoxic activity of daunorubicin or vindesin conjugated to monoclonal antibody on cultured MCF-7 breast carcinoma cells. *Biochem. Pharm. 38, 641–648.*

2 Abraham, A., Nair, M.G., Kisliuk, R.L., Gaumont, Y. and Galivan, J. (1990) Folate analogues. 33. Synthesis of folate and antifolate poly-γ-glutamates by [(9-fluorenylmethoxy)oxy] carbonyl chemistry and biological evaluation of certain methotrexate polyglutamate polylysine conjugates as inhibitors of the growth of H35 hepatoma cells. *J. Med. Chem. 33, 711–717.*

3 Adams, G.P. and Schier, R. (1999) Generating improved single-chain Fv molecules for tumor targeting. *J. Immunol. Methods 231, 249–260.*

4 Aherne, G., Piall, E. and Marks, V. (1977) Development and application of a radioimmunoassay for methotrexate. *Br. J. Cancer 36, 608–617.*

5 Al-Bassam, M., O'Sullivan, M., Bridges, J. and Marks, V. (1979) Improved double-antibody enzyme immunoassay for methotrexate. *Clin. Chem. 25, 1448–1452.*

6 Albert, R., Smith-Jones, P., Stolz, B., Simeon, C., Knecht, H., Bruns, C. and Pless, J. (1998) Direct synthesis of [DOTA-DPhe^{1}]-octreotide and [DOTA-DPhe^{1},Tyr^{3}]-octreotide (SMT487): two conjugates for systemic delivery of radiotherapeutical nuclides to somatostatin receptor positive tumors in man. *Bioorg. Med. Chem. Lett. 8, 1207–1210.*

7 Ali, S.A., Joao, H.C., Hammerschmid, F., Eder, J. and Steinkasserer, A. (1999) Transferrin trojan horses as a rational approach for the biological delivery of therapeutic peptide domains. *J. Biol. Chem. 274, 24066–24073.*

8 Allen, H.J., Ahmed, H. and DiCioccio, R.A. (1990) Metabolic correction of fucosidosis lymphoid cells by galaptin-α-L-fucosidase conjugates. *Biochem. Biophys. Res. Commun. 172, 335–340.*

9 Arano, Y., Uezono, T., Akizawa, H., Ono, M., Wakisaka, K., Nakayama, M., Sakahara, H., Konishi, J. and Yokojama, A. (1996) Reassessment of diethylenetriaminepentaacetic acid (DTPA) as a chelating agent for ^{111}In labeling of polypeptides using a newly synthesized monoreactive DTPA derivative. *J. Med. Chem. 39, 3451–3460.*

10 Arís, A., Feliu, J.X., Knight, A., Coutelle, C. and Villaverde, A. (2000) Exploiting viral cell-targeting abilities in a single polypeptide, non-infectious, recombinant vehicle for integrin-mediated DNA delivery and gene expression. *Biotechnol. Bioeng. 68, 689–696.*

11 Arnold, L.J. Jr. (1985) Polylysine-drug conjugates. *Methods Enzymol. 112, 270–285*

12 Arnon, R. and Sela, M. (1982) *In vitro* and *in vivo* efficacy of conjugates of daunomycin with antitumour antibodies. *Immunol. Rev. 62, 5–27.*

13 Arrowsmith, J., Missailidis, S. and Stevens, M.F. (1999) Antitumour imidazotetrazines. Part 37. Conjugation of the DNA major-groove alkylating imidazotetrazine mitozolomide to peptide motifs recognizing the minor groove. *Anticancer Drug Res. 14, 205–217.*

14 Atherton, F.R., Hall, M.J., Hassall, C.H., Holmes, S.W., Lambert, R.W., Lloyd, W.J. and Ringrose, P.S. (1980) Phosphonopeptide antibacterial agents related to alafosfalin: design, synthesis, and structure-activity relationships. *Antimicrob. Agents Chemother. 18, 897–905.*

15 Bagutti, C. and Eberle, A.N. (1993) Synthesis and biological properties of a biotinylated derivative of ACTH$^{1-17}$ for MSH receptor studies. *J. Recept. Res. 13, 229–244.*

16 Ballard, J.D., Doling, A.M., Beauregard, K., Collier, R.J. and Starnbach, M.N. (1998) Anthrax toxin-mediated delivery *in vivo* and *in vitro* of a cytotoxic T-lymphocyte epitope from ovalbumin. *Infect. Immun. 66, 615–619.*

17 Bard, D.R., Knight, C.G. and Page-Thomas, D.P. (1990) Targeting of a chelating derivative of a short-chain analogue

of α-melanocyte stimulating hormone to Cloudman S91 melanomas. *Biochem. Soc. Trans. 18, 882–883.*

18 Barth, R.F., Soloway, A.H., Fairchild, R.G. and Brugger, R.M. (1992) Boron neutron capture therapy for cancer. Realities and prospects. *Cancer 70, 2995–3007.*

19 Barth, R.F., Soloway, A.H., Goodman, J.H., Gahbauer, R.A., Gupta, N., Blue, T.E., Yang, W. and Tjarks, W. (1999) Boron neutron capture therapy of brain tumors: an emerging therapeutic modality. *Neurosurgery 44, 433–450.*

20 Bekkers, A.C., van der Vuurst, H., van Willigen, G., Akkerman, J.W. and Verheij, H.M. (1995) Targeting of porcine pancreatic phospholipase A_2 to human platelets: introduction of an RGD sequence by genetic engineering. *Thromb. Haemost. 74, 1138–1144.*

21 Beljaars, L., Molema, G., Weert, B., Bonnema, H., Olinga, P., Groothuis, G.M., Meijer, D.K. and Poelstra, K. (1999) Albumin modified with mannose-6-phosphate: A potential carrier for selective delivery of antifibrotic drugs to rat and human hepatic stellate cells. *Hepatology 29, 1486–1493.*

22 Beljaars, L., Molema, G., Schuppan, D., Geerts, A., De Bleser, P.J., Weert, B., Meijer, D.K. and Poelstra, K. (2000) Successful targeting to rat hepatic stellate cells using albumin modified with cyclic peptides that recognize the collagen type VI receptor. *J. Biol. Chem. 275, 12743–12751.*

23 Ben-Yedidia, T. and Arnon, R. (1997) Design of peptide and polypeptide vaccines. *Current Opinion in Biotechnology 8, 442–448.*

24 Bernstein, A., Hurwitz, E., Maron, R., Arnon, R., Sela, M. and Wilchek, M. (1978) Higher antitumor efficacy of daunomycin when linked to dextran: *in vivo* and *in vitro* studies. *J. Natl. Cancer Inst. 60, 379–384.*

25 Bonardi, M.A., French, R.R., Amlot, P., Gromo, G., Modena, D. and Glennie, M.J. (1993) Delivery of saporin to human B-cell lymphoma using bispecific antibody: targeting via CD22 but not CD19, CD37, or immunoglobulin results in efficient killing. *Cancer Res. 53, 3015–3021.*

26 Bonfils, E., Depierreux, C., Midoux, P., Thuong, N.T., Monsigny, M. and Roche, A.C. (1992) Drug targeting: synthesis and endocytosis of oligonucleotide-neoglycoprotein conjugates. *Nucleic Acids Res. 20, 4621–4629.*

27 Bongartz, J.P., Aubertin, A.M., Milhaud, P.G. and Lebleu, B. (1994) Improved biological activity of antisense oligonucleotides conjugated to a fusogenic peptide. *Nucleic Acids Res. 22, 4681–4688.*

28 Boonkitticharoen, V. and Laohathai, K. (1992) Analysis of immunoradiological properties of radiolabelled monoclonal antibody for tumour localization. *Nucl. Med. Commun. 13, 342–348.*

29 Bore, P., Rahmani, R., Cano, J., Just, S. and Barbet, J. (1984) Radioimmunoassays of 7-hydroxymethotrexate and methotrexate. *Clin. Chim. Acta 141, 135–149.*

30 Buchsbaum, D.J., Rogers, B.E., Khazaeli, M.B., Mayo, M.S., Milenic, D.E., Kashmiri, S.V., Anderson, C.J., Chappell, L.L., Brechbiel, M.W. and Curiel, D.T. (1999) Targeting strategies for cancer radiotherapy. *Clin. Cancer Res. 5, Suppl 3048s–3055s.*

31 Chakraborty, P., Bhaduri, A.N. and Das, P.K. (1990) Neoglycoproteins as carriers for receptor-mediated drug targeting in the treatment of experimental visceral leishmaniasis. *J. Protozool. 37, 358–364.*

32 Chen, J., Stickles, R.J. and Daichendt, K.A. (1994) Galactosylated histone-mediated gene transfer and Burstein expression. *Hum. Gene. Ther. 5, 429–435.*

33 Chen, J., Giblin, M.F., Wang, N., Jurisson, S.S. and Quinn, T.P. (1999) *In vivo* evaluation of ^{99m}Tc/^{188}Re-labeled linear α-melanocyte stimulating hormone analogs for specific melanoma targeting. *Nucl. Med. Biol. 26, 687–693.*

34 Clegg, J.A., Hudecz, F., Mező, G., Pimm, M.V., Szekerke, M. and Baldwin, R.W. (1990) Carrier design: biodistribution of branched polypeptides with poly-(L-lysine) backbone. *Bioconjugate Chem. 2, 425–430.*

35 Collard, M.P., Rao, K.S. and Trouet, A. (1989) Vinblastine-C4 alkyl maleoyl and amino acid maleoyl derivatives. II. Experimental antitumor activity against leukemias and human tumor xenografts. *Anticancer Res. 9, 625–630.*

36 Corvalan, J.R., Smith, W., Gore, V.A. and Brandon, D.R. (1987) Specific *in vitro* and *in vivo* drug localisation to tumour cells using a hybrid-hybrid monoclonal antibody recognising both carcinoembryonic antigen (CEA) and vinca alkaloids. *Cancer Immunol. Immunother. 24, 133–137.*

37 Czerwinski, G., Tarasova, N.I., Michejda, C.J. (1998) Cytotoxic agents directed to peptide hormone receptors: defining the requirements for a successful drug. *Proc. Natl. Acad. Sci. USA 95, 11520–11525.*

38 Decout, A., Labeur, C., Goethals, M., Brasseur, R., Vandekerckhove, J. and Rosseneu, M. (1998) Enhanced efficiency of a targeted fusogenic peptide. *Biochim. Biophys. Acta 1372, 102–116.*

39 Denmeade, S.R., Nagy, A., Gao, J., Lilja, H., Schally, A.V. and Isaacs, J.T. (1998) Enzymatic activation of a doxorubicin-peptide prodrug by prostate-specific antigen. *Cancer Res. 58, 2537–2540.*

40 Deonarain, M.P. and Epenetos, A.A. (1998) Design, characterization and anti-tumour cytotoxicity of a panel of recombinant, mammalian ribonuclease-based immunotoxins. *Br. J. Cancer 77, 537–546.*

41 Di Stefano, G., Busi, C., Mattioli, A. and Fiume, L. (1995) Selective delivery to the liver of antiviral nucleoside analogs coupled to a high molecular mass lactosaminated poly-L-lysine and administered to mice by intramuscular route. *Biochem. Pharmacol. 49, 1769–1775.*

42 Doebber, T.W., Wu, M.S., Bugianesi, R.L., Ponpipom, M., Furbish, S.F., Barranger, J.A., Brady, R.O. and Shen, T.Y. (1982) Enhanced macrophage uptake of synthetically glycosylated human placental β-glucocerebrosidase. *J. Biol. Chem. 257, 2193–2199.*

43 Dokka, S., Toledo-Velasquez, D., Shi, X., Wang, L. and Rojanasakul, Y. (1997) Cellular delivery of oligonucleotides by synthetic import peptide carrier. *Pharm. Res. 14, 1759–1764.*

44 Drew, M.D., Estrada-Correa, A., Underdown, B.J. and McDermott, M.R. (1992) Vaccination by cholera toxin conjugated to a herpes simplex virus type 2 glycoprotein D peptide. *J. Gen. Virol. 73, 2357–2366.*

45 Duncan, R., Seymour, L.C., Scarlett, L., Lloyd, J.B., Rejmanová, P. and Kopeek, J. (1986) Fate of N-(2-hydroxypropyl)methacrylamide copolymers with pendent galactosamine residues after intravenous administration to rats. *Biochim. Biophys. Acta 880, 62–71.*

46 Duncan, R., Kopeková-Rejmanová, P., Strohalm, J., Hume, I., Cable, H.C., Pohl, J., Lloyd, J.B. and Kopeek, J. (1987) Anticancer agents coupled to N-(2-hydroxypropyl)methacrylamide copolymers. I. Evaluation of daunomycin and puromycin conjugates *in vitro*. *Br. J. Cancer 55, 165–174.*

47 Duncan, R. (1992) Drug-polymer conjugates; potential for improved chemotherepy. *Anti-Cancer Drugs 3, 153–156.*

48 Edwards, D.C. (1983) Targeting potential of antibody conjugates. *Pharmacol. Ther. 23, 147–177.*

49 Erskine-Grout, M.E., Olivier, G.W., Lucas, P., Sahm, U.G., Branch, S.K., Moss, S.H., Notarianni, L.J., Pouton and, C.W. (1996) Melanocortin probes for the melanoma MC1 receptor: synthesis, receptor binding and biological activity. *Melanoma Res. 6, 89–94.*

50 Fawell, S., Seery, J., Daikh, Y., Moore, C., Chen, L.L., Pepinsky, B. and Barsoum, J. (1994) Tat-mediated delivery of heterologous proteins into cells. *Proc. Natl. Acad. Sci. USA 91, 664–668.*

51 Fiume, L., Mattiolo, A., Balboni, P.G., Tongon, M., Brabanti-Brodano, G., De Vries, J. and Wieland, Th. (1979) Enhanced inhibition of virus DNA synthesis in hepatocytes by trifluorothymidine coupled to asialofetuin. *FEBS Lett. 103, 47–51.*

52 Fiume, L., Di Stefano, G., Busi, C., Mattioli, A., Bonino, F., Torrani-Cerenzia, M., Verme, G., Rapicetta, M., Bertini, M. and Gervasi, G.B. (1997) Liver targeting of antiviral nucleoside analogues through the asialoglycoprotein receptor. *J. Viral. Hepat. 4, 363–370.*

53 Foxwell, B.M., Band, H.A., Long, J., Jeffery, W.A., Snook, D., Thorpe, P.E., Watson, G., Parker, P.J., Epenetos, A.A. and Creighton, A.M. (1988) Conjugation of monoclonal antibodies to a synthetic peptide substrate for protein kinase: a method for labelling antibodies with ^{32}P. *Br. J. Cancer 57, 489–493.*

54 Franssen, E.J., Koiter, J., Kuipers, C.A., Bruins, A.P., Moolenaar, F., de Zeeuw, D., Kruizinga, W.H., Kellogg, R.M. and Meijer, D.K. (1992) Low molecular weight proteins as carriers for renal drug targeting. Preparation of drug-protein conjugates and drug-spacer derivatives and their catabolism in renal cortex homogenates and lysosomal lysates. *J. Med. Chem. 35, 1246–1259.*

55 Fukuta, M., Okada, H., Iinuma, S., Yanai, S. and Toguchi, H. (1994) Insulin fragments as a carrier for peptide delivery across the blood brain barrier. *Pharm. Res. 11, 1681–1688.*

56 Gaál, D. and Hudecz, F. (1998) Low toxicity and high antitumour activity of daunomycin by conjugation to immunpotential amphoteric branched polypeptide. *Eur. J. Cancer 34, 155–161.*

57 Gabius, H.J., Bokemeyer, C., Hellmann, T. and Schmoll, H.J. (1987) Targeting of neoglycoprotein-drug conjugates to cultured human embryonal carcinoma cells. *J. Cancer Res. Clin. Oncol. 113, 126–130.*

58 Gac, S., Coudane, J., Boustta, M., Domurado, M. and Vert, M. (2000) Synthesis, characterisation and *in vivo* behaviour of a norfloxacin-poly(L-lysine citramide imide) conjugate bearing mannosyl residues. *J. Drug Target. 7, 393–406.*

59 Garnett, M.C., Embleton, M.J., Jacobs, E. and Baldwin, R.W. (1983) Preparation and properties of a drug-carrier-antibody conjugate showing selective antibody-directed cytotoxicity *in vitro. Int. J. Cancer 31, 661–670.*

60 Ghose, T. and Nigam, S.P. (1972) Antibody as carrier of chlorambucil. *Cancer 29, 1398–1400.*

61 Ghosh, M.K., Kildsig, D.O. and Mitra, A.K. (1989) Preparation and characterization of methotrexate-immunoglobulin conjugates. *Drug. Des. Deliver. 4, 13–25.*

62 Gijsens, A., Missiaen, L., Merlevede, W. and de Witte, P. (2000) Epidermal growth factor-mediated targeting of chlorin e6 selectively potentiates its photodynamic activity. *Cancer Res. 60, 2197–2202.*

63 Golightly, L., Brown, J.E., Mitchell, J.B. and Brown, J.R. (1988) Trypanocidal activity of free and carrier bound daunomycin. *Cell. Biol. Int. Rep. 12, 77–83.*

64 Govindan, S.V., Goldenberg, D.M., Griffiths, G.L., Leung, S.O., Losman, M.J. and Hansen, H.J. (1995) Site-specific modifications of light chain glycosylated antilymphoma (LL2) and anti-carcinoembryonic antigen (hImmu-14-N) antibody divalent fragments. *Cancer Res. 55, Suppl 5721s–5725s.*

65 Griffiths, G.L., Goldenberg, D.M., Roesch, F. and Hansen, H.J. (1999) Radiolabeling of an anti-carcinoembryonic antigen antibody Fab' fragment (CEA-Scan) with the positron-emitting radionuclide ^{94m}Tc. *Clin.Cancer Res. 5, Suppl 3001s–3003s.*

66 Haas, M., Kluppel, A.C., Wartna, E.S., Moolenaar, F., Meijer, D.K., de Jong, P.E. and de Zeeuw, D. (1997) Drug-targeting to the kidney: renal delivery and degradation of a naproxen-lysozyme conjugate *in vivo. Kidney Int. 52, 1693–1699.*

67 Hammond, P.J., Arka, A., Peters, A.M., Bloom, S.R. and Gilbey, S.G. (1994) Localization of metastatic gastroenteropancreatic tumours by somatostatin receptor scintigraphy with [^{111}In-DTPA-D-Phe^{1}]-octreotide. *Q. J. Med. 87, 83–88.*

68 Han, J.H., Oh, Y.K., Kim, D.S. and Kim, C.K. (1999*a*) Enhanced hepatocyte uptake and liver targeting of methotrexate using galactosylated albumin as a carrier. *Int. J. Pharm. 188, 39–47.*

69 Han, W.R., Murray-Segal, L.J., Gershenzon, A., Zhang, J.G., Hodder, A.N., Pietersz, G.A. and Mottram, P.L. (1999*b*) Idarubicin-145-2C11-F(ab')$_2$ promotes peripheral tolerance and reduces chronic vascular disease in mouse cardiac allografts. *Transpl. Immunol. 7, 207–213.*

70 Hansen, T., Tyndall, R. and Soll, D. (1987) Methotrexate-anticollagen conjugate inhibits *in vitro* lens cell outgrowth. *Invest. Ophthalmol. Vis. Sci. 28, 1206–1209.*

71 Hashimoto, K., Loader, J. and Kinsky, S. (1985) Synthesis and characterization of methotrexate-dimyristoylphosphatidylethanolamine derivatives and the glycerophosphorylethanolamine analogs. *Biochim. Biophys. Acta 816, 163–168.*

72 Haubner, R., Wester, H.J., Reuning, U., Senekowitsch-Schmidtke, R., Diefenbach, B., Kessler, H., Stöcklin, G. and Schwaiger, M. (1999) Radiolabeled alpha(v)beta3 integrin antagonists: a new class of tracers for tumor targeting. *J. Nucl. Med. 40, 1061–1071.*

73 Heath, T., Lopez, N., Piper, J., Montgomery, J., Stern, W. and Papahadjopoulos, D. (1986) Liposome-mediated delivery of pteridine antifolates to cells *in vitro*: potency of methotrexate, and its α and γ substituents. *Biochim. Biophys. Acta 862, 72–80.*

74 Hendel, J., Sarek, L. and Hvidberg, E. (1976) Rapid radioimmunoassay for methotrexate in biological fluids. *Clin. Chem. 22, 813–816.*

75 Hnatowich, D.J., Layne, W.W., Childs, R.L., Lanteigne, D., Davis, M.A., Griffin, T.W. and Doherty, P.W. (1983) Radioactive labeling of antibody: a simple and efficient method. *Science 220, 613–615.*

76 Hu, S., Shively, L., Raubitschek, A., Sherman, M., Williams, L.E., Wong, J.Y., Shively, J.E. and Wu, A.M. (1996) Minibody: A novel engineered anti-carcinoembryonic antigen antibody fragment (single-chain Fv-C$_H$3) which exhibits rapid, high-level targeting of xenografts. *Cancer Res. 56, 3055–3061.*

77 Hudecz, F., Votavova, H., Gaál, D., Sponar, J., Kajtár, J., Blaha, K. and Szekerke, M.(1985) Branched polypeptides with a poly(L-lysine) backbone: synthesis, conformation and immunomodulation. In: Gebelein, Ch.G. and Carraher, Ch. E. (eds.), *Polymeric Materials in Medication,* Plenum Press, New York, pp. 265–289.

78 Hudecz, F., Ross, H., Price, M.R. and Baldwin, R.W. (1990) Immunoconjugate design: a predictive approach for coupling of daunomycin to monoclonal antibodies. *Bioconjugate Chemistry 1, 197–204.*

79 Hudecz, F., Clegg, J.A., Kajtár, J., Embleton, M.J., Szekerke, M. and Baldwin, R.W. (1992*a*) Synthesis, conformation, biodistribution and *in vitro* cytotoxicity of daunomycin-branched polypeptide conjugates. *Bioconjugate Chem. 3, 49–57.*

80 Hudecz, F., Garnett, M., Khan, T. and Baldwin, R. (1992*b*) The influence of synthetic conditions on the stability of methotrexate-monoclonal antibody conjugates determined by reversed phase high performance liquid chromatography. *Biomed. Chromatogr. 6, 128–132.*

81 Hudecz, F., Clegg, J.A., Kajtár, J., Embleton, M.J., Pimm, M.V., Szekerke, M. and Baldwin, R.W. (1993) Influence of carrier on biodistribution and *in vitro* cytotoxicity of methotrexate-branched polypeptide conjugates. *Bioconjugate Chem. 4, 25–33.*

82 Hudecz, F. (1995*a*) Design of synthetic branched-chain polypeptides as carriers for bioactive molecules. *Anti-Cancer Drugs 6, 171–193.*

83 Hudecz, F. (1995*b*) Alteration of immunogenicity and antibody recognition of B-cell epitopes by synthetic branched polypeptide carriers with poly(L-lysine) Backbone. *Biomed. Peptides, Proteins and Nucleic Acids 1, 213–220.*

84 Hudecz, F., Pimm, M.V., Rajnavölgyi, É., Mező, G., Fabra, A., Gaál, D., Kovács, A.L., Horváth, A. and Szekerke, M. (1999) Carrier design: New generation of polycationic branched polypeptides containing OH groups with prolonged blood survival and diminished *in vitro* cytotoxicity. *Bioconjugate Chemistry 10, 781–790.*

85 Hurwitz, E., Levy, R., Maron, R., Wilchek, M., Arnon, R. and Sela, M. (1975) The covalent binding of daunomycin and adriamycin to antibodies with retention of both drug and antibody activities. *Cancer Res. 35, 1175–1181.*

86 Inbar, J. and Chet, I. (1997) Lectins and biocontrol. *Crit. Rev. Biotechnol. 17, 1–20.*

87 Kahán, Z., Nagy, A., Schally, A.V., Hebert, F., Sun, B., Groot, K. and Halmos, G. (1999) Inhibition of growth of MX-1, MCF-7-MIII and MDA-MB-231 human breast cancer xenografts after administration of a targeted cytotoxic analog of somatostatin, AN-238. *Int. J. Cancer 82, 592–598.*

88 Kaneko, Y. (1981) Thyrotropin-daunomycin conjugate shows receptor mediated cytotoxicity in cultured thyroid cells. *Horm. Metab. Res.* ***13,*** *110–114.*

89 Kasugai, S., Fujisawa, R., Waki, Y., Miyamoto, K. and Ohya, K. (2000) Selective drug delivery system to bone: small peptide $(Asp)_6$ conjugation. *J. Bone Miner. Res. 15, 936–943.*

90 Kiaris, H., Schally, A.V., Nagy, A., Sun, B., Armatis, P. and Szepeshazi, K. (1999) Targeted cytotoxic analogue of bombesin/ gastrin-releasing peptide inhibits the growth of H-69 human small-cell lung carcinoma in nude mice. *Br. J. Cancer 816, 966–971.*

91 Kinkade, J.M. Jr, Vogler, W.R. and Dayton, P.G. (1974) Plasma levels of methotrexate in cancer patients as studied by an improved spectrophotofluorometric method. *Biochem. Med. 10, 337–350.*

92 Kinne, R.W., Schemer, K., Behr, T., Sharkey, R.M., Palombokinne, E., Emmrich, F., Goldenberg, D.M., Wolf, F. and Becker, W. (1999) *In vivo* stability and metabolism of ^{99m}Tc-labelled anti-CD4 monoclonal antibodies and Fab' fragments. *Nucl. Med. Commun. 20, 67–75.*

93 Knight, A., Carvajal, J., Schneider, H., Coutelle, C., Chamberlain, S. and Fairweather, N. (1999) Non-viral neuronal gene delivery mediated by the HC fragment of tetanus toxin. *Eur. J. Biochem. 259, 762–769.*

94 Knowles, D.M., Sullivan, T.J., Parker, C.W. and Williams, R.C. (1973) *In vitro* antibody-enzyme conjugates with specific bactericidal activity. *J. Clin. Investigation 52, 1443–1452.*

95 Kóczán, G., Ghosh, A.K, Mookherjee, A., Ghose, A.C and Hudecz, F. (1998) Application of branched chain polymeric polypeptides for methotrexate targeting to macrophages in Leishmania donovani infection. *The Immunologist S1, 610.*

96 Kóczán, G., Ghose, A.C and Hudecz, F. (2002) Synthesis and anti Leishmania donovani effect of methotrexate-branched polypeptide conjugates *in vitro* and *in vivo. Bioconjugate Chem. 13, 518–524.*

97 Kóczán, G. and Hudecz, F. (2000) Novel RP-HPLC approach for the analysis and purification of methotrexate-polymeric branched polypeptide conjugates. *Chromatographia 55, 163–170.*

98 Köhler, G. and Milstein, C. (1975) Continuous cultures of fused cells secreting antibody of predefined specificity. *Nature 256, 495–497.*

99 Kohls, M.D. and Lappi, D.A. (2000) Mab-ZAP: a tool for evaluating antibody efficacy for use in an immunotoxin. *Biotechniques 28, 162–165.*

100 Kosmas, C., Linardou, H. and Epenetos, A.A. (1993) Review: advances in monoclonal antibody tumour targeting. *J. Drug Target. 1, 81–91.*

101 Kralovec, J., Singh, M., Mammen, M., Blair, A.H. and Ghose, T. (1989*a*) Synthesis of site-specific methotrexate-IgG conjugates. *Cancer Immunol. Immunother. 29, 293–302.*

102 Krantz, M.J., Holtzman, N.A., Stowell, C.P. and Lee, Y.C. (1976) Attachment of thioglycosides to proteins: enhancement of liver membrane binding. *Biochemistry 15, 3963–3968.*

103 Kreitman R.J. (1999) Immunotoxins in cancer therapy. *Curr. Opin. Immunol. 11, 570–578.*

104 Krinick, N.L., Rihova, B. and Ulbrich, K. (1990) Targetable photoactive drugs. *Macromol. Chem. 191, 839–856.*

105 Krüger, K., Jochum, C., Glüsenkamp, K.H., Krüsemann, C., Lorenz, P., Eberle-Adamkiewicz, G., Drosdziok, W., Beelen, D.W., Coenen, H.H. and Rajewsky, M.F. (1998) New modular delivery system for diagnostic and therapeutic pre-targeting using tautomer-specific monoclonal antibody EM-6-47 and 3-substituted adenines. *Int. J. Cancer* ***77,*** *610–619.*

106 Kulkarni, P.N., Blair, A.H. and Ghose, T.I. (1981) Covalent binding of methotrexate to immunoglobulins and the effect of antibody linked drug on tumor growth *in vivo. Cancer Res. 41, 2700–2706.*

107 Kulkarni, P., Blair, A., Ghose, T. and Mammen, M. (1985) Conjugation of methotrexate to IgG antibodies and their $F(ab)_2$ fragments and the effect of conjugated methotrexate on tumor growth *in vivo. Cancer Immunol. Immunother. 19, 211–214.*

108 Kumagai, A.K., Eisenberg, J.B. and Pardridge, W.M. (1987) Absorptive-mediated endocytosis of cationized albumin and

a β-endorphin-cationized albumin chimeric peptide by isolated brain capillaries. Model system of blood-brain barrier transport. *J. Biol. Chem. 262, 15214–15219.*

109 Lagueux, J., Pagé, M. and Delorme, F. (1984) Daunorubicin-albumin copolymer targeting to leukemic cells *in vitro* and *in vivo*. *Seminars in Oncology 11, 59–63.*

110 Latif, Z., Lozzio, B., Wust, C., Krauss, S., Aggio, M. and Lozzio, C. (1980) Evaluation of drug-antibody conjugates in the treatment of human myelosarcomas transplanted in nude mice. *Cancer 45, 1326–1333.*

111 Le Doussal, J.M., Barbet, J. and Delaage, M. (1992) Bispecific-antibody-mediated targeting of radiolabeled bivalent haptens: theoretical, experimental and clinical results. *Int. J. Cancer Suppl. 7, 58–62.*

112 Lemieux, P., Page, M. and Noel, C. (1992) *In vivo* cytotoxicity and antineoplastic activity of a transferrin-daunorubicin conjugate. *In vivo 6, 621–627.*

113 Leopold, C.S. and Friend, D.R. (1995) *In vivo* pharmacokinetic study for the assessment of poly(L-aspartic acid) as a drug carrier for colon-specific drug delivery. *J. Pharmacokinet. Biopharm. 23, 397–406.*

114 Li, S., Zhang, X.Y., Qiao, T.D., Chen, X.T., Zhang, S.Y. and Chen, L.J. (1992) Enhanced antitumor activity of daunomycin conjugated with antigastric cancer monoclonal antibody Mgb2. *Oncol. Res. 4, 447–453.*

115 Li, M., Meares, C.F., Salako, Q., Kukis, D.L., Zhong, G.R., Miers, L. and DeNardo, S.J. (1995) Prelabeling of chimeric monoclonal antibody L6 with ^{90}Y- and ^{111}In-1,4,7,10-tetraazacyclododecane-N,N′,N″,N‴-tetraacetic acid (DOTA) chelates for radioimmunodiagnosis and therapy. *Cancer Res. 55, 5726s–5728s.*

116 Lindford, J.H., Froese, G., Berczi, I. and Israels, L.G. (1974) An alkylating agent-globulin conjugate with both alkylating and antibody activity. *J. Natl. Cancer Inst. 52, 1665–1667.*

117 Liu, L., Barth, R.F., Adams, D.M., Soloway, A.H. and Reisfeld, R.A. (1996) Critical evaluation of bispecific antibodies as targeting agents for boron neutron capture therapy of brain tumors. *Anticancer Res. 16, 2581–2587.*

118 Lu, X.M., Fischman, A.J., Jyawook, S.L., Hendricks, K., Tompkins, R.G. and Yarmush, M.L. (1994) Antisense DNA delivery *in vivo*: liver targeting by receptor-mediated uptake. *J. Nucl. Med. 35, 269–275.*

119 Lu, Z.R., Gao, S.Q., Kopeková, P. and Kopeek, J. (2000) Synthesis of bioadhesive lectin-HPMA copolymer-cyclosporin conjugates. *Bioconjug. Chem. 11, 3–7.*

120 Manabe, Y., Tsubota, T., Haruta, Y., Kataoka, K., Okazaki, M., Haisa, S., Nakamura, K. and Kimura, I. (1984) Production of a monoclonal antibody-methotrexate conjugate utilizing dextran T-40 and its biological activity. *J. Lab. Clin. Med. 104, 445–454.*

121 Manfredini, S., Marastoni, M., Tomatis, R., Durini, E., Spisani, S., Pani, A., Marceddu, T., Musiu, C., Marongiu, M.E. and La Colla, P. (2000) Peptide T-araC conjugates: solid-phase synthesis and biological activity of N^4-(acylpeptidyl)-araC. *Bioorg. Med. Chem. 8, 539–547.*

122 Marcello, A., Loregian, A., Cross, A., Marsden, H., Hirst, T.R. and Palù, G. (1994) Specific inhibition of herpes virus replication by receptor-mediated entry of an antiviral peptide linked to Escherichia coli enterotoxin B subunit. *Proc. Natl. Acad. Sci. USA 91, 8994–8998.*

123 Masquelier, M., Baurain, R. and Trouet, A. (1980) Amino acid and dipeptide derivatives of daunorubicin. *J. Med. Chem. 23, 1166–1170.*

124 Mathe, G., Loc, T.B. and Bernard, J. (1958) Effet sur la leucemie 1210 de la souris d'un combinaison par diazotation d'Amethopterine et de gamma-globulines de hamsters porteurs de cette leucemie par hetirogreffe. *C.R. Acad. Sci (D) Paris 246, 1626–1632.*

125 Mező, G., Mező, I., Seprődi, A., Teplán, I., Kovács, M., Vincze, B., Pályi, I., Kajtár, J., Szekerke, M. and Hudecz, F. (1996*a*) Synthesis, conformation, biodistribution and hormon related *in vitro* anti-tumor effect of a GnRH antagonist-branched polypeptide conjugate. *Bioconjugate Chem. 7, 642–650.*

126 Mező, G., Sármay, G., Hudecz, F., Kajtár, J., Nagy, Zs., Gergely, J. and Szekerke, M. (1996*b*) Synthesis and characterization of p-borono-Phe-branched polypeptide-monoclonal antibody ternary systems for potential use in boron neutron capture therapy (BNCT). *J. Bioactive and Compatible Polymers 11, 263–285.*

127 Mező, G., Kajtár, J., Nagy, I., Majer, Zs., Szekerke, M. and Hudecz, F. (1997) Carrier design: Synthesis and conformational studies of poly[L-lysine] based branched polypeptides with hydroxyl groups. *Biopolymers 42, 719–730.*

128 Mokotoff, M., Swanson, D.P., Jonnalagadda, S.S., Epperly, M.W. and Brown, M.L. (1997) Evaluation of laminin peptide fragments labeled with ^{111}In for the potential imaging of malignant tumors. *J. Pept. Res. 49, 510–516.*

129 Monsigny, M., Kieda, C., Roche, A.C. and Delmotte, F. (1980) Preparation and biological properties of a covalent antitumor drug-arm-carrier (DAC conjugate). *FEBS Lett. 119, 181–186.*

130 Moolten, F.L. and Cooperband, S.R. (1970) Selective destruction of target cells by diphteria toxin conjugated to antibody directed against antigens on the cells. *Science 169, 68–70.*

131 Moolten, F.L., Schreiber, B.M. and Zajdel, S.H. (1982) Antibodies conjugated to potent cytotoxins as specific antitumor agents. *Immunol. Rev. 62, 47–73.*

132 Mu, Y., Kamada, H., Kodaira, H., Sato, K., Tsutsumi, Y., Maeda, M., Kawasaki, K., Nomizu, M., Yamada, Y. and Mayumi, T. (1999) Bioconjugation of laminin-related peptide YIGSR with polyvinyl pyrrolidone increases its antimetastatic effect due to a longer plasma half-life. *Biochem. Biophys. Res. Commun. 264, 763–767.*

133 Mukhopadhyay, A., Mukhopadhyay, B., Srivastava, R.K. and Basu, S.K. (1992) Scavenger receptor-mediated delivery of daunomycin elicits selective toxicity towards neoplastic cells of macrophage lineage. *Biochem. J. 284, 237–241.*

134 Murphy, J.R., Bishai, W., Williams, D., Bacha, P., Borowski, M., Parker, K., Boyd, J., Waters, C. and Strom, T.B. (1987) Genetic assembly and selective toxicity of diphtheria-toxin-related polypeptide hormone fusion proteins. *Biochem. Soc. Symp. 53, 9–23.*

135 Nagy, A., Szoke, B. and Schally, A.V. (1993) Selective coupling of methotrexate to peptide hormone carriers through a γ-carboxamide linkage of its glutamic acid moiety: benzotriazol-1-yloxytris(dimethylamino)phosphonium hexafluorophosphate activation in salt coupling. *Proc. Natl. Acad. Sci. USA 90, 6373–6376.*

136 Nagy, A., Armatis, P., Cai, R.Z., Szepeshazi, K., Halmos, G. and Schally, A.V. (1997) Design, synthesis, and *in vitro* evaluation of cytotoxic analogs of bombesin-like peptides containing

doxorubicin or its intensely potent derivative, 2-pyrrolinodoxorubicin. *Proc. Natl. Acad. Sci. USA 94, 652–656.*

137 Nagy, A., Schally, A.V., Halmos, G., Armatis, P., Cai, R.Z., Csernus, V., Kovács, M., Koppán, M., Szepesházi, K. and Kahán, Z. (1998) Synthesis and biological evaluation of cytotoxic analogs of somatostatin containing doxorubicin or its intensely potent derivative, 2-pyrrolinodoxorubicin. *Proc. Natl. Acad. Sci. USA 95, 1794–1799.*

138 Nagy, A., Plonowski, A. and Schally, A.V. (2000*a*) Stability of cytotoxic luteinizing hormone-releasing hormone conjugate (AN-152) containing doxorubicin 14-O-hemiglutarate in mouse and human serum *in vitro*: implications for the design of preclinical studies. *Proc. Natl. Acad. Sci. USA 97, 829–834.*

139 Nagy, I.B., Alsina, M.A., Haro, I., Reig, F. and Hudecz, F. (2000*b*) Phospholipid-model membrane interactions with branched polypeptide conjugates of a hepatitis a virus peptide epitope. *Bioconjugate Chemistry 11, 30–38.*

140 Nagy, I.B., Varga, I. and Hudecz, F. (2000*c*) Preparation method of In(III)-labelled DTPA-peptides purified by 8-hydroxy quinoline. *Analytical Biochemistry 287, 17–24.*

141 Narula, J., Petrov, A., Pak, K.Y., Ditlow, C., Chen, F. and Khaw, B.A. (1997) Noninvasive detection of atherosclerotic lesions by ^{99m}Tc-based immunoscintigraphic targeting of proliferating smooth muscle cells. *Chest 111, 1684–1690.*

142 Neri, D., Natali, P.G., Petrul, H., Soldani, P., Nicotra, M.R., Vola, R., Rivella, A., Creighton, A.M., Neri, P. and Mariani, M. (1996) Recombinant anti-human melanoma antibodies are versatile molecules. *J. Invest. Dermatol. 107, 164–170.*

143 Nogusa, H., Yano, T., Kajiki, M., Gonsho, A., Hamana, H. and Okuno, S. (1997) Antitumor effects and toxicities of carboxymethylpullulan-peptide-doxorubicin conjugates. *Biol. Pharm. Bull. 20, 1061–1065.*

144 Ogawara, K., Hasegawa, S., Nishikawa, M., Takakura, Y. and Hashida, M. (1999) Pharmacokinetic evaluation of mannosylated bovine serum albumin as a liver cell-specific carrier: quantitative comparison with other hepatotropic ligands. *J. Drug Target. 6, 349–360.*

145 O'Hare, K.B., Duncan, R., Strohalm, J., Ulbrich, K. and Kopeckova, P. (1993) Polymeric drug-carriers containing doxorubicin and melanocyte-stimulating hormone: *in vitro* and *in vivo* evaluation against murine melanoma. *J. Drug Target. 1, 217–229.*

146 Pasqualini, R. (1999) Vascular targeting with phage peptide libraries. *Q. J. Nucl. Med. 43, 159–162.*

147 Pató, J., Ulbrich, K., Baker, P., Mező, G. and Hudecz, F. (1999) Synthesis of macromolecular conjugates of a urokinase inhibitor. *J. Bioactive and Compatible Polymers 14, 99–121.*

148 Perkins, A.C., Frier, M., Pimm, M.V. and Hudecz, F. (1998) ^{99m}Tc-branched chain polypeptide (BCP): a potential synthetic radiopharmaceutical. *J. Labelled Compounds 41, 631–638.*

149 Pietersz, G., Cunningham, Z. and McKenzie, I. (1988) Specific *in vitro* anti-tumour activity of methotrexate-monoclonal antibody conjugates prepared using human serum albumin as an intermediary. *Immunol. Cell. Biol. 66, 43–49.*

150 Pietersz, G.A., Patrick, M.R. and Chester, K.A. (1998) Preclinical characterization and *in vivo* imaging studies of an engineered recombinant ^{99m}Tc-labeled metallothionein-containing anti-carcinoembryonic antigen single-chain antibody. *J. Nucl. Med. 39, 47–56.*

151 Pimm, M., Clegg, J., Garnett, M. and Baldwin, R. (1988) Biodistribution and tumour localization of a methotrexate-monoclonal-antibody 791T/36 conjugate in nude mice with human tumour xenografts. *Int. J. Cancer 41, 886–891.*

152 Pimm, M.V., Robins, R.A., Embleton, M.J., Jacobs, E., Markham, A.J., Charleston, A. and Baldwin, R.W. (1990) A bispecific monoclonal antibody against methotrexate and a human tumour associated antigen augments cytotoxicity of methotrexate-carrier conjugate. *Br. J. Cancer 61, 508–513.*

153 Pimm, M.V., Gribben, S.J., Mező, G. and Hudecz, F. (1995) Strategies for labelling branched polypeptides with a poly(L-lysine) backbone with radioiodines (^{123}I, ^{125}I, ^{131}I) and radiometals (^{111}In, ^{51}Cr) for biodistribution studies and radiopharmaceutical development. *J. Labelled Compounds 36, 157–172.*

154 Ponzetto, A., Fiume, L., Forzani, B., Song, S.Y., Busi, C., Mattioli, A., Spinelli, C., Marinelli, M., Smedile, A. and Chiaberge, E. (1991) Adenine arabinoside monophosphate and acyclovir monophosphate coupled to lactosaminated albumin reduce woodchuck hepatitis virus viremia at doses lower than do the unconjugated drugs. *Hepatology 14, 16–24.*

155 Poznansky, M.J. (1983) Enzyme-protein conjugates: New possibilities for enzyme therapy. *Pharmacol. Ther. 21, 53–76.*

156 Poznansky, M.J. and Juliano, R.L. (1984) Biological approaches to the controlled delivery of drugs: a critical review. *Pharmacological Reviews 36, 277–336.*

157 Poznansky, M.J., Singh, R., Singh, B. and Fantus, G. (1984) Insulin: Carrier potential for enzyme/drug therapy. *Science 223, 1304–1306.*

158 Poznansky, M.J. (1988) Soluble enzyme-albumin conjugates: new possibilities for enzyme replacement therapy. *Methods Enzymol. 137, 566–574.*

159 Proby, C.M., Ota, T., Suzuki, H., Koyasu, S., Gamou, S., Shimizu, N., Wahl, J.K., Wheelock, M.J., Nishikawa, T. and Amagai, M. (2000) Development of chimeric molecules for recognition and targeting of antigen-specific B cells in pemphigus vulgaris. *Br. J. Dermatol. 142, 321–330.*

160 Prochiantz, A. (1999) Homeodomain-derived peptides. In and out of the cells. *Ann. N. Y. Acad. Sci. 886, 172–179.*

161 Przybylski, M., Fell, E., Ringsdorf, H. and Zaharko, D.S. (1978) Synthesis and characterization of polymeric derivatives of the antitumour agent methotrexate. *Makromol. Chem. 179, 1719–1733.*

162 Rakestraw, S.L., Tompkins, R.G. and Yarmush, M.L. (1990) Antibody-targeted photolysis: *in vitro* studies with Sn(IV) chlorin e6 covalently bound to monoclonal antibodies using a modified dextran carrier. *Proc. Natl. Acad. Sci. USA 87, 4217–4221.*

163 Raso, V. and Griffin, T. (1980) Specific cytotoxicity of a human immunoglobulin-directed Fab'-ricin A chain conjugate. *J. Immunol. 125, 2610–2616.*

164 Raso, V. and Griffin, T. (1982) Hybrid antibodies with dual specificity for the delivery of ricin to immunoglobulin-bearing cells. *Cancer Res. 41, 2073–2078.*

165 Raso, V., Brown, M. and McGrath, J. (1997) Intracellular targeting with low pH-triggered bispecific antibodies. *J. Biol. Chem. 272, 27623–27628.*

166 Rékási, Z., Szöke, B., Nagy, A., Groot, K., Rékási, E.S. and Schally, A.V. (1993) Effect of luteinizing hormone-releasing

hormone analogs containing cytotoxic radicals on the function of rat pituitary cells: tests in a long term superfusion system. *Endocrinology 132, 1991–2000.*

167 Reményi, J., Tóth, S., Fabra, A., Falus, A. and Hudecz, F. (2001) Polypeptide based bioconjugates containing daunomycin isomers: synthesis, chemical stability and *in vitro* antitmour effect. In: Epton, R. (ed.), *Innovation and Perspectives in Solid Phase Synthesis & Combinatorial Libraries, 1999,* Mayflower Ltd, Birmingham, UK pp 349–350.

168 Remy, M.H. and Poznansky, M.J. (1978) Immunogenicity and antigenicity of soluble cross-linked enzyme-albumin polymers advantages for enzyme therapy. *Lancet 2, 68–70.*

169 Reubi, J.C., Waser, B., Schaer, J.C., Laederach, U., Erion, J., Srinivasan, A., Schmidt, M.A. and Bugaj, J.E. (1998) Unsulfated DTPA- and DOTA-CCK analogs as specific high-affinity ligands for CCK-B receptor-expressing human and rat tissues *in vitro* and *in vivo*. *Eur. J. Nucl. Med. 25, 481–490.*

170 Rigaudy, P., Charcosset, J.Y., Garbay-Jaureguiberry, C., Jacquemin-Sablon, A. and Roques, B.P. (1989) Attempts to target antitumor drugs toward opioid receptor-rich mouse tumor cells with enkephalin-ellipticinium conjugates. *Cancer Res. 49, 1836–1842.*

171 Rihova, B., Kopeckova, P., Strohalm, J., Rossmann, P., Vetvicka, V. and Kopecek, J. (1988) Antibody-directed affinity therapy applied to the immune system: *in vivo* effectiveness and limited toxicity of daunomycin conjugated to HPMA copolymers and targeting antibody. *Clin. Immunol. Immunopathol. 46, 100–114.*

172 Rihová, B., Jelínková, M., Strohalm, J., Subr, V., Plocová, D., Hovorka, O., Novák, M., Plundrová, D., Germano, Y., Ulbrich, K. (2000) Polymeric drugs based on conjugates of synthetic and natural macromolecules. II. Anti-cancer activity of antibody or $F(ab')_2$-targeted conjugates and combined therapy with immunomodulators. *J. Controlled Release 64, 241–261.*

173 Ringrose, P.S. (1983) Small peptides as carriers and target in human therapy. *Biochem. Soc. Transactions 11, 804–808.*

174 Roche, A.C., Midoux, P., Pimpaneau, V., Négre, E., Mayer, R. and Monsigny, M. (1990) Endocytosis mediated by monocyte and macrophage membrane lectins-application to antiviral drug targeting. *Res. Virol. 141, 243–249.*

175 Roos, C.F., Matsumoto, S., Takakura, Y., Hashida, M. and Sezaki, H. (1984) Physicochemical and antitumor characteristics of some polyamino acid prodrugs of mitomycin C. *Int. J. Pharmaceuticals 22, 75–87.*

176 Ross, W.C., Thorpe, P.E., Cumber, A.J., Edwards, D.C., Hinson, C.A. and Davies, A.J. (1980) Increased toxicity of diphtheria toxin for human lymphoblastoid cells following covalent linkage to anti-(human lymphocyte) globulin or its $F(ab')_2$ fragment. *Eur. J. Biochem. 104, 381–390.*

177 Rosowsky, A. and Yu, C.S. (1978) Methotrexate analogues. 10. Direct coupling of methotrexate and diethyl L-glutamate in the presence of peptide bond-forming ragents. *J. Med. Chem. 21, 170–175.*

178 Rousselle, C., Clair, P., Lefauconnier, J.M., Kaczorek, M., Scherrmann, J.M. and Temsamani, J. (2000) New advances in the transport of doxorubicin through the blood-brain barrier by a peptide vector-mediated strategy. *Mol. Pharmacol. 57, 679–686.*

179 Rowland, G.F., O'Neill, G.J. and Davies, D.A. (1975) Suppression of tumour growth in mice by a drug-antibody conjugate using a novel approach to linkage. *Nature 255, 487–488.*

180 Ryser, H.J.P. and Shen, W.C. (1978) Conjugation of methotrexate to poly(L-lysine) increases drug transport and overcomes drug resistance. *Proc. Natl. Acad. Sci. USA 75, 3867–3870.*

181 Sanzgiri, Y., Blaton, C.D. and Gallo, J.M. (1990) Synthesis,characterization and *in vitro* stability of chitosan-methotrexate conjugates. *Pharmaceutical Res. 7, 418–421.*

182 Schally, A.V. and Nagy, A. (1999) Cancer chemotherapy based on targeting of cytotoxic peptide conjugates to their receptors on tumors. *Eur. J. Endocrinol. 141, 1–14.*

183 Schmidt, M., Vakalopoulou, E., Schneider, D.W. and Wels, W. (1997) Construction and functional characterization of scFv(14E1)-ETA-a novel, highly potent antibody-toxin specific for the EGF receptor. *Br. J. Cancer 75, 1575–1584.*

184 Seitz, U., Neumaier, B., Glatting, G., Kotzerke, J., Bunjes, D. and Reske, S.N. (1999) Preparation and evaluation of the ^{188}Re-labelled anti-NCA antigen monoclonal antibody BW 250/183 for radioimmunotherapy of leukaemia. *Eur. J. Nucl. Med. 26, 1265–1273.*

185 Sett, R., Sarkar, K. and Das, P.K. (1993) Macrophage-directed delivery of doxorubicin conjugated to neoglycoprotein using leishmaniasis as the model disease. *J. Infect. Dis. 168, 994–999.*

186 Seymour, L.W., Ulbrich, K., Wedge, S.R., Hume, I.C., Strohalm, J. and Duncan, R. (1991) N-(2-hydroxypropyl) methacrylamide copolymers targeted to the hepatocyte galactose-receptor: pharmacokinetics in DBA2 mice. *Br. J. Cancer 63, 859–866.*

187 Sharma, S.D., Granberry, M.E., Jiang, J., Leong, S.P., Hadley, M.E. and Hruby, V.J. (1994) Multivalent melanotropic peptide and fluorescent macromolecular conjugates: new reagents for characterization of melanotropin receptors. *Bioconjug. Chem. 5, 591–601.*

188 Shen, W.C. and Ryser, H.J.P. (1981) Cis-aconityl spacer between daunomycin and macromolecular carriers. *Biochem. Biophys. Res. Commun. 102, 1048–1054.*

189 Shen, W.C., Ryser, H.J. and LaManna, L. (1985) Disulfide spacer between methotrexate and poly(D-lysine). A probe for exploring the reductive process in endocytosis. *J. Biol. Chem. 260, 10905–10908.*

190 Shier, W.T. (1979) Lectins as drug carriers. In: Gregoriadis, G., (ed.) *Drug Carriers in Biology and Medicine.* Academic Press, New York, pp. 40–70.

191 Shih, L.B., Sharkey, R.M., Primus, F.J. and Goldenberg, D.M. (1988) Site-specific linkage of methotrexate to monoclonal antibodies using an intermediate carrier. *Int. J. Cancer 41, 832–839.*

192 Shih, L.B., Goldenberg, D.M., Xuan, H., Lu, H., Sharkey, H. and Hall, T.C. (1991) Anthracycline immunoconjugates prepared by a site-specific linkage via amino-dextran intermediate carrier. *Cancer Research 51, 4192–4198.*

193 Sivolapenko, G.B., Douli, V., Pectasides, D., Skarlos, D., Sirmalis, G., Hussain, R., Cook, J., Courtenay-Luck, N.S., Merkouri, E. and Konstantinides, K. (1995) Breast cancer imaging with radiolabelled peptide from complementarity-determining region of antitumour antibody. *Lancet 346, 1662–1666.*

194 Stehle, G., Wunder, A., Sinn, H., Schrenk, H.H., Schütt, S., Frei, E., Hartung, G., Maier-Borst, W. and Heene, D.L. (1997) Pharmacokinetics of methotrexate-albumin conjugates in tumor-bearing rats. *Anticancer Drugs 8, 835–844.*

195 Stehle, G., Wunder, A., Schrenk, H.H., Hartung, G., Heene, D.L. and Sinn, H. (1999) Albumin-based drug carriers: comparison between serum albumins of different species on pharmacokinetics and tumor uptake of the conjugate. *Anticancer Drugs 10, 785–790.*

196 Stickney, D.R., Anderson, L.D., Slater, J.B., Ahlem, C.N., Kirk, G.A., Schweighardt, S.A. and Frincke, J.M. (1991) Bifunctional antibody: a binary radiopharmaceutical delivery system for imaging colorectal carcinoma. *Cancer Res. 51, 6650–6655.*

197 Szekerke, M. and Driscoll, J.S. (1977) The use of macromolecules as carriers of antitumor drugs. *Eur. J. Cancer 13, 529–537.*

198 Takakura, Y. and Hashida, M. (1995) Macromolecular drug carrier systems in cancer chemotherapy: macromolecular prodrugs. *Critical Reviews in Oncology/Haematology 18, 207–231.*

199 Thorpe, P.E. and Ross, W.C. (1982) The preparation and cytotoxic properties of antibody-toxin conjugates. *Immunol. Rev. 62, 119–158.*

200 Torchilin, V.P., Klibanov, A.L., Nossif, N.D., Slinkin, M.A., Strauss, H.W., Haber, E., Smirnov, V.N. and Khaw, B.A. (1987) Monoclonal antibody modification with chelate-linked high-molecular-weight polymers: major increases in polyvalent cation binding without loss of antigen binding. *Hybridoma 6, 229–239.*

201 Trouet, A., Masquelier, M., Baurain, R. and Deprez-De Campeneere, D. (1982) A covalent linkage between daunorubicin and proteins that is stable in serum and reversible by lysosomal hydrolases, as required for a lysosomotropic drug-carrier conjugate: *in vitro* and *in vivo* studies. *Proc. Natl. Acad. Sci. USA 79, 626–629.*

202 Tsuruo, T., Yamori, T., Tsukagoshi, S. and Sakurai, Y. (1980) Enhanced cytocidal action of methotrexate by conjugation to concanavalin A. *Int. J. Cancer 26, 655–659.*

203 Uchida, T., Yamaizumi, M., Mekada, E., Okada, Y., Tsuda, M., Kurokawa, T. and Sugino, Y. (1978) Reconstitution of hybrid toxin fragment A of diphtheria toxin and a subunit of *Wisteria fluoribunda* lectin. *J. Biol. Chem. 253, 6307–6310.*

204 Uherek, C., Fominaya, J. and Wels, W. (1998) A modular DNA carrier protein based on the structure of diphtheria toxin mediates target cell-specific gene delivery. *J. Biol. Chem. 273, 8835–8841.*

205 Umemoto, N., Kato, Y. and Hara, T. (1989*a*) Cytotoxicities of two disulfide-bond-linked conjugates of methotrexate with monoclonal anti-MM46 antibody. *Cancer Immunol. Immunother. 28, 9–16.*

206 Umemoto, N., Kato, Y., Endo, N., Takeda, Y. and Hara, T. (1989*b*) Preparation and *in vitro* toxicity of a MTX-anti-MM46 monoclonal antibody conjugate via an oligopeptide spacer. *Int. J. Cancer 43, 677–684.*

207 Vallera, D.A., Taylor, P.A., Panoskaltsis-Mortari, A. and Blazar, B.R. (1995) Therapy for ongoing graft-versus-host disease induced across the major or minor histocompatibility barrier in mice with anti-CD3F(ab′)$_2$-ricin toxin A chain immunotoxin. *Blood 86, 4367–4375.*

208 Varga, J.M., Asato, N., Lande, S. and Lerner, A.B. (1977) Melanotropin-daunomycin conjugate shows receptor-mediated cytotoxicity in cultured murine melanoma cells. *Nature 267, 56–58.*

209 Vasey, P.A., Kaye, S.B., Morrison, R., Twelves, C., Wilson, P., Duncan, R., Thomson, A.H., Murray, L.S., Hilditch, T.E., Murray, T., Burtles, S., Fraier, D., Frigerio, E., and Cassidy, J. (1999) Phase I clinical and pharmacokinetic study of PK1 [*N*-(2-hydroxypropyl)methacrylamide copolymer doxorubicin]: first member of a new class of chemotherapeutic agents – drug-polymer conjugates. *Clinical Cancer Research 5, 83–94.*

210 Vincze, B., Pályi, I., Daubner, D., Kálnay, A., Mező, G., Hudecz, F., Szekerke, M., Teplán, I. and Mező, I. (1994) Antitumor effect of a GnRH antagonist and its conjugate on human breast cancer cells and their xenografts. *J. Cancer Res. Clin. Oncol. 120, 578–584.*

211 Virgolini, I., Szilvasi, I., Kurtaran, A., Angelberger, P., Raderer, M., Havlik, E., Vorbeck, F., Bischof, C., Leimer, M., Dorner, G., Kletter, K., Niederle, B., Scheithauer, W. and Smith-Jones, P. (1998) ^{111}In-DOTA-lanreotide: biodistribution, safety and radiation absorbed dose in tumor patients. *J. Nucl. Med. 39, 1928–1936.*

212 Vitetta, E.S., Krolick, K.A., Miyama-Lanab, M., Cushley, W. and Uhr, J.W. (1983) Immunotoxins: A new approach to cancer therapy. *Science 219, 644–650.*

213 Wadhwa, M.S. and Rice, K.G. (1995) Receptor mediated glycotargeting. *J. Drug. Target. 3, 111–127.*

214 Wawrzynczak, E.J. (1992) Rational design of immunotoxins: current progress and future prospects. *Anticancer Drug Des. 7, 427–441.*

215 Willuda, J., Honegger, A., Waibel, R., Schubiger, P.A., Stahel, R., Zangemeister-Wittke, U., Plückthun, A. (1999) High thermal stability is essential for tumor targeting of antibody fragments: engineering of a humanized anti-epithelial glycoprotein-2 (epithelial cell adhesion molecule) single-chain Fv fragment. *Cancer Res. 59, 5758–5767.*

216 Wirth, M., Fuchs, A., Wolf, M., Ertl, B. and Gabor, F. (1998) Lectin-mediated drug targeting: preparation, binding characteristics, and antiproliferative activity of wheat germ agglutinin conjugated doxorubicin on Caco-2 cells. *Pharm.Res. 15, 1031–1037.*

217 Woodley, J.F. (2000) Lectins for gastrointestinal targeting-15 years on. *J. Drug Target. 7, 325–333.*

218 Wu, A.M., Chen, W., Raubitschek, A., Williams, L.E., Neumaier, M., Fischer, R., Hu, S.Z., Odom-Maryon, T., Wong, J.Y. and Shively, J.E. (1996) Tumor localization of anti-CEA single-chain Fvs: improved targeting by non-covalent dimers. *Immunotechnology 2, 21–36.*

219 Yin, W. and Cheng, P.W. (1994) Lectin conjugate-directed gene transfer to airway epithelial cells. *Biochem. Biophys. Res. Commun. 205, 826–833.*

220 Yokoyama, M., Miyauchi, M., Yamada, N., Okano, T., Sakurai, Y., Kataoka, K. and Inoue, S. (1990) Characterization and anticancer activity of the micelle-forming polymeric anticancer drug adriamycin-conjugated poly(ethylene glycol)-poly(aspartic acid) block copolymer. *Cancer Res. 50, 1693–1700.*

221 Zunino, F., Savi, G., Giuliani, F., Gambetta, R., Supino, R., Tinelli, S. and Pezzoni, G. (1984) Comparison of antitumor effects of daunorubicin covalently linked to poly-L-amino acid carriers. *Eur. J. Cancer Clin.Oncol. 20, 421–425.*

Chapter 44

Current approaches to peptidomimetics

Michael J. Kelso and David P. Fairlie

Contents

Peptides and proteins have limited usefulness as drug candidates due to their poor bioavailability and high cost of manufacture. As a consequence smaller peptidomimetic molecules are usually sought to reproduce the biological properties of peptides but with higher metabolic stability and bioavailability. There are a number of current approaches to peptidomimetics, ranging from designing conformationally constrained molecules that structurally mimic the folded backbone main chain structures adopted by peptides (e.g. helices, turns, strands, sheets) to non-peptidic scaffolds that completely replace peptide main chains but similarly position and orientate peptide side chains or their surrogates for optimal interactions with receptors. The former category represents a minimalist approach to drug design, restricting changes made to bioactive peptides to the fewest required for improving bioactivity while achieving structural and functional mimicry of a peptide surface. The latter category includes natural product peptidomimetics as well as designed compounds containing three dimensional non-peptidic scaffolds that appropriately project functionality for attaching structural or functional mimetics of peptide side chains. We illustrate these approaches ahead with a limited number of classic examples from the different categories of backbone and side chain peptidomimetics, but focus particularly on new approaches to developing molecules that mimic secondary structures of proteins.

44.1 Introduction

Endogenous proteins and peptides affect the majority of the physiological processes in the human body. They characteristically contain hydrophobic and hydrophilic surfaces folded into epitopes or clefts that selectively interact with another protein, peptide, organic molecule, or component of DNA or RNA. Proteins and peptides are perhaps the quintessential evolutionary examples of chemicals that have become optimized for a specific complementary fit with a unique binding partner. Figure 44.1 illustrates how proteins use a combination of non-covalent interactions (hydrogen bonds, pocket-filling hydrophobic and van der Waals interactions) to effect highly regioselective and enantioselective molecular recognition that defines a single, unique, and sometimes catalytic, chemical reaction. Such interactions can similarly be used by smaller molecules to compete with or mimic protein surfaces (Figure 44.1B).

The exogenous use of proteins and peptides in medicine as drugs to regulate physiological processes has been severely limited by their poor pharmacokinetic and pharmacodynamic profiles. Proteins and peptides tend to be metabolically unstable, being readily decomposed or modified by enzymes in the gut, blood and cells. Many peptides are too water soluble or too large to penetrate cell membranes via either passive diffusion or active transport mechanisms. Large molecules tend also to be immunogenic, being recognized and eliminated through host defense mechanisms. As a result conventional oral administration of protein or peptide based therapeutics has been unsuccessful to date. Proteins can be administered intravenously to people (e.g. insulin, growth factors, cytokines) but the expense, toxicity, rapid

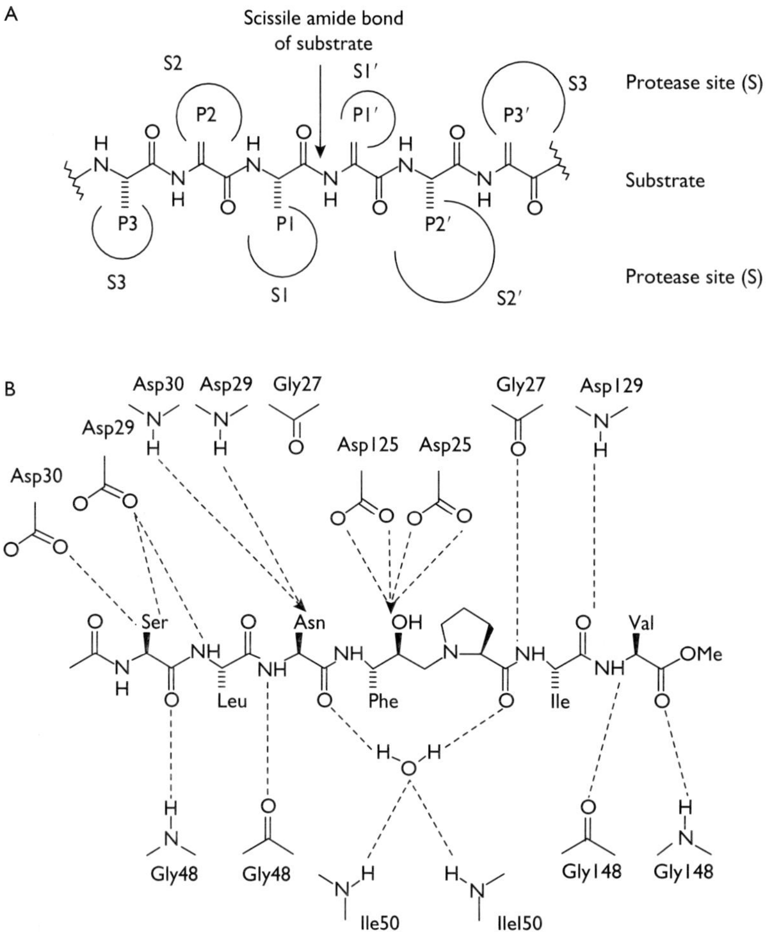

Figure 44.1 (A) Protein–protein molecular recognition as exemplified by the interactions between HIV-1 protease and a substrate. (B) Interactions between HIV-1 protease and an inhibitor[1].

degradation and/or elimination, and poor patient compliance even for fatal diseases like AIDS and cancers limit their effectiveness[1]. Although extensive efforts are being made to solve these problems, drug development will most likely continue to focus on cheaper, more bioavailable, more stable and more easily delivered small non-peptidic drugs, some 30–40 of which enter the marketplace each year as registered new medicines.

One approach to drug development is to reproduce the bioactivities of peptides and proteins with smaller "peptidomimetic" molecules. The definition of a peptidomimetic[2] is "a substance having a secondary structure or other structural feature analogous to a bioactive peptide, which allows it to displace the peptide from its target receptor or enzyme causing inhibition (antagonist) or duplication (agonist) of its activity". Either the entire surface of the bioactive region of a peptide can be mimicked (backbone peptidomimetics), or just those components of the surface that specifically interact with a receptor (side chain peptidomimetics). Both classes of peptidomimetics aim to present key functionality in appropriate positions and orientations that facilitate multiple interactions with target receptors (Figure 44.2). Backbone peptidomimetics achieve this by restraining the polypeptide backbone (shaded) to shapes that directly mimic the receptor binding conformation of the natural ligand, thereby correctly positioning and orienting side chains for receptor binding. On the other hand side chain peptidomimetics replace the peptide backbone with an alternative scaffold that can similarly maintain peptide side chains or their surrogates for receptor binding. In this review we discuss peptidomimetics from both classes but focus mainly on the more extensively documented backbone mimetics.

44.2 Backbone peptidomimetics

Bioactive surfaces of proteins result from the folding of the main chain peptide backbone into distinct "secondary structures" that include β-sheets, β-strands, α-helices, β- and

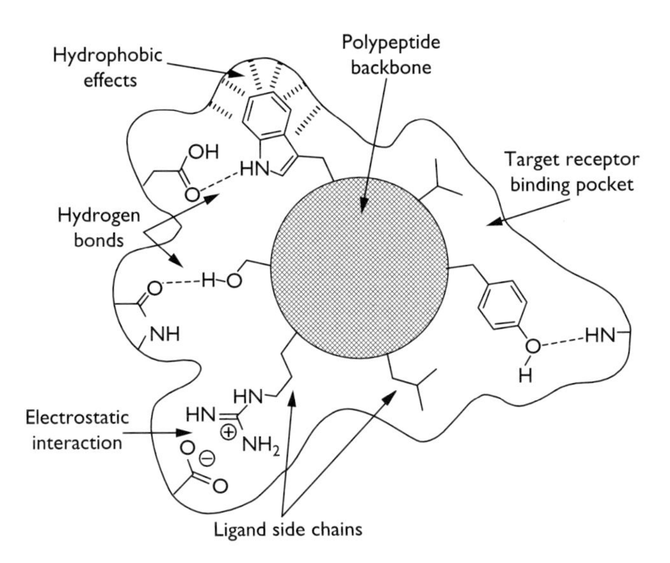

Figure 44.2 Binding interactions between peptidic side chains and a target receptor. Backbone peptidometics mimic the polypeptide backbone (shaded). Side chain mimetics contain scaffolds that replace the peptide backbone.

γ-turns, and their combinations known as "tertiary structures". These structural units position and orient component amino acids, exposing their side chains for complementary interactions with other surfaces. Unfortunately when these structural elements are removed from the stabilizing influences of the protein they usually unfold and adopt random structures in solution. The energy associated with binding of peptides to their target receptors is directly proportional to the free energy required to fold the peptide into its receptor bound conformation. Since random coil peptides must pay a considerable entropic penalty for folding into a specific conformation, they tend to be poor ligands binding with relatively low affinity to binding partners. Conformational flexibility also makes it difficult to achieve specificity for a target receptor with indiscriminant binding to multiple targets leading to toxic side effects.

Designer molecules or "protein surface mimetics" that can fix or stabilize key bioactive surface shapes on small peptide fragments can, in principle, reproduce the same biological functions as peptides and proteins[3]. When synthetic peptides are constrained to shapes that match their receptor bound conformation, 10^2–10^5 fold improvements in affinities can be achieved. Thus the goal in mimicking secondary structure with backbone mimetics is to devise molecular constraints or scaffolds that can force peptides into a bioactive conformation [3,4]. With their greatly reduced size, peptidomimetics of this type are much smaller than proteins, and can display improved pharmacokinetic/dynamic properties and superior bioavailability compared to their parent proteins. The following sections illustrate a few methods that have been used to date to mimic bioactive secondary structures of peptides.

44.2.1 β-Strand mimetics

The β-strand or extended conformation is the simplest peptide structural element in that it does not contain any intramolecular hydrogen bonding between component amino acid residues. It is a linear or saw-toothed arrangement of amino acids whose amide bonds are almost coplanar with side chains that reside alternately above and below the plane. Isolated β-strands are not common in proteins, they are usually hydrogen bonded in pairs forming β-sheet structures. Not normally considered a secondary structure, β-strands are now known to be the crucial structural element recognized by proteolytic enzymes[3, 4], MHC proteins[5] and transferases[6].

For example, proteolytic enzymes (proteases) preferentially recognize and cleave β-strand regions within a polypeptide, either directly or following unfolding through protein denaturation, forming a β-sheet network of hydrogen bonds between the substrate sequence and the enzyme active-site (Figure 44.1)[7]. Although there are thousands of protease inhibitors known[7], only a few have been deliberately constrained to mimic the β-strand conformation which is entropically favored for protease binding. The conformationally constrained macrocycles in peptidomimetics **1–3** involve a peptide side chain condensed onto the main chain of the peptide. These cycles were demonstrated[3] to structurally mimic the protease-binding conformation of tripeptide components of peptides that bind to proteases, leading to potent inhibition of HIV-1 protease (1.7, 0.6, 3 nM respectively) and also to potent inhibition of viral replication. Examples of such macrocyclic protease inhibitors have now been developed for other aspartic proteases (renin, plasmempsins, rhizopuspepsin, porcine pepsin, penicillopepsin), serine proteases (trypsin, thrombin, chymotrypsin, Hepatitis C NS3 protease) and metalloproteases (angiotensin converting enzyme, aminopeptidase, neutral endopeptidase, thermolysin, bacterial collagenase, TNFα convertase, matrix metalloproteases MMP-1, 3, 8, 9).

The advantages of designing protease inhibitors to mimic β-strand conformations were recently validated for thrombin. This serine protease participates in blood clotting by cleaving peptide bonds on the C-terminal side of arginine residues of various regulatory factors in the coagulation cascade (e.g. fibrinogen). Inhibitors of thrombin may be useful for ameliorating certain thromboembolic disease states. PPACK (D-Phe-Pro-ArgCH_2Cl) was a thrombin inhibitor that reached phase II clinical trials and spawned numerous inhibitors based on its sequence. When it was recently constrained to mimic an extended conformation by incorporating a rigid bicyclic Phe-Pro scaffold, picomolar inhibitors (**4–7**) of this protease were produced[7].

Table 44.1 β-Strand mimetics

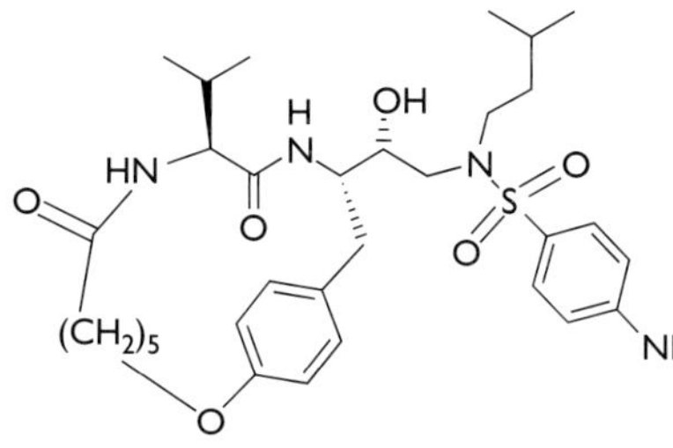

1

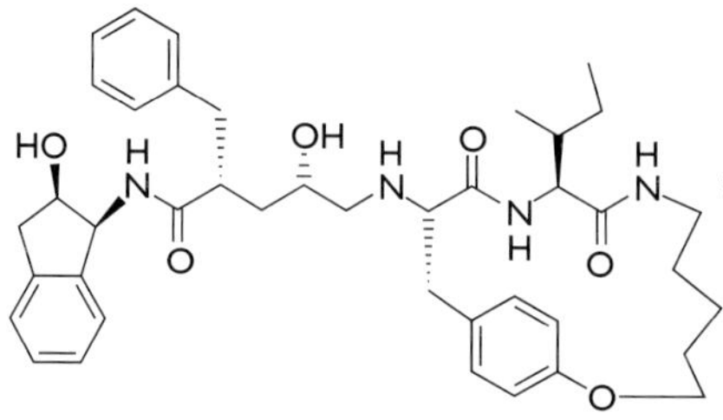

2

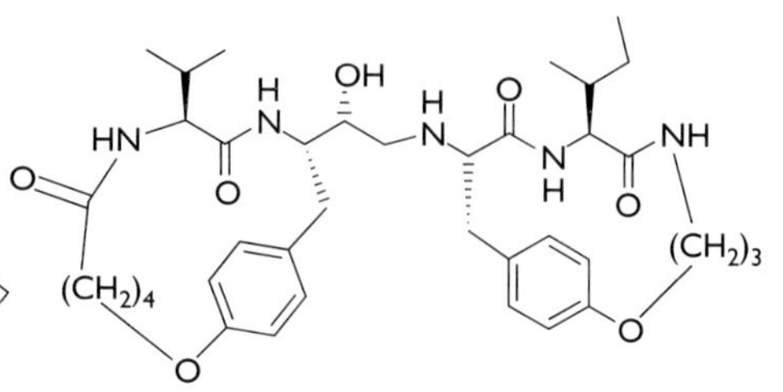

3

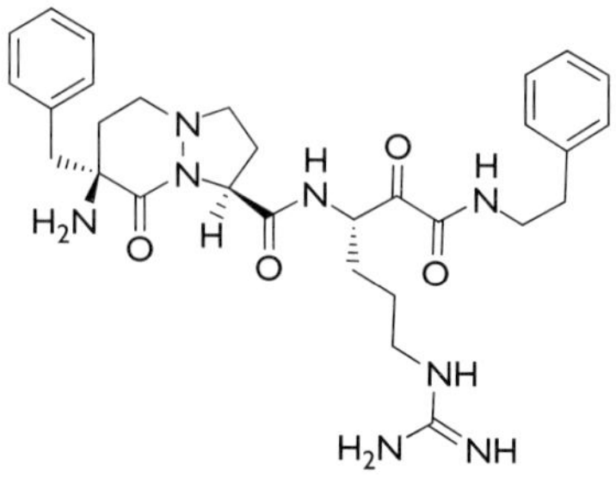

4

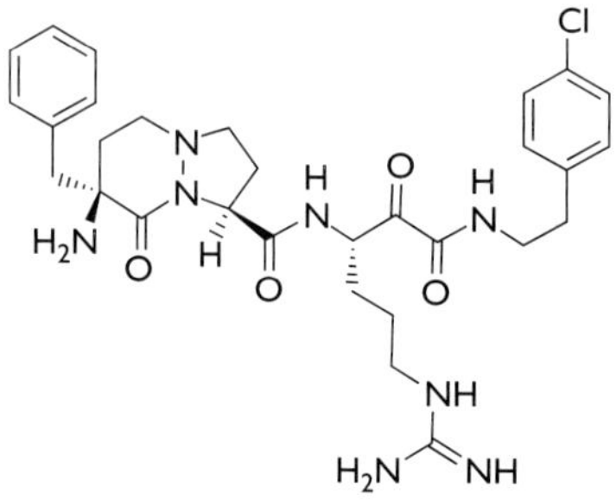

5

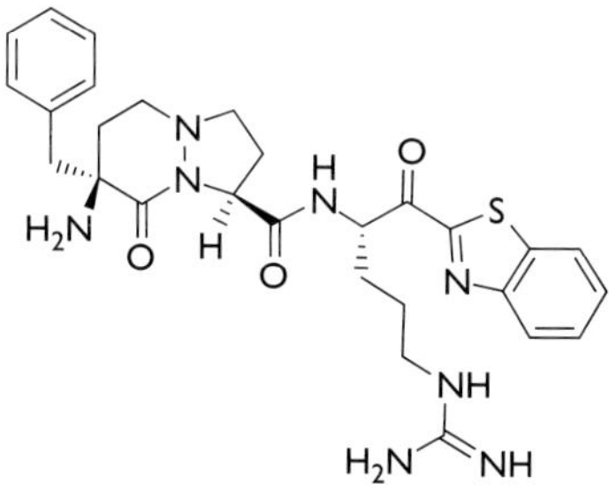

6

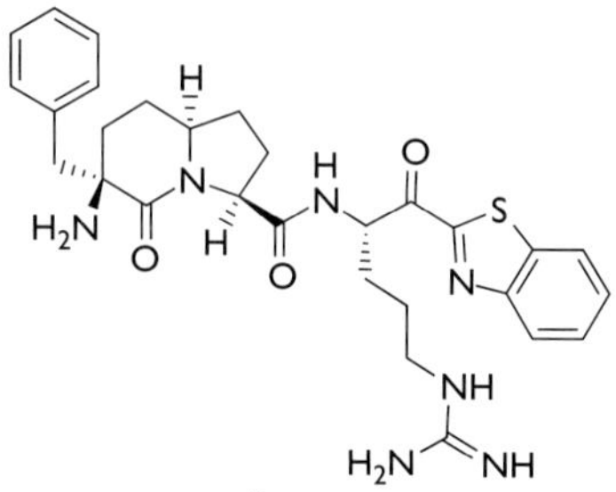

7

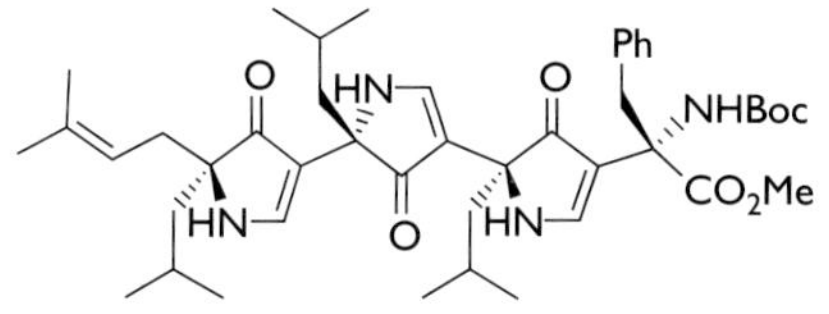

8

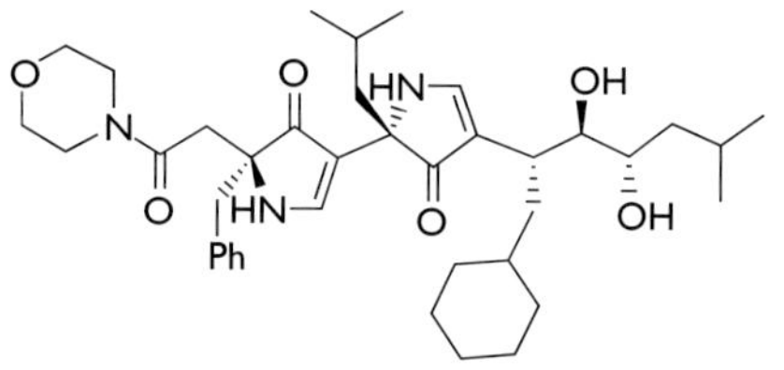

9

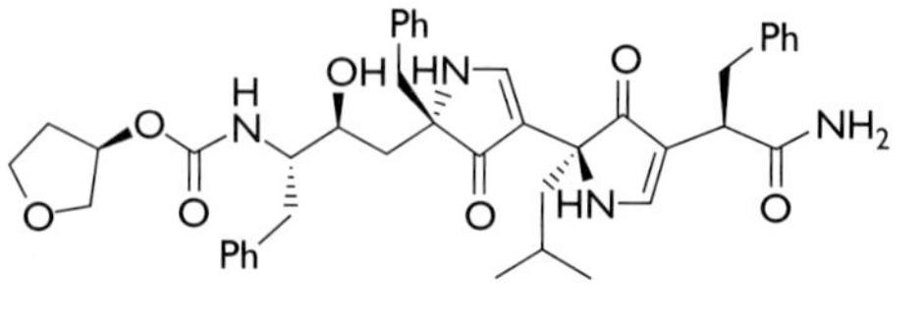

10

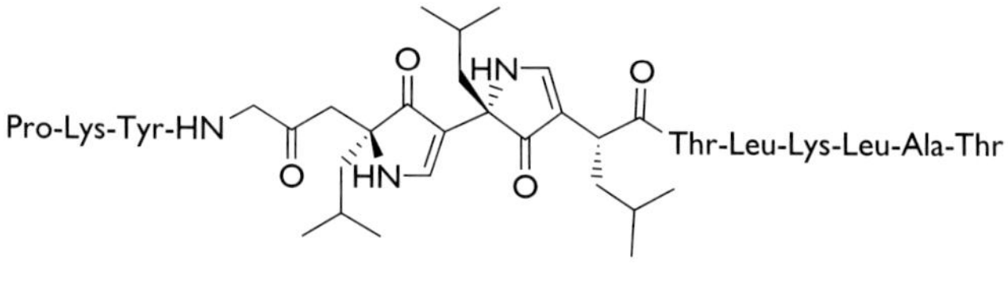

11

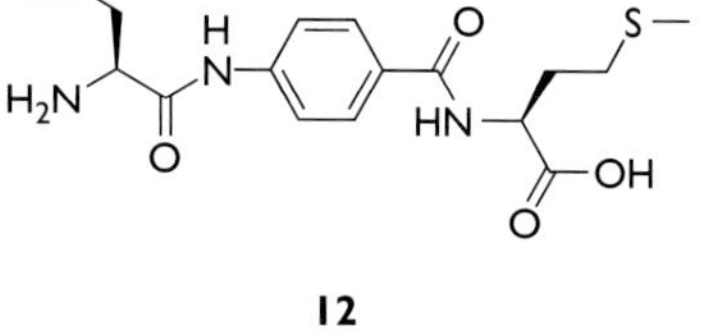

12

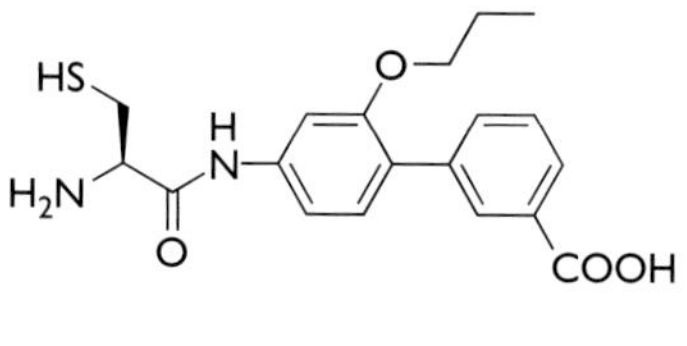

13

Oligopyrrolinones are another class of β-strand mimetics. These compounds are composed of 3,5-linked pyrrolin-4-ones (prototypical oligomer **8**) which adopt a β-strand conformation in the solid state and have shown promise as non-peptidic inhibitors of the proteases renin (**9**) and HIV-1 Protease (**10**). Compared to their peptidic counterparts, these compounds showed similar inhibitory potencies but were more resistant to proteolysis and displayed improved membrane permeabilities. Oligopyrrolinone-peptide hybrids (e.g. **11**) can inhibit the binding of peptides to class II major histocompatibility complex (MHC)[8]. This membrane-bound protein is a surface component of antigen presenting cells and helps to display antigens to T-cells for processing by the immune system. Like protease-substrate interactions, class II MHC molecules have been shown to bind their target peptides in an extended conformation[5]. Under some circumstances, normal self antigens can be inappropriately presented as class II MHC complexes leading to autoimmune diseases. The finding that molecules such as **11** can inhibit class II MHC-peptide interactions suggests that that β-strand mimetics in general warrant proper investigation as potential drugs.

Ras-farnesyltransferase (Ras-FT) is an enzyme responsible for transferring a farnesyl group to the protein Ras allowing it to partake in normal signal transduction functions. Mutated ras oncogenes occur in approximately 30% of all human tumors and for some tumors (e.g. pancreatic), the frequency can be as high as 90%. Farnesylation is both a necessary and sufficent requirement for oncogensis caused by ras mutations so inhibitors of Ras-FT have been heavily studied as potential anti-cancer agents. Ras-FT recognizes the four C-terminal residues of Ras. The "CAAX box" (C = cysteine, A = aliphatic amino acid, X = methionine or serine) as it is known, is thought to bind to Ras-FT in an extended conformation. Inhibitors designed as β-strand tetrapeptide mimetics such as **12** and **13**[6] are able to block Ras processing in a range of ras-transformed tumour cell lines in addition to inhibiting tumor growth in nude mice[9].

44.2.2 β-Sheet mimetics

Pairing of β-strands in parallel and antiparallel alignments leads to β-sheet structures which play crucial roles in many biological processes and in some notable diseases. For example, normal gene regulation by the *met* repressor requires its dimerization through β-sheet domains with the resulting sheet being recognized by the major groove of DNA. Clustering of membrane ion channel PDZ domains involves β-sheets as does the binding interaction between lymphocyte function-associated antigen-1 (LFA-1) and intercellular adhesion molecule-1 (ICAM-1). Aggregation of some proteins to form insoluble β-sheet structures is thought to be responsible for neurological disorders such as Alzheimer's, Creutzfeld-Jacob and other prion diseases. Infection of human immune cells by the HIV virus requires a β-strand mediated interaction of viral gp-120 receptors with CD4 receptors located on the surface of target cells. Strand mimetics that antagonize β-sheet formation may represent a viable therapeutic strategy towards the prevention or treatment of diseases associated with β-sheet structures. While there are relatively few examples of monomeric β-strand mimetics, probably due to their high tendency to aggregate to insoluble polymers, β-sheet mimetics are more prevalent with a variety of structurally diverse examples having been reported[10, 11]. Such β-sheet mimetics can afford useful information for the design of β-strand mimetics.

Three types of β-sheets are found in proteins and are classified as either parallel, antiparallel or mixed. Parallel β-sheets contain strands that run in the same direction and are characterized by a series of twelve membered hydrogen-bonded rings (Figure 44.3A). Antiparallel β-sheets contain strands that run in opposite directions and are characterized by an alternating series of ten and fourteen membered hydrogen-bonded rings (Figure 44.3B). Mixed β-sheets contain mixtures of both patterns.

β-sheets in proteins are often arranged as β-hairpins in which the loop connecting two hydrogen-bonded antiparallel strands is very short. Many of the β-sheet mimetics explored to date have attempted to mimic this arrangement by attaching peptide strands to a scaffold which places them in close proximity to each other and in an orientation compatible with interstrand hydrogen bonding. The first artificial β-sheet (**14**) reported in 1986 typifies this approach as do later examples by Kemp (**15**) and Kelly (**16**).

A different approach to β-sheet mimetics involves attaching peptide strands to scaffolds which are themselves

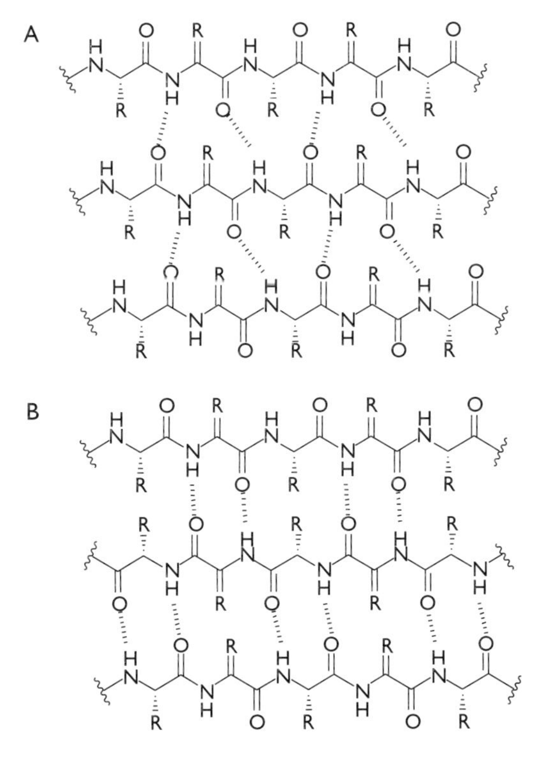

Figure 44.3 Schematic representation of (A) parallel β-sheet and (B) antiparallel β-sheet. Amino acid side chains are represented by R. Hydrogen-bonds are indicated by dashed lines.

Table 44.2 β-Sheet mimetics

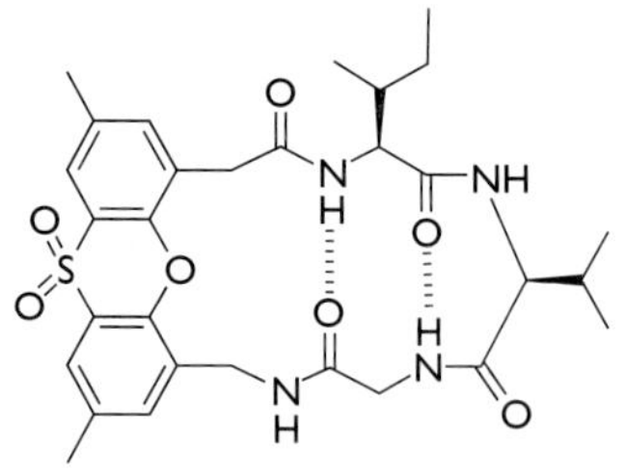

14

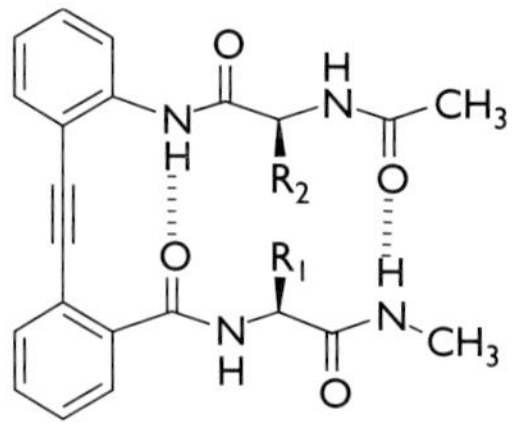

15

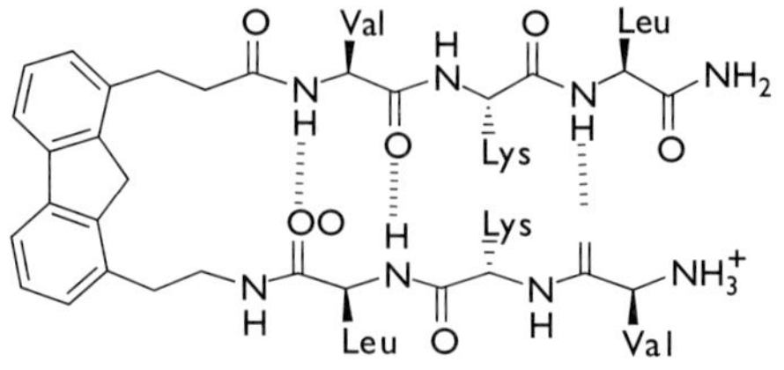

16

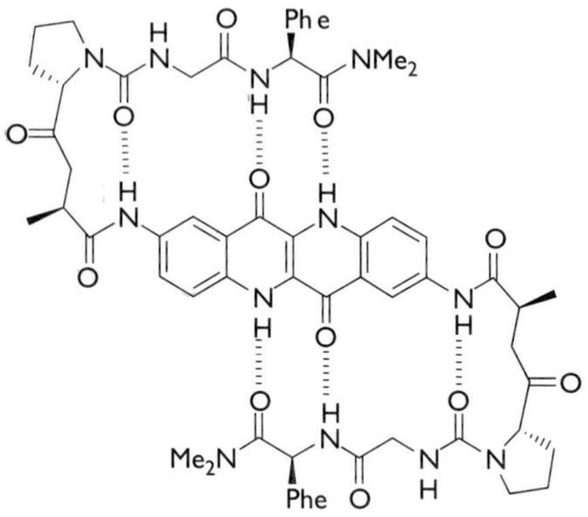

17

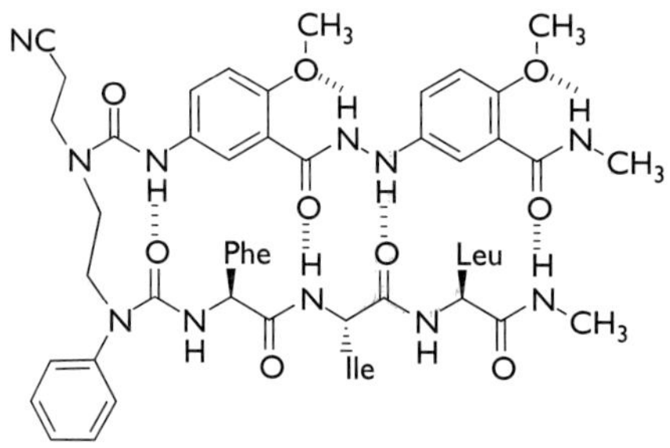

18

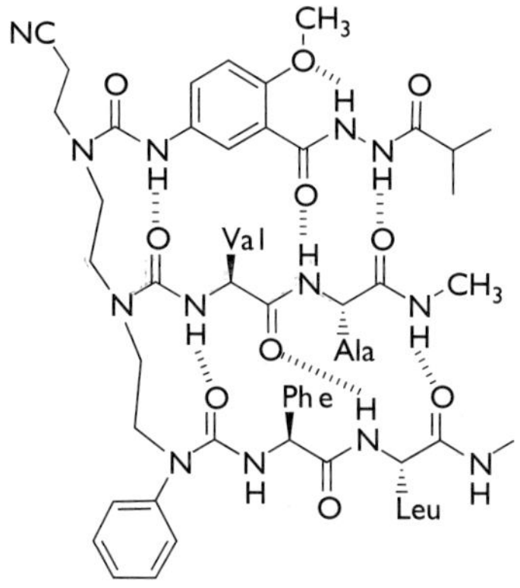

19

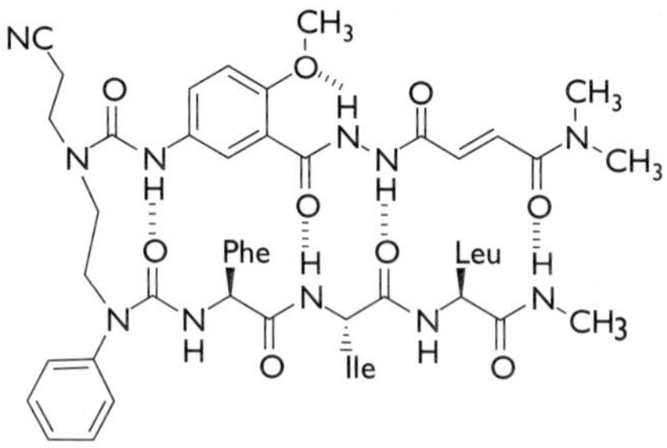

20

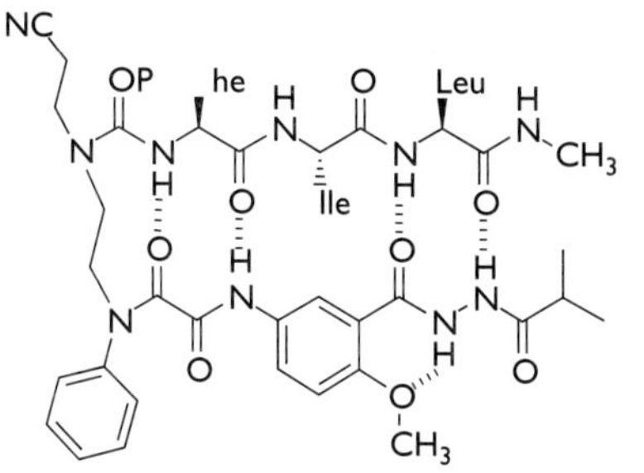

21

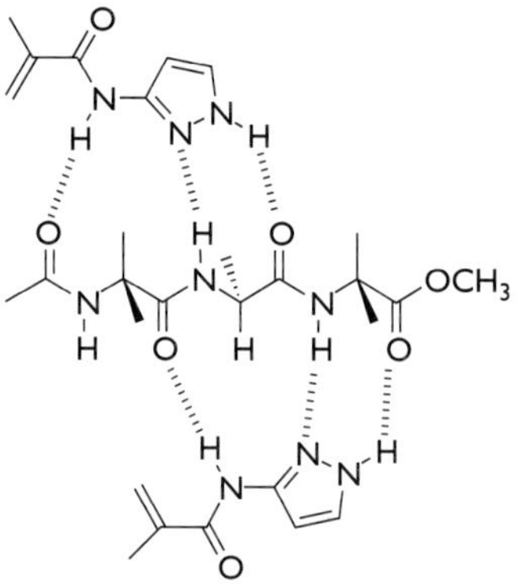

22

β-strand mimetics. These scaffolds contain hydrogen bonding functionality spatially oriented to allow β-sheet formation with the attached peptides as illustrated in **17**. This approach has more recently been investigated using 5-amino-2-methoxybenzamide derivatives as the β-strand mimicking scaffold as shown in **18–21**. When attached to an N,N-linked oligourea template to which is also appended either one or two peptide strands, the oligourea template positions the peptide strands in close proximity to the 5-amino-2-methoxybenzamide moiety inducing the whole structure to adopt the desired β-sheet conformation[12].

All of the mimetics described so far display only intramolecular β-sheet hydrogen-bonding patterns. For β-sheet mimetics to be able to mimic or antagonize a biological process they necessarily must interact with other biomolecules, as for example in Alzeimer's Disease, where native β-sheet formation appears to need preventing. If aggregation of β-amyloid peptides into the β-sheets, that are thought to ultimately form amyloid plaques, could be inhibited, it may offer a way of preventing and perhaps treating this disease. It is more likely that a β-strand rather than β-sheet mimetic, which interacts with the peptide preventing strand aggregation, will be used as β-sheet antagonists. The first reported example of a non-peptide that binds to and stabilises a peptide in a β-strand conformation purely by intermolecular interactions was a simple aminopyrazole compound **22**. This may be the forerunner to future therapeutic agents that act by preventing peptide sheet formation *in vivo*.

44.2.3 Reverse turn mimetics

Reverse turns are defined as those sites in polypeptides at which an overall reversal of chain direction occurs. Although not strictly secondary structural elements, since they display no repeating backbone torsional angles as found in β-sheets and helices, turn motifs dominate the bioactive surfaces of proteins and peptides and are key recognition motifs in peptide-protein and protein-protein interactions. Owing to their relatively small size and non-peptide properties, turn mimetics have been studied more intensively than any of the other secondary structure mimetics for their potential as pharmaceuticals. Only a few of the many examples reviewed extensively elsewhere [3, 13, 14] are illustrated ahead.

Turns may contain either three or four amino acids. Turns defined by three residues are termed γ-turns, while β-turns are defined by four residues. γ-turns contain a characteristic pseudo 7-membered hydrogen-bonded ring (Figure 44.4) and are sub-classified as either classic or inverse γ-turns depending on the backbone dihedral angles ϕ and ψ of residue $i+1$ (Table 44.3). β-turns generally, though not always, contain a pseudo 10-membered hydrogen-bonded ring between the amide carbonyl of residue i and the NH of residue $i+3$ (Figure 44.4). Sub-classification of β-turns depends on the values of four backbone dihedral angles ϕ_{i+1}, ψ_{i+1}, ϕ_{i+2}, ψ_{i+2} defining them as one of types I, I′, II, II′, III, III′ VIa1, VIa2, VIb, VIII and IV (Table 44.3). A more accurate sub-type definition for β-turns was introduced by Wilmot and Thornton[15] which classifies them according to their location in Ramachandran space (Table 44.3) however the older nomenclature still prevails in the literature of β-turn mimetics.

Table 44.3 Turn types and their sub-classification according to backbone dihedral angles

Turn type	ϕ_{i+1}	ψ_{i+1}	ϕ_{i+2}	ψ_{i+2}	*Ramachandran Nomenclature*
β-Turns					
I	−60	−30	−90	0	$\alpha R \alpha R$
I′	60	30	90	0	$\alpha L \gamma L$
II	−60	120	80	0	$\beta \gamma L$
II′	60	−120	−80	0	$\epsilon \alpha R$
III	−60	−30	−60	−30	$\alpha R \alpha R$
III′	60	30	60	30	$\alpha L \alpha L$
VIa1	−60	120	−90	0	$\beta \alpha R$
VIa2	−120	120	−60	0	$\beta \alpha R$
VIb	−135	135	−75	160	$\beta\beta$
VIII	−60	−30	−120	120	$\alpha R \beta$
IV	−60	10	−50	20	
γ-Turns					
Classic	70 to 85	−60 to −70			
Inverse	−70 to −85	60 to 70			

Neurotransmitters and hormones are naturally occurring peptides that commonly use turn motifs for recognition and binding of target receptors[4]. Some of these peptides play key roles in disease and turn mimetics functioning as agonists or antagonists may represent potential pharmaceuticals. Examples of these peptides include Somatostatin, LHRH, Bradykinin, Oxytocin, Leu-Enkephalin, Angiotensin II and Vasopressin. Most of these peptides adopt β-turns of one type or another except for vasopressin which prefers to form an inverse γ-turn.

44.2.3.1 β-Turn mimetics

β-turn mimetics aim to constrain one or more backbone torsional angles to values close to those of the desired turn conformation (i.e. β-turn Type I, II, IV etc). Ideally, they also aim to correctly position the side-chains in orientations similar to those observed in the native turn. One common approach towards achieving this has been to replace residues $i+1$ and $i+2$ of the putative β-turn with constrained cyclic dipeptide isosteres such as benzodiazepines, spiro compounds, and lactams. Of these, the Freidinger lactams[9] and their related polycyclic derivatives have probably been the most heavily investigated. They are informally defined as compounds in which the α-position of an amino-acid (residue i) is linked either by a carbon or heteroatom containing bridge to the nitrogen of the next ($i+1$) residue. Examples of such mimetics are shown for angiotensin II (**23**, **24**), LHRH (**25**), somatostatin (**26**), bradykinin (**27**) and leu-enkephalin (**28**)[4, 9, 14].

Another popular strategy for mimicking β-turns is to replace the hydrogen-bond of the pseudo ten-membered ring with a covalent linkage. This type of approach incorporating an ethylene linkage was recently used to produce a mimetic (**29**) of Leu-enkephalin (**30**). Continuing studies are

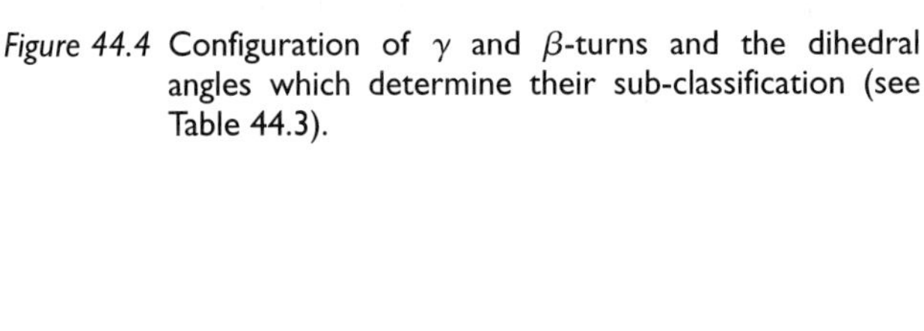

Figure 44.4 Configuration of γ and β-turns and the dihedral angles which determine their sub-classification (see Table 44.3).

Table 44.4 β-Turn mimetics

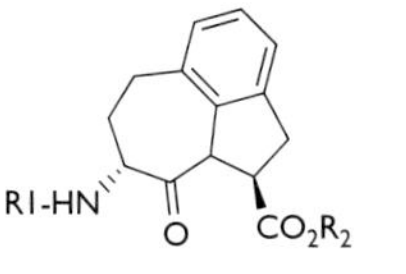

23

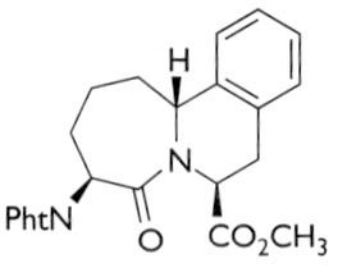

24

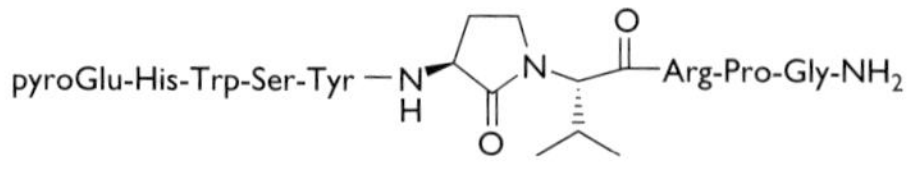

25

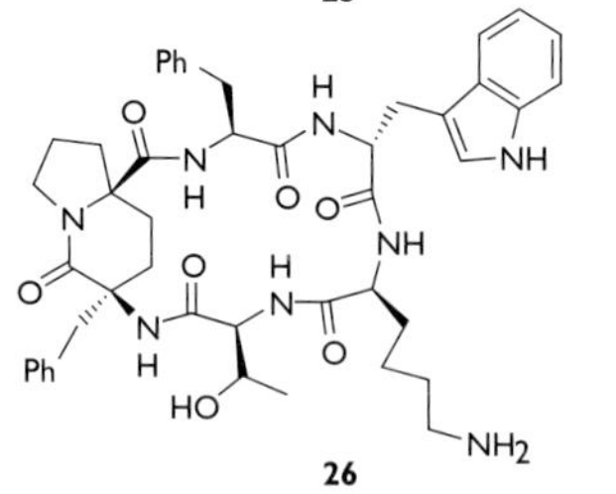

26

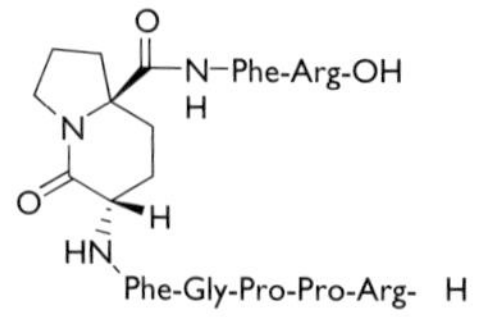

27

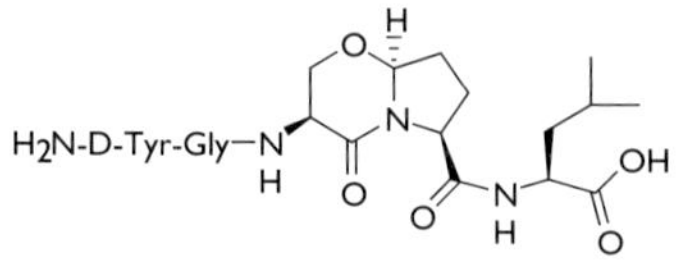

28

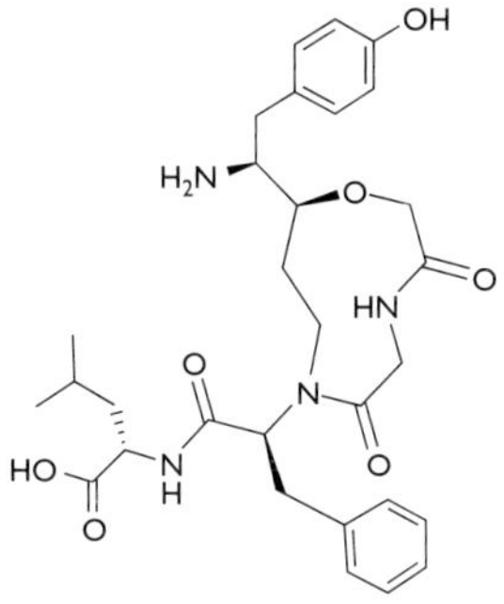

29

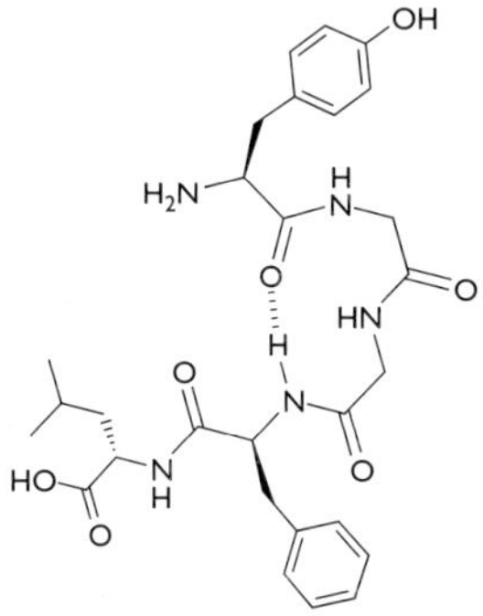

30

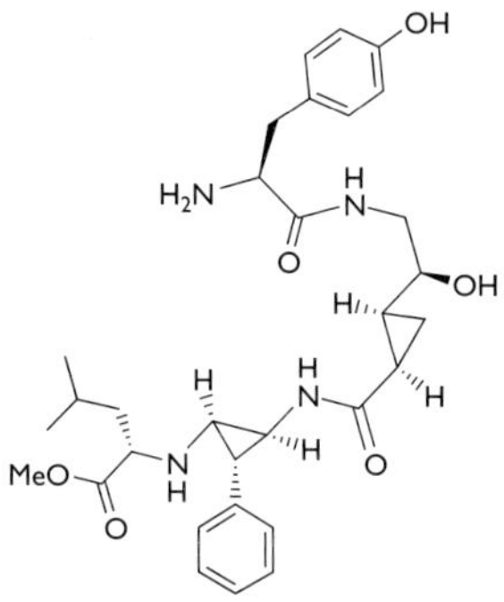

31

H-Ala-Gly-Cys-Lys-Asn-Phe-Phe-Trp
HO-Cys-Ser-Thr-Phe-Thr-Lys

32

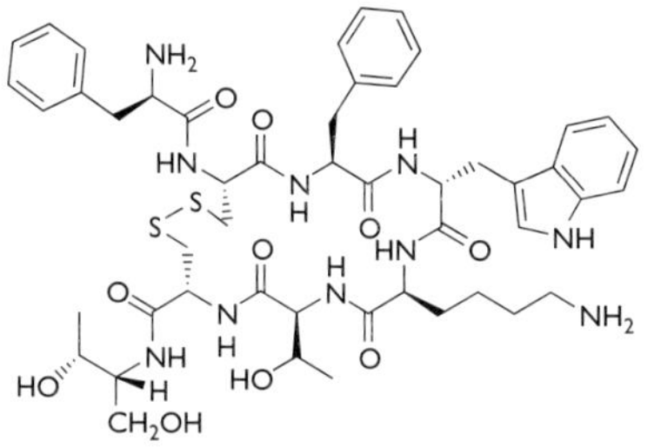

33

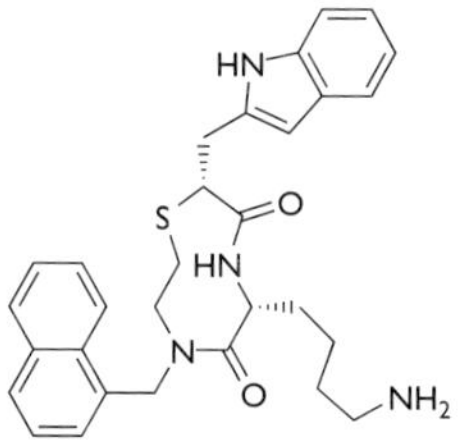

34

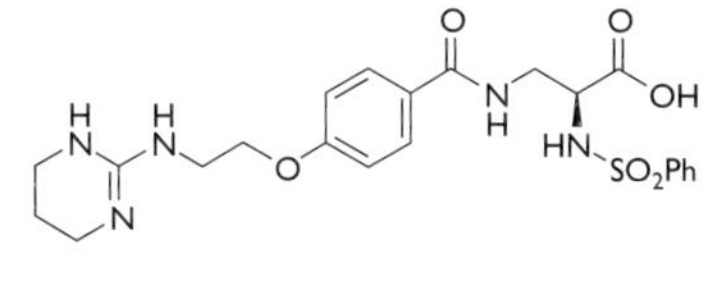

35

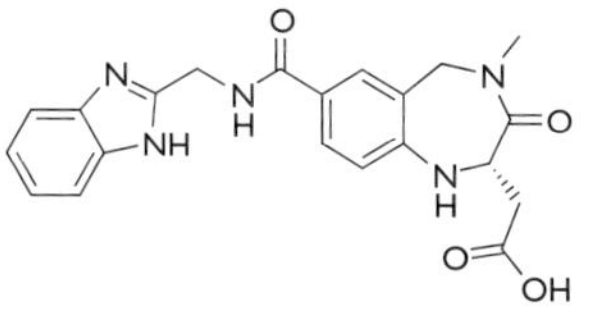

36

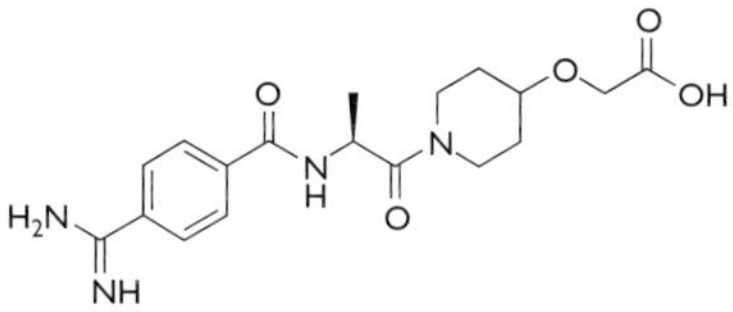

37

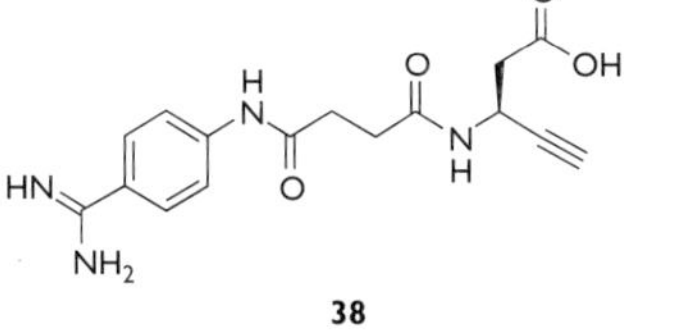

38

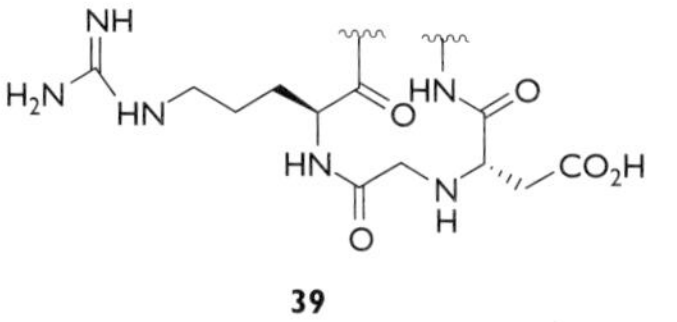

39

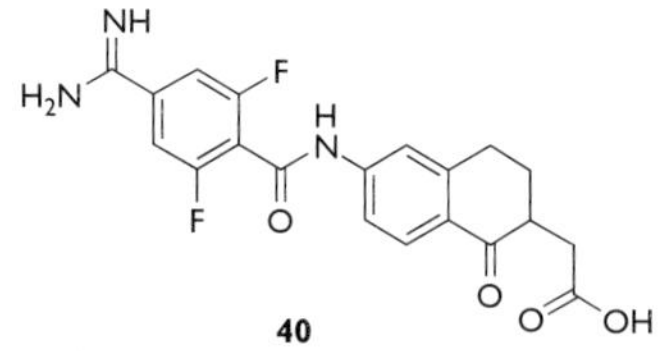

40

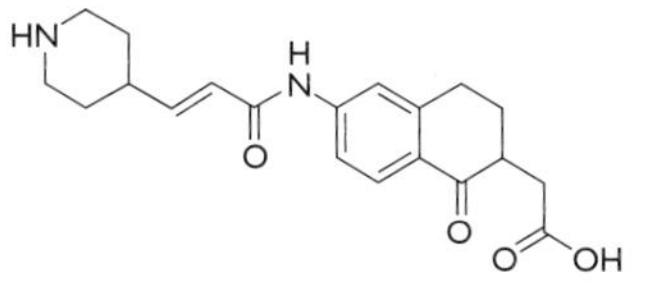

41

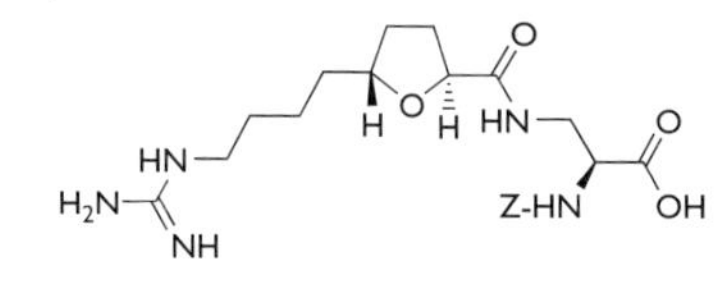

42

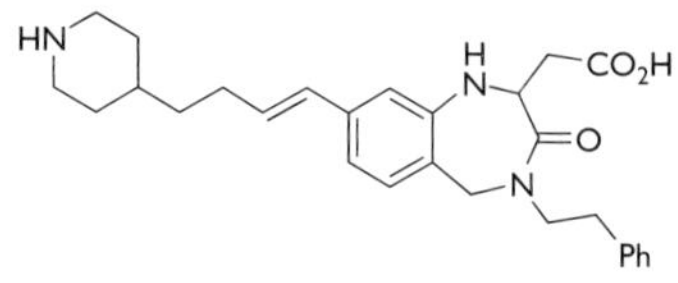

43

pursuing the conformation and bio-activity of these types of structures in addition to preparing analogs which incorporate side chains in the $i+1$ and $i+2$ positions.

Most β-turn mimetics only constrain the backbones of turn residues and very few are capable of correctly positioning the side chains. This is a significant deficiency since the side chains are the sites responsible for recognition, affinity and specificity. A recent approach attempted to counter this problem by incorporating cyclopropane containing dipeptide mimics (**31**) into the β-turn of Leu-Enkephalin (**30**). Modeling studies suggested compounds of this type should adopt the desired β-turn with correctly positioned side chains. Biological testing however showed them to be inactive, highlighting the current lack of understanding of side chain conformations and the role they play in biomolecular interactions. Perhaps the most promising approach aimed at addressing this issue is the use of combinatorial chemistry. By synthesizing large libraries of β-turn mimetics the limitations in defining the exact bioactive conformation should be more easily overcome. This was demonstrated recently in a library of β-turn mimetics for somatostatin (**32**).

Octreotide (*Sandostatin*®) (**33**) is a somatostatin agonist currently used for treating acromegaly and gastroentero-pancreatic tumors. Due to its peptidic nature it is far from an ideal drug and must be administered by injection. In searching for small orally bioavaliable substitutes, workers at Glaxo Wellcome produced a focussed library of 172 β-turn mimetics and discovered the highly potent (IC_{50} 87 nM) and $hSST_5$ selective agonist (**34**)[16], as well as potent ligands to a number of other receptor sub-types. This success demonstrates the enormous potential of combinatorial libraries for identifying mimetics of peptide ligands when relatively little is known about the peptide-receptor interaction.

β-turns from larger peptides and proteins can also be mimicked. Crystal structures of antibody-peptide complexes established that β-turns on antigens are recognized by the binding loops of antibodies. The reverse is also true with β-turns at the ends of antibody complementarity determining regions (CDRs) being used to bind antigens, and successful mimetics of these loops have been prepared [17]. β-turns can also act as recognition sites for post-translational modifications especially phosphorylation and glycosylation. The most studied beta turns in proteins for which mimetics have been developed are a class of cell surface receptors known as integrins. Integrins are heterodimeric glycoproteins that bind to extracellular adhesive molecules including fibrinogen, fibronectin, vitronectin and VCAM-1 (vascular cell adhesion molecule-1). They are critical mediators of cell-cell and cell-matrix adhesion processes and abnormal functioning has been implicated in pathological processes including angiogenesis, platelet aggregation, tumor growth and renal dysfunction. Integrin antagonists have therefore become attractive targets for treating some types of cancers, osteoporosis, thrombosis and renal diseases.

Different combinations of α- and β-subunits in integrins give rise to their different classes (e.g. $\alpha_v\beta_3$, $\alpha_{IIb}\beta3$, $\alpha_4\beta_1$) and each class has a different specificity pattern for particular extracellular adhesive proteins. Many of these target proteins bind integrins through a common recognition sequence Arg-Gly-Asp (RGD) located on their surfaces. Selectivity among ligands for the different integrins is primarily achieved through different conformations adopted by their RGD sequences including different types of β-turns (e.g. βII$'$ and βI). β-turn mimetics incorporating RGD side chains have therefore been investigated by most pharmaceutical companies in search of selective integrin antagonists. Examples are **35–38**. The underlying design principle for all of these compounds was similar and involved using a rigid core unit to separate a positively charged (Arg-isostere) group from a carboxylic acid (Asp-isostere) so as to mimic a Gly-Asp β-turn (**39**). This approach continues to be popular particularly in the search for $\alpha_{IIb}\beta3$ selective antagonists for controling thrombosis. Eli Lilly (**40, 41**)[18], Merck (**42**)[19] and SmithKline Beecham (**43**)[20] have all recently described some new $\alpha_{IIb}\beta3$ selective integrin antagonists derived from an RGD β-turn.

44.2.3.2 γ-Turn mimetics

γ-turns are less common in peptides and proteins than β-turns and there have been fewer mimetics developed. Examples are **44–51**. Vasopressin (**52**) is a nine residue cyclic peptide hormone secreted from the hypothalamus and is a primary regulator of fluid and salt retention in the kidneys. The drug desmopressin (**53**), a synthetic analog of vasopressin, has a similar antidiuretic effect but does not cause vasoconstriction or smooth muscle contraction and is useful in the treatment of diabetes insipidus, mild hemophilia A, von Willebrand's disease and thrombocyte dysfunction prior to surgery. Both vasopressin and desmopressin are now thought to interact with their target receptor through an inverse γ-turn encompassing residues Phe3-Asn5 so the search for less peptidic drugs here has moved towards γ-turn mimetics[3,4].

In a promising recent approach, one of the amide bonds of the γ-turn was replaced with an amide bond isostere, $\Psi[CH_2O]$, and the hydrogen-bond between residues i and $i+2$ was replaced with a methylene bridge to give a six membered morpholin-3-one ring (**54**). These easily synthesized compounds provide access to both classical and inverse γ-turns[21], stereoselectively incorporating all three side chains of a γ-turn into a mimetic. When incorporated into desmopressin (**55**), the morpholin-3-one ring was shown by ^1H NMR spectroscopy to successfully adopt the desired inverse γ-turn conformation. However no biological activity has yet been reported for these compounds so their value is still unproven. Nevertheless, owing to their facile and flexible synthesis and demonstrated preference for a γ-turn conformation

Table 44.5 γ-Turn mimetics

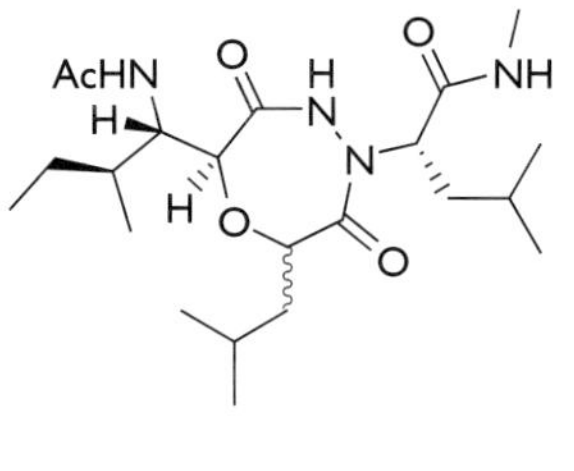

44

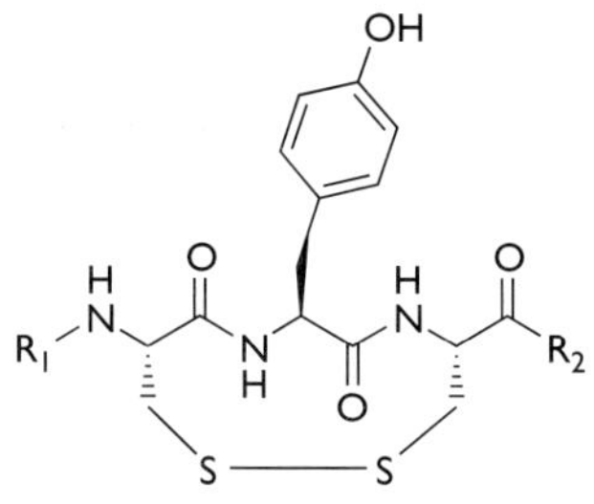

45

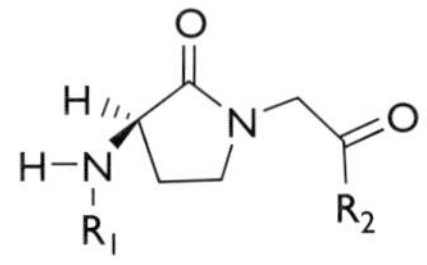

46

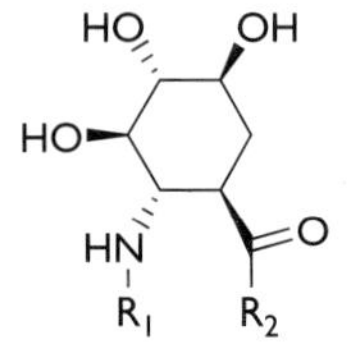

47

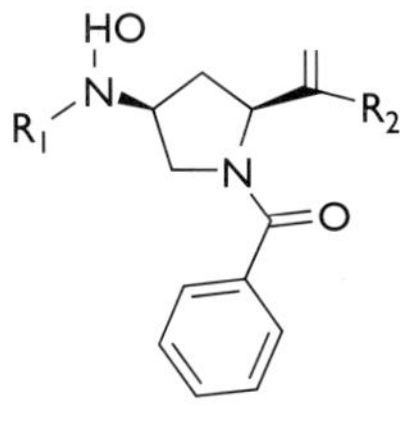

48

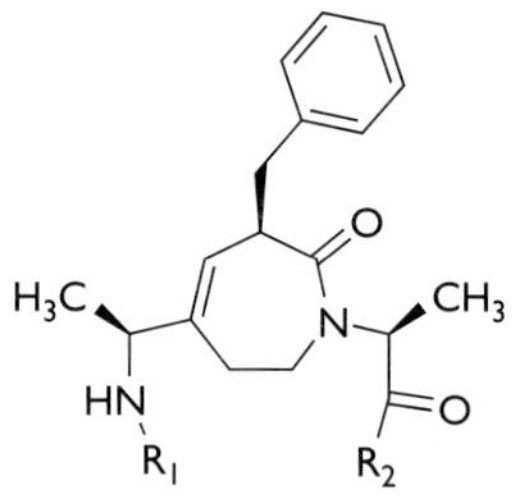

49

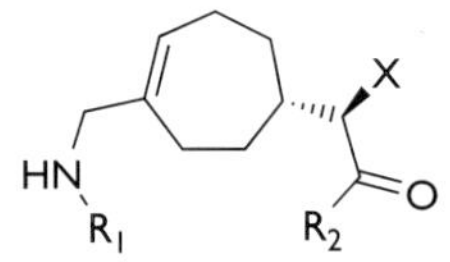

50

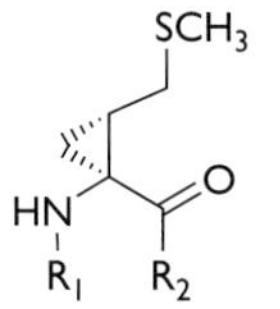

51

Cys-Tyr-Phe-Gln-Asn-Cys-Pro-D-Arg-Gly-NH_2

52

Mpa-Tyr-Phe-Gln-Asn-Cys-Pro-D-Arg-Gly-NH_2

53

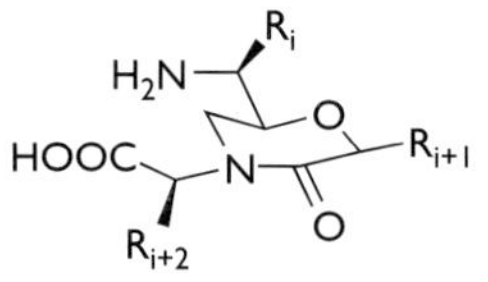

54

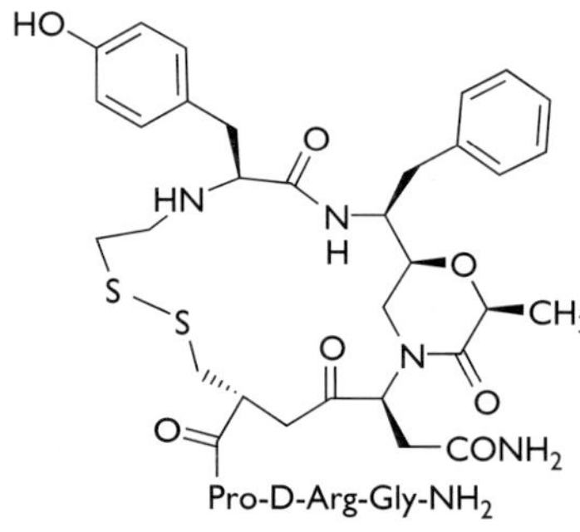

55

it is forseeable that these structures will feature in future γ-turn mimetics.

44.2.4 Helix mimetics

The α-helix is the most abundant protein secondary structure. The classical α-helix is composed of repeating units known as helical turns which interact with each other through a repetitive hydrogen-bonding network to give a characteristic rod-like structure with a radius of approximately 2.3 Å. Each turn is a pseudo 13-membered macrocyclic ring composed of 3.6 amino acid residues with the carbonyl oxygen of residue i being hydrogen bonded to the amide-NH of residue $i+4$ that begins the next turn. The optimal torsional angles for the α-helical backbone are approximately $-58°$ for ϕ and $-47°$ for Ψ. Naturally occurring α-helices are all right-handed because they are composed of L-amino acids and are typically three to four turns in length corresponding to

11–15 residues. Buried α-helices within the hydrophobic core of a protein usually function as stabilizers of tertiary structure. However, exposed helices located on a protein surface may contain crucial binding residues required for interactions with a target like DNA, RNA or another protein. These surface helices are of particular interest because molecules designed to mimic them could be agonists or antagonists of protein function.

Four examples of helix recognition motifs relevant to diseases are p53, HIV-Rev, PTHrP and vaccines. *Anticancer* agents could be developed based upon mimicking an α-helix of p53, a tumor suppressor protein that helps maintain cellular genomic integrity following DNA damage by inducing cell cycle arrest or apoptosis. Inactivation of p53 is common in neoplastic transformation via genetic alteration or competitive binding to the cellular oncoprotein MDM2. Elevated levels of MDM2 causing inactivation of p53 in tumors could, in principle, be blocked by mimetics of the p53 helix that antagonize the p53-MDM2 interaction[22]. *Antiviral* drugs could be developed by mimicking the RNA-binding α-helical domain[23] of the Rev protein. This helix is essential for replication of the Human Immunodeficiency Virus, and compounds that interfere with REV-RNA interaction prevent replication of HIV-1 in cells by preventing export of mRNA from the nucleus of HIV-infected cells.

Hypercalcemia could potentially be treated by mimetics of an α-helix of parathyroid hormone related protein (PTHrP), a protein which regulates calcium homeostasis[24] and is known to be involved in malignancy-associated hypercalcemia. Helix-mimicking antagonists of the G protein-coupled receptor for PTHrP may therefore display useful therapeutic properties. New *vaccines* can be constructed from small peptides, since antibodies raised to them can cross react with their native protein antigens. Antibodies raised to a protein surface helix epitope have been shown to bind with an affinity similar to antibodies raised only from the corresponding peptide. This was only true if the peptide was made helical by incorporating the helix stabilizing amino acid α-amino isobutyric acid (Aib) or appropriately positioned salt bridges, and helicity directly correlated with binding affinity, suggesting that helix mimetics could be promising vaccine candidates. Although comparatively little has been achieved so far towards the objective of mimicking α-helices with small molecules, a number of methods of stabilizing helices are providing important clues to the design of helix mimetics.

44.2.4.1 Helix-inducing amino acids

Unnatural amino acids such as **56–65**[3, 25] can stabilize helicity in certain short peptides. Theoretically it should be possible to examine crystal stuctures of ligand-receptor complexes involving helices and design peptide ligands which incorporate helix stabilizing amino acids at positions not important for interactions with the target. This approach was recently validated by a group attempting to prepare antagonists of the p53/MDM2 interaction[26]. The isolated α-helical segment of p53, Ac-Gln-Glu-Thr-Phe-Ser-Asp-Leu-Trp-Lys-Leu-Leu-Pro-NH2, is unstructured and a weak inhibitor of the p53/MDM2 interaction (IC_{50} 8.6 μM), but when Aib (**65**) and some other non-native amino acids (phosphonomethylphenylalanine (Pmp), 1-amino-cyclopropanecarboxylic acid (Ac3c) and 6-Cl-Trp) were incorporated into the sequence, a highly potent ($IC_{50} = 5$ nM) and helical antagonist (Ac-Phe-Met-Aib-Pmp-6-Cl-Trp-Glu-Ac3c-LeuNH2) was produced.

44.2.4.2 Helix-inducing side chain interactions

Certain side chains of amino acids in an α-helix are in close proximity because they reside on the same face of the helix. These include residue pairs positioned at i and $i+3$ or $i+4$ which are separated by a single turn as well as residue pairs separated by two turns (i, $i+7$). These pairs can be stabilized in proteins by non-covalent interactions such as ion pairs and π-stacking or hydrophobic effects. Charged side-chains can also stabilize α-helicity through the helix macrodipole. Helices have a natural dipole moment of approximately 3.2 Debye per residue due to their parallel alignment of polarized amide bonds resulting in a positive charge at the amino terminus and a negative charge at the carboxy terminus. Thus, helicity can be stabilized by incorporating negatively charged residues at the N-terminus and positively charged residues at the C-terminus.

Peptide helices can also be stabilized by covalently tethering side chains that are close together on the surface of the helix. Linkages that have been studied include lactam bridges (e.g. (i, $i+4$) (**66, 69**), (i, $i+7$) (**67, 68**)); disulphide bonds and aliphatic linkers, and metal chelates (**70–72**)[3, 25]. These methods tend to contribute only 1–2 kcal.mol^{-1} towards helix stability.

A limited number of studies have shown that these methods of helix induction can result in increased bioactivity. The use of i, $i+4$ Asp-Lys lactam bridges in a 29-residue sequence of human growth hormone releasing factor (cyclo(Lys4-Asp8)-[Ala15]-hGRF(1–29) increased both helicity and biological activity. Researchers at Rhone-Poulenc incorporated Asp-Lys lactam bridges into the human parathyroid hormone sequence (hPTH-(1–31)NH_2)[24] preparing a number of lactam analogs including mono-, bi- and tricyclic (**69**) systems, **69** being most active (EC_{50} 0.14 nM) and showing twice the potency of (hPTH-(1–31)NH_2).

44.2.4.3 Helix nucleators

One approach to stabilizing a helix is to insert a nucleating template at the N- or C- terminus of a peptide or within the sequence. For effective nucleation, templates must be

Table 44.6 Helix inducing amino acids and side chain interactions

H_2N COOH **56** H_2N COOH **57** H_2N COOH **58** H_2N COOH **59** H_2N COOH **60**

H_2N COOH **61** H_2N COOH **62** H_2N COOH **63** H_2N COOH **64** H_2N COOH **65**

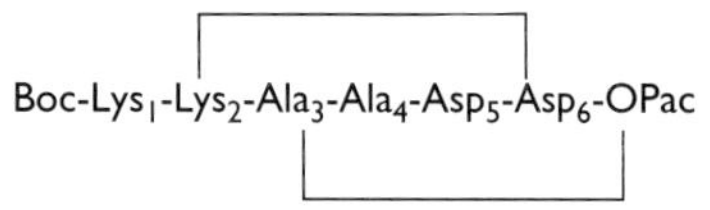

Boc-Lys_1-Lys_2-Ala_3-Ala_4-Asp_5-Asp_6-OPac

66

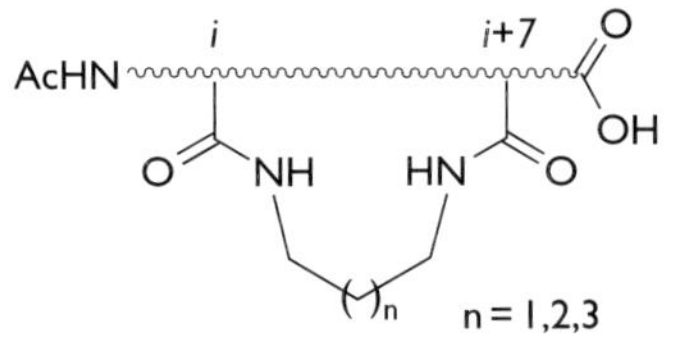

AcHN *i* *i*+7 O OH O NH HN O n = 1,2,3

67

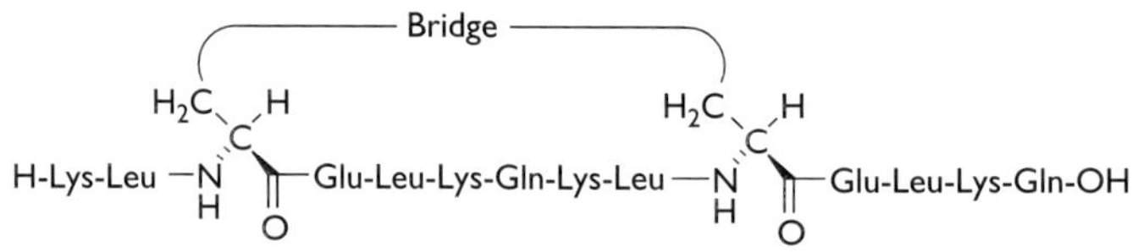

Bridge

H_2C H C H-Lys-Leu —N H O Glu-Leu-Lys-Gln-Lys-Leu —N H H_2C H C O Glu-Leu-Lys-Gln-OH

68

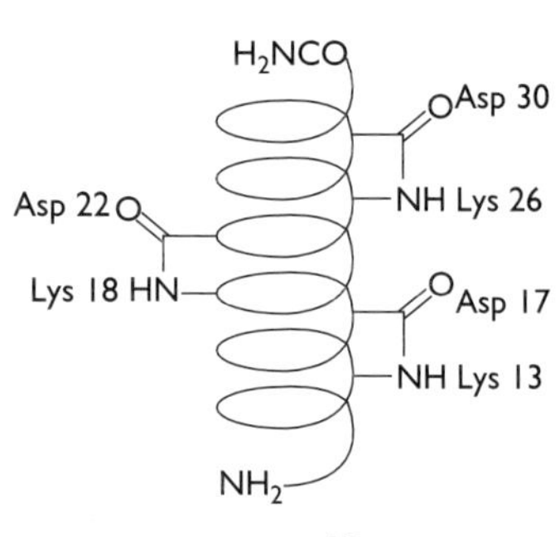

H_2NCO Asp 30 NH Lys 26 Asp 22 Lys 18 HN Asp 17 NH Lys 13 NH_2

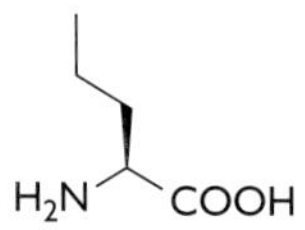

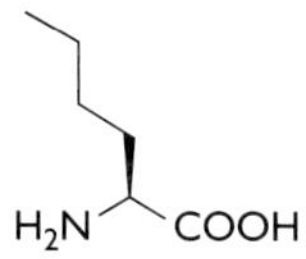

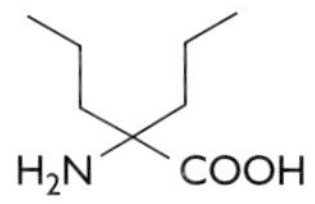

69

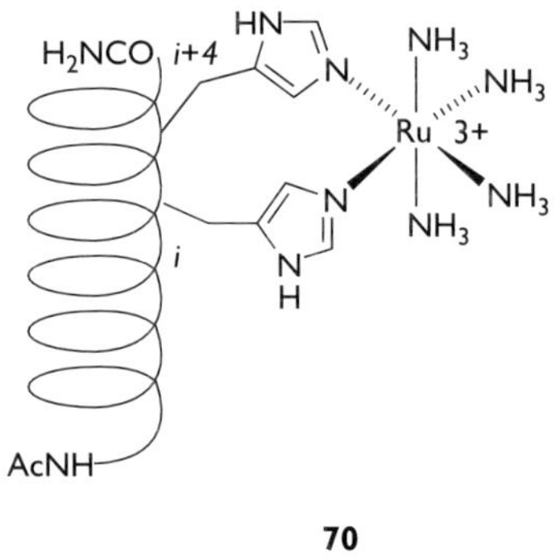

H_2NCO *i*+4 HN N NH_3 NH_3 Ru 3+ N NH_3 NH_3 *i* N H AcNH

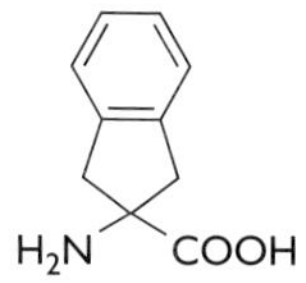

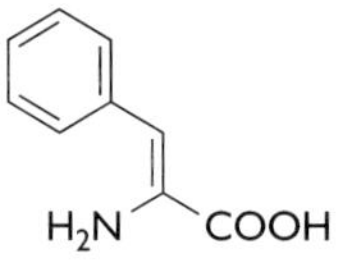

70

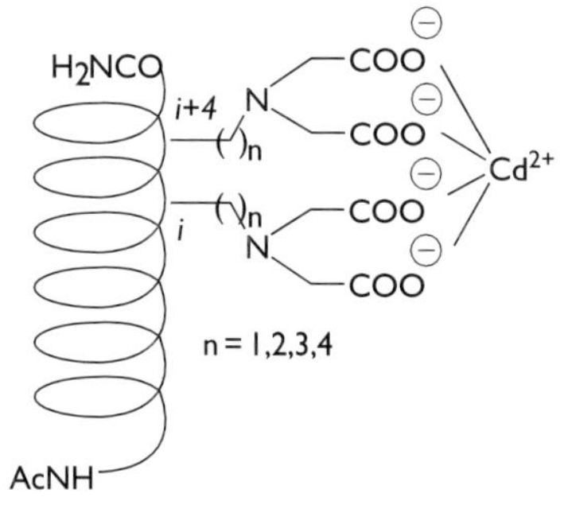

H_2NCO *i*+4 N COO COO Cd^{2+} *i* N COO COO n = 1,2,3,4 AcNH

71

H_2N O H N O H N HN O H_2 N Pd N H_2 N NH N O H N HN NH O O

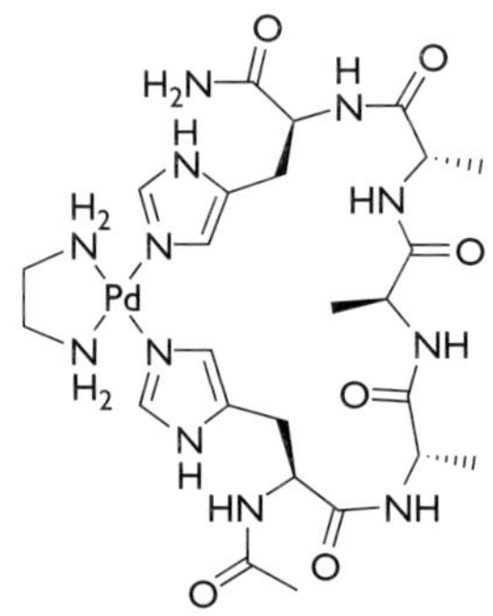

72

conformationally rigid and have appropriate hydrogen-bonding functionality fixed in the same positions as found in a turn of a helix. Most of the few helix nucleators reported to date initiate peptide helices from the N-terminus[3, 25]. Only a couple of examples are known of C-terminal or internal nucleators. N-terminal nucleators require three or four carbonyl oxgens (or some other hydrogen-bond acceptor) spatially oriented to mimic the pitch and spacing of a right-handed α-helix while C-terminal nucleators require similarly positioned hydrogen bond donors such as amide NH's.

One of three interconverting conformers of compound **73** is an N-terminal helix nucleator. Since the amount of template present in the nucleating conformation (as determined by ^{1}H NMR spectroscopy) directly correlated with the degree of helicity in attached peptides, **73** has subsequently been used as a "reporter" of peptide conformation leading to quantified *s* values for amino acids, studies of helix stabilizing effects of side chain interactions, and the use of trifluoroethanol (TFE) as a solvent. Conformational mobility in **73** and its low ability to nucleate α-helices led Kemp's group to design tri-proline templates **74** and **75**. Template **74** showed increased flexibility relative to **73** and was ineffective as a nucleator or reporter of helicity. Interestingly, a remarkably similar compound **76** prepared by an independent group showed better nucleation properties. The thioether bridged analog of **74** (i.e. **75**) was only capable of nucleating 310 helices. Kemp's triacid **77**, Muller's cage-compound **78**, Bartlett's hexahydroindol-4-one derivative **79** and modified cyclic peptide **80** are all examples N-terminal helix nucleators. Compound **80** is different however from other nucleators in that it contains a hydrazone linkage as a covalent mimic of an α-helical hydrogen bond.

Ambiguities regarding the relative helix inducing capacities of these nucleators exist because each template was studied under very different experimental conditions as summarised in Table 44.7.

The most commonly accepted technique for quantifying peptide helicity is circular dichroism (CD) spectroscopy, whereby the mean residue molar ellipticity at 222 nm ($[\theta]_{222}$) is used as a direct measure of the fraction of peptide existing in helical conformations. This relies on assumptions which are accurate only for relatively long peptides of more than about 15–20 residues. In fact Kemp has recently suggested that they may still be inaccurate even in the most ideal cases and this may be the reason that he hasn't reported per cent helicity in the presence of his templates (Table 44.7). Another limitation is that the location of helicity within a peptide sequence cannot be ascertained using this method. Nevertheless, CD continues to be the most common method of quantitating the effects of nucleators and other helix stabilizing devices. Another commonly used technique for characterizing helicity in peptides is 2D-NMR spectroscopy. Using this method the location of a helical region within a peptide can be identified by short distances between residues residing at i and $i+3$ and i and $i+4$ positions giving rise to a characteristic pattern of NOE's in a NOESY spectrum. This type of analysis can be used to generate detailed structural information at the single-residue level but cannot be used to quantify overall helicity. Thus the best way of comparing the efficencies of helix stabilizers is probably still to use a combination of both CD and NMR spectroscopy.

Table 44.7 Comparison of known N-terminal helix nucleators

Number	*Attached peptide*	*Solvent*	*NMR characterization of helicity*	*Peptide helicity by CD {FH (%)} (No nucleator)*	*Peptide helicity by CD {FH (%)} (Plus Nucleator)*
73	NH_2-(A)$_n$-OtBu (n = 1–6)	Various	Yes	N.D	N.D
74	None	$CDCl_3$	Yes	N.D	N.D
75	NH_2-(A)$_n$-OH (n = 1–3)	$CDCl_3$	Yes	N.D	N.D
76	NH_2-(A)$_2$-NHMe	$CDCl_3$	Yes	N.D	N.D
77	†gp-41(573–590) gp-41(567–580)	H_2O	No	*N.D	*N.D
78	NH_2-A-Aib(A)$_4$Aib-(A)$_2$-NHC_6H_4-I	1:1 H_2O:TFE	No	40	70
79	NH_2-EALAKA-$CONH_2$	H_2O	Yes	8	50 (23 °C) 77 (0 °C)
80	NH_2-AEAAKA-$CONH_2$	H_2O	Yes	#N.D	#N.D

Notes: FH = fractional helicity; N.D = not determined; †HIV-1 transmembrane viral coat protein gp-41; *Change in CD was not quantitated but the visual difference appeared to be negligible; #Change in CD was not quantitated but the visual difference appeared to be substantial.

Table 44.8 Helix nucleators

73

R = H **74**
= CH_3 **76**

75

77

78

79

80

81

Z-Ala–NH HN CO_2-tBu

Ac-YGGG-$(A_4K)_3$ n = 2-4

82

83

84

85

NLITRQRY-NH_2

86

QSLLSTLT-OH

87

Effective C-terminal helix nucleators have not yet been reported although a number of C-capping auxillaries are known. These are not helix nucleators because they do not contain rigidly positioned hydrogen bond donors, but instead stabilize the helix by stabilizing the macrodipole or via hydrogen-bonded capping interactions or both. Tripeptide **81** was synthesized in order to investigate whether cyclic amidines could act as C-terminal caps using both their stable positive charge and hydrogen-bond donor groups to stabilize helicity but no capping efficencies have been reported. Alkyldiamines and monoguanylated diamines have been incorporated into derivatives of poly-Ala helices **82** to investigate their C-capping propensities. Alkyldiamines provided only modest helix stabilization ($\Delta\Delta G^\circ_c = -0.2$ to -0.5 $kcal \cdot mol^{-1}$) while monoguanylated diamines were slightly better ($\Delta\Delta G^\circ_c = -0.3$ to $-0.7\,kcal \cdot mol^{-1}$). Interestingly, D-Arg carboxamide was much more efficent ($\Delta\Delta G^\circ_c = -1.2$ $kcal \cdot mol^{-1}$) and it was suggested that the extra stability arose from specific hydrogen bonds between the auxillary and C-terminal main chain carbonyls as well as macrodipole stabilization.

For nucleators to be located within a peptide sequence, rather than at one end as described above, they must contain both hydrogen-bond donors and acceptors appropriately positioned for directing helicity in both directions, as well as have functional groups at either end of the nucleator for peptide attachment. The term "position-independent nucleators (PIN)" has been used to describe these compounds. The versatility of PINs suggests they would be ideal for use as helix stabilizers in peptides derived from bioactive protein surface helices since the termini or some internal positions of the mimicked α-helix may be crucial for binding. Examples of PIN's are **83–85**. PIN **83** has been incorporated at the N-terminus, C-terminus and internal positions of hydrophobic model peptides and conformational analysis by CD, NMR and X-ray crystallography revealed helix induction in all three cases. Simpler PINs **84** and **85** reported by Kahn were respectively incorporated at the N-terminus of a mimetic of neuropeptide Y to give **86** and of the effector sequence of interleukin-2 to give **87**. Compound **86** evidently showed higher helicity and NPY antagonist activity than the native peptide, but no helicities or bioactivities were reported for **87**. Neither helix nucleating capacities nor bioactivities were reported for **84** or **85** incorporated into peptides at C-terminal or internal positions.

44.3 Side chain peptidomimetics

All of the aforementioned examples are peptidomimetics that use a peptidic or pseudo-peptidic scaffold, folded into a characteristic secondary structural unit, to support side chains that make appropriate interactions with a target receptor. By contrast, all of the following examples use a non-peptide scaffold to correctly position/orientate a peptide side chain or side chain surrogate in the same three dimensional space as achieved in a peptide. Clues to arriving at these non-peptide scaffolds, which are usually more suitable components of drugs than peptides due to greater lipophilicity and metabolic stability, have traditionally come from natural products or more recently from *de novo* design facilitated by computer methods that screen or build the most suitable molecules to best fill a required three dimensional space. Although these side chain mimetics occupy a major focus in modern peptidomimetic drug design, discovery and development there are surprisingly few general reviews on the subject[2].

44.3.1 Side chain peptidomimetics from natural products

Natural product peptidomimetics broadly fall into two categories: 1) compounds identified from random screening processes (and their analogs) that displace endogenous peptide ligands from their target receptors, and 2) compounds obtained by modifying the structures of known endogenous peptide ligands to pseudopeptides and other non-peptidic molecules. In both cases the initial lead compound is painstakingly modified using structure-activity relationships to guide the optimization process towards a potential drug candidate. Only a couple of examples are provided here to illustrate concepts.

An example of a molecule from category one is asperlicin (**88**), a natural product isolated by researchers at Merck from the fungus *Aspergillus alliaceus*[27]. This compound was found to have both *in vitro* and *in vivo* activity as an antagonist of the 33 amino acid polypeptide cholecystokinin (CCK) and was selective for peripheral CCK receptors over those located in the central nervous system. Problems associated with its low potency and poor water solubility led to extensive programs aimed at improving these attributes[28], culminating in the production of the highly potent, orally active CCK-A selective antagonist L-364,718 (**89**).

In searching for CCK-B selective antagonists, researchers at Lilly Research Labs found that the 4(3H)-quinazolinone moiety of asperlicin was a crucial feature. Following optimization studies, a highly potent (IC_{50} 9.3 nM at mouse CCK-B receptors) and selective quinazolinone (**90**) was obtained. The production of both CCK-A and CCK-B selective antagonists from asperlicin provides an excellent example of how a side chain peptidomimetic drug candidate molecule can evolve from the discovery of a natural product peptidomimetic. An excellent recent review of cholecystokinin receptor ligands and their chemistry has been published[29].

The second category of natural product peptidomimetics includes compounds obtained by modifying an endogenous peptide ligand by progressive removal of peptidic features responsible for problems associated with their use as drugs. This approach has been extremely popular and is illustrated here using the example of Substance P (**91**)[9], an 11 amino-acid endogenous neurotransmitter that binds to neurokinin 1 (NK-1) receptors located in both the central and peripheral nervous systems. Substance P plays a key role in neurogenic inflammation and pain transmission processes and neurokinin antagonists may be useful for treating diseases such as emesis, schizophrenia, asthma, migraine and rheumatoid arthritis. Typical modifications made to Substance P in the search for peptidomimetic antagonists are shown in Table 44.9.

Attempts to generate a useful drug by modifying the structure of Substance P in this way have met with limited success to date. More useful compounds have evolved from screening of natural product sources and pharmaceutical company libraries. Examples of compounds identified in this way include WIN 64821 (**99**) and CP-96345 (**100**) respectively.

Neurokinin receptors are not the only examples of receptors that have spawned natural product peptidomimetics from both of the categories described above. In fact, more often than not this is the case as exemplified by the history of the opioid family of receptors. In 1975, Hughes and Smith[30] characterized the two pentapeptides Met and Leu-enkephalin (**101**) as the endogenous ligands of opioid receptors and their findings were soon followed by the identification of other endogenous dynorphin and endorphin peptide ligands. These discoveries were seized upon and a plethora of analogs of these peptides have since been prepared in the search for receptor subtype selective non-narcotic analgesic drugs[31]. This is in addition to the massive number of analogs of morphine (**102**) and other naturally occurring opioids that have also been prepared and investigated for the same reason.

Morphine (**102**) is a classic example of a natural product peptidomimetic. It was the first bridge between peptidic and non-peptidic ligands of the same receptor. It is also a true side-chain mimetic because it acts as a rigid scaffold that correctly positions the crucial phenolic residue, needed for binding of the endogenous Leu-enkephalin (**101**) [32], in the same orientation that it is presented in the peptide to opioid receptors[33, 34].

Table 44.9 Names and structures of some peptidomimetic neurokinin receptor antagonists derived from Substance P[9].

Compound	*Structure*
Substance P (**91**)	Arg-Pro-Lys-Pro-Gln-Gln-Phe-Phe-Gly-Leu-Met-NH_2
Spantide (**92**)	(D)Arg-Pro-Lys-Pro-Gln-(D)Trp-Phe-(D)Trp-Leu-Leu-NH_2
Spantide II (**93**)	(D)Lys-Pro-β(3-pyridenyl)Ala-Pro-(4-Chloro)Phe-Asn-(SD)Trp-Phe-(D)Trp-Leu-Nle-NH_2
FR11368 (**94**)	Ac-Thr-(N-formyl)Trp-Phe-(N-Me)(N-Bzl)
Cam-104 (**95**)	Boc-Trp-Phe-NH_2
Cam-4261 (**96**)	(2-Benzofuranyl)CH_2OCO[R]α-MeTrp-NH[S]$CH(CH_3)Ph$
L-668,169 (**97**)	cyclo(Gln-D-Trp-(N-Me)-Phe-Gly-R-γ-lactam-Leu-Met)$_2$
GR71251 (**98**)	Arg-Pro-Lys-Pro-Gln-Phe-Phe- [spiro-lactam structure: N, O] N-Leu-Met-NH_2

44.3.2 Designed side chain peptidomimetics

Designed side chain peptidomimetics use an artificial non-peptidic framework to support amino acid side chains (or surrogates) and project them in an orientation that is optimal for tight binding to a receptor. Compounds of this type require scaffolds that confer significant oral bioavailability, one of the major hurdles to be overcome in designing a peptidomimetic. For this reason, most scaffolds investigated utilize molecules that are known to be absorbed when administered orally such as sugars, steroids or lipophilic units like bicyclo[2.2.2]octane. Examples of some designed side chain peptidomimetics are **103–107** with the peptide side-chains that they mimic shown in bold[35–39]. Another favorable property of designed side chain peptidomimetics is that in rigidly separating important hydrophobic side-chains they prevent the phenomenon of hydrophobic collapse, the process whereby flexible molecules containing multiple hydrophobic regions tend to fold up upon themselves into inactive conformations.

One interesting class of designer side chain peptidomimetics is the cyclic ureas[9, 40] reported as excellent mimics of tetrapeptide components of inhibitors of HIV-1 protease. In this case the two urea nitrogens and their adjacent carbon atoms each project benzylic or alkyl side chains (e.g. **108**) into the same three dimensional space as P2, P1, P1′ and P2′ side chains of peptides that are recognized by HIV-1 protease. At the same time, the cyclic urea scaffold is achiral and also substitutes for one or two water molecules that feature in the structures of most peptides bound to HIV-1 protease, thus reducing the entropic cost of inhibitor binding by 3–4 $kcal.mol^{-1}$. These latter features are important advantages conferred on the inhibitor by use of a space-filling scaffold between key receptor-binding side chains.

No designer side chain peptidomimetics are yet available as drug candidates in man but this is unquestionably a fruitful direction being adopted by most drug companies and promises significant numbers of new drugs in the next few years. The high activity of some of these compounds combined with oral bioavailability has proven the principle and it is only a matter of time before a molecule from this class advances all the way to marketable drugs.

44.4 The future for peptidomimetics

It appears to us that the future of peptidomimetic drug development lies in harnessing all of the available approaches to generate lead molecules. Rapid developments in molecular and cellular biology, macromolecular structure determination, and computing techniques have made it possible to

Table 44.10 Natural product peptidomimetics

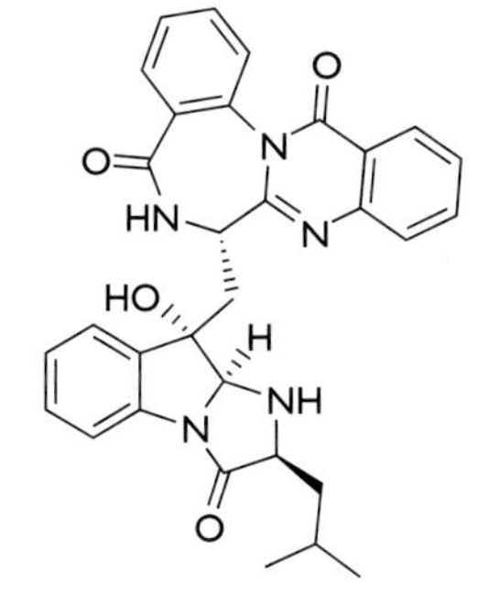

88 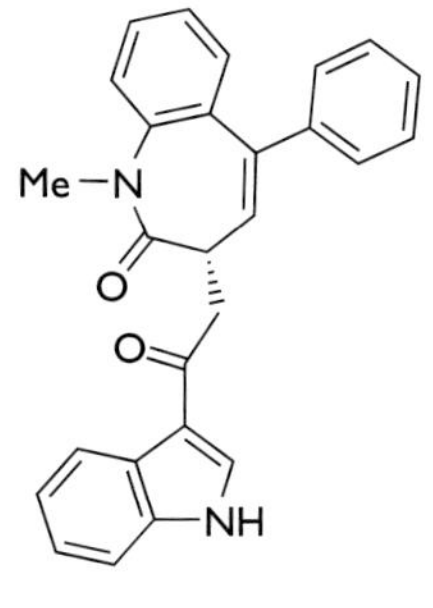 89 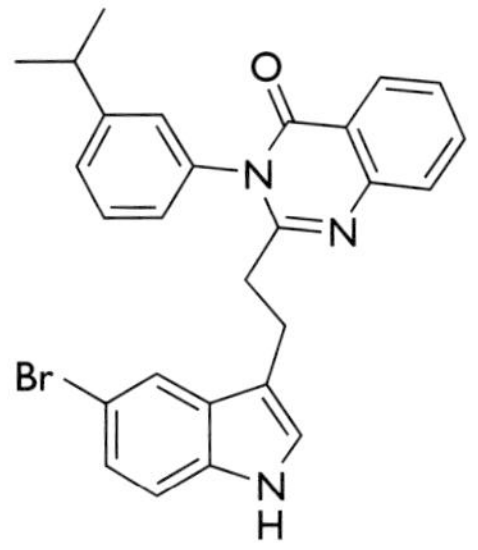 90

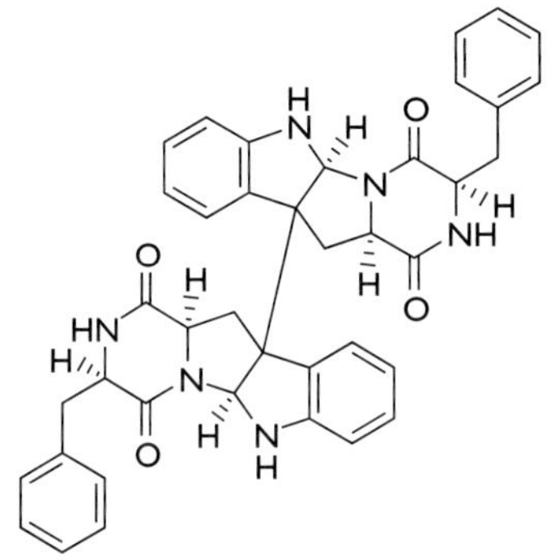

99 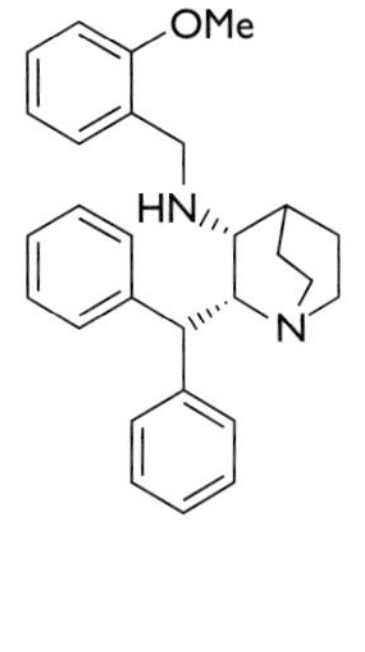 100 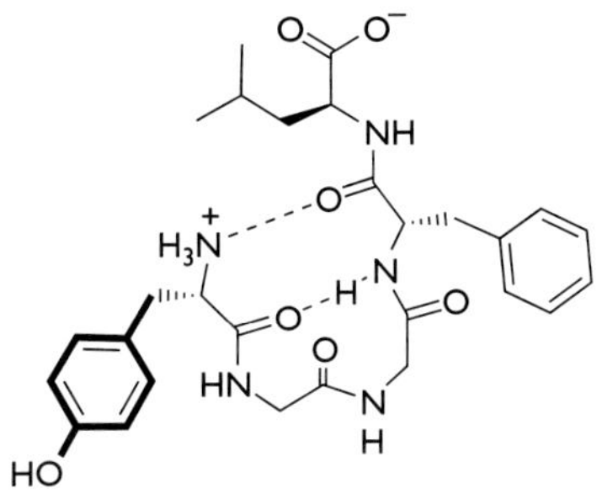 101

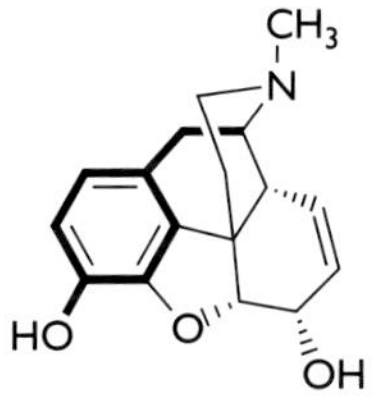

102

rationally employ structural knowledge of a target receptor in designing bioactive molecules with complementary fits for enzyme active sites or receptor pockets. Such structure-based drug design was originally heralded in the 1980s as the future for drug development, but the difficulties in predicting bioactivities by *de novo* design eventually lost many converts. The *de novo* design approach was inherently difficult and laborious because it required a detailed understanding of often poorly understood processes such as protein and peptide folding, conformational stability, and cooperativity during the "induced fit" in peptide-protein and protein-protein interactions. Structure-based drug design has, however, quickly matured during the 1990s and is now beginning to bear fruit. The recent crops of antiviral drugs used to treat AIDS are just a few examples of the potential of this approach which will likely be more routine in a few years.

It seems likely that structural techniques will increasingly facilitate these efforts in structurally reproducing specific bioactive surfaces of proteins in small bioactive molecules. While some progress has been made towards mimicking some of the simpler elements of secondary structure, this field is still in relative infancy. As it begins to mature, we can expect successful results for molecules that mimic tertiary structure as well. The focus on small molecules is, however, likely to persist for quite a few years because of the advantages over protein-based therapeutics in terms of cost, bioavailability, stability, pharmacodynamics, pharmacokinetics and immunogenicity.

Computer software packages are also rapidly evolving and one can already use coordinates of receptor-binding components of bioactive compounds to search molecular databases for matching components in different molecules. Packages

Table 44.11 Scaffold peptidomimetics and peptide being mimicked

Scaffold peptidomimetic	Peptide being mimicked
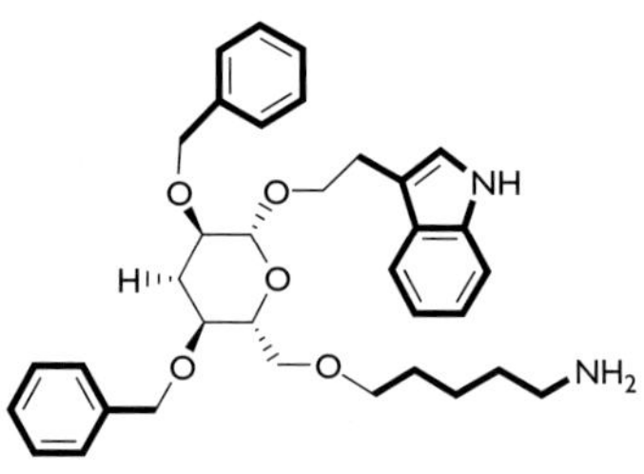 **103**	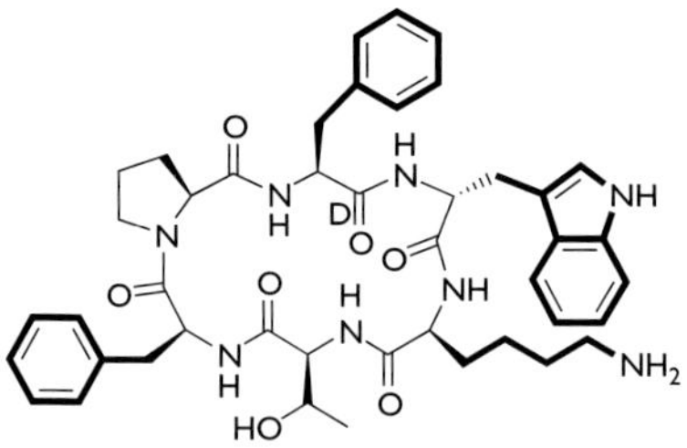 Somatostatin antagonist
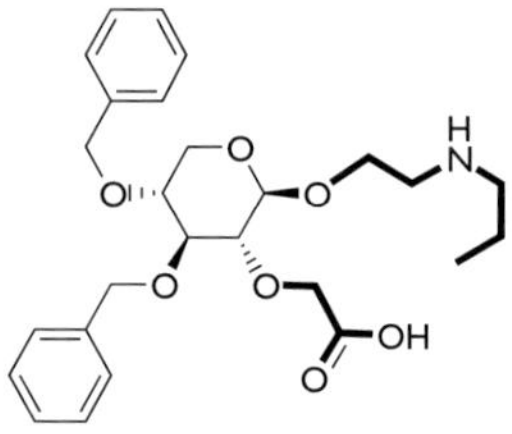 **104**	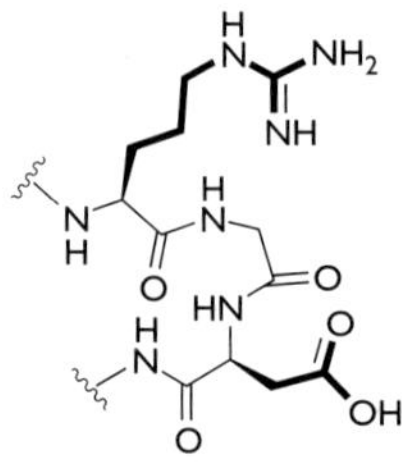 -Arg-Gly-Asp-sequence
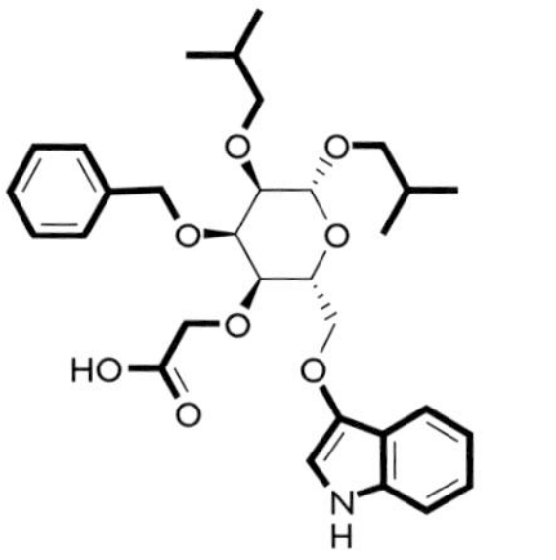 **105**	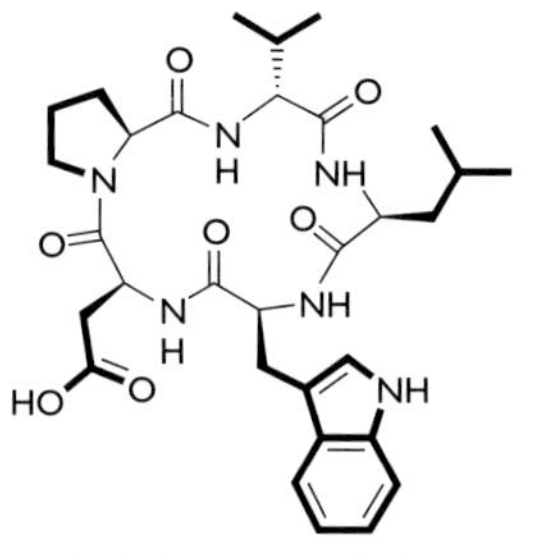 Endothelin antagonist BQ123
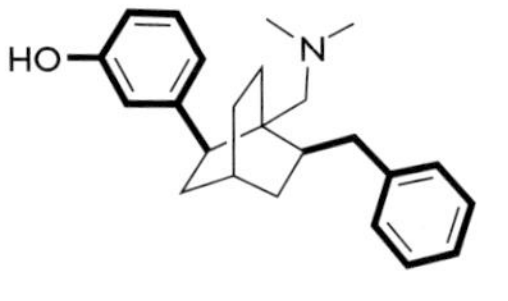 **106**	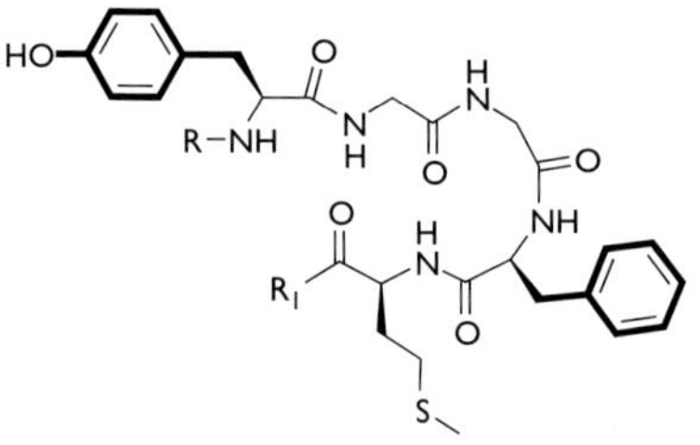 Methionine enkephalin
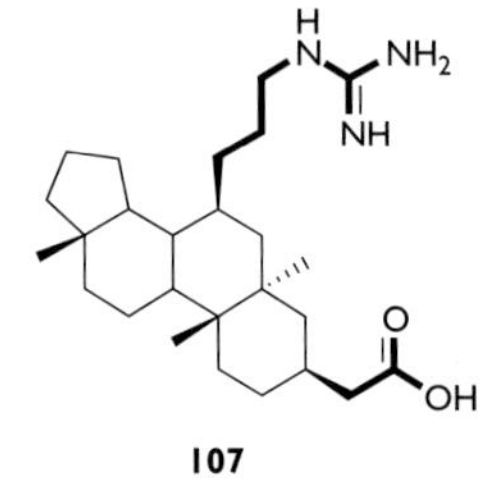 **107**	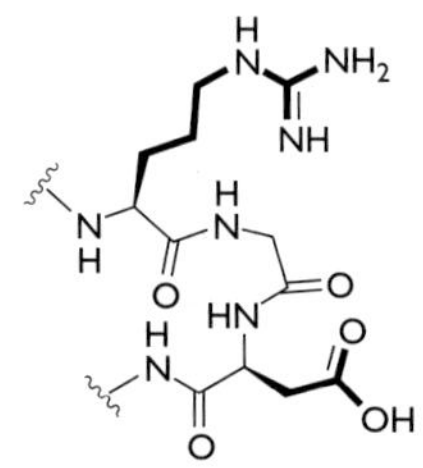 -Arg-Gly-Asp-sequence
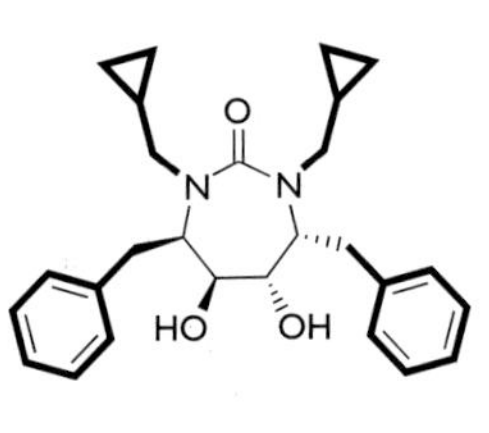 **108**	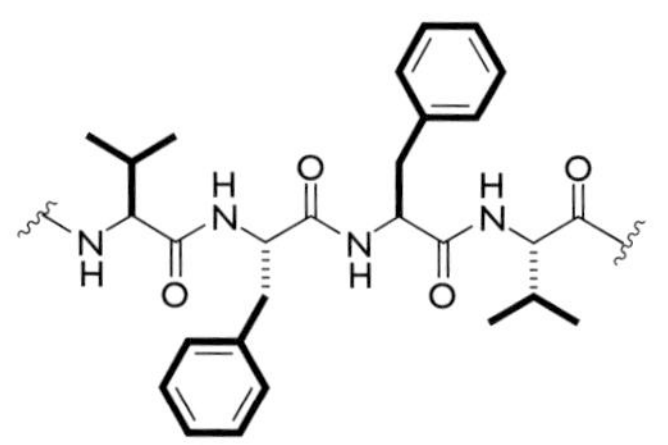 HIV-1 protease substrate

Table 44.12 Ras-FT inhibitors recently discovered from "virtual" screening approaches

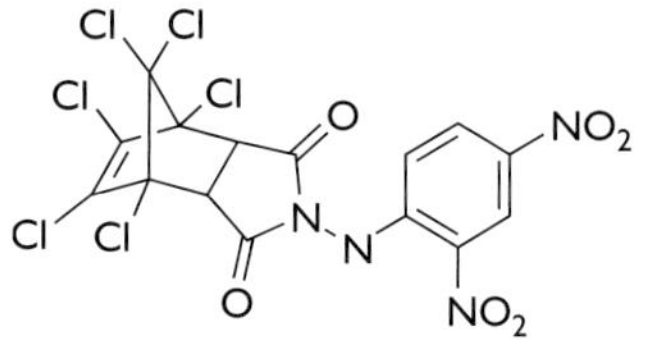

109 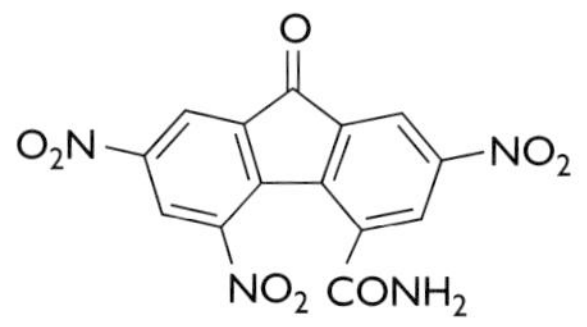 **110**

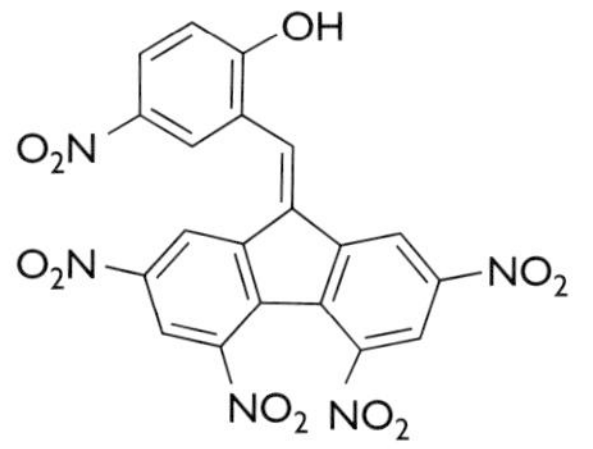

111 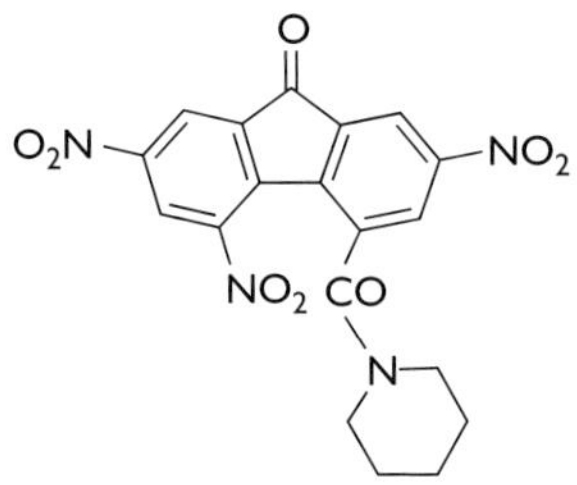 **112**

such as Cerius, Catalyst, Cybyl and EUDOC all allow "virtual" screening of databases to obtain alternative examples of pharmacophores. Successfully applied, this approach could potentially save a great deal of time in defining and optimizing pharmacophores and is particularly appropriate for designing the side chain peptidomimetics described above. A very recent example showed that it can, in fact, provide whole new classes of lead structures. By "virtually" screening the Available Chemicals Directory (ACD) with EUDOC, Pang and co-workers[41] discovered a new class of Ras-FT inhibitors. Four active compounds (**109–112**) ranging in potency from 25–100 μM were identified from 21 "virtual hits" with **109** also showing activity against human lung cancer cells. None of the compounds had previously been reported as Ras-FT inhibitors and all were significantly different in structure from the tetrapeptide mimetics (**12**,**13**).

Combinatorial chemistry has recently permitted the generation of vast libraries of compounds that can be rapidly evaluated for bioactivity using new automated high-throughput *in vitro* assay methods. New high-throughput pharmacological models and screens are also becoming available for predicting and monitoring drug development success. In the 1990s, combinatorial libraries were sought as alternatives to natural product libraries for identifying novel bioactive lead compounds. Instead of using structural knowledge of the interactions of a compound with a receptor, the focus was on directed screening of large random libraries of compounds for specific disease indications. While combinatorial approaches have been hailed as the definitive modern approach to drug design, it seems more likely that the power of this technique will be properly harnessed only when used in combination with, rather than as an alternative to, other drug design and development methodologies.

In summary, approaches to mimicking both the backbones and the side chains of bioactive protein surfaces have been successful in generating potent and selective drug candidates. Because of the inherent problems associated with delivering peptide based drugs the focus in the pharmaceutical arena has shifted during the past decade away from simply constructing peptide and protein libraries towards using the information provided by proteins and peptides to produce less peptidic or totally non-peptidic lead candidates. The recent advances described above have already illuminated the path toward faster and more effective development of peptidomimetics and we can expect more "designer" peptidomimetic drugs with high potency, selectivity and bioavailability to emerge during the next decade.

References

1 West, M. L.; Fairlie, D. P. (1995) *Trends Pharmacol. Sci. 16, 67–75.*

2 Gante, J. (1994) *Angew. Chem. Int. Ed. Engl., 33, 1699.*

3 Fairlie, D. P.; West, M. L.; Wong, A. K. (1998) *Current Medicinal Chemistry 5, 29.*

4 Fairlie, D. P.; Abbenante, G.; March, D. (1995) *Curr. Med. Chem., 2, 672.*

5 Brown, J. H.; Jardetzky, T. S.; Gorga, J. C.; Stern, L. J.; Urban, R. G.; Strominger, J. L.; Wiley, D. C. (1993) *Nature 364, 33.*

6 Quian, Y.; Blaskovich, M. A.; Saleem, M.; Seong, C. M.; Wathen, S. P.; Hamilton, A. D.; Sebti, S. M. (1994) *J. Biol. Chem. 269, 12410.*

7 Leung, D.; Abbenante, G.; Fairlie, D. P. (2000) *J. Med. Chem. 43, 305.*

8 Smith III, A. B.; Benowitz, A. B.; Sprengeler, P. A.; Barbosa, J.; Guzman, M. C.; Hirschmann, R.; Schweiger, E. J.; Bolin, D. R.; Nagy, Z.; Campbell, R. M.; Cox, D. C.; Olson, G. L. (1999) *J. Am. Chem. Soc. 121, 9286.*

9 Abell, A. (Ed.) (1997) *Advances in Amino Acid Mimetics and Peptidomimetics.* Volume 1, JAI Press Inc. London.

10 Schneider, J. P.; Kelly, J. W. (1995) *Chem. Rev. 95, 2169.*

11 Nowick, J. S.; Smith, E. M.; Parish, M. (1996) *Chem. Soc. Rev. 25, 401.*

12 Stigers, K. D.; Soth, M. J.; Nowick, J. S. (1999) *Current Opinion in Chemical Biology 3, 714.*

13 Giannis, A.; Kolter, T. (1993) *Angew. Chem. Int. Ed. Engl. 32, 124.*

14 Hanessian, S.; McNaughton-Smith, G.; Lombart, H. G.; Lubell, W. D. (1997) *Tetrahedron 53, 12789.*

15 Wilmot, C. M.; Thornton, J. M. (1990) *Protein Eng. 3, 479.*

16 Souers, A. J.; Virgilio, A. A.; Rosenquist, A.; Fenuik, W.; Ellman, J. A. (1999) *J. Am. Chem. Soc. 121, 1817.*

17 Kahn, M. (1993) *Synlett 821.*

18 Fisher, M. J.; Giese, U.; Harms, C. S.; Kinnick, M. D.; Lindstrom, T. D.; McCowan, J. R.; Mest, H-J, Morin Jr, J. M.; Mullaney, J. T.; Paal, M.; Rapp, A.; Ruhter, G.; Ruterbories, K. J.; Sall, D. J.; Scarborough, R. M.; Schotten, T.; Stenzel, W.; Towner, R. D.; Um, S. L.; Utterback, B. G.; Wyss, V. L.; Jakubowski, J. A. (2000) *Bioorg. Med. Chem. Lett. 10, 385.*

19 Osterkamp, F.; Ziemer, B.; Koert, U.; Weisner, M.; Raddatz, P.; Goodman, S. L. (2000) *Chem. Eur. J. 6(4), 666.*

20 Keenan, R. M.; Callahan, J. F.; Samanen, J. M.; Bondinell, W. E.; Calvo, R. R.; Chen, l.; DeBrosse, C.; Eggleston, D. S.; Haltiwanger, R. C.; Hwang, S. M.; Jakas, D. R.; Ku, T. W.; Miller, W. H.; Newlander, K. A.; Nichols, A.; Parker, M. F.; Southhall, L. S.; Uzinskas, I.; Vasko-Moser, J. A.; Venslavsky, J. W.; Wong, A. S.; Huffman, W. F. (1999) *J. Med. Chem. 42, 545.*

21 Brickmann, K.; Yuan, Z. Q.; Sethson, I.; Somfai, P.; Kihlberg, J. (1999) *Chem. Eur. J. 5, 2241.*

22 Kussie, P. H.; Gorina, S.; Marechal, V.; Elenbaas, B.; Moreau, J.; Levine, A. J.; Pavletich, N. P. (1996) *Science 274, 948.*

23 Scanlon, M. J.; Fairlie, D. P.; Craik, D. J.; Engelbretsen, D. R.; West, M. L. (1995) *Biochemistry 34, 8242.*

24 Condon, S. M.; Morize, I.; Darnbrough, S.; Burns, C. J.; Miller, B. E.; Uhl, J.; Burke, K.; Jariwala, N.; Locke, K.; Krolikowski, P. H.; Kumar, N. V.; Labaudiniere, R. F. (2000) *J. Am. Chem. Soc. 122, 3007.*

25 Andrews, M. J. I.; Tabor, A. B. (1999) *Tetrahedron 55, 11711.*

26 Garcia-Echeverria, C.; Chene, P.; Blommers, M. J. J.; Furet, P. (2000) *J. Med. Chem. 43, 3205.*

27 Chang, R. S. L.; Lotti, V. J.; Monaghan, R.L.; Birnbaum, J.; Stapley, E. O.; Goetz, M. A.; AlbersSchonberg, G. A.; Patchett, A.; Liesch, J. M.; Hensens, O. D.; Springer, J. P. (1985) *Science 230, 177.*

28 Bock, M. G.; DiPardo, M. G.; Rittle, K. E.; Evans, B. E.; Freidinger, R. M.; Veber, D. F.; Chang, R. S. L.; Chen, T.; Keegan, M. E.; Lotti, V. J. (1986) *J. Med. Chem. 29, 1941.*

29 De Tullio, P.; Delarge, J, Pirotte, B. (1999) *Curr. Med. Chem. 6, 433.*

30 Hughes, J.; Smith, T. W.; Kosterlitz, H. W.; Fothergill, L. A.; Morgan, B. A.; Morris, H. R. (1975) *Nature 258, 577.*

31 Hruby, V. J.; Agnes, R. S. (1999) *Biopolymers 51, 391.*

32 Bradbury, A. F.; Smyth, D. G.; Snell, C. R. (1976) *Nature 260, 165.*

33 Currie, B. L.; Krstenansky, J. L.; Lin, Z. L.; Ungwitayatorn, J.; Lee, Y. H.; del Rosario-Chow, M.; Sheu, W. S.; Johnson, M. E. (1993) *Tetrahedron 49, 3489.*

34 Su, T.; Nakanishi, H.; Xue, L.; Chen, B.; Tuladhar, S.; Johnson, M. E.; Kahn, M. (1993) *Bioorg. Med. Chem. Lett. 3, 835.*

35 Hirshmann, R.; Nicolaou, K. C.; Pietranico, S.; Salvino, J.; Leahy, E. M.; Sprengeler, G. F.; Smith, A. B. (1992) *J. Am. Chem. Soc. 114, 9217.*

36 Moitessier, N.; Minoux, H.; Maigret, B.; Chretien, F.; Chapleur, Y. (1998) *Letters in Peptide Science 5, 75.*

37 Diguarher, T. L.; Boudon, A.; Elwell, C.; Paterson, D. E.; Billington, D. C. (1996) *Bioorganic & Medicinal Chemistry Letters 6, 1983.*

38 Belanger, P. C.; Dufresne, C. (1986) *Can. J. Chem. 64, 1514.*

39 Hirschmann, R.; Sprengeler, P. A.; Kawasaki, T.; Leahy, J. W.; Shakespeare, W. C.; Smith, A. B. (1992) *J. Am. Chem. Soc. 114, 9699.*

40 Lam, P.; Ru, Y. S.; Jadhav, P. K.; Aldrich, P. E.; DeLucca, G. V.; Eyermann, C. J.; Chang, C. -H.; Emmet, G.; Holler, E. R.; Daneker, W. F.; Li, L.; Confalone, P. N.; McHugh, R. J.; Han, Q.; Li, R.; Markwalder, J. A.; Seitz, S. P.; Sharpe, T. R.; Bacheler, L. T.; Rayner, M. M.; Klabe, R. M.; Shum, L.; Winslow, D. L.; Kornhauser, D. M.; Jackson, D. A.; Erickson-Viitanen, S.; Hodge, C. N. (1996) *J. Med. Chem 39, 3514.*

41 Perola, E.; Xu, K.; Kollmeyer, T. M.; Kaufmann, S. H.; Prendergast, F. G.; Pang, Y. -P. (2000) *J. Med. Chem. 43, 401.*

Chapter 45

Drug delivery systems utilizing lipids and sugars

Allan Wong, Veronika Bender and Istvan Toth

Contents

45.1 Introduction

Successful delivery of many highly active compounds and the majority of therapeutic peptides, proteins, oligosaccharides, nucleotides and nucleosides via the oral route represents a major challenge in drug development. These drug moieties have a wide spectrum of therapeutic applications and the significance and scope of their functions have been further extended by advances in the areas of biotechnology, biochemistry, molecular biology and solid phase synthesis. Advances in these fields have led to the availability of large quantities of pure, potent and highly specific drugs, often with modified or "super-agonist" properties. However the level of progress attained in the area of drug production has not been matched by a similar rate of progress in the area of their delivery. Conventional processes to deliver drugs to their sites of action *in vivo* are inefficient and introduction of selective delivery and transport is required.

For example, peptide and protein drugs possess unique physicochemical and biological properties, however they have inherent problems for their delivery. Therapeutic peptides and proteins have short plasma half-lives, due in part to their rapid clearance by the liver or kidneys. Achieving successful delivery of therapeutic peptides and proteins via the oral route is particularly difficult, due to the formidable enzymatic and transport barriers in the gastrointestinal tract. Endopeptidases cleave internal bonds and exopeptidases cleave peptides and proteins at their N- and C-termini. An important feature of peptide degradation is that the biological activity of a peptide can be destroyed by the cleavage of just a single bond in the structure, rather than by complete hydrolysis of the molecule. Therefore merely protecting a single peptide bond from a specific enzyme is insufficient to confer metabolic protection on the peptide.

The field of drug delivery has been developed in an attempt to overcome these limitations and can be defined as a strategy using processes or devices designed to enhance the efficiency of therapeutic agents.[1] In this review the term "drug delivery" will be used in the context that therapeutic agents with improved drug delivery will have one or more of the following:

1. improved chemical stability,
2. produce positive changes in pharmacokinetics,
   a. rate of absorption/release,
   b. modify stability in the body,
   c. rate of distribution,
   d. change of metabolic parameters,
   e. modify elimination,
3. target the drug to particular areas in the body,
4. reduce toxicity,
5. reduce side effects,
6. improved patient acceptance or compliance e.g. through being easily applied or a longer lasting delivery system.

The prodrug approach is an important part of the drug delivery field and is a method of overcoming the limitations of existing drugs. It involves linking another molecule to an existing drug to form an inactive derivative (prodrug), which is then converted to the parent compound through enzymatic and/or chemical means, before or after reaching the site(s) of action.[2] In addition to this prodrug approach, derivatization of an existing drug can also produce a new entity, which by itself is stable and has improved drug delivery.

Conjugation of a drug with lipid and/or sugar units represents an important aspect of the prodrug/derivatization approach and is the focus of this review. The term "lipid" will apply to any fatty acid-like molecule with a long aliphatic chain. There are a large number of these conjugates and examples where conjugation has resulted in increased drug delivery compared to the parent drug will be highlighted. It should be kept in mind, that there are other prodrug/derivatization systems such as conjugation with aromatics, amino acids and glycols.[2]

45.2 Lipidic conjugates

Different types of lipids have been attached to drugs in an attempt to modulate their drug delivery. In terms of their structure, lipids could be attached to a drug either (a) directly to the drug or (b) via a linker moiety as shown schematically in Figure 45.1. Direct linkage of a fatty acid to a drug is limited in its ability to increase the lipophilicity, since fatty amino acids are monofunctional. On the other hand, tris and glycerol linker molecules are able to multiply the number of fatty acid lipids attached to a drug (Figure 45.2). Lipoamino acids (LAAs) are unique in their capacity to be linked through either their amine or carboxylic acid functional groups. LAAs combine the properties of amino acids (NH_2 and COOH groups) with those of lipids (hydrophobic side chains). LAAs are highly versatile molecules, for instance their length and degree of saturation, as well as their stereochemistry can be modified. They can, therefore, be specifically tailored for the transport of a wide variety of peptides or poorly absorbed drugs through the skin, and across the barriers that normally inhibit absorption from the gut or into the brain. This dual functionality allows these lipoamino acids to be linked directly to a larger number of functional groups (and hence a greater number of drugs) than the other lipids which contain only one site of attachment. Other advantages of this system include the attachment of multiple copies of the lipoamino acids together to modulate the lipophilicity, and the ability to place a LAA anywhere in a peptide sequence (Figure 45.2).

The following sections give some examples where lipid conjugation of drugs has resulted in improved drug delivery. Table 45.1 provides a summary of the different types of lipid-drug conjugates which have been reported.

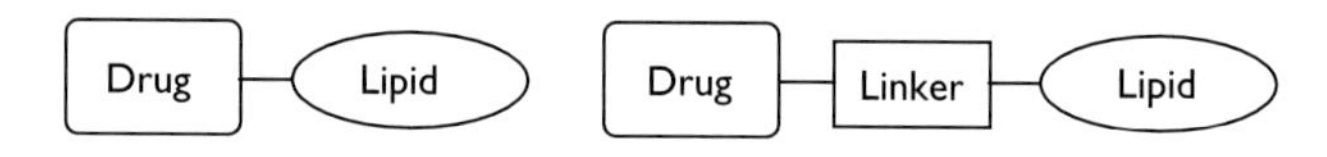

Figure 45.1 Lipidic delivery systems.

45.2.1 *Increased absorption/bioavailability*

Peptides are a major target for lipidic derivatization, since they normally exhibit poor bioavailability. For example, they have poor permeability characteristics, show instability

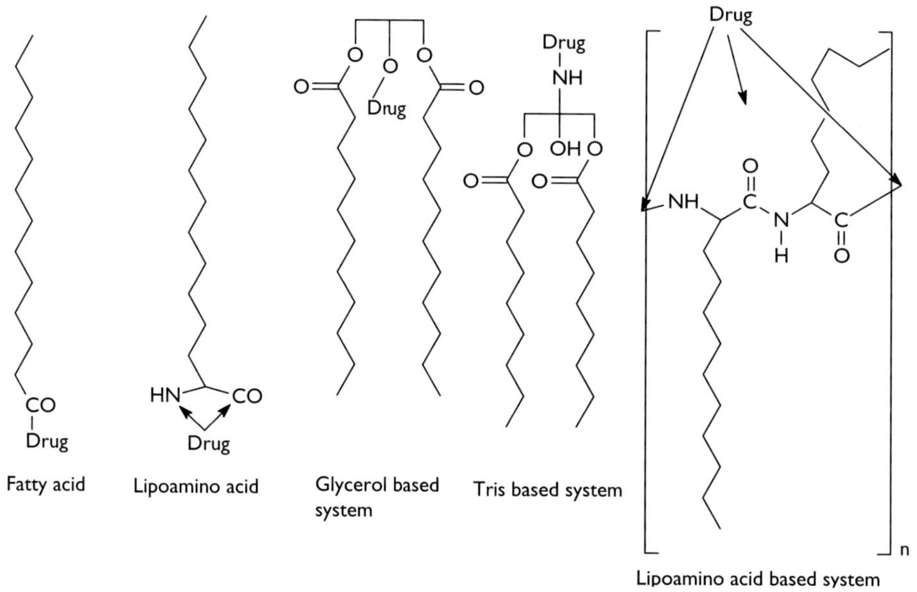

Fatty acid
Lipoamino acid
Glycerol based system
Tris based system
Lipoamino acid based system

Figure 45.2 Lipophilic structures

Table 45.1 Drug-lipid conjugates

Lipid/structure	*Drug/lipid identity*
Fatty acid	1 AZT, lipid C2 to C18.[47,#,‡]
	2 Tetragastrin, lipid C2, C4, C6, C12.[8,48–50†,‡,]
	3 Fluorouracil, lipid C2 to C7.[51,†,‡]
	4 Insulin, lipid C16, 2 × C16.[52–55,†,‡]
	5 TRH, lipid C12.[7,56,57,†]
Drug-CO–$(CH_2)n$–CH_3	6 Chlorambucil, unsaturated lipid C18:1, C20:4, C22:6.[12,*]
	7 Methotrexate, lipid C4 to C16.[58]
	8 Iododeoxyuridine, C18:1.[59,#]
	9 Trp-Leu, lipid C14.[60]
	10 Phosphonoformate, lipid C16.[61]
Tertiary amine Drug-CO-O-$(CH_2)_n$-N(R_1)(R_2)	Indomethacin, lipids R1 = C1 to C4 and R2 = C1 to C4.[14,§]
Lipoamino acids CH_3–$(CH_2)_n$–CH(NH–Drug)–C(=O)–Drug	1 Morphine, lipids (1–2) × C10, C14, C20.[62]
	2 Penicillin & cephalosporin, lipids (1–2) × C8, C14, C16.[63]
	3 Benzoquinolizine, lipids (1–3) × C10, C14, C20.[64]
	4 Chlorambucil, lipids (1–2) × C10, C14, C16.[65]
	5 LHRH & TRH, lipids (1–2) × C12.[10,#,†]
	6 AlaAlaProVal, lipids (1–3) × C14.[66,†,‡]
Glycerol, fatty acids H_2C–O–R Drug-CO_2-CH H_2C–O–R R = Fatty acid	1 Chlorambucil, lipids 2 × C16.[67,‡,II]
	2 L-Dopa, lipids 2 × C16.[9,#,‡]
	3 Phenytoin, lipids 2 × C14.[68,#,‡]
	4 AZT, lipids C16, C18:1.[69]
Modified glycerols R_2–CH(CH_2–R_1)–CH_2–O–P(=O)(O–)–X–Drug	1 1-ß-D-Arabinofuranosyl-cytosine & cytidine, R_1 = S-(C16 & C18), R_2 = C1, C2, C16, X = PO_2–O.[70,‡]
	2 AZT & DDI, R_1 = NHCO–C17, O–C16, S–C16, R_2 = C1, C2, X = O.[71,II]
Tris, fatty acids Drug-NH–C(OR_1)(OR_2)(R_3) R = H of fatty acid	1 Morphine, suc-Gly spacer, lipids 2 × C16.[72,‡]
	2 Indomethacin, Gly spacer, lipids 2 × C16.[72,Δ]
	3 Chlorambucil, Gly spacer, lipids C16.[72,†]
	4 AZT, suc-Gly spacer, lipids C16 and 2 × C16.[72,†]
	5 Methotrexate, Gly spacer, lipids C16 and 2 × C16.[72,II]
Cholesteryl group Drug–O (cholesteryl)	1 DHT-Tyr-D-Ala-Gly-Phe-D-Leu, lipid cholesterol.[73,74,‡,*]
	2 DHT-Gln-Leu-Pro-Gly, lipid cholesterol.[17,35,‡,*]
Dimyristoylphosphatidylethanol-amine (DMPE) H_3C - $(CH_2)_{12}$–C(=O)–O–CH_2 H_3C - $(CH_2)_{12}$–C(=O)–O–CH CH_2 HO–P(=O)–O Drug—NH-CH_2CH_2-O	1 MTX-γ-DMPE, MTX-α-DMPE, MTX-α,γ-di-DMPE.[76–81,†]

Abbreviations: C2 = two carbons in the conjugated lipid, C18:1 = 18 carbons with one double bond in the lipid, AZT = azidothymidine, TRH = thyrotropin releasing hormone, LHRH = luteinizing hormone releasing hormone, DDI = dideoxyinosine, MTX = methotrexate, DHT = 1,4-dihydrotrigonellyl.

Symbols: # Prodrug, † Increased activity/bioavailability *in vitro*, ‡Increased activity/bioavailability *in vivo*, II Decreased toxicity, * Increased drug targeting to sites in the body, Δ Increased skin retention.

in vivo, and can show rapid clearance, i.e. the first pass effect and excretion in the bile.[3] Attaching a lipid to a drug increases the lipophilicity of the drug and can also result in increased absorption, since lipid solubility is necessary for passive transport across intestinal mucosal membranes.[2] The conjugates represented in Table 45.1 have increased lipophilicity compared with the parent drug and should theoretically have increased passive transport ability across biological membranes.

Fatty acids have been attached to peptides such as thyrotropin-releasing hormone (TRH), tetragastrin (TG) and insulin,[4] resulting in improved intestinal delivery characteristics. Lauric acid (C12) was attached through the N-terminus of TRH to give Lau-TRH **1**. This conjugate showed significantly increased penetration across the upper small intestine compared to TRH, in an *in vitro* everted sac experiment.[5] A caproyl-tetragastrin (Cap-TG) conjugate **2** produced increased acid secretion in rats, 1.8 times higher than native TG, following intravenous injection. Sugar monolaurate (2%) solutions of TG and Cap-TG were also administered to both small and large intestine loops and the gastric acid secretion measured. In both cases, Cap-TG showed a significantly greater amount of acid secretion than TG, which was explained by the improved absorption of Cap-TG.[6]

A glycerol prodrug of L-dopa, containing two 16 carbon long fatty acid chains **3**, was tested for its anti-Parkinsonian activity in mice.[7] Mice were orally dosed with either L-dopa or the conjugate, with the result that the conjugate had a significantly longer anti-Parkinsonian activity than L-dopa. Levels of L-dopa and dopamine (a major metabolite) in the brain were also measured. For the prodrug, the levels increased less rapidly and remained at elevated levels for longer periods than L-dopa. The prodrug was also found to be selectively absorbed from the intestinal tract by the lymphatic route without any chemical or enzymatic degradation. Thus, lipid conjugation resulted in improved uptake and greater, more steady levels of L-dopa and dopamine in the brain.

Bioavailability of lipoamino acid-peptide conjugates **4** and **5** has been examined in experiments with human intestinal epithelial (Caco-2) cell homogenates.[8] The decapeptide hormone, LHRH, degraded rapidly when incubated with the Caco-2 cell homogenates, the half-life of the peptide being about 3 min. One of the diastereomers of lipoamino acid LHRH conjugate **4** showed an increased half-life of 50 min. Additionally, this conjugate was found to release the parent LHRH, therefore acting as a prodrug. The other diastereomer also showed reduced degradation, but LHRH release was not observed. The four diastereomers of LHRH conjugates with two lipoamino acids **5** ranged in their stabilities, one of them was not degraded even after 6 hrs. Thus, these lipoamino acid conjugates greatly increased the enzymatic stability of attached peptides. The mechanism for this increased enzymatic resistance was not reported, however a possible explanation could be that the bulky lipoamino acid inhibited binding of the peptide into the enzyme pocket, reducing the recognition and hence, cleavage of the attached peptide.

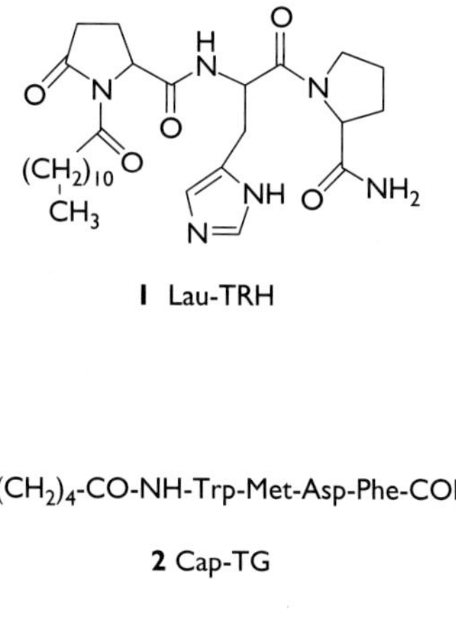

1 Lau-TRH

CH_3-$(CH_2)_4$-CO-NH-Trp-Met-Asp-Phe-$CONH_2$

2 Cap-TG

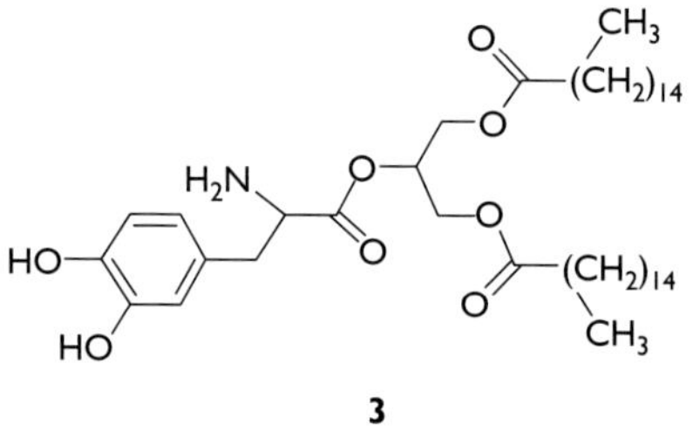

3

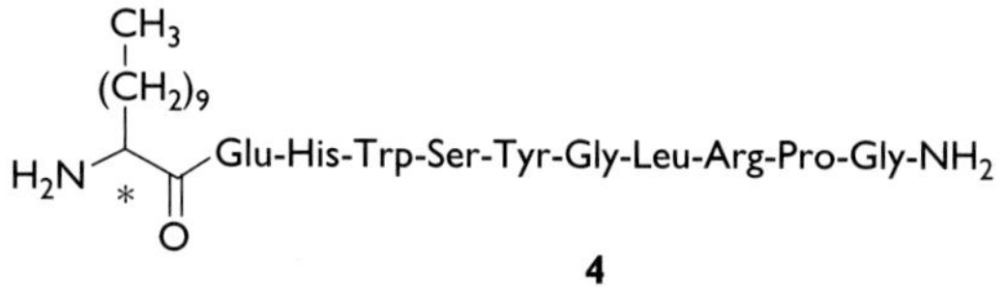

H_2N-CH($(CH_2)_9CH_3$)-CO-Glu-His-Trp-Ser-Tyr-Gly-Leu-Arg-Pro-Gly-NH_2

4

Glu-His-Trp-Ser-Tyr-Gly-Leu-Arg-Pro-Gly-NH_2

* Racemic amino acid

5

45.2.2 Decreased toxicity/increased drug targeting

Unsaturated fatty ester conjugates of chlorambucil (Chl) were tested *in vitro* upon human lymphoma cell lines for toxicity and compared against quiescent and mitogen-activated lymphocytes.[9] Chl-arachidonic acid **6** (C20:4) and

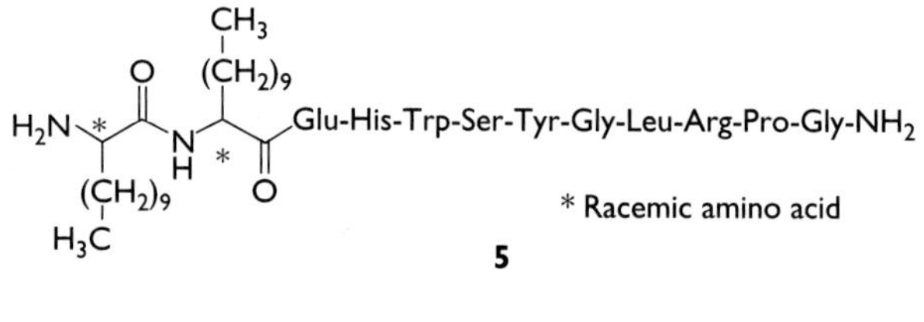

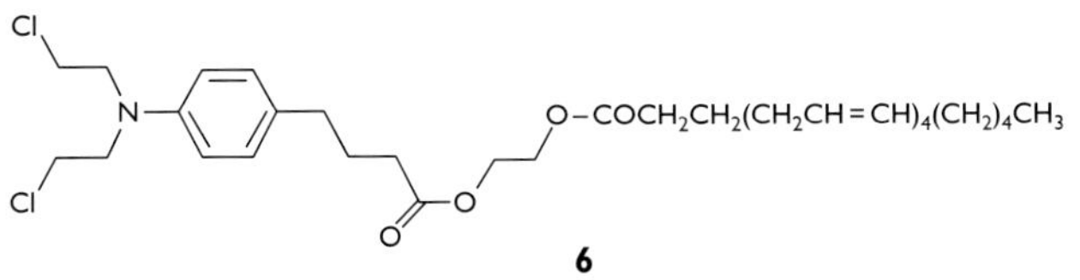

$O{-}COCH_2CH_2(CH_2CH{=}CH)_4(CH_2)_4CH_3$

6

(spacer)-Gln-Leu-Pro-Gly–O

CH_3

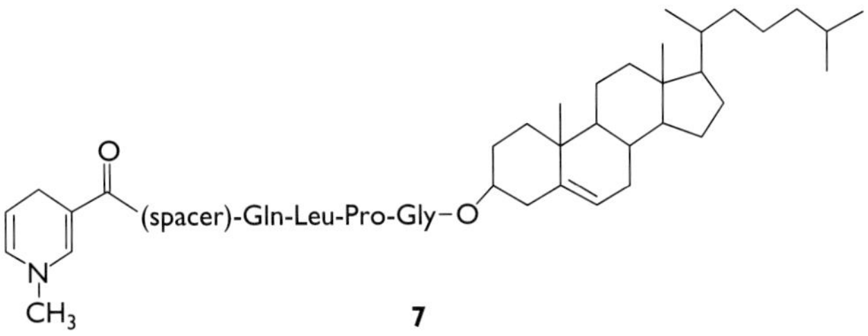

7

Chl-docosahexaenoic acid (C22:6) both showed equal or higher cell toxicity than the individual toxic potential of either chlorambucil or fatty acid, against lymphoma cells. However, compared to chlorambucil, Chl-arachidonic acid at any concentration tested, lacked toxicity towards normal non-activated lymphocytes. Thus, coupling of chlorambucil with polyunsaturated fatty acids increased the selectivity against neoplastic versus quiescent lymphocytes and the toxicity for B lymphoma cells. The increased selectivity was explained by expression of alphafetoprotein (AFP) receptor on B and T lymphoma cells which has a high affinity for polyunsaturated fatty acids.

Central nervous system (CNS) targeting was achieved by attaching a 1,4-dihydrotrigonellyl at the N-terminus and a lipidic cholesteryl group at the C-terminus of a peptide **7**. Both of these groups were hydrophobic and increase the blood-brain barrier permeability of the hydrophilic peptides. Once inside the brain, the 1,4-dihydrotrigonellyl (T) group underwent an enzymatically mediated oxidation to a hydrophilic, membrane-impermeable trigonellyl (T^+) salt, whose efflux from the brain was reduced. These two groups were then cleaved enzymatically to give the required free peptide. Compounds based on a TRH analog (Gln-Leu-Pro-Gly) were synthesized and tested *in vivo* in mice.[10] All of the modified peptides showed significant antagonism of barbiturate-induced analgesia, after i.v. injection. Some *in vitro* stability and metabolism studies were also performed on a model peptide T^+-Ala-Gln-Leu-Pro-Gly-OH. One of the key conversions of this peptide to T^+-Ala-Gln-Leu-Pro-NH_2 was shown to occur in significant amounts only in the brain.

45.2.3 *Lipids in vaccine technology*

Recent advances in molecular medicine have made it possible to identify distinct cell subsets, cell surface markers and cell products, which affect the pathologic immune-mediated inflammatory responses. A pivotal role is assigned to T-cell antigen receptors (TCR). T-cells are essential components of our immune system and are responsible for immune surveillance and recognition of foreign antigens like bacteria and viruses. Loss of "self-tolerance" and subsequent recognition of "self" antigens by T-cells leads to a large spectrum of autoimmune diseases including diabetes, allergy, arthritis, psoriasis, multiple sclerosis and rheumatoid arthritis (RA). It was shown that it is possible to inhibit this response with small peptides in a number of animal models of inflammation.[11] The mode of action is still not known but is thought to be the inhibition of assembly between subunits of the TCR by the peptide, probably by inhibiting pairwise interaction between subunits α and CD3-δ. The sequence of the nonapeptide is part of a transmembrane region of the α chain of the TCR complex and contains four hydrophobic amino acids separated by two positively charged ones, arginine and a lysine, which are essential for activity. When the C-terminal of the peptide was reacted with glycine-tris-palmitates via an active ester to give the corresponding lipopeptides with one, two and three palmitates, inhibition of IL-2 production, a measure of T-cell function following antigen recognition, was considerably enhanced when the conjugate was used. More dramatic was the difference when the effect of peptides and lipopeptides were tested for a protective effect against the induction of arthritis in an animal model (adjuvant induced arthritis in rats).

Lipopeptide conjugates have been developed as synthetic adjuvants. One of these conjugates carries three fatty acid residues, one N-linked and two O-linked palmitoyl groups (Pam_3 Cys Ser). The molecule is a synthetic analog of the N-terminal portion of the bacterial lipoprotein from *E. coli*, which is a specific B-cell mitogen. When covalently coupled to a non-immunogenic peptide, a specific antibody response was produced.[12]

Lipidated Multiple Antigenic Peptides were successfully used for mucosal immunization with an antigenic peptide containing a sequence from the V3 loop of HIV-1 gp120 $_{IIIB}$. By using a combined strategy of lipidation and micro-encapsulation of multimeric peptide antigens (**8**), induction of both systemic and mucosal antibodies was enhanced.[13]

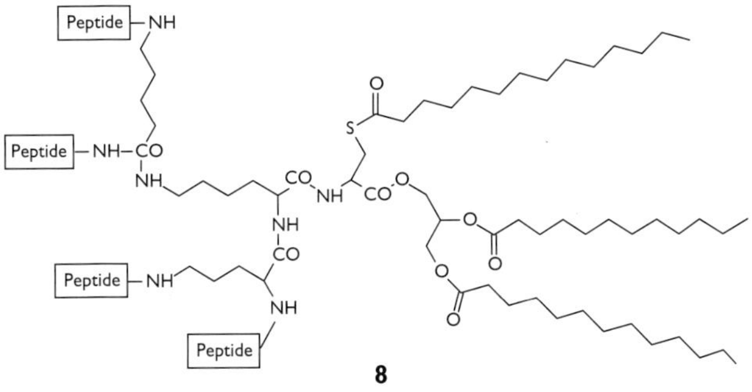

Peptide–NH

Peptide–NH–CO

Peptide–NH

Peptide

8

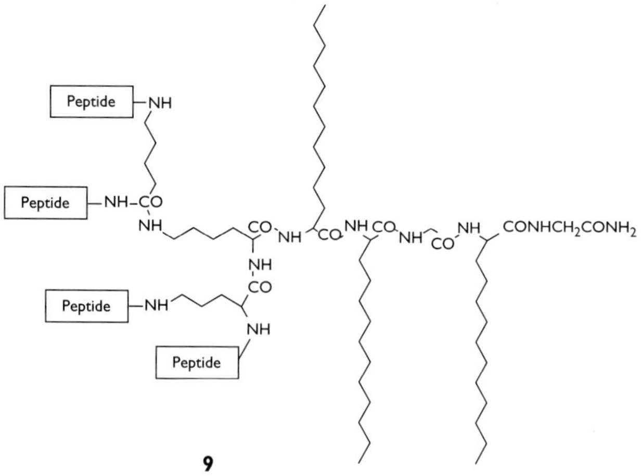

Lipoaminoacids conjugated to a branched polylysine construct (LCP) were coupled to immunodominant peptides from the outer membrane proteins of *Chlamidia trachomatis*. The LCP complexes (**9**) showed a much greater increase in antipeptide antibody titers than the non-conjugated parent peptide, administered with adjuvant.[14]

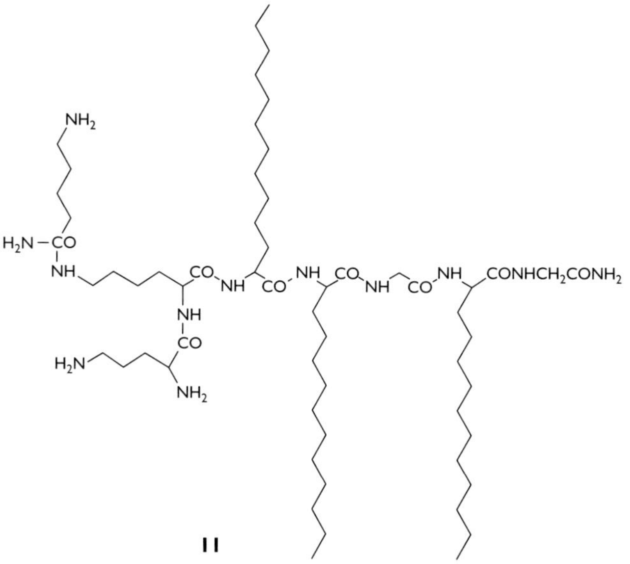

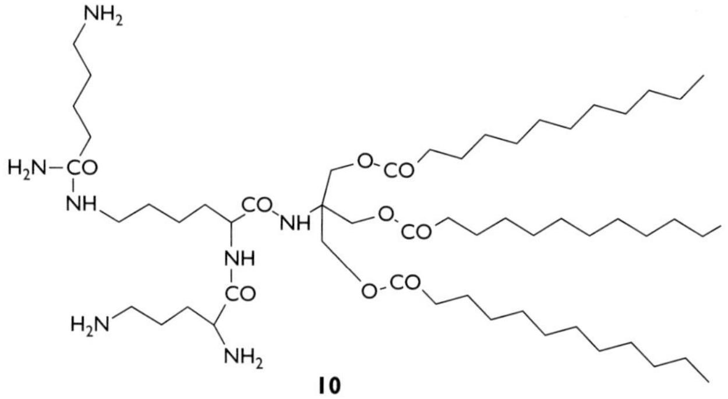

45.2.4 Cationic lipids in gene delivery

Cationic lipids have been used successfully to deliver genes and oligonucleotides into a variety of cell types *in vitro* and more recently they have been used for the delivery of therapeutic genes in a number of gene therapy clinical trials. Recently a tris-based delivery system (**10**) was used in synthesizing cationic lipid conjugates for gene delivery. The versatility of the complex allows a change in either the cationic head-group or the type and lengths of the lipid moieties. These changes have a profound effect on the rate of entry of the conjugate into cells.[15]

The efficacy of synthetic vectors for gene therapy depends to a large extent on the ease of DNA condensation and delivery to the cell nucleus. The empirical design of these vectors precludes targeted delivery and expression. A lipoamino acid-based system was also used in the rational design of novel non-viral dendritic vectors (**11**). These vectors are composites consisting of modules that bind and condense DNA and a lipid moiety for membrane lipid-bilayer transit. These vectors are versatile since a number of permutatable parameters, i.e. size and shape, surface charge, position and length of lipid components, could be altered.[16]

45.3 Sugar conjugates

Due to the poly-hydroxylated nature of sugars and the large array of sugars that are available, attachment of sugars to drugs is a more complicated process than lipid conjugation. Table 45.2 summarizes some of the drug-sugar conjugates, which have been synthesized. (Table 45.2 is grouped according to the type of linkage between the sugar and drug molecule, e.g. glycosidic bond, ester bond through the 6-O-position). Sugars can also be linked to the drug directly or via a linker moiety, as in the case of lipid-drug conjugates. The sugar conjugation can (i) utilize active transport systems, (ii) modify the physico-chemical properties of the construct, and (iii) target the compound.

45.3.1 Sugar transporters

There are a number of different types of monosaccharide transporter in different organs within the body. In terms of oral drug delivery, the transporters within the brush-border and basolateral membranes of intestinal epithelial cells are most important.[17] These transporters have been well characterized. The brush-border membranes express a sodium ion-dependent active transporter, SGLT1, and a sodium ion-independent facilitative transporter, GLUT5. In the basolateral membrane a sodium ion-independent facilitative transporter,

Table 45.2 Drug-sugar conjugates

Sugar linkage	*Drug/sugar/linker*
Glycosidic bond, 1-O-position e.g. 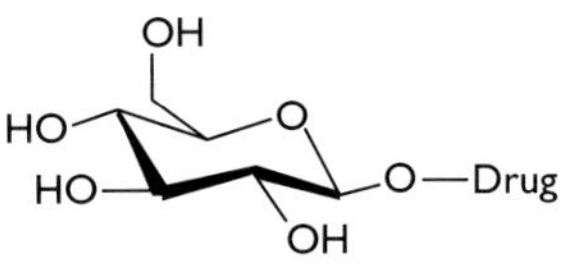	1 [D-Met 2, Hyp 5]-enkephalinamide; β-anomers Glc & Gal; attached to OH of hydroxyproline.[82,‡] 2 Enkephalin analogs; β-Glc; attached to OH of Ser.[27,‡,*] 3 Deltorphin and dermorphin analogs; β-Glc; attached to OH of Thr-4 or Thr-7.[26,‡,*] 4 Insulin; p-aminophenyl-α-D-glucopyranoside, p-aminophenyl-α-D-mannopyranoside, p-aminophenyl-α-L-arabino-pyranoside; attached to free amines through succinyl linker.[83,†,‡] 5 Taxol analog; sialic acid; polyethylene glycol linker.[84,f] 6 Arg-vasopressin (AVP) & oxytocin; β-Glc, β-Gal, β-Man, β-2-deoxy-Glc, α-Man; octamethylene spacer.[30,*] 7 9-Aminocaptothecin; β-D-glucuronylamine; benzyloxy-carbonyl labile linker.[39,#,Π,f] 8 Anthracyclines; β-D-glucuronylamine; labile linker Sugar-O-CH(OEt)-$(CH_2)_3$-NH-Drug.[85,#] 9 Anthracyclines; β-D-glucuronylamine; substituted benzyloxycarbonyl labile linker.[40,#,Π] 10 Gly-Gly-Tyr-Arg; α and β-Glc; p-(succinylamido) phenyl.[21,†]
Thioglycosidic, 1-S-position e.g. 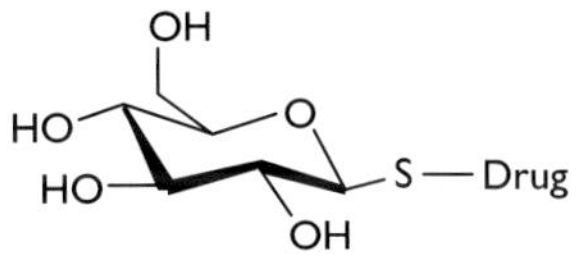	1 AVP, tryptamine, 4-nitrobenz-2-oxa-1,3-diazole; β-thio-Glc; C5, C8 & C11 linker.[31,*] 2 Vitamin K_5; thiogalactose; Sugar-CH_2-C(NH)-NH-$(CH_2)_2$-NH-poly-glutamic acid polymer.[33,‡,*]
Glycosyl amine, 1-N, 2-N, 6-N-position e.g. 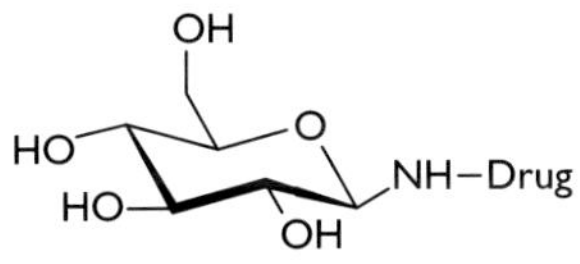	1 Leu-enkephalin; 6-amino-6-deoxy- & 2-amino-2-deoxy-D-glucopyranose & β-D-glucopyranosylamine; attached to C-terminus.[86] 2 Leu-enkephalin; β-D-glucopyranosylamine uronic acid & β-D-glucopyranosylamine uronic acid (1–6) β-D-gluco pyranosyl amine uronic acid, attached to C-terminus.[87,88,†]
3-O- and 6-O-position e.g. 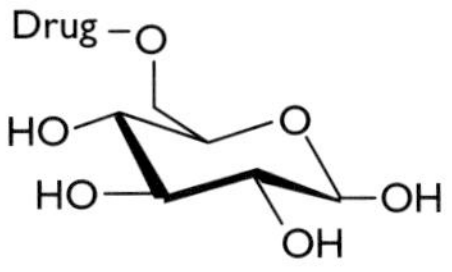	1. α-Tocopherol; Glc; succinate linker.[89,†] 2. Nipecotic acid; 6-O- and 3-O-D-Glc, 6-O-D-Gal; ester bond.[90] 3. Met-enkephalin; 6-O-D-Glc, 6-O-Glc-2-NAc, 6-O-D-$GlcAc_4$; attached to C-terminus via ester bond.[91]
Reduced sugar e.g. 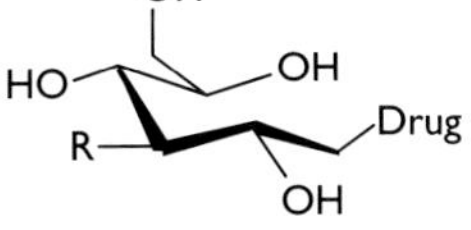	Tyr-Gly-Gly; Glc, cellobiose, maltose, lactose, gentiobiose; attached to N-terminus.[25,†]

Abbreviations: Glc = glucose, Gal = galactose, Man = mannose, Hyp = hydroxyproline, Ac = acetate.
Symbols: #Prodrug, †Increased activity/bioavailability *in vitro*, ‡Increased activity/bioavailability *in vivo*, *Π* Decreased toxicity, *Increased drug targeting to sites in the body, *f* Increased solubility.

GLUT2, has been found. SGLT1 shows specificity for D-glucose, D-galactose, 3-O-methyl-D-glucose, 1-deoxy-D-glucose and 6-deoxy-D-glucose, whereas, D-fructose, D-mannitol, 2-deoxy-D-glucose, L-glucose and L-galactose are not actively transported. The GLUT5 transporter has a major role in the intestinal absorption of D-fructose. GLUT2 has specificity for D-fructose, D-glucose and D-galactose. Thus for sugars such as D-glucose, SGLT1 and GLUT5 transport the sugars across the brush border membrane into the cell and the GLUT2 receptor then transports these sugars out of the enterocyte across the basolateral membrane into the blood.

45.3.2 Increased transport/absorption

Attachment of one of the sugars transported by GLUT5 or SGLT1 to a drug could potentially be used to transport a drug into the epithelium and then through GLUT2 into the blood stream. A number of researchers tested this hypothesis with various drugs (Table 45.2). The importance of the type of sugar and the nature of the linkage were investigated using small molecules such as p-nitrophenol[18] and 2-naphthol.[19] Intestinal absorption of p-nitrophenol conjugated with β-linked sugars showed the order of absorption in an *in vitro* everted sac assay in rats. The order of decreasing absorption clearance was glucose (4.45 μL/min/cm) > galactose (1.99 μL/min/cm) > mannose (0.811 μL/min/cm) ~ glucuronide (0.708 μL/min/cm). Only the glucose and galactose conjugates were actively transported, as evidenced by their inhibition by phloridzin, a known glucose transport carrier inhibitor and by their inhibition in the absence of sodium ions. Further work by these researchers with p-nitrophenol conjugates of glucose and galactose showed a preference of the β- over the α-anomer in Na^+ dependent intestinal absorption.[20] Interestingly, p-nitrophenol was the main compound appearing on the serosal side after absorption of the α-anomer of both glucose and galactose, suggesting the glucosides were hydrolyzed as they passed through the intestinal membrane. Thus, the α-anomers appeared to be hydrolyzed more readily than the β-anomers in this series of compounds.

Attachment of sugars to peptides was also investigated as a means of increasing absorption and peptide stability *in vivo*. Intestinal absorption of a tetrapeptide, Gly-Gly-Tyr-Arg (GGYR) was increased by glycosylation with p-(succinylamido)phenyl α- or β-D-glucopyranoside α,β-SAPG).[21] Uptake into brush border membrane vesicles (BBMVs) and transport through the intestinal membrane were examined using the rapid filtration technique and the everted sac method. The BBMV studies suggested the Na^+ dependent glucose transporter (SGLT1) played an important role in the uptake of both α- and β-SAPG-GGYR. Consistent with the results for the p-nitrophenol conjugates, the amount of β-SAPG-GGYR taken up was greater than that of α-SAPG-GGYR. Glycosylation at the N-terminus of the peptide was also shown to increase the resistance to aminopeptidase activity and inhibit its degradation. However, compared to the p-nitrophenol conjugates, α-SAPG-GGYR was found to be more stable to enzymatic degradation than β-SAPG-GGYR. Evidence of a high and low affinity Na^+ dependent glucose transporter was supported by the finding that permeation of α-SAPG-GGYR was only slightly inhibited by phloridzin, compared to a much higher inhibition of β-SAPG-GGYR. It was suggested that the α-anomer was actively transported by the Na^+ dependent D-glucose transporter with low affinity for β-D-glucoside.

The intestinal absorption of β-linked sugar-drug conjugates is greater than the α-anomers. NMR and molecular modeling studies have been carried out to examine the reasons for this β-anomeric preference.[22] It was found that the attached aglycone did not affect the conformation of the glucose ring and the crucial factor in SGLT1-mediated transport was the configuration of the aglycone at C-1 of glucose. The hypothesis proposed was that SGLT1 had a greater space available for aglycones attached to glucose in the β-conformation compared to those in the α-conformation. Thus the rate of transport also depended on the size and shape of the attached aglycone, which was supported by lower transport clearances for sugar-acetaminophen over p-nitrophenol conjugates.[22] Although the rate of absorption of sugar-drug conjugates was important, there were other factors important to the drug delivery. Factors such as the amount of desglycosylation product formed during passage through the intestinal tissues and the degree of enzymatic resistance of the conjugate are essential to a prodrug approach. With p-nitrophenol and 2-naphthol as the aglycone, more than half of the β-glycosides transported to the serosal side of a membrane were desglycosylated to the aglycone and the glucuronides.[19,20] The β-glucosides were less desglycosylated than the α-glucosides in this case.

Degradation of the disaccharide, maltose and then absorption of the resulting degradation product, glucose, from the intestine is greater than the absorption of free glucose.[23] Therefore the effect of coupling disaccharides (cellobiose, maltose and lactose) to p-nitrophenol was investigated in an attempt to increase absorption.[24] The disaccharide conjugates were not significantly absorbed, however, p-nitrophenyl β-glucoside (p-NPβglc) formed on the mucosal side and appeared on the serosal side in significant amounts. Furthermore, the absorption clearance of p-NPβglc formed by the enzymatic degradation of p-NP β-cellobioside **10** and p-NP β-lactoside **11** by lactase-phloridzin hydrolase (LPH), was much higher than that of p-NPβglc by itself. The hypothesis proposed to explain these results was that p-NPβglc was transported by the cooperation of SGLT1 with LPH from mucosal p-NP β-cellobioside or p-NP β-lactoside.

Disaccharides (cellobiose, maltose, lactose and gentiobiose) were also attached to peptides through reductive amination with peptides (Table 45.2).[25] In this case the attached reduced sugar was linear and the second sugar retained the pyranose ring structure. All of the conjugates attached to the tripeptide, TGG, could be transported from the mucosal to serosal side in rat everted small intestine and their transport was significantly decreased in the absence of Na^+, indicating involvement of the Na^+ dependent transporter. Additionally, the sugar-coupled TGG showed increased resistance to aminopeptidase compared to TGG, which was rapidly degraded.

45.3.3 *In vivo activity and targeting*

Drug sugar conjugates have also shown enhanced *in vivo* activity and ability to be targeted to particular organs within the body, as will be discussed in the following sections. The attached sugar served to target the drug to a particular sugar transporter/receptor and in some cases also increased resistance to degradation.

45.3.4 *Central nervous system targeting*

Administration of drugs to the central nervous system (CNS) is severely hampered by the blood-brain barrier (BBB), which blocks the passage of certain drugs into the brain. A number of groups have used sugar-peptide conjugates to deliver opioid peptides into the CNS to produce analgesia.[26–28] A β-O-linked glucoside attached to the side chain of serine was incorporated into an appropriate place in the peptide sequence of some enkephalin analogs.[27] The i.p. administration of two of the glycopeptides (**12** and **13**) led to dose-dependent antinociception in tail-flick and hot-plate tests, whereas the unglycosylated peptides showed no significant antinociceptive effects. As found for other CNS active drugs, the analgesia slowly increased to a maximum after 60 min and then declined over time. Additionally, the effects of one of the glycopeptides was reversible by the well known opioid antagonist, naloxone. Thus, the enkephalin analogs were not only metabolically stable for extended periods *in vivo* but were also able to cross the BBB to produce prolonged analgesia.

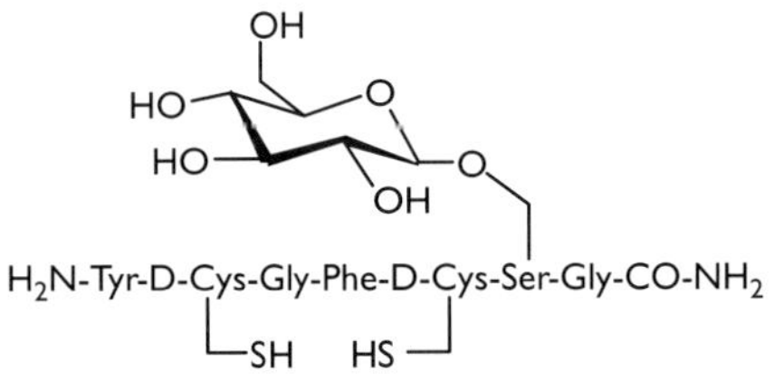

H_2N-Tyr-D-Cys-Gly-Phe-D-Cys-Ser-Gly-CO-NH_2 **12**

H_2N-Tyr-D-Cys-Gly-Phe-D-Cys-Ser-Gly-CO-NH_2 **13**

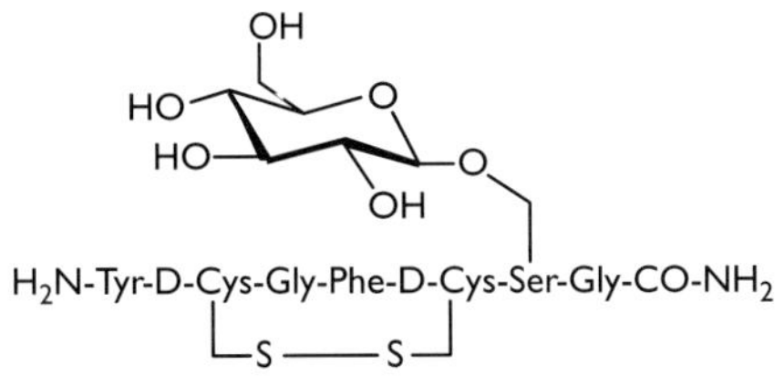

R-$(CH_2)_8$-CO-NH-Cys-Tyr-Phe-Gln-Asn-Cys-Pro-Arg-Gly-NH_2

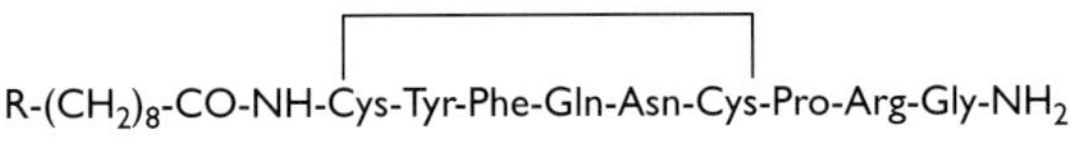

R = Glucose, Mannose, 2-deoxy-glucose

45.3.5 *Renal targeting*

Some peptide and nonpeptide-drug conjugates have been targeted to the kidneys by modification with alkylglycosides.[29–31] A model peptide Arg-vasopressin (AVP) was modified by linking it to a variety of sugars (glucose, galactose, mannose, 2-deoxy-glucose, α-mannose) via an octamethylene linker and the tissue uptake in rats was monitored after administration by i.v. injection. The glucosyl, mannosyl and 2-deoxy-glucosyl derivatives exhibited selective renal uptake and were found to be distributed in the proximal tubes of the renal cortex. It was postulated that renal uptake of the conjugates depended on the structure of the sugar, took place from the blood and involved specific binding to the renal microsomal fraction, followed by distribution in the proximal tubes.

45.4 Hepatic targeting

Drugs and proteins have been targeted to the liver through conjugation onto glycosylated polymers.[32–35] Hepatocytes recognize galactose- and N-acetygalactosamine terminated glycoproteins via the asialoglycoprotein receptor.[36,37] Additionally, the liver targeting efficiency was affected by the molecular weight, with large macromolecules (greater than 70 kDa) showing good targeting.[38] Hence, polymers containing attached sugars and drug molecules can be targeted to the liver. Polymers that were used as carriers for drugs include poly-L-glutamic acid,[33] poly-L-lysine,[34] polyacrylamide[35] and carboxymethyl-dextran.[32]

A galactosylated poly-L-glutamic acid conjugate was modified with vitamin K_5 (K_5 Gal PLGA **14**) to give 18.2 and 11.7 units of galactose and vitamin K_5, respectively per molecule.[33] This conjugate was examined for its pharmacological activity in hemorrhagic mice pretreated with warfarin.

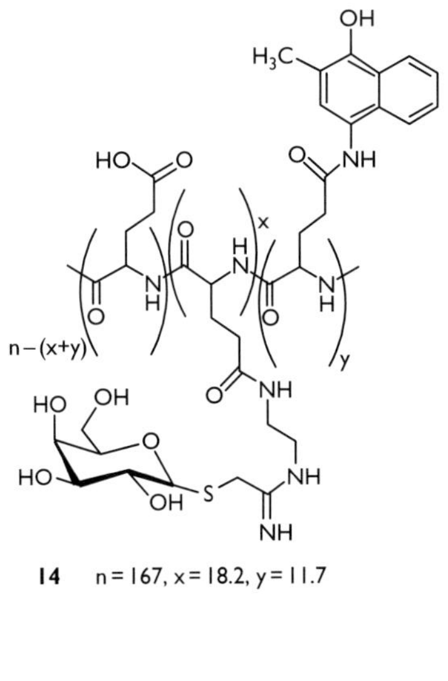

14 n = 167, x = 18.2, y = 11.7

K_5-Gal-PLGA showed anti-hemorrhagic effects 2–4 hrs after intravenous injection, whereas free vitamin K_5 was only effective at 4 hrs after administration. Liver accumulation-time course studies of ^{111}In-labelled K_5-Gal-PLGA showed the conjugate efficiently accumulated in the liver after i.v. injection, indicating the potential of these constructs for targeting the liver. One of the drawbacks with these polymer conjugates is their undefined structure, which makes quality control of the drug difficult.

45.4.1 Prodrugs with increased water solubility/decreased toxicity

Whereas lipid conjugation was used to increase the lipophilicity of an attached drug, sugar conjugation could be used to make a drug more water soluble. A glucuronide prodrug **15** of the poorly water soluble anticancer drug, 9-aminocamptothecin was found to be over 80 times more soluble than the parent drug.[39] Additionally, the prodrug was 20–80 fold less toxic than 9-aminocamptothecin and could be released to form the parent drug in the presence of β-glucuronidase. Thus, such compounds may be used for prodrug monotherapy of tumors that accumulate extracellular lysosomal β-glucuronidase, as well as for antibody-directed enzyme prodrug therapy (ADEPT) of cancer. A series of glucuronyl prodrugs of the anticancer drug doxorubicin were synthesized as candidates for the ADEPT strategy.[40] This work resulted in a glucuronyl conjugate HMR 1826 **16**, connected by a 3-nitro-benzyloxycarbonyl linker, which is almost 100-fold less cytotoxic than doxorubicin. HMR 1826 also shows tumor selectivity in *in vitro* models of cancer[41,42] and is a preclinical candidate. These results highlight the ability of sugar-drug conjugates to target particular receptors/ enzymes within the body.

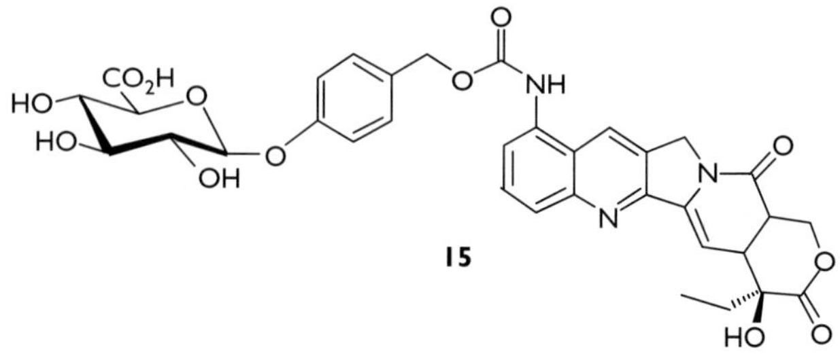

15

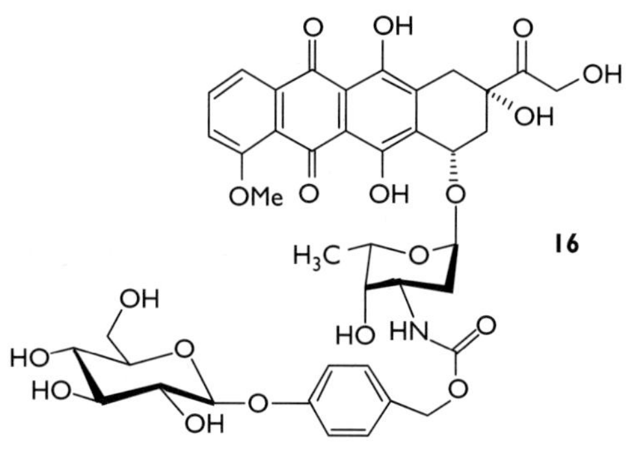

16

45.5 Lipid-sugar conjugates

Conjugation of drugs with lipids increases the passive transport across intestinal mucosal membranes, whereas sugar-drug conjugates show greater absorption through targeting the sugar transporters in the intestinal epithelium. Thus, coupling both a lipid and sugar to a drug molecule can potentially take advantage of both of these mechanisms.

Increased water solubility was obtained upon conjugation of a gulonic acid onto a lipid-cephalosporin conjugate **17**.[43] Oral uptake of cephalosporin, the glyco-lipid and lipid conjugate were examined in rats. The lipid conjugate was poorly soluble in water and was administered as a fine suspension. Cephalosporin showed no significant oral uptake, whereas both the glyco-lipid and lipid conjugates were taken up significantly over a 12 hr period. However, the water soluble glyco-lipid showed a slightly lower uptake than the lipid conjugate, indicating that decreased lipophilicity reduces the passive transport of the glyco-lipid conjugate.

Various sugar, lipid and sugar-lipid conjugates of a somatostatin analog (TT-232) have been synthesized and their *in vitro* antiproliferative effects determined on various tumor cell lines.[44] In most cases, there was no significant loss of antiproliferative activity compared to the parent peptide. Intestinal epithelial permeability of some of these compounds was determined using an *in vitro* Caco-2 cell assay, which can be correlated with oral drug absorption.[45] The sugar-lipid conjugate **18** had a high permeability coefficient, indicating that it would be well absorbed through the intestinal epithelium, whereas the lipid-conjugates that were analyzed had much lower permeability coefficients and would be poorly absorbed.

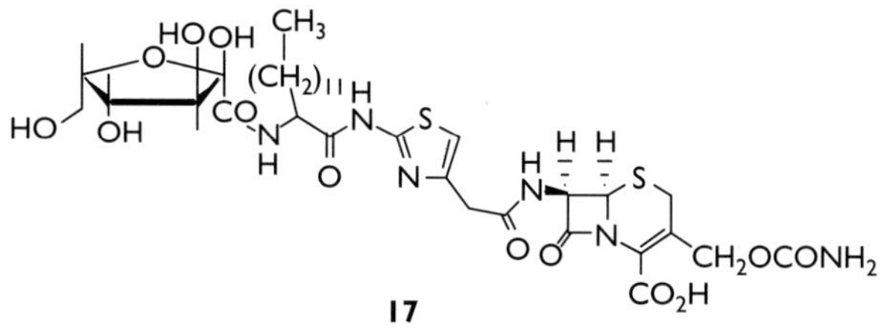

17

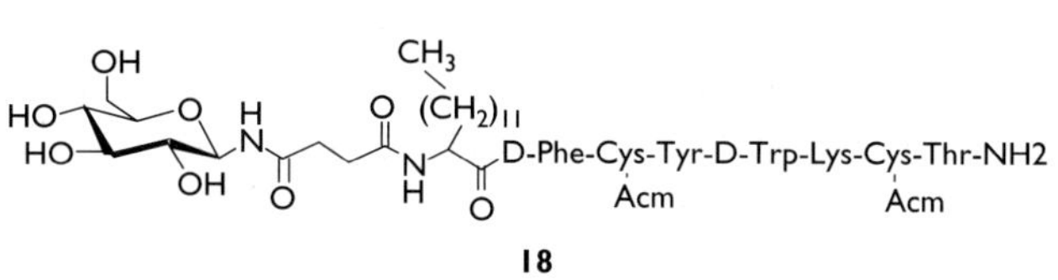

18

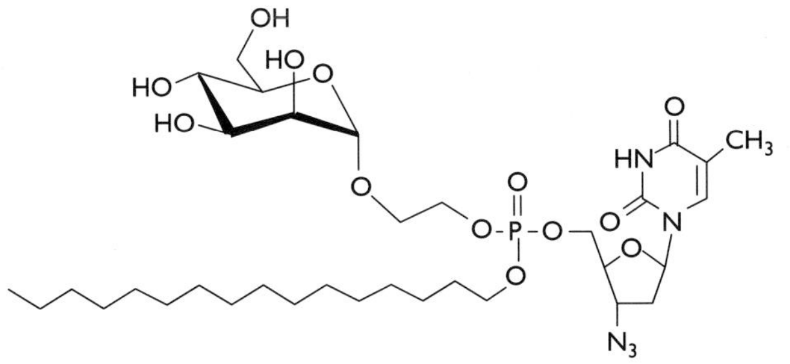

19

An extensive study on the delivery of AZT into the brain using a lipophilic glycosyl phosphotriester prodrug **19** was carried out in mice.[46] Oral administration of AZT gave a peak level of unmodified AZT after 1 hr and a total disappearance after 4 hr. However, oral administration of the phosphotriester **19** gave mainly the metabolite AZT 5′-monophophate, phosphodiester and AZT, with cerebral concentrations of AZT derivatives being 50 times higher than the minimum level needed for antiviral activity up to 48 hrs after ingestion. The phosphotriester itself could not be detected, possibly due to binding of this lipophilic derivative to lipoproteins or cellular membranes. Additionally, the concentration of AZT derivatives was much higher in the brain than in plasma indicating that its lipophilic properties allowed transfer across the BBB. Administration of the mannopyranosyl phosphodiester also gave levels of AZT derivatives higher than parent AZT, but at concentrations lower than those for phophotriester **19**.

45.5.1 Future perspectives

In this review we have discussed some of the ways in which an existing drug can be conjugated to a lipid and/or sugar molecule to improve its oral uptake, targeting, solubility, stability and other properties. Drugs with poor water solubility can be conjugated to sugars and those with poor absorption can be made more lipophilic through lipidic conjugation. However we,[43] as well as others,[27,46] have identified a major problem with lipidic conjugation, since the new entity often has poor water solubility. Conjugation with both lipids and sugars is one way of overcoming the solubility problems and will become more prominent as others experience these difficulties. Whether one can access both the sugar transporters, as well as increase passive absorption in the one sugar-lipid conjugate remains to be proven conclusively. A lack of experimentation involving oral delivery systems has also been observed as a limitation in lipid/sugar conjugate research. The oral route is still the most favored method of administration of drugs and needs to be a major focus of drug delivery researchers.

References

1 Gardner, C.R. (1987) Drug delivery-where now? In: *Drug Delivery Systems* (ed. Johnson, P., Lloyd-Jones, J. G.), p. 12. Ellis Horwood Ltd., Chichester.

2 Bundgaard, H., Hansen, A.B. (1981) Prodrugs as drug delivery systems. *Pharm. International 136–140.*

3 Vyas, S.P., Venugopalan, P., Sood, A., Mysore, N. (1997) Some approaches to improve bioavailability of peptides and proteins through oral and other mucosal routes. *Pharmazie, 52, 339–345.*

4 Muranishi, S., Murakami, M., Hashidzume, M., Yamada, K., Tajima, S., Kiso, Y. (1992) Trials of lipid modification of peptide hormones for intestinal delivery. *J. Control. Release 19, 179–188.*

5 Tanaka, K., Fujita, T., Yamamoto, Y., Murakami, M., Yamamoto, A., Muranishi, S. (1996) Enhancement of intestinal transport of thyrotropin-releasing hormone via a carrier-mediated transport system by chemical modification with lauric acid. *Biochim. Biophys. Acta 1283, 119–26.*

6 Tenma, T., Yodoya, E., Tashima, S., Fujita, T., Murakami, M., Yamamoto, A., Muranishi, S. (1993) Development of new lipophilic derivatives of tetragastrin: physicochemical characteristics and intestinal absorption of acyl-tetragastrin derivatives in rats. *Pharm. Res. 10, 1488–92.*

7 Garzon Aburbeh, A., Poupaert, J.H., Claesen, M., Dumont, P. (1986) A lymphotropic prodrug of L-dopa: synthesis, pharmacological properties, and pharmacokinetic behavior of 1,3-dihexadecanoyl-2-[(S)-2-amino-3-(3,4-dihydroxyphenyl)prop anoyl] propane-1,2,3-triol. *J. Med. Chem. 29, 687–91.*

8 Toth, I., Flinn, N., Hillery, A., Gibbons, W.A., Artursson, P. (1994) Lipidic Conjugates of Luteinizing-Hormone-Releasing Hormone (Lhrh)+ and Thyrotropin-Releasing-Hormone (Trh)+ That Release and Protect the Native Hormones in Homogenates of Human Intestinal Epithelial (Caco-2) Cells. *Int. J. Pharm. 105, 241–247.*

9 Anel, A., Halmos, T., Torres, J.M., Pineiro, A., Antonakis, K., Uriel, J. (1990) Cytotoxicity of chlorambucil and chlorambucil-fatty acid conjugates against human lymphomas and normal human peripheral blood lymphocytes. *Biochem. Pharmacol. 40, 1193 200.*

10 Prokai, L., Prokai Tatrai, K., Ouyang, X., Kim, H.S., Wu, W.M., Zharikova, A., Bodor, N. (1999) Metabolism-based brain-targeting system for a thyrotropin-releasing hormone analogue. *J. Med. Chem. 42, 4563–4571.*

11 Manolios, M., Collier, S., Taylor, J., Pollard, J., Harrison, C., Bender, V.J. (1997) T-cell antigen receptor transmembrane peptides modulate T-cell function and T-cell mediated disease. *Nature Med. 3, 84.*

12 Bessler, W.G., Suhr, B., Buhring, H.J., Muller, C.P., Wiesmuller, K.H., Becker, G., Jung, G. (1985) Specific antibodies elicited by antigen covalently linked to a synthetic adjuvant. *Immunobiol 170, 239–244.*

13 Wang, C.Y., Looney, D.J., Li, M.L., Walfield, A.M., Ye, J., Hossein, B., Tam, J.P., Wong-Staal, F. (1991) Long term higher titer neutralising activity induced by octameric synthetic HIV-1 antigen. *Science 254, 285–288;* Mora, A., Tam, J.P. (1998) Controlled lipidation and encapsulation of peptides as a useful approach to mucosal immunizations. *J. Immunol. 161, 3616–3623.*

14 Zhong, G., Toth. I., Reid, R., Brunham, (1993) R.C. Immunogenicity evaluation of lipidic amino acid based synthetic peptide vaccine for *Chlamidia trachomatis*. *J. Immunol. 151, 3728–3736.*
15 Cameron, F.H., Moghaddam, M.J., Bender, V.J., Whittaker, R.G., Lockett T.J. (1999) A transfection compound series based on a versatile tris linkage. *Biochim. Biophys. Acta 1417, 37–50.*
16 Toth, I., Sakthivel, T., Wilderspin, A.F., Bayele, H., O'Donnell, M., Perry, D.J., Pasi, K.J., Lee. C.A., Florence, A.T. (1999) Novel cationic lipidic peptide dendrimer vectors. In vitro gene delivery. *S.T.P. Pharma Sciences 9(1) 93–99.*
17 Thorens, B. (1993) Facilitated glucose transporters in epithelial cells. *Ann. Rev. Physiol. 55, 591–608*; Wright, E.M. (1993) The intestinal Na+/glucose cotransporter. *Annu. Rev. Physiol. 55, 575–589*; Tamai, I., Tsuji, A. (1996) Carrier-mediated approaches for oral drug delivery. *Adv. Drug Del. Rev. 20, 5–32.*
18 Mizuma, T., Ohta, K., Hayashi, M., Awazu, S. (1992) Intestinal active absorption of sugar-conjugated compounds by glucose transport system: implication of improvement of poorly absorbable drugs. *Biochem. Pharmacol. 43, 2037–9.*
19 Mizuma, T., Ohta, K., Awazu, S. (1994) The beta-anomeric and glucose preferences of glucose transport carrier for intestinal active absorption of monosaccharide conjugates. *Biochim. Biophys. Acta. 1200, 117–22.*
20 Mizuma, T., Ohta, K., Hayashi, M., Awazu, S. (1993) Comparative study of active absorption by the intestine and disposition of anomers of sugar-conjugated compounds. *Biochem. Pharmacol. 45, 1520–3.*
21 Nomoto, M., Yamada, K., Haga, M., Hayashi, M. (1998) Improvement of intestinal absorption of peptide drugs by glycosylation: Transport of tetrapeptide by the sodium ion-dependent D-glucose transporter. *J. Pharm. Sci. 87, 326–332.*
22 Mizuma, T., Nagamine, Y., Dobashi, A., Awazu, S. (1998) Factors that cause the beta-anomeric preference of Na+/glucose cotransporter for intestinal transport of monosaccharide conjugates. *Biochim. Biophys. Acta 1381, 340–6.*
23 Cook, G.C. (1973) Comparison of absorption rates of glucose and maltose in man in vivo. *Clin. Sci. 44, 425–428.*
24 Mizuma, T., Awazu, S. (1998) Intestinal Na+/glucose cotransporter-mediated transport of glucose conjugate formed from disaccharide conjugate. *Biochim. Biophys. Acta 1379, 1–6.*
25 Mizuma, T., Sakai, N., Awazu, S. (1994) Na(+)-dependent transport of aminopeptidase-resistant sugar-coupled tripeptides in rat intestine. *Biochem. Biophys. Res. Commun. 203, 1412–6.*
26 Tomatis, R., Marastoni, M., Balboni, G., Guerrini, R., Capasso, A., Sorrentino, L., Santagada, V., Caliendo, G., Lazarus, L.H., Salvadori, S. (1997) Synthesis and pharmacological activity of deltorphin and dermorphin-related glycopeptides. *J. Med. Chem. 40, 2948–52.*
27 Polt, R., Porreca, F., Szabo, L.Z., Bilsky, E.J., Davis, P., Abbruscato, T.J., Davis, T.P., Horvath, R., Yamamura, H.I., Hruby, V.J. (1994) Glycopeptide enkephalin analogs produce analgesia in mice – evidence for penetration of the blood-brain-barrier. *Proc. Nat. Acad. Sci. USA. 91, 7114–7118.*
28 Mizuma, T., Ohta, K., Awazu, S. (1998) Intestinal absorption and analgesic activity of aminopeptidase-resistant cellobiose-coupled leucine enkephalinamide. *Biopharm. Drug Dispos. 19, 605–10.*
29 Suzuki, K., Ando, T., Susaki, H., Mimori, K., Nakabayashi, S., Sugiyama, Y. (1999) Structural requirements for alkylglycoside-type renal targeting vector. *Pharm. Res. 16, 1026–1034.*
30 Suzuki, K., Susaki, H., Okuno, S., Yamada, H., Watanabe, H. K., Sugiyama, Y. (1999) Specific renal delivery of sugar-modified low-molecular-weight peptides. *J. Pharm. Exp. Ther. 288, 888–897.*
31 Suzuki, K., Susaki, H., Okuno, S., Sugiyama, Y. (1999) Renal drug targeting using a vector "alkylglycoside". *J. Pharmacol. Exp. Ther. 288, 57–64.*
32 Hashida, M., Nishikawa, M., Takakura, Y. (1995) Hepatic targeting of drugs and proteins by chemical modification. *J. Control. Rel. 36, 99–107.*
33 Hashida, M., Hirabayashi, H., Nishikawa, M., Takakura, Y. (1997) Targeted delivery of drugs and proteins to the liver via receptor-mediated endocytosis. *J. Control. Rel. 46, 129–137.*
34 Gonsho, A., Irie, K., Susaki, H., Iwasawa, H., Okuno, S., Sugawara, T. (1994) Tissue-targeting ability of saccharide-poly(L-lysine) conjugates. *Biol. Pharm. Bull. 17, 275–82.*
35 Kojima, S., Andre, S., Korchagina, E.Y., Bovin, N.V., Gabius, H.J. (1997) Tyramine-containing poly(4-nitrophenylacrylate) as iodinatable ligand carrier in biodistribution analysis. *Pharm. Res. 14, 879–86.*
36 Ashwell, G., Harford, J. (1982) Carbohydrate specific receptors of the liver. *Annu. Rev. Biochem. 51, 531–554.*
37 Monsigny, M., Roche, A.C., Midoux, P., Mayer, R. (1994) Glycoconjugates As Carriers For Specific Delivery of Therapeutic Drugs and Genes. *Adv. Drug Del. Rev. 14, 1–24.*
38 Takakura, Y., Fujita, T., Hashida, M., Sezaki, H. (1990) Disposition characteristics of macromolecules in tumor-bearing mice. *Pharm. Res. 7, 339–346.*
39 Leu, Y.L., Roffler, S.R., Chern, J.W. (1999) Design and synthesis of water-soluble glucuronide derivatives of camptothecin for cancer prodrug monotherapy and antibody-directed enzyme prodrug therapy (ADEPT). *J. Med. Chem. 42, 3623–3628.*
40 Florent, J.C., Dong, X., Gaudel, G., Mitaku, S., Monneret, C., Gesson, J.P., Jacquesy, J.C., Mondon, M., Renoux, B., Andrianomenjanahary, S., Michel, S., Koch, M., Tillequin, F., Gerken, M., Czech, J., Straub, R., Bosslet, K. (1998) Prodrugs of anthracyclines for use in antibody-directed enzyme prodrug therapy. *J. Med. Chem. 41, 3572–3581.*
41 Bosslet, K., Straub, R., Blumrich, M., Czech, J., Gerken, M., Sperker, B., Kroemer, H.K., Gesson, J.P., Koch, M., Monneret, C. (1998) Elucidation of the mechanism enabling tumor selective prodrug monotherapy. *Cancer Res. 58, 1195–201.*
42 Platel, D., Bonoron Adele, S., Dix, R.K., Robert, J., (1999) Preclinical evaluation of the cardiac toxicity of HMR-1826, a novel prodrug of doxorubicin. *Br. J. Cancer 81, 24–27.*
43 Toth, I., Hughes, R.A., Dekany, G., Hillery, A.M., Ward, P. (1994) Synthesis and Oral Uptake Studies of Lipidic and Glyco-Lipidic Conjugates of Beta-Lactam Antibiotics. *Liebigs Ann. Chem. 7, 685–688.*
44 Toth, I., Malkinson, J.P., Flinn, N.S., Drouillat, B., Horvath, A., Erchegyi, J., Idei, M., Venetianer, A., Artursson, P., Lazorova, L., Szende, B., Keri, G. (1999) Novel lipoamino acid- and liposaccharide-based system for peptide delivery: Application for oral administration of tumor-selective somatostatin analogues. *J. Med. Chem. 42, 4010–4013.*

45 Artursson, P., Karlsson, J. (1991) Correlation between oral drug absorption in humans and apparent drug permeability coefficients in human intestinal epithelial (Caco-2) cells. *Biochem. Biophys. Res. Commun. 175, 880–885.*

46 Namane, A., Gouyette, C., Fillion, M.P., Fillion, G., Huynh Dinh, T. (1992) Improved brain delivery of AZT using a glycosyl phosphotriester prodrug. *J. Med. Chem. 35, 3039–44.*

47 Kawaguchi, T., Ishikawa, K., Seki, T., Juni, K. (1990) Ester prodrugs of zidovudine. *J. Pharm. Sci. 79, 531–3.*

48 Yodoya, E., Uemura, K., Tenma, T., Fujita, T., Murakami, M., Yamamoto, A., Muranishi, S. (1994) Enhanced permeability of tetragastrin across the rat intestinal membrane and its reduced degradation by acylation with various fatty acids. *J. Pharmacol. Exp. Ther. 271, 1509–13.*

49 Setoh, K., Murakami, M., Araki, N., Fujita, T., Yamamoto, A., Muranishi, S. (1995) Improvement of transdermal delivery of tetragastrin by lipophilic modification with fatty acids. *J. Pharm. Pharmacol. 47, 808–11.*

50 Fujita, T., Kawahara, I., Quan, Y., Hattori, K., Takenaka, K., Muranishi, S., Yamamoto, A. (1998) Permeability characteristics of tetragastrins across intestinal membranes using the Caco-2 monolayer system: comparison between acylation and application of protease inhibitors. *Pharm. Res. 15, 1387–92.*

51 Burr, A., Bundgaard, H. (1987) Prodrugs of 5-fluorouracil. VIII. Improved rectal and oral delivery of 5-fluorouracil via various prodrugs. Structure-rectal absorption relationships. *Int. J. Pharm. 36, 41–49.*

52 Hashizume, M., Douen, T., Murakami, M., Yamamoto, A., Takada, K., Muranishi, S. (1992) Improvement of large intestinal absorption of insulin by chemical modification with palmitic acid in rats. *J. Pharm. Pharmacol. 44, 555–9.*

53 Hashimoto, M., Takada, K., Kiso, Y., Muranishi, S. (1989) Synthesis of palmitoyl derivatives of insulin and their biological activities. *Pharm. Res. 6, 171–6.*

54 Asada, H., Douen, T., Mizokoshi, Y., Fujita, T., Murakami, M., Yamamoto, A., Muranishi, S. (1994) Stability of acyl derivatives of insulin in the small intestine: relative importance of insulin association characteristics in aqueous solution. *Pharm. Res. 11, 1115–20.*

55 Asada, H., Douen, T., Waki, M., Adachi, S., Fujita, T., Yamamoto, A., Muranishi, S. (1995) Absorption characteristics of chemically modified-insulin derivatives with various fatty acids in the small and large intestine. *J. Pharm. Sci. 84, 682–7.*

56 Yamada, K., Murakami, M., Yamamoto, A., Takada, K., Muranishi, S. (1992) Improvement of intestinal absorption of thyrotropin-releasing hormone by chemical modification with lauric acid. *J. Pharm. Pharmacol. 44, 717–21.*

57 Muranishi, S., Sakai, A., Yamada, K., Murakami, M., Takada, K., Kiso, Y. (1991) Lipophilic peptides: synthesis of lauroyl thyrotropin-releasing hormone and its biological activity. *Pharm. Res. 8, 649–52.*

58 Rosowsky, A., Forsch, R.A., Yu, C.S., Lazarus, H., Beardsley, G.P. (1984) Methotrexate analogues. 21. Divergent influence of alkyl chain length on the dihydrofolate reductase affinity and cytotoxicity of methotrexate monoesters. *J. Med. Chem. 27, 605–9.*

59 Bijsterbosch, M.K., van de Bilt, H., van Berkel, T.J. (1996) Specific targeting of a lipophilic prodrug of iododeoxyuridine to parenchymal liver cells using lactosylated reconstituted high density lipoprotein particles. *Biochem. Pharmacol. 52, 113–21.*

60 Veuillez, F., GanemQuintanar, A., Deshusses, J., FalsonRieg, F., Buri, P. (1998) Comparison of the ex-vivo oral mucosal permeation of tryptophan-leucine (Trp-Leu) and its myristoyl derivative. *Int. J. Pharm. 170, 85–91.*

61 Camaioni Neto, C., Steim, J.M., Sarin, P.S., Sun, D.K., Bhongle, N.N., Piratla, R.K., Turcotte, J.G. (1990) Lipid conjugates of antiretroviral agents. II. Disodium palmityl phosphonoformate: anti-HIV activity, physical properties, and interaction with plasma proteins. *Biochem. Biophys. Res. Commun. 171, 458–64.*

62 Hughes, R.A., Toth, I., Ward, P., Ireland, S.J., Gibbons, W.A. (1991) Lipidic peptides. III: Lipidic amino acid and oligomer conjugates of morphine. *J. Pharm. Sci. 80, 1103–5.*

63 Hughes, R.A., Toth, I., Ward, P., McColm, A.M., Cox, D.M., Anderson, G.J., Gibbons, W.A. (1992) Lipidic peptides. V: Penicillin and cephalosporin acid conjugates with increased lipophilic character. *J. Pharm. Sci. 81, 845–8.*

64 Toth, I., Hughes, R.A., Munday, M.R., Murphy, C.A., Mascagni, P., Gibbons, W.A. (1991) Lipidic peptides. II. Synthesis, activity and transport of anti-inflammatory benzoquinolizine-lipidic peptide conjugates. *Int. J. Pharm. 68, 191–198.*

65 Wood, I.P., Toth, I., Holley, J.L., Gibbons, W.A. (1992) XIII:Synthesis, structure elucidation and in vitro toxicity assessment of chlorambucil conjugates with lipidic acids, lipidic amino acids and their oligomers. *Int. J. Pharm. 87, 141–147.*

66 Toth, I., Christodoulou, M., Bankowsky, K., Flinn, N., Gibbons, W.A., Godeau, G., Moczar, E., Hornebeck, W. (1995) Design of potent lipophilic-peptide inhibitors of human neutrophil elastase: In vitro and in vivo studies. *Int. J. Pharm. 125, 117–122.*

67 Garzon Aburbeh, A., Poupaert, J.H., Claesen, M., Dumont, P., Atassi, G. (1983) 1,3-dipalmitoylglycerol ester of chlorambucil as a lymphotropic, orally administrable antineoplastic agent. *J. Med. Chem. 26, 1200–3.*

68 Scriba, G.K.E., Lambert, D.M., Poupaert, J.H. (1995) Bioavailability of phenytoin following oral administration of phenytoin-lipid conjugates to rats. *J. Pharm. Pharmacol. 47, 945–948.*

69 Steim, J.M., Camaioni Neto, C., Sarin, P.S., Sun, D.K., Sehgal, R.K., Turcotte, J.G. (1990) Lipid conjugates of antiretroviral agents. I. Azidothymidine-monophosphate-diglyceride: anti-HIV activity, physical properties, and interaction with plasma proteins. *Biochem. Biophys. Res. Commun. 171, 451–7.*

70 Hong, C.I., Kirisits, A.J., Nechaev, A., Buchheit, D.J., West, C.R. (1990) Nucleoside conjugates. 11. Synthesis and antitumor activity of 1-beta-D-arabinofuranosylcytosine and cytidine conjugates of thioether lipids. *J. Med. Chem. 33, 1380–6.*

71 Piantadosi, C., Marasco, C.J.Jr., Morris Natschke, S.L., Meyer, K.L., Gumus, F., Surles, J.R., Ishaq, K.S., Kucera, L.S., Iyer, N., Wallen, C.A., *et al.* (1991) Synthesis and evaluation of novel ether lipid nucleoside conjugates for anti-HIV-1 activity. *J. Med. Chem. 34, 1408–14.*

72 Wells, X.E., Bender, V.J., Francis, C.L., He Williams, H.M., Manthey, M.K., Moghaddam, M.J., Reilly, W.G., Whittaker, R.G. (1999) Tris and the ready production of drug-fatty acyl conjugates. *Drug Dev. Res. 46, 302–308.*

73 Prokai Tatrai, K., Prokai, L., Bodor, N. (1996) Brain-targeted delivery of a leucine-enkephalin analogue by retrometabolic design. *J. Med. Chem. 39, 4775–82.*

74 Bodor, N., Prokai, L., Wu, W.M., Farag, H., Jonalagadda, S., Kawamura, M., Simpkins, J. A (1992) strategy for delivering peptides into the central nervous system by sequential metabolism. *Science 257, 1698–700.*
75 Prokai, L., Ouyang, X., ProkaiTatrai, K., Simpkins, J.W., Bodor, N. (1998) Synthesis and behavioral evaluation of a chemical brain-targeting system for a thyrotropin-releasing hormone analogue. *Eur. J. Med. Chem. 33, 879–886.*
76 Hashimoto, K., Loader, J.E., Knight, M.S., Kinsky, S.C. (1985) Inhibition of cell proliferation and dihydrofolate reductase by liposomes containing methotrexate-dimyristoylphosphatidylethanolamine derivatives and by the glycerophosphorylethanolamine analogs. *Biochim. Biophys. Acta 816, 169–78.*
77 Hashimoto, K., Loader, J.E., Kinsky, S.C. (1985) Synthesis and characterization of methotrexate-dimyristoylphosphatidylethanolamine derivatives and the glycerophosphorylethanol amine analogs. *Biochim. Biophys. Acta 816, 163–8.*
78 Kinsky, S.C., Hashimoto, K., Loader, J.E., Knight, M.S., Fernandes, D.J. (1986) Effect of liposomes sensitized with methotrexate-gamma-dimyristoylphosphatidylethanolamine on cells that are resistant to methotrexate. *Biochim. Biophys. Acta 885, 129–35.*
79 Noe, C., Hernandez Borrell, J., Kinsky, S.C., Matsuura, E., Leserman, L. (1988) Inhibition of cell proliferation with antibody-targeted liposomes containing methotrexate-gamma-dimyristoylphosphatidylethanolamine. *Biochim. Biophys. Acta 946, 253–60.*
80 Williams, A.S., Love, W.G., Williams, B.D. (1992) Synthesis of methotrexate-dimyristoylphosphatidylethanolamine analogs and characterization of methotrexate release in vitro. *Int. J. Pharm. 85, 189–197.*
81 Williams, A.S., Topley, N., Amos, N., Williams, B.D. (1994) Effect of three lipophilic methotrexate derivatives upon mediator release by lipopolysaccharide-stimulated rat peritoneal macrophages. *J. Pharm. Pharmacol. 46, 291–5.*
82 Rodriguez, R.E., Rodriguez, F.D., Sacristan, M.P., Torres, J.L., Valencia, G., Garcia, Anton J.M. (1989) New glycosylpeptides with high antinociceptive activity. *Neurosci. Lett. 101, 89–94.*
83 Haga, M., Saito, K., Shimaya, T., Maezawa, Y., Kato, Y., Kim, S. (1990) Hypoglycemic effect of intestinally administered monosaccharide-modified insulin derivatives in rats. *Chem. Pharm. Bull. 38, 1983–1986.*
84 Takahashi, T., Tsukamoto, H., Yamada, H. (1998) Design and synthesis of a water-soluble taxol analogue: Taxol-sialyl conjugate. *Bioorg. Med. Chem. Lett. 8, 113–116.*
85 Bakina, E., Wu, Z., Rosenblum, M., Farquhar, D. (1997) Intensely cytotoxic anthracycline prodrugs: Glucuronides. *J. Med. Chem. 40, 4013–4018.*
86 Varga-Defterdarovic, L., Horvat, S., Chung, N.N., Schiller, P.W. (1992) Glycoconjugates of opioid peptides. *Int. J. Pep. Prot. Res. 39, 12–17.*
87 Drouillat, B., Kellam, B., Dekany, G., Starr, M.S., Toth, I. (1997) Solid phase synthesis of C-terminal carbohydrate modified enkephalins. *Bioorg. Med. Chem. Let. 7, 2247–2250.*
88 Kellam, B., Drouillat, B., Dekany, G., Starr, M.S., Toth I. (1998) Synthesis and in vitro evaluation of lipoamino acid and carbohydrate-modified enkephalins as potential antinociceptive agents. *Int. J. Pharm., 161, 55–64.*
89 Bonina, F., Lanza, M., Montenegro, L., Salerno, L., Smeriglio, P., Trombetta, D., Saija, A. (1996) Transport of alpha-tocopherol and its derivatives through erythrocyte membranes. *Pharm. Res. 13, 1343–7.*
90 Bonina, F.P., Arenare, L., Palagiano, F., Saija, A., Nava, F., Trombetta, D., de Caprariis, P. (1999) Synthesis, stability, and pharmacological evaluation of nipecotic acid prodrugs. *J. Pharm. Sci., 88, 561–567.*
91 Horvat, S., Horvat, J., Varga-Defterdarovic, L., Pavelic, K., Chung, N.N., Schiller, P.W. (1993) Methionine-enkephalin related glycoconjugates. *Int. J. Peptide and Protein Res. 41, 399–404.*

Chapter 46

Pharmacokinetic-pharmacodynamic modeling and an example of its application to drug delivery

Edward J. Triggs

Contents

46.1 Introduction

The rational selection of a dosage regimen of a drug for treatment of any patient requires adequate knowledge of the disease process affecting the patient, the pharmacology of the drug, and the drug's pharmacokinetic and pharmacodynamic variability in the individual.

Figure 46.1 is an illustration of how pharmacokinetics (PK) and pharmacodynamics (PD) determine the pharmacological effect(s) of a drug. The relationship between drug dose and plasma concentration is most useful when it is also linked to a therapeutic (or adverse) effect that is associated with a particular concentration.

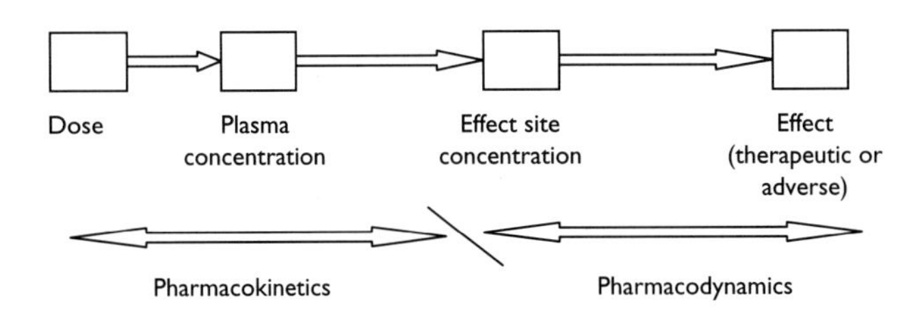

Figure 46.1 Pharmacokinetics and pharmacodynamics as determinants of the dose-effect relationship.

The relationship between plasma drug concentration and pharmacological response depends on the mechanism by which the drug exerts its effect. This effect may be direct e.g. receptor binding (muscle relaxants), where in such cases there is a direct relationship between the time course of drug concentrations (PK) and the pharmacologic effect. For other drugs, the relationship may be indirect e.g. inhibition of clotting factor synthesis (warfarin), where the relationship between PK and PD is more complex.

46.2 PK-PD modeling

A model is best viewed as a mathematical representation of a system whose purpose is to predict the response of that system after a given input. In the field of clinical pharmacology, a PK-PD model represents the time course of drug absorption and disposition linked to the pharmacologic response (effect). The input is the dose administered and the outputs are the observed plasma concentrations and the level of the response (Gabrielsson and Weiner, 1997).

46.3 Data collection

The first phase of a PK-PD study consists of an appropriate experimental design to obtain kinetic and dynamic data. In general, two different methodologies are possible, the classical and population approaches.

46.3.1 *Classical approach – experimental design*

This is often used during Phase I and Phase II trials. A PK-PD model is fitted to a set of plasma concentration and effect data measured in the same individual. In order to obtain good precision in the model parameter estimates (e.g. clearance CL, apparent volume of distribution V, bioavailability F, maximum effect EMax, etc.), the number of concentrations and effect measurements must be as large as possible for each individual since the population sample size is limited.

Such studies are normally performed at non-steady state conditions. It is theoretically possible, using optimal sampling procedures, to select the best times of measurement after drug administration, but in practice this may be difficult if the underlying PK-PD model is unknown and measurement time is often selected as for a simple PK study.

The range of measured concentrations must be large enough to examine the shape of the complete concentration-effect relationship and this may require the use of a large dose or several doses of the drug under investigation.

46.3.2 *Population approach – experimental design*

This approach is often conducted during Phase III clinical trials where the population sample is much larger and more closely represents those patients who will be receiving the drug as part of their treatment. Here, the concentration-effect relationship is modeled using only a limited number of measurements per patient.

One of the advantages of this methodology is its better relevance to the 'real' clinical situation coupled with an ability to examine the effect of covariates (age, gender, weight, serum creatinine, etc.) on PK-PD parameters and, therefore, to model inter- and intra-patient variability.

46.4 PK-PD models

Having collected the data, the PK-PD modeling can now be conducted to examine the structure of the mathematical representation between the administered dose and the resultant plasma concentration and effect measurements.

The first step in the modeling procedure is to plot the measured drug effects against the plasma concentrations in increasing time order. If the observed concentration-effect relationship does not show hysteresis, a direct PD model can usually be fitted to the data (Figure 46.2A). However, if the observed concentration-effect relationship shows a hysteresis loop (Figure 46.2B), two different effects are observed for most concentrations and either a linked PD or an indirect PD model must then be used to derive the concentration-effect relationship.

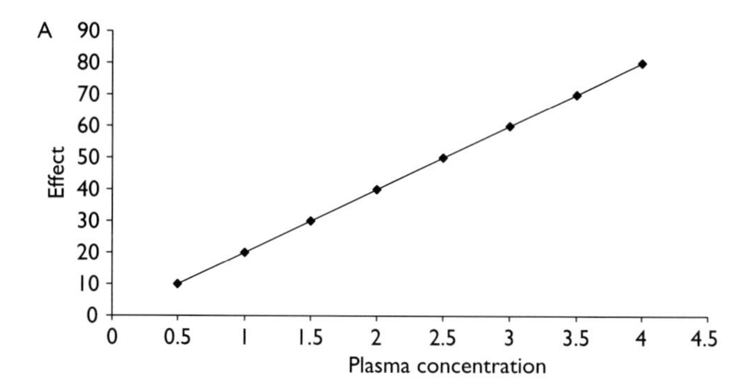

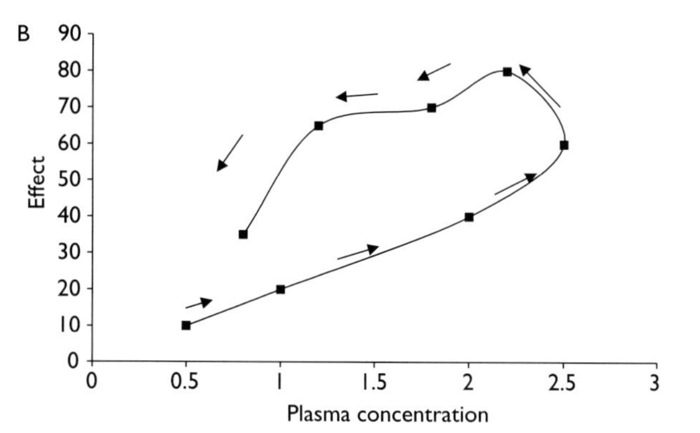

Figure 46.2 Plasma concentration-effect relationship with the data connected in order of increasing time. (A) Direct relationship. (B) Counterclockwise hysteresis.

46.4.1 *Direct PD models*

46.4.1.1 *Linear model*

This model predicts that the effect, E, of a drug is directly proportional to its concentration, C, at the effect site.

The relationship between E and C is:

$$E = S*C + E_0 \qquad \text{(Equation 46.1)}$$

Where, S is the slope and E_0 is the intercept of the effect-concentration relationship. The parameters, S and E_0, can be easily determined by linear regression. One disadvantage of this model is that it cannot be used to predict a maximum effect (EMax) nor can it be used to predict effects for concentrations above those used to estimate the parameters.

46.4.1.2 *Log-linear model*

In this model E is directly proportional to log C at the effect site.

The model can be represented as:

$$E = S * \log C + \mathrm{I} \qquad \text{(Equation 46.2)}$$

Where, S is the slope and I is the intercept of the effect-concentration relationship. Again, as with model 1, the parameters, S and I, can be estimated by linear regression.

The model cannot be used to predict the value of E when $C = 0$ nor can it be used to predict EMax and it is strictly only applicable over the range between 20% and 80% of EMax.

46.4.1.3 *Non-linear EMax model*

This model is derived from classical concepts of drug-receptor interaction and describes a hyperbolic effect-concentration relationship.

It is expressed as:

$$E = \frac{E_{Max} * C}{EC_{50} + C} + E_0 \qquad \text{(Equation 46.3)}$$

Where, EMax is the maximum effect, EC_{50} is the plasma concentration required to produce 50% of EMax and E_0 is the effect at baseline. This model requires non-linear regression techniques to estimate the model parameters but has the advantage of predicting a maximum effect.

46.4.1.4 *Non-linear sigmoid EMax model*

The Hill equation, rearranged in the form of Equation 46.4, is used to account for the sigmoidicity of the effect-concentration relationship.

It is represented as:

$$E = \frac{E_{Max} * C^{\gamma}}{EC_{50} + C^{\gamma}} + E_0 \qquad \text{(Equation 46.4)}$$

Where, γ is the sigmoidicity factor (or Hill coefficient) and other parameters are identical to those defined for the non-linear EMax model. Use of this model requires the estimation by non-linear regression of an additional parameter from that of the EMax model and has the same advantage of estimating a maximum effect.

46.4.1.5 *Time dependent PD models*

As mentioned previously, the presence of hysteresis in a plot of the effect-concentration relationship in increasing time order is indicative of a time-dependent PD model. Most often these plots will show "anticlockwise" hysteresis, suggestive of a time-delay in the kinetics of the effect as compared with the kinetics of plasma concentrations. In such cases either a link-model or an indirect effect model are used in the PK-PD modeling procedure (Levy, 1998).

If the plot should show "clockwise" hysteresis where the effect decreases quicker than plasma concentration, a PK-PD model incorporating drug tolerance may be useful to fit the data.

PK-PD LINK MODELS

By the use of a link model the time course of pharmacologic effect enables the determination of the rate of drug movement into the effect site. As plasma concentrations increase from zero to a value C and are maintained at that level, then the effect will progressively increase based on the rate of drug accumulation at the effect site. Further, if drug entry and exit from the effect site are determined by separate first-order rate constants, then the rate of onset of effect up to E at C will be determined by the rate constant (k_{e0}) describing drug loss from the effect site.

This model was initially used to describe the time-course of muscle paralysis with tubocurarine and since has found many applications in PK-PD modeling.

The exact form of the PK model may vary depending on the nature of the drug's pharmacokinetics (e.g. 1 or 2 compartmental etc.), but is immaterial as long as it adequately describes the plasma concentrations. These are then linked to a hypothetical effect compartment by a rate constant, k_{1e}. If it is assumed that k_{1e} is very small compared to all other rate constants in the PK model, then only a negligible amount of drug enters the effect compartment. Hence, the amount of drug returning to the plasma may be neglected and the effect compartment does not alter the plasma concentration time curve. Under these circumstances, the value of k_{1e} is unimportant, and the rate constant k_{e0} will characterize the equilibration time between plasma concentrations and pharmacologic effect.

This may be expressed as:

$$dC_e/dt = k_{e0} * (C - C_e) \qquad \text{(Equation 46.5)}$$

Where, C and C_e are the concentrations in the plasma and the effect compartment, respectively. The effect E is then modelled using one of the linear or non-linear models described previously and is linked to the plasma concentrations via the variable C.

INDIRECT RESPONSE MODELS

An observed delay between PK and PD can also be explained by an indirect mechanism of drug action (e.g. by stimulating or inhibiting an enzyme responsible for effect production or dissipation). In such cases, the rate of change of response in the absence of drug can be described as:

$$dR/dt = k_{in} - k_{out} * R \qquad \text{(Equation 46.6)}$$

Where, k_{in} is the zero-order rate constant for production of the response and k_{out} defines the first-order rate constant for loss of the response.

If an inhibitory effect of the drug should occur, then this may be given as:

$$I(t) = \frac{1 - C(t)}{CI_{50} + C(t)} \qquad \text{(Equation 46.7)}$$

Where, CI_{50} is the drug concentration, which produces 50% of maximum inhibition.

Similarly, a stimulatory process can be described by:

$$S(t) = 1 + \frac{C(t)}{CI_{50} + C(t)} \quad \text{(Equation 46.8)}$$

The incorporation of these two functions into Equation 46.6 leads to a family of four basic PD models, namely:

$$dR/dt = k_{in} * I(t) - k_{out} * R$$ i.e. inhibition of k_{in} (Equation 46.9)

$$dR/dt = k_{in} - k_{out} * I(t) * R$$ i.e. inhibition of k_{out} (Equation 46.10)

$$dR/dt = k_{in} * S(t) - k_{out} * R$$ i.e. stimulation of k_{in} (Equation 46.11)

$$dR/dt = k_{in} - k_{out} * S(t) * R$$ i.e. stimulation of k_{out} (Equation 46.12)

Indirect response models have been applied successfully to model the anticoagulant effects of warfarin, adrenal suppression by corticosteroids, antipyretic effects of ibuprofen and the diuretic action of frusemide.

46.4.1.6 *Tolerance models*

Several different methods have been suggested to model the development of tolerance to a drug's pharmacologic effect. For example, the non-linear EMax model can be modified to include a time-dependent exponential decrease in EMax or increase in EC_{50}. Other more complex models have been successfully used to describe the development of tolerance to nicotine on heart rate and adrenergic responsiveness both during and following withdrawal of propranolol.

46.5 Pharmacological efficiency

Recently it has been proposed that a different method of describing the time course of pharmacological effects of a drug be utilized in order to consider the cumulative effect produced by a dose. Use of the standard PD models has been criticized as only describing instantaneous drug action (Alvan *et al.*, 1999).

The pharmacological efficiency of a drug is defined as the pharmacologic effect per unit of drug concentration. Efficiency varies with drug concentration and time and, thus, the cumulative effects of a drug may vary depending on its administration profile. It has been argued that the pharmacodynamics of frusemide and other loop diuretics are best characterized by this approach and that the method is applicable to a number of other drugs including opioids and antibacterial agents.

Efficiency (Eff) is calculated as:

$$Eff = \frac{E - E_0}{C} \quad \text{(Equation 46.13)}$$

For the sigmoid EMax model for example, efficiency can then be incorporated as:

$$Eff = \frac{E_{Max} * C^{\gamma}}{EC_{50}^{\gamma} + C^{\gamma}} * \frac{1}{C} \quad \text{(Equation 46.14)}$$

or,

$$Eff = \frac{E_{Max} * C^{\gamma - 1}}{EC_{50}^{\gamma} + C^{\gamma}} \quad \text{(Equation 46.15)}$$

Total efficiency during a dosage interval can be used to compare different routes of drug administration providing any differences in bioavailability are taken into account, via the expression:

$$TotalEff = \frac{AUC(E - E_0)}{AUC} \quad \text{(Equation 46.16)}$$

Where, AUC $(E - E_0)$ is the area under the effect-time curve corrected for baseline and AUC is the area under the plasma concentration-time curve.

46.6 An example of the application of PK-PD modeling to drug delivery – target controlled infusion systems

A simple yet elegant application of knowledge of the PK-PD parameters of anaesthetic agents has lead to the development of target controlled infusion (TCI) technology. TCI is based on the assumption that by using the PK-PD model parameters of an anesthetic agent (e.g. alfentanil) (see Figure 46.3), an infusion rate for the drug can be calculated to achieve a predetermined plasma concentration suitable for the management of intraoperative conditions during surgery. A TCI system for propofol is now commercially available in several countries for clinical use (Bellissant *et al.*, 1996).

Pharmacokinetic and, more importantly, pharmacodynamic parameters of any drug are subject to considerable interindividual variability and there is also a need to target varying plasma concentrations of anesthetic agents depending on the intensity of stimulation (e.g. induction versus incision versus postoperative pain management). Such factors complicate the rational intraoperative use of these drugs making a titratable infusion rate relative to clinical requirements an attractive proposition.

TCI devices provide the anesthetist with estimates of the plasma concentration of the drug at any time and permit the target concentration to be achieved as rapidly as possible. The computerized device then alters the infusion rates after a particular target has been reached to maintain essentially stable conditions.

When using TCI in clinical practice the initial target concentration has to be defined for the drug being used, based

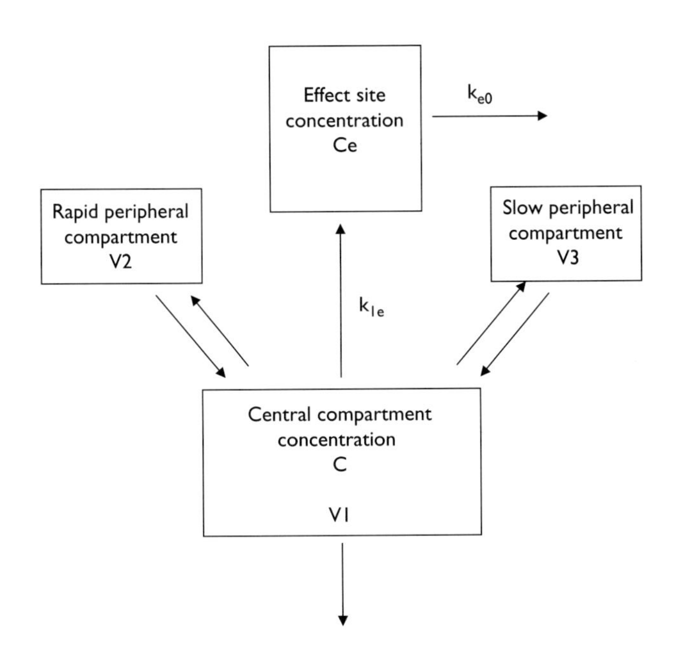

Figure 46.3 Three compartment pharmacokinetic model for alfentanil. Compartments have volumes V1, V2 and V3. The central compartment or plasma concentration is C. An effect site is then linked to the PK model via the rate constant k_{1e}. k_{e0} is the rate constant describing drug exit from the effect site. The pharmacodynamic model (e.g. EMax or sigmoid EMax model) is driven by concentrations in the effect compartment (Ce).

on its effect-concentration relationship. Safety measures also need to be programmed (in a similar fashion to patient-controlled analgesia) for a step up in target concentrations where there is inadequate effect, and a step down when adverse effects occur. Titration of the target concentration should be adjusted to obtain optimal intraoperative conditions including adjustment for rapid awakening of the patient after surgery and for postoperative analgesia.

One disadvantage of the commercially available TCI device is that an effect site compartment is not yet a feature of the device. Those devices that do incorporate an effect compartment remain experimental at present. For many anesthetic agents the target concentration correlates better with a theoretical effect site concentration rather than directly with the plasma concentration. Therefore, there will always be a delay between a change in target concentration and the corresponding clinical effect. For fentanyl for example, the time delay is about 20 minutes. With present TCI devices it is possible to adjust for such a delay by setting the initial target concentration higher than the desired concentration and then readjusting to a lower target thereafter. Addition of an effect compartment would help to reduce the need for concentration overshoot.

References

1 Lalonde, R.L. (1992) Pharmacodynamics. In: Evans, W.E., Schentag, J.J. and Jusko, W.J. (eds.) *Applied pharmacokinetics – principles of therapeutic drug monitoring*, 3rd Edition, Applied Therapeutics Inc., Vancouver, Canada, pp. 4–1 to 4–33.

2 Derendorf, H. and Hochhaus, G. (eds.) (1995) *Handbook of pharmacokinetic/pharmacodynamic correlation*, CRC Press, London, UK.

3 Gabrielsson, J. and Weiner, D. (1997) *Pharmacokinetic and pharmacodynamic data analysis – concepts and applications*, 2nd Edition, The Swedish Pharmaceutical Press, Stockholm, Sweden.

4 Castaneda-Hernandez, G., Caille, G. and du Souich, P. (1994) Influence of drug formulation on drug concentration-effect relationships. *Clin. Pharmacokin. 26: 135–143.*

5 Levy, G. (1998) Predicting effective concentrations for individual patients – determinants of pharmacodynamic variability. *Clin. Pharmacokin. 34: 323–333.*

6 Bellissant, E., Sebille, V. and Paintaud, G. (1998) Methodological issues in pharmacokinetic-pharmacodynamic modelling. *Clin. Pharmacokin. 35: 151–166.*

7 Alvan, G., Paintaud, G. and Wakelkamp, M. (1999) The efficiency concept in pharmacodynamics. *Clin. Pharmacokin. 36: 375–389.*

8 van den Nieuwenhuyzen, M.C.O., Engbers, F.H.M., Vuyk, J. and Burm, A.G.L. (2000) Target-controlled infusion systems – role in anaesthesia and analgesia. *Clin. Pharmacokin. 38: 181–190.*

Author index

Subject index